EXPANDING THE PRODUCTION AND USE OF COOL SEASON FOOD LEGUMES

Current Plant Science and Biotechnology in Agriculture

VOLUME 19

Aims and Scope
The book series is intended for readers ranging from advanced students to senior research scientists and corporate directors interested in acquiring in-depth, state-of-the-art knowledge about research findings and techniques related to plant science and biotechnology. While the subject matter will relate more particularly to agricultural applications, timely topics in basic science and biotechnology will also be explored. Some volumes will report progress in rapidly advancing disciplines through proceedings of symposia and workshops while others will detail fundamental information of an enduring nature that will be referenced repeatedly.

The titles published in this series are listed at the end of this volume.

Expanding the Production and Use of Cool Season Food Legumes

A global perspective of peristent constraints and of opportunities and strategies for further increasing the productivity and use of pea, lentil, faba bean, chickpea and grasspea in different farming systems

Proceedings of the Second International Food Legume Research Conference on pea, lentil, faba bean, chickpea, and grasspea, Cairo, Egypt, 12–16 April 1992

Edited by

F.J. MUEHLBAUER and W.J. KAISER

USDA-ARS, Washington State University, Pullman, Washington, USA

Springer-Science+Business Media, B.V.

Library of Congress Cataloging-in-Publication Data

Expanding the production and use of cool season food legumes / edited
by F.J. Muehlbauer and W.J. Kaiser.
      p.   cm. -- (Current plant science and biotechnology in
agriculture ; 19)
   Selected papers from the Second International Food Legume Research
Conference on Pea, Lentil, Faba Bean, Chickpea, and Grasspea, held
in Cairo, Egypt, April 12-16, 1992.
   Includes index.
   ISBN 978-94-010-4343-4   ISBN 978-94-011-0798-3 (eBook)
   DOI 10.1007/978-94-011-0798-3
   1. Legumes--Congresses.   2. Food crops--Congresses.   3. Legumes as
food--Congresses.   I. Muehlbauer, Fred J. (Frederick Joseph), 1940-
. II. Kaiser, W. J.   III. International Food Legume Research
Conference on Pea, Lentil, Faba Bean, Chickpea, and Grasspea (2nd :
1992 : Cairo, Egypt)   IV. Title: Cool season food legumes.
V. Series.
SB177.L45E95   1994
635'.65--dc20   93-31539

ISBN 978-94-010-4343-4

printed on acid-free paper

Table of contents

About the Conference   xiii
Institutional Hosts – Organizing Committee   xiv
International Steering Committee   xiv
Benefactors of the Conference – Industrial sponsors   xv/xvi
Contributing authors   xvii
Preface   xxvii
Acknowledgements   xxix
Editorial notes and glossary   xxxi

Keynote address

Current status and future trends in supply and demand of cool
season food legumes
P.A. Oram and M. Agcaoili   3

Processing and animal feeds

Aspects of the nutritional quality and use of cool season food
legumes in animal feed
J. Huisman and A.F.B. van der Poel   53

Nature, composition, and utilization of food legumes
J.H. Hulse   77

Diversifying use of cool season food legumes through processing
R. Jambunathan, H.L. Blain, K.S. Dhindsa, L.A. Hussein,
K. Kogure, L. Li-Juan and M.M. Youssef   98

Improving nutritional quality of cool season food legumes
P.C. Williams, R.S. Bhatty, S.S. Deshpande, L.A. Hussein and
G.P. Savage   113

Enhancing the use of cool season food legumes in different farming systems
M. Pala, M.C. Saxena, I. Papastylianou and A.A. Jaradat   130

Grasspea (*Lathyrus sativus* L.) as a potentially safe legume food crop
J. Smartt, A. Kaul, W.A. Araya, M.M. Rahman and J. Kearney   144

Climate change and biotic and abiotic stresses

Potential effects of global climate change on cool season food legume productivity
C. Grashoff, R. Rabbinge and S. Nonhebel   159

Biotic and abiotic stresses constraining productivity of cool season food legumes in Asia, Africa and Oceania
C. Johansen, B. Baldev, J.B. Brouwer, W. Erskine, W.A. Jermyn,
L. Li-Juan, B.A. Malik, A. Ahad Miah and S.N. Silim   175

Biotic and abiotic stresses of cool season food legumes in the western hemisphere
A.E. Slinkard, G. Bascur and G. Hernández-Bravo   195

Biotic and abiotic stresses of pulse crops in Europe
L. Monti, A.J. Biddle, M.T. Moreno and P. Plancquaert   204

Biotic and abiotic stresses constraining the productivity of cool season food legumes in different farming systems: specific examples
M.B. Solh, H.M. Halila, G. Hernández-Bravo, B.A. Malik, M.I. Mihov
and B. Sadri   219

Host plant resistance to manage biotic stress

Using host plant resistance to manage biotic stresses in cool season food legumes
F.J. Muehlbauer and W.J. Kaiser   233

Screening techniques and sources of resistance to foliar diseases caused by fungi and bacteria in cool season food legumes
A. Porta-Puglia, C.C. Bernier, G.J. Jellis, W.J. Kaiser and M.V. Reddy   247

Screening techniques and sources of resistance to root rots and wilts in cool season food legumes
J.M. Kraft, M.P. Haware, R.M. Jiménez-Díaz, B. Bayaa and
M. Harrabi   268

Research achievements in plant resistance to insect pests of cool season
food legumes
S.L. Clement, N. El-Din Sharaf El-Din, S. Weigand and S.S. Lateef   290

Insects in relation to virus epidemiology in cool season food legumes
L. Bos and K.M. Makkouk   305

Screening techniques and sources of resistance to parasitic angiosperms
J.I. Cubero, A.H. Pieterse, S.A. Khalil and J. Sauerborn   333

Screening techniques and sources of resistance to nematodes in cool
season food legumes
S.B. Sharma, R.A. Sikora, N. Greco, M. Di Vito and G. Caubel   346

Policy incentives

Policy incentives for expanding European pulse production
G.P. Gent   361

Effects of markets on the development of cool season food legumes,
experiences from Sudan and India
M. von Oppen, H. Faki, S. Abdelmagid and A. Hashim   367

Chickpea and lentil production in Turkey
N. Açìkgöz, M. Karaca, C. Er and K. Meyveci   388

Lentil production in Chile
J.U. Tay, G. Bascur and E. Peñaloza   399

Pea and chickpea production in Australia
*R.O. Rees, J.B. Brouwer, J.E. Mahoney, G.H. Walton, R.B. Brinsmead,
E.J. Knights and D.F. Beech*   412

Breeding methods and selection indices

Breeding methods and selection indices for improved tolerance to biotic
and abiotic stresses in cool season food legumes
R.J. Baker   429

Screening techniques and sources of tolerance to extremes of moisture
and air temperature in cool season food legumes
J. Wery, S.N. Silim, E.J. Knights, R.S. Malhotra and R. Cousin   439

Screening techniques and sources of tolerance to salinity and mineral
nutrient imbalances in cool season food legumes
*N.P. Saxena, M.C. Saxena, P. Ruckenbauer, R.S. Rana, M.M. El-Fouly
and R. Shabana*   457

Screening techniques and improved biological nitrogen fixation in cool
season food legumes
D.F. Herridge, O.P. Rupela, R. Serraj and D.P. Beck   472

Infrastructural support

Provisions for agronomic inputs for cool season food legumes in some
developing countries (discussion session)
*A.J.G. van Gastel, P.N. Bahl, H. Faki, P. Plancquaert, A.M. Nassib and
B.A. Snobar*   495

Infrastructural support to promote farmer adoption of improved
technologies
C.L.L. Gowda, D.G. Faris and A.F.M. Maniruzzaman   504

Developing and delivering mechanization for cool season food legumes
J. Diekmann, R.K. Bansal and G.E. Monroe   517

Cool season food legume breeding

Potential for wild species in cool season food legume breeding
F.J. Muehlbauer, W.J. Kaiser and C.J. Simon   531

Current status and future strategy in breeding pea to improve resistance
to biotic and abiotic stresses
S.M. Ali, B. Sharma and M.J. Ambrose   540

Current and future strategies in breeding lentil for resistance to biotic
and abiotic stresses
*W. Erskine, M. Tufail, A. Russell, M.C. Tyagi, M.M. Rahman and M.C.
Saxena*   559

Current status and future strategy in breeding chickpea for resistance to
biotic and abiotic stresses
K.B. Singh, R.S. Malhotra, M.H. Halila, E.J. Knights and M.M. Verma 572

Present status and future strategy in breeding faba beans (*vicia faba* L.)
for resistance to biotic and abiotic stresses
*D.A. Bond, G.J. Jellis, G.G. Rowland, J. Le Guen, L.D. Robertson,
S.A. Khalil and L. Li-Juan*   592

Current status and future strategy in breeding grasspea (*Lathyrus sativus*)
C.G. Campbell, R.B. Mehra, S.K. Agrawal, Y.Z. Chen, A.M. Abdel Moneim, H.I.T. Khawaja, C.R. Yadov, J.U. Tay and W.A. Araya   617

Management to control biotic and abiotic stress

Crop and soil management practices for mitigating stresses caused by extremes of soil moisture and temperature
M.C. Saxena, A. Gizaw, M.A. Rik and M. Ali   633

Integrated control of diseases of cool season food legumes
S.P.S. Beniwal and A. Trapero-Casas   642

Integrated management systems to control biotic and abiotic stresses in cool season food legumes
Y.L. Nene and W. Reed   666

Integrated control of insect pests of cool season food legumes
S. Weigand, S.S. Lateef, N. El-Din Sharaf El-Din, S.F. Mahmoud, K. Ahmed and K. Ali   679

Integrated control of the parasitic angiosperm *Orobanche* (Broomrape)
A.H. Pieterse, L. García-Torres, O.A. Al-Menoufi, K.H. Linke and S.J. ter Borg   695

Biotechnology and gene mapping

The potential of gene technology and genome analysis for cool season food legume crops: theory and practice
G. Kahl, D. Kaemmer, K. Weising, S. Kost, F. Weigand and M.C. Saxena   705

Identifying and mapping genes of economic significance
N.F. Weeden, G.M. Timmerman and J. Lu   726

Cloning *Bacillus thuringiensis* toxin genes for control of nodule-feeding insects
D.F. Bezdicek, M.A. Quinn, L. Forse, D.P. Beck and S. Weigand   738

Crop physiology and productivity

Crop physiology and productivity in the cool season food legumes:
recent advances in the measurement and prediction of photothermal
effects on flowering
R.J. Summerfield, E.H. Roberts and R.H. Ellis   755

Plant architecture, competitive ability, and crop productivity in food
legumes with particular emphasis on pea (*Pisum sativum* L.) and faba
bean (*Vicia faba* L.)
M.C. Heath, C.J. Pilbeam, B.A. McKenzie and P.D. Hebblethwaite   771

Reproductive physiology as a constraint to seed production in cool
season food legumes
G. Duc, P. Gates, B. Ney, G.G. Rowland and A. Telaye   791

Root form and function in relation to crop productivity in cool season
food legumes
P.J. Gregory, N.P. Saxena, J. Arihara and O. Ito   809

Biological nitrogen fixation: basic advances and persistent agronomic
constraints
A. Stanforth, J.I. Sprent, J. Brockwell, D.P. Beck and H. Moawad 821

VA mycorrhiza: benefits to crop plant growth and costs
E. George, S.K. Kothari, X.-L. Li, E. Weber and H. Marschner   832

Farmers' constraints and on-farm research

Experience with Ascochyta blight of chickpea in the United States
W.J. Kaiser, F.J. Muehlbauer and R.M. Hannan   849

Lygus bug on lentil in the United States
R.J. Summerfield, R.W. Short and F.J. Muehlbauer   859

Addressing farmers' constraints through on-farm research: peas in
western Canada
A.E. Slinkard, C. van Kessel, D.E. Feindel, S.T. Ali-Khan and R. Park   877

Addressing production constraints for cool season food legumes in
West Asia and North Africa through on-farm research: problems and
ways forward
J.D.H. Keatinge, I. Kusmenoglu and D. Sakar   890

Approaches to overcoming constraints to winter chickpea adoption in
Morocco, Syria and Tunisia
R.N. Tutwiler, M. Amine, M.B. Solh, S.P.S. Beniwal and M.H. Halila   899

On-farm research addressing lentil farmers' constraints in West Asia
B.A. Snobar, M. Abi Antoun, N.I. Haddad, M. Tawil and A.B. Silkine   911

Addressing farmers' constraints through on-farm research: chickpea in
Maharashtra State of India
P.W. Amin, Y.L. Nene and H.A. van Rheenen   926

Reports of seven concurrent discussion groups based on geography

North America
W.J. Kaiser and J.M. Kraft   941

Latin America
J.U. Tay and G. Hernández-Bravo   944

Europe
J. Picard and M.C. Heath   947

Africa
H.M. Halila and S.P.S. Beniwal   951

Near East
I. Kusmenoglu and N.I. Haddad   956

Asia
B. Sharma and C.L.L. Gowda   959

Oceania
R.O. Rees and J.B. Brouwer   962

Continuation of the IFLRC concept

International Food Legume Research Conference (IFLRC): concept
and continuity
R.J. Summerfield and F.J. Muehlbauer   971

Conference summary

Impressions of the second International Food Legume Research
Conference
E.H. Roberts   983

Author index   989

About the conference

The Second International Food Legume Research Conference (IFLRC-II) was held from 12 to 16 April 1992 at the Ramses Hilton Hotel, Cairo, Egypt. Five cool season food legume crops were included in the deliberations: pea (*Pisum sativum*), lentil (*Lens culinaris*), faba bean (*Vicia faba*), chickpea (*Cicer arietinum*), and grasspea (*Lathyrus sativus*). The principal objectives of IFLRC-II were to review and assess recent results from national and international research programs on cool season food legumes and to develop strategies for increasing the productivity, improving the quality and extending the use of these crops in different farming systems. Topics in both basic and applied research were addressed and multidisciplinary research efforts were emphasized.

The Conference format was based on 13 plenary sessions with invited speakers, a large contributed poster session (with different posters each day), an all-day field tour of research facilities and farmers' lands in the delta area of Egypt, and a morning session devoted to Regional Discussion Groups. Invited papers were organized according to subject areas: Processing and Animal Feeds; Climate Change and Biotic and Abiotic Stresses; Host Plant Resistance to Manage Biotic Stresses; Policy Incentives; Breeding Methods and Selection Indices; Infrastructural Support; Cool Season Food Legume Breeding; Management to Control Biotic and Abiotic Stresses; Biotechnology and Gene Mapping; Crop Physiology and Productivity; and Farmers' Constraints and On-Farm Research. Sessions devoted to Regional Discussion Groups and to the Continuation of the IFLRC Concept were also included. A Conference Summary was presented by Professor Eric Roberts.

The Conference was organized and planned by an Organizing Committee which was established during IFLRC-I. The International Steering Committee worked in close concert with the Egyptian hosts and the Local Arrangements Committee. These combined efforts culminated in a Conference program consisting of 57 invited papers co-authored by a total of 196 contributors from 37 countries.

F.J. Muehlbauer and W.J. Kaiser (eds.), Expanding the Production and Use of Cool Season Food Legumes, xiii–xvi.

Institutional Hosts

The Conference was held under the Patronage of His Excellency Prof. Dr. Yousef Wally, Deputy Prime Minister and Minister of Agriculture and Land Reclamation, Cairo, Egypt.

On behalf of His Excellency, the Conference was inaugurated by Prof. Dr. Adel El-Beltagy, Director General, Agricultural Research Center/National Agricultural Research Project, Giza, Egypt.

Organizing Committee

Dr. Saad Nasser (Host Country Chairman), Ministry Consultant and Supervisor, Foreign Agricultural Relations Department, Ministry of Agriculture and Land Reclamation, Cairo, Egypt

Dr. A. M. Nassib (Host Country Representative), Technology Transfer, National Agricultural Research Project, Agricultural Research Center (ARC), Field Crops Research Institute, P.O. Box 12619, Giza, Egypt

Dr. M. B. Solh (Local Arrangements), Regional Coordinator, Nile Valley Research Project (NVRP), International Center for Agricultural Research in the Dry Areas (ICARDA), P.O. Box 2416, Cairo, Egypt

Dr. A. E. Slinkard (Conference Chairman), Crop Development Centre, University of Saskatchewan, Saskatoon, Saskatchewan S7N 0W0, Canada

Dr. M. C. Saxena (Program Chairman), International Center for Agricultural Research in the Dry Areas (ICARDA), P.O. Box 5466, Aleppo, Syria

Dr. F. J. Muehlbauer (Conference Co-Editor), USDA-ARS, Grain Legume Genetics and Physiology Research, 303W Johnson Hall, Washington State University, Pullman, WA 99164-6434, USA

Dr. W. J. Kaiser (Conference Co-Editor), USDA-ARS, Regional Plant Introduction Station, 59 Johnson Hall, Washington State University, Pullman, WA 99164-6402, USA

Dr. G. G. Rowland (Poster Sessions Chairman), Crop Development Centre, University of Saskatchewan, Saskatoon, Saskatchewan S7N 0W0, Canada

International Steering Committee

Dr. A. E. Slinkard (Conference Chairman), Crop Development Centre, University of Saskatchewan, Saskatoon, Saskatchewan S7N 0W0, Canada

Dr. M. C. Saxena (Program Chairman), International Center for Agricultural Research in the Dry Areas (ICARDA), P.O. Box 5466, Aleppo, Syria

Dr. F. J. Muehlbauer (Conference Co-Editor), USDA-ARS, Grain Legume Genetics and Physiology Research, 303W Johnson Hall, Washington State University, Pullman, WA 99164-6434, USA

Dr. W. J. Kaiser (Conference Co-Editor), USDA-ARS, Regional Plant Introduction Station, 59 Johnson Hall, Washington State University, Pullman, WA 99164–6402, USA

Dr. M. E. Tapia, Instituto Nacional de Investigación y Promoción Agropecuaria (INIPA), Programa Nacional de Sistemas Andinos de Producción Agropecuaria, Apartado 110697, Lima 11, Peru

Prof. R. J. Summerfield, University of Reading, Department of Agriculture, Plant Environment Laboratory, Cutbush Lane, Shinfield, Reading, Berkshire RG2 9AD, UK

Dr. A. Telaye, Nile Valley Regional Project, Institute of Agricultural Research (IAR), P.O. Box 2003, Holetta Research Center, Addis Ababa, Ethiopia

Dr. A. M. Nassib, Technology Transfer, National Agricultural Research Project, Agricultural Research Center (ARC), Field Crops Research Institute, P.O. Box 12619, Giza, Egypt

Dr. W. A. Jermyn, Crop Research Division, Department of Scientific and Industrial Research (DSIR), Canterbury Agriculture and Science Centre, Private Bag, Christchurch, New Zealand

Dr. B. A. Malik, Pakistan Agricultural Research Council (PARC), National Agricultural Research Centre, P.O. National Institute of Health, National Park Road, Islamabad, Pakistan

Dr. Y. L. Nene, International Crops Research Institute for the Semi-Arid Tropics (ICRISAT), Patancheru, Andhra Pradesh 502 324, India

Dr. R. H. Lockerman, Department of Plant and Soil Science, Montana State University, Bozeman, MT 59717, USA

Dr. D. F. Bezdicek, Department of Crop and Soil Sciences, Washington State University, Pullman, WA 99164–6420, USA

Benefactors of the Conference

Egyptian Ministry of Agriculture and Land Reclamation, Cairo, Egypt
Deutsche Gesellschaft für Technische Zusammenarbeit (GTZ) GmbH, Eschborn, Germany
Federal Ministry for Economic Cooperation (BMZ), Bonn, Germany
Australia International Development Assistance Bureau, Australia
United States Agency for International Development, USA
International Development Research Centre, Canada
International Center for Agricultural Research in the Dry Areas, Syria
International Crops Research Institute for the Semi-Arid Tropics, India
Food and Agriculture Organization of the United Nations, Italy
United States Department of Agriculture – Agricultural Research Service, USA
Overseas Development Administration of the United Kingdom Foreign and Commonwealth Office, UK

Washington State University, USA
University of Saskatchewan, Canada

Industrial Sponsors

American Dry Pea and Lentil Association, USA
    Washington and Idaho Dry Pea and Lentil Commissions, USA
    Washington and Idaho Associations of Dry Pea and Lentil Producers, USA
    Saskatchewan Pulse Crop Development Board, Canada
    Alberta Pulse Growers Commission, Canada
    Manitoba Pulse Producers Association, Canada
    Canadian Special Crops Association, Canada
    Finora Canada Ltd., Canada
    International Grain Trade Canada Ltd., Canada
    Pioneer Grain Co. Ltd., Canada
    Walker Seeds Ltd., Canada
    Seedtec Ltd., Canada
    Newfield Seeds Ltd./Svalof Seeds, Canada
    Saskatchewan Department of Agriculture and Food, Canada
    Chin Ridge Seed Processors Ltd., Canada
    Parent Seed Farm, Canada
    Roy Legumex Inc., Canada
    Saskatchewan Wheat Pool, Canada
    United Grain Growers, Canada
    X-Can Grain, Canada

Contributing authors

A principal objective of IFLRC-II was to promote collaboration among participants and authors of invited papers for mutual benefit and for the advancement of the knowledge of cool season food legumes. Invited papers which comprise these proceedings were, in nearly all cases, authored by two or more contributors (85% of the papers had three or more co-authors), and representing different regions where cool season food legumes are produced. The International Steering Committee, the International Observer, the Benefactors, Industrial Sponsors and our Institutional hosts wish to thank all the authors listed below.

The special efforts made by these authors in the timely submission of the manuscripts, attention to editorial details and responses to suggested revision are gratefully acknowledged.

ABDELMAGID, S., University of Hohenheim (490), Postfach 700562, D-7000 Stuttgart 70, Germany

ABD EL MONEIM, A. M., International Center for Agricultural Research in the Dry Areas (ICARDA), P. O. Box 5466, Aleppo, Syria

ABI ANTOUN, M., Agricultural Research Organization, Tel Amara, Lebanon

AÇÌKGÖZ, N., Aegean Agricultural Research Institute, P. O. Box 9, Menemen, Izmir, Turkey

AGCAOILI, M., International Food Policy Research Institute, 1200 17th Street, Washington D.C. 20036, USA

AGRAWAL, S. K., Indira Gandhi Agricultural University, Raipur, Madra Pradesh, India

AHMED, K., National Agricultural Research Center, Islamabad, Pakistan

AL-MENOUFI, O. A., Alexandria University, Alexandria, Egypt

ALI, K., Institute of Agricultural Research, Addis Ababa, Ethiopia

ALI, M., Directorate of Pulses Research Indian Council of Agricultural Research, Kalyanpur, Kanpur 208 024, Uttar Pradesh, India

ALI, S. M., South Australian Department of Agriculture, GPO Box 1671, Adelaide, SA 5001, Australia

ALI-KHAN, S. T., Agriculture Canada Research Station, P. O. Box 3001, Morden, Manitoba R0J 1J0, Canada

F.J. Muehlbauer and W.J. Kaiser (eds.), Expanding the Production and Use of Cool Season Food Legumes, xvii–xxvi.

AMBROSE, M. J., John Innes Institute, Colney Lane, Norwich NR4 7NH, UK

AMIN, P. W., Punjabrao Krishi Vidyapeeth, Krishi Nagar, Akola, Maharashtra, India

AMINE, M., DPV, Ministry of Agriculture and Agrarian Reform, Rabat, Morocco

ARAYA, W. A., Institute of Agricultural Research, Adet Research Centre, Bahar Dar, Ethiopia

ARIHARA, J., Hokkaido National Agricultural Experiment Station, 1 Hitsujigaoka, Toyohira-Ku, Sapporo 062, Japan

BAHL, P. N., Indian Council of Agricultural Research, Krishi Bhawn, Dr. Rajendra Prasad Road, New Delhi 110 001, India

BAKER, R. J., Crop Development Centre, Department of Crop Science and Plant Ecology, University of Saskatchewan, Saskatoon, Saskatchewan S7N 0W0, Canada

BALDEV, B., Indian Agricultural Research Institute (IARI), New Delhi 110 012, India

BANSAL, R. K., Project Aridoculture, BP 290, Settat, Morocco

BASCUR, G., Instituto de Investigaciones Agropecuarias (INIA), Estación Experimental La Platina, Casilla 439, Correo 3, Santiago, Chile

BAYAA, B., International Center for Agricultural Research in the Dry Areas (ICARDA), P. O. Box 5466, Aleppo, Syria

BECK, D. P., International Center for Agricultural Research in the Dry Areas (ICARDA), P. O. Box 5466, Aleppo, Syria

BEECH, D. F., CSIRO Division of Tropical Crops and Pastures, St. Lucia, Australia

BENIWAL, S. P. S., Legume Improvement Program, International Center for Agricultural Research in the Dry Areas (ICARDA), B.P. 2335, Fes, Morocco

BERNIER, C. C., Plant Science Department, University of Manitoba, Winnipeg, Manitoba R3T 2N2, Canada

BEZDICEK, D. F., Department of Crop and Soil Sciences, Washington State University, Pullman, WA 99164–6420, USA

BHATTY, R. S., Crop Development Centre, Department of Crop Science and Plant Ecology, University of Saskatchewan, Saskatoon, Saskatchewan S7N 0W0, Canada

BIDDLE, A. J., Processors and Growers Research Organisation, Great North Road, Thornhaugh, Peterborough PE8 6HJ, UK

BLAIN, H. L., American Dry Pea and Lentil Association, 5071 Hwy 8 W., Moscow, ID 83843, USA

BOND, D. A., Plant Breeding International Cambridge, Maris Lane, Trumpington, Cambridge CB2 2LQ, UK

BOS, L., Research Institute for Plant Protection (IPO-DLO), P. O. Box 9060, 6700 GW Wageningen, The Netherlands

BRINSMEAD, R. B., Queensland Department of Primary Industries, Warwick, Australia

BROCKWELL, J., CSIRO Division of Plant Industry, GPO Box 1600, Canberra, ACT 2601, Australia

BROUWER, J. B., Victorian Institute for Dryland Agriculture, Department of Agriculture, Natimuk Road, Private Bag 260, Horsham, Victoria 3401, Australia

CAMPBELL, C. G., Agriculture Canada Research Station, Morden, Manitoba R0G 1J0, Canada

CAUBEL, G., Institut National de la Recherche Agronomique (INRA), Laboratoire de Zoologie, 35650 Le Rheu, France

CHEN, Y. Z., Lanzhou University Lanzhou, Gansu Province, China

CLEMENT, S. L., U. S. Department of Agriculture, Agricultural Research Service, Regional Plant Introduction Station, Washington State University, 59 Johnson Hall, Pullman, WA 99164–6402, USA

COUSIN, R., Institut National de la Recherche Agronomique (INRA), Route de Saint-Cyr, 78026 Versailles Cedex, France

CUBERO, J. I., Departamento de Genética, Universidad de Córdoba, Córdoba, Spain

DESHPANDE, S. S., Agriculture Canada Research Station, P. O. Box 3001, Morden, Manitoba R0G 1J0, Canada

DHINDSA, K. S., Department of Chemistry and Biochemistry, Haryana Agricultural University, Hisar 125 004, Haryana, India

DIEKMANN, J., International Center for Agricultural Research in the Dry Areas (ICARDA), P. O. Box 5466, Aleppo, Syria

DI VITO, M., Istituto di Nematologia Agraria, C.N.R., 70126, Bari, Italy

DUC, G., Institut National de la Recherche Agronomique (INRA), Station de Génétique et d'Amélioration des Plantes, BV 1540, 21034 Dijon Cédex, France

EL-DIN SHARAF EL-DIN, N., Agricultural Research Center, Wad Medani, Sudan

EL-FOULY, M. M., Botany Laboratory National Research Center, Dokki, Cairo, Egypt

ELLIS, R. H., University of Reading, Department of Agriculture, Plant Environment Laboratory, Cutbush Lane, Shinfield, Reading, Berkshire RG2 9AD, UK

ER, C., Field Crop Improvement Center, P. O. Box 226, Ulus, Ankara, Turkey

ERSKINE, W., International Center for Agricultural Research in the Dry Areas (ICARDA), P. O. Box 5466, Aleppo, Syria

FAKI, H., Agricultural Research Corporation, Wad Medani, Sudan

FARIS, D. G., Coordinator, AGLN, International Crops Research Institute for the Semi-Arid Tropics (ICRISAT), Patancheru P. O., Andhra Pradesh 502 324, India

FEINDEL, D. E., Department of Soil Science, University of Saskatchewan, Saskatoon, Saskatchewan S7N 0W0, Canada

FORSE, L., Department of Crop and Soil Sciences, Washington State University, Pullman, WA 99164–6420, USA

GARCÍA-TORRES, L., Institute of Agronomy and Plant Protection, CSIC, Córdoba, Spain

GATES, P., University of Durham, Department of Biological Sciences, South Road, Durham DH1 3LE, UK

GENT, G. P., Processors & Growers Research Organisation, Thornhaugh, Peterborough PE8 6HJ, England

GEORGE, E., Institute of Plant Nutrition, Hohenheim University, P.B. 700562, 7000 Stuttgart 70, Germany

GIZAW, A., Institute of Agriculture Research, Addis Ababa, Ethiopia

GOWDA, C. L. L., Senior Legumes Breeder, AGLN, International Crops Research Institute for the Semi-Arid Tropics (ICRISAT), Patancheru P. O., Andhra Pradesh 502 324, India

GRASHOFF, C., Centre for Agrobiological Research (CABO-DLO), P. O. Box 14, 6700 AA Wageningen, The Netherlands

GRECO, N., Istituto di Nematologia Agraria, C.N.R., 70126, Bari, Italy

GREGORY, P. J., CSIRO Dryland Crops and Soils Research Unit, Private Bag, P. O., Wembley, WA 6014, Australia

HADDAD, N., International Center for Agricultural Research in the Dry Areas (ICARDA), P. O. Box 950764, Amman, Jordan

HALILA, M. H., Food Legume Program, Institut National de la Recherche Agronomique de Tunisie (INRAT), 2080 Ariana, Tunis, Tunisia

HANNAN, R. M., U. S. Department of Agriculture, Agricultural Research Service, Regional Plant Introduction Station, Washington State University, 59 Johnson Hall, Pullman, WA 99164–6402, USA

HARRABI, M., Institut National Agronomique de Tunisie, Laboratorie de Genetique, Tunis, Tunisia

HASHIM, A., University of Hohenheim (490), Postfach 700562, D-7000 Stuttgart 70, Germany

HAWARE, M. P., International Crops Research Institute for the Semi-Arid Tropics (ICRISAT), Patancheru P. O., Andhra Pradesh 502 324, India

HEATH, M. C., ADAS Arthur Rickwood, Mepal, Ely, Cambridge CB6 2BA, England

HEBBLETHWAITE, P. D., Faculty of Agriculture and Food Science, University of Nottingham, Sutton Bonington, Loughborough, Leics LE12 5RD, UK

HERNÁNDEZ BRAVO, G., Chachalacas, Atizapan, Mexico

HERRIDGE, D. F., New South Wales Agricultural Research Centre, RMB 944, Tamworth, NSW 2340, Australia

HUISMAN, J., TNO-ILOB, Department of Animal Nutrition and Physiology, P. O. Box 15, 6700 AA Wageningen, The Netherlands

HULSE, J. H., Siemens-Hulse International Development Association, Inc., 1628 Featherston Drive, Ottawa, Ontario K1H 6P2, Canada

HUSSEIN, L. A., Department of Nutrition, National Research Centre, El-Tahrir Street, Giza, Dokki, Egypt

ITO, O., Legumes Program, International Crops Research Institute for the Semi-Arid Tropics (ICRISAT), Patancheru P. O., Andhra Pradesh 502 324, India

JAMBUNATHAN, R., Crop Quality Unit, International Crops Research Institute for the Semi-Arid Tropics (ICRISAT), Patancheru P. O., Andhra Pradesh 502 324, India

JARADAT, A. A., Jordan University of Science and Technology, Irbid, Jordan

JELLIS, G. J., Plant Breeding International Cambridge, Maris Lane, Trumpington, Cambridge CB2 2LQ, UK

JERMYN, W. A., Crop Research Division, Department of Scientific and Industrial Research (DSIR), Canterbury Agriculture and Science Centre, Private Bag, Christchurch, New Zealand

JIMÉNEZ-DÍAZ, R. M., Instituto de Agronomía y Protección Vegetal, CSIC, and Departamento de Agronomía-Patología Vegetal, ETSIAM, Universidad de Córdoba, Apdo. 3048, 14080 Córdoba, Spain

JOHANSEN, C., International Crops Research Institute for the Semi-Arid Tropics (ICRISAT), Patancheru P. O., Andhra Pradesh 502 324, India

KAEMMER, D., Plant Molecular Biology, University of Frankfurt/Main, 6000 Frankfurt/Main 1, Germany

KAHL, G., Plant Molecular Biology, Department of Biology, University of Frankfurt/Main, Slesmayerstraße 70, 6000 Frankfurt/Main 1, Germany

KAISER, W. J., U. S. Department of Agriculture, Agricultural Research Service, Regional Plant Introduction Station, Washington State University, 59 Johnson Hall, Pullman, WA 99164–6402, USA

KARACA, M., Field Crop Improvement Center, P. O. Box 226, Ulus, Ankara, Turkey

KAUL, A., Winrock International Institute for Agricultural Development, Petit Jean Mountain, Morrilton, AR 72110–9537, USA

KEARNEY, J., Department of Biology, School of Biological Sciences, The University of Southampton, Southampton S09 3TU, UK

KEATINGE, J. D. H., International Center for Agricultural Research in the Dry Areas (ICARDA), PK 39 Emek, Ankara 06511, Turkey

KHALIL, S. A., Field Crop Research Institute, Agricultural Research Center, P. O. Box 12619, Giza, Egypt

KHAWAJA, H. I. T., National Agricultural Research Centre, Islamabad, Pakistan

KNIGHTS, E. J., New South Wales Government, Department of Agriculture, Agricultural Research Centre, RMB 944, Tamworth, NSW 2340, Australia

KOGURE, K., Faculty of Agriculture, Kagawa University, 2393 Ikenobe, Miki-tyo, Japan

KOST, S., Plant Molecular Biology, University of Frankfurt/Main, 6000 Frankfurt/Main, Germany

KOTHARI, S. K., Central Institute of Medicinal and Aromatic Plants, Lucknow, India

KRAFT, J. M., U. S. Department of Agriculture, Agricultural Research Service, Washington State University, Irrigated Agricultural Research and Extension Center, Route 2, Box 2953A, Prosser, WA 99350–9687, USA

KUSMENOGLU, I., Food Legumes, Field Crops Improvement Center, Ministry of Agriculture and Rural Affairs, P. O. Box 226, Ulus, Ankara, Turkey

LATEEF, S. S., International Crops Research Institute for the Semi-Arid Tropics (ICRISAT), Patancheru P. O., Andhra Pradesh 502 324, India

LE GUEN, J., Station d'Amelioration des Plantes, Institut National de la Recherche Agronomique (INRA), Le Rheu, France

LI, X.-L., Department of Soil Science and Plant Nutrition, Beijing Agricultural University, Beijing, China

LI-JUAN, L., Faba Bean Germplasm and Breeding, Zhejiang Academy of Agricultural Sciences, Hangzhou, China

LINKE, K. H., International Center for Agricultural Research in the Dry Areas (ICARDA), P. O. Box 5466, Aleppo, Syria

LU, J., Department of Horticultural Sciences, New York State Agricultural Experiment Station, Cornell University, Geneva, NY 14456, USA. *Present Address*: Center for Viticultural Science, Florida Agricultural and Mechanical University, Tallahassee, FL 32314, USA

MAHMOUD, S. F., Agricultural Research Center, P. O. Box 12619, Giza, Egypt

MAHONEY, J. E., Victorian Institute for Dryland Agriculture, Horsham, Private Bag 260, Victoria 3401, Australia

MAKKOUK, K. M., International Center for Agricultural Research in the Dry Areas (ICARDA), P. O. Box 5466, Aleppo, Syria

MALHOTRA, R. S., International Center for Agricultural Research in the Dry Areas (ICARDA), P. O. Box 5466, Aleppo, Syria

MALIK, B. A., Pakistan Agricultural Research Council (PARC), National Agricultural Research Centre, P. O. National Institute of Health, National Park Road, Islamabad, Pakistan

MANIRUZZAMAN, A. F. M., Director Research, Bangladesh Agricultural Research Institute (BARI), Joydebpur, Bangladesh

MARSCHNER, H., Institute of Plant Nutrition, Hohenheim University, P. B. 700562, 7000 Stuttgart 70, Germany

McKENZIE, B. A., Plant Science Department, Lincoln University, Canterbury, New Zealand

MEHRA, R. B., Indian Agricultural Research Institute (IARI), New Delhi 110 012, India

MEYVECI, K., Field Crop Improvement Center, P. O. Box 226, Ulus, Ankara, Turkey

MIAH, A. A., Bangladesh Agricultural Research Institute (BARI), Joydebpur, Gazipur 1701, Bangladesh

MIHOV, M. I., Institute of Wheat and Sunflower, "Dobroudja" Gen. Toshevo, 9520, Bulgaria

MOAWAD, H., National Research Center (NRC), Dokki, Cairo, Egypt

MONTI, L., Department of Agronomy and Plant Genetics, University of Naples, Via Università 100, 80055 Portici, Italy

MONROE, G. E., MidAmerican International Agricultural Consortium (MIAC), Project Aridoculture, Settat, Morocco

MORENO, M. T., Centro de Investigacion y Desarrollo Agrario, Departamento de Meyora y Agronomia, Cordoba, Spain

MUEHLBAUER, F. J., U. S. Department of Agriculture, Agricultural Research Service, Washington State University, 303W Johnson Hall, Pullman, WA 99164–6434, USA

NASSIB, A. M., Technology Transfer, National Agricultural Research Project, Agricultural Research Center (ARC), Field Crops Research Institute, P. O. Box 12619, Giza, Egypt

NENE, Y. L., Deputy Director General, International Crops Research Institute for the Semi-Arid Tropics (ICRISAT), Patancheru P. O., Andhra Pradesh 502 324, India

NEY, B., Institut National de la Recherche Agronomique (INRA), Station d'Agronomie, BV 1540, 21034 Dijon Cédex, France

NONHEBEL, S., Department of Theoretical Production Ecology, Agricultural University, P. O. 430, 6700 AK Wageningen, The Netherlands

ORAM, P. A., International Food Policy Research Institute, 1200 17th Street, Washington D.C. 20036, USA

PALA, M. International Center for Agricultural Research in the Dry Areas (ICARDA), P. O. Box 5466, Aleppo, Syria

PAPASTYLIANOU, I., Agricultural Research Institute, Nicosia, Cyprus

PARK, R., Alberta Agriculture, Field Crops Branch, Bag Service 47, Lacombe, Alberta T0C 1S0, Canada

PEÑALOZA, E., Instituto de Investigaciones Agropecuarias (INIA), Estación Experimental Carillanca, Casilla 58-D, Temuco, Chile

PICARD, J. J. A., Union Nationale Interprofessionnelle des Proteagineux (UNIP), 12 Avenue George V, 75008 Paris, France

PIETERSE, A. H., Royal Tropical Institute, Rural Development Programme, Mauritskade 63, 1092 AD Amsterdam, The Netherlands

PILBEAM, C. J., Department of Soil Science, University of Reading, Reading RG1 5AQ, UK

PLANCQUAERT, P., Institute Technique des Céréales et des Fourrages (ITCF), 8 Avenue du Président Wilson, 75116 Paris, France

PORTA-PUGLIA, A., Istituto Sperimentale per la Patologia Vegetale, Via C. G. Bertero, 22 I-00156, Rome, Italy

QUINN, M. A., Department of Crop and Soil Sciences, Washington State University, Pullman, WA 99164–6420, USA

RABBINGE, R., Department of Theoretical Production Ecology, Agricultural University, P. O. 430, 6700 AK Wageningen, The Netherlands

RAHMAN, M. M., Bangladesh Agricultural Research Institute (BARI), Regional Agricultural Research Station, Ishurdi 6620, Pabna, Bangladesh

RANA, R. S., Indian Agricultural Research Institute (IARI) National Bureau of Plant Genetic Resources (NBPGR), New Delhi 110 012, India

REDDY, M. V., International Crops Research Institute for the Semi-Arid Tropics (ICRISAT), Patancheru P. O., Andhra Pradesh 502 324, India

REED, W., 11 Wilberforce Road, Waterside, Sherborne Street, Bourton-on-the-Water, GL54 2BY, UK

REES, R. O., Crops and Economics Section, Australian Bureau of Agricultural and Resource Economics (ABARE), Edmund Barton Building, Broughton St., Barton ACT, GPO Box 1563, Canberra ACT 2601, Australia

RIZK, M. A., Field Crops Research Institute, Agricultural Research Center, P. O. Box 12619, Giza, Egypt

ROBERTS, E. H., University of Reading, Department of Agriculture, Plant Environment Laboratory, Cutbush Lane, Shinfield, Reading, Berkshire RG2 9AD, UK

ROBERTSON, L. D., Genetic Resources Unit, International Center for Agricultural Research in the Dry Areas (ICARDA), P. O. Box 5466, Aleppo, Syria

ROWLAND, G. G., Crop Development Centre, Crop Science Department, University of Saskatchewan, Saskatoon, Saskatchewan S7N 0W0, Canada

RUCKENBAUER, P., Institute fuer Pflanzenbau und Pflanzenzuechtung, Vien, Austria

RUPELA, O. P., International Crops Research Institute for the Semi-Arid Tropics (ICRISAT), Patancheru P. O., Andhra Pradesh 502 324, India

RUSSELL, A., Department of Scientific and Industrial Research, Private Bag, Christchurch, New Zealand

SADRI, B., Food Legume Research Section, Seed & Plant Improvement Section, Ministry of Agriculture and Rural Development, Mar-Abada Avenue, Karaj, Iran

SAKAR, D., Ministry of Agriculture and Rural Affairs, P. O. Box 72, Diyarbakir 2111, Turkey

SAUERBORN, J., Institut für Pflanzenproduktion in den Tropen und Subtropen, Universität Hohenheim, Stuttgart, Germany

SAVAGE, G. P., Lincoln University, P. O. Box 84, Canterbury, New Zealand

SAXENA, M. C., Legume Program, International Center for Agricultural Research in the Dry Areas (ICARDA), P. O. Box 5466, Aleppo, Syria

SAXENA, N. P., Legumes Program, International Crops Research Institute for the Semi-Arid Tropics (ICRISAT), Patancheru P. O., Andhra Pradesh 502 324, India

SERRAJ, R., University of Marrakech, Marrakech, Morocco

SHABANA, R., Cairo University, Cairo, Egypt

SHARMA, B., Division of Genetics, Indian Agricultural Research Institute (IARI), New Delhi 110 012, India

SHARMA, S. B., Legumes Pathology, International Crops Research Institute for the Semi-Arid Tropics (ICRISAT), Patancheru P. O., Andhra Pradesh 502 324, India

SHORT, R. W., U. S. Department of Agriculture, Agricultural Research Service, Washington State University, 303W Johnson Hall, Pullman, WA 99164–6434, USA

SILIM, S. N., Eastern Africa Regional Cereals and Legumes Program (EARCAL), ICRISAT, c/o OAU/STRC, J.P. 31 SAFGRAD, P. O. Box 39063, Nairobi, Kenya

SILKINE, A. B., International Center for Agricultural Research in the Dry Areas (ICARDA), P. O. Box 5466, Aleppo, Syria

SIKORA, R. A., Institut für Pflanzenkrankheiten der Rheinischen Friedrich Wilhelms, Universitat, Nussallee 9, 5300 Bonn, Germany

SIMON, C. J., U. S. Department of Agriculture, Agricultural Research Service, Washington State University, 303W Johnson Hall, Pullman, WA 99164–6434, USA

SINGH, K. B., International Center for Agricultural Research in the Dry Areas (ICARDA), P. O. Box 5466, Aleppo, Syria

SLINKARD, A. E., Crop Development Centre, University of Saskatchewan, Saskatoon, Saskatchewan S7N 0W0, Canada

SMARTT, J., Department of Biology, School of Biological Sciences, The University of Southampton, Biomedical Sciences Building, Basset Crescent East, Southampton S09 3TU, UK

SNOBAR, B. A., Faculty of Agriculture, University of Jordan, Amman, Jordan

SOLH, M. B., Nile Valley Research Project (NVRP), International Center for Agricultural Research in the Dry Areas (ICARDA), P. O. Box 2416, Cairo, Egypt

SPRENT, J. I., University of Dundee, Department of Biological Sciences, DD1 4HN Scotland, UK

STANFORTH, A., University of Dundee, Department of Biological Sciences, DD1 4HN Scotland, UK

SUMMERFIELD, R. J., University of Reading, Department of Agriculture, Plant Environment Laboratory, Cutbush Lane, Shinfield, Reading, Berkshire RG2 9AD, UK

TAWIL, M., Directorate of Agriculture, Damascus, Syria

TAY, J. U., Instituto de Investigaciones Agropecuarias (INIA), Estación Experimental Quilamapu, Casilla 426, Chillán, Chile

TELAYE, A., Nile Valley Research Project (NVRP), Institute of Agricultural Research (IAR), P. O. Box 2003, Holetta Research Center, Addis Ababa, Ethiopia

Ter BORG, S. J., Department of Vegetation Science, Plant Ecology and Weed Science, Agricultural University, Wageningen, The Netherlands

TIMMERMAN, G. M., Crop and Food Research, Cantebury Agriculture and Science Centre, Lincoln, New Zealand

TRAPERO-CASAS, A., Departmento de Agronomía, ETSIAM, Universidad de Córdoba, Apdo. 3048, 14080 Córdoba, Spain

TUFAIL, M., Pulses Research Institute, Ayub, Faisalabad, Pakistan

TUTWILER, R. N., International Center for Agricultural Research in the Dry Areas (ICARDA), P. O. Box 5466, Aleppo, Syria

TYAGI, M. C., Genetics Department, Indian Agricultural Research Institute (IARI), New Delhi 110 012, India

Van der POEL, A. F. B., TNO-ILOB, Department of Animal Nutrition and Physiology, P. O. Box 15, 6700 AA Wageningen, The Netherlands

Van GASTEL, A. J. G., International Center for Agricultural Research in the Dry Areas (ICARDA), P. O. Box 5466, Aleppo, Syria

Van KESSEL, C., Department of Soil Science, University of Saskatchewan, Saskatoon, Saskatchewan S7N 0W0, Canada

Von OPPEN, M., University of Hohenheim (490), Postfach 700562, D-7000 Stuttgart 70, Germany

Van RHEENEN, H. A., ICRISAT, Patancheru P. O., Andhra Pradesh 502 324, India

VERMA, M. M., Department of Plant Breeding, Punjab Agricultural University, Ludhiana 141 004, Punjab, India

WALTON, G. H., Western Australian Department of Agriculture, South Perth, Australia

WEBER, E., Institute of Plant Nutrition, Hohenheim University, P. B. 700562, 7000 Stuttgart 70, Germany

WEEDEN, N. F., Department of Horticultural Sciences, New York State Agricultural Experiment Station, Cornell University, Geneva, NY 14456, USA

WEIGAND, F., Legume Program, International Center for Agricultural Research in the Dry Areas (ICARDA), P. O. Box 5466, Aleppo, Syria

WEIGAND, S., International Center for Agricultural Research in the Dry Areas (ICARDA), P. O. Box 5466, Aleppo, Syria

WEISING, K., Plant Molecular Biology, University of Frankfurt/Main, 6000 Frankfurt/Main 1, Germany

WERY, J., ENSAM Chaire de Phytotechnie et d'Amélioration des Plantes, 2 Place Viala, 34060 Montpellier Cedex 01, France

WILLIAMS, P. C., Agriculture Canada, Canadian Grain Commission, Grain Research Laboratory, 1404–303 Main Street, Winnipeg, Manitoba R3C 3G8, Canada

YADOV, C. R., Rampur Research Station, Chitwan, Nepal

YOUSSEF, M. M., Department of Agricultural Industries, Faculty of Agriculture, University of Alexandria, El-Chatby, Alexandria 21526, Egypt

Preface

The goal of the Second International Food Legume Research Conference held in Cairo, Egypt was to build on the success of the first conference held nearly 6 years earlier at Spokane, Washington, USA. It was at that first conference where the decision was made to hold the second Conference in Egypt and so near the ancestral home of these food legume crops. It has been a long held view that the cool season food legumes had their origin in the Mediterranean basin and the Near-east arc, and there is little doubt that food legumes were a staple food of the ancient Egyptian civilization.

The cool season food legumes have the reputation for producing at least some yield under adverse conditions of poor fertility and limited moisture, i.e., in circumstances where other crops are likely to fail completely. Yields of cool season food legumes are particularly poor in those regions where they are most important to local populations. The influx of more profitable crops such as wheat, maize, and soybeans have gradually relegated the food legumes to marginal areas with poor fertility and limited water which exposes them to even greater degrees of stress. In the past two decades, production of food legumes has declined in most of the developing countries while at the same time it has expanded greatly in Canada, Australia, and most notably in Turkey.

The drastic shift of production away from developing countries (where these food legumes are needed most) to developed countries (from where these crops are mostly exported) is a major concern. To reverse this trend and to expand production in those regions where these crops are most needed will require greater attention to the many biotic and abiotic stresses likely to be encountered.

The Conference organizers, realizing that stresses are major hazards, planned the program accordingly. The major challenge in the future will be to determine how to make effective use of limited resources in the production of cool season food legumes. To meet the projected future needs for these legumes in places where they are needed most will require networking of research efforts and collaboration. The Conference agenda emphasized attempts to alleviate these stresses through germplasm development, improved crop management schemes, and various other means. Many aspects of the several biotic and

F.J. Muehlbauer and W.J. Kaiser (eds.), Expanding the Production and Use of Cool Season Food Legumes, xxvii–xxviii.

abiotic stresses affecting the cool season food legumes were presented and discussed. Special emphasis was placed on reviewing current technologies used to evaluate crop plants for resistance or tolerance to stress of one sort or another.

Expanding the use of the cool season food legumes is a somewhat different matter. It will depend on policy incentives to increase availability in those regions where these commodities are needed most. Policies related to transferring improved production technology to farmers in developing regions are critically important. The elimination of antinutritional factors in grasspea is also important for expanded usage.

These Conference proceedings will provide a more complete understanding of what is required where and when in order to expand production and use of cool season food legume crops.

Acknowledgements

On behalf of the delegates to IFLRC-II we gratefully acknowledge the contributions and efforts of our Institutional hosts in Egypt for providing the meeting venue and support for the Conference activities. The benefactors and industrial sponsors listed elsewhere in these Proceedings (pp. xv–xvi) are gratefully acknowledged for providing resources, advice and counsel. In addition we gratefully acknowledge the technical and secretarial contributions of: in Egypt, Drs. A. M. Nassib, Saad Nasser, M. El-Sherbeeny and M. B. Solh, Mrs. Nagwa Lutfi, Miss Mai Elremisy, Ms. Tahany Ramez, Mr. Mustafa Abaza and Mr. Khaled Genene; in Canada, Dr. B. B. Vandenberg and Mrs. Jo-anne Relf-Eckstein; in ICARDA, Miss Mary Bogharian; in the USA, Dr. C. J. Simon, Mr. D. C. Hoyt, Mrs. Rhonda Gaylord, Mrs. Susan Chipman, Mrs. Tammi Schaumloffel, and Miss Suzie Farrow; and in the UK, Mrs. C. Hadley and Mr. A. Pilgrim.

The editors are especially grateful to Mrs. Joy Barbee (USDA-ARS) for her tireless and diligent efforts as editorial assistant in preparing the manuscripts for submission to the publisher.

Editorial notes and glossary

In the interest of continuity of the usage of various terms, the Editorial Notes and Glossary of Summerfield (1988) is included with minor revisions as follows:

(1) Nitrogen gas is correctly termed "dinitrogen" but since the colloquialism "nitrogen" is used more or less universally for the gaseous element N_2, especially in the literature pertaining to the fixation of N_2 by legumes, it is used in that context throughout this Volume.

(2) The definitions of Polhill and van der Maesen (1985) and of Barnes and Beard (1992) are adopted throughout this Volume with respect to:

accession: Sample in a gene bank; its number is unique and not re-used in cases where the sample is lost.

cultivar: Cultivated variety; an assemblage of cultivated plants clearly recognizable from other cultivars of the same species by structural features and performance.

genotype: (a) The genetic make-up of an organism (all dominant and recessive genes).
(b) A group of organisms with the same genetic make-up.

germplasm: The living substance of the cell nucleus that determines the heredity properties of organisms and that transmit these properties to the next generation.

host-plant resistance: Response of plants to infection by a disease-causing agent or an insect.

landrace: A traditional cultivar not subject to scientific selection; often a population or a mixture of closely related genotypes.

phenotype: Observable characteristics, resulting from interaction between an organism's genetic make-up and environment.

resistance: The ability of an organism to live (survive, or grow and produce) in a stress environment (e.g., drought, cold, or heat) or in the presence of a pathogen, pest, or predator through inhibition or avoidance of the cause or tolerance of its effects; e.g., frost resistance, disease resistance, insect resistance.

F.J. Muehlbauer and W.J. Kaiser (eds.), Expanding the Production and Use of Cool Season Food Legumes, xxxi–xxxii.

resistant: Able to restrict, inhibit, avoid, or tolerate the activities of a specified pest, pathogen, or environmental stress.

tolerance: Ability of an organism to perform well, or survive, despite the existence of a stress condition, pathogen, or predator within its tissues (e.g., frost tolerance, disease tolerance).

variety: Botanical variety; taxonomic level below the rank of sub-species, above the level of cultivar.

(3) There are many common (vernacular, trivial) names for each of the principal legume crops (Kay, 1979), including those five species which are the subject of this Conference. The common names pea, lentil, faba bean, chickpea, and grasspea are used throughout this Volume for *Pisum sativum* L., *Lens culinaris* Medik., *Vicia faba* L., *Cicer arietinum* L., and *Lathyrus sativus* L., respectively.

References

Barnes, R. F. and Beard, J. B. 1992. *Glossary of Crop Science terms.* Crop Science Society of America. Madison, WI. 88 pp.

Kay, D. E. 1979. *Food Legumes.* Tropical Products Institute, Crop and Product Digest No. 3. London.

Polhill, R. M. and Van der Maesen, L. J. G. 1985. In: *Grain Legume Crops*, pp. 3–36 (eds. R. J. Summerfield and E. H. Roberts). London: Collins.

Summerfield, R. J. (ed.). 1988. *World Crops: Cool Season Food Legumes*, pp. xxxvii-xxxviii. Dordrecht: Kluwer Academic Publishers.

Keynote address

Current status and future trends in supply and demand of cool season food legumes

P.A. ORAM and M. AGCAOILI

International Food Policy Research Institute, 1200 17th Street, Washington, D.C. 20036, USA

Abstract

The paper looks at the current distribution and use of the four main cool season legumes (faba bean, chickpea, lentil, and dry pea) for food and for feed, since those roles are interchangeable and in many countries they are produced for both purposes. While recognizing that *Lathyrus sativus* may be grown more extensively if neurotoxic problems can be solved, this species is not dealt with comprehensively for lack of data. Lupins and vetches are grown almost exclusively for animal feed, their role is discussed only in this context.

The cool season food legumes have become widely distributed geographically and geoclimatically over time, but in reviewing the current situation the paper notes that they remain essentially concentrated in temperate and subtropical climates, with only very limited production in the warm lowland tropics, mainly at higher altitudes.

They contribute about 60% of total world pulse production on about 40% of world area, but the share of the different species in the total is unevenly distributed as is their geographical focus. Chickpeas and lentils are predominantly produced in the developing countries, and peas in the developed countries, with faba beans being more evenly distributed. Overall, yields of food legumes in the developing countries are only about half those of the developed countries. For all four species, over 90% of their production is concentrated in only 12 countries, although those countries are not the same for all of the commodities.

In reviewing the growth of world pulse production, the paper notes that while the cool season species contributed most to overall output, and performance through the 1980s was quite good (3.5% per annum), that in the developing countries (1.5% per annum) was well below their population growth. Thus, per capita production has declined. Faba bean production fell in absolute terms; chickpea showed modest growth; lentil production expanded significantly; and peas performed excellently in the developed market economy countries with production growing at 10% annually, but declining sharply in China, the largest developing-country producer.

F.J. Muehlbauer and W.J. Kaiser (eds.), Expanding the Production and Use of Cool Season Food Legumes, 3–49.
© 1994 *Kluwer Academic Publishers.*

Consumption of cool season legumes increased by 10 million tons in developed countries, almost entirely for feed, and by 4.5 million tons in developing countries, primarily for food. Nevertheless, per capita consumption in those countries fell from 9.6 kg to 9.1 kg between 1980 and 1991.

Both imports and exports rose rapidly, almost doubling over the decade, with the proportion of production traded rising from 6.2 to nearly 11%. Nevertheless, pulse trade represents a rather thin market with prices sensitive to temporary changes in supply due to weather and the perceptions of the few major trading nations.

Looking ahead, the paper notes that there are no recent forward-looking projections of supply and demand for the food legumes, therefore it compares actual trends over the 1980s with 1986 FAO projections for that period. This assessment proved too optimistic with respect to food use and somewhat underestimated the rapid growth of feed use and imports, especially after 1985 when the actual growth of production lagged. The paper analyzes the factors likely to determine future demand for pulses and concludes that the main constraint is with supply. Drawing on these conclusions and on recent short-term assessments by Australian and North American producers, the paper concludes that while there is no cause for pessimism over the 1990s, especially with respect to demand for cool season food legumes, it seems likely that there will be some slow-down in the very rapid acceleration in their use for feed in developed countries which characterized the last half of the 1990s.

In order to maintain present levels of per capita food consumption and feed in the face of projected population growth, the paper estimates that legume production will need to rise by at least 13 million tons in the developing countries during the 1980s and the early 1990s, almost double the addition to their production in the 1980s. In exploring approaches to expanding supply, it is noted that while some countries have opportunities for increasing the area of pulses cultivated, the main thrust in many will have to be through increasing the present low levels of yield through research.

In conclusion, some issues related to research are discussed. These include:
- how to raise yields of cool season species in developing countries to narrow the gap with those now being attained by top producing countries, such as for peas in the EEC; and whether the goal should be to raise yield ceilings or to improve yield stability;
- the need to strengthen research on social sciences and to link technical and social science effort more closely so as to accelerate adoption of new technology; and
- how to make most effective use of the relatively limited global research resources likely to be available for research on the food and feed legumes.

Introduction

In preparing this paper, I have been asked to concentrate on the cool season food legumes, which are generally considered to comprise four principal species. These are *Vicia faba* (faba bean), *Cicer arietinum* (chickpea), *Lens culinaris* (lentil), and *Pisum sativum* (dry pea).

While these are the most widely grown and consumed, certain other cool season species of annual legumes are also used for human food, of which the most important in terms of its area and production is probably *Lathyrus sativus* (chickling pea or chickling vetch). There are no official statistics published by FAO on this species, nor on other *Lathyrus* species (*Lathyrus ochrus, Lathyrus cicera*, etc.), but various sources indicate an area of about a million and a quarter hectares in South Asia, principally in India, Bangladesh, and Nepal; and around a hundred thousand hectares in East Africa, mainly Ethiopia. It is also cultivated for human consumption and animal fodder in West Asia and North Africa. It would probably be more widely grown if neurotoxic problems could be solved (Kaul and Fenton, 1991). Total production is estimated at about 600,000 tons – not a negligible quantity. There may also be some human consumption of dry vetch seeds, e.g., in Iraq (Al-Anney, 1990), but no data are available on its magnitude.

In retrospect, I believe that the words "and feed" should have been added to the title of my paper, since it is difficult to discuss the current and future economic importance of the cool season food legumes without considering their use for animal feed. For dry peas, this is preeminent, and it is not negligible for faba bean. *Lathyrus sativus* is also grown as a fodder crop, as are some other *Lathyrus* species. In addition, lupins and several species of vetches are grown widely for feed as well as for forage and may compete for land in farming systems with food legumes, as well as in trade. The current strong demand for feed and fodder, in West Asia and North Africa for example, which is basically a function of high prices of animal products and of imported feeds, provides a considerable incentive to production of such crops.

Although the cool season legumes are primarily of West Asian derivation and were probably first domesticated around the Mediterranean Basin and in India, they have since spread over a much wider geographical range. Nevertheless, unlike *Phaseolus vulgaris*, which is even more widely distributed and may overlap the cool season species seasonally in countries in which both are cultivated, the latter are not adapted to the lowland humid tropics. While they may be grown at higher elevations in the tropics, for example in Ethiopia or the Yemen, they are most important agriculturally and economically in cooler and less humid climates. Analysis of national data on their production gives the following geoclimatic breakdown of the total 1989–91 cool season output of about 33 million tons.

Temperate climates: USSR, Western and Eastern Europe:	44%
Temperate climates: Canada and the United States:	2%
Temperate climates/subtropical: China:	13%

Winter rainfall subtropical (Mediterranean): Southern Europe, West
Asia, North Africa, Southern Cone South America, Mexico, Australia: 18%
Summer rainfall semi-arid subtropical/marginal tropical: South Asia: 20%
Tropical Sub-Saharan Africa/Central and South America: 3%

The distribution of the principal cool season species among these broad
geoclimatic groupings varies considerably, and in this respect an aggregate
analysis may conceal as much as it reveals. Peas are of predominant importance
in temperate regions, chickpeas, and lentils in the subtropics, and faba beans are
more evenly distributed across regions. In response to more specific ecological
niches, cultivars have evolved suited to fall, winter, and spring sowing with a
range of tolerance of cold, drought, and heat. Their existence has probably had
much to do with the development of dietary habits and consumer preferences
for particular food legumes, seed size, color, and other factors which need to be
taken into account in crop improvement policies and in trade. I shall be
referring to some of these issues in the next section of my paper, but will not be
able to deal with all local nuances in depth.

The Current Situation

The Distribution of Cool Season Legume Production

The current global distribution of the main leguminous species grown for food
and feed is shown in Table 1. It will be seen that, overall, the cool season legumes
occupy about 40% of total world pulse area, but contribute a substantially
larger share of world production, nearly 57%. The share of the developing
countries in their production is about 54%, whereas those countries represent
65% of the world area of cool season pulses.

The discrepancy between the developing countries share of total pulse area
and of production reflects the large differences between the average yields of the
various pulses in developing and developed countries. Overall, the developing
country yields of cool season pulses are less than half of those in the
industrialized countries; the difference is even larger when warm season and
tropical species are included, yields in the industrialized countries are then
almost three times those of the developing countries.

With respect to the cool season species, the largest gap is between the yields
of peas in the developed and developing economies, the latter being only 45% of
the former. For the other cool season pulses, the average yields for all
developing countries are somewhat closer to those in the developed countries,
being respectively 75% of the latter for faba bean, 72% for chickpea, 60% for
lentils, and 64% for lupins and vetches combined.

An important question which can be posed in this context is to what extent
these "yield" differences reflect improved technology in the industrialized
countries, opening up possibilities for closing the gap over time through
research and technology transfer in the developing economies; or to what extent

Table 1. Distribution of world production of major food and feed legumes (1989–91)

	Cool Season (CS) Food Legumes				C.S. Feed Legumes (Lupin + Vetch)	Total Cool Season Legumes[1]	Warm Season & Tropical Food Legumes				Total Warm Season Legumes	World Total All Species
	Faba Bean	Chickpea	:ntil	Pea			Bean[2]	Cowpea	Pigeon Pea	Other Species[3]		
Total world area (000 ha)	3,178	9,793	3,191	9,338	2,521	28,021	25,922	5,095	4,002	5,950	40,969	69,000
Share of world area (percent)	4.6	14.2	4.6	13.5	3.7	40.6	37.6	7.4	5.8	8.6	59.4	100.0
Total world production (000 MT)	4,241	7,068	2,263	16,937	2,710	33,219	15,990	2,201	3,015	3,917	25,123	58,342
Share of world production (percent)	7.3	12.1	3.9	29.0	4.6	56.9	27.4	3.8	5.2	6.7	43.1	100.0
Share of developing countries in world production (percent)	86.1	97.0	84.0	15.7	11.9	53.9	86.0	98.7	100.0	72.7	79.8	63.4
Developed-country yields (kg ha^{-1})	1.71	1.00	1.08	2.16	1.15	1.97	0.99	1.16	-	1.90	1.18	1.78
Developing-country yields (kg ha^{-1})	1.29	0.72	0.67	0.97	0.74	0.85	0.48	0.43	0.75	0.53	0.61	0.65

Source of data: Principally FAO Agrostat package and spreadsheets.

Notes:

[1] This does not include *Lathyrus sativus* for which reliable published data are not available.

[2] "Bean" is mainly composed of *Phaseolus vulgaris* which has a large ecological range, from temperate to tropical. However, other Phaseolus species, including *Phaseolus mungo* (black gram), *Phaseolus lunatos* (lima bean), etc. are sometimes reported under this heading in national statistics.

[3] Other species reported by FAO under the heading "NES" includes mainly tropical species, such as *Voandzeia subterranea* (Bambara groundnut) for which production is only reported for Malawi. It also includes a cool-season species *Lathyrus sativus* (chickling pea) which is discussed more fully in the text.

they are mainly indicative of better soils and climate – particularly more reliable and stable precipitation in the temperate latitudes where most developed countries are located. This might especially be the case with peas which are more widely grown in temperate environments and for which the yield differential is greatest (Table 2). I shall explore this issue further later in this paper.

With respect to total production of the cool season species most widely used for human food, over 80% of faba bean and lentil, and nearly all of the chickpea, compared to only 15% for peas, is grown in the developing countries. Most of their production is consumed *in situ* (which probably means that official statistics underestimate total supply), there are only about six significant developing-country exporters, and only two major ones – China for faba bean and Turkey for chickpea and lentil. There are no important exporters of peas from those countries, but Turkey exports considerable quantities of vetches. China and Thailand are significant faba bean exporters.

The share of the different species in total cool season legume production for each major region is summarized in Table 3 for 1980 and 1990. This shows that, for the world as a whole, the cool season pulses have increased their share of overall production by about 3.4% (probably mainly at the expense of *Phaseolus* beans), but that this has been due largely to the expansion of pea production in developed countries, since the share of faba beans and chickpeas has fallen by 6% and that of lentils has increased by only 1%.

Regionally, there has been a considerable expansion of pea production in the EEC countries, partly at the expense of chickpeas, lentils, and dry beans; and an increase in pea and lentil production in North America, at the expense of dry beans. In Oceania, the importance of faba bean and chickpea production has increased, possibly reflecting the increase in the share of people from countries with a Mediterranean climate in the total population of Australia. Among developing regions, the most important changes have been a noteworthy decrease of 12% in the share of dry beans within total legume production, and a 15% increase in the share of chickpea and lentils in the West Asia/North Africa region.

Production of cool season food legumes is quite highly concentrated in relatively few countries (Annex A). Over 90% of the world production of faba bean, chickpeas, lentils, and dry peas is located in only 12 countries, although those countries are not always the same for each of the four commodities.[1] Furthermore, 80% or more of the global production of each commodity originates in only six countries, with two-thirds or more in only *two* countries for faba bean, chickpea, and dry pea; and almost 60% for lentils (Table 4). The degree of concentration of vetch production is even higher, with 90% of global production in only 4 countries, while that of lupin is higher still, 94% in three

[1] The coverage of the top 12 countries for all four cool season food legumes embraces a total of 26 countries. The top six grouping involves 17 countries. China is among the top six for each commodity, India for three of them, Australia and Turkey for two, and each of the remaining 13 countries only once.

Table 2. Highest commercial average yield levels; cool-season pulses and wheat, 1978–81/1989–91 (kg ha^{-1})

Faba Bean Producer	1980	1990	Chickpea Producer	1980	1990	Lentil Producer	1980	1990	Pea Producer	1980	1990	Wheat Producer	1980	1990
France	3,063	3,755	Egypt	1,538	1,830	Egypt	2,701	2,152	France	4,284	4,972	Ireland	5,316	7,890
Germany	3,224	3,634	Australia	-	1,199	France	2,000	1,358	Netherlands	3,726	4,951	Netherlands	6,280	7,513
Egypt	2,134	2,705	Greece	1,064	1,157	Canada	822	1,356	Denmark	3,513	4,501	Zimbabwe	4,780	5,608
Sudan	1,820	1,775	Mexico	1,101	1,068	USA	1,073	1,135	Turkey	2,155	2,446	Egypt	3,192	4,967
Mexico	1,746	1,309	Turkey	1,149	952	Greece	1,313	1,080	USSR*	1,040	1,633	China**	2,047	3,063
China*	1,161	1,480	India*	627	707	India*	438	660	China**	1,583	1,295	USSR*	1,511	1,978
World Average	1,163	1,350	World Average	625	712	World Average	596	677	World Average	1,140	1,821	World Average	1,886	2,431

Source: FAO Production Tapes and Production Yearbook (1990), Vol. 44.

Notes:

* World's largest producing country.
** Largest developing-country producer.

Table 3. Share of cool season pulse species in total pulse production for each region, 1980 and 1990

Commodities/Regions	Year	Faba Bean	Dry Pea	Chickpea	Lentil	Total Cool-Season
World	1980	10.4	20.7	14.6	3.2	48.9
	1990	7.3	29.0	12.1	3.9	52.3
North America	1980	0.2	15.3	0.0	7.1	22.6
	1990	0.9	20.6	0.0	11.9	33.4
Oceania	1980	2.8	57.6	0.6	0.0	61.0
	1990	3.7	30.8	8.8	0.3	43.6
USSR	1980	0.0	84.5	0.0	0.1	84.6
	1990	0.0	85.1	0.0	0.5	85.6
EEC	1980	26.0	26.0	6.0	4.0	62.0
	1990	7.9	74.8	1.2	0.8	84.7
DME	1980	5.1	54.8	1.1	2.0	63.0
	1990	2.8	67.1	1.0	1.7	72.6
Africa Dev'g	1980	13.6	7.2	5.0	1.3	27.1
	1990	8.0	5.1	4.4	4.6	22.1
Latin America	1980	3.6	2.3	5.1	1.3	12.3
	1990	3.2	2.0	3.4	1.3	10.9
WANA	1980	18.9	2.9	20.7	18.4	60.9
	1990	16.7	2.1	29.6	24.3	72.7
Asia	1980	13.1	13.4	25.1	3.3	54.9
	1990	11.0	9.5	24.2	4.5	49.2
Dev'g Econ.	1980	12.1	10.2	18.8	3.6	44.7
	1990	9.9	7.2	18.5	5.1	40.7

countries. By comparison, the producer-country concentration of dry beans (*Phaseolus vulgaris*) which dominate warm-season pulse production is much lower, only 78% of the commodity total is in the top 12, and 67% in the top six countries. With the exception of chickpeas, the cool season legumes are more widely traded than the warm season species, a higher proportion of which (especially in the case of cowpea and pigeon pea) is consumed in the country of production (Table 5).

Trends in Production: 1980–1990 (Annex B)

Production of pulses in the world as a whole grew at a respectable rate of 3.5% annually over the decade of the 1980s, but growth was unevenly balanced between the developed and developing economies, with the former expanding faster at over 8% annually, more than doubling from 9.7 million tons in 1980 to 21.3 million tons in 1990. Both yield growth (4.5%) and area growth (3.6%) contributed to this major progress (Table 6).

By comparison, performance in the developing countries was disappointing: production expanding at 1.5%, well below the 2.2% growth of population, consequently per capita production declined (Table 7). Growth was also the result of a combination of area (0.85%) and yield growth (0.63%), but a much higher rate will be necessary to avoid an increasing gap between production and consumption, with damaging nutritional consequences to lower income

Table 4. Shares of total production of each legume represented by six largest producers, 1989–91 average (000 MT)

	Faba Bean		Chickpea		Lentil		Pea		Total Pulses	
	Producer	Production	Producer	Production	Producer	Production	Producer	Production	Producer	Production
	China	2,483	India	4,720	India	725	USSR	8,296	India	13,313
	Egypt	380	Turkey	802	Turkey	600	France	3,301	USSR	9,746
	Ethiopia	277	Pakistan	534	Canada	208	China	1,617	China	5,948
	Morocco	170	China	200	Bangladesh	157	Denmark	514	France	3,438
	Germany	124	Mexico	147	China	100	India	438	Brazil	2,173
	Italy	117	Australia	131	Nepal	79	Australia	407	United States	1,519
	Total 1st Six	3,551	Total 1st Six	6,534	Total 1st Six	1,869	Total 1st Six	14,573	Total of 1st Six	36,137
	% of World Total	84	% of World Total	92	% of World Total	83	% of World Total	64	% of World Total	62

Source: FAO production tapes (1992). See Annex A for greater detail.

Table 5. Imports and exports of legumes as a share of world pulse production and of each individual commodity's production (1988–90)

	Production (000 MT)	Imports (000 MT)	Share of Imports In			Exports (000 MT)	Share of Exports In	
			% World Pulse Production (%)	% Own Commodity Production (%)			% World Pulse Production (%)	% Own Commodity Production (%)
Dry beans	15,990	1,520	2.6	9.5		1,703	2.9	10.6
Other warm season pulses[1]	9,133	425	0.7	10.8		367	0.6	9.7
All Warm Season Food Legumes	25,123	1,945	3.3	7.7		2,070	3.5	8.2
Faba bean	4,241	599	1.0	14.1		493	0.8	11.6
Chickpea	7,068	436	0.7	6.2		501	0.9	7.1
Lentil	2,263	664	1.2	29.3		619	1.1	27.4
Dry pea	16,937	2,294	3.9	13.5		2,344	4.0	13.8
All Cool Season Food Legumes	30,509	3,993	6.8	13.1		3,957	6.8	13.0
Lupins/Vetches2	2,710	1	NS	0.1		469	0.8	40.7
World Total All Pulses	58,342	5,939	10.2	-		6,496	11.1	-

Sources of Data: FAO data sheets on production and trade.

Notes:
[1] Other species include cowpea, pigeon pea, bambara nut, lathyrus, etc.
[2] Imports of both species are insignificant; exports represent only lupins, vetches insignificant.

Table 6. Production data and growth rates for cool season and total pulses, 1979–81 to 1989–91

	World			Developed Countries			Developing Countries		
	Area	Yield	Production	Area	Yield	Production	Area	Yield	Production
	(000 ha)	(MT ha⁻¹)	(000 MT)	(000 ha)	(MT ha⁻¹)	(000 MT)	(000 ha)	(MT ha⁻¹)	(000 MT)
Faba beans, 1989-91	3,178	1.34	4,241	343	1.71	588	2,834	1.29	3,653
Growth rate, 1980-91: % per annum	-1.29	1.35	0.06	0.19	3.93	3.70	-1.48	0.89	-0.57
Chickpea, 1989-91	9,792	0.72	7,068	210	1.00	210	9,583	0.72	6,858
Growth rate, 1980-91: % per annum	-0.01	1.79	1.79	3.14	8.36	5.26	-0.06	1.71	1.66
Lentil, 1989-91	3,190	0.71	2,263	331	1.08	361	2,859	0.67	1,902
Growth rate, 1980-91: % per annum	4.01	6.36	2.35	5.45	7.40	1.79	3.85	2.37	6.20
Pea, 1989-91	9,338	1.82	16,937	6,611	2.16	14,283	2,728	0.97	2,653
Growth rate, 1980-91: % per annum	2.77	7.02	4.22	4.29	10.01	5.65	-0.43	-2.11	-2.53
Total World, 1989-91	68,999	0.85	58,342	12,112	1.76	21,274	56,887	0.65	37,068
Growth rate, 1980-91: % per annum	1.31	3.52	2.21	3.64	8.16	4.51	0.85	0.63	1.48

Source: FAO production tapes (1992). See Annex B for detailed regional distribution and 1980 data.

Table 7. Per capita production of pulses, 1980 and 1990 (Kg)

Regions	Year	Dry Bean	Faba Bean	Dry Pea	Chickpea	Lentil	Total Pulses
World	1980	3.?5	0.97	1.91	1.36	0.30	9.26
	1990	3.02	0.80	3.20	1.34	0.43	11.02
North America	1980	5.07	0.02	1.00	0.00	0.46	6.52
	1990	5.12	0.07	1.59	0.00	0.92	7.74
Oceania	1980	0.16	0.32	6.57	0.07	0.00	11.40
	1990	0.27	2.7	22.45	6.42	0.24	72.90
USSR	1980	0.24	0.00	16.09	0.00	0.02	19.04
	1990	0.38	0.00	28.75	0.00	0.17	33.77
EEC	1980	0.88	1.24	1.24	0.29	0.19	4.77
	1990	0.19	1.31	12.39	0.20	0.14	16.57
DME	1980	1.93	0.42	4.54	0.09	0.16	8.29
	1990	1.75	0.47	11.44	0.17	0.29	17.04
Africa Dev'g	1980	4.18	1.40	0.92	0.51	0.13	12.89
	1990	3.68	0.59	0.55	0.37	0.53	10.85
Latin America	1980	11.21	0.48	0.31	0.67	0.17	13.08
	1990	9.88	0.36	0.23	0.38	0.14	11.23
WANA*	1980	1.22	2.01	0.31	1.71	1.36	8.52
	1990	1.11	2.17	0.36	3.00	2.31	12.46
Asia	1980	2.38	1.14	1.17	2.19	0.29	8.72
	1990	2.56	0.90	0.78	1.97	0.37	8.16
Dev'g Econ.	1980	3.45	1.16	0.98	1.81	0.34	9.61
	1990	2.97	0.90	0.66	1.70	0.47	9.16

*Includes Algeria, Morocco, and Tunisia which are part of Africa Developing in FAO regional aggregation.

families. In the WANA region, for example, 90% of the increase in production originated in one country, Turkey, primarily for export.

Looking at the individual commodities in Annex B, it is apparent that the majority of the overall growth was in the cool season food legumes. The annual growth rate of *dry beans* (*Phaseolus sp.*), which dominate warm season pulse production, was only 0.56% of a total pulse growth rate of 3.52%. The pattern of growth for the cool season species is discussed below:

Faba bean production worldwide showed virtually no growth over the decade, despite a considerable expansion of area and yield in Australia leading to an increase in output of 50,000 tons; a notable growth of yield in the EEC, where production also increased by almost 50,000 tons despite a decrease in area; and strong growth of area and yield in West Asia and North Africa (WANA). Production there increased by 219,000 tons.

These increases were more than offset by a 25% decrease of 566,000 hectares in faba bean area in China, resulting in a decline of 133,000 tons in production despite an improvement in yield; and a large decrease in yields in Africa, where production fell by 201,000 tons in consequence. As a result, production of the developing regions as a whole fell by 145,000 tons.

Chickpea production worldwide has increased modestly by about a million tons over the decade at 1.8% annually, due both to an apparent improvement in yields, especially in the developed market economies and to area expansion. The area under crop declined in the EEC, but this was offset by a large increase in Australia, resulting in a net gain for the Developed Market Economies (DME). This combined with a 5.2% growth in yields raised production in developed countries by 107,000 tons.

Both area and yield rose in the developing countries as a whole; there was a particularly large area increase of 670,000 hectares in West Asia and North Africa, mainly but not entirely due to the fallow replacement program in Turkey, since the area also expanded in Algeria, Egypt, Morocco, Sudan, Tunisia, Iran, and Syria. Yields in that region failed to increase, but yield increases in the Far East resulted in an overall positive annual yield growth of 1.71%, and an expansion of 933,000 tons in production of the developing regions as a whole.

Lentil

Compared to faba bean, dry bean, and chickpea, lentil production gave a stellar performance, with an overall annual growth rate of 4% in area and 2.35 in yield resulting in an increase of 951,000 tons over the decade, equivalent to 72% of the *total* 1980 production of that commodity. This was the result of area and yield growth in both the developed and developing countries, leading to an increase in production of 172,000 tons in the former group and 780,000 tons in the latter. North America (+138,000 tons) and the USSR (+44,000 tons) were the main contributors to the increases in the DME group, but there was a slight decline of 17,000 tons in EEC production due to a decrease of 25,000 hectares in the area devoted to lentils.

In the developing countries, lentil production in Africa increased by 203,000 tons from a 1980 level of 45,000 tons to 248,000 tons, due mainly to a fivefold yield increase; while production in the WANA region rose by 431,000 tons due to a large area increase and despite a 25% fall in average yield. Production also rose by 4,000 tons in Latin America and 357,000 tons in Asia due mainly to yield improvement.

Dry Peas

There has been a rapid growth of production (10% annually) in the developed economies, due both to area and yield increases, with a decline in production of 2.5% annually in the developing countries. Again, WANA is the developing region with a significant growth of production (4.1% per annum). This was entirely due to increased area (which doubled), whereas yields fell by 18% over the decade. Production in China (the largest developing-country producer) fell by 527,000 tons, a fifth of total production for all of Asia. Nevertheless, because the share of the developed countries in pea production is so much higher than for the other cool season legumes, the overall growth of pea production has been strongly positive (7.0% per annum), with an increase over the decade of 8.4 million tons. This was composed of almost 9 million tons in the developed economies, partly offset by a decrease of 551,000 tons in production of the developing regions. The overall result has been to double total world dry pea output over 10 years, and to increase that of the developed countries by two-and-a-half times.

In the case of the latter, the total area devoted to dry peas rose by 42% and yields by 86%. Areas almost doubled in the United States, increased five fold in Oceania, and more than seven fold in the EEC. In the USSR (the world's largest producer), the increase was only 13% over the decade, but this represented an additional 532,000 hectares. The corresponding increases for North America, Oceania, and the EEC were respectively 114,000 ha, 316,000 ha, and 793,000 ha. The only area increases in the developing regions were 86,000 ha in WANA and 19,000 ha in Africa. However, in those regions and for the developing economies as a whole, area growth was more than offset by a sharp decline of 2.1% per annum in yields, whereas the increase in yields of peas in Western Europe and the USSR made a major contribution to the expansion of production in the developed countries.

Trends and Patterns of Consumption (Annex C)

Domestic utilization of pulses is defined here as production, plus imports and opening stocks, minus closing stocks and exports. It is normally broken down into food use, feed use, seed use, waste, and "other uses" – such as processing. "Consumption" can be equated with utilization as defined above. Current patterns of consumption and trends over the 1980–89 period are displayed in

Annex C, for total pulses, peas, beans, and other pulses.

Unfortunately, consumption is not nearly as well-documented as production, the FAO data only disaggregates dry beans, and dry peas from total pulses; faba beans, chickpeas and lentils are amalgamated with cowpeas, pigeon peas, and pulses NES in an umbrella category of "other pulses" and it was not possible to identify them separately. However, since the three cool season pulses represent 60% of world production and 80% of world trade in pulses (minus peas and beans), it is not unreasonable to assume that the "other pulse" category of consumption will be a fair approximation of the consumption patterns of the unidentified cool season pulses, at least at the regional level. Lupins and vetches are not included with the "other pulses" in the FAO accounts.

At the world level, the pattern of consumption for all pulses over the last decade is clear. Both food use and feed use show an upward trend, but the growth rate of the former is much slower, only 1.34% per annum, compared to 9.73% per annum for feed. Between 1979–81 and 1988–90, feed use rose from 18.4% of total domestic utilization to 30.8%, while food use decreased correspondingly from

Table 8. Changes in total pulse consumption by category and region, 1980 and 1989 (000 MT)

Region	Year	Food	% Total Use	Feed	% Total Use	Seed	% Total Use	Other Uses	% Total Use	Total Use
World	1980	28,362	68.7	7,600	18.4	3,387	8.2	1,957	4.7	41,307
	1989	31,891	57.2	17,189	30.8	4,001	7.2	2,675	4.8	55,755
North America	1980	703	85.9	30	3.6	83	10.1	3	0.4	819
	1989	669	79.3	59	6.9	110	13.0	7	0.8	844
Oceania	1980	25	18.6	81	60.8	21	15.6	7	5.0	133
	1989	37	5.3	506	73.2	114	16.4	36	5.1	692
USSR	1980	790	14.9	3,607	68.2	621	11.7	270	5.1	5,287
	1989	523	5.7	7,415	81.4	711	7.8	458	5.0	9,106
EEC	1980	973	38.5	916	36.2	149	5.9	492	19.4	2,529
	1989	1,172	16.5	5,102	72.0	285	4.0	532	7.5	7,090
DME	1980	3,351	34.3	5,035	51.5	980	10.0	403	4.1	9,769
	1989	3,319	17.0	14,126	72.2	1,386	7.1	732	3.7	19,562
Africa	1980	3,773	78.3	111	2.3	400	8.3	536	11.1	4,821
(Developing)	1989	4,855	78.0	152	2.4	473	7.6	742	11.9	6,222
Latin America	1980	4,136	86.1	73	1.5	385	8.0	211	4.4	4,805
	1989	4,519	87.9	33	0.5	371	7.2	218	4.2	5,140
WANA	1980	1,347	75.4	183	10.3	170	9.5	86	4.8	1,786
	1989	2,033	68.1	456	15.3	347	11.6	148	4.9	2,983
Asia	1980	15,748	78.3	2,198	10.9	1,452	7.2	720	3.6	20,117
	1989	17,154	78.6	2,423	11.1	1,424	6.5	837	3.8	21,838
Dev.g Mkt.	1980	25,011	79.3	2,565	8.1	2,408	7.6	1,553	4.9	31,537
Econ.	1989	28,572	78.9	3,064	8.5	2,615	7.2	1,944	5.4	36,194

68.7 to 57.2%. Seed use, represented 8.2% of total use in 1980, but declined by 1 to 7.2% in 1989, whereas the proportion of waste remained static (Table 8).

For the developed economies as a whole, consumption for food was almost static, whereas that in the developing world grew positively but slowly, at just under 1.5% annually, well below the level needed to match population growth. Growth of feed use exceeded 12% per annum in the industrialized countries, increasing threefold between 1980 and 1989 from approximately 5 million to 14 million tons. In 1980, the volume of feed use in those countries was only 51% of total consumption (and 1.5 times that of food use); by 1989, it was 72% of total consumption and over four times the volume of food use. The contrast with the developing economies is striking: in 1980, feed use was only 10% of the volume of food use and 8.2% of total consumption, but although feed use grew slightly faster than food use over the period it still barely reached 8.5% of total consumption by 1989.

However, there are considerable regional differences in the proportionate distribution of total pulse consumption in both the developed and developing economies. In the former, food use expanded at 4.22% per annum in Oceania and 2.9% in the EEC countries, while feed use rocketed at 21% per annum in both regions. Conversely, food use declined in North America and the USSR, while feed use – although upward – grew more slowly in those regions than in the EEC, at 8.8 and 8.3% per annum, respectively.

In the developing regions, food consumption expanded quite rapidly in Africa and in the Near East (at 2.9 and 5.7% per annum, respectively), but very slowly at under 1% per annum in Latin America and Asia/Far East. As the latter region includes India, Pakistan, Bangladesh, and China (a huge consumer of pulses for food), representing 54% of total world food use for 1988–90, slow growth there is a cause for serious concern. Although the satisfactory increase in consumption in the WANA region, which is a considerable importer, appears to be a matter for congratulation, the region represents only 6.4% of global pulse consumption for food; the smallest share of any developing region.

In fact, the "improvement" in the Near East is essentially a reflection of the outstandingly successful Turkish campaign to expand food and feed legume production in Anatolia, which masks increasing legume deficits in several other countries of the WANA region. A detailed analysis of average imports and exports among 23 countries of the region over the 1980–87 period shows that when Turkish exports of 470,000 tons of food legumes are deducted from the region's total exports, those of all other WANA countries total only 59,000 tons. Only two other countries (Syria and Morocco) were net exporters, and total net *imports* by the 22 countries other than Turkey were 381,000 tons (Oram and Belaid, 1990). This is a good example of the pitfalls of aggregate analysis at the regional level, but unfortunately time does not permit more detailed disaggregation!

Among developing regions, *feed use* expanded most rapidly (10.5% per annum) in the Near East, followed by Africa (6.4%). It fell by more than half in Latin America where feed use is small, and rose by under 1% in the Far East where total feed use of 2.4 million tons represents almost 80% of that in the developing regions and 18% of world feed consumption from legumes.

The major deficit region overall is the EEC, due principally to the rapid expansion in its requirements of legumes for animal feed, although it has also increased food use to a considerable extent over the decade. Among developing regions, Asia just about balances production and consumption during the 1980s, but the gap widens over the period in Africa and Latin America, with production falling behind utilization.

Pea consumption shows a similar pattern to that of total pulses, but with a much greater emphasis on feed use, which increases from 45% of total utilization in 1980 to 71.4% in 1989, with a corresponding decrease in food use from 41.6 to 17.5% of total utilization over the same period. This trend is evident in both the developed and the developing economies, and is in fact more marked in the latter than is the case with total consumption of pulses. Thus, by 1989, approximately 81% of pea utilization was for feed in the DME countries, and 22% in the developing regions compared to 72% in the former and only 8.5% in the latter in the case of total pulses. This suggests that to describe dry peas as a "food" legume may soon be a misnomer. However, it should be noted that the decline in pea use for food in the developing countries was not just the result of an increase in feed use, but even more seems to be due to a larger decrease in food use in Asia, partially independent of a shift to feed. Thus, food use declined by 834,000 tons, whereas feed use only rose by 62,000 tons. This seems to reflect mainly a considerable fall in production in China during the decade.

"Other pulses", as defined above, including faba beans, chickpeas, and lentils, show a pattern of increased use for feed similar to that for peas and total pulses, but on a more limited scale. Both food and feed use show significant growth over the decade (2.25% and 6.25% per annum, respectively over the 1980–89 period), but with the rapid growth of feed use being confined largely to the developed countries, and with the share of food use in total utilization being very much greater. Food use overall declines by only 5 to 65% of the total, with an equivalent rise from 17.3 to 22.6% in feed use: a much smaller shift in utilization than with peas or total pulses. In the developing regions, the increase in food use is greater than that of feed use over the decade for pulses other than peas.

Because of the failure of production to keep pace with population growth, due to a variety of factors which I will discuss later with reference to future outlook, and because of balance of payment problems restricting imports, per capita consumption is declining in many developing countries, including several of the major Asian countries and many in the WANA region. For example, reports from India (Jeswani and Baldev, 1990) and Bangladesh (Sarker *et al.*, 1991) show that the area, yield, and production of key pulses, such as lentil and chickpea, is declining despite long-standing research efforts to increase productivity, consequently the overall net daily per caput availability of pulses in India, the world's largest producer and consumer of pulses has fallen progressively from 64.6 grams during the late 1950s to 48.6 grams in the late

1960s, and 39 grams in the first half of the 1980s, and is continuing to decline. The minimum per caput daily requirement is considered in India to be 50–60 grams, and the optimum is 104 grams.

World Trade in Pulses

Recent trends in world and regional trade in pulses are summarized in Tables 9 and 10. Peas are clearly the predominant single commodity in world trade; both exports and imports exceeding those of the other three cool season food legumes combined, as well as being about 40% above those of beans – the second most traded legume. Overall, both in terms of exports and imports, peas represent nearly 40% of total pulse trade; even when lupins are vectored into imports and exports, peas still comprise 36% of world trade.[2]

Table 9. World imports of cool season pulses, 1984/85 and 1989/90 (000 MT)

Commodities/ Regions	Dry Faba Beans		Chickpea		Lentil		Dry Pea		Total C.S. Pulses	
	1984/85	1989/90	1984/85	1989/90	1984/85	1989/90	1984/85	1989/90	1984/85	1989/90
World	242,185	668,341	166,045	412,716	345,746	640,868	1,115,496	2,311,213	1,869,472	4,033,138
North America	-	1,327	10,819	15,467	1,093	8,112	13,746	26,907	25,652	51,813
Oceania	-	4	-	387	1,416	8,112	5,531	3,766	6,947	12,269
USSR	-	-	-	-	-	-	34,425	3,000	34,425	3,000
EEC	198,995	607,973	29,879	97,907	108,985	129,193	878,091	1,820,992	1,215,950	2,655,995
Dev'd Mkt. Econ.	211,994	637,252	40,698	115,717	119,575	146,051	970,545	1,912,507	1,342,899	2,811,527
% Total of Imports	88	95	25	28	35	23	87	83		
Africa	1,129	6,617	33,980	52,133	14,743	47,899	16,227	12,690	66,079	119,339
Latin America	-	-	10,120	9,493	31,602	56,638	75,801	88,843	117,523	154,974
WANA	28,933	20,060	56,741	58,201	91,080	126,954	13,713	24,136	190,467	229,351
Asia	7	8,919	24,506	177,172	88,747	263,326	39,212	273,038	152,472	722,455
Dev'g Mkt.	30,192	31,089	125,346	296,999	226,171	494,817	144,951	398,706	526,660	1,221,611
% of Total Imports	12	5	75	72	65	77	13	17		

Source: FAO trade tapes (1992).

Among developed regions, the EC is clearly the main importer; production there only represents 79% of domestic utilization. Nevertheless, although the 12 EEC countries as a whole have a *net* trade deficit in total pulses of 1.3 million tons, the average 1988–90 exports of individual EEC countries totalled about 1.73 million tons. Both imports and exports of European countries as a whole have risen rapidly during the 1980s, total cool season pulse imports having

---

[2] Lupins are incompletely reported in the FAO trade statistics, since 469,000 tons are shown as being exported from Australia, but no imports are recorded. It is apparent from Australian and Canadian data (ABARE 1991, Agriculture Canada 1989) that the majority of those exports go to EEC countries. Trade in vetches is shown to be almost zero by FAO and no other information could be obtained on this commodity.

Table 10. World exports of cool seasons pulses, 1984/85 and 1989/90 (000 MT)

Commodities/ Regions	Dry Faba Beans		Chickpea		Lentil		Dry Pea		Total C.S. Pulses	
	1984/85	1989/90	1984/85	1989/90	1984/85	1989/90	1984/85	1989/90	1984/85	1989/90
World	225,996	481,164	253,588	424,607	367,545	492,865	1,246,805	2,426,546	2,093,934	3,825,182
North America	7,341	6,002	-	3,710	71,451	157,182	230,505	325,488	309,297	513,186
Oceania	10,060	26,806	-	91,979	-	6,676	77,530	290,615	87,590	416,076
USSR	-	-	-	-	-	-	62,442	55,857	62,442	55,857
EEC	112,810	200,325	1,500	7,812	26,542	14,508	793,371	1,519,124	934,223	1,741,769
Dev'd Mkt. Econ.	130,211	233,519	1,500	104,427	97,995	183,844	1,231,828	2,404,645	1,461,534	2,946,435
% of Total Exports	58	49	1	25	27	37	99	99	70	71
Africa	18,614	17,807	3,734	8,827	4,857	7,798	508	1,394	27,713	35,836
Latin America	-	-	47,188	58,953	7,705	4,662	4,812	2,281	59,705	65,936
WANA	26,201	40,589	197,956	246,527	230,140	281,577	49	157	454,346	568,860
Asia	50,971	189,121	3,211	5,814	26,850	14,986	9,610	18,071	90,642	228,992
Dev'g Mkt. Econ.	95,786	247,645	252,089	320,180	269,551	309,021	14,977	21,901	632,400	898,747
% of Total Exports	42	51	99	75	73	63	1	1	30	23

Source: FAO trade tapes (1992).

doubled and exports risen by 86% between 1985 and 1990. Probably, a majority of those exports represent internal trade within the European Community; however, exports to Europe from Australia (lupins), Turkey and Canada (lentils), peas (Canada, Australia, USA), and chickpeas (Australia, Turkey) are important and increasing.

Australia, Canada, the United States, Argentina, China, Thailand, Turkey, Hungary, and Poland are the major exporting countries outside the EEC, and represent almost 90% of non-EC exports.

The world pulse market is quite thin, however, and exports represent only about 11% of total pulse production compared to 18% in the case of wheat. Lupins are the most widely traded commodity, with exports representing about 41% of production, followed by lentils 28%, and peas 13.8%.[3] Chickpea trade only represents 6–7% of its production, and the proportion is even lower for warm-season pulses other than dry beans (Table 5).

Consequently, export availabilities are very sensitive to price changes affecting major producers, especially in developed countries, and to year-to-year weather variability, in both developing and developed countries. Table 11 shows coefficients of variation in area, production, and yield of cool season pulses in major producing countries, and it will be seen that while yield variability tends to be higher in the developing countries, inter-annual variation in area is often greater in the developed countries.

Thus, Canada, France, the United States, and Turkey tend to rank among the most variable countries in terms of area fluctuations, but to have more stable

---
[3] Egypt, Turkey, Algeria, Morocco, Syria, Costa Rica, India, and China.

Table 11. Coefficients of variation of pulses and wheat in selected producing countries

Countries	Faba Bean			Dry Pea			Chickpea			Lentil			Wheat		
	Area	Prod.	Yield	Area	Prod.	Yield	Area	Prod.	Yield	Area	Prod.	Yield	Area	Prod.	Yield
Morocco	0.21	0.38	0.30	0.20	0.46	0.33	0.21	0.37	0.30	0.32	0.54	0.40	0.18	0.40	0.28
Tunisia	0.27	0.31	0.26	-	-	-	0.32	0.23	0.25	0.62	0.49	0.22	0.30	0.46	0.26
Syria	0.20	0.25	0.15	0.40	0.39	0.19	0.30	0.42	0.20	0.44	0.50	0.29	0.13	0.23	0.19
Turkey	0.13	0.14	0.03	0.24	0.27	0.10	0.51	0.47	0.11	0.39	0.41	0.18	0.02	0.08	0.07
Pakistan	-	-	-	0.09	0.07	0.03	0.10	0.24	0.21	0.18	0.13	0.12	0.04	0.11	0.08
India	-	-	-	0.05	0.18	0.16	0.09	0.15	0.11	0.08	0.22	0.15	0.03	0.15	0.12
China	0.12	0.11	0.11	0.04	0.19	0.16	-	1.30	-	-	1.29	-	0.02	0.17	0.15
Ethiopia	0.10	0.35	0.28	0.14	0.30	0.18	0.15	0.15	0.17	0.18	0.37	0.27	0.09	0.13	0.10
Mexico	0.12	0.38	0.31	0.29	0.51	0.24	0.04	0.10	0.10	0.27	0.30	0.08	0.16	0.18	0.06
Canada	0.64	0.61	0.12	0.66	0.53	0.16	-	-	-	0.59	0.84	0.36	0.07	0.19	0.16
USA	-	-	-	0.14	0.18	0.19	-	-	-	0.40	0.41	0.18	0.13	0.14	0.07
France	0.39	0.35	0.10	0.79	0.84	0.09	-	-	-	0.22	1.66	1.31	0.03	0.13	0.10
USSR	-	-	-	0.11	0.20	0.19	-	-	-	0.58	0.81	0.37	0.10	0.13	0.16
Australia	0.55	0.97	0.66	0.76	0.79	0.47	1.21	1.23	0.16	-	-	-	0.13	0.24	0.20

Source: International Food Policy Research Institute (IFPRI)

yields; while the Near East and North African countries and Australia, where legumes are often grown in a drought-prone environment tend to have more variable yields.

This is clear also for Pakistan, in the case of chickpea, where that crop is mainly grown under extremely stressful conditions (both drought and disease) in the Thar desert. However, chickpea yields are less variable in Turkey, which has a more stable climate despite its colder winters. Area fluctuations are relatively small in developing countries with a strong internal demand for pulses and goals of foodgrain self-sufficiency such as China, India, Pakistan, Mexico, and Ethiopia, but climatic and disease stresses render yields more variable.

It is somewhat surprising to note, looking across countries and commodities in Table 11, that area variability is frequently greater than that of yield for each of the four cool season legumes, but that in the case of wheat, area variability is lower than that of yield, although both sources of variability are lower for wheat than for any of the legumes. In the case of area, this may be due to the fact that wheat prices are subsidized or guaranteed in virtually every country, whereas in many countries legume prices are not supported by governments. Changes in relative prices of cereals, legumes, and other sources of high protein feed, and in national policies concerning subsidies, are probably the major factors influencing shifts in the *area* of pulses, especially in the major exporting countries. Price instability is probably an important disincentive to farmers in developing countries expanding their pulse production.

Lower *yield* variability of wheat compared to pulses may be due to two factors. These are *first* the much greater and much more successful research effort devoted worldwide to wheat; and *second* the fact that wheat is often grown under more favorable conditions than pulses, especially in developing countries (such as China, India, Pakistan, Syria, and Mexico) where a significant and increasing proportion of the wheat crop is irrigated. Very little of the pulse area in those countries receives irrigation, nor do pulses generally receive as much care as wheat with respect to fertilizer and weed control. Thus, the often reported fact that pulses are perceived by farmers to be more risky than cereals may to some extent be a self-fulfilling prophecy!

Future Outlook

Lessons of Recent Experience

No projections of future supply and demand for pulses, as a whole or for individual cool season species, have been undertaken since a 1986 FAO analysis (FAO, 1986). FAO has a new effort underway (Greenfield, personal communication), but it will not be completed in time for this conference, although FAO's Commodity Division has been most helpful in providing relevant current information (FAO commodity tapes, FAO Commodity Review, 1989). However, a number of forward-looking efforts have been made to assess prospects on a regional or national basis. Somel (1988) has undertaken a projections study for commodities of the WANA region from 1990 to the year 2000. The Australian Bureau of Agricultural and Resource Economics (ABARE, 1991) has also undertaken some projections of statistics on production, exports, and prices of Australian lupins and field peas (including chickpeas and cowpeas) to 1996/97. These various sources as well as reports on USA (1991), EEC, and Canadian (Longmuir, 1991) output have been taken into account in the discussion which follows. A key question which will be addressed is whether the constraint (if any) lies with demand or supply.

As a starting point, it may be illuminating to examine the conclusions of the FAO study referred to above, in the light of what has happened over the period 1980–1990 actually covered by their projections. Comparing the 1970s with the 1980s, the report states:

Over the 1970s, world pulse production fell, mainly due to better returns from other crops such as cereals and soybeans, the lack of new improved varieties, the absence of support measures in many countries, and bad weather in a number of major producing countries at the end of the decade. The fall in production led to rising prices in many countries and reduced demand, especially by the poorer sections of the population in developing countries. In addition, the higher incomes of some groups and the increased availabilities and relatively lower prices of livestock products in some countries, resulted in a shift in consumption to animal proteins.

In the early 1980s, however, these trends have been reversed. Production has risen sharply following increased support to pulse production in a number of countries, while demand for both food and feed has expanded.

The increased emphasis on raising pulse production stems from a variety of considerations, including shortages and high prices of pulses; the achievement of self-sufficiency and emergence of surplus production of cereals and low prices for these crops on export markets; the opening up of new export markets for some pulses (e.g., mung beans from Thailand; and lentils and chickpeas from Turkey); and the high prices of fertilizer, which stimulated interest in pulses as they return nitrogen to the soil. In addition, research into new varieties has been stepped up in a number of national and international centers, which along with improved extension services is expected to raise yields. These factors are expected to lead to a substantial increase in output by 1990, assuming normal weather conditions.

As a result, FAO made the following assumptions for production, consumption, and trade of pulses through 1990:

Output of pulses during the 1980s is projected to grow at 2.5% per annum in developing countries, due to an increase in both area and yields. This is in contrast to the previous decade when production expanded at only 0.2% annually. In developed countries, a substantial turnaround in production is anticipated from a fall of 2.9% a year in the 1970s to an increase of 3.7% a year in the 1980s. Overall, an annual growth of 2.8% in world production is expected, from 41 million tons in 1979–81 to 54 million tons in 1990.

In the event, these projections of *output* proved both too optimistic and too conservative. They were an overestimate with respect to the developing countries, where output increased at only 1.5 annually over the 1980s; both area and yield did increase, but at a much slower rate than expected. On the other hand, growth of production in the developed countries was significantly higher than projected, in fact more than double: output expanding at 8.16% annually due to both area and yield increases.

The overall result was a world growth of 3.5% per annum, with area contributing 1.3% and yield 2.2%. Production for 1989–91 exceeded 58 million tons, from 41 million tons in 1979–81; area rose from 61 to 69 million hectares, and yield from 0.67 to 0.85 tons ha^{-1}.

Demand for pulses was expected by FAO to grow by 2.7% annually, slightly slower than output, to reach 54 million tons by 1990, consumption exactly balancing production. Consumption data are not yet available for 1991, but the 1988–90 average was 55.7 million tons against a production of 56.6 million, indicating a slight global surplus of supply.

The FAO projected an increase of 2.1% per annum in demand for food, and of 5.3% in demand for feed through the 1980s; whereas in fact the respective growth rates of food use and feed use up to 1989 were 1.3% and 9.7%, respectively. Food use rose at 1.5% per annum in the developing countries, and showed no growth in the DME group; whereas feed use expanded at 2.0% annually in the former, and 12.2% in the latter. Thus, the growth of feed use exceeded expectations, whereas that of food consumption fell below them. The relative share of food in total use fell in all developed regions and in WANA; remaining practically constant in Africa, Latin America, and Asia. Feed use increased as a proportion of total use in all developed and developing regions, except Latin America where it declined.

In absolute terms, food use increased over the 1980s in Oceania and the EEC, possibly reflecting an influx of immigrants of Mediterranean origin; but it declined in North America and the USSR. Feed use increased in absolute terms in all regions, except Latin America. Seed use expanded quite fast in developed regions, especially in Australia, and the EEC – probably reflecting increasing seed exports.

Per capita consumption was projected to rise in the developing countries from 7.7 kg per capita to 7.9 kg per capita by 1990, but it is not clear whether that refers to food or to total use. Assuming that it refers only to food, our calculations indicate a figure of 7.6 kg per capita in 1980, falling to 7.2 kg in 1989. For the developed market economies, the levels were 2.9 kg per capita in 1980 and 2.7 kg per capita in 1990.

Equating total consumption with food, feed, seed, and other uses gives a figure for the developing countries of 9.6 kg per capita for 1980 and 9.1 kg for 1989. For the DME countries, the figures are 8.0 kg per capita and 16.4 kg per capita, respectively. A massive increase in DME utilization is almost entirely due to an expansion in the use of cool season pulses for feed; there was a small expansion of consumption for seed, but food use remains static in per capita terms.

FAO based its assumptions that per capita consumption would rise in the developing countries on expectations of a fall in relative prices of pulses. It is not clear that this has happened. Pulse prices seem to have risen in the early 1980s, then declined somewhat, followed by an increase towards the end of the decade. Relative to cereals, they have generally increased somewhat, partly because of weak international cereal prices, but not always enough to compensate for their lower yields and higher production costs. Recent data from developed-country exporting countries suggests a short-term outlook of reasonably remunerative prices relative to wheat, corn, and oilseeds; partly because of expectations of reduced EEC pulse production, and partly because more pulses are being used by their own domestic feed industries.

As far as the developed countries are concerned, the hoped for stabilization of food legume consumption due to a move towards "healthier" foods is proving slow to materialize, despite increasing public concern about the prevalence of high dietary cholesterol and fat, and wider recognition of the need for fiber in the diet. However, in some areas of the United States where strong propaganda campaigns have been launched via the media (USA Dry Pea and Lentil Council, 1990), and also in Turkey (Guler, personal communication), significant increases in pulse consumption have resulted.

Trade. FAO projected a 3% worldwide growth in pulse trade to reach 3.7 million tons by 1990. In fact, imports grew at 9.6% per annum, reaching almost 6 million tons by 1989 (1988–90 average); while exports rose at about 10.5% per annum to reach nearly 6.5 million tons (Table 5). Fifty-eight% of the imports were destined for developed countries, and 65% of the exports originated in those countries. Net imports by EEC countries averaged 1.3 million tons for 1988–90. Imports by developing countries grew at 8.7% per annum, nearly double the FAO projection, almost as fast as those of developed countries in percentage terms, although for a smaller total volume and with a larger volume for food than feed.

In fact, as Table 9 shows, imports by developing countries table of chickpeas and lentils, used primarily for food, rocketed up between 1985 and 1990 – growing at over 16% annually, while pea imports (mainly for feed) rose even faster. Imports of faba beans and chickpeas also rose rapidly in DME countries between 1985 and 1990 at about 20% per annum, while imports of dry peas rose a little more slowly at about 15% per annum. Net imports of developing countries rose by 100,000 tons over the decade, and those of developed countries by 660,000 tons. Exports of developing countries also exceeded the FAO expectations, averaging 2.3 million tons in 1988–90, compared to a projected 1.9 million tons.

It is important to note that both imports and exports increased much more rapidly between 1985 and 1990, than in the first half of the 1980s – growth rates for both developed and developing countries being nearly double in the later years (Table 12). This appears to be a response to a slowdown in production in the latter part of the decade, the reasons for which are not entirely clear.

The primary factors driving up imports are probably different in developing

Table 12. Trends in pulse production and imports, 1980–1990

Characterization	Total Production (000 MT)			Growth Rates of Production		
	1979-81	1984-86	1989-91	1980-85	1985-90	1980-90
				(% per annum)		
All Pulses						
Developed	9,685	16,128	21,274	10.7	5.6	8.3
Developing	31,506	35,050	37,068	2.2	1.1	1.6
Cool Season Pulses						
Developed	6,090	10,757	15,442	12.0	7.5	9.8
Developing	14,051	14,866	15,066	1.2	0.3	0.7
	Imports (000MT)			Growth Rates of Imports		
All Pulses	2,553	3,720	5,922	7.8	9.8	8.7
Cool Season Pulses	n.a.	1,870	4,033	n.a.	15.9	n.a.

and developed regions, with food demand being preeminent in the former and feed demand in the latter; but feed is not a negligible source of import growth in developing countries either. Concessionary imports of pulses (World Food Programme, etc.) do not play a major role.

Conclusions Concerning Demand and Supply

Factors Affecting Consumption and Demand

The comparison of the FAO projections with actual trends leads to some significant conclusions with important implications for future research and development policies for food legumes. They are:
(1) *Demand for pulses in the 1980s, particularly for the cool season legumes, has proved considerably more dynamic than might have been expected on the basis of trends in the 1970s.* It has greatly exceeded the FAO projections, which did not themselves appear particularly cautious in the light of past trends. It has also confounded a projection for the WANA region (Somel, 1988), which postulates growth rates in demand of 2.4% per annum for food and −0.04 for feed from 1983–2000 based on low income elasticities of demand of 0.17 for food and −2.6 for feed. In fact, the WANA growth rates so far since 1980 are 6.5% per annum overall, and 4.3% per annum for peas, the main feed source.
(2) *Declining per capita consumption of pulses for food does not necessarily imply declining per capita demand.* The former is largely a function of the continuing gap between production and population, which is being partly but not sufficiently closed by imports. Effective demand is a complex involving population growth; urbanization; purchasing power as determined by *per capita* income, income growth; price and income elasticity of demand, and consumer preferences (Table 13).
Income elasticities of demand for food are generally lower than those for other goods, since as incomes rise people tend to spend a smaller proportion of

Table 13A. CGIAR research expenditures by category, 1987–91 (US$ millions)

Item	1987 Actual	1987 % Total Expenditure	1991 Proposed	1991 % of Total Expenditure
Cereals	33.26	40.0	54.00	40.0
Roots and tubers	6.59	7.9	12.77	9.5
Food legumes	10.81	13.0	14.59	10.8
Livestock	15.82	19.0	29.82	22.1
Natural resources	12.57	15.1	18.95	14.0
Food policy	4.13	5.0	4.82	3.6
Total	83.18	100.0	134.95	100.0

Table 13B. Distribution of CGIAR food legume research expenditures, 1987–91 (US$ millions)

Item	1987 Actual	1987 % Total Expenditure	1991 Proposed	1991 % of Total Expenditure
Beans (*Phaseolus*)	3.59	33.2	5.01	34.3
Pigeon peas	0.75	7.0	1.07	7.3
Cowpea	1.06	9.8	2.19	15.0
Groundnut	1.99	18.4	2.71	18.6
Total warm season	7.39	68.4	10.98	75.2
Chickpea	1.69	15.6	1.86	12.8
Lentil/faba bean	1.73	16.0	1.75	12.0
Total cool-season	3.42	31.6	3.61	24.8
Total Food Legumes	10.81	100.0	14.59	100.0

Source: Consultative Group on International Agricultural Research (CGIAR) Secretariat, International Centers Week, Agenda Item 11, September 1990.

their income on food. Among food commodities elasticities tend to be lower for basic staples, and may be negative where consumption of staples is high relative to nutritional needs (as with wheat in the WANA region and potatoes in Western Europe). At that stage, population growth becomes the primary factor generating demand, and per capita consumption may decline even when supply is not a constraint because demand is saturated.

This does not seem to be the case with cool season pulses, which represent only about 5% of total caloric supply and 12% of total protein in most developing countries.

On the contrary, governments in a number of developing countries are projecting higher *per caput* as well as absolute levels of consumption of cool season food legumes for the 1990s, associated with enhanced research, price support, and other policies to expand production and reduce unit costs to consumers.[4] Rising imports of those commodities is also indicative of strong demand. Consequently, projections which assume negative demand elasticities should be treated with reserve.

(3) *For cool season food legumes in the developed countries, the constraint seems to be with supply rather than demand.* This is indicated by the rapid increase in imports when the growth of production lagged in the second half of the 1980s. Population growth is the primary source of growth of food legume consumption, and production falls well behind this.[5] This situation seems unlikely to change dramatically in the next few years, given UN projections of a 2.0% per annum growth of population in the developing countries to the year 2000.

(4) In developed countries, as a whole, per capita consumption of *food* legumes has stagnated, but two factors may change this situation in the present decade. These are:

• *Migration of people* from countries around the Mediterranean Basin to Australia and Western Europe – the only developing regions where *per capita* pulse consumption for food has risen since 1980.

• *Health-related concerns* for a diet low in fat and cholesterol and high in fiber, and a related interest in "healthy" or "organic" foods grown with low or no chemical inputs. This should favor a higher proportion of vegetable protein compared to protein of animal origin in the diet.

On the surface, this does not seem to have stimulated demand for cool season pulses in North America, the region where such concerns are most evident. However, the importance of frozen foods in diets there should not be overlooked. Data supplied by the American Frozen Food Institute (1990, 1991) shows that 1989–91 consumption of frozen peas averaged about 230,000 tons, up almost 30% from 1979–81. In addition, there are around 210,000 tons of frozen beans (78% green beans and 22% lima beans), consumption of the former has also risen by around 20% since 1982–84. Frozen black-eyed pea (cowpea) production is now around 15,000 tons, this has increased by 27% since 1979–81. The 1990 and 1991 planted area of green peas and beans averaged 111,000 hectares, of which peas represented 83,000 hectares.

---

[4] India, which imported 266,000 tons of lentils in 1988/89, is aiming to expand production of lentils and chickpeas by 50% and 21% respectively in its seventh 5-year plan. China, 70% of whose food legume production is of cool season species, is proposing a three-fold increase in food legume production, raising annual per capita consumption from 5.4 kg in 1988 to 18 kg by the year 2000.

[5] Had population in developing countries increased at only 1% per annum between 1980 and 1990, instead of over 2.0%, the FAO assumption that food consumption *per caput* would rise from 7.7 to 7.9 kg over the decade would have been correct. In the event, it fell to 7.2 kg.

The 1989–91 planted U.S. acreage of all types of dried pea (*strictu sensu*) was 91,000 hectares for an output of 170,000 tons. Thus, the frozen pea and dried pea area is approximately the same, but the production of frozen peas is about 35% higher and rising faster. Similar data were not obtained for other developed regions, but the impact of frozen foods on demand for peas and beans is clearly a factor to be taken into account. It may well explain why consumption of dry food legumes in North America has not risen in response to health concerns.

This is less likely to be significant in the short run in the developing regions, both because chickpeas, lentils, and faba beans (none of which are usually frozen) are more important among cool season legumes than peas, and because frozen food consumption is much less widespread due to low incomes and lack of refrigeration. With rising incomes, increasing urbanization, and a wish for foods of convenience, demand for frozen legumes is likely to increase there also.

(5) *In the case of legumes for feed and fodder, the driving force has been the high income elasticity of demand for meat and other livestock products, especially in the EEC, but apparently also in some higher-income developing countries.* Should demand for those products decline due to health concerns, as is becoming evident in the EEC and North America; or to rising prices of meat, as is the case in some developing countries (Oram and Belaid, 1990), demand for feed overall, including legumes, may slacken. In the latter case, however, particularly in the lower income countries, consumers are turning to legumes as a cheaper alternative source of protein.

(6) *The extent to which feed demand for cool season legumes (especially peas and lupins) will continue to increase will depend partly on technical factors affecting their suitability to animal diets, but probably even more on the ability of producers to compete in terms of price, quality, and suitability with alternative feed sources, particularly oilseeds and cereals (such as barley).* Recent changes in support price policies in the EEC, affecting all of those commodities, introduce an element of uncertainty. A ceiling of 3.5 million tons for output of pulses that qualify for the full subsidy was introduced in 1988, with cuts of 0.5% in minimum prices to producers for every 1% of output above that ceiling. This resulted in a 13% cut in prices in 1990 which has reduced the area planted in some EEC countries, but not in France (the largest producer), and has been more than offset by yield increases there and in other EEC countries (USDA Food Outlook, May 1991). Expected decreases in feed pulse prices, both absolutely and relative to cereal and oilseed prices, were expected to lead to cutbacks in area in Canada and Australia in 1990 (FAO, 1989), but the effect on production was negligible due to good growing conditions.

Since then wheat, barley, and oilseed prices have weakened and market returns on pulses have improved, especially in the case of lentils. Australia projects an increase of 271,000 hectacres in the area of lupins and 271,000 hectacres in that of peas between 1990 and 1996, with production increasing by 368,000 and 351,000 tons respectively (ABARE, 1991). Greater domestic utilization for peas in pig-feed compounds is foreseen, but for each commodity about half would be exported. Price forecasts in real terms are relatively stable at around A$ 210/ton and A$ 240/ton, respectively. A recent Canadian review of production and export prospects for legumes in Canada, Australia, and the United States (Longmuir, Market Commentary, December 1991) is also quite optimistic. It notes a strengthening of domestic feed demand and a potential for a modest rise in European imports of feed peas. However, in the case of Canada, food pea sales are expected to dominate exports, based on a feeling that lower world interest and inflation rates, and lower oil prices will increase developing-countries ability to afford food imports. The prospects for price stability have been improved by various policy interventions, including the U.S. Export Enhancement Program, the U.S. Export Credit Guarantee Scheme, the Canadian Gross Revenue Insurance Plan, and increased government intervention prices in Turkey. A danger exists, however, that subsidizers in countries with a good potential for expanding the area under pulses may lead to overproduction in the longer term with a consequent decline in market prices.

(7) Another factor which seems to be encouraging greater production of pulses in Australia and North America is the benefit conferred by the legumes via soil nitrogen supply, to yields of other crops in the rotation, especially cereals. This is certainly not a new discovery but may be of increasing benefit in the light of higher energy costs and concerns about environmental pollution. It should be of even greater importance to developing countries, but implies an enhanced research effort to raise yields of pulses and/or to reduce yield variability, as well as to identify appropriate rhizobia; backed by an extension drive to convince producers to adopt improved cultivars and to inoculate where necessary.

(8) *Both supply and demand of pulses have proved quite price-elastic*, and there appears to be scope for continuing increases in consumption for food as well as for feed over time in developing countries, as well as export opportunities, if unit costs of production can be kept down and year-to-year fluctuations in output reduced. This is a crucial task for research; since lack of successful technological innovation, relative unresponsiveness to irrigation, and high costs of weed control and harvesting are likely to place cool season legumes (especially the food legumes) at an increasing disadvantage compared to cereals and oilseeds if yields of competing crops continue to rise.

The Size of the Demand-Supply Gap

As far as food needs are concerned, the crux of the problem lies in the developing countries, but recent trends in the growth of production of cool season pulses in those countries are not encouraging. Nor are other pulses, such as beans, likely to be substitutes: not only are their growth rates equally disappointing, but there are strong consumer preferences for certain species between and within geographical regions – as is very obvious in WANA, as well as differences in ecological requirements of the different species. Thus, cross-elasticities of supply and demand for pulses are quite limited.

The current gap between supply and consumption for developing countries is around 2.5 million tons, represented by their total imports; but since there were 2.3 million tons of developing-country exports, the apparent net deficit is only about 200,000 tons. However, a majority of those exports probably went to developed countries (Japan and Europe). The gap between supply and demand is probably much larger; in several major consuming countries, such as India and China, government targets aim at per capita intakes considerably above current levels of consumption.

Assuming that population in developing countries, which was just over 4 billion in 1990 continued to grow at its projected level of about 2.0% per annum to the year 2000, another 900 million people would have to be fed. Even if per capita consumption of pulses for food remained static over the 1990s at around 7 kg, approximately another 6.3 million tons of food legumes would be needed by the year 2000 to feed those extra people, irrespective of any increase in demand due to income growth. This represents a 22% increase over current consumption. If yields did not increase above their 1990 average of 0.65 tons ha^{-1}, another 9.5 million hectares of arable land would be needed. This does not take feed, seed, and other requirements into account, currently representing only 2 kg per capita in developing countries, but rising rapidly relative to food consumption. On conservative growth assumptions of a 3 kg per capita consumption by 2000, another 6.5 million tons would be required to meet those needs by the year 2000, requiring a further 10 million hectares of land at constant yields.

On these assumptions, the minimum extra production required just to maintain constant 1990 per-capita consumption of food and other uses in the developing countries by the year 2000 would be 12.8 million tons, of which about 60% would probably have to come from cool season pulses.

This a conservative estimate, since it takes no account of income growth. It does not assume negative income elasticities of demand for food or feed, but past projections suggest that the latter are likely to be strongly positive (c: 2.0) and I have indicated earlier why I feel that at least a modest positive elasticity for food (c: 0.05) may be justified. This is the level postulated by Somel (1988) for barley in WANA, which is certainly not as widely used for human food as pulses, while he implies an income elasticity for feed barley of 2.35. In WANA, barley is the main source of feed.

To meet the overall year 2000 demand of around 49 million tons for developing countries implied by the above assumptions, at current yield levels, would require a 35% increase in the area under pulses in those countries from approximately 57 million hectares in 1989–90 to 77 million hectares. However, an increase in their average yields from the 1990 level of 0.65 tons ha^{-1} to 0.87 tons ha^{-1} by 2000 would achieve the same level of production with no increase in area. This does not seem an impossible target.

As far as the developed countries are concerned, a modest expansion of food consumption per caput can be envisaged related to health concerns, above the current level of 2.6 kg, perhaps to 3 kg; associated with a population growth of about 0.7% per annum. This would raise total DME food consumption to 4 million tons by the year 2000, an increase of 700,000 tons (20%) over 1990.

The main uncertainty lies with demand for feed (mainly peas and lupins) in the developed countries which currently represents almost three-quarters of pulse consumption in those countries, and a quarter of global pulse consumption. A projection to 2000 of their annual 1980–1990 compound growth rate of 12% would raise feed consumption in those countries to 44 million tons, the equivalent of total global pulse consumption for all purposes in 1980.

Given their low rates of population growth, such a level of domestic consumption in DME countries implies a very high income-induced increase in per capita consumption of animal products, which seems unlikely given increasing health concerns, recessionary economic trends in Japan, North America, Australia, and Western Europe, and the situation in the former Soviet Union. A scenario for continued high demand for feed legumes in DME countries might be based on their expanding use as substitutes for cereals and oilseeds in animal rations, but this would only be a partial solution if demand for livestock products declined significantly, and implies strong price, quality, nutritional, or other comparative advantages of pulses over competitive products. Rising demand for food and feed in developing countries might absorb some of the potential surplus supply of DME exporters, but exports to those countries are unlikely to exceed 6 million tons of their 12.8 million additional requirement by the year 2000, since they were able to increase total output by about 6 million tons during the 1980s.

Consequently, although without being unduly pessimistic, I conclude that the longer term outlook for feed demand is likely to be less dynamic than in the past decade with an increasing food and feed requirement in developing countries, but a slower growth of feed demand in the main developed-country markets, especially in Europe.

Closing the Supply Gap: Options and Related Research Needs

The five main options for increasing agricultural supply include:
(1) importing to supplement local production capacity;
(2) expanding the net area cultivated. . . bringing in new land;
(3) increasing cropping intensities (expanding the gross area under crop);
(4) changing the enterprise-mix; and
(5) increasing yields through technological change, including irrigation and
    drainage, . . which may facilitate all of the previous options.

Import Potential. A possible short-run solution to raising per capita intakes of pulses would be for developing countries with favorable trade balances to raise imports. If prices are right, some of those needs could probably be supplied by other developing countries, such as Turkey, Morocco, Chile, Argentina,

Myanmar, Thailand and China, which have both the land resources and the ecological conditions for expanding production of some of the cool season species (ICRISAT, 1991). Opportunities also exist for developed-country exporters, especially Australia, Canada, and the United States, to expand exports – probably by increasing the area cropped with legumes, or bringing idled land into use. Generally, they also have some yield advantage over developing-country producers, are relatively highly mechanized, and are experienced exporters, so they are likely to be cost-competitive. However, raising food imports may not be an easy option for low-income countries with adverse trade balances and high debt servicing burdens.

Expanding the Net Area under Crop. This option is almost foreclosed for most WANA, South, and Southeast Asian countries, as well as in most of Central America with the possible exception of Mexico. Scope remains in the "Southern Cone" of South America, as well as in some African countries, such as Tanzania; but these are not the main cool season pulse producers.

Increasing Cropping Intensities. This is still a feasible option in several WANA countries, mainly by eliminating annual fallow; and in South Asia and China by double-cropping irrigated land. The successful Turkish campaign to replace the cereal-fallow rotation by a cereal-legume system is an outstanding example of what can be done by a carefully planned and coordinated research, extension, and marketing effort. It raised total pulse output from 800,000 tons to 2 million tons in 10 years, with exports expanding from 165,000 tons to 750,000 tons over the same period, both through the Turkish Grain Marketing Board and through private trade, with striking benefits to the economy.

While relatively few cool season developing country producers have such scope, there is some fallow-replacement potential in WANA, estimated by Oram and Belaid (1990) to be about 6 million hectares in 11 countries with ecological situations suited to legume production. According to Kassam (1988), there may also be scope in upland areas of Eastern Asia (China and Korea) where conditions are analogous to those in Anatolia.

However, there are other crops which might be grown with wheat as alternatives to cool season food legumes – including vegetables, melons, oilseeds, forage crops in higher rainfall zones, and continuous barley at the low rainfall margins. Moreover, the comparative suitability of the various legume species varies with the ecological situation, as do producer and consumer preferences. Thus, successful exploitation of their fallow replacement and cropping systems potential will require careful agroecological characterization, as well as sociological studies, cultivar selection, and cropping systems research.

Changing the Enterprise-Mix. This may be an option linked to increasing cropping intensities, or a separate alternative through shifting patterns of land use within existing farming systems. In the latter case, the farmer must anticipate some comparative advantage from the introduction of a new crop;

reduced production costs, higher productivity, lower risk, better use of labor, dietary preference, new market opportunities, etc. The successful large scale introduction of legumes or an expansion of area under pulses in existing systems must therefore meet such requisites; presupposing both farming systems research to understand farmers' needs and constraints, and strategic or applied research, including policy and economic analysis, to find solutions to them.

Increasing Yields Through Technological Change. The competitive power of food legumes with other crops in a situation where the land constraint is likely to become increasingly severe depends heavily on improving their productivity per unit of area and time and/or on reducing their costs of production, preferably both. This implies technological change, which for most developing countries is the primary need.

Since this is what this conference is mainly about, I will not enter into detail about the constraints which must be overcome to raise the productivity of the cool season species, and indeed of all food legumes. I would, however, like to touch on three overriding issues related to the supply side. These are:

(a) Compared to wheat or barley (even under rainfed conditions), current pulse yields in most countries are both low and highly variable. In many countries, the variability is so high and the average yield so low that relatively minor weather related changes over time appear as changes in yield. In such circumstances, "yield" is a misnomer. *What is the attainable level of yields of the major legume species, as demonstrated by averages attained on a large scale by efficient producers, and what lessons (if any) can be learned from them?*

For *peas*, the 1990/91 yield average for the EEC was 4.65 tons ha^{-1} on an area of nearly a million hectares (Annex B). In some major producing countries (France, the Netherlands, and Denmark), pea yields were an astonishing 5 tons ha^{-1}. No other developed region is near those levels – the averages for peas were 2.15 tons ha^{-1} in Canada, 1.69 tons in the United States, 1.76 tons in the USSR, and 1.2 tons in Australia. Turkey achieved 3.5 tons ha^{-1}; no other developing country surpassed 2.5 tons. While these yields are mostly below those of wheat for which the 1989–91 average *world* yield was 2.4 tons ha^{-1}, and the world record was 7.9 tons ha^{-1}, they are much closer than is the case with chickpea, lentil, faba bean, or Phaseolus. The levels now being attained in EEC countries cast doubt on the contention that legumes have a physiological barrier to yields comparable to those of cereals, but do not explain why peas are so much ahead of the other major pulses, nor why the latters' yields are also low in the developed countries.

The top pea yield reported by ICARDA of the selected highest yielding entries of 348 accessions evaluated at its trial sites in Syria was 2.96 tons ha^{-1} (ICARDA, 1990). This suggests that the potential yield ceiling for peas in WANA is well in excess of the yields currently being achieved there or in other countries outside of Europe. The problem is to identify and overcome the multiple problems affecting yields in the developing countries, and to sort out the crucial factors limiting yields which are amenable to improvement.

Yields of faba beans are higher on average than those of chickpea, lentil, or other cool season species, reaching 2.5 tons ha^{-1} in Egypt, and exceeding 3.5 tons ha^{-1} in France and West Germany. In stark contrast, hardly any country (developed or developing) exceeds even 1 ton ha^{-1} in the case of chickpeas, lentils, or lupins. Is this a result of environment, lack of research compared to peas and faba beans, or both? Yields in some ICARDA trials exceed 2 tons ha^{-1}, but are highly variable from site to site. Can the intergeneric yield differences be narrowed, and by what means? Average farmer yields of lentils reported in a recent survey in Syria vary from 1,775 kg ha^{-1} in good years to 468 kg ha^{-1} in poor years (Mazid, 1990). Should research aim at narrowing the farm-level gaps and reducing instability or raising the yield ceiling? Are those goals incompatible? These are issues which merit discussion at this meeting.

(b) *Is there an imbalance between technical and socio-economic research on problems affecting supply and demand for cool season legumes, leading to inefficient use of research resources?* There are certain recognized major constraints to increasing the productivity of each of the cool season legumes, especially in the developing countries, which are reflected in low average yields, high yield instability, high costs of production, and farmers perceptions of high risk. This translates into poor competitiveness with alternative crops for land use. Apart from environmental problems related to the specific ecological niches in which each of those crops tend to predominate, those constraints include several major diseases (especially Ascochyta blight, rust, root rots, botrytis, and Fusarium wilt); serious weed infestation problems, including the parasitic *Orobanche*; poor cultivation practices; harvesting difficulties, especially with lentil, leading to losses during harvesting and high labor costs; and pests such as *Sitona* affecting nodulation, and pod borers causing crop and storage losses (ICARDA, 1990; Oram and Belaid, 1990; Malik *et al.*, 1988; etc.).

However, the incidence and severity of these problems varies between geographical regions and among countries within those regions. In some cases, there are additional local problems, minor on a world scale, but locally important; and there are also problems of local consumer preference for certain species and cultivars, and characteristics such as plant height, seed size and color, maturation time, straw quality, etc., which must be taken into account in breeding programs, if new genetic material is to be acceptable to producers. How far are producer preferences for faba bean in North Africa and chickpea in West Asia; green lentils in Europe; desi chickpeas in South Asia; and kabuli chickpeas in WANA, connected with biological and environmental factors affecting growers perceptions, and how far are they consumer-driven?

What is the most effective institutional means of dealing with the major crosscutting technical problems, while increasing understanding of local impediments which may be crucial to transferring the results to potential beneficiaries at the country level and getting them adopted? There are global studies of the adoption of improved cultivars of wheat, rice, and maize, and assessment of their impact (Dalrymple, 1986); but while ICARDA (1990) has made some estimates of winter chickpea adoption in the Mediterranean environment, and the CGIAR (Hawkes, 1985) has looked at the use of genetic material, I know of no publication which deals comprehensively with the legumes.

To answer this, and other questions I raise here, I believe that more emphasis needs to be placed on social and economic issues. There needs to be a closer marriage of technical and social sciences, not only to find solutions to technical problems, but also to provide guidance on market requirements, consumer preferences and price policies, and institutional support to producers (investment in research, extension, seed or input supply, etc.). In fairly extensive inquiries related to this paper, I found very little research on such matters, yet this is needed not only to help producers to make the right choices, but also to assist decisionmaking and effective targeting by research directors and national policymakers.

(c) What resources should be devoted to research on cool season pulses internationally and nationally, and what should be the priorities regionally and for individual species?

In this connection, a sense of perspective is necessary. Here I quote a comment by the late Dr. Frosty Hill, one of the founding fathers of the Consultative Group on International Agricultural Research (CGIAR): "There are many interesting problems: some of them are important!" Cool season food legumes are an important but relatively minor dietary staple in developed and most developing countries. Even though they have had little past research attention compared to winter cereals, it is difficult to argue that resources equivalent to the cereals should now be devoted to pulses. In a recent paper, Lipton and Longhurst (1985) argue that pulses should receive quite low priority on grounds that energy, not protein, is the primary nutritional need of the poor; and that legumes are not a very cheap source of protein anyway. These authors are particularly severe on research related to protein quality and attempts to raise the protein or amino acid content of legumes and cereals, as well as legume research on antinutritional factors, consumer preferences, and processing and culinary requirements. They claim that such research raises prices, whereas breeding for stability and quantity leads to lower costs to consumers.

While I agree with Lipton and Longhurst that attempts to improve the protein content of cereals have absorbed vast resources with little real success, it is difficult to accept their narrowly limited approach concerning the nutritional value of the legumes. Failure to raise cereal quality seems to enhance the case for research on the legumes. Their argument seems to be based largely on the substitution of pulses by wheat in India during the

1970s when the "Green Revolution" had its greatest impact and wheat prices declined significantly compared to pulses as cereal yields increased, especially in the irrigated areas where legumes are currently at greatest disadvantage (Ryan and Asokan, 1977).

While these problems of relative comparative advantage between wheat and pulses have not yet been overcome in India, demand for pulses in South Asia has improved and government policies in many countries of South Asia, WANA, and Latin America are supportive of raising their consumption both in absolute and per capita terms. Although a one-commodity diet may be the only option for the poorest of the poor, it is clearly not compatible with most human needs and preferences, as food consumption surveys show. *How adequate are international and national resources currently being devoted to research on pulse crops, and how far have they been prejudiced by arguments of the type described above?*

Data on expenditure on pulse research within the Consultative Group on International Agricultural Research are given in Table 14. Four institutes are concerned with research on chickpea, lentil, faba bean, pea, cowpea, pigeon pea, and field beans; pulse research receiving overall 10.8% of the total CGIAR research expenditure. However, it will be seen that cool season pulses are being given lower priority in recent years compared to warm-season species, one argument apparently being that the cool season crops are mainly consumed in large countries (e.g., China and India) or middle-income countries which should be in a position to support much of the necessary research themselves. Data on the allocation of scientific staff to research on pulses and other major groups of commodities by national systems in developing regions are cited in Table 15. It appears that the ratio of resources devoted to pulses overall relative to cereals and root crops by the CGIAR managers is somewhat higher than the allocations of national research directors in developing countries, although those are fairly commensurate with the contribution of pulses to agricultural GDP and dietary energy. With respect to the latter, it could be argued that pulse research merits greater support since the research problems associated with pulse improvement

Table 14. Factors affecting demand for pulses

FACTOR	DEMAND POSITIVE		DEMAND UNCERTAIN OR NEGATIVE	
SOCIAL	Population growth	+ FO/FE	Long preparation time	-FO
	Migration	+ FO	Urbanization	?-FO + FE
DIETARY	Cheap quality protein	+ FO/FE	Antinutritional factors	-FO
	"Healthy" diets	+ FO	Reduced meat consumption	-FE
ECONOMIC	Import substitution	+ FO/FE	EEC policies	?-FE
	High meat demand	+ FE	Rising incomes	-FO, + FE
	Save fertilizer	+ FO/FE		
AGRICULTURAL ENVIRONMENTAL	Good rotation crops	+ FO/FE	High production costs	-FO
	Reduced chemicals	+ FO/FE	Low yields	-FO/FE
			Risk perception	-FO

FO = Food   FE = Feed

Table 15. Share of scientific staff by crop group, developing regions, mid-1980s

Region	Pulses[1]	Cereals	Roots/Tubers	Annual Oilseeds[2]	Horticulture[3]	Other[4]
Sub-Saharan Africa	6.8	29.6	9.0	7.4	10.8	36.4
Asia	7.7	32.0	5.4	2.6	7.4	44.8
Latin America	9.5	31.5	10.7	7.4	10.9	28.3
West Asia/North Africa	7.4	23.1	2.0	4.0	34.7	28.8
Weighted Average	7.9	30.6	6.3	4.4	12.4	38.4
Total Number of Scientists	1,845	7,154	1,486	1,021	2,897	8,994

Source: Peter A. Oram, IFPRI data base for 91 countries, latest post 1980 year.

Notes:

[1] Pulses, cereals, and roots/tubers include all annual species.
[2] Oilseeds: annuals only, excludes perennials (olives, oilpalm, coconut).
[3] Horticulture comprises fruits and vegetables.
[4] "Other" includes perennial oil crops, sugar, tobacco, rubber, beverages, cotton and other fibers, and unspecified crop science research.

are more complex than with cereals because of the nitrogen metabolism and antinutritional factors, and because the research resources devoted to cereals and industrial crops, both in developed and developing countries have been so much larger. In the CGIAR case, I think that it may be difficult to argue for more resources to be allocated to legume research, but a strong case can be made for a better balance between cool and warm season species.

Given the relatively limited resources likely to be available globally, it is relevant to consider (i) which research objectives among the many possible uses of scarce scientific resources for pulses should receive priority, and for what species; and (ii) how the resources available among developed and developing countries, and international, ecoregional, and national programs, can best be coordinated and utilized so that the maximum synergism may be achieved. Among the devices which are now being employed are contract or aid-related research by developed countries on difficult problems requiring strategic research (biotechnology, etc.); collaborative research among international and national institutions in both developed and developing countries; and research networks generally among international and national institutions in developing countries.

It may be pertinent for this meeting to review some of these arrangements, to examine their potential value or actual impact and how they might be made more effective, and to consider whether information systems on available resources and ongoing research efforts devoted to legume research are adequate to achieve the increases in productivity of the major cool season food legumes which are clearly necessary.

In this context, I believe that it is important to access and make available data on all relevant research whether on food, feed, or forage legumes, cool or warm season. Research outside the cool season species *strictu sensu* may well generate new knowledge and information which could provide solutions to some of the intractible problems which are on the agenda of this conference.

References

Agriculture Canada. 1989. *Canadian Pulses Report*. Ottawa: Agriculture Canada, Crop Development Division, June.

Al-Anney, A. H. K. 1990. In: *The Role of Legumes in the Farming Systems of the Mediterranean Areas* 71–75 (eds. A. E. Osman, M. H. Ibrahim and M. A. Jones). Dordrecht: Kluwer Academic Publishers.

American Frozen Food Institute. 1990. *Frozen Food Pack Statistics*. Burlingame, California.

American Frozen Food Institute. 1991. *Statistical Bulletins*. Burlingame, California: October/November.

Australian Bureau of Agricultural Resource Economics. 1991. *Agriculture and Resources Quarterly*, Vol. 3 (4). Canberra: ABARE, December.

Dalrymple, D. G. 1986. *The Development and Spread of High-Yielding Wheat Varieties in Developing Countries*. Washington, D.C.: Bureau of Science and Technology, Agency for International Development.

Food and Agriculture Organization of the United Nations. 1986. FAO Agricultural Commodity

Projections to 1990. *FAO Economic and Social Development* Paper No. 62. Rome, Italy: FAO.

Food and Agriculture Organization of the United Nations. 1988. *Commodity Review, 1987/88 (Pulses)*. Rome, Italy: FAO.

Food and Agriculture Organization of the United Nations. 1989. *Commodity Review, 1989/90 (Pulses)*. Rome, Italy: FAO.

Food and Agriculture Organization of the United Nations. 1990a. *Trade Yearbook: Trade tapes*, Vol. 44. Rome, Italy: FAO.

Food and Agriculture Organization of the United Nations. 1990b. *Pulse, Producer Prices in National Currency, 1984–91*. Rome, Italy: FAO.

Food and Agriculture Organization of the United Nations. 1990c. *Production Yearbook: Production Tapes*, Vol. 44. Rome, Italy: FAO.

Hawkes, J. G. 1985. *Plant Genetic Resources: The Impact of the International Agricultural Research Centers*. CGIAR Study Paper No. 3. Washington, D.C.: The World Bank.

Hu Jiapeng. 1988. In: *Winter Cereals and Food Legumes in Mountainous Areas* (eds. J. P. Srivastava, M. C. Saxena, S. Varma and M. Tahir). Aleppo, Syria: ICARDA.

International Center for Maize and Wheat Improvement (CIMMYT). 1991. *1990/91 World Wheat Facts and Trends*. Prepared in collaboration with ICARDA. Mexico: CIMMYT.

International Center for Agricultural Research in the Dry Areas (ICARDA). 1990. *Food Legume Improvement Program: Annual Report for 1990*. Aleppo Syria: ICARDA.

International Crops Research Institute for the Semi-Arid Tropics (ICRISAT). 1991. In: *Research Bulletin* No. 14, pp. 77 (eds. S. M. Virmari, D. G. Faris and C. Johansen). Patancheru, Andhra Pradesh, India: ICRISAT.

Jeswani, L. M. and Baldev, B. 1990. *Advances in Pulse Production Technology*. New Delhi, India: Indian Council of Agricultural Research.

Kamel, M., Sakr, B. and Hamdaoui, F. 1989. *Food Legume Research Priorities for Dry Land Areas: 1989–1994*. Arido-culture Programme. Settat, Morocco: Institut National de la Recherche Agronomique.

Kassam, A. H. 1988. In: *Winter Cereals and Food Legumes in Mountainous Areas* (eds. J. P. Srivastava, M. C. Saxena, S. Varma and M. Tahir). Aleppo, Syria: ICARDA.

Kaul, A. K. and Fenton, M. 1991. In: *Lathyrus and Lathyrism Newsletter*, Vol. 3 (1) (eds. A. K. Kaul and M. Fenton). New York and London: Third World Medical Foundation.

Lipton, M. and Longhurst, R. 1985. *Modern Varieties, International Agricultural Research, and the Poor*. Study Paper No. 2. Washington, D.C.: Consultative Group on International Agricultural Research.

Longmuir, N. L. 1991. In: *Market Commentary*. Ottawa, Canada: Agriculture Canada, December.

Malik, B. A., Varma, M. M., Rahman, M. M. and Bhattara, A. N. 1988. In: *World Crops: Cool Season Food Legumes*, pp. 1095–1111 (ed. R. J. Summerfield). Dordrecht: Kluwer Academic Publishers.

Mazid, A. 1990. *Lens Newsletter*, Vol. 17 (No. 2). Aleppo, Syria and Saskatoon, Canada: ICARDA and University of Saskatchewan.

Oram, P. and Belaid, A. 1990. *Legumes in Farming Systems*. Aleppo, Syria: ICARDA.

Osman, A. E., Pagnotta, M., Russi, L., Cocks, P. S. and Falcinelli, M. 1990. In: *The Role of Legumes in the Farming Systems of the Mediterranean Areas*, pp. 205–216 (eds. A. E. Osman, M. H. Ibrahim and M. A. Jones). Dordrecht: Kluwer Academic Publishers.

Ryan, J. G. and Asokan, M. 1977. *Indian Journal of Agricultural Economics* 32: 8–15.

Sarker, A., Rahman, M. M., Zaman, W., Islam, M. O and Rahman, A. 1991. In: *Advances in Pulses Research in Bangladesh*. Joydebpur, Bangladesh: Bangladesh Agricultural Research Institute.

Somel, K. 1988. *Food and Agriculture in West Asia and North Africa: Projections to 2000*. Aleppo, Syria: ICARDA.

USA Dry Pea and Lentil Council. 1990. *The Highs and Lows of USA Dry Peas and Lentils*. Moscow, Idaho: USADPLC.

USA Dry Pea and Lentil Council. 1991. *USA Dry Pea and Lentil Updates*. Moscow, Idaho: USADPLC.

United States Department of Agriculture (USDA). 1991. *Food Outlook: May 1990/91 Review*. Washington, D. C.: USDA.

USDA, Economic Research Serviceweather. 1992. *World Trade in Dry Beans Rising Rapidly.* Washington, D. C.: USDA/ERS. (Courtesy of Dr. John Parker).

ANNEX A

Table 1. Area, production, and yield of top 12 producers of faba beans, 1989–91

Country	Area (ha)	% to World	Production (MT)	% to World	Yield (kg ha⁻¹)
China	1,700,000	53.49	2,483,333	58.55	1.46
Egypt	139,826	4.40	380,710	8.98	2.71
Ethiopia	325,667	10.25	277,133	6.53	0.85
Morocco	229,300	7.22	170,333	4.02	0.74
Germany, F.	34,221	1.08	124,298	2.93	3.64
Italy	106,265	3.34	116,933	2.76	1.10
France	27,987	0.88	101,109	2.38	3.66
Turkey	40,000	1.26	75,000	1.77	1.88
Australia	42,333	1.33	55,000	1.30	1.30
Spain	34,133	1.07	41,033	0.97	1.21
Mexico	31,333	0.99	41,000	0.97	1.31
Brazil	130,000	4.09	36,333	0.86	0.28
Subtotal	2,841,066	89.40	3,902,217	92.01	1.37
World	3,178,092	100.00	4,241,124	100.00	1.34

Table 2. Area, production, and yield of top 12 producers of chickpeas, 1989–91

Country	Area (ha)	% to World	Production (MT)	% to World	Yield (kg ha⁻¹)
India	6,668,233	68.10	4,720,300	66.78	0.71
Turkey	841,479	8.59	802,667	11.36	0.95
Pakistan	706,790	7.22	534,333	7.56	2.18
China	0	0.00	200,000	2.83	0.00
Mexico	143,333	1.46	146,667	2.08	1.02
Australia	109,513	1.12	131,000	1.85	1.21
Ethiopia	129,000	1.32	122,000	1.73	0.95
Myanmar	129,357	1.32	96,628	1.37	0.75
Bangladesh	101,290	1.03	67,042	0.95	0.66
Iran	110,333	1.13	60,233	0.85	0.55
Morocco	73,167	0.75	55,267	0.78	0.75
Spain	61,233	0.63	48,700	0.69	0.79
Subtotal	9,073,729	92.66	6,984,837	98.82	0.77
World	9,792,538	100.00	7,067,976	100.00	0.72

Table 3. Area, production, and yield of top 12 producers of lentils, 1989–91

Country	Area (ha)	% to World	Production (MT)	% to World	Yield (kg ha⁻¹)
India	1,098,533	34.43	724,767	32.03	0.66
Turkey	894,000	28.02	600,000	26.51	0.67
Canada	160,333	5.03	208,333	9.21	1.27
Bangladesh	214,696	6.73	157,372	6.95	0.73
China	0	0.00	100,000	4.42	0.00
Nepal	127,443	3.99	76,870	3.40	0.60
France	14,000	0.44	76,000	3.36	5.43
Syria	134,167	4.21	74,667	3.30	0.59
Iran	88,667	2.78	48,567	2.15	0.55
USSR	52,667	1.65	48,333	2.14	0.92
USA	40,940	1.28	45,983	2.03	1.14
Ethiopia	57,667	1.81	31,867	1.41	0.55
Subtotal	2,883,112	90.36	2,192,759	96.89	0.76
World	3,190,582	100.00	2,263,072	100.00	0.71

Table 4. Area, production, and yield of top 12 producers of dry peas, 1989–91

Country	Area (ha)	% to World	Production (MT)	% to World	Yield (kg ha⁻¹)
USSR	4,713,433	50.47	8,296,000	48.98	1.76
France	665,227	7.12	3,301,350	19.49	4.95
China	1,300,000	13.92	1,616,667	9.55	1.24
Denmark	114,616	1.23	514,567	3.04	4.50
India	475,000	5.09	438,133	2.59	0.92
Australia	365,333	3.91	406,667	2.40	1.11
Hungary	144,567	1.55	344,000	2.03	2.37
Canada	167,267	1.79	291,733	1.72	1.76
UK	77,433	0.83	273,000	1.61	3.54
Czechoslovakia	72,025	0.77	190,229	1.12	2.64
USA	67,807	0.73	148,043	0.87	2.17
Bangladesh	42,333	0.45	109,000	0.64	2.59
Subtotal	8,205,041	87.86	15,929,390	94.05	1.94
World	9,338,467	100.00	16,936,595	100.00	1.82

Table 5. Area, production, and yield of top 12 producers of dry beans, 1989–91

Country	Area (ha)	% to World	Production (MT)	% to World	Yield (kg ha⁻¹)
India	9,560,300	37.28	3,658,300	25.34	0.38
Brazil	5,018,449	19.57	2,137,820	14.81	0.43
China	1,417,000	5.53	1,314,000	9.10	0.93
USA	668,100	2.61	1,076,330	7.46	1.61
Mexico	1,313,024	5.12	585,952	4.06	0.45
Myanmar	300,000	1.17	209,437	1.45	0.70
Turkey	177,000	0.69	193,000	1.34	1.09
Iran	230,000	0.90	184,000	1.27	0.80
Argentina	170,900	0.67	172,300	1.19	1.01
USSR	49,500	0.19	105,000	0.73	2.12
Pakistan	184,700	0.72	96,000	0.66	0.52
Bangladesh	123,000	0.48	80,000	0.55	0.65
Subtotal	19,211,973	74.92	9,812,139	67.97	0.51
World	25,644,880	100.00	14,436,400	100.00	0.56

Table 6. Area, production, and yield of top 12 producers of total pulses, 1989–91

Country	Area (ha)	% to World	Production (MT)	% to World	Yield (kg ha⁻¹)
India	23,380,560	33.89	13,313,100	22.82	0.57
USSR	5,870,567	8.51	9,746,000	16.70	1.66
China	4,417,000	6.40	5,948,000	10.19	1.35
France	715,281	1.04	3,437,959	5.89	4.80
Brazil	5,158,449	7.48	2,172,820	3.72	0.42
USA	864,697	1.25	1,518,937	2.60	1.76
Australia	1,373,353	1.99	1,434,089	2.46	1.04
Mexico	2,016,380	2.92	1,239,805	2.13	0.60
Iran	509,500	0.74	1,206,333	2.07	2.45
Turkey	1,142,993	1.66	879,667	1.51	0.77
Pakistan	1,476,358	2.14	748,739	1.28	0.51
Ethiopia	947,333	1.37	731,633	1.25	0.77
Subtotal	47,872,470	69.38	42,377,082	72.63	0.89
World	68,999,227	100.00	58,342,512	100.00	0.85

Annex B. Area, production, and yield of pulses, 3-year averages for 1980–1990

Regions	Year	Dry Beans			Faba Beans			Dry Peas		
		Area	Production	Yield	Area	Production	Yield	Area	Production	Yield
World	1980	24,446,293	13,568,929	0.55	3,689,748	4,292,559	1.16	7,508,801	8,508,462	1.14
	1990	25,644,880	14,436,400	0.56	3,178,092	4,241,124	1.34	9,338,467	16,936,595	1.82
	G. Rate	0.47	0.56	0.09	-1.29	0.06	1.35	2.77	7.02	4.22
North America	1980	783,133	1,275,593	1.63	5,100	4,000	0.78*	121,217	250,680	2.07
	1990	716,700	1,153,430	1.61	30,000	20,000	0.66	235,073	439,777	1.87
	G. Rate	-1.12	-1.32	0.20	16.52	13.93	-2.72	8.85	7.38	-1.35
Oceania	1980	4,132	2,886	0.68	10,667	5,667	0.53	67,682	116,912	1.74
	1990	8,000	5,000	0.63	42,333	55,233	1.31	385,510	458,895	1.20
	G. Rate	8.82	7.31	-1.12	16.38	30.95	13.91	20.32	18.01	-2.97
USSR	1980	49,000	64,000	1.30	—	—	—	4,181,000	4,272,333	1.04
	1990	49,500	105,000	2.12	—	—	—	4,713,433	8,296,000	1.76
	G. Rate	0.26	2.25	2.05	—	—	—	1.62	5.95	4.22
EEC	1980	498,084	293,768	0.59	314,848	415,054	1.32	124,016	415,988	3.35
	1990	323,079	191,444	0.60	243,253	451,226	1.85	916,909	4,262,727	4.64
	G. Rate	-4.2	-4.1	0.50	-1.68	3.04	4.67	22.82	25.44	2.68
DME	1980	2,238,823	2,255,907	1.01	368,907	494,432	1.34	4,649,831	5,303,498	1.16
	1990	2,206,615	2,187,909	0.99	343,355	587,733	1.71	6,610,672	14,283,184	2.16
	G. Rate	-0.37	-0.48	-0.10	0.19	3.93	3.70	4.29	10.01	5.65
Africa Dev.'g	1980	2,255,195	1,429,797	0.63	327,710	477,128	1.46	407,800	315,819	0.77
	1990	2,603,712	1,729,544	0.66	326,311	275,911	0.85	426,733	257,701	0.60
	G. Rate	1.31	1.75	0.42	-0.04	-4.86	-4.82*	0.41	-1.83	-2.24
Latin America	1980	7,385,499	4,064,234	0.55	264,754	172,725	0.65	155,093	111,298	0.72
	1990	7,636,572	3,721,762	0.49	248,278	160,328	0.65	131,131	102,036	0.78
	G. Rate	0.49	-0.91	-1.39	-0.19	-0.56	-0.36	-1.55	-1.10	0.44
WANA**	1980	244,796	313,400	1.28	461,710	514,940	1.12	97,096	79,015	0.81
	1990	445,498	422,233	0.95	560,148	733,820	1.31	182,865	122,401	0.67
	G. Rate	5.59	2.75	-2.70	1.77	3.27	1.47*	5.92	4.06	-1.76
Asia	1980	12,321,990	5,505,591	0.45	2,266,667	2,633,333	1.16	2,198,981	2,698,832	1.23
	1990	12,752,497	6,374,952	0.50	1,700,000	2,500,000	1.47	1,987,066	2,171,280	1.09
	G. Rate	0.25	1.40	1.15	-2.82	-1.05	1.73	-0.95	-2.99	-2.06
Dev.g Econ.	1980	22,207,472	11,313,022	0.51	3,320,841	3,798,127	1.14	2,858,970	3,204,964	1.12
	1990	23,438,272	12,248,491	0.52	2,834,737	3,653,392	1.29	2,727,795	2,653,417	0.97
	G. Rate	0.55	0.75	0.19	-1.48	-0.57	0.89	-0.43	-2.53	-2.11

* Data is for 1981 in North America and 1983 for Oceania.
** WANA includes Algeria, Morocco, and Tunisia which are part of Africa developing in FAO regional aggregation.

Annex B. (Continued)

Regions	Year	Chickpeas			Lentils			Total Pulses		
		Area	Production	Yield	Area	Production	Yield	Area	Production	Yield
World	1980	9,600,725	6,027,365	0.63	2,201,841	1,312,302	0.60	61,123,333	41,190,779	0.67
	1990	9,792,538	7,067,976	0.72	3,190,582	2,263,072	0.71	68,999,227	58,342,512	0.85
	G. Rate	-0.01	1.79	1.79	4.01	6.36	2.35	1.31	3.52	2.21
North America	1980	–	–	–	116,523	116,359	1.00	1,020,874	1,642,632	1.61
	1990	–	–	–	201,273	254,317	1.25	1,278,963	2,133,403	1.66
	G. Rate	–	–	–	6.45	7.92	1.06	2.57	2.68	0.06
Oceania	1980	1,033	1,200	1.16*	–	–	–	179,996	203,005	1.14
	1990	109,513	131,233	1.22	–	4,833	–	1,391,530	1,490,417	1.07
	G. Rate	63.89	64.43	0.75	–	–	–	22.82	23.11	0.15
USSR	1980	–	–	–	9,333	4,333	0.52	5,134,800	5,055,033	1.00
	1990	–	–	–	52,667	48,333	0.92	5,870,567	9,746,000	1.66
	G. Rate	–	–	–	19.96	25.35	4.50	1.69	5.76	3.98
EEC	1980	156,146	96,052	0.61	87,282	64,134	0.73	1,416,647	1,597,053	1.13
	1990	209,582	209,853	1.00	331,111	361,326	1.08	12,111,900	21,274,368	1.09
	G. Rate	-4.36	-1.81	2.61	-1.80	-0.35	1.53	4.57	15.10	10.54
DME	1980	162,001	102,497	0.63	219,772	189,241	0.86	9,078,095	9,685,040	1.07
	1990	209,582	209,853	1.00	331,111	361,326	1.08	12,111,900	21,284,368	1.09
	G. Rate	3.14	8.36	5.26	5.45	7.40	1.79	3.64	8.16	4.51
Africa Dev.'g	1980	210,630	175,449	0.83	51,247	45,323	0.88	8,233,073	4,406,811	0.54
	1990	259,793	174,589	0.67	57,667	248,677	4.31	11,010,113	5,073,066	0.53
	G. Rate	1.93	-0.04	-1.93*	1.08	16.74	15.49	2.68	1.29	-0.19
Latin America	1980	243,332	244,155	0.98	109,203	60,563	0.55	8,313,280	4,745,038	0.57
	1990	181,590	169,076	0.93	82,940	64,969	0.78	8,970,241	5,033,984	0.56
	G. Rate	-1.27	-2.15	-0.82	-1.25	2.20	3.43	0.63	-0.08	-0.71
WANA**	1980	522,653	439,128	0.84	408,955	350,342	0.86	2,248,687	2,188,813	1.15
	1990	1,189,852	1,012,252	0.85	1,206,891	781,429	0.65	4,168,594	4,207,280	0.91
	G. Rate	7.77	7.89	0.11*	10.34	7.57	-2.51	5.77	6.12	-2.10
Asia	1980	8,462,109	5,066,135	0.60	1,412,664	666,833	0.47	33,244,875	20,161,733	0.61
	1990	7,951,720	5,502,207	0.69	1,511,973	1,023,480	0.68	32,731,525	22,748,997	0.69
	G. Rate	-1.03	0.75	1.75	0.89	5.20	4.31	-0.31	0.91	1.22
Dev.g Econ.	1980	9,438,724	5,924,868	0.63	1,982,069	1,123,061	0.57	52,045,237	31,505,739	0.61
	1990	9,582,956	6,858,123	0.72	2,859,471	1,901,746	0.67	56,887,328	37,068,149	0.65
	G. Rate	-0.06	1.66	1.71	3.85	6.20	2.37	0.85	1.48	0.63

* Data is for 1981 in North America and 1983 for Oceania.
** WANA includes Algeria, Morocco, and Tunisia which are part of Africa developing in FAO regional aggregation.

ANNEX C

Table 1. Changes in total pulse consumption by category and region, 1980 and 1989 (000 MT)

Region	Year	Food	% Total Use	Feed	% Total Use	Seed	% Total Use	Other Uses	% Total Use	Total Use
World	1980	28,362	68.7	7,600	18.4	3,387	8.2	1,957	4.7	41,307
	1989	31,891	57.2	17,189	30.8	4,001	7.2	2,675	4.8	55,755
North America	1980	703	85.9	30	3.6	83	10.1	3	0.4	819
	1989	669	79.3	59	6.9	110	13.0	7	0.8	844
Oceania	1980	25	18.6	81	60.8	21	15.6	7	5.0	133
	1989	37	5.3	506	73.2	114	16.4	36	5.1	692
USSR	1980	790	14.9	3,607	68.2	621	11.7	270	5.1	5,287
	1989	523	5.7	7,415	81.4	711	7.8	458	5.0	9,106
EEC	1980	973	38.5	916	36.2	149	5.9	492	19.4	2,529
	1989	1,172	16.5	5,102	72.0	285	4.0	532	7.5	7,090
DME	1980	3,351	34.3	5,035	51.5	980	10.0	403	4.1	9,769
	1989	3,319	17.0	14,126	72.2	1,386	7.1	732	3.7	19,562
Africa (Developing)	1980	3,773	78.3	111	2.3	400	8.3	536	11.1	4,821
	1989	4,855	78.0	152	2.4	473	7.6	742	11.9	6,222
Latin America	1980	4,136	86.1	73	1.5	385	8.0	211	4.4	4,805
	1989	4,519	87.9	33	0.6	371	7.2	218	4.2	5,140
WANA	1980	1,347	75.4	183	10.3	170	9.5	86	4.8	1,786
	1989	2,033	68.1	456	15.3	347	11.6	148	4.9	2,983
Asia	1980	15,748	78.3	2,198	10.9	1,452	7.2	720	3.6	20,117
	1989	17,154	78.6	2,423	11.1	1,424	6.5	837	3.8	21,838
Dev.g Mkt. Econ.	1980	25,011	79.3	2,565	8.1	2,408	7.6	1,553	4.9	31,537
	1989	28,572	78.9	3,064	8.5	2,615	7.2	1,944	5.4	36,194

Table 2. Pea utilization, by category, 1980 and 1990 (000 MT)

Region	Year	Food	% Total Use	Feed	% Total Use	Seed	% Total Use	Other Uses	% Total Use	Total Use
World	1980	3,622	41.6	3,925	45.0	796	9.1	371	4.3	8,715
	1989	2,701	17.5	11,003	71.4	1,107	7.2	598	3.9	15,408
North America	1980	58	61.6	17	17.6	18	18.7	2	2.1	95
	1989	61	38.2	44	27.3	49	30.7	6	3.8	160
Oceania	1980	17	27.6	26	43.6	13	21.0	5	7.7	60
	1989	17	8.1	114	55.9	59	29.1	14	6.9	203
USSR	1980	724	16.1	3,034	67.5	508	11.3	230	5.1	4,496
	1989	405	5.2	6,417	82.4	573	7.4	392	5.0	7,786
EEC	1980	230	43.9	247	47.1	42	8.1	5	1.0	525
	1989	404	9.9	3,447	84.1	182	4.4	65	1.6	4,098

Table 2. (Continued)

Region	Year	Food	% Total Use	Feed	% Total Use	Seed	% Total Use	Other Uses	% Total Use	Total Use
DME	1980	1,117	20.6	3,435	63.3	614	11.3	263	4.8	5,430
	1989	1,030	8.0	10,451	80.9	931	7.2	511	4.0	12,923
Africa	1980	313	88.4	5	1.5	18	5.1	18	5.0	354
(Developing)	1989	261	88.6	1	0.3	18	5.9	15	5.1	295
Latin America	1980	153	91.1	–	–	10	6.2	5	2.8	168
	1989	160	90.1	1	0.6	10	5.4	7	3.9	178
WANA	1980	48	87.9	–	–	5	8.5	2	3.6	55
	1989	68	86.6	–	–	8	9.6	3	3.8	79
Asia	1980	1,991	73.5	484	17.9	150	5.5	84	3.1	2,708
	1989	1,182	61.1	550	28.4	141	7.3	63	3.2	1,936
Dev.g Mkt.	1980	2,505	76.3	490	14.9	182	5.5	108	3.3	3,285
Econ.	1989	1,671	67.2	552	22.2	176	7.1	87	3.5	2,485

Table 3. Other pulse: utilization, by category, 1980 and 1989 (000 MT) [1]

Region	Year	Food	% Total Use	Feed	% Total Use	Seed	% Total Use	Other Uses	% Total Use	Total Use
World	1980	13,421	70.0	3,326	17.4	1,476	7.7	947	4.9	19,171
	1989	15,930	64.9	5,546	22.6	1,776	7.2	1,310	5.3	24,562
North America	1980	28	57.8	12	25.2	8	15.6	1	1.4	49
	1989	50	63.9	16	20.0	11	14.2	2	1.9	78
Oceania	1980	1	1.0	54	84.5	8	11.9	2	2.6	64
	1989	16	3.2	393	81.2	54	11.2	22	4.4	484
USSR	1980	4	0.6	573	79.6	106	14.8	36	5.0	720
	1989	35	2.8	998	81.3	133	10.8	62	5.0	1,227
EEC	1980	306	30.3	583	57.6	84	8.3	39	3.9	1,012
	1989	357	17.9	1,500	75.4	87	4.4	47	2.3	1,991
DME	1980	435	19.5	1,476	66.3	227	10.2	88	4.0	2,226
	1989	566	12.7	3,411	76.4	328	7.3	159	3.6	4,462
Africa	1980	2,229	75.2	106	3.6	259	8.7	370	12.5	2,964
(Developing)	1989	3,097	75.1	151	3.7	331	8.0	545	13.2	4,123
Latin America	1980	483	76.7	73	11.5	36	5.8	38	6.0	630
	1989	531	83.0	32	5.0	34	5.3	43	6.7	640
WANA	1980	990	71.3	183	13.2	144	10.4	72	5.2	1.389
	1989	1,637	65.0	456	18.1	295	11.7	132	5.2	2,519
Asia	1980	9,277	77.6	1,488	12.4	808	6.8	379	3.2	11,952
	1989	10,088	78.8	1,497	11.7	789	6.2	433	3.4	12,806
Dev.g Mkt.	1980	12,986	76.6	1,850	10.9	1,248	7.4	860	5.1	16,945
Econ.	1989	15,364	76.4	2,136	10.6	1,449	7.2	1,153	5.7	20,101

[1] Includes faba bean, chickpea, lentil, cowpea, pigeon, other pulses NES.

Processing and animal feeds

Aspects of the nutritional quality and use of cool season food legumes in animal feed

J. HUISMAN and A.F.B. VAN DER POEL
TNO-ILOB, Department of Animal Nutrition and Physiology, P.O. Box 15, 6700 AA Wageningen, The Netherlands

Abstract

Cool season food legumes are good sources of protein for pigs and poultry. Special attention has to be paid to the sulphur amino acids, threonine, and tryptophan, because they are less concentrated when compared with soybean. The amino acid profile of the storage proteins can vary considerably. This knowledge can possibly be used to improve protein quality by genetic manipulation. Large concentrations of carbohydrates can, besides flatulence, affect the small intestinal digestion. The antinutritional factors (ANFs) in peas, and tannins in faba beans, reduce protein digestibility. Chickens are less sensitive to ANFs than piglets. The native proteins of pea and faba bean are highly digestible (< 90%), indicating high nutritional quality of their proteins. Various aspects for upgrading the nutritional value of pulses are discussed, including thermal treatments, fractionation procedures, and germination.

Introduction

The major part of protein in animal diets is of vegetable origin. Apart from soybean and grains, the main protein sources in these diets are cool season food legumes seeds including pea, faba bean, chickpea, and lentil. The protein concentrations in pulses can vary between 15 and 40% while having well-balanced amino acid profiles (Bressani and Elias, 1988). A large component in food legumes is the carbohydrate fraction that represents up to 60% of total seed weight (Bressani and Elias, 1988).

The cool season grain legumes are included in animal feed mainly in raw form because the cooking or heating of faba beans and peas has until now been considered too expensive for animal nutrition. For soybean this is different because toasting is part of the oil extraction procedure.

In raw legumes various so called antinutritional factors (ANFs) are present which affect the digestion and performance of animals (Huisman, 1990). Nutritional value can be improved by using various technologies such as: heat

F.J. Muehlbauer and W.J. Kaiser (eds.), Expanding the Production and Use of Cool Season Food Legumes, 53–76.
© 1994 *Kluwer Academic Publishers.*

treatments, biotechnological processes, and plant breeding. Possibilities of improving nutritional value to increase the use of grain legumes in animal feeding, will be discussed in this paper.

Because most of the grain legumes are used in diets for monogastric animals, this paper deals only with monogastrics.

Some Aspects of the Chemical Composition

The main components in grain legumes are proteins and carbohydrates. Protein concentrations can vary distinctly between batches (Bressani and Elias, 1988; Wiseman and Cole, 1988; Singh *et al.*, 1990). Marquardt and Bell (1988) summarized data from the literature and quoted a mean protein concentration of 23.8% for peas, 24.2% for lentils, 22.3% for chickpeas, 28.6% for Canadian faba beans, and 24% for UK faba beans. An important criterion for protein quality is the amino acid composition. A comparison between soybean, pea, lentil, chickpea, and faba bean is given (Table 1). Compared with soybean, lysine concentration of the pulses tends to be higher and the other essential amino acids methionine + cystine, threonine, and tryptophan lower. Therefore, special attention needs to be paid to the amino acid profile when pulses are incorporated in the diets of pigs and poultry. This can be done by using other protein sources richer in these particular amino acids or by using biosynthetic amino acids. Lysine and methionine are relatively cheap and can be economically incorporated into animal diets. Natural sources of tryptophan are scarce and compensating for tryptophan deficiency by synthetic tryptophan in a diet is more expensive and could be uneconomical.

As pointed out by Marquardt and Bell (1988) pulses have more lysine but smaller concentrations of sulphur containing amino acids when compared to

Table 1. Amino acid profiles (grams per 100 grams protein) in soybean, lentil, chickpea, and faba bean

Amino Acid	Soybean		Pea		Lentil*		Chickpea		Faba bean		
	A	B	A	B	B	C	B	D	A	B Can.	B UK
Lysine	6.4	6.7	7.2	6.7	7.2	7.3	6.2	7.1	6.3	7.0	6.3
Methionine + Cystine	3.1	3.0	2.4	2.0	1.7	1.4	2.0	2.5	2.1	1.7	1.2
Threonine	4.2	4.1	3.9	3.9	4.0	3.5	3.4	3.5	3.7	2.7	4.0
Tryptophan	1.3	1.4	0.9	1.0	--	--	--	0.9	0.8	--	1.0

A.   Source: CVB Table, 1990. The amino acid contents in this table are related to soybean meal with a crude protein content of 423 to 476 g kg^{-1} and to pea with a crude protein content of 207 to 229 g kg^{-1}. For faba bean no distinction is made in this table between the white- and the colored-flowered cultivars, although the mean crude protein contents are indicated to be significantly different (305 g kg^{-1} and 254 g kg^{-1}, respectively).
B.   Adapted from Marquardt and Bell (1988). A distinction is made between Canadian faba beans (Can.) and faba beans from the United Kingdom (UK).
C.   Adapted from Bhatty and Christison (1984).
D.   Cordese (1990).
*   Savage (1988) summarized numerous literature reports and found the following values for lentil: lysine 5.57 to 9.60, Methionine 0.42 to 1.28, cystine 0.31 to 1.87, threonine 3.00 to 4.60 and tryptophan 0.5 to 1.1.

the cereals. In this respect, cereals and legumes are nutritionally complementary; amino acids that are deficient in one are generally adequate in the other.

The amino acid profile of the storage proteins within a given legume can vary considerably as demonstrated by Leterme *et al.* (1990) for peas. The amino acid profile of these fractions is given in Table 2. In the whole grain, the globulins represent 47.2%, the albumin 17.2%, the insoluble protein 17.7%, and the non-protein material 17.9% of the N compounds, respectively. A striking observation is that the concentration of the essential amino acids, lysine and methionine, is distinctly higher in albumins than in globulins. The globulins of pea consist of the legumin and the vicillin fraction, both varying widely in amino acid profile (Gueguen and Barbot, 1988). It is questionable whether this variability can be used to improve the amino acid profiles. The first attempts were not promising and failed for peas (Schroeder, 1982) and faba beans (Duc and Lacassagne, 1990). Genetic manipulation may be considered as a way of improving protein quality (De Lumen, 1990).

Table 2. Amino acid (AA) profile of the protein fractions of peas (grams per 16 grams nitrogen) (Leterme *et al.*, 1990)

	Whole grain	Albumins	Globulins	Insoluble proteins	Non-protein material
Essential AA					
Arginine	6.84	4.09	8.08	6.74	1.34
Histidine	2.52	2.87	2.23	2.36	0.69
Isoleucine	3.33	4.43	4.74	5.28	0.92
Leucine	6.58	5.22	8.94	8.57	1.58
Lysine	6.84	10.33	7.01	7.65	2.78
Methionine	1.03	0.97	0.61	1.20	0.26
Phenylalanine	4.19	4.55	5.49	5.64	1.09
Threonine	3.59	4.52	3.24	4.73	0.94
Tryptophan	0.94	1.23	0.97	1.08	0.0
Valine	3.89	5.21	4.86	5.83	1.29
Non-essential AA					
Alanine	4.27	5.19	4.00	5.46	1.83
Aspartic acid	10.68	11.40	11.96	10.85	4.62
Cystine	1.55	3.13	1.18	1.33	0.70
Glutamic acid	16.92	16.61	16.94	13.67	8.51
Glycine	4.32	6.41	3.84	4.92	2.57
Proline	3.76	5.38	4.47	5.28	1.72
Serine	4.79	4.11	4.87	4.92	1.16
Tyrosine	3.16	4.32	3.40	4.19	1.00
Total	89.20	99.97	96.83	99.70	33.00

The carbohydrates mainly consist of starch but oligosaccharides are also present in grain legumes (Table 3). Some examples of the carbohydrate composition of grain legumes are given.

Saini (1989) stated that oligosaccharides are present in a wide variety of grain legumes and vary in their concentration among leguminous species and within cultivars of a given legume. The oligosaccharides of the raffinose family can

Table 3. Mean carbohydrate composition of pea, lentil, chickpea, and faba bean

Component	Pea		Lentil		Chickpea		Faba bean	
	A	B	A	B	A	C	A	B
Total carbohydrates	57	--	60	--	61	--	58	--
Starch	43	39	44	42	44	--	47	44
Amylose	28	--	33	--	39	--	29	--
Glucose	--	0.1	--	0.05	--	--	--	0.04
Sucrose	--	2.0	--	1.6	--	--	--	2.1
Raffinose	--	0.8	--	0.3	--	1.0	--	0.3
Stachyose	--	2.3	--	2.6	--	2.5	--	0.9
Verbascose	--	2.8	--	1.0	--	4.0	--	1.8
Total oligo- saccharides	--	5.9	--	3.8	--	--	--	3.0
Total sugars	7.0	8.0	5.2	5.5	--	--	5.1	5.2

A   % in seed, adapted from Reddy et al. (1984).
B   % in dry matter, adapted from Bhatty and Christison (1984).
C   Adapted from Saini (1989).

induce flatulence in monogastrics. Flatulence is due to the action of the intestinal anaerobic microflora in the lower intestinal tract. Large concentrations of oligosaccharides can impair the utilization of nutrients in the grain legumes and cause flatulence symptoms such as diarrhea, nausea, cramps, and discomfort in animals (Saini, 1989).

Huisman and Le Guen (1991) demonstrated that pea carbohydrates also may affect the digestion in the small intestine of piglets. Piglets (live weight 15 kg) were fed diets containing casein + herring meal or pea protein isolate from "Finale" or pea protein isolate from "Frijaune" as the sole protein source. In these diets, two carbohydrate sources were used: either only corn carbohydrates, or a mixture of corn carbohydrates and pea carbohydrates. The piglets were cannulated and ileal chyme was collected and analyzed. The composition of the diets and the results of ileal digestibility and other criteria are summarized in Table 4.

Inclusion of pea carbohydrates did not clearly affect the ileal apparent protein digestibility. The dry matter contents of the ileal chyme were less for the piglets fed diets containing pea carbohydrates, while the amounts of ileal chyme produced was greater from diets containing no pea carbohydrates. The increased chyme production may be related to the fact that the pea carbohydrates were incompletely digested in the small intestine. As a result,

Table 4. Effects of pea carbohydrates on digestion. (Huisman and Le Guen, 1991)

	Diet I	Diet II	Diet III	Diet IV	Diet V	Diet VI
Ingredients						
Casein	12.5	12.5	--	--	--	--
Herring meal	6.9	6.9	--	--	--	--
'Finale' protein isolate	--	--	18.4	18.4	--	--
'Frijaune' protein isol.	--	--	--	--	17.9	17.9
Corn starch	42.0	30.0	42.0	30.0	42.2	30.7
Pea carbohydrates from 'Finale' a)	--	26.1	--	26.1	--	--
Pea carbohydrates from 'Frijaune' a)	--	--	--	--	--	23.9
Pea hulls	--	3.4	--	3.4	--	3.4
Dextrose	22.8	12.7	23.0	13.0	23.4	14.9
Cellulose (Arbocel 3800)	7.0	3.0	7.0	3.0	7.0	3.0
Vitamins, minerals, amino acids	8.7	5.4	9.6	8.9	9.5	6.2
Analysed contents (%):						
Crude protein	17.2	18.2	18.1	18.9	16.9	17.6
Glucose	16.3	10.6	16.9	10.3	17.2	13.2
Saccharose	<dl	1.5	<dl	1.3	<dl	0.8
Raffinose	<dl	1.1	<dl	1.1	<dl	1.3
Stachyose	<dl	0.3	<dl	0.3	<dl	0.2
Arabinose	<dl	1.7	<dl	1.4	<dl	1.8
Xylose	0.3	0.6	0.3	0.7	0.5	0.8
Results:						
Ileal digestibility						
- protein	87.1	84.8[bcde]	88.2[a]	86.7[a]	85.8	86.0[a]
- dry matter	86.1[a]	82.7[bcde]	84.1[ce]	80.8	84.5	81.9[bcd]
Dry matter content in ileal chyme	13.6	11.1[b]	12.1	10.9	14.1	13.2
Amount of chyme excreted (g 12 h^{-1})	244	409	323	424	275	347[c]

	Composite data Diets I, III, and V	Effect of pea carbohydrates Diets II, IV, and VI
Protein digestibility	87.1	85.7[a]
Dry matter content ileal chyme (%)	13.2	11.7
Amount of chyme (g 12 h^{-1})	274	393

Means with superscript that do not have a common letter in the same row, differ significantly (P < 0.05).
a) The amounts of pea carbohydrates in the diets were comparable with an dietary inclusion of 60% raw peas.

more osmotically active compounds will be present in the chyme, requiring a greater inflow of water into the intestinal lumen to maintain osmotic equilibrium between blood and lumen content. The results show that due to the inclusion of pea carbohydrates comparable with dietary inclusion of 60% raw peas, there was an increase in the flow of ileal chyme.

Constraints to the Use of Legumes

The main factor restricting the use of legumes seems to be the presence of ANFs, and to a lesser extent, as discussed in Chapter 2, some components of the carbohydrate fraction.

ANFs in Pulses

Legumes contain various ANFs (Liener, 1980, 1989; Savage, 1988; Wiseman and Cole, 1988; Savage and Deo, 1989; Huisman *et al.*, 1990d; Huisman, 1991; Huisman and Jansman, 1991; Sissons and Tolman, 1991; Huisman and Tolman, 1992). Literature reviews show that cereals generally contain less ANFs than pulses. Exceptions are tannins in some barley cultivars and in sorghum. The main ANFs in pulses are protease inhibitors, lectins, tannins, vicine/convicine, and flatulence factors. Other less relevant ANFs present in pulses are: α-amylase inhibitors, saponins, and phytate (Liener, 1989). Many legumes contain more than one type of ANF. For example, Liener (1981) mentioned the presence in soybean of trypsin inhibitors, haemagglutinins (lectins), goitrogens, anti-vitamin factors, phytates, saponins, oestrogens, flatulence factors, and allergens. Negative effects on performance may therefore be attributable to various factors.

Trypsin Inhibitors (TIs)

Related to animal nutrition there are two important families of trypsin inhibitors (TIs): the soybean trypsin inhibitor (STI) and the Bowman-Birk trypsin inhibitor (BBI) (Liener and Kakade, 1980; Birk, 1989). STI is mainly found in soybean, while the TIs in pulses seem to belong predominantly to the BBI family (Birk, 1989). STI inhibits trypsin and to a lesser extent, chymotrypsin and is inactivated by heat and gastric juice. BBI inhibits both trypsin and chymotrypsin. BBI is resistant to gastric juice and to various proteolytic enzymes. There are conflicting reports about the heat stability of BBI. Some report heat stability but others report inactivation by heat (summarized by Huisman and Jansman, 1991). The primary effect of trypsin inhibitors in the animal is related to the inactivation of (chymo)trypsin produced by the pancreas. Due to this inactivation the activity of (chymo)trypsin may be reduced and, via a negative feedback mechanism, the pancreas is stimulated to secrete more enzymes (Liener and Kakade, 1980; Gallaher and Schneeman, 1986; Birk, 1989). These effects may lead to a reduced apparent protein digestibility due to an increased loss of endogenous protein originating from the pancreas. Pancreatic enzymes are relatively rich in S-containing amino acids. Thus, when TIs are present in seeds, special attention should be paid to these amino acids (Liener and Kakade, 1980). For pulses this is very important because they are already deficient in sulfur-containing amino acids.

Lectins

The negative effect of lectins in animals is related to the binding of lectins to the mucosa of the intestinal wall. Due to this binding, the intestinal epithelial cells may be damaged which results in a decreased absorption of nutrients, a change in the activity of brush border enzymes, hypersecretion of endogenous protein due to the shedding of damaged cells, an increased production of mucins, and a loss of plasma proteins to the intestinal lumen (Jaffé, 1980; Kik *et al.*, 1989; Pusztai, 1989). These effects may cause decreased nitrogen digestibility and retention and, occasionally, scouring, resulting in a reduced weight gain and a less efficient feed conversion. There are distinct differences in the toxicity of lectins. Most lectin research has been carried out with those of *Phaseolus vulgaris*, which were exceptionally toxic (Liener, 1986; Pusztai, 1989). Reports on the toxicity of lectins from pea, lentil, and faba bean are scarce and sometimes conflicting. Pea lectins were found to be non-toxic in piglets by Bertrand *et al.* (1988), but toxic in piglets by M. J. L. Kik (personal communication), as well as in rats by Jindal *et al.* (1982). Lentil lectins were found to be toxic by Jindal *et al.* (1982). M. J. L. Kik (personal communication) compared the toxicity of various lectins in piglets and found that pea lectins and faba bean lectins caused some changes in the mucosa of the intestinal wall, but that these changes were not as severe as those caused by the lectins in *Phaseolus* beans. Jaffé (1980) reported that soybean lectins are less toxic than those of kidney bean. Results reported in the literature are difficult to interpret for various reasons:

– Most research was carried out on small laboratory animals, while Huisman *et al.* (1990a,b,c) demonstrated that piglets are distinctly more sensitive to ANFs in *Phaseolus* beans and peas than rats and chickens.
– Most research was carried out using semi-synthetic diets and the usual concentrations of lectins in these diets are distinctly higher than in practice. It is likely that at lower practical concentrations the effects are less pronounced or possibly even absent.

Tannins

Tannins are present in the colored seeds of faba beans, and other pulses. Tannins form complexes with proteins, carbohydrates, and other polymers in feed and food (Rao and Prabhavathi, 1982). Tannins form complexes more easily with proteins than with carbohydrates. Due to the formation of these complexes, protein digestibility is reduced (Liebert and Gebhardt, 1983; Wiseman and Cole, 1988; Jansman *et al.*, 1989). The mode of action in animals is not entirely clear. Besides the complexes of tannins with feed protein, the ingested tannins may form complexes with digestive enzymes, thus leading to a decreased enzyme activity that may result in decreased nutrient digestibility (Griffiths and Mosely, 1980; Marquardt, 1989). Other antinutritional effects of tannins are damage to the intestinal mucosa and interference with mineral absorption (Mitjavila *et al.*, 1977). In summary, tannins interfere with various aspects of digestion and result in reduced growth and poorer feed conversion efficiency (Marquardt, 1989).

Vicine and Convicine

Vicine and convicine are present in faba bean. Intestinal microflora hydrolyse these agents to divicine and isouramil (Frolich and Marquardt, 1983). These two factors have been implicated in haemolytic anemia (favism) in man. There are some indications that fertility may be negatively affected in non-ruminant farm animals. Marquardt (1989) stated that, among other effects, vicine and convicine could decrease fertility and hatchability of the eggs, but these effects were questioned by Wiseman and Cole (1988). Nielsen and Kruse (1974) found smaller numbers of born piglets in litters when large amounts of faba beans were fed to sows. There is thus a need for more research into the effects of vicine and convicine in pigs and poultry.

Flatulence Factors

As discussed in Chapter 2 [pp. 000–000] flatulence factors are related to oligosaccharides that are fermented by bacteria in the large intestine. There, they are broken down by bacterial α-1,6-galactosidase to monomers that are converted into volatile fatty acids, carbon dioxide, hydrogen, and methane. This results in flatulence, diarrhea, nausea, cramps, and discomfort of the animal.

Some Aspects of the Effect of ANFs on the Digestibility of Faba Bean and Pea

Faba bean

Many reports show that faba beans have a low nutritional value due to the presence of ANFs, particularly tannins. Most of this research was carried out in small laboratory animals, such as rats and chickens. It is, however, questionable whether the effects of tannins found in rats and chickens can be extrapolated to pigs. Moreover, very large amounts of faba beans or tannins are often used in the experimental diets. Jansman *et al.* (1989) evaluated four cultivars of faba bean with different tannin contents at the inclusion amount of 300 g kg^{-1} diet for piglets and chickens. In piglets, fecal and ileal apparent digestibility were measured; and, in chickens, growth and feed conversion efficiency were determined. The contents of ANFs in these faba beans are summarized in Table 5.

The four cultivars of faba beans were evaluated in a trial with piglets initially weighing 9 kg. In Table 6, the ileal and fecal digestibility of protein and N-free extract (NFE) are given. The results clearly show that in piglets the apparent ileal and fecal digestibility of crude protein decreases with increasing concentrations of tannins in faba beans. Not only is the ileal apparent digestibility of crude protein lower than the fecal apparent digestibility, but at the ileal level the differences are more pronounced than at the fecal level. From a nutritional point of view, ileal apparent digestibility of crude protein is more

Table 5. Levels of ANF's in samples of four cultivars of faba bean

Cultivar	Flower color	Tannins*	Trypsin inhibitors**	Lectins (HA)***
		g kg⁻¹	mg g⁻¹	
'Blandine'	White	0.2	1.3	5
'Herz Freya'	Colored	4.0	1.4	5
'Mythos'	Colored	9.8	1.6	5
'Alfred'	Colored	9.6	0.7	2

*   Measured according to the vanillin sulphuric acid method.
**   Expressed as mg trypsin inhibited per gram product.
*** Haemagglutination units with rabbit red blood cells; one HA
       is 1:1000 dilution step.

important than the fecal apparent digestibility because protein digested in the
large intestine does not contribute to the animal's protein deposition (Wünsche
et al., 1982). A lower apparent protein digestibility in pigs due to the presence
of tannins has also been reported by Liebert and Gebhardt (1983) for faba beans
and by Hlödversson (1987), Buraczewska *et al.* (1989), and Hauschild and
Köhler (1991) for peas.

Table 6. Ileal and fecal apparent digestibility coefficients of crude protein and N-free extract (NFE)
of cultivars of faba bean (300 g kg⁻¹ inclusion amounts) in piglet diets

Cultivar	Flower color	Crude protein		NFE	
		ileal	fecal	ileal	fecal
'Blandine'	White	85.3[a]	89.3[a]	72.9[a]	94.3[a]
'Herz Freya'	Color	75.3[b]	85.2[ab]	75.4[a]	93.0[ab]
'Mythos'	Color	74.1[b]	82.4[bc]	72.7[a]	90.7[bc]
'Alfred'	Color	68.7[b]	79.4[c]	69.1[a]	89.5[c]

Values with a different superscript within a column differ
significantly at P < 0.05.

The NFE ileal apparent digestibility was about 20% lower than the fecal
digestibility. This indicated that a substantial part of the NFE is digested in the
large intestine. There are more energy losses when carbohydrates are digested in
the large intestine than when digested in the small intestine. Only at fecal level
there were slight indications that tannins may decrease the NFE digestibility.

    The same batches of faba bean cultivars were evaluated in experiments with
broilers. The experiment started when the birds were five days old. The faba
beans were tested in experiments using isonitrogenous and isocaloric (ME)

diets, balanced for digestible amino acid content (Jansman *et al.*, 1989). The results are summarized in Table 7. The results shown in Table 7 indicate that diets had no significant effect on feed intake, weight gain, and feed conversion efficiency. This is in contrast to the observations of Marquardt *et al.* (1977), Marquardt and Ward (1979), and Laccassagne *et al.* (1988), which demonstrate negative effects on digestibility and other performance criteria with large dietary amounts of isolated tannins or tannin-rich faba beans. The results demonstrated that tannin-rich faba beans can be included up to 30% in the diet without causing negative effects on performance. This conclusion is confirmed by the work of Wiseman *et al.* (1991) who found that even 40% dietary inclusion of faba beans with large tannin concentrations did not exert negative effects in chicks, and by Jeroch *et al.* (1985) with a dietary inclusion up to 45% peas with large tannin concentrations.

Results with piglets and broilers show that piglets are more sensitive to tannins than chickens. This species-dependent sensitivity agrees with the results

Table 7. Feed intake, weight gain, and feed conversion efficiency of broilers fed diets containing faba beans differing in tannin content

Diet	Feed intake	Weight gain	Feed conversion efficiency
	--g--	--g--	kg feed per kg weight gain
Control diet	1592	1039	1.53
30% 'Blandine' in the diet	1547	1038	1.49
30% 'Herz Freya' in the diet	1594	1047	1.52
30% 'Mythos' in the diet	1600	1058	1.51
30% 'Alfred' in the diet	1589	1046	1.52

Adapted from Jansman *et al.*, 1989.

of Huisman (1990) and Huisman *et al.* (1990a,b,c) who found that the piglet is also more sensitive than chickens to feeding *Phaseolus* beans and peas.

The results of Jansman *et al.* (1989) and Wiseman *et al.* (1991) demonstrate that in chick diets, tannin amounts of up to 3 to 4 g kg^{-1} diet (based on measurements according to the vanillin sulphuric method) can be tolerated. This level is, however, too high for piglets.

Pea

Peas are a valuable source of protein for both animals and man. However, various reports have shown that the inclusion of more than 150 g kg^{-1} raw peas in diets for young piglets may exert negative effects on performance (Fekete *et al.*, 1984; Castaing and Grosjean, 1985; Grosjean and Gatel, 1986, 1989; Bengala-Freire *et al.*, 1989). These effects are not due to tannins, since modern pea cultivars are white-flowered and tannin-free. Different factors may explain the reduced performance. Many authors have discussed the possible relation of poor performance with the presence of ANFs in pea (Leterme *et al.*, 1989; Gatel and Grosjean, 1990). Peas contain large concentrations of carbohydrates (60 to 63%), consisting predominantly of starch (48 to 50%) (Grosjean and Gatel, 1986) and some α-galactosides (5 to 10%) (Savage and Deo, 1989). α-Galactosides are not broken down in the small intestine because of a lack of appropriate digestive enzymes. These components are hydrolyzed in the large intestine by bacterial fermentation. This fermentation results in energy losses and flatulence with diarrhea also may occur (Saini, 1989). Therefore, it cannot be ruled out that these carbohydrates also play a role in the negative effect on performance.

To determine which factors are responsible for the impaired performance, first the ileal apparent protein digestibility of different cultivars was measured in young piglets of 10 to 15 kg live weight. The ileal apparent protein digestibility coefficient was between 72 and 79. These values for peas are distinctly lower than those obtained with adequately heat-treated soybean flour.

The next step was to investigate which factor in pea is responsible for the low ileal apparent protein digestibility: the ANFs or the carbohydrates. To study this, Huisman and Le Guen (1991) prepared a pea protein isolate with very small concentrations of ANFs and free of pea carbohydrates, an ANF-rich concentrate, and a fraction consisting exclusively of soluble and insoluble pea carbohydrates from a batch of raw peas. These fractions were used in digestibility experiments with piglets. The effects of including pea carbohydrates on ileal protein digestibility have already been given in Table 3. These results showed that the inclusion of pea carbohydrates in the diets of piglets did not affect the apparent protein digestibility.

Table 8 gives the results of the inclusion of ANFs. The ileal apparent digestibility of crude protein of the pea protein isolate was about 14% greater than that of raw peas. When the ANFs were added to the pea protein isolate the ileal apparent protein digestibility was depressed about 7%. Thus, half the difference in ileal apparent protein digestibility between raw peas and pea protein isolate could be explained by the addition of ANFs. A point of discussion is which other factor(s) may be responsible for the lower protein digestibility in raw peas. Le Guen *et al.* (1991) indicated that pea protein may be antigenic.

Table 8. Apparent ileal protein digestibility and live weight gain measured in raw peas and pea protein isolate enriched with ANF's

	Raw peas	Pea protein isolate
Lectin content (μg g^{-1})	2301	405
TIA (mg inhibited trypsin g^{-1})	1.2	< 0.1
Ileal protein digestibility	72	86[b]

	Pea protein isolate + ANF's	Pea protein isolate
Lectin content (μg g^{-1})	2732	507
TIA (mg inhibited trypsin g^{-1})	1.2	0.1
Ileal protein digestibility	79	86[b]
Live weight gain (g d^{-1})	229	277[b]

Means in the same row that do not have the same superscript differ significantly at $P < 0.05$.

True Ileal Protein Digestibility of Peas and Faba Beans

It is not clear to what extent the low ileal apparent protein digestibility is due to a low digestibility of the natural protein in the seed or to an increased secretion of endogenous protein originating from the animal. By correcting the apparent protein digestibility for the part contributed by endogenous protein, the true protein digestibility is calculated. To study this, experiments were carried out in which the excretion of endogenous protein was measured using the radio labelled N-dilution technique. With this technique the body protein, including excreted endogenous protein, is labelled (Souffrant *et al.*, 1986). Using the data for the labelled excreted endogenous protein, a distinction can then be made between excreted undigested feed protein and excreted endogenous protein. White-flowered tannin-free "Blandine" and colored-flowered "Alfred" faba beans and two white-flowered pea cultivars, Finale and Frijaune, were tested using the N-dilution technique. Each batch of peas or faba beans was tested on three animals. Details of the experimental technique have been described by Huisman *et al.* (1992).

In Table 9 some results obtained with peas and faba beans are summarized. For faba beans, the difference between apparent and true protein digestibility was 16% for Blandine and 17% for Alfred, for peas it was 16% for Finale and 19% for Frijaune. The true protein digestibility figures show that the natural protein of faba beans and peas is almost entirely digested in the small intestine. The low apparent digestibility must therefore be attributed to an increased secretion of endogenous protein.

Table 9. Apparent and true protein digestibility of peas and faba beans

	Apparent protein digestibility	True protein digestibility
	-%-	-%-
Spring pea (low ANF's)	79[a]	95[b]
Winter pea (high ANfs)	74[a]	93[b]
Faba bean 'Blandine' (Low tannin cultivar)	79[a]	95[b]
Faba bean 'Alfred' (high tannin cultivar)	74[a]	91[b]

Means with superscript that do not have a common letter in the same row, differ significantly at P < 0.05.

Data with peas are from Huisman *et al.* (1992) and those with faba beans from A. J. M. Jansman, personal communication.

This information is important for studies aimed at increasing the nutritional value of faba beans and peas by processing and plant breeding. Clearly, processing and plant breeding objectives for upgrading the nutritional quality of peas and faba beans need to be focused on reducing factors stimulating the excretion of endogenous protein and not on the quality of the natural protein in terms of digestibility. This information is also a guide to biotechnologists in their attempts to develop enzymes to improve the protein digestibility by reducing the ANFs. For peas and faba beans, proteinases will not have a positive effect on the apparent protein digestibility because the natural protein is already highly digestible. It may be important to find enzymes that breakdown those factors causing a hypersecretion of endogenous protein.

Opportunities for Extending the Use of Cool Season Food Legume Seeds

Several approaches, including breeding and processing techniques, can be considered to extend the usage of cool season food legumes. These approaches are all aimed at improving the nutritional value of seeds, especially the protein quality. Numerous experiments have been carried out on protein quality, directed at either the apparent ileal or fecal digestibility of protein/amino acids, at the true digestibility, or the amino acid availability.

For reasons of process optimization or specific breeding objectives it is necessary to distinguish between true and apparent digestibility measurements. By knowing both true and apparent digestibility one can, for example, distinguish between a large endogenous protein secretion or a low digestibility of the protein as such. This partitioning is important for the choice of specific objectives and thus is of relevance to the improvement of nutritional protein by: (1) the removal of antinutritional factors from seeds, and/or (2) a change in storage protein quality. These two objectives may require different approaches and, subsequently, different costs can be involved.

Several methods can be used to improve the nutritional quality of legume seeds (Table 10).

Table 10. Possible treatments for upgrading the nutritional quality of legume seeds

1.	Breeding and genetic manipulation
2.	Feed formulation
	a.　selection of ingredients
	b.　supplementation with amino acids
3.	Processing
	a.　thermal treatments
	b.　enzyme treatments
	c.　fractionation procedures

Selection of Legume Seeds as a Diet Ingredient

Although some legume seeds are considered a potential component of diets for monogastric animals, it is known that performance of animals given diets containing legume seeds often falls short of what was expected from the chemical composition of the diets. Of course, this discrepancy becomes more apparent with larger inclusion amounts in these diets. For an extended use of legume seeds, small inclusion amounts because of possible negative effects of ANF is not a real option.

Of course, a further possibility lies in the selection of certain cultivars of legume seeds. Advantage can be taken of the genetic variation for ANF concentration or ANF activity by making use of breeding or genetic manipulation techniques. For example, the composition of the storage protein can be altered to produce an improved amino acid profile (Gueguen and Barbot, 1988) or a lower concentration of lectin (Osborn and Bliss, 1985). Also, the selection of tannin-free faba beans is quite possible based on the white-flowering character of reduced tannin germplasm and cultivars (Duc and Lacassagne, 1990; Wareham *et al.*, 1991; Van der Poel *et al.*, 1992a). The effects of reduced-tannin cultivars, however, may vary with the animal species used. Two near-isogenic faba bean lines were examined for apparent ileal digestibility of dry matter and nitrogen for piglets (Table 11).

Hlödversson (1987) determined the nutritional value of two white-and one dark-flowered pea cultivar in pigs (Table 12). All pea cultivars differed significantly with respect to the fecal digestibility of crude protein, but the dark-flowering cultivar had considerably lower digestibility coefficients than the other cultivars.

Providing that breeding leads only to low ANF cultivars, ways are open for the routine incorporation of larger amounts of these cultivars. In this case,

Table 11. Ileal digestibility values in piglets of low- and high-tannin (near-isogenic lines) faba beans and concentrations of antinutritional factors (Van der Poel *et al.*, 1992a)

	Faba bean cultivar	
	Low tannin	High tannin
Antinutritional factors (g kg[-1])		
Condensed tannins	<0.5	5.2
TIA[a]	1.61	1.47
Vicine + convicine	10.4	8.4
Functional lectins	13.3	11.6
Ileal digestibility (%)[b]		
Dry matter	69 ± 3	63 ± 2
Nitrogen	83 ± 3	72 ± 3

a Mg of trypsin inhibited (porcine)
b Mean and SE of mean; PVTC-methodology, 5 piglets per treatment; initial weight ca 23 kg; bean inclusion level: 20%.

Table 12. Apparent fecal digestibility[a] of nutrients and metabolizable energy content (ME) of pea cultivars (Hlödversson, 1987)

	Pea cultivar[b]		
	'Lotta' (W)	'Vreta' (W)	'Timo' (D)
Digestibility coefficients			
Organic matter	90	91	83
Crude protein	83	90	71
Crude fiber	79	62	29
Energy	85	88	76
ME content (MJ kg[-1] DM)	15.1	15.7	13.6

a W = white flowering D = dark flowering
b Growing-finishing pigs; weight range 31-55 kg; pea inclusion level 35%.

however, there should not be adverse consequences for agronomic characteristics such as resistance to disease and tolerance of adverse weather conditions. In addition, one must be certain that no other constituent of such cultivars other than ANFs (for example, storage proteins themselves) (Van der Poel, *et al.*, 1990c) will cause a significant negative effect on the feeding value before they can be safely included in diets without the need for additional processing.

Thermal Treatments

Besides changing the physical structure of pea starch (Bertrand *et al.*, 1982), the objective of thermal processing of pea for pigs also should be related to the inactivation of ANF, particularly the trypsin inhibitors (Grosjean and Gatel, 1989). For poultry, however, it has been reported that the breakdown of the cell wall of pea cotyledons is an important objective to provide an increased accessibility of nutrients to digestive enzymes (Carré et al., 1991).

Several different treatments have been reported with these purposes in mind. For example, the effects of processing conditions using twin-screw extrusion on the inactivation of ANF and on processing costs in two types of peas was investigated (Van der Poel et al., 1992b; Table 13).

Extrusion of peas showed the dependency of trypsin inhibitors and lectins on the processing variables used. For the round-seeded pea cultivar "Finale", both moisture and temperature proved to be important variables, although inactivation of TIA was complete for all temperatures under investigation (105 to 140°C). For a wrinkle-seeded pea "C306", it was temperature instead of moisture which largely inactivated the ANF.

In vivo trials should be used to evaluate improvement in the feeding value of legume seeds after processing. For early weaned piglets, Bengala Freire et al. (1991) observed a positive effect of extrusion of spring peas of 5% increase in the fecal digestibility of protein but only at large inclusion amounts of 45%. Seve et

Table 13. Effects of extrusion variables[a] on pea quality parameters (Van der Poel et al., 1992b; *with permission*)

Pea quality parameter	Raw pea	Extruded pea regression model[b]	coefficient of variation
(a) RS pea, cv Finale			
TIA (mg g⁻¹)	1.75	< 0.10	
HA (units g⁻¹)	16	$\ln Y = 6.270 - 0.077$ [T] $+ 0.145$ [M]	0.71
		$\ln Y = -0.485 + 0.174$ [PDI]	0.92
Tannins (%)	0.39	$Y = 0.513 - 0.002$ [T] $+ 0.003$ [M]	0.69
PDI (%)	70.4	$Y = 40.880 - 0.307$ [T] $+ 1.081$ [M]	0.79
(b) WS pea, cv C306			
TIA (mg g⁻¹)	3.56	$Y = 6.674 - 0.0473$ [T]	0.77
HA (units g⁻¹)	10	$\ln Y = 19.730 - 0.1410$ [T]	0.75
Tannins (%)	0.77	no significant correlations found	
PDI (%)	84.4	$Y = 33.690 - 0.2320$ [T] $+ 1.16$ [M]	0.89

[a]   All variables $\dot{P} < 0.05$
[b]   [T] = Temperature, [M] = Moisture; n = 14

al. (1985) did not observe a positive effect on the digestibility of amylolytic enzymes added to dehulled spring peas.

Veldman (1988) observed positive effects of pelleting or steam heating of faba beans before their inclusion into diets for growing pigs (Table 14). Besides the legume genotype, treatment and conditions of treatment, it should be emphasized that the diet form also may affect feeding value. Experimental evidence indicates that for pigs the digestibility of legume seeds also will be increased by their inclusion as pellets in the diets.

Poultry seem less sensitive to ANF in peas and *Phaseolus* beans (Huisman *et*

Table 14. Digestibility coefficients of mash diets with processed faba bean (Veldman, 1988)[a]

	Digestibility coefficient	
	Crude protein	Crude fat
Diet + raw faba bean	61	61
Diet + pelleted faba bean	64	68
Diet + steam heated faba bean	70	70
Lsd (P < 0.05)	42	84

a   Bean inclusion level: 50%

al., 1990a,b,c; Van der Poel *et al.*, 1990a,b). From the literature it can be concluded that the effect of technological treatments of legumes may vary with the type of legume and cultivar, the type of treatment, and the age of the birds. A short overview of these effects in poultry is reported by Gatel (1992) and the reader is referred to this publication.

There is no doubt that thermal treatments are effective in decreasing the level of proteinaceous ANF or even antigenicity of legume proteins. In addition, thermomechanical treatments such as pelleting may have an additional positive effect due to the breakdown of cell wall material.

Fractionation Procedures

Constituents of legume seeds have a different size or density that make them suitable for mechanical fractionation after certain pretreatments. These mechanical separation procedures such as dehulling techniques or milling and subsequent air classification, is much simpler than complex technologies involved in extractions with solvents.

Dehulling, for example, fractionates legumes into the seedcoat and the cotyledons. Air-classification after fine milling fractionates legumes into their main constituents, protein bodies and starch granules. This means that the yield and composition of the different fractions produced by these techniques are primarily fixed by the relative proportions of the constituents in these particular legume seeds. For feeding, utilization of these fractions will depend on the relative contribution of the unwanted components in a particular fraction to the negative effects observed with animals fed the raw seeds.

Dehulling of faba beans for example is a method by which the concentration of condensed tannins (primarily located in the seedcoat) is decreased. These treatments also may affect other components which therefore need to be validated in *in vivo* experiments. For faba bean, dehulling increases the nutritional value for pigs (Eggum, 1980; Wiseman and Cole, 1988; Van der Poel *et al.*, 1992c). The effect of dehulling of the faba bean Alfred on apparent ileal digestibility in pigs is shown in Table 15.

The results of this experiment confirm results of Eggum (1980) and of Wiseman and Cole (1988) that the dehulling of faba beans increases their nutritional value compared to that of whole faba beans. However, there are still factors hampering a high digestibility, since the protein digestibility of Blandine (Table 6) was not reached. Moreover, the true protein digestibility was above 90% (Table 9). This means that there is still a significant endogenous protein fraction passing out of the terminal ileum. It is further noted that an important feature associated with dehulling is the disposal of the hulls and that the improved nutritional quality of faba beans has to compensate for the extra costs associated with dehulling.

In flour milling of cereals it was discovered that the finest flour particles contain the largest protein concentrations. This information led to the development of commercial procedures for protein enrichment in fractions and to developments in air-classification of legume flours into protein-rich and starch-rich fractions. These fractions represent new raw materials that can be used as ingredients in industrial applications or in animal feeds depending on the chemical composition and the nutritional value of the crude fractions (Table 16). These nutritional characteristics, however, must include the known

Table 15. Effect of dehulling and dehulling/heat processing of faba bean on the level of ANF and apparent digestibility in pigs[a] (Van der Poel *et al.*, 1992c)

	Faba bean		
	Untreated	Dehulled	Dehulled + steam processed[b]
Antinutritional factors			
Tannins	6.7	< 0.01	< 0.01
TIA	0.64	0.81	0.54
Ileal digestibility			
Dry matter	63	72	71
Nitrogen	67	75	76
Starch	95	98	94

[a]   Pigs, 35 kg liveweight; faba bean inclusion level 30%
[b]   Autoclaving: 103°C for 20 minutes

Table 16. Yield and protein concentrations of whole and dehulled seeds (M) from pea and faba bean and their air-classified starch (S) and protein (P) fractions[a] (Vose *et al.*, 1976)

		Whole seed		Dehulled seed	
		Yield	Protein	Yield	protein
		-%-	-%-	-%-	-%-
				Pea	
Fraction	M	100	25.7	100	28.8
	P1	25	63.5	31	60.5
	S1	75	13.3	69	14.5
	P2	19	48.3	17	40.0
	S2	56	5.6	52	4.8
				Faba bean	
Fraction	M	100	27.9	100	31.9
	P1	28	66.1	30	69.0
	S1	72	15.1	70	16.5
	P2	14	51.1	18	49.6
	S2	58	5.2	52	4.2

[a] For yield: M = P1 + S1 and S1 = P2 + S2; p = protein rich and s = starch rich fraction

antinutritional factors. It has been established that the protein fraction of *Phaseolus* beans (Elkowicz and Sosulski, 1982) and peas (Van der Poel *et al.*, 1989) are rich in constituents such as trypsin inhibitors, lectins, saponins, and phytic acid. Pitz *et al.* (1980) found that faba bean fines fractions were rich in vicine and convicine.

Unless additional pretreatments are used for this fractionation procedure, this phenomenon represents a serious drawback to the use of these fractions in diets for monogastric animals. As mentioned above, utilization of the fines fractions in feeding practice will depend on the relative contribution of the ANF to the negative effects observed in *in vivo* trials. Thermal processing or extraction of these fractions or heat treatment before milling and air-classification are ways to establish and eliminate or reduce ANF. Dry heat, for example, will reduce proteinaceous ANF but also will reduce the moisture level of seeds which is important in the efficiency of the air-classification procedure (Reichert, 1982). As reported previously (Van der Poel *et al.*, 1989) for basic nutritional studies (e.g., the determination of threshold levels of ANF in diets) air-classification may provide seed fractions with varying and large concentrations of ANF.

Germination

During the germination of legume seeds, enzymes become active so as to degrade starch, storage proteins, and several ANF. Degradation of storage proteins is necessary to make peptides and amino acids available for seed growth and early plant growth. During germination, ANF are degraded to a lower level of activity by the action of several enzymes (Savelkoul et al., 1992). Isolation and characterization of the enzyme(s) responsible for this ANF inactivation, could lead to the treatment of cool season food legume seeds with this enzyme, and can be a future step in the upgrading of legume seeds for feeding purposes where thermal processing may either fail or result in protein damage.

Concluding Remarks and Recommendations for Future Research

The pulses have a good amino acid profile but the content of essential amino acids is less well-balanced than in soybean. Thus, when soybean is replaced by pulses, special attention should be paid to the amino acid profile. This can be done by combining other feed ingredients with a complementary amino acid profile or by adding synthetic amino acids. Albumins in pea have a better amino acid profile than the globulins (Table 2). Although the first attempts to improve the protein profile were not successful, this information may be important for plant breeders, especially in research focussed on improvement by gene manipulation. It has been shown that the inclusion of large concentrations of pea carbohydrates causes an increase in small intestinal chyme flow, which is most likely due to the presence of the oligosaccharides. Further studies are required to identify the precise factor. Special attention should be given to the maximum dietary inclusion level of pea and other pulse carbohydrates in relation to excess small intestinal chyme production.

As discussed, pulses contain various ANFs. The results in Tables 6 and 8 show that indeed trypsin inhibitors plus lectins in pea and tannins in faba bean have negative effects on performance of piglets. However, it is largely unknown to what extent each ANF is responsible for the negative effects in the target farm animal under practical conditions. It was shown that there were no negative effects when practical amounts of tannin-rich faba beans were included in the diets of chickens. It could be calculated that 3 to 4 g tannins (measured according to the vanillin sulphuric assay) per kg diet can be tolerated in diets of chickens. Clearly there is a need for more data about threshold concentrations of ANFs to be used in practical diet formulation. The results of Jansman et al. (1989) and Huisman et al. (1990a,b,c) showed that pigs are more sensitive than chickens, implying the necessity for threshold concentrations for each animal species. The lack of such information restricts the increase of the use of pulses in animal diets.

Results concerning the ileal true protein digestibility show that the natural protein of faba bean and pea is highly digested in the small intestine and that the quality of protein in terms of true digestibility is high. The low apparent protein digestibility is due to an increase in the excretion of endogenous protein. It is important to do more research into the causative factors for the hypersecretion of endogenous protein. In this respect, improving the nutritional value of protein should be focused on reducing ANFs and possibly repressing of the antigenicity of the protein. It is not clear from an economic point of view which direction is the best: plant breeding or technology.

From the possibilities for upgrading legume seed quality by means of several approaches described above it is concluded that extending the use of these seeds in the nutrition of animals is possible in a number of ways. The various ways of treatment (see Table 11) have their specific advantages and disadvantages.

Breeding and genetic manipulation of plants, for example, are long term efforts to remove ANFs while improving the nutritional quality of raw beans (Osborne and Bliss, 1985). In addition, one should consider the implications for possible simultaneous consequences in agronomic characteristics such as resistance to adverse weather conditions and resistance to disease and predators. Moreover, yield of beans may be affected by changes in ANF (Bond and Smith, 1989).

The use of enzymes to change the composition of seed proteins or to decrease the level of ANF (as feed additive or by the use of inherent enzymes through, for example, germination) implies a rather specific application relative to a specific objective. This application may be limited by the frequent presence of different ANFs in a particular seed. Moreover, the increased use of thermal treatments in the routine procedures for animal feed manufacturing may limit the use of feed enzymes in dry complete feeds.

It has been well established that the protein nutritional value of vegetable protein is improved by processing. Many efforts have been made to define the processing conditions in some legume seeds. These approaches are largely based on thermal treatment, but other ways, such as fractionation, have been explored. Heat may render seed protein more digestible through denaturation (storage proteins; proteinaceous ANF). However, overheating in terms of temperature and/or exposure time may adversely affect the amount and, particularly, the availability of essential amino acids. It may also depress digestibility and cause a slower release of amino acids from the protein (De Wet, 1982).

In addition to this general discussion about upgrading procedures, it should be remembered that for each particular seed it must be established whether the digestibility of protein needs to be increased by the removal of ANFs or by enhancing the digestibility of the protein itself. In the latter case, for example the seeds of the *Phaseolus* family (Van der Poel *et al.*, 1991), the inactivation of ANF gives no guarantee of a high nutritional value and the threshold amount for ANF cannot be used to calculate the contribution of this particular legume seed to the protein value of the diet.

If ANFs contribute to the negative nutritional effects observed in animals, a choice has to be made between the possibilities described on the basis of improvement in the nutritional value which then has to compensate for the extra cost associated with breeding, enzyme treatment, or processing.

References

Bengala-Freire, J., Aumaitre, A. and Peiniau, J. 1991. *Journal of Animal Physiology and Animal Nutriton* 65: 154–164.

Bengala-Freire, J. P., Hulin, J. C., Peiniau, J. and Aumaitre, A. 1989. *Journées Recherche Procine en France* 21: 75–82.

Bertrand, D., Delort-Laval, J., Melcion, J. P. and Valdebouze, P. 1982. *Sciences des Aliments*, HS II, pp. 197–202.

Bertrand, G., Séve, B., Gallant, B. and Tomé, R. 1988. *Sciences des Aliments* 8: 187–212.

Bhatty, R. S. and Christison, G. I. 1984. *Plant Foods for Human Nutrition* 34: 41–51.

Birk, Y. 1989. In: *Recent Advances of Research in Antinutritional Factors in Legume Seeds*, pp. 239–250 (eds. J. Huisman, A. F. B. van der Poel and I. E. Liener). Wageningen, The Netherlands: Pudoc.

Bond, D. A. and Smith, D. B. 1989. In: *Recent Advances of Research on Antinutritional Factors in Legume Seeds*, pp. 285–296 (eds. J. Huisman, A. F. B. van der Poel and I. E. Liener). Wageningen, The Netherlands: Pudoc.

Bressani, R. and Elias, L. G. 1988. In: *World crops: Cool Season Food Legumes*, pp. 381–404 (ed. R. J. Summerfield). Dordrecht: Kluwer Academic Publishers.

Buraczewska, L., Gdala, J. and Grala, W. 1989. In: *Recent Advances of Research in Antinutritional Factors in Legume Seeds*, pp. 181–184 (eds. J. Huisman, A. F. B. van der Poel, and I. E. Liener). Wageningen, The Netherlands: Pudoc.

Carré, B., Beaufils, E. and Melcion, J. P. 1991. *Journal of Agricultural Food Chemistry* 39: 468–472.

Castaing, J. and Grosjean, F. 1985. *Journées Recherche Porcine en France* 17: 407–418.

Cordese, R. 1990. *Options Méditerranéennes – Série Séminaires* 9: 127–131.

CVB Table. 1990. Veevoedertabel: Gegevens over voederwaarde, verteerbaarheid en samen telling. Centraal Veevoederbureau in Nederland, Lelystad.

De Lumen, B. O. 1990. *Journal of Agricultural Food Chemistry* 38: 1779–1788.

De Wet, P. J. 1982. In: *Handbook of Nutritive Value of Processed Food*, pp. 321–341 (ed. M. Rechcigl Jr.). Vol. II. Boca Raton, FL, USA: CRC-Press, Inc.

Duc, G. and Lacassagne, L. 1990. In: *Symposium Qualité des Céréales, des Oléagineux et des Protéagineux Francais pour l'Alimentation Animale*, pp. 93–95 (eds. AGPM, CETIOM, ITCF, ONIDOL and UNIP). Toulouse, France.

Eggum, B. O. 1980. In: *Vicia faba, Feeding Value, Processing and Viruses*, pp. 107–123 (ed. D. A. Bond). Den Haag, The Netherlands: Martinus Nijhoff.

Elkowicz, K. and Sosulski, F. W. 1982. *Journal of the Science of Food and Agriculture* 47: 1301–1304.

Fekete, J., Castaing, J., Lavorel, O., Leuillet, M. and Quemere, P. 1984. *Journées Recherche Porcine en France* 16: 393–400.

Frolich, A. A. and Marquardt, R. R. 1983. *Journal of the Science of Food and Agriculture* 34: 153–163.

Gallaher D. and Schneeman, B. O. 1986. In: *Nutritional and Toxicological Significance of Enzyme Inhibitors in Foods*, pp. 167–185 (ed. M. Friedman). New York: Plenum Press.

Gatel, F. 1992. Protein quality for monogastric animals. *Proceedings 1st European Conference on Grain Legumes*, pp. 461–473. June 1–3, Angers, France.

Gatel, F. and Grosjean, F. 1990. *Livestock Production Science* 26: 155–175.

Griffiths, D. W. and Mosely, G. 1980. *Journal of the Science of Food and Agriculture* 31: 255–259.

Grosjean, F. and Gatel, F. 1986. *Pig News and Information* 4: 443–448.

Grosjean, F. and Gatel, F. 1989. In: *Recent Advances of Research in Antinutritional Factors in Legume Seeds*, pp. 239–242 (eds. J. Huisman, A. F. B. van der Poel and I. E. Liener). Wageningen, The Netherlands: Pudoc.

Gueguen, J. and Barbot, J. 1988. *Journal of the Science of Food and Agriculture* 42: 209–224.

Hauschild, A. and Köhler, R. 1991. In: *Proceedings of the 6th International Symposium on Protein Metabolism and Nutrition*, pp. 21–23 (eds. B. O. Eggum, S. Boisen, C. Borsting, A. Danfaer and T. Hvelplund). Herning, Denmark: EAAP-publication No. 59.

Hlödversson, R. 1987. *Animal Feed Science & Technology* 17: 245–255.

Huisman, J. 1990. *Antinutritional Effects of Legume Seeds in Piglets, Rats and Chickens*, 149 pp. Ph.D. Thesis, Agriculture University, Wageningen, The Netherlands.

Huisman, J. 1991. *Aspects of Applied Biology* 27: 11–22.

Huisman, J., Heinz, Th., Van der Poel, A. F. B., Van Leeuwen, P., Souffrant, W. B. and Verstegen, M. W. A. 1992. True protein digestibility and amounts of endogenous protein measured with the 15N dilution technique in piglets fed peas and common beans. *British Journal of Nutrition* (In press)

Huisman, J. and Jansman, A. J. M. 1991. *Nutriton Abstracts and Reviews, Series B: Livestock Feeds and Feeding* 61: 902–921.

Huisman, J. and Le Guen, M. P. 1991. In: *Digestive Physiology of the Pig*, pp. 60–66 (eds. M. W. A. Verstegen, J. Huisman and L. A. den Hartog). Wageningen, The Netherlands: Pudoc.

Huisman, J. and Tolman, G. H. 1992. Antinutritional factors in plant proteins of diets for non-ruminants. In: *Recent Advances in Animal Nutrition*, Vol. 68, No. 1, pp. 101–110 (eds. D. J. A. Cole and P. C. Garnsworthy). London: Butterworths.

Huisman, J., Van der Poel, A. F. B., Kik, M. J. L. and Mouwen, J. M. V. M. 1990a. *Journal of Animal Nutrition and Animal Physiology* 63: 273–279.

Huisman J., Van der Poel, A. F. B., Muwen. J. M. V. M. and Van Weerden, E. J. 1990c. *British Journal of Nutrition* 64: 755–764.

Huisman J., Van der Poel, A. F. B., Van Leeuwen, P. and Verstegen, M. W. A. 1990b. *British Journal of Nutrition* 64: 743–753.

Huisman, J., Van der Poel, A. F. B., Verstegen, M. W. A. and Van Weerden, E. J. 1990d. *World Review of Animal Production* 2: 77–82.

Jaffé, W. G. 1980. In: *Toxic Constituents of Plant Foodstuffs*, pp. 73–102 (ed. I. E. Liener). New York: Academic Press.

Jansman, A. J. M., Huisman, J. and Van der Poel, A. F. B. 1989. In: *Recent Advances in Research of Antinutritional Factors in Legume Seeds*, pp. 176–180 (eds. J. Huisman, A. F. B. van der Poel and I. E. Liener). Wageningen, The Netherlands: Pudoc

Jeroch, H., Berger, H. and Gebhardt, G. 1985. *Archiv für Tierernährung* 35: 817–821.

Jindal, S., Soni, G. L. and Singh, R. 1982. *Journal of Plant Foods* 4: 95–103.

Kik, M. J. L., Rojer, M., Mouwen, J. M. V. M., Konirkx, J. F. J. G., Van Dijk, J. E. and Van der Hage, M. H. 1989. *The Veterinary Quaterly* 11: 108–115.

Lacassagne, L., Francesh, M., Carré, B. and Melcion, J. 1988. *Animal Feed Science and Technology* 20: 59–68.

Le Guen, M. P., Tolman, G. H. and Huisman, J. 1991. In: *Digestive Physiology of the Pig*, pp. 99–103 (eds. M. W. A. Verstegen, J. Huisman and L. A. den Hartog). Wageningen, The Netherlands: Pudoc.

Leterme, P., Beckers, Y. and Thewis, A. 1989. In: *Recent Advances of Research in Antinutritional Factors in Legume Seeds*, pp. 121–124 (eds. J. Huisman, A. F. B. van der Poel and I. E. Liener). Wageningen, The Netherlands: Pudoc.

Leterme, P., Monmart, T. and Baudart, E. 1990. *Journal of the Science of Food and Agriculture* 53: 107–110.

Liebert, F. and Gebhardt, G. 1983. *Archiv für Tierernährung* 33: 47–56.

Liener, I. E. (ed.). 1980. *Toxic Constituents of Plant Foodstuffs*, 502 pp. New York: Academic Press.

Liener, I. E. 1981. *Journal of the American Oil Chemists' Society* 58: 406–415.

Liener, I. E. 1986. In: *The Lectins*, pp. 527–552 (eds. I. E. Liener, N. Sharon and I. J. Goldstein). New York: Academic Press.

Liener, I. E. 1989. In: *Recent Advances of Research in Antinutritional Factors in Legume Seeds*, pp. 6–13 (eds. J. Huisman, A. F. B. van der Poel and I. E. Liener). Wageningen, The Netherlands: Pudoc.

Liener, I. E. and Kakade, M. L. 1980. In: *Toxic Constituents of Plant Foodstuffs*, pp. 7–71 (ed. I. E. Liener). New York: Academic Press.

Marquardt, R. R. 1989. In: *Recent Advances in Research in Antinutritional Factors in Legume Seeds*, pp. 141–155 (J. Huisman, A. F. B. van der Poel and I. E. Liener). Wageningen, The Netherlands: Pudoc.

Marquardt, R. R. and Bell, J. M. 1988. In: *World Crops: Cool Season Food Legumes*, pp. 421–444 (ed. R. J. Summerfield). Dordrecht: Kluwer Academic Publishers.

Marquardt, R. R. and Ward, A. T. 1979. *Canadian Journal of Animal Science* 59: 781–789.

Marquardt, R. R., Ward, A. T., Campbell, L. D. and Cransfield, P. E. 1977. *Journal of Nutrition* 107: 1313–1324.

Mitjavila, S., Lacombe, C., Carrera, G. and Derache, R. 1977. *Journal of Nutrition* 107: 2113–2121.

Nielsen, H. E. and Kruse, P. E. 1974. *Livestock Production Science* 1: 179–185.

Osborne, T. C. and Bliss, F. A. 1985. *Journal of the American Society of Horticultural Science* 110: 484–488.

Pitz, W. J., Sosulski, F. W. and Hogge, L. R. 1980. *Canadian Institute of Food Science and Technology*, 13: 35.

Pusztai, A. 1989. In: *Recent Advances of Research in Antinutritonal Factors in Legume Seeds*, pp. 17–29 (eds. J. Huisman, A. F. B. van der Poel and I. E. Liener). Wageningen, The Netherlands: Pudoc.

Rao, B. S. N. and Prabhavathi, J. 1982. *Journal of the Science of Food and Agriculture* 33: 89–96.

Reddy, N. R., Pierson, M. D., Sathe, S. K. and Salunkhe, D. K. 1984. *Food Chemistry* 13: 25–68.

Reichert, R. D. 1982. *Journal of Food Science* 47: 1263–1267,1271.

Saini, H. S. 1989. In: *Recent Advances of Antinutritional Factors in Legume Seeds*, pp. 329–341 (eds. J. Huisman, A. F. B. van der Poel and I. E. Liener). Wageningen, The Netherlands: Pudoc.

Savage, G. P. 1988. *Nutrition Abstracts and Reviews (Series A)* 58: 320–344.

Savage, G. P. and Deo, S. 1989. *Nutrition Abstracts and Reviews (Series A)* 59: 66–68.

Savelkoul, F. H. M. G., Van der Poel, A. F. B. and Tamminga, S. 1992. *Plant Foods for Human Nutrition* 42: 71–85.

Schroeder, H. E. 1982. *Journal of the Science of Food and Agriculture* 33: 623–633.

Seve, B., Aumaitre, A. and Bouchez, P. 1985. *Science Alimentaire* 5: 119–126.

Singh, K. B., Wlliams, P. C. and Nakkoul, H. 1990. *Journal of the Science of Food and Agriculture* 53: 429–441.

Sissons, J. W. and Tolman, G. H. 1991. In: *Proceedings of the Second Spring Conference Edinburgh*, pp. 66–85 (eds. J. P. F. D. Mello and C. M. Duffus). Edinburgh, Scotland: Scottish Agricultural College.

Souffrant, W. B., Darcy-Vrillon, B., Corring, T., Laplace, J. P., Köhler, R., Gebhardt, G. and Rerat, A. 1986. *Archiv für Tierernährung* 36: 269–274.

Van der Poel, A. F. B., Aarts, H. L. M. and Stolp, W. 1989. *Netherlands Journal of Agricultural Science* 37: 273–278.

Van der Poel, A. F. B., Blonk, J., Huisman, J. and Den Hartog, L. A. 1991. *Livestock Production Science* 28: 305–319.

Van der Poel, A. F. B., Blonk, J., Van Zuilichem, D. J. and Van Oort, M. G. 1990c. *Journal of the Science of Food and Agriculture* 53: 215–228.

Van der Poel, A. F. B., Dellaert, L. M. W., Van Norel, A. and Helsper, J. P. F. G. 1992a. The digestibility in piglets of near isogenic lines of faba beans (*Vicia faba* L.) differing in levels of condensed tannins. *British Journal of Nutrition* (In press).

Van der Poel, A. F. B., Gravendeel, S., Van Kleef, D. J., Jansman, A. J. M. and Kemp, B. 1992c. *Animal Feed Science Technology* 36: 205–214.

Van der Poel, A. F. B., Liener, I. E., Mollee, P. W. and Huisman, J. 1990b. *Livestock Production Science* 25: 137–150.

Van der Poel, A. F. B., Mollee, P. W., Huisman, J. and Liener, I. E. 1990a. *Livestock Production Science* 25: 121–135.

Van der Poel, A. F. B, Stolp, W. and Van Zuilichem, D. J. 1992b. *Journal of the Science of Food and Agriculture* 58: 83–87.

Veldman, B. 1988. *Veevoedkundige mededelingen*, CLO-Instituut voor de Veevoeding "De Schothorst", 1–7.

Vose, J. R., Basterrechea, M. J, Gorin, P. A. J., Finlayson, A. J. and Youngs, C. G. 1976. *Cereal Chemistry* 53: 928–936.

Wareham, C. N., Wiseman, J. and Cole, D. J. A. 1991. Proceedings Winter Meeting British Society of Animal Production, paper no. 22, 18–20 March, Scarborough, United Kingdom.

Wiseman, J. and Cole, D. J. A. 1988. In: *Recent Advances in Animal Nutrition*, pp. 13–37 (eds. W. Haresign and D. J. A. Cole). London: Butterworths.

Wiseman, J., Wareham, C. N. and Cole, D. J. A. 1991. *Aspects of Applied Biology* 27: 23–30.

Wünsche, J., Hennig, U., Meinl, K., Kreienbring, F. and Bock, H. D. 1982. *Archiv für Tierernährung* 32: 337–348.

Nature, composition, and utilization of food legumes

J.H. HULSE
Siemens-Hulse International Development Association, Inc., 1628 Featherston Drive, Ottawa, Ontario, K1H 6P2 Canada

Abstract

The *Leguminosae* family comprises almost 700 genera and 1800 species. The most widely cultivated are the oilseeds, soybeans (*Glycine max*), and groundnuts (*Arachis hypogaea*), which provide 50 and 16% of world grain legume harvests, respectively. Apart from the oilseeds, scientists have given more attention to protein nutritional quality than to useful functional properties. Traditional and modern processes of transformation and fractionation of high and low lipid legumes are reviewed and the useful biochemical and biophysical properties of various components and derivatives are discussed. Protein properties examined include water activity in relation to solubility, dispersibility, viscosity, gel, and foam stabilities. Practical systems of protein fractionation and the properties of derived fractions are described. Particular properties of legume lipids and carbohydrates are examined. Attention is given to sources and control of mycotoxins and to quality implications related to the progenies of wide transgenic crosses. Priorities for future research and opportunities for utilization are suggested.

Introduction

Food legumes are broadly classified as those with small and those with large lipid concentrations; the latter being the oilseeds which are the most abundantly cultivated. World production and yield patterns over the past decade indicate greater increases for dry pulses when compared to the cereal grains (Table 1). World cereal production and yields both increased by about 25%, with greater increases noted for the Asian countries. World production of pulse crops increased by 44% and yields by 28%. These data suggest that the growth in production of cereals was almost entirely the result of higher yields, while the increased production of pulse crops was partly due to an increase in area planted. Both global and developing regions' production and yield of chickpea rose by about 15%. World production of both dry peas and lentils doubled,

F.J. Muehlbauer and W.J. Kaiser (eds.), Expanding the Production and Use of Cool Season Food Legumes, 77–97.
© 1994 *Kluwer Academic Publishers.*

Table 1. World production and yields of food grains, 1980 and 1990 compared*

	Production			Yield		
	1980	1990	Increase	1980	1990	Increase
Cereals	------tons------		-%-	---kg ha^{-1}---		-%-
World	1590	1955	+23	2210	2765	+25
LDCs	776	1021	+32	1900	2405	+26
Asia	640	860	+34	2105	2770	+32
Pulses						
World	41	59	+44	670	860	+28
LDCs	32	38	+19	605	665	+ 9
Asia	22	26	+20	625	725	+16
Pea (Dry)						
World	8.5	17.5	+106	1140	1890	+66
LDCs	3.2	2.7		1120	970	
Chickpea						
World	6.0	6.9	+15	625	720	+15
LDCs	5.9	6.7	+14	625	710	+14
Lentil						
World	1.3	2.7	+108	595	845	+42
LDCs	0.7	1.1	+57	470	750	+60

*Source: FAO Production Year Book, 1990

while yields showed a two-thirds increase. Production and yields of lentils in developing countries grew by 60%.

Legume Composition

Legume seed composition is influenced by genetic background, conditions of crop cultivation, state of maturity at harvest, method of harvest, and storage conditions. Composition can vary among seeds from the same crop harvest and among seeds from the same plant (Sjodin, 1982; Matthews and Arthur, 1985). Total amino acid analyses show that N content of legume seeds is greater than

17.5%, and that the N-to-protein conversion factor for most cereals and legumes lies between 5.3 and 5.7 (Tkachuk, 1977; Sosulski and Holt, 1980).

The nutritional evaluation of protein and amino acids is discussed by Pellett and Young (1980) and Young and Pellett (1990). They state that daily protein requirements vary among age groups as follows: 85 g kg^{-1} of body mass for infants 3 to 6 months of age, 0.8 g kg^{-1} for 10- to 12-year-olds, and 0.57 g kg^{-1} for healthy adults. Their conclusion is that protein quality requirements remain about the same throughout the life span. Protein quality is important in regions where two-thirds or more of the protein is derived from cereals. Such diets probably require additional lysine, for example, from legumes.

Singh *et al.* (1968) give analyses of seedcoat, cotyledon, and embryo fractions of several legumes. Seedcoats varied from 8 to 15.5%, cotyledons from 83 to 90%, and embryos from 0.99 to 2.3% of total seed weight. Cotyledons contain over 90% of the total protein and ether extracts in the seed, and over 85% of the total mineral and N-free extract, while seedcoats contain between 70 and 98% of the crude fiber. Reddy *et al.* (1984) present data on carbohydrate contents in legumes (Table 2). Protein and lipid contents and estimates of total fiber summarized from various sources are given in Table 3. Ranges for lysine and methionine in chickpea, lentil, and faba bean in comparison with sorghum indicate considerably larger concentrations for the pulses (Table 4).

Singh and Jambunathan (1982) give data on chickpea and others describe protein fraction distribution among seed components. It is difficult to make comparisons among authors since various distinctly different methods and

Table 2. Ranges of carbohydrate concentrations (%) of selected cool season food legume seeds*

	Starch	Amylose	Sucrose	Raffinose	Stachyose	Verbascose
Smooth Pea	37–49	24–33	2.3–2.4	0.3–0.9	2.2–2.9	1.7–3.2
Wrinkled Pea	24–37	63–66	2.3–4.2	1.2–1.6	2.9–5.5	2.2–4.2
Chickpea	37–50	32–46	0.7–2.9	0.7–2.4	2.1–2.6	0.4–4.5
Faba bean	41–53	22–35	1.4–2.7	1.1–0.5	0.5–2.4	1.6–2.1
Lentil	35–53	21–45	1.8–2.5	1.4–1.0	1.9–2.7	1.0–3.1

*Source: Reddy *et al.*, 1984

Table 3. Protein, lipid, and crude fiber concentrations of cool season food legume seeds

	Protein	Total lipids Approx. mean	Total Crude fiber
	--%--	--%--	--%--
Chickpea	15 – 30	5.0	1.2 – 13.5
Pea	21 – 33	2.4	4.6 – 7.0
Faba bean	23 – 39	1.6	5.0 – 8.5
Lentil	20 – 31	1.2	3.8 – 4.6

Table 4. Lysine and methionine concentrations (mg 100 gN^{-1}) of chickpea, lentil, and faba bean as compared to sorghum*

	Lysine	Methionine
Chickpea	394 - 437	75 - 97
Lentil	391 - 471	37 - 76
Faba bean	391 - 660	44 - 57
Sorghum	71 - 212	39 - 172

*Sources: Hulse *et al.* 1980; Salunkhe and Kadam, 1989

variations of methods of analysis were employed. The classical Osborne method (Osborne, 1924) uses four solvents in sequence: (1) Water (to extract the albumins), (2) NaCl (to extract the globulins), (3) Ethanol (to extract the prolamines), and (4) NaOH (to extract the glutelins). Landry and Moureaux (1970) employed a sequence of five solvents: (1) NaCl, (2) 70% Isopropanol, (3) Isopropanol + 2-Mercaptoethanol, (4) Borate buffer + 2-Mercaptoethanol, and (5) Borate buffer plus sodium lauryl sulphate + 2-Mercaptoethanol. Kanamori *et al.* (1982) used yet a different sequence: (1) NaCl, (2) 30% Isopropanol, (3) 4% lactic acid, and (4) 0.5% KOH.

Non-protein nitrogen often accounts for 10 to 15% of total legume seed nitrogen and is found in the form of free amino acids, amines, purine and pyrimidine bases, nucleic acids, lipoproteins, and other complex substances (Adsule and Kadam, 1989). Mineral concentrations of legume seeds range from 2.5 to 4.5%. Phosphorus is in the highest proportion and averages close to 300 mg 100 g^{-1} of seed, with between 300 and 500 mg 100 g^{-1} in the cotyledons.

Chickpea

About 95% of the world's chickpea crop is produced in developing countries. Chickpea provides over half the pulse harvest of India, and about 1 million tonnes are grown annually in West Asia, the Near East, and North Africa. Both kabuli and desi types are produced on the Indian sub-continent, while kabuli types predominate in the Near East. Although there are exceptions, kabuli types have relatively large seeds with thin testa while the desi types are characterized as having smaller, more angular seeds with thicker testa. The nutritional quality of chickpea has been discussed by Rao (1969), Rama Rao (1974), and Singh (1985) among others.

Williams *et al.* (1991) have recommended that breeders consider quality factors that include: medium to large, uniform, ram-head shaped seeds which are bright beige in color; low proportion of testa; greater concentrations of protein and methionine; reduced oligosaccharides; and shorter cooking times. At ICARDA, quality traits were affected by genotype, place of cultivation, sowing time, season, and available water. In those studies, seed size varied from 8 to 67 g 100 seeds^{-1}, protein content ranged from 14 to 27%, and cooking times ranged from 50 to 300 minutes. Location had the greatest influence on seed size and protein while samples from irrigated fields had smaller seed sizes, reduced protein concentrations, and longer cooking times.

Hulse (1976) recalculated chickpea data from several sources into a standard format and described analytical variations among samples. Data for whole chickpeas and dehulled chickpeas (in parentheses) were as follows: ether extract from 3.9 to 6.2% (4.6 to 6.9%), protein from 20.8 to 25.9% (25.3 to 28.9%), soluble carbohydrates from 60 to 63% (63 to 65%), and crude fiber from 8.0 to 8.7% (1.0 to 1.5%). Analyses showed significant variations between inner and outer segments of the cotyledons. The inner fractions of the cotyledons which accounted for 25% of whole grain, contained on average 19.4% protein, while the outer fraction, or 65% of the whole grain, contained 25.7% crude protein. The outer fraction of the chickpea cotyledons generally had greater concentrations of methionine and lysine when compared to the inner fraction.

Variation among 125 desi chickpea genotypes (ICRISAT, personal communication) indicated that 100 seed weight ranged from 10.4 to 36.7 g, protein 14.5 to 29%, carbohydrate from 52.4 to 71%; lipid from 3.8 to 10.2%, cooking time for the whole seed from 52 to 98 minutes, and from 26 to 46 minutes for dhal (decorticated and split seeds). Increased fertilizer applications increased total and non-protein nitrogen in the seeds. Desi types, having a greater proportion of seedcoat, were higher in total fiber, cellulose, and hemicellulose when compared to kabuli types. A study (Swaminathan and Jain, 1975) of 16 cultivars grown at 12 locations indicated a range of protein concentrations from 12.4 to 28.1%, with a mean of 19.5%. Previously, Hawtin *et al.* (1977) reported a range of 18.7 to 28.4% in protein concentration among 2667 germplasm accessions.

Hulse (1976) reviewed several publications and found significant variation (52 to 78) in reported Biological Values and coefficients of digestibility (76 to 92) (see Pellett and Young, 1980). It is not clear to what degree these variations reflect either true differences or experimental inconsistencies.

Pea

Pea is one of the four most cultivated legumes, only soybeans, groundnuts, and beans (*Phaseolus vulgaris*) being grown in greater quantities. Peas are grown extensively in northern Europe, the USA, Canada, Russia, and China. Traditionally the pea crop was harvested to be eaten fresh or field dried before storage and later rehydration. During the past 40 years, a significant proportion has been grown and harvested immature for commercial freezing. For this purpose wrinkled cultivars are usually preferred.

Some early taxonomic classifications have been questioned. For example, Ben-Ze'ed and Zohary (1973) intercrossed cultivated *P. sativum* with *P. elatius*,

P. humile, and *P. fulvum* and concluded that there is no cytogenetic basis for classifying the first three as other than members of the species *P. sativum*.

Though excavations in Neolithic settlements (ca 7000 BC) in the Near East and Europe brought to light carbonized pea seeds, neither a wild progenitor nor the early history of *Pisum* has been discovered. The external characters of the carbonized relics suggest that peas have been cultivated for at least as long as wheat and barley.

Vavilov (1949) believed the centers of origin to be Ethiopia, the Mediterranean, and Central Asia.

Most modern pea cultivars differ from wild types in having larger seeds and a shorter, more compact habit of growth, though at present times both tall and small-seeded plants are cultivated. Though peas were culinary ingredients in Greece and in Rome over 2000 years ago, no varieties were described until the sixteenth century. During the nineteenth century, breeders in Britain, who developed several classic varieties, appear to have been aware of Mendel's research without understanding the essential implications.

With the expansion of canning and freezing, pea production diversified from being a small-scale horticultural crop to intensive commercial cultivation. Among the industrialized countries, many pea crops are grown under contract, the processor specifying the cultivars and conditions of cultivation.

Analyses of peas (dry weight basis) in the former USSR (Makasheva, 1973) indicated protein concentrations ranged from 18 to 35%, starch concentrations from 20 to 50%, soluble sugar concentrations from 4 to 10%, cellulose concentrations from 2 to 10%, and mineral matter from 2 to 4%. Amylose in starch from round-seeded cultivars varied from 37.6 to 41.6% and in wrinkled-seeded types from 65.5 to 72.9%. On the average, wrinkled seeds accumulated 55 g of starch per 1000 seeds as compared to over 100 g per 1000 round seeds.

Faba Bean

Three botanical types of the subspecies *V. faba* eu-faba have been described (Chavan *et al.*, 1989) and include: *V. faba* var. minor Beck having thick pods with small seeds; *V. faba* var. equina Pers. having pods with medium-sized seeds; and *V. faba* var. major Harz., having broad pods with large seeds. Though cultivated on all continents, over 60% of world production of faba bean is in the People's Republic of China. World plantings and production have fallen significantly over the past decade.

Cotyledons of faba bean represent, on average, about 87% of the seed weight which ranges from 0.4 to 1.8 g among accessions (Chavan *et al.*, 1989). Proximate analyses of faba bean (Eden, 1968; Chavan *et al.*, 1989) indicate wide variation among accessions and cultivars for nearly all traits, but especially for crude protein concentrations (Table 5). Likewise, Hawtin *et al.* (1977) also reported wide variation (22.3 to 37.1%) in protein concentrations among 500 faba bean accessions.

Table 5. Proximate analyses (ranges) of faba beans from two sources

	Chavan et al. (1989)	Eden (1968)	
		Spring sown	Fall sown
	--%--	--%--	--%--
Crude protein	20.3 - 41.0	25.5 - 35.4	24.3 - 29.9
True protein		22.3 - 31.7	22.0 - 27.0
Ether extract		1.2 - 2.0	1.3 - 1.7
Crude fiber	5.0 - 8.5	6.0 - 10.6	7.5 - 10.6
Mineral matter	2.7 - 3.7	3.1 - 4.8	3.4 - 4.4
N-free extractives		49.9 - 59.4	55.5 - 59.8
Total carbohydrate	50.9 - 67.9		
Crude lipid	1.0 - 1.6		

Amino acid composition reported as mg g^{-1} of nitrogen ranged from 36 to 69 mg for methionine, from 44 to 94 mg for cystine, and from 343 to 400 mg for lysine (Chavan *et al.*, 1989).

Purified faba bean globulins contain two main proteins which are characterized by their sedimentation coefficients and molecular weights. Faba bean protein fractions, as determined by two extraction procedures, are shown in Table 6. Legumin is the predominant globulin with a molecular weight of 300,000 to 450,000, while the molecular weight of vicilin ranged from 150,000 to 250,000. Legumin has larger concentrations of arginine, threonine, and tryptophan; while vicilin has larger concentrations of lysine, leucine, and phenylalanine. Protein digestibility varies from 82 to 92%, biological value from 45 to 55%, and utilizable protein from 14.8 to 15.5%.

Table 6. Protein fractions of faba bean as determined by two methods

Osborne method	protein concentration	Kanamori method	protein concentration
	--%--		--%--
Albumin	8.6	NaCl 2%	69.6
Globulin	42.0	Isopropanol 30%	5.6
Glutelin	30.0	Lactic acid 4%	5.9
		KOH 0.5%	5.6

*from Chavan *et al.*, 1989

Faba bean starch granules are largely spherical or oval and can vary between 6 and 40 microns in diameter though the majority generally lie in the 20 to 40 micron range. Cerning *et al.* (1975) analyzed a series of faba bean samples grown in various parts of France from seed brought from Asia, North Africa, and European countries. Total carbohydrate, oligosaccharides, and sugars fell within the ranges given above, but starch concentrations varied from 30 to 42%. The seeds varied in size from 5 mm × 7 mm to 10 mm × 20 mm, and included ovoid and flat seeds, both either smooth or wrinkled. Ranges of carbohydrate fractions are shown in Table 7.

Table 7. Faba bean carbohydrate concentrations

Carbohydrate	Concentration		
Total CHO	51.0	–	68.0
Starch	41.0	–	53.0
Amylose	34.0	–	44.0
Sugars	3.0	–	7.0
Sucrose	1.4	–	2.7
Raffinose	0.1	–	0.5
Stachyose	0.5	–	2.4
Verbascose	1.6	–	2.6
Cellulose	1.0	–	4.8
Hemicellulose	4.0	–	6.0

Pitz and Sosulski (1980) determined concentrations of vicine and convicine in *V. faba* major, *V. faba* minor, and *V. narbonensis*. The two haemolytic toxins were significantly higher in the two *V. faba* varieties when compared to *V. narbonensis*, and also in fresh green beans when compared to dried stored samples. Both vicine and convicine are located in the protein bodies and have been concentrated in the protein fractions of air-classified faba bean flour. Neither glycoside was found in chickpea, lentil, pea, navy bean, lima bean, or cowpea. Marquardt *et al.* (1975) reported significant variability in haemagglutinin activity among faba bean cultivars in Canada and considerably smaller concentrations of trypsin inhibitors in faba beans than in soybeans. Chavan *et al.* (1989) quote lectin concentrations which are similar to those reported by Pitz and Sosulski (1980); but they also refer to significant variability. They recommend more research on lectins and unusual amino acids such as dihydroxy phenylalanine (DOPA) since faba beans find their way into various processed foods.

Lentil

Protein concentrations in lentil reportedly range from 22 to 31% (Adsule *et al.*, 1989). Similar ranges of protein concentrations have been reported by Swaminathan and Jain (1975), Hawtin *et al.* (1977), and Bhatty (1984). Ether extracts of lentil accessions from several sources (Bhatty, 1988) range from 0.7 to 3.8%, mineral matter from 2.4 to 4.2%, P from 0.28 to 0.63%, Ca from 0.04

to 0.16%, and K from 0.88 to 1.44%. Bhatty (1988) stated that 93% of P, 60% of Ca, and 90% of the lipids are found in the cotyledons. Lentil lipids have received scant attention and little is known about their possible role in off-flavor development during storage. Oleic, palmitic, and linoleic are the dominant fatty acids.

About 90% of lentil protein is in the cotyledons. Albumins and globulins are the dominant fractions. The globulins 2S, 7S (vicilin), and 11S (legumin) proportions account for 14, 5, and 35% of the protein, respectively. Analyses of lysine and methionine from different authors vary between 356 to 456 and 37 to 65 mg g^{-1} of nitrogen, respectively. No heritable high methionine germplasm has been reported in either cultivated or wild *Lens* species (Bhatty, 1988).

Digestibility coefficients for lentil are relatively high and range from 78 to 93%, while biological values reportedly range from 32 to 58% (Adsule *et al.*, 1989).

Starch concentrations in lentil seeds range from 35 to 53% (Reddy *et al.*, 1984). Estimates of amylose in the starch vary from 20.7 to 38.5%. Starch granules are elipsoid or kidney-shaped and vary in size from 15 to 30 microns long and from 10 to 25 microns wide. Under polarized light the starch granules show a typical birefringence. Raffinose ranges from 0.39 to 1.0, stachyose 1.5 to 3.1, and verbascose 0.5 to 3.1 as percent of seed weight (Bhatty, 1988).

Starch pasting properties are similar to yellow peas and chickpeas and gelatinization temperatures range from 64 to 74°C.

Lentil haemagglutinins appear to be homologous with pea and faba bean lectins; they bind to glucose and mannose and are inactivated during autoclaving. Trypsin inhibition is reported as about one-tenth that of soybean (Bhatty, 1977).

General Comment

Summaries of legume analyses from many authors are given in Salunkhe and Kadam (1989). There is much variability in composition among species and genotypes, among different systems and seasons of cultivation, among plants from the same crop harvest, and among seeds from the same plant and from the same pod. There seems little to be gained from intensive pursuit of inheritable high concentrations of lysine or sulfur containing amino acids. In general, consumption and production would be better stimulated by systematic studies of useful functional properties and how these may be enhanced by genetic modification and cultivation practices.

The diversity of products from soybean and groundnut illustrates the progress that is possible from determinations of functional properties and the chemical and physical technologies by which these properties can be modified and diversified.

Functional Properties

Functional properties determine for what purposes and products a biological material can be adapted and transformed. Factors which affect functional properties of the seed components of cereals and legumes include: source (species, genotype, and cultivation), intrinsic nature (biochemical composition, molecular size, structure, and conformation), and post-harvest modifications (storage, processing, chemical, physical, and biological treatments and interactions).

Proteins

Functional properties of legume proteins have been reported (Hermansson, 1975, 1979; Kinsella, 1979; Martinez, 1979; Hulse, 1991). The important functional properties of the dry storage proteins of seeds appear after they have been hydrated, and are affected by composition, conformation, charge distribution, inter- and intra-molecular bonding, temperature, and pH. In globular proteins, polar amino acids tend to orient towards the surface which makes for better hydration. Proteins which unfold at aqueous-lipid interfaces assist emulsification. Long coiled polypeptides which unfold when heated are conducive to gel formation. Electrostatic interactions and bonding are affected by ionic concentrations and covalent bonds such as disulphide linkages and contribute to gels and fibrillar structures.

Water-Binding Capacity

Water-binding relates to sorption (bound), retention (entrapped H_2O), solubility, viscosity, and gel formation. In general, at low ERH – low water activity (AW) – water binds tightly with high absorption energy. At higher AW levels, absorption energy decreases and water is progressively less tightly bound. Absorption energy at different AW levels characterizes a protein and influences solubility, swelling, and apparent viscosity. Highest absorption energies are where polar amino acids (lysine, tyrosine, aspartic and glutamic acids) predominate, and lowest where non-polar amino acids (alanine, glycine, and leucine methionine) are found.

Legume proteins display various degrees of water-holding capacity and the water is held in the interstices in a relatively free, unbound state. Capacity to retain water is influenced by temperature, ionic strength, and pH, and affects apparent viscosity of aqueous-protein dispersions.

Solubility

True solubility of proteins is difficult to measure. It is convenient to rely on NSI (nitrogen solubility index) or PDI (protein dispersibiliy index) (Johnson, 1970).

Solubility, important to surface activity and other functional properties, is affected by post-harvest treatment and processes of extraction and separation. Moist heat markedly reduces solubility which suggests an advantage for pulses over soybean and groundnut since frictional heat in expellers and steam for solvent stripping are not involved.

Solubility is affected by pH though results from acid and alkali treatment may be misleading in that protein molecules may disintegrate into smaller units which appear more soluble. In some cases NSI falls during storage, particularly under hot and moist conditions.

Swelling and Viscosity

As proteins take up water, they swell, which is important when the ingredient is required as a thickener and to contribute to viscosity. As with most other functional properties, swelling is influenced by pH, temperature, protein concentration, NSI, and PDI. Apparent viscosity is important to flow properties and product consistency and is influenced by size and aggregation of dispersed particles. Being non-Newtonian, apparent viscosities of aqueous-protein systems are non-equilibrium properties influenced by the method of dispersion and determination, particularly by shear rates in viscosimeters. High rates of shear cause disaggregation and reduced apparent viscosity.

Gels

Protein gels are three-dimensional matrices of interlinked, partially associated polypeptides which entrap free liquid. Gels may form at high apparent viscosities and by heating concentrated protein solutions. Gel strengths are affected by protein concentrations, by conditions of heating and cooling, and by pH and electrolytes present. Gels can be formed by heating protein dispersions in the presence of bivalent Ca and/or Mg cations, calcium sulphate being commonly used to produce gels and curds in traditional and commercial food products. Protein gels may be reversibly thermosetting or irreversibly thermoplastic and can display thixotrophy. Empirical methods of assessing gel strength variously depend on relative resistances to compression, penetration, shear and other deforming forces, and by susceptibility to syneresis. No method seems to have been objectively standardized and officially adopted.

Surface Activity

The balance between hydrophilic (polar) and hydrophobic (non-polar) amino acids and the relative protein solubility critically influence surfactant properties, and the ability to orient, reduce surface tension and act as a protective colloid at oil-and-water interfaces. Strongest attachments with lipids are at hydrophobic sites where non-polar amino acids are concentrated. These in turn affect lipid binding and ability to stabilize oil-in-water and mixed emulsions.

Surfactant properties enable legume proteins to stabilize foams. Protein foams consist of gas bubbles encapsulated in impervious films. Foam formation and stability depend on proteins that are soluble and that concentrate at air-liquid interfaces to form cohesive continuous films sufficiently strong and elastic to resist rupture and coalescence. Foams are usually formed by whipping air into protein solutions. Foam stability is empirically determined by measuring rate of collapse and liquid released by syneresis. With foam stability being dependent on protein solubility and concentration, protein isolates and air-classified fractions generally produce more stable foams than legume flours, or chemically derived concentrates. Other factors that affect protein foams include pH, the maximum effect being close to the iso-electric point, and presence of other surfactants (Cumper, 1953).

The functional properties of legume proteins depend upon their colloidal nature and other characters described above. It must be emphasized that the information given is drawn almost entirely from studies of soybean and other oilseed proteins, comparatively little being published on the functional properties of proteins derived from pulses. The experience gained with soybean and groundnut can be valuable in refining methodologies but results will, in general, not be practically transferrable.

Carbohydrates

Starch is a water-insoluble storage polysaccharide that exists in granules which vary in shape, size distribution, and composition, their morphology and molecular organization being typical of species and governed by each plant's physiology and biochemistry (French, 1984). Legume and cereal starch granules consist, in varying proportions, of the linear amylose and branched amylopectin polymers. When heated in water, starch granules swell and eventually gelatinize. Above a critical temperature, granules swell irreversibly and amylose leaches out into the intergranular aqueous phase. Dependent on conditions, including concentration, composition, and temperature, these changes lead to a rise in viscosity and gelatinization. Heated hydrated starch is also more susceptible to enzyme degradation. Temperatures of gelatinization and the character and strength of starch gels are important functional properties. Amyloses tend to form stiffer more opaque gels than those from amylopectins (Dengate, 1984; Lund, 1984).

Various instruments follow changes with temperature in starch-water systems. The Brabender Amylograph plots profiles of viscosity changes as the temperature of starch-water is gradually raised, held close to 100°C, then reduced. Schoch and Maywald (1968) studied hot paste viscosities of several legume starches and stated that:

Hot paste viscosity patterns of various starches appear to be determined by two factors: 1) the extent of swelling of the starch granules, 2) the resistance of the swollen granules to dissolution by heat or fragmentation by shear. The viscosity patterns can be roughly classified into four types:

A. High-swelling starches...the Brabender shows a high pasting peak followed by rapid thinning during cooling.
B. Moderate-swelling starches (typical of many cereals) which show a lower pasting peak and less thinning during cooling.
C. Restricted-swelling (typical of cross-linked CHO polymers) which show no pasting peak but a viscosity which stays constant or increases.
D. Highly-restricted-swelling (typical of high amylose starches) which do not produce a hot viscous paste at normal concentrations.

Schoch and Maywald (1968) reported that starch from yellow pea, lentil, and chickpea showed type C Brabender patterns with no pasting peak and either constant or rising viscosity at 95°C. Doublier (1987) also reported legume starches to behave like cross-linked cereal starches, showing restricted swelling and low solubility at 95°C. Suggestions that legume starches might be substituted for chemically cross-linked polymers must be regarded with reservation since in other respects, legume starch pastes and gels are different. No single assay or determination can assess or predict the composite functional properties of any biological polymer in a food material.

Lineback and Ke (1975) describe the granules and gelatinization of starch from chickpea and faba bean. Chickpea granules were mainly oval, 17 to 29 μm in diameter. Faba bean granules were mainly oval or irregular and 17 to 31 μm in size. Gelatinization temperature ranges were, for faba bean 61–63.5–70°C, and for chickpea 63.5–65–69°C. The latter were determined by a Kofler hot-stage on a polarizing microscope and correspond to loss of birefringence by 2, 50, and 98%, respectively, of granules observed. Hot paste Viscograph patterns were similar to those reported by Schoch and Maywald (1968). Naivikul and D'Appalonia (1979) described faba bean granules that ranged from 12 to 24 μm wide and from 20 to 48 μm long and lentil granules that ranged from 16 to 28 μm wide and from 16 to 36 μm long. Amylose concentrations in faba bean and lentil were estimated at 24 and 21%, respectively.

Colonna *et al.* (1981) reported amylose concentrations in smooth pea and faba bean to range from 32 to 34% with gelatinization temperatures of 44–65–86°C and 48–61–80°C for faba bean and pea, respectively. Pullulanase debranching before or after beta-amylolysis indicated amylopectin structures similar to those of common cereal starches.

Naivikul and D'Appalonia (1979) also studied water soluble non-starch polysaccharide (WSNP) and water insoluble non-starch polysaccharide (WINP) fractions from faba bean and lentil. Total water solubles and WSNP from faba bean were 42.2 and 7.3%, respectively; and from lentil 33.6 and 2.2%, respectively. Crude WINP from faba bean was 3.5% and from lentil 3.8%. Faba bean WSNP was composed mainly of glucose, but also contained mannose, arabinose, xylose, and galactose. WINP from faba bean and lentil contained mainly glucose with traces of arabinose and xylose.

The isolation of pure starch from many legumes is made difficult by the presence of fine cell wall fiber. This fiber displays a high water absorption capacity which impedes any clean separation by washing and centrifuging or settling.

The removal of the outer seedcoats from chickpea is made difficult by the adhesive gum which holds the outer husk to the cotyledons. The gum is believed to be a polysaccharide composed mainly of gluco- and galacto-mannans. Evidence suggests that the higher the proportion of uronic acids, the stronger is the adhesion. Stanley and Aguilera (1985) state that legume seedcoats contain 20% polyuronides (pectin substances), 40 to 75% cellulose and hemicellulose, 0.5 to 1.0% lipids, 1 to 11% lignin, 2 to 8% mineral matter, and 5% protein. They also suggest that legume cell walls are composed of approximately 25% pectic substances with hemicellulose, cellulose, and lignin, together with 5 to 10% of polyphenols as polysaccharide-protein-polyphenol complexes.

Relatively little attention has been given to the polysaccharides present in legumes or to their specific functional properties and how these are influenced by genetic diversity. Legume starch properties are clearly different from those of cereals. They differ among species and conceivably among genotypes within legume species.

Starch and other carbohydrates vary in composition and properties among pulses, and legume starches are significantly different from their cereal counterparts. The useful functional properties of starch from legumes have not been extensively studied. A Canadian company in cooperation with a university has demonstrated how commercial benefit can be derived from research on carbohydrate properties. Yellow mustard is used mainly for its pungent flavor. By suppression of the myrosinase present, the flavor does not develop. The polysaccharides present were found to possess unique colloidal properties valuable as stabilizers and binders in various food products. The company that stimulated the research now exports the flavorless mustard polysaccharides to more than 60 countries. Though mustard is not a legume, the example is valid since such research as has been undertaken indicates unique properties in certain polysaccharides derived from pulses. (Private communication from an industrial food processor.)

Legume Transformation and Utilization

Traditionally, legumes are cooked either as whole seeds or after dehulling. Parpia (1973) estimated that at least 80% of chickpea and other pulses in India are dehulled, split, and consumed as dhal. Domestic and small village mills give yields after dehulling of about 75% for chickpea and 68% for pigeon pea. An improved process, in which after incipient heating the legume is held in a tempering bin, gives yields of ca 82%. Parpia (1973) and Hawtin (1981) describe many traditional foods from legumes, and Milner (1981) reviews legume foods used in international aid programs.

Cooking Legumes

Stanley and Aguilera (1985) present an extensive review of legume composition, structure, and texture and how they affect and are affected by cooking. Others who have discussed cooking quality include: Rockland and Metzler (1967), Bressani and Elias (1974), Kon (1979), Bueno and Narasimha (1980), Simpson (1980, 1983), Kon and Sanshuck (1981), and International Development Research Centre (IDRC) (1992).

The length of time to cook pulses increases with age after harvest and cooking time is adversely affected by high temperature and humidity in storage. The IDRC Bean Research Network has developed a software package named McBean to predict hardening, cooking time, and economic loss under different storage conditions (IDRC, 1992). No single method correlates with all properties that influence cooking quality. Changes that take place during storage and lead to increased cooking time are often designated "bean hardening". The literature indicates two interrelated phenomena: 1) hardness of the seedcoats which do not imbibe water and soften when soaked; and 2) hard-to-cook pulses which, after soaking, require long cooking and/or do not become tender after long cooking. Hardening may result from cross-linking of pectates in the middle lamella. As in calcium pectate, cross-linking is facilitated by a bivalent cation. The cross-linked pectates remain intact after cooking. Calcium is chelated by phytic acid demonstrably present in beans that soften when cooked. Storage at high temperature and humidity may stimulate hydrolysis of phytates by phytase thus preventing chelation and permitting cross-linking to take place.

Stanley and Aguilera (1985) implicated phenolic compounds in bean hardening and postulated that hard-to-cook beans result from a failure of cotyledon cells to separate during cooking, a condition caused by lignification of the cell walls. Chemical and microscopic examinations indicate that condensed tannins migrate from testa to cotyledons where they bind to form macromolecules in cell walls that resist water penetration and restrict cell separation and softening during cooking.

In extreme cases, beans harden to the point of being useless for food or feed. Various means of transforming hard beans to a useful state have been proposed. Feldberg and Fritsche (1956) recommended soaking beans for 8 hours before cooking, freezing, and hot air-drying, a technically feasible but commercially uneconomical process. In whole beans, water penetration was accelerated by treatment with various salts: $NaHCO_3$, Na_2CO_3, $NaHPO_4$, and $NaCl$ among others (Rockland and Metzler, 1967; IDRC, 1992). Dry and wet dehulling reduce cooking times. The utility of hard bean flour can be improved by extrusion and roller-drum cooking.

Processing of Legume Flour

The simplest processed products are mixtures of legume flour with other finely ground particulate materials. Milner (1981) describes some which achieved commercial acceptance. Hulse *et al.* (1981) reviewed various combinations of sorghum and millets with legume flours. Particles of different size and effective mass tended to separate during transport and storage. Homogeneity can be realized by agglomeration; the surface of each particle is wetted with a film of water sufficient to cause particles to form discrete agglomerates which, after drying in warm air, form stable porous granules.

Also simple and inexpensive is low-pressure extrusion, a process used to make pasta. Legume flour composites are mixed with water to form smooth, stiff doughs which can be extruded and dried. Pasta made from cereal-legume composites may be cooked as is or ground into flour; the latter process is used in UNICEF's infant food supramine (Milner, 1981). More complex are the single and twin screw extrusion cookers in which legume flours and other compatible raw materials are cooked under pressure.

An extrusion process recently developed in Canada processes whole dry grains at ultra high temperature and pressure. Dry split legumes are fully cooked in 2 to 3 seconds, rapidly cooled, and ground into flour. Green pea, yellow pea, lentil, chickpea, and faba bean have been processed into precooked flours that absorb 5 to 8 times their weight in water (Table 8).

Table 8. Properties of HTST extruded legume flour

	Protein	H_2O	Dietary fiber	Absorption
	--%--	--%--	--%--	ml 100 g^{-1}
Lentil	25.5	5.5	8.0	700 - 800
Green Pea	23.5	6.0	4.0	600 - 700
Yellow Pea	23.5	4.0	5.0	700 - 800
Chickpea	21.5	5.0	8.5	350 - 450

Protein Concentration

Legume protein concentrates containing 60 to 70% protein are made by removing most soluble carbohydrates. Alternative solvents include:
1) Hot water (70 to 90°C) at pH 5.5 to 7.5;
2) Food grade acid pH ca 4.5: close to the iso-electric point;
3) Aqueous ethanol (EtOH:H_2O 70:30 v/v).

Isolates containing ca 90% protein can be extracted with dilute alkali. After centrifuging insoluble carbohydrate and other material, the dissolved protein is precipitated by food grade acid at the iso-electric pH. The liquid fraction is comparable to the whey produced when casein is precipitated from milk. The dissolved legume protein may be alternatively precipitated and extruded through textile spinnerets into long filaments which can be woven and texturized to resemble meat or fish.

Though extruded pulse flours are effective as fat and water binders in ground meat, most texturized products are derived from soybean protein.

Air-Classification

Legume cotyledon structures are conducive to protein concentration by ultra-fine grinding and air-classification (Hulse, 1991). Legume starch granules, being relatively larger and more uniform, are more easily detached from protein matrices than cereal starch granules when subjected to high speed shearing in a pin mill. Air-classification is achieved by opposing a centrifugal force with a centripetal air-drag, the two being counterbalanced to cause separation of particles according to their effective mass. The heavier starch flows in one direction, and the lighter protein fragments in the other direction. Youngs (1975), Sosulski and Youngs (1979), Vose (1980), Simpson (1983), Cloutt *et al.* (1987), and Sosulski and McCurdy (1987) have reported various results from air-classification of legume flour. Sosulski and Youngs (1979) reported the following protein increases, the first value being protein concentration in original flour and the second value being protein concentration after air-classification: chickpea 19.5 to 28.9%, faba bean 23 to 47.7%, and lentil 23.9 to 57.9%. The high protein fractions showed high oil absorption, good emulsification, and foam stability; high starch fractions showed high water absorption. Chickpea starch gave both high cold and hot paste (Amylograph) peak viscosities.

Table 9. Comparison of air-classified fractions with large protein concentrations with protein isolates from pea and faba bean

	Protein	NSI	H$_2$O absn (g/g)	Oil absn.	Foal vol Ml
Pea					
Flour	25.0	80.3	1.02	0.34	300
AC fract	47.2	65.1	1.65	0.72	565
Isolate	80.3	38.1	2.97	0.98	315
Faba bean					
Flour	29.2	85.9	1.22	0.40	220
AC fract	63.3	64.2	1.53	0.72	440
Isolate	86.3	40.0	2.34	1.78	200

Sosulski and McCurdy (1987) compared air-classified fractions with protein isolates from field pea and faba bean (Table 9). The data demonstrated the superior foaming properties of air-classified protein fractions, probably because air-classification imposes less biochemical change and reduced solubility than wet processes.

Cloutt *et al.* (1987) reported variations in protein concentration among cultivars of faba beans and other pulses. One faba bean cultivar gave a higher protein fractionation percentage than all others tested, an increase from 29.1% (flour) to 62.7% (high protein fraction).

Legume Product Development

There are many food products made from legumes; but of those produced on a sizable commercial scale, most are derived from soybean or from groundnut. Apart from those based on soybean or groundnut oil, a range of products of comparable diversity could be developed from pulses by physical, biological, and chemical transformations. Hydrolized vegetable protein, made mainly by HCL or fungal protease treatment of soybean flour, could be derived from other legumes. Pulse protein concentrates and isolates can be pre-cooked or texturized by various techniques.

There are three demonstrated routes to commercial food product development. The first is to start with a widely accepted traditional domestic product, determine the limiting quality parameters, then design and construct a technological system to produce and distribute it in marketable quantity. The second is to invent and construct a processing device – a grain mill, a particle classifier, an extruder-cooker – then empirically to discover the products that emerge as various materials are fed in and different processing conditions are imposed. The third is to begin with an established but underutilized crop and, by using the many scientific tools available, determine the structures, biochemical and biophysical properties of its constituents, how these are genetically controlled, what their functional potential is, and how they may be modified to enhance and diversify their functional utility.

General Comment

Published literature shows little evidence of cooperation between food and plant scientists in exploring how the functional properties of legumes other than soybean and groundnut vary with genetic diversity. Much more time and effort have been devoted to demonstrating that rats gain weight faster on diets of cereals supplemented with pulse flour than on cereals alone. Those who have baked bread from composites of cereal and pulse flour have generally evinced greater interest in the quasi-nutritional consequences than with functional compatibility. Few appear to have given serious attention to the degrees of disparity among such important characters as water activity, hydration, pasting, and gelatinization properties of the constituent carbohydrates and proteins of the cereals and legumes. There is relatively little in the published literature on the biophysical, rheological, and functional consequences of mixing pulse flours with cereal flours; whether or not some genotypes are more functionally compatible than others.

It is not suggested that International Agricultural Research Centers such as ICARDA significantly expand their capacities for food science and technology. It does appear desirable that plant breeding objectives be as much concerned with functional utility as with agronomic traits. This calls for collaboration between centers of plant breeding and agronomy on the one side, and of institutes of food science on the other. Equally necessary is the involvement of food processing and distributing industries. Governmental and academic research institutes tend to be too remote from consumer markets to make economically sound decisions on food product design and development. Consequently, for food legume utilization to expand and diversify in a commercially rational and reliable pattern requires orderly and coordinated cooperation in which plant scientists, food scientists, industrial food processors, and distributors are equal and effective partners.

The remarkable diversity of products available from soybean, groundnut, and maize came through such patterns of cooperation. Plant breeders produced cultivars with diverse functional properties. The nature of these properties and how they may be modified by chemical and/or physical treatments were determined by food scientists, biochemists, and biophysicists. The development of marketable food products and how best to use by-products was the responsibility of food industries. Through similar patterns of collaboration, the utilization of many pulse crops could be remarkably expanded and diversified with immense benefits to farmers, industries, and consumers.

References

Adsule, R. N. and Kadam, S. S. 1989. Proteins. In: *CRC Handbook of World Food Legumes*, 75–97 (eds. D. K. Salunkhe and S. S. Kadam). Boca Raton, Florida, USA: CRC Press.

Adsule, R. N., Kadam, S. S. and Leung, H. K. 1989. Lentil. In: *CRC Handbook of World Food Legumes* (eds. D. K. Salunkehe and S. S. Kadam). Boca Raton, Florida, USA: CRC Press.

Ben-Ze'ed, N. and Zohary, D. 1973. *Israel Journal of Botany* 22: 73–91.

Bhatty, R. S. 1977. *Canadian Journal of Plant Science* 57: 979–983.

Bhatty, R. S. 1984. *Journal of Agriculture Food Chemistry* 32: 1161.

Bhatty, R. S. 1988. *Journal of Canadian Institute of Food Science Technology* 21: 144–160.

Bressani, R. and Elias, L. G. 1974. In: *New Protein Foods*, pp. 231–297 (ed. A. Altschul). New York and London: Academic Press.

Bueno, E. C. and Narashimha. 1980. *Journal of Food Science Technology* (Mysore) 7: 235–240.

Cerning, J., Saponik, A. and Guilbot, A. 1975. *Cereal Chemists* 52: 125–138.

Chavan, J. K., Kute, L. S. and Kadam, S. S. 1989. In: *CRC Handbook of World Legumes*, pp. 223–245 (eds. D. K. Salunkhe and S. S. Kadam). Boca Raton, Florida, USA: CRC Press.

Clarke, H. E. 1970. *Proceedings of the Nutrition Society* 29: 64–73.

Cloutt, P., Walker, A. F. and Pike, D. J. 1987. *Journal of Food Science Agriculture* 38: 177–186.

Colonna, P., Buleon, A. and Mercier, C. 1981. *Journal of Food Science* 46: 88–93.

Cumper, C. W. N. 1953. *Transactions of the Faraday Society* 49: 1360–1365.

Dengate, H. N. 1984. In: *Advances in Cereal Science and Technology*, pp. 49–82 (ed. Y. Pomeranz). St. Paul, Minnesota, USA: American Association of Cereal Chemists.

Doublier, J. L. 1987. *Journal of Cereal Science* 5: 247–262.

Eden, A. 1968. *Journal of Agricultural Science* (Cambridge). 70: 299–301.

FAO. 1990. *Production Year Book*. Rome: FAO.

Feldberg, C. and Fritsche, H. W. 1956. *Food Technology* 10: 523–527.

French, D. 1984. In: *Starch: Chemistry and Technology, Second Edition*, pp. 184–247 (eds. R. L. Whistler, J. N. BeMiller and E. F. Paschall). Orlando, Florida, USA: Academic Press.

Hawtin, L. C. 1981. *Faba Bean Cook Book*. Aleppo, Syria: ICARDA.

Hawtin, G. C., Rachie, K. O. and Green, J. M. 1977. In: *Nutritional Standards and Methods of Evaluation for Food Legume Breeders*, pp. 43–50 (eds. J. H. Hulse, K. P. Rachie and L. W. Billingsley). Ottawa, Canada: International Development Research Centre.

Hermansson, A. M. 1975. *Journal of Food Science* 40: 603–607.

Hermansson, A. M. 1979. *Journal of American Oil Chemistry Society* 56: 272–279.

Hulse, J. H. 1975. In: *International Workshop on Grain Legumes*, pp. 189–207. Hyderabad, India: ICRISAT.

Hulse, J. H. 1980. In: *Utilization of Protein Resources*, pp. 1–17 (eds. D. W. Stanley and D. M. Murray). Westport, Connecticut, USA: Food and Nutrition Press.

Hulse, J. H. 1991. In: *Uses of Tropical Grain Legumes*, pp. 11–27 (ed. R. Jambunathan). Patancheru, India: ICRISAT.

Hulse, J. H., Laing, E. M. and Pearson, O. E. 1981. *Sorghum and the Millets: Their Composition and Nutritional Value*. London: Academic Press.

IDRC. 1992. *IDRC Bean Research Network*. Ottawa, Canada: International Development Research Centre (In press).

Johnson, D. W. 1970. *Journal of American Oil Chemistry Society* 47: 402–406.

Kanamori, M., Ikeuchi, T. and Ibuki, F. 1982. *Journal of Food Science* 47: 1991–1994.

Kinsella, J. E. 1979. *Journal of American Oil Chemistry Society* 56: 242–257.

Kon, S. 1979. *Journal of Food Science* 44: 1329–1334.

Kon, S. and Sanshuck, D. W. 1981. *Journal of Food Processing Preservation* 5: 169–178.

Landry, J. and Moureaux, Г. 1970. *Bulletin of the Society of Chemical Biology* 52: 1021–1037.

Lineback, D. R. and Ke, C. H. 1975. *Cereal Chemists* 52: 334–347.

Lund, D. 1984. *Critical Review Food Science Nutrition* 20: 249–261. New York and London: Academic Press.

Makasheva, R. K. 1973. *The Pea*. Leningrad: Kolos Publishers. Translated from Russian, New Delhi, India: Amerind Publishing Co., 1983. 267 pp.

Marquardt, R. R., McKirdy, J. A., Ward, T. and Campbell, L. D. 1975. *Canadian Journal of Animal Science* 55: 421–429.

Martinez, W. H. 1979. *Journal of American Oil Chemistry Society* 56: 281–283.

Matthews, P. and Arthur E. 1985. In: *Pea Crop*, pp. 369–381 (ed. P. D. Hebblethwaite). London: Butterworth.

Milner, M. 1981. In: *Utilization of Protein Resources*, pp. 18–31 (eds. D. W. Stanley and E. D. Murray). Westport, Connecticut, USA: Food and Nutrition Press.

Naivikul, O. and D'Appalonia, B. L. 1979. *Cereal Chemists* 56: 24–28, 45–49.

Osborne, T. B. 1924. *The Vegetable Proteins*. London: Longsman, Green and Co. 154 pp.

Parpia, H. A. B. 1973. In: *Nutritional Improvement of Legumes by Breeding*, pp. 281–295 (ed. M. Milner). New York and London: Wiley-Intersciences.

Pellett, P. L. and Young, V. R. 1980. *Nutritional Evaluation of Protein Foods*. Tokyo: United Nations University.

Pitz, W. J. and Sosulski, F. W. 1980. *Journal of Canadian Institute of Food Science Technology* 13: 35–39.

Rama Rao, G. 1974. *Indian Journal of Nutrition and Dietetics* 11: 268–273.

Rao, P. S. 1969. *Indian Journal of Medical Research* 57: 2151–2157.

Reddy, N. R., Pierson, M. D., Sathe, S. K. and Salunkhe, D. K. 1984. *Food Chemistry* 13: 25.

Rockland, L. B. and Metzler, E. A. 1967. *Food Technology* 21: 344–348.

Salunkhe, D. K. and Kadam, S. S. (eds.). 1989. *CRC Handbook of World Food Legumes*. Boca Raton, Florida, USA: CRC Press.

Schoch, T. J. and Maywald, E. C. 1968. *Cereal Chemists* 52: 564–573.

Simpson, A. D. F. 1980. In: *Vicia Faba, Feed Value, Processing and Viruses*, pp. 257–272 (ed. D. A. Bond). Dordrecht: Martinus Nijhoff.

Simpson, A. D. F. 1983. In: *The Faba Bean: A Basis for Improvement*, pp. 535–552 (ed. P. D. Hebblethwaite). London: Butterworths.

Singh, S., Singh, H. D. and Sikka, K. C. 1968. *Cereal Chemists* 45: 13–17.

Singh, U. 1985. *Qualitas Planatarum: Plant Foods for Human Nutrition* 35: 339–351.

Singh, U. and Jambunathan, R. 1982. *Qualitas Plantarum: Plant Foods for Human Nutrition* 31: 347–354.

Sjodin, J. 1982. In: *Faba Bean Improvement*, pp. 319–331. (eds. G. Hawtin and C. Webb). Dordrecht: Martinus Nijhoff.

Sosulski, F. W. and Holt, N. W. 1980. *Canadian Journal of Plant Science* 60: 1327–1331.

Sosulski, F. W. and McCurdy, J. 1987. *Journal of Food Science* 52: 1010–1014.

Sosulski, F. W. and Youngs, C. G. 1979. *Journal of American Oil Chemistry Society* 56: 292–295.

Stanley, D. W. and Aguilera, J. M. 1985. *Journal of Food Biochemistry* 9: 277–323.

Swaminathan, M. S. and Jain, H. K. 1975. In: *Nutritional Improvement of Food Legumes by Breeding*, pp. 69–82 (ed. M. Milner). New York and London: Wiley-Interscience.

Tkachuk, R. 1977. In: *Nutritional Standards and Methods of Evaluation for Food Legume Breeders*, pp. 78–82 (eds. J. H. Hulse, K. O. Rachie and L. W. Billingsley). Ottawa, Canada: International Development Research Centre.

Vavilov, V. I. 1949. *Chronicle of Botany* 13: 1–54.

Vose, J. R. 1980. *Cereal Chemists* 57: 406–410.

Williams, P. C., Singh, K. B. and Saxena, M. C. 1991. In: *Uses of Tropical Grain Legumes*, pp. 55–61 (ed. R. Jambunathan). Patancheru, India: ICRISAT.

Young, V. and Pellett, P. 1990. In: *Sorghum Nutritional Quality*, pp. 25–40 (ed. Gebisa Ejeta). Lafayette, Indiana, USA: Purdue University.

Youngs, C. G. 1975. In: *Oilseed and Pulse Crops in Western Canada*, pp. 617–632 (ed. J. T. Harapiak). Calgary, Alberta, Canada: Western Co-operative Fertilizers Ltd.

Diversifying use of cool season food legumes through processing

R. JAMBUNATHAN[1], H.L. BLAIN[2], K.S. DHINDSA[3], L.A. HUSSEIN[4], K. KOGURE[5], L. LI-JUAN[6] and M. M. YOUSSEF[7]

[1] *Crop Quality Unit, ICRISAT, Patancheru P.O., Andhra Pradesh 502 324, India;*
[2] *American Dry Pea and Lentil Association, 5071 Hwy 8 W, Moscow, Idaho 83843, USA;*
[3] *Department of Chemistry and Biochemistry, Haryana Agricultural University, Hisar 125 004, Haryana, India;*
[4] *Department of Nutrition, National Research Centre, El-Tahrir Street, Giza, Dokki, Egypt;*
[5] *Faculty of Agriculture, Kagawa University, 2393 Ikenobe, Miki-tyo, Japan;*
[6] *Faba Bean Germplasm and Breeding, Zhejiang Academy of Agricultural Sciences, Hangzhou, China, and*
[7] *Department of Agricultural Industries, Faculty of Agriculture, University of Alexandria, El-Chatby, Alexandria 21526, Egypt*

Abstract

Traditional methods of processing food legumes such as decortication, boiling, roasting, frying, puffing, germinating, and fermenting are used in one form or the other in preparing legume products for consumption in different regions of the world. These processing methods contribute to improved product quality as well as nutritional quality. Extrusion cooking, textured vegetable product, quick-cooking products, weaning foods, and beverages are some of the technologies that have good potential in the improved utilization of cool season food legumes. There is also a growing trend to market legume products as value-added snack items. To expand the markets worldwide for these new products, it is necessary to provide products that are more attractive, convenient, and are nutritionally equal or superior to other established popular products in the market. Other less explored but potentially promising avenues include the use of components of legumes as therapeutics, in dairy industry, and as food preservatives. Leaf protein concentrate also offers scope for use as a protein source for livestock and humans, and in cell culture.

Introduction

The last decade witnessed the development of many products originating from grain legumes with diverse functional, nutritional, and biological activities. Production will depend on development of markets for such products, as well as the balance between costs of raw materials and processing, and the return generated by the recovery and functionality of proteins and starch from dry and wet processing of grain legumes.

Increasing the uses and diverse applications of legumes other than soybean

F.J. Muehlbauer and W.J. Kaiser (eds.), Expanding the Production and Use of Cool Season Food Legumes, 98–112.
© *1994 Kluwer Academic Publishers.*

has been enhanced by processing them into protein-rich and starch-rich fractions. Convenience and appearance are the major reasons for additional processing. Food legumes traditionally have been consumed as staple items in many parts of the world. Removing the seedcoat, splitting, and polishing the legumes not only makes a more attractive product but reduces the cooking time. This is a process that has been used for hundreds of years.

In today's economy there is a growing trend to market many food legumes as value-added products. This is particularly true in countries such as Japan, Taiwan, Thailand, and the USA where the economies are strong. These value added snack items are becoming increasingly popular with the younger generation whose tastes are changing and who now have the resources to purchase these items. Increased advertising through magazines and television have also been a factor in creating an increased demand and acceptance for snack items.

Faba Bean

Faba bean (*Vicia faba* L.) can be fractionated using air- classification or by a wet process (Bramsnaes and Olsen, 1978; Tyler *et al.*, 1981; Gueguen 1983; Sosulski and Sosulski, 1986). The efficiencies of protein (70 to 80%) and starch (88 to 93%) recoveries from faba bean and field pea (*Pisum sativum* L.) were found to be higher by the dry process than respective recovery figures (73 to 79%) obtained from wet processing (Sosulski and Sosulski, 1986). However, wet processing was reported to be more effective in removing toxic constituents (Arntfield *et al.*, 1985). Pilot scale processes have been developed by Bramsnaes and Olsen (1978), Murray *et al.* (1981), Gueguen (1983), and McCurdy and Knipfel (1990).

Faba bean and pea concentrate prepared by air classification has been reported to reduce the cooking losses to negligible values for boiled meat patties where 30% of the meat was substituted by legume protein concentrate (Vaisey *et al.*, 1975). The air-classified products contain about 20 to 25% starch which might be responsible for a significant part of the water-and-fat-binding properties of these products.

By sensory evaluation of the legume beaf-patties, Vaisey *et al.* (1975) found that dried pea flavor and bitter after taste were the dominant flavor characteristics and were attributed to the presence of lipoxygenase activity in faba bean. The mechanisms of both the faba bean and pea flavor have been suggested to be similar to that of the soybean (*Glycine max* L.) and could be solved by partial enzymatic hydrolysis of the protein extract with microbial proteolytic enzyme, isolated from *Penicillium du ponti* (Vieth *et al.*, 1983). Mixtures of whey skim milk and faba bean protein in which faba bean grain made up to 15 to 45% by weight, proved to have good emulsifying capacity, foam stability and could form gels (Fayed and Morshed, 1990).

Hydrolyzed Vegetable Proteins (HVP)

Hydrolyzed Vegetable or Plant Proteins (HVP or HPP) are defined as mixtures composed of amino acids and peptides, which are obtained by hydrolysis of vegetable proteins and frequently other substances, such as salt (Olsman, 1978). The industrial interest in hydrolyzed vegetable proteins (HVP) grew sharply after Ikeda's discovery in 1908 of monosodium glutamate (MSG) as the major ingredient and flavor compound in HVPs. This know-how was the start for its commercial production in Europe, and proved to be a commercial success in bouillon cubes.

On an industrial scale, HVP are prepared either enzymatically or by acid hydrolysis. Either process converts proteins into peptides and amino acids, whereas the carbohydrates are converted into sugars which degrade to a large extent into products like hydroxymethylfurfural and levulinic acid (Olsman, 1978). Under the conditions of HVP production, or during the concentration process, part of the amino acids, sugars or sugar degradation products are converted in nonenzymic browning reactions.

Enzymatic hydrolysis offers an attractive way of increasing the solubility of vegetable proteins. However, the bitter peptides, identified as those fractions having leucine at the termini, are not decomposed. Vieth *et al.* (1978) used a microbial thermophilic enzyme, isolated from *Penicillium du ponti*, with specificity similar to pepsin for hydrolysis.

The legal position of HVPs has been discussed by Codex Alimentarius Commission, which considered HVP as food ingredient and as food additive. There is no evidence from the available information to indicate that HVPs, applied as flavoring agents at their current levels, pose a public hazard. However, the commission stated that evidence is insufficient to determine the reported adverse effects (lesions in the central nervous system) and HVP's are not deleterious to infants when added as flavoring agents to infant or baby foods. The industry had already decided to abandon the use of HVP as flavor agents in these foods. Murata *et al.* (1988) reported that acidic, neutral, and alkaline proteinases originating from microorganisms and plants were capable of coagulating faba bean milk protein. The curd was made up of 73% of the faba bean milk protein. However, pea milk-protein did not coagulate under these experimental conditions.

Chemical Modifications of Grain Legumes

Acylation is the most extensively studied modification of legumes. Schmandke *et al.* (1982) tested acylated faba bean proteins, with different degree of substitutions (0 to 78%) and recommended their use to increase the viscosity of gelatine solutions. Treatment of faba bean protein with succinic anhydride (succinylation) is reported to increase its water and oil absorption capacity by 25 and 40%, respectively. Moderately succinilated (27%) faba bean protein improved substantially its emulsifying activity, emulsifying capacity, and emulsion stability.

Isolation of Compounds with Biological Activities

Up to 80% of the total protein in grain of faba bean and pea is in the cotyledons as a non-metabolic reserve. This globulin protein contains legumin and vicilin in the ratio of 2.3:1. Arntfield and Murray (1985), and Gueguen and Schaeffer (1984) described appropriate methods for the isolation of legumin and vicilin.

Hypocholesteremic Compounds

Spadoni *et al.* (1981) reported that rats fed faba bean concentrate in a diet with high fat content reduced the total plasma cholesterol significantly and increased the bile excretion. The authors suggested that protein concentrate of faba bean had a hypocholesteremic effect through modifying the pattern of bile acid excretion. The identification of the hypocholestremic factor and its isolation in purified form is needed.

Faba Bean Products

The most popular dishes made from faba beans are *Medamis* (stewed beans), *Falafel* or *Ta'meya* (deep fried cotyledons paste with some vegetables and spices), *Bissara* (cotyledons paste poured into plates), and *Nabet* soup (boiled germinated beans). Processing methods of these products and their nutritive values have been described previously (Youssef *et al.* 1986, 1987; Shekib *et al.*, 1989; Ziena *et al.*, 1991). *Medamis* and *Falafel* taken with bread are very popular breakfast food and snack sandwiches for the majority of Egyptians (Ragab, 1988). A less popular product *Fool Mekalley*, is made from faba beans by roasting and consumed as a snack.

China has the largest area of faba bean in the world. Excluding faba bean crops grown for green manure, the current production area is about 1×10^6 ha, output is close to 2×10^6 t and average yields are about 1700 kg per hectare. Faba bean is an important winter and spring legume in China. Faba bean grains are an important item in the daily food of the Chinese people as a nutritious food, rich in protein (24 to 34%) and amino acids. They are made into many kinds of traditional foods. Faba bean has been utilized as staple and non-staple foods in different styles. An agronomist G. Q. Xu (AD 1562–1633) in Ming dynasty evaluated faba bean as a versatile foodstuff and so did scientists in Qing dynasty (AD 1616–1911) (Li, 1987).

For a long time, farmers in northwestern China have commonly mixed faba bean flour with other flours such as maize flour to make meals. Due to improved national economy and better living standards of people, there was an increasing demand for foods with desirable quality. Food processing methods of faba bean have also changed considerably and products are expected to have superior color, smell, and taste. At present, non-staple foods of faba bean have a ready market in China and elsewhere. Faba bean foods are divided into three classes based on their processing and cooking methods.

Fried Products

Fried products of faba bean are used as popular refreshments and made by simple processes (Zhang, 1987). Salty faba bean is made by boiling with salt, crisp faba bean by frying with sand or salt. Fragrant faba bean products such as "aniseed faba bean", "spiced faba bean", and "unusual aromatic bean" are prepared by adding varied flavoring agents. Among them, fried "orchid bean" is the most popular one. Cakes and pastes of faba bean are also commonly consumed. A majority of such products is made in individual families for their own use and some also made by small factories.

The orchid faba bean is made by the following process. Faba beans are cooked in boiling water until they can be easily pricked through by a needle. They are dried and each bean is cut both vertically and horizontally to make a cross. The beans are dried in air to remove surface wetness and fried in oil under high flame. When the top splits and the hulls change color from yellow to red, the beans are removed and cooled. Salt could be added for taste before they are eaten.

The spiced faba bean (Wu Xiang Dou) is made by the following process. Selected faba beans are washed and boiled in water. Salt, Chinese prickly ash, staranise, aniseed, and Cassia bark cinnamon are added and cooked under low flame. The quantity of addition of these ingredients are based on the amount of faba bean used and individual preferences. When the shape of a bean can be changed by a gentle bite (to indicate the texture), beans are removed and dried in air. A second procedure is to fry the beans until the hull splits a bit, licorice powder is then added and fried to dry the beans.

Brewed Products

Brewing industry in China has a long history with an outstanding record (Jiang, 1988). As a result of the development of science and technology, improvements have been made in the brewing industry of China. Being rich in proteins and various amino acids, faba bean has been used to brew different kinds of sauces. They are made by mixing faba bean with flour, salt, and water using a special process. Faba bean pastes with specific flavor such as chili, sesame, chicken, ham, beef, and "huoguo paste" are prepared by adding specific flavoring ingredients to faba bean paste in Sichuan and Anhui provinces. "Juancheng" brand faba bean paste made in Pi country, Sichuan province and "Anqing" chili faba bean paste made in Anhui province are famous. Sauce is one of the essential factors for making delicious Chinese dishes. Proper proportion of different flavoring ingredients is also very important besides the superb cooking skills needed in preparing these Chinese products.

Starch Foods

Starch has variable usage as it can be used directly as food or processed into varied non-staple foods (Huang, 1987). Starch extracted from faba bean can be further processed to make high-grade bean-starch vermicelli and noodles sheet jelly. Products of faba bean starch have been used in regular meals and they have similar quality as mung bean product and much better quality than other starch products. Cooked, hulled, and mashed faba bean has also been used to produce dumpling and steamed bread by adding oil, sugar, sweet osmanthus and orange skin. Sometimes, sesame and sugar are added to make refreshments such as bean sweet and sesame bean sweet. In summary, faba bean has a great prospect in making traditional, popular, high-grade, and instant foods.

Beans (*Phaseolus vulgaris* L.), peas, and black eye peas (*Vigna unguiculata* L.) are cooked or canned with chunks of beef and tomato sauce. According to El-Ashway *et al.* (1985) and El-Hashmy *et al.* (1985), cooking of the afore-mentioned leguminous seeds by both traditional and pressure cooking methods improved their sensory properties as well as protein quality. Until only four decades ago, faba bean has been an important part of the Japanese diet in various forms such as the main item of food, the subsidiary food item, protein curd, and fermented sauce and paste. Today, this bean has become very popular among most Japanese, especially young people, as snack items. Kagawa-ken (a state in Japan) is famous for making them. Some of the popular products are, fried beans, processed fried beans with sugar, sesame, ginger, red pepper, green laver, and curry. Press cakes are also prepared with faba bean flour and sugar and cut into different shapes for consumption.

Utilization of Faba Bean Plant

The efficient utilization of whole green crop or plants is important for meeting the future world demand for food, especially protein for humans and livestock (Pirie, 1978). The effect of fertilizer on the distribution of different fractions of plant harvested at three stages are reported (Kogure and Ohshima, 1991a,b).

Faba bean cultivar "Boshu-wase" was grown under zero level (I), standard level (II) (28 kg ammonium sulfate, 45 kg superphosphate, 18 kg potassium chloride), and three times the level (III) of fertilizer. Fertilizer was applied at the beginning of flowering stage. Samples were collected at 0 days (start of flowering), 15 days (end of flowering), and 30 days (pod-developing). The top portion of the plant material was cut and disintegrated in a pulper. It was then squeezed and fractions of green juice (GJ) and fibrous residue (FR) were obtained. GJ was heated (70°C) after adjusting it to pH 4 with HCl. The coagulated leaf protein concentrates (LPC) and brown juice (BJ) were separated by centrifugation (Kogure and Ohshima, 1991a,b). FR was placed in bottles and ensiled for about six months (Ohshima and Kogure, 1984) (Figure 1). Total non-

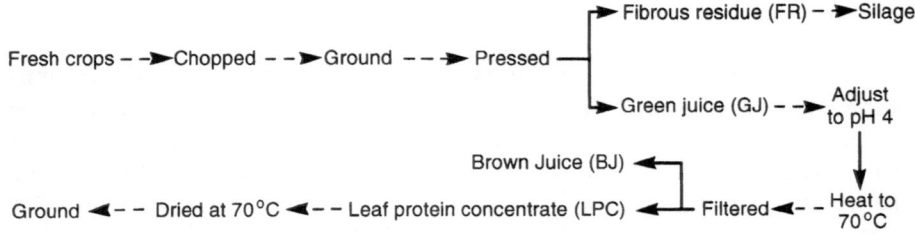

Figure 1. A flow diagram for the laboratory fractionation of green crops.

constructive carbohydrate (TNC) was determined by Somogyi-Nelson method and total nitrogen (N) by Kjeldahl method.

Fertilizer accelerated the growth and development of plants and increased the biomass and yield of raw material for fractionation (Table 1) (Kogure *et al.*, 1977). The concentration of carbohydrate in raw material increased in contrast to the amount of nitrogen and ash with increased applications of fertilizer. The carbohydrate concentration in BJ was two-to-three fold greater than the corresponding LPC and FR in early stages of growth, but it increased in LPC and FR at the pod-developing stage, especially in LPC. Nitrogen concentration in LPC, which was remarkably high, declined rapidly towards the pod-developing stage. Nitrogen concentration in FR and BJ was similar at various sampling times and at different fertilizer amounts. The ash concentration was high in BJ throughout the three stages while it declined in LPC and FR, especially in FR at the later stages. The quality of silage prepared from FR at each stage of sampling of plant and at different fertilizer amounts was good.

Fertilizer dressing caused the vigorous growth of plants and increased the carbohydrate concentration of raw material. Due to this, the nutritional value of LPC was improved by changing the balance of carbohydrate to nitrogen concentration. This observation would be useful for the utilization of the plant as a human food in the near future. The fractionation process resulted in the uniform distribution of carbohydrate content in FR and in removal of the detrimental elements for better lactic acid fermentation during the ensiling process of FR (Ohshima and Kogure, 1984). Soluble carbohydrates and ash containing the detrimental elements were separated into BJ. It was concluded that faba bean plants grown with high amounts of fertilizer and cut at the pod-development stage can be fractionated and utilized. This procedure results in high yield of chemical components, value-added LPC, good quality of FR silage, and valuable BJ.

Dry peas are a good example of a growing snack market made from legumes. Two major types of value-added snack items currently are made from peas. The usual process is to soak peas overnight and fry them in hot oil. Palm oil is most commonly used. Sometimes peas are coated with other materials such as rice flour before frying to provide different flavoring. The product is then seasoned and packaged. The larger marrowfat type pea is preferred by most consumers for this process.

Table 1. Characteristics of raw material, three fractions and FR-silage

Sampling time	0 day		15 days		30 days		
Fertilizer level	I	I	II	III	I	II	III
Raw material (RM) Yield (g m⁻²)	149.0	200.0	219.0	221.0	219.0	310.0	418.0
Total Non-constructive Carbohydrate (TNC)							
RM (%)	14.7	13.9	11.2	24.1	16.6	21.5	24.8
LPC (%)	15.5	16.1	17.4	16.6	27.5	28.4	36.5
FR (%)	7.6	7.4	7.5	9.1	12.0	10.7	12.5
BJ (%)	61.0	51.8	41.9	61.7	61.0	54.9	58.9
Nitrogen (N)							
RM (%)	3.6	3.5	3.5	3.3	2.9	3.1	2.9
LPC (%)	8.0	8.6	8.8	7.4	5.2	6.2	5.4
FR (%)	3.1	2.4	2.7	2.7	2.6	2.6	2.5
BJ (%)	2.2	2.5	2.5	2.3	2.4	2.5	2.1
Ash (%)							
RM (%)	11.8	9.6	13.2	8.0	7.5	9.0	7.4
LPC (%)	9.4	6.3	10.0	6.8	6.9	6.0	7.1
FR (%)	12.2	12.3	12.4	8.7	6.4	6.9	6.6
BJ (%)	12.0	12.8	15.9	10.1	10.0	12.4	10.7
FR-silage							
pH	3.84	3.82	3.89	3.74	3.79	3.83	3.79
Lactic acid (%)	2.91	2.59	2.65	2.83	2.41	2.32	2.47
TVFA (%)	1.39	1.15	1.55	1.26	1.11	1.05	1.13
NH₄-N (%TN)[2]	4.4	5.2	5.6	4.2	6.2	5.8	6.0

[1]Total Volatile Fatty Acid.
[2]Percent of Total Nitrogen. Fertilizer level I, II, III (see text)

The second major value-added snack product from peas is made by grinding the peas into a fine flour that is then forced under pressure through an extruder to create different shapes. The extruded shapes are fried, seasoned, and packaged.

Lentil (*Lens Culinaris* L.)

Legumes contain antinutritional factors including trypsin inhibitors, hemagglutinins, and flatulence causing oligosaccharides (Liener, 1980). Heating and germination have been found to be effective in reducing the concentration of antinutritional factors in lentil (Batra *et al.*, 1986; Batra, 1987; Batra and Dhindsa, 1989), pigeonpea (Vasishta *et al.*, 1986), and chickpea (Bansal *et al.*, 1988).

Seven genotypes of lentil *viz.* L-9-12, L-82-3, L-82-4, L-82-6, L-82-7, LH-21, and LH-311 were studied. Trypsin inhibitor activity (TIA), haemagglutinin activity (HA), and oligosaccharides were estimated in dry heated, autoclaved, and boiled samples of lentil (Batra *et al.*, 1986; Vasishta *et al.*, 1986; Batra and Dhindsa, 1989). All the genotypes possessed TIA (Table 2). Autoclaving of lentil seeds for 20 minutes or heating in boiling water for 10 min inactivated TIA completely. Moist heating was more effective than dry heating. Purified trypsin inhibitors are, in general, resistant to heat (Taukamote *et al.*, 1983). Trypsin in horse gram appears to be thermostable even during cooking (Ghorpade *et al.*, 1986). Soaking of lentil seeds in water and subsequent germination resulted in progressive loss of TIA. Limited proteolysis of the inhibitor protein may be the basis for such a loss in activity (Wilson and Tan-Wilson, 1983).

Table 2. Effect of heating and germination on trypsin inhibitor activity[1] and phytohaemagglutinin activity[2] in lentil seeds/flour[3]

Treatments	Trypsin inhibitor activity	Phytohemaglutinin activity
Control	672	2326
Dry heating		
1 h	119	14
2 h	59	3
Autoclaving		
10 minutes	90	155
20 minutes	Nil	14
Heating in boiling water		29
10 minutes	Nil	13
20 minutes	---	
Soaking		
24 h	254	969
Germination		
3 days	230	330
6 days	84	90

[1]μmol of tyrosine released per g material.
[2]Expressed in terms of maximum dilution of the seed extract in which agglutination could be observed.
[3]Each value in this table is an average of seven values representing seven genotypes and each value for single genotype is based on four determinations (duplicate extract for each sample and duplicate estimation for each extract).

Heating in boiling water is the most effective means of destroying HA in lentil (Table 2). More than half of HA was eliminated when seeds were soaked in water for 24 h, and the decrease in HA continued with time up to 6 days of germination. In cereals, HA has been associated with several protein fractions (Newburg and Concon, 1985). Assuming a similar situation in legumes, differences in HA and its response to heat and germination may be due to differences in amount and proportion of different protein fractions contributing to haemagglutinin activity.

Oligosaccharide concentration in lentil increased irrespective of the heating procedure (Table 3). Even though oligosaccharide concentrations in the seeds did not change much on heating in boiling water for 10 minutes, the increase was evident when taking into account the oligosaccharides leached out into the surrounding water. The observed increase might be due to non-enzymatic hydrolysis or to the release of oligosaccharides from bound macromolecules including higher molecular weight alpha-galactosides which may also be protein in nature.

Table 3. Effect of heating and germination on oligosaccharides concentration[1] (g 100^{-1} g dry weight) in lentil seeds/flour

	Oligosaccharides			
Treatments	Sucrose	Raffinose	Stachyose	Total
Control	1.71	1.11	0.83	3.65
Dry heating				
10 minutes	2.46	2.34	2.21	7.01
20 minutes	1.59	1.02	1.17	3.78
Autoclaving				
10 minutes	2.20	1.29	1.06	4.55
20 minutes	2.33	1.69	1.04	5.06
Heating in boiling water				
10 minutes	1.27	0.95	1.38	3.60
Seeds	0.35	0.11	0.22	0.68
Medium	1.62	1.06	1.60	4.28
Total				
20 minutes				
Seeds	1.69	1.48	1.77	4.94
Medium	0.63	0.54	0.34	1.51
Total	2.32	2.02	2.11	6.45
Soaking 24 h				
Seeds	1.81	1.36	1.05	4.22
Medium	0.20	0.04	0.04	0.28
Total	2.01	1.40	1.09	4.50
Germination				
3 days	1.14	0.40	0.32	1.86
6 days	1.74	0.00	0.00	1.74

[1]Each value in this table is an average of values representing seven genotypes and each value for single genotype is based on four determinations (duplicate extract for each sample and duplicate estimation for each extract).

Although a slight increase in total as well as individual oligosaccharides occurred at 24 h soaking of seeds in water, all oligosaccharides decreased 3 days after germination. At 6 days after germination, raffinose and stachyose had disappeared completely, while sucrose showed a proportionate increase.

Increase in sucrose at the expense of stachyose and raffinose at 3 days after germination, further strengthens the view that raffinose and stachyose are hydrolyzed to produce sucrose during germination. Germination of lentil seeds for 6 days is perhaps the most reliable means for complete elimination of raffinose and stachyose, the most gas forming sugars, and therefore, may be used in the preparation of lentil based food products.

Lentil seeds are widely consumed in Mideastern countries in the following four forms: 1) whole seeds are cooked with tomato sauce, 2) soup, 3) paste prepared from decorticated seeds with rice and carrots, and 4) *Koshary*. *Koshary* is a very popular dish especially in Cairo, Egypt. It is prepared by blending rice and whole or decorticated lentil seeds in the ratio of 2:1 (w/v). Chemical composition and nutritive value of *Koshary* have been extensively investigated (Shekib *et al.*, 1985; Shekib *et al.*, 1986, 1987). It was observed that proteins of *Koshary* exhibited moderate concentrations of lysine and sulfur-containing amino acids that are the first limiting amino acids in rice and lentils, respectively.

Chickpea

Chickpea is a major food legume in many countries including Algeria, Myanmar, Ethiopia, Iran, India, Malawi, Morocco, Pakistan, Spain, Syria, Tanzania, Tunisia, and Turkey. Desi and kabuli are two types of chickpea that are grown in the world. However, more than 80% of the world production of chickpea is of the desi type. The use and versatility of chickpea has been well recognized for many centuries. Various aspects of chickpea including production, nutritional quality, postharvest technology, utilization, and marketing have been reported in detail (Saxena, 1987). The Indian subcontinent is the largest producer of chickpea in the world. It also accounts for a much larger variety of usage of chickpea than any other region in the world. A meeting held at the International Crops Research Institute for the Semi-Arid Tropics (ICRISAT, 1991) discussed specifically the utilization of tropical legumes including chickpea. Scientists from 21 countries including Bangladesh, India, Spain, Pakistan, Turkey, Ethiopia, Sudan, West Asia, and North Africa region participated in this meeting. A brief mention of the products reported by these participants is given below:

In the Indian subcontinent, desi chickpea is usually consumed in the form of *dhal* (decorticated split cotyledons) or dhal flour (*besan*). In Bangladesh, most of the chickpea produced is consumed in the form of *dhal*, followed by dhal flour. Chickpea flour is also mixed with wheat flour to make *roti*. These preparations are also common in India, Pakistan, and Nepal.

A popular Egyptian dish, *Lokmet El-kadi* is prepared from chickpea wheat flour and other ingredients. In Ethiopia, legumes are eaten in the form of sauce to supplement the cereal-based staple diet. Some of the products in which chickpea is used are *nifro* (boiled and served by itself or mixed with other cereals), *kollo* (roasted), and *dabo* (fermented wheat and chickpea are mixed and baked).

In India, in addition to *dhal* and *besan*, whole dehulled grain, sprouted grain, immature pods, seeds, and mature green seeds are some of the other forms in which chickpea is consumed. The secondary processing of dhal may involve dry or moist heating, e.g., roasting, steaming, and frying. Puffed chickpea production is a cottage industry in India.

Processing methods of chickpea have certain built in advantages both from nutrition and convenience aspects. Soaking of chickpea reduces cooking time considerably and also reduces the trypsin inhibitor, haemagglutinating activity, and flatulence- inducing sugars as some of them are leaked out. Although the concentration of these inhibitors and antinutritional factors are not high enough to cause major concern, the processing method reduces the concentration further.

During the roasting of dhal, the chickpea becomes brown due to Maillard reaction and the aroma and quality of seed improves. On puffing, the seed becomes light from shrinkage of the endosperm and loss of water. The seed starch is thus dextrinized.

In Spain, a chickpea product *Cocido* (boiled chickpea) is quite popular. The price of chickpea also varies from 2 to US$ 3 kg^{-1} in some towns. Canning of chickpea could offer new opportunities for consumers. In the Sudan, chickpea is consumed as *balilah* (boiled chickpea with salt and sesame oil, an energy food eaten especially during fasting period of Ramadan) and *tamia* (soaked dhal ground to a paste, which, after addition of spices, is deep fried).

In Turkey, chickpea is added to improve the taste of many dishes e.g., *nohutlu pilav* (*pulao*), *nohutlu kabak* (sweet squash dish), *asure* (dessert), and *eksili corba* (soup). Roasted white and yellow chickpea are eaten as plain, salted, or sugared nuts. Fermented product and canning of chickpea offer scope for future utilization though at present it is used in small quantities.

In the Mediterranean region of West Asia and North Africa, 75% of all kabuli chickpea grown is consumed as three products: *Tisqieh* (boiled, mixed with soaked bread, olive oil, and yogurt). *Falafel* (mashed with peppers and herbs, and deep fried) and *Homos biteheneh* (*Mousabaha*) [boiled mashed chickpea, mixed with pulverized sesame (tiheneh), olive oil, lemon, and herbs]. Chickpea noodles are made in Myanmar (Burma) through an elaborate process and are quite popular.

Diversifying the use of chickpea

Mathur *et al.* (1964) reported that an epidemiological survey in Agra, India, revealed lower levels of serum cholesterol and a lower incidence of ischemic heart disease in people who consumed chickpea as a staple diet than those who did not consume chickpea. Two isoflavones, biochanin and formononetin, isolated from chickpea have been reported to reduce the concentration of cholesterol (Siddiqui and Siddiqi, 1976). However the results of therapeutic effects of chickpea on human beings are not conclusive and require further investigation. Any confirmed effect in humans would open new avenues of utilization of chickpea.

The process of dry extrusion generates heat by friction and is capable of cooking, partial sterilization, expansion, partial dehydration, enzyme inactivation, and shaping of the product. Products like these can yield convenience foods of high nutrient density, and have good scope for diversifying the use in expanded snack foods and breakfast foods.

Quick-cooking whole seed and dhal can be popularized. Treatments with salt solutions and enzymes have shown promising results and are advantageous in that they diversify chickpea usage. Sprouted chickpea (called *Quanty* in Nepal) improves the availability of vitamins and minerals and also enhances the flavor. Similarly, fermented products like *tempeh*, *natto*, and *kinema* can be prepared from chickpea and popularized.

Chickpea starch is used in textile industry and in the manufacture of plywood. An indigo-like dye is obtained from chickpea leaves. The stems and leaves have high concentration of malic, malonic, citric, and oxalic acids that are used in medicine. There is further scope in utilizing the chickpea in industry.

Chickpea straw contains almost twice the amount of protein per kg when compared with cereal straw. If biological value is considered, 8 t of cereal straw will be equal to 1 t of legume straw. Seedcoats obtained during dehulling is also being used as animal feed. Although chickpea seed may not be used for animal feed for economical reasons, the seed and the whole plant can be used in the animal feed industry.

Nutrition is becoming an important factor in the acceptance of value-added snack items throughout the world. Snack foods have long been labeled as junk food. The snack industry is a complicated mix of radically different companies ranging from large companies with new and modern factories to tiny family-owned and operated businesses selling a single snack item. However, consumers are demanding that nutrition information be provided on the package due to the fact that consumers are becoming more nutrition conscious. Dietary guidelines today suggest we should avoid excessive intake of total fat, saturated fat, cholesterol, sugar, and sodium, and we should increase our consumption of complex carbohydrates. By nature, food legumes are high carbohydrate, high fiber, high protein, low fat foods. Fiber-rich foods such as cool season food legumes play a significant role in treating and preventing obesity, cardiovascular disease, hypertension, diabetes, and cancer. Being able to provide a product that is more attractive, convenient, and nutritionally superior is a must if food legumes are to compete in the value-added snack market. Food legumes have all the necessary ingredients to increase their share of this expanding market.

References

Arntfield, S. D., Ismond, M. A. H. and Murray, E. D. 1985. *Canadian Institute of Food Science and Technology Journal* 18: 137–143.
Bansal, K. K., Dhindsa, K. S. and Batra, V. I. P. 1988. *Journal of Food Sciences and Technology* 25: 46–48.

Batra, V. I. P. 1987. *Indian Journal of Nutrition and Dietetics* 24: 15–19.

Batra, V. I. P. and Dhindsa, K. S. 1989. *Indian Journal of Nutrition and Dietetics* 26: 75–83.

Bramsnaes, F. and Olsen, H. S. 1978. *Journal of American Oil Chemists' Society* 56: 450–454.

El-Ashway, E. T., Abdallah, N. M., El-Hashmy, F. S. A. and Hassan, E. M. 1985. *Egyptian Journal of Food Science* 13: 11–22.

El-Hashmy, F. S. A., El-Ashway, E. T., Abdallah, N. M. and Hassan, E. M. 1985. *Egyptian Journal of Food Science* 13: 113–128.

Fayed, H. H. and Morshed, M. A. 1990. *Egyptian Journal of Dairy Science* 18: 75–83.

Ghorpade, V. M., Kadam, S. S. and Salunkhe, D. K. 1986. *Journal of Food Science and Technology* 23: 164–165.

Gueguen, J. 1983. *Qualitas Plantarum Plant Foods for Human Nutrition* 32: 267.

Gueguen, J. and Schaeffer, F. 1984. *Journal of the Science of Food and Agriculture* 35: 1024–1033.

Huang, H. D. 1987. *Food Science* 7: 62–64.

ICRISAT (International Crops Research Institute for the Semi-Arid Tropics). 1991. In: *Uses of Tropical Grain Legumes: Proceedings of a Consultants' Meeting*, pp. 350. 27–30 March 1989, ICRISAT Center, Patancheru, Andhra Pradesh, India.

Jiang, Y. 1988. In: *Handbook for Processing Industry in Family*, pp. 4–5. China: Scientific Publishing House of Hubei.

Kogure, K., Asanuma, K. and Naka, J. 1977. *Technical Bulletin of the Faculty of Agriculture, Kagawa University* 29: 1–9.

Kogure, K. and Ohshima, M. 1991a. *Technical Bulletin of the Faculty of Agriculture, Kagawa University* 43: 1–10.

Kogure, K. and Ohshima, M. 1991b. *Technical Bulletin of the Faculty of Agriculture, Kagawa University* 42: 97–109.

Li, Q. L. 1987. In: *Techniques of Grain Food Processing*, pp. 294–299. Beijing, China, Food Publishing House of China.

Liener, I. E. (ed.). 1980. *Toxic Constituents of Plant Foodstuffs*. New York: Academic Press. 502 pp.

Mathur, K. S., Singhal, S. S. and Sharma, R. D. 1964. *Journal of Nutrition* 84: 201–204.

McCurdy, S. M. and Knipfel, J. E. 1990. *Journal of Food Science* 55: 1093–1101.

Murata, K., Kusakabe, J., Kobayashi, H., Akaike, M. and Murakami, M. 1988. *Agricultural and Biological Chemistry* 52: 1317–1318.

Murray, E. D., Myers, C. D., Barker, L. D. and Maurice, T. J. 1981. In: *Utilization of Protein Resources*, pp. 158–176 (eds. D. W. Stanley, E. D. Murray and D. H. Lees). Westport, Connecticut, USA: Food and Nutrition Press.

Newburg, D. S. and Concon, J. M. 1985. *Journal of Agriculture and Food Chemistry* 33: 685–687.

Ohshima, M. and Kogure, K. 1984. *Journal of Japanese Grassland Science* 30: 178–183.

Olsman, H. 1978. *Journal of American Oil Chemists' Society* 56: 375–376.

Pirie, N. W. 1978. *Leaf Protein and Other Aspects of Fodder Fractionation*. Cambridge: Cambridge University Press. 183 pp.

Ragab, M. H. 1988. *Traditional foods in Egypt and Sudan*. A report prepared by the Food and Agriculture Organization of the United Nations (FAO), Rome: FAO.

Saxena, M. C. and Singh, K. B. (eds.). 1987. *The Chickpea*, 409 pp. Wallingford, Oxon, UK: CAB International.

Schmandke, H., Bottcher, H., Plaschnik, D. and Kuhrt, A. 1982. *Die Nahrung* 26: 385–390.

Shekib, L. A., Youssef, M. M., Zoueil, M. E. and Mohamed, M. S. 1987. *Alexandria Journal of Agricultural Research* 33: 192–201.

Shekib, L. A., Zoueil, M. E., Youssef, M. M. and Mohamed, M. S. 1985. *Food Chemistry* 18: 161–168.

Shekib, L. A., Zoueil, M. E., Youssef, M. M. and Mohamed, M. S. 1986. *Food Chemistry* 20: 61–67.

Siddiqui, M. T. and Siddiqi, M. 1976. *Lipids* 11: 243–246.

Sosulski, F. W. and Sosulski, K. 1986. In: *Plant Proteins Applications: Biological Effects and Chemistry*, pp. 176–186 (ed. R. Ory). Washington, D.C.: American Chemical Society.

Spadoni, M. A., Mengheri, E., Scarino, M. L. and Vignolini, F. 1981. In: *Proceedings of the European Congress on Plant Proteins for Human Food*, p. 19 (eds. C. E. Bodwell and L. Petit). The Hague, Boston: Martinus Nijhoff/Dr. W. Junk.

Taukamoto, I., Miyoski, N. and Hamaguchi, Y. 1983. *Cereal Chemistry* 60: 281–286.

Tyler, R. J., Youngs, C. G. and Sosulski, F. W. 1981. *Cereal Chemistry* 58: 144.

Vasishta, R. and Dhindsa, K. S. 1986. *Journal of Food Science and Technology* 23: 260–263.

Vasishta, R., Dhindsa, K. S. and Batra, V. I. P. 1986. *Current Science* 55: 1236–1237.

Vieth, W. R., Constantinides, A. and Bernath, F. R. 1978. In: *Proceedings of the National Science Foundation Grantee-Users Conference*, pp. 43–62. Washington, D. C., USA: NSF.

Wilson, K. A. and Tan-Wilson, A. L. 1983. *Acta Biochemica* (Poland) 30: 139.

Youssef, M. M., Hamza, M. A., Abdel-Aal, M. H., Shekib, L. A. and El-Banna, A. A. 1986. *Food Chemistry* 22: 225–233.

Zhang, Y. J. 1987. In: *Food Processing in Family*, pp. 134–146. Shanghai, China: Scientific Publishing House of Shanghai.

Ziena, H. M., Youssef, M. M. and El-Mahdy, A. R. 1991. *Journal of Food Science* 56: 1347–1349, 1352.

Improving nutritional quality of cool season food legumes

P.C. WILLIAMS[1], R.S. BHATTY[2], S.S. DESHPANDE[3], L.A. HUSSEIN[4] and G.P. SAVAGE[5]

[1] *Agriculture Canada, Canadian Grain Commission, Grain Research Laboratory, 1404–303 Main Street, Winnipeg, Manitoba, R3C 3G8 Canada;*
[2] *Crop Development Centre, Department of Crop Science and Plant Ecology, University of Saskatchewan, Saskatoon, Saskatchewan, S7N 0W0 Canada;*
[3] *Agriculture Canada Research Station, P.O. Box 3001, Morden, Manitoba, R0G 1J0 Canada;*
[4] *Department of Nutrition, National Research Centre, El-Tahrir Street, Giza, Dokki, Egypt, and*
[5] *P.O. Box 84, Lincoln University, Canterbury, New Zealand*

Abstract

Nutritional quality of food legumes includes the composition and functionality of the seeds. Composition is the main factor affecting nutritional value (contributions to energy and protein requirements, amino acid balance, digestibility, antinutritional factors, etc.). Functionality embraces preliminary preparation steps, digestibility, and cooking quality itself (physical aspects of food preparation, flavor, appearance, and acceptability).

Factors affecting nutritional quality include genetic make-up; growing environment, including location and season, storage, insect infestations (in the field and during storage); and other factors. Nutritional quality parameters of cool season food legumes are tabulated. The improvement of nutritional quality will be considered in the light of the heritability of the chemical and physical factors which affect it. Strategies for improvement in nutritional quality are presented.

Introduction

Cool season food legumes (CSFLs) considered include the kabuli type of chickpea (*Cicer arietinum*), dry pea (*Pisum sativum* or *P. arvense*), faba bean (*Vicia faba*), grasspea or chickling vetch (*Lathyrus sativa*), and lentil (*Lens culinaris*). Interest in the nutritional quality of foods has grown during the past two decades due partly to rising costs of red meats of all types, a concurrent increase in awareness of the benefits to human health of "white" meats (mainly fish and poultry), and the benefits of using food legumes as a protein source.

Nutritional quality embraces all factors essential for people to maintain a condition of healthy living conducive to productive work in terms of manual, mental, and athletic effort. It includes antinutritional as well as nutritional factors. But nutritional quality should also accommodate parameters such as

F.J. Muehlbauer and W.J. Kaiser (eds.), Expanding the Production and Use of Cool Season Food Legumes, 113–129.
© 1994 *Kluwer Academic Publishers.*

flavor, biting texture, and voluntary intake. If something doesn't taste good, people won't like it. If they don't like it, they won't eat it, so without voluntary intake the food won't be very useful, no matter how "nutritious" it appears to be on the basis of its composition.

This paper outlines the composition of five common cool season food legumes. The main emphasis is to define the main parameters which contribute to the nutritive quality of CSFLs, and to suggest strategies for improvements in their acceptability by people. Acceptability of CSFLs by people is probably the end-point in our crusade. How well we have progressed in this is illustrated in Table 1, which summarizes trends in production since 1950.

Table 1. Trends in world production of cool season food legumes, 1950–1990*

Legume	1950	1960	1970	1980	1990
Chickpea	5.4+	7.0	7.0	6.0	6.8
Dry pea	5.8	10.7	10.2	8.5	17.5
Faba bean	4.4	4.6	5.2	4.3	4.3
Lentil	0.8	1.0	1.1	1.3	2.7
Others**	2.8	3.3	3.2	7.5	11.8
Total CSFL	19.2	25.6	26.7	27.6	43.1

*Source was FAO Crop Production Yearbooks.
**Includes cowpea, grasspea, pigeon pea, etc.
+Million tons.

Dry pea heads the list with an annual production of about 17.5 million tonnes (Mt) and a healthy increase in production over the past 10 years. Chickpea follows, with about 7 Mt, then faba bean at a little over 4 Mt, and lentil (about 3 Mt). Other legumes combined total about 12 Mt, and their production is also climbing. These other legumes include cowpea, grasspea (chickling vetch or lathyrus), and pigeon pea. Production of chickpea and faba bean have been more or less stable over the last 30 years, while lentil production is increasing, having more than doubled over the past 20 years. The increase in dry pea production may be misleading to the nutritionist, since a significant proportion is processed industrially for fiber, starch, and protein, and not consumed directly as food.

Nutritional Quality Parameters of Food Legumes

What makes up nutritional quality? The most important aspects are summarized in Table 2.

Food legumes are considered to be good sources of protein, lysine, and some minerals, particularly calcium and magnesium. They are favored as a

Table 2. Principal aspects of nutritional quality in cool season food legumes

```
Protein concentration

Protein digestibility

Amino acid composition

Amino acid availability

Fiber concentration

Calorific value

Cooking quality

    Appearance

    Biting texture/"mouth feel"

    Flavor

    Cooking time

Consumer acceptability (voluntary intake)

Antinutritional factors
```

replacement for meat, particularly by vegetarians and vegans, and serve as a satisfactory complement to cereals, which are generally low in lysine, but higher in sulphur-containing amino acids. In developing a scheme for improvement of food legume quality, all of the factors which comprise "quality" must be considered in order that one aspect of quality is not improved at the expense of another. The important aspects of legume quality are listed in Table 3.

Functionality factors affect mainly marketing and industrial processing, but also consumer acceptability, since long cooking times are not acceptable to the housewife. In developing and developed countries, housewives are seldom reluctant to hide their feelings concerning what they consider to be inferior goods. The retailer, the wholesaler, and the entire chain of agencies down to the breeder will come to learn of any serious defects in their produce. Appearance, biting texture and flavor (see Table 1), while they are apparently functionality elements, can be considered as nutritional factors, since they affect voluntary intake. The "true" nutritional items include protein concentration and digestibility, amino acid balance, husk percentage (fiber concentration), mineral concentration, and antinutritional factors. "Stability" pertains to all factors. It encompasses heritability and stability to environmental conditions.

Table 3. Aspects of food legume quality

Seed size*	Seed shape*
Seed size uniformity*	Seed texture*
Decortication loss*	Husk percentage**
Cooking quality*	Protein content**
Protein quality**	Antinutritional factors**
Digestibility	Beta-oxalyl amino-alanine
Essential and	Chymo-trypsin inhibitors
limiting amino	Cyanogens
acids	Favogens
Haemagglutinins	
Oligosaccharides	
Phytic acid	
Color and appearance**	Tannins
Mineral content**	Trypsin inhibitors
Water absorption*	Stability**

* = functionality factors
** = nutritional factors

Protein and Amino Acids

Protein concentration is probably the most fundamental of nutritional considerations (Table 4).

While food legumes are regarded as natural substitutes for meat, in practice they fall far short of providing the same amount of protein per unit weight of food intake. Cooked meats usually contain 45 to 60% moisture, but on a moisture-free basis they contain upwards of 75% protein. This gives the as-eaten cooked product a protein concentration of about 32 to 40%. Chickpea and dry peas, as eaten, do not offer much more protein than wheat, even though on the face of things they contain twice as much protein.

Bread, including flat bread, is the form in which wheat is usually eaten and contains about 30 to 35% moisture. The original wheat contains about 10.5 to 12% protein, so that the product consumed contains only about 7 to 8% protein. Chickpea and dry pea contain about 20 to 22% protein, as harvested; however, they are eaten after soaking and boiling, and the product as served contains over 60% moisture which reduces the functional protein concentration to about 7.0

Table 4. Protein concentrations of some common food legumes (total N × 6.25, moisture-free basis)

Legume	Mean	High	Low
Chickpea[a]	20.2	27.0	14.3
Dry pea	22.5[b]	39.7[c]	15.5[c]
Faba bean[d]	28.4	37.8	18.6
Grasspea[e]	28.2	34.6	18.2
Lentil[f]	28.0	32.8	20.2

a-f refer to sources of data:
[a]Singh *et al.* (1990)
[b]Williams and Singh (1988)
[c]Bressani and Elias (1988)
[d]El Sayed *et al.* (1982)
[e]Deshpande, personal communication (1992)
[f]Erskine and Witcombe (1984)

to 7.5% (Saxena *et al.*, 1991). To obtain the same protein intake as there is in 25 g of cooked chicken one would need to eat 120 g of cooked chickpea. On the other hand, for countries where the staple is rice, CSFLs provide much more protein than the cereal of choice, since cooked rice contains only about 2.2 to 2.5% protein.

Differences in digestibility further reduce the value of the food legume proteins relative to that of meat and fish. Digestibility is affected by factors such as: the fiber components of the seed, amino acid composition, the physical nature of the proteins, antinutritional constituents, and other factors.

Protein digestibility is an area of food legume technology which has been much discussed in scientific papers but about which little has been done in terms of defining practical methods of improvement in the inherent digestibility of the protein of the raw seeds. Legume proteins are generally lower in digestibility than are the proteins of the cereal they are intended to supplement. Furthermore, the determination of protein digestibility is at best a slow process, involving balance experiments, usually with animals rather than humans. The question has often been raised as to the transferability of digestibility (or any nutrition-related) data from – for example, rat experiments – to what would happen if humans of different ages were subjected to the same diets.

Composition of the proteins of CSFLs in terms of the ten or so amino acids which are regarded as essential to human and monogastric nutrition is summarized in Table 5, which also illustrates the degree to which the individual legumes conform to the FAO reference protein.

The data in Table 5 indicate areas where useful improvements can be made in nutritive quality. With the exceptions of arginine, lysine, and leucine, most CSFLs are deficient in essential amino acids, relative to the FAO standard. All

Table 5. Essential amino acid composition of proteins of cool season food legumes

Amino acid*	Chickpea	Dry pea	Faba bean	Grasspea	Lentil	FAO Reference egg protein
Arginine	8.8	9.6	9.6	7.8	6.9	6.7
Cystine	1.2	1.7	1.4	1.3	1.9	--
Isoleucine	4.4	4.1	4.1	6.7	3.8	6.8
Leucine	7.6	7.0	7.6	6.6	7.1	7.8
Lysine	7.2	7.2	6.4	7.4	7.3	7.2
Methionine	1.4	1.0	0.7	0.6	1.6	3.4**
Ph-alanine	6.6	4.6	4.2	4.2	4.6	5.8
Threonine	3.5	3.8	3.6	2.3	3.5	5.2
Tryptophan	0.8	0.8	0.9	0.4	0.9	1.5
Valine	4.6	4.6	4.6	4.7	4.2	7.4

*   Source: Deshpande and Demodaran (1990); data represent amino acids per gram of seed protein.
**   Includes total sulphur-containing amino acids.

CSFLs are rich in arginine. On an individual basis, chickpea is satisfactory in leucine, lysine, and phenylalanine, but deficient in all of the others. The amino acid complement of chickpea is also plagued by the fact that it is the lowest of the CSFLs in total protein concentration. Dry pea is adequate in lysine, leucine, and phenylalanine, but deficient to very deficient in all other amino acids. Faba bean is adequate in arginine and leucine, but deficient in all of the other seven. Grasspea is satisfactory only in lysine and arginine. Lentil is well-endowed with lysine, adequate in arginine, reasonably high in leucine, and low in all others except the sulphur-containing amino acids, in which it appears to be the richest of the five.

Amino acids which may be present in the diet at levels low enough to inhibit normal metabolism and growth are called Limiting Amino Acids. These may vary depending on the animal, and an amino acid which is limiting to one species may not affect another, and vice versa. The amounts of what are considered to be the most important of the limiting amino acids in CSFLs are given in Table 6.

The estimated percentages of the limiting amino acids in the as-processed

Table 6. Limiting amino acid concentrations of food legumes

Legume	Lysine + Cystine	Methionine	Threonine	Tryptophan
Chickpea	1.45*	0.52	0.71	0.16
Dry pea	1.62	0.61	0.86	0.18
Faba bean	1.82	0.60	1.02	0.26
Grasspea	2.09	0.54	0.65	0.11
Lentil	2.04	0.98	0.98	0.25

*   Percent of whole seed, based on amino acid data of Deshpande and Demodaran (1990), and protein data of Table 1.

seeds indicate that chickpea is lowest of the CSFLs in lysine and the sulphur amino acids, and second lowest in threonine and tryptophan (Table 6). Grasspea is lowest in threonine and tryptophan, and second lowest in sulphur amino acids, but richest in lysine. Lentil appears to be rather more well-balanced in limiting amino acid concentrations than the other four CSFLs.

Antinutritional Factors

Cool season food legumes vary widely in the amounts of various antinutritional factors (Table 7).

Table 7. Common antinutritional factors in cool season food legumes

Factor	Approximate Concn. in Seeds	Effect of Domestic Cooking
Beta-oxalyl amino-alanine	0.01-0.9%	P[a]
Cyanogens	Traces	P
Favogens[b]	0-1%	P
Haemagglutinins	0-400 HU/mg[c]	D
Oligosaccharides	0-9%	P
Phytic acid	0.1-1%	P
Proanthocyanidins/tannins	1-8%	P
Trypsin inhibitors	0-14.5 TIU/mg	D
Saponins	0.4-0.56%	

[a]D = mainly destroyed; P = partially destroyed
[b]Mainly aglycones of convicine and vicine
[c]HU = International haemagglutinin unit
TIU = International trypsin inhibitor unit

   The antinutritional factors, listed in Table 7, do not occur in all CSFLs, and where they occur in more than one species, they may not be present in similar concentrations. Beta-oxalyl amino-alanine (BOAA) occurs in the grasspea. It is a neuro-toxin which causes paralysis, particularly of the lower limbs. Trypsin and chymotrypsin inhibitors occur in several widely-consumed species and they cause difficulties in digestibility of proteins, but are largely destroyed during cooking (particularly by boiling). Cyanogens can produce highly toxic substances under certain circumstances.
   Favogens include the aglycones of convicine and vicine, respectively, divicine, isouramil, and dihydroxyphenylalanine (DOPA) (Marquardt, 1982).

They occur in faba bean, are only partially destroyed during cooking, and cause haemolytic anaemia. The main cause of the disease has been traced to a deficiency in glucose-6-phosphate dehydrogenase (Mager *et al.*, 1965).

Haemagglutinins (lectins) cause a different type of blood disorder. They are also found in the faba bean, which appears to contain many troublesome substances. Fortunately they are largely destroyed during normal domestic food preparation. Oligosaccharides, mainly stachyose, raffinose, and verbascose, are found in many CSFLs, and are particularly prevalent in chickpea and faba bean. Oligosaccharides contain galactose as well as glucose residues. They do not break down via the normal sugar metabolism pathway in the digestive tract of monogastrics, and are fermented by bacteria in the lower intestine to produce methane and other gases. This causes discomfort and abdominal pains, which can become severe. Oligosaccharides are not extensively destroyed during cooking, and persist into the food.

Phytic acid or inositol hexa-phosphoric acid and its salts (mainly calcium and magnesium phytates) are the forms in which developing grains and seeds store phosphorus which is required by the developing embryo during germination. The amount of phytate present in a seed is more a function of the phosphorus status of the soil than an inherent genetic trait. Phytate is not extensively broken down during normal food preparation, although prolonged boiling or simmering, encountered, for example, in the preparation of "Ful", and "Ful medames" from faba beans, is likely to cause appreciable degradation. The principal antinutritional action of phytates is to combine with calcium and magnesium, thereby effectively removing them from the metabolic pathway. This can cause bone disorders such as rickets.

Saponins cause bitter flavors, and thereby reduce voluntary intake. Although their actual physiological role has not been clearly established, they are nevertheless considered to be antinutritional factors. Finally the tannins, including condensed tannins, are present to a certain extent in all CSFLs. They occur mainly in the seedcoat and persist after cooking. Their principal modus operandum is to cause disorders in protein digestion, mainly by combining with the proteins, thereby reducing their availability for digestion. This causes stomach cramps and diarrhea.

In terms of the relative proclivity of CSFLs to contain these objectionable substances, the faba bean leads the field in concentrations of antinutritional factors, while lentil is low in all of them (Bhatty, 1988). (All of the factors enumerated in Table 7 except BOAA occur in faba bean.) The only antinutritional factors which occur to a significant degree in lentil are tannins. These are concentrated in the seedcoat, which is removed during most lentil foodprocessing. The distribution of some antinutritional factors among CSFLs is given in Table 8.

Table 8. Distribution of antinutritional factors in cool season food legumes

Factor[a]	Legume				
	Chickpea	Dry Pea	Faba Bean	Grasspea	Lentil
BOAA –	–	–	0.02–0.75%[b]	–	
Favogens	–	–	0.3–0.7[c]	–	–
Haemagglutinins	180HU[d]	100–400[c]	640[f]	–	2–8[g]
Oligosaccs	3–5%[h,i]	5–7%[j]	3–4.3%[j]	–	2–7%[g]
Phytate	–	0.2–0.7[k]	1–1.2%[l]	–	0.1–0.[g]
Tripsin Inhibitor	–	3–11TIU[c]	3–6[c]	6–8mg/[b]	2–5[g]
Tannins	0.2–0.6%[m]	0.2–8%[k]	0–6%[n,p]	0–0.4%[b]	0–0.7%[f]

Notes:[a] Haemagglutinin concentration is reported in International Haemagglutinin units (HU/mg.); Trypsin Inhibitor concentration is reported in International Trypsin Inhibitor units (TIU/mg).

b-p refer to sources of data:

[b] Deshpande, personal communication (1992)
[c] Cordeiro and Williams, previously unpublished data (1986)
[d] Rea et al. (1985)
[e] Valdebouze et al. (1980)
[f] Hussein et al. (1974)
[g] Savage (1988)
[h] Lineback and Ke (1975)
[i] Rao and Belavady (1978)
[j] Cerning-Beroard and Filiatre (1976)
[k] Savage and Deo (1989)
[l] Griffiths and Thomas (1981)
[m] Singh (1984)
[n] Cansfield et al. (1980)
[p] The Cansfield et al. (1980) paper reported tannin content in faba bean testa only, and not intact seed.

Other Nutritional Parameters

Included in other nutritional parameters are starch, the fiber components, and oil or fat. CSFLs contain 30 to 50% starch. Some legume starches, e.g., dry pea starch, possess unique properties which have an industrial value greater than their food or feed value. Fiber components include cellulose and other non-starch carbohydrate polymers, which are present to the extent of about 10 to 20% in CSFLs. With the exception of chickpea, which contains up to 8% oil, CSFLs contain only about 1 to 3% oil.

Functionality includes cooking quality, hard seedcoats, and associated factors. Cooking quality includes cooking time, appearance, flavor, odor, and texture or "mouth-feel" of the final processed food. Methods for determination of cooking times vary from manual methods (e.g., Williams et al., 1983) to methods involving commercial equipment such as the Kramer Shear-press

(Bhatty, 1989). CSFLs may vary in the time it takes for their texture to become acceptable; some CSFLs may take over 6 hours boiling under reflux. Cooking time has been shown to be related to seed size (Williams *et al.*, 1983; Erskine *et al.*, 1985). Seed size is highly heritable and breeders may have to make a choice between desirable seed size and cooking time. In practice, most CSFLs are soaked in water for several hours or overnight to greatly reduce cooking times (Erskine *et al.*, 1985); therefore, cooking times may actually be of less significance than seed size. Price differentials often favor larger seeds within a CSFL type.

Color and biting texture are aspects of cooking quality which affect consumer acceptability and these are best established by taste panels. The emergence of computerized equipment to assist taste panelists, such as that offered by Tecator AB, underline the importance of flavor and texture in food evaluation.

The hard seedcoat characteristic has attracted much attention in recent years. Some legumes, such as dry pea and lentil, may contain a high proportion of seeds which do not imbibe water as fast as normal seeds. These are referred to as having hard seedcoats and are characterized by exceptionally long cooking times. Several causal reasons have been proposed including lignification during storage (of faba beans) and the relationship between phytate and hard-seed-coatedness described by Bhatty and Slinkard (1989). The condition is likely to be affected by growing environment; however, the degree of heritability of hard seedcoats remains to be clarified.

The Strategy for Improvement of Nutritional Quality

The early sections of this chapter outlined the most important aspects of the nutritional quality of CSFLs and where improvements can and should be made.

Table 9. Strategy for improvement of nutritional quality of cool season food legumes

---

1. Increase:

    a) Protein content
    b) Protein quality
        i)   Improve amino acid balance
        ii)   Improve digestibility

2. Improve acceptability:

    a) Appearance
    b) Solubility
    c) Aroma
    d) Flavor

3. Reduce antinutritional factors

4. Maintain yield

A suitable strategy for improvement in the nutritional quality of CSFLs is summarized in Table 9. This strategy is simple and self-explanatory, and covers the most important aspects of nutritional quality in CSFLs.

Steps in Improvement of Nutritional Quality

There are several steps in the implementation of such a strategy.

The first step in the improvement of the nutritional status of a CSFL is to determine its present nutritional status with respect to the above nutritional factors. This first step will identify factors which are either too low, in the case of desirable factors, or too high, in the case of things like antinutritional factors. The investigation should include a separate study of samples drawn from throughout the region where the CSFL is expected to be grown and ideally over at least two growing seasons. The standard deviations and coefficient of variability should be determined for as many individual lines as possible. If the standard deviations are low, the substance or parameter is likely to be very stable over environments. If the standard deviations are high, the reasons underlying the variability should be determined.

Table 10. Steps in improvement of nutritional quality

1. Determine:

    a) Present nutritional status
    b) Range and standard deviation
    c) Desired range
    d) Heritability
    e) Effect of environment on nutritional status

2. Identify parental material

3. Evaluate:

    a) Progeny under field conditions over region
    b) Impacts

Next, the desired range should be selected with reasonable consideration given to practicable limits. The range experienced in the study of variability can serve as guidelines, and the objective can be set to raise or lower the mean depending on the constituent or parameter. It is important to determine the heritability of the constituent or parameter, since years of labor and large amounts of resources can be wasted in trying to change the level of a constituent with low heritability. Heritability can be determined as broad-sense or narrow-sense heritability, or as heritability in standard units.

The first two methods require an analysis of variance, and trials must be grown at two or more locations, and ideally over two or more seasons, to enable differentiation between genetic and environmental variance. The trials should also be replicated. Heritability in standard units is less exacting, since it does not call for replication, but trials still need to be grown over at least two and preferably more locations. The levels of the constituent, or the functionality parameter, are determined in all lines.

Coefficients of correlation among the lines at the different locations indicate the degree of heritability in standard units (Frey and Horner, 1957). Broad or narrow-sense heritabilities carry a higher degree of scientific credibility than does heritability in standard units; however, if heritability is high by any method, it is appropriate to proceed with a breeding program to improve the status of the CSFL with respect to that characteristic. There have been few estimates of heritability of quality factors in CSFLs (Table 11).

Table 11. Heritability of some quality factors in lentil and chickpea

Factor	Lentil	Kabuli Chickpea	Faba bean
Protein %	0.71[a]	0.37[b]	0.36[c,d]
Seed size	0.98[a]	0.96[b]	–
Cooking time	0.82[a]	0.68[b]	–

[a]Erskine *et al.* (1985)
[b]Singh *et al.* (1990)
[c]Robertson *et al.* (1985)
[d]Heritability in standard units in this study

Determination of environmental effects coincides with the determination of heritability. This is not the case with standard unit heritability estimates, but if this value is poor, it is still an indication that the influences of the growing location and/or season are greater than the genetic influence. Knowledge of the environmental effects on composition and functionality is essential in developing a breeding program for improvement in nutritional status. It is unavailing to use as parents lines believed to be high in desirable (or low in undesirable) characteristics if a change in growing location, or an unusually cool or hot season can change the composition or functionality to the extent that the lines may no longer effect the improvement.

Having determined in which direction a breeding program should be orientated, the next step is to select the parental material carrying the necessary genes. This is best achieved by analysis of as wide a population as practicable for the desired characteristics. Catalogs which provide information on protein concentrations for entries in the international nurseries of lentil, chickpea, and faba bean are available from ICARDA.

Time and expense have been serious deterrents to the screening of very large populations for parameters such as essential or limiting amino acids. The new technique of near-infrared reflectance (NIR) spectroscopy has opened the door to screening for characteristics hitherto believed to cost too much in terms of both time and money (Williams *et al.*, 1978). For example, a catalog of barley germplasm has recently been published (by ICARDA) which contained lysine data on several thousand lines of barley. The lysine tests were completed in a few weeks by NIR analysis at very low cost, using unskilled labor.

There are several ways to introduce new characteristics to plants. In addition to the time-honored methods of pedigree breeding, backcrossing, and other well-tried forms of applied genetics. modern biotechnology offers procedures for by-passing and short-cutting some of the steps involved in the traditional approaches.

No matter how they have been introduced, genes are chemical in nature and, as chemicals react with each other. so do genes. To determine the effect of introducing new characteristics to a plant type on those which are there already, new genotypes must be grown under field conditions. This should also be carried out over the range of growing conditions anticipated throughout the region for which the plant is intended, and preferably for more than one successive growing season. If the region is large enough (e.g., the WANA region), two seasons may provide sufficient variance in climatic conditions to enable confident field evaluation.

As a result of painstaking steps in plant improvement, the impact of changes in genetic constitution will become apparent. For example, the introduction of genes to reduce the amounts of an antinutritional factor may reduce drought-tolerance or increase susceptibility to attack by diseases or insects. Physical characteristics of the plants may be changed, including attributes such as seed shape or color. Changes in composition with respect to protein content or one or more antinutritional factors may result in changes in seed color or texture which can affect cooking time or palatability. It is most important to ensure that yield per hectare has not been reduced to the extent that farmers will not grow the new cultivar.

The conflict in plant improvement by breeding is that if nutritional or functional quality (in particular) are reduced in the quest for increased yield, the consumer won't buy the product; whereas, if quality is improved at the expense of a notable reduction in yield, farmers won't grow it. Compromise can be affected only up to a certain extent by paying premiums for higher quality.

Success in breeding for improved nutritional quality is evaluated by screening early generation material (up to F_4), identifying lines most likely to lead to success, followed by a more thorough examination of genotypes which have been advanced on the basis of both quality and agronomic attributes. Techniques for use in screening are summarized in Table 12.

The classical methods of analysis, such as the Kjeldahl method for protein analysis, ion-exchange chromatography, and the protein quality tests are described or cited in the IDRC Technical Bulletin No. TS7e. High performance liquid chromatography is a well-documented technique, and methods for its application are described in the manufacturers' manuals. Combustion analysis for total nitrogen and protein content is a relatively recent introduction made practicable only by increases in the sample weight which the instruments are

Table 12. Analytical procedures for nutritional quality improvement

Parameter	Method
Amino acids	ICE*
Protein %	Kjeldahl, Combustion
Antinutritional factors	HPLC
Cooking quality	Boiling, Organoleptic
Protein quality	Amino acid score
	PER, NPR, etc.
	Near-infrared techniques

*IEC = ion exchange chromatography; HPLC = high performance liquid chromatography; PER = protein efficiency ratio; NPR = net protein ratio.

able to process. The technique is based on the Dumas method, which predates the Kjeldahl test by about 60 years (e.g., Sweeney and Rexroad, 1987).

During the past decade near-infrared (NIR) instruments have improved in accuracy, reproducibility, and durability to the extent that the technology can be regarded as equal or superior to reference methods (Kjeldahl protein, extraction methods for fiber and oil, etc.). Calibrations can be transferred among some models of instruments so that a calibration developed in one laboratory can be installed via diskette or modem to another instrument of the same type anywhere in the world. Screening large populations of germplasm for parameters such as antinutritional factors or essential amino acids, a step which was impractical for reasons of time and expense up until recently, can be accomplished in a few weeks or even days with NIR techniques. Accuracy can be determined by reference analysis of a few selected samples, which can then themselves be used to extend the calibration. A strategy for NIR screening is summarized in Table 13.

Re-prediction of results can be achieved in a few minutes by computer. Results for NIR prediction of some antinutritional factors in faba bean, chickpea, and grasspea are given in Table 14, while Figure 1 illustrates the relationship between NIR and HPLC results for the determination of BOAA in *Lathyrus*.

Table 13. Strategy for NIR screening

1. Develop calibration using up to 100 samples.

2. Analyze all genetic material using the calibration.

3. Verify accuracy of 5-10% by reference tests.

4. Update calibration by adding verification samples.

5. Apply slope/bias correction if necessary, and re-predict all results.

Table 14. NIR prediction of antinutritional factors

Factor	Crop	Mean	r*	SEP
		%		
Vicine	Faba bean	0.68	0.85	0.10
Oligosaccs	Chickpea	3.10	0.85	0.42
BOAA	Grasspea	0.30	0.91	0.08

*r = coefficient of correlation; SEP = standard error of prediction; BOAA - beta-oxalyl amino-alanine.

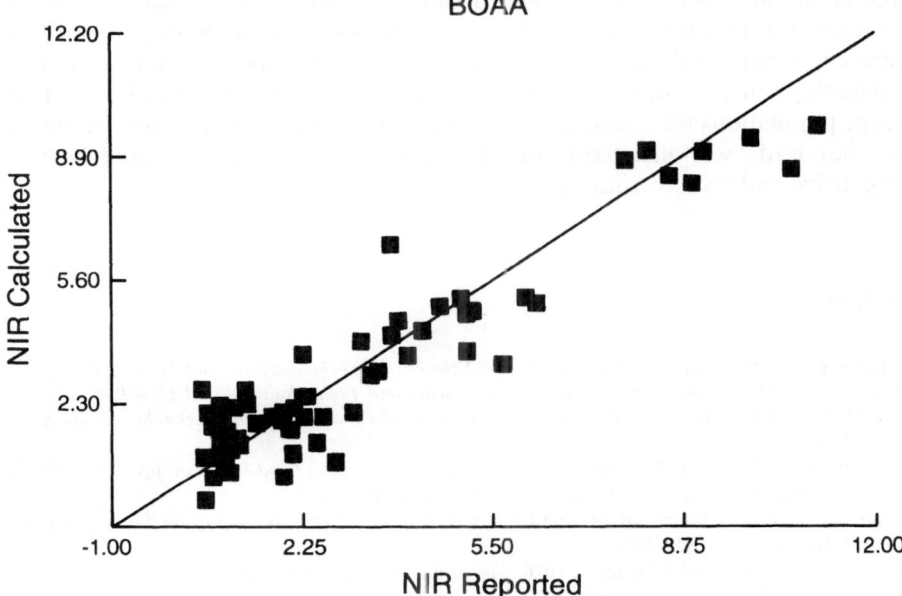

Figure 1. Relationship between NIR and HPLC determination of BOAA in grasspea.

A final word in the quest to improve CSFL acceptability by people concerns promotion. The mature, dry seeds take a long time to cook and tend to be bland in flavor; most of the flavors which are appealing in the final foods have been enhanced by the use of herbs and condiments. The variety of foods which can be prepared from CSFLs leaves little to be desired, and the appearance, texture, and flavor of many of the dishes served on Indian and Pakistani tables can tantalize even the most discerning palate. Preparation is painstaking, and may be a lengthy process, but the results justify the effort.

At present, commercially-prepared, ready-to-eat foods are limited to snacks such as falafel, and dips such as hommos. More attention paid to improved methods of food preparation, including the introduction of commercial preparations of the most popular dishes, coupled with enhanced advertising and popularizing campaigns to complement the efforts of plant breeders would likely result in a healthy increase in consumption of these valuable and inexpensive sources of proteins.

Concluding Remarks

Nutritional quality and its chief components are defined as applicable to cool season food legumes. Data from the literature are presented to illustrate protein and essential amino acid contents and levels of antinutritional factors present in the common CSFLs. Factors which can be improved are discussed, and a strategy for affecting improvements in nutritional status is described. The strategy is applicable to any parameter. Attention is drawn to the role of NIR technology in providing a rapid, accurate, and low-cost means of screening large populations for constituents and parameters, the determination of which by standard "wet chemistry" or physicochemical techniques would be too expensive and time-consuming.

References

Bhatty, R. S. 1988. *Canadian Institute of Food Science and Technology Journal* 21: 144–160.
Bhatty, R. S. 1989. *Canadian Institute of Food Science and Technology Journal* 22: 450–455.
Bhatty, R. S. and Slinkard, A. 1989. *Canadian Institute of Food Science and Technology Journal* 22: 137–142.
Bressani, R. and Elias, L. G. 1988. In: *World Crops: Cool Season Food Legumes*, pp. 381–404 (ed. R. J. Summerfield). Dordrecht: Kluwer Academic Publishers.
Cansfield, P. E., Marquardt, R. R. and Campbell, L. D. 1980. *The Journal of the Science of Food and Agriculture* 31: 802–812.
Cerning-Beroard, J. and Filiatre, A. 1976. *Cereal Chemistry* 53: 968–978.
Deshpande, S. S. and Damodaran, S. 1990. *Advances in Cereal Science and Technology* 10: 147–241.
El Sayed, F., Nakkoul, H. and Williams, P. C. 1982. *FABIS Newsletter* 5: 37.
Erskine, W., Williams, P. C. and Nakkoul, H. 1985. *Field Crops Research* 12: 153–161.
Erskine, W. and Witcombe, J. R. 1984. *Lentil Germplasm Catalog*, 363 pp. The International Center for Agricultural Research in the Dry Areas, Aleppo, Syria.
Frey, K. J. and Horner, T. 1957. *Agronomy Journal* 49: 59–62.
Griffiths, K. W. and Thomas, T. A. 1981. *The Journal of the Science of Food and Agriculture* 33: 187–192.
Hussein, L., Gabrial, G. J. and Marcos, S. R. 1974. *The Journal of the Science of Food and Agriculture* 25: 1433–1440.
Lineback, D. R. and Ke, C. H. 1975. *Cereal Chemistry* 52: 334–347.
Mager, J., Glaser, G. N., Razin, A., Bien, S. and Noam, M. 1965. *Biochemical and Biophysical Research Communications* 20: 235–240.
Marquardt, R. R. 1982. In: *World Crops: Production, Utilization, Description, Volume 9. Faba Bean Improvement*, pp. 343–353 (eds. G. C. Hawtin and C. Webb). The Hague: Martinus Nijhoff.

Rao, P. U. and Belavady, B. 1978. *The Journal of Agricultural and Food Chemistry* 26: 316–319.

Rea, R. L., Thompson, L. U. and Jenkins, D. J. A. 1985. *Nutrition Research* 5: 919–929.

Robertson, L. D., Nakkoul, H. and Williams, P. C. 1985. *FABIS Newsletter* 11: 17–18.

Savage, G. P. 1988. *Nutrition Abstracts and Reviews, A (Human and Experimental)* 58: 319–343.

Savage, G. P. and Deo, S. 1989. *Nutrition Abstracts and Reviews (Series A)* 59: 66–88.

Saxena, M. C., Singh, K. B. and Williams, P. C. 1991. Utilization of chickpea in West Asia and North Africa. In: *Uses of Tropical Grain Legumes: Proceedings of a Consultants' Meeting*, pp. 63–67. The International Crops Research Institute for the Semi-Arid Tropics, Patancheru, Andhra Pradesh, India.

Singh, U. 1984. *Nutrition Reports International* 29: 745–753.

Singh, K. B., Williams, P. C. and Nakkoul, H. 1990. *The Journal of the Science of Food and Agriculture* 53: 429–441.

Sweeney, R. A. and Rexroad, P. R. 1987. *The Journal of the Association of Official Agricultural Chemists* 70: 1028–1030.

Valdebouze, P., Bergeron, E., Gaborit, T. and Dehort-Laval, J. 1980. *Canadian Journal of Plant Science* 60: 695–701.

Williams, P. C. and Singh, U. 1988. In: *World Crops: Cool Season Food Legumes*, pp. 445–457 (ed. R. J. Summerfield). Dordrecht: Kluwer Academic Publishers.

Williams, P. C., Singh, K. B. and Nakkoul, H. 1983. *The Journal of the Science of Food and Agriculture* 34: 492–496.

Williams, P. C., Stevenson, S. G., Starchy, P. M. and Hawtin, G. C. 1978. *The Journal of the Science of Food and Agriculture* 29: 285–292.

Enhancing the use of cool season food legumes in different farming systems

M. PALA[1], M.C. SAXENA[1], I. PAPASTYLIANOU[2] and A.A. JARADAT[3]

[1] ICARDA, P. O. Box 5466, Aleppo, Syria
[2] Agricultural Research Institute, Nicosia, Cyprus, and
[3] Jordan University of Science and Technology, Irbid, Jordan

Abstract

Cool season food legumes currently occupy nearly 40% of the world's pulse area. They contribute to sustainable farming systems in low-input and marginal production areas through an influx of biologically fixed nitrogen, increased availability of other mineral nutrients, improved water-use efficiency, and decreased build-up of pests and diseases in cereal-dominated cropping systems. The low yield potential of existing landraces, yield instability due to various biotic and abiotic stress factors, poor production techniques, high cost of labor for weeding and harvesting, and inadequate pricing policies and marketing systems are the major constraints to the use of these crops in different farming systems. There is, however, great scope for expansion of their use through fallow replacement, relay cropping, and intercropping. Development of well-adapted cultivars and practices that enable these crops to fit in the niches available in major cropping systems, and promotion of appropriate pricing and marketing policies and extension support, would enhance the use of cool season food legumes in world agriculture.

Introduction

Agricultural production is limited by availability of land and water resources. About 85% of the total arable land in the world is rain-fed, with large temporal and spatial variation in rainfall, resulting in unstable production. Growing population pressure is threatening the long-term capacity of the world's agricultural system to continue to provide for increasing food needs. A large part of the world's population growth is taking place in the developing countries, necessitating substantial increase in food production from the resources available there. As there is little scope for bringing additional area under cultivation, increases in crop production can only come about by increasing yields and cropping intensity. These two sources were projected to account, respectively, for 72 and 21% of the total crop production increases

F.J. Muehlbauer and W.J. Kaiser (eds.), Expanding the Production and Use of Cool Season Food Legumes, 130–143.
© 1994 Kluwer Academic Publishers.

needed in the WANA region by the year 2000 (Somel, 1988). This implies both higher use of improved technologies and more intensive use of land, particularly the fallow areas. The major challenge facing the farmers in the stressful environment is how to raise productivity of their drylands. The issues which merit greatest attention are (1) increasing the biological efficiency of the cropping systems, (2) improving the management of soils to conserve moisture and fertility, and (3) effecting a better integration of livestock and crop production.

The cool season food legumes (CSFL) such as chickpea, lentil, faba bean, and dry pea play an important role in sustaining the productivity of the farming systems. Their ability to fix atmospheric nitrogen, and to improve soil structure in the cereal-dominated cropping systems are key to the systems' sustainability (Beck and Materon, 1988; Plancquaert and Haggar, 1989; Osman *et al.*, 1990). Their high protein concentrations makes them a valuable yet cheap substitute for meat and other high-protein animal products in the developing countries. The crop residues of these legumes provide an important livestock feed (Jaradat, 1989). Although evolved in the Fertile Crescent, they are now cultivated all over the world. They account for nearly 40% of current total world pulse area and 50% of the pulse production (Table 1). About 70% of the current world area under CSFL is in the developing countries, providing only 50% of the world total CSFL production. Ninety percent of both area and production of CSFL in the developing world is in West Asia and North Africa (WANA), India and China. The world area under these crops increased by only 10% during the last decade, mainly through the increases in developed countries and Turkey. Unfortunately, in the developing countries as a whole there was no change in the area in the last decade, and the production increased by only 5% because of a

Table 1. Area and production of total pulses and cool season food legumes in the world

	Pulses			Cool season food legumes			
	1979/81	1988/90	% of world in 1988/90	1979/81	1988/90	% of world in 1988/90	% of total pulses in 1988/90
Area (million ha)							36.9
World	61.1	68.8	100.0	23.0	25.4	100.0	
Developed	69.1	12.7	18.5	5.4	7.8	30.7	61.4
Developing	52.0	56.1	81.5	17.6	17.6	69.3	31.4
-WANA	4.7	6.4	9.3	3.5	4.9	19.3	76.6
-India	22.8	22.9	33.3	8.5	7.9	31.1	34.5
-China	5.4	4.4	6.4	3.7	3.0	11.8	68.2
Production (million t)							
World	41.2	56.9	100.0	20.2	29.2	100.0	52.5
Developed	9.7	20.7	36.4	6.1	15.1	50.5	72.9
Developing	31.5	36.2	63.6	14.1	14.8	49.5	40.9
-WANA	3.7	5.3	9.3	2.7	4.0	13.4	75.5
-India	10.5	12.7	22.3	4.9	5.4	18.1	42.5
-China	6.6	5.8	10.2	4.9	4.2	14.0	72.4

Source: FAO Production Yearbook, 1990, vol. 44

Table 2. World area, production, mean yield, and countries having major production of chickpea, pea, faba bean, and lentil, average of 1988/90

Crop	Area (million ha)	Production (million ha)	Yield (t ha⁻¹)	Major countries (production, million t)	
Chickpea	9.3	6.7	0.7	India (4.3) Turkey (0.8)	Pakistan (0.5) Mexico (0.2) Ethiopia (0.1)
Pea	9.6	16.4	1.7	USSR (8.1) France (3.0)	China (1.5) India (0.4) Denmark (0.5) Hungary (0.4) UK (0.3)
Faba bean	3.2	4.3	1.3	China (2.4)	Egypt (0.4) Ethiopia (0.3) Morocco (0.2) Italy (0.1)
Lentil	3.2	2.5	0.8	India (0.7) Turkey (0.8)	Bangladesh (0.2) Canada (0.1) Iran (0.1)

Source: FAO Production Yearbook, 1990, vol. 44

small increase in yield levels. World production and major producing areas of the cool season food legumes are given in Table 2.

The reduction in the area of CSFL in some developing countries (FAO, 1990) has mainly occurred because of the introduction of mechanized production practices for wheat and barley, which have made these crops more reliable and profitable than food legumes. The higher price of legumes has not made them more remunerative than cereals because of the higher cost of their production owing to high use of manual labor. Nevertheless, and in spite of a number of socioeconomic, physical and institutional constraints, opportunities for increasing CSFL production in the developing world do exist. This paper outlines the role of the CSFL in different farming systems, and discusses the constraints and possible solutions to their production in the developing world.

Place in Different Farming Systems

The place of cool season food legumes in different farming systems has been discussed in several publications (Beck and Materon, 1988; Summerfield, 1988; Plancquaert and Haggar, 1989; Osman *et al.*, 1990). In the subtropical and warm temperate climates they are grown during winter, either on water conserved from preceding monsoon rains or on residual moisture after the main crop of cereals (maize, sorghum, or millet). Winter rains are scant and moisture usually limits production irrespective of farming system (Papendick *et al.*, 1988).

The West Asia and North Africa (WANA) region, which is one of the most important legume production areas of the developing world, has a Mediterranean-type climate characterized by hot, dry summers and wet, mild winters except for high-plateau areas where winters are severe. The CSFL are grown mostly under rain-fed conditions throughout the region, which fit well into the farming systems because they permit optimizing the use of resources and balancing risks (Saxena, 1988). The rain-fed farming systems in the dry areas of the region are dominated by the cereal livestock system. The intensity of cropping may range from 33% to more than 100%, depending on precipitation and soil type. Faba bean, lentil, and dry pea are grown as winter crops and chickpea as a spring crop in the lowlands of the region. However, all are grown as spring crops in the high-altitude areas because of severe winters. Yields are often limited by lack of water and also by the short growing period. Chickpea and lentil are used mainly in rotation with wheat or barley and summer crops in either two- or three-course rotations depending on the locality (Saxena, 1988; Nassib *et al.*, 1988). Ecologically, chickpea is a major component of the wetter side of the rainfall range and lentil is better adapted to the drier areas. Faba bean and dry pea are rotated with cereals, cotton, or sugar beet in the coastal areas (Nassib *et al.*, 1988; Buddenhagen, 1990).

China is an important producer of faba bean and dry pea among the developing countries. The crops are generally autumn-sown in the southern provinces. Faba bean is planted after rice or intercropped with cotton or maize in October and harvested in May or June. In the northern provinces faba bean is usually grown in rotation with spring-sown winter wheat and harvested in July or August (Papendick *et al.*, 1988).

Advantages of Cool Season Food Legumes in the Farming Systems

The effects of incorporating CSFL in the farming systems can be assessed in terms of water and nitrogen balance in comparison with "fallow-cereal" or "continuous cereal" systems.

Water Use Efficiency

The use of fallow in rotation with cereals has been widely advocated in rain-fed agriculture in various regions of the world, mainly for conserving water in soil, accumulating nutrients in available form, assisting seed-bed preparation, and limiting the build-up of pests and pathogens to increase yields (Oram and Belaid, 1990; Pala, 1992). Current estimates are that about 20 million ha are kept fallow in WANA, which makes the region's cropping intensity low (Pala, 1992). The fallow cereal rotations in the region do not conserve water as efficiently as is generally thought. On the Anatolian plateau of Turkey, with relatively mild evaporative conditions in spring and summer, fallow efficiencies of 23 to 31% were reported by Durutan *et al.* (1989). A soil depth of > 90 cm is required for an efficient fallow (Guler and Karaca, 1988). In the lowlands of the region, low fallow efficiency was reported in areas with less than 300 mm annual rainfall because moisture penetration into the soil profile below 70 to 90 cm was negligible, and traditional cultural practices did not permit much moisture conservation in the fallow year (Jaradat, 1988). Similarly, at a dry site in

Table 3. Effect of different preceding crops in a two course rotation on the yield and water use of crops and of subsequent wheat, Turkey 1983–88 (Karaca *et al.*, 1991)

Rotation	Crop yield (kg ha⁻¹)	ET⁺ for crop (mm)	WUE⁺ of crop (kg ha⁻¹ mm⁻¹)	Soil moisture (mm/120 cm)
Sunflower – wheat	840	354 ab	2.0 cd	203 d
Winter lentil – wheat	1050	310 cd	3.7 bcd	245 bc
Spring lentil – wheat	940	346 b	2.9 cd	210 d
H. vetch – wheat	1860	299 d	6.4 a	255 b
Safflower – wheat	1020	369 a	2.7 cd	191 e
Wheat – wheat	1760	345 b	5.7 ab	209 d
Cumin – wheat	530	322 c	1.7 d	235 c
Chickpea – wheat	1120	348 b	4.0 bc	209 d
Fallow – wheat	–	–	–	284 a
LSD (P ≤ 0.05)	–	15	2.0	10

⁺ ET = evapotranspiration; WUE = water–use efficiency.
‡ Moisture storage efficiency (%) = (net gain in soil water, cm/precipitation, cm) x 100.
Means in columns followed by the same letter are not significantly different at the 0.05 level.

northern Syria, with long-term mean annual rainfall of 280 mm, Harris (1989) found that by the beginning of the cereal season, less than 10% of the rain that fell during the fallow season remained in the soil profile, implying a very low fallow efficiency.

Use of food legumes to replace inefficient fallow can increase crop water-use efficiency and result in increased productivity of the system. Karaca *et al.* (1991) reported that the wheat crop grown after food legumes had a higher water-use efficiency than wheat after fallow in the Central Anatolian Plateau of Turkey (Table 3). In addition, it must be borne in mind that since the land is cropped only every second year in the fallow cereal rotation the productivity per unit amount of rainfall received in the whole period of rotation will be lower.

Table 3. (Continued)

Moisture storage efficiency ‡ (%)	Wheat yield (kg ha⁻¹)	ET for wheat (mm)	WUE of wheat (kg ha⁻¹ mm⁻¹)	WUE compared with fallow (%)
-5.2	2910 d	320 cd	9.5 abc	+6.7
7.8	3190 bc	358 b	9.8 ab	+10.1
-2.6	3040 cd	329 c	10.0 a	+12.4
9.8	3300 ab	366 b	9.7 ab	+9.0
-9.2	2500 e	309 d	8.3 c	-6.7
-2.2	1890 f	326 c	6.4 d	-28.1
3.8	3140 bc	355 b	9.3 abc	+4.5
-3.4	3180 bc	326 c	10.2 a	+14.6
19.4	3410 a	399 a	8.9 bc	0.0
-	168	13	1.2	-

Nitrogen Fixation

Another advantage of including legumes in the cropping system is that they reduce the dependence of the system on nitrogen fertilizer because of their ability to fix atmospheric nitrogen in association with effective rhizobia. The amount of N symbiotically fixed by food legumes may exceed 100 kg ha⁻¹ year⁻¹ (Saxena, 1988; White, 1989; Beck *et al.*, 1991). Biological nitrogen fixation is an important and low-cost source of N added to the system, especially in N-deficient environments and where the use of N fertilizer is low because of uncertain rainfall and high costs. However, the amount of N that is contributed by legumes depends on management practices and method of harvest. The damage by the larvae of *Sitona* weevil to nodules of lentils can adversely affect the nitrogen-fixation ability of lentil plants. Studies on control methods in Syria (Weigand *et al.*, 1992) and Jordan (AGRODEV, 1989; Jaradat and Badarneh, unpublished) showed that *Sitona* damage is significantly reduced by application of Carbofuran insecticide or seed treatment with Promet. Effective *Rhizobium* inoculation and *Sitona* control resulted in yield increases of lentils over control by 47% in arid and by 100% in semiarid regions of Jordan (Jaradat and Badarneh, unpublished). The potential for, and limitations of, N₂ fixation by CSFL have been reported in detail elsewhere (Beck and Materon, 1988; Herridge *et al.*, this volume; Stanforth *et al.*, this volume).

Effects on Subsequent Cereals

Growing cereals continuously in low-input dryland conditions is likely to be an unsustainable system in the long run, with a high probability of declining profit margins due to depletion in soil fertility and development of unfavorable soil physical and biological conditions (Jones, 1989; Harris, 1990; Karaca et al., 1991). An incorporation of fallow or a legume crop in the rotation can reduce the requirement for the use of fertilizer nitrogen in the system in contrast to a continuous cereal rotation (Papastylianou, 1987; Keatinge et al., 1988). Introduction of legumes in the system could improve the productivity of cereals (Saxena, 1988; Harris, 1990), not only because of reduced depletion of soil nitrogen by legume crops in contrast to the cereal, but also because of some other beneficial effects. Benbella (1987) attributed the beneficial effects of faba bean on a subsequent wheat crop, compared with continuous wheat, in part to the nitrogen enrichment of the soil and in part to a change in the soil physical and biological properties. Sharma et al. (1985) also reported that legumes encouraged the development of a deeper and more extensive rooting system in the subsequent wheat, enabling the crop to take more water and nutrients. Grain legumes are also valuable in reducing disease and pest infestations of cereals that follow by serving as a break crop (White, 1989). The effect of different crops in a two-course rotation on a succeeding wheat crop, and availability of soil nitrogen to it, was studied at ICARDA. The soil after legumes had better nitrogen status as revealed by "A" value of soil nitrogen in contrast to that after cereals or even fallow (Table 4). The wheat following legumes gave high total dry matter as well as nitrogen yield in contrast to wheat after wheat, reflecting the beneficial effect of legumes.

Table 4. Residual effect of 1982/83 rotational treatments on wheat crop grown in 1983/84 season at Tel Hadya, Syria

Cropping treatment in 1982/83	%N15 atom excess in wheat N (a)	Arc sine trans- formed values of a	%N derived from fertilizer (b)	Arc sine trans- formed values of b	A value for soil N (kg ha^{-1})	Total dry matter yield of wheat (kg ha^{-1})	Total N yield in wheat shoot (kg ha^{-1})
Lentil without Carbofuran	1.25	6.42	25.05	30.02	60.19	2708	21.46
Lentil + Carbofuran	1.23	6.36	24.66	29.73	62.38	2322	15.78
Lentil harvested as hay	1.08	5.96	21.55	27.66	73.03	4003	30.11
Winter chickpea	0.90	5.44	17.64	24.81	94.42	2163	12.51
Spring chickpea	1.37	6.72	27.48	31.60	53.46	1756	10.77
Faba bean	0.97	5.64	19.90	26.44	82.29	4117	23.33
Dry pea	1.29	6.54	25.95	30.62	57.13	3294	19.84
Winter chickpea + wheat	1.26	6.44	25.15	30.09	59.72	2718	22.11
Spring chickpea + wheat	1.36	6.70	27.28	31.47	53.85	2183	13.10
Wheat at 20 kg N ha^{-1}	1.75	7.58	34.95	36.20	38.23	1212	10.47
Wheat at 60 kg N ha^{-1}	1.44	6.90	28.90	32.50	49.61	1597	9.35
Clean fallow	1.40	6.80	28.10	32.01	51.26	2818	15.25
LSD (5%)		0.52	2.5		12.55	342	3.35
CV %		5.60	5.85		14.23	9.24	13.68

Production Constraints and Possible Solutions

Physical Stresses

Food legumes are more sensitive than cereals to drought and extremes of temperatures. In the rain-fed environments of the WANA region, drought and temperature stresses are common and they make the risk for failure higher with cool season food legumes than with cereals. Local landraces of food legumes adapted to these abiotic stresses are often low yielding. Efforts have been made by the national programs in WANA, in collaboration with ICARDA and other international research organizations, to develop cold- and disease-tolerant high-yielding cultivars of chickpea, lentil, and faba bean and some have been recently released (FLIP, 1991). Breeding drought-resistant cultivars is a difficult task, but the stress can be managed through the use of improved production practices such as appropriate land preparation, early sowing with optimum inter- and intra-row spacings, use of appropriate fertilizers, and adequate control of pests and diseases. For example, winter sowing of an Ascochyta blight-resistant and cold-tolerant chickpea cultivar results in a considerably increased yield and water-use efficiency in contrast to a spring-sown crop, partly by releasing the drought stress (Saxena, 1987; Pala and Mazid, 1992a).

Similarly, drought and high temperatures reduce lentil yields considerably when the crop is sown late in the season (Silim *et al.*, 1991). Advancing the date of sowing, from late to early winter in lowlands, or from spring to winter in cold highlands, results in substantial yield increases and water-use efficiency in lentil through a rapid and large canopy development, ensuring better interception of photosynthetically active radiation and a substantial reduction in evaporation (Silim *et al.*, 1991; Pala and Mazid, 1992b).

Biotic Stresses

While early sowing of cool season food legumes in WANA invariably improves water-use efficiency by reducing drought and heat stress, it can accentuate a weed problem. Therefore, to get the maximum benefit from early sowing of food legumes, optimum control of weeds is necessary. Marked increases in yield of early sown chickpea and lentil crops through weed control have been reported by several workers (Basler, 1981; Solh and Pala, 1990; Pala and Mazid, 1992a,b). Rising costs of labor impose limitations on hand-weeding (Jaradat, 1988); thus chemical weed control merits consideration. Several herbicides have been tested in different countries, as well as at ICARDA research stations, for their effectiveness and crop tolerance. A pre-emergence application of cyanazine (0.5 kg ha^{-1}) and terbutryne (2.0 kg ha^{-1}), alone or in combination with pronamide (0.5 kg ha^{-1}), proved promising for broad-spectrum control of weeds in lentil and chickpea, respectively. Adoption of a pre-emergence herbicide by farmers, however, necessitates good extension efforts. Availability of the chemicals in the local market can also become a limitation. Mechanical weed control is a feasible alternative. In fact, farmers in Algeria and Morocco control weeds by sowing in wide (1 to 2 m) rows and doing inter-row cultivation, but such row spacing often results in low yields because of inadequate plant populations. Effective mechanical weed control requires careful setting of planting and inter-row cultivation equipment. A good potential for weed control by inter-row cultivation has been shown in a recent study on winter-sown chickpea in northern Syria (Pala, 1991).

The cool season food legumes are also affected by several diseases and insects which can substantially reduce crop yields (Van Emden *et al.*, 1988). Their control requires an integrated approach based on the use of resistant/tolerant cultivars, strategic use of safe chemicals, and proper management. Although several examples of successful solutions to these problems are available, more work is needed for developing economic and environmentally safe control measures to enhance the use of CSFL in different farming systems.

Mechanization

One of the major constraints in expanding the use of cool season food legumes in the farming systems of WANA and several other developing countries is the low degree of mechanization of these crops. The whole process of food legume production is labor intensive, resulting in high production costs. For this reason the economic competitiveness of cool season food legumes is lower compared with fully mechanized cereals, which may explain the decline in their production area in several countries (AGRODEV, 1990; Buddenhagen, 1990; Oram and Belaid, 1990; Osman *et al.*, 1990).

Drill-seeding for food legumes is not as widespread in the developing world as is the case with cereals. Proper placement of seed in soil is essential for satisfactory germination and seedling emergence, especially where moisture is limited. Thus, management for higher yields requires the use of mechanical seeders with good control of sowing depth, seeding rate, and row spacing. Drills designed for cereals are generally satisfactory for sowing legumes provided some adjustments are made (Papendick *et al.*, 1988).

Mechanization of harvest is another major constraint for the production, especially in the broadcast method of sowing, which causes heterogeneous emergence and stand establishment, and uneven soil leveling leads to impediments in mechanical harvest. Widespread use of landrace cultivars characterized by a short stature, lodging habit, and susceptibility to shattering due to pod dehiscence and drop (Saxena *et al.*, 1987) further accentuates the problem. Therefore, achieving a drill sowing on a well-leveled seed bed with improved cultivars will facilitate mechanical harvest.

Various methods for mechanical harvest of the most problematic crop, lentil, have been tested in some countries of the WANA region. These include use of single- and double-knife cutter-bars, puller, swather, and combine harvester

(Saxena *et al.*, 1987; Snobar and Haddad, 1987; AGRODEV, 1989, 1990; Friedrich *et al.*, 1989; Erskine *et al.*, 1991). However, due to high losses in seed and/or straw, some of these harvesting methods are only of limited use (Snobar *et al.*, 1990). It is worth noting that these losses are largely attributed to poor management practices (AGRODEV, 1989) and the plant growth characteristics of landraces mentioned earlier.

Mechanization of chickpea harvest presents fewer problems. With the tall cultivars which are now available, the traditional grain combines can be used by adjusting them for the seed size and other parameters of the crop (Saxena *et al.*, 1987). The spread of winter sowing in the lowlands of WANA will further encourage the mechanical harvest because of the gain in plant height in contrast to the spring-sown crop.

Sound improvement in mechanization of food legumes is considered one of the most important factors for enhancing their production in WANA (Osman *et al.*, 1990). For example, a large increase in lentil area in Turkey during the 1980s accentuated the labor shortage for harvest and encouraged research on mechanization of lentil harvesting. This resulted in the development of a suitable cutting and threshing system. A combine harvester was reported to give highest field efficiency at the lowest cost, but its use would only be applicable for large fields. Although hand harvesting is still done in some parts of Turkey where cheap labor is available, combine harvesters are being used extensively in the country (Guler, 1990).

Socioeconomic Factors

The expansion of cool season food legume production can occur with increasing local demand and more profitable prices. The demands for legumes in developing countries in recent years is indeed increasing at a rate faster than projected by the World Bank. For example, the per capita consumption of pulses in WANA was projected to reach 6.1 kg per annum by 1990; it reached 6.6 kg already by 1987 (Oram and Belaid, 1990). However, local production did not increase in several WANA countries and the increased demand was met through increased imports. Therefore, the major factor responsible for holding back the expansion of local production in the developing countries seems to be the lower profitability of food legumes because of higher production costs and low yields compared with other crops. There is no guaranteed floor prices for legumes in the Indian subcontinent and most of the countries of WANA although such floor for wheat exists (Guler, 1990; Snobar *et al.*, 1990). This leads to instability of legume prices and contributes to a perception among farmers that legumes are a more risky enterprise (Oram and Belaid, 1990). Government policy change that provides for guaranteed floor price for legumes would certainly contribute to an expansion in the production of legumes as has occurred, for example, in Jordan, Syria and Turkey.

The cost of production of cool season food legumes in many countries of

WANA is high because of growing shortage and escalating wages of labor required for such operations as manual weeding and harvest. Collaborative research with ICARDA done by national programs in Algeria, Egypt, Jordan, Lebanon, Morocco, Syria, Tunisia and Turkey has shown that it is possible to reduce production costs by replacing manual weeding with the use of selective herbicides. Similarly, mechanization has reduced harvest costs in Algeria, Jordan, Lebanon, Morocco, Syria and Turkey. In order to benefit from these developments, the farmers in these countries need government support for making herbicides and harvest machinery available at reasonable costs. Also more extension efforts are needed for rapid transfer of these technologies to farmers.

Subsidizing the prices of legumes can also contribute to enhanced local production in countries where demands are currently met through imports. Because of lower international prices the government of Algeria and Jordan, for example, have preferred in the past years to import lentil and chickpea, respectively, than to subsidize the prices of local produce at high levels. A change in government's economic policy there can encourage domestic production.

Much of the legume production in the developing countries is done by subsistence farmers who produce the crops for their own use and for the local market. Expansion of production through greater commercialization would necessitate infrastructural support not only for making the needed inputs available, but also for handling, storage, marketing and distribution of surplus produce. Government and private sector interventions could play a key role in this as has been so successfully done in Turkey.

The prevailing succession laws in several WANA countries encourage fragmentation of holdings as the land is divided among all successors when ownership passes from one generation to another. Greater commercialization of legume production in such countries would be possible if holdings could be consolidated into more efficient production units on which improved inputs and mechanized paractices could be used. Encouraging the development of local cooperatives and establishment of custom service for farm machinery would make collective management of contiguous holdings possible despite the fact that each parcel of land may belong to a different owner.

Concluding Remarks

Expansion of cool season food legumes in different farming systems is limited by several technological and socioeconomic constraints. These can, however, be alleviated through increased efforts in (1) research on development of appropriate cultivars and production technology to suit the needs of specific farming systems, (2) transferring this technology to the farmers, and (3) encouraging development of supportive government policies.

Systems-oriented research on various aspects of production agronomy is needed to enhance yield and water-use efficiency of these legumes in major farming systems in different agroecological conditions. Such studies will be particularly necessary for the crops that will be introduced to replace fallow and may be very location-specific. Replacement of fallow offers one major avenue for expanding the production of cool season food legumes in the WANA region, as has already been occurring in Turkey. Introduction of cool season food legumes as intercrops in olive plantations and fruit orchards can further augment production in WANA. Relay and intercropping of chickpea, lentil, and grasspea with other crops is common in the tropics where land is scarce, cropping intensity is already high, and labor is available. Research is needed to develop systems that optimize the use of limited resources in such systems and maximize profits.

Availability of seeds of improved cultivars of food legumes has been another major constraint. Government policies that may facilitate this are needed, including the development of a strong private sector to produce seeds. The production of cool season food legumes will have to undergo a change, evolving from subsistence farming to a market-oriented enterprise, through consolidation of small holdings into larger fields. This will allow more efficient husbandry operations and better marketing could be organized. Government policies that ensure guaranteed floor prices and assured procurement of produce from farmers would make cool season legumes more competitive with other crops and would additionally contribute to their expanded production.

References

AGRODEV. 1989. *Lentil Mechanization Project, Jordan. Year 1 Report*, p. 79. AGRODEV Canada Inc.

AGRODEV. 1990. *Lentil Mechanization Project, Jordan. Year 2 Report*, p. 68. AGRODEV Canada Inc.

Basler, F. 1981. In: *Lentils*, pp. 143–154 (eds. C. Webb and G. Hawtin). Wallingford, UK: CAB International.

Beck, D. P. and Materon, L. A. (eds.) 1988. *Nitrogen Fixation by Legumes in Mediterranean Agriculture*. Proceedings of the Workshop, 14–17 April, 1986, ICARDA, Aleppo, Syria. 379 p. Dordrecht: Martinus Nijhoff.

Beck, D. P., Wery, J., Saxena, M. C. and Ayadi, A. 1991. *Agronomy Journal* 83: 334–341.

Benbella, M. 1987. Effect of fallow, wheat (*Triticum aestivum* L.), broad bean (*Vicia faba* L.) and nitrogen fertilizer on subsequent wheat. Communication presented at the National Workshop on Food Legumes in Morocco, 7–9 April, 1987, Settat.

Buddenhagen, I. W. 1990. In: *The Role of Legumes in the Farming Systems of the Mediterranean Areas*, pp. 3–29 (eds. A. E. Osman, M. H. Ibrahim and M. A. Jones). London: Kluwer Academic Publishers.

Durutan, N., Pala, M., Karaca, M. and Yesilsoy, M. S. 1989. In: *Soil, Water, and Crop/Livestock Management Systems for Rainfed Agriculture in the Near East Region*, pp. 60–77 (eds. C. E. Whitman, J. F. Parr, R. I. Papendick and R. E. Meyer). Washington, DC, USA: USDA.

Erskine, W., Diekmann, J., Jegatheeswaran, P., Salkini, A., Saxena, M. C., Ghanaim, A. and El Ashkar, F. 1991. *Journal of Agricultural Science* (Cambridge) 117: 333–338.

FAO. 1990. In: *Production Yearbook*, Vol. 44: 98–106. Rome: FAO.

FLIP. 1991. In: *Food Legume Improvement Program, 1990 Annual Report*, p. 17, 117, 154. Aleppo, Syria: ICARDA.

Friedrich, T., Diekmann, J. and Erskine, W. 1989. In: *FLIP Annual Report for 1988*, pp. 91–94.

ICARDA-146 En. Aleppo, Syria: ICARDA.

Guler, M. 1990. In: *The Role of Legumes in the Farming Systems of the Mediterranean Areas*, pp. 131–139 (eds. A. E. Osman, M. H. Ibrahim and M. A. Jones). London: Kluwer Academic Publishers.

Guler, M. and Karaca, M. 1988. In: *Winter Cereals and Food Legumes in Mountainous Areas*, pp. 41–49 (eds. J. P. Srivastava, M. C. Saxena, S. Varma and M. Tahir). ICARDA-136 En. Aleppo, Syria: ICARDA.

Harris, H. 1989. In: *FRMP Annual Report for 1988*, pp. 36–64. ICARDA-142 En. Aleppo, Syria: ICARDA.

Harris, H. 1990. In: *FRMP Annual Report for 1989*, pp. 137–166. ICARDA-162 En. Aleppo, Syria: ICARDA.

Herridge, D. F., Rupela, O. P., Serraj, R. and Beck, D. P. 1994. In: *Expanding the Production and Use of Cool Season Food Legumes*, pp. 472–492 (eds. F. J. Muehlbauer and W. J. Kaiser). Dordrecht: Kluwer Academic Publishers.

Jaradat, A. A. 1988. In: *An Assessment of Research Needs and Priority for Rainfed Agriculture in Jordan*, Published by USAID, 415 p. Irbid, Jordan: Al-Hurria Press.

Jaradat A. A. 1989. *A Progress Report on Cropping Systems Research in Jordan*, p. 63. Irbid, Jordan: ACSAD.

Jones, M. J. 1989. In: *Barley Rotation Trials at Tel Hadya and Breda Stations – A Summary of Biological Yield Data, 1981–1987*, p. 30. ICARDA-140 En. Aleppo, Syria: ICARDA.

Karaca, M., Guler, M., Durutan, N., Meyveci, K., Avci, M., Eyuboglu, H. and Avcin, A. 1991. In: *Proceedings of International Workshop on Soil and Crop Management Practices for Improved Water Use Efficiencies in Rainfed Areas*, pp. 251–259 (eds. H. C. Harris, P. J. M. Cooper and M. Pala). Aleppo, Syria: ICARDA.

Keatinge, J. D. H., Chapanian, N. and Saxena, M. C. 1988. *Journal of Agricultural Science* (Cambridge) 110: 651–659.

Nassib, A. M., Sakar, D., Solh, M. and Salih, F. A. 1988. In: *World Crops: Cool Season Food Legumes*, pp. 1081–1094. (ed. R. J. Summerfield). London: Kluwer Academic Publishers.

Oram, P. and Belaid, A. 1990. *Legumes in Farming Systems*. A joint ICARDA/IFPRI Report, ICARDA-160, p. 206. Aleppo, Syria: ICARDA.

Osman, A. E., Ibrahim, M. H. and Jones, M. A. (eds.). 1990. *The Role of Legumes in the Farming Systems of the Mediterranean Areas*, p. 310. Dordrecht: Kluwer Academic Publishers.

Pala, M. 1991. In: *FRMP Annual Report for 1990*, pp. 41–47. ICARDA-221. Aleppo, Syria: ICARDA.

Pala, M. 1992. *Sustainability of fallow replacement*. Paper presented in Seminar on Natural Resources and Environmental Management in Dry Areas sponsored by EDI/World Bank in collaboration with ICARDA and AOAD, 16–27 February, 1992, Aleppo, Syria.

Pala, M. and Mazid, A. 1992a. *Experimental Agriculture* 28: 175–184.

Pala, M. and Mazid, A. 1992b. *Experimental Agriculture* 28: 185–193.

Papastylianou, I. 1987. *Journal of Agricultural Science* (Cambridge) 108: 623–626.

Papendick, R. I., Chowdhury, S. L. and Johansen, C. 1988. In: *World Crops: Cool Season Food Legumes*, pp. 237–255 (ed. R. J. Summerfield). London: Kluwer Academic Publishers.

Plancquaert, P. and Haggar, R. (eds.). 1989. In: *Legumes in Farming Systems*, p. 181. Dordrecht: Kluwer Academic Publishers.

Saxena, M. C. 1987. In: *The Chickpea*, pp. 207–232 (eds. M. C. Saxena and K. B. Singh). Wallingford, UK: CAB/ICARDA.

Saxena, M. C., Diekmann, J., Erskine, W. and Singh, K. B. 1987. In: *Mechanization of Field Experiments in Semi-arid Areas*, pp. 211–228 (ed. D. Karlsson). ICARDA-115 En. Aleppo, Syria: ICARDA.

Saxena, M. C. 1988. In: *Nitrogen Fixation by Legumes in Mediterranean Agriculture*, pp. 11–24 (eds. D. P. Beck and L. A. Materon). Dordrecht: Martinus Nijhoff.

Sharma, K. N., Bhandari, Kapur, M. L. and Rana, D. S. 1985. *Journal of Agricultural Science* (Cambridge) 104: 609–613.

Silim, S. N., Saxena, M. C. and Erskine, W. 1991. *Experimental Agriculture* 27: 145–153.

Snobar, B. A. and Haddad, N. I. 1987. In: *Mechanization of Field Experiments in Semi-arid Areas*, pp. 202–210 (ed. D. Karlsson). ICARDA-115 En. Aleppo, Syria: ICARDA.

Snobar, B., Masadeh, A. and Haddad, N. 1990. *Food Legume Improvement and Mechanization Project Report for 1989/90 and Workplan for 1990/91*, IDRC, UJ, oa, P. 160. Amman: University of Jordan.

Solh, M. B. and Pala, M. 1990. In: *Present Status and Future Prospects of Chickpea Crop Production and Improvement in the Mediterranean Countries*, pp. 93–99 (eds. M. C. Saxena, J. I. Cubero and J. Wery). Zaragoza, Spain: CIHEAM.

Somel, K. 1988. In: *FRMP Annual Report for 1987*, pp 122–143. ICARDA-131 En. Aleppo, Syria: ICARDA.

Stanforth, A., Sprent, J. I., Brockwell, J., Beck, D. P. and Moawad, H. 1994. In: *Expanding the Production and Use of Cool Season Food Legumes*, pp. 823–833 (eds. F. J. Muehlbauer and W. J. Kaiser). Dordrecht: Kluwer Academic Publishers.

Summerfield, R. J. (ed.). 1988. *World Crops: Cool Season Food Legumes*, p. 1179. London: Kluwer Academic Publishers.

Van Emden, H. F., Ball, S. L. and Rao, M. R. 1988. In: *World Crops: Cool Season Food Legumes*, pp. 519–534 (ed. R. J. Summerfield). London: Kluwer Academic Publishers.

Weigand, S., Pala, M. and Saxena, M. C. 1992. *Journal of Plant Diseases and Plant Protection* 99: 174–181.

White, J. G. H. 1989. In: *Grain Legumes: National Symposium and Workshop*, pp. 109–115 (eds. G. D. Hill and G. P. Savage). Agronomy Society of New Zealand Special Publication No. 7. Lincoln, New Zealand: Agronomy Society

Grasspea (*Lathyrus sativus* L.) as a potentially safe legume food crop

J. SMARTT[1], A. KAUL[2], WOLDE AMLAK ARAYA[3], M.M. RAHMAN[4] and J. KEARNEY[1]

[1] *Department of Biology, School of Biological Sciences, The University of Southampton, Biomedical Sciences Building, Bassett Crescent East, Southampton, S09 3TU UK;*
[2] *Winrock International Institute for Agricultural Development, Petit Jean Mountain, Morrilton, Arkansas 72110–9537, USA;*
[3] *Adet Research Centre, Bahar Dar, Ethiopia, and*
[4] *Bangladesh Agricultural Research Institute (BARI), Regional Agricultural Research Station, Ishurdi 6620, Pabna, Bangladesh*

Abstract

The grasspea is potentially an extremely valuable pulse crop. It is a quick maturing annual which is agronomically undemanding and is tolerant of both drought and waterlogging. It is well adapted to production in the cool season (winter) of warm temperate and subtropical areas (Mediterranean basin, India, Pakistan, Bangladesh, and Ethiopia). As a pulse it is palatable and nutritious (ca. 27% protein) but it contains a toxic non-protein amino acid ODAP (β-N-oxalyl-L-$\alpha\beta$ diamino propionic-acid) or BOAA (β-N-oxalylamino-L-alanine). This interferes with the normal functioning of nerves controlling leg movement and induces an irreversible paralysis. The apparent solution of this problem lies in the development of stable low ODAP containing lines whose consumption does not evoke the paralytic syndrome of neurolathyrism.

Introduction

The grasspea is an enigmatic crop. It is of very ancient cultivation but has acquired a very bad reputation since from classical times its consumption has been associated with the development of an irreversible paralysis of the lower limbs – neurolathyrism. In spite of this, and attempts to discourage and even prohibit its cultivation, it is still grown and in some countries such as Bangladesh it may still be a very significant pulse crop and source of dietary protein.

The reason for its persistence in cultivation is that not only is it a palatable food but it is a remarkably hardy plant in its optimal climate zone. It is essentially a cool season crop of warm-temperate (Mediterranean) and subtropical regions in which it is grown successfully because it is tolerant of both drought conditions and waterlogging. Therefore it is equally at home in drought prone areas of Ethiopia as it is in Bangladesh where floods and waterlogging are problems.

F.J. Muehlbauer and W.J. Kaiser (eds.), Expanding the Production and Use of Cool Season Food Legumes, 144–155.
© 1994 *Kluwer Academic Publishers.*

The source of the problem in its use and the incitant of the condition of neurolathyrism is the non-protein amino acid variously known as BOAA, beta-N-oxalylamino-L-alanine; ODAP (or oxdabro) beta-N-oxalyl-L-alpha, beta-diaminopropionic acid; or OAP, L-3-oxalylamino-2-aminopropionic acid. As is the case with certain non-protein amino acids they behave as antagonists or mimics of particular protein amino acids with often very profound consequences. The seriousness of neurolathyrism lies in the fact that irreversible neurological damage is inflicted upon people who are usually economically disadvantaged and unable to secure less hazardous food supplies. Neurolathyrism is therefore a serious public health problem which has socio-economic roots and is clearly related to poverty. Since, as the New Testament tells us, the poor are always with us and unfortunately are likely to remain so in the foreseeable future it clearly behooves us to do whatever is humanly possible to develop safe foods for the poor and prevent the misery of poverty being compounded with the handicap of paralysis.

The purpose of this review is to consider in a broad context, the nature of the grasspea itself, the condition of neurolathyrism and its causation and strategies for improving the situation which in our present state of knowledge give the greatest hope of success. In the past decade the Third World Medical Foundation has encouraged study of the grasspea and the associated problem of lathyrism through its support of the International Network for the Improvement of *Lathyrus sativus* and the Eradication of Lathyrism (INILSEL). Two international meetings have been held in France and England the reports of which have been published (Kaul and Combes, 1986; Spencer, 1989) which together summarize a great deal of our present knowledge of *Lathyrus sativus* and lathyrism.

Biosystematics

The grasspea is a rather delicate looking vetch-like herb which is much less vigorous in its growth than the congeneric sweetpea *Lathyrus odoratus* and the everlasting pea *L. latifolius*. It is a member of the tribe Vicieae which includes the major pulse crops of the classical Mediterranean civilizations, the faba bean (*Vicia faba*), the common pea (*Pisum sativum*), and the lentil (*Lens culinaris*). The chickpea (*Cicer arietinum*) formerly also included in the tribe has been transferred to the monogeneric tribe Cicereae (Kupicha, 1981). From available evidence (Renfrew, 1973) it is clear that the grasspea has been cultivated since ancient times. Its development as a crop plant and to some extent its dissemination have been inhibited by its association with lathyrism. The removal of this hazard could provide the incentive for plant breeders by selection to accelerate the development of this species into a truly modern and productive crop plant.

The genus *Lathyrus* to which the grasspea belongs, is quite large containing about 150 species (Kupicha, 1983). The genus comprises some 13 sections of which section *Lathyrus* includes the grasspea itself and some popular garden subjects such as the sweetpea and everlasting pea. The relationship between *L. sativus* and its closest relatives in section *Lathyrus* have been studied by Yunus (1990) and Yunus and Jackson (1991). Experimental hybridizations were attempted between 15 species of section *Lathyrus* and the grasspea of which only two, *L. amphicarpos* and *L. cicera* gave viable F_1 hybrids. Six other species produced pods in experimental crosses but the seeds within them either aborted and were shrivelled or if fully developed failed to germinate or produced inviable seedlings.

These results give us some measure of the extent of the genetic resources which can be mobilized in the improvement of the grasspea. In terms of Harland and De Wet's (1971) gene pool concept, the primary gene pool (GP1), the most accessible to the breeder, comprises both wild (or feral) and cultivated populations of the grasspea. There are no clearcut discontinuities between wild and cultivated forms of the grasspea as there are in other pulse species however, so no subdivision of the GP1 into cultivated (A) and wild (B) segments serves any useful purpose. The secondary gene pool (GP2), which can be exploited by interspecific hybridization, consists only of *L. cicera* and *L. amphicarpos* as far as is known at present. The remaining species of the genus comprise the tertiary gene pool (GP3) which can only be tapped by techniques of bridging species crosses and biotechnological procedures.

Recent cytogenetic studies in section *Lathyrus* are of interest in this context. The chromosome complement $2n = 2x = 14$ is found in the vast majority of species. There is some variation in karyotype but the majority of chromosomes are submetacentric, in *L. sativus* all seven pairs are submetacentric while both cross-compatible species (*L. cicera* and *L. amphicarpos*) have 1 pr. metacentric and 6 prs. submetacentric. This implies that some chromosome structural differentiation has occurred between genomes of different species. From meiotic studies of interspecific hybrids it would seem that *L. amphicarpos* is structurally more differentiated from *L. sativus* than is *L. cicera*. In the F_1 hybrid *L. cicera* × *L. sativus* the configurations observed were 6II + 2I and 7II. In the hybrid *L. amphicarpos* × *L. sativus* multivalents were frequently observed suggesting that translocational changes had occurred (Yunus, 1990; Yunus and Jackson, 1991).

The evolution of the grasspea as a crop did not apparently conform very closely to the general pattern of grain legumes evolution suggested by Smartt (1990) in that some distinctive characters of cultigens, such as gigantism of seeds, were poorly developed. It was suggested that this was due to the dual purpose use of the grasspea both for forage and as a pulse. Large seed would be disadvantageous in a forage crop and if the same populations were used for both purposes, selection pressures imposed by dual usage would cancel each other out. In addition, the practice of undersowing grasspea in standing crops of rice for example favors the use of a relatively large number of small seed rather than the same biomass of large seed. Recently Kearney (unpublished) has shown that in peripheral areas of the crop's distribution, the evolutionary progression actually conforms to the general model quite closely. This is noteworthy in some important seed characters. There are accessions from the former USSR which have appreciably enhanced seed size (5 to 10 fold) and which also show reductions and even loss of the phenolic testa pigments. While these compounds are fungistatic and protect the seed to some extent, they interact with seed proteins rendering them less "cookable". Although an aid to seed survival and seedling establishment, the increase in cooking times these phenolics produce is most undesirable where cooking fuels are in short supply. This drawback may be acceptable in areas where the crop is undersown without seedbed preparation and the presence of phenolics may contribute significantly to crop establishment.

Such conflicting selection pressures could in effect cancel each other out and a primitive *status quo* maintained. Where such constraints have been removed, there are indications that in terms of development of advanced attributes, such as large, easily cooked seeds, the grasspea may yet prove to be the equal of the other pulses of Mediterranean origin.

Utilization

Undoubtedly the most important constraint in the development of the grasspea as a crop and its dissemination is the occurrence in its seed of the toxic, non-protein amino acid ODAP. The condition it incites, neurolathyrism, has been known from the time of Hippocrates and other ancient writers notably from India who observed that eating certain pulses "causes a man to become lame – it cripples and irritates nerves" (Barrow et al., 1974). Such knowledge over the ages must have provided a powerful incentive to the adoption of alternative pulses wherever possible. This has led to the present situation where the crop tends to be produced in areas where few if any alternative pulses can be grown reliably. Its wide range of ecological tolerance explains its cultivation in agriculturally difficult environments such as drought prone areas of Northern India, Pakistan, and Ethiopia as well as areas of Bangladesh and Nepal subject to flooding and waterlogging.

The grasspea can be prepared and utilized in a number of ways, which have implications in the development of the paralysis syndrome. There is an appreciation among consumers that certain grasspea products are more likely to produce lathyrism than others. The reason for this is that a certain level of detoxification is produced in some modes of preparation. If seed is soaked and boiled a major part of the toxin may be eluted in the water in which it is steeped and cooked. Obviously the effect is enhanced if water is changed after soaking and during cooking. However when the seed is ground to produce a meal or flour, which is then used in cooking or baking, no toxin may be lost (Haque and Mannan, 1989; Pushpamma, 1989) and symptoms may develop more rapidly or in a more severe form.

Epidemiology and Etiology of Lathyrism

The circumstances in which neurolathyrism develops are well understood. In those areas in which it is recorded its incidence is not constant, tending to reach peaks in times of famine. It has been suggested that if it comprises more than 25% of the food intake over a period of 45–180 days the condition may develop. Higher levels of consumption can produce symptoms in only 20 days (Rutter and Percy, 1984; Dwivedi, 1989). The incidence of the disease was in times past increased by the practice of paying laborers in kind with grasspea seed, a practice which is now illegal. Attempts at controlling the disease by banning cultivation and trade have not succeeded. This was first attempted in Europe in the 17th century, the duke Georg of Wurtemburg issued an edict banning it in 1671 (Spencer *et al.*, 1986), and subsequently similar attempts have been made in the Indian subcontinent. These have been unsuccessful in the absence of satisfactory alternative crops. Without these, those dependent on the crop for survival have a stark choice of death through starvation or the possibility of survival with an attendant higher risk of lower limb paralysis.

Socio-economic factors are clearly important in controlling the incidence of lathyrism. Dwivedi (1989) clearly demonstrated that in social classes I (richest) to V (poorest) the syndrome did not occur in classes I and II and in a season of high incidence (1981) increased rapidly through social classes III (5.65%) and IV (26.5%) to an incidence of 67.8% in one sample of social class V. A similar distribution by occupation was also apparent with the highest incidence (72.31%) among unskilled laborers. There is also a very marked difference in incidence between the sexes, males are more susceptible than females, who do not apparently develop the condition between puberty and the menopause. Among males the incidence declines markedly after 35 years of age.

The onset of the disease may be insidious (chronic) or sudden (acute). It can be progressive but the progression from mild to more severe forms is not rapid (Kaul *et al.*, 1989). Patients can be classified in four categories, depending on the extent of the disability and resultant loss of mobility, these are (1) "non-stick", (2) "one-stick", (3) "two-stick", and (4) "crawler" (Dwivedi, 1989). The condition of patients who seek treatment in the majority of cases does not improve, but in about 25% there may be no further deterioration in their condition. While some also reported improvement in their condition none have been reported as making a complete recovery. The onset of the disease is indicated by a feeling of heaviness and pain in the lower limbs. Although that part of the nervous system controlling the movement of the legs is impaired other parts of the nervous system are apparently not affected; there is, for example, no loss of mental faculties.

The pathological action of ODAP arises from its excitatory effects on parts of the nervous system (MacDonald *et al.*, 1986; Ross and Spencer, 1989) which severely disrupt neurotransmission. ODAP acts as an antagonist or mimic of the normal glutamate moiety, its effects being to overexcite neurons, disrupt the restoration of normal ionic balance after transmission, and induce imbibition of excess water producing an oedematous state and subsequent degeneration. The latter provides some explanation for the irreversibility of the paralysis caused by "an upper-motor neuron disease producing corticospinal dysfunction" (Ross and Spencer, 1989).

The Toxin and its Biochemistry

ODAP is a non-protein amino acid which can be regarded as a derivative of alanine (a protein amino acid) itself a derivative of propionic acid. The actual biosynthetic pathway producing it has only recently been elucidated (Lambein and Kuo, 1991) although the structure of ODAP itself had been independently established by two groups in India (Murti and Seshadri, 1964; Rao *et al.*, 1964). The critical missing link is apparently beta-(isoxazolin-5-on-2-yl)-alanine (BIA) which gives rise to diamino propionic acid (DAPRO) which by oxalylation produces ODAP (Lambein *et al.*, 1990). ODAP is known to exist in two isomeric forms, alpha and beta.

```
COOH — CO              CO — COOH
        |              |
NH₂    NH              NH    NH
 |      |              |      |
CH₂ —— CH              CH₂ — CH
        |                     |
      COOH                  COOH

   alpha-isomer           beta isomer
```

Isomerization occurs spontaneously at room temperatures producing an isomeric mixture of 5% alpha : 95% beta-isomer. The alpha-isomer appears to have feeble if any neurotoxic activity (Nunn, 1989).

Potential for Improvement

Genetic Control of Toxin Level

The first priority in the improvement of the grasspea crop is a reduction in the seed content of ODAP. Since physical detoxification procedures, apparently simple though they are (Pushpamma, 1989), have not been extensively adopted, then genetic detoxification seems to be the most feasible solution to the neurotoxin problem. There are encouraging precedents in other grain legumes such as lima beans and lupins, where contents of cyanogenic compounds and alkaloids have been reduced to acceptable levels. In both lima beans and lupins there are polymorphisms for toxin concentrations in the seed and it was a logical step to explore the possibility of a similar polymorphism in the grasspea. Such an exploration depends critically on the development of suitable techniques for measurement of ODAP contents in seeds (Barat *et al.*, 1989). These have enabled surveys to be made of local landrace collections, international collections and experimental breeding lines and selections with very useful results.

Kaul *et al.* (1986) screened a collection of germplasm lines in Bangladesh for seed contents of ODAP and found this ranged between 450 and 1,400 mg ODAP per 100 g of seed (0.45 to 1.40%). This clearly indicated that there is adequate scope for selecting genotypes with reduced ODAP contents. Later work of Rathod (1989) in India reported concentrations in some samples as low as 0.07%. Barat *et al.* (1989), also in India, reported a range of ODAP contents within the range 0.10% to greater than 1.0%. The effective range of concentrations appears to be from approximately 0.10 to 1.40% considering all these reports. It is interesting to note that it was possible to detect ODAP in all samples. It may therefore be unrealistic to hope that ODAP-free genotypes can be established but concentrations reliably below 0.5% in the first instance reducing subsequently to below 0.25% and even lower are probably not unrealistic. In the lima bean, stable low-cyanide genotypes have proven satisfactory over the years. Total elimination of the toxin may even be considered undesirable as it is thought to contribute to the flavor of the grasspea.

When ODAP contents of accessions have been measured in successive seasons, these have in some instances been found to vary. Kaul *et al.* (1986) measured toxin concentrations in seeds of 18 strains in two successive seasons, these were all relatively high in ODAP content. The strain Pahartoli produced one of the highest contents in 1980 (1.06%) but the lowest (0.68%) in 1981. Variation in other strains was less and in some apparent consistency was shown in the two seasons. Similar observations have been made in Ethiopia (Araya, 1992, unpublished). This is a cause for some concern since stability of low toxin levels is essential, while absolute constancy could probably not be ensured, the potential range of variation needs to be established reliably.

Barat *et al.* (1989) report ODAP levels in seed produced at three sites in India over three seasons (1971–73). Levels at Delhi were consistent for three seasons in seven strains. At Rewa one strain (P678) was consistent over all three seasons (0.24, 0.22, 0.24%, respectively) but others were less so (P101 at 0.41, 0.22, 0.32). Contents at the third site Tancha showed little consistency overall, they were uniformly high in '71, lower in '72, and lowest in '73. This clearly shows that consistency of ODAP levels between locations and seasons cannot be taken for granted. It also needs to be established whether ODAP content is correlated with stress during the growing season. The results of Rathod (1989) present a generally similar situation.

Heritability studies of ODAP contents and other attributes have been reported by Dahiya (1986) from India and Quader *et al.* (1989) from Bangladesh. Heritability estimates obtained from parent/offspring regression are in the range 0.49–0.65 indicating that there is a substantial environmental effect expressed in this study as differences between locations. Genotype-environment interactions are also high.

There is clearly a need to identify specific environmental factors which can influence toxin content and also to identify genotypes with the greatest consistency in producing low levels of ODAP regardless of season and location. Stability of seed toxin content could also be influenced by another set of circumstances. Dr. C. G. Campbell (pers. comm.) has observed an apparent linkage of male sterility with low ODAP content in one accession, this clearly imposes a higher level of outcrossing that might otherwise be the case. Since the toxin is laid down in the embryo the genetic constitution of incoming pollen could directly affect the level of ODAP production in the cotyledonary tissue. It is of the greatest importance to establish as strict self-pollination as possible in low ODAP grasspea strains in order to prevent undesirable gene introgression. Such a breeding system has ensured stability of low cyanide lines of *Phaseolus lunatus* while appreciable outcrossing in lupins poses problems in maintaining low alkaloid status in the sweet forms.

The measurement of ODAP content does create problems for the breeder in carrying out his selection program and as a result correlations have been sought with readily observable phenotypic characters such as flower and seed coat colors. It was thought that pale flowers and light testa colors (cream-white) might be associated with low seed contents of ODAP. In their investigations Quader *et al.* (1986, 1989) found no support for this hypothesis, which seems to have arisen from the small number of samples initially studied. Nevertheless, in some circumstances flower and seed coat colors could be used as genetic markers to monitor possible introgression. Were a low ODAP strain with white flowers and pale testa color to be cultivated among populations characterized by blue flowers and dark testae, its low toxin status could probably be safeguarded by appropriate roguing at flowering and after harvest.

While considerable progress has been achieved in the understanding of the genetic control of toxin level, there is clearly scope for some further investigation of environmental effects upon it. Some experiments under controlled environment conditions could be very instructive. Further studies of genotype-environment interactions and heritabilities on genotypes showing low and stable ODAP levels in their seeds are also indicated. These would require initial study of single plant selections maintained by strict self-pollination and monitored over several generations for toxin level before crosses were made. This would obviate apparently equivocal results obtained in the past arising from the sampling of heterogeneous populations, the occurrence of cross-pollination, and the erratic response to variable environmental conditions.

Other Aspects of Improvement

The approach to other aspects of grasspea improvement is likely to be determined by changes which can be made in husbandry practices. Many possible avenues of improvement are ruled out if the current practices such as undersowing rice crops with grasspea in Bangladesh continue. In such circumstances reduction of ODAP levels is clearly desirable, but most of the present characteristics of the landraces used could only be changed to the detriment of this system of production. Dark testa colors, high testa phenolic content, and relatively small seed size serve on the one hand to minimize seed predation, spread its risk and give some protection to the seedling in the course of establishment. In such a farming system, large pale seeds would be visible and readily predated, and achievement of adequate plant populations would be costly in terms of the investment in seed sown. The seedlings would also be relatively unprotected from pest and pathogen attack. In addition, the improvement of pest and pathogen resistance generally would be a valid and sensible objective. Bharati and Neupane (1989) have found that pod borer attacks and that of *Botrytis* inhibit cultivation in some areas of Nepal. Lal *et al.* (1986) report rust (*Uromyces fabae*) and powdery mildew, Campbell (1989) reported that powdery mildew (*Erysiphe polygoni*) and downy mildew (*Peronospora lathi-palustris*) are major fungal pathogens, while thrips can be a serious pest. There are indications of resistance in some lines to the mildews.

If there are reasonable prospects of developing improved cultural practices for the grasspea crop then there would be a real incentive to initiate wider ranging plant breeding programs and the development of one or more grasspea ideotypes. At the present time the grasspea is a rather delicate vetch-like plant which does not individually carry a particularly heavy load of pods and seed. Selection for higher yielding capacity would imply selection also for a more sturdy stem and branch system and a denser canopy. Some basic studies of the kind necessary for the development of ideotypes have already been carried out (Kaul *et al.*, 1986). These would clearly have to be developed further if for example it was envisaged that the grasspea were to be produced in a conventional cropping system with a high standard of seedbed preparation and cultivation. It would not be unreasonable to develop an ideotype perhaps resembling in form a modern dwarf garden pea cultivar.

Breeding Strategies in Improvement

It would seem that when the evaluation of grasspea genetic resources has reached a reasonably definitive state, there will be no shortage of suitable parental material with which to launch traditional plant breeding programs. There appear to be no particular or specific difficulties in the handling of the grasspea in the making of controlled crosses. In such programs it is well to be prepared for the unexpected particularly when material of diverse geographic origin is brought together. It may also be necessary to produce very large segregating progenies and to institute multilocation selection, testing, and evaluation at the earliest opportunity. Production of F_1 hybrid material could be centralized and F_2 material produced in quantity. This could be disseminated for selection in a range of locations, seasons, and environments and the products evaluated for local and more general utility.

The work of Yunus (1990) and Yunus and Jackson (1991) has indicated that interspecific hybridization could be used to augment, if only to a limited extent, the gene pool available for grasspea improvement. Until the currently accessible genetic resources are exhausted there is no present incentive to explore the wider gene pool of the remainder of the genus *Lathyrus*. Circumstances may however change and justify the higher levels of investment and expenditure entailed in state-of-the-art biotechnology.

Concluding Remarks

The laudable objective of developing the grasspea as a safe crop must look not only to the present state of agriculture and socio-economic conditions in the developing world but attempt to anticipate intelligently the probable progressive changes likely to occur in these societies. A phased program of improvement must be initiated and safe cultivars must be developed to meet the requirements of those whose need is greatest which can be produced successfully in farming systems currently practiced. These are by and large minimal input systems. If socio-economic development plans bear fruit and higher levels of inputs become feasible then there is scope for the development of improved cultivars responsive to higher standards of crop husbandry. One probably useful line of development could be selection of distinct forage and grain cultivars. It is commonplace at the present time for the crop to be grazed at least once and a grain crop harvested from the aftermath. This is a flexible system of utilizing the crop but not necessarily the most productive when agricultural conditions improve. [The human population incidentally would not be the sole beneficiary of improvement, livestock in famine conditions fed on high ODAP forage also develop pathological symptoms (Spencer *et al.*, 1986).] Specialized cultivars (for forage and grain production) with the common characteristic of low seed ODAP contents could usefully be developed at the present time: the current demand for them, though small at present, could be expected to expand.

These developments must be considered in their likely socio-economic context. Lathyrism, though regarded as a problem of the Indian sub-continent and Ethiopia at the present time, has been experienced in other areas and even in Europe during World War II. There is a wider awareness of such problems internationally and international response to famine crises can alleviate the most extreme famine conditions even in remote geographic areas. Where political conditions permit, food aid is likely to reach lathyrism-prone areas before the disease reaches epidemic proportions. However shortages of other grain crops, especially pulses such as chickpea, can lead to adulteration of their products by the grasspea which is used as filler or extender. This practice has been viewed with some concern as being in effect an exporting of the lathyrism problem to areas in which it was not endemic. This has provided some impetus to those seeking to ban its production and trade. While clearly undesirable in principle in practice it is not a problem provided that adulteration is not excessive.

The hazard of lathyrism can be exacerbated by the disruption caused by civil disturbance and war. In these circumstances farmers may be prevented from producing crops which require a high level of husbandry and become more dependent on those such as grasspea which are less exacting. Such a situation could increase the relative availability of the grasspea and hence the risk of lathyrism. Poverty pure and simple can be instrumental in promoting lathyrism independently of famine conditions. When foods have to be purchased by the poverty stricken, the cheapest foods (including grasspea) would be favored even when less hazardous alternatives were available.

The first step of producing safe grasspea cultivars, to be fully effective, needs to be carried out not piecemeal but in a coordinated way under the aegis of dedicated organizations such as INILSEL and the Third World Medical Research Foundation with the cooperation of IBPGR, FAO, and collaborating organizations. In order to make the effort most effective, the closest collaboration between interested and involved national organizations must be developed with the freest possible exchange of information, materials, and personnel. In this way the grasspea could become an unqualified boon to many of the poorest in the Third World and not a potential bane to those at risk of developing lathyrism.

References

Barat, G. K., Ghose, C. and Singh, J. 1989. In: *The Grasspea – Threat and Promise*, pp. 122–127 (ed. P. S. Spencer). New York: Third World Medical Reserch Foundation.

Barrow, M. W., Simpson, C. F. and Miller, E. J. 1974. *The Quarterly Review of Biology* 49: 101–128.

Bharati, M. P. and Neupane, R. K. 1989. In: *The Grasspea – Threat and Promise*, pp. 159–167 (ed. P. S. Spencer). New York: Third World Medical Research Foundation.

Campbell, C. 1989. In: *The Grasspea – Threat and Promise*, pp. 139–146 (ed. P. S. Spencer). New York: Third World Medical Research Foundation.

Dahiya, B. S. 1986. In: *Lathyrus and Lathyrism*, pp. 161–168 (eds. A. K. Kaul and D. Combes). New York: Third World Medical Research Foundation.

Dwivedi, M. P. 1989. In: *The Grasspea – Threat and Promise*, pp. 1–26 (ed. P. S. Spencer). New York: Third World Medical Research Foundation.

Haque, A. and Mannan, M. A. 1989. In: *The Grasspea – Threat and Promise*, pp. 27–35 (ed. P. S. Spencer). New York: Third World Medical Research Foundation.

Harlan, J. R. and De Wet, J. M. J. 1971. *Taxon* 20: 509–517.

Kaul, A. K. and Combes, D. (eds.). 1986. *Lathyrus and Lathyrism*, 334 pp. New York: Third World Medical Research Foundation.

Kaul, A. K., Hamid, M. A. and Akanda, R. U. 1989. In: *The Grasspea – Threat and Promise*, pp. 41–54 (ed. P. S. Spencer). New York: Third World Medical Research Foundation.

Kaul, A. K., Islam, M. Q. and Hamid, A. 1986. In: *Lathyrus and Lathyrism*, pp. 130–141 (eds. A. K. Kaul and D. Combes). New York: Third World Medical Research Foundation.

Kupicha, F. K. 1981. In: *Advances in Legume Systematics*, pp. 377–381 (eds. R. M. Polhill and P. H. Raven). London: HMSO.

Kupicha, F. K. 1983. *Notes from the Royal Botanic Garden Edinburgh* 41: 287–326.

Lal, M. S., Agrawal, I. and Chitale, M. W. 1986. In: *Lathyrus and Lathyrism*, pp. 146–160 (eds. A. K. Kaul and D. Combes). New York: Third World Medical Research Foundation.

Lambein, F. and Kuo, Yu-Haey. 1991. The biosynthesis of BOAA finally resolved. *INILSEL Newsletter*, Spring 1991.

Lambein, F., Ongena, G. and Kuo, Y. H. 1990. *Phytochemistry* 29: 2793–2796.

MacDonald, J. F., Morris, M. E. and Miljkovic, Z. 1986. In: *Lathyrus and Lathyrism*, pp. 306–314 (eds. A. K. Kaul and D. Combes). New York: Third World Medical Research Foundation.

Murti, V. V. S. and Seshadri, T. R. 1964. *Current Science* 33: 323–329.

Nunn, P. B. 1989. In: *The Grasspea – Threat and Promise*, pp. 89–96 (ed. P. S. Spencer). New York: Third World Medical Research Foundation.

Pushpamma, P. 1989. In: *The Grasspea – Threat and Promise*, pp. 198–204 (ed. P. S. Spencer). New York: Third World Medical Research Foundation.

Quader, M., Ahad Miah, M. A., Wahiduzzaman, Md. and Rahman S. 1989. In: *The Grasspea – Threat and Promise*, pp. 152–158 (ed. P. S. Spencer). New York: Third World Medical Research Foundation.

Quader, M., Ramanujam, S. and Barat, G. K. 1986. In: *Lathyrus and Lathyrism*, pp. 93–97 (eds. A. K. Kaul and D. Combes). New York: Third World Medical Research Foundation.

Rao, S. L. N., Adiga, P. R. and Sharma, P. S. 1964. *Biochemistry* 3: 432–436.

Rathod, K. L. 1989. In: *The Grasspea – Threat and Promise*, pp. 168–174 (ed. P. S. Spencer). New York: Third World Medical Research Foundation.

Renfrew, J. M. 1973. *Palaeoethnobotany*, 248 pp. London: Methuen.

Ross, S. M. and Spencer, P. S. 1989. In: *The Grasspea – Threat and Promise*, pp. 97–108 (ed. P. S. Spencer). New York: Third World Medical Research Foundation.

Rutter, J. and Percy, S. 1984. *New Scientist* 23: 22–23.

Smartt, J. (ed.). 1990. In: *Grain Legumes – Evolution and Genetic Resources*, 379 pp. Cambridge, UK: Cambridge University Press.

Spencer, P. S. (ed.). 1989. *The Grasspea – Threat and Promise*. New York: Third World Medical Research Foundation.

Spencer, P. S., Roy, D. N., Palmer, V. S. and Dwivedi, M. P. 1986. In: *The Grasspea – Threat and Promise*, pp. 297–305 (ed. P. S. Spencer). New York: Third World Medical Research Foundation.

Yunus, A. G. 1990. Biosystematics of *Lathyrus* section *Lathyrus* with special reference to the grasspea, *L. sativus* L. Ph.D. thesis, University of Birmingham, UK.

Yunus, A. G. and Jackson, M. T. 1991. *Plant Breeding* 106: 319–328.

Climate change and biotic and abiotic stresses

Potential effects of global climate change on cool season food legume productivity

C. GRASHOFF[1], R. RABBINGE[2] and S. NONHEBEL[2]

[1] Centre for Agrobiological Research, (CABO-DLO), P. O. Box 14, 6700 AA Wageningen, The Netherlands, and
[2] Department of Theoretical, Production Ecology, Agricultural University, P. O. Box 430, 6700 AK Wageningen, The Netherlands

Abstract

A simulation study was done with a well-tested and validated model for crop growth and production of faba bean to evaluate the consequences of some aspects of climate change on yield and yield variability. The model used was a version of SUCROS87, including a water balance.

For three locations differing in climate (Tel Hadya, Syria; Migda, Israel; Wageningen, Netherlands) at least 8 years with detailed weather data were used to simulate the consequences of temperature rise and increase of atmospheric CO_2 (based on assessments of the Intergovernmental Panel on Climate Change [IPCC]), separately and combined. It appears that temperature rise causes a decrease in seed yield of rain-fed crops in Wageningen and Migda, due to a shortening of the growing season. At Tel Hadya, seed yield of rain-fed crops increases, due to an accelerated start of the reproductive phase and consequently an "escape" from water shortage later in the season. For fully irrigated crops, temperature rise causes, at all locations, a decrease of seed yield which is greatest at Migda and smallest at Tel Hadya. CO_2-enrichment causes, in all situations, an increase in growth and production of faba bean, which compensates for the decrease from temperature rise. The effects are not completely additive at all locations. Yield increases due to CO_2-enrichment are much higher than the yield decrease due to temperature rise. At Wageningen, Tel Hadya, and Migda the positive net effect of a CO_2 concentration to 460 ppm and a temperature increase of 1.7°C was respectively, 12, 68, and 28% for rain-fed crops and 5, 16, and 13% for fully irrigated crops. Fully irrigated crops show remarkably smaller yield variability than rain-fed crops in all these assessments. In rain-fed crops, the variation in yield over the years remains the same or is somewhat reduced due to reduced sensitivity to water shortage. Thus the net effects on productivity and stability due to the scenarios used for global climate change are at all locations positive. Other effects, such as, for example, morphological effects, may overrule these physiological effects. Such effects are not taken into account in this simulation study.

F.J. Muehlbauer and W.J. Kaiser (eds.), Expanding the Production and Use of Cool Season Food Legumes, 159–174.

Introduction

The increasing presence of atmospheric trace gases such as CO_2, CH_4, and N_2O due mainly to human activity, directly or indirectly, may influence the earth's climate by transmitting incoming solar radiation, while partly blocking outgoing terrestial black body radiation. The increased "greenhouse" effect may cause temperature rise. This may affect the functioning of various agro-ecosystems in general and faba bean production more specifically.

Different processes are influenced by the various factors that are affected by climate change. CO_2-increase affects the stomatal conductance and increases photosynthesis rate (Lemon, 1983; Cure and Acock, 1986) and water use efficiency (Gifford, 1979; Sionit *et al.*, 1980). Temperature rise may increase developmental rate of the crop, and adversely effect crop production. Evaluation of the effects which work in contrary directions with direct qualitative or quantitative methods is difficult. Crop growth simulation models may be used for such an evaluation as the causal relations between rate variables and forcing variables is present in such models. The consequences of CO_2-increase and temperature rise may be evaluated with these models.

Climate change may have strong effects on faba bean production, as this crop is very sensitive to water shortage and has high yield variability when grown in the present climate (Dantuma *et al.*, 1983; Grashoff, 1990a, 1990b). A feasibility study of effects of climate change on growth and production of faba bean is described.

Consistent Climate Change Assessments

With respect to simulation of the future climate, the same procedure as described in Nonhebel (1993) was used. Based on the Report of Working Group I to the Intergovernmental Panel on Climate Change (IPCC) (Houghton *et al.*, 1990) two scenarios were considered: for the year 2030 with a CO_2 concentration of 460 ppm and a temperature of 1.7°C above the present level, and for the year 2080 with a CO_2 concentration of 700 ppm and a temperature rise of 3°C. The changes in precipitation as estimated by the General Circulation Models (GCMs) are low (<10%) in comparison with the present inter-annual variability in precipitation, and were not taken into account.

Model Description

The simplified structure of the model[1] is shown in Figure 1.

Simulation of crop growth under potential growth situations is done with a

---

[1] A complete, documented listing of the simulation model and a quantification of the parameters and functions specific for faba bean are available from the authors.

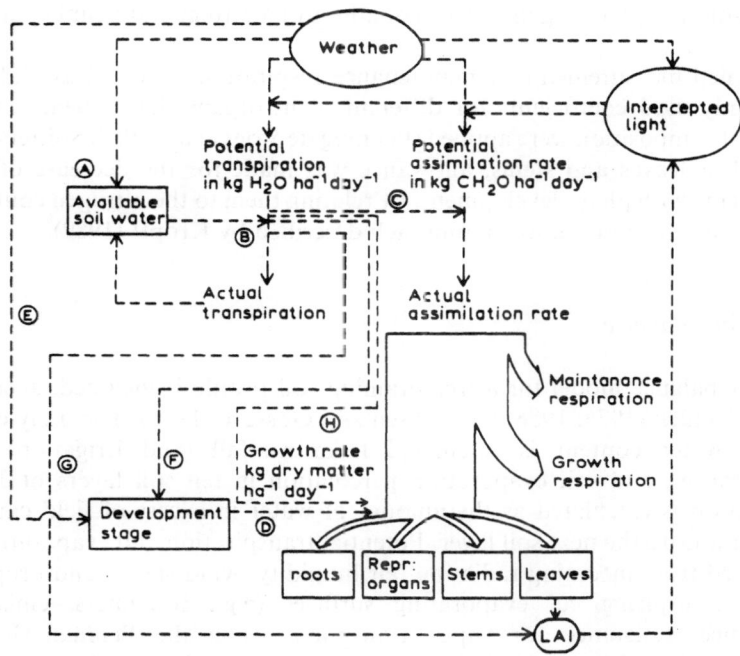

Figure 1. Schematic representation of faba bean growth. Effects of TRAN/TRP: on rate of photosynthesis (C); on developmental rate (F); on death rate of leaves and other plant organs (G); on dry matter partitioning (H). NB: the scheme provides an overview, not a relation diagram.

general crop growth simulator (Spitters *et al.*, 1989; Penning de Vries *et al.*, 1989). The developmental stage of the crop is simulated through integration of the rate of development, which is influenced by temperature. Assimilation rate is calculated from the incident amount of photosynthetically active radiation (PAR), the fraction of PAR intercepted by the canopy, and the photosynthesis/ light response curve of individual leaves. The total amount of carbohydrates available for growth is calculated from the assimilation rate, the reallocation and redistribution of dry matter, and losses due to respiration of the plant organs ("maintenance respiration"). Dry matter growth of the plant organs is determined by the total net carbohydrate production, the fraction of carbohydrates allocated to the plant organs and the values for conversion of carbohydrates into dry matter ("growth respiration"). The fraction of carbohydrates allocated to the organs is affected by the developmental stage of the crop (cf. Van Heemst, 1986). Leaf area growth is calculated from the dry matter allocated to the leaves and the specific leaf weight, as influenced by the developmental stage of the crop.

The parameters and functions specific for faba bean were mainly derived from experiments with faba bean in The Netherlands (Grashoff, 1990a,b). The parameters for growth respiration were calculated from analysis of chemical composition of plant organs, using equations of Vertregt and Penning de Vries (1987).

Detailed measurements of maintenance respiration were not available, so commonly used coefficients for different plant organs that depend on their chemical composition were applied (Penning de Vries *et al.*, 1989; Spitters *et al.*, 1989). For leaves and stems, allowance was made for the decrease of these coefficients with plant development, by relating them to the nitrogen content of these organs, according to an approach described by Kropff (1989).

Soil Water Balance

A water balance model for a free draining soil profile is included, according to Van Keulen (1975, 1986) and Jansen and Gosseye (1986). The daily change in soil water content is calculated from rainfall (and irrigation), crop transpiration, and soil evaporation percolation in ten soil layers of 10 cm. Percolation is calculated as the amount of water in excess of field capacity, which drains to the next soil layer. Potential transpiration and evaporation are calculated from incoming radiation, air humidity, wind speed, and properties of the transpiring cq. evaporating surfaces (e.g., roughness, emissivity, reflectance, resistance for vapor transport), using the Penman/Monteith equation (Monteith, 1965). Soil evaporation is reduced when the upper layer of the soil dries out. The model does not calculate capillary rise, however the relative contribution of each soil layer to the total evaporation is calculated following a mimic procedure (Van Keulen, 1975). In this concept, the relative contribution of each soil layer to the total evaporation decreases exponentially with the depth of the layer. The relative contribution of each soil layer to the total transpiration is based on the rooted length in each layer and the maximum water uptake per rooted length.

Crop-Water Relations

When the soil dries out, stomata close and transpiration falls below its potential value. The ratio between actual and potential transpiration (TRAN/TRP) is supposed to decrease linearly with soil water content from unity at a critical water content, to zero at wilting point (Feddes *et al.*, 1978). According to Doorenbos and Kassam (1979) and Driessen (1986), the critical soil water content at which water uptake is reduced, is typical for a group of species. This critical content is also affected by TRP and the leaf area index (LAI) and in this way, the rate of water uptake at a certain soil water content depends on the actual leaf area of the crop and on the evaporative demand. Based on the sensitivity of faba bean to water shortage, the crop belongs to the group with a rather "weak" extracting capacity, like clover and carrot (Driessen, 1986). The ratio TRAN/TRP linearly affects the actual assimilation rate (Figure 1), assuming a constant water use efficiency. From a physiological point of view, the ratio TRAN/TRP indicates the water availability to the crop.

Two other crop-water relations were included in the simulation model. First, water shortage accelerates the development of the faba bean crop. With water shortage, pod development begins earlier in the season and maturity also is reached earlier (Grashoff, 1990a). This accelerating effect was modelled by multiplying the developmental rate of the crop by a factor proportional to 1/(TRAN/TRP) (Grashoff and Stokkers, 1992). Second, a mild water shortage results in an increase of the relative part of assimilates which flows towards the reproductive organs (Grashoff, 1990a,b; Grashoff and Verkerke, 1991). This is included in the following way. The total daily assimilation decreases *linearly* with decreasing TRAN/TRP, but the daily relative part of assimilates which flows to vegetative organs decreases *exponentially* with decreasing TRAN/TRP. Hence, as water shortage results in a decrease of the ratio TRAN/TRP, an increasing part of assimilates cannot be incorporated in vegetative organs. This part is used for the developing pods and seeds (Grashoff and Stokkers, 1992).

Input Data

The input variables for the model are standard data of daily solar radiation, temperature, rainfall, air humidity, and wind speed; date of crop emergence and values of crop dry matter and leaf area index at emergence; rootable depth; some characteristics of soil water retention, such as total pore space, volumetric soil water content at field capacity, at wilting point, and at air-dry soil; and reflection coefficient for total radiation of the soil, average clod height, and extinction coefficient for soil evaporation.

Model Calibration, Validation, and Testing

The crop growth model was calibrated with a series of field experiments with varying water supply patterns in The Netherlands in 1977, 1980–1982, and 1988. Then, the reliability of the model was tested in a comparison of simulated yields with measured yields. For this test, we used a data set consisting of faba bean yields collected over a 14-year period (1975–1988) at one experimental farm at Wageningen, The Netherlands. Secondly, we used a data set which was collected over a two-year period (1985, 1986) at nine locations of the so-called "EC-Joint Faba Bean Trials" (Roskilde in Denmark, Dundee, Nottingham, and Cambridge in the United Kingdom, Wageningen in The Netherlands, Gottingen and Hohenheim in Germany, Vienna in Austria, and Dijon in France). For both data sets, the average seed yields over the years and/or locations and the standard deviation (used as a measure of variability) were simulated correctly. Moreover, linear regressions of measured versus simulated yields fitted through the origin, had slopes of almost one and accounted for up to 80% of the yield variation (Grashoff and Stokkers, 1992). An example of the model performance for the Wageningen data set 1975–1988 is presented in Figure 2. The model was tested also for autumn-sown faba bean under Mediterranean climate conditions, using weather conditions for Giza (Egypt) and with three levels of irrigation (Kropff and Schippers, 1986). The simulated seed yields varied between 2 t ha^{-1} (rain-fed) to 7 t ha^{-1} (fully irrigated). These yields are realistic when compared with results of field trials (Nassib *et al.*, 1984; Zahran, 1982). However, an evaluation of the model results in comparison with detailed data sets from field experiments in the Mediterranean region was not yet possible.

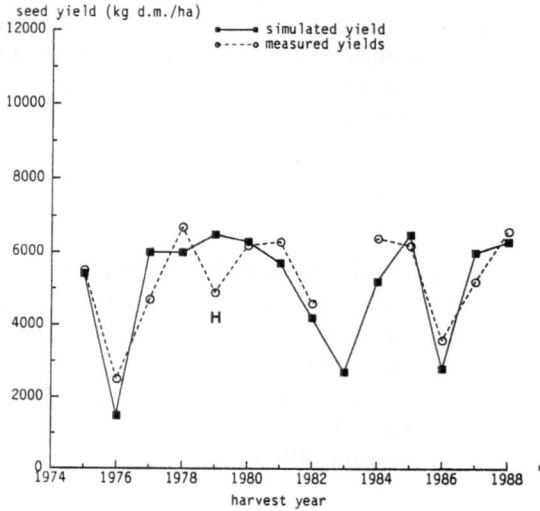

Figure 2. Simulated and measured seed yields at Wageningen 1975–1988. H = hail damage.

Robustness of the Model

In the above sections, some details about the reliability of the model were given. Besides this, with small changes of model parameters and/or input, simulation models should give consistent results. This is called the *robustness* of the model. Like all the models, derived from the SUCROS-stem, the faba bean model contains many negative feed-backs, which prevents the model from being hyper-sensitive to small changes in input. A point of attention is the effect of environmental factors on specific leaf weight (SLW), which is not yet included in the model. Lieth *et al.* (1986) found an increase in SLW of 9% with a doubling of the CO_2 concentration in soybean.

Adaptations of the Model for the Feasibility Study on Climatic Change

For the present study on the effects of climatic change, adaptations to the model had to be made for increased CO_2 concentration and increased temperatures as follows:

The Effect of Increased CO_2 Concentration

The effect of higher CO_2 levels on leaf photosynthesis was simulated according to Goudriaan and Unsworth (1990). The initial light use efficiency (EFF) and the maximum rate of photosynthesis (AMAX) are affected by the CO_2 concentration. At an average temperature of 20°C, a doubling of the CO_2 concentration results in an increase of EFF by 15% and a doubling of AMAX. A higher CO_2 concentration also can affect stomatal resistance and consequently transpiration. Based on Goudriaan and Unsworth (1990) it is assumed that faba bean follows the most common response for C3 plants. This means that the ratio of intracellular to external CO_2 concentration is stabilized. With a doubling of external CO_2 concentration, the stomatal aperture is reduced very little, combined with a strong response of assimilation. Typically in C3 plants transpiration will be reduced by 10 to 20% and assimilation stimulated by 40%. By consequence, the water use efficiency (WUE) is considerably stimulated.

The Effect of Increased Temperature

This effect was simulated by adding the estimated temperature rise to the daily input data on minimum and maximum temperature. The vapor pressure was adjusted in such a way that the relative humidity of the air was kept at the original value (Nonhebel, 1993).

Simulation Runs

For each of the scenarios 2030 and 2080, two conditions were distinguished: "rain-fed" crops and "fully irrigated" crops (which means no water shortage during the whole growing season). For rain-fed crops, the following simulation runs were made: one with the present weather situation, one run with the temperature rise only, one with the increased CO_2 concentration only, and one with both effects combined. Next, the same runs were made for fully irrigated crops. Other weather variables (radiation, windspeed) were not changed.

Historic Weather Data as Basis for the Feasibility Study

As input for the following simulation studies, weather and soil data sets for three locations were used. Set 1: 14 years (1975–1988) for Wageningen, Netherlands (Meteorological Station of the Agricultural University) combined with a clay soil data set for the experimental farm "De Bouwing" of the Centre for Agrobiological Research (CABO-DLO), Wageningen. Set 2: 8 years (1979/80–1985/86, 1989/90) for Tel Hadya, Syria (Meteorological Station of ICARDA) combined with an estimated soil data set for the heavy clay soil of the ICARDA Experimental Farm at Tel Hadya (Harmsen *et al.*,1983; Penning de Vries *et al.*, 1989). Set 3: 13 years (1962/63, 1963/64, 1965/66, 1966/67, 1969/70–1973/74, 1976/77–1979/80) for Migda, Israel (Gilat Meteorological Station) combined with a sandy soil data set for the Migda Experimental Farm (Van Keulen and Seligman, 1987).

All simulation runs were started at crop emergence, with the soil assumed to be at field capacity. For Wageningen, measured dates of emergence were used as input. On average, the date of emergence of the spring-sown crops in Wageningen is day 125 (5 May). As measured dates were not available for Tel Hadya and Migda, an estimated and fixed date of emergence for these autumn-sown crops was used: day 330 (26 November) in all the runs. This agrees with earlier simulation studies made by Kropff and Schippers (1986).

Results of the Feasibility Study

Effects of Temperature Rise and CO_2-Enrichment on Average Yields

Temperature Rise

Table 1 presents the general results of the simulation runs. In fully irrigated crops, a temperature increase of 3°C resulted in a general decrease of average seed yield at all locations. The strongest reduction was found in Migda (23%), the smallest in Tel Hadya (17%). This general reduction is mainly due to the fact that higher temperatures cause a 15 to 25 day shortening of the growing season. In rain-fed crops, the effect is more complicated: at Wageningen and Migda, seed yield was reduced 25 and 21%, respectively, but at Tel Hadya, seed yield increased 29%. This is due to the fact that the present temperature in Tel Hadya during the phase of vegetative growth of the crop (in autumn/winter) is about 5°C lower than in the same period at Migda and up to 8°C lower than at the comparable period for spring-sown crops at Wageningen (Figure 3). Due to this low temperature, pod filling starts much later at Tel Hadya than at Wageningen and Migda (Figure 3). Due to rapidly developing water shortage, the period of pod filling is short. Temperature increases of 1.7 and 3.0°C enhance the start of pod filling by 15 and 30 days, respectively. Consequently, the total period of pod filling will be longer than in the present situation, which causes an increase of

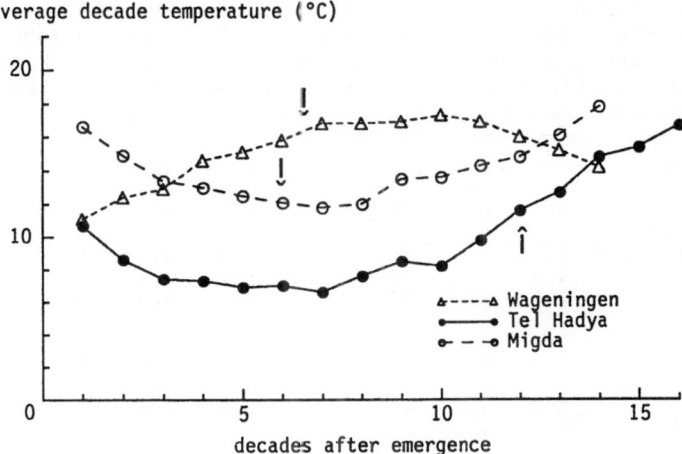

Figure 3. Comparison of present temperatures after crop emergence at Wageningen, Tel Hadya, and Migda. The arrows (→; ↓) in the Figure indicate the start of pod filling. NB: in the figure, "decade" is used for a 10-day period.

seed yields. However, the physiological parameters and functions used in the model are derived from experimental data of the cultivar Minica, one of the most productive West European cultivars and may be different for cultivars used in the Mediterranean region.

CO_2-Enrichment

Increased CO_2 enhanced average seed yields of the fully irrigated crops at Wageningen, Tel Hadya, and Migda by 48, 54, and 72%, respectively (Table 1). In these simulations, the total transpiration of the crops was reduced by 10% to 20%. This agrees with results of Goudriaan and Unsworth (1990). The increase in assimilation, with a concomitant decrease in transpiration, resulted in a large increase of water use efficiency (WUE). This explains why the effects of CO_2 increase are greater in the rain-fed crops: at Tel Hadya and Migda the rain-fed yields were more than doubled (Table 1) while at Wageningen, a 61% yield increase was found.

Combined Effects of Warming and CO_2

When temperature and CO_2 increase were included together in the simulations, a completely consistent increase of average yields of rain-fed and fully irrigated crops resulted. At Wageningen, Tel Hadya, and Migda the increase was 12, 68, and 28%, respectively, for rain-fed crops and 5, 16, and 13%, respectively, for fully irrigated crops in scenario 2030. In scenario 2080, the increments at these locations were 31, 164, and 72%, respectively, for rain-fed crops and 18, 38, and

168   *C. Grashoff* et al.

Table 1. Average seed yields (y) in t ha^{-1}, absolute standard deviation (s) in t ha^{-1} and relative standard deviation s (%) in %, for standard simulations, simulations with temperature rise (T + 1.7 and T + 3.0), simulations with increased CO_2 concentration (C460 and C700), and simulations with temperature rise and CO_2-increase combined in two scenarios for 2030 and 2080 (Sc 2030 and Sc 2080). All simulations are presented for rain-fed and fully irrigated crops

Locations and	rain-fed			fully irrigated		
	y	s	s(%)	y	s	s(%)
De Bouwing						
'standard sim'	5.1	1.6	32	6.1	0.5	9
T +1.7	4.4	1.4	32	5.2	0.3	7
T +3.0	3.8	1.2	32	4.7	0.3	5
C 460	6.4	1.8	28	7.4	0.7	9
C 700	8.2	2.0	24	9.0	0.8	9
Sc 2030	5.7	1.7	29	6.4	0.4	7
Sc 2080	6.7	1.5	22	7.2	0.4	6
Tel Hadya						
'standard sim'	2.8	1.1	41	6.9	0.8	11
T +1.7	3.5	1.3	37	6.4	0.4	7
T +3.0	3.6	1.2	32	5.7	0.3	6
C 460	3.6	1.5	43	8.5	1.1	12
C 700	6.2	2.5	40	10.6	1.5	14
Sc 2030	4.7	1.7	35	8.0	0.6	8
Sc 2080	7.4	1.6	22	9.5	0.6	6
Migda						
'standard sim'	3.9	1.5	39	6.4	0.6	9
T +1.7	3.5	1.5	42	5.2	0.6	12
T +3.0	3.1	1.3	41	4.3	0.6	14
C 460	5.4	2.1	40	8.4	0.6	7
C 700	7.9	2.8	36	11.0	0.7	7
Sc 2030	5.0	2.0	41	7.2	0.7	10
Sc 2080	6.7	2.4	35	8.5	0.9	11

33%, respectively, for fully irrigated crops. The negative effect of temperature rise is more than compensated by the positive effect of CO_2 increase, both for the 2030 and for the 2080 scenario (Table 1). The net effect of temperature and CO_2 agrees almost completely with the addition of the separate effects (Table 1).

Effects of Temperature Rise and CO_2-Enrichment on Stability in Yield Responses

As indicators for yield stability, the absolute standard deviation and the relative standard deviation of the simulated yield series are used. The relative standard deviation is defined as the absolute standard deviation, expressed as a percentage of the average yield. Stable yields require both an absolute and a relative standard deviation that are small.

Effects of Climate Change on Stability of Rain-Fed Crops

With temperature rise, the absolute standard deviations decrease with about 0.3 t ha^{-1} (at Wageningen and Migda) or stay the same (at Tel Hadya) (Table 1). At Tel Hadya, this unchanged absolute standard deviation and the increased average yield result in a clear decrease of 9% in the relative standard deviation (Table 1). At Wageningen and Migda, the combination of decreased average yields and decreased absolute standard deviations results in almost unchanged relative standard deviations. The effect of increased temperature for individual years at Wageningen is illustrated in Figure 4. It clearly shows that temperature rise reduces yields in the "high-yielding years" (the years with adequate precipitation). The reduction is due to a shortening of the growing season. In the "low yielding years" (the years without adequate precipitation) the negative effect is smaller. In those years, a part of the negative effect is compensated by an earlier start of pod filling and a consequent escape from water shortage later in the season. In a year such as 1976, the net effect of temperature rise will be positive. Figure 4 shows that the stabilizing effect of temperature rise is a stabilization in the "wrong direction" as it reduces the larger yields.

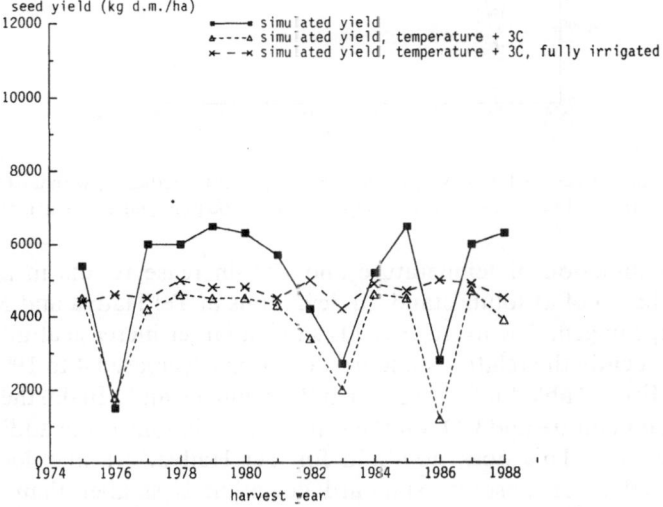

Figure 4. Simulated seed yields at Wageningen with 1) present weather, 2) with a temperature rise of 3.0°C and a rain-fed situation, 3) with a temperature rise of 3°C and a fully irrigated situation.

With CO_2-increase, the absolute standard deviations increase (with 0.4–1.4 t ha^{-1}) at all locations. At Wageningen and Migda, this increase is smaller than the increase in average yields. Thus, the relative standard deviations decrease about 9% at these locations. At Tel Hadya, the relative standard deviation remains unchanged (Table 1). The effect of increased CO_2 for the individual years at Wageningen is illustrated in Figure 5. The large positive effect on seed yield is present even in drier years such as 1976, 1983, and 1986. This is due to the increased water use efficiency with CO_2 increase. Figure 5 shows that, in the present situation, the yields at Wageningen vary from 1.5 t ha^{-1}–6.5 t ha^{-1}. With CO_2 increase, seed yields vary from 2.5–10.5 t ha^{-1}, which means a range of 8 t ha^{-1}. Figure 5 shows that, although the relative standard deviation decreases with CO_2-increase (Table 1), the absolute variation in seed yields increases.

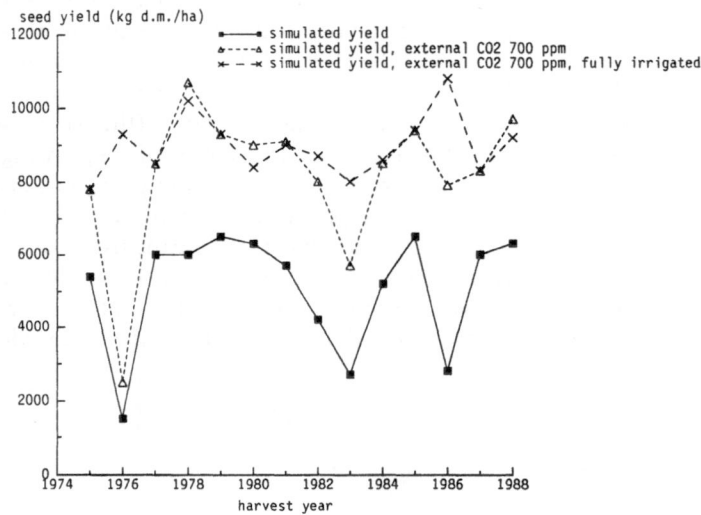

Figure 5. Simulated seed yields at Wageningen with 1) present weather, 2) with a CO_2-increase to 700 ppm and a rain-fed situation, 3) with a CO_2-increase to 700 ppm and a fully irrigated situation.

The combination of temperature and CO_2-increase results in an increase (0.5–0.9 t ha^{-1}) of absolute standard deviations at Tel Hadya and Migda, but not at Wageningen. The average yields show a larger increase at all locations, and consequently the relative standard deviations decreased 4 to 19% at all the locations. From Table 1 it follows that at Wageningen and Migda, the combined effect of temperature and CO_2 on the standard deviations is the addition of the separate effects. This does not hold for Tel Hadya. At that location, the combined effect on absolute standard deviation is smaller than the added separate effects. In contrast, the combined effect on relative standard deviation is much larger than the added separate effects.

In general, the explored climate changes in Table 1 stabilize relative variations in seed yields. It is uncertain if this can be seen as a real stabilizing effect. Figure 6 illustrates this for the individual years at Wageningen: although the relative standard deviation is reduced, the relatively small yields in dry years e.g., 1976, will remain in the rain-fed crops. In the present situation, yields vary from 1.5–6.5 t ha^{-1}, which is a range of 5 t ha^{-1}. In the 2080 simulation, the yields of the rain-fed crops vary from 3.0–8.5 t ha^{-1}, which is a range of 5.5 t ha^{-1} (Figure 6). This small increase in yield range and the increase in absolute standard deviation indicate that absolute yield variability slightly increases under rain-fed conditions; however, the relative standard deviation decreases.

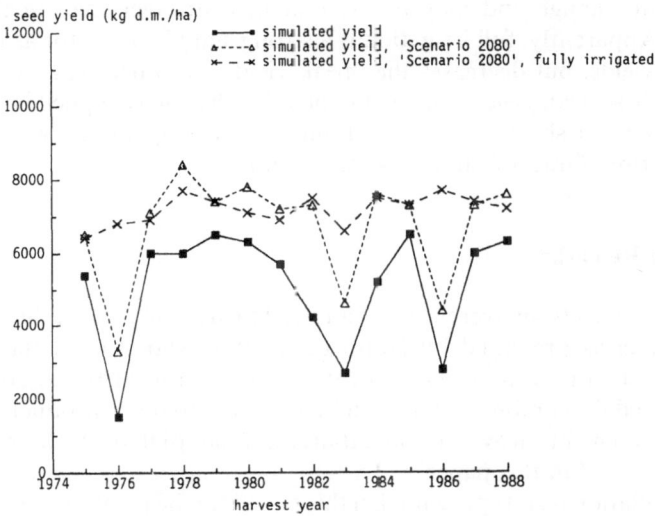

Figure 6. Simulated seed yields at Wageningen with 1) present weather, 2) with a scenario 2080 and a rain-fed situation, 3) with scenario 2080 and a fully irrigated situation.

Effects of Climate Change on Stability of Fully Irrigated Crops

The effects of temperature and CO_2 on yield variation in fully irrigated crops are of minor importance. In these crops, the separate effects of temperature rise, increase of CO_2, and the combination result only in minor increases or decreases of absolute and relative standard deviations (Table 1). In fact, the most important factor is the stabilizing effect of irrigation itself. Over all the simulations for fully irrigated crops, the absolute standard deviations (0.3 to 1.5 t ha^{-1}) are much lower than for the rain-fed crops (1.1 to 2.8 t ha^{-1}). Also the relative standard deviations (9 to 14%) are much lower than for the rain-fed crops (22 to 43%) (Table 1). This shows the general stabilizing effect of enhanced water supply on faba bean yields and is in agreement with earlier experimental and simulation studies (Grashoff, 1990a,b; Grashoff and Verkerke, 1991; Grashoff and Stokkers, 1992). Apparently, this stabilizing effect of water supply remains very large in the assessments dealing with climate change.

The stabilizing effect of irrigation in the changed climate is illustrated in Figure 6 for the individual years at Wageningen. On one hand, it shows that yields of fully irrigated crops in dry years (e.g., 1976 and 1986) will be much higher than those of rain-fed crops, due to the higher availability of water. On the other hand, it shows that yields of fully irrigated crops will be slightly lower than those of the rain-fed crops in years such as 1978, 1980, 1981, and 1988. This is due to the fact that faba bean needs a mild water shortage during flowering for optimum dry matter partitioning and optimum seed yields (Grashoff and Stokkers, 1992). In years with such a pattern of precipitation, this effect will stay with climate change and may be even slightly stronger than in the present situation. Apparently, full irrigation is a strong stabilizing factor as it increases the lower yields, but decreases the "peak yields" in other years, which is not efficient. These peak yields can be obtained if irrigation is applied in such a way that a mild water shortage is allowed during flowering and no water shortage after flowering (Grashoff and Stokkers, 1992).

Concluding Remarks

The assessed effects on average yield and yield variability are in the same order of magnitude as presented by Nonhebel (1993), who studied the combined climate effects for spring wheat with the scenarios for 2030 and 2080. Jansen (1990) studied the combined climate effects on potential rice production in Asia. He found increments in average potential seed yields of 0 to 10% for comparable scenarios as used in this paper.

The simulation results presented in this paper are the results of a survey of the sensitivity of the crop/weather system to changes in temperature and CO_2 concentration. CO_2-increase and temperature rise are regarded as the most important aspects of climate change. This paper shows that the direct CO_2 effect on assimilation and thus on crop production is substantial and thus essential in assessments of crop production under climate change. When only the temperature effects were studied, it is concluded that the climate change induced by the greenhouse gases would result in a yield decline in most cases. However, the combined effect of temperature rise on development and of CO_2-increase on assimilation is resulting in a yield increase on all sites. This shows that simulation models which include a synthesis of the most important physiological processes on crop growth are a valuable tool to explore the possible effects of climate changes on crop growth, yields and yield stability. It emphasizes also, the need for validation of these models with experiments of crop growth under increased atmospheric CO_2 concentrations. In this validation, two aspects have to be considered. Firstly, the positive effect of CO_2-increase on AMAX (maximum rate of photosynthesis), which we used, is a maximum possible effect (J. Goudriaan, pers. comm.). Measured positive effects on AMAX and thus on photosynthesis may be smaller. Secondly, attention has to be paid to the possible effects of CO_2-increase on morphological characters such as specific leaf weight (SLW). These effects are neglected in the present study. Usually SLW increases with CO_2-increase, which results in a slower increase of LAI. This effect is small in experiments with comparable crops such as soybean (Lieth *et al.*, 1986). However, when CO_2-experiments with faba bean show smaller positive effects on AMAX and important increments of SLW, this will result in a tendency towards smaller positive effects of CO_2-increase on crop growth than assessed in this paper.

Climate change may also affect other weather variables such as global radiation, precipitation, vapor pressure, and wind speed. Based on the IPCC reports, it is assumed in this paper that changes in radiation, relative humidity, precipitation, and wind speed are of minor importance in comparison with the effects of CO_2-increase and temperature rise. However, precipitation has a strong regional character, and when the GCMs are able to give reliable assessments with respect to possible changes in (regional) precipitation in the future (Houghton *et al.*, 1990), a detailed analysis is needed. As soon as these studies might assess major changes in precipitation, radiation, or windspeed in individual regions, the results can be included in the presented type of crop growth models to continue the study of regional aspects of climate change on crop growth and yields. Once more, this emphasizes the usefulness of these types of models in assessments of future yields.

From this paper the following conclusions are drawn. In the first place, the positive effect of CO_2-increase on average yields of faba bean more than compensates for the negative effect of temperature rise. Based on these effects, a general increase of future average seed yields is assessed. Secondly, the effects of climate change on yield variability are not completely equal at all locations, but for both scenarios, 2030 and 2080, the relative variability shows a tendency to decrease. Nevertheless, in the scenarios for 2030 and 2080, faba bean remains relatively sensitive to water shortage, and irrigation will remain the most effective factor in stabilizing yields.

Acknowledgements

The authors wish to thank Dr. Ir. J. Goudriaan (TPE-LUW) and Dr. Ir. A. J. Haverkort (CABO-DLO) for helpful criticism of the manuscript.

References

Cure, J. D. and Acock, B. 1986. *Agriculture and Forestry Meteorology* 38: 127–145.
Dantuma, G., Von Kittlitz, E., Frauen, M. and Bond, D. A. 1983. *Zeitschrift für Pflanzenzüchtung* 90: 85–105.
Doorenbos, J. and Kassam, A. H. 1979. *Actual evapotranspiration*, 193 pp. FAO irrigation and drainage paper 33. Rome, Italy: FAO.
Driessen, P. M. 1986. In: *Modelling of Agricultural Production: Weather, Soils and Crops*, pp. 182–193 (eds. H. van Keulen and J. Wolf). *Simulation Monographs*. Wageningen, The Netherlands: Pudoc.
Feddes, R. A., Kowalik, P. J. and Zaradny, H. 1978. In: *Simulation of Field Water Use and Crop Yield*, 189 pp. *Simulation Monographs*. Wageningen, The Netherlands: Pudoc.

Gifford, R. M. 1979. *Australian Journal of Plant Physiology* 6: 367–378.
Goudriaan, J. and Unsworth, M. H. 1990. In: *Impact of Carbon Dioxide, Trace Gases, and Climate Change on Global Agriculture*, pp. 111–129. ASA special publication no 53. Madison, Wisconsin, USA: American Society of Agronomy.
Grashoff, C. 1990a. *Netherlands Journal of Agricultural Science* 38: 21–44.
Grashoff, C. 1990b. *Netherlands Journal of Agricultural Science* 38: 131–143.
Grashoff, C. and Stokkers, R. 1992. *Netherlands Journal of Agricultural Science* 40: 447–468.
Grashoff, C. and Verkerke, D. R. 1991. *Netherlands Journal of Agricultural Science* 39: 247–262.
Harmsen, K., Shepherd, K. D. and Allan, A. Y. 1983. In: *Proceedings of the 17th Colloquium of the International Potash Institute*, pp. 223–248. Bern, Switzerland: The Institute.
Houghton, J. T., Jenkins, G. J. and Ephraums, J. J. 1990. In: *Climate Change, The IPCC Scientific Assessment*. Cambridge, UK: Cambridge University Press.
Jansen, D. M. 1990. *Netherlands Journal of Agricultural Science* 38: 661–680.
Jansen, D. M. and Gosseye, P. 1986. *Simulation Reports CABO-TT* 10. 108 pp. Wageningen, The Netherlands.
Kropff, M. J. 1989. *Quantification of SO₂ Effects on Physiological Processes, Plant Growth and Crop Protection*, 210 pp. Ph.D. Thesis, Agricultural University, Wageningen, The Netherlands.
Kropff, M. J. and Schippers, P. 1986. In: *Biology and Control of Orobanche. Proceedings of a Workshop in Wageningen, The Netherlands*, pp. 70–79 (ed. S. J. ter Borg). Wageningen, The Netherlands: LH/VPO.
Lemon, E. R. 1983. *CO₂ and Plants, The Response of Plants to Rising Levels of Atmospheric Carbon Dioxide*, 280 pp. Boulder, Colorado, USA: Westview Press.
Lieth, J. H., Reynolds, J. F. and Rogers, H. H. 1986. *Field Crops Research* 13: 193–203.
Monteith, J. L. 1965. *Proceedings of Symposiums of the Society of Experimental Biology* 19: 205–234.
Nassib, A. M., Hussein, A. H. A. and El Rayes, F. M. 1984. *FABIS Newsletter* 10: 11–15.
Nonhebel, S. 1993. *The Importance of Weather Data in Crop Growth Simulation Models and Assessment of Climatic Change Effects*, 144 pp. Ph.D. Thesis, Agricultural University of Wageningen, The Netherlands.
Penning de Vries, F. W. T., Jansen, D. M., Ten Berge, H. F. M. and Bakema, A. 1989. *Simulation monographs*, 271 pp. Wageningen, The Netherlands: Pudoc.
Sionit, N., Hellmers, H. and Strain, B. R. 1980. *Crop Science* 20: 456–458.
Spitters, C. J. T., Van Keulen, H. and Van Kraalingen, D. W. G. 1989. In: *Simulation and Systems Management in Crop Protection*, pp. 147–181 (eds. R. Rabbinge, S. A. Ward and H. H. van Laar). *Simulation Monographs*. Wageningen, The Netherlands: Pudoc.
Van Heemst, H. D. J. 1986. *Potato Research* 29: 55–66.
Van Keulen, H. 1975. *Simulation Monographs*, 175 pp. Wageningen, The Netherlands: Pudoc.
Van Keulen, H. 1986. In: *Modelling of Agricultural Production: Weather, Soils and Crops*, pp. 130–152 (eds. H. van Keulen and J. Wolf). *Simulation Monographs*. Wageningen, The Netherlands: Pudoc.
Van Keulen, H. and Seligman, N. G. 1987. *Simulation monographs*, 310 pp. Wageningen, The Netherlands: Pudoc.
Vertregt, N. and Penning de Vries, F. W. T. 1987. *Journal of Theoretical Biology* 128: 109–119.
Zahran, M. K. 1982. In: *Faba Bean Improvement*, pp. 191–198 (eds. G. Hawtin and G. Webb). The Hague, The Netherlands: Martinus Nijhoff.

Biotic and abiotic stresses constraining productivity of cool season food legumes in Asia, Africa and Oceania

C. JOHANSEN[1], B. BALDEV[2], J.B. BROUWER[3], W. ERSKINE[4], W.A. JERMYN[5], L. LI-JUAN[6], B.A. MALIK[7], A. AHAD MIAH[8] and S.N. SILIM[9]

[1] ICRISAT, Patancheru P. O. Andhra Pradesh 502 324, India;
[2] Indian Agricultural Research Institute (IARI), New Delhi 110 012, India;
[3] Victorian Institute for Dryland Agriculture, Department of Agriculture, Natimuk Road, Private Bag 260, Horsham, Victoria 3401, Australia;
[4] ICARDA, P.O. Box 5466, Aleppo, Syria;
[5] Crop Research Division, Department of Scientific and Industrial Research (DSIR), Canterbury Agriculture and Science Centre, Private Bag, Christchurch, New Zealand;
[6] Faba Bean Germplasm and Breeding, Zhejiang Academy of Agricultural Sciences, Hangzhou, People's Republic of China;
[7] Pakistan Agricultural Research Council (PARC), National Agricultural Research Centre, P.O. National Institute of Health, National Park Road, Islamabad, Pakistan;
[8] Bangladesh Agricultural Research Institute (BARI), Joydebpur, Gazipur 1701, Bangladesh, and
[9] Eastern Africa Regional Cereals and Legumes Program (EARCAL), ICRISAT, c/o OAU/STRC, J.P. 31 SAFGRAD, P.O. Box 39063, Nairobi, Kenya

Abstract

We attempt to categorize on the basis of current knowledge, the most important biotic and abiotic stresses faced by chickpea, lentil, pea, faba bean, and grasspea in their major production regions in Asia, Africa, and Oceania. Foliar diseases rank high overall as major contributors to yield loss, whereas root diseases tend to assume greater importance at lower latitudes. Despite the obvious signs of damage caused by various insect pests, their impact on yield is less. Among abiotic stresses, drought is a major factor in reducing yield. The cool season food legumes are also very sensitive to even minor imbalances in soil nutrients. For each stress we considered the prospects for alleviating the problem. Although availability of genetic resistance to the major constraints of foliar diseases and drought have so far proven to be limited, we propose that they offer attractive opportunities for research investment. Redesign of canopies such that they are less conducive to establishment and growth of fungal diseases, the use of short-duration cultivars to escape drought, and the incorporation of particular drought-resistant traits are some of the options worth pursuing. The grasspea appears well-adapted to a range of biotic and abiotic stress conditions and we suggest that a greater understanding of this crop's adaptive traits would give indications of how best to improve stress tolerance of the other cool season food legumes.

F.J. Muehlbauer and W.J. Kaiser (eds.), Expanding the Production and Use of Cool Season Food Legumes, 175–194.
© 1994 Kluwer Academic Publishers.

Introduction

Grain yields in farmers' fields of the legumes under consideration, chickpea (*Cicer arietinum* L.), lentil (*Lens culinaris* Medik.), pea (*Pisum sativum* L.), faba bean (*Vicia faba* L.), and grasspea or lathyrus (*Lathyrus sativus* L.), are usually well below their established potential yields. This can invariably be attributed to their sensitivity to particular biotic and abiotic stresses. In Asia and Africa, these legumes are mainly grown as subsistence crops, with their areas of cultivation declining over time in many places, despite increasing demand for their grain. The unpredictable nature of many of the stresses affecting these legumes discourages their cultivation. On the other hand, rapid increases in commercial production of several of these legumes have been recorded in some regions, such as in Australia and New Zealand, often revealing new types of constraints where the crop was previously not cultivated.

This paper attempts to summarize our current knowledge of the main biotic and abiotic constraints facing each of these legumes in the major producing countries of Asia, Africa, and Oceania. We try not only to estimate the severity of each constraint but also to suggest the potential for alleviation of the stress. Of course, much of our knowledge is incomplete and subjective judgements have been attempted. Nevertheless, we hope that such a compilation will provide guidance in setting priorities for research and development in the major production regions for these legumes.

Methodology

The biotic and abiotic constraints that we consider are listed in Table 1. We have also suggested the appropriate means of alleviation of these constraints; genetic, management, or a combination of these.

In the subsequent tables, we have ranked the degree of limitation posed by each constraint as follows:

1 = Severe yield reducer – greater than 50% yield loss from crop potential in some years at least.

2 = Moderate yield reducer – 15 to 50% yield loss across years.

3 = Minor yield reducer – less than 15% yield loss in any year.

? = Problem suspected but status unknown.

– = Known to be not a problem.

Blank space = No knowledge concerning the problem.

We have tried to estimate the potential for alleviation of each constraint, through a concerted research and extension effort, according to the following designations:

A = High – regional production breakthrough probable in the medium term (e.g., 3 to 7 years).

B = Moderate – production breakthrough possible over the longer term (e.g., > 7 years).

C = Low – only marginal improvement expected ever or substantial improvement only after a decade or more.

We have only considered constraints in the major producing countries for these legumes, according to FAO statistics (FAO, 1991), to give a broad picture for different regions. We recognize that lesser production levels may be of particular economic and social importance in specific countries and that a different constraint scenario than the one we generally present may apply in these cases. We also consider only constraints to cultivars generally adapted to the region of concern; for example, cultivars having the appropriate photoperiod response and phenology.

Table 1. Biotic and abiotic constraints to the cool season food legumes considered, their abbreviation, and proposed means of alleviation

Constraint	Abbrev-iation	Means of alleviation[1]
Biotic		
Ascochyta blight (*Ascochyta* spp.)	ASC	g/m
Botrytis gray mold (*Botrytis* spp.)	BOT	g/m
Chocolate spot (*Botrytis fabae*)	CHS	g/m
Stemphylium blight (*Stemphylium* spp.)	STE	m/g
Powdery mildew (*Erysiphe polygoni* in pea and lentil)	PMI	g/m
Downy mildew (*Peronospora* spp.)	DMI	g/m
Rust (*Uromyces* spp.)	RUS	g
Anthracnose	ANT	g
Phoma stem blight (*Phoma medicaginis* var. *pinodella*)	PSB	m/g
Septoria blotch (*Septoria pisi*)	SEP	m/g
Sclerotinia rot (*Sclerotinia sclerotiorum*)	SCT	g/m
Fusarium wilt (*Fusarium oxysporum*)	FUS	g/m
Sclerotium rolfsii	SCL	g/m
Dry root rot (*Rhizoctonia bataticola*)	DRR	g/m
Root rot (*Aphanomyces euteiches*)	AMY	g/m
Damping off (*Pythium* and *Rhizoctonia* spp.)	DOF	m
Phytophthora root rot (*Phytophthora megasperma* f.sp. *medicaginis*)	PHY	g/m
Viral diseases	VIR	g/m
Nematodes	NEM	g/m
Helicoverpa pod borer (*Helicoverpa* spp.)	HEL	m/g
Leaf miner (*Liriomyza cicerina* in chickpea	LMI	g/m
Aphids (e.g. *Aphis craccivora*)	APH	m/g
Thrips	THR	m/g
Sitona weevil (*Sitona lineatus*)	SIT	g/m
Cutworm (*Agrostis* spp.)	CUT	g/m
Army worm (*Spodoptera* spp.)	ARW	m
Pea stem fly (*Ophiomya phaseoli*)	PST	m
Pea pod borer (*Etiella zinckenella*)	ETZ	m
Red-legged earth mite (*Halotydus destructor*)	RLE	m
Seed storage pests (e.g. *Callosobruchus* spp.)	SSP	m
Mammalian pests (e.g. rodents)	MAM	m
Weeds	WDS	m
Orobanche spp.	ORO	g/m

Table 1. (Continued)

Constraint	Abbrev-iation	Means of alleviation[1]
Abiotic		
Drought	DRO	g/m
Waterlogging	WLG	m/g
High temperature	HTP	g
Low temperature	LTP	g
Wind or hail (lodging)	WIN	g/m
Alkaline soil problems	HPH	m/g
Acid soil problems	LPH	m/g
Salinity	SAL	m/g
Nitrogen fixation	N	m/g
Phosphorus deficiency	P	m
Sulphur deficiency	S	m
Iron deficiency	FE	g/m
Boron deficiency	B	m
Boron toxicity	BT	g/m

[1] g   – Genetic improvement offers best prospects
   m   – Management improvement offers best prospects
   g/m – Combined g and m improvement options with g predominant
   m/g – Combined g and m improvement options with m predominant

Our estimates are based on cited published data where available but, in many cases, we rely on unpublished survey information and personal experience of the authors. We recognize that some of these estimates are predominantly judgmental, reflecting personal assessments of individuals, and perhaps based on different perceptions between regions. Indeed, this exercise has made us aware that we have much less quantified knowledge than we initially had thought.

Chickpea

A major yield constraint to chickpea in Asia, Africa, and Australia is posed by the foliar diseases, Ascochyta blight and Botrytis gray mold (Tables 2 and 3). But these diseases are confined to regions having the appropriate temperature and humidity conditions to permit disease development. In South Asia, for example, they are only found in specific areas in the sub-tropics and their incidence diminishes to insignificance at lower latitudes, with warmer and drier winters. Despite intensive screening efforts it is difficult to find substantial levels of host plant resistance, especially across biotypes of the pathogen (Nene and Reddy, 1987). However, progress in genetic enhancement of host plant resistance to Ascochyta blight in West Asia has been made (Reddy and Singh, 1992). These foliar diseases can be controlled by fungicides, but their widespread use in the largely subsistence farming systems of Asia and Africa where chickpea is grown is not a viable option.

Table 2. Ranking of constraints of chickpea and their potential for alleviation in the major producing countries of Asia[1]. Area and yield estimates (FAO, 1991) for 1990 are indicated

Production/ constraint	W. Asia			S. Asia					E. Asia
	Iran	Syria	Turkey	Bangla-desh	India	Myan-mar	Nepal	Pak-istan	China[2]
Area ('000 ha)	112	55	800	100	6495	134	28	1002	–
Yield (kg ha⁻¹)	723	660	1075	650	652	752	590	536	–
Biotic									
ASC		1B	1B	–	2B	–	–	1B	2B
BOT		–	–	1C	3C	–	2C	3C	
FUS		2B	2B	2A	1A	2A	2A	2A	2B
SCL				2B	3B	1B	–	?B	
DRR		3B		2A	2A	3A	3A	2A	3B
VIR		3C		3C	2B	–	–	3C	3C
NEM		2C		?C	3C	–	?C	3C	
HEL		2C	2C	2B	2B	2B	2B	2B	
LMI		2C	2C	–	–	3B	–	–	
APH		3C	3C	–	–	–	–	3B	
SSP		3A	3A	2A	3A	3A	3A	3A	
WDS		2A	3A	2B	3A	3B	3B	3A	
Abiotic									
DRO	1C	1B	3C	2B	1B	1C	3C	2B	2B
WLG				3C					
HTP	3C	3C	–	2C	2C	3C	–	3C	3C
LTP	3A	2A	2A	3C	2B	–	2C	2C	3C
WIN	–	–	–	3C	3C	?C	–	3C	
HPH	–	–	–	–	3C	–	–	3C	3C
LPH	–	–	–	3C	–	?C	3C	–	
SAL	2C	3C	–	–	2C	–	–	2C	3C
N	?A	3A	3A	3A	3A	3A	3A	3A	2A
P	3A	3A	3A	2A	3A	3A	3A	3A	
S				2A					
B				3A					

[1]   Only countries with area of chickpea exceeding 10,000 ha are considered.
[2]   Area and yield data not available but production of 250,000 t reported for China in 1990 (FAO, 1991). Constraint estimates based on 1991 tour report by Dr. Jagdish Kumar, ICRISAT.

Sources: Smithson *et al.*, 1985; Saxena, 1987; Saxena and Singh, 1987; Baldev *et al.* 1988; Summerfield, 1988; ICRISAT, 1990; BARI, 1991.

Root fungal diseases, particularly Fusarium wilt, are also universal yield reducers of chickpea (Tables 2 and 3). They assume greater importance at lower latitudes, where foliar diseases are not a problem. In this case, good sources of host plant resistance have been identified and resistant cultivars bred (Nene and Reddy, 1987). These diseases can also, to some extent, be minimized by management options, such as appropriate crop rotations to reduce soil inoculum level and application of fungicides to seeds at sowing.

In Australia, where chickpea is a recent introduction and is rapidly increasing in area under cultivation, a different suite of root diseases are found compared with those in traditional chickpea growing regions (Table 3).

Table 3. Ranking of constraints of chickpea and their potential for alleviation in the major producing countries of Africa and Oceania[1]. Area and yield estimates (FAO, 1991) for 1990 are indicated

Production/ constraint	N. Africa				E.Africa				Oceania
	Alg- eria	Egypt	Mor- occo	Tun- isia	Eth- iopia	Mal- awi	Tanz- ania	Uga- nda	Aust- ralia
Area ('000 ha)	60	8	77	45	130	50	75	7	94
Yield (kg ha⁻¹)	333	1750	766	622	962	500	333	500	1138
Biotic									
ASC	1B	–	1B	2B	3B	–	1B	1B	
BOT									2B
SCT									2B
FUS	3A	3A	3A	3A	2A	2A	1A	1A	2B
SCL					3C	?B	3B	3B	
DRR		3A			2A		3A	3A	2B
PHY									1A
DOF									2B
VIR					3C	?C	3C	3C	?C
NEM									2C
HEL					2C				2A
LMI	3B	–	3B	3B	–	–	–	–	
APH	?B	–	?B	?B	3C				
RLE									3A
SSP	3A	?A	3A	3A	?A	3A	3A	3A	
WDS	2A	?A	2A	2A	?A	3A	3A	3A	2B
Abiotic									
DRO	1B	–	1B	1A	3B	2B	1B	3B	2B
WLG	–	?A	–	–	?B	–	–	?B	2C
HTP	2C	2C	2C	2C	–	–	–	–	3C
LTP	3B	–	3B	3B	–	–	–	–	2B
WIN									3C
HPH	3C	2C	3C	3C	?C	–	–	–	
LPH	–	–	–	–	?C	?C	?C	2C	1C
N	3A	3A	3A	3A	?A	3A	3A	3A	3B
P									2A
FE									3B

[1] Only countries with area of chickpea exceeding 7,000 ha are considered.

Sources: Smithson *et al.*, 1985; Hamblin, 1987; Saxena, 1987; Saxena and Singh, 1987; Summerfield, 1988; Ryley and Irwin, 1989; Walton, 1989; ICRISAT, 1990; Lamb and Poddar, 1990; Wolde Amlak *et al.*, 1990.

Phytophthora root rot can be devastating in subtropical regions (Ryley and Irwin, 1989) as can Phoma blight in southern Australia (Lamb and Poddar, 1990).

Viral diseases, commonly referred to as "chickpea stunt", can be moderate yield reducers of chickpea (Tables 2 and 3).

Acid exudation by leaves, stems, and fruits of chickpea acts as a general deterrent to insect attack (Reed *et al.*, 1987). Nevertheless, *Helicoverpa* pod borer is a universal yield reducer of chickpea, and can be quite serious in parts of South Asia (Tables 2 and 3). Leaf miner is the major insect pest of chickpea in the West Asia-North Africa (WANA) region. Seed storage pests are of universal occurrence but are controllable with relatively simple management technology. Uniquely in southern Australia, red-legged earth mites are a threat to chickpea (Table 3).

Weeds assume greater importance as yield reducers of chickpea where soil conditions remain moist during early growth stages (Tables 2 and 3). Thus in South Asia for example, weed problems increase with latitude.

Among abiotic stresses, drought stands out as by far the major yield reducer of chickpea in all regions (Tables 2 and 3). This is because the crop is normally grown in a receding soil moisture environment and is thus exposed to terminal drought stress. Indeed, at higher latitudes, outside of the tropics, good soil moisture conditions (near field capacity), can induce excessive vegetative growth with consequent lodging and foliar disease attack. There are some genetic options for making substantial improvements here, through use of shorter duration cultivars (to escape terminal drought) and exploiting drought resistance traits such as prolific rooting ability (Saxena *et al.*, 1993). Also, management techniques to better conserve soil moisture and maximize crop transpiration versus soil evaporation, provide scope to reduce drought effects on chickpea.

Terminal drought stress is normally accompanied by increasing temperature towards maturity, often to levels (e.g., > 30°C) that may interfere with pod filling (Tables 2 and 3). Low temperatures, causing either freezing injury or mortality in WANA, or delaying onset of podding in subtropical South Asia, have also been identified as yield reducers (Tables 2 and 3). Only genetic options are available to better adapt chickpea to unfavorable temperature regimes. This is yet to be exploited in the case of high temperature stress but tolerance to low temperature has been found (Singh, 1987; Saxena *et al.*, 1988).

There are several reports of nutrient imbalances adversely affecting chickpea; but, apart from P deficiency, effects are quite localized (Tables 2 and 3). It is suspected that a more thorough diagnosis of such possible problems is needed and that nutrient imbalances will become more apparent when major yield limiting factors are corrected. In traditional chickpea growing areas, native strains of chickpea *Rhizobium* normally ensure effective nodulation, provided other environmental conditions are conducive. However, when chickpea is introduced into new areas, its host-specific *Rhizobium* also needs to be introduced through inoculation.

Lentil

The most serious biotic constraints facing lentil appear to be the foliar diseases, such as Ascochyta blight, rust, Stemphylium blight, and Botrytis gray mold (Tables 4 and 5). Rust is a key yield reducer in Morocco and Ethiopia, and Ascochyta and Stemphylium blights are important in wetter parts of South Asia. Fusarium wilt and Sclerotinia root rot are also important universally; the former being more prominent in drier areas less prone to foliar diseases. There appear to be greater opportunities for using host plant resistance to control root diseases rather than foliar diseases (Khare, 1981).

Among insect pests, *Sitona* weevils, the larvae of which feed on nodules, are

Table 4. Ranking of constraints of lentil and their potential for alleviation in the major producing countries of Asia[1]. Area and yield estimates (FAO, 1991) for 1990 are indicated

Production/ constraint	W. Asia					S. Asia			
	Iran	Jor- dan	Leb- anon	Syria	Tur- key	Bangla- desh	India	Nepal	Pak- istan
Area ('000 ha)	107	9	4	144	900	210	1095	122	80
Yield (kg ha⁻¹)	477	703	965	701	1000	771	641	625	475
Biotic									
ASC	–	–	–	–	?C	–	2C	?C	2C
BOT	–	–	–	–	–	2C	?C	3B	3B
STE	–	–	–	–	–	1B	–	–	–
PMI	–	–	–	–	–	–	3B	–	–
DMI	3B	–	–	–	–	–	–	–	–
RUS	–	–	–	–	–	3C	2C	3C	2C
FUS	?B	3B	3B	2A	3A	2B	2B	2B	2A
SCL	–	–	–	–	–	3C	2C	2C	3C
ANT									3C
SCT									3C
VIR	3C	?C	?C	3C	?C	?C	–	?C	?C
NEM	?C	?C	?C	3C	?C	?C	–	–	?C
HEL	–	–	–	–	–	–	3B	?B	–
APH	?C	3B	3B	3B	–	3B	2B	3B	3B
SIT	?A	2A	2A	2A	3A	–	–	–	–
CUT	–	–	–			–	–	3B	
SSP	?A	3A	3A	3A	3A	2A	3A	3A	3A
MAM						2B			
WDS						2B	3B	3B	
ORO	?C	?C	?C	3B	–	–	–	–	–
Abiotic									
DRO	2B	1B	2B	1B	2B	2B	2B	3B	3B
HTP	–	3C	3C	3C	3C	3B	3C	–	–
LTP	3C	–	–	–	3B	–	–	–	–
WIN						2C			
SAL	?C	–	–	–	–	–	–	–	3C
N						3B			
P						3A	3A		

[1]Only countries with area of lentil of 4,000 ha or more are considered.

Sources: Webb and Hawtin, 1981; Malik *et al.*, 1983; Muehlbauer *et al.*, 1985; Baldev *et al.*, 1988; Summerfield, 1988; BARI, 1991.

a key pest of lentil in WANA. Aphids are the main insect pest in South Asia and in Ethiopia and they may also be important in some years in WANA. Lentil seeds are also attractive to seed storage pests and precautions in storage are mandatory.

Lentil is susceptible to weed competition, especially in wetter areas, and necessary precautions are required to minimize yield loss (Tables 2 and 3). *Orobanche* (broomrape), a parasitic weed, is an important pest of lentil in the WANA region and is difficult to control by either management or genetic means.

As lentil production is predominantly rainfed, drought stress is the most important abiotic stress in all regions (Tables 4 and 5). Appropriate matching of crop duration to probable soil moisture availability pattern offers the best scope for yield improvement in this regard. This also requires matching of photoperiod response to target environments (Webb and Hawtin, 1981).

Table 5. Ranking of constraints of lentil and their potential for alleviation in the major producing countries of Africa[1]. Area and yield estimates (FAO, 1991) for 1990 are indicated

Production/ constraint	N. Africa				E. Africa
	Algeria	Egypt	Morocco	Tunisia	Ethiopia
Area ('000 ha)	6	6	63	4	45
Yield (kg ha⁻¹)	333	270⁻	523	516	498
Biotic					
ASC	–	–	3B	–	3B
RUS	3B	–	1A	–	2A
FUS	3B	3B	3B	3B	3B
DRR	–	2A	–	–	2A
ANT					2B
APH	?B	3B	?B	?B	2B
SIT	3A	–	3A	?B	–
SSP	2A	?A	3A	3A	3A
WDS	2A	?A	2A	2A	–
ORO	–	?B	2B	–	–
Abiotic					
DRO	2B	–	2B	2B	2B
WLG	–	?A	–	–	?B
HTP	3C	3C	2B	3B	–
LTP	3B	–	3B	3B	–
HPH	3C	2C	3C	3C	–
LPH	–	–	–	–	?C
N	?A	3A	3A	?A	?A

[1]only countries with area of lentil of 4,000 ha or more are considered.

Sources: Webb and Hawtin, 1981; Muehlbauer *et al.*, 1985; Summerfield, 1988.

Nutrient imbalances perhaps rank next as abiotic yield reducers but problems are usually very location specific.

Although China is reported to have produced 150,000 t of lentil in 1990 (FAO, 1991), area and yield data are not available, and we have little information on constraints of this crop in China.

Faba Bean

Faba bean is usually grown in higher rainfall areas or with irrigation. Thus, with the possible exception of Australia and northwestern China, the crop is not regularly exposed to drought, to which it is particularly susceptible compared to chickpea or lentil, for example (Bond *et al.*, 1985). Therefore biotic stresses loom as the major yield reducers of faba bean (Tables 6 and 7).

Again, foliar diseases predominate, with chocolate spot being the most serious yield reducer (Tables 6 and 7). Ascochyta blight and rust are the other major foliar diseases of widespread distribution. These three diseases are particularly predominant in China, by far the major producer of faba bean in the world (Lang and Zheng, 1988). However, they are much more severe in southern China, where the crop is sown in autumn, than in the spring-sown crop of northern China. In southern China, rainfall, humidity, and temperature conditions during March to May are conducive to these foliar diseases (Liu,

Table 6. Ranking of constraints of faba bean and their potential for alleviation in the major producing countries of Asia and Oceania[1]. Area and yield estimates (FAO, 1991) for 1990 are indicated

Production/ Constraint	W. Asia		E. Asia	Oceania
	Syria	Turkey	China	Australia
Area ('000 ha)	12	42	1700	44
Yield (kg ha⁻¹)	1667	1867	1529	977
Biotic				
CHS	3B	3B	1B	1A
ASC	3B	3B	2C	2A
BOT				2B
RUS			1B	2B
FUS			3B	?C
SCL				?C
VIR	3C	3C		?C
NEM	3C	3C		3C
HEL				2A
APH	3B	3B		
RLE				3A
SSP	3B	3B		
WDS				2A
ORO	2A	2A		
Abiotic				
DRO	3C	3C	2B	1C
WLG			2B	3C
HTP				2C
LTP			1C	3C
WIN				3C
LPH				2C
N				3B
P				2C
FE				2C

[1]Only countries with area of faba bean of 10,000 ha or more are considered.

Sources: Hawtin and Webb, 1982; Hebblethwaite et al., 1984; Liu, 1984; Bond et al., 1985; Hamblin, 1987; Lang and Zheng, 1988; Summerfield, 1988.

1984). Lines with broad-based resistance to chocolate spot have been identified but progress in developing cultivars resistant to Ascochyta blight and rust is slower (Nene *et al.*, 1988). Globally, root diseases are minor yield reducers compared to foliar diseases. However, root rot (*Fusarium solani*) can cause major yield losses in southern China (Yunnan Province in particular) (Liu, 1984). *Orobanche* is a particularly widespread problem in WANA. Virus diseases and stem nematodes are other biotic constraints of widespread occurrence, but the extent to which they cause yield losses is not well-documented.

High temperatures can reduce yield of faba bean in Ethiopia (lower altitudes), Sudan, and Egypt (Table 7) even if the crop is well-watered. In southern China, frosts and low temperatures (0 to 5°C), largely during January, can reduce yield (Table 6). Low temperatures from late January to March are particularly damaging as this period coincides with flowering and early pod development stages; in severe circumstances, the entire crop may be lost (Liu, 1984). Terminal drought stress in July can affect spring-sown faba bean in northwestern China, with yield reductions of 31 to 36%. On the other hand,

Table 7. Ranking of constraints of faba bean and their potential for alleviation in the major producing countries of Africa[1]. Area and yield estimates (FAO, 1991) for 1990 are indicated

Production/ Constraint	N. Africa				E. Africa	
	Algeria	Egypt	Morocco	Tunisia	Ethiopia	Sudan
Area ('000 ha)	63	142	233	52	330	15
Yield (kg ha⁻¹)	317	2641	713	519	848	1867
Biotic						
CHS	1B	3B	2B	2B	2B	–
ASC	2B	–	2B	2B		
PMI	–	–	–	–	3B	3A
RUS	2B	2B	2B	2B	2B	–
FUS	–	–	–	–	–	2B
DRR	–	–	–	–	–	2B
VIR	3C	–	3C	3C	3C	–
NEM	3C	?C	3C	3C	?C	–
HEL	–	–	–	–	2C	3C
LMI	–	–	–	–	–	3B
APH	?B	3B	?B	?B	2B	
SIT	–	–	3B	3B	–	–
ARW	–	–	–	–	–	3B
SSP	3A	?A	3A	3A	3A	2A
WDS	3A	?A	3A	3A	?B	3B
ORO	2B	3B	1B	2B	–	–
Abiotic						
DRO	1C	3C	1C	3C	3B	–
WLG	–	?A	–		?B	–
HTP	3C	2B	3C		2C	1C
LTP	3A	–	3A		3C	–
HPH	3C	3C	3C		–	–
LPH	–	–	–		?C	–
N	?A	3A	3A		?A	–

[1]Only countries with area of faba bean exceeding 10,000 ha are considered.

Sources: Hawtin and Webb, 1982; Hebblethwaite *et al.*, 1984; Bond *et al.*, 1985; Hamblin, 1987; Summerfield, 1988.

waterlogging is a common problem for the autumn-sown crop of southern China, following excessive rains during March and April. In Australia, faba bean is often exposed to drought, acid soil problems, and phosphorus deficiency. Here also, *Helicoverpa* poses a problem, whereas in other major producing areas insect pests seem to be of relatively minor importance (Table 6).

Pea

Powdery mildew stands out as the major yield reducer in pea (Tables 8 and 9). There are also several other foliar diseases of widespread occurrence, such as downy mildew, Ascochyta blight, Botrytis leaf spot, and rust. Of the root diseases, Fusarium wilt is most widespread (Tables 8 and 9). Although good progress has been made in developing pea cultivars resistant to powdery mildew, disease control could be enhanced by combining resistance with reduced foliage characteristics (Hagedorn, 1985). Such progress is not apparent for the other foliar diseases. Fusarium wilt resistant cultivars of pea are

Table 8. Ranking of constraints of pea and their potential for alleviation in the major producing areas of Asia and Oceania[1]. Area and yield estimates (FAO, 1991) for dry (D) and green (G) pea in 1990 are indicated

Production/ Constraint		W.Asia Iran	S.Asia Bangla- desh	India	Myan- mar	Pak- istan	E.Asia China	Oceania Aust- ralia	New Zealand
Area	D	77	45	488	20	142	1300	331	23
('000 ha)	G			96			59	13²	7
Yield	D	727	667	902	672	527	1231	1184	3130
(kg ha⁻¹)	G			2741			5559	3940²	4615
Biotic									
PMI			2B	1B	2B	2B	1B	2B	1B
DMI				2C	3C		2C	2B	3B
ASC				3C		3C	3B	2B	-
BOT				?C				3B	2B
RUS				3B			3B		
SEP								3B	
PSB								2B	2B
SCT				•				3C	2B
FUS				3B			2B	3B	
AMY								?C	2B
SCL		3C							-
DOF								3A	3A
DRR								3A	
VIR				2A			2B	?C	2C
NEM									?C
HEL								2A	2A
RLE								3A	3A
LMI				2B	3B	3B	3B		
PST				3C					
ETZ				2B					
ARW				3B					
APH							2B		
SSP			2A	2A	2A	2A	2A	2A	-
MAM			2B	3B					-
WDS			3A	2A	?A	3A	-	2A	2A

available and thus host plant resistance offers the best prospect for control of this disease (Hagedorn, 1985).

In Australia, where the crop has recently expanded, there appear to be additional disease problems, including bacterial blight (Table 8).

Insect pests rate as relatively minor yield reducers of pea (Tables 8 and 9). However, bruchids can be devastating to developing and maturing seed and stored grain if adequate management precautions are not taken. *Bruchus pisorum*, the pea weevil, is the most important insect pest of pea in Australia; it is a bruchid that invades the developing pod and seed of pea and continues to develop and grow inside the pea seed even after maturity. In southern Australia, the red-legged earth mite can also be a serious pest of pea.

As pea is normally grown in moister environments, weed problems can be severe at early growth stages (Tables 8 and 9). *Orobanche* reduces yield of pea in North Africa (Table 9).

Like faba bean, pea is particularly sensitive to drought and high temperature

Table 8. (Continued)

Production/ Constraint	W.Asia Iran	S.Asia Bangla-desh	India	Myan-mar	Pak-istan	E.Asia China	Oceania Aust-ralia	New Zealand
Abiotic								
DRO		3B	2B	2B	3B	2B	2B	2B
WLG		3B	3B		3C		3C	
HTP		2C	2C	2C			2C	
LTP			3C		2C	2C	3C	
WIN		2C	−				2C	
HPH				−	3C	−	−	
LPH			−	3C	−	3C	3C	
N			3A				3B	
P		3A	3A		3A		2A	
FE			−				3A	
BT			−				3C	

[1]Only countries with area of dry pea production exceeding 10,000 ha are considered.

[2]Values from Australian Bureau of Statistics - 1989/90 Summary of Crops in Australia; as there appear to be inconsistencies in FAO data.

Sources: Davies *et al.*, 1985; Hebblethwaite *et al.*, 1985; Hamblin, 1987; Baldev *et al.*, 1988; Summerfield, 1988; Armstrong *et al.*, 1989; Walton, 1989; BARI, 1991; Zong Xuxiao, 1992.

(Tables 8 and 9). At higher latitudes, cold stress (i.e., frost damage) can be a problem. Peas can face waterlogging damage in excessively wet soils, as in the sub-montane soils of South Asia. There appears to be scope for genetic improvement of pea to be better able to face these extremes of temperature and soil moisture (Hebblethwaite *et al.*, 1985).

Grasspea

Grasspea seems to be able to withstand both biotic and abiotic stresses much better than the other legumes considered here. Indeed, it is normally grown in marginal lands considered unsuitable for cultivation of the other legumes, such as under late-sown conditions in rice fallows in South Asia. Even under temperature and humidity conditions highly conducive to foliar diseases in chickpea and lentil (e.g., as in northern Bangladesh), relatively little foliar disease is recorded in grasspea. Such disease is usually powdery or downy mildew (Table 10). The lesser disease incidence may be related to the open canopy of grasspea, with its "grass-like" leaves and canopy habit, allowing light penetration and air circulation even at high plant densities and with prolific plant growth.

Root diseases and insect pests also appear to cause only minor yield reductions (Table 10). Grasspea effectively smothers weeds, and even uses more erect ones to support its vining habit.

Table 9. Ranking of constraints of pea and their potential for alleviation in the major producing areas of Africa[1]. Area and yield estimates (FAO, 1991) for dry (D) and green (G) pea in 1990 are indicated

Production/ Constraint		N. Africa			Central and E. Africa	
		Algeria	Egypt	Morocco	Burundi	Ethiopia
Area	D	16	4	70	52	150
('000 ha)	G	19	9	4		
Yield	D	313	2143	900	519	727
(kg ha⁻¹)	G	1895	10941	7500		
Biotic						
PMI		3B		2B	2B	2B
ASC					2C	3C
BOT				2B		2B
SEP					2B	
FUS		3B				3B
HEL						2C
APH		?B		?B	2B	
SSP		3A	?A	3A		
WDS		3A	?A	3A	2B	
ORO				2B	2B	
Abiotic						
DRO		2B	–	2B		3B
WLG		–	3A	–	3B	?B
HTP		3B	2B	3B	2C	–
LTP		3A	–	3A		3C
HPH		3C	3C	3C		–
LPH		–	–	–	2C	?C
N		–	3A	3A		?B
P					2A	

[1]Only countries with area of dry pea production exceeding 10,000 ha or green pea exceeding 9,000 ha are considered. Zaire, Tanzania, Rwanda, and Uganda fit these criteria, with respective areas for dry pea of 91,000, 80,000, 35,000, and 25,000 ha and yields of 659, 313, 714, and 540 kg ha⁻¹, but these countries have been excluded from this table as we have been unable to find constraint information for them.

Sources: Davies *et al.*, 1985; Hebblethwaite *et al.*, 1985;   Summerfield, 1988.

Grasspea establishes well in waterlogged soils, even when relay-sown into rice, and can grow to maturity in waterlogged soils. Grasspea has good drought resistance and produces more reliably than other legumes when sown on drought-prone upland soils. In South Asia, grasspea nodulates prolifically, presumably satisfying its own nitrogen needs and also contributing nitrogen to the cropping system as a whole. However, in Ethiopia, ineffective nodulation of grasspea has been reported (Wolde Amlak *et al.*, 1991).

Cultivation of grasspea is generally officially discouraged because of its neurotoxin content (β-N-oxalylamino-L-alanine [BOAA]) which can cause lathyrism (Kaul and Combes, 1986). Breeding programs to reduce levels of this toxin or appropriate education of consumers of grasspea on how to avoid ingestion of toxic quantities of the neurotoxin would permit more confident and widespread use of a pulse crop already having a high level of resistance to biotic and abiotic stresses in general.

Table 10. Ranking of constraints of lathyrus and their potential for alleviation in the major producing areas of Asia and Africa. Area and yield estimates are also indicated

Production/ Constraint	Asia				Africa
	Bangladesh	India	Nepal	Pakistan	Ethiopia
Area ('000 ha)	234[1]	1190[2]	49[3]	143[4]	60[5]
Yield (kg ha⁻¹)	715[1]	406[2]	380[3]	497[4]	704[5]
Biotic					
PMI	3B	3B	3C	?C	2C
DMI		3C			
RUS		–			2C
FUS	?B	3B	3B		3B
SCL	3C	3C			–
HEL	–	3C			3C
APH		2C	3C		2C
THR	3C	2C			3C
SSP	3A				
WDS	–	3A			
Abiotic					
DRO	–	2C		3C	
HTP	3C	3C		3C	
LPH	–	3B	3C		3B
SAL				3C	
N	–	3B			2B
P	–	3A	3A		

[1]Five year mean for 1983/84 to 1987/88. Source: Statistics available with Bangladesh Agricultural Research Institute

[2]For 1982/83. Source: Chitale (1993)

[3]For 1983/84. Source: Bharati (1986)

[4]Source: Anonymous (1991)

[5]Mean for 1987 and 1988. Source: Wolde Amlak et al. (1991)

Sources: Bharati, 1986; Kaul and Combes, 1986; Campbell, 1989; BARI, 1991; Wolde Amlak et al., 1991; Chitale, 1993

National average grain yields of grasspea still remain low (<1 t ha⁻¹), when it is considered that yields exceeding 3 t ha⁻¹ are possible (Wolde Amlak *et al.*, 1991). In South Asia at least, this can largely be attributed to low seeding and establishment rates under relay sowing. Also, planting is usually done after the optimum time. Nevertheless, there must be other constraints which are limiting yield, but they probably have not been adequately diagnosed and evaluated as yet due to the relative advantage under stress that grasspea maintains over comparable legumes.

Considerations Across Crops

For simplicity of presentation, we have so far only considered each constraint in isolation, and not the possible interactions between them. Further, we have considered constraints to crops sown at the normal time for a given region and using commonly followed agronomic practices. Each of these legumes are particularly sensitive to sowing time, showing marked yield declines with deviation from the optimum sowing time. This is because of their sensitivity to environmental factors such as photoperiod, temperature, and soil moisture status. Increasing the flexibility and adaptation of these legumes, such that their yield is less sensitive to sowing date, would require less sensitivity to deviations in these environmental factors.

Biotic and abiotic factors usually intimately interact to produce an ultimate effect on yield. A particular example is the interaction between fungal root diseases and soil moisture status. The interactions between foliar diseases and environmental factors can be very subtle. Disease outbreak requires humid conditions within the canopy and a favorable temperature regime, specific for the fungus. Rainfall, or otherwise high soil moisture during reproductive growth stages of chickpea, lentil, faba bean, and pea, favor further vegetative growth because of their indeterminate growth habits. This results in a closed canopy conducive to creating a humid microenvironment within the canopy suitable for fungal proliferation.

The results of the present analysis indicate that foliar diseases and drought are the major yield reducers of chickpea, lentil, faba bean, and pea in the growing regions considered here. Alleviation of these stresses by management means, such as fungicidal sprays for foliar diseases and irrigation for drought, is possible but impractical on a wide scale as these crops are largely grown as subsistence crops. These stresses are characterized by relatively slow rates of genetic progress in development of resistances to combat them. In the case of foliar diseases, this can be attributed to the low levels of resistances found, their usually monogenic nature (leading to rapid breakdown of resistance), and diverse and rapidly evolving pathotypes (Bernier *et al.*, 1988). However, in view of the large yield losses caused by foliar diseases and drought, it is suggested that greatest returns on investment in research may be made by a renewed and concerted attack on them, perhaps by adopting modified approaches to the problems.

Attempts to improve the ability of grain legumes to cope with foliar diseases have focussed on improvement of host plant resistance, or the ability of plant tissue to resist penetration by fungal hyphae. An additional and complementary approach would be to initially discourage potential for fungal growth in the canopy. Grasspea may give a clue in this regard in that, in South Asia at least, its naturally open canopy is only ever marginally affected by foliar disease in situations where the closed canopies of other legumes often grown adjacent, such as chickpea and lentil, are devastated by foliar diseases. However, the relative contributions of host plant resistance and open canopy structure to disease incidence need to be established for grasspea. Development of leafless and semileafless pea has permitted a more open canopy and less tendency to lodge, which mitigates against development of foliar disease (Heath and Hebblethwaite, 1985). It is suggested that canopies of chickpea, lentil, and faba bean be redesigned so as to maximize light penetration and air circulation without unduly jeopardizing photosynthetic efficiency. Erect plant types with fewer and smaller leaves and a more determinate habit are thus needed. Definitive experiments are required to establish the degree of advantage, and possible associated disadvantages, of such open canopies. Close collaboration between physiologists, pathologists, and breeders is required to achieve this.

Although a seemingly intractable problem, renewed efforts are needed to develop cultivars that are better able to produce under drought-prone situations. The first line of attack on terminal drought stress is use of short-duration cultivars, which can reach maturity within the probable period of soil moisture availability (Saxena *et al.*, 1993). Additional genetic gains can be expected by incorporating traits shown to be beneficial to yield in drought environments, such as a prolific root system in the case of chickpea (Saxena *et al.*, 1993).

Grasspea needs to be more intensively investigated as to how it copes with several different types of stress, compared with the other four legumes discussed here. This may give clues as to what traits may be improved to increase stress resistance in the other legumes, as already discussed in terms of foliar disease and canopy structure. If the neurotoxin problem in grasspea can be satisfactorily dealt with, then further knowledge on mechanisms of stress resistance in this crop would also permit enhancement of its ability to produce under various stress conditions.

Constraint Analysis

The exercise conducted for this paper is a rather general one, on a continental scale. Within countries, even covering such small geographical areas as Bangladesh, there are marked spatial variations in incidence and extent of a particular constraint. For most countries we have simply "averaged" the effect of a constraint across different agro-ecological zones where the crop is grown. However, we suggest that further detailed analysis on a regional or country basis is a very worthwhile pursuit in allowing prioritization of research on these crops, an increasing necessity in view of a generally declining research funding trend.

We have only broadly ranked yield losses from a given constraint. More precise estimates of yield loss would allow calculation of expected financial gains from a successful research input to alleviate a particular stress. The time lags involved between the research phase and actual adoption in farmers' fields must be taken into account. It should thus be possible to estimate returns on investment to particular constraint-oriented research inputs. This process now seems advantageous, if not necessary, in attracting funds for research. The process can be enhanced, and permit presentation in an attractive form by use of geographic information systems (GIS) technology. However, the credibility of the process depends on obtaining better constraint survey and yield loss data than we now generally have for these legumes.

Concluding Remarks

From the incomplete information available to us, foliar diseases and drought rank as the most widespread and severe yield reducers of chickpea, lentil, faba bean, and pea. Although genetic resistance to these constraints is usually disappointingly minimal, it is suggested that greater research emphasis be directed towards these constraints. This is because of large potential gains on research investment and possible viable alternative approaches to these hitherto difficult problems. It is strongly suggested that better knowledge of the mechanisms by which grasspea withstands a range of stresses would contribute to understanding of how to increase stress tolerance of all legumes. More detailed constraint analysis than the general overview attempted here would permit improved research prioritization and possibly attract research support.

Acknowledgements

We thank Dr. R. S. Malhotra and Dr. S. Weigand, ICARDA, for advice on pea and insect pests, respectively. We are also grateful to Dr. M. P. Haware, ICRISAT for his suggestions on improving the manuscript.

References

Anonymous. 1991. Agricultural Statistics of Pakistan. Islamabad: Planning Unit, Ministry of Food, Agriculture and Cooperatives.

Armstrong, E. L., Butler, B. and Fisher, S. G. 1989. In: *Proceedings of the Fifth Australian Agronomy Conference*, p. 499 (ed. G.P. Ayling). Perth, 24–28 Sep 1989. Australian Society of Agronomy.

Baldev, B., Ramanujam, S. and Jain, H. K. 1988. In: *Pulse Crops*, 626 pp. New Delhi: Oxford and IBH.

BARI (Bangladesh Agricultural Research Institute). 1991. In: *Advances in Pulses Research in Bangladesh: Proceedings of the Second National Workshop on Pulses*, p. 254 (ed. J. Kumar). 6–8 June 1989, Joydebpur, Bangladesh. Patancheru, Andhra Pradesh, India: ICRISAT.

Bernier, C. L., Bijiga, G., Nene, Y. L. and Cousin, R. 1988. In: *World Crops: Cool Season Food Legumes*, pp. 97–106 (ed. R. J. Summerfield). Dordrecht: Kluwer Academic Publishers.

Bharati, M. P. 1986. In: *Lathyrus and Lathyrism*, pp. 142–145 (eds. A. K. Kaul and D. Combes). New York: Third World Medical Research Foundation.

Bond, D. A., Lawes, D. A., Hawtin, G. C., Saxena, M. C. and Stephens, J. H. 1985. In: *Grain Legume Crops*, pp. 199–265 (eds. R. J. Summerfield and E. H. Roberts). London: Collins.

Campbell, C. G. 1989. In: *The Grasspea, Threat and Promise. Proceedings of the International Network for the Improvement of Lathyrus sativus and the Eradication of Lathyrism*, pp. 139–146 (ed. P. S. Spencer). New York: Third World Medical Research Foundation.

Chitale, M. W. 1993. *Lathyrus* (grasspea). In: *Proceedings of the RAS/89/040 Workshop on Unexploited and Potential Food Legumes*, 31 Oct – 3 Nov 1990, Chiang Mai, Thailand. Bangkok: FAO (In press).

Davies, D. R., Berry, G. J., Heath, M. C. and Dawkins, T. C. K. 1985. In: *Grain Legume Crops*, pp. 147–198 (eds. R. J. Summerfield and E. H. Roberts). London: Collins.

FAO (Food and Agriculture Organization of the United Nations). 1991. *FAO Yearbook*. Production. 1990. Vol. 44. Rome, Italy: FAO.

Hagedorn, D. J. 1985. In: *The Pea Crop. A Basis for Improvement*, pp. 205–213 (eds. P. D. Hebblethwaite, M. C. Heath and T. C. K. Dawkins). London: Butterworths.

Hamblin, J. 1987. In: *Proceedings of the Fourth Australian Agronomy Conference*, pp. 65–82 (ed. T. J. Reeves, Parkville, Victoria, Australia). Melbourne, Australia, Aug. 1987. Australian Society of Agronomy.

Hawtin, G. and Webb, C. (eds.). 1982. *Faba Bean Improvement*. The Hague, The Netherlands: Martinus Nijhoff.

Heath, M. C. and Hebblethwaite, P. D. 1985. In: *The Pea Crop. A Basis for Improvement*, pp. 19–29 (eds. P. D. Hebblethwaite, M. C. Heath and T. C. K. Dawkins). London: Butterworths.

Hebblethwaite, P. D., Dawkins, T. C. K., Heath, M. C. and Lockwood, G. (eds.) 1984. *Vicia faba: Agronomy, Physiology and Breeding*, 333 pp. The Hague, The Netherlands: Martinus Nijhoff/Dr. W. Junk Publishers.

Hebblethwaite, P. D., Heath, M. C. and Dawkins, T. C. K. (eds.). 1985. *The Pea Crop. A Basis for Improvement*, 486 pp. London: Butterworths.

ICRISAT (International Crops Research Institute for the Semi-Arid Tropics). 1990. In: *Chickpea in the Nineties*, 403 pp. (eds. H. A. van Rheenen, M. C. Saxena, B. J. Walby and S. D. Hall). Patancheru, Andhra Pradesh, India: ICRISAT.

Kaul, A. K. and Combes, D. (eds.). 1986. *Lathyrus and Lathyrism*. New York: Third World Medical Research Foundation.

Khare, M. N. 1981. In: *Lentils*, pp. 163–172 (eds. C. Webb and G. Hawtin). Farnham Royal, UK: Commonwealth Agricultural Bureaux. Aleppo, Syria: ICARDA.

Lamb, J. and Poddar, A. 1990. Grain Legume Handbook. South Australian Peagrowers Co-operative Ltd.

Li-juan, L. and Zheng, Z. 1988. In: *World Crops: Cool Season Food Legumes*, pp. 1135–1152 (ed. R. J. Summerfield). Dordrecht: Kluwer Academic Publishers.

Liu, Qing-Fong. (ed.). 1984. *Faba Bean (Vicia faba) Cultivation*, 186 pp. Yunnan, China: People's Publishing House.

Malik, B. A., Huqqani, A. M. and Bashir, M. 1983. *Lens* 10: 15–16.

Muehlbauer, F. J., Cubero, J. I. and Summerfield, R. J. 1985. In: *Grain Legume Crops*, pp. 266–311 (eds. R. J. Summerfield and E. H. Roberts). London: Collins.

Nene, Y. L., Hanounik, S. B., Qureshi, S. H. and Sen, B. 1988. In: *World Crops: Cool Season Food Legumes*, pp. 577–589 (ed. R. J. Summerfield). Dordrecht: Kluwer Academic Publishers.

Nene, Y. L. and Reddy, M. V. 1987. In: *The Chickpea*, pp. 233–270 (eds. M. C. Saxena and K. B. Singh). Wallingford: CAB International.

Reddy, M. V. and Singh, K. B. 1992. *Crop Science* 32: 1079–1080.

Reed, W., Cardona, C., Sithanantham, S. and Lateef, S. S. 1987. In: *The Chickpea*, pp. 283–318 (eds. M. C. Saxena and K. B. Singh). Wallingford: CAB International.

Ryley, M. J. and Irwin, J. A. G. 1989. In: *Australian Chickpea Workshop Proceedings*, pp. 100–109 (eds. R. B. Brinsmead and E. J. Knights). Melbourne: Australian Institute of Agricultural Science.

Saxena, M. C. 1987. In: *The Chickpea*, pp. 207–232 (eds. M. C. Saxena and K. B. Singh). Wallingford, UK: CAB International.

Saxena, M. C. and Singh, K. B. (eds.). 1987. *The Chickpea*, 409 pp. Wallingford, UK: CAB International.

Saxena, N. P., Johansen, C., Saxena, M. C. and Silim, S. N. 1993. In: *Breeding for Stress Tolerance in Cool Season Food Legumes*, pp. 245–270 (eds. K. B. Singh and M. C. Saxena) Chichester, UK: John Wiley & Sons.

Saxena, N. P., Johansen, C., Sethi, S. C., Talwar, H. S. and Krishnamurthy, L. 1988. *International Chickpea Newsletter* 19: 17–29.

Singh, K. B. 1987. In: *The Chickpea*, pp. 127–162 (eds. M. C. Saxena and K. B. Singh). Wallingford, UK: CAB International.

Smithson, J. B., Thompson, J. A. and Summerfield, R. J. 1985. In: *Grain Legume Crops*, pp. 312–390 (eds. R. J. Summerfield and E. H. Roberts). London: Collins.

Summerfield, R. J. (ed.). 1988. *World Crops: Cool Season Food Legumes*, 1179 pp. Dordrecht:

Kluwer Academic Publishers.

Walton, G. H. 1989. Technical Report No. 19. Division of Plant Industries, Western Australian Department of Agriculture.

Webb, C. and Hawtin, G. (eds.). 1981. *Lentils.* Farnham Royal, UK: Commonwealth Agricultural Bureaux. Aleppo, Syria: ICARDA.

Wolde Amlak Araya, Asgelil Dibabe, Bekele Hundie, Regassa Ensermu, Wasie Haile and Yeshanew Ashagsie. 1991. Presentation at Third Triennial Colloquim of INILSEL, Dhaka, Bangladesh, 30 Nov – 3 Dec, 1991.

Wolde Amlak Araya, Seifa Tsegay and van Rheenen, H. A. 1990. *International Chickpea Newsletter* 23: 27–29.

Zong Xuxiao. 1993. Sweet pea. In: *Proceedings of the RAS/89/040 Workshop on Unexploited and Potential Food Legumes*, 31 Oct – 3 Nov 1990, Chiang Mai, Thailand pp. 103–108 (eds. N. Chomchalow, C. L. L. Gowda and P. Laosuwan). Bangkok: FAO.

Biotic and abiotic stresses of cool season food legumes in the western hemisphere

A.E. SLINKHARD[1], G. BASCUR[2] and G. HERNÁNDEZ-BRAVO[3]

[1] Crop Development Centre, University of Saskatchewan, Saskatoon, Saskatchewan, S7N OWO Canada;
[2] Instituto de Investigaciones Agropecuarias, Estación Experimental La Platina, Casilla 439, Correo 3, Santiago, Chile, and
[3] Chachalacas, Atizapan, Mexico

Abstract

Pea, lentil, faba bean, and chickpea suffer from many of the same biotic and abiotic stresses throughout the world. The recent large increase in production of pea and lentil in western Canada has been relatively free of biotic stresses. However, as production increases further and as these crops are grown with increasing frequency in the same fields, increased problems with biotic stresses will occur. The most severe biotic stresses will be the soilborne diseases and the Ascochyta complex of foliage diseases. Fortunately, genetic differences occur in response to most of these biotic stresses and are being incorporated into new cultivars.

Abiotic stresses are largely a function of the climate and the area. Thus, they change slowly, although extreme variations can occur among years especially in areas with a continental type climate. Additional research is needed to identify genetic differences in abiotic stress tolerance and incorporate them into improved cultivars.

Biotic Stresses

Each cool season food legume is best adapted to a specific environment and is affected by different biotic stresses, although they have many common biotic stresses (Table 1). Certain areas produce large volumes of a specific crop and thus biotic stresses are more important there and more research has been done on them. However, many of these biotic stresses occur in other areas and are not included in Table 1 because of lack of information or their relative insignificance in a given area.

F.J. Muehlbauer and W.J. Kaiser (eds.), Expanding the Production and Use of Cool Season Food Legumes, 195–203.
© 1994 Kluwer Academic Publishers.

Table 1. Biotic constraints to the production of pea, lentil, chickpea, and faba bean in North America, Central America, and South America

Constraint and causal organism	North America	Central America	South America
Seed, seedling and root rots			
Aphanomyces root rot (*A. euteiches*)	P*	P	P,F
Fusarium root rot (*F. solani*)	P,L,F,C	P,L,F,C	P,L,F,C
Fusarium wilt (*F. oxysporum*)	P,L,F,C	P,L,F,C	P,L,F,C
Pythium seed rot, root rot and damping-off (*P. ultimum*)	P,L,F,C	P,L,F,C	P,L,F,C
Rhizoctonia seedling rot (*R. solani*)	P,L	F	P,L,F,C
Bacterial diseases			
Bacterial blight (*Psuedomonas syringae* pv. *pisi*)	P		P
Fungal foliage diseases			
Ascochyta blight of lentil (*A. fabae* f. sp. *lentis*)	L	L	L
Ascochyta blight of faba bean (*A. fabae*)	F	F	F
Ascochyta foot rot of pea (*A. pinodella* = *Phoma medicaginis* var. *pinodella*)	P	P	P
Ascochyta blight of pea (*A. pinodes* = *Mycosphaerella pinodes*)	P	P	P
Ascochyta leaf and pod spot of pea (*A. pisi*)	P	P	P
Ascochyta blight of chickpea (*A. rabiei*)	C	C	C
Gray mold (*Botrytis cinerea*)	L,F,C	F,C	L,F,C
Anthracnose of lentil (*Colletotrichum truncatum*)	L		L
Rust (*Uromyces* sp.)		F,C	L,F
Powdery mildew (*Erysiphe* sp.)	P,L,F,	P	P,L
Downy mildew (*Peronospora viciae*)	P,F	P	P,L

Table 1. (Continued)

Constraint and causal organism	North America	Central America	South America
Sclerotinia rot (white mold) (*S. sclerotiorum*)	P,L	P	P,F,C
Virus diseases			
Pea enation mosaic	P,L,F		
Bean yellow mosaic virus	F	F	F
Bean (pea) leaf roll	P,L,F		
Pea seedborne mosaic	P,L	P	P
Pea streak	P,F		F
Pea stunt	P,F		
Faba bean stain			F
Necrotic and mottle faba bean virus		F	
Nematodes			
Pea cyst nematode (*Heterodera goettengiana*)	P,L,F,C		
Root-knot nematode (*Meloidogyne hapla*)	P	C	L,C
Root-lesion nematode (*Pratylenchus penetrans*)	P		L,C
Stem nematode (*Ditylenchus dipsaci*)	P P		
Insects			
Grasshoppers (many species)	L		
Leaf miners (*Melanognomyza* sp. and *Liriomyza* sp.)		F,L,C	F,L,C
Lygus bug	L		
Aphids (*Acyrthosiphon* sp.)	P,L,F	P,L,F	F,L
Budworm and pod borer (*Heliothis* sp.)		C	L,F,C
Seed weevil (*Bruchus pisorum*)	P		P
Leaf weevil (*Sitona* sp.)	P		
Blister beetle (*Epicauta* sp.)	F	F	F
Slug (*Deroicerus* sp. and *Limax* sp.)			L
Armyworm (*Spodoptera* sp.)		P,C	

* P,L,F,C refer to pea, lentil, faba bean, and chickpea, respectively.

Seed, Seedling, and Root Rots

These types of diseases are usually caused by soilborne fungi (Kaiser, 1981; Hagedorn, 1984; Kraft *et al.*, 1988) and commonly occur in fields that have grown a specific legume or a closely related one over a prolonged period of time, e.g., Aphanomyces root rot of pea in fields surrounding a vining pea processing plant. Another common soilborne disease of legumes is seed rot, root rot, and damping-off caused by *Pythium ultimum* Trow (and other *P.* species). It is found nearly everywhere legumes are grown due to its extensive host range, but is usually only a problem in poorly drained soils or in kabuli chickpeas and wrinkle-seeded peas where integrity of the seed coat is limiting or imbibitional injury is induced by excessively rapid water uptake, especially at low temperatures.

At least an intermediate level of resistance is available for most of these soilborne diseases, except for *Pythium* where use of well drained soils and seed treatment with a fungicide such as metalaxyl are fairly effective. The use of crop rotations with a susceptible crop only once in four years is commonly recommended as a means of reducing the rate of build-up of these diseases in the soil.

Bacterial Diseases

Bacterial blight of pea (*Pisum sativum* L.) (Nene *et al.*, 1988) is the most important and widespread bacterial disease of pea (Table 1). It is seedborne and most countries require field inspection and phytosanitary certificates for pea seed imports in an attempt to prevent importation of infected seed. It is not a very serious disease in dry areas as it is spread by raindrop splash, especially so in a driving rain or hail storm.

Fungal Foliage Diseases

The most serious and widespread fungal foliage diseases of cool season food legumes are caused by the Ascochyta complex (Nene *et al.*, 1988). All species of *Ascochyta* are seedborne and survive for a year or more on infested crop residue. These diseases require free water on the leaf surface for infection and have a low optimum temperature for growth and development. Accordingly, these diseases are most important in cool areas characterized by rainy spells during late vegetative growth and pod fill.

Each food legume is attacked by a specific species of *Ascochyta* (Table 1). The most serious Ascochyta disease of pea is Mycosphaerella blight, caused by *A. pinodes* L. K. Jones (teleomorph [perfect stage] *Mycosphaerella pinodes* [Berk. & Bloxam] Vestergr.). The most resistant pea lines are at best only moderately resistant and, since ratings of foliage infection in the moderately resistant range are highly subjective and have a low level of repeatability, little progress has been made in breeding for Mycosphaerella blight resistance.

Researchers at the University of Saskatchewan recently initiated a breeding program for Mycosphaerella blight resistance that should be effective in increasing the level of resistance. Different sources of Mycosphaerella blight resistance, as identified by Dr. J. M. Kraft (USDA at Prosser, WA), will be crossed with adapted cultivars and the resulting F_2-derived families selected for yield in replicated yield trails grown in a Mycosphaerella blight nursery. Since early infection by Mycosphaerella blight can cause yield losses up to 50% (Wallen, 1974), the highest yielding lines will have a genetically determined, moderately resistant reaction. Then the highest yielding lines from different crosses will be intercrossed and subjected to another cycle of selection for yield in the Mycosphaerella blight nursery. In this way the additive effect of different genes for Mycosphaerella blight resistance (or tolerance) can be accumulated into one or a few lines. Selected lines can be double-checked for Mycosphaerella blight resistance by growing them in paired plots (sprayed vs. unsprayed for blight control). The success of this approach is related to the greatly increased efficiency of the F_2-derived family method of breeding relative to other methods of breeding self-pollinated crops (Slinkard, 1990). The F_2-derived family method develops its efficiency by using selection for yield among F_2-derived families in replicated trials in F_3, F_4, F_5, and F_6 with concurrent elimination of low-yielding crosses and F_2-derived families within the few remaining superior crosses. Efficiency in this case refers to the resources and number of generations required to select the highest yielding line from a given number of crosses.

Ascochyta blight of chickpea (*Cicer arietinum* L.) is the most devastating disease of the Ascochyta complex. Completely susceptible plants can be killed in about two weeks under heavy inoculum pressure as in an Ascochyta nursery. More and more resistant lines are being reported every year (Singh and Reddy, 1991), but breeding for resistance is compounded by the presence of different races.

Ascochyta blight of lentil (*Lens culinaris* Medik.) first became widespread in Canada in 1978 (Morrall and Shepherd, 1981). Excellent progress is being made in breeding for Ascochyta blight resistance in lentil at the University of Saskatchewan. Laird lentil, registered in 1978 and grown on over 200,000 ha in western Canada in 1992, is moderately resistant in the vegetative stage, but becomes very susceptible as the plant starts to senesce. A second type of resistance from ILL 5588, active throughout the growth cycle, was obtained from the International Centre for Agricultural Research in the Dry Areas (ICARDA), Aleppo, Syria and is rapidly being incorporated into the breeding material. The first resistant cultivar should be available in 1994.

Anthracnose of lentil causes severe losses in areas of Canada with frequent showers during the late vegetative and reproductive stages of lentil. An excellent level of resistance has been found and is being incorporated into the breeding program. Resistance to both Ascochyta and anthracnose has been recovered in some breeding lines.

Lentil rust is a serious problem in South America, but resistance to these races is available in Laird lentil. Several resistant cultivars have been developed recently that are well adapted to Chile and Argentina.

Powdery mildew of pea occurs widely and serious yield losses occur if infection occurs before flowering. However, resistance is widely available and simply inherited. Few cultivars are resistant at the present time, but breeders are currently incorporating powdery mildew resistance into many of their newer cultivars.

Sclerotinia rot has a wide host range and is widely distributed. Various species of plants differ in their general level of susceptibility, but genetic variability for resistance is limited within a species. Spore production of the fruiting bodies is promoted by prolonged wet conditions under a dense canopy of foliage. These spores light on petals on the soil surface and germinate, producing mycelia. However, infection of pea and lentil stems does not occur unless the dense foliage lodges and the stems come in contact with the mycelia. Thus, the severity of Sclerotinia infection can be reduced by the use of semi-leafless pea cultivars that have a more open canopy and lodge later in development of the crop.

Virus Diseases

Some 44 viruses have been reported in these cool season food legumes (Bos *et al.*, 1988). The more common ones (Hagedorn, 1974) are reported in Table 1. Genetic resistance to these viruses has been reported, but few of these resistance genes have been incorporated into new cultivars. Many viruses survive on perennial legumes such as alfalfa (*Medicago sativa* L.) and are transmitted to food legumes by aphids. Thus, virus diseases are most serious in areas of high aphid populations or where aphids are blown into the area early in the season. Thus, viruses are rarely of any significance in western Canada with its severe winters and short growing season.

Several viruses are seedborne and pea seedborne mosaic virus on pea is the most serious one. A pedigreed seed program with close monitoring for the virus is fairly effective in most areas. However, the symptoms are masked at low temperatures, making visual detection difficult in cool season areas. Resistance to pea seedborne mosaic virus is present in pea (Hagedorn and Gritton, 1973) and lentil (Haddad *et al.*, 1978).

Nematodes

Plant parasitic nematodes attack many food legumes and may build up in fields when legumes are grown frequently. Most nematodes occur on the roots where researchers rarely look except occasionally to see if nodules are present. The presence of nematodes and their identification requires their extraction from the roots and few people are trained in this technique. Above ground symptoms are nondescript, such as reduced growth rate and slight yellowing of the foliage. Nodulation is often reduced, but again this can be caused by many things. Presumably nematodes would be reported with increasing frequency if more people were specifically looking for them.

Insects

Food legumes are attacked by many insects in the western hemisphere. Various grasshoppers attack lentil in western Canada, but they eat only the flower buds. In severe cases the grasshoppers can practically eliminate lentil seed production unless they are sprayed with an insecticide. The pea leaf weevil and the pea seed weevil are serious insect pests of pea in the Pacific Northwest of the USA, but are not yet a problem in western Canada, possibly because they cannot survive the severe winters.

Leaf miners occur with a high frequency on faba bean (*Vicia faba* L.) in the Andean Region, but the severity of their damage has not been determined. Aphids are ubiquitous and may require chemical control in isolated areas throughout the western hemisphere.

Blister beetles attack the buds and growing points of faba bean plants in Canada and Chile. They concentrate in local areas of the field and are usually only a serious problem in small fields (less than 2 hectares). Budworms and pod borers are important pests of chickpea in local areas of South and Central America.

Abiotic Stresses

Cool season food legumes have poor tolerance to saline soils and thus saline areas should be avoided. Food legumes have varying levels of drought tolerance and can be ranked from the most tolerant to the least tolerant as chickpea > lentil H pea > faba bean. Chickpea gets its drought tolerance primarily from its deep, extensive root system which enables it to extract water from a greater volume of soil than is possible by the other three. Lentil plants are small and produce little dry matter until just before flowering, thus deferring heavy water use. In addition lentil plants have a strongly indeterminate growth habit and are able to respond rapidly to available water during the first half of the flowering and pod filling period. Consequently, chickpea and lentil respond poorly to irrigation other than to have the soil profile full at seeding time and one or two light (5 to 7 cm) irrigations at first flower and perhaps again three weeks later during mid-pod fill. Excess water is conducive to root rot, especially in chickpea.

Peas have an intermediate level of drought tolerance and respond well to irrigation as long as the soil does not become water logged. The Canadian pea cultivar Express apparently has a slightly higher level of drought tolerance than other cultivars, but this needs further confirmation.

Faba bean has poor drought tolerance and responds well to irrigation. A slight drought stress prior at flowering will help prevent excess plant height, but water demand is very high during pod fill. Any drought stress during pod fill will markedly reduce seed yield.

All four of these cool season food legumes are winter annuals and have good cold tolerance as seedlings. They will tolerate $-2°C$, but top growth may be damaged at $-6°C$. If the top growth is "burned off" by frost in the early seedling stage, regrowth usually will occur from a dormant bud at the uppermost scale node at or below the soil surface. However, if these food legumes are grown as winter annuals, the larger seedlings may be damaged severely by late frosts and the buds at the two scale nodes may no longer be capable of regenerating a new shoot. "Winter" peas are capable of withstanding $-10°C$ when completely exposed and they will tolerate short periods of as low as $-40°C$ if they are protected by a deep layer of snow.

These cool season food legumes have a low optimum temperature for vegetative growth and even a lower optimum temperature for reproductive growth. Thus, Stanfield *et al.* (1966) reported that the optimum temperature for pea was 21/16°C (day/night) for vegetative growth and 16/10 °C for reproductive growth. These crops are most sensitive to high temperatures during flowering and pod fill as exhibited by flower abortion, early pod abortion, and even abortion of seeds within pods.

Food legumes are poor weed competitors. The seeds germinate and emerge rapidly, but seedlings develop slowly between emergence and formation of a complete canopy. Severe weed competition and high yield losses can occur if the weeds are not controlled, either culturally or chemically. Yield losses can exceed 80% in extreme cases.

These four species originated in the Near East and thus are classed as long-day plants. However, some short-day accessions have evolved, especially lentils in southern India.

All legumes have a high phosphorus requirement and respond well to seed-placed phosphate fertilizer. However, pea roots are very sensitive to seed-placed fertilizers. Thus, low phosphate rates must be used on peas if the phosphate fertilizer is placed in a narrow band with the seed.

References

Bos, L., Hampton, R. O. and Makkouk, K. M. 1988. In: *World Crops: Cool Season Food Legumes*, pp. 591–615 (ed. R. J. Summerfield). Dordrecht: Kluwer Academic Publishers.

Haddad, N. I., Muehlbauer, F. J. and Hampton, R. O. 1978. *Crop Science* 18: 613–615.

Hagedorn, D. J. 1974. *Virus diseases of pea, Pisum sativum. Monograph No. 9*, 47 pp. American Phytopathological Society. St. Paul, MN, USA.

Hagedorn, D. J. (ed.). 1984. *Compendium of Pea Diseases*, 57 pp. St. Paul, MN, USA: American Phytopathological Society.

Hagedorn, D. J. and Gritton, E. T. 1973. *Phytopathology* 63: 1130–1133.

Kaiser, W. J. 1981. *Economic Botany* 55: 300–320.

Kraft, J. M., Haware, M. P. and Hussein, M. M. 1988. In: *World Crops: Cool Season Food Legumes*, pp. 565–575 (ed. R. J. Summerfield). Dordrecht: Kluwer Academic Publishers.

Morrall, R. A. A. and Shepherd, J. W. 1981. *Canadian Plant Disease Survey* 61: 7–13.

Nene, Y. L., Hanounik, S. B., Qureshi, S. H. and Sen, B. 1988. In: *World Crops: Cool Season Food Legumes*, pp. 577–589 (ed. R. J. Summerfield). Dordrecht: Kluwer Academic Publishers.

Singh, K. B. and Reddy, M. V. 1991. *Advances in Agronomy* 45: 191–222.

Slinkard, A. E. 1993. In: *Breeding for Stress Tolerance in Cool Season Food Legumes*, pp. 429–438 (eds. M. C. Saxena and K. B. Singh). Chichester, UK: John Wiley & Sons.

Stanfield, B., Ormrod, D. P. and Fletcher H. F. 1966. *Canadian Journal of Plant Science* 46: 195–203.

Wallen, V. R. 1974. *Canadian Plant Disease Survey* 54: 86–90.

Biotic and abiotic stresses of pulse crops in Europe

L. MONTI[1], A.J. BIDDLE[2], M.T. MORENO[3] and P. PLANCQUAERT[4]

[1] Dept. of Agronomy and Plant Genetics, University of Naples, Via Università 100, 80055 Portici, Italy;
[2] Processors and Growers Research Organisation, Great North Road, Thornhaugh, Peterborough, PE8 6HJ UK;
[3] Centro de Investigación y Desarrollo Agrario, Departamento de Mejora y Agronomía, Córdoba, Spain, and
[4] Institute Technique des Céréales et des Fourrages (ITCF), 8 Avenue du Président Wilson, 75116 Paris, France

Abstract

Biotic and abiotic stresses seriously reduce seed yields of pulses in Europe and are major causes of yield instability.

Many chemicals have proved to be effective in controlling biotic stresses, but several are not cost-efficient. Healthy seeds, seed treatment with chemicals, and the use of resistant and tolerant cultivars are now used to control several diseases. On the other hand, some constraints can only be controlled by avoiding infested soil. Pest control is increasingly being carried out by appropriate use of chemicals and by the use of monitoring systems. Weeds are becoming the most serious problem, and merit considerable scientific attention.

Regarding abiotic stresses, cold and drought are the most important. Control of drought stress in southern Europe is obtained by winter sowing of cold-tolerant cultivars which are able to escape drought that often occurs in late spring and summer.

Introduction

The full genetic potential of pulse crops is limited by poor management and by natural constraints. The identification of biotic and abiotic constraints which affect these crops in Europe is difficult due to the differences in agro-climatic conditions in which these species are grown and because these crops can be sown both in spring and in autumn/winter in some regions (Mediterranean area). Different stresses predominate in the various agroclimatic zones: for instance, *Fusarium* predominates in warm climates, and *Verticillium* in cooler regions; *Erysiphe* predominates in dry conditions, and *Peronospora* in cold and humid areas. Viruses cause more damage to spring sown crops than to winter-sown crops. Waterlogging and frost are absent in southern Europe.

Biotic constraints due to viruses, bacteria, fungi, insects, and nematodes that are important in pulse crops in Europe, together with precise control measures to be adopted and an indication of resistant or tolerant cultivars are reviewed.

F.J. Muehlbauer and W.J. Kaiser (eds.), Expanding the Production and Use of Cool Season Food Legumes, 204–218.
© 1994 *Kluwer Academic Publishers.*

Regarding abiotic stresses, only the most common natural stresses are considered, namely the lack of water and unfavorable temperatures.

Data are given only for pea, chickpea, faba bean and lentil. Grasspea (*Lathyrus sativus*) and *L. cicera* are grown in very small restricted areas (e.g., Maseta in Spain). Grasspea is considered a rather primitive crop and very tolerant of harsh climatic conditions; however, the production could expand provided that cultivars free of neurotoxins become available (Franco-Jubete, 1991).

Biotic Constraints

Viral and Bacterial Diseases

The major viral and bacterial diseases which affect pulse crops in Europe are listed in Table 1. Pea enation mosaic virus (PEMV) is common in peas during seasons which encourage pea aphids; other viruses such as pea streak (Hagedorn, 1984) are also found in such years. Until there is widespread availability of PEMV resistant cultivars, effective control of pea aphid remains as the only measure to reduce damage from this virus.

Serious yield losses due to pea seedborne mosaic virus (PSbMV) have been reported and a survey has shown that several cultivars of vining and combining peas are susceptible. Although initially seedborne, PSbMV is spread rapidly through the crop by aphid vectors; however, controlling the spread of PSbMV with aphicides has been ruled out (McKeown and Biddle, 1991). Resistant breeding lines of peas are available, but until there is widespread introduction of resistant cultivars, the use of healthy seed is of prime importance. An ELISA test has been developed for use with whole seeds (Ding *et al.*, 1992), and screening of commercial seed stocks for the presence of PSbMV is currently used in the UK (McKeown and Biddle, 1991).

Viruses of faba bean, chickpea, and lentil rarely reach epidemic proportions. Seedborne viruses such as broad bean stain and broad bean true mosaic virus are relatively uncommon at high levels, but bean yellow mosaic and bean leafroll virus are more frequently encountered in seasons when pea aphids are numerous.

As far as bacterial diseases are concerned, only pea bacterial blight (*Pseudomonas syringae* pv. *pisi*) is found regularly in seed stocks (Roberts *et al.*, 1991). Yield losses, however, have been confined so far to autumn-planted peas, where it is considered the most important constraint. However, in the UK, the disease is controlled by a statutory certification scheme for field pea seed and by a voluntary code of practice for vining pea seed producers. A sensitive seed test using a combination of serology and host inoculation is used by seed testing laboratories (Ball and Reeves, 1991). These workers have also reported on the development of a DNA probe-based test and Candlish *et al.* (1988) have developed an Mab-based test for rapid screening of infected seed.

Table 1. Main viral and bacterial diseases of pulses in Europe

Plant Species	Disease	Control Measures
Pea	Pea seedborne mosaic (PSbMV)	Resistance; Healthy seed
	Pea enation mosaic (PEMV)	Aphid control; Resistance
	Pea early browning (PEBV)	Healthy seed; Avoid vector; infested soil
	Bean leaf roll (BLRV)	Resistance; Aphid control
	Bean yellow mosaic (BYMV)	Resistance
	Bacterial blight (*Pseudomosan syringae* pv. *pisi*)	Resistance; Healthy seed
Chickpea	Stunt (BLRV)	Healthy seed
Faba bean	Broad bean true mosaic (BBTMV)	Healthy seed; Aphid control
	Broad bean stain (BBSV)	Healthy seed; Aphid control
	Bean yellow mosaic (BYMV)	Resistance
	Pea seedborne mosaic (PSbMV)	Healthy seed
	Bean leaf roll (BLRV)	Aphid control
Lentil	Pea seedborne mosaic (PSbMV)	Resistance

As the majority of viral and bacterial diseases are seed-transmitted, the use of pathogen-free seeds is very important. Currently, the use of healthy seed and resistant cultivars is the most common measure for controlling these diseases in several European countries.

Fungal Diseases

Fungal diseases which affect pulse crops in Europe are listed in Table 2. In pea, downy mildew (*Peronospora viciae*) is very commonly found in crops grown on land with a history of pea growing. The oospores are very long-lasting in the soil and under cool conditions, emerging seedlings can become systemically infected early in the season (Pegg and Mence, 1970). Crop rotation is of little value in reducing the risk of infection. In faba bean, the same pathogen causes disease during flowering.

Table 2. Main fungal diseases of pulses in Europe and their control

Plant Species	Disease	Control Measures
Pea	Leaf and pod spot (*Ascochyta pisi*)	Plant treatment; Seed treatment
	Ascochyta blight (*Mycosphaerella pinodes*)	Plant treatment; Seed treatment
	Grey mold (*Botryotinia fuckeliana*)	Plant treatment
	Foot rot (*Fusarium solani* f. sp. *pisi Phoma medicaginis* var. *pinodella*)	Resistance; Avoid infested soil
	Root rot (*Aphanomyces euteiches*)	Avoid infested soil
	Fusarium wilt (*Fusarium oxysporum* f. sp. *pisi*)	Resistance
	Stem rot (*Sclerotinia sclerotiorum*)	Avoid infested soil; Plant treatment
	Powdery mildew (*Erysiphe pisi*)	Plant treatment; Resistance
	Downy mildew (*Peronospora viciae*)	Seed treatment; Resistance
Chickpea	Fusarium wilt (*Fusarium oxysporum* f. sp. *ciceris*)	Resistance
	Blight (*Ascochyta rabiei*)	Healthy seed; Crop rotation; Resistance
	Grey mold (*Botryotinia fuckeliana*)	Healthy seed
Faba bean	Chocolate spot (*Botrytis fabae*)	Resistance; Plant treatment
	Down mildew (*Peronospora viciae*)	Plant treatment
	Blight (*Ascochyta fabae*)	Healthy seed; Seed treatment
	Rust (*Uromyces viciae-fabae*)	Plant treatment; Resistance
	Foot root (*Phoma medicaginis*)	Avoid infested soil
Lentil	Vascular wilt (*Fusarium oxysporum* f. sp. *lentis*)	Avoid infested soil
	Stem rot (*Sclerotinia sclerotiorum*)	Avoid infested soil
	Root rot (*Rhizoctonia* spp.)	Crop rotation
	Blight (*Ascochyta fabae* f. sp. *lentis*)	Healthy seed; Crop rotation
	Rust (*Uromyces fabae*)	Resistance; Plant treatment

Protection of emerging pea seedlings is afforded by systemic seed treatments such as metalaxyl (Miller and De Whalley, 1981) or fosetyl aluminium. Foliar sprays with fungicides are not effective. Pea cultivars exhibit a wide range of resistance to infection (Dixon, 1981). Little work has been done on the use of seed treatments for faba bean, but foliar applications of fosetyl aluminium or metalaxyl during flowering have produced significant yield improvements (King, 1981).

Powdery mildew in pea is caused by *Erysiphe pisi* which negatively affects not only yields, but also grain quality. This fungus is important in hot and dry areas, particularly in autumn-sown material, where the first symptoms may appear as early as April. The fungus is seedborne and survives on crop residues. Wind, high temperatures associated with low humidity, and lack of water on leaf surfaces are all conditions that positively influence the spread of the disease. Several fungicides are known to control the disease. Two genes have been identified that afford resistance; and, in Italy, a cultivar has been bred which is resistant to one of the two races known to exist (Ciccarese and Cirulli, 1985). Also, many lines resistant to both races are at an advanced selection stage (Monti and Frusciante, in press).

Botryotinia fuckeliana can infect pea pods during flowering when the weather is wet and flower petals adhere to the developing pods. Fungicides (which contain iprodione or vinclozolin) applied during flowering and pod set have been shown to improve yield by preventing disease development in pea (Biddle and Yeatman, 1986; Plancquaert, 1991). *Mycosphaerella pinodes* can also cause severe yield loss especially in dry pea during wet summers in northern temperate areas. The fungus is both seedborne and soilborne and although effective seed treatments are available, the onset of disease in late summer cannot be prevented by this means. Sprays containing mixtures of chlorothalonil and flutriafol are currently used during periods of disease risk (Dobson and Heath, 1991; PGRO, 1991; Plancquaert, 1991). There is little evidence that any commercial cultivars are resistant to *Mycosphaerella*, but they differ in their tolerance to both *Mycosphaerella pinodes* and *Ascochyta pisi* (Bretag, 1989).

In faba bean, chocolate spot is troublesome in wet seasons by reducing the effective photosynthetic area and encouraging premature defoliation. Sprays at early flowering and repeated after three weeks are beneficial (Dobson and Giltrap, 1991). Rust (*Uromyces fabae*) is also a problem in both the spring and autumn-planted faba bean crop but is effectively controlled by fenpropimorph. The development of the disease is favored by hot conditions. Faba bean cultivars differ in their susceptibility to both chocolate spot and rust. There is some evidence that susceptibility to rust is associated with short-stature cultivars (Thomas and Sweet, 1991). *Ascochyta fabae* is not effectively controlled by either seed treatment or foliar sprays and healthy seed is the only means of control. Chickpea cultivars resistant to Ascochyta blight are now available (Crinò *et al.*, 1985): control of this disease has allowed winter-sown chickpea in the Mediterranean region to achieve good yields.

Stem rot caused by *Sclerotinia sclerotiorum* is an occasional disease problem in the pea crop. The sclerotia are quite long-lasting in the soil, but the disease is worse in areas of intensive vegetable growing and in wet summers. Crop rotation and the use of fungicides are recommended where the problem is recurrent (Davies, 1991).

Soilborne root infecting fungi are often found in crops with a long history of legume production and for many of them the only means of control is to avoid infested soil. Foot rot caused by *Fusarium solani* f. sp. *pisi* and *Phoma medicaginis* var. *pinodella* occurs as a complex in many soils (Biddle, 1983). The development of a foot rot prediction test which picks out those fields with a potential for disease has been established in the UK since 1983 (Biddle, 1984). Breeding lines resistant to *F. solani* are currently being selected (Kraft, 1989). Foot rot is also induced by poor soil conditions (Allmaras *et al.*, 1988) and Biddle (1988) has demonstrated the effects of cultivation practice on the severity of the disease. Recent work has shown that faba bean is often infected with *P. medicaginis* and currently, their susceptibility to a range of isolates from pea are being tested (PGRO, 1991).

Root rot caused by *Aphanomyces euteiches* is common in moisture retentive soils and occurs more frequently in wet years. Breeding lines are also being screened for tolerance (Kraft, 1989), and prediction tests like that used for foot rot are being used in the USA and in Scandinavia.

Because of the cross infectivity of some root-infecting fungi between pea, *Phaseolus*, and faba bean, legume crops in the UK are no longer grown in rotation separated by less than four non-legume crops (Gent *et al.*, 1988).

Fusarium wilt (*Fusarium oxysporum* f. sp. *pisi*) of pea is present in many soils with a long history of pea crops. Races 1 and 2 have been recorded, but other types of *F. oxysporum* which have yet to be race-typed have been recently isolated from pea. All cultivars of field pea grown in the UK are tested for their resistance to Race 1, and many resistant cultivars are grown commercially (Thomas and Sweet, 1991). Nearly all available vining pea cultivars are resistant to Race 1 and therefore the disease is not as common as in the past. Race 2 is not often found, although many of the common cultivars are thought to be susceptible. In chickpea, Fusarium wilt has become the most important disease in Spain (Cubero *et al.*, 1990).

Other diseases affecting the four cool season food legume crops are listed in Table 2. Among these four species, pea is the one for which research has obtained the best results in avoiding and controlling many diseases. Pea wilt is now a very minor problem in several countries, and seed treatments have effectively controlled *Ascochyta* (Biddle, 1981) and downy mildew (Miller and De Whalley, 1981); crop rotation and the use of healthy seeds have reduced diseases caused by other pathogens considerably.

Pests

The main insect pests affecting pulses in Europe are listed in Table 3. There are relatively few insect pests which seriously affect yield, although several cause indirect losses through quality spoilage. For insects and nematodes, no resistant cultivars are available; the main controls are performed through the use of insecticides on plants and through appropriate crop rotation to avoid infested soils.

Of the more important insects, aphids remain the most damaging by the direct feeding effects and by virus transmission; the use of aphicides remains the only possible control. In the UK, pea aphids do not show resistance to any of the commonly used aphicides, including the extensively used pirimicarb. Recent work on dry pea has shown that the highest significant yield responses to aphid control occurred when applications were made on infested crops before and during flowering (Lane and Walters, 1991). Work to determine threshold numbers of aphids and the optimum timing of sprays in vining pea is currently in progress (PGRO, 1991).

Black bean aphid (*Aphis fabae*) colonizes faba bean and also has a wide host range. This aphid species covers the growing point and flower trusses and reduces yields while transmitting damaging viruses. Insecticides such as pirimicarb are routinely used.

Pea midge (*Contarinia pisi*) causes yield losses where the production areas are restricted to a small geographical area which allows the development of large midge populations, as occur in some localized areas of the UK. In Sweden, control of midges is effected by large scale rotational practices where the main pea production area is moved regularly (Jonsson, 1988). In other European countries, especially in the more northerly temperate areas, insecticide use is the only practical means of control, but the timing of applications is also critical. Often sprays are mis-timed or applied prophylactically (Biddle, 1985). A female sex-pheromone has been demonstrated by Wall *et al.* (1985) and work is currently in progress to synthesize the pheromone for its possible use in a monitoring system (PGRO, 1990, 1991).

Pea and faba bean weevil (*Sitona lineatus*) can be damaging to newly emerged seedlings of pea, faba bean, and lentil. The larvae feed below ground on the root nodules thereby reducing nitrogen availability and subsequent yield. Chemical control by sprays is aimed at controlling the adults early in the spring (Dore and Bouthier, 1991). However, this spray is often mis-timed and the use of granular insecticides such as aldicarb or phorate applied to the seedbed before planting has resulted in more effective control (Biddle, 1985). There is potential for seed treatment with insecticides (Baughan *et al.*, 1985), and the development of a pheromone-based monitoring system (Blight *et al.*, 1991) may also aid the timing and necessity of insecticidal application.

The larva of the pea moth (*Cydia nigricana*) is more likely to spoil product quality than to directly reduce yield, but an effective monitoring and spraying program has been in use in the UK for several years (Biddle *et al.* 1983). Early maturing pea cultivars escape damage by pea moth larvae.

Table 3. Main insect pests of pulses in Europe and their control

Plant Species	Pest	Control Measures
Pea	Pea aphid (*Acyrthosiphon pisum*)	Plant treatment
	Pea weevil (*Sitona lineatus*)	Plant treatment
	Pea midge (*Contarinia pisi*)	Crop rotation; Plant treatment
	Pea moth (*Cydia nigricana*)	Early cv; Plant treatment/monitoring system
	Bean seed fly (*Delia platura*)	Avoid infested soil
	Pea seed beetle (*Bruchus pisorum*)	Seed treatment; Plant treatment
	Adzuki bean seed beetle (*Callosobruchus chinensis*)	Seed treatment; Plant treatment
	Pea cyst nematode (*Heterodera goettingiana, Pratylenchus* spp.)	Avoid infested soil
	Root knot nematode (*Meloidogyne javanica*)	Avoid infested soil
Chickpea	Cowpea aphid (*Aphis craccivora*)	Plant treatment
	Gram pod borer (*Helicoverpa armigera*)	Plant treatment
	Leaf miner (*Liriomyza cicerina*)	Plant treatment
	Seed beetle (*Callosobruchus* spp.)	Plant treatment
	Root knot nematode (*Meloidogyne artiellia*)	Avoid infested soil
Faba bean	Black bean aphid (*Aphis fabae*)	Plant treatment
	Cowpea aphid (*Aphis craccivora*)	Plant treatment
	Bean weevil (*Sitona lineatus*)	Plant treatment
	Bean seed fly (*Delia platura*)	Avoid infested soil
	Stem nematode (*Ditylenchus dipsaci*)	Avoid infested soil; Healthy seed
	Bean seed beetle (*Bruchus rufimanus*)	Seed treatment; Plant treatment
Lentil	Cowpea aphid (*Aphis craccivora*)	Plant treatment
	Lentil seed beetle (*Bruchus lentis*)	Plant treatment
	Bean weevil (*Sitona lineatus*)	Plant treatment

Chickpea fly (*Liriomyza cicerina*) is spread in southern Europe, but damages have not been evaluated.

Pea cyst nematode (*Heterodera gottingiana*) is very damaging to pea crops; it also infests faba bean, but without affecting yields. Rotations which include close cropping of pea and faba bean are avoided to prevent the development of infestations. Chemical control with oxamyl is possible (Whitehead *et al.*, 1974) but is very costly; the best means of control is to avoid pea growing in fields known to be infested.

Stem nematode (*Ditylenchus dipsaci*) is common as a free-living nematode in many soils, but a giant race is particularly damaging to faba bean causing severe distortion of the plant and loss of yield as well as infesting the seed (Hooper, 1971; Hanounik and Bisri, 1991). There is no economically viable chemical means of control, and rotation and the use of non-infested seed are the only effective methods (Hooper, 1991).

Faba bean seed fly larvae (*Delia platura*) often infest late-sown legumes including pea and faba bean, and late planting in fields containing crop residues is avoided where possible.

Serious damage during storage, but less than in the past, is caused by *Bruchus* spp. and *Callosobruchus* spp. (Weigand and Bishara, 1991). The bruchid larvae feed and develop inside the seeds. Treating fields at the early flowering stage is the most effective control measure. Seeds stored for sowing are treated with chemicals.

Weeds

In order to achieve the full economic potential of intensively grown crops, weeds must be controlled to remove plant competition. There are several weeds that infest pulses in Europe which differ according to region.

Peculiar to the Mediterranean region, where it is endemic, is broomrape (*Orobanche* spp.), a parasitic weed. *Orobanche crenata*, in particular, attacks faba bean, lentil, and pea. The fruits contain an average of 150,000 seeds which can remain viable when released into soil for 15–20 years. Most of the damage to the host is done during the first two weeks after emergence. Hand-pulling of parasitic shoots before seed formation is the traditional practice to control the weed, but it has limited effectiveness because of the persistance of seeds in the soil. The only way to prevent damage is to avoid infested fields. Various chemicals have been tested but none is as yet cost-effective. Studies are needed on the chemical substances that can stimulate or inhibit seed germination (Cubero *et al.*, 1988). Some satisfactory source of resistance has been found and resistant breeding lines were obtained in Spain that, when cultivated in combination with the application of glyphosate, produce completely clean fields (Cubero and Hernández, 1991; Moreno and Martínez, 1991).

As for other weeds, pea is the crop which suffers most from these constraints. In this crop, weeds can severely reduce yield, and tall species which shade the crop plants are particularly damaging (Knott, 1990). Vining pea crops are sometimes rejected by the processor where there is a risk of contamination with poisonous berries of *Solanum nigrum* and volunteer potatoes. Faba bean crops offer more competition to weeds and suppress low-growing species in most situations. Lentil crops are poor competitors with weeds, probably because growth rates are slow in the early/initial stages.

In Europe, the main means of weed control is with herbicides (Knott, 1985). For pea crops, narrow row widths of 200 mm or less preclude inter-row cultivations and, although faba bean crops are sown with precision drills on wider rows of 450 mm, machine hoeing is an option which is seldom used except on soils which are unsuitable for residual herbicide application.

There are several recommendations for residual triazine herbicides although these cannot be used on highly organic soils and sands. For most food legume crops, broad-leaved weeds are controlled with pre-emergence treatments. There are a few foliar-acting herbicides for post-emergence application to pea crops. Faba bean crops are sensitive to "hormone" herbicides (which cause severe epinasty) and to most contact-acting herbicides, since cuticular leaf wax is insufficient to avoid phytotoxicity (Knott, 1990), although bentazone is sometimes used. In most crops, one herbicide application for broad-leaved weeds is sufficient. Infestations of *Elymus repens* in pea and faba bean in the UK have declined largely due to eradication with glyphosate elsewhere in the rotation. Post-emergence graminicides which suppress weeds in pea and faba bean are often applied too late to prevent yield reduction (Knott, 1982). Lentil crops are more sensitive to herbicide toxicities than other pulses.

Though a wide range of herbicide products is used, new herbicides are not produced specifically for pulses since these legumes are considered "minor crops" by the agrochemical manufacturers. There are few new developments and there is the possibility of withdrawal of approval for older chemicals which are under review. Besides the problem of *Orobanche* mentioned above, the trend of cutting weed control costs in cereals in rotation, the ban on straw burning, the policy of "set-aside", and the increasing use of semi-leafless pea cultivars with reduced smothering capacity (King, 1978) are also likely to increase weed problems in legume crops.

Abiotic Constraints

In northern Europe, the four cool season food legume crops are all spring-sown and excess soil moisture at the initial vegetative stage and drought at the reproductive stages are the main constraints to be considered in connection with some diseases and insect pests. In most seasons, pea and faba bean do not require additional water, but in some areas, free-draining soils suffer moisture deficit and irrigation is often applied. Work has shown that yield responses in pea could be obtained from applications made during flowering and not during the vegetative stage (Salter, 1962, 1963). Experiments in France have shown that whenever possible it is better to irrigate little and often and that irrigation before flowering can improve yield; some cultivars were also identified which appear tolerant to drought or respond better to irrigation (Plancquaert, 1991; Deumier *et al.*, 1991). Semi-leafless pea genotypes characterized by substantial leaf reduction proved in Italy to be more tolerant in water stress conditions (Monti *et al.*, in press) as a result of a higher CER (CO_2 exchange rate) per unit of leaf surface, lower stomatal resistance, and lower canopy temperature (Leone *et al.*, 1987; Alvino and Leone, in press).

Late-planted faba bean crops often suffer moisture stress and produce reduced yields. Similar findings in spring-sown faba bean were reported by Hebblethwaite *et al.* (1991). Hebblethwaite *et al.* (1991) showed that spring-sown faba bean responded to irrigation all through the growing season, although the willingness of growers to irrigate faba bean in preference to other more economically responsive crops is limited. Water is essential during the vegetative phase to obtain a normal differentiation of flower buds and after flowering to supply fruits with assimilates through a good canopy (Karamanos and Gimenez, 1991).

In Mediterranean areas, faba bean is traditionally winter-sown, while pea, lentil, and chickpea are sown at the beginning of spring. In these regions, temperatures during the reproductive period may rise above optimum, leading to heat and drought stress (Saxena *et al.*, 1988) which can reduce yields. In the last few decades, a large output of literature has highlighted the considerable advantage gained by earlier-sown pea and chickpea in southern Europe. Autumn sowing results in vigorous winter growth which render the crops more susceptible to frost damage and leaf diseases due to *Ascochyta* spp. or to *Botrytis* spp. It is therefore necessary to use cultivars with genetic resistance to cold temperatures and leaf diseases. Autumn sowing dates should also be chosen in each region in such a way that during cold spells the plants have a strong and deep root system, a slow growth rate, and a small above-ground vegetative structure with few leaves (Murray *et al.*, 1988). Several pea cultivars have been bred which are adapted to winter sowing (Cousin *et al.*, 1985). In France, sowings of winter pea in November are performed with cultivars in which so much time elapses between emergence and the flower initiation stage that in cold weather conditions, plants are still in the vegetative phase (Plancquaert, 1991). Experiments carried out in Italy have shown that winter sowings can also extend the pea protein crop to non-irrigated areas of southern Italy in rotation with durum wheat. A large breeding program is under way in Italy to introduce cold tolerance to plants that are at the flowering stage during February-March, and to incorporate powdery mildew resistance as well: several advanced lines have already been selected (Monti, 1988; Monti *et al.*, in press).

Chickpea represents one of the best examples of the effectiveness of plant breeding in crop improvement. In Italy, France, and Spain (Saxena, 1990; Saccardo and Calcagno, 1990; Moreno, 1984), starting from material selected at ICARDA, cultivars were bred which were able to withstand very low temperatures ($-12.5°C$ with no snow cover on the crop) and therefore adapted to autumn-winter sowings. Chickpea sown in autumn can utilize winter rainfall and have vigorous growth and large yields. The new cultivars are also resistant to Ascochyta blight which is very aggressive on winter-sown crops.

Concluding Remarks

Estimates of the losses due to biotic stresses in pulses are vague and contradictory, mainly because there is a lack of data on the potential yield of these crops in the experimental sites to be compared with the yield obtained under stress (Cook, 1988). An overall yield loss of about 25% can be estimated in Europe, due to diseases, insect pests, and weeds.

For almost all the biotic constraints, methods of detection, isolation and identification are well-established. Because of the poor consideration into which these crops are taken, cost-efficient methods of controlling diseases and insect pests are needed. The selection of seeds on the basis of health tests and the use of seed dressings with chemicals now control several diseases arising from fungi, viruses and bacteria. Pest control is also becoming increasingly efficient through a more appropriate use of insecticides and the use of monitoring systems.

Resistance to many diseases has been obtained through breeding. Among the four crops we have dealt with, the greatest breeding successes have been achieved with pea. In this species, resistance or tolerance has been reported for nine of the 15 major diseases which are known to attack the crop. Weed control has been inadequately studied in pulses, and weeds may well become, as a consequence, one of the most important constraints to pulse production in Europe in future years.

As far as the abiotic stresses are concerned, success may be achieved through a more thorough knowledge of the physiological mechanisms underlying plant-stress interactions (Monti, 1987; Monti *et al.*, 1991), and more effective collaboration among legume scientists (Monti and Venezian, 1990).

References

Allmaras, R. R., Kraft, J. M. and Miller, D. E. 1988. *Annual Review of Phytopathology* 26: 219–243.

Alvino, A. and Leone, A. Response to low soil water potential in pea genotypes (*Pisum sativum* L.) with different leaf morphology. *Scientia Horticulturae* (In press).

Ball, S. and Reeves, J. C. 1991. In: *Techniques for the Rapid Detection of Plant Pathogens*, pp. 92–207 (eds. J. M. Duncan and L. Torrance). London: Blackwell.

Baughan, P. J., Biddle, A. J., Blackett, J. A. and Toms, A. M. 1985. In: *Symposium on Application and Biology*, pp. 97–106 (ed. E. S. E. Southcombe). British Crop Protection Council Monogram No. 28. Brighton, England, UK: British Crop Protection Council.

Biddle, A. J. 1981. Tests of Agrochemicals and Cultivars. *Annals of Applied Biology* (Supplement No. 2) 97: 34–35.

Biddle, A. J. 1983. In: *Proceedings of the 10th International Congress of Plant Protection*, p. 117. Vol. 1. Brighton, England, UK: The British Crop Protection Council.

Biddle, A. J. 1984. In: *Proceedings of the British Crop Production Conference – Pest and Diseases*, pp. 773–777. Vol. 2. Brighton, England, UK: The British Crop Protection Council.

Biddle, A. J. 1985. In: *The Pea Crop – A Basis for Improvement*, pp. 257–266 (eds. P. D. Hebblethwaite, M. C. Heath and T. C. K. Dawkins). London: Butterworths.

Biddle, A. J. 1988. In: *Proceedings of 78th Annual Meeting of the Western Washington Horticultural Association*, pp. 54–55. Puyallup, WA, USA.

Biddle, A. J., Blood Smyth, J., Cochrane, J., Emmett, B., Garthwaite, D. G., Graham, J. C., Greenway, A. R., Lewis, T., Macaulay, E. D. M., Perry, J. N., Smith, M. C., Sturgeon, D. M. and Wall, C. 1983. In: *Proceedings of the 10th International Congress of Plant Protection*, Vol. 1, pp. 161. Brighton, England, UK: The British Crop Protection Council.

Biddle, A. J. and Yeatman, C. J. 1986. In: *Proceedings of the 1986 British Crop Protection Conference – Pests and Diseases*, Vol. 3, pp. 1021–1025. Brighton, England, UK: The British Crop Protection Council.

Blight, M. M., Dawson, G. W., Pickett, J. A. and Wadhams, L. J. 1991. *Aspects of Applied Biology 27, Production and Protection of Legumes*, pp. 137–142. Association of Applied Biologists, National Vegetable Research Station, Wellesbourne, UK.

Bretag, T. W. 1989. *Annals of Applied Biology* 114: 156–157.

Candlish, A. A. G., Taylor, J. D. and Cameron, J. 1988. In: *Proceedings of the 1988 Brighton Crop Protection Conference – Pests and Disease – British Crop Protection Council*, pp. 787–794. Brighton, England, UK: The British Crop Protection Council.

Ciccarese, F. and Cirulli, M. 1985. In: *Proceedings of the EUCARPIA Meeting on Pea Breeding*, pp. 135–136. Sorrento, Italy, 10–13 June 1985.

Cook, R. J. 1988. In: *World Crops: Cool Season Food Legumes*, pp. 649–660 (ed. R. J. Summerfield). Dordrecht: Kluwer Academic Publishers.

Cousin, R., Vingere, A. and Eteve, G. 1985. In: *Proceedings of the EUCARPIA Meeting on Pea Breeding*, pp. 109–117. Sorrento, Italy, 10–13 June 1985.

Crinò, P., Porta-Puglia, A. and Saccardo, F. 1985. *Genetica Agraria* 39: 319–330.

Cubero, J. I. and Hernández, L. 1991. *Options Méditerranéennes – Série Séminaires* 10: 51–57.

Cubero, J. I., Moreno, M. T. and Gil, J. 1990. *Options Méditerranéennes – Série Séminaires* 10: 157–161.

Cubero, J. I., Pieterse, A., Saghir, A. R. and Borg, S. T. 1988. In: *World Crops: Cool Season Food Legumes*, pp. 549–563 (ed. R. J. Summerfield). Dordrecht: Kluwer Academic Publishers.

Davies, J. M. L. 1991. *Aspects of Applied Biology 27, Production and Protection of Legumes*, pp. 351–354. Association of Applied Biologists, National Vegetable Research Station, Wellesbourne, UK.

Deumier, J. M., Bouthier, A. and Gillet, J. P. 1991. *Perspectives Agricoles* 143.

Ding, X. S., Cockbain, A. J. and Govier, D. A. 1992. Annals of Applied Biology 121: 75–83.

Dixon, G. R. 1981. In: *The Downy Mildews*, pp. 487–514 (ed. D. M. Spencer). London: Academic Press.

Dobson, S. C. and Giltrap, N. J. 1991. *Aspects of Applied Biology 27, Production and Protection of Legumes*, pp. 111–116. Association of Applied Biologists, National Vegetable Research Station, Wellesbourne, UK.

Dobson, S. C. and Heath, M. C. 1991. *Aspects of Applied Biology 27, Production and Protection of Legumes*, pp. 343–346. Association of Applied Biologists, National Vegetable Research Station, Wellesbourne, UK.

Doré, T. and Bouthier, A. 1991. Perspectives Agricoles 160.

Franco-Jubete, F. 1991. Los titarros. El cultivo de Lathyrus en Castilla y León. Junta de Castilla y León, Valladolid, Spain.

Gent, G. P., Biddle, A. J. and Knott, C. M. 1988. *The Pea Growing Handbook*. Processors and Growers Research Organisation, Peterborough, UK.

Hagedorn, D. G. (ed.). 1984. *Compendium of Pea Diseases*, 57 pp. St. Paul, Minnesota, USA: American Phytopathological Society.

Hanounik, S. B. and Bisri, M. 1991. *Options Méditerranéennes – Série Séminaires* 10: 59–66.

Hebblethwaite, P. D., Batts, G. R. and Kantar, F. 1991. *Aspects of Applied Biology 27, Production and Protection of Legumes*, pp. 77–84. Association of Applied Biologists, National Vegetable Research Station, Wellesbourne, UK.

Hooper, D. J. 1971. *Plant Pathology* 20: 25–27.

Hooper, D. J. 1991. *Aspects of Applied Biology* 27, *Production and Protection of Legumes*, pp. 143–149. Association of Applied Biologists, National Vegetable Research Station, Wellesbourne, UK.

Jonsson, B. G. 1988. An ecological approach to management of the pea midge, *Contarinia pisi* (Winn.) in vining peas. Waxtskyddsrapporter, Avhandlingar 17 Uppsala, Sweden.

Karamanos, A. J. and Gimenez, G. 1991. *Options Méditerranéennes – Série Séminaires* 10: 79–90.

King, J. M. 1978. In: *Proceedings of the British Crop Protection Conference, Weeds*, pp. 511–517. Brighton, UK. Association of Applied Biologists, National Vegetable Research Station, Wellesbourne, UK.

King, J. M. 1981. *Tests of Agrochemicals and Cultivars. Annals of Applied Biology*, (Supplement 2) 97: 32–33.

Knott, C. M. 1982. In: *Proceedings of the British Crop Protection Conference*, Weeds, Brighton, UK, pp. 835–842. Association of Applied Biologists, National Vegetable Research Station, Wellesbourne, UK.

Knott, C. M. 1985. In: *The Pea Crop – A Basis for Improvement*, pp. 267–275 (eds. P. D. Hebblethwaite, M. C. Heath and T. C. K. Dawkins). London: Butterworths.

Knott, C. M. 1990. In: *Weed Control Handbook – Principles*, pp. 329–366 (eds. R. J. Hance and K. Holly). Oxford, UK: Blackwell Scientific Publications.

Kraft, J. M. 1989. *Pisum Newsletter* 21: 82–85.

Lane, A. and Walters, K. F. A. 1991. *Aspects of Applied Biology* 27, *Production and Protection of Legumes*, pp. 363–368. Association of Applied Biologists, National Vegetable Research Station, Wellesbourne, UK.

Leone, A., Alvino, V., Magliulo, V. and Zerbi, G. 1987. In: *Proceedings of the EEC Meeting on Drought Resistance in Plants: Genetic and Physiological Aspects*, pp. 297–310. Amalfi, Italy, 20–23 October 1986.

McKeown, B. M. and Biddle, A. J. 1991. *Aspects of Applied Biology* 27, *Production and Protection of Legumes*, pp. 333–338. Association of Applied Biologists, National Vegetable Research Station, Wellesbourne, UK.

Miller, M. W. and De Whalley, C. V. 1981. In: *Proceedings of the 1981 British Crop Protection Conference – Pests and Disease*, Vol. 1, pp. 341–348. Brighton, England, UK: The British Crop Protection Council.

Monti, L. M. 1987. In: *Proceedings of the EEC Meeting on Drought Resistance in Plants: Genetic and Physiological Aspects*, pp. 1–8. Amalfi, Italy, 20–23 October 1986.

Monti, L. 1988. *Agricoltura Ricerca* 85: 15–18.

Monti, L. M., De Pace, C. and Scarascia Mugnozza, G. T. 1991. *Options Méditerranéennes – Série Séminaires* 10: 143–151.

Monti, L. and Frusciante, L. In: *Proceedings of the Workshop on "Fondamenti dello Sviluppo di Germoplasma Resistente a Stress Biotici"*, Rome, 19–23 Feb. 1990.

Monti, L., Frusciante, L. and Romano, R. 1993. In: *Breeding for Stress Tolerance in Cool Season Food Legumes*, pp. 43–73. (eds. K. B. Singh and M. C. Saxena). Chichester, UK: John Wiley & Sons.

Monti, L. M. and Venezian, M. E. 1990. In: *Proceedings of The First Congress of the European Society of Agronomy*, 5–7 December 1990.

Moreno, M. T. 1984. *Cortijo del Cuarto* 51: 4–7.

Moreno, M. T. and Martínez, A. 1991. *Options Méditerranéennes – Série Séminaires* 10: 165–167.

Murray, G. A., Eser, D., Gusta, L. V. and Eteve, G. 1988. In: *World Crops: Cool Season Food Legumes*, pp. 831–843 (ed. R. J. Summerfield). Dordrecht: Kluwer Academic Publishers.

Pegg, G. F. and Mence, M. J. 1970. *Annals on Applied Biology* 71: 19–31.

PGRO. 1990. *Annual Report of the Processors and Growers Research Organisation*. Peterborough, UK.

PGRO. 1991. *Annual Report of the Processors and Growers Research Organisation*. Peterborough, UK.

Plancquaert, P. 1991. *Aspects of Applied Biology* 27, *Production and Protection of Legumes*: 189–197. Association of Applied Biologists, National Vegetable Research Station, Wellesbourne, UK.

Roberts, S. J., Reeves, J. C., Biddle, A. J., Taylor, J. D. and Higgins, P. 1991. *Aspects of Applied Biology* 27, *Production and Protection of Legumes*, pp. 327–332. Association of Applied Biologists, National Vegetable Research Station, Wellesbourne, UK.

Saccardo, F. and Calcagno, F. 1990. *Options Méditerranéennes – Série Séminaires* 9: 35–41.

Salter, P. J. 1962. *Journal of Horticultural Science* 37: 141–149.

Salter, P. J. 1963. *Journal of Horticultural Science* 38: 321–334.

Saxena, M. C. 1990. *Chickpea in the Nineties*, pp. 13–25. (eds. H. A. van Rheenen, M. C. Saxena, B. J. Walby and S. D. Hall). Patancheru, Andhra Pradesh, India: ICRISAT.

Saxena, M. C., Saxena, N. P. and Mohamed, A. K. 1988. In: *World Crops: Cool Season Food Legumes*, pp. 845–856 (ed. R. J. Summerfield). Dordrecht: Kluwer Academic Publishers.

Thomas, J. E. and Sweet, J. B. 1991. *Aspects of Applied Biology* 27, *Production and Protection of Legumes*, pp. 321–326. Association of Applied Biologists, National Vegetable Research Station, Wellesbourne, UK.

Wall, C., Pickett, J. A., Garthwaite, D. G. and Morris, N. 1985. *Entomologia Experimentalis et Applicata* 39: 11–14.

Weigand, S. and Bishara, S. I. 1991. *Options Méditerranéennes – Série Séminaires* 10: 67–74.

Whitehead, A. G., Tite, D. J., Fraser, J. R. and French, E. M. 1974. *Annals of Applied Biology* 78: 331–335.

Biotic and abiotic stresses constraining the productivity of cool season food legumes in different farming systems: specific examples

M.B. SOLH[1], H.M. HALILA[2], G. HERNÁNDEZ-BRAVO[3],
B.A. MALIK[4], M.I. MIHOV[5] and B. SADRI[6]

[1] Nile Valley Research Project (NVRP), ICARDA, P.O. Box 2416, Cairo, Egypt;
[2] Food Legume Program, Institut National de la Recherche, Agronomique de Tunisie (INRAT), 2080 Ariana, Tunis, Tunisia;
[3] Chachalacas, Atizapan, Mexico;
[4] Pakistan Agricultural Research Council (PARC), National Agricultural Research Centre, P. O. National Institute of Health, National Park Road, Islamabad, Pakistan;
[5] Institute of Wheat and Sunflower, "Dobroudja" near General Toshevo, Bulgaria, and
[6] Food Legume Research Section, Seed and Plant Improvement Section, Ministry of Agriculture and Rural Development (MARA), Mar-Abada Avenue, Karaj, Iran

Abstract

Productivity of chickpea (*Cicer arietinum*), faba bean (*Vicia faba*), grasspea (*Lathyrus sativus*), pea (*Pisum sativum*), and lentil (*Lens culinaris*) is constrained by various abiotic and biotic stresses in different farming systems. The yield potential of these crops is seldom achieved due to unsuitable cultivars and inadequate crop management to cope with these stresses. Major biotic stresses include diseases, insect pests, and weeds (parasitic and non-parasitic) while major abiotic stresses include extremes of soil moisture (drought or water logging), temperature extremes (heat stress or cold temperature/frost), imbalance in soil fertility (nutrient deficiencies or toxicity including salinity). Specific examples of biotic stresses include: Ascochyta blight and Fusarium wilt of chickpea in South Asia, Mexico, and North Africa; and faba bean necrotic yellows virus (FBNYV); and *Orobanche* on faba bean in Egypt. Specific examples of abiotic stresses include: drought and heat stress of chickpea in West Asia and North Africa, drought stress of lentil in Bulgaria, water logging of lentil in Ethiopia, and high temperature/heat stress on food legumes in Bulgaria and Sudan.

Introduction

Productivity of chickpea (*Cicer arietinum* L.), faba bean (*Vicia faba* L.), grasspea (*Lathyrus sativus* L.), pea (*Pisum sativum* L.), and lentil (*Lens culinaris* Medik.) is constrained by various abiotic and biotic stresses in different farming systems. Yield potential of these crops is seldom achieved due to unsuitable cultivars and crop management that is inadequate to cope with these stresses.

F.J. Muehlbauer and W.J. Kaiser (eds.), Expanding the Production and Use of Cool Season Food Legumes, 219–230.
© 1994 *Kluwer Academic Publishers.*

Major biotic stresses include diseases, insect pests, and weeds (parasitic and non-parasitic) while major abiotic stresses include extremes of soil moisture (drought or water logging), temperature (heat stress or cold frost), and soil fertility imbalances (nutrient deficiencies, toxicity and salinity). The seriousness of these stresses is often associated with particular farming systems characterized by specific agro-climatic and edaphic conditions.

Association of Stresses with Farming Systems

Cool season food legumes are grown in different farming systems throughout the world because of the differences in the adaptation of these crops to various agro-climatic conditions. For example, lentil, faba bean, pea, and grasspea are more winterhardy than chickpea (Murray *et al.*, 1988). Critical temperatures for freezing injury appear to be higher for chickpea than lentil, pea, or faba bean (Saxena *et al.*, 1988). Field rankings for drought tolerance are lentil > chickpea > pea > faba bean. These crops also differ in their susceptibility to diseases, insect pests, and parasitic weeds. Food legumes are generally poor competitors with weeds, with lentil being the least competitive.

In farming systems with favorable rainfall (above 450 mm annually) and where cool season food legumes are grown as winter crops, high moisture and humidity favor both foliar and root rot diseases. Major biotic stresses include Ascochyta blight (caused by *Ascochyta rabiei, A. fabae* f. sp. *lentis, A. pisi*, or *A. fabae*), chocolate spot and Botrytis gray mold (*Botrytis fabae, B. cinerea*), rusts (*Uromyces fabae, U. pisi*), collar and root rots (*Sclerotium rolfsii, Rhizoctonia bataticola, Fusarium solani*), wilts (*Fusarium oxysporum* f. sp. *ciceris, Verticillium albo-atrum*), stem rots (*Sclerotinia* spp.), *Stemphylium* spp., stem nematodes, stunt and bean leaf roll virus (BLRV).

The seriousness of these diseases depends on agro-climatic (mostly temperature, rainfall, and humidity) and soil moisture conditions that prevail during the cropping season. Rainfall during the growing season with relatively cool temperatures (5 to 15°C), encourages both development and spread of Ascochyta blight on winter and spring chickpea, lentil, pea, and faba bean. In extreme cases, yield losses in winter or spring chickpea can be up to 100% irrespective of chemical control measures, particularly when rainfall extends to the pod filling stage. In severe winter areas in northern latitudes (Europe, North America, and North Asia), rust is less important than Botrytis gray mold and Ascochyta blight. Fusarium wilt is associated with dark/black vertisols in North Africa, the Nile Valley, and the Indian sub-continent.

In farming systems under low rainfall and relatively low humidity, foliar diseases are usually less important than soilborne diseases. These farming systems, in addition to moisture stress, are usually characterized by higher temperatures where insect pests become more important production constraints. Common insect pests include aphids, pod borers (*Helicoverpa armigera*), stem borers (*Lixus* spp.), *Apion* spp., leaf miner (*Liriomyza* spp.),

Sitona spp., *Spodoptera* spp., white flies, cutworms (*Agrotis* spp.), and seed beetles (*Bruchus* spp. and *Callosobruchus* spp.). Viruses are more important in these areas because of the high activity of aphids as vectors.

In irrigated farming systems (mostly flood and furrow irrigation) and hot dry environments such as in Egypt and Sudan, foliar diseases are generally less important in cool season food legumes except in areas of high relative humidity and wetness, as is the case in the North Delta of Egypt. Botrytis gray mold and rust and to a much lesser extent Ascochyta blight are constraints to high production of chickpea in certain seasons. In these hot environments, aphids and other insect pests are of major importance due to the favorable warm environments. Viruses are of a major concern because of high aphid activity. In 1992, Egypt lost about 40% (about 60,000 ha) of its faba bean crop due to faba bean yellows necrotic virus (FBYNV), which is a new aphid-transmitted virus of faba bean.

In both irrigated and high rainfall areas, broomrape (*Orobanche* spp.) is a serious parasitic weed of faba bean, lentil, pea, and, to a lesser extent, winter chickpea. Broomrape is a serious parasitic weed throughout the Mediterranean region where three species are known to parasitize cool season food legumes. The most common species is *O. crenata*, particularly in Egypt, Morocco, Tunisia, Syria, and southern Spain, followed by *O. aegyptiaca* and *O. faetida* which was identified recently and is found in Beja area in Tunisia. Although *O. crenata* attacks most cool season food legumes, it is most serious on faba bean in Egypt and Morocco where yield losses of 100% are commonly observed in heavily infested areas. In Morocco, it is also serious on pea and lentil.

Abiotic stresses are strongly associated with the agro-climatic and edaphic conditions of the various farming systems. In cropping systems practiced in dry areas (minimal rainfall), drought and moisture stress are major production constraints, particularly in spring-grown crops and rice-based cropping systems where food legumes are raised on residual moisture. High temperature is another major constraint in cropping systems in dry areas under rainfed and irrigated conditions, especially when high temperatures prevail during the reproductive stage. Sudan is an extreme case where high temperatures at both the early and the late crop growth stages of the short winter season (100 days) is a major constraint to increased food legume production. In systems under high rainfall (above 450 mm annually), water-logging that causes anaerobic conditions is a major constraint in Europe and the highlands of Ethiopia. In areas with severe winters and continental climates, low temperatures and frosts are major concerns. Lentil, pea, vetch, and faba bean show adequate winterhardiness (and can endure temperatures up to -8 and $-15°C$) compared to chickpea which is more susceptible to cold/frost damage.

The following specific examples cover serious biotic and abiotic stresses in different farming systems at various agro-climatic and geographical regions of the world. Research efforts have succeeded with the stresses described below.

Specific Examples of Biotic Stresses

Ascochyta Blight of Chickpea in South Asia

Ascochyta blight is the most devastating disease of chickpea in South Asia where about 80 to 85% of the world's chickpea production is grown. In this region, chickpea is grown in three different cropping systems:
1. Rainfed cropping: where chickpea, as other pulses, is grown as a sole crop, in rotation with cereals or *Brassica* spp. or as a relay crop.
2. Rice-based cropping: where chickpea is grown after rice on residual moisture.
3. Irrigated cropping: where chickpea is grown on a limited area only.

Several serious Ascochyta blight epiphytotics have occurred in India and Pakistan, the two largest chickpea producing countries of the world. Production of chickpea in Pakistan dropped by 46 to 48% annually due to Ascochyta blight epiphytotics in three consecutive crop seasons (1979/80, 1980/81, 1981/82). Production declined in both the rainfed and the rice-based cropping systems in three zones lying above 28 to 30°N latitude. The chickpea growers of Pakistan suffered losses valued at US$ 158 million. The shortfall in the domestic chickpea supply due to blight damage were met through imports of 282,000 tons of pulses worth US$ 90 million. The chickpea growers of North India also suffered losses and India had to import large quantities of pulses to meet the domestic demand.

Because of the complex nature of *A. rabiei* and its epidemiology, integrated control of the disease was and continues to be the only reliable strategy to control or reduce the damage of Ascochyta blight in South Asia. In Pakistan, the integrated approach involves blight resistant cultivars, management of infested crop debris, crop rotation (in certain areas) and the use of clean seed (disease-free). This approach resulted in the restoration of an annual average production of 500,000 to 600,000 metric tons of chickpea compared to 300,000 tons during blight seasons.

Fusarium Wilt of Chickpea in Mexico

Fusarium wilt caused by *F. oxysporum* f. sp. *ciceris* is the major biotic stress to chickpea production in Mexico where it is grown on 70,000 ha (50,000 ha kabuli, 20,000 ha desi). The kabuli chickpea is grown as a winter crop (sown in fall) in the northwest region of Mexico in the Culiacan Valley (Sinaloa), Hermosillo Valley (Sonora) and Santo Domingo Valley (Baja, California). Most of the chickpea is grown under rainfed conditions except in relatively dry areas where two to three supplementary irrigations are applied during the growing season. Large seed size is critical for the export market which accounts for 90% of total production. Thus, research efforts by the National Research Institute of Mexico developed several chickpea cultivars with large seeds and resistance to Fusarium wilt: "Surutato 77", "Santo Domingo 82", and

"Mocorito 88". More recently, a new cultivar "Blanco Sinaloa 92" was developed with resistance to Fusarium wilt, very large seed (65 grams per 100 seeds), white salmon color, rugose seed surface, 160 days to harvest, and an average yield of 1.7 t ha^{-1}.

Because of the persistence of Fusarium wilt in the soil, control measures involve an integrated approach. Besides using resistant cultivars, farmers usually avoid planting on heavily infested fields where land is not limited. Crop rotations are not effective in reducing wilt incidence since the Fusarium wilt pathogen has the ability to survive in the soil for long periods. Other measures include the use of high-quality, disease-free seeds and seed dressing with Benlate T (30% Benomyl + 30 Thiram) at 1.5 g kg^{-1} of seed to eradicate seedborne inoculum.

Fusarium Wilt and Ascochyta Blight in North Africa

Wilt of chickpea is a serious biotic stress on both spring and winter sown chickpea in North Africa where the crop is grown exclusively under rainfed conditions in rotation with cereals and oil crops. Ascochyta blight is another serious biotic stress, particularly in recently introduced winter chickpea and in traditional spring chickpea, when rainfall occurs late in the cropping season (April/May). Yield losses from both diseases can be up to 100% when conditions are favorable for disease development. Disease reactions and yields of two Tunisian cultivars with resistant/tolerant and susceptible reactions to either one of the two diseases is shown in Table 1.

Table 1. Yield performance of two Tunisian chickpea cultivars in the presence of Ascochyta blight and Fusarium wilt compared to disease free situations

Cultivar	Resistance reactions		Yield[1] (Kg ha⁻¹)		
	Ascochyta	Wilt	Disease Free	Ascochyta Cond. Wilt Free	Wilt Cond. Asco Free
Kasseb	Tolerant	100%	3884	2090	0
Amdoun 1	Susceptible	0%	2867	0	2566

[1]   Tunisia in the 1990/91 crop season.

Considering the seriousness of both diseases, the breeding program in Tunisia is aimed at developing cultivars with dual resistance to Ascochyta blight and Fusarium wilt diseases using the shuttle breeding method (Harrabi and Halila, 1992). Already, advanced breeding lines with dual resistance are under agronomic testing anticipating release. However, as in other parts of the world, the complexity of Ascochyta blight necessitates an integrated approach to control the disease and maintain durable resistance. For this purpose, recently released resistant cultivars (FLIP 83-46C, FLIP 84-79C, and FLIP 84-92C) are used in combination with an effective fungicide (Bravo 500 or chlorothalonil) as shown in Table 2. Fungicide usage on susceptible chickpea cultivars (i.e.,

Amdoun 1 in Tunisia) is ineffective in controlling Ascochyta blight (Table 2). To control Fusarium wilt, farmers are advised to use resistant cultivars, avoid planting in heavily infested fields, use disease-free seeds, and apply seed treatments (e.g., Benlate T).

Table 2. Seed yield of chickpea under integrated control of Ascochyta blight in Tunisia using resistant cultivars and fungicides, 1991/92

			Yield (Kg ha⁻¹)		
	Rate	Susceptible	Resistant Cultivars (FLIP)		
Fungicide[1]	a.i. ha⁻¹	Amdoun[1]	83-46C	84-79C	84-92C
Bravo	1.5 l	142	1943	2343	1676
Peltar	500 g	47	1295	2076	1619
Bavistiu	120 g	0	1085	723[2]	1371
Quinolate Pro	120 g	0	1295	1371	1866
Check	–	0	1103	1307	1271

[1]   Three applications (seedling, mid-vegetative, and pod setting).
[2]   Heavily attacked by virus diseases.

Source: Halila (1992)

Botrytis and Rust on Faba Bean in the Delta Area of Egypt

Chocolate spot (*B. fabae*) and rust (*U. fabae*) are among the most serious yield-limiting biotic stresses of faba bean in the Delta area of Egypt and in the Mediterranean basin. Losses can be devastating when a virulent *Botrytis* pathogen and a susceptible host are brought together in an environment that favors disease development (Hannounik and Bisri, 1991). Even when faba bean is grown under irrigation in the Delta with very little rainfall, yield losses as high as 50% were reported in faba bean production under chocolate spot epiphytotic conditions and 5 to 20% yield losses were estimated annually. This is mainly due to the high relative humidity (80 to 90%) and favorable daily temperature (around 15 to 20°C) which prevails in the Delta in winter and early spring. Late rains in certain seasons increase relative humidity and wetness which leads to epiphytotic conditions. With the increase in temperature in spring and high relative humidity, rust prevails also and becomes serious. Because of such conditions in 1987/88 and 1990/91, chocolate spot and rust epiphytotics reduced faba bean production in the Delta by 50%. The integrated disease control package demonstrated to farmers since 1990/91 by the Agricultural Research Center, included resistant cultivars (e.g., "Giza 461"), an effective fungicide (Dimetheme M45, 250 g 100 l⁻¹) and an appropriate date of planting (around mid-November) increased seed yield by an average of 20 to 41% and more than 50% in certain locations (Nassib *et al.*, 1991). Currently, this practice is being adopted by most farmers in the Delta as a standard practice because of its effectiveness.

Faba Bean Necrotic Yellows Virus on Faba Bean in Egypt

In 1991/92, a virus disease epidemic destroyed the irrigated faba bean crop on about 60,000 ha in two major faba bean-producing Governorates (Minya and Beni Suif) in Upper Egypt. As a result, faba bean production was reduced by 40%. The epidemic was associated with an excessively cold winter and an outbreak of aphid vectors. Preliminary tests conducted in 1991/92 suggested the involvement of faba bean necrotic yellows virus (FBNYV), a recently identified virus of faba bean. In 1992/93, yield losses due to viral diseases was about 20% since conditions were not favorable to aphids which transmit FBNYV. In 1992/93, two systematic surveys (in early February and mid-March) on diseases with emphasis on virus diseases confirmed FBNYV as the main casual organism for the 1991/92 epidemic. The results of the two surveys indicated the presence of six viruses on faba bean: FBNYV (49.3%), bean yellow mosaic (BYMV) (24.3%), broadbean wilt (BBWV) (4.2%), cucumber mosaic (CMV) (0.4%), alfalfa mosaic (AMV) (0.3%), and bean leaf roll (BLRV) (0.2%) with varying frequencies (Makkouk, 1993; Rizkallah *et al.*, 1993). FBNYV causes chlorosis, thick leaves, leaf rolling, stunting, necrosis, and complete killing of the faba bean plants which then turn black. The virus was observed on food legumes in West Asia and North Africa in 1987/88 (Katul *et al.*, 1993). Fortunately, FBNYV is not seed transmitted since it is a phloem-limited plant virus which is transmitted by aphids. The potential danger of such a virus is considerable in North Africa and West Asia where the virus has been detected (Katul *et al.*, 1993). Resistance of faba bean to FBNYV has not been detected thus far and very little information is available about this virus. The Egyptian national program has started concentrated efforts to cope with this virus problem. Other virus diseases, such as BYMV which is seed transmitted, deserve special attention in Egypt.

Broomrape on Faba Bean in Egypt

Broomrape (mostly *O. crenata*) is the most serious biotic stress of faba bean in Egypt. After a long history of concentrated efforts by the Agricultural Research Center in Egypt, an integrated control package was developed and transferred to farmers in Middle and Upper Egypt where *O. crenata* is most serious. The first package included:
- A cultivar tolerant to *Orobanche*: "Giza 402"
- Date of sowing: first half of November
- Fertilization: 36 kg N ha^{-1} (starter) + 71 kg P$_2$O$_5$ ha^{-1}
- Chemical control: glyphosate (lancer) at 64 g a.i. 500 l^{-1} water ha^{-1}.
   Results of the adoption of this package over the last five years indicated that *Orobanche* infestation was reduced considerably (up to 96% in some farmers' fields in 1989/90) and yield was increased up to 292% in 1988/89 (Nassib *et al.*, 1988). More recently, the integrated package was modified to replace the high

rate of glyphosate by a reduced rate of 34 g a.i. ha^{-1} combined with NPK (1:1:2) with an increase in yield up to 182% (Nassib *et al.*, 1990). Economic analysis indicated that one Egyptian pound (L.E.) invested in reduced and high rate packages increased farmers gain by 27 and 10 L.E., respectively. A new chemical, Pursuit, is showing better results on faba bean than glyphosate, particularly in the crop tolerance margin which will make chemical control, as one component of the integrated control package, easier to use by farmers.

Examples of Abiotic Stresses

Drought and Heat Stress on Chickpea in West Asia and North Africa

Drought is the major abiotic constraint facing chickpea in West Asia and North Africa (WANA). Chickpea in WANA is traditionally grown as a spring-sown crop on conserved or residual soil moisture and is, therefore, exposed to both terminal drought and heat stress during the reproductive stage (Saxena, 1992). The considerable fluctuations in the amount and distribution of winter rains, which are common in the region, limits moisture supply and reduces the productivity of rain-fed chickpea. Yield losses observed in certain areas of the WANA region reached up to 100% occasionally as was the case in the past two seasons in Morocco. However, losses of 30 to 50% are commonly observed due to terminal drought and heat stress. ICARDA's strategy to overcome these constraints has been to shift sowing date from spring (March/April) to early fall (November/December) to be grown as a winter crop. Such a shift would make various phenological stages of crop development match with appropriate environmental conditions (Saxena, 1992), thus adequate moisture would be available and relatively cool temperatures would prevail at the reproductive stage. However, the requirements for winter chickpea in the lowlands of WANA include resistance to Ascochyta blight, cold tolerance, and better management of weed infestations. ICARDA, in collaboration with national programs, has worked on these requirements since 1977 and considerable achievements have been made in the development of winter chickpea with resistance to Ascochyta blight and drought tolerance (Singh, 1990). Productivity of chickpea was increased by nearly 100% by adopting winter sowing in place of the traditional spring sowing. On-farm results obtained by ICARDA and national programs throughout WANA confirmed the superiority of winter sowing. More than 25 winter chickpea cultivars have been released in WANA, and winter chickpea is grown on a large commercial scale in many countries of the region.

Drought Stress on Lentil in Bulgaria

The drought which affects all grain legumes in Bulgaria is a common phenomenon for the moderate continental climate of the country. Fall planting of the grain legumes often coincides with long drought periods which cause delayed and non-uniform emergence. The unhardened crop plants will then be exposed to severe winter temperature at a critical stage of crop growth and consequently more than 50% of the plants are killed by freezing. During the past 20 years, three very serious droughts were experienced in Bulgaria (1974, 1983, and 1985) and caused significant yield reductions in grain legumes and winter wheat (Tsenov, 1984; Mihov, 1986).

The assessment of drought damage which occurred during 1983 (Mihov, 1985, 1986) indicated that the direct and indirect damage reduced crop yields by 14 to 88% for the different regions of the country. Yields of different lentil cultivars were reduced by 22 to 64% in 1983, while in 1985 yield reductions were from 15 to 52%. The lentil cultivars showed differential reaction to drought severity and late maturing cultivars were damaged the most. Drought also affected seed quality of lentil by increasing the time required for cooking by three to seven times.

To cope with drought stress, breeding work in Bulgaria has aimed at developing earlier maturing cultivars with a more vigorous root system and rapid initial growth. On peas, increasing wax coatings on the leaves was a very important selection criteria because it decreased transpiration and created a better micro-environment to tolerate dry weather conditions.

High Temperatures in Bulgaria and Heat Stress in Sudan

In Eastern Europe, cool season food legumes are grown under rainfed conditions as spring crops. High temperatures (above 30°C) are a serious abiotic constraint to increased production of cool season food legumes. In Bulgaria, high temperatures (over 30°C) occur in June, July, and August when most of the food legume crops are in the reproductive or pod filling stages at which high temperature becomes critical to pod set and seed filling. High temperatures, which may also prevail at vegetative stages in early spring, cause withering, burning of lower leaves, and desiccation of poorly developed plants. In addition, plants will be stunted with lower pod-bearing nodes and fewer branches. The dry hot wind in combination with drought during flowering, pod set, and pod filling cause great yield losses. Such conditions reduce the period of flowering and pod filling and cause flower abortion, poor pod set, and poor seed filling.

The breeding program in Bulgaria has developed cultivars adapted to high-temperatures and drought conditions. Selection criteria involved early maturity through shorter vegetative stage and more rapid rate of growth to achieve early flowering and efficiency in pod filling. Stronger pod retention is another important selection criteria to avoid pod shedding as a result of rainfall at the crop maturity stage which is typical of some regions in Bulgaria.

In Sudan, where food legumes are grown under irrigation, high temperatures are a major constraint to high production, particularly in central Sudan. The growing season for food legumes is extremely short (100 to 110 days) and is characterized by high temperature stress (35°C or above) at early and late crop stages. There is no confounding effect between drought and heat since the crop is under irrigation. As a result of breeding efforts by the Agricultural Research Corporation, all faba bean, chickpea, and lentil cultivars were developed in Sudan to mature in about 100 days and to tolerate high temperatures at germination, crop establishment, pod set, and seed filling. Even with such a short season, yields of up to 2.2, 3.0, and 3.5 tons ha^{-1} can be obtained in lentil, chickpea, and faba bean, respectively.

The 1991/92 season in Sudan was exceptionally cool and consequently the growing season was longer compared to a normal year. Temperature differences between the 1991/92 and 1990/91 crop seasons were up to 6°C in December and similar for the rest of the season. As a result of the reduction in temperature, productivity of faba bean and lentil was increased by 88% and 31%, respectively, in some areas (Salih, 1992). The increase in chickpea productivity was less dramatic.

Water Logging of Lentil in Ethiopia

Lentil requires proper soil and water management since it is highly sensitive to excess soil moisture. On heavy soils of low infiltration and unlevelled fields, water logging, as a result of heavy rains, can cause serious damage to the lentil crop growth, development and yield. Such conditions which prevail in the vertisols of the highlands of Ethiopia make water logging the most serious constraint to increased production of lentil. To solve this problem, a broad-bed-furrow (BBF) animal-drawn implement was developed to facilitate drainage in vertisols through raised seedbeds. The implement was developed through collaborative efforts on vertisols management among the Ethiopian National

Table 3. Grain and straw yields of lentil in response to improved drainage of vertisols using broad-bed-furrow (BBF) and ridge/furrows at three locations in Ethiopia over two seasons, 1990 and 1991

Location	Seedbed	Grain yield t ha⁻¹	Increase %	Straw yield t ha⁻¹	Increase %
Akaki	BBF	2.32 a	59	4.65 a	28
	Ridge and Furrows				
	LSD (0.01)	0.78			
	CV (%)	9.50			
Dibandiba	BBF	1.30 a	102	4.40 a	27
	Ridge and Furrows	0.20 b		3.40 b	
	LSD (0.01)	0.80		0.60	
	CV (%)	8.00		0.50	
Keteba	BBF	1.60 a	99	4.90 a	59
	Ridge and Furrows	0.80 b		3.10 b	
	LSD (0.01)	0.30		1.40	
	CV (%)	8.00		1.00	

For each location, mean values within the same variable (grain or straw) followed by the same letter are not significantly different at P < 0.01.

Program, ICRISAT, and ILCA. Testing the BBF over several years confirmed its superiority and yields increased from 59 to 102% and 27 to 59% in seed and straw yields, respectively, over two seasons, 1989/90 and 1990/91 (Table 3, Mamo *et al.*, 1991). The implement is being made available to Ethiopian farmers after being demonstrated successfully.

Concluding Remarks

The productivity of cool season food legumes is constrained by particular biotic and abiotic stresses in different farming systems. The seriousness of these losses depends on the agro-climatic conditions prevailing in the crop season. Research efforts to minimize or control the damage of these stresses have mostly used an integrated management approach involving genetic resistance/tolerance of the crop, cultural practices, and chemical control in cases of biotic stresses. However, specific examples of these stresses in different farming systems need to be well documented. Documentation of specific examples of biotic and abiotic stresses deserves further attention to learn from past experience.

References

Halila, H. 1992. In: *Tunisian Progress Report 1991/92 of the UNDP Project RAB 91/007.*

Hanounik, S. B. and Bisri, M. 1991. In: *Proceedings of the CIHEAM, EEC and ICARDA Seminar on Present Status and Future Projects of Faba Bean Production and Improvement in the Mediterranean countries*, 27–29 June, Zaragoza, Spain, pp. 59–66 (eds. J. I. Cubero, and M. C. Saxena). Zaragoza, Spain: CIHEAM. Options Médit.

Harrabi, M. and Halila, H. 1992. In: *Disease Resistance Breeding in Chickpea*, pp. 98–109 (eds. K. B. Singh and M. C. Saxena). Aleppo, Syria: ICARDA.

Katul, L., Vetten, H. J., Maiss, E., Makkouk, K. M., Lessmann, D. E. and Casper, R. 1993. *Annals of Aphid Biology* (Submitted).

Makkouk, M. 1993. *Survey for Faba Bean Disease in Egypt with Special Emphasis on Viruses*, February 1–10, 1993, Aleppo, Syria: ICARDA.

Mamo, T., Duffera, M., Abebe, M. and Kidanu, S. 1991. In: *1990/91 Annual Report, The Nile Valley Regional Program on Cool Season Food Legumes*, pp. 71–72. Aleppo, Syria: ICARDA.

Mihov, M. 1985. *Plant Science* 22: 32–36.

Mihov, M. 1986. *Agricultural Science* 24: 45–54.

Murray, G. A., Eser, D., Gusta, L. V. and Eteve, G. 1988. In: *World Crops: Cool Season Food Legumes*, pp. 831–843 (ed. R. J. Summerfield). Dordrecht, The Netherlands: Kluwer Academic Publishers.

Nassib, A. M., Hussein, A. H. A., and El-Deeb, M. A. 1989. In: *Nile Valley Regional Program on Cool Season Food Legumes and Cereals*, pp. 14–16. Aleppo, Syria: ICARDA.

Nassib, A. M., Hussein, A. H. A., El-Deeb, M. A. and Mosa, M. 1990. In: *Nile Valley Regional Program on Cool Season Food Legumes and Cereals*, pp. 5–7. Aleppo, Syria: ICARDA.

Nassib, A. M., Khalil, S. A., El-Borai, M. A. and Radi, M. M. 1991. In: *Nile Valley Regional Program on Cool Season Food Legumes and Cereals. 1990/91 Annual Report, Egypt*. Cairo, Egypt: ARC/ICARDA.

Rizkallah, L., El-Sherbeeny, M., Solh, M. B. and Kumari, S. 1993. *Survey for Faba Bean Virus Diseases in Egypt*. ARC/ICARDA, Cairo, Egypt.

Salih, H. Salih. 1992. Preface. In: *Nile Valley Regional Program on Cool Season Food Legumes and Cereals*. Food Legumes Annual National Coordination Meeting, 6–10, 1992. Wad Medani, Sudan: Agricultural Research Corporation.

Saxena, M. C., Saxena, N. P. and Mohamed, A. K. 1988. In: *World Crops: Cool Season Food Legumes*, pp. 845–856 (ed. R. J. Summerfield). Dordrecht, The Netherlands: Kluwer Academic Publishers.

Saxena, M. C. 1992. In: *Disease Resistance Breeding in Chickpea*, pp. 1–10 (eds. K. B. Singh and M. C. Saxena). Aleppo, Syria: ICARDA.

Singh, K. B. 1990. In: *Proceedings of the CIHEAM, EEC (AGRIMED), and ICARDA Conference on Present Status and Future Prospects of Chickpea Crop Production and Improvement in the Mediterranean Countries*, 11–13 July 1988, Zaragoza, Spain, pp. 25–34 (eds. J. I. Cubero, M. C. Saxena and J. Wery). Zaragoza, Spain: CIHEAM.

Tsenov, A. and Petrova, D. 1984. *Plant Science* 21: 77–86.

Host plant resistance to manage biotic stress

Using host plant resistance to manage biotic stresses in cool season food legumes

F.J. MUEHLBAUER[1] and W.J. KAISER[2]

[1] *United States Department of Agriculture, Agricultural Research Service, 303 W Johnson Hall, Washington State University, Pullman, WA 99164–6434 USA, and*
[2] *United States Department of Agriculture, Agricultural Research Service, Regional Plant Introduction Station, 59 Johnson Hall, Washington State University, Pullman, WA 99164–6402 USA*

Abstract

The cool season food legumes are seriously affected by diseases and pests that collectively cause yield reductions variously estimated at over 50% on a world wide basis. The use of host plant resistance to increase and stabilize yields depends on a well planned and coordinated program of germplasm evaluation, hybridization with otherwise adapted material, and screening and selection methods that efficiently identify segregants with combined resistance to multiple diseases and insect pests. Sequential and simultaneous screening has successfully combined resistance to Ascochyta blight and Fusarium wilt of chickpea; Fusarium wilt, powdery mildew, and viruses of pea; and Fusarium wilt and rust of lentil.

Resistance has generally been durable for the soilborne diseases; however, resistance to Ascochyta blight has often been overcome by new pathotypes. Resistance to powdery mildew, pea enation mosaic and other viruses has been durable. Some of the most serious biotic stresses of the cool season food legumes remain as chronic production constraints. These include Aphanomyces root rot of pea, rust and Ascochyta blight of lentil, root rot of chickpea, chocolate spot of faba bean, and *Orobanche* ssp. that parasitize all of the cool season food legumes.

The use of host plant resistance to control insect pests is almost non-existent; however, resistance to *Heliocoverpa* and leaf miner of chickpea has been identified and work is underway toward developing resistant cultivars. The control of *Bruchus* spp. and *Sitona* spp. through host plant resistance remains as a remote possibility.

Cultivars which are resistant or tolerant to one or more biotic stresses are a critical component of integrated pest management. Decisions as to crop rotations, monitoring of field populations of pathogens or insects, pesticides or biological control agents, tillage, planting dates, method of planting, and other factors can all be critical to reducing the effects of biotic stresses. Successful production of cool season food legumes appears to depend on the creation of cultivars with genetic resistance to one or more pests followed by management

F.J. Muehlbauer and W.J. Kaiser (eds.), Expanding the Production and Use of Cool Season Food Legumes, 233–246.
© 1994 *Kluwer Academic Publishers.*

decisions designed to delay development of pathotypes or biotypes capable of overcoming the available resistance.

Introduction

The cool season food legumes are often severely damaged by disease and insect pests that individually and collectively reduce yields and crop quality. Biotic stresses have been estimated to be responsible for 5 to 15% yield reductions in the temperate regions and 50 to 100% yield losses in the tropics (Van Emden *et al.*, 1988). Van Emden *et al.* (1988) listed the biotic stresses affecting the cool season food legumes. In addition, broomrape (*Orobanche* spp.) has proved to be a debilitating parasite of all cool season food legumes, particularly of faba bean in the Mediterranean region.

Host plant resistance along with suitable crop rotations and other cultural practices offer the most economical, long term, and environmentally acceptable means of controlling disease and insect pests of the cool season food legumes. Until very recently, nearly the entire world's crop was produced from plantings of indigenous land races. Those land races having evolved in the presence of pathogens, insects, and parasites are able to not only survive but produce valuable products for local populations. Legumes have been and continue to be completely intertwined in farming systems of many developing countries. The replacement of those land races by homogeneous cultivars can entail a significant loss of genetic variation for resistance to pests. Improved cultivars, usually selected for yield and quality, may not possess the combined resistances already present in the land races they are intended to replace. Fortunately, germplasm collections, which include many ancient landraces, have been established and are being preserved in gene banks. These collections provide breeders with genetic variation not found in available cultivars or enhanced germplasm.

While crop improvement efforts have largely focused on yield and quality, progress is also being made to incorporate disease and pest resistance. Incorporation of resistance to biotic stresses is of high priority to stabilize yields and quality, the lack of which is often cited as a major limitation to more widespread use of grain legumes in farming systems. Breeding for and using host plant resistance to manage biotic stresses in the cool season food legumes and particularly its place in pest management systems is reviewed and discussed.

Host Plant Resistance

Host plant resistance can range from single gene resistance to multigenic resistance (Parlevliet, 1981; Bernier *et al.*, 1988). Numerous examples are available in the cool season food legumes. Resistance to the different races of the Fusarium wilt fungus in pea and chickpea is controlled by single genes.

Likewise, resistance to pea enation mosaic and bean leaf roll viruses of pea and pea seedborne mosaic virus of pea and lentil is controlled by single genes in all cases. Resistance to powdery mildew of pea is still unclear but there is general agreement that two recessive genes control resistance. Resistance to Ascochyta blight of chickpea, lentil, and pea appears to be under multigenic control. Resistance to Fusarium and Pythium root rot of pea and common root rot of pea (incited by *Aphanomyces*) is considered to be multigenic. Resistance to aphids and weevils in pea and lentil is incomplete and considered to be multigenic.

Resistance to pests may entail avoidance and/or resistance mechanisms (Parlevliet, 1981). Avoidance such as brought about by glandular hairs on plant surfaces, secretion of volatile repellant compounds that effectively ward off insect pests, or growth out of synchronization with the pest are prime examples (see Clement *et al.*, this volume, for a more complete treatment of this area). Actual resistance may be in the form of chemical compounds in the host that are either naturally occurring (e.g., pre-formed phenolics in seed) or are induced during contact of the pathogen with the host (e.g., phytoalexins). Generally, resistance is the ability of the host to limit infection and disease spread.

Escape, although not considered resistance *per se*, can take on many forms in the cool season food legumes. A prime example is the semi-leafless trait of pea with its reduced leaf area and increased tendril number which in turn keeps the plants and canopy more erect, allows more air movement, and thus creates an environment less conducive to foliar fungal pathogens which thrive in humid environments (e.g., *Ascochyta, Sclerotinia*).

The erect plant habit of chickpea also contributes to foliar disease control, especially Ascochyta blight, by providing greater air movement through the canopy and reducing relative humidity to a range less favorable to the pathogen. As an example, chickpea lines resistant to Ascochyta blight in the Mediterranean region are tall, erect, and late maturing (Singh and Reddy, 1991).

Major Germplasm Collections

Major collections of cool season food legumes are maintained in the USA by the National Plant Germplasm System, at the International Centers in Syria and India, and at facilities of national programs in several countries (Table 1).

Germplasm collections of pea are available from several sources and include accessions which have resistance to important diseases. The Weibullsholm genetic stocks collection maintained by the Nordic Gene Bank at Alnarp, Sweden is the World *Pisum* germplasm collection as designated by the International Board of Plant Genetic Resources (IBPGR) (Bettencourt *et al.*, 1989). This World collection is the work of the late Professor Herbert Lamprecht and more recently, Dr. Stig Blixt. The collection is particularly well characterized genetically and samples of accessions are available on request.

Another well-defined genetic stocks collection was developed by the late Dr. Gerry A. Marx at Geneva, NY. This collection includes cultivars, Plant Introduction accessions, mutants, and enhanced and well characterized genetic stocks. Much of the genetic stocks collection has been made available to basic and applied researchers. A catalog describing the G. A. Marx genetic stock collection is being prepared by Dr. James McFerson of the USDA-ARS Northeast Regional Plant Introduction Station, Geneva, NY. Other genetically well-defined *Pisum* collections are maintained at the John Innes Institute at Norwich, England, at Wiatrowo, Poland by Dr. W. K. Swiecicki, and at Hobart (Tasmania), Australia by Dr. Ian Murfet.

Important *Pisum* germplasm collections are also maintained at Edinburgh, Scotland, UK; the Vavilov Institute at St. Petersburg, Russia; Gatersleben, Germany; and at Bari, Italy. The USA *Pisum* collection is currently maintained by the USDA-ARS Western Regional Plant Introduction Station, Pullman,

Table 1. Major collections of cool season food legumes (from van der Maesen *et al.*, 1988, revised)

Institute and Location	Species conserved				
	Pisum	Lens	Cicer	Vicia sect. faba	Lathyrus
1. Ege Agric. Res. Introd. Centre Menemen, Turkey	2000				
2. Ethiopian Genebank Addis Ababa, Ethiopia	1860	413	717	1298	
3. Marx Genetic Stocks Collection Geneva, NY, USA	>500				
4. Germplasm Laboratory Bari, Italy	5000			2000*	
5. ICARDA Aleppo, Syria		6000	4500	5000	100
6. ICRISAT Patancheru, India			14400		
7. INIA Mexico City, Mexico			1600		
8. John Innes Institute Norwich, England	2000				
9. NBPGR New Delhi, India	1400				
10. Netherlands Genebank Wageningen, The Netherlands	800			700	
11. Nordic Genebank Alnarp, Sweden	5000				
12. Nat. Seed Storage Lab. Fort Collins, CO, USA	2213	702	2698	18	
13. Pakistan Agr. Res. Council Islamabad, Pakistan	10	144	626	13	
14. Reg. Plant Introduction Station Pullman, WA, USA	2800	1973	3431	295	330
15. Vavilov Inst. Pl. Industry St. Petersburg, USSR	5550	2470	1685	2525*	
16. ZG Kulturpfl. Gatersleben, Germany	2000	160	40	1300	
17. Inst. of Crop Germplasm Resources Beijing, Peoples Republic of China	1677	336	23	1999	

Washington. The collection numbers over 2800 accessions and includes cultivars, land races, and wild and primitive forms.

By far, the largest and most diverse collection of lentil germplasm is maintained by the International Center for Agricultural Research in the Dry Areas (ICARDA), Aleppo, Syria and numbers 4700 accessions (Table 1). This collection includes many accessions from the national collections of India, Turkey, Ethiopia, and the USA. The USA collection contains over 2250 accessions from 40 countries. The USA and ICARDA collections now include wild *Lens* species that were recently (since 1980) collected in the centers of origin. Parts of the USA collection have been evaluated for resistance to pea enation mosaic and pea seedborne mosaic viruses. The ICARDA collection contains accessions known to have resistance to rust and *Ascochyta*.

The International Crops Research Institute for the Semi-Arid Tropics (ICRISAT), Patancheru, India and ICARDA have extensive collections of chickpea (Table 1). The initial material used to establish these collections was collected by the former USDA Regional Pulse Improvement Project centered in Iran and India in the late 1960s. That project assembled 4177 accessions of cultivated types that were then made available to the ICARDA and ICRISAT collections. Most of these accessions were also added to the *Cicer* collection maintained at the USDA-ARS Regional Plant Introduction Station at Pullman, WA. The USDA-ARS collection currently numbers over 3800 accessions of cultivated and wild forms from over 20 countries. Many of the wild forms in the collection were obtained from the ICRISAT collection, and were largely the result of collection efforts of Dr. L. J. G. van der Maesen and Gideon Ladizinsky. Also included among the wild *Cicer* species in the USDA collection are the collections made by F. J. Muehlbauer, W. J. Kaiser and C. R. Sperling in Turkey in 1985 and 1989.

An extensive collection of faba bean is maintained at ICARDA (Table 1). Other centers have sizeable collections of faba bean, the most important of which are the collections maintained at Bari, Italy; St. Petersburg, Russia; and Beijing, Peoples Republic of China. The origin of faba bean is still unknown and no extant wild species have been found which are cross compatible with the cultigen. Ladizinsky *et al.* (1988) speculated that the origin of faba bean might be northern Pakistan.

The USDA-ARS *Lathyrus* collection maintained at Pullman, WA contains 330 accessions of 13 species (Table 1). Little information is available concerning other *Lathyrus* collections, but sizeable collections are maintained at ICRISAT and ICARDA. There is also a collection of *Lathyrus* maintained by Agriculture Canada at Saskatoon, Saskatchewan, Canada (and see Campbell *et al.*, this volume). These collections represent important sources of germplasm for development of this drought tolerant crop.

Multiple pest resistance is an obvious goal where the crop is known to confront a number of pest problems throughout its growth cycle. Multiple disease resistance as pointed out by Nene (1988) is a major task for breeders because it requires identifying sources of resistance to pathogens which may also have a number of divergent races or pathotypes. The genes for resistance must then be combined in an otherwise acceptable genetic background. Screening methods which are efficient, practical and eliminate or greatly reduce the number of escapes are required.

Disease resistance in many crops is known to be overcome by the development of new and more virulent pathotypes. Cases where this has taken place in cool season food legumes are still very limited. This is likely because breeding for disease resistance in cool season food legumes other than pea is a rather recent undertaking. Generally, pathogens which have a perfect stage produce large numbers of airborne spores and are windblown, are likely to develop new pathotypes when a resistant host is planted extensively. As an example, evidence for the development of new races of *Ascochyta rabiei* in response to resistant cultivars of chickpea has been reported (Porta-Puglia, 1992; Singh and Reddy, 1993). However, powdery mildew resistance in peas caused by *Erysiphe pisi* is an exception, and host resistance has been stable for over 50 years (Hagedorn, 1984).

Resistance to soilborne diseases is much more stable when compared to diseases disseminated by airborne spores. For example, resistance to the different races of *Fusarium oxysporum*, the wilt fungus, in pea and chickpea has remained stable. Virus resistance has also remained stable possibly because most viruses that affect cool season food legumes have alternative hosts such as alfalfa or other perennial forage legumes.

Methods of Breeding/Testing

Development of improved cultivars with resistance to a single pathogen is often straight forward if a good source of resistance is available and an efficient, easily controlled, and practical screening procedure exists that provides good selection pressure. The task of recombining resistance to two or more pathogens is considerably more difficult and time consuming. Many breeders use a sequential approach with cool season food legumes in which segregating material is subjected to a single pathogen in one generation and the resistant, segregating progeny are screened for resistance to a different pathogen in succeeding generations. However, if sufficient seed is available simultaneous testing against several pathogens is often attempted. Regardless of the approach taken, success in obtaining recombinants with multiple disease resistance depends on the screening procedures used, the selection pressure, and the ability to obtain seed from selected individuals. In our work with dry peas, we use simultaneous selection for resistance to common root rot, Fusarium wilt race 1, powdery mildew, and tolerance to certain viruses. In nearly all seasons, critical evaluations have been made for root rot and wilt resistance. In addition, by inoculating the nursery with infested powdery mildew plants we are able to produce a sufficient epidemic to critically evaluate selections for that disease. Virus resistance testing in the field is often erratic and dependent upon the severity of the previous winter and the effect on populations of aphid vectors.

Breeding populations being screened should be exposed to the pathogen as uniformly as possible and under conditions favorable for disease development. It is often difficult to do in foliar disease nurseries because wind currents can alter spore movement and distribution leaving certain areas inadequately inoculated. The use of spreader rows, infested crop debris, sprinkler irrigation to maintain high relative humidity, multiple inoculations, and other procedures have been employed to obtain uniform disease in nurseries (Singh *et al.*, 1981). Most often uniform disease development is dependent upon weather conditions at inoculation time and afterwards. In our experience with Ascochyta blight of chickpea, the use of infested debris from previous chickpea crops for inoculation and the use of spreader rows every five rows has been very successful in obtaining uniform infection among the plots. Sprinkler irrigation for short periods in the evenings is used to maintain a moist environment conducive to disease development and spread.

Screening in artificial environments for foliar pathogens such as Ascochyta blight is considered unreliable because of the difficulty in obtaining environments similar to those in the field and because plant growth and disease development tends to be atypical. Space limitations in controlled environments so restrict population sizes that only small numbers of plants or selections can be accurately screened. Consequently, it is our opinion that screening against Ascochyta blight and other foliar diseases be carried out in the field and that controlled environments be used only as a supplement.

The lack of uniformity of virus disease incidence in field nurseries and the fact that several viruses may be present simultaneously suggests that screening against specific viruses sequentially in controlled environments would be desirable. Susceptibility to one virus often masks resistance to other viruses in field screening trials. In order to utilize all the sources of virus resistance, controlled inoculations in which the germplasm is tested against the important viruses sequentially is necessary. For many viruses, mechanical inoculation by leaf rubbing is a rapid and efficient means of virus transfer. After multiple virus resistant selections have been identified, field screening to verify results from controlled inoculations should be used.

Sources of Resistance to Biotic Stresses

Sources of resistance or tolerance to the major diseases of the cool season food legumes have been reported (Table 2). In addition, the international centers coordinate and distribute disease resistance nurseries that contain resistant or tolerant materials. Availability of resistance or tolerance to insects is extensively reviewed by Clement *et al.* (this volume). Sources of tolerance to nematodes was recently reviewed by Singh *et al.* (1989), and Cubero *et al.* (1988) has reviewed sources of resistance to parasitic weeds.

Table 2. Reports of resistance to biotic stresses in cool season food legumes

Diseases		References
Chickpea	Ascochyta blight	Nene and Reddy (1987); Reddy et al. (1990)
	Fusarium wilt	Haware et al. (1990); Nene and Reddy (1987)
	Botrytis gray mold	Nene and Reddy (1987); Reddy et al. (1990)
	Bean leaf roll virus	Nene and Reddy (1987)
Faba bean	Ascochyta blight	ICARDA (1989); Nene et al. (1988)
	Chocolate spot	ICARDA (1989); Nene et al. (1988)
	Rust	ICARDA (1989); Nene et al. (1988)
	Stem nematode	ICARDA (1989)
Lentil	Ascochyta blight	Khare et al. (1990); Nene et al. (1988)
	Rust	Khare et al. (1990); Nene et al. (1988); Muehlbauer (1992)
	Pea enation mosaic virus	Muehlbauer (1992)
	Pea seedborne mosaic virus	Muehlbauer (1992)
Pea	Aphanomyces root rot	Hagedorn (1984); Kraft and Kaiser (1993); Muehlbauer (1992)
	Fusarium wilt and root rot	Hagedorn (1984); Kraft and Kaiser (1993); Muehlbauer (1992)
	Pythium and Rhizoctonia seedling disease	Hagedorn (1984); Kraft and Kaiser (1993); Muehlbauer (1992)
	Powdery mildew	Hagedorn (1984); Kraft and Kaiser (1993); Muehlbauer (1992)
	Pea enation mosaic virus	Kraft and Kaiser (1990); Muehlbauer (1992)
	Pea seedborne mosaic virus	Kraft and Kaiser (1990); Muehlbauer (1992)

Insects		Reference
Chickpea	Pod borers (Helicoverpa armigera)	Clement et al. (this volume); Reed et al. (1987); Singh et al. (1990)
	Leafminer	Clement et al. (this volume); Reed et al. (1987); Singh et al. (1990)
	Weevils (Bruchus and Callosobruchus spp.)	Clement et al. (this volume); Reed et al. (1987)
	Aphids	Clement et al. (this volume)
Faba bean	Aphids	Clement et al. (this volume)
	Weevils	Clement et al. (this volume)
Lentil	Aphids	Clement et al. (this volume)
	Weevils	Clement et al. (this volume)
Pea	Aphids	Clement et al. (this volume)
	Weevils	Clement et al. (this volume)

Nematodes and Parasitic Weeds		
Chickpea	Heterodera ciceri	Singh et al. (1989)
Faba bean	Orobanche crenata	Cubero et al. (1988)
Pea	Orobanche crenata	Cubero et al. (1988)

Combined resistance to two or more biotic stresses have been reported for chickpea, faba bean, lentil, and pea (Table 3). The lines listed in Table 3, while not intended to be a complete listing, represent examples of useful germplasm for breeding purposes. There are a number of examples where resistance to certain biotic stresses of cool season food legumes do not occur in commercial cultivars, although resistance can be found in wild species, germplasm accessions, or breeding lines. Resistance to Aphanomyces root rot of pea incited by Aphanomyces euteiches has been found in germplasm lines. Progress is now being made to transfer this resistance into commercially acceptable pea cultivars (Kraft and Kaiser, 1993). Resistance to virus diseases in cultivars of most cool season food legumes is generally lacking. Bean leaf roll virus (BLRV) is but one of several viruses that affect these crops in different countries. It is an important,

Table 3. Multiple resistance to biotic stresses in cool season food legumes

Crop	Cultivars/germplasm lines	Biotic stresses	Reference
Chickpea	FLIP 82-78-C, FLIP 84-43-C, FLIP 84-130-C	Ascochyta blight and Fusarium wilt	Singh *et al.* (1990)
	ICC 1069	Ascochyta blight, Fusarium wilt, and Botrytis gray mold	Nene (1988)
	ICC 12237, ICC 12269	Fusarium wilt, dry root rot, and black root rot	Nene (1988)
	ICCL 86111	Fusarium wilt and pod borer	Singh *et al.* (1990)
	ICC 858, ICC 959, ICC 4918, ICC 8933, ICC 9001	Fusarium wilt and Sclerotinia stem rot	Nene (1988)
Faba bean	L 82005, L 82009, BPL 266	Ascochyta blight, chocolate spot, and rust	Nene (1988)
	BPL 266	Ascochyta blight, chocolate spot, rust, Stemphylium blight, and Alternaria blight	Nene (1988)
Lentil	JL 599, JL 632, JL 674, JL 105, Bombay 18, L 9-12	Fusarium wilt and rust	Nene (1988); Nene *et al.* (1975)
	ILL 4605	Ascochyta blight and rust	W. Erskine (personal communication)
	Pant L 406	Ascochyta blight, rust, and wilt	Pandya *et al.* (1980)
	Coll. 10066, 10463, 10496, 10498, 10509, 10518, 10520, 10528, 10534, 10536, 10537	Powdery mildew and rust	Khare *et al.* (1990)
Pea	74SN3	Foot and root rot and wilt	Muehlbauer (1992)
	Garfield	Fusarium and Pythium root rot	Muehlbauer (1992)
	C-12, Wisconsin, PI 207508	Powdery mildew and rust	Nene (1988)
	Glacier	Foot rot and wilt	Nene (1988)
	OSU 559-6, OSU 564-3, OSU 584-16, OSU 589-12	Bean leaf roll, bean yellow mosaic, pea enation mosaic and pea seedborne mosaic viruses, and wilt	Baggett and Kean (1988)
	PI 173052, Cobri, Rondo, Sun Valley	Ascochyta complex	Kraft and Kaiser (1993)
	VR 74-410-2, VR 74-1492-1	Fusarium wilt, pea seedborne mosaic virus, and Fusarium root rot	Kraft and Kaiser (1993)
	86-2236	Fusarium wilt races 1 and 2 and Aphanomyces root rot	Kraft and Kaiser (1993)
	79-2022	Fusarium wilt races 1 and 2, Fusarium root rot, and Aphanomyces root rot	Kraft and Kaiser (1993)
	90-2131	Fusarium wilt races 1 and 2, Fusarium root rot, and Aphanomyces root rot	Kraft (1992)

and at times, a devastating virus disease of chickpea, faba bean, lentil, and pea. In peas, commercial cultivars resistant or immune to BLRV have been identified (Kraft and Kaiser, 1993). The resistance of chickpea, faba bean, and lentil to BLRV is poorly understood and no BLRV-resistant cultivars of these crops are currently being grown.

Sources of resistance to several important field and storage insect pests of the cool season food legumes, including pod borers (*Helicoverpa* spp.), leafminers (*Liriomyza* spp.), weevils (*Bruchus* and *Sitona* spp.), and aphids (*Aphis craccivora* and *Acyrthosiphon pisum*) have been identified (see Clement *et al.*, this volume). At present, no insect resistant cultivars of these crops have been released to growers.

In some production areas, nematodes adversely affect plant growth and yields of the four crops under discussion. In chickpeas, the most important nematodes are root knot (*Meliodogyne* spp.), cyst (*Heterodera* spp.), and lesion (*Pratylenchus* spp.) (Greco and Sharma, 1990). Sources of resistance to most parasitic nematodes infecting cool season food legumes are limited. For example, resistance in chickpea to the cyst nematode, *H. ciceri*, was found only in accessions of the wild chickpea, *C. bijugum* (Singh *et al.*, 1989).

Broomrape (*Orobanche* spp.) is the most important parasitic weed of cool season food legumes, and is particularly damaging to faba bean crops in the Mediterranean region (Cubero *et al.*, 1988). Resistance found in Giza 402, a faba bean breeding line from Egypt, is being incorporated into improved cultivars (Cubero *et al.*, 1988).

Standardized Screening Procedures

Over the years, researchers in different countries have used various procedures to screen germplasm accessions, breeding lines, and cultivars of the different cool season food legumes for resistance to many biotic stresses. Researchers at ICARDA and ICRISAT have been instrumental in developing reliable and efficient inoculation techniques for screening germplasm of chickpea (ICARDA and ICRISAT), faba bean (ICARDA), and lentil (ICARDA) for resistance to different biotic stresses, including soilborne and foliar diseases, insects, nematodes, and parasitic weeds.

Since establishment of ICARDA and ICRISAT in the 1970s, there has been a trend toward standardization of inoculation techniques and disease rating scales with different biotic stresses affecting their mandate crops. For example, researchers in most countries now follow techniques developed by ICARDA and ICRISAT scientists to screen chickpea germplasm for resistance to Ascochyta blight (Singh and Reddy, 1991). Readings for disease severity are made at periodic intervals and are based on a visual 9-point scale, where 1 = no infection (immune) and 9 = plants killed.

Further testing of chickpea lines exhibiting resistance to Ascochyta blight under field conditions is often conducted in the greenhouse where various environmental conditions, (e.g., temperature and relative humidity) can be controlled more precisely. Additionally, the plants can be inoculated with an inoculum of one or more isolates of the blight fungus. Inoculated plants are incubated in a moist chamber for several days where the relative humidity is maintained near 100%. Disease severity readings usually are made 14 to 21 days after inoculation using the 9-point scale.

ICARDA and ICRISAT have provided a very useful and valuable service of providing germplasm of chickpea, faba bean, and lentil to researchers worldwide, and also by distributing international yield, adaptation, and disease and insect resistance nurseries for testing in many countries. These multilocation trials also provide valuable information on the resistance of this germplasm to different biotic stresses in different areas and on the incidence and distribution of new races, strains, and pathotypes.

Management of Biotic Stresses

Management decisions affecting newly developed pest resistant materials should focus on reducing opportunities for the pathogen to infect the host and to develop new and more virulent pathotypes. Management practices that delay development of new pathotypes might include: clean seed sources where the pathogen is seedborne, fungicide or biological seed treatments to reduce or eliminate seed infection, sanitation (especially with regard to crop debris), elimination of alternative hosts, and possibly chemical control. Crop rotations, tillage methods, and planting time and method also influence disease development and spread.

Development of cultivars with resistance to a particular biotic stress often leads to the expectation on the part of producers that the stress has been completely alleviated. In many cases where one or two genes control resistance such as in the case for Fusarium wilt, pea enation mosaic virus, and powdery mildew, the stress is substantially reduced. In other cases, such as Ascochyta blight of chickpea, resistance does not completely eliminate the stress. It should always be emphasized that host plant resistance cannot be expected to do it all. As part of pest management schemes, host plant resistance varies from being the primary control component to being almost of secondary importance. The following are several examples of the use of host plant resistance to manage disease stresses in the cool season food legumes.

Fusarium Wilt Resistance in Pea

Race 1 wilt of pea caused by *F. oxysporum* f. sp. *pisi*, has been controlled for over 65 years through the use of a single dominant gene for resistance. While this resistance has been stable in all regions of the world where race 1 is a problem, occurrences of races 5 and 6 in an intensively cropped pea growing region of western Washington, USA (Haglund and Kraft, 1970, 1979) clearly indicates that this pathogen can overcome single gene resistance in the host. Continuous cropping in that region with cultivars resistant to race 1 of the pathogen perhaps provided sufficient selection pressure for development of new races. In other areas, peas are generally grown in rotations which may be the reason new races of the pathogen have not appeared.

Ascochyta Blight Resistance in Chickpea

Ascochyta blight of chickpea caused by *Ascochyta rabiei* can be controlled through the use of resistant cultivars. Nevertheless, other practices can augment resistance, reduce primary infection and disease spread by reducing primary inoculum. The use of pathogen free seed properly treated with fungicides (benomyl or thiabendazole) can substantially reduce or eliminate the number of primary infection sites in chickpea fields. Although there are foliar fungicides available for control of blight, their use is generally not economical. Crop residues are known to harbor the fungus and, after an overwintering period, produce the sexual stage of the fungus which can be devastating to succeeding chickpea crops. Management of infested crop residues to eliminate or reduce that source of primary inoculum is crucial to minimize infection of chickpea fields in the vicinity. The residues may be collected and fed to livestock or buried by plowing or other means. Crop rotations should be used to reduce the frequency of chickpea crops in individual fields.

Aphanomyces Root Rot of Pea

Common root rot of pea incited by *Aphanomyces euteiches* can be severe in fields with wet and heavy soils and a history of frequent pea crops. The disease is especially severe where soil organic matter is depleted and in cases of soil compaction from tillage implements. Breeding for resistance to *Aphanomyces* has been partially successful (see Kraft *et al.*, this volume); and control may only be accomplished through careful attention to cultural practices, use of resistant/tolerant cultivars, and careful monitoring of soil populations of the pathogen in order that highly infested fields might be identified and avoided for pea crops.

Cultural practices considered essential to control common root rot include: crop rotations, careful tillage to avoid compaction, fungicidal seed treatments, and tolerant/resistant cultivars. The most important factor in the control of common root rot may be crop rotation in which a pea crop is not planted more than once in every five growing seasons. Alfalfa, clover, faba bean, and common bean are all susceptible and cannot be used in rotation to control common root rot. The recent registration of Tachigaren has real potential for control of this disease as do other biocontrol agents. Soil compaction can be avoided through proper tillage operations that are performed when soil moisture is ideal.

Concluding Remarks

There is reason to be optimistic about the prospect of developing multiple pest resistant cultivars of the cool season food legumes for use in developed and developing countries. Significant progress has been made in developing enhanced germplasm with measurable resistance to *A. euteiches*, multiple wilt resistance, Fusarium root rot of pea, powdery mildew of pea, virus resistance in pea, *Ascochyta* and wilt of chickpea, rust and *Ascochyta* of lentil, *Botyritis* of faba bean, and *Orobanche* of faba bean Critical needs for cool season food legumes include better resistance or tolerance to *Aphanomyces euteiches* in pea and lentil, better resistance to Ascochyta blight and the root rot/wilt complex of chickpea, better resistance to *Orobanche* spp, and better resistance to insect pests. Germplasm collections have a wide array of cultivated accessions and wild species that are available to breeders. The wild species represent an untapped resource for multiple pest resistance breeding; however, evaluations need to be made and procedures to transfer important genes from the secondary and tertiary gene pools need to be developed and refined. Nevertheless, for every serious disease thus far studied and germplasm screened, resistance has been found.

Multiple pest resistance is not expected to do everything. Sound management practices with the judicious use of pesticides can augment gains made by breeding multiple pest resistant cultivars and substantially increase and stabilize yields of the cool season food legumes.

References

Baggett, J. R., and Kean, D. 1988. *HortScience* 23: 630–631.

Bernier, C. C., Bijiga, G., Nene, Y. L. and Cousin, R. 1988. In: *World Crops: Cool Season Food Legumes*, pp. 97–106 (ed. R. J. Summerfield). Dordrecht: Kluwer Academic Publishers.

Bettencourt, E., Konopka, J. and Damania, A. B. 1989. *Directory of Germplasm Collections, 1.1 Food Legumes, International Board for Plant Genetic Resources*, 190 pp. Rome.

Campbell, C. G., Mehra, R. B., Agrawal, S. K., Chen, Y. Z., Abd El Moneim,A. M., Khawaja, H. I. T., Yadov, C. R., Tay, J. U. and Araya W. A. 1994. Expanding the Production and Use of Cool Season Food Legumes, pp. 617–630 (eds. F. J. Muehlbauer and W. J. Kaiser). Dordrecht: Kluwer Academic Publishers.

Clement, S. L., El Din Sharafeldin, N., Weigand, S. and Lateef. S. S. 1994. Expanding the Production and Use of Cool Season Food Legumes, pp. 290–304 (eds. F. J. Muehlbauer and W. J. Kaiser). Dordrecht: Kluwer Academic Publishers.

Cubero, J. I., Pieterse, A., Saghir, A. R., and ter Borg, S. 1988. In: *World Crops: Cool Season Food Legumes*, pp. 549–563 (ed. R. J. Summerfield). Dordrecht: Kluwer Academic Publishers.

Greco, N. and Sharma, S. B. 1990. In: *Chickpea in the Nineties: Proceedings of the Second International Workshop on Chickpea Improvement*, pp. 135–137 (eds. H. A. van Rheenen, M. C. Saxena, B. J. Walby and S. D. Hall). Patancheru, India: ICRISAT.

Hagedorn, D. J. (ed.). 1984. *Compendium of Pea Diseases*. American Phytopathological Society, St. Paul, MN.

Haglund, W. A. and Kraft, J. M. 1970. *Phytopathology* 60: 1861–1862.

Haglund, W. A. and Kraft, J. M. 1979. *Phytopathology* 69: 818–820.

Haware, M. P., Jiménez-Díaz, R. M., Amin, K. S., Phillips, J. C. and Halila, H. 1990. In: *Chickpea in the Nineties: Proceedings of the Second International Workshop on Chickpea Improvement*, pp. 129–133 (eds. H. A. van Rheenen, M. C. Saxena, B. J. Walby and S. D. Hall). Patancheru, India: ICRISAT.

ICARDA. 1989. *Food Legume Improvement Program Annual Report for 1989*. Aleppo, Syria: ICARDA.

Khare, M. N., Bayaa, B., and Beniwal, S. P. S. 1993. In: *Breeding for Stress Tolerance in Cool-Season Food Legumes*, pp. 107–121. (eds. K. B. Singh and M. C. Saxena). Chichester, UK: John Wiley & Sons.

Kraft, J. M. 1992. Crop Science 32: 1076.

Kraft, J. M., and Kaiser, W. J. 1993. In: *Breeding for Stress Tolerance in Cool-Season Food Legumes*, pp. 123–144. (eds. K. B. Singh and M. C. Saxena). Chichester, UK: John Wiley & Sons.

Kraft, J. M., Haware, M. P., Jiménez-Díaz, R. M. and Bayaa, B. 1994. Expanding the Production and Use of Cool Season Food Legumes, pp. 268–289 (eds. F. J. Muehlbauer and W. J. Kaiser). Dordrecht: Kluwer Academic Publishers.

Ladizinsky, G., Pickersgill, B. and Yamamoto, K. 1988. In: *World Crops: Cool Season Food Legumes*, pp. 967–978 (ed. R. J. Summerfield). Dordrecht: Kluwer Academic Publishers.

Muehlbauer, F. J. 1992. In: *Use of Plant Introductions in Cultivar Development, Part 2*, pp. 49–73 (eds. H. L. Shands and L. E. Weisner). Crop Science Society of America: Madison, WI.

Nene, Y. L. 1988. *Annual Review of Phytopathology* 26: 203–217.

Nene, Y. L., Kannaiyan, J. and Saxena, G. C. 1975. *Indian Journal of Agricultural Science* 45: 177–178.

Nene, Y. L., Hanounik, S. B., Qureshi, S. H. and Sen, B. 1988. In: *World Crops: Cool Season Food Legumes*, pp. 577–589 (ed. R. J. Summerfield). Dordrecht: Kluwer Academic Publishers.

Nene, Y. L. and Reddy, M. V. 1987. In: *The Chickpea*, pp. 233–270 (eds. M. C. Saxena and K. B. Singh). Wallingford: CAB International.

Pandya, B. P., Pandey, M. P. and Singh, J. P. 1980. *Lentil Experimental News Service* 7: 34–37.

Parlevliet, J. E. 1981. Chapter 9. In: *Plant Breeding II*, pp. 309–364 (ed. K. J. Frey). Ames: Iowa University Press.

Porta-Puglia, A. 1992. In: *Disease Resistance Breeding in Chickpea*, pp. 135–143 (eds. K. B. Singh and M. C. Saxena). Aleppo, Syria: ICARDA.

Reddy, M. V., Nene, Y. L., Singh, G. and Bashir, M. 1990. In: *Chickpea in the Nineties: Proceedings of the Second International Workshop on Chickpea Improvement*, pp. 117–127 (eds. H. A. van Rheenen, M. C. Saxena, B. J. Walby and S. D. Hall). Patancheru, India: ICRISAT.

Reed, W., Cardona, C., Sithantham, S. and Lateef, S. S. 1987. In: *The Chickpea*, pp. 283–316 (eds. M. C. Saxena and K. B. Singh). Wallingford: CAB International.

Singh, K. B., Hawtin, G. C., Nene, Y. L. and Reddy, M. V. 1981. *Plant Disease* 65: 586–587.

Singh, K. B., Di Vito, M., Greco, N. and Saxena, M. C. 1989. *Nematologia Mediterranea* 17: 113–114.

Singh, K. B., Kumar, J., Haware, M. P. and Lateef, S. S. 1990. In: *Chickpea in the Nineties: Proceedings of the Second International Workshop on Chickpea Improvement*, pp. 233–238 (eds. H. A. van Rheenen, M. C. Saxena, B. J. Walby and S. D. Hall). Patancheru, India: ICRISAT.

Singh, K. B. and Reddy, M. V. 1991. *Advances in Agronomy* 46: 191–222.

Singh, K. B. and Reddy, M. V. 1993. *Crop Science* 33: 186–189.

van der Maesen, L. J. G., Kaiser, W. J., Marx, G. A. and Worede, M. 1988. In: *World Crops: Cool Season Food Legumes*, pp. 55–66 (ed. R. J. Summerfield). Dordrecht: Kluwer Academic Publishers.

van Emden, H. F., Ball, S. L. and Rao, M. R. 1988. In: *World Crops: Cool Season Food Legumes*, pp. 519–534 (ed. R. J. Summerfield). Dordrecht: Kluwer Academic Publishers.

Screening techniques and sources of resistance to foliar diseases caused by fungi and bacteria in cool season food legumes

A. PORTA-PUGLIA[1], C.C. BERNIER[2], G.J. JELLIS[3], W.J. KAISER[4] and M.V. REDDY[5]

[1] *Istituto Sperimentale per la Patologia Vegetale, Via C. G. Bertero, 22, I-00156 Rome, Italy;*
[2] *Plant Science Department, University of Manitoba, Winnipeg, Manitoba, R3T 2N2 Canada;*
[3] *Plant Breeding International, Cambridge, Maris Lane, Trumptington, Cambridge CB2 2LQ UK;*
[4] *U. S. Department of Agriculture, Agricultural Research Service, Regional Plant Introduction Station, 59 Johnson Hall, Washington State University, Pullman, WA 99164-6402 USA, and*
[5] *ICRISAT, Legume Program, Patancheru P. O., Andhra Pradesh 502 324, India*

Abstract

Screening techniques are an important component of the overall strategy of breeding for resistance to diseases in cool season food legumes. Suitable screening methods have been developed for several major foliar diseases of chickpea, pea, faba bean, and lentil, and sources of resistance have been identified. International cooperation plays an important role in promoting research and keeping collections of cultivated species and their wild relatives. New biotechnological approaches are promising for enhancing the practical use of genes for resistance.

Introduction

Chickpea (*Cicer arietinum* L.), faba bean (*Vicia faba* L.), lentils (*Lens culinaris* Medik.), and pea (*Pisum sativum* L.) are very important sources of proteins for the human diet, or for animal feed, in several countries. Acreage and production are second only to cereals, although research efforts have been discontinuous and, globally, rather poor (Hawtin *et al.*, 1988). Considering the studies related to disease resistance in the four crops, a comparatively greater amount of research has been devoted to pea and to faba bean, probably because developed countries are more interested in these crops. In more recent years, national governments of developing countries and international organizations have put more impetus on research on chickpea, faba bean, and lentil. Also some countries in the northern hemisphere have been involved in such studies, in programs supporting less developed countries abroad, or marginal areas inside their own borders, and our knowledge of these crops is progressively increasing, along with their yield. Nevertheless, a lot remains to be done, and resistance to diseases has a major role to play in the improvement of food legumes, both in terms of increasing and stabilizing production.

F.J. Muehlbauer and W.J. Kaiser (eds.), Expanding the Production and Use of Cool Season Food Legumes, 247–267.
© *1994 Kluwer Academic Publishers.*

The crops considered in this paper suffer from a number of diseases caused mainly by viruses, bacteria, fungi, and nematodes, that can affect one or more organs of the plant. Major diseases are systemic virus infections, fungal root rots, fungal wilts, and fungal and bacterial diseases of leaves and stems. This paper deals with diseases caused by fungi and bacteria that affect the above ground tissues of the plant.

Damages caused by foliar pathogens affecting supply and translocation of photosynthates are of primary importance. Nevertheless, other effects, e.g., on growth regulators and water relations, influence disease impact on yield (Griffiths, 1984). The yield reductions caused by bacterial and fungal diseases of the aerial parts of plants can be severe. For example, losses caused by Ascochyta blight in chickpea, reported by diverse authors in different years and locations, were more than 40% (Nene and Reddy, 1987). Most of the pathogens causing foliar diseases often attack pods and infect or infest seeds; and, by this means, several pathogens are able to survive and disseminate, even through the small amount of seeds exchanged among scientists (Frison *et al.*, 1990). Although soilborne pathogens affecting roots or the vascular system can be, under some conditions, seed transmitted, transmission of foliar diseases by seed is far more common.

Resistant cultivars provide an efficient means of controlling foliar diseases. The judicious choice of appropriate methods of screening should make it possible to identify genotypes with suitable resistance in germplasm collections, in segregating populations, or in advanced breeding lines. This paper deals with the rationale behind this approach and gives practical examples of screening for resistance and known resistance sources to major foliar diseases of cool season food legumes.

Strategies and Methods of Screening for Resistance

The prerequisites to obtain resistant cultivars are: (1) the knowledge of the pathogenic variability of the disease incitant, (2) the development of a screening method able to mimic the conditions met by the plants when exposed to natural sources of inoculum in diverse field environments, and (3) the availability of usable sources of resistance.

Screening methods for disease resistance should be developed within the framework of a general strategy for resistance. The changes in the frequency of virulence genes among the populations of pathogens inciting disease of the above-ground parts of plants are very frequent. Populations of such pathogens vary in time and space because of the airborne or seedborne nature of inoculum which facilitates long distance dispersal of their variants. As a result of these situations, breeding for resistance to foliar pathogens is, in general, more difficult than in the case of less mobile pathogens, e.g., soilborne fungi which are, therefore, more stable.

Screening as part of a strategy for developing resistant cultivars requires good planning and understanding of the processes involved in resistance. A screening program should be initiated with a clear statement of the type of resistance which is sought, i.e., complete resistance or partial resistance, or both, and with at least some knowledge of pathogenicity and virulence patterns in the pathogen. Considerable progress has been made in the last decade regarding the nature and durability of resistance, and on effective methods of evaluating plant material for resistance to a number of pathogens. A clear understanding of the concepts and terminology involved is essential to the success of any screening program. *Host resistance* is defined as the ability of the host to hinder the growth and/or development of the pathogen, *complete resistance* is used when the sporulation of the pathogen is prevented. *Incomplete resistance* refers to resistance that allows some sporulation. *Partial resistance* is used when the host is susceptible to infection but spore production is reduced (Parlevliet, 1979; Bernier *et al.*, 1988). The term *durable resistance* has generated some confusion because of its sometimes inappropriate use. The term is descriptive and does not explain the underlying causes. The durability of resistance can be practically tested only when the resistant cultivar is widely used in space and time (Johnson, 1984). Multilocation cultivar testing or the challenge of cultivars with a large collection of pathogen variants can help to verify resistance and give timely warning of the possibility of resistance breakdown, but cannot actually be considered as a test for durability of resistance.

Care must be taken in interpreting results of glasshouse or laboratory tests, as the expression of resistance in the field may be considerably modified because of interaction between microorganisms and between pathogens and environmental conditions.

For foliar pathogens, the plant material must be adequately challenged with a single race or pathotype at a realistic inoculum dose to allow disease development, but at the same time, not obscure minor differences in host response required to identify partial resistance. Use of inoculum composed of a mixture of races or naturally infested crop debris of unknown pathotype composition will not be adequate to achieve this objective. The rationale behind this approach has been reviewed by Parlevliet (1979, 1983). To illustrate, three cultivars each having a single gene for complete resistance to a given race would be identified only when inoculated singly with each isolate but not if the isolates were used in a mixture. Parlevliet also concluded that using a single race provides the best conditions for the selection of partial resistance in the presence of complete resistance, and that the selected race should have the broadest possible virulence spectrum to suppress the expression of as many complete resistance genes as possible. Genotypes with resistance to one virulent race should then be systematically tested to a collection of other isolates.

The identification of cultivars with complete resistance is but a first step in the development of effective, durable resistance and genetic analysis of resistant reactions is essential to reveal similarities and differences in the gene(s) that confer resistance in each genotype. The information is then used to recombine, in a single cultivar, several genes known to be effective against a given race and genes effective against all prevalent races in a production region.

In the last 20 years, there has been a tremendous increase in the number of countries that have established germplasm banks. Many of these countries maintain germplasm collections of varying sizes of one or more of the cool season food legumes (van der Maesen *et al.*, 1988). Two of the international agriculture research centers, ICARDA in Syria and ICRISAT in India, maintain large and diverse collections of chickpea (ICARDA and ICRISAT), faba bean (ICARDA), and lentil (ICARDA). For several years, both centers have had excellent programs for distributing germplasm of their mandate crops to researchers worldwide. One particularly useful and valuable service of ICARDA and ICRISAT is the distribution of international yield, adaptation and disease and insect resistance nurseries to cooperators in different countries, i.e., ICARDA's Chickpea International Ascochyta Blight Nursery and Lentil Rust and Ascochyta Blight Nurseries. These and other multilocation tests, promoted by international centers help also in standardization of inoculation techniques and rating scales.

Screening Techniques in the Field

Screening for resistance in the field provides a comparatively cheap means for testing, during the whole plant cycle, a large number of individuals under conditions similar to those in which the resistant cultivars are expected to perform. On the other hand, the environmental conditions, the nature of the inoculum available in the area, and the interactions with other organisms, can affect the expression of resistance to such an extent that screening can be effective only during epidemic years. There is also some risks of confusing escape with resistance. Some of these constraints can be controlled by appropriate techniques. For example, relative humidity can be increased by sprinkler or perfo-irrigation. Artificial inoculation can be applied and epidemics can be encouraged by interplanting rows of susceptible genotypes (spreaders).

Screening Techniques in the Glasshouse

Screening germplasm of cool season food legumes for resistance to foliar diseases under glasshouse conditions has some advantages over screening in the field. It is often easier to perform than field screening because various environmental factors, such as temperature and relative humidity, can be regulated to favor infection and disease development. Plants can be inoculated at different stages of development and with varying concentrations of inoculum of one or more purified isolates of the pathogen. Additionally, the inoculum can be distributed in such a way that it will result in more uniform infection. When a high relative humidity is required for infection, inoculated plants can be incubated in a chamber covered by plastic or cloth where the relative humidity can be maintained at or near 100% by various methods. One major disadvantage of glasshouse screening is that space is often a limiting constraint. As cool season food legumes are field crops, results obtained in the greenhouse need to be validated in the field.

Screening Techniques in the Laboratory

In circumstances when conditions in the field are not favorable for infection and disease development and when it is not possible to increase the leaf wetness duration period by irrigating or misting the plots or by covering them with polyethylene sheets, detached leaves can be used to assess host reactions in the laboratory. The method has been used successfully to assess reactions of faba bean genotypes to chocolate spot and also to test a large number of isolates of the fungus for pathogenicity (Hanounik and Maliha, 1986; Hanounik and Robertson, 1988).

In several crops, culture filtrates or purified phytotoxins have been tested, particularly *in vitro*, as agents for selection. There is a lack of research regarding the possible use of such methods in cool season food legumes. The results obtained with other species indicate that these methods may be useful for selection only when a host-specific toxin is involved in pathogenesis. In the case of nonhost-specific toxins, the methods can be of some help in a few host-pathogen systems (Buiatti and Ingram, 1991; van den Bulk, 1991).

Rating Scales

For rapid evaluation of a large number of lines in the field, it is necessary to have a simple rating scale. But for more precise studies, such as components of resistance, genetics of resistance, pathogenic variability, etc., it is desirable to have a more detailed scale taking into consideration disease severity and sporulation of the fungus. A 9-point scale based on visual judgement of disease severity in chickpea Ascochyta blight has been found to be very useful (Reddy et al., 1984). The advantages of the scale are that it is rapid, repeatable, and covers a wider range of disease severities. The 9-point scale is suitable for scoring progeny rows and yield plots. Similar scales can be developed for other foliar diseases.

It is also important to establish relationships between disease severity scores and extent of yield loss. This information is essential in selecting lines in resistance breeding programs. In case of chickpea Ascochyta blight, the relationship between 1 to 9 scores and yield loss was estimated (Reddy and Singh, 1990). The loss in yield in lines scored 2 to 4 was less than 10% and in those with a 5 score, it was 16%. The yield loss in lines with a 6 to 7 score was 26 to 27%, while in those with an 8 to 9 score, 81 to 98%.

A more precise 9-point scale can be devised taking into consideration the extent of damage to the crop in the form of defoliation, stem blighting, pod infection, lesion size on stem and pods, and the extent of sporulation in the lesions. Generally, a correlation is observed between these characters. The major problem observed in devising a precise rating scale (e.g., for Ascochyta blight of chickpea) is variation in disease severity between the plants of the same line and variation in the lesion size and sporulation on the same plant. Diverse scales can be developed for evaluating material in the greenhouse and laboratory.

Testing Related and Linked Characters

Many of the testing procedures for disease resistance are rather complex and time consuming, and so breeders have looked for easily selectable markers linked to the resistance.

Unfortunately, disease resistance is often associated with traits which the breeder is trying to select against, such as late maturity [e.g., resistance to *Mycosphaerella pinodes* (Berk. & Bloxam) Vestergr. in pea, see Lawyer, 1984], high content of phenolics in seed (e.g., resistance to various pathogens affecting establishment in pea, chickpea, and faba bean, see Muehlbauer and Kraft, 1978; Knights and Mailer, 1989; Pascual Villalobos and Jellis, 1990), and long straw (e.g., resistance to *Ascochyta fabae* Spegazzini in beans, see Jellis *et al.*, 1985).

Recently, there has been much interest in using isozymes and restriction fragment length polymorphisms (RFLPs) as selectable markers. These can identify the presence of resistance genes without depending on phenotypic testing.

The use of isozymes depends on a close association between resistance and a specific enzyme banding pattern. An example of a case where this has been used very successfully is in wheat, where an endopeptinase gene *Ep-Dlb* is tightly linked to the gene *Pch1* which confers high resistance to eyespot [*Pseudocercosporella herpotrichoides (Fron) Deighton*] (Summers *et al.*, 1988). Gaur and Slinkard (1991) have recently described an isozyme gene map for chickpea. Linkage of morphological markers and isozymes has also been studied in lentil (Vaillancourt, 1989). In pea, a considerable amount of work has been done on linkage relationships of isozyme loci (Mahmoud *et al.*, 1984; Weeden and Marx, 1987). Recently, Weeden *et al.* (1992) have shown a close linkage between the peroxidase gene *Prx-3* and the gene conferring resistance to strain P1 of pea seedborne mosaic virus. Such developments may lead to the successful development of isozyme marker assisted selection for resistance to pea diseases in the future.

With the development of molecular biology, there is currently considerable interest in using RFLPs as markers (see Tanksley *et al.*, 1989, for a general review). RFLP linkage maps are currently being constructed for pea (Davies, 1990) and other legumes and we can expect rapid developments in this field.

Screening for Resistance in the Major Cool Season Food Legume Crops

Chickpea

One of the major reasons for slow progress of foliar disease resistant work in chickpea in the past has been the lack of efficient field inoculation techniques. Techniques developed only recently for Ascochyta blight caused by *Ascochyta rabiei* (Pass.) Labrousse.

Ascochyta Blight

In view of the polycyclic nature of Ascochyta blight (and other foliar diseases) and association between age of the plant and disease susceptibility, field evaluation of the lines exposing all stages of crop to the disease is necessary. It has been found to be necessary to expose the materials to disease even after the susceptible checks were killed as the disease progresses with time in the resistant materials until maturity. At ICARDA, efficient techniques to evaluate large amounts of chickpea germplasm against Ascochyta blight in up to 8.5 ha fields have been developed (Reddy *et al.*, 1984).

Temperature and humidity are critical factors for blight development. Planting the crop in a period when the average minimum and maximum temperatures are between 10 and 20°C is essential. Relative humidity can be increased by sprinkler or perfo-irrigation, if needed. Inoculations are done either by spraying plants with spore suspensions of the fungus multiplied in the laboratory or by scattering diseased debris in the field. The advantage of the diseased debris method is that inoculations can be done any time and blight develops when conditions become favorable for the disease. With the spore suspension method, it is necessary to inoculate plants when natural conditions are favorable for disease development or by providing favorable conditions after inoculation. The success rate of this method is low and requires repeated inoculations. Instead of using diseased debris, the fungus can be multiplied on chickpea seed or chickpea dextrose broth and the dried seed or mycelial mats can be spread in the field. In the Mediterranean region, planting the crop in the winter season exposes it to high disease pressure. The other advantages of the diseased debris method are that it is closer to the natural mode of spread of the disease and it can be used where laboratory facilities are lacking.

Systematic evaluation of chickpea for Ascochyta blight resistance started with the initiation of the ICRISAT-ICARDA kabuli chickpea program in 1978. The entire world collection of germplasm of over 15,000 accessions comprising both desi and kabuli types was evaluated at ICARDA in Syria and the accessions resistant at this site (Table 1) were evaluated at other blight endemic locations in the world (Reddy and Singh, 1984). A few accessions of wild *Cicer* species were also found promising against blight at ICARDA (Table 2) (K. B. Singh and M. V. Reddy, unpublished). Many cultivated chickpea lines that were resistant in the vegetative stage showed severe infection in the pod stage. The reaction of the lines varied with the location. In general, the lines showed higher disease severity in India and Pakistan than in West Asia, North Africa, and Southern European countries. The variable reaction of the lines was attributed to physiologic specialization in the blight pathogen (Reddy and Kabbabeh, 1985). On the other hand, other workers obtained results suggesting that the isolates differ only in the degree of virulence (Gowen *et al.*, 1989). More information is needed about these aspects to improve the efficacy of screening methods for disease resistance.

Table 1. Desi and kabuli chickpea germplasm accessions resistant or moderately resistant in field and greenhouse evaluation trials (on a 1 to 9 scale) to *Ascochyta rabiei* in Syria (ICARDA)

	Field evaluation (1979 to 1991)[b]		Greenhouse evaluation
Accession No.[a]	Average blight score (range)	No. of years of evaluation	Blight score (1990)
Desi			
ICC 3606	3.0 (2-4)	2	5.0
ICC 4286	4.0 (4-5)	2	5.0
ICC 4475	3.5 (3-4)	4	4.0
ICC 4828	4.0 (3-5)	2	5.0
ICC 6328	3.5 (3-4)	2	4.0
ICC 8540	4.0 (4-4)	2	5.0
ICC 8566	4.0 (4-4)	2	5.0
ICC 9584	4.0 (4-4)	2	5.0
ICC 12004	3.0 (3-3)	3	4.0
Kabuli			
ILC 187	3.0 (2-4)	9	5.0
ILC 200	3.0 (2-4)	9	3.5
ILC 3856	3.0 (3-4)	6	5.0
ILC 5913	3.0 (2-4)	2	5.0
ILC 6482	4.0 (4-4)	2	2.0

[a]   ICC = ICRISAT Chickpea assigned to ICRISAT germplasm accessions and ILC = International Legume Chickpea assigned to ICARDA germplasm accessions.
[b]   Evaluations during 1985, 1987 and 1990 were not effective.

A few lines, such as ILC 72, ILC 182, ILC 201, ILC 202, ILC 2380, ILC 2956, ILC 3279, ILC 3868, ILC 3870, and ILC 4421 showed resistance across locations (Singh *et al.*, 1984). Most of these were found to be either kabuli or intermediate types. They were also found to be tall and late maturing. A few breeding lines, such as FLIP 82-191C, FLIP 83-46C, FLIP 83-49C, FLIP 83-72C, FLIP 83-97C, FLIP 84-83C, FLIP 84-93C that were recently developed at ICARDA also showed resistance to blight at multiple locations (Reddy *et al.*, 1992).

Table 2. Accessions of wild annual *Cicer* species resistant or moderately resistant to Ascochyta blight, Tel Hadya, Syria, 1988 to 1991

ICARDA legume wild accession no.	Ascochyta blight score on 1 to 9 scale					
	Field tests			Greenhouse tests		
	1988	1989	1991	1989	1990	1991
C. judaicum						
ILWC 4-2	3	2	3	5	5	4
ILWC 29/S-9	3	2	3	5	4	4
ILWC 31-2	3	2	4	4	4	2
ILWC 47/2	3	2	3	5	5	5
C. pinnatifidum						
ILWC 7-5	3	3	4	5	5	4
ILWC 9/2	3	2	3	4	5	5
ILWC 29/S-10	3	2	4	5	5	5
ILWC 30-1	2	2	3	5	5	4
ILWC 30-3	3	2	3	5	5	4
ILWC 30/S-2	3	2	3	4	4	2
ILWC 49/1	3	2	5	5	5	5
C. arietinum						
ILC 1929 (Susceptible check)	9	9	9	9	9	9

Other Diseases

Satisfactory screening techniques for the other foliar diseases have not been developed (Reddy *et al.*, 1990). "Hot-spot" locations for Botrytis gray mold, Stemphylium blight, and rust are known. For gray mold, Northeast India, Nepal, and Bangladesh are endemic areas. Pantnagar in India, Rampur in Nepal, and Ishurdi in Bangladesh are gray mold hot-spot locations. Closer spacing, early sowing, establishing a good stand and good canopy through good agronomic practices, and increasing humidity by perfo- or flood irrigation helps in obtaining higher disease pressure.

For Stemphylium blight, Dholi, India and Ishurdi, Bangladesh are hot-spot locations and the conditions that favor Botrytis grey mold are also favorable for Stemphylium blight. Chickpeas sown in the summer (July to October) at Terbol, Lebanon were found to develop severe rust infections. The evaluation of chickpea for resistance to Botrytis gray mold, Alternaria blight, Stemphylium blight, and rust has been very limited, and mainly confined to field tests. The experience with Botrytis gray mold at ICRISAT indicates that it may be difficult to get high levels of resistance to the disease. A few lines that showed some promise with moderate disease pressure at Pantnagar, India were susceptible at Rampur, Nepal, a hot-spot location for the disease. Again, the tall types with compact canopy were found to be more resistant to the disease than the conventional spreading types (Reddy et al., 1990). The work on the host-plant resistance to foliar diseases in chickpea other than Ascochyta blight is very limited. Frequent loss of resistance and lack of stability across locations was observed.

Faba Bean

Faba bean germplasm and breeding lines have been evaluated for resistance to three major diseases, namely rust [(Uromyces fabae (Grev.) Fuckel = U. viciae-fabae Pers.:Pers.) J. Schröt], Ascochyta spot (blight) (Ascochyta fabae), and chocolate spot (Botrytis fabae Sardina).

Rust

In the case of rust, evaluations were first conducted indoors to identify resistance and to provide information on the race composition of rust isolates from cultivated and wild legume hosts in Manitoba, Canada (Conner and Bernier, 1982a). Single pustule isolates were used as inoculum, and single plants of four cultivars were assessed for reaction on the basis of infection type (IT) of 0 to 4, where 0 = highly resistant (no sporulation), 1 and 2 = resistant (small pustules, about 0.5 mm in diameter), and 3 and 4 = susceptible. The four cultivars were heterogeneous for rust reactions but a few plants in each were found to be resistant to each of two rust isolates. When seven faba bean inbred lines (some derived from single resistant plants), and 12 pea cultivars were used to test 17 rust isolates, 11 races were identified. The rust reactions remained the same when the lines and cultivars were tested in the field, indicating that as with other rust fungi, evaluations can be conducted in the greenhouse as well as in the field.

Crosses involving 11 faba bean inbred lines revealed the presence of several genes conditioning resistance to two rust races (Rashid and Bernier, 1986a). The study also showed that IT 2 was controlled by a single gene in several inbred lines, and that gene(s) controlling resistance to race 3 appear to condition IT 0 in line 1, IT 1 in line 3, and IT 3 in line 4. These results indicate that selection for intermediate IT would not lead to quantitative resistance in this host-pathogen system, and confirm results obtained with rust on cereals.

In view of the existence of numerous races in faba bean rust, open-pollinated faba bean accessions were evaluated for their ability to retard rust development in order to identify more durable and quantitative resistance. Some 252 accessions were tested in the field over four years as single or double-row plots 2.5 m long (Rashid and Bernier, 1986b). Two rows of a rust susceptible accession were planted at right angles to the plots, at both ends, to act as rust-spreader rows. The spreader rows were inoculated 3 to 4 weeks after emergence with a mixture of two virulent races. Since the accessions were generally heterogeneous in their disease reactions, mass selection (MS) was performed every year by eliminating the resistant plants (IT 0 to 2) from accessions with a predominance of IT 3 and 4. Rust development was assessed using keys of 1, 5, and 10% leaf area with sporulating pustules. The values of leaf area infected over time were summarized as area under the disease progress curve (AUDPC) values. Eight populations consistently had low AUDPC values and were considered slow rusters. Rust development and spread were also evaluated in isolated test plots using one slow-, one moderate- and one fast-rusting MS population. Populations were found to be similarly ranked in both adjacent and isolated plots, confirming the adequacy of using small adjacent plots in evaluating large number of accessions for slow-rusting.

More recently, it was shown that rust caused yield losses of only 1 to 2% in three slow-rusting populations, whereas in other populations with equal AUDPC values, losses ranged from 6 to 43% (Rashid and Bernier, 1991). The results of this work agree with the conclusions reached by Buddenhagen (1981) that it is essential to assess yielding ability as well as disease severity in order to identify populations with tolerance. Clearly, slow-rusting capability may not reflect the yield potential under epidemic conditions. Evaluation of genotypes for slow-rusting and tolerance must be done concurrently over several years of field testing to ensure that genotypes with both traits are identified.

Table 3. Inventory of genetic resistance to major faba bean diseases available in germplasm and breeding lines

Disease	Pathogen	Resistance		Patho-types	References
		Type	Nature of Inheritance		
Rust	*Uromyces viciae-fabae*	SR rr	Monogenic dominant --	yes	Conner & Bernier (1982a,b); Rashid & Bernier (1984, 1986a); Khalil *et al.* (1985);
		tol.	--		Rashid & Bernier (1991)
Ascochyta spot (blight)	*Ascochyta fabae*	SR rr	Monogenic dominant --	yes	Kharanda & Bernier (1980); Hanounik & Maliha (1984); Rashid *et al.*, (1991a,b)
Chocolate spot	*Botrytis fabae*	SR	Monogenic?	yes?	Hanounik (1983); Hanounik & Maliha (1986); Elliott & Whittington (1979); Khalil *et al.* (1984); Hanounik & Maliha (1986); Hanounik & Robertson (1988)

SR = strong resistance
rr = rate reducing resistance
tol. = tolerance
-- Data not available

Leaf Spots

Two necrotrophic diseases of faba bean, chocolate spot, and Ascochyta spot, provide good examples of successful screening for resistance where the work was conducted first in the field and then confirmed in greenhouse tests (Table 3). For both diseases, genotypes were evaluated by artificially inoculating micro plots in the evening and covering them with polyethylene sheets supported by metal or wood frames to maintain leaf wetness overnight (Hanounik and Robertson, 1988; Rashid and Bernier, 1991).

Chocolate Spot

Since the extent of pathogenic variability in *Ascochyta fabae* was not known, a mixture of 20 isolates was used as inoculum in a first evaluation (Hanounik and Robertson, 1988). Some of the plants in this test developed a few susceptible lesions which were believed to be due to highly virulent forms of the fungus. To provide more rigorous testing, 20 isolates of the fungus were obtained from such lesions and used to retest 53 resistant progenies from the first season. Fourteen of the 53 progenies remained resistant in the second testing. These were also resistant when tested on detached leaves in the laboratory. Further comparisons of the efficiency of the resistant populations were made by growing 19 lines in Syria, Egypt, England, and the Netherlands. Only three lines remained resistant at all locations, and 19 and 16 were resistant in Egypt and Syria, respectively (Hanounik and Maliha, 1986). The results indicate that more virulent pathotypes of the fungus exist in Europe than in the Middle East, or that the environmental conditions in the former were more favorable for disease development.

Finally, it is worth mentioning that some hybrid bulk populations tested in Egypt under field conditions showed multi-resistance to chocolate spot, rust, and Alternaria leaf spot (Khalil *et al.*, 1986).

Ascochyta Spot

A slightly different approach was used to evaluate faba bean lines for resistance to Ascochyta spot in Manitoba since isolates of known virulence were available (Kharbanda and Bernier, 1980). Open-pollinated and mass-selected populations were tested in the field over a 3-year period using two virulent isolates (Isolate A and Y1) each at a separate location (Rashid *et al.*, 1991a). The plots were inoculated and covered with polyethylene as described above for rust, and rated for reaction on a 0 to 5 scale where 0 = no infection, 1 = flecking and 2 = localized lesions without pycnidia, i.e., no sporulation, 3 = localized lesions with pycnidia, 4 = spreading lesions with pycnidia, and 5 = coalescing lesions with pycnidia. Classes 0 to 2 were resistant and 3 to 5 susceptible. To improve the homogeneity for disease reaction, all susceptible plants were removed from a given population. The number of populations tested was reduced from 370 in the first year to 50 and 23 in subsequent years. In the last 2 years, plants were sequentially inoculated with each isolate. Isolate A was applied first and isolate Y1 applied to new growth 8 days later.

After three cycles of testing and mass-selection, the level of heterogeneity was reduced but no population was homogenous for resistance. A total of 18 and 11 populations were identified with > 80% plants resistant to isolates A and Y1, respectively, and two populations had > 80% plants resistant to both isolates. These populations were then tested and selfed in the growth room where, after four cycles, seven and eight inbred lines were homogenous to isolates A and Y1, respectively, and five were homogenous for resistance to both isolates. The inbred lines were then used to differentiate 10 isolates originating from various regions of the world into seven races. Such a high number of races can now be explained by the recent first report of the teleomorph (sexual stage) of *A. fabae* in the UK (Jellis and Punithalingam, 1991). In Europe, selection has not been made within partially inbred lines in the hope that a more heterogeneous population may provide a more durable resistance.

In a recent study, the genetics of resistance to five isolates of *A. fabae* was investigated in 19 faba bean inbred lines (Rashid *et al.*, 1991b). Seven genes for resistance to specific isolates were identified. Single genes, or in some cases two genes, controlled resistance to a given isolate. Some genes appear to confer resistance to more than one isolate of the fungus.

It is noteworthy that both leaf spot disease resistant reactions were effectively scored on the basis of infection types, that high levels of resistance (no sporulation) were identified, and that considerable variation was observed in pathogenicity and virulence of each pathogen.

Pea

The pea crop is attacked by a number of fungal and bacterial pathogens that affect foliar growth, seed quality, and yield. Resistance to many of these diseases has been sought and found, and breeding programs have been developed to incorporate these resistances. Many of the testing procedures described in the literature rely on glasshouse inoculation techniques and not all have been verified by field performance.

Some of the foliar diseases, particularly those caused by *Mycosphaerella pinodes* and *Phoma medicaginis* Malbr. & Roum. var. *pinodella* (L. K. Jones) Boerema affect the base of the stem as well as the foliage. Tests for resistance involve inoculating seed (e.g., Sakar *et al.*, 1982), or infesting the growing medium with inoculum either after planting or after plant establishment (e.g., Kraft and Roberts, 1970).

Assessment of resistance in the foliage to these two species and *Ascochyta pisi* Lib. has been performed in the glasshouse by inoculating with conidia, incubating for 48 h in a mist chamber and then placing the plants on a glasshouse bench and assessing symptoms after 7 to 14 days (Ali *et al.*, 1978). No single sources of resistance to all three species was found. Furthermore, there is no correlation between resistance to the footrot and foliage phases of the disease caused by *P. medicaginis* var. *pinodella* according to Sakar *et al.* (1982). Resistance to *A. pisi* is complicated by the existence of distinct races (Darby *et al.*, 1986) and differential interactions have also been reported for *M. pinodes* (Clulow *et al.*, 1991). The situation is further complicated by a strong correlation between maturity and infection by *M. pinodes*, so care must be taken in interpreting results.

Screening for resistance to downy mildew [*Peronospora viciae* (Berk.) Casp.] in the glasshouse or growthroom can be done either by inoculating seed with oospores or pregerminated seed with sporangia before planting, when plants become systemically infected (Ryan, 1971; Hubbeling, 1975), or by spraying seedlings with a sporangial suspension. Differential interactions between host and pathogen have been reported (Hubbeling, 1975).

Selection for resistance to powdery mildew [*Erysiphe pisi* Syd. (= E. polygoni DC.)] normally relies on natural attacks in the field in areas where the disease is known to be severe. Infections are most damaging on late maturing crops (Thomas and Sweet, 1990), so sowing trials later than usual can be an advantage.

Resistance to bacterial blight (*Pseudomonas syringae* Van Hall pv. *pisi*) is race specific. Six races have been characterized and one partially characterized at present (Taylor *et al.*, 1989). Pathogenicity tests are usually made on the stem of pea seedlings. Bacteria are scraped from the surface of 24 to 48 h cultures and stabbed into the main stem at the junction with the stipules at the two youngest nodes and plant reaction recorded after 5 to 10 days. Resistant lines show a localized necrotic reaction and susceptible ones, an area of water soaking that spreads from the site of inoculation.

Table 4. Genes reported to control resistance to some pea diseases

Disease	Pathogen	Resistance gene(s) (where known)	Some key references to sources of resistance
Ascochyta complex			
Leaf and pod spot	*Ascochyta pisi* race 1	*rap-1*	Ali *et al.* (1978)
	2	*Rap-2*	
	3	*Rap-3*	Darby *et al.* (1986)
	4	*Rap-4*	
Mycosphaerella blight	*Mycosphaerella pinodes*	*Rmps-1*	Ali *et al.* (1978)
		Rmps-2	Clulow *et al.* (1991)
		Rmps-3	
		Rmps-4	
		Rmpf-1	
		Rmpf-2	
		Rmpf-3	
		Rmpf-4	
Root rot	*Phoma medicaginis* var. *pinodella*	polygenic	Ali *et al.* (1978)
			Sakar *et al.* (1982)
Downy mildew	*Peronospora pisi*	*rpv-1*	Hubbeling (1975)
		rpv-2	Stegmark (1988)
Powdery mildew	*Erysiphe pisi*	*er-1*	Gritton & Hagedorn
		er-2	(1971)
Bacterial blight	*Pseudomonas syringae* pv. *pisi*	5 dominant genes	Taylor *et al.* (1989)

These techniques have proved very useful in identifying sources of resistance to specific pathogens. Details of resistant lines can be found in the references cited in Table 4.

Phoma medicaginis var. *pinodella* and, in particular, *Mycosphaerella pinodes* affect the foliage as well as the base of the plant. Assessment of resistance in the foliage to these two species and *Ascochyta pisi* has been performed in the glasshouse by inoculating with conidia, incubating for 48 h in a mist chamber and then placing the plants on a glasshouse bench and assessing symptoms after 7 to 14 days (Ali *et al.*, 1978).

Lentil

Two of the most important and devastating foliar diseases of lentil in many countries are rust caused by *Uromyces fabae* (= *U. viciae-fabae*) and Ascochyta blight incited by *Ascochyta fabae* Speg. f. sp. *lentis* Gossen *et al.* (= *A. lentis* Vassiljevsky). Rust and blight affect all aerial parts of the lentil plant (Khare, 1981). Infection and disease development and spread are favored by cool, wet weather (Khare, 1981; Nene *et al.*, 1988). Sources of resistance in lentil germplasm to rust and blight have been identified.

Rust

The rust pathogen, which is autoecious, infects several legumes, including species of *Lathyrus*, *Lens*, *Pisum*, and *Vicia* (Laundon and Waterston, 1965). In India, pathotypes of *U. fabae* from pea were identified which varied in their virulence on a set of differential lentil hosts (Singh and Sokhi, 1980).

Most screening of lentils for resistance to rust has been done in the field (Khare, 1981; Khare *et al.*, 1990). The results of field screening often vary from season to season and location to location depending on environmental conditions and source of viable inoculum. Field screening needs to be done in conjunction with screening in the glasshouse where environmental conditions and inoculation with known isolates of the pathogen can be regulated more precisely. Kramm and Tay (1984) refined glasshouse techniques for inoculating lentils with rust. However, information is lacking on the pathogenic variability within the lentil rust population. Additional information in this area would help to improve the effectiveness and reliability of screening in the glasshouse and field, and in selection of more durable and improved sources of resistance.

Ascochyta Blight

The blight fungus is a seedborne pathogen with a host range confined to lentils. The teleomorph (sexual stage) of the lentil blight fungus was identified recently as *Didymella* sp. (W. J. Kaiser, unpublished data). The pathogen survives from one season to the next in infected seeds and infested crop residues (Nene *et al.*, 1988).

A number of sources of resistance to the blight pathogen have been identified in different countries. Most of the screening has been done under field conditions (Khare *et al.*, 1990), but there is a need to corroborate the results of field screening with those in the glasshouse where controlled inoculations can be done with single or combined isolates of the pathogen. Isolates of *A. fabae* f. sp. *lentis* from different countries vary greatly in cultural characteristics (W. J. Kaiser, unpublished). However, little is known concerning the existence of races or pathotypes of the blight pathogen (Nene *et al.*, 1988). Information on pathogenic variability of the fungus could lead to a more efficient screening of lentil germplasm in the field and glasshouse and in development of lentil cultivars with resistance to one or more diseases.

Researchers in different countries have identified lentil germplasm with resistance or tolerance to lentil rust and Ascochyta blight. Some lentil accessions have been identified that have resistance or tolerance to more than one disease, such as Fusarium wilt and rust (Nene *et al.*, 1975), Ascochyta blight and rust (W. Erskine, personal communication), Ascochyta blight, rust, and wilt (Pandya *et al.*, 1980), and powdery mildew and rust (Khare *et al.*, in press). Recent screening studies at ICARDA have identified several sources of resistance to Ascochyta blight in wild lentil germplasm (Bayaa *et al.*, 1991). Observations on the reaction of wild lentil germplasm is needed for other diseases of lentil, including rust. Lentil germplasm accessions, breeding lines, and cultivars reported to be resistant or tolerant to rust and Ascochyta blight are listed in Table 5.

Future Research Needs and Prospects

The increase in international exchanges of plant germplasm and the involvement of researchers in different countries in the breeding efforts to improve food legume resistance demand the standardization of the techniques utilized in different centers devoted to research and development of resistant plants. Standardization is particularly needed in the study of pathogenic variability of the pathogens and in laboratory tests to evaluate germplasm for resistance. The testing methods must be standardized to give reproducible results and produce disease intensity sufficiently high to allow appropriate selection, but not so severe that plants with some resistance are graded as susceptible (Dhingra and Sinclair, 1985). In fact, the nature of inoculum, its concentration and distribution on the host plants, the age of the host to be tested, and the environmental conditions under which the testing is performed are of paramount importance. They should be accurately defined, if possible as a result of comparative international testing, and then constantly used. This approach implies a coordinated effort to be shared by diverse institutions and specialists. Cooperation between plant pathologists and breeders should be a must from the very beginning of a screening program.

Sources of resistance from different institutions need to be pooled and

Table 5. Lentil germplasm accessions, breeding lines, and cultivars with resistance or tolerance to rust (*Uromyces viciae-fabae*) and Ascochyta blight (*Ascochyta fabae* f. sp. *lentis*)

RUST

India				Ecuador	Morocco
Bombay 18	LG 8	LWS 43	10465	INIAP 406	ILL 215
BC 10	LG 12	LWS 81	10475	(FLIP 84-	ILL 234
C 31	LG 41	NP 47	10495	94L)	ILL 255
EC 10	LG 60	Pant L 406	10502		ILL 275
HPL 5	LL 3	Pant L 620	10506	Ethiopia	ILL 277
HY1-1	LL 48	Pant L 638	10507	ILL 358	ILL 857
JL 599	LL 56	Pant L 639	10511	ILL 857	ILL 4605
JL 632	LL 71	PL 5	10526		(Precoz)
JL 642	LL 82	PL 8		Pakistan	ILL 5883
JL 648	LL 83	PL 538	Chile	Manserha 89	ILL 6002
JL 673	LL 103	PL 539	Araucana-	(ILL 4605)	ILL 6209
JL 674	LL 133	PL 620	INIA		ILL 6212
JL 676	LL 178	PL 640	Centinela-		ILL 6471
JL 688	LP 338	PL 642	INIA		
JL 1004	LP 409	PLMA 183	Laird		
JL 1005	LP 846	Pusa 10	Tekoa		
K 75	LWS 30	RR 25			
L 9-12	LWS 38	T 36			
L 1278	LWS 39	UPL 172			
L 2895	LWS 42	UPL 175			

ASCOCHYTA BLIGHT

India		Pakistan	Morocco	Canada
HPL 5	LG 201	FLIP 84-27L	ILL 5698	ILL 358
L 442	LG 209	FLIP 84-43L	ILL 5700	ILL 5588
L 448	LG 217	FLIP 84-55L	ILL 5883	ILL 5684
LG 169	LG 218	FLIP 84-85L	ILL 6212	Laird
LG 170	LG 219	FLIP 86-9L		
LG 171	LG 221	FLIP 86-12L	Syria	Chile
LG 172	LG 223	ILL 358	ILL 857	ILL 358
LG 173	LG 225	ILL 858	ILL 4605	ILL 4605
LG 174	LG 231	ILL 5588	ILL 5244	
LG 176	LG 232	ILL 5684	ILL 5588	Ethiopia
LG 177	LG 236	ILL 6024	ILL 5562	ILL 358
LG 179	89 S 26013	78 S 26018	ILL 5590	ILL 857
LG 195	Pant L406	78 S 26052	ILL 5593	
		Manserha 89	ILL 5684	
		(ILL 4605)		

References:
Rust: Bascur & Sepúlveda (1989); W. Erskine (personal communication, 1991); Khare (1981); Khare *et al.* (1979); Khare *et al.* (1990); Mishra *et al.* (1985); Nene *et al.* (1988); Pandya *et al.* (1980); Singh & Sandhu (1988).
Ascochyta blight: W. Erskine (personal communication, 1991); Iqbal *et al.* (1990); Kapoor *et al.* (1990); Khare *et al.* (1990); Pandya *et al.* (1980); Singh *et al.* (1982); Slinkard *et al.* (1983); Tay (1989); Tay *et al.* (1981).

evaluated in different disease endemic locations. This helps in both distribution of resistant materials to all those interested and in obtaining information on stability of resistance. The activities of international organizations in this field should be continuously supported. Furthermore, international institutions can contribute in coordinating the work on variability in the major pathogens by identifying suitable institutes and financially supporting research directly or through donors. The source of some newly released disease resistant cultivars in several countries has been from these international nurseries, such as rust and Ascochyta blight resistant lentil accession ILL 4605 that was released recently in Pakistan as "Manserha 89" (W. Erskine, personal communication).

Support given to gene banks worldwide by the International Board of Plant Genetic Resources (IBPGR) in Rome, Italy, is to be encouraged. IBPGR is also instrumental in organizing and supporting germplasm collecting expeditions to different regions of the world. Wild relatives of the cool season food legumes are often poorly represented in most gene banks. There is a continuing need to collect additional germplasm of the wild species of these crops in the centers of diversity. Little is known concerning the resistance of the wild species of cool season food legumes to different diseases. An important step in this direction is the recent report by Bayaa *et al.* (1991) on the resistance of wild lentil germplasm to Fusarium wilt and Ascochyta blight.

Screening techniques can benefit from innovative approaches. At present, there is considerable interest in the development of techniques facilitating indirect selection for disease resistance using isozymes and RFLP markers, and linkage maps for pea, lentil, faba bean, and chickpea are being developed. As the maps become more complete, the opportunities for using linkage in effective screening programs will increase. A major goal will be the marking of specific chromosome segments involved in quantitative traits.

Quantitative assessment of pathogens in host tissue using enzyme-linked immunosorbent assay (ELISA) systems is another area where rapid progress is being made. Both polyclonal and monoclonal antisera have been produced to many fungal pathogens and these have been used to quantify mycelial growth. In host-pathogen systems where assessment is largely subjective, applications of the ELISA system may play a valuable role in the future (Harrison *et al.*, 1991).

The introduction of foreign genes by genetic engineering is another innovative approach commanding much attention at the present time. In order to be successful, transformation requires a system for delivering foreign DNA into rapidly dividing cells. *Agrobacterium*-mediated transformation is often the favored system in dicotyledonous crops but regeneration of transformed tissues has proved to be problematical in legumes. Success has been achieved in transforming soybean (Hinchee *et al.*, 1988), but the technique is dependent on a large-scale tissue culture effort and regeneration of many elite lines may prove difficult (Chee *et al.*, 1989). Other techniques, designed to overcome the regeneration problem, which have been successfully employed are the infection of germinating seeds with *Agrobacterium tumefaciens* (Smith and Towsend) Conn containing a binary vector (Chee *et al.*, 1989) and the penetration of meristematic cells with DNA-coated microprojectiles (McCabe *et al.*, 1988).

In pea, Nauerby *et al.* (1991) have described a successful system using nodal thin cell layer segments as explants. Furthermore, preliminary experiments demonstrated that transformation of pea with *A. tumefaciens* was possible using this system. With further refinement, transgenic legumes may soon be a commonplace as transgenic tobacco and potato plants are today and future developments in gene construction and expression will be available to these important crops.

Concluding Remarks

Methods suitable for screening for resistance to foliar diseases in cool season food legumes have been developed for a number of major diseases. In the last decade, international organizations and national governments and institutions have augmented and coordinated their efforts aimed at improving the efficiency of research projects and other initiatives concerning breeding for resistance in the four crops taken into account in this paper. The results are encouraging, nevertheless more impetus should be given to improve the exchange of knowledge and – with due quarantine precautions – plant materials among scientists and professionals involved in breeding for resistance. Collections of germplasm should also be increased and properly characterized to find sources of resistance suitable for breeding programs.

Future achievements in screening techniques and practical use of sources of resistance also can be expected from biotechnological approaches for cool season food legumes. Nevertheless, as it appears difficult to obtain suitable levels of resistance to the most important diseases of each species, the importance of an integrated disease management program should not be underestimated.

References

Ali, S. M., Nitschke, L. F., Dube, A. J., Krause, M. R. and Cameron, B. 1978. *Australian Journal of Agricultural Research* 29: 841–849.

Bascur B. G. and Sepúlveda R. P. 1989. *LENS Newsletter* 16: 24–26.

Bayaa, B., Erskine, W. and Hamdi, A. 1991. Screening wild lentil for resistance to vascular wilt and Ascochyta blight disease. In: *Proceedings of the Fourth Arab Congress of Plant Protection*, Cairo, Egypt.

Bernier, C. C., Bijiga, G., Nene, Y. L. and Cousin, R. 1988. In: *World Crops: Cool Season Food Legumes*, pp. 97–106 (ed. R. J. Summerfield). Dordrecht: Kluwer Academic Publishers.

Buddenhagen, I. W. 1981. In: *Plant Disease Control: Resistance and Susceptibility*, pp. 221–234 (eds. R. C. Staples and G. H. Toenniessen). New York: John Wiley & Sons.

Buiatti, M. and Ingram, D. S. 1991. *Experientia* 47: 811–819.

Chee, P. P., Fober, K. A. and Slightom, J. L. 1989. *Plant Physiology* 91: 1212–1218.

Clulow, S. A., Lewis, B. G. and Matthews, P. 1991. *Journal of Phytopathology* 131: 322–332.

Conner, R. L. and Bernier, C. C. 1982a. *Canadian Journal of Plant Pathology* 4: 157–160.

Conner, R. L. and Bernier, C. C. 1982b. *Canadian Journal of Plant Pathology* 4: 263–265.

Darby, P., Lewis, B. G. and Matthews, P. 1986. *Plant Pathology* 35: 214–223.

Davies, D. R. 1990. *The Pisum Newsletter* 22: 87–88.

Dhingra, O. D. and Sinclair, J. B. 1985. *Basic Plant Pathology Methods*, 355 pp. Boca Raton, Florida: CRC Press.

Elliot. J. E. M. and Whittington, W. J. 1979. *Journal of Agricultural Science* 93: 411–417.

Frison, E. A., Bos, L., Hamilton, R. I., Mathur, S. B. and Taylor, J. D. 1990. In: *FAO/IBPGR*

Technical Guidelines for the Safe Movement of Legume Germplasm, 88 pp. Rome, Italy: FAO and IBPGR.

Gaur, P. M. and Slinkard, A. E. 1991. *International Chickpea Newsletter* 24: 25–28.

Gowen, S. R., Orton, M., Thurley, B. and White, A. 1989. *Tropical Pest Management* 35: 180–186.

Griffiths, E. 1984. In: *Plant Diseases. Infection, Damage and Loss*, pp. 149–159 (eds. R. K. S. Wood and G. J. Jellis). Oxford, UK: Blackwell Scientific Publications.

Gritton, E. T. and Hagedorn, D. J. 1971. *Crop Science* 11: 941.

Hanounik, S. 1983. *FABIS Newsletter* 5: 24–26.

Hanounik, S. B. and Maliha, N. F. 1984. *FABIS Newsletter* 9: 33–36.

Hanounik, S. B. and Maliha, N. F. 1986. *Plant Disease* 70: 770–773.

Hanounik, S. B. and Robertson, L. D. 1988. *Plant Disease* 72: 696–698.

Harrison, J. G., Lowe, R. and Duncan, J. M. 1991. *Plant Pathology* 40: 431–435.

Hawtin, G. C., Muehlbauer F. J., Slinkard, A. E. and Singh, K. B. 1988. In: *World Crops: Cool-Season Food Legumes*, pp. 67–80 (ed. R. J. Summerfield). Dordrecht: Kluwer Academic Publishers.

Hinchee, M. A. W., Connor-Ward, D. V., Newell, C. A., McDonnell, R. E., Sato, S. J., Gasser, C. S., Fischhoff, D. A., Re, D. B., Fraley, R. T. and Horsch, R. B. 1988. *Bio/Technology* 6: 915–922.

Hubbeling, N. 1975. *Mededelingen van de Faculteit Landbouwwetenschappen Rijksuniversiteit te Gent* 40: 539–543.

Iqbal, S. M., Bakhsh, A. and Malik, R. A. 1990. *LENS Newsletter* 17: 26–27.

Jellis, G. J., Lockwood, G. and Aubury, R. G. 1985. *Plant Pathology* 34: 347–352.

Jellis, G. J. and Punithalingam, E. 1991. *Plant Pathology* 40: 150–157.

Johnson, R. 1984. *Annual Review of Phytopathology* 22: 309–330.

Kapoor, S., Singh, G., Gill, A. S. and Singh, B. 1990. *LENS Newsletter* 17: 26–28.

Khalil, S. A., Nassib, A. M. and Abou Zeid, N. M. 1986. *Biologisches Zentralblatt* 105: 1–7.

Khalil, S. A., Nassib, A. M. and Mohammed, H. A. 1985. *FABIS* 11: 18–20.

Khalil, S. A., Nassib, A. M., Mohammed, H. A. and Habib, W. F. 1984. *FABIS* 10: 18–21.

Kharbanda, P. D. and Bernier, C. C. 1980. *Canadian Journal of Plant Pathology* 2: 139–142.

Khare, M. N. 1981. In: *Lentils*, pp. 163–172 (eds. C. Webb and G. Hawtin). Slough, UK: CAB International/ICARDA.

Khare, M. N., Agrawal, S. C. and Jain, A. C. 1979. *Diseases of lentil and their control*, 46 pp. Technical Bulletin JNKVV, Jabalpur, M.P., India.

Khare, M. N., Bayaa, B. and Beniwal, S. P. S. 1993. In: *Breeding for Stress Tolerance in Cool-Season Food Legumes*, pp. 107–121 (eds. K. B. Singh and M. C. Saxena). Chichester, UK: John Wiley & Sons.

Knights, E. J. and Mailer, R. J. 1989. *Journal of Agricultural Science, Cambridge* 113: 325–330.

Kraft, J. M. and Roberts, D. D. 1970. *Phytopathology* 60: 1814–1817.

Kramm, M. V. and Tay, J. U. 1984. *LENS Newsletter* 11: 24.

Laundon, G. F. and Waterston, J. M. 1965. *Uromyces viciae-fabae*. Descriptions of Pathogenic Fungi and Bacteria, No. 60, 2 pp. Commonwealth Mycological Institute, Kew Surrey, England.

Lawyer, A. S. 1984. In: *Compendium of Pea Diseases*, pp. 11–15 (ed. D. J. Hagedorn). St. Paul, Minnesota: APS Press.

Mahmoud, S. H., Gatehouse, J. A. and Boulter, D. 1984. *Theoretical and Applied Genetics* 68: 559–566.

McCabe, D. E., Swain, W. F., Martinell, B. J. and Christou, P. 1988. *Bio/Technology* 6: 923–926.

Mishra, R. P., Kotasthane, S. R., Khare, M. N., Gupta, O. and Tiwari, S. P. 1985. *LENS Newsletter* 12: 25–26.

Muehlbauer, F. J. and Kraft, J. M. 1978. *Crop Science* 18: 321–323.

Nauerby, B., Madsen, M., Christiansen, J. and Wyndaele, R. 1991. *Plant Cell Reports* 9: 676–679.

Nene, Y. L., Hanounik, S. B., Qureshi, S. H. and Sen, B. 1988. In: *World Crops: Cool-Season Food Legumes*, pp. 67–80 (ed. R. J. Summerfield). Dordrecht: Kluwer Academic Publishers.

Nene, Y. L., Kannaiyan, J. and Saxena, G. C. 1975. *Indian Journal of Agricultural Science* 45: 177–178.

Nene, Y. L. and Reddy, M. V. 1987. In: *The Chickpea*, pp. 233–270 (eds. M. C. Saxena and K. B.

Pandya, B. P., Pandey, M. P. and Singh, J. P. 1980. *LENS* 7: 34–37.
Parlevliet, J. E. 1979. *Annual Review of Phytopathology* 17: 203–222.
Parlevliet, J. E. 1983. *Phytopathology* 73: 379.
Pascual Villalobos, M. J. and Jellis, G. J. 1990. *Journal of Agricultural Science, Cambridge* 115: 57–62.
Rashid, K. Y. and Bernier, C. C. 1984. *Plant Disease* 68: 16–18.
Rashid, K. Y. and Bernier, C. C. 1986a. *Canadian Journal of Plant Pathology* 8: 317–322.
Rashid, K. Y. and Bernier, C. C. 1986b. *Crop Protection* 5: 218–224.
Rashid, K. Y. and Bernier, C. C. 1991. *Canadian Journal of Plant Science* 71: 967–972.
Rashid, K. Y., Bernier, C. C. and Conner, R. L. 1991a. *Plant Disease* 75: 852–855.
Rashid, K. Y., Bernier, C. C. and Conner, R. L. 1991b. *Canadian Journal of Plant Pathology* 13: 218–225.
Reddy, M. V. and Kabbabeh, S. 1985. *Phytopathologia Mediterranea* 24: 265–266.
Reddy, M. V., Nene Y. L., Singh, G. and Bashir, M. 1990. In: *Chickpea in the Nineties*, pp. 117–128 (eds. H. A. van Rheenen, M. C. Saxena, B. J. Walby and S. D. Hall). Patancheru, Andhra Pradesh, India: ICRISAT.
Reddy, M. V. and Singh, K. B. 1984. *Plant Disease* 68: 900–901.
Reddy, M. V. and Singh, K. B. 1990. *Phytopathologia Mediterranea* 29: 32–38.
Reddy, M. V., Singh, K. B. and Malhotra, R. S. 1992. *Phytopathologia Mediterranea* 31: 59–66.
Reddy, M. V., Singh, K. B. and Nene Y. L. 1984. In: *Ascochyta Blight and Winter Sowing of Chickpeas*, pp. 45–54 (eds. M. C. Saxena and K. B. Singh). The Hague, The Netherlands: Martinus Nijhoff/Dr. W. Junk Publishers.
Ryan, E. W. 1971. *Irish Journal of Agricultural Research* 10: 315–322.
Sakar, D., Muehlbauer, F. J. and Kraft, J. M. 1982. *Crop Science* 22: 988–992.
Singh, G., Singh, K., Gill, A. S. and Brar, J. S. 1982. *Indian Phytopathology* 35: 678–679.
Singh, K. B., Reddy, M. V. and Nene, Y. L. 1984. *Plant Disease* 68: 782–784.
Singh, K. and Sandhu, T. S. 1988. *LENS Newsletter* 15: 28–29.
Singh, S. J. and Sokhi, S. S. 1980. *Plant Disease* 64: 671–672.
Slinkard, A. E., Morrall, R. A. A. and Gossen, B. 1983. *LENS Newsletter* 10: 31.
Stegmark, R. 1988. *Acta Agriculturae Scandinavica* 38: 373–379.
Summers, R. W., Koebner, R. M. D., Hollins, T. W., Förster, J. and Macartney, D. P. 1988. In: *Proceedings of the Seventh International Wheat Genetics Symposium*, pp. 1195–1197. Cambridge, UK.
Tanksley, S. D., Young, N. D., Paterson, A. H. and Bonierbale, M. V. 1989. *Bio/Technology* 7: 257–264.
Tay, J. 1989. *Inheritance of resistance to Ascochyta blight in lentil*, 70 pp. M.S. Thesis, University of Saskatchewan, Saskatoon, Sask., Canada.
Tay, J., Paredes, M. and Kramm, V. 1981. *LENS* 8: 30.
Taylor, J. D., Bevan, J. R., Crute, I. R. and Reader, S. L. 1989. *Plant Pathology* 38: 364–375.
Thomas, J. E. and Sweet, J. B. 1990. *Diseases of Peas and Beans*, 51 pp. Cambridge, UK: National Institute of Agricultural Botany.
Tiwari, S. P. 1985. *LENS Newsletter* 12: 25–26.
Vaillancourt, R. 1989. *Inheritance and linkage of morphological markers and isozymes in lentil*, 143 pp. Ph. D. Thesis. University of Saskatchewan, Saskatoon, Sask., Canada.
van den Bulk, R. W. 1991. *Euphytica* 56: 269–285.
van der Maesen, L. J. G., Kaiser, W. J., Marx, G. A. and Worede, M. 1988. In: *World Crops: Cool Season Food Legumes*, pp. 55–66 (ed. R. J. Summerfield). Dordrecht: Kluwer Academic Publishers.
Weeden, N. F. and Marx, G. A. 1987. *The Journal of Heredity* 78: 153–159.
Weeden, N. F., Provvidenti R. and Wolco, B. 1992. *Pisum Genetics* 23: 42–43.

Screening techniques and sources of resistance to root rots and wilts in cool season food legumes

J.M. KRAFT[1], M.P. HAWARE[2], R.M. JIMÉNEZ-DÍAZ[3], B. BAYAA[4] and
M. HARRABI[5]

[1] U. S. Department of Agriculture, Agricultural Research Service, WSU-IAREC, Route 2 Box
2953A, Prosser, Washington, 99350 USA;
[2] ICRISAT, Patancheru P. O., Andhra Pradesh 502 324, India;
[3] Instituto de Agronomía y Protección Vegetal, CSIC, and Departamento de Agronomia-Patologia
Vegetal, ETSIAM, Universidad de Córdoba, Apdo. 3048, 14080 Córdoba, Spain;
[4] ICARDA, P.O. Box 5466, Aleppo, Syria, and
[5] Institut National Agronomique de Tunisie, Laboratorie de Genetique, Tunis, Tunisia

Abstract

Soilborne, fungal pathogens of cool season food legumes, including seed and
seedling blights, rot rots, and wilts are described. Seed and seedling diseases are
caused primarily by *Pythium* and *Rhizoctonia* spp. The most important fungi
causing root rots include *Aphanomyces euteiches*, *Fusarium solani*, *Pythium* spp.,
Sclerotium rolfsii, and *Macrophomina phaseolina*. Wilt is caused primarily by
various host-specific forms of *Fusarium oxysporum*. This paper discusses these
diseases and screening procedures that emphasize standardization of inoculum
levels, maintenance of virulent pathogen cultures, inoculum growth media,
environmental conditions, and host plant age. Sources of resistance to these
diseases are discussed.

Seedling Diseases

Any environmental or physiological factors which delay emergence or result in
uneven stands, such as: a) poor seed vigor; b) cold wet soil; c) poor seedbed
preparation; d) herbicide injury; or e) crusting of soil after planting can
predispose developing plants to seedling disease. Worldwide, seedling diseases
of peas (*Pisum sativum* L.), chickpeas (*Cicer arietinum* L.), faba beans (*Vicia
faba* L.), and lentils (*Lens culinaris* Medik.) are caused primarily by *Pythium*
spp. and *Rhizoctonia solani* Kuehn.

Pythium Seed and Seedling Rot

Pythium ultimum Trow and sporangial forms resembling *P. ultimum* are often
described as seed and seedling pathogens of both chickpeas and peas. Other
species, such as *P. splendens* Braun and *P. irregulare* Buisman (Harman, 1984;

*F.J. Muehlbauer and W.J. Kaiser (eds.), Expanding the Production and Use of Cool Season Food
Legumes, 268–289.*
© 1994 *Kluwer Academic Publishers.*

van der Plaats-Niterink, 1981), have also been reported as pathogenic to peas. In studies conducted at Prosser, Washington, Plant Introduction (PI) accessions resistant to *P. ultimum* were also resistant to other *Pythium* species (Kraft, unpublished data). In addition, zoospore inoculum, produced by such species as *P. irregulare*, was equal in pathogenicity to mycelial inoculum produced by *P. ultimum*.

Stasz and Harman (1980) reported that resistance in peas to Pythium seed and seedling rot was due to differences in numbers of infections occurring at a given inoculum level. They also reported that aging seeds prior to planting increased exudation, decreased vigor, and increased disease severity. Woyke (1987) reported that large seed was more vigorous and resistant to seedling attack by *Pythium* than small seed. Pea seed with the dominant *A* gene for anthocyanin production was more resistant than lines possessing the recessive *a* gene for lack of anthocyanin. Resistance is due to the presence of delphinidin, an anthocyanin (anthocyanin-aglycone) pigment in the testae, which is fungistatic to a number of seed and root pathogens. However, peas with an *A* gene can be susceptible in the seedling stage, despite the presence of delphinidin, if they exude sufficient amounts of reducing sugars (Kraft, 1977). Similar to peas, kabuli chickpea seeds, lacking pigmentation in the testae, are extremely susceptible to *Pythium* attack, as compared to pigmented, desi types (Kaiser and Hannan, 1983).

Methods for screening peas for resistance to *Pythium* include infesting soil with: cornmeal-sand inoculum (Perry, 1973); inoculum produced on medium-saturated vermiculite (Kraft and Roberts, 1969); oospore inoculum (Stasz and Harman, 1980); or soaking seed in a hyphal suspension prior to planting (Ohn *et al.*, 1978). Soil infested with inoculum of *P. ultimum*, grown on vermiculite saturated with a basal medium, is used at Prosser. Infested soil is air-dried to induce the formation of thick-walled sporangia and oospores of *P. ultimum*. This soil is then mixed with sufficient quantities of non-infested soil so that test lines are exposed to a population of 200 to 500 propagules of *P. ultimum* g^{-1} of air-dry soil (Mircetich and Kraft, 1973).

Several PI accessions, including 257593, 140165, 166159 and 140295 with the dominant *A* gene for pigmented testae, were found resistant and were used in breeding for resistance at Prosser (Kraft and Roberts, 1970). However, more advanced pea breeding lines with the recessive *a* gene have been developed, which are resistant to *Pythium* seed and seedling attack. Early generation lines (F_3-F_6) are planted in the field with and without seed treatment fungicides to screen for resistance to *Pythium* pre- and post-emergence damping-off.

Rhizoctonia Seed and Seedling Rot

The imperfect stage classification of *Rhizoctonia solani* is based on the anastomosis grouping concept (AG). The pathogen responsible for seed and seedling disease is classified in AG4 (Anderson, 1982). *Rhizoctonia solani* (AG4) can attack pea seedlings whenever environmental conditions are favorable. For seedling infection to occur, the sclerotium or hyphal fragment must germinate or resume growth. The resulting hyphae may grow through soil for several millimeters to infect the epicotyl, seed, or hypocotyl of a seedling host. The pathogen prefers well aerated areas at the soil surface and is most aggressive under warm, moist (24 to 30°C) conditions.

Screening tests for resistance to *R. solani* have utilized the following procedures: a) mycelial discs from potato-dextrose agar (PDA), V8 juice agar, or synthetic media (McCoy and Kraft, 1984); b) corn kernel inoculum (Shehata *et al.*, 1981, 1984) placed at the base of seedling stems; and c) sclerotia infested soil (McCoy and Kraft, 1984). McCoy and Kraft (1984) reported that resistance to *R. solani* was positively correlated with epicotyl thickness when test lines were grown in soil infested with 20 sclerotia g^{-1} of soil. Shehata *et al.* (1983) reported that pea lines resistant to Fusarium root rot were susceptible when grown in soil infested with both *R. solani* and *Fusarium solani* (Mart.) Sacc. f. sp. *pisi* (Jones) Snyd. & Hans. The type of nutrient medium used to produce primary inoculum of *R. solani* can affect overall disease severity ratings. For example, mycelial discs of *R. solani* incubated on PDA or V8 juice agar caused lower disease ratings than mycelial-disc inoculum from dextrose-asparagine agar. However, the overall relationship among test lines remained similar (McCoy and Kraft, 1984). Table 1 lists those pea lines reported resistant to *R. solani* AG4.

Table 1. Pea lines resistant to *Rhizoctonia solani* AG4

Pea Line #	Reference
Dark Skin Perfection	McCoy and Kraft, 1984
	Shehata *et al.*, 1981
B77-634-4	McCoy and Kraft, 1984
	Shehata *et al.*, 1981
PI 189171	McCoy and Kraft, 1984
	Shehata *et al.*, 1981
PI 197990	McCoy and Kraft, 1984
	Shehata *et al.*, 1981
74SN3	Shehata *et al.*, 1981
PI 257593	Shehata *et al.*, 1981

In India, *R. solani* can sometimes cause heavy damage to seedling and adult lentil plants depending upon the time of infection. In both cases, affected plants exhibit yellowing of leaves progressing from lower to upper plant parts. The roots of the affected plants become reddish brown with a clear constriction at the collar region or below (Shukla *et al.*, 1972; Kannaiyan and Nene, 1973; Shatla *et al.*, 1974). The main root below the constriction remains healthy in the early stages of infection but may become infected at a later date. In such cases, the affected seedlings/plants are easily pulled out of the soil, but some break at the constriction point. When infection occurs in the late seedling stage, the plants can survive for some time but exhibit a progressive yellowing of leaves from the base upwards.

To screen lentil lines for resistance to *R. solani* in the greenhouse, the fungus is grown on PDA and the mycelial mat mixed thoroughly with the planting medium (Kannaiyan and Nene, 1973). Lentil seedlings are grown in sterilized soil for 10 days after which the top 2.5 cm soil is removed and a uniform amount of fungus inoculum placed near the collar region of each seedling. Then the soil is replaced. Mortality counts are taken every 5 days. None of the 158 lentil lines/accessions tested through a pot-screening method in the greenhouse were found resistant; however, line UPL 172 showed the lowest mortality of 30% followed by 40% in line UPL 288 (Kannaiyan, 1974).

Collar rot of lentil, caused by *Sclerotium rolfsii* Sacc., is omnipresent and can cause heavy losses at the early stages of crop growth. Infected plants droop and ultimately exhibit damping-off symptoms. In the final stages of the disease cycle, white strands of the fungus and sclerotia are formed around the collar region. The disease is more severe under sufficient soil moisture conditions and ambient temperatures of around 28 to 30°C (Aycock, 1966; Mathur and Deshpande, 1968; Khare, 1980).

In the greenhouse, screening for collar rot resistance is conducted either in trays or large pots (Kannaiyan and Nene, 1976; Khare, 1980). Pot screening is conducted similarly to the method described for wet root rot (Kannaiyan and Nene, 1973). In another method, *S. rolfsii* is multiplied on soil-maize medium (Kannaiyan and Nene, 1976); or sand-oatmeal (Claudius and Mehrotra, 1973) for 12 days. Screening trials are conducted in 30-cm-diameter pots using 200 g inoculum in 5 kg^{-1} of soil. Infested pots are covered with moist jute bags for 48 h and then planted with 50 seeds per pot. Another technique involves inoculating each plant with two mature 21-day-old sclerotia at the hypocotyl region and covering with moistened soil (Mohammad and Kumar, 1986). After a 14 to 21 day incubation, test lines are rated for percentage damping-off.

Of several hundred lentil lines and accessions tested for resistance to collar rot in the greenhouse, 12 (JL 678, JL 719, JL 727, JL 828, LP 18, LP 288, LP 338, LP 379, Pant 370 P 23, Pant 638, Pusa 1, and Pusa 3) were found resistant with less than 10% mortality (Kannaiyan, 1974; Mohammad and Kumar, 1986), respectively.

Root Diseases

Fusarium Root Rot

Fusarium root rot of peas is caused by *F. solani* f. sp. *pisi* (Kraft *et al.*, 1981). This is a serious disease of peas in all USA pea-producing areas. This pathogen is now also recognized as a serious root pathogen of chickpeas, especially in warm growing conditions (Bhatti and Kraft, 1992a). Fusarium root rot is distinct from Fusarium wilt and usually occurs in conjunction with other diseases of peas and chickpeas (Kraft *et al.*, 1981; Bhatti and Kraft, 1992a,b). This pathogen can reduce pea yields from a constant 10% to as high as 50%. *Fusarium solani* f. sp. *pisi* usually invades the cotyledonary attachment area of both peas and chickpeas. Initial symptoms on primary and secondary roots consist of reddish-brown streaks that later coalesce. The external root color becomes dark reddish-brown to black, especially at the soil line and in the cotyledonary attachment area. Above ground symptoms of severely infected chickpea and pea plants include stunting, graying, yellowing, and necrosis of lower foliage. Chlamydospores are the naturally occurring survival structure in field soil.

Whalley (1984) developed a rapid test for screening peas for resistance to *F. solani* f. sp. *pisi in vitro*. Test seeds are surface disinfested and germinated on moist filter paper until the plumules are 30 mm long. Resultant seedlings are then transferred to test tubes and suspended in 0.1% water agar containing 1 × 10^6 conidia ml^{-1}. Peas are incubated for 14 days in a growth chamber set at 24°C. Lockwood (1960) screened PI accessions in pure culture by pipetting a 1 × 10^6 ml^{-1} conidial suspension onto seed planted in autoclaved quartz sand. Plants were dug and read 23 days later and scored on a 0 to 9 scale. An infested soil technique was developed at Prosser where seeds are planted in soil infested with 20,000 to 40,000 colony forming units (cfu) g^{-1} soil in a controlled environment (24°C day, 15°C night, 6480 lux maximum illumination with a 16 h day) (Kraft, 1975). Plants are harvested 14 to 21 days after emergence and each plant is scored on a 0 to 5 scale with 5 indicating a completely rotted root. For segregating material, plants with more healthy root systems (i.e., less epicotyl and root necrosis than the susceptible control) are transplanted into an autoclaved potting mixture. Transplants that survive are grown to seed set.

Good seed vigor is an important consideration in comparing one line to another. A line with poor seed vigor may be susceptible to Fusarium root rot when in fact it is resistant (Kraft, 1986), which is similar to an earlier report on Pythium seed rot of peas (Stasz and Harman, 1980). Recently, we have tested a technique first developed by Dr. Simon Menzies, Plant Pathologist, DSIR, Auckland, New Zealand (retired), to screen peas for resistance to Fusarium root rot. Untreated seeds of a test line are soaked overnight at room temperature, in a conidial suspension of *F. solani* f. sp. *pisi* adjusted to 1 × 10^6 spores ml^{-1}. Inoculated seed are then planted into coarse-grade perlite in plastic flats and incubated in a growth chamber at a constant 24°C with a 16 h photoperiod at

6480 lux. The perlite is kept moist throughout the 14 day incubation period by watering with micropore filtered water (0.45 μ). Resultant plants are scored for disease severity using the 0 to 5 scale and segregating plants are saved for seed production as mentioned previously.

Fusarium Root Rot of Chickpea

There are few reports describing *F. solani* f. sp. *pisi* as an important pathogen of chickpea (Kraft, 1969; Grewal *et al.,* 1974; Westerlund *et al.,* 1974). However, *F. solani* is often cited as a serious pathogen of chickpeas (Viswakarma and Chaudhary, 1981; Nain and Agnihotri, 1984; Mani and Sethi, 1985; Mario and Carolina, 1987; Nene, 1987), but it is unknown if this pathogen is actually *F. solani* f. sp. *pisi.*

Several chickpea lines from ICRISAT and ICARDA were evaluated for resistance to *F. solani* f. sp. *pisi* (Bhatti and Kraft, 1992b) using the technique described earlier (Kraft, 1975). Only four lines of 39 tested exhibited some resistance when the incubation temperature varied from 22 ± 3°C. However, when the incubation temperature was 25 ± 3°C no lines tested were resistant.

Progress has been made and will continue in developing peas with quantifiable resistance to Fusarium root rot and with acceptable horticultural attributes. Table 2 lists a number of PI accessions and breeding lines of peas which are resistant/tolerant to *F. solani* f. sp. *pisi.*

Dry Root Rot of Chickpea

Because disease development is highly influenced by temperature and soil moisture, screening for resistance to dry root rot of chickpeas caused by *Rhizoctonia bataticola* (Tabenhaus) E. J. Butler [syn. *Macrophomina phaseolina* (Tassi) Goidanich] in the field is not practical. High temperatures and dry soil conditions at flowering and podding stage can dramatically increase dry root rot severity. However, Fusarium wilt screening plots at ICRISAT center became infested with the dry root rot pathogen, and were used in eliminating chickpea breeding lines susceptible to dry root rot. There are no reported cases of uniform and effective dry root rot disease screening plots being developed for chickpeas. Pot culture techniques for greenhouse screening and a paper towel technique for laboratory screening were developed (Nene *et al.,* 1981). Further improvement of pure culture screening techniques is needed to correspond more closely with results obtained in the field.

Dry root rot of lentil, caused by *R. bataticola*, is becoming an increasingly important lentil disease in India under both dry and humid climates. The affected plant exhibits sudden drooping of top leaves and drying without showing any yellowing. The roots turn ashy to ashy-brown and desiccate. Black sclerotial bodies of variable size and shape develop on the surface as well as within the root. The pathogen is also responsible for pre- and post-emergence seed and seedling rots which reduce plant stands.

Table 2. Pea lines resistant/tolerant to *Fusarium. solani* f. sp. *pisi*

Pea Line	Reference
PI 140165	Kraft, 1975
PI 164417	King *et al.*, 1960
PI 164837	King *et al.*, 1960
PI 164971	King *et al.*, 1960
PI 165577	King *et al.*, 1960
PI 165965	King *et al.*, 1960
PI 166082	King *et al.*, 1960
PI 166084	King *et al.*, 1960
PI 169606	King *et al.*, 1960
PI 171816	King *et al.*, 1960
PI 173057	King *et al.*, 1960
PI 174921	Kraft, 1975
PI 174922	Kraft, 1975
PI 179969	Kraft, 1975
PI 196013	Kraft, 1975
PI 196021	Kraft, 1975
PI 196022	Kraft, 1975
PI 242028	Kraft, 1975
PI 257593	Kraft, 1975
VR-410-2	Kraft and Giles, 1978
VR-1492-1	Kraft and Giles, 1978
RR-1178	Kraft, 1984
WR-1167	Kraft, 1984
792022	Kraft, 1981

Evaluation of lentils for resistance to dry root rot is usually conducted in the laboratory (Kannaiyan and Nene, 1976) using the paper towel method (Deshkar *et al.*, 1973). The inoculum is raised in Richard's liquid medium (Deshkar *et al.*, 1973) or PDA (van Rheenen *et al.*, 1989) for 6 days. The resultant mycelial mat, separated by filtration, is macerated in a sterilized blender for 30 seconds with sterilized water to get a suspension (10:100, v/v). The mycelial suspension is transferred to an autoclaved enamel tray. Paper towels (30 × 30 cm) are first autoclaved at 15 psi for 15 minutes at 110°C, and then immersed in the fungal suspension and removed. Seeds are placed at constant intervals on the towel which is then folded and kept for 3 days in a moist chamber at 25–28°C. The numbers of diseased seeds and seedlings are recorded and diseased seedlings discarded, while the healthy ones are transferred to pots containing soil infested with *R. bataticola* for further screening. The optimum temperature for disease development appears to be 30–35°C (Singh and Nema, 1987; van Rheenen *et al.*, 1989). Unfortunately, using the techniques described above, no sources of resistance to dry root rot of lentils, caused by *R. bataticola*, have been identified.

Common Root Rot of Peas

Common root rot of peas, caused by *Aphanomyces euteiches* Drechs., occurs in most pea growing areas of North America, Europe, Australia, New Zealand, and Japan (Pfender, 1984). This disease can infect peas at any age and plant symptoms can appear as early as 10 days after emergence, if the inoculum level is high and soil moisture levels are conducive for disease development. Straw-colored lesions spread through the root cortical tissues, which become soft and darken as secondary organisms invade the colonized tissues. Microscopic examination of infected cortical tissue will reveal typical oospores of the pathogen. The oospores are readily recognized and are sufficiently distinct to permit quick identification. They are large (25 to 30 μ) and contain a unique, large oil globule.

Oospores can persist in soil for years. Upon germination, oospores form hyphae or sporangia. Sporangia, which are undifferentiated from hyphae, produce asexual primary zoospores which quickly encyst. Soon secondary zoospores emerge from these cysts and swim to a susceptible host where they encyst, germinate, and infect.

All published reports on screening peas for resistance to *A. euteiches* in pure culture have utilized zoospores as the primary inoculum source. Zoospore inoculum has been produced on corn kernel broth (Haglund and King, 1961), on maltose-peptone broth (Carmen and Lockwood, 1960), and on pea seed broth (Kraft, 1988). At Prosser, 10 g of pea seed are placed in 200 ml of glass distilled water and autoclaved for 0.5 h. An agar disc of *A. euteiches*, from a colony margin, grown for 5 days on 2% cornmeal agar is aseptically placed in each flask. Inoculated flasks are incubated for 5 to 7 days in the dark at room temperature. Resultant mycelial mats are washed three times in sterile tap water and drained. Mats are then placed in a mineral salts solution (Carmen and Lockwood, 1960) and aerated overnight with filtered air. Seven mats are placed in 250 ml of mineral salts solution. The mats are harvested by swirling several times in the salt solution and removed. The resultant zoospore suspension is then counted with a hemacytometer. Zoospore counts usually range from 300,000 to 800,000 ml^{-1}. Zoospore numbers are adjusted to 200,000 per ml^{-1}.

Five-day-old seedlings, germinated in coarse grade perlite, are removed and dipped in the zoospore suspension for 1 minute and transplanted back into the perlite. Inoculated plants are compared to an uninoculated control of the same line after a 14 day incubation period under greenhouse conditions (25 to 30°C daytime temperatures). Any test line which produces at least 70% of the fresh weight of that line's uninoculated control is considered resistant.

Marx *et al.* (1972) reported that tolerance to Aphanomyces root rot was associated with dominant, undesirable alleles at three unlinked marker loci (*le*-long internodes; *A*-anthocyanin pigment production; and *Pl*-pigmented hilum). Substitution of recessive alleles which express horticulturally desirable traits at each of these loci resulted in a reduction in tolerance. However, Kraft (1988) reported that resistance to *A. euteiches* was recovered in breeding lines with desirable horticultural traits. Resistance did not break down. Resistance was expressed as less disease and fewer oospores produced in resistant plant roots than in susceptible roots. Lewis and Gritton (1988) reported that resistance to *A. euteiches* appears to be quantitatively inherited with low inheritability. A recurrent selection program where disease pressure is intense was used to increase resistance to common root rot in horticulturally acceptable types. The recent release of several public germplasm lines with resistance to *A. euteiches* greatly improves the prospect of developing commercial cultivars with improved resistance/tolerance to this important disease (Davis *et al.*, 1976; Kraft, 1986, 1989, 1992; Gritton 1990). Table 3 lists PI accessions and breeding lines resistant to *A. euteiches*.

Fusarium Wilt

Pea Wilt

Wilt of peas, caused by *Fusarium oxysporum* Schl. f. sp. *pisi* (van Hall) Snyd. & Hans. race 1, was first described in 1924 (Linford, 1928). Resistance to this disease was determined to be inherited by a single dominant gene (Wade, 1929; Walker, 1931). Race 2 was found when race 1 resistant cultivars were developed and again attributed to a separate, dominant gene (Hare *et al.*, 1949). Races 3 and 4 found in the Netherlands and Canada, respectively, are thought to be more virulent strains of race 2 (Hubbeling, 1974). In addition, the genetic basis for resistance in the host to races 3 and 4 was not defined. In 1970, race 5 was detected and described from northwestern Washington state. All cultivars known to be resistant to races 1 and 2 were susceptible (Haglund and Kraft, 1970). Resistance to race 5 was first found in USDA PI accessions and was also attributable to a single dominant gene. In 1979, race 6 was also described from northwestern Washington state (Haglund and Kraft, 1979). This new race was pathogenic on cultivars and PI accessions resistant to races 1, 2, and 5, and resistance was again reported to be governed by a single, dominant gene.

Techniques to screen peas for resistance to wilt include: 1) pruning roots

Table 3. Pea lines resistant/tolerant to *Aphanomyces euteiches*

Pea Line	Reference
PI 166159	Lockwood, 1960
PI 167250	Lockwood, 1960
PI 169604	Lockwood, 1960
PI 175227 (sel)	Marx et al., 1972
PI 176721	Lockwood, 1960
PI 180693	Lockwood, 1960
PI 180702	Lockwood, 1960
PI 180868	Lockwood, 1960
PI 184129	Lockwood, 1960
Minn 108	Davis et al., 1976
Minn 494A-1	Haglund and King, 1961
792022	Kraft, 1981
75-786	Kraft and Tuck, 1986
84-1638	Kraft and Tuck, 1986
84-1930	Kraft and Tuck, 1986
86-2236	Kraft, 1989
90-2079	Kraft, 1992
90-2131	Kraft, 1992
90-2322	Kraft, 1992
Wis 8901-RR	Gritton, 1990
Wis 8902-RR	Gritton, 1990
Wis 8903-RR	Gritton, 1990
Wis 8904-RR	Gritton, 1990
Wis 8905-RR	Gritton, 1990

while submersed in a conidial suspension (Wells *et al.*, 1949; Haglund, 1989); 2) pouring mycelial fragments and conidia into a trough adjacent to seedling roots growing in sand (Armstrong and Armstrong, 1974); and 3) pouring conidia and hyphal fragments into holes punched into potting soil with a pointed rod to wound roots (Doling, 1963). In Washington state, resistance to race 1 wilt is determined under field conditions at Pullman, Washington, where a race 1 field nursery was established by Dr. F. J. Muehlbauer, USDA/ARS. As described by Wade (1929), elimination of race 1 susceptible cultivars is complete when inoculum levels are high. Resistance to races 2, 5, and 6 is determined in the greenhouse under pure culture conditions. Tests to screen for race 2 resistance are conducted when ambient greenhouse temperatures range from 20 to 24°C (Wells *et al.*, 1949). Tests for race 5 and 6 resistance are conducted in the winter months when ambient greenhouse temperatures range from 15 to 21°C.

All cultures of races 2, 5, and 6 used as primary inoculum are derived from single spores on 2% water agar (Toussoun and Nelson, 1976) and increased on fresh PDA under fluorescent light with a 12 h photoperiod. Only colonies appearing to be representative of the wild type are maintained in soil tubes (2 ml conidial suspension placed in 10 g of autoclaved soil mix in a test tube) (Toussoun and Nelson, 1976). Primary inoculum of a test isolate is produced by dispersing a small amount of infested soil on a PCNB plate (Nash and Snyder, 1962) and a resulting colony is selected, which is representative of the wild type for each race. A small agar plug from the colony margin of a 5-day-old culture is placed in 50 ml of liquid Kerr's medium (Kerr, 1963). Inoculated flasks are incubated for 5 days on a rotary shaker at 120 cycles per minute with constant fluorescent light on a laboratory bench. Spore concentrations for each isolate are determined with a hemacytometer. Usually, three separate isolates of each race are combined so that the fungal spore concentration is 1×10^6 conidia ml^{-1}.

Seeds of each test line are surface disinfested with a 0.53% NaOCl solution before planting in coarse, autoclaved perlite. All seedlings are inoculated in the third to fourth node by carefully removing each plant, dipping and pruning one-half of the root system of each plant with a razor blade while immersed in a conidial suspension. Inoculated seedlings are transplanted back into the perlite and incubated on a greenhouse bench until wilt symptoms are evident and/or known susceptible controls are severely wilted or dead. Wilt symptoms consist of stunting, yellowing, dying of lower leaves, curling of leaf margins, and usually death of the plant (Kraft and Haglund, 1978). The pathogen should be readily isolated from the above ground stem of any susceptible, inoculated plant when whole plant symptoms are evident. Suggested pea lines to use as differentials to distinguish races 1, 2, 5, and 6 are shown in Table 4.

Lentil Wilt

The disease is a serious threat to lentil production in many parts of the world and is caused by *Fusarium oxysporum* f. sp. *lentis*. This pathogen is responsible for severe grain losses (Fleischmann, 1937; Khare, 1980; Bayaa *et al.*, 1986). The disease appears either in the early stage of crop growth (seedling wilt) or during reproductive growth (adult plant wilt) (Khare, 1981). Moderately high soil temperatures (20 to 25°C) which favor fungal growth, and sunlight, which enhances transpiration, seem to be the key factors determining symptom

Table 4. Suggested pea lines to differentiate races 1, 2, 5, and 6 of the Fusarium wilt fungus

	Wilt Reaction[a]				Reference
Pea Line	**R 1**	**R 2**	**R 5**	**R 6**	
M410	S	S	S	S	Brotherton Seed Co.
Vantage	R	S	S	S	Brotherton Seed Co.
Mini	S	R	S	S	Asgrow Seed Co.
Mini 93	R	R	S	S	Asgrow Seed Co.
Sundance II	R	S	R	S	Pure Line Seed Co.
Grant	R	S	S	R	Brotherton Seed Co.
WSU 23	R	R	R	S	W. A. Haglund
WSU 28	R	S	R	R	W. A. Haglund
74SN5	R	R	R	R	J. M. Kraft

[a]R = resistant; S = susceptible.

[b]Brotherton Seed Co., Inc., Moses Lake, WA 98837; Asgrow Seed Co., Twin Falls, ID 83303; Pure Line Seeds, Inc., Moscow, ID 83843; Dr. W. A. Haglund, NW Washington Research & Extension Unit, Washington State University, Mt. Vernon, WA 98273.

expression. In Syria, wilt usually appears in April/May (Erskine *et al.*, 1990). However, seedling wilt was very widespread in the wilt sick plot at Tel Hadya, Syria, in November 1991, on 1-month-old seedlings of a susceptible cultivar, probably because of abnormally high temperatures in October (Bayaa and Erskine, unpublished).

Typical symptoms of wilt are first seen as sudden drooping of leaflets starting at the plant top and progressing downward. Leaflets close and do not shed prematurely, turning dull green. Finally the whole plant wilts. Wilting may be unilateral and confined to individual branches. When wilt appears at flowering no seeds are produced. When wilt occurs in mid-late pod filling, yield is drastically reduced and resultant seeds are often shrivelled. Vascular discoloration of infected plants sometimes occurs.

Lentil Wilt Screening Procedures

To screen under field conditions, the wilt nursery is infested by repeated cultivation of a susceptible cultivar and the incorporation of wilted plant material (Bayaa and Erskine, unpublished; Khare *et al.*, 1990). In order to increase disease pressure, lentil wilt inoculum can be increased on sterilized lentil seeds and uniformly spread in the nursery before sowing a susceptible cultivar. Screening is initiated when a uniform and high level of wilt damage is observed. Information on appropriate inoculum levels of *F. oxysporum* f. sp. *lentis* for field evaluation is lacking. This could be determined by estimating the number of propagules g^{-1} of soil using plate dilution with selective medium for *Fusarium* (Nash and Snyder, 1962; Komada, 1975). Test lines should be interplanted with a susceptible check, repeated every two to four test rows. Disease incidence should be recorded regularly during the season (Khare *et al.*, 1990). The wilt-sick plot should be monitored for possible interactions with other soil microorganisms.

Laboratory and Greenhouse Pure Culture Screening

Pure cultures of the lentil pathogen are maintained in autoclaved soil as mentioned previously (Toussoun and Nelson, 1976). Primary inoculum of *F. oxysporum* f. sp. *lentis* can be increased on any suitable medium (Kerr, 1963) and then mixed with the planting soil according to procedures previously described (Kannaiyan and Nene, 1976). Various growth media to increase inoculum are reported including PDA, lentil extract, dextrose agar (LD), and Richard's solution as the most commonly used media (Kannaiyan and Nene, 1978; Khare, 1980; Bayaa and Erskine, 1990).

The water culture technique, originally described by Wensley and McKeen (1962) and Roberts and Kraft (1971), was modified by Omar *et al.* (1988), and used to screen lentil lines for resistance to wilt. The roots of 10-day-old seedlings, grown previously in sterilized sand, were dipped in a spore suspension of 10^5 spores ml^{-1}. The seedling reaction was rated on a 0 to 7 scale described by Dixon and Doodson (1971) approximately 7 to 10 days later. Using the water culture technique of Roberts and Kraft (1971), the optimum spore concentration to differentiate between susceptible and resistant lines in 7 days was found to be 1.5×10^5 to 2×10^5 or 6.5×10^5 spores ml^{-1} (Bayaa and Erskine, unpublished; Kamboj *et al.*, 1990).

Seeds may also be sown in autoclaved soil and inoculum, grown in a liquid medium for 14 days, applied 14 days after sowing (Bayaa and Erskine, 1990). At ICARDA, a spore concentration of 2.5×10^6 microconidia ml^{-1} was used to inoculate sterilized field soil in the greenhouse to screen for wilt. The resultant inoculum concentration was 5.0×10^4 microconidia g^{-1} soil (Bayaa and Erskine, unpublished). Other reports have described the use of oatmeal-sand inoculum mixed with planting soil. However, the authors believe that it is impossible to quantify inoculum produced on oatmeal-sand and incorporated into soil. The ability to standardize inoculum levels is extremely important to interpret results from one evaluation test to the next.

Wilt severity readings are recorded either at 56 days after planting for seedling wilt or during pod development for adult plant wilt (Bayaa and Erskine, unpublished). The observations on seedling/plant mortality are converted to percent wilt for use on a 1 to 9 scale where: 1 = 1% or less plants wilted; 3 = 1 to 10% plants wilted; and 9 = 51% or more plants wilted. Observations on the intensity of wilting are also recorded to assess symptom development on test lines (Bayaa and Erskine, 1990).

The effects of various environmental factors on pathogen growth and disease expression have been studied (Dhingra *et al.*, 1974; Khare, 1980; Saxena and Khare, 1988; Erskine *et al.*, 1990). Lentil suffered more damage (48%) in sandy loam soil than in clay soil (22%). The mortality of lentil plants increased with soil pH up to 7.5 beyond which it declined. The optimum soil temperature for disease development was found to be 20 to 25°C. The optimum soil moisture level for disease expression is 25% in the soil used at ICARDA.

Kannaiyan and Nene (1976) reported that 32 out of 158 lines were found immune under glasshouse conditions. However, none of them was immune, resistant or tolerant under field conditions, which indicated that their inoculation procedure was not adequate, or different races or strains were prevalent at their test site. Khare (1980) identified 25 out of 440 lines as resistant. Of these, JL 80, JL 500, JL 674, Pusa-3, and Pant 234 were highly promising. At ICARDA, sources of wilt resistance have been identified. These are distributed in the Lentil International Fusarium Wilt Nursery (LIFWN) for testing under field conditions at different locations. Their resistance was confirmed in Egypt (Hamdi *et al.*, 1991). Mihov *et al.*, (1987) reported four cultivars with resistance under field conditions in Bulgaria. Seven lines out of 100 were found resistant to wilt in Bangladesh (Hossain *et al.*, 1985). Five breeding lines (JL 599, JL 632, JL 674 JL 1005, and L 406), which combined immunity to rust and resistance to lentil wilt, have been reported (Nene *et al.*, 1975; Pandya *et al.*, 1980). Resistance to wilt was also found among wild relatives of lentil (Bayaa *et al.*, 1991).

In reporting resistance, it is very important to mention the crop growth stage of the screening. Most lentil lines, exhibiting resistance at the seedling stage, lose their resistance at the adult stage (ICARDA, 1990).

Chickpea Wilt

Fusarium wilt, caused by *Fusarium oxysporum* Schlecht emd.:Fr. f. sp. *ciceris* (Padwick) Matuo & K. Sato, is an important soilborne disease of chickpea. This disease has been reported from most all the chickpea growing regions in the world (Nene and Reddy, 1987). Although no precise information on yield losses from this disease is available, a rough estimate of about 10% yield loss was reported from India (Singh and Dahiya, 1973) and Spain (Trapero-Casas and Jiménez-Díaz, 1985), and up to 40% in Tunisia (Bouslama, 1980). The disease can appear at any stage of plant growth. The chickpea wilt pathogen can penetrate young seedling roots directly. Penetration occurs mainly through the cotyledons, or close to the cotyledons, and to a lesser extent in the zone of elongation and maturation (Jiménez-Díaz *et al.*, 1989a). Symptoms can develop in a highly susceptible cultivar within 25 days after sowing, but can also occur up to podding stage. Early wilting causes more loss than late wilting, but seeds from late-wilted plants are lighter, rougher, and duller than those from healthy plants (Haware and Nene, 1980).

The pathogen is both soilborne and seed transmitted. Seedborne inocula can be eradicated by seed dressing with commercial rates of Benlate T (30% benomyl + 30% thiram) (Haware *et al.*, 1978). *Fusarium oxysporum* f. sp. *ciceris* is pathogenic to *Cicer* spp. and can invade roots of other grain legumes (Haware and Nene, 1982a), melon, potato, sugarbeet, vetch, white lupine, *Amaranthus retroflexus*, and *Chenopodium album* (Trapero-Casas and Jiménez-Díaz, 1985; Cabrera de la Colina *et al.*, 1987) without causing external symptoms. Crop rotations to control chickpea wilt are not feasible because *F. oxysporum* f. sp. *ciceris* can reproduce on symptomless carriers and chlamydospores can survive for at least 6 yr in the soil. The most practical and economical control of Fusarium wilt of chickpea is resistant cultivars (Nene and Reddy, 1987).

The existence of pathogenic races in populations of *F. oxysporum* f. sp. *ciceris* is well established. Haware and Nene (1982b) first identified races 1, 2, 3, and 4 in India based on the differential reactions of ten chickpea lines. Later, three additional races, namely races 0, 5, and 6, were identified in Spain (Jiménez-Díaz *et al.*, 1989b). Race 0 is the least virulent of the seven races described. It is not pathogenic to desi line JG 62, which is susceptible to all other races. All race 0 isolates tested so far induce a progressive foliar yellowing as compared to the severe leaf chlorosis, flaccidity, and early wilt induced by races 1–6. Race 0 is widespread in southern Spain and seems to occur in Tunisia (Nene and Sheila, 1986). Italian isolates of *F. oxysporum* f. sp. *ciceris* have been identified as belonging to race 0, although the disease reaction of differential lines correspond to race 1 (Frisullo *et al.*, 1989). Recently, Phillips (1988) described a race 6 in California based on the disease reaction of the ten differential lines of Haware and Nene (1982b) in naturally infested field plots. No difference can be established as yet between race 6 from Spain and California until cross inoculation studies are made under standardized conditions. Furthermore, the possibility that differential host-interactions between more than one race in the field cannot be ruled out. The race status of *F. oxysporum* f. sp. *ciceris* in California seems to be more complex than that reported by Phillips (1988). Buddenhagen and Workneh (1988) concluded that wilt isolates from three central coast counties in California could be classified into at least two pathogenicity groups. Jiménez-Díaz (unpublished) carried out cross inoculation studies of differential lines with races 0 and 5 from Spain and with four isolates of *F. oxysproum* f. sp. *ciceris* provided by Buddenhagen and Workneh. Three of the isolates were similar to races 0, 1, or 5, respectively, and the fourth one resembled race 6.

Screening Techniques

Efficient field, greenhouse, and laboratory procedures to evaluate chickpea lines for resistance to Fusarium wilt have been developed and standardized at ICRISAT (Nene *et al.*, 1981). Haware and Nene (1980, 1982b) reported on screening under controlled conditions using a pot-culture inoculation method in artificially infested soil. Stock isolates of *F. oxysporum* f. sp. *ciceris* should not be maintained by repeated transfers on growth media because loss of virulence through mutation can occur (Toussoun and Nelson, 1976; Burnett, 1984). We recommend that virulent isolates be stored as dormant cultures in either sterile soil (Toussoun and Nelson, 1976), silica gel (Windels *et al.*, 1988), or sterile filter paper (Correll *et al.*, 1986).

Recent work at ICRISAT (1989) indicates that a threshold of 483 propagules of race 1 g^{-1} (ppg) of soil is required for a 100% wilt incidence of JG 62. The late wilting cultivar K850 had a resistant reaction with that inoculum density, but wilt incidence increased with an increased inoculum level of 3283 ppg of soil. No wilt occurred with the resistant cultivar WR 315 at any inoculum level tested. Plants inoculated by the pot-culture method are grown in the greenhouse or in a growth chamber adjusted to a 14 h photoperiod of fluorescent light at 252 μE m^{-2} s^{-1} (Jiménez-Díaz *et al.*, 1989b, 1991). Chauhan (1963) reported that the optimum temperature for infection was 25°C, whereas disease developed with low severity at 35°C and no disease occurred at 15°C. However, Bhatti and Kraft (1992a) found that in soil artificially infested with 500 and 1000 ppg of *F. oxysporum* f. sp. *ciceris* wilt was severe at 25 and 30°C, but moderately severe at 15 and 20°C. No disease developed at 10°C even with an inoculum density of 5000 ppg.

Screening of a large number of germplasm accessions is more efficiently accomplished in the field. Experience at ICRISAT indicates that the development of a wilt-sick plot for field screening is relatively easy (Nene and Haware, 1980). Initially, a test site is selected where wilted chickpea plants are frequent and the soil is a slightly alkaline vertisol. Next, a highly susceptible cultivar is grown at this site for 2 to 3 seasons and infected plant debris is incorporated into the soil. Since 1980, wilt-sick plots have been used for resistance screening at many research centers including ICRISAT Center, Bihar, and Ludhiana (India), El Bajio and Culiacan (Mexico), Faisalabad (Pakistan), Beja (Tunisia), Santaella (Spain), and the Central Valley of California, USA.

Screening for resistance in a wilt-sick plot may present difficulties such as: a) the occurrence of other soilborne pathogens of chickpea; b) uneven distribution of the pathogen(s); c) the existence of more than one race of *F. oxysporum* f. sp. *ciceris*. Additionally, the yearly use of a wilt-sick plot for screening of different germplasm may lead to an uneven shift in the population and/or race distribution of the pathogen in soil. Screening chickpeas for resistance to Fusarium wilt in Santaella, Spain illustrates some of the points raised above. This wilt-sick plot was established in a naturally infested field in 1981 and since then has been continuously used for wilt resistance screening of germplasm from ICARDA, ICRISAT, and the regional breeding program. In 1983, a uniform wilt reaction occurred in highly susceptible lines, but cultivar JG 62 had a resistant reaction indicating that race 0 was prevalent in the plot. Results from field screening in 1989 indicated that a race shift may have occurred in the plot and at least two pathogenicity groups are now present in the plot (Jiménez-Díaz et al., 1991). Furthermore, large differences occurred in the disease reaction of selected lines in the field. These same lines exhibited resistant to moderately resistant reactions when they were inoculated with isolates representative of the two pathogenicity groups and with races 0 and 5 of the pathogen in the greenhouse.

Development of Wilt Resistant Sources and Cultivars

Good progress has been made in identifying sources of resistance and development of wilt resistant, high-yielding cultivars (Singh, 1987; Singh and Reddy, 1991). At ICRISAT over 150 wilt resistant desi and kabuli germplasm lines are available (Haware et al., 1989), and additional sources of resistance to wilt and root rots have been identified (Reddy et al., 1990a,b). Multilocation testing has shown that a few lines have broad-based resistance, such as ICC-2862, ICC-9023, ICC-9033, ICC-10803, ICC-11550, and ICC-11551, or broad based and stable resistance, such as ICC-267, ICC-858, and ICC-8933 (Nene et al., 1989).

At Santaella, Spain, a total of 2702 kabuli lines have been screened for resistance to Fusarium wilt in collaboration with ICARDA, including 713 FLIP lines in 1987, 991 ILC lines in 1989, and 196 FLIP and 802 ILC lines in 1990. The most promising resistant lines identified in 1987 were screened again in 1989, and these together with resistant lines identified in 1989 were also screened in 1990. Four lines [FLIP 84–43 C (ILC-5411), FLIP 85–20 C, FLIP 85–29 C, and FLIP 85–30 C] had a resistant reaction, and six additional lines had a resistant to moderately resistant reaction, in replicated tests in 1987 and 1989. These lines also tolerate cold and Ascochyta blight (Jiménez-Díaz et al., 1991). In the 1990 replicated test, lines FLIP 85–20 C, FLIP 85–30 C, and ILC 219 were resistant, while the other lines showed a moderately resistant reaction. In addition, FLIP 87–26 C, FLIP 87–78 C, and FLIP 87–82 C, and ILC-267, ILC-1278, and ILC-1300 were resistant among those screened in 1990 (Jiménez-Díaz et al., unpublished).

Fifteen accessions of 11 wild *Cicer* spp. and nine accessions of *C. pinnatifidum* were screened in pot culture inoculations for resistance to races 0 and 5 of *F. oxysporum* f. sp. *ciceris* in collaboration with the USDA-ARS Western Regional Plant Introduction Station, Pullman, Washington (Kaiser and Jiménez-Díaz, unpublished). *Cicer bijugum*, *C. cuneatum*, and *C. judaicum* were resistant to both races, and *C. canariense* and *C. chorassanicum* were resistant to race 0 but susceptible to race 5. All accessions of *C. pinnatifidum* were susceptible to race 5 but PI 458555, PI 458556, and PI 510654 were resistant to race 0.

Resistance to Fusarium wilt of chickpeas has been successfully incorporated into high yielding desi and kabuli backgrounds, including ICCV 2, 3, 4, and 5 (Kumar *et al.*, 1985) and ICCV 6 (Ghanekar *et al.*, 1990) from ICRISAT; "Surutato 77", "Sonora 80", "Gavilan", "Kino", and "Tubutama" from Mexico (Morales, 1986); "UC 15" and "UC 27" from California (Buddenhagen *et al.*, 1988); and "Andoum 1" (Halila and Harrabi, 1990) from Tunisia. However, in most cases, wilt resistant kabuli cultivars are susceptible to Ascochyta blight and cultivars resistant to Ascochyta blight are susceptible to wilt (Halila and Harrabi, 1990; Jiménez-Díaz and Trapero-Casas, 1990).

Concluding Remarks

Significant progress has been made in developing resistance to *Pythium* seed and seedling diseases in peas. Kabuli chickpeas are extremely sensitive to *Pythium* seed rot and preemergence damping-off, whereas desi chickpeas are resistant. Most likely desi chickpeas are resistant to *Pythium* due to thickened testae which reduce nutrient leakage during germination and the presence of fungistatic, anthocyanin compounds. More effort is needed to develop kabuli chickpeas with *Pythium* resistance. Resistance to Rhizoctonia seed and seedling rot in peas and lentils has been identified but much more work is needed to develop commercial cultivars with significant *Rhizoctonia* resistance. Techniques have been developed to screen lentils for resistance to collar rot caused by *S. rolfsii* and sources of resistance have been identified. However, the best sources of lentils with resistance to collar rot still exhibited a 10% mortality rate. Good resistance to Fusarium root rot of peas has been reported. This resistance is available in lines with acceptable horticultural type and significant progress is being made to develop commercial pea cultivars with resistance to this pathogen. However, the threat of new races of *F. oxysporum* f. sp. *pisi* developing exists in areas where short rotations are common. Resistance to Fusarium root rot of chickpea is not stable at high temperatures (30°C) and additional research is needed to identify resistant, chickpea germplasm. In addition, the *formae speciales* of *F. solani*, attacking chickpeas in the Middle East and North Africa needs to be determined.

Significant progress has been made in developing resistance to Aphanomyces root rot of peas with acceptable horticultural type. However, as is the case with most root diseases, this resistance is not complete and will break down with high inoculum levels and/or climatic conditions favorable to disease development. Good single gene resistance exists to the four races of the pea wilt pathogen. Apparently, resistance to all four economically important races of Fusarium wilt in peas is governed by single, independent, dominant genes. Screening techniques are described and are reproducible from one location to another around the world. Resistance to lentil wilt, caused by *F. oxysporum* f. sp. *lentis*, has been reported at ICARDA, ICRISAT, Bangladesh, and Bulgaria. Unfortunately, lentil breeding lines or cultivars exhibiting wilt resistance in the seedling stage are often not resistant in the adult stage. More work is needed to find mature plant resistance to lentil wilt and in developing accurate screening procedures for this pathogen. Significant progress has been made in finding resistance to chickpea wilt, caused by *F. oxysporum* f. sp. *ciceris*, and this resistance has been incorporated into high yielding kabuli and desi cultivars. Currently, seven races (0, 1, 2, 3, 4, 5, and 6) of *F. oxysporum* f. sp. *ciceris* are recognized. More work needs to be conducted on determining the genes responsible for resistance to these races and to find linkages that can been used as markers in a breeding program. Lastly, more effort is needed to combine Fusarium wilt resistance with Ascochyta blight resistance in chickpeas.

References

Anderson, N. A. 1982. *Annual Review of Phytopathology* 20: 329–347.

Armstrong, G. M. and Armstrong, J. K. 1974. *Phytopathology* 64: 849–857.

Aycock, R. 1966. *North Carolina Experiment Station Technical Bulletin* 174: 202 pp.

Bayaa, B. and Erskine, W. 1990. *Arab Journal of Plant Protection* 8: 30–33.

Bayaa, B., Erskine, W. and Hamdi, A. 1991. Fourth Arab Congress of Plant Protection, Cairo, Egypt, December 1–5, 1991.

Bayaa, B., Erskine, W. and Khoury, L. 1986. *Arab Journal of Plant Protection* 4: 118–119.

Bhatti, M. A. and Kraft, J. M. 1992a. *Plant Disease* 76: 50–54.

Bhatti, M. A. and Kraft, J. M. 1992b. *Plant Disease* 76: 54–56.

Bouslama, M. 1980. In: *Proceedings International Workshop Chickpea Improvement*, pp. 277–280 (eds. J. M. Green, Y. L. Nene and J. B. Smithson). Hyderabad, India: ICRISAT.

Buddenhagen, I. W. and Workneh, F. 1988. *Phytopathology* 78: 1563.

Buddenhagen, I. W., Workneh, F. and Bosque-Pérez, N. 1988. *International Chickpea Newsletter* 19: 9–10.

Burnett, J. H. 1984. In: *The Mycology of Fusarium*, pp. 39–69. (eds. O. M. Moss and J. E. Smith). Cambridge, England, UK: Cambridge University Press.

Cabrera de la Colina, J., Trapero-Casas, A. and Jiménez-Díaz, R. M. 1987. In: *Proceedings of the 7th Congress of the Mediterranean Phytopathological Union*, p. 109, SEF/MPU, Sept. 20–26, Granada, Spain, Dirección General de Investigación y Extensión Agrarias, Sevilla, Spain.

Carmen, L. M. and Lockwood, J. L. 1960. *Phytopathology* 50: 826–830.

Chauhan, S. K. 1963. *Proceeding Indian Academy of Science B* 33: 552–554.

Claudius, G. R. and Mehrotra, R. S. 1973. *Indian Phytopathology* 26: 268–273.

Correll, J. C., Puhalla, J. E. and Schneider, R. W. 1986. *Phytopathology* 76: 396–400.

Davis, D. W., Shehata, M. A. and Bissonnette, H. L. 1976. *HortScience* 11: 434.

Deshkar, M. V., Khare, M. N. and Joshi, L. K. 1973. *JNKVV Research Journal* 7: 300–301.

Dhingra, O. D., Agrawal, S. C., Khare, M. N. and Kushwaha, L. S. 1974. *Indian Phytopathology* 27: 408–410.

Dixon, G. R. and Doodson, J. K. 1971. *Journal of the National Institute of Agricultural Botany* 12: 299–307.

Doling, D. A. 1963. *Transactions of the British Mycological Society* 46: 577–584.

Erskine, W., Bayaa, B. and Dolli, M. 1990. *Arab Journal of Plant Protection* 8: 34–37.

Fleischmann, R. 1937. *Pflanzenbau* 14: 49–65.

Frisullo, S., Ciccarese, F., Amenduni, M. and Zamani, H. R. 1989. In Italian. *La difesa delle piante* 12: 181–186.

Ghanekar, A. M., Sethi, S. C., Jalali, B. L. and Nene, Y. L. 1990. *International Chickpea Newsletter* 23: 22–23.

Grewal, J. S., Pal, M. and Kullshrestha, D. D. 1974. *Indian Journal Genetics and Plant Breeding* 34: 242–246.

Gritton, E. T. 1990. *Crop Science* 30: 1166–1167.

Haglund, W. A. 1989. *Plant Disease* 73: 457–458.

Haglund, W. A. and King, T. H. 1961. *Phytopathology* 51: 800–802.

Haglund, W. A. and Kraft, J. M. 1970. *Phytopathology* 60: 1861–1862.

Haglund, W. A. and Kraft, J. M. 1979. *Phytopathology* 69: 818–820.

Halila, H. M. and Harrabi, M. M. 1990. *Options Méditerranéennes Série Séminaires* 9: 163–166.

Hamdi, A., Omar, S. A. M. and Amer, M. L. 1991. *Egyptian Journal of Applied Science* 6: 18–29.

Hare, W. W., Walker, J. C. and Delwiche, E. J. 1949. *Journal of Agricultural Research* 78: 239–250.

Harman, G. E. 1984. In: *Compendium of Pea Diseases*, pp. 5–6 (ed. D. J. Hagedorn). St. Paul, Minnesota, USA: American Phytopathological Society.

Haware, M. P., Jiménez-Díaz, R. M., Amin, K. S., Phillips, J. C. and Halila, M. H. 1990. In: *Chickpea in the Nineties*, pp. 129–133 (eds. H. A. van Rheenen, M. C. Saxena, B. J. Walby and S. D. Hall). Patancheru, Andhra Pradesh, India: ICRISAT.

Haware, M. P. and Nene, Y. L. 1980. *Tropical Grain Legume Bulletin* 19: 38–44.

Haware, M. P. and Nene, Y. L. 1982a. *Plant Disease* 66: 250–251.

Haware, M. P. and Nene, Y. L. 1982b. *Plant Disease* 66: 809–810.

Haware, M. P., Nene, Y. L. and Rajeshari, R. 1978. *Phytopathology* 68: 1364–1367.

Hossain, M. A., Auyb, A. and Ahmed, H. I. 1985. Bangladesh Agricultural Research Institute. Abstracts of First National Plant Pathology Conference, Joydebpur, Bangladesh.

Hubbeling, N. 1974. *Overdruk VIT: Mendelingen Fakulteit Landbouwwetenschappen Gent* 39: 991–1000.

ICARDA. 1990. Food Legume Improvement Program, Annual Report for 1990, 129 pp. Aleppo, Syria: ICARDA.

ICRISAT (International Crops Research Institute for the Semi-Arid Tropics). 1989. Annual Report 1988, pp. 63–64. Patancheru, Andhra Pradesh, India: ICRISAT.

Jiménez-Díaz, R. M., Basallote Ureba, M. J. and Rappoport, H. 1989a. In: *Vascular Wilt Diseases of Plants, Vol. H28*, pp. 113–121 (eds. E. C. Tjamos and C. Beckman). Berlin, Germany: Springer-Verlag.

Jiménez-Díaz, R. M., Singh, K. B., Trapero-Casas, A. and Trapero-Casas, J. L. 1991. *Plant Disease* 75: 914–918.

Jiménez-Díaz, R. M. and Trapero-Casas, A. 1990. *Options Mediterranéennes Série Seminaires* 9: 65–72.

Jiménez-Díaz, R. M., Trapero-Casas, A. and Cabrera de la Colina, J. 1989b. In: *Vascular Wilt Diseases of Plants, Vol. H28*, pp. 515–520 (eds E. C. Tjamos and C. Beckman). Berlin, Germany: Springer-Verlag.

Kaiser, W. J. and Hannan, R. M. 1983. *Plant Disease* 67: 77–81.

Kamboj, R. K., Pandey, M. P. and Chaube, H. S. 1990. *Euphytica* 50: 113–117.

Kannaiyan, J. 1974. *Studies on the Control of Lentil Wilt*, 93 pp. Ph.D. Thesis, G. B. Pant University of Agriculture and Technology, Pantnager, India.

Kannaiyan, J. and Nene, Y. L. 1973. *Current Science* 42: 257.

Kannaiyan, J. and Nene, Y. L. 1976. *Indian Journal of Agricultural Science* 46: 165–167.

Kannaiyan, J. and Nene, Y. L. 1978. *LENS Newsletter* 5: 8–10.

Kerr, A. 1963. *Australian Journal of Biological Sciences* 16: 55–69.

Khare, M. N. 1980. *Wilt of Lentil*, 155 pp. Jabalpur, M.P., India: JNKVV.

Khare, M. N. 1981. In: *Lentils*, pp. 163–172 (eds. C. Webb and G. Hawtin). UK: ICARDA/CAB.

Khare M. N., Bayaa, B. and Beniwal, S. P. S. 1993. In: *Breeding for Stress Tolerance in Cool-Season Food Legumes*, pp. 107–121 (eds. K. B. Singh and M. C. Saxena). Chichester, UK: John Wiley & Sons.

King, T. H., Johnson, H. G., Bissonnette, H. and Haglund, W. A. 1960. *Proceedings of the American Society for Horticultural Science* 75: 510–516.

Komada, H. 1975. *Review of Plant Protection Research* 8: 114–124.

Kraft, J. M. 1969. *Plant Disease Reporter* 53: 110–111.

Kraft, J. M. 1975. *Plant Disease Reporter* 59: 1007–1011.

Kraft, J. M. 1977. *Phytopathology* 67: 1057–1061.

Kraft, J. M. 1981. *Crop Science* 21: 352–353.

Kraft, J. M. 1984. *Crop Science* 24: 389.

Kraft, J. M. 1986. *Plant Disease* 70: 743–745.

Kraft, J. M. 1988. *Phytopathology* 78: 1545.

Kraft, J. M. 1989. *Crop Science* 29: 494–495.

Kraft, J. M. 1992. *Crop Science* 30: 1076.

Kraft, J. M., Burke, D. W. and Haglund, W. A. 1981. In: *Fusarium: Diseases, Biology, and Taxonomy*, pp. 142–156 (eds. P. E. Nelson, T. A. Toussoun and R. J. Cook). University Park, Pennsylvania, USA: Pennsylvania State University Press.

Kraft, J. M. and Giles, R. A. 1978. *Crop Science* 18: 1098.

Kraft, J. M. and Haglund, W. A. 1978. *Phytopathology* 68: 273–275.

Kraft, J. M. and Roberts, D. D. 1969. *Phytopathology* 59: 149–152.

Kraft, J. M. and Roberts, D. D. 1970. *Phytopathology* 60: 1814–1817.

Kraft, J. M. and Tuck, J. A. 1986. *Crop Science* 26: 1262–1263.

Kumar, J., Haware, M. P. and Smithson, J. B. 1985. *Crop Science* 25: 576–577.

Lewis, M. E. and Gritton, E. T. 1988. *Pisum Newsletter* 20: 20–21.

Linford, M. B. 1928. *Research Bulletin Wisconsin Agricultural Experiment Station 85*, 44 pp.

Lockwood, J. L. 1960. *Quarterly Bulletin Michigan Agricultural Experiment Station* 43: 358–366.

Mani, A. and Sethi, C. L. 1985. *Indian Phytopathology* 38: 542–543.

Mario, A. A. and Carolina, B. G. 1987. *Agricultura Técnica* 47: 78–79.

Marx, G. A., Schroeder, W. T., Provvidenti, R. and Mishanec, W. 1972. *Journal of the American Society for Horticultural Science* 97: 619–621.

Mathur, B. N. and Deshpande, A. L. 1968. *Indian Phytopathology* 21: 455–456.

McCoy, R. J. and Kraft, J. M. 1984. *Plant Disease* 68: 491–493.

Mihov, M. I., Stoeva, I. and Ivanov, P. 1987. *Rasteniev dri-Nauki* 24: 45–51.

Mircetich, S. M. and Kraft, J. M. 1973. *Mycopathologia et Mycologia Applicata* 50: 151–161.

Mohammad, A. and Kumar, U. 1986. *Indian Phytopathology* 39: 93–95.

Morales, G. J. A. 1986. *International Chickpea Newsletter* 15: 11–12.

Nain, V. S. and Agnihotri, J. P. 1984. *International Chickpea Newsletter* 11: 43–35.

Nash, S. M. and Snyder, W. C. 1962. *Phytopathology* 52: 567–572.

Nene, Y. L. and Haware, M. P. 1980. *Plant Disease* 64: 379–380.

Nene, Y. L., Haware, M. P. and Reddy, M. V. 1981. *Information Bulletin 10*, 11 pp. Patancheru, Andhra Pradesh, India: ICRISAT.

Nene, Y. L., Haware, M. P., Reddy, M. V., Phillips, J. C., Castro, E. L., Kotasthane, S. R., Gupta, D., Singh, G., Ahukla, P. and Sah, R. P. 1989. *Indian Phytopathology* 42: 499–505.

Nene, Y. L., Kannaiyan, J. and Saxena, G. C. 1975. *Indian Journal of Agricultural Science* 45: 177–178.

Nene, Y. L. and Reddy, M. V. 1987. In: *The Chickpea*, pp. 233–270 (eds. M. C. Saxena and K. B. Singh). Oxon, UK: CAB International Publishers.

Nene, Y. L. and Sheila, V. K. 1986. Fusarium wilt of chickpea. *Fusarium Notes*, Vol. 5 (November).

Ohn, S. H., King, T. H. and Kommedahl, T. 1978. *Phytopathology* 68: 1644–1649.

Omar, S. A. M., Salem, D. E. and Rizik, M. A. 1988. *LENS Newsletter* 15: 37–39.

Pandya, B. P., Pandey, M. P. and Singh, J. P. 1980. *LENS Newsletter* 7: 34–37.

Perry, D. A. 1973. *Transactions of the British Mycological Society* 61: 135–144.

Pfender, W. F. 1984. In: *Compendium of Pea Diseases*, pp. 25–28 (ed. D. J. Hagedorn). St. Paul, Minnesota, USA: American Phytopathological Society.

Phillips, J. C. 1988. *International Chickpea Newsletter* 18: 19–21.

Reddy, M. V., Raju, T. N. and Nene, Y. L. 1990a. *International Chickpea Newsletter* 23: 22.

Reddy, M. V., Raju, T. N. and Pundir, R. P. S. 1990b. *International Chickpea Newsletter* 22: 36–38.

Roberts, D. D. and Kraft, J. M. 1971. *Phytopathology* 61: 324–323.

Saxena, D. R. and Khare, M. N. 1988. *Indian Phytopathology* 41: 69–74.

Shatla, M. N., Kamal, K. and Shanwani, M. Z. 1974. *Archiv für Phytopathologie and Pflanzenschutz* 10: 333–339.

Shehata, M. A., Davis, D. W. and Anderson, N. A. 1981. *Plant Disease* 65: 417–419.

Shehata, M. A., Davis, D. W. and Anderson, N. A. 1984. *Plant Disease* 68: 22–24.

Shehata, M. A., Pfleger, F. L. and Davis, D. W. 1983. *Plant Disease* 67: 1146–1148.

Shukla, T. N., Ahmed, Z. U. and Garg, S. K. 1972. *Indian Phytopathology* 25: 584–585.

Singh, D. and Nema, K. G. 1987. *International Chickpea Newsletter* 17: 23–25.

Singh, K. B. 1987. In: *The Chickpea*, pp. 127–162 (eds. M. C. Saxena and K. B. Singh). Oxon, UK: CAB International Publishers.

Singh, K. B. and Dahiya, K. B. 1973. In: *Symposium on Wilt Problems and Breeding for Wilt Resistance in Bengal Gram*, pp. 13–14. New Delhi, India: Indian Agricultural Research Institute. (Abstract).

Singh, K. B. and Reddy, M. V. 1991. *Advances in Agronomy* 45: 191–222.

Stasz, T. E. and Harman, G. E. 1980. *Phytopathology* 70: 27–31.

Toussoun, T. A. and Nelson, P. E. 1976. *Fusarium: A Pictorial Guide to the Identification of Fusarium Species*, 43 pp. University Park, Pennsylvania, USA: Pennsylvania State University Press.

Trapero-Casas, A. and Jiménez-Díaz, R. M. 1985. *Phytopathology* 75: 1146–1151.

van der Plaats-Niterink. 1981. *Studies in Mycology*, 21, 242 pp. Centraalbureau voor Schimmelcultures, Baarn, The Netherlands

van Rheenen, H. A., Reddy, M. V., Kumar, J. and Haware M. P. 1989. Consultancy Meeting on *Breeding for Disease Resistance in Kabuli Chickpeas*, ICARDA, Syria, March 6–8, 1989.

Viswakarma, S. N. and Chaudhary, B. C. 1981. *Proceedings of the Indian National Science Academy* B47, No. 5, 748–750.

Wade, B. L. 1929. *Research Bulletin Wisconsin Agricultural Experiment Station* No. 97, Madison, Wisconsin, USA. 32 pp.

Walker, J. C. 1931. *Research Bulletin Wisconsin Agricultural Experiment Station* No. 107, Madison, Wisconsin, USA. 15 pp.

Wells, D. G., Hare, W. W. and Walker, J. C. 1949. *Phytopathology* 39: 771–779.

Wensley, R. N. and McKeen, C. D. 1962. *Canadian Journal of Microbiology* 8: 818–819.

Westerlund, F. V., Campbell, R. N. and Kimble, K. A. 1974. *Phytopathology* 64: 432–436.

Whalley, W. M. 1984. *Annals of Applied Biology* 104: 118–119.

Windels, C. E., Burnes, P. M. and Kommedahl, T. 1988. *Phytopathology* 78: 107–109.

Woyke, H. W. 1987. *Acta Horticulturae* 215: 77–81.

Research achievements in plant resistance to insect pests of cool season food legumes

S.L. CLEMENT[1], N. EL-DIN SHARAF EL-DIN[2], S. WEIGAND[3] and
S.S. LATEEF[4]

[1] U. S. Department of Agriculture, Agricultural Research Service, Regional Plant Introduction
Station, 59 Johnson Hall, Washington State University, Pullman, WA 99164–6402 USA;
[2] Agricultural Research Center, Wad Medani, Sudan;
[3] ICARDA, P. O. Box 5466, Aleppo, Syria, and
[4] ICRISAT, Legume Program, Patancheru P. O., Andhra Pradesh 502 324, India

Abstract

Plant resistance to at least seventeen field and storage insect pests of cool season food legumes has been identified. For the most part, this resistance was located in the primary gene pools of grain legumes via conventional laboratory, greenhouse, and field screening methods. The use of analytical techniques (i.e., capillary gas chromatography) to characterize plant chemicals that mediate the host selection behavior of pest insects offers promise as a new, more rapid way to differentiate between insect-resistant and susceptible plant material. Examples of research achievements in mechanisms of resistance and host-plant resistance within the context of integrated control programs are discussed. Accelerating the development and subsequent releases of insect-resistant cultivars to pulse farmers requires more involvement from interdisciplinary teams of plant breeders, entomologists, plant pathologists, plant chemists, molecular biologists, and other scientists.

Introduction

Entomologists and plant breeders have located sources of plant resistance to several of the most important insect pests of cool season food legumes (Horber, 1978; Reed et al., 1988; Weigand and Pimbert, in press). However, the transfer of resistance-conferring genes from this material to regionally adapted lines has been constrained by several biological and technological factors, including but not limited to: a lack of sufficient information about the chemical and physical nature and genetic bases of insect resistance in plants; the need for breakthrough research and technology to overcome barriers to the development of cultivars with multiple insect and disease resistance; and the requirement for new and improved technology to overcome barriers to inter-specific hybridization so resistance genes can be transferred from nonadapted to adapted backgrounds. Moreover, pest resistance research and breeding has been "undervalued and underfunded" (Reed et al., 1988). Overcoming these barriers, and expediting the

F.J. Muehlbauer and W.J. Kaiser (eds.), Expanding the Production and Use of Cool Season Food
Legumes, 290–304.
© 1994 Kluwer Academic Publishers.

development and first releases of chickpea (*Cicer arietinum* L.), faba bean (*Vicia faba* L.), dry pea (*Pisum sativum* L.), lentil (*Lens culinaris* Medik.), and grasspea (*Lathyrus sativus* L.) cultivars with insect resistance or with the ability to tolerate more insect damage than normally sensitive cultivars, will require much more involvement from interdisciplinary teams of plant breeders, entomologists, plant pathologists, plant chemists, molecular biologists, and other scientists. The reader is referred to Reed *et al.* (1988) and Singh *et al.* (1990) for indepth discussions on constraints to breeding for insect resistance in cool season food legumes (grain legumes).

This chapter reviews research achievements in plant resistance to insect pests of grain legumes. Although emphasis is on progress since the first International Food Legume Research Conference in 1986, some pre-1986 literature and work overlooked in previous reviews is highlighted to provide a comprehensive review of the topic. After reviewing known cases of plant resistance to insect pests of grain legumes, we summarize and discuss the screening methods and evaluation criteria that researchers have used to separate susceptible from resistant germplasm. Next, we address mechanisms and levels of plant resistance, citing examples from the body of literature on plant resistance to insect pests of grain legumes. These aspects warrant consideration because they relate to the longterm durability of insect-resistant crop cultivars (Kennedy *et al.*, 1987). We briefly discuss host-plant resistance as a complementary pest control strategy before ending with comments on the prospects for breeding insect-resistant cultivars of grain legumes.

Insect Resistance in Grain Legumes

Through the efforts of several researchers, sources of plant resistance to at least seventeen of the most important field and storage insect pests of chickpeas, faba beans, dry peas, and lentils have been located (Table 1). These searches for resistance have involved as few as two, normally 6 to 140, and at times more than 14,000 accessions or entries. As is normally the result when searches for insect-resistant plant material are undertaken, grain legume workers have found low frequencies of resistance among plant materials examined (Table 2). The reader is referred to the citations in Tables 1 and 2 for listings of specific insect-resistant plant genotypes, plant introductions, accession numbers, and breeding lines. We are unaware of any reports of plant resistance to insect pests of grasspea.

For the most part, insect resistance has been located in the primary gene pools of grain legumes. Rarely have the secondary (i.e., species that will cross with crop but gene transfer often difficult) and tertiary (i.e., species related to crop; however, gene transfer not possible or requiring radical techniques) (definitions according to Harlan and De Wet, 1971) gene pools been examined for insect resistance. The only evaluations of wild and related species of grain legumes for insect resistance have involved wild species of *Cicer* against

Table 1. Reports of plant resistance to field and storage insect pests of grain legumes

Crop and insects	Field or storage pest	Plant taxa evaluated	References
Chickpea			
Aphis craccivora (Koch)[a]	Field	*Cicer arietinum*	Weigand and Tahhan (1990)
Callosobruchus chinensis (L.)[c]	Storage	*C. arietinum*	ICRISAT (1976)
		Wild *Cicer* species	Weigand and Tahhan (1990); Weigand and Pimbert (in press)
Callosobruchus maculatus (F.)[c]	Storage	*C. arietinum*	Salunkhe and Jadhav (1982); Ahmed et al. (1989)
Helicoverpa armigera (Hüb.)[f]	Field	*C. arietinum*[k]	Rembold (1981); Lateef (1985, 1990); Lateef et al. (1985); Ahmed et al. (1990); Lateef and Pimbert (1990); Lateef and Sachan (1990); Pimbert (1990); Rembold et al. (1990a,b); Weigand and Pimbert (in press)
		Wild *Cicer* species	ICRISAT (1987)
Liriomyza cicerina (Rondani)[h]	Field	*C. arietinum*	Weigand (1990); Weigand and Tahhan (1990)
Faba bean			
Aphis craccivora (Koch)[a]	Field	*Vicia faba*	ICARDA (1989; El-Defrawi et al. (1991)
Aphis fabae Scop.[a]	Field	*V. faba* and wild species	Holt (1983); Birch and Wratten (1984); Holt and Birch (1984); Birch (1985); ICARDA (1989)
Acyrthosiphon pisum (Harris)[a]	Field	*V. faba* and wild species	Birch and Wratten (1984); Holt and Birch (1984)
Megoura viciae (Buckt.)[a]	Field	*V. faba* and wild species	Birch and Wratten (1984); Holt and Birch (1984)
Empoasca fabae (Harris)[b]	Field	*V. faba*	Wolfenbarger and Sleesman (1963)
Bruchus dentipes Baudi[c]	Field[j]	*V. faba*	Tahhan (1986); Tahhan and van Emden (1989)
Callosobruchus chinensis (L.)[c]	Storage	*V. faba*	Ishii (1952)
Callosobruchus maculatus (F.)[c]	Storage	*V. faba*	Fam and El-Sayed Ahmed (1985)
Epilachna varivestris Muls.[d]	Field	*V. faba*	Wolfenbarger and Sleesman (1961)
Liriomyza trifolii (Burgess)[h]	Field	*V. faba*	El-Din Sharaf El-Din (unpublished data)
Pea			
Acyrthosiphon pisum (Harris)[a,i]	Field	*Pisum sativum*	Semenova (1990); Soroka and MacKay (1990a,b,c); Soroka and MacKay (1991)
Bruchus pisorum (L.)[c]	Field[j]	*P. sativum*	Vilkova and Kolensichenko (1973); Aleksandrova (1977); Pesho et al. (1977); Sokolov (1977); Annis (1983); Pillsbury (1986); Clement (unpublished data)
		P. sativum ssp. *humile*	Hardie (1990)
		Lathyrus species	Annis and O'Keeffe (1984)
Sitona lineatus (L.)[c]	Field	*P. sativum*	Nouri-Ghanbalani (1974); Nouri-Ghanbalani (1977); Auld et al. (1980)
Cydia nigricana (F.)[g,i]	Field	*P. sativum*	Wright et al. (1951); Bingefors et al. (1964); Wnuk (1968)
Chromatomyia horticola (Goureau)[h]	Field	*P. sativum*	Sehgal et al. (1987)

Table 1. (Continued)

Crop and insects	Field or storage pest	Plant taxa evaluated	References
Lentil			
Aphis craccivora (Koch)[a]	Field	*Lens culinaris*	Weigand and Pimbert (in press)
Bruchus lentis Froel.[c]	Field[j]	*L. culinaris*	Chopra and Pajni (1987)
Sitona spp.[e]	Field	*L. culinaris*	Sedivy (1972)

[a]Homoptera:Aphididae; [b]Homoptera:Cicadellidae; [c]Coleoptera:Bruchidae;
[d]Coleoptera:Coccinellidae; [e]Coleoptera:Curculionidae; [f]Lepidoptera:Noctuidae;
[g]Lepidoptera:Tortricidae; [h]Diptera:Agromyzidae.

[i]Resistance found in dry and/or green pea cultivars.

[j]Infestation starts in the field as eggs on green pods but larval feeding damage is manifested in stored seed.

[k]Resistance found mainly in desi and to some extent in kabuli (Mediterranean) types.

Helicoverpa armigera (Hüb.) (ICRISAT, 1987) and the storage pest *Callosobruchus chinensis* (L.) (Weigand and Tahhan, 1990), wild *Vicia* against the aphids *Aphis fabae* Scop., *Acyrthosiphon pisum* (Harris), and *Megoura viciae* (Buckt.) (Birch and Wratten, 1984; Holt and Birch, 1984; Birch, 1985), and *Lathyrus sativus* and *L. tingitanus* L. against *Bruchus pisorum* (L.) (Annis and O'Keeffe, 1984) (Table 1). The latter study was conducted in conjunction with research into the mechanisms of plant resistance to *B. pisorum*, which is not a pest of grasspea but is a major, worldwide pest of peas (Clement, 1992). That these few evaluations led to the discovery of insect-resistant plant materials suggests a need for more evaluations of the secondary and tertiary gene pools of grain legumes.

Screening Methods and Measurement of Resistance

Field Pests

Usually, grain legume researchers have relied upon conventional methods such as open-field tests, field confinement techniques, and laboratory assays to search for differences in the ability of plants to serve as hosts for insect pests and to withstand attacks and recover from injury. Tingey (1986) and Smith (1989) are useful general references on screening methods and evaluation criteria currently used in host-plant resistance.

Open-field tests have been used to successfully segregate chickpea (Lateef, 1985; Weigand and Tahhan, 1990), faba bean (Sharaf El-Din, unpubl. data; Wolfenbarger and Sleesman, 1961, 1963; Tahhan and Van Emden, 1989), pea (Wright *et al.*, 1951; Nouri-Ghanbalani, 1974, 1977; Pesho *et al.*, 1977; Nouri-Ghanbalani *et al.*, 1978; Sehgal *et al.*, 1987; Soroka and Mackay 1990a), and lentil (Chopra and Pajni, 1987) germplasm for resistance to attack by pod borer (*H. armigera*); leafminers (*Liriomyza cicerina* [Rondani], *L. trifolii* [Burgess],

Table 2. Mass screening of grain legumes and frequency of insect-resistant genotypes among screened material

Insect	Plant taxa evaluated	Approximate no. of entries		References
		in mass screenings	showing antixenosis, antibiosis and/or tolerance after screening and re-testing	
Callosobruchus chinensis (L.)	Cicer arietinum	6,697	0	Weigand and Pimbert (in press)
Helicoverpa armigera (Hüb.)	C. arietinum	14,800	21	Lateef and Pimbert (1990)
Liriomyza cicerina (Rondani)	C. arietinum	6,800	10	Weigand and Tahhan (1990)
Aphis craccivora (Koch)	Vicia fabae	7,156	114	El-Defrawi et al. (1991)
Bruchus dentipes Baudi	V. fabae	1,000	0[a]	Tahhan and van Emden (1989)
Bruchus pisorum (L.)	Pisum sativum	1,571	10	Annis (1983); Pesho et al. (1977)
Sitona lineatus (L.)	P. sativum	2,074	2	Auld et al. (1980); Nouri-Ghanbalani (1977)

[a]Phenological resistance related to late flowering and pod setting was reported in one accession.

Chromatomyia horticoloa [Goureau]); weevils (*Bruchus pisorum* [L.], *B. dentipes* Baudi, *B. lentis* Froel., *Sitona lineatus* [L.]); aphids (*Aphis craccivora* [Koch], *Acyrthosiphon pisum* [Harris]); potato leafhopper (*Empoasca fabae* [Harris]); pea moth (*Cydia nigricana* [F.]); and Mexican bean beetle (*Epilachna varivestris* [Muls.]). These searches for resistance often employed small plots without replication to quickly eliminate susceptible plant genotypes. These trials have sometimes been followed by larger field plots containing standard checks and promising lines from initial screenings, all replicated and grouped in plots according to similar maturities (Lateef, 1985; Lateef and Sachan, 1990). Since 1980, multilocational testing of promising *H. armigera* resistant selections in replicated field plots has become part of the chickpea entomology program at the International Crops Research Institute for the Semi-Arid Tropics (ICRISAT), Patancheru, India (Lateef and Sachan, 1990; Pimbert, 1990). While multilocational testing helps chickpea breeders determine the agronomic performance of promising lines across several agroecological zones, it provides entomologists with a mechanism to assess variation in the virulence of allopatric pod borer populations on resistant lines. Only Smith *et al.* (1982), in a study involving the screening of pea lines against *B. pisorum*, seem to have rigorously addressed the selection of appropriate experimental and statistical designs for use in open-field screening trials.

To compensate for low insect populations during field evaluations, some researchers caged laboratory-reared insects on test plants (Birch, 1985) while others released laboratory-reared insects into plots (Lateef, 1985). Laboratory tests conducted alone or in concert with field studies and utilizing caged insects on plant material have also proven useful for evaluation of insect resistance in grain legumes, such as aphid (*A. craccivora, A. faba, A. pisum, M. viciae*) resistance in the genus *Vicia* (Birch and Wratten, 1984; Holt and Birch, 1984; El-Defrawi *et al.*, 1991) and weevil (*B. pisorum, S. lineatus*) resistance in peas (Nouri-Ghanbalani, 1977; Annis, 1983; Pillsbury, 1986).

Plant resistance workers normally separate susceptible from resistant plant materials during screening and evaluation programs by measuring the deleterious effects of plant resistance traits on insect development, population dynamics, and behavior and/or measuring the effect of insects on plant yield and quality (Tingey, 1986; Smith, 1989). These general approaches have been used by grain legume researchers, as well. For example, resistance has been evaluated in terms of insect feeding and oviposition preferences (Clement, unpubl. data; Pesho *et al.*, 1977; Pillsbury, 1986), insect infestation levels (Wolfenbarger and Sleesman, 1961, 1963; Lateef, 1985; El-Defrawi *et al.*, 1991), and through the effects of plants on insect development, survival, and fecundity (Birch and Wratten, 1984; Holt and Birch, 1984; Sehgal *et al.*, 1987; Soroka and Mackay, 1991). Visual rating scales based on percentages or numerical ratings of damage have been used routinely to measure plant susceptibility to insect attack (Wolfenbarger and Sleesman, 1961, 1963; Nouri-Ghanbalani, 1974, 1977; Lateef and Reed, 1985; Semenova, 1990; Weigand, 1990; El-Din Sharaf El-Din, unpubl.; Weigand and Pimbert, in press). Resistance also has been expressed in terms of the effect of insect injury on plant development, yield, and seed quality (Nouri-Ghanbalani, 1974, 1977; Pesho et al., 1977; Lateef, 1985; Chopra and Pajni, 1987; Tahhan and van Emden, 1989). Under field conditions, these researchers have used a variety of methods to measure and compare insect population levels on plants, namely direct observation, sweepnet and vacuum sampling, and trapping. The specific sampling method used depended upon the insect species and crop plant (including growth stages) being sampled and other factors such as available resources and the amount of material being evaluated.

Microanalytical methods like capillary gas chromatography-mass spectrometry offer promise as a more rapid way to differentiate between insect-resistant and susceptible plant material. However, before researchers can use this method to screen germplasm and breeding lines for insect susceptibility they must have knowledge of the specific phytochemical stimuli that mediate the behavior of a target pest. These biochemical determinants of resistance are usually identified via basic studies on the host-selection behavior of insect pests. While much has been written about the importance of such research in host-plant resistance work (e.g., Beck and Schoonhoven, 1980; Kogan, 1986), little attention has been given to this area of research by grain legume researchers. Indeed, we know of only one case in which insect resistance in grain legumes has been correlated with specific phytochemicals. This information emerged from collaborative work by entomologists at ICRISAT and chemists at the Max-Planck Institute for Biochemistry, Munich, Germany, on the host-selection behavior of H. armigera and the biochemical basis of resistance in chickpea germplasm to this pest (Rembold, 1981; Rembold et al., 1990a,b). These investigators related H. armigera resistance in chickpeas to relatively high amounts of malic and oxalic acids (Rembold et al., 1990b). Research is now underway at the International Center for Agricultural Research in the Dry Areas (ICARDA), Aleppo, Syria, to assess the role of malic acid in chickpea resistance to the leafminer, L. cicerina (Weigand, 1990). More recently, capillary gas chromatography revealed the presence of major volatile compounds in the headspace surrounding pea flowers, some of which may be unique to flowers from genotypes varying in their susceptibility to pea weevil (B. pisorum) attack (Clement et al., 1991; Fellman and Clement, unpubl.). With more research, this preliminary work may lead to methodology for the rapid quantitative and qualitative screening of pea germplasm for phytochemicals that mediate the host-selection behavior of B. pisorum.

Storage Pests

Laboratory tests have been conducted several times to assess variation in susceptibility of seed of cultivated chickpea and wild Cicer spp. to the weevils Callosobruchus chinensis (L.) and C. maculatus (F.) (Raina, 1971; Schalk et al., 1973; ICRISAT, 1976; Salunkhe and Jadhav, 1982; Ahmed et al., 1989; Weigand and Tahhan, 1990; Weigand and Pimbert, in press). On the other hand, there have been few searches for *C. chinensis* (Ishii, 1952) and *C. maculatus* (Fam and El-Sayad, 1985) resistance in seed of faba bean germplasm.

Researchers usually differentiated between resistant and susceptible chickpea and faba bean seed on the basis of ovipositional preference, adult emergence, percentage of damaged or weevil infested seed, insect developmental periods, and/or reproductive capacity of females exposed to seed of different cultivars. Several workers have reported that chickpea cultivars with rough, hard, and wrinkled seedcoats were least preferred by *Callosobruchus* weevils (Raina, 1971; Schalk *et al.*, 1973; ICRISAT, 1976; Salunkhe and Jadhav, 1982; Ahmed *et al.*, 1989; Weigand and Pimbert, in press). However, such "unsightly" seeds may be unacceptable to consumers (Weigand and Pimbert, in press).

Mechanisms and Stability of Resistance

There is now ample evidence that pest populations have the ability to evolve and overcome specific plant resistance factors. When this happens, pest-resistant crops will lose their ability to resist insect attack. To slow pest evolution and thus prolong the useful life of insect-resistant cultivars, some entomologists have suggested (Kennedy *et al.*, 1987; Smith, 1989; Gould, 1991) that resistance breeding programs place more emphasis on: the breeding of insect-resistant cultivars with more than one type of resistance; the deployment of crop cultivars with partial resistance to insect pests; and the development and use of tolerant crop cultivars. For example, a new cultivar with genes conferring resistance at both the behavioral (antixenosis) and physiological (antibiosis) levels might last much longer in the field than a cultivar possessing only one type of resistance. Intuitively, exposure of pest insects to plants exhibiting strong antibiosis and antixenosis resistance would subject them to intense selection pressure, with subsequent development of resistance-breaking insect biotypes (Smith, 1989); therefore, the effect of partial resistance in cultivars and deployment of tolerant crop cultivars would be less selection pressure on pest populations (Lamberti *et al.*, 1983; van Emden, 1991).

Using specific assays to monitor the effects of particular physical and chemical plant traits on insect behavior and physiology, as well as inferences drawn from the results of initial screenings and evaluations, researchers have differentiated between the antixenosis, antibiosis, and tolerance categories of plant resistance to insect pests of grain legumes. To date, however, more antibiosis than antixenosis or tolerance has been reported in grain legumes. There are also documented cases in which pulse genotypes avoided insect attack or suffered less damage than other entries because of phenological asynchrony, i.e., ecological resistance as defined by Kogan (1982) (Table 3).

Where multiple types of resistance (tolerance, antixenosis, antibiosis) are reputed to be associated with pulse resistance to insects, breeders may be able to circumvent the breakdown of plant resistance by releasing cultivars with multiple types of insect resistance. However, this strategy might not work against the pea aphid (*P. pisum*) because of its ability to develop resistance-breaking biotypes (Reed *et al.*, 1988). On the other hand, the breeding of chickpea cultivars with polygenic resistance combining insect repellency (antixenosis), toxicity (antibiosis), and tolerance would likely slow the breakdown of plant resistance to *H. armigera*, and possibly to other chickpea insect pests as well (Pimbert, 1990). Moreover, Reed *et al.* (1988) were of the view that resistance to *H. armigera* in chickpea is likely to be stable, in part because of the polygenic nature of the resistance. This polygenic resistance is based on the discovery of all three types of genetic resistance in chickpea (Table 3),

Table 3. Status of types of plant resistance reportedly involved with insect resistance in grain legumes

| Crop and insects | Ecological resistance[b] | Categories of genetic resistance[b] | | |
		Antixenosis	Antibiosis	Tolerance
Chickpea				
Callosobruchus maculatus(F.)		+	+	
Helicoverpa armigera (Hüb.)	+	+	+	+
Faba bean				
Aphis craccivora (Koch)			+	
Aphis fabae Scop.			+	
Acyrthosiphon pisum (Harris)			+	
Megoura viciae (Buckt.)			+	
Empoasca fabae (Harris)		+	+	
Bruchus dentipes Bandi	+			
Callosobruchus chinensis (L.)			+	
Callosobruchus maculatus (F.)			+	
Epilachna varivestris Muls.		+	+	
Pea				
Acyrthosiphon pisum (Harris)[c]		+	+	+
Bruchus pisorum (L.)		+	+	
Sitona lineatus (L.)			+	+
Cydia nigricana (F.)[c]	+			
Chromatomyia horticola (Goureau)		+	+	
Lentil				
Bruchus lentis Froel.	+			

[a]Information compiled from references listed in Table 1.

[b]Types of genetic resistance as defined by Kogan (1982).

[c]Resistance found in dry and/or green pea cultivars.

which gives breeders the option of creating combinations of resistance factors in a single cultivar.

In addition, both antixenosis and antibiosis resistance have been detected in germplasm evaluated against the field pests *E. fabae, E. varivestris, C. horticola,* and *B. pisorum* (Table 3). However, until more details about the nature of plant resistance to the first three species are forthcoming there is little reason to discuss the deployment of different resistance modalities in resistance breeding. Pesho *et al.* (1977) detected antixenosis resistance in peas to *B. pisorum* in the United States; however, this resistance did not hold up under field conditions in Chile and Australia (Clement, unpublished information; Hardie, 1990). Apparently, the effects of chemical antixenosis were not strong enough to substantially decrease weevil oviposition on pods of the nonpreferred pea lines. There is, however, room for optimism concerning the use of plant resistance against *B. pisorum* and it is based on Hardie's (1990) recent discovery that a wild line of *Pisum sativum* ssp. *humile* (= ssp. *elatius* var. *pumilio* [van der Maesen *et al.,* 1988]) responded to the presence of pea weevil eggs on pods by forming callus. If it can be shown that this pod callus inhibits the development of eggs or impedes larval penetration of the pod wall, Hardie's (1990) discovery may represent a new type of antibiosis-based resistance against pea weevil. A similar reaction against pea weevil oviposition was first reported by Annis and O'Keeffe (1984) for pods of *Lathyrus* spp. Efforts to increase levels of plant tolerance and antibiosis resistance in peas to *S. lineatus* (Table 3) were not always successful (Nouri-Ghanbalani *et al.,* 1978; Auld *et al.,* 1980), leading to a cessation of breeding efforts against this pest in the western United States.

Only antibiosis resistance, and some of it in the form of partial resistance, has been found in faba bean cultivars and related *Vicia* species against the aphids *A. craccivora, A. fabae, A. pisum,* and *M. viciae* (Table 3). Although Holt and Birch (1984) considered the usefulness of partial resistance to aphid pests of faba beans, they viewed the incorporation of high levels of antibiosis from wild *Vicia* species into faba bean cultivars as a longer term solution to the development of virulent, resistance-breaking aphid biotypes.

Although sources of antixenotic- and antibiotic-based resistance to storage pests in the genus *Callosobruchus* have been found in chickpea and faba bean seed (Table 3), some researchers (Bushara, 1988; Reed *et al.* 1988; Pimbert, 1990) have not expressed confidence in host-plant resistance as a feasible strategy to control these weevils. Their reservations have centered around the fact that relatively few sources of weevil resistance in pulse seeds have been found, despite the many attempts made. Rather than aggressively pursuing weevil resistance in pulse seeds, Reed *et al.* (1988) and Pimbert (1990) suggested it may be more productive to work towards improving seed storage conditions and improving other control methods for storage pests. We would only add that the recent discovery of *Callosobruchus* resistance in seed of wild *Cicer* (Weigand and Tahhan, 1990; Weigand and Pimbert, in press) suggests the need for more evaluations of secondary and tertiary gene pools for seed resistance to storage pests.

Host-Plant Resistance in Pest Management

Host-plant resistance can serve both as a principal pest management method and as a complementary pest control method in integrated pest management systems (Kogan, 1982). The latter approach clearly has been embraced by chickpea entomologists at ICARDA and ICRISAT (Reed *et al.*, 1987; Lateef, 1990; Lateef and Pimbert, 1990; Pimbert, 1990). In addition, statements in the literature (Holt and Birch, 1984) and workshops on specific pests (National Pea Weevil Workshop, Victoria, Australia; Smith, 1990) indicate that entomologists working on insect pests of other grain legumes plan to deploy host-plant resistance as part of integrated control programs. Indeed, traditional methods of pest control, such as the use of insecticides, are often impractical and uneconomical for grain legume producers, especially in the developing countries (Singh *et al.*, 1990). Moreover, "other factors such as toxicity, environmental pollution, the extermination of natural enemies, and eventually, build-up of insecticide resistance in the pests make chemical control a risky strategy" (Lateef, 1990). Hence the need for more sustainable approaches to managing insect pests of grain legumes.

The potential interactive role of plant resistance and classical biological control in managing insect pests of grain legumes has been addressed by some researchers. For example, Annis and O'Keeffe (1987) investigated the influence of pea genotypes on parasitization of the pea weevil (*B. pisorum*) by a pteromalid wasp in the western United States. Other investigators (Kareiva and Sahakian, 1990) studied the interaction of plant resistance and biological control in peas by assessing the effect of plant morphology on the population growth of pea aphids (*A. pisum*) in the presence and absence of coccinellid beetle predators. What they found was that the predators were more effective at controlling aphid populations on leafless as opposed to normal-leafed peas. Soroka and Mackay (1990a) also found fewer pea aphids on more architecturally simple pea plants but they attributed their findings to the increased vulnerability of aphids on semi-leafless plants to adverse weather and to the reduction of leaflets, which allowed for less preferred space for aphid population development. The work of Karieva and Sahakian (1990) and Soroka and Mackay (1990a,b,c) in the United States and Canada suggests it would be prudent to consider the effects of plant morphology on insect predators and pea aphid populations if breeding efforts are directed towards the development of semi-leafless or leafless types. More examples of research on the integration of plant resistance with biological control can be found in Weigand *et al.* (1994, this volume). These researchers also addressed the potential interplay of plant resistance with cultural and chemical control methods in the development of integrated control programs in grain legumes.

Largely unexplored by pulse entomologists are the effects that different types and levels of plant resistance could have on the success or failure of chemical and biological control methods. The importance of this aspect in breeding for insect resistance in crops was pointed out by van Emden (1991), Kennedy *et al.* (1987) and Smith (1989).

Prospects

This chapter is testimony to the many advances made in plant resistance to insect pests of cool season food legumes by entomologists and plant breeders, who through their interests and energy have developed plant screening methods, located insect resistance in germplasm, and characterized mechanisms of resistance. With new progress by interdisciplinary, mission-oriented research teams at ICARDA and ICRISAT, we have reason to be optimistic about the future development of insect-resistant grain legumes, especially chickpeas for the developing countries. For example, entomologists and chemists have learned much about the biochemical bases of resistance in *Cicer* to *H. armigera* and *L. cicerina* and the factors governing the host-selection behavior of these major insect pests (Pimbert, 1990; Rembold *et al.*, 1990a,b). Moreover, research begun by ICRISAT breeders and entomologists after the discovery that most *Helicoverpa* resistant chickpea lines were highly susceptible to Fusarium wilt and Ascochyta blight has led to the successful combination of pod borer and Fusarium wilt resistance in chickpea lines ICCL 8611 and ICPX-730020-11-1-1H (Lateef, 1985, 1990; Lateef and Sachan, 1990; Singh *et al.*, 1990). Team research like this must continue. Moreover, it must be expanded to include molecular biologists who can apply new biotechnological innovations to the development of insect-resistant cultivars of cool season food legumes.

References

Ahmed, K., Khalique, F., Afzal, M., Tahir, M. and Malik, B. A. 1989. *Journal Stored Products Research* 25: 97–99.

Ahmed, K., Lal, S. S., Morris, H., Khalique, F. and Malik, B. A. 1990. In: *Chickpea in the Nineties*, pp. 165–168 (eds. H. A. van Rheenen, M. C. Saxena, B. J. Walby and S. D. Hall). Patancheru, Andhra Pradesh, India: ICRISAT.

Aleksandrova, E. A. 1977. *Selektsiya i Semenovodstvo* 1: 46–47.

Annis, B. A. 1983. *Mechanisms of host plant resistance to pea weevil in peas*, 89 pp. Ph.D. Thesis, University of Idaho, Moscow, Idaho, USA.

Annis, B. and O'Keeffe, L. E. 1984. *Entomologia Experimentalis et Applicata* 35: 83–87.

Annis, B. and O'Keeffe, L. E. 1987. *Environmental Entomology* 16: 653–655.

Auld, D. L., O'Keeffe, L. E., Murray, G. A. and Smith, J. H. 1980. *Crop Science* 20: 760–766.

Beck, S. D. and Schoonhoven, L. M. 1980. In: *Breeding Plants Resistant to Insects*, pp. 115–135 (eds. F. G. Maxwell and P. R. Jennings). New York, USA: John Wiley & Sons.

Bingefors, S., Johanson, N. and Wiklund, K. 1964. Losses caused by pea moth in variety trials at the Swedish seed association (in Swedish). *Medd. St. Vaxtskydds. anst. Stockh.* 12: 413–432. (Abstract in *Review Applied Entomology, Series A* 55: 620).

Birch, N. 1985. *Annals Applied Biology* 106: 561–569.

Birch, N. and Wratten, S. D. 1984. *Annals Applied Biology* 104: 327–338.

Bushara, A. G. 1988. In: *World Crops: Cool Season Food Legumes*, pp. 367–378 (ed. R. J. Summerfield). Dordrecht, The Netherlands: Kluwer Academic Publishers.

Chopra, N. and Pajni, H. R. 1987. *Lens Newsletter* 14: 23–26.

Clement, S. L. 1992. *Entomologia Experimentalis et Applicata* 63: 115–121.

Clement, S. L., Pike, K. S. and Fellman, J. K. 1991. In: *Program and Abstracts, Biennial Meeting of The National Pea Improvement Association*, p. 21. Lincoln, Nebraska, USA.

El-Defrawi, G., El-Gantiry, A. M., Weigand, S. and Khalil, S. A. 1991. *Arab Journal of Plant Protection* 9: 138–141.

Fam, Z. E. and El-Sayed, S. A. 1985. *FABIS Newsletter* 13: 30–31.

Gould, F. 1991. *American Scientist* 79: 496–507.

Harlan, J. R. and de Wet, J. M. J. 1971. *Taxon* 20: 509–517.

Hardie, D. 1990. In: *Proceedings of National Pea Weevil Workshop*, pp. 72–79 (ed. A. M. Smith). Melbourne, Victoria, Australia: Department of Agriculture and Rural Affairs.

Holt, J. 1983. *Aphid resistance in faba beans*. Ph.D. Thesis, University of Southampton, Southampton, UK.

Holt, J. and Birch, N. 1984. *Annals Applied Biology* 105: 547–556.

Horber, E. 1978. In: *Pests of Grain Legumes: Ecology and Control*, pp. 281–295 (eds. S. R. Singh, H. F. van Emden and T. A. Taylor). London, UK: Academic Press.

ICARDA. 1989. *Annual Report*. Aleppo, Syria.

ICRISAT. 1976. *Annual Report*, 1975–1976. Hyderabad, India.

ICRISAT. 1987. *Annual Report*, 1986. Patancheru, India.

Ishii, S. 1952. *National Institute Agricultural Sciences, Series C* 1: 185–256.

Kareiva, P. and Sahakian, R. 1990. *Nature* 345: 433–434.

Kennedy, G. G., Gould, F., de Ponti, O. M. B. and Stinner, R. E. 1987. *Environmental Entomology* 16: 327–338.

Kogan, M. 1982. In: *Introduction to Insect Pest Management*, pp. 93–134 (eds. R. L. Metcalf and W. H. Luckmann). New York, USA: John Wiley & Sons.

Kogan, M. 1986. *Iowa State Journal of Research* 60: 501–527.

Lamberti, F., Walker, J. M. and van der Graaff, N. A. 1983. *Durable Resistance in Crops*, 454 pp. New York, USA: Plenum.

Lateef, S. S. 1985. *Agriculture, Ecosystems and Environment* 14: 95–102.

Lateef, S. S. 1990. In: *First National Workshop on Heliothis* Management : Current Status and Future Strategies, pp. 129–140 (ed. J. N. Sachan). Kanpur, India: Directorate of Pulses Research.

Lateef, S. S., Bhagwat, V. R. and Reed, W. 1985. *International Chickpea Newsletter* 13: 29–32.

Lateef, S. S. and Pimbert, M. P. 1990. In: *Proceedings of First Consultative Group on Host Selection Behavior of Helicoverpa armigera*, pp. 14–18. Patancheru, Andhra Pradesh, India: ICRISAT.

Lateef, S. S. and Reed, W. 1985. In: *National Seminar on Breeding Crop Plants for Resistance to Pests and Diseases*, pp. 127–131. Tamil Nadu, India: Coimbatore, Tamil Nadu Agricultural University.

Lateef, S. S. and Sachan, J. N. 1990. In: *Chickpea in the Nineties*, pp. 181–189 (eds. H. A. van Rheenen, M. C. Saxena, B. J. Walby and S. D. Hall). Patancheru, Andhra Pradesh, India: ICRISAT.

Nouri-Ghanbalani, G. 1974. *Plant resistance in peas, Pisum spp., to the pea leaf weevil, Sitona lineatus (L.) (Coleoptera:Curculionidae)*, 43 pp. M.S. Thesis, University of Idaho, Moscow, Idaho, USA.

Nouri-Ghanbalani, G. 1977. *Host plant resistance to the pea leaf weevil, Sitona lineatus (L.), in pea (Pisum sativum L.) and its inheritance*, 125 pp. Ph.D. Thesis, University of Idaho, Moscow, Idaho, USA.

Nouri-Ghanbalani, G., Auld, D. L., O'Keeffe, L. E. and Campbell, A. R. 1978. *Crop Science* 18: 858–860.

Pesho, G. R., Muehlbauer, F. J. and Harberts, W. H. 1977. *Journal of Economic Entomology* 70: 30–33.

Pillsbury, B. P. 1986. *Development of a laboratory assay to study the resistance of peas to the pea weevil, Bruchus pisorum L.*, 63 pp. M.S. Thesis, University of Idaho, Moscow, Idaho, USA.

Pimbert, M. P. 1990. In: *Chickpea in the Nineties*, pp. 151–163 (eds. H. A. van Rheenen, M. C. Saxena, B. J. Walby and S. D. Hall). Patancheru, Andhra Pradesh, India: ICRISAT.

Raina, A. K. 1971. *Journal of Stored Products Research* 1: 213–216.

Reed, W., Cardona, C., Lateef, S. S. and Bishara, S. I. 1988. In: *World Crops: Cool Season Food Legumes*, pp. 107–115 (ed. R. J. Summerfield). Dordrecht, The Netherlands: Kluwer Academic Publishers.

Reed, W., Cardona, C., Sithanantham, S. and Lateef, S. S. 1987. In: *The Chickpea*, pp. 283–318 (eds. M. C. Saxena and K. B. Singh). UK: CAB International.

Rembold, H. 1981. *International Chickpea Newsletter* 4: 18–19.

Rembold, H., Schroth, A., Lateef, S. S. and Weigner, Ch. 1990a. In: *Proceedings of the First Consultative Group Meeting on Host Selection Behavior of Helicoverpa armigera*, pp. 23–26. Patancheru, Andhra Pradesh, India: ICRISAT.

Rembold, H., Walner, P., Köhne, A., Lateef, S. S., Grüne, M. and Weigner, Ch. 1990b. In: *Chickpea in the Nineties*, pp. 191–194 (eds. H. A. van Rheenen, M. C. Saxena, B. J. Walby and S. D. Hall). Patancheru, Andhra Pradesh, India: ICRISAT.

Salunkhe, V. S. and Jadhav, I. D. 1982. *Legume Research* 5: 45–48.

Schalk, J. M., Evans, K. H. and Kaiser, W. J. 1973. *FAO Plant Protection Bulletin* 21: 126–131.

Sedivy, J. 1972. *Arch Pflanzenschutz* 8: 209–217.

Sehgal, V. K., Sen, A. and Singh, K. V. 1987. In: *ACIAR Proceedings Series 18*, p. 299. Canberra, Australia: Australian Centre for International Agricultural Research.

Semenova, A. G. 1990. *Sbornik Nauchnykh Trudov Po Prikladnoi Botanike, Genetik I Selektsii* 132: 78–82.

Singh, K. B., Kumar, J., Haware, M. P. and Lateef, S. S. 1990. In: *Chickpea in the Nineties*, pp. 233–238 (eds. H. A. van Rheenen, M. C. Saxena, B. J. Walby and S. D. Hall). Patancheru, Andhra Pradesh, India: ICRISAT.

Smith, A. M. (ed.) 1990. In: *Workshop Proceedings: National Pea Weevil Workshop*. Melbourne, Victoria, Australia: Victoria Department of Agriculture and Rural Affairs.

Smith, C. M. 1989. *Plant Resistance to Insects: A Fundamental Approach*, 286 pp. New York, USA: John Wiley & Sons.

Smith, J. H., O'Keeffe, L. E. and Muehlbauer, F. J. 1982. *Journal of Economic Entomology* 75: 530–534.

Sokolov, Yu A. 1977. *Zashchita Rastenii* 10: 34.

Soroka, J. J. and Mackay, P. A. 1990a. *Canadian Entomologist* 122: 503–513.

Soroka, J. J. and Mackay, P. A. 1990b. *Canadian Entomologist* 122: 1193–1199.

Soroka, J. J. and Mackay, P. A. 1990c. *Canadian Entomologist* 122: 1201–1210.

Soroka, J. J. and Mackay, P. A. 1991. *Journal of Economic Entomology* 84: 1951–1956.

Tahhan, O. 1986. *Bionomics of Bruchus dentipes Baudi and varietal resistance in Vicia faba L.* Ph.D. Thesis, University of Reading. Reading, UK.

Tahhan, O. and Van Emden, H. F. 1989. *Bulletin of Entomological Research* 79: 211–218.

Tingey, W. M. 1986. In: *Plant Insect Interactions*, pp. 251–284 (eds. J. A. Miller and T. A. Miller). New York, USA: Springer.

van der Maesen, L. J. G., Kaiser, W. J., Marx, G. A. and Worede, M. 1988. In: *World Crops: Cool Season Food Legumes*, pp. 55–66 (ed. R. J. Summerfield). Dordrecht: Kluwer Academic Publishers.

van Emden, H. F. 1991. *Bulletin of Entomological Research* 81: 123–126.

Vilkova, N. A. and Kolesnichenko, L. I. 1973. *Zashchity Rastenii* 37: 164–171.

Weigand, S. 1990. *Options Méditerranéennes, Série Séminaires* 9: 73–76.

Weigand, S., Lateef, S. S., El-Din Saraf El-Din, N., Mahmoud, S. F., Ahmed, K. and Kemal, Ali, 1994. *Expanding the Production and Use of Cool-Season Food Legumes*, pp. 680–695 (eds. F. J. Muehlbauer and W. J. Kaiser). Dordrecht: Kluwer Academic Publishers.

Weigand, S. and Pimbert, M. P. 1993. In: *Breeding for Stress Tolerance in Cool-Season Food Legumes*, pp. 145–156 (eds. K. B. Singh and M. C. Saxena). Chichester, UK: John Wiley & Sons.

Weigand, S. and Tahhan, O. 1990. In: *Chickpea in the Nineties*, pp. 169–175 (eds. H. A. van Rheenen, M. C. Saxena, B. J. Walby and S. D. Hall). Patancheru, Andhra Pradesh, India: ICRISAT.

Wnuk, A. 1968. *Polskie Pismo Ent.* 38: 453–461 (Abstract in *Review of Applied Entomology, Series A* 59: 931).

Wolfenbarger, D. A. and Sleesman, J. P. 1961. *Journal of Economic Entomology* 54: 1018–1022.
Wolfenbarger, D. A. and Sleesman, J. P. 1963. *Journal of Economic Entomology* 56: 895–897.
Wright, D. W., Geering, Q. A. and Dunn, J. A. 1951. *Bulletin of Entomological Research* 41: 663–677.

Insects in relation to virus epidemiology in cool season food legumes

L. BOS[1] and K.M. MAKKOUK[2]

[1] Research Institute for Plant Protection (IPO-DLO), P.O. Box 9060, 6700 GW Wageningen, The Netherlands, and
[2] ICARDA, P.O. Box 5466, Aleppo, Syria

Abstract

All but two of 44 viruses of cool season food legumes (chickpea, faba bean, lentil, and pea) are known to be spread by vectors, and their majority (37) by insects. One of the insect-borne viruses is transmitted by leafhoppers, two are by thrips, five by beetles, and 29 by aphids. The viruses have different, more or less intimate relationships with their insect vectors, and all have different ecologies largely determined by those of their vectors. Opportunities for epidemic development of the resulting virus diseases, therefore, differ considerably, as do possibilities for control of these diseases by vector control. There is no single effective approach; breeding for resistance to vectors of the viruses is only one of a range of control measures, and new problems caused by insect-vectored viruses are likely to emerge with time.

Introduction

Emphasis in this volume is on further improvement of the cool season food legumes (chickpea, faba bean, lentil, and pea). Crop improvement involves (1) improvement of cropping practices, (2) genetic crop upgrading, and (3) better crop protection. These aspects are linked in various ways. Crop protection is increasingly through adapted crop management and by choice of resistant genotypes or by breeding for resistance. The grower's choice of improved crop genotypes also is a matter of better exploitation of natural resources, and thus of improved crop management.

Breeders, as in International Agricultural Research Centers, often claim exclusive rights with respect to the use of the term crop improvement, which they think equivalent to genetic crop upgrading. Breeding for resistance to biotic constraints, including viruses and their vectors, is the subject of this section of the volume. Resistance to vectors is sometimes suggested to be "the" solution to virus problems in crops. However, reality is not that simple, and genetic crop upgrading covers only part of the scene. That is why this

F.J. Muehlbauer and W.J. Kaiser (eds.), Expanding the Production and Use of Cool Season Food Legumes, 305–332.
© 1994 *Kluwer Academic Publishers.*

contribution, which is a literature review rather than a research report, goes slightly beyond the subject of this section of the volume.

An earlier contribution on the viruses and virus diseases of cool season food legumes briefly included information on vectors, including fungi, nematodes, and man himself (Bos *et al.*, 1988). Insects are particularly important vectors because they are usually highly mobile and may travel long distances. This paper will further elaborate their role.

Ecology and Epidemiology

Whether a virus will reach a crop and to what extent, depends upon a range of factors (Figure 1), which may interact in various ways. Qualitative analysis of these interactions is called "ecology". Their quantification is called "epidemiology". This aims at forecasting whether and when crop damage will result, and tries to determine whether and when interference is needed to prevent eventual economic damage. Much information on the often decisive role of insects in the epidemiology of virus diseases in crops is provided by proceedings of international conferences, which have been organized at regular intervals since 1981 by the Plant Virus Epidemiology Committee of the International Society for Plant Pathology (Plumb and Thresh, 1983; McLean *et al.*, 1986).

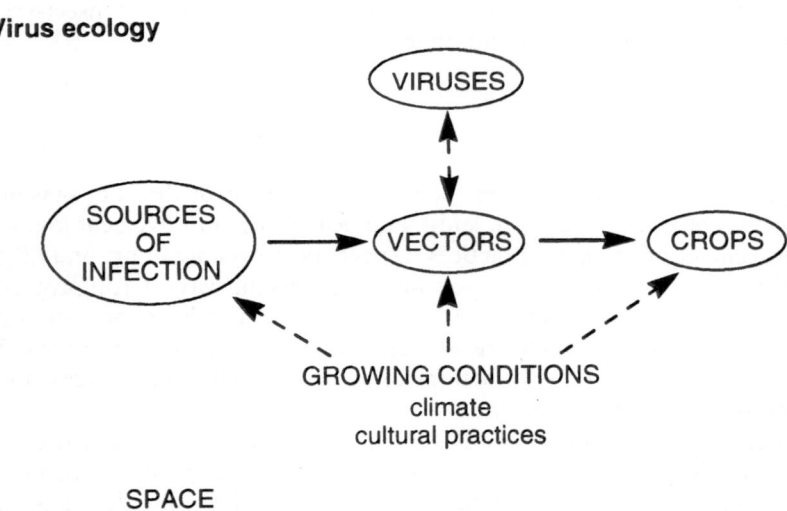

Virus ecology

Figure 1. Factors involved in the ecology of plant viruses, indicating a central role of vectors. (Simplified after Bos, 1983.)

Most of the information on the role of insect vectors in the epidemiology of virus diseases is derived from research on crops other than legumes, particularly potatoes where the determination of roguing dates and certification require quantitative data, and by far most of the data concerns aphids (Harris and Maramorosch, 1977). Such information, however, complements the limited information obtained with legumes and will help explain the epidemiology of virus diseases of food legumes.

Viruses

Final target of any control measure are the viruses. Some 44 were listed for the four cool season food legume crops by Bos *et al.* (1988). Several of them are important constraints in the cultivation of these crops. Surveys conducted in the region of outreach of the International Center of Agricultural Research in the Dry Areas (ICARDA) in Aleppo, Syria, have revealed the occurrence of at least 12 different viruses with usually wide distribution and often high incidence in the cool season food legume crops (Makkouk *et al.*, 1988a; Fortass and Bos, 1991). A number were found unexpectedly widespread, and most are of great potential importance for crop improvement programs because of their transmission in seed. They must be seriously taken into account in international breeding programs and in systems of commercial seed production.

For example, broad bean mottle bromovirus, until recently only of academic interest as one of a few bromoviruses with an interesting molecular biology, is now known to occur in Syria, Egypt, Sudan, Tunisia, Morocco (Makkouk *et al.*, 1988b), and Algeria (Ouffroukh, 1985). A more systematic and detailed survey in Morocco revealed its occurrence in 56% of the fields inspected, with a maximum incidence of 33% recorded in one field (Fortass and Bos, 1991). The virus is beetle- and seed-transmitted and may well cause serious, but still overlooked damage in chickpea, lentil, and especially pea (Makkouk *et al.*, 1988b).

Incompletely identified luteoviruses, including bean leafroll luteovirus (BLRV), have in recent years been found to be very prevalent in faba bean in the Middle East and North Africa (Makkouk *et al.*, 1988a; Fortass and Bos, 1991). The diseases they cause in several legumes have been and continue to be overlooked in many places, or their symptoms mistaken for nutrient deficiency or the effect of drought. They also attract increasing attention elsewhere. BLRV, known from pea and faba bean in Western Europe since 1954 (Quantz and Völk, 1954), caused a major epidemic in pea in southern Idaho in 1980 where nearby crops of lentil and chickpea developed severe chlorosis and stunting, leading, in chickpea, to premature death (Hampton, 1983b). During the 1990 growing season, a severe epidemic, predominantly caused by pea enation mosaic enamovirus and BLRV developed on grain legumes grown in the Palouse region of eastern Washington and northern Idaho, and incidences in chickpea fields approached 100% (Klein *et al.*, 1991). The legume yellows diseases resemble the global barley yellow dwarf problem of cereals (Plumb, 1983), now also known to be due to a complex of luteoviruses, and will undoubtedly prove to be of similar, yet incompletely appreciated importance. In fact, the luteoviruses seem to form a continuum of viruses with serological, cytological, and epidemiological similarities to beet western yellows luteovirus which has the widest host range of all luteoviruses (Duffus *et al.*, 1990).

A recent shift of chickpea culture from summer to winter sowing in California led to a tremendous increase in infection of the crop by three luteoviruses (beet western yellows, legume yellows, and subterranean clover red leaf) and three other viruses (cucumber mosaic cucumovirus, alfalfa mosaic alfamovirus, and lettuce mosaic potyvirus). In field trials at Davis, Salinas, and in the San Joaquin Valley, their incidence was 60 to 100% (Bosque-Pérez and Buddenhagen, 1990). The chickpea disease is likely the one already reported there by Snyder *et al.* (1956).

Meanwhile, nature may have much more in store. A hitherto unknown, ungrouped non-luteovirus, faba bean necrotic yellows virus, transmitted persistently by aphids, was recently detected in Syria and other countries (Katul, 1992). The single-stranded DNA-containing virus with small isometric particles of ca 18 to 19 nm is related to the ungrouped subterranean clover stunt virus in New Zealand, which causes similar symptoms in clovers, faba bean, French bean, and other legumes (Grylls and Butler, 1959). Another newly detected virus, the leafhopper-transmitted chickpea chlorotic dwarf geminivirus (Horn *et al.*, 1992), which causes symptoms in chickpea similar to those of luteoviruses, is prevalent in the north of India. It is able to infect several other legumes and may well turn out to be of great economic importance. In "yellows" diseases, different viruses often occur in mixtures, and such infections are often difficult to study. Viruses currently considered single entities may finally prove to be mixtures.

Despite extensive laboratory testing of samples from commercial seed lots in the US Pacific Northwest (Mink and Parsons, 1978) and a cleanup of germplasm collections (Hampton, 1983a), the aphid-transmitted pea seedborne mosaic potyvirus now appears to have settled there. The virus is suspected to have spread worldwide with germplasm (Khetarpal and Maury, 1987). Seed-transmitted viruses in general are causing increasing concern. Efforts to contain their spread with germplasm have been intensified, as by the development of FAO/IBPGR Guidelines for the Safe Movement of Legume Germplasm (Frison *et al.*, 1990), but fool-proof systems may be impracticable (Bos, 1989). One reason is the sensible attempt at liberalization of international trade, including plant propagation material, as through recent General Agreement on Trade and Tariffs (GATT) negotiations. Absolute virus-freedom of commercial seed can never be guaranteed by testing and certification. Viruses remain hard to detect in seed, and usually cannot be removed from them once they are in the embryo. Thus, in, and as a result of, endeavors to further improve cool season legumes, viruses are causing increasing, rather than just continuing concern.

In the ecology of a virus, it is the virus itself that largely determines the type of interaction with the vector (Tables 1 and 2) and whether virus can be acquired and transferred during short probes by migrant insects that are in search of palatable sources of food, or whether an actual feeding process of longer duration is required, as for the luteoviruses which are limited to the phloem tissues of the vascular bundles of their hosts. The type of virus and its behavior (aggressiveness) in the host determine its multiplication and thus its availability for transmission. Figure 2 illustrates how efficiency of transmission depends upon the concentration of the virus in the plant on which the insect feeds or probes. The concentration of cucumber mosaic virus declines much more rapidly in infected plants soon after a peak than does that of a potyvirus. Aphids, therefore, less readily acquire the first virus, and its spread from plants infected by that virus is less. Of course, the concentration of the virus available for transmission also depends on the source plant and on its susceptibility.

Table 1. Aphid-transmitted viruses of cool season food legumes

(1) Nonpersistently transmitted viruses

Alfamoviruses
alfalfa mosaic virus[*]

Carlaviruses
pea streak virus
red clover vein mosaic virus

Cucumoviruses
cucumber mosaic virus[*]
peanut stunt virus[*]

Fabaviruses
broad bean wilt virus[*]

Potyviruses
bean common mosaic virus[*]
bean yellow mosaic virus[*]
beet mosaic virus
clover yellow vein virus
lettuce mosaic virus[*]
pea seedborne mosaic virus[*]
peanut mottle virus[*]
soybean mosaic virus[*]
turnip mosaic virus
watermelon mosaic virus 2

Non-grouped viruses
broad bean mild mosaic virus

(2) Persistently transmitted viruses

Enamoviruses
pea enation mosaic virus

Luteoviruses
bean (pea) leafroll virus
beet western yellows virus
legume yellows virus
milk vetch dwarf virus
soybean dwarf virus
(= subterranean clover
red leaf virus)

Rhabdoviruses
lettuce necrotic yellows virus
lucerne enation (papillosity)
virus

Non-grouped viruses
bean yellow vein-banding virus
faba-bean necrotic yellows virus
subterranean clover stunt virus

(3) Semipersistently transmitted viruses

Closteroviruses
clover yellows virus

[*] *virus names marked with an asterisk are seed transmissible.*

Table 2. Beetle-, leafhopper-, and thrips-transmitted viruses of cool season food legumes

Vector group Virus group Virus	Vector
Beetle-transmitted viruses	
Bromoviruses	
broad bean mottle virus*	*Apion arrogans*
	Apion vorax
	Colaspis flavida
	Diabrotica undecipunctata
	Sitona lineatus var. *viridifrons*
bean yellow stipple virus	*Cerotoma ruficornis rogersi*
	Diabrotica balteata
	Diabrotica adelpha
Comoviruses	
broad bean stain virus*	*Apion vorax*
broad bean true mosaic virus*	*Apion vorax*
cowpea mosaic virus*	*Acalymma vittatum*
	Cerotoma spp.
	Diabrotica spp.
	Nematocerus acerbus
	Ootheca mutabilis
	Paraluperodes quaternus
Leafhopper-transmitted viruses	
Geminiviruses	
chickpea chlorotic dwarf virus	*Orosius orientalis*
Thrips-transmitted viruses	
Ilarviruses	
tobacco streak virus	*Frankliniella* spp.
	Thrips tabaci
Tospoviruses	
tomato spotted wilt virus	*Frankliniella* spp.
	Thrips tabaci

* *virus names marked with an asterisk are seed transmissible.*

In general, viruses belonging to different groups are likely to differ in ecology because of differences in vectors and in relationships with their vectors. On the other hand, viruses belonging to different taxonomic groups sometimes have similar ecologies. This is the case with faba bean necrotic yellows virus and subterranean clover stunt virus which differ considerably from luteoviruses but are transmitted by almost the same aphid species, are phloem-limited in their plant hosts, cause almost identical symptoms, and none of them is seed-transmitted.

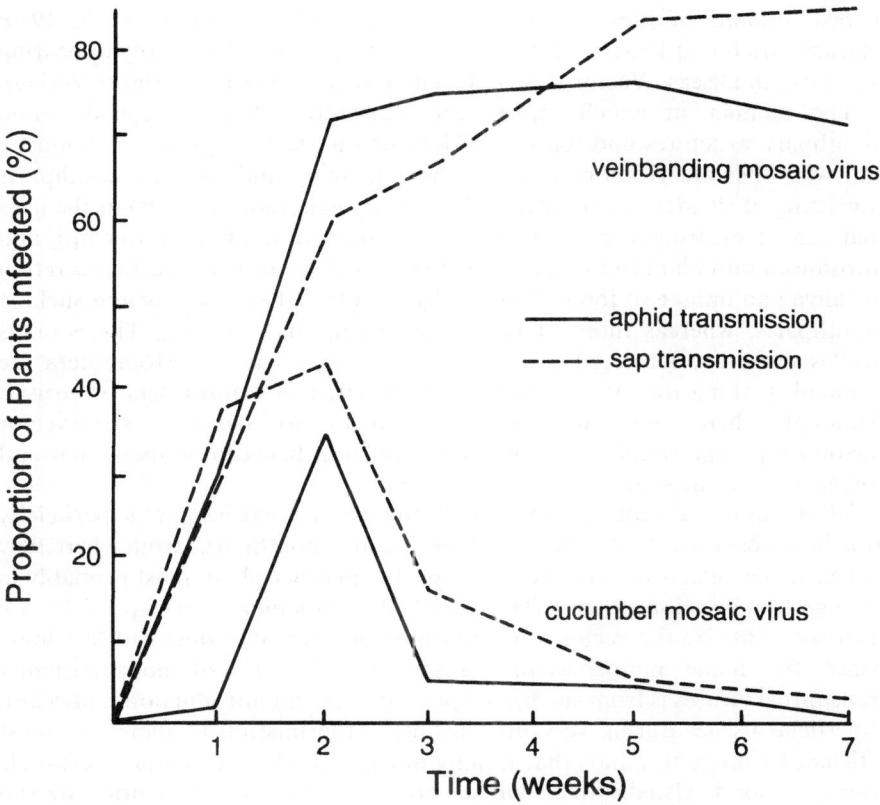

Figure 2. Effect of virus concentration in source plant (*Capsicum frutescens*) as determined by sapinoculation (interrupted line) on proportion of plants infected after transmission by aphids (solid line). (After Simons, 1958.)

Insects

Out of the 44 viruses of cool season legumes that were listed in the earlier survey (Table 1 in Bos *et al.*, 1988), all but two are known to be spread by vectors, and the majority (37) by insects. These viruses can be grouped according to the type of insects that transmit them, i.e., aphids (29) (Table 1), beetles (5), leafhoppers (1), and thrips (2) (Table 2). Aphid transmission appears to be by far the most important way of aerial spread of viruses of cool season legumes. Whitefly transmission, which is of rapidly increasing importance in the tropics and subtropics, is so far of no known importance in cool season legumes. This situation, however, may change with the movement of *Bemisia tabaci*, or of a special biotype of this vector, into cooler regions.

Comprehension of the ecology of insect-vectored viruses requires profound knowledge of the relationships between virus and vector. This field is covered by a vast amount of literature (e.g., Harris and Maramorosch, 1977, 1980; Maramorosch and Harris, 1979; and an ongoing series of annually appearing Advances in Disease Vector Research edited by K. F. Harris, Springer Verlag).

The manner in which viruses are transmitted largely depends upon mouthpart structures and the way of feeding on plants. Aphids, leafhoppers, and whiteflies are Homoptera which have piercing and sucking mouthparts consisting of slender mandibular and maxillary extensions (stylets) of the jaws that can be protruded from the beak or rostrum (a modified lower lip), and introduced into plant tissue. The stylets surround separate canals for secretion of saliva and uptake of food. Thrips (Thysanoptera) have rasping and sucking mouthparts, whereas those of beetles are biting and chewing. The wounds beetles make are large, whereas the wounds caused by the Homoptera are minimal, making the latter group of insects efficient transmitters of viruses. Although there are different terminologies to describe virus-vector relationships, the simplest remains the distinction between nonpersistent and persistent transmission.

With "nonpersistent transmission", the virus particles are superficially attached (absorbed) to the interior of the vector's mouthparts, from where they can easily be detached when the vector probes another plant, most probably in an ingestion-egestion process (Harris, 1977). Contaminant virus particles are then soon lost by the vector, and transmission capability does not last long; hence the name nonpersistent transmission. Uptake of nonpersistently transmitted viruses is from superficial plant tissues and introduction is also into superficial tissues during very brief probes. Transmission is, therefore, most efficient by migrant aphids that rapidly move around in search of a palatable source of food. Usually these viruses also can be easily transmitted by sap inoculation, using an abrasive for making minute superficial wounds. Conversely, mechanically transmissible viruses, when also transmitted by insects, are in the nonpersistent manner, but pea enation mosaic virus is an exception to this rule.

In the case of "semipersistent transmission", as with closteroviruses transmitted by aphids (Table 1), the relationships are slightly more intimate, in that deeper surfaces of the insect's food canal are involved. There is no latency period, but persistence may last for hours if not days. The viruses concerned are absorbed on inner surfaces of the pharingeal area of the food canal and foregut, hence deeper along the alimentary tract than nonpersistently transmitted viruses, but they also are lost during moulting, and transmission may be according to a similar ingestion-egestion mechanism (Harris, 1983). In plants, most of these viruses are more or less concentrated in the phloem, and they may also be transmitted mechanically but with difficulty.

Pea seedborne mosaic virus is one of a few viruses with "bimodal transmission", that is, both nonpersistently and semipersistently. The virus can be acquired during short probes but also during long feeding. When acquired during short probes, the virus is retained only briefly, and when acquired during long feeds, for extended periods of time. One biotype of *Myzus persicae* transmitted the virus bimodally and another only nonpersistently (Lim and Hagedorn, 1977).

Most viruses that are characterized by "persistent transmission" are so-called phloem-limited in plants and cannot be transmitted artificially by sap inoculation. They are taken up by their vectors while feeding on phloem contents, then "circulate" through the body of the vector until they reach the salivary glands. After the latency period, elapsing since the time of uptake, they can be introduced into healthy plants upon which the vector may happen to feed. Such viruses are not transmitted in brief probes because the vector needs time to search for and reach the phloem with its mouthparts. Most viruses will get digested in the insect's alimentary canal, some very stable ones may pass unharmed, and only few can penetrate the gut wall and get all the way into the salivary glands undamaged. The latter are the ones that may be transmitted. Depending upon the dose of virus ingested by the vector, transmission capability may then last "persist" for days or weeks. Such transmission is merely "circulative". Some viruses even multiply in vector tissues, such as the lettuce necrotic yellows rhabdovirus of Table 1 in *Hyperomyzus lactucae* (O'Loughlin and Chambers, 1967); their transmission is "propagative".

Virus transmission by beetles with biting mouthparts is of a unique nature. Acquisition is very rapid, as is inoculation, and there seems to be no latency period. Efficiency of transmission increases with longer feeding, and retention may be for days. Specificity of transmission is determined by virus ability to resist inactivation by RNase activity in the regurgitant (Gergerich and Scott, 1991). Thrips transmission of tomato spotted wilt tospovirus is special in that only the larvae can acquire the virus. Transmission is in a persistent manner, although thrips feed by sucking the contents of subepidermal cells.

Nonpersistent transmission is largely non-specific, and single viruses may be transmitted by many different aphid species. Soybean mosaic potyvirus, for example, can be transmitted by at least 23 aphid species (Halbert *et al.*, 1981). That is why in Table 1 no special aphids have been mentioned and why nonpersistent transmission was long thought to be a mere mechanical process. Contact between virus and vector is superficial (internal stylet contamination), but there is some degree of specificity in such transmission. Specificity is high in the case of persistence. So there are specific interactions between virus and vector. For example, aphid-transmitted viruses are not transmitted by other types of vector and vice versa. Luteoviruses can often be distinguished by the different aphid species that transmit them, and this may even hold for the distinction between strains of barley yellow dwarf virus. For example, the RPV form of the latter virus is transmitted efficiently by *Rhopalosiphon padi* but not by *Sitobion avenae*. For another form, MAV, the reverse is the case. Bean leafroll virus is readily transmitted by *Acyrthosiphon pisum* but poorly by *Myzus persicae*, the vector of the related beet western yellows luteovirus (Waterhouse *et al.*, 1988). Although many, if not most, aphid species can transmit a virus that is aphid- transmitted in the nonpersistent manner, not all do so with the same efficiency, as especially studied for potato viruses (van Harten, 1983; De Bokx and Piron, 1990).

Specificity is likely to occur at two levels, *viz.* (1) at the surface where nonpersistently transmitted virus particles absorb and from which they become detached, and (2) at the membranes through which the persistently transmitted viruses pass while circulating through the body of their vector. Rapidly augmenting information on structure and function of virus particles, and of the proteins they code for, is providing new insights into the underlying molecular mechanisms of specific interaction between virus particles and vector tissues.

Potyviruses often lose aphid transmissibility during purification and storage because of loss of N-terminal parts of their coat proteins, while remaining sap transmissible. This suggests that transmission specificity resides in the N-terminal parts of the coat protein. Loss of aphid transmissibility may take place also during passage to plants by mechanical transmission, and this has been found for a number of potyviruses to result from one point mutation in the RNA (Harrison and Robinson, 1988). The genome of potyviruses also codes for a nonstructural protein, the so-called helper protein, that assists attachment of the virus particles to vector tissues with a certain degree of specificity (Harrison and Murant, 1984).

For the persistently aphid-transmitted luteoviruses, a nontransmissible virus or "strain", e.g., barley yellow dwarf virus, may become transmissible through heterologous encapsidation with coat protein of the transmissible virus or strain (for discussion, see Harrison and Murant, 1984). Luteovirus capsid protein can also encapsidate nucleic acid of some totally different viruses, thus making them transmissible by the vector of the luteovirus. A similar dependent aphid transmission has been claimed for the incompletely described ungrouped bean yellow vein-banding virus, found occasionally in southern England in mixed infections with pea enation mosaic virus, which it needs for aphid transmission (Cockbain *et al.*, 1986). Recent experiments with a panel of monoclonal antibodies (MAbs which react to very specific amino acid configurations or epitopes on the coat protein) and poorly and readily aphid-transmitted strains of the potato leafroll luteovirus have shown that aphid transmissibility is correlated with reactivity to special MAbs, and that these can block transmission in mixtures with the virus. Antibodies raised against the specific MAbs, when fed to *Myzus persicae* prior to virus acquisition, were found to considerably reduce transmission (van den Heuvel, 1991). These investigations suggest the presence of virus-specific receptors on the membranes in the vector responsible for physical attachment of virus particles (as for nonpersistent transmission) and their passage (as for persistent transmission).

Of course, each vector usually has its own extremely complex ecology, population dynamics, and dependence upon climate and farming practices, and this has been documented in much detail for aphids (Carter and Harrington, 1991). Virus transmission by aphids is largely determined by their behavior in crops, and this differentially affects persistent and nonpersistent transmission. With respect to aphids as virus vectors, distinction must be made between visiting noncolonizing (= migrating), settling noncolonizing, and colonizing aphid species. Population densities and vector activity may vary greatly according to conditions, and to location, time of the year, and year. Tremendous variations in temperature and rainfall in capricious climates account for often erratic and unpredictable epidemic development. Wind and wind direction greatly affect insect movement, hence the rate, direction, and distance of virus transfer (see also under Distance).

The part of the vector in virus transmission may be quantified by "vector pressure" (VP), determined by flight activities (number of insects monitored, N) and the so-called relative efficiency factor (REF) to be determined for each species by laboratory comparison with an efficient vector (van Harten, 1983; De Bokx and Piron, 1990). Vector pressure can be determined daily by aphid trapping and calculating VP according to the formula VP = Z (N × REF). Accumulated daily vector pressure for the entire growing season would then indicate the spread of a given virus during that season at a given inoculum potential.

Sources of Infection

Each virus has its own range of hosts, which may be extremely wide and include several nonlegumes, such as of cucumber mosaic virus, broad bean wilt fabavirus, and tomato spotted wilt virus. Perennial legumes (clovers) are highly important in regions where annual legumes are grown, and such perennials develop in or near crops as weeds (Bos *et al.*, 1988). Fifteen of the insect-transmitted viruses of Tables 1 and 2 are seedborne and most of them (10) are further spread by aphids. Then, seedlings developing from infected seed are important within-crop sources of further spread.

The efficiency of virus acquisition by vectors is a function of virus concentration in the source plant. Watermelon mosaic potyvirus 2 was, therefore, found to be less efficiently transmitted by *Aphis gossypii* and *Myzus persicae* from virus resistant melon plants (Romanow *et al.*, 1986). This may also explain why the pea aphid, *Acyrthosiphon pisum*, acquired pea streak carlavirus twice as easily from pea and faba bean than from lucerne (alfalfa, *Medicago sativa*) (Hampton and Weber, 1983a). The virus concentration in an infected source plant may also vary according to the season. Alfalfa mosaic virus has been reported to be poorly spread in alfalfa fields by aphids during summer because of its low concentration during that season (Matisová, 1971). ELISA-measurable concentration of bean leafroll virus in lucerne plants was found to gradually increase to a maximum level in the spring, when the reproduction level of the pea aphid on lucerne also increases, and to decrease with the onset of daytime temperatures above 30°C. The virus was sometimes not ELISA-detectable in infected lucerne plants during July through September (Hampton, 1983b).

The amount of virus available at a given site for spread ("inoculum pressure" or "inoculum potential") not only depends on virus aggressiveness and on source plant susceptibility, but also largely on the number of sources of infection. A nearby infected crop may, therefore, provide much more inoculum potential than will a few infected weed plants. Hence the extremely important role of perennial legume crops (clovers).

For seedborne viruses, the extent of seed infestation at the time of sowing is the determinant for final disease incidence and damage in the ensuing crop. A rate of infestation of soybean seed with bean pod mottle comovirus of 0.013% leads to 32 infected plants in one ha of ca. 250,000 plants, and this is considered sufficient for final economic damage if vector beetle population densities are high (Ross, 1986). Whether final damage results, and which level of seed infestation may be tolerated, thus also depends upon vector pressure. At low vector pressures, infestation of soybean seed with soybean mosaic virus is likely to decrease when farmers save their own seed for planting the next year (Ruesink and Irwin, 1986). In Australia, cucumber mosaic virus spread from seed-infected seedlings was slow in periods of drought, and the percentage of infected seed harvested was 3 to 5 times less than that in the seed originally sown (Jones and Proudlove, 1991). The same was found for broad bean stain and broad bean true mosaic comoviruses in Scotland where *Sitona* weevils were relatively scarce and the main vector *Apion vorax* was absent (Jones, 1978).

Final "infection pressure" to which a given crop is subject, of course, is determined by inoculum potential and vector pressure.

Crops

Whether or not crop infection results from virus dissemination by viruliferous insects finally depends upon whether and to what extent the crop will attract and host the vector. Host palatability will determine whether the vector, once it has alighted, will only probe and move on, or will probe, settle, and colonize. Both, crop attractiveness to insect vectors, depending upon color and other physical and chemical stimuli, and palatability are determined by crop genotype.

Müller (1964) found that brown-colored lettuce cultivars were colonized by aphids to a much lesser extent and were less frequently infected by lettuce mosaic virus than green cultivars. A silver-leaved breeding line of summer squash (*Cucurbita pepo*) was substantially less infected in the open by cucumber mosaic virus and clover yellow vein potyvirus than a green-leaved cultivar, although by mechanical inoculation there was no difference in susceptibility. The difference was ascribed to escape from virus infection because of lower incidence of aphid visitation due the light reflectance from the silvery leaf surface (Davis and Shifriss, 1983). Here, aphids were even prevented from probing, and this explains the effect on the nonpersistently transmitted viruses. Spread of nonpersistently transmitted viruses by vectors does not require crop palatability. Migrant vectors making short probes and immediately thereafter moving on, may be highly efficient in transmitting viruses to and from such crops and even within them.

Evans (1954a,b) was one of the first to notice the effect of true resistance to a virus vector. Incidence of groundnut rosette virus (ungrouped virus) was less in crops of the Mwitunde-type of groundnut (peanut), resistant to *Aphis craccivora*, partially because the aphid could not acquire the virus so readily from it, but also because restricted aphid multiplication resulted in further population reduction by predator control before secondary dispersal. Reduction in aphid population density on a resistant cultivar may also explain why fields of red clover "Dollard", which in comparison to "Wegener" is less preferred and less suitable for multiplication of *Acyrthosiphon pisum*, were much less infected by pea mosaic virus (= bean yellow mosaic potyvirus) (10% versus 90%), although both cultivars were equally susceptible when tested by mechanical inoculation (Wilcoxson and Peterson, 1960). Consistently lower incidences of groundnut bud necrosis tospovirus (formerly tomato spotted wilt virus, Sreenivasulu *et al.*, 1991) in a number of groundnut genotypes resulted from resistance to the vector *Thrips palmi* (Reddy *et al.*, 1991). Genotypes that are susceptible to the virus after mechanical inoculation may escape infection in the field because of resistance to the vector.

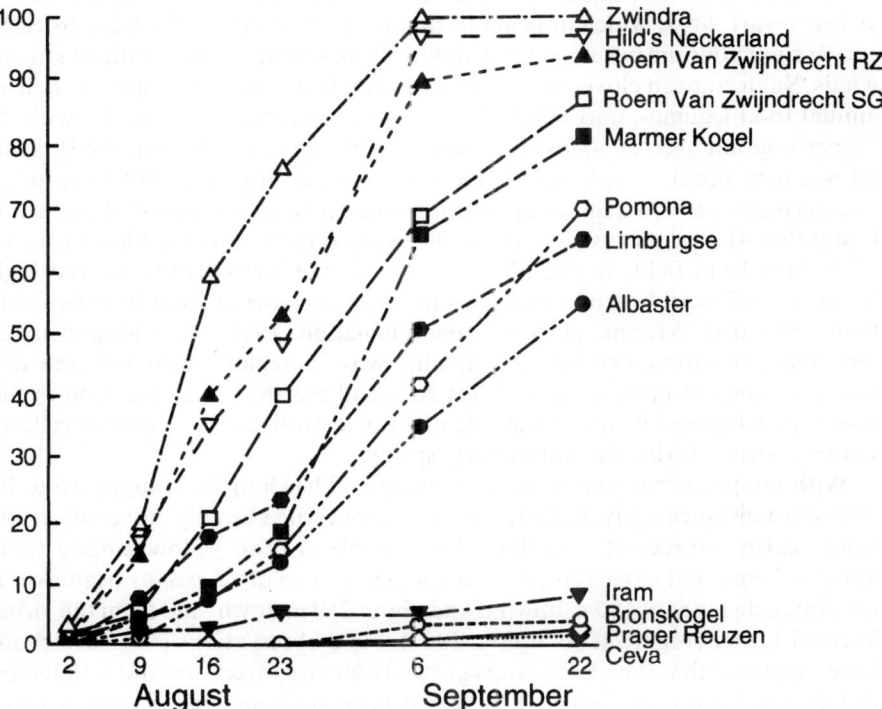

Figure 3. Differences in epidemic buildup in cultivars with differences in resistance to a nonpersistently aphid-transmitted virus (celery mosaic potyvirus in celeriac, *Apium graveolens*). (After Bos *et al.*, 1989.)

Crop plants initially infected by virus from outside sources are mostly sources of further within-crop spread. The extent of secondary spread is greatly determined by the susceptibility of the given crop genotype (irrespective its sensitivity). Resistant plants are poorer and more slowly available secondary sources of infection. Epidemic buildup (increase in disease incidence) is slower in resistant cultivars than in susceptible ones (as found for celery mosaic potyvirus in celeriac: Figure 3; Bos et al., 1989).

Distance

Whether and to what extent an insect-vectored virus will finally reach a susceptible crop depends upon the distance of this crop from the source of infection. Distance also determines spatial spread within a crop. The distance to be covered by insects while remaining viruliferous depends on the type of relationship between virus and vector, as well as on whether alatae (winged) or wingless insects (apterae) are involved and on the type of insects.

Leafhoppers are strong flyers, but even aphids can cover long distances while carrying persistently transmitted viruses. The latter has been shown for barley yellow dwarf virus, which appears in the northern states of the USA in early spring soon after the influx of cereal aphids from southern states with jet stream winds. Subterranean clover stunt virus, which infects pasture legumes as well as annual food legumes, may similarly cover long distances in Australia with its "super-migrant" vector Aphis craccivora (for literature see Thresh, 1983). Such viruses may occur in aphids at high incidences. Ashby et al. (1979) trapped Aulacorthum solani in a mixed agricultural area in New Zealand for 4 years and found that 41% of the aphids carried the subterranean clover red leaf virus. In 1989, faba bean fields in the Flevo Polders in the Netherlands showed high incidences of bean leafroll virus, and more so downwind from lucerne fields than upwind (J. Meems, personal communication, 1991). This long-distance movement of circulative viruses explains why in regions with hot and dry summers, such as many areas of West Asia and North Africa, infection of the cool season legumes by luteoviruses (e.g., bean leafroll and beet western yellows viruses) occurs during the fall or early spring.

With nonpersistent transmission, virus spread has long been supposed to be over short distances only. In fact, the viruses concerned usually move into crops from nearby sources of infection. An example is bean yellow mosaic virus moving from a red clover crop into an adjacent crop of Phaseolus bean over a maximum distance of 100 m upwind and about 250 m downwind from the virus-infected source (Figure 4; Hampton, 1967). Field observations in Great Britain have suggested that faba bean crops grown from virus-free seed and isolated by 250–500 m from crops infected by broad bean stain and true mosaic viruses remain free of infection (Cockbain et al., 1975). With the above aphid-borne viruses, incidence decreases according to a gradient from the source of infection, as was also found for pea streak virus and alfalfa mosaic virus in pea, in contrast to the persistently transmitted bean leafroll virus (Hampton, 1983b). In recent years, however, information is accumulating about the retention of nonpersistently transmitted viruses in aphid vectors for hours during long-distance flights. Hampton and Weber (1983a) found pea streak virus to retain infectivity for at least 2 hours of post-acquisition fasting. These results supported field observations of spread of the virus from lucerne fields to pea over a distance of 12.2 km. Skaf and Makkouk (1988) showed that *Aphis craccivora* remained viruliferous when starved for 6 hours after acquisition of a Syrian isolate of bean yellow mosaic virus. For the related maize dwarf mosaic potyvirus, such retention for over 21 h has been reported for *Schizaphis graminum* (Berger *et al.*, 1987). For many years, this virus was confined to southern states of the USA, and a sudden and dramatic epidemic in sweet corn in Minnesota in 1977 was associated with a weather pattern allowing long-distance aphid transport from south to north (Zeyen *et al.*, 1987).

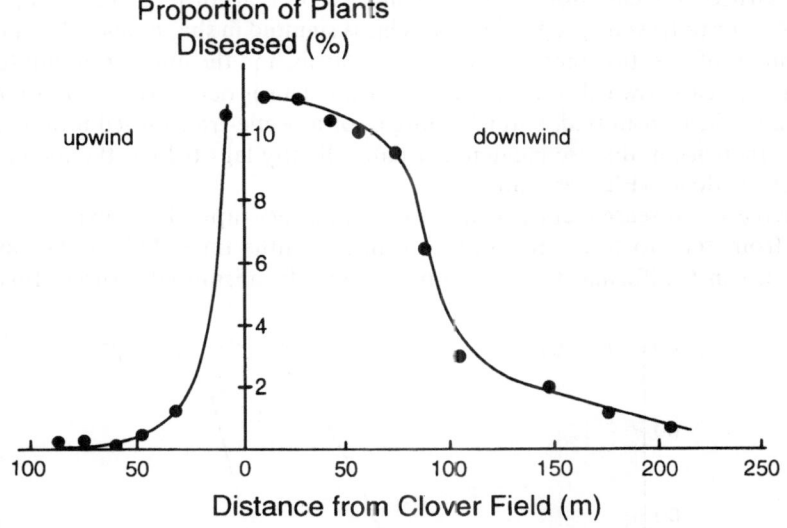

Figure 4. Short-distance spread of nonpersistently aphid-transmitted bean yellow mosaic potyvirus from red clover into a *Phaseolus* bean field as influenced by prevailing wind direction. (After Hampton, 1967.)

Although movement of virus into a crop along a gradient is usually associated with nonpersistent transmission (e.g., Hampton, 1967, 1983b), persistently transmitted viruses also may move in from the edge, if infection comes from an adjacent source, as was the case with subterranean clover stunt virus in Australia moving from faba bean into French bean. This was ascribed to large numbers of aphids walking onto bean plants while these were germinating (Garrett and Mclean, 1983). Erratic distribution of initially infected plants throughout a crop with no obvious gradient of infection is a good indication of introduction from a distant source, irrespective of the type of transmission. With both types, secondary spread often is along gradients in clusters, most likely by wingless aphids that cover short distances only. In Australian experiments with faba bean, the nonpersistently transmitted bean yellow mosaic virus and the persistently transmitted subterranean clover red leaf virus were both found to spread at similar rates from a central source, but they differed in their dependence upon vector colonization on the virus source plants. Spread of the first virus was attributed mainly to migrating alate aphids, and of the second virus by local movement of colonizing aphids (Jayasena and Randles, 1984).

Time

In crops, incidence of infected and diseased plants increases with time. With most viruses, increase during the growing season is according to a "compound interest" or polycyclic progress curve. This is sigmoid in shape, since the number of sources of infection increase during the season, and the number of uninfected plants decrease towards the end of the season. This is demonstrated in Figure 5 for an aphid-transmitted, and in Figure 6 for a beetle-transmitted food legume virus. Increase in disease incidence usually slightly lags behind the increase in vector incidence (Figures 5 and 6).

   Increase in disease incidence in crops may be very rapid, if not explosive, and lead from zero to total crop infection in 3 months time. This is the case in chickpea in California with a number of aphid-transmitted viruses (Bosque-

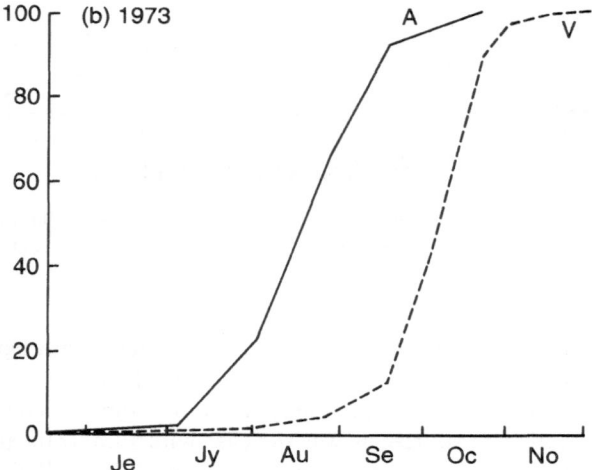

Figure 5. Increase in incidence of subterranean clover red leaf luteovirus (*V*) and its aphid vector (*A*) *Aulacorthum solani* in a faba bean crop in Tasmania. (After Johnstone and Rapley, 1979, redrawn by Thresh, 1983.)

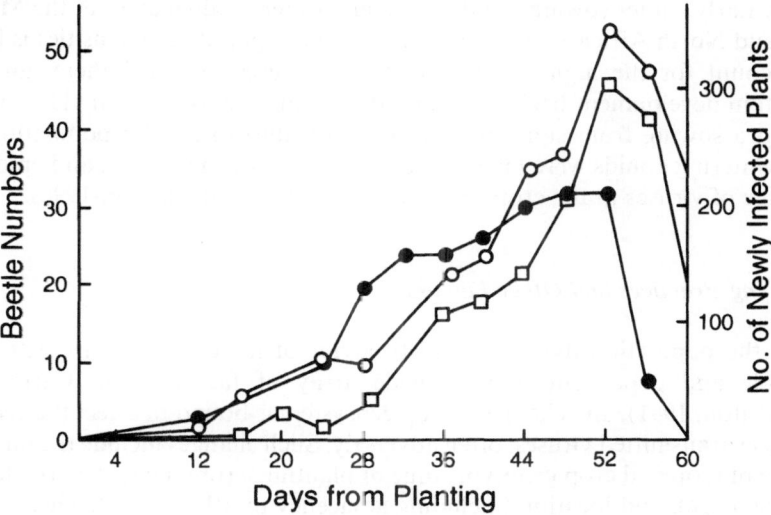

Figure 6. Increase in incidence of cowpea mosaic comovirus as indicated by number of newly infected plants (open squares) and its beetle vectors *Diabrotica balteata* (open rounds) and *Cerotoma ruficornis rogersi* (filled rounds) in cowpea. (After Valverde *et al.*, 1982, from Gámez and Moreno, 1983.)

Pérez and Buddenhagen, 1990), but also holds for beetle-transmitted viruses (Figure 6). Speed of epidemic build-up in a crop greatly depends upon crop susceptibility determining the build-up of inoculum pressure. Perennial crops, such as fodder (forage) legumes that may host viruses of annual legumes, often do so without symptoms, and this may be due to some degree of resistance impeding virus multiplication. In such crops, incidence may build up more slowly than in highly susceptible annual legumes, and will do so through the years. In the northwestern USA pea production areas, pea streak and alfalfa mosaic viruses were not detected in 2 year old stands of lucerne, but all 5 year-old fields contained both viruses, with at least 40% of the plants infected with the first virus. Lucerne fields, particularly in eastern Washington, USA, remain productive for 15 to 20 years (Hampton and Weber, 1983b).

Time of cropping may greatly determine the risk of infection with respect to the presence of both sources of virus infection and vector infestation. Annual food legume crops sown in spring in areas with perennial fodder legumes, especially lucerne, are bound to be subject to invasion by large numbers of aphids from the perennial legume early during the development of the annual crop, particularly when the aphids are forced to move by early cutting of the fodder crop (Hampton, 1983b). In Germany, pea enation mosaic virus was found to move in the spring into pea and faba bean fields from overwintering crimson clover and vetch, and thence to late crops of both, and to sweet pea, from which winter legumes were infected in turn to complete the cycle (Quantz,

1961). Early winter sowing of faba bean and other food legumes in the Middle East and North Africa at times of still high aphid population densities is likely to account for the high incidences of luteoviruses recorded there, and for symptom development beginning early during plant development. The shift of chickpea sowing from summer to winter in California, when populations of overwintering aphids were high, has been blamed for the sudden and epidemic upsurge of viruses in the crop (Bosque-Pérez and Buddenhagen, 1990).

Cropping Practices and Other Factors

Since the population dynamics and behavior of insect vectors are extremely variable and dependent upon a wide array of factors (e.g., Carter and Harrington, 1991), any change in crop ecosystem is bound to affect the ecology of insect-transmitted viruses, often adversely. Such changes include the farmer's choice of crop and crop genotype, time of planting, cropping system (including crop acreage), and location (including adjacency to other crops). Their effects on individual epidemiological factors do not need discussion here.

Mention should be made briefly of the consequences of the introduction of new vector insects or biotypes of them at places where they did not occur before. Vectors may move long distances naturally, as in jet stream air currents. They may come along also in or on transport vehicles, including airplanes and with cargo, especially planting materials such as ornamentals, and may become established in new environments. The introduction of the western flower thrips, *Frankliniella occidentalis*, into many new regions, such as the northern USA and Canada and several northern European countries where it has established itself well in greenhouses, and southern European countries where it is already widespread in the open, is causing serious concern. It has led to an upsurge of the extremely polyphagous tomato spotted wilt virus in many crops (Cho *et al.*, 1989). The natural host range of this virus includes legumes, such as faba bean, in which it was already quite common in Australia prior to 1947 (Stubbs, 1947), pea in Brazil (Reifschneider *et al.*, 1989), and white clover in Canada (Paliwal, 1974). The green pea aphid, *Acyrthosiphon pisum*, was introduced into Tasmania in 1980 and has displaced *Aulacorthum solani*, as the major component of the aphid fauna there and as a vector of soybean dwarf luteovirus (Johnstone *et al.*, 1984).

The complexity of the ecology of insect-transmitted viruses is demonstrated by the finding in cage experiments that a natural aphid enemy, *Coccinella californica*, made the green pea aphid move more between plants, which resulted in a higher incidence of bean yellow mosaic virus (Roitberg and Myers, 1978). Forecasting of epidemic developments, therefore, remains risky.

Control

Disease control is the final goal of ecological and epidemiological studies of pathogens. Apart from those measures more directly addressing the virus or its sources of infection (Bos *et al.*, 1988), the measures directed towards insect vectors concern (1) avoidance of vectors or reduction of their incidence, (2) breeding for vector resistance, or, (3) integrated approaches.

Avoidance of Vectors or Reduction of their Incidence

Important overwintering sources of virus infection and of vector infestation, such as clover fields, can be avoided through spatial isolation. With nonpersistent viruses, such as bean yellow mosaic virus, a few hundred meters may suffice (Hampton, 1967), but 1000 m is better (Schmidt *et al.*, 1979). However, in eastern Washington, severe attack by pea streak virus, although nonpersistently aphid-borne, followed massive pea aphid invasions of unknown origin, in one instance lucerne fields as far away as 12.2 km (Hampton and Weber, 1983a). Persistently transmitted viruses can be carried from clover, such as bean leafroll virus from lucerne, over very long distances, making it difficult to evade the source of vector infestation and virus infection.

Manipulation of sowing dates may help avoid peaks of insect flights. In the Netherlands, nonpersistently aphid-transmitted viruses seldom create problems in annual legume crops, unless sown late to extend the harvesting period for industrial processing. In Australia, yield losses in faba bean due to subterranean clover red leaf virus in 1972/73 were 21, 30, 61, and 8% in plots sown in May, July, September, and November, respectively. The increases in disease incidence were parallelled by increases in the infestation by *Aulacorthum solani*, the aphid vector of the virus. Yield losses were least after sowing in November when seedlings emerged after the peak of the aphid flight had declined (Johnstone and Rapley, 1979). In East Germany (former DDR), incidence of insect-transmitted viruses in faba bean was considerably less in fields sown in February and early March, than in those sown in late March or April (Schmidt *et al.*, 1979).

Crops grown in dense stands are generally claimed to have less infection by insect-transmitted viruses than more open crops. However, Way and Heathcote (1966) found more winged *Aphis fabae* on narrowly spaced faba bean plants, and in 1 year also found more bean leafroll virus in narrow rows. Aphid landing rates were reduced in soybean by increasing crop canopy closure. Higher crop densities were advised as part of an integrated control of the virus, but some aphid species were caught more abundantly in a closed soybean canopy than in an open one (Halbert and Irwin, 1981).

With nonpersistently transmitted viruses, taller barrier crops, on which incoming aphids tend to alight first and rid their mouthparts of contaminant virus, have been found to be effective. Introduction of bean yellow mosaic virus into yellow lupin was considerably less when plots were surrounded by narrow strips of oats (Corbett and Edwardson, 1957), but this may be impracticable for large scale cropping. Barrier crops and intercropping of a legume (cowpea) with a taller crop, such as maize or cassava, also have been found effective with cowpea severe mosaic comovirus, a beetle-transmitted legume virus in the tropics (for literature see Gámez and Moreno, 1983). In Kenya, mixed cropping of *Phaseolus* bean with maize showed less incidence of bean common mosaic potyvirus (van Rheenen *et al.*, 1981).

Mulching of the soil around crop plants, as with grey or white plastic sheets or aluminum strips, has been found higly effective in repelling aphids from vegetable crops and limiting virus incidence (Loebenstein *et al.*, 1975; Loebenstein and Raccah, 1980; Lecoq and Pitrat, 1983). A straw mulch repelled whiteflies and limited infection of tomato by tomato yellow leafcurl geminivirus (Cohen *et al.*, 1974), but mulches are expensive and their effect tends to decrease as the plants cover the mulch. More promising are polypropylene sealed fiber fleeces like Lutrasil to protect crops from probing and feeding by aphids, and, in potato, complete protection against transmission of potato virus Y potyvirus and potato leafroll virus was obtained (Harrewijn *et al.*, 1991). Their application is likely to be limited to high value crops other than legumes, or to experimental plots only.

Use of insecticides to directly reduce vector populations has had a limited effect on virus transmission. This effect greatly depends upon the type of transmission and whether infection is from outside or within crop sources.

Systemic insecticides will not reduce the introduction of persistently transmitted viruses into crops because they have already been introduced into plants before the insects are killed. Secondary spread of such viruses, however, from infected plants within a treated crop will be reduced. In Tasmania, incidence of infection of faba bean by subterranean clover red leaf virus at harvest varied from 13 to 23% in plots sprayed 3 to 5 times with demeton-S-methyl compared with 31 to 84% in unsprayed plots (Johnstone and Rapley, 1981). Pyrethroids, such as deltamethrin which has a rapid knock-down effect on insects, were later shown to also reduce primary infections by incoming viruliferous aphids in faba bean crops (Johnstone, 1984).

Common insecticides have been found to have an adverse effect on the spread of nonpersistently transmitted viruses by causing restless feeding behavior of vectors. Large area aphid control, leading to an overall reduction in aphid population density, however, may reduce infections by the nonpersistently transmitted viruses. Three year trials over large areas with insecticide application by aircraft reduced the average infection of faba bean crops in East Germany (former DDR) by bean yellow mosaic, pea enation mosaic, and bean leafroll viruses by 63, 72, and 71%, respectively (Schmidt *et al.*, 1977). Promising results in the control of aphid-transmitted viruses have more recently been obtained in nonlegume crops with pyrethroids (Gibson *et al.*, 1982; Asjes, 1985), but the widespread use of such broad-spectrum insecticides has caused concern about possible hazardous effects on beneficial insects and possible selection for insecticide resistance. Promising results have also been achieved in the laboratory with aphid alarm pheromone derivatives, which inhibit acquisition as well as inoculation of nonpersistently transmitted potato virus Y and semipersistently transmitted beet yellows closterovirus (Gibson *et al.*, 1984). They have more recently also been found effective in field tests against barley yellow dwarf virus (Dawson *et al.*, 1988).

Within-crop spread by beetle vectors of broad bean stain and true mosaic viruses in faba bean crops grown from infested seed lots can be checked by applying insecticides, such as fenitrothion and malathion (Cockbain, 1980). In cowpea with the beetle-transmitted cowpea severe mosaic virus, carbofuran applied to the seed at sowing time was less effective than the use of barrier crops (for references see: Gámez and Moreno, 1983). The spread of broad bean stain and true mosaic viruses by weevil vectors also can be checked with fenitrothion and malathion, but it was found to be more satisfactory to ensure that crops are grown from seed that is free, or as free as possible, from infection (Cockbain, 1980).

Spraying with mineral oils to prevent the introduction of virus into plants once they have landed on them has only found limited application for nonfood crops. Application is costly, reduces yield, and requires repeated application to extend protection to newly developing young sprouts.

Breeding for Insect Vector Resistance

Currently, resistance to insects is an important aim in breeding programs to reduce insecticide application for preventing or reducing insect feeding damage to crops. Much attention has been paid to resistance to aphids (Gibson and Plumb, 1977; Klingauf, 1982). Furthermore, it is often assumed that plants resistant to a viruliferous insect would be less subject to infection by the vectored virus. Some examples have already been given of reduced virus disease incidence in legume cultivars resistant to the vectors. However, deliberate breeding for vector resistance with the aim of reducing virus spread (Kennedy, 1976; Jones, 1987) is not yet widely practiced, and in this respect relatively little attention has been paid to legume crops. The objective of breeding for vector resistance is to reduce the incidence of infected plants rather than to diminish the effect of the virus on infected plants.

Breeding for vector resistance based on reduced host palatability [as expressed by nonpreference (antixenosis) or even antibiosis] to the vector is likely to have more effect on persistently transmitted viruses than on nonpersistently transmitted ones. Aphids may equally probe unpalatable and host genotypes. Hence, they may transfer nonpersistently transmitted viruses to and from both equally well, but they may not be able to do so with persistently transmitted viruses that require feeding on phloem tissue. It is often noncolonizing aphids that spread the nonpersistent viruses. Resistance in plants to virus vectors may even prematurely lead to vector dispersal and thereby influence secondary virus spread (Kennedy and Kishaba, 1977). In fact, blackeye cowpea mosaic potyvirus has been found to be more readily transmitted on cowpea genotypes resistant to *Aphis craccivora* than on nonresistant ones. On resistant genotypes, they were more restless and probed more frequently and for shorter durations (Atiri *et al.*, 1984). There are several crop species which are poorly colonized by aphids, but virus spread in them may be extensive. However, plant characteristics that decrease aphid population development are likely to reduce the rate of virus spread within crops if there are relatively few incoming aphids. This was found for fields of red clover Dollard, which is less preferred than Wegener and less suitable for multiplication of *Acyrthosiphon pisum* (Wilcoxson and Peterson, 1960). It suggests the need to include perennial fodder legumes in a program of breeding for vector resistance to control virus diseases in annual legumes.

Some plant characteristics, other than reduced palatability, that might be exploited in breeding for vector resistance, are (1) leaf color, (2) leaf glandularity, including the exudation of sticky material, and (3) dense leaf pubescence. Such genotype characteristics that deter probing or reduce movement are likely to be more effective for the control of nonpersistently transmitted viruses than of persistently transmitted ones. Despite well-documented responses of insects to color, no information exists as to the exploitation of leaf color in breeding of legumes for vector resistance. Heavy leaf pubescence reduced field spread of soybean mosaic virus in soybean (Gunasinghe *et al.*, 1988). Tomato genotypes with glandular hairs, such as *Lycopersicon pennellii*, *L. hirsutum*, and *L. hirsutum* forma *glabratum*, were less infected by tomato yellow leafcurl virus transmitted by whiteflies (Berlinger and Dahan, 1987), but in California, five chickpea cultivars, that differed considerably in amount of glandular hairs and in aphid colonization on them, showed only little difference in incidence of virus infection. All aphid species were able to survive on chickpea cultivars with abundant glandular exudate for 24 to 36 h, sufficient for transmission to occur (Bosque-Pérez and Buddenhagen, 1990). Wild potato was found to repel aphids by the release of aphid alarm pheromone (Gibson and Pickett, 1983). One might speculate about the possibility of protecting plants against viruliferous aphids by transferring to them the genes for producing the aphid alarm pheromone.

Breeding for resistance to specific aphid vectors of nonpersistently transmitted viruses may not be effective in nature since these viruses can be transmitted by a wide range of aphid species, unless mechanisms of resistance, such as glandularity operating against all of them, can be exploited. For example, the resistance of muskmelon in France to *Aphis gossypii*, also preventing its colonization, is ineffective against other aphid species, including *Myzus persicae*, which are efficient vectors of cucumber mosaic virus (Lecoq *et al.*, 1980).

Search is underway to find novel approaches of controlling virus diseases by use of molecular techniques. One way could be the obstruction of luteovirus circulation in an aphid vector by virus-specific anti-idiotypic antibodies (that is, antibodies having surface structures identical to vector membrane-specific epitopes of the virus) acquired by the vector together with the virus from plants genetically engineered for the production of such antibodies (van den Heuvel, 1991). Other attempts a long way off the beaten track concern modification of the helper component (HC) protein of potyviruses into an inactive form and making plants transgenic for it.

Integrated Approaches

With a few exceptions, there is no single way to solve virus problems and effectively control the diseases they cause. Vector control is only part of the overall strategy to deal with viruses in crops and reduce their adverse effects. Vector control in itself provides a variety of approaches (Maelzer, 1986), often needed in combination to obtain acceptable results. Control of cucumber mosaic virus in muskmelon in the south of France could neither be acceptably achieved with resistant crop genotypes alone because resistance was only partial, nor by cultural hygiene meant to reduce aphid movement into the crops by mulching and to reduce the number of virus-infected weeds near plantings. However, adequate protection was obtained by a combination of all three measures (Figure 7; Lecoq and Pitrat, 1983). In the USA, it was found that in melons slight vector resistance, although not sufficient alone, could, in combination with (slight) virus resistance, significantly improve virus control (Moyer *et al.*, 1985; Romanow *et al.*, 1986).

Management of virus diseases in food legumes must include measures outside these crops, even when vectors are being addressed. The partial resistance of

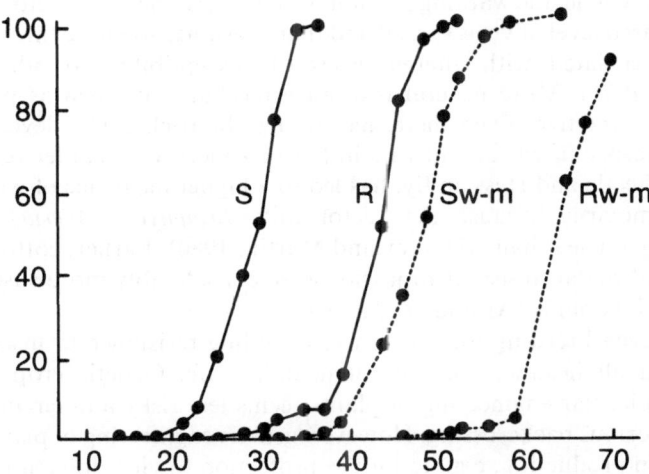

Figure 7. Effect of breeding for (partial) resistance to cucumber mosaic cucumovirus and of resistance in combination with mulching and weed control in melon. *S*, susceptible cultivar; *R*, resistant cultivar; *w*, weeding; *m*, mulching. (After Lecoq and Pitrat, 1983.)

Dollard red clover to the pea aphid and the ensuing low incidence of pea mosaic virus in the crop (Wilcoxson and Peterson, 1960), suggest the feasibility of deliberate breeding of perennial legumes for such vector resistance to reduce population densities of virus vectors of annual legumes and to restrict infection pressure in an area. This would result from the combined effect of reducing inoculum and vector pressures. Swenson and Hagedorn (1974) have used this example and endeavors to breed for aphid resistance in lucerne as a possible solution to the pea aphid problem in pea crops, to suggest more emphasis on breeding for vector resistance in clovers as a probable management of aphid-borne viruses of annual legumes. Such resistance would not have to be complete to be useful. Partial resistance would decrease the developmental rate of the aphid but not that of the natural enemies, permitting these to overtake the aphid population sooner on resistant than on susceptible cultivars. Methods which aim at reducing infection pressure in the crop environment rather than at introducing extreme genetic crop resistance may also reduce selection pressure on the vector and virus to develop new vector biotypes and new strains of the virus.

Concluding Remarks

Since the ecology of viruses, including that of their vectors, is highly complex, we should be aware of the fact that any change in the ecosystem, even when meant to control a special virus or other pathogen, may create niches for others to come to the fore. Developments are hard if not impossible to predict. Since breeding for resistance, including resistance to insects that may act as vectors of viruses, is in vogue, the warning still holds that "resistance to a vector may result in an increased level of virus spread and that resistance to one arthropod species may be associated with altered levels of susceptibility to other species" (Kennedy, 1976). More general resistance mechanisms, such as pubescence, seem most attractive. Even there, nature may be tricky. The development of such pubescence in wheat cultivars in North America, to confer resistance to cereal leaf beetle and Hessian fly, has led to a higher incidence of wheat streak mosaic rymovirus because the vector mite *Eriophyes (Aceria) tulipae* is attracted by the leaf hairs (Harvey and Martin, 1980). Earlier, cotton cultivars in India and Sudan resistant to jassids were considerably more susceptible to whiteflies and aphids (Arnold *et al.*, 1976).

Hence, even breeding for resistance, including resistance to insect vectors, may not in all instances be environmentally safe. Genetic crop upgrading through molecular engineering of plants seems less risky and circumvents the need of laborious backcrossing. However, genetic engineering of plants for viral coat-protein production, e.g., to induce protection to virus infection, may also entail risks. A plant-produced luteovirus protein may heterologously encapsidate the nucleic acid of other luteoviruses against which it does not protect the plant, but which thence may be spread by aphids that previously were not able to do so or were poor vectors. Such futuristic approaches therefore have to be dealt with carefully.

Finally, agricultural practice continues to exhibit immense complexity of agro-ecosystems, including viruses and their vectors, and continues to show that these systems are highly dynamic and will remain so. Insect vectors are only part of the system. Vector control and genetic upgrading of crops by breeding and other more sophisticated techniques are only two of a series of measures of crop improvement. Continuing collaboration between virologists, entomologists, mycologists, nematologists, plant breeders, and agronomists is therefore needed to improve disease and pest control for the sake of crop improvement and to cope with new problems as they emerge.

References

Arnold, M. H., Innes, N. L. and Brown, S. J. 1976. In: *Agricultural Research for Development*, pp. 175–195 (ed. M. H. Arnold). Cambridge, UK: Cambridge University Press.

Ashby, J. W., Teh, P. B. and Close, R. C. 1979. *New Zealand Journal of Agricultural Research* 22: 361–365.

Asjes, C. J. 1985. *Crop Protection* 4: 485–493.

Atiri, G. I., Ekpo, E. J. A. and Thottappilly, G. 1984. *Annals of Applied Biology* 104: 339–346.

Berger, P. H., Zeyen, R. J. and Groth, J. V. 1987. *Annals of Applied Biology* 111: 337–344.

Berlinger, M. J. and Dahan, R. 1987. *Insect Science and Its Application* 8: 783–784.

Bos, L. 1983. *Introduction to Plant Virology*, 160 pp. London/New York: Pudoc, Wageningen/Longman

Bos, L. 1989. In: *Introduction of Germplasm and Plant Quarantine Procedures*, pp. 19–29 (eds. Abdul Wahid Jalil *et al.*). PLANTI Proceedings No. 4. Serdang, Selangor, Malaysia: PLANTI.

Bos, L., Hampton, R. O. and Makkouk, K. M 1988. In: *World Crops: Cool Season Food Legumes*, pp. 591–615 (ed. R. J. Summerfield). Dordrecht: Kluwer Academic Publishers.

Bos, L., Mandersloot, H. J., Vader, F. and Steenbergen, B. 1989. *Netherlands Journal of Plant Pathology* 95: 225–240.

Bosque-Pérez, N. A. and Buddenhagen, I. W. 1990. *Plant Disease* 74: 372–378.

Carter, N. and Harrington R. 1991. *Advances in Disease Vector Research* 7: 19–51.

Cho, J. J., Mau, R. F. L., German, T. L., Hartmann, R. W., Yudin, L. S., Gonsalves, D. and Provvidenti, R. 1989. *Plant Disease* 73: 375–383.

Cockbain, A. J. 1980. In: *Vicia faba. Feeding Value, Processing and Viruses*, pp. 297–308 (ed. D. A. Bond). Brussels-Luxembourg: ECSC, EEC, EAEC.

Cockbain, A. J., Cook, S. M. and Bowen, R. 1975. *Annals of Applied Biology* 81: 331–339.

Cockbain, A. J., Jones P. and Woods, R. D. 1986. *Annals of Applied Biology* 108: 59–69.

Cohen, S., Melamed-Madjar, V. and Hameiri, J. 1974. *Bulletin of Entomological Research* 64: 193–197.

Corbett, M. K. and Edwardson, J. R. 1957. *Proceedings Soil & Crop Science Society of Florida* 17: 294–301.

Davis, R. F. and Shifriss, O. 1983. *Plant Disease* 67: 379–380.

Dawson, G. W., Griffiths, D. C., Pickett, J. A., Plumb, R. T., Woodcock, C. M. and Zhong-Ning, Z. 1988. *Pesticide Science* 22: 17–30.

De Bokx, J. A. and Piron P. G. M. 1990. *Netherlands Journal of Plant Pathology* 96: 237–246.

Duffus, J. E., Falk, B. W. and Johnstone, G. R. 1990. In: *World Perspectives on Barley Yellow Dwarf*, pp. 86–104 (ed. P. A. Burnett). Mexico, D. F.: CIMMYT.

Evans, A. C. 1954a. *Annals of Applied Biology* 41: 189–206.

Evans, A. C. 1954b. *Nature*, London 173: 1242–1243.

Fortass, M. and Bos, L. 1991. *Netherlands Journal of Plant Pathology* 97: 369–380.
Frison, E. A., Bos, L., Hamilton, R. I., Mathur, S. B. and Taylor, J. D. (eds.). 1990. *FAO/IBPGR Technical Guidelines for the Safe Movement of Legume Germplasm*, 88 pp. Rome: FAO/IBPGR.
Gámez, R. and Moreno, R. A. 1983. In: *Plant Virus Epidemiology*, pp. 103–113 (eds. R. T. Plumb and J. M. Thresh). Oxford, UK: Blackwell Scientific Publishers.
Garrett, R. G. and McLean, G. D. 1983. In: *Plant Virus Epidemiology*, pp. 199–209 (eds. R. T. Plumb and J. M. Thresh). Oxford, UK: Blackwell Scientific Publishers.
Gergerich, R. C. and Scott., H. A. 1991. *Advances in Disease Vector Research* 8: 1–13.
Gibson, R. W. and Pickett, J. A. 1983. *Nature*, London 302: 608–609.
Gibson, R. W., Pickett, J. A., Dawson, G. W., Rice, A. D. and Stribley, M. F. 1984. *Annals of Applied Biology* 104: 203–209.
Gibson, R. W. and Plumb, R. T. 1977. In: *Aphids as Virus Vectors*, pp. 473–500 (eds. K. F. Harris and K. Maramorosch). New York: Academic Press.
Gibson, R. W., Rice, A. D. and Sawicki, R. M. 1982. *Annals of Applied Biology* 100: 49–54.
Grylls, N. E. and Butler, F. C. 1959. *Australian Journal of Agricultural Research* 10: 145–159.
Gunasinghe, W. B., Irwin, M. E. and Kampincier, G. E. 1988. *Annals of Applied Biology* 112: 259–272.
Halbert, S. E. and Irwin, M. E. 1981. *Annals of Applied Biology* 98: 15–19.
Halbert, S. E., Irwin, M. E. and Goodman, R. M. 1981. *Annals of Applied Biology* 97: 1–9.
Hampton, R. O. 1967. *Phytopathology* 57: 476–481.
Hampton, R. O. 1983a. *Seed Science & Technology* 11: 535–546.
Hampton, R. O. 1983b. *Plant Disease* 67: 1306–1310.
Hampton, R. O. and Weber, K. A. 1983a. *Plant Disease* 67: 305–307.
Hampton, R. O. and Weber, K. A. 1983b. *Plant Disease* 67: 308–310.
Harrewijn, P., Den Ouden, H. and Piron, P. G. M. 1991. *Entomologia Experimentalis et Applicata* 58: 101–107.
Harris, K. F. 1977. In: *Aphids as Virus Vectors*, pp. 165–220 (eds. K. F. Harris and K. Maramorosch). New York: Academic Press.
Harris, K. F. 1983. *Advances in Virus Research* 28: 113–140.
Harris, K. F. and Maramorosch, K. (eds.). 1977. *Aphids as Virus Vectors*, 559 pp. New York: Academic Press.
Harris, K. F. and Maramorosch, K. (eds.). 1980. *Vectors of Plant Pathogens*, 467 pp. New York: Academic Press.
Harrison, B. D. and Murant, A. F. 1984. In: *Vectors in Virus Biology*, pp. 1–36 (eds. M. A. Mayo and K. A. Harrap). London: Academic Press.
Harrison, B. D. and Robinson, D. J. 1988. *Philosophical Transactions Royal Society, London B* 321: 447–462.
Harvey, T. L. and Martin, T. J. 1980. *Journal of Economic Entomology* 73: 225–227.
Horn, N. M., Reddy, S. V., Roberts, I. M. and Reddy, D. V. R. 1992. Chickpea chlorotic dwarf virus, a new leafhopper-transmitted geminivirus of chickpea in India (In preparation).
Jayasena, K. W. and Randles, J. W. 1984. *Annals of Applied Biology* 104: 249–260.
Johnstone, G. R. 1984. *Australasian Plant Pathology* 13: 55–56.
Johnstone, G. R. and Rapley, P. E. L. 1979. *Annals of Applied Biology* 91: 345–351.
Johnstone, G. R. and Rapley, P. E. L. 1981. *Annals of Applied Biology* 99: 135–141.
Johnstone, G. R., Ashby, J. W., Gibbs, A. J., Duffus, J. E., Thottappilly, G. and Fletcher, J. D. 1984. *Netherlands Journal of Plant Pathology* 90: 107–115.
Jones, A. T. 1978. *Annals of Applied Biology* 88: 137–144.
Jones, A. T. 1987. *Annals of Applied Biology* 111: 745–772.
Jones, R. A. C. and Proudlove, W. 1991. *Annals of Applied Biology* 118: 319–329.
Katul, L. 1992. Serologische und molekularbiologische Charakterisierung des bean leaf roll virus (BLRV) und des faba bean necrotic yellows virus (FBNYV), 115 pp. *Dissertation Georg-August-Universität, Göttingen / Biologische Bundesanstalt für Land und Forstwitschaft, Braunschweig.*
Kennedy, G. G. 1976. *Environmental Entomology* 5: 827–832.
Kennedy, G. G. and Kishaba, A. N. 1977. *Journal of Economic Entomology* 70: 407–410.

Khetarpal, R. K. and Maury, Y. 1987. *Agronomie* 7: 215–224.
Klein, R. E., Larsen, R. C. and Kaiser, W. J. 1991. *Plant Disease* 75: 1186.
Klingauf, F. A. J. 1982. In: *Faba Bean Improvement*, pp. 285–295 (eds. G. Hawtin and C. Webb). The Hague, The Netherlands: Martinus Nijhoff.
Lecoq, H., Labonne, G. and Pitrat, M. 1980. *Annales de Phytopathologie* 12: 139–144.
Lecoq, H. and Pitrat, M. 1983. In: *Plant Virus Epidemiology*, pp. 169–176 (eds. R. T. Plumb and J. M. Thresh). Oxford, UK: Blackwell Scientific Publishers.
Lim, W. L. and Hagedorn, D. J. 1977. In: *Aphids as Virus Vectors*, pp. 237–251 (eds. K. F. Harris and K. Maramorosch) New York: Academic Press.
Loebenstein, G., Alper, M., Levy, S., Palevitch, D. and Menagem, E. 1975. *Phytoparasitica* 3: 43–53.
Loebenstein, G. and Raccah, B. 1980. *Phytoparasitica* 3: 221–235.
Maelzer, D. A. 1986. In: *Plant Virus Epidemics: Monitoring, Modelling and Predicting Outbreaks*, pp. 483–512 (eds. G. D. McLean, R. G. Garrett, and W. G. Reesink). Australia: Academic Press.
Makkouk, K. M., Bos, L., Azzam, O. I., Koumari, S. and Rizkallah, A. 1988a. *Arab Journal of Plant Protection* 6: 61–53.
Makkouk, K. M., Bos, L., Rizkallah, A., Azzam, O. I. and Katul, L. 1988b. *Netherlands Journal of Plant Pathology* 94: 195–212.
Maramorosch, K. and Harris, K. F. (eds.). 1979. *Leafhopper Vectors and Plant Disease Agents*, 654 pp. New York: Academic Press.
Matisová, J. 1971. *Acta Virologica* 15: 411–420.
McLean, G. D., Garrett, R. G. and Ruesink, W. G. (eds.). 1986. *Plant Virus Epidemics: Monitoring, Modelling and Predicting Outbreaks*, 550 pp. Sidney: Academic Press.
Mink, G. I. and Parsons, J. L. 1978. *Plant Disease Reporter* 62: 249–253.
Moyer, J. W., Kennedy, G. G. and Romanow, L. R. 1985. *Phytopathology* 75: 201–205.
Müller, H. J. 1964. *Entomologia Experimentalis et Applicata* 7: 85–104.
O'Loughlin, G. T. and Chambers, T. C. 1967. *Virology* 33: 262–271.
Ouffroukh, A. 1985. Contribution à la connaissance des viroses des plantes en Algérie. Inventaire des virus présents chez des legumineuses à longue cosse. Étude approfondie de deux maladies isolées de fève et de haricot, 91 pp. *Thèse Université Pierre et Marie Curie, Paris VI*.
Paliwal, Y. C. 1974. *Canadian Journal of Botany* 52: 1177–1182.
Plumb, R. T. 1983. In: *Plant Virus Epidemiology*, pp. 185–198 (eds. R. T. Plumb and J. M. Thresh). Oxford, UK: Blackwell Scientific Publishers.
Plumb, R. T. and Tresh, J. M. (eds.). 1983. *Plant Virus Epidemiology; the Spread and Control of Insect-borne Viruses*, 377 pp. Oxford, UK: Blackwell Scientific Publishers.
Quantz, L. 1961. *Mitteilungen aus der Biologischen Bundesanstalt für Land- und Forstwirtschaft Berlin-Dahlem* 104: 121–127.
Quantz, L. and Völk, J. 1954. *Nachrichtenblatt des Deutschen Pflanzenschutzdienstes (Braunschweig)* 6: 177–182.
Reddy, D. V. R., Wightman, J. A., Beshear, R. J., Highland, B., Black, M., Sreenivasulu, P., Dwivedi, S. L., Demski, J. W., McDonald, D., Smith Jr., J. W. and Smith, D. H. 1991. *ICRISAT Information Bulletin No.* 31, 20 pp.
Reifschneider, F. J. B., Cafe, A. C., Dusi, A. N. and Kitajima, E. W. 1989. *Tropical Pest Management* 35: 304–306.
Roitberg, B. D. and Myers, J. H. 1978. *Journal of Applied Ecology* 15: 775–779.
Romanow, L. R., Moyer, J. W. and Kennedy, G. G. 1986. *Phytopathology* 76: 1276–1281.
Ross, J. P. 1986. *Plant Disease* 70: 222–224.
Ruesink, W. G. and Irwin, M. E. 1986. In: *Plant Virus Epidemics: Monitoring, Modelling and Predicting Outbreaks*, pp. 295–313 (eds. G. D. McLean, R. G. Garrett and W. G. Ruesink). Australia: Academic Press.
Schmidt, H. E., Dubnik, H., Karl, E., Schmidt, H. B. and Kamann, H. 1977. *Nachrichtenblatt für den Pflanzenschutz in der DDR* 31: 247–250.
Schmidt, H. E., Karl, E., Rollwitz, W., Klein, W. and Kästner, H. F. 1979. *Archiv für Acker- und Pflanzenbau und Bodenkunde, Berlin* 23: 389–396.

Simons, J. N. 1958. *Phytopathology* 48: 265–268.

Skaff, J. S. and Makkouk, K. M. 1988. *Phytopathologia Mediterranea* 27: 133–137.

Snyder, W. C., Paulus, A. O. and Gold, A. H. 1956. *Phytopathology* 46: 27.

Sreenivasulu, P., Demski, J. W., Reddy, D. V. R., Naidu, R. A. and Ratna, A. S. 1991. *Plant Pathology* 40: 503–507.

Stubbs, L. L. 1947. *Journal Department of Agriculture Victoria* 45: 323–332.

Swenson, K. G. and Hagedorn, D. J. 1974. *Agricultural Experiment Station, Oregon State University, Corvallis, Station Bulletin* 615, 15 pp.

Thresh, J. M. 1983. *Philosophical Transactions of the Royal Society London B* 302: 497–528.

Valverde, R., Moreno, R. and Gámez, R. 1982. *Turrialba* 32: 29–32.

van den Heuvel, J. F. J. M. 1991. *Acquisition and Transmission of Potato Leafroll Virus by Myzus persicae; Quantitative Aspects.* Ph.D. Thesis, Wageningen, The Netherlands.

van Harten, A. 1983. *Potato Research* 26: 73–78.

van Rheenen, H. A., Hasselbach, O. E. and Muigai, S. G. S. 1981. *Netherlands Journal of Plant Pathology* 87: 193–199.

Waterhouse, P. M., Gildow, F. E. and Johnstone, G. R. 1988. *AAB Descriptions of Plant Viruses* 339, 9 pp.

Way, M. J. and Heathcote, G. D. 1966. *Annals of Applied Biology* 57: 409–423.

Wilcoxson, R. D. and Peterson, A. G. 1960. *Journal of Economic Entomology* 53: 863–864.

Zeyen, R. J., Stromberg, E. L. and Kuehnast, E. L. 1987. *Annals of Applied Biology* 111: 325–336.

Screening techniques and sources of resistance to parasitic angiosperms

J.I. CUBERO[1], A.H. PIETERSE[2], S.A. KHALIL[3] and J. SAUERBORN[4]

[1] Departamento de Genética, Universidad de Córdoba, Córdoba, Spain;
[2] Royal Tropical Institute, Rural Development Programme, Mauritskade 63, 1092 AD Amsterdam, The Netherlands;
[3] Field Crop Research Institute, Agricultural Research Center, Giza, Egypt, and
[4] Institut für Pflanzenproduktion in den Tropen und Subtropen, Universität Hohenheim, Stuttgart, Germany

Abstract

Parasitic angiosperms cause great losses in many important crops under different climatic conditions and soil types. The most widespread and important parasitic angiosperms belong to the genera *Orobanche*, *Striga*, and *Cuscuta*. The most important economical hosts belong to the Poaceae, Asteraceae, Solanaceae, Cucurbitaceae, and Fabaceae. Although some resistant cultivars have been identified in several crops, great gaps exist in our knowledge of the parasites and the genetic basis of the resistance, as well as the availability of *in vitro* screening techniques. Screening techniques are based on reactions of the host root or foliage. *In vitro* or greenhouse screening methods based on the reaction of root and/or foliar tissues are usually superior to field screenings and can be used with many species. To utilize them in plant breeding, it is necessary to demonstrate a strong correlation between *in vitro* and field data. The correlation should be calculated for every environment in which selection is practiced. Using biochemical analysis as a screening technique has had limited success. The reason seems to be the complex host-parasite interactions which lead to germination, rhizotropism, infection, and growth of the parasite. Germination results from chemicals produced by the host. Resistance is only available in a small group of crops. Resistance has been found in cultivated, primitive and wild forms, depending on the specific host-parasite system. An additional problem is the existence of pathotypes in the parasites. Inheritance of host resistance is usually polygenic and its transfer is slow and tedious. Molecular techniques have yet to be used to locate resistance to parasitic angiosperms. While intensifying the search for genes that control resistance to specific parasitic angiosperms, the best strategy to screen for resistance is to improve the already existing *in vitro* or greenhouse screening techniques.

F.J. Muehlbauer and W.J. Kaiser (eds.), Expanding the Production and Use of Cool Season Food Legumes, 333–345.
© 1994 *Kluwer Academic Publishers.*

Introduction

Parasitic angiosperms belong to several families. Their parasitism obviously has different origins and different biological mechanisms which enable them to extract nutrients from their hosts (Kuijt, 1969). Most of them are regarded more as "botanical curiosities" than authentic pests. Among the best known parasitic angiosperms are broomrapes (Orobanchaceae), witchweeds (*Striga* species in the Scrophulariaceae), dodders (*Cuscuta* species in the Convolvulaceae), mistletoes (Loranthaceae), and dwarf mistletoes (Viscaceae). There are others which are of no interest in agronomy at the present time, hence in plant breeding, since they either parasitize wild plants (e.g., the Hydnoraceae) or they do not attack important cultivated crops (*Tapinanthus* spp.). Certain parasites which attack cultivated crops are *still* not important on a global scale (e.g., *Alectra* and *Osyris* spp.).

There appears to be more interest in parasitic angiosperms because of their *weedy* nature than for their *parasitic* way of life. However, even though they are authentic weeds, they also may be considered authentic diseases because their tissues become interconnected with those of their hosts. Consequently, they should be included under the realm of plant pathology. Literature reviews confirm this dual biological nature. Their status shifts from being regarded as weeds by weed scientists to diseases by plant breeders and plant pathologists.

As they share characteristics of both weeds and pathogens, there is no incongruency in trying to control them by *cultural practices* (especially using herbicides), as well as by *genetic resistance*. More studies have been conducted on their weedy aspect than on their pathogenic nature. Control by agronomic practices have been reviewed (Cubero and Moreno, 1979; Ramaiah, 1987; Parker, 1991) and they continue to be studied by several research and extension groups worldwide. Sources of resistance and screening methods were discussed in previous papers (Cubero, 1986, 1991).

Within the scope of the present paper, *resistance* will be referred to as a protective action by the host which may be either passive or active. It will not be considered synonymous with immunity which was a common practice in Eastern European literature. *Resistance* is related to attacks and symptoms, whereas *immunity* implies the absence of symptoms. Another argument not to use *immunity* is its technical meaning in words such as *immumology* and *immunogenetics*. The term *tolerance* will be used when a host thrives and reproduces satisfactorily, even though it is severely infested. To accentuate the differences between microorganisms and parasitic angiosperms, the term *biotype, pathogenic biotypes*, or *pathotypes* will be used instead of *physiological race* for growth forms of a particular species varying in aggressiveness to a particular host plant.

Screening Methods

Characteristics of a Good Screening Index

A good index to screen for resistance would have to reveal a high correlation with field resistance and should be easy to record in and out of season (e.g., greenhouse). Its expression should be related to at least one important host resistance barrier. Ultimately, it should be based on "qualitative observations" (counts) rather than on "quantitative measurements".

Screening in "crops other than" cool season food legumes

Screening for resistance in hosts other than cool season food legumes was usually conducted under field or greenhouse conditions. The usual screening index was the number of broomrape shoots attached (either emerged or the total number) to individual host plants. Other indices included the total dry matter of the parasitic plants per host plant, per sown row or per unit area (Cubero, 1986, 1991).

Although several horticultural crops such as tomato and tobacco have been screened for resistance to *O. ramosa*, the best example is provided by sunflower/*O. cernua*. Selection of resistant cultivars of sunflower was made possible by Pustovoit's (1976) work in the former USSR at the beginning of this century. At the beginning of the program, host plants free from broomrape were selected. These plants were cultivated in naturally infested field plots whose infestation level was high by periodically adding 250,000 to 400,000 seeds of broomrape m^{-2}. In the off season, sunflower was cultivated in greenhouses and the roots were observed for *Orobanche* shoots after two to three months. By adjusting light, moisture and temperature, attached parasitic shoots could be counted within a month after emergence of the sunflower seedlings. The time required for the parasite to penetrate the roots of susceptible sunflower cultivars was 12 to 14 days after germination. Infection resulted in development of round yellow swellings which were not produced on roots of resistant plants.

The method described above was simple and very efficient. Its efficiency may be attributed to several factors: a systematic long term research program initiated in 1910; an excellent collaboration among different agricultural research stations, coordinated by Pustovoit (taking advantage of the rigid centralism of the former USSR); and last but not least, a shift from a quantitative (the number of broomrapes per host plant) to a qualitative index (the existence of root swellings produced by the penetration of parasitic haustoria).

An excellent example of a qualitative way (i.e., *yes or no*) of classifying resistant host plants is provided by cowpea/witchweed (*Vigna unguiculata/ Striga gesnerioides*). Work at IITA had produced some resistant cowpea lines, one of them (B301), discovered in Botswana, was resistant to all strains of *Striga*

(Aggarwal *et al.*, 1984, 1986; Parker and Polniaszek, 1990). Lane *et al.* (1991) developed an *in vitro* growth system for investigating attachment of *Striga gesnerioides* to cowpea roots. Roots of young cowpea plants were spread over glass fibre filter paper in a shallow plastic tray and germinated *Striga* seeds were then placed on the surface of the cowpea roots. An additional advantage of this system was that after testing, the plants could be transferred to soil and grown to maturity. Using the resistant lines and the technique just described, two different mechanisms of resistance were identified (Lane and Bailey, 1991; Lane *et al.*, 1991). Resistance was expressed in both cases after the parasite penetrated the host root. Resistance was not associated with reduced germination of the parasite's seeds. In cowpea line 58–57, a hypersensitive host reaction resulted in the death of the parasite within 2 to 3 days. In line B301, small tubercles formed by the parasite after penetration of host roots failed to grow larger than 1 to 2 mm in diameter. Resistance of cowpea line B359 to *Alectra* was also associated with this mechanism (Lane *et al.*, 1991). The experimental method used by these authors could be applied to temperate food legumes, as decribed below.

A hypersensitive reaction also was described to demonstrate the resistance of *Lycopersicon esculentum* var. *cerasiforme* to *Cuscuta campestris* (Al-Menoufi and Ashton, 1991). A similar mechanism has been suggested for green gram (*Vigna radiata*) lines resistant to *Cuscuta chinensis* (Rao and Rao, 1991).

Thus far, the use of host root stimulants to test for host resistance has been unsuccessful. Numerous reasons may explain this phenomenon. There are many substances involved in germination of seed of the parasite. These include stimulants as well as inhibitors which are present in the host root and seed of the parasite. In addition, the possible influence of endophytes and other soil microorganisms should be taken into account (Cezard, 1965, 1973a,b,c,d; Hiron, 1973; Whitney, 1979, 1986). Seeds of the parasite are affected by many chemicals and by conditioning treatments (Pavlista *et al.*, 1979a). Some substances produce antagonistic effects on parasite seed germination (Pavlista *et al.*, 1979b). Besides, crude root extracts (likely containing stimulants as well as inhibitors) or artificial stimulants (very probably not produced by the host, e.g., strigol was purified from cotton which is not susceptible to broomrape or witchweed) were used. Using root stimulants, two sorghum lines out of 14,000 tested at ICRISAT, India were highly resistant to *Striga*. One of these lines was already known to be resistant to *Striga*, while and the other produced large amounts of root stimulants.

Methods of screening plants under greenhouse conditions or *in vitro* in an environmental chamber were described by various authors. Parker *et al.* (1977) devised a method for *Striga*. An improved host-screening technique for testing sorghum for resistance to *Striga* was described by Vasudeva Rao *et al.* (1983). Sorghum plants grown in shallow pans developed a denser growth of roots which favored a higher frequency of *Striga* attachments. Subsequently, Parker and Dixon (1983) developed the so-called polybag technique. Host plants were grown on a layer of filter paper in a flattened, polyethylene bag. It was used to test sorghum for resistance to *Striga hermonthica*. The polybag technique was successful in identifying resistant lines. A new germination stimulant produced by sorghum roots, called "sorgoleone", was recently identified and isolated (Chang *et al.*, 1986; Netzly and Butler, 1986). Sorgoleone was used for *in vitro* sorghum screening by Hess *et al.* (1991). They found that sorgoleone production by different sorghum cultivars was rather uniform and that sorgoleone production by host roots did not correlate with the reaction of the host to *Striga* under field conditions. Sorghum seeds germinating on water agar containing preconditioned *Striga* seeds produced a much better correlation with field resistance. The interval between host root and the most distant witchweed seed that germinated was more rapidly measured (only 72 hours were needed) than the percentage of total germination of *Striga* seeds. This screening method measures many interactions between the host root and *conditioned* seeds of the parasite. To explain the failure when using sorgoleone *alone*, other stimulants must be present in sorghum roots (Hess *et al.*, 1991). Olivier *et al.* (1991) also worked with production of sorgoleone by sorghum roots and suggested that this substance was responsible for the resistance of some sorghum lines to *Striga* but not for others, such as IS-7777, where resistance seems to be the result of active defense reactions (successive layers of cellulose-rich material and perhaps phenolic compounds).

The main reason for setting up a rapid screening method in sunflower, sorghum and cowpea was the identification of highly resistant genotypes, which would permit further studies on mechanisms of resistance. Lack of identifying resistant genotypes is one of the reasons why there have been so many failures in setting up efficient screening methods for most host species.

Other host/parasite systems reveal the complexity of the resistance barriers. Sallé *et al.* (1991a) found that the mechanism of resistance in poplar (*Populus* spp.) to mistletoe (*Viscum album* L.) involves flavonoids as well as mechanical barriers. Sallé *et al.* (1991b) found that mechanical barriers were mainly responsible for resistance of three species of *Quercus* to mistletoe. Resistance was described in other species of herbs (reviews in Cubero, 1986, 1991) and trees (Scharpf and Roth, 1991), but mechanisms of resistance and, hence, reliable screening methods other than field data, are not described.

Screening Methods Used for Cool Season Food Legumes

The favorite index for resistance to broomrape, the primary parasitic weed of temperate food legumes, is the number of emerged shoots per host plant. In faba bean, there is a high correlation between the total number of attached broomrapes and those that have emerged (Cubero, 1986). Other indices are the number of broomrapes per unit area, the total weight of broomrapes per host plant, the dry matter of parasitic plants per host plant, and the height of the tallest parasitic shoot (Cubero and Hernández, 1991). Hanounik and Bisri (1991) suggested the rate of broomrape reproduction as an index of resistance. Plant pathologists would find this to be a suitable index of resistance in faba bean. In *Vicia sativa/O. crenata*, the number of emerged broomrape shoots per host plant was the most stable index compared to the number of broomrape shoots, and the weight and the height of parasitic plants per host plant (Gil *et al.*, 1982, 1984, 1987). Cubero (1983, 1991) and Cubero and Hernández (1991) obtained the same result for faba bean/*O. crenata*.

Most screening tests to select faba beans resistant to broomrape have been conducted under field conditions. The number of emerged shoots or total dry matter of the parasitic plants per host plant, per row, or per plot, as an index of infestation, are advantageous in that it gives an indication of field resistance and is relatively simple. A problem, however, is that cross breeding lines may be less tolerant to factors such as low temperature, drought, and disease. Consequently, plants may die in the field before resistance can be observed. Other disadvantages of field testing include heterogeneity in spread of the seed of the parasite in the soil, dependence on a growing season, and difficulty in simultaneous data recording of many genotypes. Even when all these factors are controlled, these indices cannot be used easily in greenhouse or growth chamber studies. Additionally, one generation per year is required, or even more if progeny testing is conducted.

Kukula and Masri (1985) who used polyethylene bags and attapulgus clay artificially infested with broomrape seeds counted the number of attached broomrapes as well as the number of germinated seeds of the parasite after 65 days. Pieterse *et al.* (1986) reported that with *Orobanche* the polybag technique as used for sorghum (see above) or modifications of this technique cannot be used successfully with faba bean. Faba bean plants died within a few days after germination possibly due to the lack of aeration. Better results were obtained with lentil whose roots appeared less susceptible to oxygen depletion. However, growth of lentil plants under these conditions was not optimal and most plants died after initial development of *Orobanche*. Consequently, other methods were tested. For faba bean and lentil, growing plants in vermiculite in nylon gauze bags provided encouraging results. There were, however, some disadvantages with this method. Firstly, host plant roots may grow through the gauze into adjacent bags and it was not possible to directly observe development of haustoria of the pathogen.

Linke *et al.* (1991) used a modification of this technique (plastic pots instead of nylon gauze bags) to determine the optimal broomrape seed density to develop a greenhouse screening method for chickpea. The only disadvantage of this method was the large greenhouse space required for extensive screening.

A method primarily aimed at testing seed viability also could be used to determine resistance (Aalders and Pieters, 1987). It involved growing faba bean plants on agar in glass test tubes that contained *Orobanche* seed. There were no problems with aeration of the root system which seems to be in contradiction with the results of Pieterse *et al.* (1986). Sauerborn *et al.* (1987), subsequently developed a technique for screening lentil genotypes using Petri dishes filled with clay. With the Petri dish technique, lentil seedlings were grown between glass filter paper that were sprinkled with *Orobanche* seed. Unfortunately, this method could not be used with faba bean. Finally, Linke and Vogt (1987) designed special plexiglass vessels for studying early development of *Orobanche* and *Striga* species on roots of host plants. The parasite/host interaction (up to a depth of 18 cm) could be observed for several weeks.

There are advantages and disadvantages to the various methods that use *in vitro* techniques. Host plant species differ considerably in their ability to grow under artificial conditions. Faba bean, in particular, is difficult to grow *in vitro* due to susceptibility of the roots to oxygen depletion. Theoretically, mass screening in the greenhouse or environmental chamber would be an ideal solution as changes in environmental factors would be negligible. On the other hand, the method should be relatively easy and reliable. Presently, it seems more advisable to begin with mass screening in the field, preferably in different regions. If promising genotypes are identified, additional screening could be conducted in the greenhouse or environmental chamber. *In vitro* techniques may be particularly helpful to clarify resistance mechanisms, as was demonstrated by Lane and Bailey (1991) and Lane *et al.* (1991) with the cowpea/*Striga gesnerioides* interaction.

Special attention needs to be given to the correlation between *in vitro* techniques and field resistance data. Data on sorghum screening using root exudates were discussed above. A similar example with faba bean was presented by Aalders and Pieters (1986, 1987). These authors found that faba bean lines showing acceptable field resistance did not exhibit differences in germination of seed of the parasite when compared to susceptible lines in *in vitro* trials. On the contrary, in a line (BPL 2210) without much field resistance, there was a low level of germination of *Orobanche* seed in *in vitro* tests. Research is needed to determine whether lines, such as BPL 2210, have specific genes for resistance. Faba bean roots stimulated up to 100% germination of *O. cernua* (which is a sunflower parasite) seeds when they were suspended in agar. However, under these *in vitro* conditions, faba bean roots induced less than 40% germination of *O. crenata* seeds, although sunflower broomrape does not parasitize faba beans under field conditions (Cubero *et al.*, 1992a). *O. cernua* does *not* parasitize faba bean under field conditions; it seems that to become a parsite requires something more than just to germinate under *in vitro* conditions.

The Search for a Screening Method

Some characteristics required for a reliable resistance index have been mentioned above. Swellings or limited necrotic areas on the root where the haustorium is penetrating, as well as phytoalexin production in roots after broomrape infection (as in sunflower) are important data for setting up a rapid, field resistance screening method. These qualitative reactions are usually associated with Mendelian segregations which would permit the detection of a qualitative rather than a quantitative trait.

In two temperate food legumes, faba bean and common vetch, the inheritance of resistance seems to be quantitative and not Mendelian as in cowpea. The resistance mechanism may be complex. Cubero *et al.* (1993) referred to different mechanisms as follows:

(1) Stimulating or inhibiting *Orobanche* seed germination;
(2) Attracting or rejecting parasitic haustoria;
(3) Permitting or disturbing haustorial penetration, even by mechanical means (i.e., radicular cortex structure);
(4) Permitting or preventing vascular connection between vessels of the host and parasite;
(5) Permitting or disturbing the internal flow of nutrients to the young parasitic tubercles; and
(6) Supplying of nonessential substances to the parasitic shoot.

As discussed above, mechanisms (1) and (3) seem to be present in sorghum with resistance to *Striga hermonthica*; (3) and (5), or (6), or both, were described for cowpea with resistance to *S. gesnerioides*; and (3) or (4) for sunflower with resistance to *O. cernua*. On the contrary, anatomical sections (3) of the faba bean resistant line VF 1071 and highly susceptible cultivar "Prothabon", as well as root sections from segregating materials obtained from crosses between these lines, did not reveal differences in the number and size of cells in the different radicular zones, at least sufficient to explain the marked difference in field resistance shown by these two lines. A similar situation was described for Solanaceae and Cucurbitaceae attacked by *O. ramosa* or *O. aegyptiaca* (Cubero, 1991). However, these results are different from those reported by Khalaf and El-Bastawesy (1989) and Zaitun *et al.* (1991) regarding the level of tolerance of "Giza 402" when compared with "Aguadulce" or related cultivars. Further research is needed to resolve this problem.

It is worth mentioning that mechanism (1) referred to above, does not seem to be the cause of resistance of cowpea to *Striga* (Lane and Bailey, 1991; Lane *et al.*, 1991). It is not the cause of differences detected between faba bean cultivars Giza 402 and Aguadulce, although differences in root structure were detected. The complexity of biochemical reactions leading to seed germination of the parasite and/or attraction by the host (Cezard, 1973a,b,c,d; Hiron, 1973; Whitney, 1979, 1986) seems to suggest that assays using root extracts are not reliable screening methods. Khalaf *et al.* (1991) reported that there is more than one natural stimulant in faba bean and flax root exudates. This is in agreement with the results of Hess *et al.* (1991) regarding sorghum. Studies by Matthews *et al.* (1991) also indicate the complexity of the *in vitro* germination of seeds of the parasite.

Thus, the use of root exudates as a screening method has many disadvantages due to the complexity of the reactions involved in germination of the parasite's seed, as well as its rhizotropism. The almost general lack of correlation with field resistance is understandable. Better results could be obtained if a certain compound producing a *qualitative* response was discovered and its correlation with field data demonstrated. In this context, the role of flavonoid and other molecules in the germination and rhizotropism of seed of the parasite may provide a new "avenue" of research (Riopel, 1986; Wegman, 1986), although it is an "avenue" full of obstacles. El-Ghamrawy *et al.* (1990), showed that gibberellin activity is broadly correlated with *Orobanche* seed germination, and it is related to *gibberellin-like* substances not yet identified. Flavonoids were identified in faba bean root exudates for the first time (Tomas-Lorente *et al.*, 1990), but their role in stimulating germination of broomrape seeds has yet to be clarified. They stimulate haustorium initiation in *Agalinis purpurea* and *Striga* (Riopel, 1986).

Not much attention has been given to mechanism (4) as a screening method even though there is much experimental data on the subject (Whitney, 1973; Abou-Raya *et al.*, 1973; Leonard, 1973; Wolswinkel, 1973, 1979, 1984; Fer, 1984). To be of use, it would be necessary to develop a *rapid* method for extensive screening.

Even when a mechanism of resistance is found, other factors involved in resistance expression should be taken into consideration (Cubero *et al.*, 1993), such as:
(a) The role of mycorrhiza in relation to germination of broomrape seed;
(b) Root morphology (superficial or deep); and
(c) The environment *sensu lato* (humidity, soil reaction, etc.).

Both (a) and (b) have been suggested as factors *affecting* resistance of sorghum to *Striga hermonthica* (Parker and Dixon, 1983), and (b) is present in some *Vicia sativa* resistant lines, even though it was discarded as a factor in resistance after greenhouse experiments in which seeds of the parasite were placed on the roots (Gil *et al.*, 1982).

Sources of Resistance

Screening for resistance to broomrape (*Orobanche* spp.), the only important parasitic weed of cool season food legumes, was successful in faba bean but not in lentil. Available data suggests that it will not be difficult to select resistant chickpea lines. Observations of different pea germplasm collections suggest the existence of some resistant lines in heavily infested fields. Selection in common vetch also produced resistant lines (Gil *et al.*, 1982, 1984, 1987). Observations indicate that screening for resistance to narbonne vetch (*V. narbonensis*) and lathyrus peas (*Lathyrus* spp.) will not be easy. Differences among close relatives also were found in Asiatic *Vigna* spp. (e.g., resistance in green gram (*Vigna radiata*), but only tolerance in black gram (*Vigna mungo*) (Rao and Rao, 1991). The difficulty in identifying resistant lines is usually not related to the importance of the crop. For example, there are as yet no sources of resistance in maize to *Striga*, in spite of the crop's importance and the research conducted to date (Kim and Winslow, 1991).

To date, the most successful screening program was developed for faba bean (Cubero, 1973, 1986; Nassib *et al.*, 1982; Cubero *et al.*, 1992a). Cubero (1991), Cubero and Hernández (1991), and Cubero *et al.* (1992a) identified promising faba bean material for Western Mediterranean countries. Line VF1071, selected from the Egyptian cultivar Giza 402 under field conditions in southern Spain, was crossed with the Spanish cultivars Alameda and Brocal to produce the cultivar Baraca after two cycles of recurrent selection. Baraca shows a higher degree of field resistance than VF1071, very likely because some genes from Alameda, known to be a tolerant cultivar, were transferred during the selection process in a heavily infested plot. Experimental lines leading to Baraca proved to be stable under different conditions in southern Spain, Lattakia, Syria and Morocco. It does not appear to be resistant to a Tunisian population of *Orobanche foetida*.

Baraca shows a somewhat higher degree of resistance than the resistant parent line. As the aim was to recover the VF1071 genes and not to produce a transgressive segregation, the importance of incorporating *any* kind of resistant *and/or* tolerant material in a breeding program is stressed. Even lines showing a lesser *in vitro* degree of parasitic seed germination should be included in a crossing program as they may carry genes controlling certain steps in a long chain leading to field resistance. Lines, such as ICARDA BPL 2210 (Aalders and Pieters, 1986, 1987), VF172 (Cubero, 1973) or landraces, such as Locale di Castellano (Perrino *et al.*, 1988) should be regarded as sources of useful, although minor, genes.

Concluding Remarks

Resistant cultivars were obtained in sunflower/*O. cernua*, cowpea/*S. gesnerioides*, and sorghum/*S. hermonthica* by extensive field screenings. Development of these resistant cultivars permitted the testing of different laboratory and/or greenhouse screening methods with an optimum degree of confidence. The screening methods developed in this way allowed for intensive and reliable screenings which have been applied successfully to other parasites, such as *Alectra vogelii* on cowpea. These techniques make it possible to study the mechanisms of resistance and this knowledge will lead to development of new and more efficient screening methods.

Research on the cool season food legumes has not been as successful. There is resistance to broomrape in both faba bean and common vetch, and its presence in chickpea appears likely, but not in pea. In other species, such as lentil, resistance has yet to be found.

In conclusion, some lessons can be learned from the experience of other host/parasite systems. First, there is a need to coordinate laboratory and field experiments. Second, more research is needed on the host rather than on the complex environment that exists between the host root and the parasite's seed. Resistance by avoiding or preventing germination of seed of the parasite cannot be excluded, but the best known cases suggest that resistance in the host is an active response following contact with the haustorium of the parasite. In fact, a completely resistant cultivar from an agronomic point of view could be one showing an extremely strong barrier (chemical, physical, or both) and at the same time inducing germination of seeds of the parasite thereby helping rid the soil of the parasite's seeds.

And the third lesson is to have patience and hope.

References

Aalders, A. J. G. and Pieters, R. 1986. In: *Biology and Control of Orobanche, Proceedings of a Workshop on Biology and Control of Orobanche*, pp. 140–144 (ed. S. ter Borg). Wageningen, The Netherlands: Agricultural University.

Aalders, A. J. G. and Pieters, R. 1987. *Euphytica* 36: 227–236.

Abou-Raya, M. A., Radi, A. F. and Darwish Heikal, M. M. 1973. In: *Anonymous, Symposium on parasitic weeds*, pp. 167–176. Wageningen, The Netherlands: Agricultural University.

Aggarwal, V. D., Haley, S. D. and Brockman, F. E. 1986. In: *Biology and Control of Orobanche, Proceedings of a Workshop on Biology and Control of Orobanche*, pp. 176–180 (ed. S. ter Borg). Wageningen, The Netherlands: Agricultural University.

Aggarwal, V. D., Muleba, N., Drabo, I., Souma, J. and Mbewe, M. 1984. In: *Proceedings of the Third International Symposium on Parasitic Weeds*, pp. 143–147 (eds. C. Parker, L. J. Musselman, R. Polhill and A. K. Wilson). Aleppo, Syria: ICARDA.

Al-Menoufi, O. A. and Ashton, F. M. 1991. In: *Proceedings of the 5th International Symposium of Parasitic Weeds*, pp. 293–297 (eds. J. K. Ransom, L. J. Musselman, A. D. Worsham and C. Parker). Nairobi, Kenya: CIMMYT.

Cézard, R. 1965. *Bulletin de lÉcole Nationale Supérieure Agronomique de Nancy* 7: 153–168.

Cézard, R. 1973a. *Bulletin Académie et Société Lorraine des Sciences* 12: 97–120.

Cézard, R. 1973b. *Bulletin Académie et Société Lorraine des Sciences* 12: 121–139.

Cézard, R. 1973c. *Bulletin Académie et Société Lorraine des Sciences* 12: 269–288.

Cézard, R. 1973d. In: *Anonymous, Symposium on Parasitic Weeds*, pp. 55–67. Wageningen, The Netherlands: Agricultural University.

Chang, M., Netzly, D. H., Butler, L. G. and Lynn, D. G. 1986. *Journal of the American Chemical Society* 108: 7858–7860.

Cubero, J. I. 1973. In: *Anonymous, Symposium on Parasitic Weeds*, pp. 205–217. Wageningen, The Netherlands: Agricultural University.

Cubero, J. I. 1983. In: *The Faba Bean*, pp. 493–521 (ed. P. D. Hebblethwaite). London, UK: Butterworths.

Cubero, J. I. 1986. In: *Biology and Control of Orobanche, Proceedings of a Workshop on Biology and Control of Orobanche*, pp. 127–139 (ed. S. ter Borg). Wageningen, The Netherlands: Agricultural University.

Cubero, J. I. 1991. In: *Progress in Orobanche Research Proceedings of the International Workshop on Orobanche Research*, pp. 257–277 (eds. K. Wegmann and L. J. Musselman). Tubingen, Germany: Erberhard-Karls-Universität.

Cubero, J. I. and Hernández, L. 1991. In: *Present Status and Future Prospects of Faba Bean Production and Improvement in the Mediterranean Countries*, pp. 51–57 (eds. J. I. Cubero and M. C. Saxena). Zaragoza, Spain.

Cubero, J. I., Khalil, S. and Moreno, M. T. 1993. In: *Breeding for Stress Tolerance in Cool-Season Food Legumes*, pp. 167–178 (eds. K. B. Singh and M. C. Saxena). Chichester, UK: John Wiley & Sons.

Cubero, J. I. and Moreno, M. T. 1979. In: *Some Current Research on Vicia faba in Western Europe*, pp. 41–80 (eds. D. A. Bond, G. T. Scarascia-Mugnozza and M. H. Poulsen). Luxemburg: Commission of the European Communities.

Cubero, J. I., Moreno, M. T. and Hernández, L. 1992a. In: *1st European Conference on Grain Legumes*, pp. 41–42. 1–3 June 1992, Angers, France.

El-Ghamrawy, N., Salem, S. M. and Neumann, K. H. 1990. *Faba Bean Abstracts* 11: 216.

Fer, A. 1984. In: *Proceedings of the Third International Symposium on Parasitic Weeds*, pp. 164–174 (eds. C. L. Parker, J. Musselman, R. M. Polhill and A. K. Wilson). Aleppo, Syria: ICARDA.

Gil, J., Martín, L. M. and Cubero, J. I. 1982. *Anales del Instituto Nacional de Investigaciones Agrarias, Serie Agrícola* 21: 175–181.

Gil, J., Martín, L. M. and Cubero, J. I. 1984. In: *Proceedings of the Third International Symposium on Parasitic Weeds*, pp. 221–237 (eds. C. L. Parker, J. Musselman, R. M. Polhill and A. K. Wilson). Aleppo, Syria: ICARDA.

Gil, J., Martín, L. M. and Cubero, J. I. 1987. *Plant Breeding* 99: 134–143.

Hanounik, S. B. and Bisri, M. 1991. In: *Present Status and Future Prospects of Faba Bean Production and Improvement in the Mediterranean Countries*, pp. 59–66 (eds. J. I. Cubero and M. C. Saxena). Zaragoza, Spain.

Hess, D. E., Ejeta, G. and Butler, L. G. 1991. In: *Proceedings of the 5th International Symposium of Parasitic Weeds*, pp. 217–222 (eds. J. K. Ransom, L. J. Musselman, A. D. Worsham and C. Parker). Nairobi, Kenya: CIMMYT.

Hiron, R. W. P. 1973. In: *Anonymous, Symposium on Parasitic Weeds*, pp. 76–88. Wageningen, The Netherlands: Agricultural University.

Khalaf, K. A., Ali, A. M. and El-Masry, R. R. 1991. In: *Proceedings of the 5th International Symposium of Parasitic Weeds*, pp. 83–89 (eds. J. K. Ransom, L. J. Musselman, A. D. Worsham and C. Parker). Nairobi, Kenya: CIMMYT.

Khalaf, K. K. and El-Bastawesy, F. I. 1989. *FABIS Newsletter* 25: 5–9.

Kim, S. K. and Winslow, M. D. 1991. In: *Proceedings of the 5th International Symposium of Parasitic Weeds*, pp. 494–497 (eds. J. K. Ransom, L. J. Musselman, A. D. Worsham and C. Parker). Nairobi, Kenya: CIMMYT.

Kuijt, J. 1969. *The Biology of Parasitic Flowering Plants*, 246 pp. Berkeley, California, USA: University of California Press.

Kukula, S. and Masri, H. 1985. *FABIS Newsletter* 12: 20–23.

Lane, J. A. and Bailey, J. A. 1991. In: *Progress in Orobanche Research, Proceedings of the International Workshop on Orobanche Research*, pp. 344–350 (eds. K. Wegmann and L. J. Musselman). Tübingen, Germany: Erberhard-Karls-Universität.

Lane, J. A., Kershaw, M. J., Moore, T. H. M., Child, D. V., Terry, P. S. and Bailey, J. A. 1991. In: *Proceedings of the 5th International Symposium of Parasitic Weeds*, pp. 237–240 (eds. J. K. Ransom, L. J. Musselman, A. D. Worsham and C. Parker). Nairobi, Kenya: CIMMYT.

Leonard, O. A. 1973. In: *Anonymous, Symposium on Parasitic Weeds*, pp. 188–193. Wageningen, The Netherlands: Agricultural University.

Linke, K. H., Singh, K. B. and Saxena, M. C. 1991. *Chickpea Newsletter* 24: 32–34.

Linke, K. H. and Vogt, W. 1987. In: *Parasitic Flowering Plants, Proceedings of the 4th Symposium on Parasitic Flowering Plants*, pp. 501–509 (eds. H. C. Weber and W. Forstreuter). Marburg, Germany: Phillipps University.

.Matthews, D. E., Joel, D. M. and Steffens, J. C. 1991. In: *Proceedings of the 5th International Symposium of Parasitic Weeds*, pp. 428–434 (eds. J. K. Ransom, L. J. Musselman, A. D. Worsham and C. Parker). Nairobi, Kenya: CIMMYT.

Nassib, A. M., Ibrahim, A. A. and Khalil, S. A. 1982. In: *Faba Bean Improvement*, pp. 199–206 (eds. G. Hawtin and C. Webb). The Hague, The Netherlands: Martinus Nijhoff.

Netzly, D. H. and Butler, L. G. 1986. *Crop Science* 26: 775–778.

Olivier, A., Benhamou, N. and Leroux, G. D. 1991. In: *Proceedings of the 5th International Symposium of Parasitic Weeds*, pp. 127–136 (eds. J. K. Ransom, L. J. Musselman, A. D. Worsham and C. Parker). Nairobi, Kenya: CIMMYT.

Parker, C. 1991. *Crop Protection* 10: 6–22.

Parker, C. and Dixon, N. 1983. *Annals of Applied Biology* 103: 485–488.

Parker, C., Hitchcock, A. M. and Ramaiah, K. V. 1977. In: *Anonymous, Proceedings, 6th Asian Pacific Weed Science Society Conference*, pp. 67–74. 11–17 July 1977, Djakarta, Indonesia.

Parker, C. and Poliniaszek, T. I. 1990. *Annals of Applied Biology* 116: 305–311.

Pavlista, A. D., Worsham, A. D. and Moreland, D. E. 1979a. In: *Second International Symposium on Parasitic Weeds*, pp. 219–227 (eds. L. J. Musselman, A. D. Worsham and R. E. Eplee). Raleigh, NC, USA: North Carolina State University Press.

Pavlista, A. D., Worsham, A. D. and Moreland, D. E. 1979b. In: *Second International Symposium on Parasitic Weeds*, pp. 228–237 (eds. L. J. Musselman, A. D. Worsham and R. E. Eplee). Raleigh, NC, USA: North Carolina State University Press.

Perrino, P., Pace, M. S. and Polignano, G. B. 1988. *FABIS Newsletter* 20: 40–44.

Pieterse, A. H., Roorda, F. A. and Wiselius, S. I. 1986. In: *Biology and Control of Orobanche, Proceedings of a Workshop on Biology and Control of Orobanche*, pp. 150–153 (ed. S. ter Borg). Wageningen, The Netherlands: Agricultural University.

Pustovoit, V. S. 1976. Selection, seed culture and some agrotechnical problems in sunflower. Indian National Scientific Documentation Centre, New Delhi, India.

Ramaiah, K. V. 1987. In: *Parasitic Flowering Plants, Proceedings of the 4th Symposium on Parasitic Flowering Plants*, pp. 637–664 (eds. H. C. Weber and W. Forstreuter). Marburg, Germany: Phillipps University.

Rao, K. N. and Rao, R. S. N. 1991. In: *Proceedings of the 5th International Symposium of Parasitic Weeds*, pp. 170–175 (eds. J. K. Ransom, L. J. Musselman, A. D. Worsham and C. Parker). Nairobi, Kenya: CIMMYT.

Riopel, J. L. 1986. In: *Biology and Control of Orobanche, Proceedings of a Workshop on Biology and Control of Orobanche*, pp. 52–56 (ed. S. ter Borg). Wageningen, The Netherlands: Agricultural University.

Sallé, G., Hariri, E. B. and Andary, C. 1991a. In: *Proceedings of the 5th International Symposium of Parasitic Weeds*, pp. 270–278 (eds. J. K. Ransom, L. J. Musselman, A. D. Worsham and C. Parker). Nairobi, Kenya: CIMMYT.

Sallé, G., Hariri, E. B., Jeune, B. and Urech, K. 1991b. In: *Proceedings of the 5th International Symposium of Parasitic Weeds*, pp. 525–526 (eds. J. K. Ransom, L. J. Musselman, A. D. Worsham and C. Parker). Nairobi, Kenya: CIMMYT.

Sauerborn, J., Masri, H., Saxena, M. C. and Erskine, W. 1987. *LENS Newsletter* 14: 15–16.

Scharpf, R. F. and Roth, L. F. 1991. In: *Proceedings of the 5th International Symposium of Parasitic Weeds*, pp. 519–522 (eds. J. K. Ransom, L. J. Musselman, A. D. Worsham and C. Parker). Nairobi, Kenya: CIMMYT.

Tomas-Lorente, F., Garcia-Grau, M. M. and Tomas-Barberan, M. M. 1990. *Zeitschrift für Naturforschung* 45: 1070–1072.

Vasudeva Rao, M. J., Chidley, V. L., Ramaiah, K. V. and House, L. R. 1983. In: *Proceedings of the Second International Workshop on Striga*, pp. 61–70. Ouagadougou, Mali: IDRC/ICRISAT.

Wegman. 1986. In: *Biology and Control of Orobanche, Proceedings of a Workshop on Biology and Control of Orobanche*, pp. 107–113 (ed. S. ter Borg). Wageningen, The Netherlands.

Whitney, P. J. 1973. In: *Anonymous, Symposium on Parasitic Weeds*, pp. 154–166. Wageningen, The Netherlands: Agricultural University.

Whitney, P. J. 1979. In: *Proceedings of the 2nd International Symposium on Parasitic Weeds*, pp. 182–192 (eds. L. J. Musselman, A. D. Worsham and R. E. Eplee). Raleigh, NC, USA: North Carolina State University Press.

Whitney, P. J. 1986. In: *Biology and Control of Orobanche, Proceedings of a Workshop on Biology and Control of Orobanche*, pp. 42–49 (ed. S. ter Borg). Wageningen, The Netherlands: Agricultural University.

Wolswinkel, L. 1973. In: *Anonymous, Symposium on Parasitic Weeds*, pp. 177–187. Wageningen, The Netherlands: Agricultural University.

Wolswinkel, L. 1979. In: *Proceedings of the 2nd Symposium on Parasitic Weeds*, pp. 156–164 (eds. L. J. Musselman, A. D. Worsham and R. E. Eplee). Raleigh, NC, USA: North Carolina State University.

Wolswinkel, L. 1984. In: *Proceedings of the 3rd International Symposium on Parasitic Weeds*, pp. 156–163 (eds. C. L. Parker, J. Musselman, R. M. Polhill and A. K. Wilson). Aleppo, Syria: ICARDA.

Zaitun, F. M. F., Al-Menoufi, O. A. and Weber, H. C. 1991. In: *Proceedings of the 5th International Symposium of Parasitic Weeds*, pp. 195–207 (eds. J. K Ransom, L. J. Musselman, A. D. Worsham and C. Parker). Nairobi, Kenya: CIMMYT.

Screening techniques and sources of resistance to nematodes in cool season food legumes

S.B. SHARMA[1], R.A. SIKORA[2], N. GRECO[3], M. DI VITO[3] and
G. CAUBEL[4]

[1] Legumes Pathology, ICRISAT, Patancheru P. O., Andhra Pradesh 502 324, India;
[2] Institut für Pflanzenkrankheiten der Rheinischen Friedrich Wilhelms Universität, Nussallee 9, 5300
Bonn, Germany;
[3] Istituto di Nematologia Agraria, C.N.R., 70126, Bari, Italy, and
[4] INRA, Laboratoire de Zoologie, 35650 Le Rheu, France

Abstract

Identification of sources of resistance in cool season legumes to cyst (*Heterodera* spp.), root-knot (*Meloidogyne* spp.), and stem nematode (*Ditylenchus dipsaci*) is generally based on number of cysts on roots, root-knot nematode induced gall index, and stem nematode reproduction in shoot tissue, respectively. Various levels of resistance to cyst nematodes have been detected in chickpea and pea. Resistance has also been identified in chickpea, faba bean, and pea to the root-knot nematodes. Broad based durable sources of resistance to plant parasitic nematodes are required. Basic research is needed to develop transgenic plants with resistance based on hatch stimulants, inhibitors, toxins, or repellents found in antagonistic rhizosphere microorganisms. Selection of genotypes that favor development of beneficial rhizosphere microorganisms or root endophytes that increase the plant resistance to nematode infection deserves attention.

Introduction

Many species of plant parasitic nematodes attack chickpea (*Cicer arietinum* L.), faba bean (*Vicia faba* L.), lentil (*Lens culinaris* Medik.), and pea (*Pisum sativum* L.). Cyst, root-knot, and stem nematodes are considered to be of widespread economic importance (Sikora and Greco, 1990). These nematodes cause vascular disorders and suppress *Rhizobium* nodulation. Damage is amplified by synergistic interrelationships with soilborne diseases. Most growers, however, are not sufficiently aware of the magnitude of damage caused by nematodes to these crops. Control of nematodes with soil application of nematicides and soil solarization is effective but expensive (Sharma and Nene, 1990; Di Vito *et al.*, 1991). Seed treatment is not always effective and crop rotation is complicated due to a wide host range of root-knot and stem nematodes. For these reasons, exploitation of host plant resistance to important nematode pests has priority and has great potential. Cultivation of nematode-resistant cultivars is a simple and economical way to prevent nematode-caused damage and to avoid

F.J. Muehlbauer and W.J. Kaiser (eds.), Expanding the Production and Use of Cool Season Food Legumes, 346–358.

environmental pollution due to improper use of pesticides. Unfortunately, limited efforts have been made to identify sources of resistance and breed for nematode resistance in chickpea, faba bean, lentil, and pea. The first requirement in any program designed to breed for resistance to nematodes is to develop practicable and simple host-plant resistance screening methods that will result in reliable selection of resistant genotypes. Identification of host-plant resistance to nematodes is generally based on either measurement of reproduction of nematodes and in some cases such as the root-knot nematodes and stem nematodes on symptoms produced (root galls, swelling of the stem, etc.).

This paper concentrates on important nematode pests of chickpea, faba bean, lentil, and pea, techniques for identification of host plant resistance, identified sources of resistance, and perspectives for future research.

Chickpea

Root-knot nematodes, *Meloidogyne incognita* (Kofoid and White) Chitwood, *M. javanica* (Treub) Chitwood, *M. artiellia* Franklin and chickpea cyst nematode, *Heterodera ciceri* Vovlas, Greco and Di Vito (Vovlas *et al.*, 1985) are economically important nematode pests of chickpea (Sikora and Greco, 1990).

Meloidogyne incognita and M. javanica

These two species of root-knot nematodes are the most important nematode pests of chickpea in many countries, particularly India, Nepal, and Pakistan (Greco and Sharma, 1990; Sharma and McDonald, 1990; Sharma *et al.*, 1990). Nematode infection does not produce any characteristic symptoms on aerial parts but reduces plant vigor, delays flowering, and induces early senescence – symptoms that are often confused with poor soil fertility and low moisture. Root galls are the most characteristic symptom of nematode infection. The galls are easily seen with the unaided eye. The tolerance limit (number of nematodes a plant can host without any measurable damage) of chickpea to these nematodes ranges from 0.2 to 2.0 eggs and juveniles cm^{-3} soil at sowing. These nematode species interfere with nitrogen fixation and increase the incidence of Fusarium wilt.

Screening Techniques

Evaluation of host-plant resistance to *M. javanica* and *M. incognita* is largely based on the number of galls produced on roots of a plant. To screen for resistance, chickpea genotypes are evaluated either in a greenhouse or in nematode-sick fields.

Greenhouse Screening

Population densities of the root-knot nematodes are increased on susceptible host plant cultivars (e.g., tomato, chickpea) in sandy soils in large pots. The nematode-infested soil obtained from these pots is added to 15-cm-diameter pots and seeds of chickpea genotypes are sown in pots. Alternatively, the infected roots of tomato and chickpea are chopped and thoroughly mixed with the soil before filling in the pots; or the infected roots of about 60-day old plants are washed free of soil, cut into 1–2 cm long segments, the nematode egg masses dissolved in 1.0% sodium hypochlorite, and the eggs released are counted (Hussey and Barker, 1973). The egg suspension can be stored at 10°C for two weeks until used. Storage for a longer period, although possible, is not recommended. Seeds of chickpea genotypes are sown in 10 to 15-cm-diameter pots and then 5- to 10-day old seedlings are inoculated with 5,000 to 10,000 eggs per plant by pouring the egg suspension into four 3 to 5 cm deep holes, around the stem base. After inoculation, pots are irrigated lightly to assure survival and even distribution of the nematode inoculum. Pots are kept at $25 \pm 5°C$ in a greenhouse.

Field Screening

Root-knot nematode infested fields are identified and the population density of the nematode species (*M. javanica* or *M. incognita*) is enhanced by growing susceptible cultivars. Chickpea genotypes are sown in 4 m rows with 10 cm plant-to-plant spacing and 30 cm between the rows. Two rows of highly susceptible chickpea genotypes (e.g., ICCC 4 and Dhanush) are sown after every 10 test entries. Each test entry is replicated at least three times and all the genotypes are evaluated for number of galls or egg masses per plant generally near crop maturity (Manandhar *et al.*, 1989). Root-knot or egg mass indices of each test genotype are compared with those of the checks.

Resistance Evaluation

For determining resistance, genotypes are evaluated for number of galls and number of egg masses per root after 60 days on a 0 to 5 or 1 to 5 or 1 to 9 scale (Table 1).

Chickpea genotypes are also evaluated for tolerance to root- knot nematodes by comparing plant growth in nematode-infested soil with plant growth in nematode-free soil. The differences in biomass of promising tolerant plants in infested and nematode-free soil are not statistically significant ($P \leq 0.05$).

Table 1. Resistance evaluation scales used for screening chickpea genotypes for resistance to *Meloidogyne incognita* and *M. javanica*

Gall index or egg mass index	Number of galls or egg masses per plant		Reaction
1(1)	0	(0)	Immune/Highly resistant
2(3)	1-10	(1-10)	Resistant
3(5)	11-30	(11-30)	Moderately resistant
4(7)	31-100	(31-50)	Susceptible
5(9)	>100	(>50)	Highly susceptible

1 to 5 and 1 to 9 (in parentheses) rating scales are generally used; 0 to 5 scale (as given for *M. artiellia*) can also be used. A 1 to 5 scale based on % galling of roots: 1 = no galling, 2 = 1 to 25% galling; 3 = 26 to 50% galling, 4 = 51 to 75% galling, and 5 = 76 to 100% galling is also used.

Sources of Resistance

Resistance in chickpea germplasm to *M. javanica* and *M. incognita* has been reported in India (Sandhu *et al.*, 1981; Hasan, 1983; Handa *et al.*, 1985; Sharma and Mathur, 1985; Gupta and Verma, 1989; Pandey and Singh, 1990). However, level of resistance in these sources needs reconfirmation. Manandhar *et al.*, (1989) screened 267 promising chickpea genotypes in a field naturally infested with *M. incognita* race 2 and *M. javanica*. Many of the test genotypes had been reported resistant to root-knot nematodes in greenhouse tests in India. All the genotypes were susceptible and only two genotypes (ICC 5875 and ICC 6371) were found to be "less susceptible" with fewer than 50 galls per plant. At ICRISAT, 1000 chickpea genotypes and 35 accessions of wild species of *Cicer* have been evaluated for resistance to *M. javanica*. Bheema, N 31, N 59 and ICCC 42 were tolerant and the other genotypes were susceptible (S. B. Sharma, unpublished).

Meloidogyne artiellia

This root-knot nematode was first described in England and has gradually spread in the Mediterranean basin (Sikora and Greco, 1990). It causes severe damage to chickpea in northern Syria, Italy, and Spain. Spring chickpea suffers more damage than winter chickpea. The tolerance limit is 0.01 eggs and juveniles cm^{-3} soil in spring and 0.14 eggs and juveniles cm^{-3} soil in winter (Di Vito and Greco, 1988). Legumes, cereals and crucifers are good hosts of the nematode (Di Vito *et al.*, 1985).

Screening Techniques

Meliodogyne artiellia reproduces well on chickpea and durum wheat at 20 ± 5°C. Greenhouse and field screening procedures are essentially similar to those described for *M. incognita* and *M. javanica*. *M. artiellia* prefers lower temperatures than the other species of *Meloidogyne*, with an optimum between 15 and 25°C for development, activity and reproduction; therefore, pots are kept at 20 ± 5°C in the greenhouse. The genotypes are evaluated for number of galls and egg masses per root 60 days later on a 0 to 5 scale, where 0 = 0 galls or egg masses, 1 = 1 to 2, 2 = 3 to 10, 3 = 11 to 30, 4 = 31 to 100, and 5 = more than 100 galls and/or egg masses (Taylor and Sasser, 1978). Galls caused by *M. artiellia* are either very small (as with Syrian populations) or absent (as with Italian populations). Therefore, evaluation on the basis of number of egg masses is more reliable (Table 1).

Sources of Resistance

Chickpea cultivars with resistance to *M. artiellia* have not been detected. However, evaluation of accessions of *Cicer* spp. have indicated availability of resistance in *C. judaicum*, *C. pinnatifidum*, *C. chorassanicum*, and *C. cuneatum* (Di Vito, unpublished).

Heterodera ciceri

The chickpea cyst nematode, *H. ciceri*, is widespread in northern Syria (Greco *et al.*, 1984) and causes severe damage when its population at sowing exceeds 1 egg g^{-1} soil (Greco *et al.*, 1988). Yield losses of 20 and 50% have been estimated in fields infested with 8 and 16 eggs g^{-1} soil, respectively. Complete crop failure may occur at ≥ 32 eggs g^{-1} soil (Greco *et al.*, 1988).

Screening Technique

Inoculum of *H. ciceri* is produced in high number on susceptible host plants. To evaluate chickpea for resistance to *H. ciceri*, test genotypes are sown in pots filled with steam-sterilized soil containing 80% sand, 15% loam, and 5% clay that are artificially infested with 20 or more eggs and juveniles g^{-1} soil. Size of the pots and number of seedlings per pot may vary, but pots must be at least 10 cm in diameter to contain one plant per pot. Pots containing about 3,000 cm^3 soil and four chickpea plants have been found to give reliable results. Pots are maintained in a greenhouse at 20 ± 5°C.

Resistance Evaluation

Evaluation of plant reaction to infection is based on number of females and/or cysts per root. Roots of test genotypes are rated on a 0 to 5 scale (where 0 = no females, 1 = 1 to 2 females, 2 = 3 to 5 females, 3 = 6 to 20 females, 4 = 21 to 50 females, and 5 = more than 50 females per root) 45 days after seedling emergence (Di Vito *et al.*, 1988). Genotypes with an average rating of ≤ 2 are considered resistant. If there is a need to obtain seeds from the resistant plants, seeds are sown in thin-walled 6 to 7-cm-diameter black plastic pots, containing noninfested soil, and having 4 to 6 holes at the bottom (Wyatt and Fassuliotis, 1979). These thin-walled pots are then buried in larger pots containing nematode-infested soil. After 60 days, the small pots are gently removed from the larger pots to recover roots growing in the infested soil of the large pots. These roots are washed free of adhering soil and scored for number of nematode females. Resistant plants are then transplanted in sterilized soil. Roots of these plants have not been seriously injured and the plants are able to produce seeds.

Sources of Resistance

More than 9,000 chickpea lines have been screened and most of them have been rated as 4 and 5 (Di Vito *et al.*, 1988, 1992). A small number of lines were rated as 3 and were considered as moderately resistant. No line has been found to be resistant. However, resistance is available in *Cicer bijugum*, *C. reticulatum*, and *C. pinnatifidum* (Singh *et al.*, 1989; Di Vito *et al.*, 1992). Resistance in accessions of *C. reticulatum* also is promising because this species can be crossed easily with *C. arietinum* and therefore can be useful in future breeding programs.

Faba Bean

The most important nematode pests of faba bean are the stem nematode, *Ditylenchus dipsaci* (Kuhn) Filipjev, and the pea cyst nematode, *Heterodera goettingiana* Liebscher, and to a lesser extent species of *Meloidogyne* (Sikora and Greco, 1990).

Ditylenchus dipsaci

Stem nematode is an important pest in most regions where faba bean is grown. Air temperatures of 15° to 20°C, high humidity, rain, fog, dew and sprinkler irrigation favor nematode infection, and disease development. The nematode withstands desiccation for many years and can survive for years in soil in the absence of a host plant or in seed in storage. Nematode infection causes severe distortions, swelling of the stem, and necrosis of all aerial parts. Disease development is greatly influenced by environmental conditions. In the Mediterranean region, where the nematode is widespread, these conditions occur from late fall to early spring. Several races of the nematode have been identified and faba beans are attacked by "giant" and "normal" races in winter and spring. The giant race, common in North Africa, is the most damaging to faba bean, especially when nematode-infested seed is sown.

Screening Technique

Screening procedures have been developed to estimate resistance in faba bean in pots under controlled environment conditions (Sturhan, 1975; Hooper, 1983; Caubel and Leclercq, 1989a) and in nematode-infested fields (Hanounik *et al.*, 1986).

Greenhouse Screening

The nematode reproduces on susceptible faba bean cultivars (Hooper, 1983) or on callus tissue (Riedel and Foster, 1970) and can be easily extracted with the incubation (Young, 1954) or centrifuge (Coolen, 1979) methods. Nematodes in water are stored at 10°C until used and are concentrated to obtain about 100 nematodes in a drop of water. Seeds of test genotypes are sown on perlite at 23°C and 5-day old seedlings are transplanted into 15-cm-long and 2.5-cm-diameter glass tubes or in small pots filled with organic sterilized compost (Ait Ighil, 1983). After a week, seedlings are inoculated with the nematode suspension in the leaf axil (Sturhan, 1975). The inoculated seedlings are placed at 15 to 20°C in a growth chamber. High humidity is required for several days to favor nematode infection.

Field Screening

For screening in the field, large quantities of *D. dipsaci*-infected stems are collected from infested fields, cut into 2-cm segments, and mixed thoroughly with soil in a ratio of 1:1 (v:v). After two weeks, the infested soil is diluted with a nematode-free soil until a population density of about 300 larvae 1000 cm^{-3} soil is obtained. Seeds are sown in rows 1 m long and 50 cm apart. A susceptible cultivar row is planted after every five test rows. All seeds are covered with infested soil to a depth of 15 cm. Data are recorded, using a scale based on symptoms (Table 2), at about 80% podding when symptoms on the stem are well developed (Hanounik *et al.*, 1986).

Table 2. Resistance evaluation scale used for field screening of faba bean genotypes for resistance to *Ditylenchus dipsaci*

Score	Symptoms	Infected plants (%)
1	No infection or very small stem swellings	0
3	Few stem infections	<20
5	Stem and leaf swellings	21-50
7	Stunting, elongated necrotic stem infections and moderate defoliation	51-75
9	Severe stunting, giant necrotic swellings and severe defoliation	>75

Resistance Evaluation

The nematode produces characteristic symptoms of swelling or necrosis near the inoculation site generally 8 weeks after inoculation. These symptoms are useful in differentiating susceptible and resistant plants. Plant reaction can be confirmed by measuring nematode reproduction 12 weeks after inoculation. All faba bean genotypes showing a reproduction rate of less than one are considered resistant. Reproduction rate in susceptible plants is high and may reach more than 100. A good correlation exists between symptoms on the stem and reproduction rates of the nematode.

Sources of Resistance

Resistance to *D. dipsaci* in faba bean has been identified. Schreiber (1977) reported a Moroccan faba bean cultivar (Souk el Arba du Rharb) having resistance to *D. dipsaci*. Abbad *et al.* (1990) found resistance in local Moroccan lines and in germplasm lines of faba bean. Hanounik *et al.* (1986) found 11 germplasm lines resistant to Tunisian populations of *D. dipsaci* and 12 lines resistant to Syrian populations. Field beans (*V. faba* ssp. *minor*) are generally poorer hosts than broad beans (*V. faba* ssp. *faba*). The field bean cultivar "Diana" is not a good host in comparison with broad bean cultivar "Hedosa" (Augustin and Sikora, 1989). Caubel and Leclercq (1989b) found faba bean line INRA 29H resistant to "giant" races of the nematode. No information is so far available on the genes conferring resistance to *D. dipsaci*. Limited information is available on the reaction of other Vicia spp. to *D. dipsaci*. Accessions of *V. cracca*, *V. sativa*, and *V. villosa* are susceptible. Two accessions of *V. narbonensis* have resistance to the nematode.

Heterodera goettingiana

The pea cyst nematode is an important pest of faba bean in many temperate regions and is a limiting factor in the cool-growing season in some countries. Greco *et al.* (1991) found that the tolerance limit of faba bean to *H. goettingiana* is 0.8 eggs g^{-1} soil, and yield losses of 20 and 50% are expected in soils infested with five and 15 eggs g^{-1} soil, respectively, while complete crop failure occurs at $\geq$ 64 eggs g^{-1} soil. *Heterodera goettingiana* is widespread in Europe and in the Mediterranean basin (Di Vito and Greco, 1986).

Screening Technique

The nematode reproduces on faba bean, pea, grasspea, and vetch and its population can be increased by rearing it on host plants in pots kept in greenhouses at about 20°C or outdoors (from mid-autumn to mid-spring under Mediterranean climates). Faba bean germplasm can be evaluated for resistance to *H. goettingiana* using the screening procedure suggested for chickpea to *H. ciceri*. Number of cysts on roots can be counted at flowering to early podding stage. Soil and environmental conditions for development and reproduction of the nematode are similar to that of *H. ciceri*. More details on host reaction can be obtained by determining the reproduction rate of the nematode. All germplasm lines giving nematode reproduction rates of ≤ 1 are considered as resistant.

Sources of Resistance

So far no faba bean cultivar has been reported as resistant to *H. goettingiana*, nor is there any germplasm screening in progress.

Lentil

Heterodera ciceri is one of the important nematode pests of lentil in northern Syria (Greco *et al.*, 1984; Vovlas *et al.*, 1985). However, lentil is less susceptible to *H. ciceri* than chickpea. The tolerance limit for lentil is 2.5 eggs g^{-1} soil (Greco *et al.*, 1988). Yield losses of 30 and 50% occur in fields infested with 32 and 128 eggs g^{-1} soil, respectively.

Screening Technique

The screening techniques described for the evaluation of chickpea genotypes for resistance to this nematode are also used for lentil.

Sources of Resistance

Screening of lentil germplasm in Syria (ICARDA, 1985) demonstrated that although lentil lines differed in their reaction to this nematode, none were resistant.

Pea

Pea is a good host of many nematode species (Sikora and Greco, 1990). However, under field conditions only the pea cyst nematode, *H. goettingiana* and the root-knot nematodes, *Meloidogyne* spp. are important. There have been few reports of root-knot nematodes causing severe damage to pea when grown in the cool season. Resistance in pea to root-knot nematodes has been reported and is extremely important wherever peas are grown as a summer crop (Tanveer and Saad, 1971). Microplot experiments have shown that the tolerance limit of pea to *H. goettingiana* is 0.5 eggs g^{-1} soil (Greco *et al.*, 1991). Yield losses of 20 and 50% as well as complete crop failure are expected when pea is cultivated in soils with two, eight, and 32 eggs of *H. goettingiana* g^{-1} soil, respectively, (Greco *et al.*, 1991). Stem nematode is considered as a minor pest of pea and nematode infection causes discoloration of the stem.

Screening Technique

Nematode-infested sandy soil containing 20 eggs and juveniles is filled in pots and seeds of test genotypes are sown in November in a greenhouse at 15°C. Roots are evaluated for number of females and cysts after two months. Number of cysts on a test genotype is compared with number of cysts on a check cultivar (e.g., "Verdone Fulminate", "Progress 9") and genotypes with cysts numbering 1 to 5% of the check are considered as resistant, and those with cysts numbering 6 to 15% of the check are considered to be moderately resistant (Di Vito and Perrino, 1978).

Sources of Resistance

Although cultivars with resistance to *H. goettingiana* have not been marketed commercially, accessions of *Pisum sativum* ssp. *abyssinicum* (A. Braun)

Govorov, *P. sativum* ssp. *sativum* var. *arvense* (L.) Poir, and *P. sativum* ssp. *elatius* (M. Bieb.) Aschers & Graebn. var. *elatius* (M. Bieb.) Alef have been shown to be moderately resistant to *H. goettingiana* (Di Vito and Perrino, 1978). Interspecific crosses between susceptible *P. sativum* cultivar Progress 9 and resistant *P. sativum* ssp. *abyssinicum* (MG 101791) are moderately resistant to the nematode in the F_2 generation indicating that resistance is recessive (Di Vito and Greco, 1986).

Concluding Remarks

Research on nematode pests of chickpea, faba bean, lentil, and pea has not received due attention and a limited number of nematologists are working on identification of host-plant resistance in these crops to important nematode species. This is largely because of lack of awareness about the economic importance of nematode-caused damage to these crops. In addition, there are not enough trained nematologists in the developing countries to identify, investigate, and demonstrate nematode disease and losses in farmers' fields. Availability of trained staff in the national programs of developing countries is essential for tangible progress in management of losses caused by nematodes using host-plant resistance. At present, identification of resistant sources is largely done in the greenhouse. Efforts should be made to develop field screening methods. Screening in greenhouses should be used as a supplement to field screening. Once resistant genotypes are identified, multilocational testing of these genotypes at hot spot locations is essential to evaluate their resistance against different populations (races) of the nematodes and to investigate the agronomic performance of the promising cultivars under different environmental situations. Use of nematode tolerant cultivars to limit economic losses in nematode-infested areas should be encouraged. Tolerant cultivars have one advantage over resistant cultivars in that they reduce yield losses from nematodes without providing selection pressure on the nematode for development of more aggressive races. More fundamental studies are required on biodiversity and coevolution to understand variability in races of nematodes, evolution of virulence genes in nematode populations, and mechanisms of resistance at biochemical and molecular levels. In addition to the on-going research on post-infectional resistance (based on galls, cysts, etc.), basic research is needed to identify genotypes with morphological and physiological preinfectional resistance. This will help in developing transgenic plants with resistance based on egg hatch inhibition, toxins, or repellents. A number of major advances have been made in developing techniques for screening rhizosphere bacteria for toxic as well as microbial metabolites that adversely affect nematode behavior (Sikora, 1991). In both cases, the presence of these organisms leads to increased plant health by either protection of the root surface from infection or through direct toxicity to the nematode parasite. A number of bacterial antagonists, for example, have been detected which are able to reduce hatch and mobility, as well as root recognition and penetration processes (Oostendorp and Sikora, 1989). These organisms could serve as a basis for future engineered plants. Fungal antagonists which produce nematicidal protein based compounds that degrade egg wall material have also been detected (Sikora *et al.*, 1990). In addition, fungal endophytes which live within the tissue of roots have been shown to prevent nematode penetration (Pedersen *et al.*, 1988), whereas others inactivate endoparasitic nematodes through specific metabolites. These fungi also could serve as a bank for future resistance breeding programs.

It should be mentioned that the development of gene engineered plants will not necessarily solve the problems associated with the selection of pathotypes out of naturally occurring nematode populations. Therefore, judicious use of resistant plants whatever their origin will be necessary.

References

Abbad, F. A., Ammati, M. and Alami, R. 1990. In *Proceedings of 8th Congress of the Mediterranean Phytopathological Union*, pp. 347–349. Agadir, Morocco.

Ait Ighil, M. 1983. *Variabilite physiologique de deux races, normale et geante, de Ditylenchus dipsaci (Kuhn) Fil. (Nematoda : Tylenchida), parasites de Vicia faba L*, 106 pp. These Ingenieur Docteur, ENSA, Universite des Sciences Rennes.

Augustine, B. and Sikora, R. A. 1989. *Gesunde Pflanzen* 5: 189–192.

Caubel, G. and Leclercq, D. 1989a. *Nematologica* 35: 216–224.

Caubel, G. and Leclercq, D. 1989b. *Fabis* 25: 45–47.

Coolen, W. A. 1979. In: *Root-knot Nematodes (Meloidogyne species) Systematics, Biology and Control*, pp. 317–329 (eds. F. Lamberti and C. E. Taylor). London, UK: Academic Press.

Di Vito, M. and Greco, N. 1986. In: *Cyst Nematodes*, pp. 321–332 (eds. F. Lamberti and C. E. Taylor). New York, USA: Plenum Press.

Di Vito, M. and Greco, N. 1988. *Nematologia Mediterranea* 16: 163–166.

Di Vito, M., Greco, N. and Saxena M. C. 1991. *Nematologia Mediterranea* 19: 109–111.

Di Vito, M., Greco, N., Singh, K. B. and Saxena. M. C. 1988. *Nematologia Mediterranea* 16: 17–18.

Di Vito, M., Greco, N., Singh, K. B. and Saxena, M. C. 1992. 21st International Nematology Symposium of ESN, Albufeira, Portugal, 11–17 April, 1992.

Di Vito, M., Greco, N. and Zaccheo, G. 1985. *Nematologia Mediterranea* 13: 207–212.

Di Vito, M. and Perrino, P. 1978. *Nematologia Mediterranea* 6: 113–118.

Greco, N., Di Vito, M., Reddy, M. V. and Saxena, M. C. 1984. *Nematologia Mediterranea* 12: 87–93.

Greco, N., Di Vito, M., Saxena, M. C. and Reddy, M. V. 1988. *Nematologica* 34: 98–114.

Greco, N., Ferris, H. and Brandonisio, A. 1991. *Revue de Nematologie* 14: 619–624.

Greco, N. and Sharma, S. B. 1990. In: *Chickpea in the Nineties*, pp. 135–137 (eds. H. A. van Rheenen, M. C. Saxena, B. J. Walby and S. D. Hall). Patancheru, Andhra Pradesh, India: ICRISAT.

Gupta, D. C. and Verma, K. K. 1989. *Haryana Agricultural University Journal of Research* 19: 318–320.

Handa, D. K., Jain, K. K., Mishra, A. and Bhatnagar, S. M. 1985. *International Chickpea Newsletter* 12: 33.

Hanounik, S. B., Halila, H. and Harrabi, M. 1986. *Fabis* 16: 37–39.

Hasan, A. 1983. *International Chickpea Newsletter* 8: 26–27.

Hooper, D. J. 1983. *Rothamsted Report* Part 2: 239–260.

Hussey, R. S. and Barker, K. R. 1973. *Plant Disease Reporter* 57: 1025–1028.

ICARDA. 1985. Studies on nematodes of food legumes, 25 pp. Progress Report 1984/1985. ICARDA, Aleppo, Syria.

Manandhar, H. K., Sharma, S. B. and Singh, O. 1989. *International Chickpea Newsletter* 20: 14.

Ostendorp, M. and Sikora, R. A. 1989. *Revue de Nematologie* 12: 77–83.

Pandey, G. and Singh, R. B. 1990. *Current Nematology* 1: 71–72.

Pedersen, J. F., Rodriguez-Kabana, R. and Shelby, R. A. 1988. *Agronomy Journal* 88: 811–814.

Riedel, R. M. and Foster, J. G. 1970. *Plant Disease Reporter* 54: 251–254.

Sandhu, T. S., Kooner, B. S., Singh, I. and Singh, K. 1981. *Indian Journal of Nematology* 11: 86–87.

Schreiber, E. R. 1977. *Lebensweise, Bedeutung und Bekampfungsmoglichkeiten von Ditylenchus dipsaci (Kuhn) Filipjev an Ackerbohne (Vicia faba L.) in Marokko.* Dissertation der Technischen Universitat, Berlin. D83 No. 79, 150 pp.

Sharma, G. L. and Mathur, B. N. 1985. *International Chickpea Newsletter* 12: 31–32.

Sharma, S. B. and McDonald, D. 1990. *Crop Protection* 9: 453–458.

Sharma, S. B. and Nene, Y. L. 1990. *Journal of Nematology* 22S: 579–584.

Sharma, S. B., Saha, R. P., Singh, O. and van Rheenen, H. A. 1990. *Tropical Pest Management* 36: 327–328.

Sikora, R. A. 1991. In: *Plant Growth-Promoting Rhizobacteria – Progress and Prospects*, pp. 3–10 (eds. C. Keel, B. Koller and G. Defago). Interlaken, Switzerland.

Sikora, R. A. and Greco, N. 1990. In: *Plant-parasitic Nematodes in Subtropical and Tropical Agriculture*, pp. 181–235 (eds. M. Luc, R. A. Sikora and J. Bridge). Wallingford, UK: CAB International Publishers.

Sikora, R. A., Hiemer, M. and Schuster, R. P. 1990. *Mededelingen van de Faculteit Landbouwwetenschappen Rijksuniversiteit Gent* 55: 699–712.

Singh, K. B., Di Vito, M., Greco, N. and Saxena, M. C. 1989. *Nematologia Mediterranea* 17: 113–114.

Sturhan, D. 1975. *Mededelingen van de Faculteit Landbonarwatenschappen Rijksuniversiteit Gent* 40: 443–450.

Tanveer, M. and Saad, A. T. 1971. *Plant Disease Reporter* 55: 1082–1084.

Taylor, A. L. and Sasser, J. N. 1978. Biology, identification and control of root-knot nematodes (*Meloidogyne* species), 111 pp. North Carolina State University Graphic, Raleigh, NC, USA.

Vovlas, N., Greco, N. and Di Vito, M. 1985. *Nematologia Mediterranea* 13: 239–252.

Wyatt, J. E. and Fassuliotis, G. 1979. *HortScience* 14: 27–28.

Young, T. W. 1954. *Plant Disease Reporter* 38: 794–795.

Policy incentives

Policy incentives for expanding European pulse production

G.P. GENT

Processors & Growers Research Organisation, Great North Road, Thornhaugh, Peterborough, PE8 6HJ UK

Abstract

For many years, peas and faba beans have been grown in Europe for human consumption and other specialist markets, but their main use is now for animal feeds.

European animal feed manufacturers have traditionally imported vegetable proteins for their compound rations, with soybean meal from the USA as their preferred product. However, in 1973, the US Government placed an embargo on soybean exports and the EEC decided that supplies of home-grown vegetable protein should be increased. Schemes were then introduced to encourage the production of protein crops, primarily peas, faba beans, and soybeans.

Initially the output of community produced peas and faba beans was small, but from the mid-1980s there have been surpluses of cereals and the introduction of pulse crops allows growers to reduce their cereal output while maintaining farm profitability.

Pea and faba bean production has expanded to about 5 million tons per annum with the key element being community payments which bridge the gap between the value of the produce to the compounder and the price the farmer needs for peas and faba beans to be competitive with other arable crops.

Recently, escalating costs of the support program have led to changes in the financial arrangements, but the principle of maintaining the profitability of the crops to the farmer and encouraging use by compounders remains.

Introduction

Dried pulse crops have been an important part of human diets in most European countries since recorded history and while primitive types had many undesirable characteristics, the availability of plentiful hand labor allowed their production and harvesting. From the earliest times of cultivation, the selection of types that had desirable culinary and agronomic characteristics changed the plants, but

F.J. Muehlbauer and W.J. Kaiser (eds.), Expanding the Production and Use of Cool Season Food Legumes, 361–366.
© 1994 *Kluwer Academic Publishers.*

systematic improvements only came in the 19th Century after the establishment of scientific plant breeding.

Despite the Mendelian experiments which used peas as a model, relatively little progress was made on improving this or other pulse species until the 1980s. Prior to this time, all pulse crops had a number of characteristics that restricted their agricultural usefulness and therefore also limited the potential for crop expansion. However, during the 1980s the combination of economic incentives from expanded crop areas and the appreciation of the value of novel plant characteristics led to substantial progress with new pea and faba bean cultivars and this has been a key element in allowing farmers to expand their output. The introduction of afila (semi-leafless) peas and dwarf, early-maturing faba beans were particularly important in this respect.

Major Crop Species

Pea (Pisum sativum L.) is the most important pulse crop in the EEC. They give produce that can be used for human consumption or inclusion in animal feeding rations and they can be successfully produced in northern temperate areas. The major weakness of the crop is its lax plant habit, but the best afila cultivars have greatly improved this characteristic and also improved both total and harvestable yield.

Faba bean (Vicia faba L.) is a major crop in the UK and traditionally suffered from poor and erratic plant characteristics and late maturity. Produce is used for horse and cattle feeding and for human consumption in middle eastern export markets. The introduction of tannin-free types has improved product quality, while early maturing types have improved agricultural adaptability.

Dry bean (Phaseolus vulgaris L.) is a major crop in southern Europe, but often the scale of production is small and output has declined over the past 20 years. Produce is very varied in size, color and shape and is best used for human consumption. Recently mechanized growing and harvesting has been improved by the development of tall, erect, early maturing cultivars.

Lupin (Lupinus spp.). Considerable development work is being undertaken on a range of species including *L. albus*, *L. multiflorus*, and *L. angustifolius*. The produce of low alkaloid lupins (usually *L. albus*) has a high protein content (about 33%) and only very low concentrations of toxins, but despite this, community production is very small.

In the community, soybean (*Glycine max*) is regarded as an oil-seed crop. Early systems of price support failed to stimulate production, so a new system was introduced in 1980 and for the marketing year 1987/88 a Maximum Guaranteed Quantity (MGQ) was introduced so that aid payments could be reduced as output increased. This has led to the production of about 2 million tons per annum, with Italy being by far the largest producer. Treated meal from this output can be used as a high protein animal feeding supplement in competition with that from imported soybeans and community produced pulses. Other pulses, including lentil (*Lens culinaris*) and chickpea (*Cicer arietinum*) are grown locally for human consumption use.

European Vegetable Protein Requirements

The EEC is a major importer of vegetable protein for use in animal feeding rations. These rations are a key element in intensive livestock husbandry for they allow milk and egg production to be maintained and animals to be fattened through the winter. The feeding value of rations varies according to the type of livestock, but in each case the protein element is most important. All ingredients have a level of protein, but in most cases it would be too low for a fully balanced and productive feed ration, so high protein supplements have to be included. Traditionally, soybean meal, manioc, fish meal, and animal offals have been used for this purpose. Most of these products are imported into the community.

Table 1. Pulse production, major EEC countries

	Production ('000 tons)						
	78	80	82	84	86	88	90
Denmark	22	7	38	255	514	497	542
UK	232	260	211	301	559	959	812
Germany	58	50	62	67	69	59	–
France	171	358	567	865	1262	2654	(3756)
Italy	352	313	274	290	239	211	–
Total *	–	–	–	–	–	4822	5529

* All countries excluding the former DDR; peas and faba beans only

( ) Figure in brackets, is an estimate

The community's vulnerability through its dependence upon an imported strategic raw material was made clear in 1973 when the USA placed an embargo on the export of soybeans and processed soybean products. Community measures to reduce this dependence on imported vegetable protein were then introduced from 1974–78 and have led to the expansion of pulse production from 200,000 tons in 1978 to about 5 million tons in 1990.

Total EEC requirements for vegetable protein can only be estimated but are probably equivalent to about 10 million tons of pulses. However, production at this level would only be desirable if the use of all other protein supplements was suspended.

EEC Production Incentives

Inevitably these have taken the form of production subsidies.

Pea and faba bean production is aided in the Community under the dried peas, beans, and lupins support scheme. This scheme was introduced in 1978 and originally only covered peas and beans for use in animal feed. It was extended in 1982 to cover peas and beans for human consumption and again in 1984 to cover sweet lupins for animal feed use.

The aim of the regime is to keep Community produced peas and beans competitive with third country products, notably soybean, for inclusion in animal feeds. This is achieved through aid payments to the end-users (compounders for example) of Community peas and beans. The payment of the aid is conditional on the grower receiving at least the minimum price.

Every year as part of the annual price fixing procedure the Agriculture Council sets an activating price, a guide price, and a minimum price. These are the basic prices for calculation of the aid. The activating price is the lowest price to which soybean cake can fall before peas and beans cease to be comptetitive. Aid for peas and beans for animal feed is 45% of the difference between the activating price and the world market price for soybean cake. Aid for peas and beans for human consumption is equal to the difference between the guide price and the world price for similar products. Monthly increments, intended to help orderly marketing, apply to the minimum, activating, and guide price. The aid rates are calculated twice a month, or more often if there are significant changes in world prices.

The minimum price and aid rate are set with reference to a standard quality for 14% moisture and impurities. Where the quality differs from the standard, the minimum price, and the aid are increased or decreased accordingly.

Due to a significant increase in budgetary expenditure in the peas and beans sector a Maximum Guaranteed Quantity system was introduced from the beginning of the 1988/89 marketing year. MGQ of 3.5 million tons of production eligible for aid has been set for the period 1988/89 to 1991/92. For every 1% of production over the MGQ a 0.5% penalty is applied to the guide price. The minimum price and the activating price are reduced by the same actual amount. Estimates of production for the coming year, and determination of actual production the previous year are adopted by the Management Committee before the end of the second month (August) in the marketing year. If actual production differs from the estimate, the necessary adjustment is made to the prices in the following year. A 16.5% price cut is in effect for 1991/92 under the MGQ system. (In 1990/91 a 20% cut was applied with 8.9% in 1989/90 and 9% in 1988/89). The MGQ arrangements for the 1992/93 marketing year have yet to be determined. It is likely that the current arrangements will continue because the proposed new scheme for the arable crops sector under the CAP reform production would not be implemented until 1993/94.

Possible Changes

The Commission have issued proposals to change the support systems to a common regime for all the main arable crops. If agreed, the new system would apply to the peas and beans sector from 1 July 1993. The current system of aid payments to the end-user would be replaced by a system of direct payments to the farmer on a per hectare basis. The aid payment to the grower would be a flat rate of aid of 55 ecu per ton varied regionally on the basis of historical cereal yields, so that higher yielding areas would receive higher payments than the average EC aid. In order to receive this payment a grower would be required to set aside a proportion (initial figure is 15%) of his arable land. Outlines of these proposals have only recently been issued by the Commission and have yet to be discussed in any detail.

Crop Inter-Relationships

All major EEC farming activities receive some form of support so the level and form of subsidy for pulse crops is partly to allow them to compete with other arable crops. Indications to date are that despite the fact that pulses are in production deficit in the Community and other sectors, such as cereals are in surplus, there will not be any dramatic changes in the ratio, or the areas, of the major crops.

One aspect, however, that could favor pulses, is their "environmentally friendly" image, being low input crops and providing the ideal entry for a high yielding winter wheat.

National Production, Major EEC Countries

1. Denmark. Output is almost wholly pea with annual production now being about 0.5 million tons.
2. U K. Faba bean is the major crop with both winter and spring types being grown. Peas are grown successfully as far north as central Scotland (latitude 56° N) but lack of true frost hardiness and late maturity restricts faba beans to the south of the country.
3. Germany. Statistics relate to the old FDR where production of pea and faba bean was very small. The incorporation of the former DDR could increase output dramatically on the very large former co-operative and state farms.
4. France. The dominant feature of European pulse production has been the rapid expansion of pea growing. Small tonnages of lentils, dried beans, soybeans, and faba beans are also grown but these have only local adaptability. In contrast, peas are a major crop in northern France and a small area (about 20,000 ha) of winter peas are grown in south and west central France.

5. Italy. All the main pulse crops are grown, including a small area of chickpeas. Faba bean is the major crop and, exceptionally, the area of dry beans exceeds that of peas. Italy also produces about 1.6 million tons of the community production of 2 million tons of soybeans.

Acknowledgements

The essential information in this paper on the EEC dried peas, faba beans and lupins support scheme has been provided by the EEC commission and their UK agents, the Ministry of Agriculture, Fisheries and Food (MAFF). Grateful thanks are therefore extended to Mr. Russell Mildon, the commission head of oilseeds and proteins and to Mr. A. Bastion, the MAFF head of sugar, tobacco, oilseeds and protein branch. The help of Miss J. Trinder and Mr. M. J. Pemberton at MAFF is also acknowledged.

Effects of markets on the development of cool season food legumes, experiences from Sudan and India

M. VON OPPEN[1], H. FAKI[2], S. ABDELMAGID[1] and A. HASHIM[1]

[1] University of Hohenheim (490), Postfach 700562, D-7000 Stuttgart 70, Germany, and
[2] Agricultural Research Corporation, Wad Medani, Sudan

Abstract

World production of cool season food legumes has increased over the past decade, in line with the total pulse production in the world. However the lion's share of 90% of this increase occurred in developed countries with a tripling of pea production, while developing countries contributed only 10% due to an increase in chickpea and lentil production, and a stagnation in faba bean. While technological innovation for chickpea and lentil could be observed in India and Sudan; for faba bean, such innovations are still being implemented. Institutional constraints such as inefficient marketing and credit systems may be partly responsible for weak performance in the Sudan. On the other hand, the tendency toward removal of trade restrictions in both countries and the observable trend of an increasing international trade in pulses will increasingly generate new patterns of production and trade flows which will sustain further growth of cool season food legume production in the world.

Development at the World Level

World production of grain legumes over the past decade has shown a remarkably steady increase. The declining trend that was prevailing up to around 1980 has been reversed and production of pulses is increasing at a faster rate than cereals (Figure 1).

Within the pulses, cool season food legumes (dry pea, chickpea, lentil, and faba bean) have increased even faster than other crops and they presently contribute more than half the world production (Table 1). Individual crops contributed differently to this rapid increase in world production of cool season food legumes. While dry pea and lentil doubled, chickpea increased only by 15% in production and faba bean remained stagnant.

These differences in performance are even more visible if the production of cool season food legumes is compared between developed and developing countries (Table 2). About 90% of the increase in cool season food legume

F.J. Muehlbauer and W.J. Kaiser (eds.), Expanding the Production and Use of Cool Season Food Legumes, 367–387.

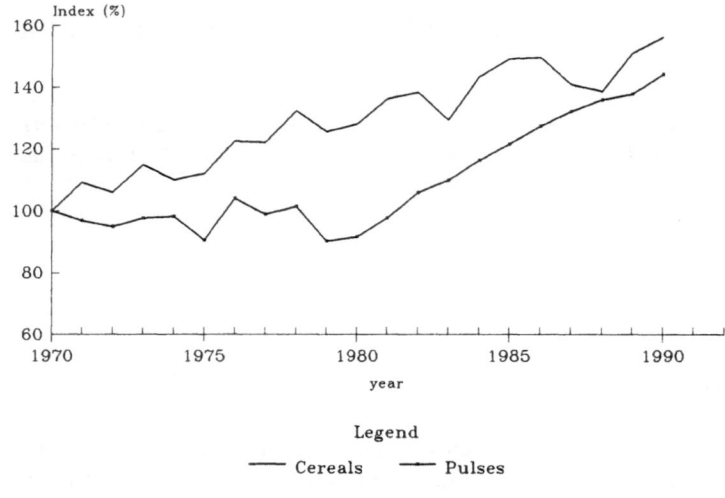

Source: *FAO production yearbook*

Figure 1. Production of cereals and pulses (world without China).

Table 1. World production of total pulses and cool season food legumes (in million tons)

Year	Pulses Total	All	Dry pea	Chickpea	Lentil	Faba bean
			Cool Season Food Legumes			
1979–81	41	20.1	8.5	6.0	1.3	4.3
1983	46	23.2	10.6	7.0	1.6	4.0
1984	48	23.0	10.7	6.5	1.5	4.3
1985	51	25.2	12.8	6.4	1.8	4.2
1986	53	26.8	12.3	8.0	2.3	4.2
1987	53	28.5	14.5	6.9	2.6	4.5
1988	56	28.8	15.7	5.9	2.7	4.5
1989	55	29.2	15.9	7.3	2.1	3.9
1990	60	31.4	17.5	6.9	2.7	4.3

Source: FAO Production Yearbooks, 1980, 1985, 1991

production occurred in developed countries, while developing countries contributed only about 10%.

This rapid increase in developed countries is primarily due to an increase in dry pea by nearly three times, against their decline by 10% in developing countries. Lentil production doubled in both groups of countries and production of faba bean stagnated in both. Chickpea, even though produced to

Table 2. Production of pulses and cool season food legumes in developed and developing countries (in million tons)

Years	Developed Countries	Developing Countries	World Total
Pulses			
1979–81	10	31	41
1990	22	38	60
Dry pea			
1979–81	5.3	3.2	8.5
1990	14.8	2.7	17.5
Chickpea			
1979–81	0.1	5.9	6.0
1990	0.2	6.7	6.9
Lentil			
1979–81	0.2	1.1	1.3
1990	0.4	2.3	2.7
Faba bean			
1979–81	0.5	3.8	4.3
1990	0.5	3.8	4.3
All Cool Season Food Legumes			
1979–81	6.1	14.0	20.1
1990	15.9	15.5	31.4

Source: FAO Production Yearbooks, 1980, 1985, 1991

a very limited extent in developed countries, doubled in those countries but production increased less than 15% in the developing countries.

Technological progress and market induced changes are responsible for these developments. World market prices for selected cereals and cool season food legumes showed different trends over the past decade (Figure 2). While cereal prices remained constant, prices for legumes increased.

Prices for dry pea were about three times those for sorghum and maize in the early 1980s but rose to four times around 1985 and seven times in 1986/87. This implies a doubling of relative prices of dry pea in world markets. In Europe, where two-thirds of the dry pea crop is produced, substantial subsidies for their production led to an expansion in production, especially in France. These forces, i.e., increases in relative prices in world markets and subsidies for protein feeds within Europe, explain the extremely rapid increases in production of dry pea until 1990.

With regard to lentil, the picture is not so clear. There were price peaks in 1980/81, 1985/86, and 1990 and troughs in between which reveal no significant change in overall trends of prices relative to cereals. Therefore, the expansion of

Fig. 2a: Prices for Selected Cereals and Pulses
        (in US $ / MT)

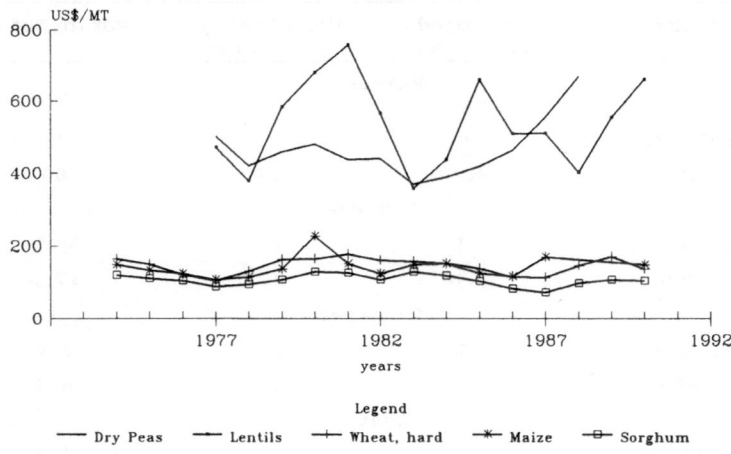

Source: FAO Production Yearbook

Fig. 2b: Relative Price betw. Cereals and Pulses
        (in US $ / MT)

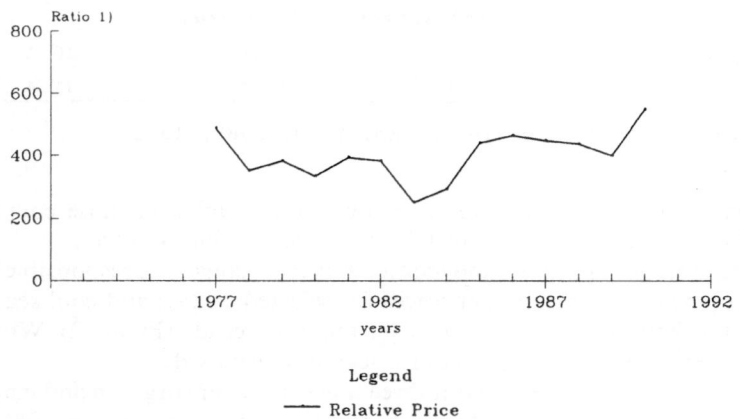

Source: *FAO Production Yearbook*
1) ratio of average of pulse prices over
average of cereal prices (for 1989 and 1990 lentil prices only)

Figure 2. (a) Prices for selected cereals and pulses (in US$/MT); (b) Relative price between cereals and pulses (in US$/MT).

lentil production in developing countries, and also in developed countries, cannot be easily explained by price changes. Improvements in production technologies and market access for lentil producers within individual countries such as India or Sudan will explain the increase in lentil production (see below).

World production of chickpea is dominated by developments within its major producing country, India (see below). However, there are some interesting developments in rapid expansion of chickpea production in Turkey and Australia which indicate a gradually increasing effect on world trade of chickpea. For instance, nearly all of Australia's production is being exported to India. The stagnation of faba bean production in developing countries such as Sudan will be discussed below.

Development in the Sudan

Consumption Patterns and Production Expansion Prospects

Faba bean, chickpea, and lentil are the cool season food legumes mostly consumed in the Sudan, of which faba bean is the most important. Faba bean consumption has been expanding within the last three decades and is increasingly accepted by the rural population. It is predominantly used as boiled grain that may be mixed with various other ingredients and eaten with wheat bread. The crop is consumed by almost all income groups in urban centers. This is shown in Table 3 for the town of Wad Medani (Abdelmagid, 1992).

Table 3. Per family faba bean consumption by income class, Wad Medani, Sudan, 1988

Income Class (LS per Month)	Quantity (kg per Month)	Expenditure (% of Food Expenses)
250 - 500	5.20	15.5
501 - 800	9.70	17.7
801 - 1150	16.44	22.1
1151 - 1450	13.62	13.1
> 1450	12.40	8.7
All Classes	11.48	15.5

Abdelmagid (1992)

However, its consumption increases with rising incomes and is highest within middle-income classes. On average, families spend more than 15% of their food budget on faba bean. The lesser quantity consumed by low-income groups is due to their lower incomes but their relative expenditure is the same as the average of all income classes.

Chickpea consumption is much lower when compared to faba bean.

Although large quantities of it are consumed as boiled grain during Ramadan, the Muslim fasting month, it is normally used in limited quantities in the preparation of many local dishes and some sweets (Nourai, 1987).

Lentil cultivation is a recent venture in the Sudan, although that crop has been consumed for a long time as an alternative dish to faba bean or as part of other meals.

Positive income/demand elasticities were recorded for pulse crops in various urban and semi-urban areas of the Sudan that ranged between 0.85 and 1.48 (Yousif, 1988). This provides prospects for high prices for those crops, in view of expected rising incomes and population growth.

Further, the demand for pulse crops is relatively inelastic. This could be shown for the case of faba bean, for which the price elasticity of demand was as low as 0.49 (Abdelmagid, 1992). In spite of that, the per capita consumption of food legumes is very low by all standards. For cool season food legumes, which are the most commonly consumed pulses, it averaged only about 1.9 kg per annum, while the average for Africa of pulse crops is 9 kg (Agostini and Khan, 1988). Consumption in some neighboring countries is much higher than in the Sudan. For example, the per capita figures for Egypt and Ethiopia are 6.5 and 19.7 kg per annum, respectively.

Most pulse consumption in the Sudan is concentrated in urban and some rural areas along the Nile. In the big urban centers of Khartoum and Wad Medani, per capita consumption was estimated in two different studies to vary from 23.6 to 27.2 kg per annum (Yousif, 1988; Abdelmagid, 1992). People in remote regions in the west, east and south are deprived of rich protein sources of food. One reason is that they are not accustomed to faba bean consumption and another is the lack of nutritional awareness in those areas. Moreover, being unaccustomed to wheat consumption, people rely on sorghum or millet bread which is not very suitable to eat with faba bean.

The complementarity between faba bean and wheat seems to affect the consumption patterns of the former. Both crops have been traditionally more expensive to purchase than other food crops, and their use has accordingly been very low in poor rural areas. Development of alternative recipes in faba bean dish preparation independent of wheat bread use may enhance faba bean use in low-consumption areas. Further increase in consumption would be possible if retail prices could be reduced through savings in marketing costs, particularly those of transport and storage. Such savings might also be reflected as higher prices to producers, conducive to increased supply.

Production Aspects and Market Response

The northern part of Sudan dominates the production of cool season food legumes in the country. Almost all faba bean, all of the limited amount of lentil, and most of the chickpea are grown in that region because of its relatively cool winters. The three crops occupy an area of about 26,000 ha, which is about 21% of the total area cultivated in northern Sudan. Although this represents only a small fraction of the total area under cropping in the country, it supplies most of the domestic needs for those crops. Faba bean is the most important of the three crops, occupying about 96% of the total production area.

The development in areas and production of the three crops within the past 10 years is depicted in Table 4. Both area and production of faba bean were highly variable and without a clear trend. A more clearly increasing trend is reflected by chickpea, where increases of over 300% were realized in both area and production over the last 10 years. However, variations in its area and production are also apparent. This is because most of the chickpea crop is produced on land with residual moisture from the receding Nile flood, the area of which varies from one season to the next.

Table 4. Area, production and average yields of faba bean, chickpea, and lentil in the Sudan, 1981/82–1990/91

Season	Area (000' ha)			Production (000' ton)			Yield (kg ha⁻¹)		
	Faba bean	Chickpea	Lentil	Faba bean	Chickpea	Lentil	Faba bean	Chickpea	Lentil
1981/82	14.07	0.46		26.97	0.62		1917	1347	-
1982/83	17.23	0.37		28.72	0.43		1667	1162	-
1983/84	34.76	0.51		74.50	0.78		2143	1529	-
1984/85	17.35	0.33		27.80	0.49		1602	1484	-
1985/86	17.47	1.07		33.65	1.28		1926	1196	-
1986/87	21.55	0.63	0.13	35.92			1713	-	-
1987/88	22.90	0.84	0.34	36.64			1600	-	-
1988/89	14.71	0.92	0.42	27.5	1.54	0.34	1969	1673	809
1989/90	18.91	1.47	0.54	36.00	2.80	0.61	1903	1905	1130
1990/91	24.37	1.89	0.59	41.43	2.70	0.59	1700	1428	1000

Sources: 1) Regional Ministry of Agriculture, Northern Region, Sudan
2) Nourai (1987)

Lentil, which used to be produced on a small scale, has been recently reintroduced as a self-sufficiency crop. Its area has been gradually increasing due to government support for lentil farmers in terms of credits for inputs such as seeds, fertilizers, and decortication equipment. Also, improved production technology has been developed through on-farm research (Faki and Nourai, 1991). The product is purchased by the government at determined prices. Otherwise, most of the Sudan's lentil needs are imported. Within the last 30 years, imports averaged about 5000 t, but with a clearly increasing trend. Recent import figures have been more than 7000 t annually (Bank of Sudan, 1989).

The farming system and the characteristics of each crop, coupled with the prevailing market conditions, shape farmers' decisions on areas and production of various cool season food legumes. Those crops are produced under predominantly small schemes that are irrigated from the Nile.

Scarcity of cultivable land limits the expansion of crop area. High value crops of comparable profitability, predominantly legumes, spices, fruits, and subsistence cereals, compete for the available land in those farming systems. In many areas, farmers traditionally produce certain crop mixes. Much of the yield variability is attributable to weather factors, especially winter temperatures and the incidence of crop pests and diseases. In spite of some of these factors, which may limit maneuvers in crop area allocations, farmers are responsive to the market conditions. The degree of farmer response is examined for the effect of faba bean price change on its area and production over time. Judged by the sharp increases in nominal faba bean prices, large expansions in its area and production should have been expected. However, the response to price change was very weak. But when those prices were deflated by the inflation rates to reflect real prices, a considerable level of lagged response could be computed over seven years in the period 1983/84–1990/91 as follows:

$$A = 16177 + 4810 \, P \, (R^2 = 0.18)$$
$$Q = 28401 + 7744 \, P \, (R^2 = 0.20)$$

where A and Q denote faba bean area (in ha) and production (in t), respectively, and P the deflated faba bean farm-gate price (LS per kg) of the preceding season. The P coefficients were respectively significant at the 20 and 15% levels, respectively. Due to the scarcity of data, figures on areas and production used in the calculation were only aggregate values of all producing areas, while faba bean prices were those prevailing in selected schemes where field surveys were conducted. Also, influencing factors other than price were not possible to consider. Accepting those limitations, the estimated relationships reflect area and production elasticities of 0.25 and 0.23, respectively, which indicate sufficient farmers' reaction to price signals.

Based on encouraging results of on-farm trials, faba bean could be successfully produced in large irrigation schemes in central and eastern Sudan. In spite of their less favorable climatic conditions, those schemes possess abundant land, cheap water supply, and are located in important consumption areas. Those areas provide a large potential for increasing the supply of faba bean in the country.

The question arises whether the right signals are reflected. The prevailing marketing channels, prices, and the distribution of market shares are examined as indicators to constraints in the reflection of price signals to producers.

Marketing Channels and Commodity Flows

Cool season food legumes are mainly produced for the market. As much as 86% of faba bean production is disposed of for sale and about 10% is retained as seeds (Gibreel and Faki, 1984). Consumption by producers is limited to the dominated by sorghum and wheat consumption. Farmers' need for cash may constitute another factor behind large sales.

The faba bean marketing channels sketched in Figure 3 show that various agents are engaged in the commodity movement from producers to consumers. The largest faba bean dealers are wholesalers in Khartoum who usually have agents in the producing areas buy and sometimes store the crop in those areas. However, buyers vary from a farmer who collects small quantities from others to lorry merchants who engage in short term speculative transactions from production areas to big towns. Big wholesalers in Khartoum channel the commodity to distant towns in southern and western Sudan. Consumers get

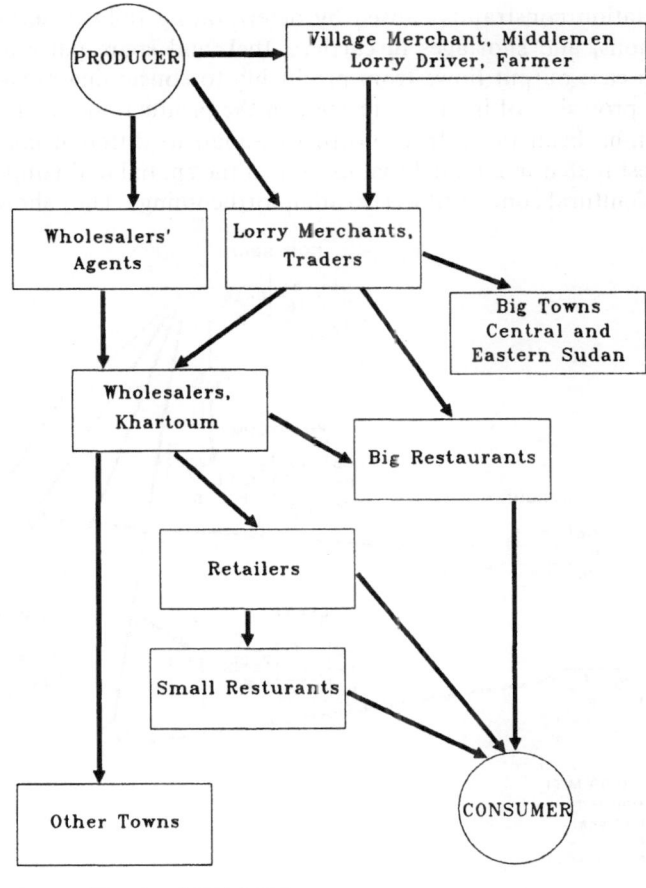

Source: Gibreel and Faki (1984).

Figure 3. Marketing channels of faba bean in Sudan.

remaining 4%. This may be partly due to the lower population densities in the rural producing areas, but also because of habits in those areas which are their supplies from retailers, although a large portion of them buy cooked faba bean from restaurants.

There are no specialized traders in faba bean, but transactions are usually made by the existing traders in a variety of grains and legumes. The vast majority of farmers do not have market access further than a periodic (usually weekly) local market.

The low amounts of faba bean consumption in many producing areas precludes farmers' knowledge about prices that prevail in distant important consumption centers. Moreover, most of the product leaves farmers' hands shortly after harvest when the prices are at very low levels. Rapid farmers' sales are dictated by their need for cash after harvest, since no credit facilities are available to them.

Transportation constraints caused by deteriorating railway services, poor road conditions, and shortages in carriers, fuel, and spare parts need to be addressed to ease output flows from producing to consuming areas and also improve the provision of inputs to farmers in the production locations.

Optimal faba bean flows from northern Sudan to different consumption regions are estimated in a model that examined the spatial and temporal flows of major agricultural commodities (Hashim, forthcoming). They show quantity

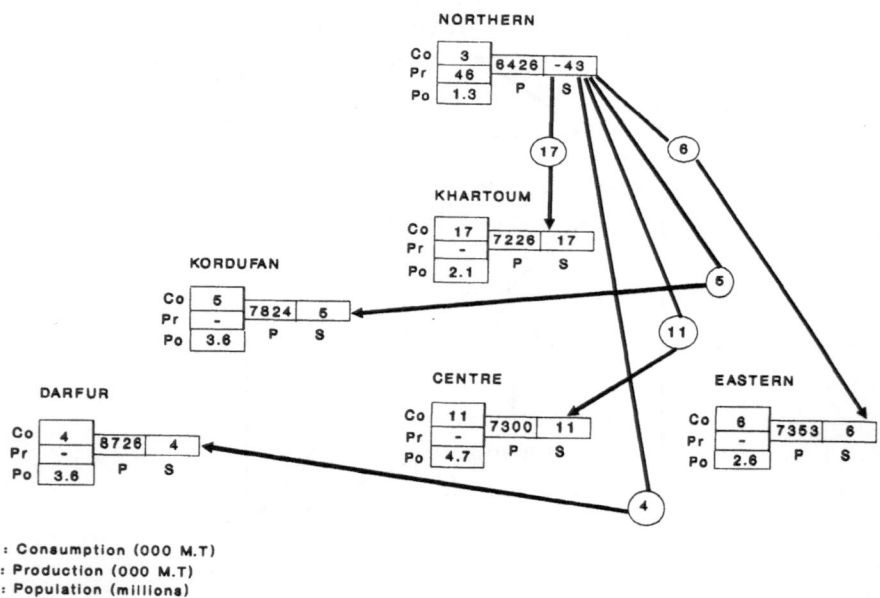

Co : Consumption (000 M.T)
Pr : Production (000 M.T)
Po : Population (millions)
P : Price (Ls/ton)
S : Spatial flows (000 M.T)

Figure 4. Spatial flows of faba beans in the first year of three planning periods (Sudan; based on data of season 1988/89).

shipments and equilibrium prices that reflect transportation costs of a uniform market (Figure 4). The quantities are a function of population densities, transportation costs, and consumption habits. The differences in optimal prices between the exporting and importing regions are considerably less than the actual differences and may be caused by high traders' profits. This reflects some level of distortion in the market. On the other hand, it is doubtful whether remote areas such as Darfur and Kordofan receive the quantities they demand, reflecting another level of distortion.

Marketing Costs and Distribution of Marketing Margins

Data from earlier studies (Gibreel and Faki, 1984; Yousif, 1985) and recent information from the Wad Medani market are used to estimate current costs and returns of marketing of faba bean produced in the Nile province (Table 5). Crop losses in store account for most of the marketing costs. Cleaning faba bean from debris is another expensive marketing activity which is performed by retailers and accounts for losses estimated at 13%. Transport costs represent a sizeable share, 7.3%, of marketing costs.

Table 5. Estimates of average returns for faba bean in Sudan

Item	LS t⁻¹	% of Retail Price
Retail Price	83333	100.0
Marketing Margins	47333	56.8
Farm-Gate Price	36000	43.2
Marketing Costs	27350	32,8
Traders' Profit	19983	24.0

Abdelmagid (1992)

The impact of market costs reductions is apparent from the effect of different levels of transport costs (Hashim, forthcoming). Table 6 shows the effect for faba bean. When transportation costs are doubled, total faba bean supply is reduced by more than 4%. A 50% reduction on the other hand is associated with about 1% increased supply. Faba bean prices are also affected by these changes. Since the value of storage losses is by far more than the transportation costs, it is expected that reductions in the latter will have a greater impact on the supply and prices of the crop.

Storage losses are primarily caused by bruchids (*Bruchus incernatus*) in the store. The degree of attack by that pest was estimated at 17% of the amount stored (Gibreel and Faki, 1984), which is assumed to result in a reduction in the faba bean retail price by the same percentage. Certain simple store treatment

Table 6. Effect of transportation costs on faba bean production in Sudan, results of a normative planning model over three years (thousand tons)

Year		Levels of Transportation Costs	
	Existing	100 % Increase	50 % Decrease
First	45.7	44.0	46.5
Second	45.9	44.7	46.5
Third	46.7	44.8	47.5
All Years	139.3	133.5	140.5

methods have been developed that proved to be technically and economically feasible for faba bean (Bushara *et al.*, 1984, 1985, 1986). Institutional support, in terms of provision of the necessary inputs, is crucial for wide implementation of the technology.

The prices received by farmers amounts only to 43% of the retail price, as influenced by the relatively high marketing margins of about 57% (Table 5). This reflects location and transport problems, besides the many marketing agents who reap high profits. In comparison, farmers in the Gezira Scheme, where faba bean is being introduced, get as much as 61% of the retail price due to their proximity to major consumption areas. This results in lower marketing margins (Abdelmagid, 1992).

Traders' profits of 24% are also high. Such profits are earned by exploiting the high price seasonality in faba bean. Five months after harvest prices were more than double the farm-gate prices (Gibreel and Faki, 1984). The wide price variation is caused by the crop moving to the hands of traders directly after harvest, but partly by the fact that pulses are produced only once a year while its consumption extends over the whole year.

Government Intervention

Production and marketing activities of faba bean and chickpea are almost completely handled by the private sector with hardly any government intervention. The government role is significant in boosting lentil production by enhancing input availability to farmers undertaking its cultivation. However, some policy measures have an effect on cool season food legumes. For example, the encouragement of wheat production to reach self-sufficiency affected the areas allocated to faba bean and accordingly its supply. This led to sharp increases in its price in 1990/91. Under such conditions, a compensating action could be an institutional support for the provision of some important inputs, especially improved seed and pesticides, to boost the supply of cool season legumes through yield improvement. Additionally, the current institutional support for technology development related to faba bean integration in irrigated farming in central and eastern Sudan could be extended to include extension and input supply. This represents a large potential for increased supply to meet increases in demand, especially if consumption in rural areas is widely promoted.

On the marketing side, there is hardly any tangible government role. However, imports of lentil are facilitated and those of faba bean are encouraged in years of short supply. The low producer prices caused by the sharp price seasonality and the farmers' need for cash after harvest call for policy legislation to guarantee encouraging prices and institutional credit support for farmers.

Development in India

The marketing system of food grains in India has been relatively efficient (von Oppen and Rao, 1987). The system of regulated markets assures competitive price formation even if market access could still be improved. Metric measures and weights facilitate efficient price formation; market information systems assure comparability of prices across the country. The existing infrastructure of railways and roads provides a functional transportation network so that interregional trade is reasonably well developed even though capacity constraints remain a limiting factor. The general liberalization of earlier trade restrictions after abolition of food zoning in the 1980s has led to an increasing specialization of crop production, thus mobilizing agricultural resources through regional specialization according to comparative advantage.

Market Channels

The marketing system of chickpea in India is shown in Figure 5. Chickpea is sold in primary wholesale markets directly by the producer or through a broker or commission agent. The bulk of the chickpea travels from the primary market to the nearby mills where it is processed and converted into splits called dhal. From the mills, a large quantity of dahl moves to secondary markets and reaches the consumer via retailers. Small amounts of chickpea dhal are further processed into flour.

In India the major part of the chickpea crop sold by farmers is offered in the regulated market yards. Farmers retain 55% of their chickpea production on the farm while the remaining 45% is marketed (Raju and von Oppen, 1980; 1982).

Supply and Demand Relations

In rural areas of India, income elasticity of demand for the lowest four expenditure groups is around 0.5 for chickpea and greater than 1.0 for other

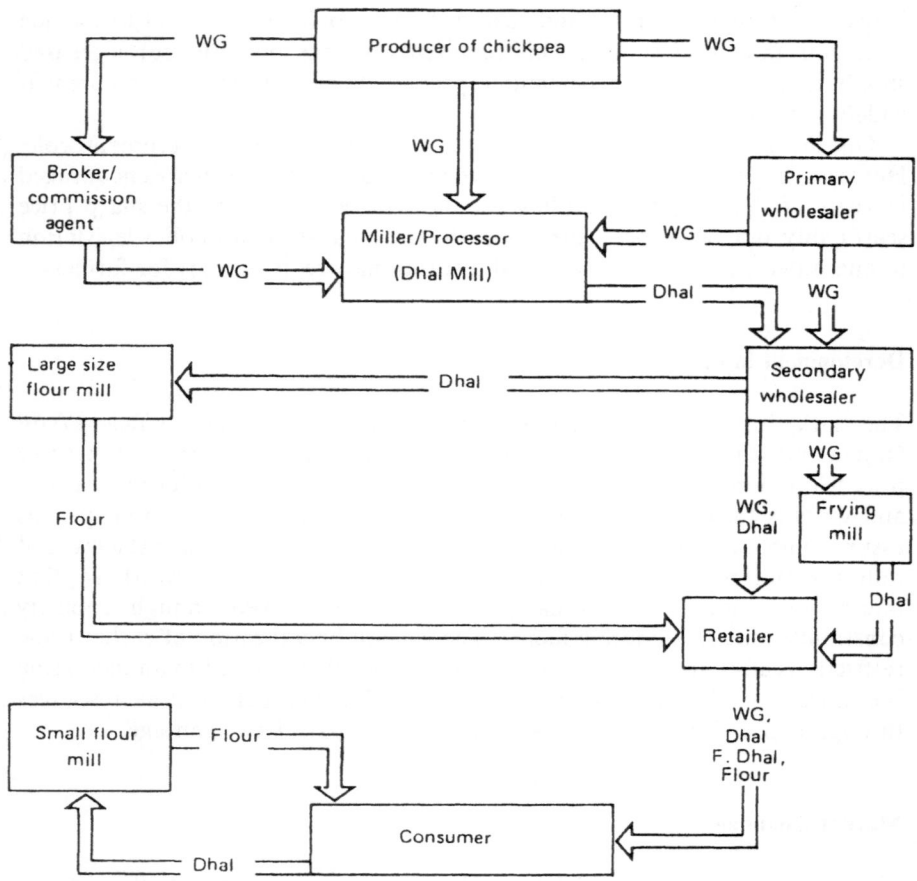

WG = Whole Gram; Dhal = Split Gram; Flour = Split Gram Flour; F. Dhal = Fried Dhal.

Figure 5. Chickpea marketing channels in India.

pulses (Table 7). Accordingly, low income consumers will increase their consumption of other pulses compared to chickpea with income rise. In urban areas, income elasticities of demand for both chickpea and pulses decline for higher expenditure groups.

Chickpea consumers, in both rural and urban areas are highly responsive to changes in chickpea prices. This implies that if production of grain legumes remains the same while population and income continue to increase at the same rate as in the past, demand for pulses will increase and prices will rise but demand for chickpea will not increase as fast.

Given the high absolute total price elasticities for the lower income groups, this section of the population will considerably reduce consumption of pulses

Table 7. Income and price elasticities of demand for chickpea in rural and urban areas of India, by expenditure groups (in ascending order of expenditure)

	Expenditure Groups				
	1	2	3	4	5
RURAL INDIA					
Own Price Elasticity					
Chickpea	-1.03	-1.61	-0.81	-1.06	-0.20
Other pulses	-1.43	-0.91	-0.63	-0.36	-0.05
Income Elasticity[1]					
Chickpea	0.50	0.79	0.47	0.47	0.07
Other pulses	1.82	1.02	1.03	0.53	0.46
URBAN INDIA					
Own Price Elasticity					
Chickpea	-2.90	-2.89	-1.01	-1.00	-0.15
Other pulses	-1.07	-0.67	-0.59	-0.38	-0.29
Income Elasticity[1]					
Chickpea	1.26	0.99	0.25	0.07	0.01
Other pulses	1.47	0.96	0.72	0.44	0.14

[1] Mean expenditures were used to approximate income.

Source: Murty, K.N. and von Oppen, M. (1985)

while the higher income groups, who exhibit low absolute price elasticity, will continue to consume pulses though in reduced quantities.

The output supply elasticity for chickpea was estimated to be around 0.5 (Bapna *et al.*, 1981); for total pulses for all the SAT region the elasticity was 0.43.

The price elasticities of demand for chickpea are about twice as high as the supply elasticities. Therefore, it could be expected that market prices should have a tendency to stabilize quickly; changes in supply would tend to be absorbed by the elastic demand.

Producers' Share in Consumers' Rupee

The producer share in the consumer rupee spent on chickpea is about 78% in Warangal market and about 81% in Tandur market (Table 8). This share is relatively high, indicating that market channels in India appear to function quite efficiently.

Table 8. Estimates of marketing costs and margins for chickpea in two markets of Andhra Pradesh, 1975/76

Item	Warangal		Tandur	
	Rs/q	% to consumer's price	Rs/q	% to consumer's price
Farmer's sale price	194.80	79.5	207.59	82.1
Farmer's market charges	4.13	1.7	3.26	1.3
Primary wholesale traders'	10.66	4.3	10.51	4.2
Millers'	9.60	3.9	7.55	3.0
Retailers'	4.40	1.8	3.06	1.2
Consumers' purchase price of dhal	245.17	100	252.77	100
Farmers' share in consumers' rupee		77.77		80.83

Source: compiled from Raju, V.T. and von Oppen, M. (1982)

Overall Development

Against this background it was reasonable to expect that, with markets becoming more accessible, productivity and welfare would increase and price signal should provide clearer information to producers. Thus, producers should be induced to shift production according to comparative advantages and consumers should be induced to choose consumption patterns according to comparative utility of food items.

The overall result has been a significant increase in per capita food production in India over the last 10 years by nearly 20% (FAO, 1991). However, this does not necessarily imply that all crops change at the same rates nor even participate in this increase. Depending upon the comparative advantage of a particular crop and upon its comparative utility, production and consumption will change individually. In India, an interesting comparison can be made between three legumes, i.e., chickpea, soybean, and lentil competing for the same resources (Table 9).

Chickpea production declined because farmers allocated less land to chickpea; and, even though yield increased slightly, this technology improvement was not sufficient to offset the area decline. Thus, production of chickpea in India declined. Soybean on the other hand, which was introduced in India in the early 1970s, has expanded to more than 2 million ha today. Benefiting from significant yield increases, production of soybean with over 2 million tons today amounts to half the quantity of chickpea produced.

Soybean is generally cultivated on the same land as chickpea (vertisols in

Table 9. Area, yield and production of selected legume crops in India

Year	lentil			chickpea			soybean		
	$p^{1)}$	$y^{2)}$	$a^{3)}$	p	y	a	p	y	a
69 - 71	0.4	505	0.8	5.0	663	7.5	0	545	0
79 - 81	0.4	438	0.9	4.5	627	7.1	0.4	679	0.5
82									
83	0.5	490	1.0	5.3	715	7.4	0.6	757	0.8
84	0.5	566	0.9	4.8	663	7.2	0.7	734	1.2
85	0.5	557	1.0	4.6	661	6.9	1.0	970	1.3
86	0.7	607	1.1	5.8	742	7.8	0.8	978	1.4
87	0.7	621	1.1	4.5	657	6.8	0.9	990	1.4
88	0.7	635	1.0	3.6	629	5.8	1.5	892	1.7
89	0.7	679	1.1	5.1	753	6.8	1.7	804	2.1
90	0.7	641	1.1	4.2	652	6.5	2.2	1000	2.2

1) p = production in 1,000 tons
2) y = yield in kg ha^{-1}
3) a = area in 1,000 ha

Government of India Agriculture in Brief

central India), but during the rainy season while chickpea is planted after the rains on residual moisture. Chickpea also competes with wheat as a winter season crop. Soybean is an ideal crop to precede the planting of wheat which benefits from the soybean's nitrogen as long as good rains or irrigation assure sufficient soil moisture. In this context, it is not surprising that chickpea – unless new production technologies can boost their yields – is losing out in this competition.

Area under lentils also expanded so that, with increasing yields, production nearly doubled in India. Lentil in India is also being grown in the same regions of chickpea production, but generally on poorer land.

Government Intervention

The explanation for the difference in development of chickpea vs. soybean production in India can be found in price policies. A study by Gulati and Sharma (1991) analyzed the degree of protection which different crops receive through government policies in India. The degree of protection against the world market was measured in two ways: (a) by a nominal protection coefficient (NPC), i.e., the ratio of farm gate price over the price the farmer would receive under free-trade, and (b) an effective subsidy coefficient (ESC), i.e., the ratio of value added at international prices of traded inputs (e.g., fertilizer, pesticides) as well as non-traded inputs (e.g., irrigation).

Comparison of the development of NPCs and ESCs for chickpea and soybean over the years 1980–81 to 1986–87 show that for soybean the average NPC was 1.14 indicating a price protection of 14% above world market levels and the ESC was 1.21, thus indicating that including subsidy protection the protection rate is 21% above world market levels. For chickpeas the protection rate over these years was near unity (average NPC = .99 and average ESC = 1.02). These figures imply that in India chickpea is being produced at world market conditions while soybean enjoys considerable price as well as input subsidies.

The study also points out for pulses that "exports are banned... Imports are permitted freely under open general license, since 1978–79, subject to registration with National Agricultural Cooperative Marketing Federation (NAFED) which monitors imports of pulses. There have been no import duties for 4 to 5 years after 1980–81. However, a duty of 10% was imposed in 1985 which was raised to 35% in March 1989, but lowered to 10% in November 1989, and has remained unchanged since then. Import of pulses which was only 173,000 tons... in 1980–81 reached a peak of 825,000 tons in 1988–89. During 1990–91, import of pulses up to December was already 592,000 tons. Since market prices rule much above support prices there are obligatory domestic purchases by government or NAFED. However, NAFED makes commercial purchases of various pulses from major producing states and also directly imports some. These are sold to various civil supplies corporations and state marketing federations" (Gulati and Sharma, 1991).

Comparison Between the Sudan and India

The observed increase in production of cool season food legumes in the world was up to 90% caused by a tripling in pea production in the developed countries, while in the developing countries the increase in chickpea and lentil contributed only 10% and faba bean remained stagnant. The weak performance of cool season food legumes in the developing countries can be attributed to various changes in technological innovation and institutional constraints. The two countries, India and Sudan which were selected to assess the situation, lead to certain conclusions.

Technological Innovation

A comparison of the findings for these countries shows for India that lentil yields increased from 450 to about 650 kg ha^{-1}, and chickpea yields rose from 650 to about 700 kg ha^{-1}; however, for soybean, a newly introduced crop competing with lentil and chickpea, yields increased from about 600 to over 800 kg ha^{-1}.

In the Sudan, chickpea yields increased from 1,250 to about 1,700 kg ha^{-1} and lentil is appearing as a new crop with yields around 1,000 kg ha^{-1}. Only yields of faba bean in the Sudan fluctuated around 1,800 kg ha^{-1}. Over the last 10 years, increases for chickpea and lentil of 10 to 40% were obtained in both countries. Technological innovations have caused yield increases for the two crops in India and for lentil in the Sudan, although for the latter the role of institutional support was also substantial. For chickpea in the Sudan, there are indications that price increases were the main reasons behind the yield rise.

New technologies have been developed for faba bean that have been verified to increase productivity and farmers' incomes in northern Sudan. Their adoption, although taking place in some areas, is not yet widely spread. Extension services and technology's input-provision are expected to enhance technology use which will reflect yield and supply of faba bean. Other technology developments have facilitated successful growing of faba bean in large irrigation schemes in central and eastern Sudan. The large potential in those areas could be exploited to increase the supply of the crop.

Institutional Constraints

The marketing systems for chickpea in India indicate that despite the processing into dhal the overall marketing margin is only about 20%, thus allowing the producer to get about 80% of the price paid by the consumer. In comparison, the faba bean producer in the Sudan receives only about 43% of the consumer's price. This is to some extent caused by the high costs of transport from the producing area in the north to the consumption centers in other parts of the country. However, it is obvious that, compared to India, considerable deficiencies in the marketing system are likely to exist in the Sudan. Complex marketing channels, involving many agents, are partly causing such deficiencies. Moreover, liquidity constraints after harvest are forcing farmers to sell at low prices, despite high consumer prices.

In India, the relatively well-functioning system of agricultural markets has helped the introduction and expansion of soybean as a new crop, at the expense of chickpea. Also market price policies in India provide a subsidy of about 21% above world markets to soybean producers. On the other hand, chickpea producers operate very near to world market levels as India is increasingly importing chickpea and other pulses from outside, e.g., from Australia.

In the Sudan, pulses are also imported from outside at increasing quantities, probably keeping the level of prices in Sudan for faba beans and lentils at world market levels. Given a unit value cool season food legumes of two to three times that of cereals, an expansion in international trade of these pulses is a logical consequence which will generate considerable gains in overall economic welfare. Over the last decade the quantities traded in world markets of grain legumes increased from 6 to over 9% of production, while cereals are traded at constant rates of 12 to 13%. The expansion of world trade in pulses is very likely to continue, especially if, as a consequence of the GATT negotiations, Europe should be obliged to open its markets and import much of those cool season food legumes which, up to now, have been increasingly produced within its own, up to now, well-protected borders.

References

Abdelmagid, S. A. 1992. *The economic impact of faba bean introduction in small holdings, a case study of the Gezira Scheme, Sudan.* Ph.D. thesis, University of Hohenheim, Germany.

Agostini, B. B. and Khan, D. 1988. In: *World Crops: Cool Season Food Legumes,* pp. 461–482 (ed. R. J. Summerfield). Dordrecht: Kluwer Academic Publishers.

Bank of Sudan. 1989. *Annual report, 1988.* Khartoum, Sudan.

Bushara, A. and Faki, H. 1986. On-farm evaluation of chemical control of bruchids. In: *Report of the 5th Annual Coordination Meeting of the ICARDA/IFAD Nile Valley Project on Faba Beans,* 9–13 Sept., 1985, Cairo, Egypt: ICARDA.

Bushara, A. and Faki, H. 1987. On-farm evaluation of chemical control of bruchids. In: *Report of the 5th Annual Coordination Meeting of the ICARDA/IFAD Nile Valley Project on Faba Beans,* 23–27 Sept., 1986, Cairo, Egypt: ICARDA.

Faki, H. and Nourai, A. H. 1991. *Quarterly Journal of International Agriculture* 30: 236–247.

FAO (Food and Agriculture Organization). *Production Yearbook,* various issues, Rome. 1980, 1985, 1991. Rome, Italy: FAO.

FAO (Food and Agriculture Organization). *Trade Yearbook,* various issues, Rome. 1980, 1985, 1991. Rome, Italy: FAO.

Gibreel, M. and Faki, H. 1984. Marketing of faba bean: Zeidab crop, 1983/84. Report presented in the 5th Annual Coordination Meeting of the ICARDA/IFAD Nile Valley Project on Faba Beans, 13–17 Sept., 1986, Cairo, Egypt.

Government of India. *Agriculture in Brief,* various issues, Ministry of Agriculture, New Delhi, India.

Government of India. *Bulletin on Food Statistics,* various issues, Ministry of Agriculture, New Delhi, India.

Gulati, A. and Sharma, P. K. 1991. *Journal of Indian School of Political Economy* 3: 205–237.

Hashim, A. A. *Impact of Market Access on Crop Productivity in Sudan.* Ph.D. thesis. University of Hohenheim, Germany (In press).

Murty, K. N. and Von Oppen, M. 1985. In: *Proceedings of the International Workshop in Aricultural Markets in the Semi-Arid Tropics,* pp. 179–200, 24–28 Oct. 1983, ICRISAT Center, India: ICRISAT.

Nourai, A. H. 1987. In: *Summary Proceedings of the Consultative Group Meetings for Eastern and Central African Regional Research on Grain Legumes,* pp. 59–64, 8–10 Dec. 1986, Addis Ababa, Ethiopia. ICRISAT Center, India: ICRISAT.

Parthasarathy Rao, P. and von Oppen, M. 1987. In: *Food Legume Improvement for Asian Farming Systems: Proceedings of an International Workshop,* pp. 54–63 (eds. E. S. Wallis and D. E. Byth). 1–5 Sept. 1986, Khon Kaen, Thailand, ACIAR Proceedings no. 18. Canberra, Australia: Australian Centre for International Agricultural Research.

Raju, V. T. and von Oppen, M. 1982. *ICRISAT Economics Program Progress Report 32.* ICRISAT Center, India: ICRISAT.

von Oppen, M. 1990. In: *Proceedings of the Second International Workshop on Chickpea Improvement,* pp. 31–39 (eds. H. A. van Rheenen and M. C. Saxena). 4–8 Dec. 1989, ICRISAT Center, India: ICRISAT.

von Oppen, M. and Parthasarathy Rao, P. 1987. In: *The Chickpea,* pp. 383–397 (eds. M. C. Saxena and K. B. Singh). Wallingford, Oxfordshire, UK: CAB International.

von Oppen, M. and Parthasarathy Rao, P. 1988. In: *World Crops: Cool Season Food Legumes*, pp. 487–500 (ed. R. J. Summerfield,). Dordrecht: Kluwer Academic Publishers.
Yousif, F. S. 1988. *Faba Bean Marketing and Markets in Sudan*. M.Sc. thesis. University of Gezira, Sudan.

Chickpea and lentil production in Turkey

N. AÇİKGÖZ[1], M. KARACA[2], C. ER[2] and K. MEYVECI[2]

[1] *Aegean Agricultural Research Institute, P. O. Box 9, Menemen, Izmir, Turkey, and*
[2] *Field Crop Improvement Center, P. O. Box 226 Ulus, Ankara, Turkey*

Abstract

Chickpea (*Cicer arietinum*) and lentil (*Lens culinaris*) are important and widely cultivated food legumes in Turkey. Extensive research on food legumes started in 1975 with the initiation of the "National Food Legumes Research Project". Since then, four research institutes, in particular, have carried out research on breeding and agronomic aspects of these crops, which has resulted in the release and registration of a number of chickpea and lentil cultivars.

The "Utilization of Fallow Areas Project" (UFAP) was initiated in 1982 with the goal of reducing the amount of land area devoted to fallow. The first phase of the UFAP was completed in 1986 and 1.4 million hectares which had been devoted to fallow were converted to annual cropping. The second phase of the project began in 1987 and an additional 1.7 million hectares were converted to annual cropping. Lentil and chickpea are now produced on 36.9 and 38%, respectively, of the land previously devoted to fallow.

Since traditional methods are still common in chickpea and lentil cultivation, there have been no significant increases in yield. Sowing, weeding, and harvesting are still performed manually. Diseases and weeds are the most important limiting factors in production. A seed production program was established recently which is responsible for producing seed of all the registered chickpea cultivars and four registered lentil cultivars.

Introduction

Food legumes are produced on about 11% of the area of over 25 million hectares devoted to field crops in Turkey (TSAS, 1989). Fallowing is practiced on about 5.2 million hectares annually and is most often used in the production of wheat or barley. In a usual 24 month period, more than 8 million hectares are devoted to fallow-cereal (wheat or barley) rotations which has been a common practice on the Anatolian Plateau for centuries. In a typical field, a cereal is planted in November and harvested in August. After harvest, the field is fallowed for

F.J. Muehlbauer and W.J. Kaiser (eds.), Expanding the Production and Use of Cool Season Food Legumes, 388–398.

about 14 months and planted to a cereal crop again.

Fallowing has been a traditional practice where annual precipitation is limited and variable; however, recent research designed to decrease such areas has produced some promising results. The first attempts to utilize fallow areas for crop production were made in the late seventies by the "Çorum – Çankìrì Rural Development Project" in the northern transitional area. Before the implementation of that project, fallow-wheat and fallow-barley were the main cropping sequences. Food legumes, such as chickpea and lentil, were produced on many farms of that region for their own use. By making credits available, legume production was made more attractive to farmers as an alternative to fallow. Over a five-year-period, the legume-wheat rotation system was largely adopted and resulted in a large increase from 1975 to 1991 in the area devoted to chickpea and lentil (Tables 1 and 2).

The success of the "Çorum – Çankìrì Project" served as a basis for the Utilization of Fallow Areas Project (UFAP) which was directed toward the utilization of fallow areas by alternative crops. The first phase of the project (1982–1986) proposes to utilize 1.4 million hectares of the fallowing area and in the second phase (1987–1991) proposes 1.7 million hectares of the fallowing area additionally. Incentives of seed and fertilizer credits were offered to producers in order to reach this amount of land area. These incentives hastened

Table 1. Area and production of chickpea (1973–1991)

Year	Area	Production	Yield
	000 ha	000 tons	kg ha^{-1}
1973	184	185	1006
1974	175	195	1114
1975	139	171	1228
1976	137	170	1232
1977	138	130	1304
1978	168	205	1220
1979	200	225	1125
1980	240	275	1146
1981	200	195	1023
1982	245	280	1143
1983	334	290	867
1984	345	335	971
1985	399	400	1003
1986	534	630	1180
1987	665	725	1106
1988	778	1040	1336
1989	818	520	858
1990	890	860	965
1991	883	895	1013

Source: TSAS, Agricultural Structure and Production (1973–1991)

Table 2. Area and production of lentil (1973–1991)

Year	Area	Production	Yield
	000 ha	000 tons	kg ha⁻¹
1973	85	672	790
1974	116	120	1035
1975	124	135	1085
1976	186	210	1129
1977	240	260	1083
1978	177	180	1015
1979	175	180	1029
1980	191	195	1023
1981	255	280	1098
1982	622	550	883
1983	650	650	1000
1984	620	570	919
1985	597	618	1035
1986	750	850	1133
1987	916	925	1010
1988	983	1040	1062
1989	997	520	522
1990	906	846	934
1991	781	727	929

Source: TSAS, Agricultural Structure and Production (1973-1991)

Table 3. Changes of field crops and fallow area during the Utilization of Fallow Areas Project (1981–1988)

	Year		
	1981	1986	1988
Field Crops area (000 ha)	16711	18149	18995
Fallow area (000 ha)	8204	5771	5179
Lentil area (000 ha)	255	750	983
Chickpea area (000 ha)	200	534	778

the conversion of fallow areas to crop production.

The success of the project resulted in a 37% decrease (3 million hectares) in land area devoted to fallow from 1981–88. Of that area, 2.3 million hectares were utilized for the production of field crops; and, of that total 43 and 34% was used for the production of lentil and chickpea, respectively (Table 3).

Table 4. World chickpea production (1990)

Countries	Production
	000 tons
India	4232
Turkey	890
Pakistan	562
China	250
Mexico	170
Spain	59
Syria	36
Tunisia	28
World total	6908

Source: FAO Production Year Book (1991)

Table 5. Lentil production in the world (1990)

Countries	Production
	000 tons
Turkey	900
India	703
Canada	219
Bangladesh	162
China	150
Syria	101
Nepal	76
Iran	51
USSR	50
USA	40
Pakistan	38
World	2692

Source: FAO Production Year Book (1991)

Turkey has become a major producer of chickpea and lentil and the leading exporter of these crops in the world (Tables 4 and 5). Chickpea is produced in almost all regions of the country (Table 6); whereas, lentil is more concentrated in central Anatolia and in the southeast (Table 7). Kabuli type chickpeas predominate production in Turkey mainly because of consumer preference.

Large seeded green (var. macrosperma) and small seeded red (var. microsperma) are the predominant types of lentil produced in Turkey, and account for approximately 30 and 70% of the production, respectively. Nearly the entire green lentil crop is produced in Central Anatolia; whereas, red lentil is produced primarily in the southeast.

Increased area devoted to irrigated crops in the southeast is expected to result in reduced lentil production as a result of expected changes in cropping patterns

Table 6. Area and production of chickpea by regions (1991)

Regions	Area sown	Production	Yield
	-- ha --	-- ton --	-- kg ha^{-1} --
C. North	180347	193733	1074
Aegean	126103	129170	1024
Marmara	4715	5071	1075
Mediterranean	172687	154260	893
Northeast	6979	4542	650
Southeast	121577	122667	1008
Black Sea	2570	1977	769
Central East	88446	103003	1164
C. South	179982	180949	1005
Total	883506	895372	1013

Source: Turkish State Agricultural Statistics Estimates, 1991
(unpublished)

Table 7. Area and production of lentil by regions (1991)

Regions	Area sown		Production		Yield	
	green	red	green	red	green	red
	-- ha --		-- ton --		-- kg ha^{-1} --	
C. North	184736	2750	155857	3475	843	1263
Aegean	3805	19	3832	19	1007	1000
Marmara	859	211	1201	221	1398	1047
Mediter.	1754	48277	2088	45477	1190	942
Northeast	2403	50	2088	51	868	1002
Southeast	1470	448460	1133	433953	770	967
Black Sea	908	20	892	20	982	1000
C. East	15943	31273	16112	27534	1010	880
C. South	41300	879	36316	355	879	887
Total	253178	528915	219519	507765	867	960

Source: Turkish State Agricultural Statistics Estimates, 1991
(unpublished)

in the future. It appears that the area devoted to red lentil production has already begun to decrease (Table 2).

Constraints and Factors for Low Yield

Chickpea and lentil production increases (Tables 1 and 2) have resulted from an expansion of area planted and not to yield increases. Environmental, agronomic, and biotic factors appear responsible for the comparatively small yields of chickpea and lentil.

Environmental Constraints

Chickpea and lentil are grown in dry areas without the benefit of irrigation. Farmers prefer to plant chickpea in spring, usually before the end of May, in order that the crop can escape from Ascochyta blight [*Ascochyta rabiei* (Pass.) Lab]. Ascochyta blight can be devastating to chickpea crops planted in autumn or early spring. Autumn or early spring planting would be possible if Ascochyta blight resistant, cold tolerant cultivars were developed.

Lentil is seeded in spring in the central and eastern regions and in late autumn in the southeast. Limited soil moisture and increasing drought stress during the growing period generally reduce yields of both lentil and chickpea.

Mediterranean-type precipitation prevails throughout the country with approximately 65% of the total annual precipitation occurring in winter and spring. Central Anatolia and certain areas of the southeastern region receive 300 to 500 mm of annual precipitation. The dry season usually begins in June and continues until the end of October during which the active growing period for lentil and chickpea takes place.

Agronomic Constraints

Although there are some released cultivars, farmers still rely on unimproved land races which generally have limited yield potential. The land races have been systematically collected throughout the country and have been preserved in gene banks for use in crop improvement programs.

Poor seedbed preparation is generally regarded as a major factor limiting yields of lentil and chickpea. Farmers broadcast the seed on the soil surface and then cover the seeds by plowing. Seed drills for lentil are used in the southeastern region where fields are large, usually level and mostly free of stones. Seeding with drills is not common for chickpea except on the state farms. Chickpea is sown in the transitional areas with a simple machine consisting of a seed hopper, three seed cells, fluted force feeds, three seed tubes and a mechanical drive wheel. This apparatus is mounted on a plow or cultivator (Açìkgöz, 1987). Farmers rarely use fertilizers and insecticides are very seldom applied for insects. Herbicide use is also rare. Harvesting is still performed manually; however, combines are used on the state farms and on some of the larger farms.

Biotic Constraints

Ascochyta blight is the most serious disease of chickpea in Turkey and can destroy entire fields under epiphytotic conditions. Resistant cultivars is the only satisfactory means of controlling the disease.

Fortunately, lentil is relatively free of disease problems. However, diseases such as wilt/root rot complex (*Fusarium solani*, *F. oxysporum*, *F. acuminatum*), downy mildew (*Peronospora lentis*), and *Ascochyta* (*Phoma medicaginis* var. *pinodella*) have been found in southeastern Anatolia.

Pod borer (*Helicoverpa* spp.), seed beetles (*Bruchus* spp.), and leaf miners (*Liriomyza* spp.) are common pests of chickpea and lentil. *Apion* sp. can also reduce yields of lentil.

Weeds are a major yield limiting factor for chickpea and lentil. Hand weeding is done by family members on small farms, but weeding is not economic and large fields are often left weedy because of the high labor costs. Herbicide use on grassy weeds is not a common practice. Effective herbicides are not available for broadleaf weeds in legumes.

Although lentil yields can be increased by winter planting, competition from weeds, not winter hardiness, is a major factor preventing a more widespread use of the practice.

Research Activities

Extensive research on food legumes began in 1975 with the "National Food Legumes Research Project". In 1982, additional studies were initiated under the UFAP.

Breeding Programs

Chickpea and lentil breeding programs have involved the creation of genetic variation, selection for desirable plants with disease resistance, and the evaluation of the selected lines for commercial production. Genetic variation for use in selection has been obtained through the direct use of cultivars and segregating material from other countries or from within the country. Genetic variation has also been obtained through hybridization of promising selections and germplasm lines and from mutagen treated material.

Breeding materials either introduced or developed through hybridization have been entered in observation nurseries, preliminary yield trials, advanced yield trials, regional yield trials, and yield trials for registration. Promising selections are registered as new cultivars followed by seed multiplication and distribution.

Chickpea Breeding Program

The main objectives of the chickpea breeding program include; resistance to Ascochyta blight, large seed size, improved yield potential, cold and drought tolerance, and suitability for mechanization.

In general, chickpea breeding activities of the agricultural research institutions are based on introduction, hybridization, and selection. The Aegean Agricultural Research Institute (AARI) and the Field Crops Improvement Center (FCIC) initiated a hybridization program in the early 1980s. Pedigree, bulk-pedigree and backcross-pedigree methods are used for the development of large seeded, Ascochyta blight resistant or tolerant cultivars. Some basic studies on Ascochyta blight have been done by various researchers and the inheritance of Ascochyta blight resistance was determined. A single recessive gene was found to be conferring resistance in lines 72012, ILC 195, and Nec 138–1; a single dominant gene was found controlling resistance in lines ILC 207 and ILC 202. Seven virulence patterns were determined in the *Ascochyta rabiei* isolates collected from the Aegean region (Açikgöz, 1983).

Screening programs for Ascochyta blight resistance in chickpea have been conducted by AARI and FCIC, both in the greenhouse and in the field, since 1980.

Screening for cold tolerance in chickpea is carried out by AARI and FCIC. No lines were sufficiently cold tolerant to survive the winter at Bozdağ, Izmir (1200 m altitude), but a number of lines survived winter field trials at Haymana, Ankara (1030 m altitude) (AARI, 1991; FCIC, 1991).

To date, four chickpea cultivars have been registered and two lines have been licensed for production (Table 8). In addition to the above mentioned cultivars, two or three other lines are planned for release in 1992 by AARI.

Chickpea seed production program is given in Table 9.

Table 8. Released chickpea cultivars

Cultivar	Institute	Year	Specific feature
Canitez-87	TARI	1987	HY,LS,MT,WA,HSB
Eser-87	AUAF	1987	HY,MS,MT,SB,ER
ILC 195/2*	AARI	1987	HY,SS,T,WA,TC,RB
ILC 482	SEAARI	1991	HY,MS,WA,RB
Akçin-91	FCIC	1991	HY,MS,TSB
87AK7112*	FCIC	1991	T,MS,RB

*: registered temporarily HY: high yield; LS: large-seed; MT: mid-tall; WA: wide adaption; MS: medium seed; T: tall; HSB: highly susceptible to blight; SB: susceptible to blight; ER: early; SS: small seed; TC: cold tolerant; RB: resistant to blight; TSB: tolerant susceptible to blight; SEAARI: South Eastern Anatolia Agricultural Research Institute; AUAF: Ankara University Agricultural Faculty; TARI: Transitional Agricultural Research Institute; FCIC: Field Crop Improvement Center; AARI: Aegean Agricultural Research Institute

Table 9. Seed production of chickpea cultivars (1991)

Cultivars	Foundation	Certified	Controlled
	-- tons --	-- tons --	-- tons --
Canitez-87	-	12	-
Eser-87	-	14	-
ILC/195-2	8.5	54	-
ILC 482	-	20	-
Akçin-91	-	-	-
87 AK 71114	-	30	-
87 AK 71112	0.1	0.8	-
Isp. population	-	-	191

Lentil Breeding Program

Lentil breeding is conducted by FCIC, Transitional Agricultural Research Institute (TARI) and South Eastern Anatolia Agricultural Research Institute (SEAARI), with the objectives of improving yielding ability, improving resistance to cold and drought, developing simultaneous maturity, and improving suitability for mechanized harvesting.

In contrast to the chickpea breeding program, lentil breeding is entirely based on germplasm introduction and selection. Research objectives of FCIC are to develop cold tolerant material which would be adapted to higher elevations of Turkey. Initial studies conducted in 1979–80 in cooperation with the International Center for Agricultural Research in the Dry Areas (ICARDA) were successful in identifying a number of cold tolerant lines (Erskine *et al.*, 1981). As a result of that initial success, collaboration has been established between FCIC, ICARDA and the US Department of Agriculture, Agricultural Research Service to develop cold tolerant cultivars that are specifically adapted to Central Anatolia.

Thusfar, eight winter and four spring lentil cultivars have been released for use in Turkey (Table 10). Seed production of new lentil cultivars has been limited and only four cultivars were included in the seed production program for 1991.

Agronomy Programs

Current research priorities include the development of improved agronomic practices specific to improved cultivars. With the initiation of the UFAP, special emphasis has been placed on crop rotation research. Detailed research on the long term effects of various systems on crop yield and soil condition has been conducted by FCIC and TARI. Studies have included sowing dates, plant populations, row spacings, methods and rates of fertilizer application, and weed control.

Table 10. Released lentil cultivars

Name	Type	Institute
Kişlik pul-11	red	AUAF
Kişlik yeşil-21	green	AUAF
Kişlik yeşil-31	green	AUAF
Kişlik kirmizi-51	red	AUAF
Sultan-1	green	TARI
Emre-20	red	TARI
Kayi-91	green	TARI
Sazak-91	red	TARI
Kişlik yerli kir.	red	SEAARI
Firat-87	red	SEAARI
Malazgirt-89	red	EAARI
Erzurum-89	green	EAARI

AUAF: Ankara University Agricultural Faculty
TARI: Transitional Agricultural Research Institute
SEAARI: South Eastern Anatolia Agricultural Research Institute
EAARI: East Anatolia Agricultural Research Institute

The results of research conducted by FCIC have shown that total nitrogen fixed and made available by legumes was at least equal to the amount accumulated by fallow at seeding time for wheat (Meyveci, 1988). Results have also shown that fallow was significantly better than other treatments in terms of moisture conservation. Winter legumes had an advantage over summer crops in rotation systems. Among the summer crops, legumes performed better than oil crops and barley, and chickpea utilized soil moisture more efficiently when compared to other grain crops (Karaca *et al.*, 1989a).

In rotation systems of winter legume-wheat and spring lentil-wheat, unweeded crops not only had decreased grain yield by 30 to 65% depending on tillage methods, they also had a greater negative effect on succeeding crops (Karaca *et al.*, 1989b; Durutan *et al.*, 1990).

FCIC placed special emphasis on weed control, particularly for the winter planted legumes. An integrated approach to weed control has been adopted and includes treatments in every management component to determine their effectiveness (Durutan *et al.*, 1989).

Numerous experiments are underway on the effects of rate and date of sowing for spring and winter lentil. Studies on the effects of seeding dates, depths of planting, and weed control for chickpea are underway by several research institutes.

References

AARI. 1991. *Annual Report of Food Legumes*. Aegean Agricultural Research Institute, Menemen, Izmir, Turkey (Turkish).
Açìkgöz, N. 1983. Ege Böl. Zir. Ar. Ens. Y. no: 29, Menemen, Izmir, Turkey.

Açìkgöz, N. 1987. In: *Proceedings of IAMFE/ICARDA Conference.* Aleppo, Syria, 23–27 May 1987. Aleppo, Syria: ICARDA.

Durutan, N., Güler, M., Karaca, M., Meyveci, K., Avçin, A. and Eyüboğlu, H. 1989. International Workshop on Soil and Crop Management for Improved Water Use Efficiency. 15–19 May, 1989. Ankara, Turkey.

Durutan, N., Meyveci, K., Karaca, M., Avcì, M. and Eyüboğlu, H. 1990. In: *A Case Study. The Role of Legumes in Dry Farming Systems of the Mediterranean Areas,* pp. 239–255 (eds. A. E. Osman et al.). Aleppo, Syria: ICARDA.

Erskine, W., Meyveci, K. and Izgin, N. 1981. *Lens* 8: 5–9.

FAO. 1991. *Production Yearbook Vol. 45.* FAO Statistics Series No. 104. 1992. Rome, Italy.

FCIC. 1991. *Annual Report of Food Legumes.* Field Crop Improvement Center, Ankara, Turkey (Turkish).

Karaca, M., Güler, M., Durutan, N., Meyveci, K., Avcì, M., Eyüboğlu, H. and Avçin, A. 1989a. International Workshop on Soil and Crop Management for Improved Water Use Efficiency. 15–19 May, Ankara, Turkey.

Karaca, M., Güler, M., Durutan, N., Meyveci, K., Avcì, M., Eyüboğlu, H. and Avçin, A. 1989b. 11th National Soil Science Congress. 1–4 Kasim 1989. Antalya, Turkey.

Meyveci, K. 1988. Orta Anadolu ko şullarinda ikili ekim nöbeti sisteminde toprakta nem ve inorganik azot formlarinin belirlenmesi. AÜZF Doktora Tezi (Turkish).

TSAS. 1989. Turkish State Agricultural Statistics. Ankara, Turkey.

Lentil production in Chile

J.U. TAY[1], G. BASCUR[2] and E. PEÑALOZA[3]

[1] Instituto de Investigaciones Agropecuarias (INIA), Estación Experimental Quilamapu, Casilla 426, Chillán, Chile;
[2] Instituto de Investigaciones Agropecuarias (INIA), Estación Experimental La Platina, Casilla 439, Correo 3, Santiago, Chile, and
[3] Instituto de Investigaciones Agropecuarias (INIA), Estación Experimental Carillanca, Casilla 58-D, Temuco, Chile

Abstract

Chile has the longest lentil crop history in the New World since it was introduced by the Spanish in the sixteenth century. Lentil is grown between 29°56' S and 38°41' S latitude, and under different agroecological conditions. The crop is sown during April-June and harvested by November-December in the north and south central areas; whereas, sowing is done between August-September in the southern-most areas, and harvested by January. The crop is grown under rain-fed conditions, where annual rainfall ranges from 800 to 1500 mm, depending on the growing area. "Araucana-INIA" is the most widely used cultivar in Chile, and one of the main features is its large seed size. The cultivated area in recent years has averaged 33,000 ha with mean seed yields of 610 kg ha^{-1}. Yields by farmers vary from 400 kg ha^{-1}, in low rainfall areas, to over 1200 kg ha^{-1} in high rainfall areas, when adequate technology is applied.

Lentil research in Chile began in 1959, and it is conducted by the "Instituto de Investigaciones Agropecuarias" (INIA) through its Grain Legume Program at three Experimental Stations (La Platina, Quilamapu and Carillanca), covering all the areas where edible grain legumes are grown. This paper examines the major constraints to the productivity of lentil in Chile and the main achievements of research programs to date.

Introduction

Dry bean (*Phaseolus vulgaris* L.), lentil (*Lens culinaris* Medik.), chickpea (*Cicer arietinum* L.), pea (*Pisum sativum* L.), lupin (*Lupinus* sp.), and, to a lesser extent, faba bean (*Vicia faba* L.) and grasspea (*Lathyrus sativus* L.) are the main food legumes cultivated in Chile. They occupy an important place in the diet of the population, and are also considered as traditional products in the export market. Among pulses, lentil ranks second after dry beans in the area cultivated.

Although lentils seem to be identified with small scale farmers, it is an

F.J. Muehlbauer and W.J. Kaiser (eds.), Expanding the Production and Use of Cool Season Food Legumes, 399–411.
© 1994 Kluwer Academic Publishers.

attractive crop for high input agriculture. In these cases, the area planted by a farmer can be as high as 150 ha, thus making mechinization of the crop possible.

Area, Distribution, and Production

The area sown with lentil has decreased sharply during the last 5 years (Table 1). This decrease is attributed to the dramatic changes experienced by Chilean agriculture in the last decade, with South Central Chile being affected the most. In this zone, an important number of medium to large farms that traditionally raise wheat, rape seed, and lentil are being planted with Pine (*Pinus radiata* D. Don) and Eucalyptus (*Eucalyptus* spp.). At present, the forest industry in Chile is having a dramatic impact on the export market, with cellulose and timber being the main products.

Table 1. Area, average seed yield and production of lentil in Chile

Years	Area	Yield	Production
	-ha-	-kg ha^{-1}-	-tons-
1980-81	52950	510	26840
1981-82	47660	370	17690
1982-83	38860	410	15820
1983-84	23050	600	13840
1984-85	36360	680	24725
1985-86	37270	780	29071
1986-87	46330	530	24555
1987-88	32750	610	20100
1988-89	14690	530	7786
1989-90	13930	590	8219
1990-91	14870	800	11896
1991-92	16320	-*	-*

*   : Data not available.
Source: National Institute of Statistics, Chile.

Due to the existence of more profitable crops, it is likely that the area planted to lentils in Chile may stabilize around 15,000 to 20,000 ha in the near future. Thus, increase in productivity is the only way to counteract the reduction in cultivated land. It is expected that technology transfer programs that are being undertaken by the government will contribute to increased yields.

The lentil growing areas in the country are concentrated between 29°56' and 38°41' S latitude, as illustrated in Figure 1. The crop is grown in sandy to heavy clay soils in areas with an annual rainfall ranging from 475 to 1500 mm. Figure

Figure 1. Lentil growing areas in Chile.

2 shows a transverse profile of the Central-South Zone, the most important lentil growing area in Chile. From east to west, the following topographic sections can be distinguished:

– *Andean foothill.* Situated in the piedmont of the Andean Mountains, it is an unirrigated area characterized by low hills with an elevation of 400 to 600 m, annual precipitation of about 1500 mm and heavy frosts between May and November.

– *Central Valley.* A flat land devoted mostly to highly profitable crops on which

irrigated agriculture takes place. Lentil is sown in this zone as a rainfed crop
but only when prices are high.

– *Coastal Mountains.* A lower chain of mountains on which inner and outer
(coastal) dryland areas can be distinguished. Its main climatic features are an
annual precipitation of about 800 mm and its moderate (frost-free)
temperatures. It is the most traditional lentil growing area in Chile, cultivated
mostly by small farmers of low input technology. The inner dryland area is
characterized by an intensive cultivation of wheat and lentil, with inadequate
soil management practices, which have resulted in soil erosion and loss of
fertility. The coastal dryland region is more productive because of its natural
soil fertility status; however, its marine climate, sometimes makes lentil
production risky because of the eventual occurrence of lentil rust.
Nevertheless, the coastal dryland area is where the lentil crop grows best and
where seed yields are better than 3000 kg ha^{-1} (Peñaloza, 1991).

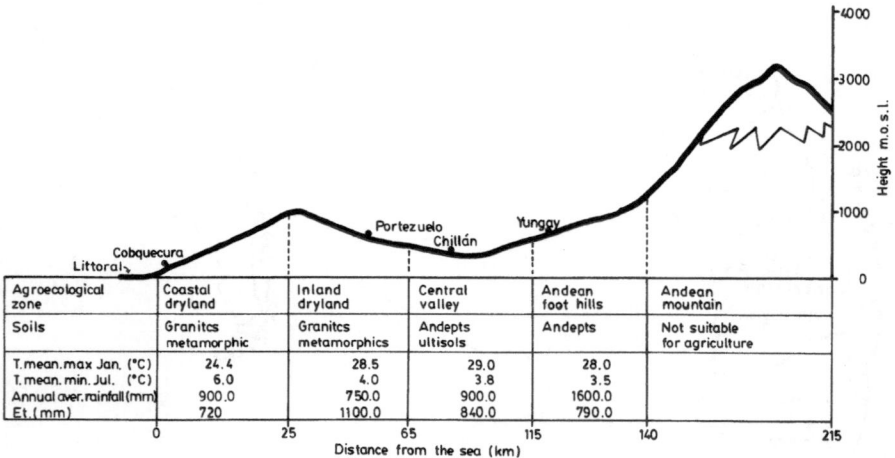

Agroecological zone	Coastal dryland	Inland dryland	Central valley	Andean foot hills	Andean mountain
Soils	Granites metamorphic	Granites metamorphics	Andepts ultisols	Andepts	Not suitable for agriculture
T.mean.max Jan. (°C)	24.4	28.5	29.0	28.0	
T.mean. min. Jul. (°C)	6.0	4.0	3.8	3.5	
Annual aver. rainfall(mm)	900.0	750.0	900.0	1600.0	
Et.(mm)	720	1100.0	840.0	790.0	

Figure 2. Transversal profile of the South-Central Chile.

Uses and Marketing

Chile produces premium quality lentils, mostly for export, with the remaining
being used in the domestic market. Between 1965 and 1985, national
consumption fluctuated between 0.32 and 0.96 kg per capita (Soto, 1991).
Rather than a change in eating habits, the large variation in consumption is
explained by the variation in the availability of lentils throughout the year,
which is determined by the demand in the export market.

The most important markets for Chilean lentils are the Latin American
countries (Brazil, Columbia, Ecuador, Perú, and Venezuela) and European
countries (Germany, Spain, France, Belgium, and The Netherlands).
Considering the cycle 1977–1989, the quantities exported have fluctuated

between 17 and 75% of the national production (Soto, 1991). Concerning the diameter of seed exported, about 50% of the total volume has a diameter of 6 mm, followed by 5 mm, not specified and 7 mm. Prices are normally higher for seeds of larger diameters (Table 2).

Table 2. Evolution of the export prices (FOB) of Chilean lentil according to its diameter during the 1980–1989 period (in US$ kg^{-1})

Year	Seed diameter (mm)			
	5	6	7	Not specified
1980	0.99	1.02	1.13	0.93
1981	0.84	0.92	0.97	0.90
1982	0.49	0.67	0.77	0.64
1983	0.48	0.60	0.65	0.61
1984	0.59	0.75	0.88	0.71
1985	0.65	0.72	0.85	0.65
1986	0.77	0.85	1.00	0.81
1987	0.48	0.60	0.75	0.50
1988	0.36	0.45	0.58	0.46
1989	0.50	0.60	0.74	0.55

From: Soto, 1991.

Germplasm Resources, Evaluation and Uses

Chilean lentil germplasm maintained by INIA is characterized by its large average seed size. This characteristic has been, without doubt, the result of selection made by farmers since the time lentil was introduced by the Spanish in the sixteenth century. From studies on the effect of planting different sized seed, the practice of seed grading has been recommended when heterogenous landrace populations are used (Krarup, 1983; Paredes *et al.*, 1987).

Systematic germplasm evaluation and utilization began in 1954 with the main effort being selection within landraces and introduction of foreign germplasm (Anónimo, 1959). Since 1978, a large number of genotypes have been introduced from ICARDA. At present, the lentil germplasm collection comprises 996 accessions, with 355 of Chilean origin and 641 from 37 other countries (Table 3). Its preservation and maintenance will ensure the maintenance of genetic variability for the future.

The germplasm has been evaluated for its adaptation, seed yields, seed size, and diseases in the most important growing areas of the country. As a result, cultivars such as "Constitución" and "Araucana-INIA" have been released, and

Table 3. Origin of the lentil germplasm resources maintained by the Food Legume Program of the Instituto de Investigaciones Agropecuarias (INIA), Chile

Country of origin	Number of lines	%
Chile	355	35.7
Iran	97	9.7
Greece	83	8.3
Russia	57	5.7
Turkey	55	5.5
Syria	34	3.4
India	33	3.3
Spain	18	1.8
Portugal	12	1.2
Other countries	126	12.7
Unknown	126	12.7
Total	996	100

From: Paredes, 1981

many accessions with good yield, large seed, and disease resistance have been identified (Table 4).

Accessions of Chilean germplasm have been introduced by other countries. The large-seeded Chilean lentil known as "lentejón" was widely used in Argentina until 1970. This cultivar was later replaced by the small-seeded cultivar "Precoz", which has a short growth cycle (Riva, 1980). According to Jiménez (1972), Chilean lentils also were introduced to Algeria.

Cultivars

The appearance of rust in 1957 promoted the introduction and evaluation of foreign germplasm, most of which came from Russia. Several lines with tolerance or resistence to rust, such as "Algeria 211", "Larisa 01", both small-seeded types, and "Penzeskaja-14", medium-seeded type, were selected from this germplasm (Anónimo, 1959; Caglevic and Kushel, 1969). However, only Penzeskaja-14 was released as a commercial cultivar (INIA, 1967). Subsequently, this cultivar was discarded and replaced by Chilean types which were likely to be more affected by rust, but with larger seed size.

In 1970, the Faculty of Agronomy of the University of Concepción, released the large-seeded cultivar Constitución which is tolerant to rust. In 1978, the

Table 4. Lentil genotypes selected in Chile for different characteristics

Cultivars	Origin	Distinguishing trait	Reference
Algeria 215	Algeria	Resistant to rust	Anonymous, 1959
Larisa 01	Greece	Resistant to rust	Anonymous, 1959
PI 209858	Greece	Resistant to rust	Anonymous, 1959
L-3067	Chile	Resistant to rust	Torrens, 1965
L-4002	Russia	Resistant to rust	Torrens, 1965
L-4007	Russia	Resistant to rust	Torrens, 1965
L-4008	Germany	Resistant to rust	Torrens, 1965
Penzeskaja*	Russia	Resistant to rust	Torrens, 1965; Jiménez, 1972; Anonymous, 1959
Palo Colorado*	Chile	Large-seeded	Jiménez, 1972
Tekoa**	USA	Resistant to rust	Bascur, 1978
Constitución**	Chile	Large-seeded, high yield	Del Villar, 1972 (unpublished)
6008	Germany	Resistant to root rot	Paredes, 1981
6006	Germany	Resistant to root rot	Paredes, 1981
3057	France	Resistant to root rot	Paredes, 1981
Araucana-INIA**	Chile	Large seeded, high yield	Tay et al., 1981
Laird	Canada	Immune to rust	Sepúlveda and Tapia, 1982
Centinela	ICARDA	Resistant to rust	Bascur and Sepúlveda, 1989

* Old cultivars, discontinued.
** Commercial cultivars, actually in use

cultivar "Tekoa" was introduced from the USA (Bascur, 1978). Tekoa is resistant to rust and is a medium-seeded type. It is recommended only for coastal drylands which have a high incidence of the disease. At present, Tekoa is being replaced by a new cultivar, "Centinela-INIA", introduced from ICARDA, with similar characteristics as Tekoa, but with larger seed size (Bascur and Sepúlveda, 1989).

At present, the most widely used cultivar grown in Chile is Araucana-INIA, released in 1981. This cultivar was obtained by mass selection from line 1284, a Chilean germplasm accession (Tay *et al.*, 1981). It is characterized by having a large seed size, large seed yields, yellow cotyledons, and a light brown seedcoat. Araucana-INIA is susceptible to rust; thus it must be protected with foliar fungicides when planted in areas of with a high incidence of rust (Sepúlveda and Alvarez, 1989; Sepúlveda and Tay, 1991). Tables 5 and 6 show the seed yield and seed size of the cultivars now grown in Chile.

Table 5. Seed yield (kg ha^{-1}) of improved lentil cultivars at four agroecological zones in Chile

		Seed yield (kg ha^{-1})		
Agroecological zone	Location	Araucana-INIA	Constitución	Tekoa
Andean foothills	Yungay [1]	1550	1590	1070
Inland dryland	Portezuelo [2]	1430	1310	1110
Coastal dryland	Chanco [3]	1810	1680	2240
Coastal dryland	Carahue [4]	2360	−*	−*

*:   Data not available
[1]:   Average yield for six years
[2]:   Average yield for three years
[3]:   Average yield for five years
[4]:   Average yield for four years

Table 6. Seed size fractions (expressed as percentage of seeds retained on sieves with hole diameters of 7, 6 and 5 mm) of selected cultivars at four agroecological zones in Chile

		Seed size fraction (%)								
		Araucana-INIA			Constitución			Tekoa		
Agroecological zone	Location	7mm	6mm	5mm	7mm	6mm	5mm	7mm	6mm	5mm
Andean Foothills	Yungay[1]	29	65	4	15	70	14	2	54	42
Inland dryland	Portezuelo[2]	26	63	7	19	60	17	3	60	37
Coastal dryland	Chanco[3]	31	65	4	17	68	15	7	62	30
Coastal dryland	Carahue[4]	60	35	5	−*	−	−	−	−	−

*:   Data not available
[1]:   Average yield for six years
[2]:   Average yield for three years
[3]:   Average yield for five years
[4]:   Average yield for four years

Because of the high price of certified seed (approximately US$ 1 to 1.5 kg^{-1}), equivalent to three to four times the price paid for the product at the farm, farmers usually use their own seeds which is generally of unknown origin. In order to make certified seed accessible to any farmer, the government started a program to deliver Araucana-INIA seed under a system known as 2 × 1. Under this system the farmer receives 100 kg of certified seed with the understanding to return 200 kg of that seed after harvest. So far this program has been susccessful and it is expected that improved cultivars will be accessible to most of the lentil farmers in the near future.

Agronomy of Lentil

Lentils which are grown in pure stands are generally planted after wheat and are usually cultivated under rainfed conditions. Soils devoted to the crop range from sandy to heavy-clay with seed yield being more dependent on rainfall than soil type.

Date of Sowing

Optimum dates of planting for most growing areas have been defined on the basis of rainfall distribution, mean temperature, disease incidence [*Ascochyta fabae* Speg. f. sp. *lentis* Gossen *et al.*, *Uromyces fabae* (Grev.) Fackel (= *U. viciae-fabae* (Pers.) Schroet.], and soil type (Tay *et al.*, 1980, 1987; Peñaloza and Mera, 1984; INIA, 1985). Depending on the above factors, lentil is sown from April to September and harvested between November and January.

Fertilization and *Rhizobium* Inoculation

Experiments conducted in different regions of the country have not shown any consistent response to applied nitrogen or *Rhizobium* inoculation in most soils. The lack of response is attributed to the presence of good populations of native *Rhizobium*. Thus, seed inoculation is not commonly practiced by farmers. Seed yield increases have been reported with phosphorus application in soils with P lower than 16 ppm. When applied, P fertilizers are most commonly used by farmers at rates between 40 to 80 kg P_2O_5 ha^{-1}, with the highest rates being used in volcanic P-fixed soils (Andisol). Similar to nitrogen, potassium fertilizer has not shown any consistent response in experimental trials, probably because of the high K status in most Chilean soils; nevertheless, applications in amounts sufficient to supply the K absorbed by the crop are being recommended for soils with K lower than 0.3 meq 100 g^{-1} (Peyrelongue, 1991).

Planting Methods

Although broadcast seeding is still practiced in some lentil-growing areas, the method most commonly practiced by farmers is to plant the crop in rows. When planting machines are not available, i.e., small farms, furrows are opened with a plow pulled by a horse or bullock and seeds are dribbled by hand into furrows at distances of 20 to 50 cm with five to 15 seeds each time. Field studies carried out to compare planting methods have shown seed yield advantages by planting seeds in rows rather than broadcasting seeds, with differences between planting machines or seeding by hand being dependent of environmental conditions (Paredes *et al.*, 1987; Peñaloza, 1987). On the farm, however, drilling is not practiced as recommended and, therefore, yields are drastically reduced mainly because of inadequate spatial arrangements of plants which results in higher losses on cultivated land.

Plant Population and Row Spacing

Plant density studies in the country have revealed that about 70 to 80 plants m^{-2} are required to get higher seed yields when large-seeded cultivars are used. Depending on seed weight, seed rates adequate to obtain the target population varies from 60 to 80 kg ha^{-1}, considering an average of 15% seed losses from planting to harvest (Tapia and Almarza, 1981; Paredes *et al.*, 1982, 1989; Bascur *et al.*, 1984; INIA, 1985; Peñaloza and Mera, 1986; Peñaloza, 1991).

Although experimental results have demonstrated the advantage of using narrow row spacing (Peñaloza and García, 1991), actual row spacings used by farmers depend on whether weed control is done by hand, with chemicals, or by using a cultivator pulled by horses.

Weed Control

Weeds are by far the most important factor limiting lentil production in Chile, especially in the southernmost humid areas. Traditional methods of weed control commonly used by small farmers include hand weeding or cultivators pulled by horses. Weed control is not practiced when seeds are broadcast. The use of herbicides is practiced mostly in large cultivated areas. Broad spectrum preemergence herbicides, such as prometrina, linuron, and metabeuztiazuron, are recommended (Ramírez *et al.*, 1968; Espinoza, 1991), as well as postemergence gramicides are being recommended (Díaz and Espinoza, 1991). No suitable postemergence herbicides for broadleaf weed control have been found so far.

Harvest

Lentil is usually harvested by pulling up plants by hand and leaving them in the field to dry. Alternatively, a cutter bar is used. Once seed moisture has dropped below 18%, plants are either threshed in the same field with a combine or transported to a threshing area were they are threshed by a stationary machine powered by a belt drive from a tractor. When no mechanization is available, threshing is done by using horses. Recently, farmers have started to harvest the crop directly by using a combine harvester, with losses that can reach as high as 20% depending on crop lodging and pod shattering.

Diseases

Lentil rust caused by *Uromyces fabae* is the most serious disease of lentil in Chile, particularly in the inland and coastal dryland areas (Anónimo, 1959; Caglevic, 1961, 1969; Torrens, 1965; France *et al.*, 1989; Sepúlveda, 1984, 1985, 1987b). Susceptible cultivars are severely affected by the fungus which may cause yield losses of more than 70% (Sepúlveda and Alvarez, 1989). Cultural practices such as crop rotation and foliar fungicides are recommended to control the disease (Sepúlveda and Alvarez, 1989; Sepúlveda and Tay, 1991).

Another important disease is Ascochyta blight, caused by *Ascochyta fabae* f. sp. *lentis* which was reported in Chile in 1982 (Sepúlveda and Alvarez, 1982). The fungus is seedborne and can infect plants at any stage of growth. The most severe symptoms appear when the fungus infects the seed. Seed treatment with fungicides is recommended to reduce or eliminate the primary seedborne inoculum (France *et al.*, 1987; Guerrero, 1987). However, the most effective and economic control measure is the use of resistant cultivars. The main objective of the lentil breeding program of INIA is to develop, mainly by hybridization, cultivars with resistance to these important diseases.

There are other foliar diseases that can be found in lentil fields, such as powdery mildew caused by *Erysiphe polygoni* DC., septoriosis caused by *Septoria* sp. (Sepúlveda, 1987; France, 1986), and *Botrytis cinerea* Pers.:Fr. (France *et al.*, 1988). Among the diseases that affect the roots and stem, *Sclerotinia sclerotiorum* (Lib.) deBary, *Fusarium solani* (Mart.) Sacc. and *Rhizoctonia solani* Kühn have been reported to cause yellowing and wilting of plants.

Pests

No serious pests affecting lentil have been detected so far in Chile. Only slugs (*Derocerus reticulatus* and *Limax agrotis*), especially in some wet areas of the coastal dryland in the South Central Zone cause some damage to the crop (Quiroz, 1983). Two aphids, *Acyrthosiphon pisum* (Harris) and *A. kondoi* Shinji & K., are frequently found on the crop. The populations of aphids are not sufficient to cause damage, probably because of the abundance of natural enemies (Gerding *et al.*, 1986).

References

Anónimo. 1959. *Agricultura y Ganadería* 5: 5–6. Santiago, Chile.
Bascur, B. G. 1978. *Estación Experimental La Platina*. Santiago, Chile. Informativo N° 16, 2 pp.
Bascur, B. G. and Sepúlveda, R. P. 1989. *Agricultura Técnica (Chile)* 49: 366–369.
Bascur, B. G., Tapia, F. F. and Sepúlveda, R. P. 1984. In: *Mejoramiento de la productividad del Secano Costero VI Región*, pp. 25–31. Estación Experimental La Platina, Santiago (Chile): INIA.
Caglevic, M. 1961. *Actas de la V Reunión Latinoamericana de Fitotécnia*, Tomo II, p. 451. Buenos Aires, Argentina.

Caglevic, M. 1969. *Simiente (Chile)* 34: 14.

Caglevic, M. and Kushel, R. 1969. *Agroinformativo* N° 72, 4 pp. Santiago, Chile: Estación Experimental La Platina (INIA).

Díaz, S. J. and Espinoza, N. N. 1991. *Investigación y Progreso Agropecuario, Carillanca, Temuco* (Chile) 10: 17–21.

Espinoza, N. N. 1991. In: *Producción de lenteja en la IX Región. Boletín Técnico N°144*, pp. 33–39 (eds. E. Peñaloza and E. Kehr). Temuco, Chile: Estación Experimental Carillanca.

France, I. A. 1986. *El Campesino (Chile)* 7: 34–42.

France, I. A., Paredes, C. M., Tay, U. J. and Cortez, A. M. 1989. *Investigación y Progreso Agropecuario, Quilamapu, Chillán (Chile)* 40: 14–18.

France, I. A., Sepúlveda, R. P. and Tay, U. J. 1988. *Agricultura Técnica (Chile)* 48: 158–160.

France, I. A., Tay, J. U. and Cortez, A. M. 1987. *Investigación Progreso Agropecuario, Quilamapu (Chile)* 31: 22–26.

Gerding, P. M., Zúñiga, S. E. and Kramm, M. V. 1986. *Investigación y Progreso Agropecuario, Quilamapu, Chillán (Chile)* 29: 3–6.

Guerrero, C. J. 1987. *Agricultura Técnica (Chile)* 47: 101–107.

INIA. 1967. *Agroinformativo* N° 3, 2 pp. Santiago, Chile: Estación Experimental La Platina.

INIA. 1985. *Boletín Divulgativo* N° 101, 78 pp. Santiago, Chile: Estación Experimental La Platina.

Jiménez, T. A. 1972. *Investigación y Progreso Agrícola, Santiago (Chile)* 4: 11–14.

Krarup, H. A. 1983. *Agro Sur (Chile)* 11: 57–60.

Paredes, C. M. 1981. Antecedentes generales y análisis del Programa de Mejoramiento de lentejas en Chile 1958/59–1980/81. Reunión Nacional Programa Leguminosas de Granos INIA. (Documento interno).

Paredes, C. M., Tay, J. U. and France, I. A. 1989. Instituto de Investigaciones Agropecuarias, Estación Experimental Quilamapu. *Chillán (Chile) Serie Quilamapu* N° 13, 19 pp.

Paredes, C. M., Tay, J. U. and Kramm, M. V. 1982. *Investigación y Progreso Agropecuario, Quilamapu, Chillán (Chile)* 14: 14–15.

Paredes, C. M., Tay, J. U. and Parra, R. C. 1987. *Investigación y Progreso Agropecuario, Quilamapu, Chillán (Chile)* 31: 27–29.

Peñaloza, H. E. and Mera, K. M. 1984. *Investigación y Progreso Agropecuario, Carillanca, Temuco (Chile)* 3: 11–14.

Peñaloza, H. E. and Mera, K. M. 1986. *Agricultura Técnica (Chile)* 46: 231–236.

Peñaloza, H. E. 1987. *Investigación y Progreso Agropecuario, Carillanca, Temuco (Chile)* 6: 2–4.

Peñaloza, H. E. 1991. In: *Producción de lenteja en la IX Región. Boletín Técnico N° 144*, pp. 18–24 (eds. E. Peñaloza and E. Kehr). Estación Experimental Carillanca, Temuco, Chile.

Peñaloza, H. E. and Garcia, D. J. C. 1991. *Investigación y Progreso Agropecuario, Carillanca, Temuco (Chile)* 10: 21–24.

Peyrelongue, C. M. 1991. In: *Producción de lenteja en la IX Región. Boletín técnico N° 144*, pp. 25–32 (eds. E. Peñaloza and E. Kehr). Estación Experimental Carillanca, Temuco, Chile.

Quiroz, E. C. 1983. In: *II. Seminario de Leguminosas de Granos*, pp. 132–133. Instituto de Investigaciones Agropecuarias, Estación Experimental Quilamapu, Chillán, Chile.

Ramírez, A., Bofarull, L. N., del Pozo, P. G. and López, V. M. 1968. Control Químico de Malezas. *Manual Técnico* N° 1, p. 37. Instituto de Investigaciones Agropecuarias. Ministerio de Agricultura. Santiago, Chile.

Riva, A. E. 1980. *Lens* 7: 5–7.

Sepúlveda, R. P. 1984. *Investigación y Progreso Agropecuario, La Platina, Santiago (Chile)* 26: 21–22.

Sepúlveda, R. P. 1985. *Agricultura Técnica (Chile)* 45: 335–339.

Sepúlveda, R. P. 1987a. *Investigación y Progreso Agropecuario, La Platina, Santiago (Chile)* 39: 29.

Sepúlveda, R. P. 1987b. *Investigación y Progreso Agropecuario, La Platina, Santiago (Chile)* 43: 14–17.

Sepúlveda, R. P. and Alvarez, A. M. 1982. *Agricultura Técnica (Chile)* 42: 351–353.

Sepúlveda, R. P. and Alvarez, M. A. 1989. *Agricultura Técnica (Chile)* 49: 309–313.

Sepúlveda, R. P. and Tapia, F. F. 1982. *Investigación y Progreso Agropecuario, La Platina, Santiago (Chile)* 109: 37–38.

Sepúlveda, R. P. and Tay, U. J. 1991. *Estacticn Experimental Quilamapu, Chilián, Serie Quilamapu* N° 29, 3 pp.

Soto, A. M. A. 1991. In: *Producción de lenteja en la IX Región. Boletín Técnico N° 144*, pp. 65–72 (eds. E. Peñaloza and E. Kehr). Temuco, Chile: Estación Experimental Carillanca.

Tapia, F. F. and Almarza, D. P. 1981. *Investigación y Progreso Agropecuario, La Platina, Santiago (Chile)* 4: 22–25.

Tay, J. U., France, I. A. and Paredes, C. M. 1987. *Investigación y Progreso Agropecuario, Quilamapu, Chillán (Chile)* 31: 19–21.

Tay, J. U., Kramm, M. V. and Paredes, C. M. 1980. *Investigación y Progreso Agropecuario, Quilamapu, Chillán (Chile)* 1: 25–27.

Tay, J. U., Paredes, C. M. and Kramm, M. V. 1981. *Agricultura Técnica (Chile)* 41: 170.

Torrens, R. C. 1965. *Reacción de 360 variedades y líneas delentejas a la roya (Uromyces fabae P.)*, 104 pp. Universidad de Chile, Santiago Chile. Tesis Ingeniero Agrónomo.

Pea and chickpea production in Australia

R.O. REES[1], J.B. BROUWER[2], J.E. MAHONEY[2], G.H. WALTON[3],
R.B. BRINSMEAD[4], E.J. KNIGHTS[5] and D.F. BEECH[6]

[1] Crops Economics Section, Australian Bureau of Agricultural and Resource Economics (ABARE),
Edmund Barton Building, GPO Box 1563, Canberra ACT, Australia 2601;
[2] Victorian Institute for Dryland Agriculture, Private Bag 260, Horsham, Victoria 2401, Australia;
[3] Western Australian Department of Agriculture, South Perth, Australia;
[4] Queensland Department of Primary Industries, Warwick, Australia;
[5] Agricultural Research Centre, RMB 944, Tamworth, New South Wales 2340, Australia, and
[6] CSIRO, Division of Tropical Crops and Pastures, St Lucia, Australia

Abstract

The significant increase in the areas sown to field pea, chickpea, and lupin in
Australia has been due principally to policy incentives favoring the production
of field pea and lupin for use as stock feed in the European Community (EC),
and the relaxation of import restrictions in India which in turn have led to
increased imports of Australian pulses for feed and food uses.

Australia's ability to respond rapidly to demand from the EC and India was
due in part to basic pulse research policies being in place. These were designed
primarily to meet fundamental crop issues, particularly the introduction of
wheat quotas in the early 1970s and the need to find alternative crops.
Specifically in Western Australia there was a need for a crop for large areas of
acid, sandy and gravelly soils. For example, lupin proved to be very satisfactory
for those conditions. In Queensland, chickpea was introduced as one option in
a comprehensive management strategy to reduce soil erosion.

The emphasis of agronomic research and extension in Australia during the
1970s and early 1980s was to recommend suitable rotation crops to reduce
cereal diseases, provide stable yields, add inexpensive nitrogen to the soil,
increase yields of cereal that follow, and to provide suitable cropping
alternatives to wheat and wool.

The establishment of a national Grain Legume Research Council in 1985, the
introduction of pulse crop levies, the availability of more diverse funding
sources, stronger state funding, and extension efforts in recent years have
contributed to a greater awareness by producers of the market potential for a
broad range of pulses at a time when wool and wheat prices have declined.

Introduction

The rapid expansion of the area sown to pea, lupin, and chickpea in Australia
during the 1980s, was due to a combination of factors including policy
initiatives in a number of countries, relative profitabilities of pulse crops

F.J. Muehlbauer and W.J. Kaiser (eds.), Expanding the Production and Use of Cool Season Food
Legumes, 412–425.

compared to traditional enterprises such as wheat and wool, and the increasing significance of pulses in the crop rotations used in Australia.

The paper outlines the trends in production and exports of field pea and chickpea, and traces the policy developments in the EC and India as two major buyers, the development of the Australian stock feed market and examines the profitability of pulses from an economic and agronomic viewpoint. Lupins have been included in the paper because of its significant role in Western Australia and as a pulse crop suitable for acid soils. The development and progress of pulse research programs in each of the major producing states is summarized.

No attempt has been made to quantify future production potential or production limitations because of the constraints of the paper. However, expansion of field pea, chickpea and lupin will be dictated more by international demand and their relative prices compared to traditional enterprises such as wheat and wool rather than by an inability to expand production to meet that demand.

Trends in Field Pea, Chickpea, and Lupin Production in Australia

The area sown to field pea and lupin in Australia peaked in 1988/89, fell for the next 2 years but recovered strongly in 1991/92. The area sown to chickpea, mainly desi, has risen rapidly since 1984/85 despite a drought in Queensland, the major production area, in 1991/92 (Table 1).

Table 1. Trends in area sown to field pea, chickpea and lupin in Australia, 1971/72–1991/92

Years	Field Pea '000 ha	Chickpea '000 ha	Lupin '000 ha
1971/72	24	–	33
1982/83	114	n.a.	257
1984/85	140	6	594
1988/89	456	70	850
1989/90	326	93	802
1990/91	309	167	789
1991/92	445	215	925

Source: Australian Bureau of Agricultural and Resource Economics (ABARE)
Commodity Statistical Bulletin 1991, Australian Bureau of Statistics
ABARE: Crop Report No. 68, 22 October 1991

Relatively high wool prices in 1987/88 and 1988/89 and wheat prices in 1988/89 and 1989/90 respectively, led to an expansion in wool and wheat production. On the other hand, three successive years of chickpea prices in excess of A$ 300 a ton would have led to a significant increase in the area sown to chickpea in Queensland in 1991/92 but grower intentions were curtailed by drought.

Production of field pea and lupin peaked in 1988/89 with lupin production falling just short of 1 million tons and field pea production exceeding half a million tons for the first time. Chickpea production has risen from a negligible amount in 1984/85 to slightly above 200 Kt in 1991/92 (Table 2).

Table 2. Trends in production of field pea, chickpea and lupin in Australia, 1971/72 to 1991/92

Years	Field Pea '000 Kt	Chickpea '000 Kt	Lupin '000 Kt
1971/72	39	NIL	24
1982/83	30	n.a.	199
1984/85	164	6	581
1988/89	523	89	930
1989/90	388	108	773
1990/91	309	196	952
1991/92	509	206	885

Source: ABARE: Commodity Statistical Bulletin 1991,
                   Australian Bureau of Statistics
           ABARE: Crop Report No. 68, 22 October 1991

The states of Victoria and South Australia, predominate in pea production, while Western Australia and Queensland are the major producers of lupin and chickpea, respectively. However, Victoria and New South Wales have increased their area sown to chickpea (Table 3).

Table 3. Area sown to field pea, chickpea, and lupin by state: 1990/91, 1991/92

State	Field pea '000 ha		Chickpea '000 ha		Lupin '000 ha	
	1990/91	1991/92	1990/91	1991/92	1990/91	1991/92
Victoria	154	200	40	75	31	32
South Australia	103	150	9	19	48	65
Queensland	–	–	80	50	–	–
New South Wales	28	45	37	70	52	46
Western Australia	24	50	1	1	665	780
Australia	309	445	167	215	789	925

Source: ABARE: Commodity Statistical Bulletin 1991.

Trends in Pea, Chickpea, and Lupin Exports from Australia

Australian exports of field pea were destined for the European Community as a major market in 1986/87 (244 kt) but exports have declined due to the emergence of India and Bangladesh as major users of Australian field pea and desi chickpea for human consumption. In recent years, Australia has exported most of its chickpea crop to these two countries (Table 4).

Table 4. Australian exports of field pea, chickpea and lupin by destination 1986/87 and 1990/91

Country	Field Pea '000 Kt 1986/87	Field Pea '000 Kt 1990/91	Chickpea '000 Kt 1986/87	Chickpea '000 Kt 1990/91	Lupin '000 Kt 1986/87	Lupin '000 Kt 1990/91
Asia						
– India	94	198		95	–	–
– Bangladesh	–	91		24	–	–
– Malaysia	8	13		2	–	–
– Japan	–	–		–	82	120
– Korea, Republic	–	–		–	11	98
– Taiwan	4	6		–	–	–
European Community	244	8			297	100
– Netherlands	136	–		–	250	61
– Portugal	–	–		0.2	–	20
– Spain	–	–		–	25	19
TOTAL	371	363	N.A.	146	396	330

Source: Australian Bureau of Statistics

Lupin, in contrast has not achieved acceptance by importing countries as a high protein food grain with qualities similar to soybean. Consequently, the Grain Pool of Western Australia (WA), the sole marketer of WA lupins, has attempted to expand its outlets by concentrating on penetrating the Japanese and South Korean markets for ruminant feed, which because of domestic policy factors, are able to pay higher for lupin (Coffey, 1992).

Increased chickpea exports compete with exports of field peas into India and Bangladesh. Chickpea, which is preferred to field pea on the Indian subcontinent, may soon force Australian field pea into the lower priced domestic stock feed markets, with a return to the European stock feed market and/or other alternative stock feed markets if production continues to expand (Table 4).

Policy Initiatives and Other Factors Leading to Production Increases in Australia of Field Pea, Chickpea, and Lupin During the 1980s

World trade in pulses doubled between 1982 and 1988 due principally to significant increases in imports by the EC and by India (Table 5). The policy initiatives taken by these two countries are detailed below.

EC Policy Incentives

Production of field pea in the EC-12 has risen rapidly since 1984 from 1.1 Mt to 4.5 Mt with production of field pea rising in France from 0.6 Mt in 1984 to 3.3 Mt in 1990 (Toepfer, 1990/91).

The European Community produced 100 Mt of compound feed in 1989 for the major livestock categories. Imports of feed stuffs in that year totalled 36.7

Table 5. World imports of pulses, by selected countries: 1982, 1985 and 1989

	1982 '000 Kt	1985 '000 Kt	1989 '000 Kt
Western Europe			
Belgium–			
Luxembourg	99	209	403
Netherlands	285	574	709
Spain	87	64	175
UK	160	176	182
Asia			
India	48	307	620
China	124	89	85
Japan	207	155	180
TOTAL	2902	3935	5537

Source: Food and Agriculture Organisation of the United
Nations: Trade Yearbook, Vol 43, 1989.

Mt and included 0.3 Mt field pea, 0.3 Mt faba bean and 0.2 Mt lupin. In order to reduce its dependence on imported protein rich feeds and partly because of an embargo on soybean imports, the European Community has encouraged domestic production and consumption of oilmeals and pulses. Soybean meal imports averaged 13 Mt for 3 years from 1989/90 to 1991/92.

A special program was introduced in 1978/79 which consisted of a minimum price paid to growers of pulses and a subsidy paid to users of domestic feed pulses. Feed pulses have become serious competitors in the EC for subsidized feed grains and soybean meal.

Any agreement at the General Agreement on Trade and Tariffs (GATT) negotiations on reduction of domestic price subsidies is likely to reduce the competitive ability of imported feed pulses. However, the introduction of annual maximum guarantee quantities for pulses produced in the EC is likely to place a ceiling on production which should increase the possibility of Australian imports, but at lower prices.

Indian Policy Incentives

The major production policies of the Indian Government are directed at increasing wheat and oilseed production in preference to pulses. As a consequence, a significant increase in the area sown to wheat has pushed the production of the major pulse, chickpea, into marginal areas despite the payment of minimum support prices (Rees, 1992).

To ensure that domestic prices do not rise excessively, the Indian Government allows imports of pulses under an open general licensing scheme (OGLC). Variable tariff levels are used to either restrict or increase the volume of imports. Better than average pulse production in 1986 and relatively large import volumes in the same year led to 25% tariffs. Reduced production in the following two years led to reduced tariff of 10% (Rees, 1992).

Despite a number of economic problems for the Indian Government during 1991, the importance of pulses to India is reflected in the maintenance of the OGLC system at the current tariff of 10% and exemption of pulses from an import cash deposit scheme (Rees, 1992).

The introduction of Exxim Scrip which is a tradeable certificate enabling Indian exporters to obtain import credits, has important ramifications for the Australian pulse industry. The re-export of pulses, imported in bulk from Australia, and re-exported in processed form (mainly as split product) to Pakistan, Sri Lanka and the Middle East could assist the future volume of exports from Australia. However, it seems inconceivable that a country whose already large, mostly vegetarian population which is increasing at 2.2% a year or by 200 million by the year 2000 can continue to export pulses in exchange for hard currency despite the short term attractiveness of the scheme to the Indian economy (Rees, 1992).

It is estimated that the demand for pulses in India by the year 2000 will be 20 to 25 million tons (Rees, 1988). At current levels, production would have to nearly double. Although there is ample genetic improvement possible for yields of chickpea and pigeonpea to increase in India, it will also require pulses to be grown on more additional wheat and oilseeds areas where yields are better. The ready availability of relatively cheap bulk shipments of field pea and chickpea from Australia is likely to reduce the incentive for Indian producers to alter their production patterns.

Australian Stock Feed Market

The initiative of State Departments of Agriculture in testing pulses for use in stock feed rations has led to a rapid rise in use. As production has become more assured, field pea has become preferred to lupin for use in monogastric rations because of the relative indigestibility of lupin seed. One of the problems in using lupin seed is the high concentrations of undegraded dietary protein relative to wheat or field pea; and the reduced availability of lysine in lupin seed when compared to lupin kernel meal, dry pea or faba bean (Rees, 1988).

From an economic viewpoint it is the relative prices in terms of the values of energy and amino acids from pulses and alternative feeds which are critical when contemplating their use in feed rations (Rees, 1988). Field pea is a good source of energy with a digestible energy value of 13.5 Mg kg^{-1} which is less than wheat but more than barley. However, its protein concentration is less (22%) relative to other protein sources such as sweet lupin seed (30%) and meat meal (50%). Peas have a relatively large concentration of lysine, small concentrations of sulphur amino acids and are deficient in calcium and phosphorus (Rees, 1988).

The development of the "sweet", low alkaloid cultivars of the narrowleaf lupin (*Lupinus angustifolius*) in Australia has led to the rapid acceptance of lupin seed in ruminant rations, particularly for sheep (which can digest the crude fiber of the seedcoat). However, the toughness of the outer layer, relatively poor digestibility and smaller lysine concentrations when compared with field pea and soybean, reduces the usage of lupin seed in monogastric rations. Dehulling of lupin seed is more expensive but the removal of the testa from the seed leaves the kernel which is superior to the whole seed for pig feeding (Rees, 1988).

Chickpea, like field pea is a good energy source with large concentrations of lysine. Chickpea was used successfully for stock feeding in Queensland during the 1991/92 drought period.

The size of the domestic stock feed market in Australia rose sharply in the last half of the 1980s as rising numbers of sheep and cattle and increases in poultry and pig production (Table 6) created more demand for alternative combinations of protein and energy sources, and in particular for relatively cheap pulses such as pea, lupin, and faba bean. There is also a growing feed-lot industry which should assist demand for lupins and other pulses.

Table 6. Livestock numbers in Australia

	1984 '000	1989 '000	1991 '000
Cattle, Calves and Dairy Cattle	22,161	22,400	23,347
Sheep and Lambs	135,137	164,888	161,092
Pigs	2,527	2,671	2,531
Poultry Production	315 Kt	432 Kt	1990 445 Kt

Source: ABARE: Commodity Statistical Bulletin 1991

Profitability of Pulses in Alternative, Intensive Rotation Systems

Pulse production were minor enterprises on Australian farms during the 1970s (Rees and Presser, 1991). Traditional farming methods were relatively profitable for cereals and livestock and there was limited technological information available on the production, marketing, and whole farm effects of growing pulses. However, a number of factors have combined to change the emphasis from traditional practices to more intensive but sustainable rotations which include pulses as grain or green manure. These factors include: declining profits from wheat and wool, the benefits of pulses for reducing soil disease problems, increased cereal yields following pulse crops, the opportunity to use effective herbicides in pulses to control grassy weeds which are difficult or expensive to control in cereal crops, the profitable use of pulse stubble and grain which has lodged for feeding, the lower wheat protein levels using traditional rotation methods, the relative profitability of pulses in their own right, and ease of mechanical operations for sowing, spraying, and harvesting.

Until the wool price boom of 1987/88, wool returns had lagged significantly behind wheat returns in the major wheat-sheep zones of Australia. However, the rapid buildup in sheep numbers and decline in wool purchases, particularly from China, led to surplus in wool stocks and the subsequent removal of the wool floor price scheme by the Federal Government. Until stocks are cleared, there will be relatively small returns for wool compared to grains for the next few years. In addition, other economic pressures such as high interest rates, drought, and low wheat prices caused farmers to implement more intensive cropping programs.

Gross margins for a range of crop and livestock activities in the wheat-sheep belt of SA in 1990/91 demonstrate the marked difference in profitability between wheat, wool and pulses. Field pea returns were forecast at A\$ 189 per ha compared with A\$ 171 per ha for wheat and A\$ 50 to A\$ 60 per ha return from livestock. If the additional agronomic benefits of pulses with their complementary role in weed and disease control in intensive cropping rotations are taken into account, there is little wonder that many farmers are substituting pulses for pasture legumes.

As a consequence of the development of and focus on cropping technology, there was investment in larger machinery in the 1980s which has meant higher overhead costs and the need to make far greater use of equipment and land.

Australian wheat board statistics show that there has been a decline in grain quality in recent years with protein concentrations, in particular, declining in many parts of Australia, protein concentrations in wheat are dependent on the amount of available nitrogen in the soil in the cereal-pasture system which is used as a means of increasing soil fertility. However, the ingress of new pasture pests and weeds, and the increase in herbicide resistant grasses have made it more difficult for farmers to maintain quality, legume-dominant pastures. This problem has been compounded with more intensive cropping, increased grazing pressures, and possibly poor grazing management as a result of economic pressures on farmers to increase production. However, the economics of pasture improvement, i.e., to establish a vigorous legume growth in a pasture ley are not favorable given current returns from sheep and cereals, although they are improving. As a consequence, legume pasture seed numbers are down significantly. A preliminary study by CSIRO has shown that rotations containing pasture legumes in the previous year rather than cereal or grain legumes increased wheat grain protein.

The economic incentives which are required for farmers to increase the cropping intensity can be illustrated by an analysis of the profitability of a number of alternative rotations (Table 7).

A pulse crop in the rotation every 3 years has been the most profitable over the last few years. There are however, several agronomic issues, such as the long

Table 7. Rotation gross margins: South Australia

	Crop Rotation			Crop Intensity %	Average Annual Gross Margin $ ha⁻¹
	Year 1	Year 2	Year 3	Year 4		
1.	Wheat	Barley	Peas	Wheat	100	133
2.	Wheat	Pasture	Barley	Pasture	50	77
3.	Wheat	Barley	Pasture	Pasture	50	62
4.	Wheat	Peas	Barley	Pasture	60	106
	Year 5					
	Pasture					
5.	Wheat	Pasture	Pasture	Wheat	33	77
6.	Wheat	Barley	Peas	Wheat	100	98
	Year 5	Year 6				
	Barley	Peas as green manure				

term effects on soil fertility, resistance of weeds to commonly used herbicides and soilborne diseases which may affect this situation.

Australian farmers are becoming more flexible in their rotation systems. There are several crop and pasture rotations which allow them to respond to market signals while maintaining sensible and ecologically sound systems.

A History of Agronomic Research and Extension Programs in Australia

Production increases to satisfy expanding market requirements can only be made if suitable agronomic programs and extension packages are in place to convince farmers that pulses were worth the production risks. For quite different reasons, each Australian State has put in place policies for production and to a lesser extent research which provided the basis for the rapid development of the pulse industry during the latter half of the 1980s.

For many years there had been a very small pulse production in the southern states. In 1969, wheat quotas were introduced which provided the initial stimulus to find alternative crops. In time there was an increasing need for diversification as a means of managing risk which is inherently associated with the production of commodities. Prices of wheat were declining and there were problems with pasture production and farmers were looking for profitable cash crops.

All states, through their Departments of Agriculture, have individually provided funds for research into the production of pulses. There were additional funds available nationally through the auspices of the Reserve Bank and then by the Wheat Industry Research Council and State Committees using money raised by a levy on wheat. In 1985, in recognition of the increasing economic and agronomic contribution of pulses, levies were imposed and the Grain Legume Research Council was formed to oversee pulse research throughout Australia.

Research has generally been carried out by the individual states in response to particular production problems such as disease, insect pests, weed control and integration into the particular rotation systems. In Western Australia, traditional pulses were not suitable to the light acid soils, consequently, lupin which is naturally adapted to this type of environment was targeted and now there is a substantial industry. There is always the ongoing challenge of improving the adaptation of the various species to the production environment as well as to the demands of the market place.

More recently there has been an increasing effort directed at market identification and development. This has been successfully carried out in Western Australia by the WA Grain Pool which is a state statutory marketing authority and without its efforts the lupin industry would not have developed so rapidly.

For South Australia, the successful introduction of field peas into compound feed rations for pigs and the first shipment of bulk field pea to Europe in 1985, boosted confidence in the field pea industry.

The emphasis of extension efforts generally in the 80s was to reduce cereal diseases by the introduction of break crops such as pulses, to add inexpensive nitrogen to the soil, to increase subsequent wheat yields and to add alternative profitable crops to the system.

In 1987, the Victorian Government developed an economic strategy (Victoria–the Next Decade) which targeted diversification of the cropping base as a high priority. Pulses were identified as suitable for expansion and a program was developed for the coordinated development of the pulse industry which involved both government and private companies. This program has encouraged the strong growth and stability of the grain legume industry in Victoria.

Chickpea has emerged as an economically attractive crop in Queensland and northern NSW because it is adapted to significant areas of alkaline clay soils and stored moisture, it is more tolerant than other pulses of higher temperatures during flowering, it is particularly attractive economically, and it has been assisted greatly by the opening of the Indian subcontinent market to bulk shipments.

Grain Legume Research Council 1985

The pulse industry recognized in the early 80s that while the market outlook for pulses appeared promising, for the full potential of pulses to be realized there needed to be increased research efforts into the production, processing, storage, transport and marketing of these crops.

A proposal made to the Minister for Primary Industries and Energy by the National Oilseeds and Protein Crops Committee of the Australian Wheat Growers Federation (now the Grains Council of Australia), and a joint national industry pulses research scheme was established in late 1985.

An initial levy of 75 cents per ton was applied to lupin and field pea payable by growers and received by grain handling authorities or through direct purchases. Subsequently, levies on faba bean (1986), chickpea, mung bean, pigeon pea and peanut (1988) have been collected.

In 1986/87 the Grain Legumes Research Program totalled A$ 491,000 with 15 projects funded during the first year of funding (Grain Legumes Research Council Annual Report, 1986–87). The funds were allocated for research on: species evaluation and cultivar improvement; pests and diseases; machinery, soil management, and farming practices; handling and marketing; and supporting extension, travel, and seminar activities.

Future Policy Directions

Grain Legumes Research Council (Now Incorporated into the Grains Research and Development Corporation)

The dramatic growth in the production of pulses since the formation of the council in 1985, coupled with an increase in levy, has seen a five-fold increase in the funds available for allocation.

However, most of the expertise in research on the production and processing of pulses is within the State Departments of Agriculture, CSIRO and the Universities. Government policies have encouraged these organizations to seek a greater proportion of their operating funds from external sources. These policies of reducing the provision of Government funds have meant that the Council has been required to pay a greater proportion of the total cost of a research project.

The council has established coordinated national breeding programs for each of the crops within the scheme. The introduction of plant variety rights and the desire of public sector institutions to commercialize outcomes have required the council to develop procedures to facilitate germplasm exchange.

The council has developed a number of research and development objectives. These have been formulated as follows:
1. To improve pulses productivity in sustainable production systems.
2. To improve the efficiency of harvesting, handling and storage of pulses.
3. To add to knowledge of pulse markets and end products with a view to increasing returns to growers by broadening the market base.
4. To allocate funding to individual research projects and categories commensurate with the expected benefits to growers.
5. To effectively promote the adoption of research results within the industry.
6. To efficiently administer the Council's responsibilities under the Act.

7. To meet the necessary standards of accountability to Parliament and to growers.
8. To encourage the development or specialist scientific skills in regard to pulse research.

These objectives have been largely maintained by the new Grains Research and Development Corporation (GRDC).

National Grain Legumes Consultative Committee

This industry body was established several years ago as a joint State Government, Grains Council of Australia and industry initiative to coordinate policy initiatives developed at a state level and to recommend appropriate courses of action to be taken for issues associated with the pulse industry.

A major initiative taken by the committee was to apply to the Federal Government to establish an Australian Grain Legumes Corporation whose role would be to facilitate the promotion and development of the pulse industry but not to become directly involved in marketing. The work of the Association would be complementary to the GRDC and would have some similarities with the work of the US Pea and Lentil Council.

Marketing Skills Program

The Department of Primary Industries and Energy in 1991 funded a Marketing Skills Program for the pulse industry.

The objectives of the program were to bring together a number of pulse industry people, including processors, who could act as catalysts for the industry in understanding the marketing requirements to successfully develop more profitable market niches, examine value adding opportunities and understand the competition the industry faces. A market study mission was sent to a number of countries in September/October 1991 with the objective of developing a number of market strategies for the pulse industry.

State Market Development Programs

All states are acutely aware of the importance of pulses not only in farm management strategies but also of the market potential of a large number of winter and summer pulse crops that could be grown.

It is clear that future development and funding will be focused mainly on the activities of the GRDC. However, each state is becoming increasingly more involved in its own market development programs. Austrade (the Commonwealth Department which facilitates Australian trade) is moving to coordinate the states marketing efforts particularly into markets where a state is perceived to have a comparative advantage or to coordinate state efforts where there are overlap problems and overall coordination could result in better national returns and more efficient research and development efforts.

Concluding Remarks

Future growth of the Australian pulse industry will be influenced by the GATT. Any decision to reduce grain subsidies in the European Community will mean lower prices for Australian pulses into the EC.

Provided the Indian Government continues the Open General Licensing System for pulse imports and maintains tariffs at reasonable levels, there are considerable long term opportunities for Australian pulse exports to India. With a population likely to increase by 200 million in the next decade, it will be difficult for Indian production to keep pace. As the pulse industry now has a well developed research and development infrastructure and farmers have experience with pulses, Australia will be better able to respond to any increase in world demand.

For the future, the Australian pulse industry will need to establish facilities for product development and processing opportunities, place greater emphasis on niche marketing and crop diversification, arrange for international testing of Australian cultivars for color, taste, cooking time, growth inhibitors and other impediments to acceptance of the Australian product and obtain more substantial market information.

While there are a number of production constraints including diseases, insect pests, weeds, water and wind erosion, and frost risk, there is a significant potential area available for the sowing of a large range of pulses in Australia.

There is considerable potential for new pulse crops. For example, recent evaluations of lentil germplasm is likely to lead to the release of selections that are much higher yielding than existing cultivars. The possibility to promote the vetch (*Vicia sativa* – c.v. Blanche fleur) as a crop for more marginal areas may have a significant impact. However, the undesirable concentrations of antinutritional substances in vetch has led to the banning of imports by India and Egypt. For the future it is more likely that lentil will replace vetch in higher and lower rainfall areas. Vetch and grasspea will become useful forage crops and used in hay mixtures with cereals. The development of low toxin cultivars of vetch and grasspea is some time away. Similarly, the development of kabuli chickpea, blue pea, and faba bean crops will allow pulse producers, particularly in southern Australia, to choose from a wide diversity of cropping options. Provided there is adequate agronomic and technical research support, the market opportunities will create significant changes to Australian cropping patterns.

References

Australian Bureau of Agricultural and Resource Economics. 1991. *Commodity Statistical Bulletin.*

Australian Bureau of Agricultural and Resource Economics. 1991. *Crop Report* No. 68.

Coffey, R. 1992. Export market prospects for lupins: *National Agricultural and Resources Outlook Conference*, Australia. Australian Bureau of Agricultural and Resource Economics.

FAO. 1989. *Trade Yearbook*, Vol. 43.

Grain Legumes Research Council Annual Reports, 1986-87.

Rees, R. 1988. *Grain Legumes Research Council*, Chapter 7, pp. 5-40.

Rees, R. and Presser, J. 1991. *Economics of pasture and grain legumes: the role of grain legumes in cereal rotations*. Waite Agricultural Research Institute, April 1991.

Rees, R. 1992. The Indian and Middle East Markets for Australian Grain Legumes. National Agricultural and Resources Outlook Conference 1992, Australia.

Toepfer International: Statistische Informationen 1990/91.

Breeding methods and selection indices

Breeding methods and selection indices for improved tolerance to biotic and abiotic stresses in cool season food legumes

R.J. BAKER

Department of Crop Science and Plant Ecology, University of Saskatchewan, Saskatoon, Saskatchewan, Canada S7N 0W0

Abstract

The objective of breeding for stress tolerance is to improve productivity for a target level of stress. If tolerance is viewed as resistance to change in productivity with increasing stress, productivity under stress depends not only on stress tolerance, but also on maximum productivity. Index selection theory indicates that selection in non-stress environments will be more effective than direct selection for productivity under stress whenever the correlation between the two types of environments exceeds the heritability of productivity under stress. With high genetic correlation, selection should be conducted within a level of stress that maximizes heritability. In cases where heritability under non-stress is much higher than under stress, an index combining data from stress and non-stress environments is expected to be more efficient than selection based on evaluation only within stress environments.

Secondary traits will be useful in breeding for productivity under stress whenever they have high heritability and high genetic correlation with productivity under stress. For some abiotic stresses and many biotic stresses, heritability will be highest in the presence of stress and indirect or index selection will be of limited value.

Definition of Stress Tolerance

Discussions of stress tolerance in food crops will often include consideration of mechanisms for avoidance or tolerance. Knowledge of physiological phenomena may be used to support arguments that certain physiological, morphological or phenological characteristics will provide some level of tolerance. From a plant breeding perspective, it is important to recognize that the ultimate objective is to improve productivity in environments that are characterized by a range of generally sub-optimal environments.

In applied terms, tolerance to abiotic or biotic stress factors must always be assessed in terms of productivity under stress conditions expected in field

F.J. Muehlbauer and W.J. Kaiser (eds.), Expanding the Production and Use of Cool Season Food Legumes, 429–438.

production. Most researchers will view tolerance in terms of differing patterns of response to increasing levels of stress; some researchers will fail to distinguish between tolerance and productivity under stress.

To define tolerance in a way that is suitable for use of selection indices, I have chosen to approximate response to stress as a linear function of increasing stress. Using a linear function is justified for two reasons. First, many response curves are nearly linear over a realistic range of stress. For example, a logistic response curve can be approximated by a straight line over the range of 5 to 95% of maximum with coefficient of determination exceeding 0.98. Likewise, cubic polynomial response curves can be described by straight lines from maximum to minimum productivity with very little loss of information. The second reason for approximating stress response curves with straight lines is that experiments measuring response in productivity are rarely precise enough to distinguish between different response curves.

Using a linear response curve, productivity (Y_j) at any level of stress (S_j) can be represented by a linear regression equation such as

$$Y_j = M - TS_j,$$

where M represents maximum productivity in the absence of stress, T is a measure of tolerance, and S_j is a quantitative measure of the level of stress. Note that high tolerance will be represented by low values of T. High tolerance (low T) will result in little change in productivity with changes in stress level.

Using two levels of stress, a low level S_l and a high level S_h, a decrease in productivity will measure $T(S_l - S_h)$. Small differences in productivity between non-stress and stress environments are thus indicators of increased tolerance. Small change with increasing stress is similar to lower variability in a sample of different stress levels.

In plant breeding, emphasis is on developing genotypes with high productivity under stress conditions, that is, high $M - TS_h$. Note that productivity under stress conditions depends upon two components, stress tolerance and maximum producitivty. If two genotypes have the same productivity in the absence of stress, the one with the greater tolerance (lower T) will be more productive under stress conditions. If two genotypes have similar tolerance (equal T), the one with higher productivity in the absence of stress will also be the one with high productivity in the presence of stress.

A critical question in studying stress tolerance concerns the possibility that high maximum productivity will be correlated with low tolerance. If productivity under stress is proportional to productivity in the absence of stress, highly productive cultivars will have low tolerance because of their greater differences in productivity between low and high stress conditions. In such cases, cultivars that perform well in the absence of stress will also perform well in its presence, and it may be more effective to select for high productivity in the absence of stress.

In an evaluation of 39 cultivars of bean (A. Vandenberg, personal communication) under dryland and irrigated conditions in Saskatchewan in

1991, the differences between yields in the irrigated and dryland tests was highly correlated ($r = 0.72$) with the yield of the irrigated test. If one uses resistance to change as a definition of tolerance, then there was a very large negative relationship between tolerance and yield potential in this study. In this study, genotypes that yielded well in the dryland experiment also yielded well in the irrigated experiment. A large component of performance in the stressful dryland environment was due to overall performance rather than tolerance *per se.*

When productivity in stress environments is highly correlated with productivity in non-stress environments, one needs to question the definition of stress tolerance. Apparent tolerance may merely reflect poor productivity in non-stress environments. This type of relationship between productivity in stress and non-stress environments does occur whenever productivity under stress is a constant proportion of productivity in non-stress conditions. Rosielle and Hamblin (1981) presented a theoretical discussion of this type of problem. They noted that tolerance and mean productivity are often negatively related and concluded that selection for tolerance will usually result in reduced yields in the absence of stress and reduced average yield over all levels of stress. Selection for mean productivity, on the other hand, would be expected to increase yield in both stress and non-stress environments.

Selection Indices for Productivity Under Stress

Blum (1988) emphasized the need for defining a target stress environment and noted that the objective of a breeding program is to improve yield in that target environment. In breeding for increased productivity under target levels of stress, legume breeders may consider using direct selection for productivity under stress, indirect selection based on productivity under non-stress, and combined selection based on productivity measured under both stress and non-stress conditions. The index selection methods reviewed by Baker (1986) can be used to identify the conditions that make one approach more efficient than another.

Let Y represent productivity measured under stress and let X represent productivity measured under non-stress conditions. The objective is to increase average value of Y by selecting for Y directly, by selecting for X, or by selecting for an index

$$I = b_1X + b_2Y.$$

For direct selection, response to truncation selection is expected to be

$$R = i_Y h_Y \sigma_{GY},$$

where i_Y is the standardized selection differential, h_Y is the square root of heritability, and σ_{GY} is the genotypic standard deviation for productivity under stress. According to Baker (1986), correlated response in Y due to indirect selection for X is expected to be

$$CR = i_X h_X \sigma_{GY} r_G,$$

where subscripts X refer to productivity under non-stress and r_G represents the genotypic correlation between productivity under stress and productivity under non-stress.

Indirect selection is expected to be more efficient than direct selection under stress only if correlated response exceeds direct response ($CR > R$ or $CR/R > 1$). When standardized selection differentials are equal for selection under stress and under non-stress, correlated response is expected to exceed direct response whenever $h_X r_G$ exceeds h_Y. That is, indirect selection will be more efficient than direct selection only if the genetic correlation between productivity under stress and under non-stress is close to 1.0 and heritability under non-stress is higher than heritability under stress.

In the absence of correlation of environmental deviations within stress environments with environment deviations within non-stress environments,

$$r_G = r_P/[h_X h_Y]$$

where r_P is the phenotypic correlation between productivity in stress and non-stress environments. Under these conditions, indirect selection under non-stress should be more effective than direct selection under stress whenever the phenotypic correlation between two environments exceeds the heritability under stress (i.e. $r_P > h^2_X$).

From a theoretical point of view, the decision to use direct or indirect selection for productivity under stress should be based on a comparison of heritability under stress with the correlation between the two types of environments. Whenever the correlation exceeds heritability under stress, selection under non-stress will be most efficient; otherwise it will be more effective to select under stress.

As an example, consider yield data for ten cultivars of *Vicia faba* grown in 1975 and 1977 at Dijon, France (Le Guen and Berthelem, 1981). For the ten cultivars, average yield in the less stressful 1975 season was 5.85 t ha^{-1} and average yield in the more stressful 1977 season was 2.37 t ha^{-1}. Using the pooled error reported by these authors, I estimate that the within-season heritability (on a mean of five replications) was 0.82 for the 1975 season and 0.62 for the 1977 season. The correlation between mean yields in the two contrasting seasons was 0.68.

Based on these estimates, one would conclude that selection in 1975 would be only slightly more effective for improving grain yield in the 1977 season than would direct selection in the 1977 season. The correlation between the two seasons was slightly greater than the heritability within the more stressful environment. While this calculation serves as a demonstration of how to evaluate the potential for indirect selection, the results provide some confidence in being able to breed for improved yield at a single location even though the level of unspecified environmental stress changes dramatically from one season to another.

If response to stress is such that, for all genotypes, productivity under stress is a constant proportion of productivity under non-stress, the genotypic correlation will be 1.0 and indirect selection will be more effective than direct selection whenever heritability is higher in non-stress than in stress environments. With high genetic correlations between stress and non-stress environments, selection will be most effective if carried out in the environment that shows the greatest heritability for yield. In an evaluation of drought tolerance in rapeseed, Richards (1978) observed that heritability of seed yield was generally higher under dryland than under irrigated conditions. He concluded that selection would be most effective in the stress environment. In contrast, Allen *et al.* (1978) concluded that selection for average yield would be more effective in wheat and soybean if carried out in high-yielding environments. Results were inconsistent for barley, oat and flax.

The third option for increasing productivity under stress is that of using combined selection (using data from stress and non-stress environments). For an optimum index, Baker (1986) has indicated that, with equal selection intensities, index selection will be more effective than single trait selection whenever the genetic correlation between the index and the single trait exceeds the square root of the heritability for the single trait. In the present context, the single trait is productivity under stress while the index is some linear function of productivity under stress and productivity under non-stress.

Methods described by Baker (1986, pp. 97–103) were used to estimate efficiency of an optimum index relative to direct selection under stress. Table 1 shows estimated efficiencies for three levels of heritability in each type of environment and six different genetic correlations between stress and non-stress environments. Result are independent of relative amounts of genetic variability in the two types of environments. Identical results would be expected for negative genetic correlations of the same magnitude (the index would put negative weight on productivity in the non-stress environment). Since these estimates assume that genetic variances and covariances are known, they are biased in favor of selection indices. Nevertheless, they serve to highlight the nature of information required to assess the problem.

The data in Table 1 show that index selection, using data from both stress and non-stress environments, will be more effective than selection in the stress environment only if there is a genetic correlation between the two types of environments. The calculations also show that index methodology will not result in much improvement if heritability is high in the stress environment. With moderately low heritabilities and moderate to high genetic correlations, considerable advantage can be expected from using data from both stress and non-stress environments. For example, if heritability under non-stress is 50%, heritability under stress is 30%, and the genetic correlation is 0.8, the expected efficiency of index selection is 126% of that expected from direct selection in stress environments at the same selection intensity.

Again, to determine whether or not a selection index would be useful in selecting for productivity in stress environments, one must have information on

Table 1. Predicted efficiency (%) of index selection relative to direct selection for productivity under stress

Genotypic correlation

h^2_1	h^2_2	0.0	0.2	0.4	0.6	0.8	1.0
30	30	100	101	104	109	115	124
30	50	100	100	101	103	105	109
30	80	100	100	100	100	101	101
50	30	100	102	107	115	126	140
50	50	100	101	102	105	109	116
50	80	100	100	100	100	101	102
80	30	100	103	110	123	141	165
80	50	100	101	103	108	116	129
80	80	100	100	100	101	102	105

h^2_1 = heritability (%) in non-stress environment.

h^2_2 = heritability (%) in stress environment.

the heritability of productivity in each type of environment as well as knowledge of the genotypic correlation between the two environments. To demonstrate the calculations, I consider data from a diverse sample of 39 bean cultivars evaluated under dryland and irrigated conditions in 1991 (A. Vandenberg, personal communication).

The two trials were located about 100 km apart in the province of Saskatchewan, Canada. In the dryland test, average yield was 1.47 t ha^{-1} and within-test heritability (based on means of four replications) was estimated as 0.91. For the irrigated test, mean yield was 2.98 t ha^{-1} and heritability was 0.93. The phenotypic correlation between the two tests was 0.74. Assuming that environmental deviations were not correlated, the estimated genotypic correlation between the two sites is $0.74/\sqrt{[(0.91)(0.93)]} = 0.80$. From Table 1, we can see that despite the high genetic correlation, the expected increase in efficiency for a selection index is likely to be less than 2% because of the high heritability in the dryland environment. Based on theoretical considerations and estimates of heritabilities within, and correlations between, tests at different levels of stress, we would conclude that index selection based on data from both dryland and irrigated tests would not be useful for improving yield in dryland sites.

Evaluating Components of Stress Tolerance

Physiological, morphological and phenological characteristics may be considered as important components of stress tolerance or productivity in stress environments. From a plant breeding perspective, such components must be considered to be secondary traits that are important only if they can be used to improve the efficiency with which improved productivity can be attained in stress environments.

Component traits may be correlated with stress tolerance because of pleiotropic effects of genes controlling these traits on productivity under stress conditions, or because of linkage between genes controlling these traits and genes controlling stress tolerance. The latter is expected to cause transient correlations between a component and stress tolerance. Pleiotropy is expected to give a more consistent relationship between stress tolerance and components that can be used in different populations in different breeding programs.

For a component trait to be useful, it must possess three characteristics; it must be related to productivity under stress, it must be amenable to genetic modification, and it must be economical to assess. In quantitative genetic terminology, a component must have a high genetic correlation with productivity under stress and it must be heritable. From earlier discussion, it is clear that indirect selection for a component trait will be useful only if the product of the heritability of the component trait and its genetic correlation with productivity under stress exceed the heritability of productivity under stress. It is also suggested that a component trait should be used in a selection index only if it has a moderate to high genetic correlation with productivity under stress. In addition to genetic considerations, one must also evaluate the practical aspects of the economics of measuring a component trait for use in indirect selection or an index.

I have chosen an example from the soybean literature (Bouslama and Schapaugh, 1984) to demonstrate some of the considerations that are required to decide when a secondary trait should be included in a program of breeding for stress tolerance. These authors investigated several laboratory measures for assessment of drought tolerance in soybean. Measurements included seedling germination in 0 and -0.6 MPa osmotic potential (using polyethylene glycol), seedling growth in 0 and -0.6 MPa osmotic potential nutrient solution, and thermostability measured on leaf disks incubated at 25 and 50°C. A subset of cultivars had been tested for four years in five irrigated and five dryland sites in Kansas.

The authors demonstrated that there was significant genetic variation among 20 cultivars for germination stress index, dry matter stress index, plant height stress index and heat tolerance. The authors did not report any quantitative measure of heritabilities of these components and it is not possible to calculate such values from the information supplied.

A subset of seven cultivars were tested in dryland and irrigated tests over a four-year period. Field drought tolerance was measured as the ratio of yield under dryland to yield under irrigation, and was found to vary from 65 to 78%. While the number of observations was too few to obtain good measures of relationship, the authors were able to show positive correlations between hydroponic seedling tests and field drought tolerance. Because of the reproducibility and consistency of cultivar differences, and because of the positive correlation with field drought tolerance, the authors concluded that hydroponic seedling tests could be useful in screening for drought tolerance in soybean.

The authors did not comment on the relationship between their measure of field drought tolerance and productivity under both dryland and irrigated environments (Table 2). The cultivar with the highest measure of drought tolerance had the lowest yield under both dryland and irrigated conditions. The cultivar with the lowest drought tolerance index was intermediate in yield under dryland environments and second highest in the irrigated tests. These observations emphasis that selection for improved performance under stress conditions requires consideration of average productivity in addition to tolerance.

Table 2. Yield and tolerance index of seven cultivars of soybean from ten locations in Kansas (1978–1981)

Cultivar	Yield		Tolerance index
	Dryland	Irrigated	
	- - - - kg ha⁻¹ - - - -		%
Elf	1888	2406	78
Douglas	2426	3333	73
Williams	2137	2997	71
Cumberland	2238	3158	71
Crawford	2043	2910	70
Calland	2137	3165	67
DeSoto	2164	3320	65

Modified from Bouslama and Schapaugh (1984).

For tolerance-related traits that are quantitative in nature, whether or not they will be useful in a crop improvement program depends upon how easy they are to measure, how heritable they are, and on the strength of their genetic correlation with productivity under stress. These three questions apply equally well to morphological characteristics such as root:shoot ratio, to phenological characteristics such as early maturity, and to physiological characteristics such as osmoregulation or carbon isotope discrimination. An additional factor that needs to be assessed is how consistent the genetic correlation is from one genetic population to another.

Measurements of heritabilities and genetic correlations require large sample sizes. Information from fewer than 20 genotypes will be of very limited value to plant breeders. Evaluation of fewer genotypes will result in imprecise estimates of heritability and genetic correlation and may well lead to erroneous conclusions about the value of candidate component.

Choosing Parents

The limit of stress performance in a breeding program is set once the parental material for the program has been chosen. If a component of stress tolerance is controlled by several to many genes, it is likely that there will be genetic variation for that trait in most crosses. In this case, it is better to choose parents so that the cross mean is as close as possible to the breeding objective than to choose parents to assure maximum genetic variation. However, if a component is rather simply inherited, it may be important to choose parents carefully, even if this component has too low a heritability or too low a genetic correlation to make it a useful secondary trait in a selection index. If a simply inherited physiological trait can be proven to have an effect of stress productivity, however small, the breeder should attempt to make certain that at least one (and preferably all) of the parents possess this desirable component.

Biotic Stresses

The discussion to this point has been directed primarily at abiotic stresses that show more or less continuous variability. Among the biotic stresses, disease resistance is paramount. Depending upon the nature of the disease, the level of disease stress may be quantitative or qualitative. For quantitative biotic stresses, it seems reasonable to suggest that the same principles that pertain to quantitative abiotic stress should apply. To improve productivity in the presence of the stress, one may select directly in the presence of the stress, indirectly in the absence of the stress if the correlation exceeds the heritability under stress, or use an index when the heritability under stress is much less than the heritability in the absence of stress. I believe the same principles apply to biotic stresses that are more qualitative in nature.

Biotic stresses tend to differ from many abiotic stresses in that heritability is often higher in the presence of the stress than in the absence of the stress. Theoretical considerations discussed earlier would then dictate that selection for improved productivity in the presence of the biotic stress will be most effective with direct selection under stress. Indirect selection will not be effective and index selection will not be any more efficient than direct selection.

For qualitative biotic stresses, tolerance is replaced by resistance. Resistance is selected for in the presence of the stress and productivity may be evaluated either in the presence or absence of the biotic stress. The combination of high mean productivity and resistance (high tolerance) will assure maximum productivity under stress.

Concluding Remarks

Many abiotic and biotic stresses vary in intensity from season to season and even during the course of a growing season. Many stresses exhibit quantitative variation in intensity and are, to some extent, unpredictable. However, regions can usually be classified as being subject to some average level of a predominant stress and the plant breeder must try to produce genotypes that will be as productive as possible within that target class of environments.

Quantitative genetics methods, including the principles of index selection, provide some guidance in answering questions that are critical to efficient plant breeding. Should selection be carried out only in the stress environment? Which is the best environment for selection? What secondary traits will be useful in improving productivity under stress? Should indirect selection or index selection be used in a breeding program? Answers to these and other important questions depend upon having good estimates of heritabilities within the different classes of environments as well as estimates of genetic correlations between different classes of environments. My impression is that the literature contains rather few of these types of estimates for food legumes.

References

Allen, F. L., Comstock, R. E. and Rasmussen, D. C. 1978. *Crop Science* 18: 747–751.

Baker, R. J. 1986. *Selection indices in plant breeding*, 218 pp. Boca Raton, FL, USA: CRC Press.

Blum, A. 1988. *Plant Breeding for stress environments*, 223 pp. Boca Raton, FL, USA: CRC Press.

Bouslama, M. and Schapaugh, Jr., W. T. 1984. *Crop Science* 24: 933–937.

Le Guen, J. and Berthelem, P. 1981. In: *Vicia faba: Physiology and Breeding*, pp. 80–98 (ed. R. Thompson). The Hague, The Netherlands: Martinus Nijhoff.

Richards, R. A. 1978. *Euphytica* 27: 609–615.

Rosielle, A. A. and Hamblin, J. 1981. *Crop Science* 21: 943–946.

Screening techniques and sources of tolerance to extremes of moisture and air temperature in cool season food legumes

J. WERY[1], S.N. SILIM[2], E.J. KNIGHTS[3], R.S. MALHOTRA[4] and
R. COUSIN[5]

[1] ENSAM, Chaire de Phytotechnie et d'Amélioration des Plantes, 2 place Viala, 34060 Montpellier Cedex 01, France;
[2] ICRISAT Legume Program, Patancheru P.O., Andhra Pradesh 502 324, India;
[3] NSW Agriculture, Agricultural Research Center, RMB 944, Tamworth, NSW 2340, Australia;
[4] ICARDA Legume Program, P.O. Box 5466, Aleppo, Syria, and
[5] INRA Route de Saint-Cyr, 78026 Versailles Cedex, France

Abstract

Breeding for resistance to extremes of temperature and moisture in cool season food legumes is limited by the lack of adequate screening techniques. The success of each technique depends upon the representativeness and reproducibility of the type of stress created. Descriptions of successful techniques are presented for frost and terminal drought. Development of new screening tests designed to select for specific adaptive traits require a better knowledge of the mechanisms of resistance in these crops, especially to drought. Rooting depth, early vigor, reduced branching, and osmotic adjustment are discussed. Other mechanisms of resistance to drought, heat, freezing, or chilling have been proposed but need to be studied jointly by crop physiologists and plant breeders.

Introduction

The level and stability of grain production and nitrogen fixation of cool season food legumes are frequently impaired by abiotic stress (Saxena et al., 1990; Wery et al., 1990). In the Semi-Arid Tropics (SAT) and Mediterranean environments, particularly in West Asia and North Africa (wANA), the major abiotic constraint is drought because these crops, with the exception of faba bean, are generally grown on poor soil without irrigation. Chickpea, in particular, experiences drought stress because of late sowing. In the Mediterranean region and in Australia, increasing drought stress from flowering to maturity is very common for chickpea. Historically, crops have escaped drought stress when sowing time is shifted from spring to fall. The latter is the normal sowing time for lentil, faba bean, and pea in Mediterranean conditions, and is being extended to chickpea (Wery, 1990). In temperate regions, faba bean and pea are sometimes sown in fall or winter. This means that cold (chilling or freezing) is also a major stress experienced by cool season food legumes.

F.J. Muehlbauer and W.J. Kaiser (eds.), Expanding the Production and Use of Cool Season Food Legumes, 439–456.
© 1994 Kluwer Academic Publishers.

The third stress is high temperatures after flowering, especially for chickpea in North Africa and West Asia. Waterlogging is not common for these crops and will not be discussed here.

Although the development of cultivars tolerant to these stresses is essential for the improvement of cool season food legumes, the breeding effort has been limited as compared to cereals, especially regarding resistance to drought. A considerable amount of work has been done for frost tolerance in pea, faba bean, and chickpea. Research on drought resistance in chickpea and lentil, and chilling resistance in chickpea have been initiated recently.

The major limitation in breeding for stress resistance is the lack of screening techniques adapted to these crops. Most breeding programs are based on visual scoring or yield measurement in the target environment. Significant progress can be made in this direction and in the development of more specific screening techniques in controlled conditions, using knowledge obtained on one of the four crops, and extending it to the others.

General Characteristics of the Screening Techniques

The efficiency of the screening technique depends on its ability to reproduce the most probable conditions of development of the stress in the target environment. It requires characterization of this most probable stress in its actual position in the plant cycle (Wery *et al.*, 1990) and its reproduction in conditions where screening of a large number of genotypes can be made. These two steps are essential for the representativeness and the reproducibility of screening techniques.

The target environment is often defined by a combination of two types of stress, for example heat and drought at the end of the plant growth cycle. Screening can be done separately for the two stresses in controlled conditions, or simultaneously in the target environment.

Nitrogen fixation is more susceptible to water and thermic stresses than nitrate assimilation and carbon metabolism (Wery, 1987). Screening for tolerance to these stresses must be done in conditions representative of the actual pattern of nitrogen nutrition, which for cool season food legumes means a soil with a low nitrogen content and a sufficient number of rhizobia.

Before defining the criteria used in the ranking of genotypes, it is essential to identify the mechanism, or combination of mechanisms, which are the most efficient to ensure escape, avoidance, or tolerance. For example, drought resistance can be obtained with different types of mechanisms which may have antagonistic effects on plant yield (Wery *et al.*, 1990). The breeder should give priority to the adaptive mechanisms which confer on the crop the ability to yield well in the best years or soils.

Direct measurement of the physiological process involved in tolerance is promising if it is possible to use it on a large number of genotypes suffering from the same stress.

Screening Techniques and Sources of Resistance to Drought

Drought Stress in Cool Season Food Legumes

Drought stress is the condition whereby the amount of water available to the plant is not sufficient for undisturbed growth and development, and consequently results in yield losses. Drought is extremely unpredictable because it depends on many factors including: rainfall and its distribution, evaporative demand of the atmosphere, and the capacity of the soil to store moisture. Two types of drought stress are commonly experienced: intermittent drought caused by breaks in rainfall, and terminal drought which occurs at the end of the growing season due to depletion of stored soil moisture. The most probable drought experienced by cool season food legumes is terminal drought, especially for lentil, because it is cultivated in the driest areas, and chickpea, because it is spring-sown and has late maturity (Saxena *et al.*, 1990). Drought is frequently associated with heat stress, especially in the Mediterranean basin (Wery *et al.*, 1990).

Resistance to drought can be obtained with various types of mechanisms, including: escape, dehydration avoidance, or dehydration tolerance (Blum, 1988; Ludlow and Muchow, 1990). Breeding programs for resistance to drought have been limited in cool season food legumes. It is important to define the traits required for the various environments before screening tests can be developed.

Drought Escape

The ability of a crop to complete its life cycle before serious soil water deficit develops is referred to as drought escape. This mechanism has been extensively used by the breeders, especially through selection of early flowering genotypes (Saxena *et al.*, 1990). Yield potential and drought escape through early flowering are the two major components of drought resistance for lentil (Silim *et al.*, 1993a), chickpea (Silim and Saxena, 1993a), and pea (Silim, unpublished). For example, flowering date accounts for 37 to 69% (in a dry year) of the genotypic differences in yield of chickpea (Silim and Saxena, 1993a). When yield potential (as yield in irrigated conditions) is taken into account, the two criteria explain 70 to 88% of the variability (Table 1).

Progress towards earliness may be limited by the susceptibility of flowers to chilling and the negative correlation between early flowering and yield potential. In seasons with higher rainfall, short duration cultivars of lentil and chickpea left potentially transpirable water in the soil and incurred a yield penalty when compared to medium-late maturing cultivars (Silim *et al.*, 1993b; Silim and Saxena, 1993a).

Other plant characters might be used to obtain drought escape and include:
a. The ability to emerge in dry soil may accelerate and synchronize the plant growth cycle in the field in semi-arid regions (Saxena *et al.*, 1990).

Table 1. Partitioning of sum of squares for unirrigated yield, among time to flowering (FL), yield potential (YFI), and drought response index (DRI) in chickpea (adapted from Silim and Saxena, 1993a)

Year	Rainfall	FL	YFI	DRI	Residual
	-mm-	-%-	-%-	-%-	-%-
1986-87	359	41	47	4	8
1987-88	504	37	37	17	9
1988-89	234	69	1	17	13

b. Accelerated phenology, and particularly the adaptation of flowering duration to the amount of water available. This mechanism, already identified in pea, can also be used in the other cool season food legume crops (Wery et al., 1990). Accelerated phenology should be more efficient than early flowering because it is more adaptive with no antagonistic effects on yield potential. Kabuli chickpea accession ILC 262, was early to flower, showed plasticity for maturity, and gave large yield under high moisture supply (Silim and Saxena, 1993b). Nevertheless, no screening of pea lines and no evaluation of the other crops have been done for this character.

c. Rapid establishment of the seed-filling sink can also be obtained with large seed size or doublepoddedness in chickpea (Saxena and Johansen, 1991).

Screening for phenological plasticity in the field involves growing a set of genotypes both under rainfed conditions (with terminal drought) and assured moisture supply.

The response of genotypes independent of escape and potential yield is called drought response index (DRI) (Bidinger et al., 1987). A line-source sprinkler irrigation system was used for screening lentil (Silim et al., 1993b) and chickpea (Silim and Saxena, 1993b) genotypes for yield under drought stress and response to increased moisture supply. DRI was then used to identify genotypes that were tolerant to drought. Lentil lines ILL 6035 and ILL 6011 (Silim et al., 1993a), and chickpea lines ILC 1919, FLIP 83-2C and Annigeri (Silim and Saxena, 1993b) were identified as drought tolerant. Correlations of traits that are significantly associated with drought tolerance, yield, time to flower, and DRI provide an indication as to whether the trait is associated with drought escape or tolerance. In lentil, early growth vigor, high percentage of ground cover, and high biomass were traits associated with drought tolerance. In chickpea, a deep root system and high leaf water potential were associated with drought tolerance. A high harvest index, a large number of pods per unit area, and a large grain mass were associated with drought escape (Table 2).

Table 2. Correlation coefficients of yield and traits related to yield in unirrigated chickpea cultivars, with time to flower (FL) and drought response index (DRI) (adapted from Silim and Saxena, 1993a)

Correlate	FL			DRI		
	1986/87	1987/88	1988/89	1986/87	1987/88	1988/89
Biomass	0.464	0.167	-0.255	0.106	0.028	0.081
Straw yield	0.641***	0.499*	0.365	-0.174	-0.239	-0.215
Harvest index	-0.631***	-0.709**	-0.816***	0.423	0.559*	0.650**
Grain yield	-0.173	-0.431	-0.842***	0.587*	0.561*	0.701***
Days to mature	0.823***	0.813***	0.863***	0.244	-0.263	-0.222
Max. rooting depth	na	na	0.075	na	na	0.657***
LWP (at 77 DAS)[1]	na	na	0.237	na	na	-0.482*
LWP (at 84 DAS)[1]	na	na	0.041	na	na	-0.626**

1 : LWP = Leaf Water Potential
    DAS = Days After Sowing

Significance : * = P < 0.05 ; ** = P < 0.01 ; *** = P < 0.001

Dehydration Avoidance

The objective is to develop mechanisms which enable avoidance of a decrease in plant water content (Blum, 1988) in order to maintain its productive functions, especially photosynthesis, growth, and nitrogen fixation.

Tolerance with High Potential

Increased Transpiration

There are some reports of genotypic differences in chickpea for root density and rooting depth, or for the number and diameter of xylem vessels (Saxena *et al.*, 1990). Rooting depth is the most interesting characteristic because it is less influenced by the soil conditions and it can reduce the amount of water left by the plant in the bottom of the profile (Wery *et al.*, 1990). A deeper rooting depth could delay the final drop of plant water potential and ensure better carbon and nitrogen metabolism during seed filling. Saxena *et al.* (1990) reviewed the literature on genetic differences in root length and spread in cool season food legumes. They reported that genotypes with greater root growth and branching were resistant to drought stress. As genetic differences for these characteristics were established early in the season, they screened seedlings using a sand culture technique for greater root growth. In the field, direct evaluation of this character is time consuming and is not adapted to a large number of genotypes. Nevertheless, Silim and Saxena (1993b) used a neutron probe to determine genotypic variations in effective rooting depth (Table 3) and ability to extract

Table 3. Grain yield and pre-dawn leaf water potential (LWP) of chickpea cultivars with various effective rooting depth (ERD) (adapted from Silim and Saxena, 1993a, 1993b)

Cultivar	ERD Unirrigated (a)	LWP Unirrigated (a)	Yield	
			Unirrig.	Irrigated
	-m-	-MPa-	- - - t ha⁻¹ - - -	
ILC 262	0.75	-2.21	0.395	2.252
ILC 1930	0.89	-2.06	0.083	2.180
ILC 1919	0.97	-1.82	0.257	2.082
ICC 4958	1.04	-1.36	0.635	2.708
Annigeri	1.13	-1.65	0.658	2.058

a : ERD and LWP are the average of the measurements made

   during the seed filling period.

water in spring-sown chickpea in a Mediterranean environment. A similar methodology was also used for lentil (ICARDA, 1987) and faba bean (Silim and Saxena, 1993c). In all cases, genotypes with a deeper root system produced large grain yields under limited soil moisture conditions (an example is given for chickpea in Table 3).

In chickpea, Silim and Saxena (1993a) reported that a genotype with a deep root system had a high leaf water potential (Table 3). Since screening for differences in root length and density is laborious and time consuming, Silim and Saxena (1992a) suggested that an indirect way of selecting for drought tolerance would be to screen for high leaf water potential during reproductive stage.

In a Mediterranean environment, early growth vigor is an important adaptive trait for terminal drought stress in winter-sown crops such as lentil (Silim *et al.*, 1993b). As solar radiation is important in winter, early growth vigor of chickpea (Saxena and Johansen, 1991) will save water for post-flowering transpiration through a significant reduction of soil water evaporation (Siddique and Sedgley, 1987). In these regions, the pattern of water consumption during the crop cycle is more important than the total amount of water used (Siddique and Sedgley, 1987). A visual score for early growth vigor on a 1 to 5 scale was highly correlated with measurements such as dry matter per unit area in lentil and was thus recommended because it is simple, swift and enables the breeder to assess a large number of genotypes (Silim *et al.*, 1993a).

In lentil, the genotypes ILL 6004 and ILL 6035 exhibited early growth vigor and gave high yield under drought. The trait is exhibited independent of moisture supply.

When rainfall is higher (above 300 to 400 mm) lines with large average yields and a good response to water supply can be used in lentil (Silim *et al.*, 1993b) and chickpea (Silim and Saxena, 1993b). This ideotype is in fact a combination of various characters including: large yield potential, earliness to flower, high harvest index, and deep rooting. Evaluation of the value of this combination of traits requires large field trials under low rainfall, with lines sown across a moisture gradient produced by a line-source sprinkler irrigation system.

Maintenance of water uptake and transpiration can be evaluated indirectly with infrared thermometry with the methodology previously used for cereals (Blum, 1988). Direct measurement of soil water consumption with a neutron probe is not adapted to screening of large numbers of lines.

Reduced Transpiration

Characters leading to reduced transpiration must be restricted to the driest situations (areas with low rainfall and shallow soils). The objective is to use morphological attributes to limit plant water consumption especially in the vegetative phase.

Reduction of leaf area is an important adaptive mechanism for drought stress and is usually the first strategy a plant adopts when water becomes limiting. Examples of reduced leaf area include the small leaflets of certain chickpea lines (Saxena and Johansen, 1991) and tendrils of semileafless peas. Work with conventional-leafed, semileafless and leafless peas in Syria showed that under drought stress the semi-leafless pea cultivar "Baf" produced the largest grain yield (Silim and Saxena, 1990). However, the usefulness of this trait has yet to be evaluated in a breeding program. Significant genotypic variations in stomata size and frequency have been recorded in a number of crops, but their importance in drought tolerance has not been fully established (Ludlow and Muchow, 1990).

The plants under drought stress are usually exposed to an abundance of sunshine. Without transpiration, temperatures of the exposed plant parts rise. A reduction of the amount of radiation received reduces the need for transpiration to keep the plant cool. Leaf rolling, leaf movement, and increased reflectance are some of the mechanisms reported in cereals for reducing water loss (Ludlow and Muchow, 1990). A chickpea line with multipinnate leaf form has a higher level of reflectance thus reducing the amount of energy stored by the leaf (Saxena and Johansen, 1991).

Similar biological yield and higher harvest index were obtained from chickpea phenotypes with reduced branching (Siddique and Sedgley, 1985). The pattern of water consumption was modified in the debranched crop with a reduction before flowering and an increase during the reproductive phase when compared to the control. The available genotypic variability for this character, estimated at between 2.3 to 16 primary branches, in chickpea should be exploited.

Reduced leaf area or branching are amenable to breeding because they are generally simply inherited and direct selection can be applied. Nevertheless, they should be limited to environments with consistently low rainfall because they frequently lead to reduced productivity. For example, the multipinnate leaf mutant with increased reflectance has significantly lower yield potential than that of its near isogenic conventional-leafed line (Knights, submitted).

Tolerance with Low Potential

Water stress in the later phases of the growth period can reduce yield potential through a lowering of biomass and harvest index. These influences can be partially negated through osmoregulation, whereby partial or full regulation of cellular turgor pressure and hydration is achieved by increases in the amount of intra-cellular solutes. This mechanism is used by field grown chickpea during its reproductive period when drought is established progressively (Lecoeur *et al.*, 1992). Genotypic differences in osmoregulation have been demonstrated in this species (Morgan *et al.*, 1991). In this study, osmoregulation was determined from measurements of osmotic potentials and relative water content made on leaves of glasshouse-grown plants. The genotypes (half-sibs) were grouped into two well-defined classes on the basis of osmoregulation, and the effect of this mechanism determined from yield trials conducted over a range of environments. The yield increases attributable to osmoregulation ranged from approximately zero in low water-deficit environments (site mean yield 3.5 t ha^{-1}) to approximately 20% in high water-deficit environments (site mean yield 1.3 t ha^{-1}). It is significant that one of the four high osmoregulating genotypes was the cultivar "Tyson", which is also known as "C235" in northern India. Despite being released 30 years ago, its consistently high yields have resulted in its retention as a popular cultivar there.

With present technology, screening for osmoregulation is very time and resource consuming. Consequently, screening for high osmoregulation should be delayed until the final stages of a breeding project or used on potential parental lines. Another means of investigation could be to screen for the ability to accumulate the main solutes involved in osmoregulation (Lecoeur *et al.*, 1992). Nevertheless, improvement in this mechanism is probably associated with field screening for yield under terminal drought.

Dehydration Tolerance

This is represented by mechanisms allowing the plant to withstand dehydration, especially at the end of the reproductive phase. The physiological and biochemical bases are not clearly established, but simple screening tests have been proposed for soybean (Bouslama and Schapaugh, 1984) and cereals (Blum and Ebercon, 1981). They are similar to those proposed for heat tolerance.

Screening Techniques and Sources of Tolerance to Heat

No breeding program has been developed for this character in cool season food legumes (Malhotra and Saxena, 1990), probably because genotypic and seasonal effects make it difficult to reproduce the field screening conditions at a specific phenological stage. Nevertheless, tolerance to this stress is probably taken into account in the WANA region when screening for terminal drought is conducted in the field. Breeding specifically for this trait is desirable, and progress could be made with screening techniques developed for other crops, such as wheat (Blum and Ebercon, 1981) and soybean (Bouslama and Schapaugh, 1984). These techniques are based on indirect measurement of plasmic membrane stability and can be used to screen both for heat and dehydration tolerance. The same methodology has also been used for freezing tolerance in alfalfa (Sulc *et al.*, 1991). The plant organs, generally leaflets or leaf-discs, are subjected to extreme temperature or dehydration stress before being immersed in distilled water. The amount of solute leakage from the leaves of each genotype are measured indirectly by electrical conductivity (Blum and Ebercon, 1981; Bouslama and Schapaugh, 1984) or UV light absorbance at a given wavelength (Sulc *et al.*, 1991). The genotypes are ranked on the basis of the ratio between this measurement and a measurement made after destruction of the organ with reference to a non-stressed sample (Blum and Ebercon, 1981). Flowers are probably the most susceptible organs (Wery *et al.*, 1990). Nevertheless, these procedures can be applied to leaves because the target ideotype is a plant with leaves able to ensure the end of seed filling, at a time when flowering has ceased. Evaluation of its potential use for screening cool season food legumes will require preliminary studies, especially on the best pattern of temperature increase during the test. Tolerance to heat seems to be better when hardening has been achieved with a preliminary short heat shock (Blum, 1988).

Breeders could also use mechanisms of escape (from accelerated phenology) and avoidance to heat (e.g., increased transpiration), when a global approach of terminal drought and heat is used (Wery *et al.*, 1990).

Screening Techniques and Sources of Tolerance to Chilling

Tolerance to Chilling Injury at Emergence

Chilling injury in the field at emergence is manifested by poor establishment and low vigor in cold and often wet soils. It is very frequent for winter sown crops, especially chickpea, and the resulting poor stand counteracts the beneficial effects of early sowing. Powell and Matthews (1978) postulated that chilling injury and imbibition damage (in peas) are the same phenomenon, with embryos being more sensitive to imbibition damage at low temperatures. Tully *et al.* (1981) refer to the phenomenon as imbibitional chilling injury. The poor emergence and low vigor of seedlings has been ascribed to a rapid uptake of water which damages cells on the cotyledonary surface, resulting in tissue death and high solute leakage (Powell, 1989). Imbibitional damage has also been implicated in poor establishment of some chickpea genotypes in cold, wet soils (Knights and Mailer, 1989).

Tully *et al.* (1981) contend that imbibitional chilling injury is mediated by the rapidity of imbibition, and this factor is controlled principally by the seedcoat. Seedcoat color is associated with establishment, and presumably the level of imbibitional damage. In peas, white-seeded lines imbibed more rapidly than their colored-seeded isolines (Powell, 1989). In Pythium-infested soil the white-seeded lines also had a higher level of cotyledon and embryo infection. These differences were attributed to differences between white- and colored-isolines in the adherence of the seedcoat to the cotyledons. When white- and brown-seeded near-isolines of kabuli chickpeas were sown into cold and wet soil, the establishment of white-seeded lines was also inferior (Knights and Mailer, 1989). This difference was associated with a higher level of tannins, presumably possessing fungistatic properties, in the colored seeds. As a matter of fact the difference in susceptibility to chilling during emergence between a black-seeded desi type and white-seeded kabuli types was reduced by seed-dressing with fungicides (Knights, 1980; Wery *et al.*, 1990). However, in the near-isogenic lines used by Knights and Mailer (1989), the kabulis also had much thinner seedcoats. The associated reduced cracking in the desi lines was assumed to be a factor in limiting the rate of water intake, and hence imbibitional damage. Differential establishment for desi and kabuli lines has also been reported for early-spring sowings in the Palouse region of the USA (Auld *et al.*, 1988) and for winter sowings in southern France (Wery *et al.*, 1990). Establishment differences between kabuli genotypes have also been reported (Auld *et al.*, 1988; Knights and Mailer, 1989) and indicate the potential for ultimately developing kabuli cultivars which establish satisfactorily without fungicidal protection in cold, disease-infested soils. Differences in establishment potential between desi genotypes have also been observed in Australia (Knights, unpub. data). Cultivars reported as tolerant to chilling at emergence are: "CPI 562896-b" (Knights, 1980), "Semsen" (Knights, unpublished) and "Sombrero" (Wery *et al.*, 1990).

Screening can be done either in the field or in controlled conditions by sowing unprotected seeds into naturally or artificially infested soil when soil temperature is low. For example these conditions are almost systematically met with December sowing in the Mediterranean plains. The logical time to apply this test is in early-generation segregating populations when some mass selection for improved establishment potential can be effected.

Tolerance to Chilling Injury at Flowering

Cold-induced flower drop is a major cause of yield loss in autumn or early-winter sown chickpeas. Savithri and Ganapathy (1980) showed that failure to produce seed at low temperatures was due to a prevention of fertilization, and not to abortion of the embryo. The same authors reported that an EMS-derived mutant set seed at lower temperatures than its source genotype. The improved seed set of the mutant at low temperatures was associated with higher pollen germination and faster pollen tube growth. Genotypes have also been developed which set seed under low temperatures (ICRISAT, 1988). These genotypes set some seed when minimum temperatures ranged from -1 to 7°C. The most interesting genotypes are M450 (Savithri and Ganapathy, 1980), ICCV 88501, ICCV 88502, and ICCV 88503 (Sethi, personal communication).

The screening technique is based on early sowing so that flowering commences in late winter/early spring. Individual plants are then tagged on the day of anthesis. At maturity, the percentage pod-set can be related to the daily minimum and maximum temperatures. Chilling injury could also be the cause of early abortion of flowering buds in fall- or winter-sown chickpeas (Wery *et al.*, 1990), but no screening has been done for this trait. Nevertheless, its evaluation could simply be based on the number of nodes between the first reproductive node (always following the highest secondary branch) and the first node having borne a pod.

Although some selection in the field may be possible in early generations, progress is likely to be impeded by micro-environmental variation. Hence, field evaluation for this character is best done on advanced selections in replicated trials with the test plants well buffered by neighboring plants to simulate a field situation.

Screening Techniques and Sources of Resistance to Frost

Mechanisms of Resistance

Escape mechanisms have frequently been used by breeders, especially vernalization requirements or rosette-type morphology (Wery *et al.*, 1990) in pea (Açìkgöz, 1982; Cousin *et al.*, 1990). Avoidance can occur via supercooling (associated with the lack of ice nucleation particles) or osmotic adjustment (Wery *et al.*, 1990). Nevertheless, there is no evidence that breeders could screen specifically for these mechanisms.

Although the physiological bases of the mechanisms of frost tolerance are not well established, breeding work has been successful on pea, faba bean, and chickpea. Freezing tolerance is often associated with mechanisms at the cellular level, including increased membrane fluidity and osmotic adjustment (Wery *et al.*, 1990). This can explain why simple screening tests, even in the field, have enabled the development of "new crops" such as winter peas in Europe and winter chickpeas in the Mediterranean basin. The nature of the mechanisms involved in frost tolerance emphasizes the frequently reported role of the hardening process. The ability to harden, i.e., the progressive adaptation to a decreasing temperature, is essential to frost resistance of the cool season food legumes. This ability seems to decrease with plant age and requires a small growth rate (Wery *et al.*, 1990). The success of the screening test will depend largely on the opportunity for plant hardening.

Screening Techniques

Screening Conditions

In the Field

Extensive screening of pea and chickpea for cold tolerance has been done using similar techniques. The main problem is to ensure the target range of low temperatures each year. Screening at high elevations has been used extensively in chickpeas. In 1979/80, ICARDA screened 3158 kabuli accessions near Ankara, Turkey (altitude 1055 m) (Singh *et al.*, 1984). But in those conditions, the plants were frequently protected by snow cover or the temperature was much lower than in the normal growing conditions (Wery, 1990). The best solution is probably to screen at lower altitudes in regions with days of $-10°C$ for chickpeas and $-15°C$ or lower for peas, without snow cover. The most susceptible stage of the plant should be matched with the period of frost and while allowing the plants enough time to emerge and harden before the first frost (Wery, 1990). This can be obtained with very early fall sowing in the Mediterranean regions. That time of sowing was determined by a date of sowing trial, conducted at Tel Hadya, Syria (altitude 82 m) which includes 100 chickpea accessions and nine sowing dates that varied from October to March (Singh *et al.*, 1989). It is generally safer to sow at two dates spaced one month apart to more likely meet the correct conditions each year of testing (Wery, 1990; Cousin *et al.*, 1990). The same type of field screening has been used on a large scale for other crops, for example, on progenies from crosses between garden peas and forage peas (Cousin, 1976; Auld *et al.*, 1983a), on pure lines of faba bean (Robertson and Sherbeeny, 1988) and on lentil accessions (Erskine *et al.*, 1981).

Field trials for cold tolerance require no specific equipment and allow for screening of thousands of genotypes. However, the correct conditions of cold must be verified through the use of appropriate checks. It is usually sufficient to sow the same susceptible check after every nine test genotypes (Erskine *et al.*, 1981; Singh *et al.*, 1989). The rating of cultivars is made only after the susceptible check suffers 100% mortality. The presentation of the results must be made with consideration of daily minimum air temperatures, the amount and importance of snow cover, and the ratings of known lines covering the scale of susceptibility (Wery, 1990). The screening test must be repeated successfully in another place or during a second year.

In Controlled Environments

The method presented here has been developed for peas (Cousin *et al.*, 1990) but could easily be adapted to the other cool season food legumes. Similar tests have been used for pea (Auld *et al.*, 1983b) and faba bean (Herzog, 1988). The pattern of temperature evolution during the test must be determined for each species. The plants, before exposure to the required minimum temperature, are subjected to a progressive decrease in temperature over a period of 2 to 3 weeks so that they become hardened. After they have been subjected to sub-zero temperatures for several days or weeks, the temperature is gradually increased. This increase is started when about 50% of plants show frost damage. When the lines show a large range of cold resistance, it is better to conduct the test at two levels of cold (− 6 and − 8°C). The ranking of lines is better at − 6°C when their level of tolerance is low, and at − 8°C when this level of tolerance is high (Table 4). This test at two temperatures is used officially in the registration process of winter pea cultivars in France. The ranking given by this test is in good agreement with field screening (Table 5). The main advantage of this type of screening in controlled conditions is the efficiency and the reproducibility of the

Table 4. Screening for cold tolerance of pea lines under two controlled environmental conditions

Minimum temperature −6 °C			Minimum temperature −8 °C		
Lines	Cold Tolerance	Newman-Keuls test (5%)	Lines	Cold Tolerance	Newman-Keuls test (5%)
1	6.0	*	1	5.5	*
Kazar	5.9	*	Kazar	4.7	*
15	5.8	*	2	4.0	* *
11	5.7	*	11	4.0	* *
Froidure	5.4	*	6	3.9	*
6	5.3	*	Froidure	3.6	*
2	5.3	*	15	3.4	*
14	4.4	*	14	2.5	*
Rafale	4.3	*	12	2.5	*
12	4.3	*	Frilene	2.2	* *
Frilene	4.2	*	Frijaune	2.2	* *
9	4.2	*	Rafale	2.0	* *
16	3.7	* *	9	1.7	* * *
Frijaune	3.5	* *	16	1.5	* *
7	3.2	*	7	1.5	* *
13	3.1	*	13	1.3	* *
10	2.1	*	10	1.1	*
Marik	1.9	*	Marik	1.1	*

Cold tolerance is assessed on a 0 (susceptible) to 6 (tolerant) scale.

Table 5. Comparative ranking of pea cultivars for cold tolerance in the field and under controlled environmental conditions

Cultivar	Controlled conditions			Field	
	1989 (a)	1990 (a)	1991 (a)	Versailles (a)	Brion (b)
					%
Kazar	5.6	5.0	5.9	5.0	98
Monitor	5.0	na	na	5.0	96
Froidure	na	na	5.4	na	93
Frilene	3.0	2.6	4.2	4.0	66
Rafale	na	na	4.3	3.0	54
Booster	na	4.4	na	2.5	78
Frijaune	2.3	2.8	3.1	3.0	27
Santon	2.8	na	na	na	38
Oscar	2.8	na	na	3.5	13
Lazer	2.5	3.4	na	3.0	33
Amac	2.3	3.0	na	2.5	16
Frisson	2.0	2.9	na	na	50
Carla	2.0	na	na	2.0	38
Marik	1.8	2.0	1.3	1.0	0

a : Cold tolerance is assessed on a 0 (susceptible) to 6 (tolerant) scale.

b : % of plants surviving after winter.

test with regard to the hardening conditions and the intensity and duration of the frost period. The main limitation is the cost of buying and maintaining cold rooms to be used for the tests. It is necessary to complement this test with field screening in order to obtain more general information on winterhardiness which includes a sum of resistance to frost, diseases and sometimes waterlogging.

Rating of Genotypes

The simplest rating is based on the number of plants killed during the winter, which requires only a count of the number of plants per row at emergence and after the thawing period. In chickpea, Wery (1990) used a "Frost Resistance Ratio" (number of plants at harvest/number of plants emerged). This ratio was used to split the genotypes into three classes ("adapted to fall, winter or spring sowing") on the basis of proximity to the ratio of known lines. Similarly, in lentil, Erskine *et al.* (1981) used a 1 to 3 scale with 1 (tolerant) = 75% surviving undamaged, 2 (moderately tolerant) = between 25 and 75%, and 3 (susceptible) = less than 25% surviving. This rating system can be sufficient if the frost is systematically severe in every test. A more precise 1 to 9 scale has been used on chickpea, using a combination of percent plants killed and visual damage on leaflets and branches (Singh *et al.*, 1989). A similar kind of scale (1 to 5 based on percent leaf area necrosis) has been used on pea (Auld *et al.*, 1983a). Other criteria, probably related to the hardening capability of the genotypes, have been used: Water/Dry Matter ratio of faba bean in a growth chamber (Herzog, 1988; Herzog and Saxena, 1988), or the accumulation of specific solutes involved in osmotic adjustment. For example, correlations between the accumulation of soluble sugars and resistance to cold have been reported in peas (Açìkgöz, 1982; Cousin *et al.*, 1990). The breeder could rank genotypes on the basis of variation in sugar content before and at the end of a regular decrease in temperature.

Sources of Resistance

Chickpea

From the evaluation of 2526 germplasm accessions and 750 breeding lines done from 1981 to 1987, Singh *et al.* (1989) identified 21 cold tolerant lines. This field screening has been extended by ICARDA to 7100 accessions and breeding lines (Singh *et al.*, 1991). Singh *et al.* (1990) evaluated 137 accessions of eight wild annual *Cicer* species. Cold tolerance was better for most of accessions of *C. bijugum* K.H. Rech., *C. reticulatum* Ladiz., and *C. echinospermum* P.H. Davis than for the cultivated species in. Some lines of *C. pinnatifidum* Jaub. & Spach showed tolerant reactions. *Cicer bijugum* had the highest degree of cold tolerance among the annual wild species tested. These two species are also resistant to important chickpea diseases such as Ascochyta blight (Singh *et al.*, 1990). Crosses have been made at ICARDA between the cultigen and the tolerant accessions of *C. reticulatum* to improve the cold tolerance in the cultigen.

Pea

The plant breeding station of INRA Versailles (France) has identified 23 genotypes of forage pea and three cultivars of dry pea with good frost resistance. These lines are able to resist temperatures of −20°C without snow cover (Cousin *et al.*, 1993). Auld *et al.* (1983b) identified 14 lines of *P. sativum* and four lines of *P. sativum* var. *arvense* with good winter survival at Moscow (Idaho). Winterhardiness (resistance to cold + other attributes) is a quantitatively inherited trait in *Pisum*, and it is possible to conduct recurrent selection using crosses between wild accessions and winter and spring cultivated types (Cousin *et al.*, 1990). The afila gene (conferring semi-leafless type) is associated with reduced frost tolerance. Linkages of cold tolerance with undesirable seed quality traits such as tannins and trypsin inhibitors have been found but have not been broken (Cousin *et al.*, 1993).

Lentil

From 3592 accessions of lentil germplasm evaluated by Erskine *et al.* (1981), 6.62% were tolerant to cold. These lines were mainly from Chile, Iran, Syria, Turkey, and Greece. In the Lentil International Cold Tolerance Nursery only two lines (ILL 632 and ILL 662) were tolerant (ICARDA, 1990).

Faba bean

From 840 faba bean pure lines (BPL) evaluated for cold tolerance by Robertson and Sherbeeny (1988), a large number had zero mortality. In *Vicia* species, accessions of *V. johannis* Tamamschjan and *V. narbonensis* L. were found to be resistant to black bean aphid and possessed improved frost tolerance (Birch, 1985). However, incompatibility of these related species with *Vicia faba* mean that this germplasm is not useful at the present time for improving cold tolerance of faba bean cultivars.

References

Açkìgöz, E. 1982. *Zeitschrift für Pflanzenzüchtung* 88: 118–126.
Auld, D. L., Adams, K. J., Swensen, J. B. and Murray, G. A. 1983a. *Crop Science* 23: 763–766.
Auld, D. L., Ditterline, R. L., Murray, G. A. and Swensen, J. B. 1983b. *Crop Science* 23: 85–88.
Auld, D. L., Lettis, B. L., Crock, J. E. and Kephart, K. D. 1988. *Agronomy Journal* 80: 909–914.
Bidinger, F. R., Mahalakshmi, V. and Rao, G. D. P. 1987. *Australian Journal of Agricultural Research* 38: 49–59.
Birch, A. N. E., Tithecott, M. T. and Bisby, F. A. 1985. *Economic Botany* 39: 177–190.
Blum, A. 1988. *Plant Breeding for Stress Environments*, 223 pp. Boca Raton, FL, USA: CRC Press Inc.
Blum, A. and Ebercon, A. 1981. *Crop Science* 21: 43–47.
Bouslama, M. and Schapaugh, W. T. 1984. *Crop Science* 24: 933–937.
Cousin, R. 1976. *Plantes* 28: 235–263.

Cousin, R., Burghoffer, A., Marget, P., Vingere, A. and Eteve, G. 1993. In: *Breeding for Stress Tolerance in Cool-Season Food Legumes,* pp. 311–320 (eds. K. B. Singh and M. C. Saxena). Chichester, UK: John Wiley & Sons.

Erskine, W., Meyveci, K. and Izgin, N. 1981. *LENS* 8: 5–9.

Herzog, H. 1988. *Plant Breeding* 101: 269–276.

Herzog, H. and Saxena, M. C. 1988. *FABIS Newsletter* 21: 19–25.

ICARDA. 1987. In: *Food Legume Improvement Program, Annual Report for 1986*, pp. 177–186. Aleppo, Syria: ICARDA.

ICARDA. 1990. *Food Legume International Nursery Report 1987/88*. Aleppo, Syria.

ICRISAT. 1988. In: *ICRISAT Annual Report*, pp. 60–61. Hyderabad, India: ICRISAT.

Knights, E. J. 1980. In: *Proceedings of the International Chickpea Workshop*, pp. 70–74 (ed. J. M. Green). Hyderabad, India: ICRISAT.

Knights, E. J. *International Chickpea Newsletter* 25: 16–17.

Knights, E. J. and Mailer, R. J. 1989. *Journal of Agricultural Science, Cambridge* 113: 325–330.

Lecoeur, J., Wery, J. and Turc, O. 1992. *Plant and Soil* 144: 177–189.

Ludlow, M. M. and Muchow, R. C. 1990. *Advances in Agronomy* 43: 107–153.

Malhotra, R. S. and Saxena, M. C. 1993. In: *Breeding for Stress Tolerance in Cool-Season Food Legumes*, pp. 227–244 (eds. K. B. Singh and M. C. Saxena). Chichester, UK: John Wiley & Sons.

Malhotra, R. S. and Singh, K. B. 1990. *Journal of Genetics and Breeding* 44: 227–230.

Malhotra, R. S. and Singh, K. B. 1991. *Theoretical and Applied Genetics* 82: 598–601.

Morgan, J. M., Rodríguez-Maribona, B. and Knights, E. J. 1991. *Field Crops Research* 27: 61–70.

Powell, A. A. 1989. *Annals of Botany* 63: 169–175.

Powell, A. A. and Matthews, S. 1978. *Journal of Experimental Botany* 29: 1215–1229.

Robertson, L. D. and Sherbeeny, M. 1988. *Faba bean germplasm catalog. Pure line collection*, 140 pp. Aleppo, Syria: ICARDA.

Savithri, K. S. and Ganapathy, P. S. 1980. *Journal of Experimental Botany* 31: 475–481.

Saxena, N. P. and Johansen, C. 1991. In: *Chickpea in the Nineties*, pp. 81–85 (eds. H. A. van Rheenen, M. C. Saxena, B. J. Walby and S. D. Hall). Patancheru, Andhra Pradesh, India: ICRISAT.

Saxena, N. P., Johansen, C., Saxena, M. C. and Silim, S. N. 1993. In: *Breeding for Stress Tolerance in Cool-Season Food Legumes*, pp. 245–270 (eds. K. B. Singh and M. C. Saxena). Chichester, UK: John Wiley & Sons.

Siddique, K. H. M. and Sedgley, R. H. 1985. *Field Crops Research* 12: 251–269.

Siddique, K. H. M. and Sedgley, R. H. 1987. *Australian Journal of Agricultural Research* 37: 599–610.

Silim, S. N. and Saxena, M. C. 1990. In: *Food Legume Improvement Program Annual Report for 1989*, pp. 238–240. Aleppo, Syria: ICARDA.

Silim, S. N. and Saxena, M. C. 1993a. Adaptation of spring-sown chickpea to the mediterranean basin. II Factors influencing yield under drought. *Field Crops Research* (In press).

Silim, S. N. and Saxena, M. C. 1993b. Adaptation of spring-sown chickpea to the mediterranean basin. I. Response to moisture supply. *Field Crops Research* (In press).

Silim, S. N. and Saxena, M. C. 1993c. Yield and water use efficiency of faba bean sown at two row spacing and densities. *Field Crops Research* (In press).

Silim, S. N., Saxena, M. C. and Erskine, W. 1993a. *Experimental Agriculture* 29: 9–19.

Silim, S. N., Saxena, M. C. and Erskine, W. 1993b. *Experimental Agriculture* 29: 21–28.

Singh, K. B., Holly, L. and Bejiga, G. 1991. *A Catalog of Kabuli Chickpea Germplasm*, 398 pp. Aleppo, Syria: ICARDA.

Singh, K. B., Saxena, M. C. and Gridley, H. E. 1984. In: *Ascochyta Blight and Winter Sowing of Chickpea*, pp. 167–177 (eds. M. C. Saxena and K. B. Singh). The Hague, The Netherlands: Martinus Nijhoff.

Singh, K. B., Malhotra, R. S. and Saxena, M. C. 1989. *Crop Science* 29: 282–285.

Singh, K. B., Malhotra, R. S. and Saxena, M. C. 1990. *Crop Science* 30: 1136–1138.

Sulc, R. M., Albrecht, K. A. and Duke, S. H. 1991. *Crop Science* 31: 430–435.

Tully, R. E., Musgrave, M. E. and Leopold, A. C. 1981. *Crop Science* 21: 312–317.

Wery, J. 1987. In: *Drought Resistance in Plants: Physiological and Genetic Aspects*, pp. 179–202. (eds. L. Monti and E. Porceddu). Brussels: CEE.

Wery, J. 1990. In: *Present Status and Future Prospects of Chickpea Crop Production and Improvement in the Mediterranean Countries*, pp. 77–85 (eds. M. C. Saxena, J. I. Cubero and J. Wery). Options Méditerranéennes – Série Séminaires – N° 9 – CIHEAM Paris.

Wery, J., Turc, O. and Lecoeur, J. 1993. In: *Breeding for Stress Tolerance in Cool-Season Food Legumes*, pp. 271–291 (eds. K. B. Singh and M. C. Saxena). Chichester, UK: John Wiley & Sons.

Screening techniques and sources of tolerance to salinity and mineral nutrient imbalances in cool season food legumes

N.P. SAXENA[1], M.C. SAXENA[2], P. RUCKENBAUER[3], R.S. RANA[4], M.M. EL-FOULY[5] and R. SHABANA[6]

[1] ICRISAT, Patancheru P. O., Andhra Pradesh 502 324, India;
[2] ICARDA, P. O. Box 5466, Aleppo, Syria;
[3] Institute für Pflanzenbau und Pflanzenzuechtung, Vien, Austria;
[4] Indian Agricultural Research Institute, Bureau of Plant Genetic Resources (NBPGR), Pusa Complex, New Delhi 110 012, India;
[5] Botany Laboratory, National Research Center, Dokki, Cairo, Egypt, and
[6] Cairo University, Cairo, Egypt

Abstract

A large global land area is affected by saline, alkali (sodic), and acid soil conditions. Cool season food legumes are important crops in many countries with such adverse soils. Tolerant genotypes have been identified in many crops, including legumes. However, very little has been published on selection of tolerant cool season food legume crops. The inadequate knowledge and understanding of the responses of cool season food legume crops to these abiotic stresses, necessitates action by a collaborative network of interdisciplinary teams to make rapid progress in identifying tolerant germplasm and developing cultivars better adapted to unfavorable soil conditions.

Introduction

Soils affected by mineral stress problems (2.9×10^9 ha) rank only second to drought (3.7×10^9 ha) compared with a relatively small area (1.4×10^9 ha) of stress-free global land (Christiansen, 1982). The total global area and distribution of saline, saline-alkali (9.0×10^8 ha; Szabolcs, 1989), and acid (nearly 1×10^9 ha; Van Wambeke, 1976) soils is known. One infers from national crop areas, production statistics of cool season food legumes (FAO, 1990), and the distribution of problem soils that adverse soil conditions are an important constraint to cool season food legume production in some countries (in Pakistan, 40% of arable land is saline; Mohammad, 1978) and a potential threat in others.

Extensive research in the past 50 years on tolerance of many crops to adverse soil conditions was summarized recently by Blum (1988) and Epstein (1985). General inferences possible from the published literature are:
1. Problems of adverse soil conditions, in particular salinity, are more variable

F.J. Muehlbauer and W.J. Kaiser (eds.), Expanding the Production and Use of Cool Season Food Legumes, 457–471.
© 1994 *Kluwer Academic Publishers.*

over space and time, and perhaps more complex than drought in crop x soil
x climate interactions (Figure 1), and

2. Both basic and applied aspects have received research attention and
   contributed to a better understanding of crop responses to these soil
   conditions.

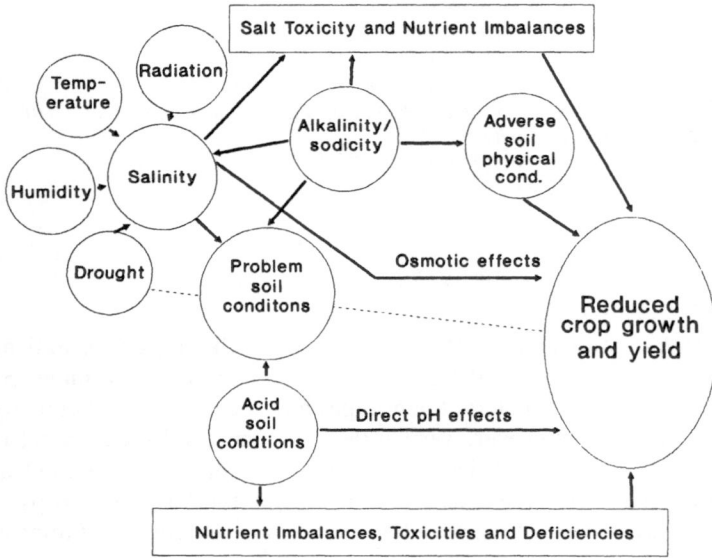

Figure 1. Interactions of factors of adverse soil conditions with other soil and climate factors.

Cool season food legumes are sensitive to salinity (Saxena *et al.*, in press);
thus, responses and cultivar differences in the salinity tolerance of chickpea and
faba bean have been studied extensively. There are more publications on
chickpea than on lentil and faba bean, and on salinity and Fe uptake than on
other constraints (Table 1).

Since the scope of the subject is fairly broad, it is necessary to establish crop
and constraint priorities for research on screening techniques and identifying
tolerant genotypes in these legumes. A realistic basis is constraint analysis maps,
prepared by overlays of crop area maps on maps of problem soils using
techniques such as Geographic Information System (GIS). In the absence of
such maps, a greater research emphasis on chickpea (9.58 × 10^6 ha) and dry
peas (9.27 × 10^6 ha) than on lentil and faba bean (both around 3.2 × 10^6 ha)
seems warranted on the basis of crop area statistics (FAO, 1990).

This review aims to summarize our current understanding of salinity
tolerance and nutrient imbalances in cool season food legumes with respect to:
screening techniques in use or of potential use, criteria for crop selection, and
sources of tolerant germplasm.

Table 1. Research papers (n = 80) on cool season food legumes found in literature searches

Crop	Percent	Constraint	Percent
Chickpea	43	Salinity	57
Faba bean	24	Nutrient imbalances	39
Lentil	21	Fe	26
Pea	11	P	7
Grasspea	1	Mn	4
		Zn	1
		B	1
		Acid soil conditions	4

Salinity

The subject of salinity tolerance in cool season food legumes was recently reviewed (Saxena *et al.*, in press). Many selection criteria and screening methods on salinity tolerance are described in the literature. Relevance of the method chosen to the target crop and environment is, however, essential because of the known interactions of salinity with many factors. For example, the osmotic and specific ion effects of salinity are well recognized (USDA, 1954) as are the differential responses to Cl^- and SO_4^- ions in chickpea (Lauter and Munns, 1986; Manchanda and Sharma, 1989) and a greater inhibitive effect of Mg SO_4 salt over NaCl and Na_2SO_4 in lentil (Jana and Slinkard, 1976). Environmental factors, such as low relative humidity and high evaporative demand in faba bean (Prisc'o and O'Leary, 1973; El-Karouri, 1976; Salim, 1987), high temperatures in chickpea (Lauter and Munns, 1986) and faba bean (El-Karouri, 1976), and low light intensity in faba bean (Helal and Mengel, 1981), aggravate salinity effects. Effects of salinity are moderated/alleviated by irrigation (USDA, 1954).

Crop response to salinity also changes with crop age. For example, lentil (Jana, 1979) and faba bean (Hamid and Talibuddin, 1976) are more sensitive at germination than at subsequent growth stages and the converse is true for chickpea (Kumar, 1985). Integrated effects of salinity over the whole crop growing period are important in determining seed yield. A careful consideration of these factors is necessary before applying a screening method to a test crop or environment.

Criteria for Selection

Osmotic Effects

There is a progressive delay in germination with increasing levels of soil salinity in chickpea (Chandra, 1980; Jaiwal *et al.*, 1983) and faba bean (Ayoub, 1970/71; Kamel, 1986). This is attributed to the osmotic effects. Growth of faba bean on non-sodic soils (low in exchangeable sodium percentage – ESP) also was affected more by the osmotic than by the toxic component (Heipko and Kauffman, 1974) of salinity. Osmotic effects of salinity can be quantified with techniques such as measurement of shoot water potential and canopy temperature, routinely used in drought research. At low levels of salinity (3.0 dS m^{-1}) but high ESP, no big differences in shoot water potentials were found in chickpea seedlings (Kumar *et al.*, 1983).

Proline accumulation, as an osmotic adjustment response to salinity stress, has been reported in pea (Bar-Num and Pojakoff-Mayber, 1979) and chickpea (Pandey and Ganpathy, 1985); but, in chickpea, Chandra (1980) found no consistent relationship between genotypic differences in salinity tolerance and accumulation of free proline. In other legume crops, evidence relating proline content with adaptation to saline environments seems to be equally conflicting, and mostly circumstantial.

Toxicity and Specific Ion Effects

Expression of characteristic symptoms in response to stress conditions is useful in field screening methods. Toxicity symptoms of salinity have been described for chickpea (Saxena, 1987; Chandra, 1980) and faba bean (El-Karouri, 1979). A rating scale of 1 to 5 (van Rheenen and Sethi, 1990) or 1 to 9 (Saxena, 1987) has been used in field screening of chickpea on naturally occurring saline fields. Results of such visual scoring were validated in pot experiments (Saxena, 1987).

Chemical analysis of leaf and root tissues has been used to identify specific-ion retention/exclusion mechanisms, such as for Na or Cl in cool season legumes, and the ratios of Na/K in other legumes. In chickpea, which is not a very efficient salt-excluding crop, relative genotypic differences in salt tolerance were associated with differences in Na accumulation in tops (Lauter and Munns, 1986). Selective retention of Na^+ in the roots and stems of faba bean also has been reported (Jacoby, 1964; Salim, 1987).

Seed germination has been used as a criterion and large genotypic differences in lentil (Jana and Slinkard, 1976) and chickpea (Chandra, 1980) have been identified.

Growth and yield are sensitive to both osmotic and specific-ion effects as shown in pea by Cerda *et al.* (1982). The threshold salinity level (EC at which reductions in growth and yield commence) is considered an important index for salt-sensitive crops such as chickpea. EC or EC_e at which 50% reduction in shoot mass or seed yield occurs also has been used as a selection criterion in chickpea (Chandra, 1980; Saxena, 1987; Johansen *et al.*, 1990) and faba bean (El-Karouri, 1979).

Other selection criteria used in chickpea are survival of cells and growth of calli in saline solutions (Pandey and Ganpathy, 1984).

Screening Methods

A screening method is readily acceptable if it is based on simple criteria for selection, provides rapid screening of large numbers, is reproducible, and, most important, is relevant to field performance. A number of screening methods, ranging from field to cell biology techniques, are available to screen cool season food legumes for salt tolerance.

Field Methods

Since the aim of screening genotypes is to identify those that will perform well in the field, a field screening method should be most appropriate. However, there are many practical limitations in the use of field screening methods. A major one is the degree of control and reproducibility of field environments. Salinity, in particular, poses a problem in this respect because of the spatial variation of salinity on the surface and down the profile at any one given site, differences in the composition of salts, and interactions with environmental conditions. It is difficult to interpret results of such screening, which often are not repeatable.

Natural heterogeneous distribution of salinity has been exploited to screen faba bean (El-Karouri, 1979) and chickpea (Saxena, 1987; van Rheenen and Sethi, 1990) genotypes. While van Rheenen and Sethi (1990) used visual scores, seed number and seed yield of test and control cultivars (dibbled in a common hole) for computing tolerance indices, El-Karouri (1979) and Saxena (1987) quantified salinity of the test spots, selected on the basis of visual differences, and related it with growth and yield reductions. Quantifying salinity by measuring EC in the laboratory is considered laborious, but soil probes are now available which overcome this difficulty and permit instantaneous *in situ* measurements of salinity and soil moisture. Combined with visual symptoms, such quantification of the magnitude of salinity should be more meaningful, cheap, and attractive to use in field screening.

Soil moisture gradients created by a line-source sprinkler irrigation method have been used in screening chickpea and lentil genotypes (ICARDA, 1990) for drought resistance. Using a technique of parallel line-source sprinkler irrigation systems, one with saline and the other with non-saline water, it should be possible to create controlled salinity and drought gradients in field experiments to study responses to each factor and their interactions. A layout of such a system is shown in Figure 2. Lauter (personal communication) intended to use such a method to screen chickpea genotypes.

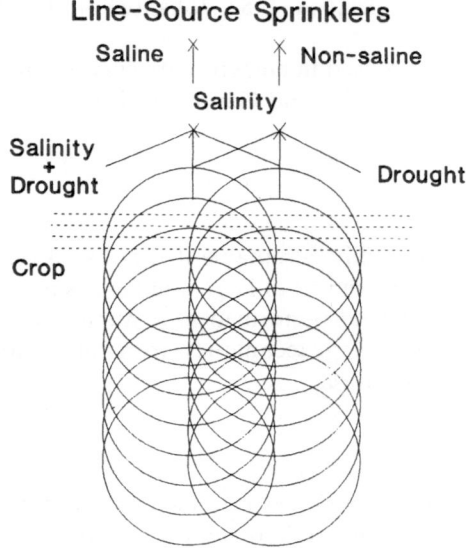

Figure 2. A parallel line-source sprinkler irrigation system for simultaneous screening of drought, salinity, and the interaction between the two.

Partially Controlled Conditions

Microplot Techniques

Microplots of salinized soil of known EC_e and composition of salts have been used to screen salt-tolerant chickpea and lentil lines (Chandra, 1980). The advantage is that there is good control on the intensity and type of salinity and screening conditions can be standardized to mimic field environments.

Lysimeter/Pot Methods Using Saline Soil and/or Salt Solutions

Lysimeters filled with calcareous soil and irrigated with saline irrigation water were used to study responses of pea to salinity (Cerda *et al.*, 1982). Graded levels of salinity treatments in pots were used to screen chickpea germplasm (Johansen *et al.*, 1990) and to study the response of chickpea to specific ion effects (Manchanda and Sharma, 1989). Using such pot methods, Saxena (1987) studied genotypic differences in salinity tolerance of chickpea and the interaction with soil moisture (at field capacity and 50% of field capacity).

Saline nutrient solutions in large tanks (700 liters capacity), have been used to screen chickpea germplasm (Lauter and Munns, 1986). This method offers good control of salinity and aerial environment.

Saline solutions of different kinds of salts and their combinations, to create desired ionic compositions, also have been used extensively in seed germination studies and as a rapid method of screening germplasm (Jana and Slinkard, 1976). Genotypic differences in germination are important in determining initial plant stand establishment, which is often suboptimal in saline conditions. However, the correlation between genotypic ranking for germination and final yield performance is often very weak, which is not surprising because the effects of salinity on crop growth are integrated over time.

Cell Tolerance/Cell Suspension

Cell cultures have been used to study the response of chickpea calli to salinity (Pandey and Ganpathy, 1984). Whole plants, regenerated from salt-tolerant cell line cultures, with a greater salinity tolerance as has been demonstrated in tobacco (Nabors *et al.*, 1980), have not been produced.

Crop and Cultivar Differences

It is difficult to make valid comparisons and draw conclusions about the relative salt tolerance of different cool season food legumes on the basis of the available data, because results were generated in separate experiments using different criteria and methods of measurements.

With regard to seed germination, faba bean seems to be most tolerant of salinity (germination occurring at 16.9 dS m^{-1}, with progressive delay at higher salinity) (Kamel, 1986), followed by lentil (satisfactory germination up to 5 dS m^{-1} and a 50% reduction at 20 dS m^{-1}) (Ayoub, 1974/75), and chickpea the most sensitive (large genotypic differences at 5.8 dS m^{-1}, ranging from total germination failure to partial effects) (Chandra, 1980).

Faba bean has the highest threshold of salinity tolerance among cool season food legumes (Kamel, 1986) compared with a lack of a threshold salinity level in chickpea (Chandra, 1980; Lauter and Munns, 1986; Johansen *et al.*, 1990) and lentil and pea (Ayoub, 1975/76). Chickpea and lentil seem equally sensitive to salinity based on this criterion.

In field experiments with faba bean, a 50% reduction in shoot mass at 10.5 dS m^{-1} and 50% reduction in seed yield at 9 dS m^{-1} was observed on a soil in which Na_2SO_4 was dominant (El-Karouri, 1979). Using a fine sandy loam soil tilled in drums, and a 1:1 mixture of $NaCl$ and $CaCl_2$ salts, there was a 50% reduction in seed yield at 6 to 7 dS m^{-1} (Ayers and Eberhard, 1960).

In pot and field experiments with lentil, a 50% reduction in seed yield occurred at EC_e of 3 to 4 dS m^{-1} and a huge reduction beyond 5 dS m^{-1} (Ayoub, 1975/76; Jana and Slinkard, 1976).

In chickpea, if the $EC_{1:2}$ (soil:water extract) values at which 50% reduction in growth occurs in pot (Johansen *et al.*, 1990) and field (Saxena, 1987) experiments, are recalculated to EC_e using the equation given by Johansen *et al.* (1990), a 50% shoot mass reduction occurs at an EC_e of 5 to 6 dS m^{-1}, which is similar to the value for lentil. Comparing the results of screening across experiments conducted in the field, in pots and in nutrient culture, "L 550" was the most tolerant chickpea cultivar. Among the cool season food legumes, responses of wild species have been studied only in chickpea, and it is discouraging that the wild relatives of chickpea are more sensitive to salinity than the cultivated types (Johansen et al., 1990).

Although no strict comparison among crops can be made on the basis of the available evidence, faba bean seems to be the most tolerant of the cool season food legumes, followed by lentil and chickpea which show similar responses. Very little information is available to allow comment on the response of pea to salinity.

Extremes of pH (Alkalinity and Acidity) and Nutrient Imbalances

Nearly 1.5×10^9 ha of global arable land are affected by extremes of pH and a third of it is in the alkaline phase (Szabolcs, 1989). Nutrient availability can change dramatically with small changes in soil pH, particularly in mineral soils low in organic carbon (Clark, 1982). On such soils, toxicity of Al, Fe, and Mn could result with a drop in pH from 5.5 to 4.5. Deficiencies of Fe, Zn and Mn occur at pH above 8.0.

High pH

Micronutrient deficiencies are easily recognized because of their characteristic appearance and have been described for many elements in both chickpea and lentil (Agarwala and Sharma, 1979). There are many reports of Fe deficiency in chickpea (Saxena and Sheldrake, 1980) and lentil (Saxena and Singh, 1977; Singh et al., 1985; Erskine, ICARDA, personal communication); Zn in pea and chickpea (Saxena and Singh, 1970) and lentil (Saxena and Singh, 1970; Gangawar and Singh, 1986); B in lentil (Sakal et al., 1988), and Mn in chickpea (Rashid et al., 1990).

Iron deficiency is by far the most commonly studied nutrient disorder on high pH calcareous soils across many crops, including cool season food legumes. Losses in yield of highly susceptible genotypes range from 22 to 50%, both in chickpea and lentil. In field experiments on a sandy loam soil, neutral to alkaline in reaction, and low in available Zn, more than a 60% reduction in seed yield was observed in lentil (Gangawar and Singh, 1986).

Iron deficiency symptoms correlate better with the available (Fe^{2+}) rather than the total iron (Fe^{2+3}) concentration of the shoot tissue (Sakal et al., 1984). Iron concentration of the seeds, which is generally sufficient to produce healthy green seedlings, can be immobilized by the high pH of the surrounding medium and thereby induce deficiency symptoms (Saxena et al., 1971).

Appearance of Fe-deficiency symptoms is highly transient across crops (chickpea: Saxena and Sheldrake, 1980; Saxena et al., 1990); (lentil: Erskine,

ICARDA, personal communication), appearing under wet (above field capacity) soil moisture conditions and at low temperatures.

Criteria for Selection

In chickpea, the characteristic appearance of Fe-deficiency in susceptible and lack of the incipient deficiency in tolerant genotypes (Saxena and Sheldrake, 1980) permits selection of tolerant material in germplasm on the basis of visible symptoms and the rejection of susceptible types in the segregating population, because the trait is governed by a single recessive gene (Gowda and Smithson, 1980; Saxena *et al.*, 1990). Genotypic differences in Fe-deficiency tolerance in chickpea have been rated on a scale of 1 to 9 (Gowda and Smithson, 1980; Saxena *et al.*, 1990) and 1 to 5 by Hamz'e *et al.* (1987).

Critical values of Fe in leaves of lentil (Sakal *et al.*, 1984) and for most of the elements in chickpea (Agarwala and Sharma, 1979) are available and can be used as criteria, if screening conditions are standardized, but such methods will be laborious and costly.

Root-induced acidification of the rhizosphere has been demonstrated in chickpea (Marschner and Romheld, 1983; Ae *et al.*, 1991). Genotypic differences in the ability to acidify the rhizosphere are reported in lentil (Kannan, 1983) but not yet in chickpea, a crop with strong root-acidifying characteristics.

Selective immobilization of ^{59}Fe isotope in the roots of Fe-inefficient genotypes of chickpea (Malewar *et al.*, 1982) and lentil (Singh *et al.*, 1985) has been shown. The technique seems more useful in studying the mechanism of occurrence of Fe-deficiency than for screening.

Biochemical probes, such as peroxidase activity for efficiency in Fe utilization, are suggested in the literature but have not yet been tested in cool season food legumes.

Screening Methods

Screening germplasm grown on high pH (more than 8.1) calcareous soils has been very effective in identifying genotypic differences in susceptibility to Fe-deficiency, both in chickpea and lentil (Gowda and Smithson, 1980; Saxena and Sheldrake, 1980; Saxena *et al.*, 1990; Erskine, ICARDA, personal communication). Gowda and Smithson (1980) used irrigation to magnify genotypic differences in response.

A pot screening method has been developed to screen soybean cultivars for differences in Fe-deficiency using an appropriate combination of HCO_3^- and a wet soil moisture regime (Coulombe *et al.*, 1984). This method is worth exploring with cool season food legumes. Screening conditions can be set up to screen crop and cultivars for tolerance to any of the nutrient deficiencies by using soil or nutrient solutions deficient in that particular element.

Laboratory Methods

Genotypic differences in root-acidifying power, with pH sensitive indicators such as bromocresol green (Ae *et al.*, 1991), can be used for routine screening of selective germplasm.

Crop and Cultivar Differences

There are many reports of genotypic differences in Fe-deficiency in chickpea and lentil. Chickpea cultivars highly resistant to lime-induced chlorosis at 63% $CaCO_3$ content but succumbing to 82% $CaCO_3$ have been reported (Hamz'e *et al.*, 1987). Chickpea and lentil germplasm originating from West Asia (Saxena *et al.*, 1990; Erskine, personal communication) was more tolerant than germplasm from the Indian subcontinent. In chickpea, the problem occurred more frequently in the late rather than in early genotypes (Gowda and Smithson, 1980).

Crop (Marschner and Romheld, 1983; Ae *et al.*, 1991) and cultivar (Kannan, 1983) differences in root acidification of the rhizosphere are known but these have not been related to differences in tolerance or susceptibility to Fe-deficiency.

Differences in the susceptibility of lentil cultivars to boron deficiency were noted by Sakal *et al.* (1988).

Low pH

Although acid soils comprise nearly 40% of the world's arable land (Hang, 1984), very little has been published on the responses of cool season food legumes to those conditions. On acid soils in Chile, a problem called *Marea Negra* in lentil was diagnosed as being due to toxic concentrations of Mn in the tissue (Andre's France and Tay, 1986; Ange'lica Sadzwaka, 1987).

In northern Idaho on acid soils not dominant in Al^{3+}, forage and food legume crops were more severely affected than cereals. This was associated with a greater distribution of legume roots in the top 0.5 m compared with 1.5 m in cereals (Mahler and McDole, 1987).

Toxicity of Fe has been reported in a pea mutant, E-107 (Kneen *et al.*, 1990; Welch and LaRue, 1990) which was not associated with toxic concentrations of Fe in the growth medium. The toxicity developed because of an impaired metabolism associated with the pleiotropic effect of a gene conferring low nodulation and enhanced redox activity (Gursak *et al.*, 1990).

Yield Reduction

According to the data of Mahler and McDole (1987), a drop in pH by one unit below the threshold value (pH 5.5) caused a greater than 86% reduction in seed yield of lentil and a 93% reduction in pea. This clearly shows that cool season food legumes are extremely sensitive in their reaction to acid conditions.

Selection Criteria and Screening Methods

There are not many publications on the screening of cool season food legumes for acid soil conditions and their associated constraints. Symbiosis and N_2 fixation have been studied in some detail. Simple methods to screen cultivars for tolerance to acid soil conditions, in particular Al^{3+} toxicity, are available and could be used with modification, if needed.

Other Nutrients

Among the major nutrients, responses of P have been inadequately studied considering its relevance to increasing food legume production. Yield of lentil and chickpea was only 17 and 42% of the P-fertilized treatments on a calcareous highly P-deficient soil (Rashid *et al.*, 1990) in Pakistan. On the other hand, on Vertisols characterized as low in available P (2 to 5 mg kg^{-1} soil by Olsen's method), chickpea did not respond to soil or foliar P applications and mobilized 0.6% P in the shoot of 30-day-old control plants grown on a native soil. This P concentration is high compared with other legumes (Saxena *et al.*, 1988). This and other evidence show that chickpea is highly efficient in mobilizing soil P (Ae *et al.*, 1991).

Genotypic differences in efficiency of P uptake, using yield and yield components as criteria in treatments involving P and no P treatments, were identified in a chickpea field experiment on a Calcic Rhodoxeralf, low in available P, at ICARDA (Bambach, 1990). Rhizosphere acidification in relation to uptake of native soil nutrients, discussed under Fe nutrition, is relevant to P nutrition as well and needs to be more intensively researched and pursued. References to phosphatase activity as an enzyme probe are made in the literature, but in pea the high efficiency of phosphate absorption was found to be independent of cell wall properties of the roots (Lefebvre and Clarkson, 1987). This may be a fruitful area of further research.

Except for these isolated pieces of information, very little has been published on genotypic variation in P utilization. We agree with the conclusion of Johansen and Sahrawat (1991) that at this stage the knowledge of genotypic difference in P-uptake is so limited it is difficult to opine about the practical use of this option to exploit soil-available P for increasing productivity.

Lacunae in our Present Knowledge

Current research emphasis does not seem to reflect the real crop constraint and research thrust priorities. Among cool season food legumes the greatest research emphasis has been on chickpea and in constraints on salinity followed by Fe-deficiency. It is difficult to decide whether this emphasis is realistic or has been because of the marked and dramatic responses of cool season legumes to these two constraints. Are the losses in yield due to deficiencies of Zn, Mn, and to B, unnoticed? A major lacuna is on genotypic differences in efficiency of P uptake in cool season legumes.

Screening methods do not seem to be a constraint but the levels of tolerance and genotypic differences in tolerance seem to be low for salinity and there is very little knowledge about responses to acid conditions.

Future Thrust Areas of Research

A redeploying of the available resources may be necessary for responding to the issues raised. Increasing resource constraints dictate that multidisciplinary, cooperative, and collaborative teams are constituted as shown in Figure 3 to utilize resources and expertise available across institutes and crops through networks developed for the most important cool season legume(s) and constraint(s).

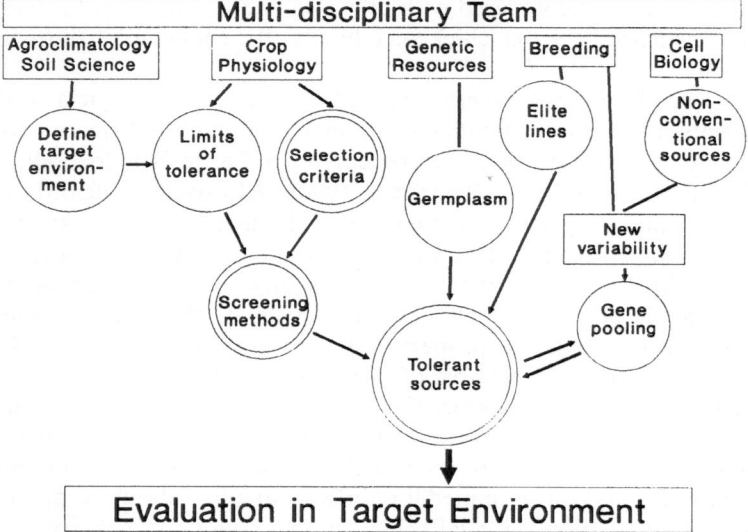

Figure 3. A proposal of a multidisciplinary team for genetic enhancement of salinity tolerance. The locations of criteria for selection, screening techniques and tolerant sources (the theme of this paper) are shown with two concentric circles.

References

Ae, N., Arihara, J. and Okada, K. 1991. In: *Phosphorus Nutrition of Grain Legumes in the Semi-Arid Tropics*, pp. 33–41 (eds. C. Johansen, K. K. Lee and K. L. Sahrawat). Patancheru, Andhra Pradesh, India: ICRISAT.

Agarwala, S. C. and Sharma, C. P. 1979. *Recognizing Micronutrient Disorders of Crop Plants on the Basis of Visible Symptoms and Plant Analysis*. Lucknow, U.P., India: Department of Botany, Lucknow University.

Ange'lica Sadzwaka, R. 1987. *Agricultura Técnica* (Chile) 47: 350–354.

Ayers, A. D. and Eberhard, D. L. 1960. *Agronomy Journal* 52: 110–111.

Ayoub, A. 1970/71. In: *Annual Report of the Hudeiba Research Station*, pp. 60–80. Democratic Republic of the Sudan: Agricultural Research Corporation, Ministry of Agriculture.

Ayoub, A. T. 1974/75. In: *Annual Report of the Hudeiba Research Station*, pp. 53–58. Democratic Republic of the Sudan: Agricultural Research Corporation, Ministry of Agriculture.

Ayoub, A. T. 1975/76. In: *Annual Report of the Hudeiba Research Station*, pp. 57–68. Democratic Republic of the Sudan: Agricultural Research Corporation, Ministry of Agriculture.

Bambach, T. 1990. *Zuchterische Untersuchungen zur Phosphateffizien bei Kichererbsen (Cicer arietinum L.) in Syrien*. Dissertation Institut für Pflanzenzuchtung, Saatgutforschung und Populationsgenetik. Ph.D. Dissertation. Stuttgart, Germany: der Universitat Hohenheim.

Bar-Num, N. and Pojakoff-Mayber, A. 1979. *Annals of Botany* 44: 309–312.

Blum, A. 1988. *Plant Breeding for Stress Environments*, 223 pp. Boca Raton, FL, USA: CRC Press.

Cerda, A., Caro, M. and Fernández, F. G. 1982. *Agronomy Journal* 74: 796–798.

Chandra, S. 1980. In: *Proceedings of the International Workshop on Chickpea Improvement*, pp. 97–105 (eds. J. M. Green, Y. L. Nene and J. B. Smithson). Patancheru, Andhra Pradesh, India: ICRISAT.

Christiansen, M. N. 1982. In: *Breeding Plants for Less Favourable Environments*, pp. 1–12 (eds. M. N. Christiensen and C. F. Lewis). New York, USA: John Wiley & Sons.

Clark, R. B. 1982. In: *Breeding Plants for Less Favourable Environments*, pp. 71–142 (eds. M. N. Christiensen and C. F. Lewis). New York, USA: John Wiley & Sons.

Coulombe, B. A., Chaney, R. L. and Wiebold, W. J. 1984. *Journal of Plant Nutrition* 7: 411–425.

El-Karouri, M. O. H. 1976. *Annual Reports* Soba Saline Research Station, ARC, Sudan.

El-Karouri, M. O. H. 1979. *Experimental Agriculture* 15: 59–63.

Epstein, E. 1985. *Plant and Soil* 89: 187–198.

FAO (Food and Agricultural Organization of the United Nations). 1990. *FAO Yearbook. Production 1990. Volume 44*. Rome, Italy: FAO.

France, I. A. and Tay, J. U. 1986. *Agricultura Técnica* (Chile) 46: 379–383.

Gangawar, K. S. and Singh, N. P. 1986. *Lens* 1: 17–20.

Gowda, C. L. L. and Smithson, J. B. 1980. *International Chickpea Newsletter* 3: 10.

Gursak, M. A., Welch, R. M. and Kochean L. V. 1990. *Plant Physiology* 93: 976–981.

Hamid, A. and Talibuddin, O. 1976. *Journal of Agricultural Science*, Cambridge 86: 49–56.

Hamz'e, M., Ryan, J., Mikadashi, R. and Solh, M. 1987. *Journal of Plant Nutrition* 10: 1031–1039.

Hang, A. 1984. *Critical Reviews in Plant Science* 1: 345–373.

Heipko and Kauffman. 1974. *Annual Report of the Hudeiba Research Station*, Agricultural Research Corporation, Ministry of Agriculture, Democratic Republic of the Sudan.

Helal, H. M. and Mengel, K. 1981. *Plant Physiology* 67: 999–1002.

ICARDA (International Centre for Agricultural Research in Dry Areas). 1990. In: *Food Legume Improvement Program, Annual Report for 1989*, pp. 107–116; 238–241. Aleppo, Syria: ICARDA.

Jacoby, B. 1964. *Plant Physiology* 39: 445–449.

Jaiwal, P. K., Bhambie, S. and Kuldip Mehta. 1983. *International Chickpea Newsletter* 9: 15–16.

Jana, M. K. 1979. *Harvester* 21: 32–38.

Jana, M. K. and Slinkard, A. E. 1976. Screening for salt tolerance in lentil. *Lens* 6: 25–27.

Johansen, C. and Sahrawat, K. L. 1991. In: *Phosphorus Nutrition of Grain Legumes in the Semi-*

Arid Tropics, pp. 227–241 (eds. C. Johansen, K. K. Lee and K. L. Sahrawat). Patancheru, Andhra Pradesh, India: ICRISAT.

Johansen, C., Saxena, N. P., Chauhan, Y. S., Subbrao, G. V., Pundir, R. P. S., Kumar Rao, J. V. D. K. and Jana, M. K. 1990. In: *Proceedings of the International Congress of Plant Physiology, Volume 2*, pp. 977–983 (eds. S. K. Sinha, P. V. Sane, S. C. Bhargava and P. K. Agrawal). New Delhi, India: Society of Plant Physiology and Biochemistry.

Kamel, M. W. 1986. *FABIS* 14: 26–27.

Kannan, S. 1983. *Journal of Plant Nutrition* 6: 1025–1031.

Kneen, B. E., La Rue, T. A., Welch, R. M. and Weeden, N. F. 1990. *Plant Physiology* 93: 717–722.

Kumar, D. 1985. *Annals of Arid Zone* 24: 334–340.

Kumar, D., Singh, M. P. and Buttar, B. S. 1983. *International Chickpea Newsletter* 8: 15–17.

Lauter, D. J. and Munns, D. N. 1986. *Plant and Soil* 95: 271–279.

Lefebvre, D. O. and Clarkson, D. T. 1987. *Canadian Journal of Botany* 65: 68–86.

Mahler, R. L. and McDole, R. E. 1987. *Agronomy Journal* 79: 751–755.

Malewar, G. V., Jadhav, D. K. and Ghonsikar, C. P. 1982. *International Chickpea Newsletter* 6: 13–14.

Manchanda, H. R. and Sharma, S. K. 1989. *Journal of Agricultural Science*, Cambridge 113: 407–410.

Marschner, H. and Romheld, V. 1983. *Plant Physiology* 111: 241–251.

Mohammad, S. 1978. In: *Membrane Biophysics and Salt Tolerance in Plants*, pp. 47–64 (eds. R. H. Qureshi, S. Mohammad and A. Aslam). Faislabad, Pakistan: Pakistan University of Agriculture Press.

Nabors, M. W., Gibbs, E., Bernstein, C. S. and Meis, M. E. 1980. *Zeitschrift für Pflanzenphysiologie* 97: 13–17.

Pandey, R. and Ganpathy, P. S. 1984. *Journal of Experimental Botany* 35: 1194–1199.

Pandey, R. and Ganpathy, P. S. 1985. *Plant Science* 40: 13–17.

Prisc'o, J. T. and O'Leary, J. W. 1973. *Plant and Soil* 39: 263–276.

Rashid, A., Jalal-ud-Din and Bashir, A. 1990. *International Chickpea Newsletter* 22: 21–22.

Sakal, R., Singh, A. P. and Singh, R. B. 1988. *Lens* 15: 27–29.

Sakal, R., Singh, B. P. and Singh, A. P. 1984. *Plant and Soil* 82: 141–148.

Salim, M. 1987. *FABIS Newsletter* 18: 11–13.

Saxena, M. C. and Singh, H. P. 1977. Chapter III In: Research on Winter Pulses. *Experiment Station Technical Bulletin* 101: 43–50. Directorate of Experiment Station. Govind Ballabh Pant University of Agriculture and Technology, Pantnagar 263 145, India.

Saxena, M. C., Singh, K. B. and Malhotra, R. S. 1990. *Plant and Soil* 23: 251–254.

Saxena, M. C. and Singh, Y. 1970. *Relative susceptibility of important varieties of some pulses and soybean to zinc deficiency*. Paper presented at III Annual Workshop of Coordinated Scheme on Micronutrients of Soils held at Ludhiana, India, Oct. 5–7, 1970.

Saxena, N. P. 1987. In: *Adaptation of Chickpea and Pigeonpea to Abiotic Stresses*, pp. 63–76 (eds. N. P. Saxena and C. Johansen). Patancheru, Andhra Pradesh, India: ICRISAT.

Saxena, N. P., Jaganmohan Rao, M. and Sakai, H. 1971. In: *Proceedings of the International Symposium on Soil Fertility Evaluation, Vol. 1*, pp. 797–804. New Delhi, India.

Saxena, N. P., Johansen, C., Saxena, M. C. and Silim, S. N. 1993. In: *Breeding for Stress Tolerance in Cool-Season Food Legumes*, pp. 245–270 (eds. K. B. Singh and M. C. Saxena). Chichester, UK: John Wiley & Sons.

Saxena, N. P., Krishnamurthy, L. and Sheldrake, A. R. 1988. In: *Phosphorus in Indian Vertisols: Summary Proceedings of a Workshop*, pp. 23–32. Patancheru, Andhra Pradesh, India: ICRISAT.

Saxena, N. P. and Sheldrake, A. R. 1980. *Field Crops Research* 3: 211–214.

Singh, B. P., Sakal, R. and Singh, A. P. 1985. *Indian Journal of Agricultural Science* 55: 56–58.

Szabolcs, I. 1989. *Salt Affected Soils*, 274 pp. Boca Raton, FL, USA: CRC Press.

USDA (United States Department of Agriculture). 1954. Diagnosis and Improvement of Saline and Alkali Soils. *Agricultural Handbook No. 60*. Washington D.C.: Soil and Water Conservation Research Branch, Agricultural Research Service.

van Rheenen, H. A. and Sethi, S. C. 1990. *International Chickpea Newsletter* 22: 25–27.

van Wambeke, A. 1976. In: *Proceedings of Workshop in Plant Adaptation to Mineral Stress in Problem Soils*, pp. 15–24 (ed. M. J. Wright). Ithaca, New York, USA: Cornell University, Agricultural Experiment Station.
Welch, R. M. and LaRue, A. T. 1990. *Plant Physiology* 93: 723–729.

Screening techniques and improved biological nitrogen fixation in cool season food legumes

D.F. HERRIDGE[1], O.P. RUPELA[2], R. SERRAJ[3] and D.P. BECK[4]

[1] NSW Agricultural Research Centre, RMB 944, Tamworth, NSW 2340 Australia;
[2] ICRISAT, Legume Program, Patancheru P. O., Andhra Pradesh 502 324, India;
[3] University Marrakech, Marrakech, Morocco, and
[4] ICARDA, P. O. Box 5466, Aleppo, Syria

Abstract

Dinitrogen fixation and legume productivity are greatly influenced through the interactions of legume host Rhizobium, and the above- and below-ground environment. The benefits of improving legume N_2 fixation include reduced reliance on soil N, leading to more sustainable agricultural systems and reduced requirements for fertilizer N, enhanced residual benefits to subsequent crops, and increased legume crop yields. Most of the gains in N_2 fixation to date have been derived from management of legume cropping systems and through inoculation of legume seed with competitive and symbiotically effective rhizobia. Further gains are possible by developing plant cultivars with tolerance to soil abiotic factors, increased plant yield, and a broader and more effective matching of plant host and rhizobia. Techniques for screening material for superior N_2 fixation and examples of programs to increase fixed N, with attention to the major abiotic stresses, are discussed.

Introduction

Research into legume N_2 fixation has two broad objectives. Most often the goal is greater legume productivity, but efforts are increasingly being directed toward maximizing the legume contribution to soil N in a farming systems context. Legumes do not inevitably nodulate and fix N_2. Nodulation requires the presence of sufficient numbers of the appropriate rhizobia in the root zone. Even where suitable rhizobia are present, other factors such as soil fertility or water availability may interfere with the processes of nodulation and N_2 fixation. Grain legumes also vary in their inherent capacity to nodulate and/or fix N_2, both under stress-free conditions (Attewell and Bliss, 1985) and when stresses such as nitrate, acidity, or drought are present (Harper and Gibson, 1984; Gibson and Harper, 1985; Beck, 1990). Thus, commercial legume crops can be poorly nodulated, resulting in loss of production (Herridge et al., 1987). Nitrogen balance studies show that legumes frequently fix considerably less N_2

F.J. Muehlbauer and W.J. Kaiser (eds.), Expanding the Production and Use of Cool Season Food Legumes, 472–492.

than the quantity removed in harvested seed (Beck *et al.*, 1991), resulting in a decline rather than an increase in soil N fertility. These limitations must be redressed if legumes are to be fully exploited in farming systems.

There are two possible strategies for improving legume N_2 fixation: management of the legume crop to minimize stresses and maximize yield, and breeding legumes or rhizobia (or combinations) with enhanced capacity for N_2 fixation. The potential for improving N_2 fixation through breeding has been considered for some time (e.g., Phillips *et al.*, 1971) but progress has been slow in part because of the limited attention given to an "operational breeding approach" (Mytton, 1983). In addition, there is a shortage of basic information on the genes that are important in the development and functioning of the legume-*Rhizobium* and -*Bradyrhizobium* symbioses. To further confound the problem, our knowledge of the bacterial genes involved in the symbioses has expanded at a far greater rate than that of the corresponding plant genes (e.g., Vance *et al.*, 1988).

The pressing objectives of cropping systems improvement are too important to wait upon possible future developments in knowledge of the host-rhizobia-environment genetics. This paper presents information on recent efforts to improve N_2 fixation in the grain legumes, and suggests methods for selection of legumes with enhanced N_2 fixing ability.

Amounts of N_2 Fixed and Factors Limiting N_2 Fixation

Published estimates of N_2 fixed by the cool season food legumes (Brunner and Zapata, 1984; Jensen, 1986; Rennie and Dubetz, 1986; Evans and Herridge, 1987; Smith *et al.*, 1987; Beck *et al.*, 1991) reveal a large range of values and reflect both the genetic capacities of the legume crops to yield and fix N_2, the environmental constraints on those capacities and the effects of cultural practices on them. Amounts of N_2 fixed were: chickpea, 3 to 141 kg N ha^{-1}; lentil, 10 to 192 kg N ha^{-1}; pea, 17 to 244 kg N ha^{-1} and faba bean, 78 to 330 kg N ha^{-1}. Ranges of values for the proportion of crop N derived from fixation (*P*fix) were: chickpea, 8 to 82%; lentil, 39 to 87%; pea, 23 to 83% and faba bean, 59 to 92%. We conclude from these figures that biomass or plant yield is a greater constraint to N_2 fixation in chickpea and, to a lesser extent, lentil than with pea and faba bean.

Data presented in Table 1 re-enforce the importance of legume yield and soil nitrate status to N_2 fixation by the cool season food legumes. At the Tel Hadya site, growth of all four species and, as a consequence, N_2 fixation was restricted by lack of water, especially in the drier season (low values). Values of N_2 fixation in winter-sown chickpea were high, whereas *P*fix values in spring-sown chickpea, where drought limited growth as early as anthesis, were negligible. In contrast, differences between N_2 fixed in spring- and winter-sown chickpea in France were relatively small because of extended moisture availability through the later stages of plant growth. High concentrations of soil nitrate in

Montpellier depressed N_2 fixation in chickpea, but had little effect on the other crops. At the Dromolaxia site, soil nitrate reduced N_2 fixation by the three species and it is probable too that genetic and environmental constraints on plant yield, together with the lack of the appropriate rhizobia for chickpea, represented further constraints in chickpea and pea.

Table 1. Estimates of N_2 fixation by the cool season food legumes and factors responsible for reduced activity

Location Species	Total crop N	N_2 fixed	Pfix	Soil N use	Major factors restricting Nfix
	kg ha^{-1}	kg ha^{-1}	%	kg ha^{-1}+	
Dromolaxia, Cyprus[1]					
	61	25	41	36	Soil N, rhizobia
Chickpea	194–220	122–176	63–80	44–72	Soil N, H$_2$O
Faba bean	64–144	23–91	36–63	41–53	Yield, soil N
Pea					
Montpillier, France					
	113–134	49–74	44–55	44–54	Soil N, rhizobia
Chickpea	199	126	63	65	Soil N, yield
Pea	210–213	133–147	63–69	59–70	Yield
Lentil	187–196	165–181	88–92	(02)–16	Yield
Faba bean					
Tel Hadya, Syria[2]					
	21–142	3–115	8–81	11–28	Yield through H$_2$O
Chickpea++	48–86	34–62	70–72	12–22	Yield through H$_2$O
Pea	124–143	88–108	68–75	31–37	Yield through H$_2$O
Lentil	117–184	78–133	63–76	17–56	Yield through H$_2$O
Faba bean					

[1]Papastylianou (1988)
[2]Beck et al. (1991)
+with removal of total above ground dry matter
++wide range due to winter- and spring-sown treatments in chickpea

In chickpea, N_2 fixation seems to be more sensitive to stresses such as high temperatures (Rawsthorne et al., 1985) and drought (Wery et al., 1988) than seed production and N assimilation. In field studies with chickpea and lentil in Syria (Beck, unpublished data), drought stress depressed Pfix more than N uptake (Figure 1). In these trials, conducted using a line-source sprinkler over two seasons with six cultivars each of chickpea and lentil, N uptake from soil in both crops remained fairly stable until moisture became sufficient for maximum fixation, at which time soil N uptake increased. Values for Pfix in different cultivars at a the lower moisture levels ranged from 15 to 38% in chickpea and 27 to 60% in lentil. These data may partly explain why N fertilization improves yield of non-irrigated chickpea in low N soils, but does not affect the irrigated crop yield (ICRISAT, 1982).

The evidence for species differences in tolerance to nitrate is conflicting, although a number of studies indicate that faba bean, pea, and the narrow-leafed lupin can continue to fix N_2 in the presence of moderate to high levels of nitrate in the soil (Richards and Soper, 1979; Evans et al., 1989; Beck et al.,

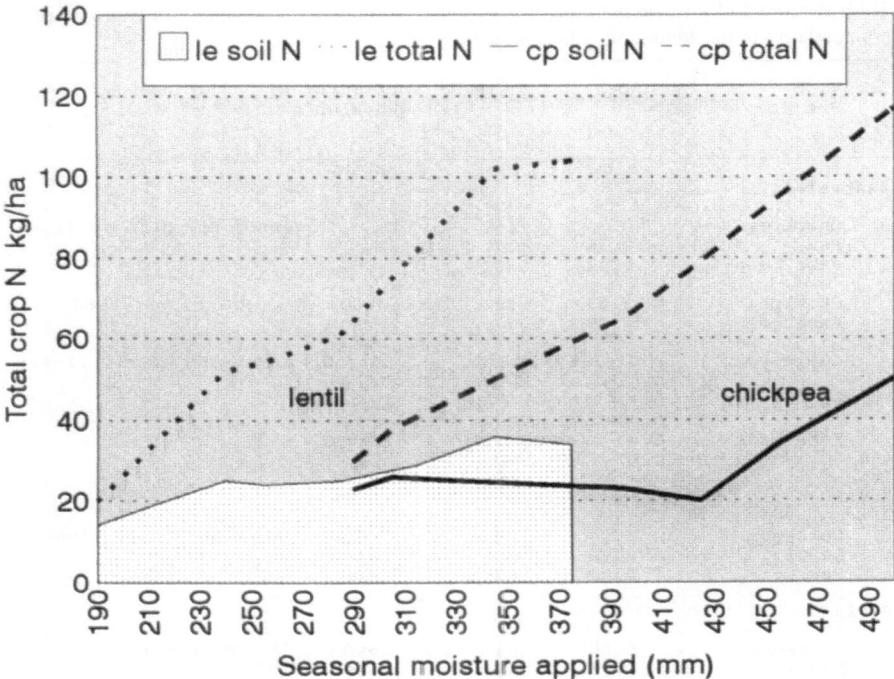

Figure 1. Total crop nitrogen as affected by moisture availability in lentil and chickpea cultivars, with portions of total derived from soil and from fixation indicated. Tel Hadya, Syria. Beck, unpublished data.

1991; Hardarson *et al.*, 1991). Evans *et al.*, (1989) quantified the relationship between *P*fix and soil nitrate (using uptake of N by an adjacent cereal crop as an index) for the narrow-leafed lupin and field pea. The functions accounted for 35 and 79% of the variance for the two species, respectively.

Contributions to the Soil N Pool and Associated Rotational Benefits

Results from legume-based rotation experiments in rainfed cropping areas of a number of countries have been published in recent years. These experiments reflect the growing concern of scientists and farmers in those areas to declining levels of N fertility in soil and reduced production of cereal grain and protein yields. In all cases, wheat following grain legumes yielded more than wheat following wheat, irrespective of the species of legume (Table 2).

A typical set of experiments are those of Armstrong *et al.* (1987). Beneficial effects of faba bean, pea, and the narrow-leafed lupin on two subsequent wheat crops were recorded at two sites in eastern Australia. Increases in the absence of fertilizer-N were between 0.17 and 2.46 t ha^{-1} (5 and 140%) (crop 1) and

Table 2. Effects of a single preceding food legume crop on yields of subsequently-grown wheat grain, relative to the continuous wheat sequence

Location Pre-crop		Increase in yield		Fertilizer N equivalence	Reference
		t ha^{-1}	-%-	kg ha^{-1}	
Australia					
Chickpea		0.31	15	–	Jessop and Mahoney (1985)
Pea		0.36	17	–	"
Faba bean		0.42	20	–	"
Chickpea		1.43	88	>50	Marcellos (1984)
Faba bean		1.25	77	>50	"
Lupin-year	1	0.91–1.82	28–103	60–>80	Armstrong *et al.* (1987)
	2	0.51–0.92	33–48	–	"
Pea -year	1	0.17–2.46	5–140	15–>80	"
	2	0.48–0.93	17–88	–	"
Faba bean	1	0.40–1.93	13–110	30–>80	"
	2	0.57–0.68	24–54	–	"
India					
Chickpea		1.04	77	<60	Ahlawat *et al.* (1981)
Lentil		0.90	66	<60	"
Pea		0.81	60	<60	"
Syria					
Chickpea		0.65	53	>90	Harris (1989)
Lentil		0.78	65	>90	"
United Kingdom					
Faba bean		1.04	47	<48	Dyke and Prew (1983)

between 0.48 and 0.93 t ha^{-1} (17 and 88%) for the second crop. Responses were generally greater at one of the sites because of the additional effects of the legumes in lessening the incidence and impact of take-all (*Gaeumannomyces graminis* var. *tritici*); in the continuous wheat plots at this site, the disease caused substantial yield reductions.

Differences in rotational effects among the cool season legume species appear to vary for particular sites and/or seasons (Harbison *et al.*, 1986; Strong *et al.*, 1986a; Harris, 1989, 1990). Strong *et al.* (1986b) detected no consistent effect of prior legume crop on levels of total soil N. This finding is perhaps not surprising because of the following:

• Soils are normally cultivated in the fallow period between the legume crop and the following (wheat) crop. Where adequate moisture is available, transformation of the N contained in the legume (organic) residues to the mineral forms will occur thereby lowering levels of organic N.

• In virtually all of the studies of grain legume – cereal rotations, the legumes are grown for a single season. It is unreasonable to expect large and detectable increases in total soil N following just one crop of a grain legume.

- Grain legumes are usually grown in soils that have been cultivated and fallowed and/or fertilized at high levels for cereal production. Consequently, nitrate levels can be high. Legume nodulation and N_2 fixation are both sensitive to soil nitrate concentration and will be depressed by the presence of soil nitrate.
- Legumes such as soybean, faba bean, common bean, pea, and chickpea, having high N harvest indices (60 to 80%), are not likely to add appreciable N to the soil through the non-harvested residue. In addition, in some areas legume crop residues are entirely removed from the field for use as valuable animal feed.

There may be more appropriate indices of the benefits of legumes on the soil N pool than measurement of total soil N. In a long-term two-course rotational trial in Syria, soil N levels were measured following 6 years of rotation. Total soil N in the surface 40 cm of the chickpea-wheat rotation receiving no fertilizer N did not differ significantly from that measured in the continuous wheat and fallow-wheat rotation (Harris and Matar, unpublished data). Soil organic C levels in the three rotational treatments were also similar (0.9 to 1.0%), but incubation measurements of the N mineralization potential (Stanford and Smith, 1972; Matar *et al.*, 1991) showed large differences between rotational treatments. Mineralization potentials in continuous wheat and fallow-wheat soils were similar, with 75 and 61 mg N kg^{-1} soil, respectively. In the chickpea-wheat soil, however, mineralization potential was 118 mg N kg^{-1} soil, indicating an increased capacity to supply plant-available N from the total N pool. These data are supported by studies on soils at 40 northern Syrian sites under different crop rotations, where mineralization potential measurements gave the best indication of N uptake in wheat under legume-cereal rotations (Matar *et al.*, 1989).

Is Breeding for Improved N_2 Fixation in Legumes Beneficial?

The simple answer to the question is "yes". Improving legume N_2 fixation has several potential benefits, including: reduced reliance on soil N, leading to more sustainable agricultural systems and reduced requirements for fertilizer N inputs; increased yields of grain and other harvestable products, e.g. straw; enhanced residual benefits to subsequent crops; and increased seed quality and quantity through efficient conversion of fixed N_2 to seed N. The importance of breeding for enhanced N_2 fixation in improved cultivars may be of particular relevance because plant breeders, when selecting for yield, tend to use sites with high soil N status. In such soils, the legume's capacity for N_2 fixation may not be a critical factor.

Increasing the capacity of the legume through selection and breeding to fix N_2 can be achieved by:

- Maximizing legume yield within the constraints imposed by agronomic and environmental conditions. Plant breeders actively select for yield within the constraints of local environments; yield largely determines the amount of N_2 that is fixed by the crop, particularly in low N soils (Hardarson *et al.*, 1984; Kumar Rao and Dart, 1987; Duc *et al.*, 1988). Those breeders who operate for the most part in low N soils will tend to select for material with good capacity for N_2 fixation.

- Active selection and breeding for symbiotic characteristics in legumes. Examples are nitrate tolerance, i.e., the ability of the plant to nodulate and fix N_2 in the presence of soil nitrate, the capacity to fix N_2 in dry soil and general nodulation capacity. Indications are that chickpea and lentil cultivars selected for drought tolerance vary in their capacity for N_2 fixation under drought stress (Beck, unpublished data). Natural variations for nitrate tolerance and nodulation capacity also exist and have been created too using plant mutagenesis (Jacobsen and Feenstra, 1984; Carroll *et al.*, 1985; Herridge and Betts, 1985; Park and Buttery, 1988). It may be impossible to produce a legume that is dependent solely upon N_2 for growth and cannot assimilate nitrate, but there is undoubted scope to improve *P*fix in the presence of nitrate for the majority of legumes. One argument is that legumes should be able to use both atmospheric and soil N sources so they can scavenge nitrate from the soil which would otherwise be lost through leaching and denitrification. Two counter arguments are that in many soils, nitrate is relatively stable over time and can be considered as a stable pool of N. Secondly, if depletion of soil nitrate was considered necessary, it would make more sense to use a cereal crop with a higher demand for N and often greater economic return.

- Optimizing the numbers and effectiveness of rhizobia in the rooting zone through strain selection and inoculation techniques and through plant breeding for promiscuous or selective nodulation (Devine, 1984; Kueneman *et al.,* 1984; Cregan and Keyser, 1986). Continued improvements are needed in the effectiveness of strains of rhizobia used as legume inoculants and in the process of inoculation. There is scope to exploit strains of rhizobia for specific cultivars in some species (Mytton, *et al.*, 1977; Beck, 1993) or environmental niches, e.g. acid tolerant strains for acidic soils (Howieson and Ewing, 1986).

Screening Techniques for Improved N_2 Fixation

There is no single method universally appropriate for measuring N_2 fixation by legumes. None of the four most widely-used methods (N difference [N yield], ^{15}N, acetylene reduction, xylem solute [ureide]) can be relied upon to provide an accurate measure of N_2 fixation for every legume species grown under all possible variations of soil type and cropping environment. Each method has relative advantages and specific limitations. However, some of the methods are more likely than others to provide reliable and quantitative estimates of N_2 fixation.

N difference

A true measure of N_2 fixation based on legume N yield can only be obtained when the contribution of soil N to total biomass N is determined. This is usually achieved by growing a non N_2-fixing crop concurrently in the same soil. The difference in total N accumulated by the legume (N_{leg}) and non-fixing control (N_{nonfix}) is regarded as the amount of N_2 fixed. Thus:

$$N_2 \text{ fixed} = N_{leg} - N_{nonfix}$$

The principal assumption is that the legume and non-fixing control take up identical amounts of N from the soil, and so the choice of the control is of utmost importance. Ideally, the legume and control should explore the same rooting volume, have the same ability to extract and utilize soil mineral-N, and have similar patterns of N uptake. The non-fixing control may be a non-legume or an unnodulated or non-nodulating legume (preferably an isoline of the test legume). Unfortunately, there are often substantial differences between N_2-fixing and non-fixing plants in their capacities to use soil N. Even when a supposed ideal non-fixing control is used, e.g., a non-nodulating isoline, erroneous estimates of N_2 fixation may still result from differences in root morphologies (Boddey *et al.*, 1984).

The observation that levels of soil mineral-N were invariably higher following a legume crop than after a non-legume (Reeves *et al.*, 1984; Evans *et al.*, 1985) led Evans and Taylor (1987) to propose a modification of the N difference equation to account for differences in the utilization of soil mineral-N by the legume and non-legume (non-fixing control). Additional measurements are made of the amounts of soil mineral-N in the root zones of the two crops at maturity. Then:

$$N_2 \text{ fixed} = (N_{leg} - N_{nonfix}) + (\text{SoilN}_{eg} - \text{SoilN}_{nonfix})$$

With both the N difference and modified N difference methods, greater accuracy will always be achieved when plant-available soil N is low and when legume biomass-N is high.

15N Methods

The stable isotope ^{15}N occurs in atmospheric N_2 at a constant 0.3663 atom %^{15}N. If ^{15}N enrichment (abundance) in plant-available soil N is different from that in atmospheric N_2, then the proportion of legume N derived from each source can be measured by the isotopic abundances in the legume and in a non-fixing control wholly dependent on the same soil N. In many cases, the very small differences in the natural abundance of ^{15}N between plant-available soil N and atmospheric N_2 can be used, provided that samples can be analyzed using a very precise mass spectrometer. More usually, the difference between the ^{15}N enrichment of the soil and atmosphere is expanded by incorporating ^{15}N-labelled materials in the soil.

The *15N enrichment* method is generally regarded as the standard method for estimating legume N_2 fixation. Its use has greatly increased over the past decade, a fact that is reflected in the extensive list of reviews on the method (e.g., Chalk, 1985; Danso, 1988; Ledgard and Peoples, 1988; Witty et al., 1988). However, the expense of instrumentation to measure ^{15}N plus the expense of the ^{15}N-labelled materials are serious constraints to even greater use of the method. Its main advantage is that it provides a time-averaged estimate of *P*fix, integrated for the period of plant growth to the time of harvest. Thus:

$$Pfix = 1 - \frac{(\text{atom } \%^{15}N \text{ excess}_{legume})}{(\text{atom } \%^{15}N \text{ excess}_{nonfix})}$$

where atom $\%^{15}N$ excess = (atom $\%^{15}N_{sample}$) − (atom $\%^{15}N_{air}$ N_2), and atom $\%^{15}N$ of air N_2 = 0.3663. The estimate of *P*fix is often considered to be independent of yield, although it is necessary to measure dry matter and N yield to estimate the amount of N_2 fixed.

The major assumption of both the ^{15}N enriched and natural ^{15}N abundance methods is that the legume and non-fixing reference plants utilize soil N with the same isotopic composition. With the enriched system, this assumption translates into the legume and non-fixing reference plants utilizing the same relative amounts of N from added ^{15}N and endogenous soil N. This may not always occur and is the major weakness of the method (Witty et al., 1988). Therefore, the choice of non-fixing reference plant is of critical importance (Witty, 1983). It may, however, be very difficult in practice to assure identical patterns of uptake in fixing and reference crops.

Natural 15N Abundance

Almost all transformations in soil result in isotopic fractionation. The net effect is often a small increase in the ^{15}N abundance of soil N compared with atmospheric N_2 (Shearer and Kohl, 1986). Although the principles of the natural abundance method are similar to those of ^{15}N enrichment, the major limitations are quite different. An isotope ratio mass spectrometer capable of measuring accurately differences of 0.1 part per thousand (about 0.00004 atom $\%^{15}N$) is needed. Great care is necessary in sample preparation to avoid isotopic fractionation or contamination (Bergersen et al., 1988). The accuracy of the method will depend ultimately on the levels and spatial and temporal uniformity of the ^{15}N in the soil. Levels of soil $\delta^{15}N > 6.0$ are preferable. Values below this are often found in pasture and plantation soils and in natural forest systems (Peoples et al., 1989). In soils that are regularly cultivated, $\delta^{15}N$ values tend to range between 6.0 and 16.0, and are relatively constant with time and depth (Peoples and Herridge, 1990). Therefore, the major limitation of ^{15}N enrichment, i.e., choice of appropriate non-fixing reference plant, is less critical.

Acetylene Reduction

The acetylene reduction assay (ARA) arose from observations in the 1960s that the N_2-fixing enzyme, nitrogenase, catalyzed the reduction of acetylene (C_2H_2) to ethylene (C_2H_4). Since that time, ARA has been used to detect N_2 fixation activity for a wide range of biological systems and has played a major role in N_2 fixation studies because of its rapidity, simplicity, high sensitivity, and inexpensive equipment and resource costs. The standard ARA method involves enclosing detached nodules or nodulated root systems in airtight containers and exposing them to an atmosphere containing C_2H_2. After an incubation period, gas samples are collected and analyzed for C_2H_4 using gas chromatography (Turner and Gibson, 1980).

The ARA method eventually lost favor because of a number of major problems. These included: the difficulty in recovering nodules; the need to interpolate between single, short-term measurements to obtain time-integrated measurements; the need to determine correct conversion ratios between C_2H_4 produced and N_2 fixed, which can vary according to environmental, diurnal, and host plant effects acting independently on N_2 and C_2H_2 assimilation and reduction; non-linearity in the rate of C_2H_2 reduction over the period of the assay; effects of nodule removal and decapitation of plants and difficulties in sampling, particularly in hard-setting soils. The magnitude of the problems and the resultant errors in estimating C_2H_2 reduction (N_2 fixation) activity suggest that even simple comparisons of material in a breeding program may be invalid. Readers are referred to Turner and Gibson (1980) and Witty and Minchin (1988) for discussion of the problems of the ARA.

Xylem N Solutes

Collection of xylem sap and analysis of its contents has been used widely in assessments of the nutritional status of field-grown plants, and, since the late 1970s, to measure legume N_2 fixation (McClure et al., 1980; Herridge et al., 1984, 1990; Herridge and Betts, 1985; Rerkasem et al., 1988; Peoples et al., 1989). The principal underlying this method is that the composition of N solutes in xylem sap changes from one dominated by ureide compounds (allantoin and allantoic acid) in N_2-dependent plants to a profile dominated by nitrate and amino-N in plants utilizing soil N. In calibration experiments, correlations between the relative abundance of ureide-N in xylem sap and the proportion of plant N derived from N_2 fixation (*P*fix) was extremely strong with regression coefficients close to unity (e.g., Herridge and Peoples, 1990).

Use of this method, however, is limited as not all legumes export fixed N_2 as ureides. With the cool season food legumes, around 80% of fixed N_2 is exported from the nodules as the amides, asparagine, and glutamine, the remainder as amino acids (Herridge et al., 1988). It would be extremely useful though to extend the principal of the ureide assay of N_2 fixation activity to the amide exporters. Unfortunately, with these species, none of the readily-measured N solutes appears to be specifically associated with N_2 fixation. There may be some scope to measure shifts in asparagine:glutamine ratios, or the relative proportion of nitrate in xylem sap. Calibration experiments involving chickpea, faba bean, lentil, and pea have been reported (Peoples *et al.*, 1987) but validation of the relations under field conditions have not yet been attempted. Groupings of the important food and oilseed legumes according to the principal forms of xylem N transport products are presented in Table 3.

Table 3. Principal N solutes in xylem sap of N_2-dependent food and oilseed legumes

Amides (asparagine, glutamine)	Ureides (allantoin, allantoic acid)
chickpea (*Cicer arietinum*)	soybean (*Glycine max*)
lentil (*Lens culinaris*)	pigeonpea (*Cajanus cajan*)
pea (*Pisum sativum*)	mung bean (*Vigna radiata*)
faba bean (*Vicia faba*)	black gram (*Vigna mungo*)
white lupin (*Lupinus albus*)	common bean (*Phaseolus vulgaris*)
narrow-leafed lupin (*L. angustifolius*)	cowpea (*Vigna unguiculata*)
peanut (*Arachis hypogaea*)	winged bean (*Psophocarpus tetragonolobus*)

Plant Improvement Strategies

There is an apparent consensus that enhanced N_2 fixation by the grain legumes will result from selection and breeding for N yield, nitrate tolerance, and specific nodulation requirements (Mytton, 1983; Beringer *et al.*, 1988; Bliss and Miller, 1988). This section examines programs that aim to develop cultivars of legumes that incorporate one or more of these characteristics. Where possible, programs involving the cool season food legumes are featured; examples of other legumes are also presented to illustrate methodologies.

Legume Yield

Agronomic and environmental considerations often limit the biomass yield of a legume crop and therefore the capacity of that crop to fix N_2. Yield will also be determined genetically. Duc *et al.* (1988) evaluated 21 genotypes of faba bean at two sites near Dijon, France, over two years. Their data show clearly the strong correlation between biomass N yield and N_2 fixation (Figure 2). Where comparisons between genotypes could be made over sites and years, they suggest a strong genetic basis for N yield and N_2 fixation.

With some species, low N yield may be a typical characteristic. In studies over a range of environments and agronomic practices, N yield and N_2 fixation by chickpea were consistently less than for the other cool season food legumes (Rennie and Dubetz, 1986; Evans and Herridge, 1987; Smith *et al.*, 1987; Beck *et al.*, 1991). Average values for N yield were 100 (chickpea), 185 (field pea), 196

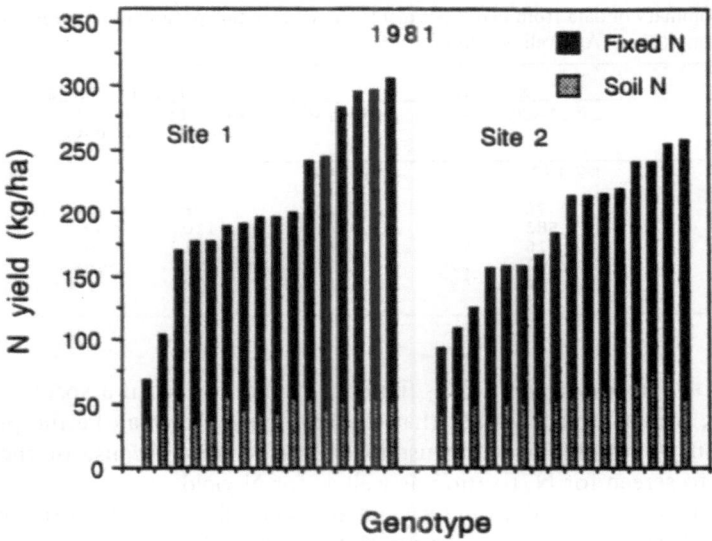

Figure 2. Contributions of soil N and fixed N_2 to total N yield of 16 genotypes of faba bean, grown at two sites near Dijon, France, in 1981.

(lentil), and 200 kg N ha^{-1} (faba bean). These studies did not indicate that the inherent capacity of chickpea for either nodulation or N_2 fixation was less than for the other species. We conclude, then, that increasing N yield of chickpea may result in increased N_2 fixation.

The antithesis of this situation involves the common bean, in which low N yield is the result of low N_2 fixation capacity, rather than *vice versa* (Attewell and Bliss, 1985). A breeding program by Bliss and his co-workers at the University of Wisconsin has produced new genotypes of common bean with higher levels of N_2 fixation resulting in increased plant vigor and improved N yields (Table 4, Expt. 1). For all of the hybrids, however, N_2 fixation capacity was substantially less than that of Puebla 152, the high-fixing donor. In a second experiment, the hybrid line 24–21 displayed higher rates of growth and N accumulation (+36%) and more total growth (+79%) than the commercial parent while retaining desirable growth characteristics. In later experiments utilizing the inbred backcross breeding method, favorable N_2 fixation alleles, independent of maturity, were recovered in line 24–21 indicating that N_2 fixation was heritable (Wolyn *et al.*, 1991).

Initial assessments of N_2 fixation were based on the ARA. More recently, ^{15}N methods (enriched and depleted) were used (Pereira *et al.*, 1989; Wolyn *et al.*, 1991). Breeding material was, of necessity, evaluated under low soil mineral-N, because the plant's capacity for N_2 fixation and not the capacity for growth (N yield) was of principal importance. In a soil with moderate to high levels of N, the capacity of the plant to fix N_2 will not be expressed to the same extent

Table 4. Summary of data from two experiments from a breeding program to increase N_2 fixation by the common bean (Attewell and Bliss, 1985)

Parent, cultivar or line	Experiment 1			Experiment 2		
	N_2 fixation	Pfix	Seed yld	Maturity (days)	Deter-minate	N yield
	mg plant^{-1}		g plant^{-1}			mg plant^{-1}
Sanilac	76	12	18	85	yes	591
24 - 17	583	48	31	110	no	1068
24 - 21	216	25	19	91	yes	1045
24 - 55	192	22	23	94	yes	668
Puebla	852	57	38	120	no	1429

because of the suppression of N_2 fixation by the soil N. In a species such as chickpea, where N yield rather than N_2 fixation *per se* may be the problem, evaluation would still be more useful in low N soils because of the added capacity to screen for N_2 fixation as well as for N yield.

Because N yield and dry matter yield are generally strongly correlated (e.g., Mytton, 1983), a program to enhance N_2 fixation in chickpea might involve the following sequence:

1. Screen a large and diverse germplasm (500 to 1000 genotypes) of chickpea, inoculated with highly effective rhizobia, for production of dry matter under low N conditions (preferably in the field, but could also be done in a glasshouse).
2. Select superior genotypes (e.g., top 10%) for further evaluation. The second round of screening is ideally done in the field on a low N-fertility soil, again with a mixture of highly-effective rhizobia. Assessments should include seed yield and total N yield.
3. Compare elite genotypes over a range of edaphic (particularly soil N fertility) and environmental (including diverse rhizobial population) conditions for seed yield, N yield and N_2 fixation, the latter using ^{15}N methods.

Genotypes identified at this stage to be superior in all three attributes and adapted to the soils and environments for which they were likely to be used would have immediate commercial application. High N_2-fixing genotypes that produced low seed yields or seed of low quality could be used as donor parents in a breeding program (e.g., as with "Puebla" in the common bean program, Table 4). There may be little scope to adopt this protocol for species such as field pea and faba bean that are already capable of producing high N yields under field conditions, i.e., > 300 kg N ha^{-1}, and fixing substantial amounts of N_2 (see Brunner and Zapata, 1984; Jensen, 1986).

It is also important in a program such as this to remove the confounding effect of crop duration on N yield. Increased N yield due to elevated rates of growth and assimilation is more useful because it can theoretically be expressed in any environment; increased crop N due to longer crop duration can only be expressed if the length of the season in a particular environment or cropping system is sufficiently long. In commercial agriculture, individual crops must fit into cropping systems which are determined by seasonal changes in temperature, moisture availability, radiation, availability of land and resources to grow and harvest the crop, marketing arrangements, and farmer priorities. The optimum duration of any crop is therefore determined by a combination of very different factors, the least of which is N yield or N_2 fixation.

Brunner and Zapata (1984) compared 19 mutant lines of faba bean with the parent cultivar for seed yield, dry matter and N yield, and N_2 fixation. One line (II-18) was clearly superior in all characteristics, but it was not clear whether the superiority of the line was because of higher growth and N-assimilation rates or because of a longer crop duration. Nevertheless, its superior performance over a range of environments suggests that growth and assimilation rate had been improved by mutagenesis. The authors also concluded that N_2 fixation was determined by N yield, rather than Pfix. In a similar study involving pigeonpea, Kumar Rao and Dart (1987) also showed the strong relation between N yield and N_2 fixation. In that instance, however, N yield appeared to be linked to crop duration.

Nodule Function and Specific Nodule Traits

Although sometimes used as criteria for increased N_2 fixation, selection of genotypes on the basis of specific traits that are either directly or indirectly associated with nodulation or nodule (N_2 fixing) activity appears to be of limited value. Heichel *et al.* (1989) concluded that the selection of lines of alfalfa (*Medicago sativa* L.) for activity of various nodular enzymes, including nitrogenase, resulted in experimental populations with enhanced or reduced enzyme activity, but did not result in populations that were different in N_2 fixation. The authors proposed several explanations for their inconclusive findings, including: (1) short-term measurements of nitrogenase activity using the ARA may not predict activity over a longer growth period; (2) the limitations of the ARA may be too great to even compare treatments on a relative, let alone a quantitative, basis; (3) performance of seedlings in a glasshouse may not predict performance in the field over one or several seasons; and (4) enhanced activities of single or several nodular enzymes may be countered by inadequate or even normal activity of other equally-important enzymes, resulting in no net change in the rate of N_2 fixation.

Controlled Nodulation

Legume inoculation with rhizobia and bradyrhizobia is a long-established and successful practice, especially with particular crops in the more technically-advanced countries. Vincent (1965) and others have argued that it is a desirable practice in most agricultural soils throughout the world although Date (1977) cautioned that the need to inoculate was not universal and should be carefully determined for each individual situation. Lacking such evaluations and because of the exacting technology required for the production, distribution, and use of inoculants, the practice of inoculation remains the exception rather than the rule in many parts of the world.

In the developing countries in particular, there are substantial advantages in growing legumes that nodulate with effective rhizobia already present in the soil. Nangju (1980) described how soybean genotypes from Southeast Asia nodulated successfully with the indigenous rhizobia in Nigeria, resulting in effective symbioses. The USA bred cultivars nodulated poorly without inoculation. Hybridization of the Asian and American cultivars has given high yielding lines capable of fixing large amounts of N_2 without inoculation (Kueneman *et al.*, 1984).

In other environments where inoculation is an option, populations of infective rhizobia often exist in large numbers in the soil and represent a formidable barrier to the introduction of more effective strains (Devine, 1984; Singleton and Tavares, 1986; Thies *et al.*, 1991). It is possible, however, to select rhizobial strains for competitiveness (Amarger and Lobreau, 1982), and there are indications that selected, highly effective, and competitive strains can be used to increase N_2 fixation and yields where native rhizobial populations are low or ineffective (Arsac and Cleyet-Marel, 1986; Beck, 1990, 1992). Successful nodulation by inoculant strains is often site-specific depending on soil factors and the composition of the indigenous rhizobial population.

In the USA, large populations of the soybean bradyrhizobia have developed with cropping and so typically less than 10% of nodules on soybean are formed by the inoculant and yield responses to inoculation are rare (Devine, 1984; Halliday, 1985). In response, research programs in several laboratories (e.g., Devine, 1984; Cregan and Keyser, 1986) have aimed to produce cultivars of the host that bypass the resident rhizobia in the soil to become nodulated by more effective inoculant strains. This strategy has been applied also to pea. Lie (1978, 1981) found that the primitive pea cultivar "Afghanistan" was not nodulated by strains of rhizobia isolated from temperate soils in Europe but did form nodules with strain Tom from Turkey. This resistance to nodulation has been ascribed to the recessive gene, *sym-2* (Holl, 1975). In competition experiments, the European strains caused differential blocking of nodulation by Tom (Lie *et al.*, 1988). Subsequent identification of rhizobial isolates that could nodulate Afghanistan as well as cultivars bred for European conditions indicated the presence in the rhizobia of a specific genetic region, termed *nodX*. Fobert *et al.* (1991) suggest that the combination of the plant gene, *sym-2*, and the rhizobial gene, *nodX*, may allow ultimate control of nodulation and provide a mechanism for enhancing N_2 fixation, even in the presence of large populations of indigenous rhizobia. This approach assumes that N_2 fixation is limited by the effectiveness of the rhizobia, which may not always be the case.

Nitrate Tolerance

Grain legumes can utilize substantial amounts of soil nitrogen (usually as nitrate) during growth. Development of host legume-rhizobia symbioses where

*P*fix is maintained at near maximum values in the presence of high soil nitrate, i.e., nitrate tolerance, could provide a significant advance in the improvement of legume N_2 fixation. It is most probable that nitrate tolerance will be achieved through manipulation of the host legume, rather than the rhizobial strain (Harper and Gibson, 1984; Gibson and Harper, 1985), though differences have been observed in rhizobial competitiveness and effectiveness at high and low nitrate concentrations (Somasegaran and Bohlool, 1990).

A mutant phenotype capable of copious nodule production under high levels of nitrate would appear to offer the means of obtaining truly nitrate tolerant (and therefore higher fixing) legumes. Ethyl methanesulphonate (EMS)-induced hypernodulating mutants have now been developed in a range of legumes including soybean (Carrol *et al.*, 1985; Gremaud and Harper, 1989), common bean (Park and Buttery, 1988), pea, and faba bean (Duc and Messager, 1989; Duc *et al.*, 1989).

Assessments of N_2 fixation activity of EMS-induced hypernodulating mutants have relied almost exclusively on the ARA. Results indicated that the mutants fixed substantially more N_2 than the wild-types in the absence as well as in the presence of nitrate (Jacobsen and Feenstra, 1984; Carroll *et al.*, 1985; Gremaud and Harper, 1989; Rosendahl *et al.*, 1989; Wu and Harper, 1991). Subsequent assessments of N_2 fixation using ^{15}N and ureide analysis indicated that the mutants rarely fixed more N than the wild-type parents (Hansen *et al.*, 1989; Herridge *et al.*, 1990; Wu and Harper, 1991). In most cases, higher or similar *P*fix values could not compensate fully for reduced plant yields. The greatest potential for exploiting hypernodulation will most likely come through backcrossing with high yielding, normally-nodulating material (e.g., J.E. Harper [USA], R. Boerma [USA], L. Song [Australia], personal communications), where the principal objective is to recover the plant yield lost during mutagenesis while, at the same time, maintaining the hypernodulating phenotype.

Variation in tolerance to nitrate in natural populations of soybean has been exploited in a breeding program in Australia. Nearly 500 genotypes were screened for nodulation and N_2 fixation in the presence of nitrate (Herridge and Betts, 1985, 1988; Betts and Herridge, 1987). Nitrogen fixation was assessed using the solute (ureide) technique. Results indicated that genotypes of Korean origin displayed higher-than-average levels of symbiotic activity in the presence of nitrate (Table 5). These genotypes had similar shoot yields to the commercial cultivars Bragg and Davis, suggesting that the higher levels of fixation by the Korean lines reduced their use of soil N. Nitrate-tolerant lines were therefore used as high-fixing donor parents in a breeding program with selection for both seed yield and N_2 fixation capacity. After crossing with commercial cultivars, F_2 combinations were sown into a high nitrate field soil and assessed during growth for plant and seed characteristics, growth habit and agronomic type and N_2 fixation using a non-destructive xylem ureide assay (Herridge and Peoples, 1990).

Evaluation of F_6 and F_7 material for seed yield and N_2 fixation on high

Table 5. Measurements of nodulation and N_2 fixation by, and growth and yield of, nitrate tolerant and commercial genotypes of soybean in a high nitrate field soil

Genotype	Nodule/plant weight	Nodule/plant number	Rel.ureide N	Rel.ureide (Pfix)[1]	Shoot yield	Seed yield	Residual soil nitrate[2]
	mg		%		g plant⁻¹	t ha⁻¹	kg ha⁻¹
Nitrate tolerant lines							
Korean 466	376	34.5	36	(31)	45.9	1.56	64
" 468	254	16.8	27	(18)	43.3	1.74	79
" 469	176	19.5	30	(22)	41.6	1.43	76
Commercial cultivars							
Bragg	24	2.0	11	(0)	39.7	2.23	45
Davis	40	1.3	15	(0)	48.5	2.17	na

1. Estimating *Pfix* using: *Pfix* = 1.56(Rel ureide-N – 15.9) (Herridge and Peoples, 1990)
2. To a depth of 1.2m

nitrate soils indicated that a number of the genotypes displayed nitrate tolerance by maintaining substantially higher levels of N_2 fixation activity than commercial cultivars (e.g., Forrest) and were equal to or better than the high-fixing Korean parent (K468) (Table 6). However, seed yields were substantially lower than in commercial types. The next step will be to backcross the most promising F_6 and F_7 selections with high yielding commercial cultivars, or to intercross selections in the hope of improving seed yield within this high-fixing germplasm.

Table 6. Seed yields and relative ureide-N values for superior lines of soybean in the F6 and F7 generations and for the commercial and high fixing controls. Data from D. F. Herridge and I. A. Rose, unpublished data

Genotype	Original cross	Flowering (days)	Seed yield, mean of 3 sites (t/ha) F₆	Pfix (%)[1] high nitrate sites F₆	Pfix (%)[1] high nitrate sites F₇
	Forrest	50	2.67	27	36
D22-8	Valder x K468	46	2.08	48	58
A82-3	K464 x Valder	52	2.00	49	50
K78-1	K468 x Bossier	57	1.91	46	50
A46-4	K464 x Valder	58	2.24	51	52
Korean 468		43	0.95	47	53

[1] Assessed during late vegetative growth/early flowering using the xylem ureide technique (Herridge and Peoples, 1990).

Concluding Remarks

Legumes and their capacity for symbiotic N_2 fixation have long been recognized and exploited in agriculture. Most of the gains have been derived through management of the various agricultural systems and through inoculation of legume seed with effective rhizobia. Further gains are possible by developing plant cultivars with enhanced capacity for N_2 fixation. This extra capability for legume N_2 fixation will result from (1) increased plant yield, (2) tolerance of the suppressive effects of soil nitrate, or (3) more effective matching of the plant and rhizobia.

There are essentially two options in the choice of a screening protocol. The method used in any single program will depend on legume species and whether or not the selection is for a plant type (where indigenous rhizobia are compatible and effective) or for the plant-rhizobia symbiosis. One option is to select for plant yield in a low-N soil, then for *P*fix in a moderate to high-N soil (most appropriate for species that have low plant vigor, e.g., chickpea, and for selection of plant-rhizobia symbioses). The alternate option is to select for *P*fix in a moderate- to high-N soil (e.g., nitrate tolerance in pea). The material identified may have sufficient yield and desirable agronomic characteristics for immediate release as a cultivar, or may be used as a source of genes in a crossing program.

References

Ahlawat, I. P. S., Singh, A. and Saraf, C. S. 1981. *Experimental Agriculture* 17: 57–62.

Amarger, N. and Lobreau, J. P. 1982. *Applied and Environmental Microbiology* 44: 583–588.

Armstrong, E., Vere, D. and Dear, B. 1987. In: *Proceedings of the Thirtieth Southern Cereal Improvement Conference*, pp. 108–112. New South Wales: Department of Agriculture.

Arsac, J. F. and Cleyet-Marel, J. C. 1986. *Plant and Soil* 94: 411–423.

Attewell, J. and Bliss, F. A. 1985. In: *Nitrogen Fixation Research Progress*, pp. 3–9 (eds. H. J. Evans, P. J. Bottomley and W. E. Newton). Dordrecht: Martinus Nijhoff.

Beck, D. P. 1990. In: *Food Legume Improvement Program, Annual Report*, pp. 150–155. Aleppo, Syria: ICARDA.

Beck, D. P. 1993. Yield and N_2 fixation response of chickpea cultivars to inoculation with selected *Rhizobium* strains. *Agronomy Journal* (In press).

Beck, D. P., Wery, J., Saxena, M. C. and Ayadi, A. 1991. *Agronomy Journal* 83: 334–341.

Bergersen, F. J., Peoples, M. B. and Turner, G. L. 1988. *Australian Journal Plant Physiology* 15: 407–420.

Beringer, J. E., Bisseling, T. A. and LaRue, T. A. 1988. In: *World Crops: Cool Season Food Legumes*, pp. 691–702 (ed. R. J. Summerfield). London: Kluwer Academic Publishers.

Betts, J. H. and Herridge, D. F. 1987. *Crop Science* 27: 1156–1161.

Bliss, F. A. and Miller, J. C., Jr. 1988. In: *World Crops: Cool Season Food Legumes*, pp. 1001–1012 (ed. R. J. Summerfield). London: Kluwer Academic Publishers.

Boddey, R. M., Chalk, P. M., Victoria, R. L. and Matsui, E. 1984. *Soil Biology Biochemistry* 16: 583–588.

Brunner, H. and Zapata, F. 1984. *Plant and Soil* 82: 407–413.

Carroll, B. J., McNeil, D. L. and Gresshoff, P. M. 1985. *Plant Physiology* 78: 34–40.

Chalk, P. M. 1985. *Soil Biology Biochemistry* 17: 389–410.

Cregan, P. B. and Keyser, H. H. 1986. *Crop Science* 26: 911–916.

Danso, S. K. A. 1988. In: *Nitrogen Fixation by Legumes in Mediterranean Agriculture*, pp. 345–358 (eds. D. P. Beck and L. A. Materon). Dordrecht: Martinus Nijhoff.

Date, R. A. 1977. In: *Biological Nitrogen Fixation in Farming Systems of the Tropics*, pp. 169–180 (eds. A. N. Ayanaba and P. J. Dart). Chichester: John Wiley.

Devine, T. E. 1984. In: *Biological Nitrogen Fixation*, pp. 127–154 (ed. M. Alexander). New York: Plenum.

Duc, G., Mariotti, A. and Amarger, N. 1988. *Plant and Soil* 106: 269–276.

Duc, G. and Messager, A. 1989. *Plant Science* 60: 207–213.

Duc, G., Trouvelot, A., Gianinazzi-Pearson, V. and Gianinazzi, S. 1989. *Plant Science* 60: 215–222.

Dyke, G. V. and Prew, R. D. 1983. In: *The Faba Bean (Vicia faba L.)*, pp. 263–269 (ed. P. D. Hebblethwaite). Cambridge: University Press.

Evans, J. and Herridge, D. F. 1987. In: *Nitrogen Cycling in Temperate Agricultural Systems*, pp. 14–43 (eds. P. E. Bacon, J. Evans, R. R. Storrier and A. C. Taylor). Wagga Wagga: Australian Society of Soil Science.

Evans, J., O'Connor, G. E., Turner, G. L., Coventry, D. R., Fettell, N., Mahoney, J., Armstrong, E. and Walsgott, D. N. 1989. *Australian Journal of Agricultural Research* 40: 791–805.

Evans, J. and Taylor, A. 1987. *Journal of Australian Institute Agricultural Science* 53: 78–82.

Evans, J., Turner, G. L., O'Conner, G. and Bergersen, F. J. 1985. In: *Nitrogen Fixation Research Progress*, p. 690 (eds. H. J. Evans, P. J. Bottomley and W. E. Newton). Boston: Martinus Nijhoff.

Fobert, P. R., Roy, N., Nash, J. H. E. and Iyer, V. N. 1991. *Applied and Environmental Microbiology* 57: 1590–1594.

Gibson, A. H. and Harper, J. E. 1985. *Crop Science* 25: 497–501.

Gremaud, M. F. and Harper, J. E. 1989. *Plant Physiology* 89: 169–173.

Halliday, J. 1985. In: *Nitrogen Fixation Research Progress*, pp. 675–681 (eds. H. J. Evans, P. J. Bottomley and W. E. Newton). Boston: Martinus Nijhoff.

Hansen, A. P., Peoples, M. B., Gresshoff, P. M., Atkins, C. A., Pate, J. S. and Carrol, B. J. 1989. *Journal of Experimental Botany* 40: 715–724.

Harbison, J., Hall, B. D., Nielson, R. G. H. and Strong, W. M. 1986. *Australian Journal of Experimental Agriculture* 26: 339–346.

Hardarson, G., Danso, S. K. A., Zapata, F. and Reichardt, K. 1991. *Plant and Soil* 131: 161–168.

Hardarson, G., Zapata, F. and Danso, S. K. A. 1984. *Plant and Soil* 82: 397–405.

Harper, J. E. and Gibson, A. H. 1984. *Crop Science* 24: 797–801.

Harris, H. 1989. In: *Farm Resource Program Annual Report*, pp. 137–157. Aleppo, Syria: ICARDA.

Harris, H. 1990. In: *Farm Resource Program Annual Report*, pp. 28–40. Aleppo, Syria: ICARDA.

Heichel, G. H., Barnes, D. K., Vance, C. P. and Sheaffer, C. C. 1989. *Journal of Production Agriculture* 2: 24–32.

Herridge, D. F, Bergersen, F. J. and Peoples, M. B. 1990. *Plant Physiology* 93: 708–716.

Herridge, D. F and Betts, J. H. 1985. In: *Nitrogen Fixation Research Progress*, p. 32 (eds. H. J. Evans, P. J. Bottomley and W. E. Newton). Boston: Martinus Nijhoff.

Herridge, D. F. and Betts, J. H. 1988. *Plant and Soil* 110: 129–135.

Herridge, D. F. and Peoples, M. B. 1990. *Plant Physiology* 93: 495–503.

Herridge, D. F., Roughley, R. J. and Brockwell, J. 1984. *Australian Journal of Agricultural Research* 35: 149–161.

Herridge, D. F., Roughley, R. J. and Brockwell, J. 1987. *Australian Journal of Agricultural Research* 38: 75–82.

Herridge, D. F., Sudin, M. N., Pate, J. S. and Peoples, M. B. 1988. In: *World Crops: Cool Season Food Legumes*, pp. 1001–1012 (ed. R. J. Summerfield). London: Kluwer Academic Publishers.

Holl, J. 1975. *Euphytica* 24: 767–770.

Howieson, J. G. and Ewing, M. A. 1986. *Australian Journal of Agricultural Research* 37: 55–64.

ICRISAT (International Crops Research Institute for the Semi-Arid Tropics). 1982. *1981 Annual Report*. Hyderabad: ICRISAT.

Jacobsen, E. and Feenstra, W. J. 1984. *Plant Science Letters* 33: 337–344.

Jensen, E. S. 1986. *Plant and Soil* 92: 3–13.

Jessop, R. S. and Mahoney, J. 1985. *Journal of Agricultural Science, Cambridge* 105: 231–236.

Kueneman, E. A., Root, W. R., Dashiell, W. E. and Hohenberg, J. 1984. *Plant and Soil* 82: 387–396.

Kumar Rao, J. V. D. K. and Dart, P. J. 1987. *Plant and Soil* 99: 255–266.

Ledgard, S. F. and Peoples, M. B. 1988. In: *Advances in Nitrogen Cycling in Agricultural Ecosystems*, pp. 351–367. (ed. J. R. Wilson). Wallingford: CAB International.

Lie, T. A. 1978. *Annals of Applied Biology* 88: 462–465.

Lie, T. A. 1981. *Plant and Soil* 61: 125–134.

Lie, T. A., Pijnenborg, J. and Timmermans, P. C. J. M. 1988. In: *Nitrogen Fixation by Legumes in Mediterranean Agriculture*, pp. 93–100 (eds. D. P. Beck and L. A. Materon). Dordrecht: Martinus Nijhoff.

Marcellos, H. 1984. *Journal of Australian Institute Agricultural Science* 50: 111–113.

Matar, A. E., Beck, D. P., Pala, M. and Garabet, S. 1989. *Farm Resource Program Annual Report* pp. 89–93. Aleppo, Syria: ICARDA.

Matar, A. E., Beck, D. P., Pala, M. and Garabet, S. 1991. *Communications Soil Plant Analysis* 22: 23–36.

McClure, P. R., Israel, D. W. and Volk, R. J. 1980. *Plant Physiology* 66: 720–725.

Mytton, L. R. 1983. In: *The Physiology, Genetics and Nodulation of Temperate Legumes*, pp. 373–393 (eds. D. G. Jones and D. R. Davies). London: Pitman.

Mytton, L. R., El-Sherbeeny, M. H. and Lawes, D. A. 1977. *Euphytica* 26: 785–591.

Nangju, D. 1980. *Agronomy Journal* 72: 403–406.

Papastylianou, I. 1988. In: *Nitrogen Fixation by Legumes in Mediterranean Agriculture*, pp. 55–64 (eds. D. P. Beck and L. A. Materon). Dordrecht: Martinus Nijhoff.

Park, S. J. and Buttery, B. R. 1988. *Canadian Journal of Plant Science* 68: 199–202.

Peoples, M. B., Faizah, A. W., Rerkasem, B. and Herridge, D. F. 1989. *Methods for Evaluating Nitrogen Fixation by Nodulated Legumes in the Field*. Canberra: ACIAR Monograph No. 11.

Peoples, M. B. and Herridge, D. F. 1990. *Advances in Agronomy* 44: 155–223.

Peoples, M. B., Sudin, M. Noor and Herridge, D. F. 1987. *Journal of Experimental Botany* 38: 567–579.

Pereira, P. A. A., Burris, R. H. and Bliss, F. A. 1989. *Plant and Soil* 120: 171–179.

Phillips, D. A., Torrey, J. G. and Burris, R. H. 1971. *Science* 174: 169–171.

Rawsthorne, S., Hadley, P., Roberts, E. H. and Summerfield, R. J. 1985. *Plant and Soil* 83: 279–293.

Reeves, T. J., Ellington, A. and Brooke, H. D. 1984. *Australian Journal of Experimental Agriculture Animal Husbandry* 24: 595–600.

Rennie, R. J. and Dubetz, S. 1986. *Agronomy Journal* 78: 654–660.

Rerkasem, B., Rerkasem, K., Peoples, M. B., Herridge, D. F. and Bergersen, F. J. 1988. *Plant and Soil* 108: 125–135.

Richards, J. E. and Soper, R. J. 1979. *Agronomy Journal* 71: 807–811.

Rosendahl, L., Vance, C. P., Miller, S. S. and Jacobsen, E. 1989. *Physiologia Plantarum* 77: 606–612.

Shearer, G. and Kohl, D. H. 1986. *Australian Journal of Plant Physiology* 13: 699–756.

Singleton, P. W. and Tavares, J. W. 1986. *Applied Environmental Microbiology* 51: 1013–1018.

Smith, S. C., Bezdicek, D. F., Turco, R. F. and Cheng, H. H. 1987. *Plant and Soil* 97: 3–13.

Somasegaran, P. and Bohlool, B. B. 1990. *Applied Environmental Microbiology* 56: 3298–3303.

Stanford, G. and Smith, S. J. 1972. *Soil Science Society America Proceedings* 38: 103–107.

Strong, W. M., Harbison, J., Nielsen, R. G. H., Hall, B. D. and Best, E. K. 1986a. *Australian Journal of Experimental Agriculture* 26: 353–359.

Strong, W. M., Harbison, J., Nielsen, R. G. H., Hall, B. D. and Best, E. K. 1986b. *Australian Journal of Experimental Agriculture* 26: 347–351.

Thies, J. E., Singleton, P. W. and Bohlool, B. B. 1991. *Applied Environmental Microbiology* 57: 19–28.

Turner, G. L. and Gibson, A. H. 1980. In: *Methods for Evaluating Biological Nitrogen Fixation*, pp. 111–138 (ed. F. J. Bergersen). Chichester: John Wiley.

Vance, C. P., Egli, M. A., Griffith, S. M. and Miller, S. S. 1988. *Plant, Cell Environment* 11: 413–427.

Vincent, J. M. 1965. In: *Soil Nitrogen*, pp. 384–435 (eds. M. V. Bartholomew and F. E. Clark). Madison, Wisconsin, USA: American Society Agronomy.

Wery, J., Deschamps, M. and Leger-Cresson, N. 1988. In: *Nitrogen Fixation by Legumes in Mediterranean Agriculture*, pp. 287–301 (eds. D. P. Beck and L. A. Materon). Dordrecht: Martinus Nijhoff.

Witty, J. F. 1983. *Soil Biology Biochemistry* 15: 631–639.

Witty, J. F. and Minchin, F. R. 1988. In: *Nitrogen Fixation by Legumes in Mediterranean Agriculture*, pp. 331–344 (eds. D. P. Beck and L. A. Materon). Dordrecht: Martinus Nijhoff.

Witty, J. F., Rennie, R. J. and Atkins, C. A. 1988. In: *World Crops: Cool Season Food Legumes*, pp. 715–730 (ed. R. J. Summerfield). London: Kluwer Academic Publishers.

Wolyn, D. J., St. Clair, D. A., DuBois, J., Rosas, J. C., Burris, R. H. and Bliss, F. A. 1991. *Plant and Soil* 138: 303–311.

Wu, S. and Harper, J. E. 1991. *Crop Science* 31: 1233–1240.

Infrastructural support

Provisions for agronomic inputs for cool season food legumes in some developing countries (discussion session)

A.J.G. VAN GASTEL[1], P.N. BAHL[2], H. FAKI[3], P. PLANCQUAERT[4], A.M. NASSIB[5] and B.A. SNOBAR[6]

[1] ICARDA, P. O. Box 5466, Aleppo, Syria;
[2] Indian Council of Agricultural Research, Krishi Bhawn, Dr. Rajendra Prasad Road, New Delhi 110 001, India;
[3] Agricultural Research Corporation, Wad Medani, Sudan;
[4] Institute Technique des Céréales et des Fourrages (ITCF), 8 Avenue du Président Wilson, 75116 Paris, France;
[5] Agricultural Research Center, Technology Transfer, National Agricultural Research Project, Field Crops Research Institute, P. O. Box 12619, Giza, Egypt, and
[6] Faculty of Agriculture, University of Jordan, Amman, Jordan

Introduction

Agronomic inputs are a key factor for the successful introduction of crops and cultivars. In many cases genetically improved cultivars are available but the genetic potential of the cultivars cannot be attained because farmers do not have access to the necessary inputs. By using the correct agronomic practices and inputs a significant improvement of profits can be obtained.

Several agronomic inputs are recommended to maximize yield and net profit from the production of cool season food legumes. These inputs include seed of improved cultivars, fertilizers, herbicides, and equipment for land and seedbed preparation, spraying, seeding, harvesting, threshing, etc.

Not only these inputs but also processing equipment and an efficient distribution system that is able to reach farmers in all areas is essential. If inputs are to be adopted by farmers, they have to be made available to farmers in their villages at reasonable cost and at the right time. The quality and quantities of the products should be geared towards the farming systems used.

Whose responsibility should it be to provide inputs; the public or the private sector? This is a matter of controversy. It can be argued that it is the responsibility of the public sector since it conducts research and supports the extension services and should thus have a keen interest in ensuring that the fruits of their research efforts reach and are used by the farming communities. In many developing countries, the introduction of improved technology in the agricultural sector, is carried out through public activities starting from conducting research to distribution of inputs and price subsidies. However, efficiency of the public sector is often questioned and the private sector, which is usually more efficient and effective in distribution and providing extension services to farmers, becomes involved. Because private sector operations are

F.J. Muehlbauer and W.J. Kaiser (eds.), Expanding the Production and Use of Cool Season Food Legumes, 495–503.
© 1994 Kluwer Academic Publishers.

governed by profitability, public sector has to ensure that inputs are made available to all farmers.

With regard to handling of farm products, it is important that the losses are minimized. In order to minimize losses and maximize profits, the produce should be properly handled, processed and stored. Quick delivery of the produce to the processing location is important; containers and transportation facilities should be made available to guarantee quick delivery of farm produce to the processor. Processing the product through threshing, cleaning and separating should help to increase the quality of the produce and minimize losses.

Distribution of the processed products to the different markets is also important; this is assured if appropriate transport facilities and a network of distributors is made available.

This paper discusses experiences in India, France, and the Sudan with regard to the provisions of agronomic input and the handling of farm products.

Provision of Agronomic Inputs: India

In the Indian sub-continent, chickpea, lentil, lathyrus, and pea are the major cool season food legumes. They form an indispensable component of dryland farming. Lathyrus is considered to be the most drought tolerant of all. Present levels of agronomic managements of food legumes are more suited for subsistence agriculture. However, agriculture in the Indian sub-continent has now reached a stage where improved agronomic management of dryland farming and post-harvest handling practices of food legumes have become increasingly more important. There appears to be considerable scope for stepping up the yield levels vertically through improved management practices, if resource-poor legume growers are provided with input-support services.

To find out the impact of improved cultivars and agronomic inputs in chickpea, analysis of yield gaps and constraints was undertaken, using data of: (1) all India coordinated trials, (2) mini-kit trials, and (3) state mean (1976–77) and front line demonstrations (1990–91). The difference in mean yields of improved cultivars recorded in the coordinated trials and those obtained in mini-kit trials conducted in farmers' fields was 931 kg ha^{-1} (Gap 1). Improved cultivars outyielded the local cultivars by a margin of 360 kg ha^{-1} when both were tested in farmers' fields (Gap 2). The superiority of improved cultivars is apparent in this gap. The difference in yields of local cultivars in the mini-kit trials and those obtained by the farmers was 501 kg/ha (Gap 3). This gap can be narrowed if improved management is followed by the farmers. Gaps 2 and 3 together, with a ratio of 2.22, indicate the yield potential of the improved cultivars under good management conditions.

To demonstrate the yield potential of improved cultivars that can be achieved, research scientists conducted 642 front line demonstration in farmers' fields during 1990–91. In these demonstrations, conventional farmers' practices in case of local cultivars and recommended package of practices in case of improved cultivars were followed. Taking the average of all demonstrations, improved cultivars yielded 1452 kg ha^{-1} compared to 841 kg ha^{-1} for local cultivars. These data clearly show that yields can be substantially increased if seeds of improved cultivars and agronomic inputs are made available to the farmers.

Agronomic inputs must be in tune with the diverse cropping patterns and agro-ecosystems in which cool season food legume crops are grown. *Lathyrus* and lentil in this regard are considered to be particularly important in the Indian sub-continent. These crops are quite often seeded following a practice, locally known as "Utera" or "Paira". In this practice, which is a kind of relay cropping, the legume crop is raised in rice fields on residual moisture and fertility after the harvest of the rice crop. Provision of appropriate agronomic inputs for such diverse cropping patterns may open up possibilities of extending the area of cultivation of cool season food legumes.

Handling of Farm Products: India

Efforts to increase production of food legumes should run concurrently with reduction of losses during some of the operations like threshing, transport, processing, and storage. Total losses during these operations in India are estimated to be around 10% and out of this about 80% occurs during storage. It has been observed that losses in storage can be reduced substantially if legumes are stored as "dal" (dehulled and split) rather than whole grains which are more prone to attack by storage insect pests.

Post-Harvest Operations

The majority of the small and marginal farmers carry out the post-harvest operations manually by use of traditional processes. Availability of simple and easy to operate post-harvest equipment like mechanical threshers and dryers for community use at the village level can bring in much needed modernization of outdated technology.

Processing and Utilization

In India, more than 75% of the total food legume production is converted into decorticated and split form commonly known as "dal". The conventional dal milling processes result in approximately 20% wastage, though a portion of it is retrieved as dal powder. There is urgent need for modernization of processing machines so that losses during milling could be mitigated.

Marketing, Prices, and Government Policy

Legume growers tend to dispose of marketable surplus food legumes almost immediately after the harvest because they are subject to insect damage during storage. Establishing of small scale dal milling units at the village level can help to reduce storage losses and bring economic benefit to the farmers.

There is considerable variation between the prices at farm gate and prices for the consumers. This paradox must be removed by market intervention by public agencies so that farmers are enthused to take up food legume cultivation, not for subsistence but for commercial purposes.

The available potential for increasing production and productivity of food legumes can be exploited by providing various support measures through government policies, plans, and programs. Recently, the government of India has formulated a Technology Mission on Pulses in which special research and development efforts will be made to accomplish yield and production advances in some of the important food legumes grown in India. A number of departments and agencies will collaborate under the aegis of the Mission through four mini-missions, covering: (1) crop production and protection technology, (2) post-harvest technology, (3) input support services to the farmers, and (4) price support, storage, processing, and marketing.

Provision of Agronomic Inputs: Sudan

The production of cool season food legumes in the Sudan has largely kept its traditional, labor-intensive, cropping approach. Important exceptions are the use of modern diesel pumps and some machinery for arduous cultivation operations. Many important inputs are not available to farmers.

Seeds, irrigation water, and pesticides are the most important inputs. Provision of these inputs is likely to result in a significant increase in productivity. Other important inputs include fertilizers and harvest- and post-harvest equipment.

Seeds

Because farmers have no access to improved cultivars they usually keep their own seeds. Farmers' saved seeds have several disadvantages. Since they are taken from the bulk harvest, the seeds for next seasons planting do not originate from any selection procedure. Moreover, in the absence of seed treatment, stored seeds are attacked and lose viability. Due to high prices prior to sowing, farmers may sell their intact seeds and buy cheaper seeds that are often insect damaged; they believe that viability is not seriously impaired. Higher seeding rates are used to compensate for the lower quality. In bad years there is a risk of seed shortage when farmers sell all their produce because they need cash immediately after harvest. The planting method (broadcasting, followed by ridging), which is widely practiced, also necessitates high seeding rates, but the method is not labor demanding.

The provision of good seed from improved cultivars will increase productivity substantially. Some recommended cultivars are available and yield increases due to improved cultivars are about 15% for faba bean and range from 20 to 30% for chickpea.

Seed multiplication of improved cultivars is slow and seed multiplication institutions should be strengthened to achieve a reliable seed supply. The seed industry should ensure a guaranteed seed supply to farmers through seed exchange or credit at the time of sowing.

Improving farmers' capabilities to manage their own seed supply, through seed cleaning, treatment, and good storage is another option which will reduce the risk for farmers.

Irrigation water

Irrigation water is the most important input in Northern Sudan. Expensive pumping practices are used and water availability at pumping sites is unreliable. Depending on locality and season, the amount of irrigation water has a large effect on crop yields. One irrigation may increase faba bean yield by an average of 288 kg ha^{-1} (calculated from on-farm data in Northern Sudan), which was equivalent to gross benefits of 10,000 Sudanese Pounds ha^{-1}. This amount far exceeds the flat water rates levied in the area. The yield of chickpea was found to decrease by 34% if watering intervals are increased from 7 to 14 days, while a 21-day interval leads to a 72% yield reduction.

Farmers in many areas operate small pumps which can provide water for approximately 0.75 ha d^{-1}. With about 8 ha per scheme, the average irrigation intervals would be 12 days. There are also crops that demand more frequent irrigation (onions, fruits, spices, and alfalfa) and maintenance, equipment break-downs, and delay in fuel provision result in prolonged watering intervals.

The allocation of water is expected to follow the level of marginal returns to irrigation water. With limited capacity for water provision in relation to land, the marginal value productivity of the latter is higher than that of irrigation water and it will be more profitable to irrigate more acreage than to maximize returns per unit of water applied. Farmers currently apply around six waterings to faba bean as compared to 11 recommended irrigations. The key action is to increase water supply to a level at which the marginal value productivity of water exceeds that of land.

More area would be grown, if provision of fuel, oil, and spare parts was better organized and if expansion of the present pump size was encouraged by providing credit. The last season has witnessed large strides in the supply of credit facilities through the Agricultural Bank of Sudan.

A continuation of this policy coupled with the availability of the required inputs will lead to a reduction of the problems regarding the pumping of water. However, water availability at pumping sites will still continue to be a problem and calls for technology development to improve pumping as the river subsides.

Pesticides

The most important pests are *Helicoverpa* and aphids in faba bean, pod borers in chickpea, and birds in lentil. Large losses were incurred by *Heliothis* in the 1984–85 season in the Nile province when yield reductions between 20 to 45% were recorded. Chemical treatment is the most reliable method for control. However, chemicals and spraying equipment are in short supply and an increased farmers' know-how about the use of chemicals is necessary. Farmers regard pest control as an important management practice but hazardous application and misuse of chemicals is often observed. The sale of chemicals should be confined to approved chemicals; legislation is necessary.

Other important inputs are fertilizers, such as nitrogen and phosphorus, and seed inoculants and decortication equipment for lentil.

In general, the provision of inputs will have limited benefits if the introduction is not associated with an efficient extension program.

Product Handling: Sudan

The handling of products has the following characteristics. The cool season food legumes are predominantly used in urban areas. The direction of trade is one-way; from the production sites in the North to other parts of the country. Crops are usually disposed with hardly any processing; storage is seasonal and mainly practiced by traders.

Transport problems, deterioration during storage, lack of product grading, numerous traders between producers and consumers, limited farmers' access to markets and market information, and farmers' need for cash are all serious constraints facing cool season food legumes in Sudan.

Infrastructural problems and shortage of transport facilities lead to high prices of cool season food legume products for consumers. Products have to be moved over long distances. An improved railroad network would lead to a reduction in marketing costs, which may be reflected in lower prices to consumers and higher prices to producers. Transport improvement will also facilitate easy movement of inputs to producing areas.

Crop processing affects the quality and use of the product. This is especially important for lentil. Furthermore, the narrow use of faba bean and chickpea in terms of recipe preparation, in addition to their high prices, limits their use in rural areas where income is low and consumption habits different.

Storage losses need to be addressed through provision of the required inputs and extension services.

Mixing of various grades of the same product, especially in faba bean, reduces the efficiency of price formation and does not provide the right price signals to producers. This affects resource allocation. Some sorting based on storage pest damage is carried out for chickpea.

Agronomic Inputs: France

Several research projects have been undertaken by ITCF and INRA in France. These projects emphasize the reduction of production costs while maintaining an acceptable yield, germination and nutritional value and on handling of grains to reduce losses and damage.

Operational costs vary greatly from farm to farm and region to region. In spring peas, because of high 1,000-grain weight and sowing density used, the seed alone can represent between 40 and 50% of the total operating costs. In the case of winter peas, handing is less costly, but costs for plant protection purposes are higher (Table 1).

Table 1. Operating costs for pea in French francs ha^{-1} (ITCF/UNIP 1991)

Inputs	Spring Peas	Winter Peas
Seed	1,200	700
Fertilizers	500	500
Herbicides	450	500
Fungicides	350	500
Insecticides	100	100
	2,600	2,300

Fertilizer treatments (P, K, Mg, and Ca) represent the second highest expenditure and insecticides, which are not costly, are often used by farmers as an insurance.

Seeds

Efforts are made to reduce the costs through reductions in seeding rates. Peas sown with precision drills rarely give better yields than those sown with conventional drills. A precision drill allows seed rate reductions but, unfortunately, only 7% of farmers in France use precision drills. Moreover, seeding rates may be adjusted as a function of the sowing date, and the type of cultivar i.e., the later the sowing date, the higher the required seeding rate. Research on the optimum rate for specific cultivars is ongoing.

Bearing the cost of seed in mind, results of experiments on the yield/plant density relationship have been used to calculate gross margins. The optimum gross margins are obtained when slightly lower planting densities than those required to maximize yield are used.

Herbicides and Fungicides

Preemergence post-planting herbicide application, using products based on toluidines (Treplik S, Tribunil, Winner), results in the best weed control. When *Galium* or *Lolium* is present other herbicides are used (Challenge 600 and Dinograne SP). Since several problems with preemergence treatments have been envisaged, some post-emergence herbicides (Vulkan T and Dribble) are often used.

Because fungicides have a preventive action rather than a curative one, treatment for *Ascochyta* and *Botrytis* have to be carried out before an infection is detected. Research into the mode of action of fungicides has been initiated; the objective is to discover the cause of poor action against principal diseases.

There is also a cultivar x fungicide interaction. Under high disease pressure differences between treated and non-treated crops can vary from 15 (a yield difference of up to 1.1 t ha^{-1}) to 25% (2.0 t ha^{-1}). Research into cultivar $\times$ treatment interaction is pursued to obtain more effective treatment.

Handling of Farm Products: France

In most regions grains can be harvested sufficiently dry (14 to 15% moisture content). However, it is recommended to cut the crop at 17 to 20% moisture to reduce cracking and shedding.

Artificial Lodging

Artificial lodging is used to simplify harvesting. Naturally, a pea crop lodges in one direction and to avoid losses the combine cuts the crop in one direction. Lodging the crop artificially enables the combine to work up and down, thus reducing harvesting time. Artificial lodging, when carried out correctly, has no effect on yield. To avoid yield losses, artificial lodging must be carried out before the crop lodges naturally, i.e., not later than 10 d after the beginning of flowering. Artificial lodging is carried out using a tube fitted beneath the boom of a sprayer or with a metal tube (round cross-section) which runs on wheels and is mounted either in front of or behind the tractor. The crop only needs to be slightly inclined; not more than 10 cm below crop height to avoid plant damage.

Harvesting, Handling, and Cleaning

All harvesting operations should be carried in such a way that losses are minimized both in quantity, and in the case of peas for seed, quality. Yield loss prevention starts at planting and continues up to harvest. Essential are a well-leveled sowing bed, careful crop management, proper cultivar choice, etc. Furthermore, combine setting is important. Cutter bars should be equipped with crop lifters.

Quality losses mainly result from damage inflicted on seed because the seed of pea is very fragile. The threshing mechanism needs to be carefully set and seed should not fall over long distances. When emptying the grain tank, reducing height, and using flow sleeves and rubber bottoms in tank and trailers is recommended.

Handling is often done with screw-type devices, using large diameter screws and reduced rotation speed. Bucket-type elevators cause less damage than screw-type elevators. It is desirable to reduce speed and height from which the peas may fall. Peas must be cleaned before storage because broken grains, green grains, and clay can block ventilation and interfere with storage.

Storage, Conservation

Drying peas harvested at 14 to 15% moisture content requires the same rules as with cereal crops.

If peas are harvested at a moisture content between 16 and 17%, ventilation must start immediately after harvest and continue day and night until the moisture has fallen to 15%. When moisture content is between 17 and 20% drying with heated dry air must be undertaken; above 20% moisture, it is recommended to use a commercial drying unit.

Drying management also depends on the use of the crop. For food it is necessary to dry at temperatures below 90°C in order not to increase the fragility of the grains, nor to alter the food value. For seed, peas should be dried more slowly. It is advisable to dry with warm air having a relative humidity below 70% and with air flow above 100 m^3 h^{-1} m^{-3} of grain.

Infrastructural support to promote farmer adoption of improved technologies

C.L.L. GOWDA[1], D.G. FARIS[1] and A.F.M. MANIRUZZAMAN[2]
[1] *ICRISAT, Legume Program, Patancheru P. O., Andhra Pradesh 502 324, India, and*
[2] *Director Research, Bangladesh Agricultural Research Institute (BARI), Joydebpur, Bangladesh*

Abstract

Adoption of technology is faster with resourceful farmers in the more productive and homogenous core areas with irrigation or assured rainfall. However, the pace of adoption is slower with the resource-poor farmers in less productive and diverse hinterlands. A decentralized research strategy is needed to develop technology suitable to diverse conditions of hinterlands. Use of farmers' indigenous knowledge and their involvement in developing and testing of technology to suit local conditions is essential. Hence, farmer-participatory on-farm research constitutes the foundation for technology adoption. Environmental factors influence adoption, but are beyond our control. Therefore, the infrastructure for adoption will be comprised of technological, institutional, socioeconomic, and human factors. Case studies from Bangladesh are given to illustrate the beneficial effects of some of the infrastructural support systems on technology adoption.

Introduction

Technology is defined as a "scientific method of achieving a practical purpose". In agriculture, improved technology usually refers to one or more crop management practices that are an improvement over traditional practices, and can produce higher yields and profits, while maintaining stability of production. Usually, most improvements are first developed in research stations, and then tested for applicability in farmers' fields. Beneficial sets of technologies are then "adopted" by farmers.

Technology transfer has been particularly successful in the green revolution era, in the highly productive and homogeneous areas called "core areas" where production is guaranteed by irrigation or assured rainfall. The physical environment in core areas is similar to that at research stations, and hence the new technologies from research stations were easily duplicated by the resource endowed farmers in these regions. At the same time adoption of technology has

F.J. Muehlbauer and W.J. Kaiser (eds.), Expanding the Production and Use of Cool Season Food Legumes, 504–516.
© 1994 *Kluwer Academic Publishers.*

been slow in the more diverse, less productive, and more heterogeneous areas called "hinterlands" (Rambo and Sajise, 1985). The highly diverse conditions and the limited resources of farmers in the hinterlands reduced the chances of applicability of improved technologies developed on well endowed research stations. This means that a single comprehensive package of improved technologies can be designed to fit the core areas, while no single package can possibly be applicable to the diverse hinterlands. In addition, resource endowed farmers in core areas could influence the research strategies to their benefit while resource poor farmers had no means of expressing their demand for suitable technology, and consequently their needs were not addressed adequately (Chambers and Jiggins, 1986; Farrington, 1988; Fujisaka, 1989; Merrill-Sands *et al.*, 1989). This paper considers the various factors responsible for technology adoption, and suggests the infrastructures needed to hasten adoption.

Farmer participation in technology adoption

The "top-down" approach of transfer of technology (TOT) was successful in the more fertile and homogeneous areas, but not the diverse, less productive and risk-prone areas (Chambers and Jiggins, 1986). In the early 1960s farmers' ignorance and the inability of the extension service to educate farmers were cited as reasons for non-adoption of technology. In the 1970s, socioeconomic factors were advanced as the main reasons for slow adoption. In the 1980s, a more challenging interpretation was put forward which stated that the problem was not the farmer, nor the farm, but the technology itself and the priorities and processes of its generation (Chambers *et al.*, 1989; Toulmin and Chambers, 1990). The farmer-participatory approach envisages that new technology should be technically feasible, economically viable, socially acceptable, and environmentally safe and sustainable for use by the small, resource poor farmers (Batugal *et al.*, 1985). In the hinterland the agroecological complexities and socio-cultural diversities of each area require the technology to be tested and modified (adapted) to suit individual farmer or farm situations. The centrality of the farmer in the whole process becomes vital and this leads to "demand driven pull" for technology in contrast to the "technology push" in the TOT approach. Farmers have been increasingly recognized as sources of indigenous knowledge and technology and hence must be involved in the technology adoption research process (Chambers *et al.*, 1989). This involves assessing the farmer's problems, priorities, and indigenous knowledge; and use with available scientific knowledge to address identified problems – the "bottom up" approach (Norman, 1980).

On-farm adaptive research should be farmer-participatory research (Farrington and Martin, 1987) to develop a basket of technology options from which the farmer can choose practices useful and affordable (Toulmin and Chambers, 1990). As aptly put by Raintree and Hoskins (1988), "both scientists and farmers have unique areas of expertise which collectively produce a better basis for development than either alone". Therefore, for the purpose of this paper, we will consider the farmer-participatory approach in technology adoption as the foundation to build the infrastructure needed to facilitate adoption.

Infrastructural Support for Adoption of Technology

Adoption of technology is a continuous process and takes places at varying degrees depending on the crop, technology, and the locations where adoption is occurring (Feder *et al.*, 1981). Grigg (1984) reported that during the 18th century new technologies in Europe spread at about a mile (1.6 km) each year. In the 20th century with improved transport facilities and communication links we would expect a faster adoption rate. However, several studies on innovation adoption among farmers have shown that there is a considerable lag period (5 to 8 years) between the first adoption and majority adoption (Jones, 1967).

Infrastructure refers to the "basic framework of a system or organization". We have already indicated that the farmer-participatory approach of technology adoption forms the foundation to build the infrastructure. In the present context we will consider infrastructure to be the organizational structures and policies needed to facilitate adoption of improved technology. In this paper we will attempt to ascertain the factors responsible for the lag period in adoption and suggest ways to improve the conditions to hasten adoption.

Factors Affecting Adoption of Technology

Farmers make choices to exploit available technology based on environment and the resources. Farmers who make correct choices will survive and prosper, and over time their successful production practices will become institutionalized (Rambo, 1983). A knowledge of the factors responsible for adoption of technology will enable agricultural scientists and planners to identify strategies to overcome some of the bottlenecks slowing adoption. These factors can be grouped into: environmental, technological, socioeconomic, institutional, and human.

Environmental or climatic factors include topography, soil, rainfall, temperature, and natural hazards. The resource poor farmer has little, if any, control over these factors even though they have a major influence on adoption in diverse, risk-prone, and unpredictable environments (Chambers and Ghildyal, 1985; Chambers and Jiggins, 1986; Rambo and Sajise, 1985). Despite their importance and although they must be kept in mind we will not discuss environmental factors as they go beyond the scope of this paper. We will discuss the technological, socioeconomic, institutional, and human factors as they form the framework for the infrastructural support.

Technological Factors

Technology development is a continuous process, and is essential for success of any adoption process. Components of technology include:

Indigenous Knowledge Base

Farmers are always experimenting and making small changes in the course of their normal agricultural activities. Farmers therefore naturally can contribute their indigenous knowledge in planning on-farm trials designed for adoption of improved technology in their area. Incorporation of results, experiences, and suggestions from these farmers will make improved technology more responsive to local conditions and also gain tacit support of the farming community who feel they own the new technology (Maurya, 1989; Toulmin and Chambers, 1990).

Research Support

Each country should have a dynamic backup research program to hasten and assist the adoption process by supporting (1) technology generation, (2) technology adaptation, (3) technology verification, and (4) technology dissemination. All these four stages of technology adoption are interlinked and need a concerted effort by scientists to provide the research backup. Countries with a poorly developed research program are likely to take more time for adoption than countries with a strong research base.

Systems Approach

Integration of disciplinary research and commodity research into a systems approach is a prerequisite to ensure the success of a technology adoption program. Scientists in individual disciplines tend to emphasize the importance of their own area of research, and are sometimes intolerant of the importance of other disciplines. Establishing a systems perspective can eliminate these biases and allow scientists to work as a team. Integration of on-station and on-farm research improves the capacity to respond to the client group, i.e., the resource poor farmer (Merrill-Sands and McAllister, 1989; Holden and Joseph, 1991). Policy changes (as discussed elsewhere in the paper) by governments are needed to get the systems perspective incorporated into the existing system.

Monitoring and Evaluation

Monitoring and evaluating the adoption process ensures that field staff and extension personnel get moral support, and provides the direction needed for making adjustments. Evaluation provides feedback that can be used to modify the strategy or technology so that they better suit the farmers' conditions.

Information Exchange

A strong information network can help in the exchange of knowledge and proven technologies between farmers, researchers, and extension staff in a region, and in other regions (Biggs and Farrington, 1991).

Socioeconomic Factors

Studies on technology adoption have emphasized the importance of socioeconomic factors. An apparently profitable technology may be rejected because of high input costs or because of increased risks associated with the new technology (Rambo, 1983). Following are some socioeconomic considerations:

Resources

Resource-poor farmers must make difficult decisions about how they allocate their resources which include land, labor, draft-power, and equipment. Security of tenure or land ownership is a necessary pre-condition before farmers feel safe to invest in long term management practices needed for adopting many new technologies (Chambers *et al.*, 1989) because the share of profit to a tenant is low. Most improved technologies require added labor, which may conflict with other essential farm operations. Similarly, the draft power required may be more than the farmer can provide. In such cases technology must be designed to reallocate labor and draft power efficiently. Equipment (planter, sprayer, etc.) needed for new technologies can be made available through policies which enable the hiring or cooperative ownership of equipment (Duwayri *et al.*, 1988; Grenoble *et al.*, 1990).

Inputs

Inputs such as seeds, fertilizers, pesticides, and fungicides must be available for many improved technologies (Dei, 1981; Feder *et al.*, 1981; Chambers and Ghildyal, 1985; Chitnis and Bhilegaonkar, 1987; Heinrich, 1991). In such cases non-availability and high cost of inputs can slow down their adoption. Necessary inputs can be made available when needed and at reasonable prices through government organized agriculture input and service organizations. Farmer cooperatives and non-government organizations (NGO) have arranged the supply of essential inputs which has lead to success in adoption of technologies dependent on inputs (Montemayor, 1987). Quality of inputs (e.g., genuineness of pesticides) should be guaranteed for the success of a technology dependent on the input.

Finances

Cash for inputs, equipment, machinery, and other services, beyond the resources of the farmer can be a major bottleneck to adoption (Conteh, 1986; Chitnis and Bhilegaonkar, 1987; Cruz, 1987; Duwayri *et al.*, 1988). Low-rate credit facilities through banks, cooperatives, and financial institutions can help farmers buy the needed inputs (Krause *et al.*, 1990). Repayment should be made easy. Subsidizing input costs is practiced in some countries but is not favored in others because of the negative effects it can have when the subsidy is withdrawn.

Markets

Market intervention can help speed up adoption (Miller and Trolley, 1989). Encouraging small and large scale industries to increase the demand for a product has accelerated adoption of technology associated with that product. Support price, guaranteed procurement, and other government policies that ensure remunerative prices to farmers can provide strong motivation for adoption (Cruz, 1987). This must be coupled with the establishment of rural markets and a good network of roads to ensure accessibility to markets. For example, in Nepal the area under groundnut has not increased despite heavy demand for groundnut oil because of poor roads and few markets for selling the crop.

Insurance

Risk and uncertainty substantially affect farmers' decisions to adopt improved technology (Feder *et al.*, 1981; Nygaard and Basheer, 1981; Kelley and Walker, 1991). Hence, reducing risk through crop insurance schemes can provide the needed moral support for farmers to accept a new technology (Knight *et al.*, 1989).

Institutional Factors

Institutional factors can have a greater effect than resource constraints on technology adoption (Merrill-Sands *et al.*, 1989). These are discussed below:

Policy

Success of technology adoption in a country depends largely on the policies adopted by the government. Some of these have been discussed above under technology and socioeconomic factors. Senior administrators and policy makers should be knowledgeable about on-farm adaptive research, show commitment towards such projects, and provide leadership and support through suitable policies. A national coordination committee or council to plan, coordinate, and provide guidance has been found effective (Merrill-Sands and McAllister, 1988; Axinn, 1991). Farmer-participatory research and farming systems perspectives need to be emphasized in formal agricultural education (Axinn, 1991).

Resources

Allocation of resources for on-farm research can facilitate adoption. Funds also are needed to recruit new staff, provide them with vehicles and fuel to visit and monitor trials to ensure that farmers are provided with timely advice and assistance in adopting the technology. Often this includes training new staff and providing new facilities in areas targeted for adoption. Sustained involvement of social scientists at all stages of adoption brings a social science perspective (Merrill-Sands *et al.*, 1989).

On-Farm, On-Station Links

There can be conflicts of interests between on-farm and on-station researchers. To prevent these conflicts, opportunities for collegial interaction should be given to allow frank discussion and free flow of information (Axinn, 1991). The on-farm researcher who is posted in remote areas can feel neglected, while the on-station researcher can feel superior. Joint planning and allocation of responsibilities can lead to a feeling of mutual success (Merrill-Sands *et al.*, 1989). A system of monetary incentives can be set up to reward those that use the farmer participatory adoption process (Axinn, 1991).

Technology Adoption System

Among countries there is a wide variety of systems for technology transfer and adoption. Most countries have an extension service, and some have on-farm research groups. The basic objective is to assist farmers to increase their agricultural production and income. The World Bank has financed the training and visit system of extension workers in about 40 countries. This system has been effective by increasing the farmer orientation of research through feedback to the system (Benor, 1987). However, the links among the different agencies need to be strengthened for the agencies to be effective in serving the farmers. McDermott (1987) suggests that there should be no clear distinction between extension and research, and that extension should provide the technical liaison between the farmer and researcher needed to support adoption activities. Some reorganization and reorientation of policies and procedures are needed. For example, in countries that have used cooperative farming, a sudden shift to individual farming can lead to unclear job goals for the extension staff. Formerly the extension system was paid by the cooperatives they were helping. With the new system such payments no longer exist, and the expected goals of extension system become unclear.

Authority

Decentralization of power and delegation of decision making can strengthen the effectiveness of adaptive on-farm research (Axinn, 1991). A rigid and inflexible bureaucratic setup can negatively effect the adoption processes because scientists or extension staff located in remote villages cannot always wait for decisions to be made by the headquarters (Merrill-Sands *et al.*, 1989). Although the local staff is usually better able to judge the situation, and make appropriate and timely decisions, such decentralization sometimes attracts local political pressures, and can lead to disruption of planned programs. Effective mechanisms to link with NGO's and farmers' organizations can also help the adoption process (Axinn, 1991).

Communication

Better communication links to assist in contacts between field staff and senior administrators can assist the adoption process.

Feedback

An effective feedback system from farmers, extension staff, and on-farm research scientists can help develop a meaningful research agenda to sustain the adoption process. Direct interaction between the farmer and scientist is better than indirect contact through the extension service (Chambers and Jiggins, 1986). Feedback systems need strong support from senior management and institutions (Merrill-Sands and McAllister, 1988; Axinn, 1991).

Human Interactions

The human interactions are basic to technology adoption. Those in the technology adoption process – the research scientist, the extension worker, and the farmer – should interact as joint partners. The level of interaction and linkage affect the adoption process. Collegial researcher-farmer interactions can improve adoption (Lightfoot *et al.*, 1989). To ensure that new technology is appropriate, scientists should avoid the resource-rich farmer and work with the representative groups of resource-poor farmers (Chambers and Jiggins, 1986). However, resource-rich farmers are in a better position to test and adopt new technologies which means that they could act as "models" to disseminate the new technologies (Feder *et al.*, 1981). Involving farmers' groups and rural leaders in the decision to adopt a technology can provide a firm basis for the adoption of that technology (Kean, 1988; Norman *et al.*, 1988). Farmer to farmer interaction then becomes important (Dequito and Abansi, 1987; Chambers *et al.*, 1989).

Sustaining the farmer's involvement in the adoption process helps research address the needs of the resource-poor farmer. For example, identification of priorities is an interactive process and can be best accomplished by continuous interchange between researcher and farmer (Merrill-Sands *et al.*, 1989). Communication between scientists and farmers is an art requiring an expertise which many biological scientists do not have (Rhoades *et al.*, 1985). Experience indicates that maintaining the involvement of social scientists improves communication between farmers and research scientists. Another approach is for national programs to train their scientists to have a social scientist's perspective.

The role that women play in participatory on-farm research and technology adoption has been overlooked. Most technology adoption programs have a male-bias, although several activities involve women. The gender issue becomes particularly critical when the decision maker or user of technology is not consulted. Recent studies have re-emphasized the important role women can have in the adoption of technology (Chambers *et al.*, 1989; Axinn, 1991).

All the factors discussed above affect the pace of technology adoption. Therefore, the infrastructure needed must be designed to facilitate the removal of bottlenecks, and provide for the congenial conditions necessary for rapid adoption. Many of these are interrelated, and sometimes interdependent. A marginal positive change in any factor can hasten the adoption process substantially.

Technology Adoption in Bangladesh

Case studies from Bangladesh are given below to illustrate how technology adoption can be successfully achieved. The examples given below illustrate the positive effects of some of the infrastructural systems described above in enhancing adoption.

Importance of Indigenous Knowledge

The Bangladesh Innovative Farmers' Workshop (BIFW) is an example of the collegiate participation where scientists work with farmers to strengthen farmers' own informal research and development systems. The methods used to identify farmer's innovation include field visits, unstructured surveys, farmers' meetings, and field days. The BIFW organized by ERP (Extension and Research Project-the forerunner of On-Farm Research Division [OFRD], of the Bangladesh Agricultural Research Institute [BARI]) met to learn about the value of new processes, spread new technological innovations widely among potential users, and plan further research for refining these technologies. Started in 1982, with farmers as resource persons, these workshops are now common. So far a total of 43 innovations have been presented. These workshops have helped the ERP improve its research agenda and also provided a forum for informal exchange of information. For example, farmers participating in the Innovative Watermelon Farmers' Workshop provided valuable ideas for the quick sprouting of watermelon seeds. In the Innovative Wheat Farmers' Workshop scientists were told about relay cropping of wheat with transplanted aman rice (main rice crop – which is usually harvested during the first half of December), which helped OFRD scientists design experiments for wheat production under minimum tillage (Jabbar and Abedin, 1989). This reversal of the roles that research and extension workers normally have with farmers has encouraged interaction and improved the relation between the participants, thereby strengthening the technology adoption process. This arrangement has also been an effective means to hasten the transfer process.

Successful On-Farm Research Program in Bangladesh: Homestead Vegetable Gardening

Homesteads occupy about 5% of the total cultivated land of Bangladesh. About 56% of the households have around 400 m^2 of land. Although there are about 13 million homesteads in rural Bangladesh, the consumption of vegetables is only 28 grams per person per day. As a result, an estimated 30,000 children are becoming blind due to vitamin A deficiency each year. If homestead areas could be properly utilized, vegetable production would increase and malnutrition decrease. The Homestead Vegetable Production and Utilization Research Project was initiated in 1985 with USAID assistance at the Farming Systems Research Site, Kalikapur, Ishurdi in north Bangladesh (Hossain *et al.*, 1990).

Technology Generation

Improved vegetable production technology consisted of five vegetable cropping patterns with 14 vegetable crop combinations. About 200 kg of fresh vegetables could be produced each year in a 6 m × 6 m plot which provided the total vitamins A and C and iron needed by a family of five. In addition the homestead garden generated cash income for resource poor farmers. A unique feature of this technology was the participation of women of landless households and marginal farms.

Adoption of Technology

The program for the adoption of technology to permit homestead vegetable production (HVP) was initiated in October 1989. It was coordinated in 20 upazilas (sub-districts) by the Bangladesh Agricultural Research Council (BARC) in collaboration with BARI and the Department of Agricultural Extension (DAE) with USAID funding.

On-farm research was conducted with landless and marginal farmers, with practical training on vegetable production, pest management, and water management. Each trainee was provided with an instruction manual and seeds of vegetable cultivars. Monitoring and evaluation of homestead gardens by researchers and extension personnel facilitated their interaction with farmers and aided the diffusion of technology. The technology was so successful that the program was extended to 100 upazilas in 1990.

Lessons Learned

The activities of the homestead vegetable production (HVP) is in its second year. Its initial success has led to plans to cover most of the districts in Bangladesh within the next 5 years (BARC, 1991). Success of the HVP technology may be attributed to:

i)   Initiatives taken to strengthen extension-research links achieved through District Technical Committees, Regional Technical Committees, Internal Review Workshops, joint field visits by researchers and extension workers, and participation by farmers.
ii) Sharing by administrators, researchers, extension staff, and farmers of the common objective to improve homestead vegetable production.
iii) Elaborate planning, adequate monitoring, and supervision of the program.
iv) Clear institutional policy and donor support in a well defined area of research and development.
v)   Inclusion of women as participants which enhanced the effectiveness of adoption.

It can be seen from these examples that an infrastructure to support successful farmer adoption of improved technologies requires input and committment at all levels from policy makers to administrators, scientists, extension workers, and farmers.

Concluding Remarks

Farmer adoption of technology has been generally good in the more productive and homogeneous "core lands," but slow in the less productive, diverse, and risk prone "hinterlands". To improve adoption in the hinterlands, the emphasis of administrators and scientists in each country should be to decentralize the research process to permit development of location specific technologies. Advantages can be gained by combining indigenous technical knowledge with scientific innovations to develop appropriate technologies for the resource-poor farmer. The centrality of the farmer in on-farm research and adoption of technology is of great significance. Therefore, the farmer participatory approach forms the foundation for the infrastructure needed for technology adoption. Infrastructural support itself involves the technology base, socioeconomic factors, institutional policies, and human interactions. A strong technology base is needed in each country to provide a research backup to support technology generation, testing, adaptation, and be responsive to feedback from farmers. Institutional changes and policy decisions act as vehicles, and socioeconomic factors provide the needed pull for adoption. Among all the factors, improved human interactions play the most important role in adoption of technologies. National governments and administrators should endeavor to provide these infrastructural supports needed to enhance technology adoption.

References

Axinn, G. H. 1991. *Journal of the Asian Farming Systems Association* 1: 69–78.

BARC (Bangladesh Agricultural Research Council) 1991. *Homestead Vegetable Gardening: Progress Report*, 50 pp. Dhaka: Bangladesh Agricultural Research Council.

Benor, D. 1987. In: *Agricultural Extension Worldwide*, pp. 137–148 (eds. W. M. Rivera and S. G. Schram). New York: Croom Helm Ltd.

Biggs, S. D. and Farrington, J. 1991. *Journal of the Asian Farming Systems Association* 1: 113–132.

Batugal, P. A., Acasio, R. F. and Khwaja, A. M. 1985. *Philippine Journal of Crop Science* 10: 107–112.

Chambers, R. and Ghildyal, B. P. 1985. *Agricultural Administration* 20: 1–30.

Chambers, R. and Jiggins, J. 1986. *Agricultural research for resource-poor farmers: A parsimonious paradigm*. Discussion paper 220, 38 pp. Brighton, Sussex, UK: Institute of Development Studies, University of Sussex.

Chambers, R., Pacey, A. and Thrupp, L. A. (eds.). 1989. *Farmer First: Farmer Innovations and Agricultural Research*, 218 pp. London: Intermediate Technology Publications Ltd.

Chitnis, D. H. and Bhilegaonkar, M. G. 1987. *Journal of Maharashtra Agricultural Universities* 12: 84–88.

Conteh, M. 1986. *The green revolution and rural development in Sierra Leone*, 9 pp. Annual Conference of the Sierra Leone Agricultural Society, Kenema, Sierra Leone, 1–5 Oct 1986.

Cruz, L. T. 1987. *Technical Report 1985–86*, Cotton Research and Development Institute, Batac, Ilocos Norte, Philippines.

Dei, K. O. 1981. Testing a model to help small-scale farmers in Central Ashanti (Ghana) – a case study, 53 pp. University of Science and Technology, Kumashi (Ghana).

Dequito, N. L. and Abansi, C. L. 1987. *Technical report 1985–86*. Cotton Research and Development Institute, Batac, Ilocos Norte. Philippines.

Duwayri, M., Estman, C., Fanex, N. and Baqquain, A. A. 1988. *Dirasat* (Jordan) 15: 7–34.

Farrington, J. 1988. *Experimental Agriculture* 24: 269–279.

Farrington, J. and Martin, A. 1987. *Farmer participatory research: A review of concepts and practices*. Discussion paper 19. London, UK: ODI Agricultural Administration Network.

Feder, G., Just, R. and Zilberman, D. 1981. *Adoption of agricultural innovations in developing countries: a survey*. World Bank Staff Working Paper no. 542, 67 pp. Washington, D.C., USA: The World Bank.

Fujisaka, S. 1989. *Experimental Agriculture* 25 423–433.

Grenoble, D. W., Daum, D. R. and Gama, D. M. 1990. *Applied Agricultural Research* 5: 235–240.

Grigg, D. B. 1984. *Understanding Green Revolution*, 384 pp. Cambridge: Cambridge University Press.

Heinrich, G. M. 1991. *ILEIA Newsletter*, May issue: 38–39.

Holden, S. T. and Joseph, L. O. 1991. *Agricultural Systems* 36: 173–189.

Hossain, S. M. M., Mallik, R. N., Abedin, M. Z. and Khan, M. H. 1990. Paper presented at the 10th Annual FSR/E Symposium, Michigan State University, East Lansing, Michigan, USA.

Jabbar, M. A. and Abedin, M. Z. 1989. *Bangladesh: A case study of the evolution and significance of on-farm and farming systems research in the Bangladesh Agricultural Research Institute*. OFCOR

Case Study No. 4. The Hague, The Netherlands: International Service for National Agricultural Research (ISNAR).

Jones, G. E. 1967. *Agricultural Economics and Rural Sociology Abstracts* 9: 1–34.

Kean, S. A. 1988. *Experimental Agriculture* 24: 289–299.

Kelley, T. G. and Walker, T. S. 1991. Paper presented at the 11th Annual FSR/E Symposium, 14–18 Oct 1991, Michigan State University, East Lansing, Michigan, USA.

Knight, T. O., Lovell, A. C., Rister, M. E. and Coble, K. H. 1989. *Southern Journal of Agricultural Economics* 21: 21–33.

Krause, M. A., Dellson, R. R., Baker, T. G., Preckel, P. V., Lowenberg-DeBoer, J., Reddy, K. C. and Maliki, K. 1990. *American Journal of Agricultural Economics* 72: 911–922.

Lightfoot, C., De Guia Jr., O., Aliman, A. and Ocado, F. 1989. In: *Farmer First: Farmer Innovation and Agricultural Research*, pp. 93–100 (eds. R. Chambers, A. Pacey and L. A. Thrupp). London: Intermediate Technology Publications Ltd.

Maurya, D. M. 1989. In: *Farmer First: Farmer Innovation and Agricultural Research*, pp. 9–13 (eds. R. Chambers, A. Pacey and L. A. Thrupp). London: Intermediate Technology Publications Ltd.

McDermott, J. K. 1987. In: *Agricultural Extension Worldwide*, pp. 89–102 (eds. W. M. Rivera and S. G. Schram). New York: Croom Helm Ltd.

Merrill-Sands, D. 1986. *Experimental Agriculture* 22: 87–104.

Merrill-Sands, D., Ewell, P., Biggs, S. and McAllister, J. 1989. *Quarterly Journal of International Agriculture* 28: 279–300.

Merrill-Sands, D. and McAllister, J. 1988. *Strengthening the integration of on-farm client-oriented research and experiment station research in National Agricultural Research Systems: Management lesions from nine country case studies.* OFCOR Comparative Study Paper No. 1. The Hague, The Netherlands: International Service for National Agricultural Research (ISNAR).

Miller, T. and Trolley, G. 1989. *American Journal of Agricultural Economics* 71: 847–857.

Montemayor, L. Q. 1987. *Philippine Journal of Crops Science* 12: 23–24.

Norman, D. W. 1980. *MSU Rural Development Paper no. 5.* East Lansing, Michigan, USA: Michigan State University.

Norman, D. W., Baker, D., Heinrich, G. and Worman, F. 1988. *Experimental Agriculture* 24: 321–331.

Nygaard, D. and Basheer, M. M. 1981. In: *Faba Bean Improvement*, pp. 297–308 (eds. G. Hawtin and C. Webb). The Hague, Netherlands: Martinus Nijhoff.

Raintree, J. B. and Hoskins, M. W. 1988. *Appropriate R&D support for forestry extension.* Paper for the FAO Consultation on Organization of Forestry Extension, 7–11 March 1988, Bangkok, Thailand.

Rambo, A. T. 1983. *Conceptual approaches to human ecology.* Research Report no. 14. Honolulu, Hawaii, USA: East-West Environment and Policy Institute.

Rambo, A. T. and Sajise, P. E. 1985. *The Environmental Professional* 7: 289–298.

Rhoades, R. E., Booth, R., Shaw, R. and Werge, R. 1985. *Appropriate Technology* 11: 11–13.

Toulmin, C. and Chambers, R. 1990. *Farmer-First: Achieving sustainable dryland development in Africa.* Paper no. 19, 12 pp. London, UK: International Institute for Environment and Development.

Developing and delivering mechanization for cool season food legumes

J. DIEKMANN[1], R.K. BANSAL[2] and G.E. MONROE[2]
[1] ICARDA, P. O. Box 5466, Aleppo, Syria, and
[2] MidAmerican International Agricultural Consortium, Project Aridoculture, B. P. 290, Settat, Morocco

Abstract

Successful mechanization of cool season food legume crops begins with smooth and level seedbeds which can be prepared using equipment such as leveling bars, rotary harrows and rollers. Small-seeded legumes can be planted with cereal seeders; however, chickpeas and faba beans are better sown with precision planters.

In the eastern Mediterranean region, lentil harvest includes the collection of straw. For two-stage harvest, a double-knife cutter bar or a lentil puller have been developed and made available. In a one-stage harvest by combine, losses at the pick-up point are reduced with a modified reel, and the straw and chaff are collected by an integrated box at the rear. Legumes with a medium seed size can be harvested with adjusted unmodified combines. Larger-seeded faba beans require a larger opening between the concaves and sieves. Breakage of large-seeded legumes can be reduced with various threshing drums, such as a flail finger drum.

In Morocco, food legumes are often grown on small farms because of their high labor demand and lack of mechanization. A comparison of hand sowing vs. seed-drill sowing showed no significant yield increases for machine use at the same seeding rates. Row spacing between 12 and 72 cm had no effect on yield. However, it is easier to control weeds with wider row spacings.

Chickpea harvesting worked well with a modified Kubota cereal grain reaper. For lentils, a double-knife cutter bar with two different attachments for swathing were tested. Local manufacturing of new equipment is difficult, as large manufacturers are not interested and small companies often do not have adequate facilities.

Introduction

Farmers' mechanization of food legume production is often much less developed than mechanization of cereals. An important prerequisite for

F.J. Muehlbauer and W.J. Kaiser (eds.), Expanding the Production and Use of Cool Season Food Legumes, 517–528.
© 1994 *Kluwer Academic Publishers.*

machine harvesting is use of cultivars with non-lodging, non-shattering growth habits, and a minimum height of 25 cm. The method of seedbed preparation is as important as the choice of cultivar. Problems related to seedbed preparation are more difficult to overcome in animal draft systems than under tractor mechanized conditions, particularly in stony soil. Most seedbeds are far too uneven for full mechanization.

Soil Tillage and Seedbed Preparation

Tillage operations aim at:
1. Weed control;
2. Restoration of soil structure in the arable layer; and
3. Preparation of level seedbeds.
   The shorter the crop or the higher the risk of lodging at harvesting, the more important it is to have a levelled surface, both perpendicular and parallel to the working direction, so the stubble length is kept to a minimum.

Seedbed preparation tools should be selected for good levelling ability. If tine implements, like ducksfoot cultivators, are used, a levelling bar is the minimum requirement for preparation of an acceptable seedbed.

If a seeder is used, a bar should be attached behind as well. In stony conditions the levelling bar must be replaced by a crosskill or Cambridge roller if the soil is still dry at the time of operation.

Spike tooth harrows, either animal-drawn or tractor- operated, will improve the seedbed. Working tractors in different track helps avoid soil compaction, as does using the largest and widest tires possible or twin wheels at the lowest permitted air inflation rates.

A spike tooth harrow used in combination with a wire cage roller improves the crumbling and levelling effect and reduces the number of required passes to one or two. One of the best tools for seedbed preparation, particularly on clay soils, is a rotary power harrow with a heavy toothed roller. This tool ensures the highest amount of lateral levelling, controlled working depth, and clod breaking.

Planting

Mechanized planting mainly controls seed depth, reduces seed waste, and ensures a flat surface finish.

Small-Seeded Legumes

Lentil (*Lens culinaris* Medik.), dry pea (*Pisum sativum* L.), chickling or grasspea (*Lathyrus* spp.), and small-seeded faba beans (*Vicia faba* L.) can be planted with existing, unmodified cereal seeders. While row widths of 15 to 40 cm are common, extremely wide rows of up to 70 cm are reported in North Africa, and from a technical point of view row spacings down to 10 cm are possible. Silim *et al.* (1990) reported that yields tend to increase with reduced row distances, as small as 20 cm.

In an experiment in Morocco, a comparison was made between hand planting and machine planting. Hand sowing 50 kg ha^{-1} behind an animal-drawn plow was compared with machine planting at seed rates of 34, 55, and 70 kg ha^{-1}. The row spacing with hand planting was 60 to 70 cm, while that of the seeder was 60 cm. The result showed no negative effect of the seeder on yield; the seed rate did not significantly increase yield. In this trial, the main advantage of using a seeder was found to be easier weed control.

Table 1 shows results of a weed control experiment in lentil with row spacings of 12 to 60 cm that was machine planted. Row spacing had little effect on yield, but weed control was decisive for increased lentil yields.

Table 1. Effect of row spacings and weed control on lentil grain yield, 1989–90

Row spacings	Weed free	No weed control	Average yields	Yield difference
	kg ha^{-1}	kg ha^{-1}	kg ha^{-1}	%
12 cm	613.2	403.2	508.2	52
24 cm	582.4	291.2	436.8	100
36 cm	733.6	411.6	572.6	78
60 cm	677.6	473.2	575.4	43
Average	651.7	394.8		68
LSD (5%) Row spacing	129.6			
LSD (5%) Weed control	91.6			

Advantages of a seeder are its ability to place seeds at a more consistent depth in less time than hand planting, and that seeding rates can be lower than with hand planting. A seeder also leaves a more level seedbed than hand planting.

Chickpeas

Chickpeas (*Cicer arietinum* L.) can be planted with cereal seed drills. However, cultivars with larger seed size (30 grams per 100 seeds) may benefit from precision planters, the use of which is well established in maize and sugarbeet production.

Mechanical precision planters are solid, simple, and reliable, but require matching seed size to the scoop wheel. Pneumatic precision planters do not require as much seed size calibration, but are more complex and expensive. The main advantages of precision planters with regard to chickpea are:
– less seed breakage, and
– lower seed rates per ha.

Row spacings of 40 to 50 cm enable mechanical weed control, but the agronomical optimum is sometimes down to 20 cm (Saxena, 1987).

The use of a planter for chickpeas in Morocco was found so effective that a machine, currently being tested, was developed for this purpose.

At ICARDA, chickpeas for production are planted on rows spaced 30 cm with commercially available mechanical precision planters.

Faba Beans

Vicia faba ssp. *minor* can be planted with a standard cereal planter, whereas seeds of other subspecies (*Vicia faba* ssp. *major* and *Vicia faba* ssp. *equina*) are often too large and require precision seeders.

Rolling

After planting, the surface of the field can best be levelled by rolling the soil with a Cambridge or crosskill roller. It helps to flatten the field surface and push stones into the soil, which make machine cutting much easier. Flat rollers are not recommended, as they may increase the risk of soil erosion. Cambridge or crosskill rollers are serrated and avoid this problem. The soil surface must be dry at the time of rolling, which should be done immediately after planting when plants reach a height of approximately 10 cm for lentils, vetches (*Vicia* spp.), and peas. No information is available on postemergence rolling of chickpea and faba bean fields.

Weed Control

If seeders are used for planting, row width is best determined by site-specific conditions. Narrow rows of less than 30 cm distance may be difficult to weed mechanically, but could be recommended when an early leaf canopy cover is needed. Wider rows may be required if the farming system depends on mechanical weed control.

Chemical Weed Control

Herbicides used for weed control can be non-selective and systemic, e.g., glyphosate. Herbicides are most commonly sprayed with tractor-mounted standard boom sprayers. Another application method is to use a weed wiper, either hand held or tractor mounted.

The main advantage weed wipers have over sprayers is that the herbicide is applied directly on the plants to be controlled, thereby requiring substantially less herbicide than in sprayer application. The main disadvantages of weed wipers are that the weeds must be taller than the crop and the operating time required several hours per ha for hand-held weed wipers compared to 2 to 4 ha per hour with a 10 to 12 m boom sprayer (KTBL, 1984).

Hand-held wipers are easier to operate than tractor-mounted versions. With working widths of 20 to 70 cm, hand-held weed wipers are useful in controlling patches of weeds.

Although tractor-mounted weed wipers have a greater working area than the hand-held version, i.e., approximately 3 to 6 m, they require an even soil surface and a minimum of 20 cm difference in heights of the weeds and the crop.

Mechanical Weed Control

Interrow cultivation is the most common form of mechanical weed control. It requires machine planting, so that rows are parallel to allow precise guidance of the tools. The most common system uses ducksfoot type blades. It can be operated with row distances down to 30 cm. Narrow tractor tires may be required. Interrow cultivator operations benefit from units being either mid-mounted or front-mounted on the tractor. Both positions enable the operator to watch the work carefully. The mid-mounted version has the advantage of being easily controlled by the tractor steering. Rear-mounted versions may require a second operator for steering or wider bands of unworked surfaces on each side of the row. Early weeding is possible if the crop rows are protected by discs or similar devices that prevent soil from covering the small plants.

Other systems such as weed brushes are also available, but are not important in production schemes because of their higher price and possibly lower capacity.

Non-interrow mechanical weed control devices being tested at ICARDA are weeding harrows, like those used in Canadian lentil production areas. They have the advantage of not leaving unworked strips, since they have narrow tine distances of 2.5 or 3 cm across the whole working width. A problem arises because this harrow has to be operated while weeds are rather small, as their main action is to cover small plants with soil rather than to uproot them. Weeding harrows will not bring up stones as much as ducksfoot cultivators. ICARDA's tests should clarify whether they are suitable for Mediterranean conditions and heavy clay soils.

In many cases a combination of pre-emergence herbicides and later interrow cultivation will provide the best weed control. In Morocco, Zimdahl and Brahli (1990) obtained the highest lentil grain yield from a treatment with Bladex pre-emergence herbicide and later interrow cultivation with an animal-drawn cultivator. Therefore, tractor operated interrow cultivators are now included in their evaluation work.

Harvest

Lack of harvest mechanization is often said to be the major production constraint for food legumes in many parts of the world. The quality of field surface levelling is one of the most important prerequisites for mechanization of all crops that bear pods close to the soil surface, regardless of whether this is caused by short plants (lentils), pods present on lower parts of the stem (some faba beans), or the likelihood of lodging (peas; Saxena *et al.*, 1987).

Lentils

The height of lentils varies between 20 and 40 cm. Traditional cultivars tend to lodge and problems with pod dehiscence and pod drop exist with all cultivars. In the eastern Mediterranean region, straw may be as valuable as grain and requires special attention when selecting harvesting techniques.

Two-Stage Harvest

The classical two-stage harvest is hand pulling at the half green stage, leaving the material for a period of drying, and a separate threshing operation. Both machine cutting and swathing, and machine pulling and swathing are two-stage harvests as well.

The influence of sowing method and seedbed surface finish on losses with two-stage harvest by cutter bar are shown in Table 2.

Under hand-harvesting conditions no alternative to a two-stage technique existed. With mechanization, the two-stage option may be chosen if fields do not ripen equally or fields are not levelled enough for combine harvesting or other problems occur during the cutting and threshing operation. In the USA, cutting and swathing at the half green stage is often done to avoid staining of the grain by late rains. Uneven fields are more easily handled either by cutter bars or a lentil puller, with approximately a 2 m working width.

Time of harvest is critical for all methods. Two-stage harvest is recommended when approximately half the pods are still green. Any later stage increases pod losses, even when done by hand or machine (Friedrich *et al.*, 1987). Both recommended methods, cutting and swathing and machine pulling and swathing, require an adapted planting and seedbed preparation system.

The current research on harvesting equipment at the Centre Regional de la

Table 2. Effect of method of seeding and harvesting on the seed yield of Syrian Local Large lentil (ILL 4400) at Tel Hadya, northern Syria, 1978/79 (Saxena *et al.*, 1987)

Method of seeding	Seed yield Hand harvest	Cutter bar	Loss with mechanical harvest
	kg ha⁻¹		%
Hand broadcast + covering by disc harrow	400.8 +/- 38.7	349.0 +/- 40.2	13.0
Hand broadcast + covering by spike-tooth harrow	348.3 +/- 31.8	298.8 +/- 50.1	14.2
Seeding by fertilizer spinner + covering by disc harrow	431.6 +/- 29.1	379.8 +/- 45.0	12.0
Seeding by fertilizer spinner + covering by spike-tooth harrow	454.9 +/- 30.5	384.1 +/- 48.5	15.6
Seeding with local seed drill (Kashashian)	489.6 +/- 22.6	439.4 +/- 35.0	10.3
Seeding with Amazone seed drill	510.9 +/- 28.8	471.5 +/- 31.8	7.8

Recherche Agronomique (CRRA), Settat, Morocco is primarily focused on lentils. The following machinery options have been evaluated:
1. Double-knife mower, followed by a hay rake for windrowing;
2. Modified walk-behind reaper for combined cutting and windrowing;
3. Set of steel straps attached for windrowing behind a double-knife cutter bar; and
4. A set of rolling disc wheels for windrowing behind a double-knife cutter bar.

The first option was evaluated in lentils that were still partially green and the hayrake worked with relatively low losses. Under these conditions 10% losses were estimated (Bashford *et al.*, 1990). Additional losses occur at the time of picking up the crop for transport.

The other three options combine cutting and swathing. The second option tested was a Kubota AR 120 vertical – conveyor reaper, modified for lentils and chickpeas. This equipment was chosen because of its light-weight design. It has a working width of 120 cm and a cutter bar stroke of 50 mm. The lower and upper crop conveyor lugs are positioned relatively low. The machine was operated by one man and was found suitable for use in small fields. It was able to cut off plants close to the soil surface. Modifications were made by:
– lowering the nylon star-wheels and spring wires from the upper lug height to the lower lug height, and
– adding lifters at the front end of the four crop dividers.

Table 3. Grain and straw losses from harvesting lentil with a modified vertical conveyor reaper (Morocco, 1990)

Row spacing	Potential yields[a] Grain	Straw	Average losses Grain	Straw
	- - -kg ha^{-1}- - -		- - - - % - - - -	
30 cm	629.9	2015.7	8.3 (4.14)[b]	18.4 (6.45)
40 cm	643.3	1867.3	25.4 (10.12)	42.3 (22.72)
60 cm	635.7	1936.7	16.8 (6.67)	29.5 (14.37)
LSD (5%)	176.55	495.13		

[a] Potential yield was determined by hand harvesting of samples.

[b] Figures in parentheses are standard deviations based on 5 observations.

Tests in 1990 were carried out on 30, 40 and 60 cm spaced rows of lentils. The best results were obtained with 30 cm spaced, machine-planted rows (Table 3).

Machinery options 3 and 4 were compared in 1991, but were done too late in the cropping season for optimum performance. It is reportedly easier to swath lentils with the rolling discs rather than the steel straps.

Machinery options 2, 3, and 4 will be tested further. In addition, the use of a pick up on a combine harvester will be tested in Morocco.

One-Stage Combine Harvest

Under optimum conditions, i.e., level seedbeds and non-lodging crops of a minimum of 20 cm height, combines can be used for direct harvest. Close attention to the cutting table is required to avoid losses.

The plant must be collected on the table once it is touched for cutting. Therefore, a reel replacement seems the most needed modification. A pneumatic device is available from Harvestair in Australia, and is used widely in Canada. It consists of a blower, powered from the existing reel drive, supplying pressurized air to angled tubes, positioned over the cutting knife, with approximately 25 cm distance across the working width. The unit is adjustable in such a way that the crop is blown towards the table at the time it is cut. In dense stands no reel or blower is required, but these conditions occur only rarely. In stony fields, a replacement of the standard cutter bar by a double-knife cutter bar may have advantages, as it cannot be obstructed by small stones squeezed between two fingers. The combine working width is recommended to be kept low, e.g., 3 m for a medium-sized combine, so the cutting table can follow ground undulations best. A problem often reported with the combine's use is grain breakage. This can be overcome if the drum is adjusted to a low circumferencial speed of below 13 m sec^{-1}. Breakage also depends on grain moisture, grain size and the machine load factor. With lentils, most combines are only used to half or less of their capacity, which makes it difficult to adjust them for either clean threshing or low breakage rates.

With some combines the standard drum speed adjustment may not offer the required low speed, but a set of exchange sprockets for the drum drive is available for many models. Wind and sieve openings are to be adjusted to the specific harvesting conditions. No constructive modifications are necessary.

When farmers want to collect all non-grain material, a device can be fitted to the rear of the combine to collect all material coming from the shakers. With some modifications, it is also possible to collect the material coming off the sieves. At present, ICARDA is using a modified wheat straw collector as a straw and chaff collector. This device is commercially available for only the Class Dominator 68 combine with straw collector. Straw collectors that do not collect material off the sieves are available, e.g., for John Deere and Fortschritt combines.

Often farmers request the straw to be chopped at the same time which requires bagging. Syrian and Lebanese farmers have modified some combines for their own use. The operation of the straw bagging device creates considerable dust, so working conditions are far from ideal. In addition chopping and bagging may limit the working capacity as it requires additional attention and time. Farmers often consider straw on the ground as lost, but it could be used for sheep grazing.

Chickpeas

Compared to lentils, kabuli chickpeas (*Cicer arietinum*) create fewer harvesting problems, particularly with winter-planted cultivars that are upright and tall.

A minimum of 30 cm height is required for problem-free mechanization. Lodging rarely occurs and losses are minor. Winter-sown chickpeas are an average of 40 cm high, while spring-sown chickpeas have an average height of 25 to 30 cm (Saxena *et al.*, 1987).

Hand planting of chickpeas results in uneven seedbeds. Machine-planted fields that are levelled with a bar are easier to harvest. If stones are present, it is strongly recommended to roll the field after planting. Two-stage harvest with a cutter bar, swathing, possibly drying, and later threshing is possible with many types of cutters.

One-stage harvest by combine is easier with chickpeas than for lentils. Few if any modifications are required for the cutting table. A standard reel is sufficient, but care must be taken to adjust the reel speed to match forward speed, so that the crop is not "raked" or "combed". Tolerable stubble length is between 5 and 10 cm, depending on the lowest pods.

The adjustments for combine threshing are:
Drum type: As for cereals or corn

Drum circumferencial speed : 15 m sec^{-1}
Concave   : 8 to 12 mm between bars
Concave opening to drum   : 2 × seed diameter
Sieves   : According to seed size
Wind   : Stronger than for cereals

In Morocco, farmers started using combines on chickpeas, but in addition to losses, seed breakage is a problem. The reaper, initially used for lentils, also was tried on chickpeas. The reaper harvested 1 ha of chickpeas in seven hours, while hand pulling required 37 man hours.

Faba Beans

This crop consists of several subspecies with different seed sizes. In the dry stage, the lowest pods of *V. faba* ssp. *major* may be either close to the ground or on the ground, which makes mechanization of harvest difficult. Therefore, level seedbeds are required to reduce losses at cutting. Two-stage harvesting using a hand-held cutter bar and removal of the crop for threshing separately is common. In irrigated crops, particularly in surface systems, direct combining is rarely used. At threshing, there may be a problem of seed breakage, particularly with large-seeded cultivars. A special set of concaves and sieves is required:
Concave opening : 18 mm
Sieve, nose type : 22 mm
Sieve, adjustable : 18 to 20 mm
Drum speed   : Slow
Small-seeded faba beans are combine-harvested in parts of Europe.

A flail finger threshing drum was developed at ICARDA in 1982 for use when other drums cause high seed breakage. (Diekmann *et al.*, 1983). The drum uses round-shaped aluminum flail fingers. The fingers are 12 mm in diameter and 95 mm long. Twenty-two fingers are mounted on a 600 mm long bar and six bars are mounted in the 350 mm diameter drum. The operational speed is between 700 and 1000 rpm. The fingers can be covered with rubber, e.g., from garden hoses, for extremely brittle seeds. The concave is closed and coated with a mildly profiled rubber, e.g., from conveyor belt material. This principle works successfully in several stationary threshers.

Dry Peas

Since peas (*Pisum sativum*) are likely to be lodged at the time of harvest, well-levelled seedbeds are required for mechanization. Crop lifters may be attached to the cutter bar to minimize stubble length and pod loss. Use of a vertical cutting device close to the unworked strip to contend with the tendrilous growth eases operations.

Cutting and swathing may be a good technique for uneven ripening conditions, but the swath may best be followed by a combine with a pick up. Hand pick-up is difficult because of the tendrilous growth habit of peas. If seedbeds are level, the crop can be single-stage harvested by combine without further machine modifications. The concave may be the standard version with a slow drum speed and relatively strong wind.

Chickling (grasspea)

Experience with *Lathyrus* spp. (chickling or grasspea) is limited to crops grown at ICARDA's Tel Hadya Station under trial conditions.

Lathyrus spp. grow 20 to 35 cm high, develop tendrils, may lodge, and also have a risk of pod shattering. Level seedbeds are, therefore, as critical as for lentils and peas.

For threshing, an 8 mm standard concave, slow drum speed and standard sieves are recommended. At ICARDA, plot combines create 15 to 20% losses with normal growth, but up to 50% under poor conditions, mainly due to crop height and pod shattering.

Special Conditions, such as Irrigated Farming, Extremely Small Plots, or Zero-Till Conditions

Mechanization of surface-irrigated legumes is difficult because of furrows and/or ditches. Moreover, well-irrigated crops tend to lodge more frequently. Sprinkler systems make mechanization easier. Extremely small plots are difficult to mechanize but may be harvested in two-stages.

Zero-till conditions can create difficulties of uneven surfaces, presence of stones on the surface, or plant residues of the previous year obstructing the harvest. Rolling after planting is, therefore, recommended.

Technology Transfer

There are three steps in transferring technology related to mechanization:
1. Working with farmers during the development of equipment;
2. Develop linkages with extension services to conduct on-farm demonstrations; and
3. Make new equipment available to farmers from a local source.

Working with farmers at an early stage of development creates a better understanding of their problems and constraints. Involving the extension services aligns research with farmers' needs and motivates the extension service.

The third step, finding a local source to manufacture the new equipment, has been a weak point of the Settat project.

ICARDA has no mandate for research on mechanization. ICARDA will, however, continue to adapt existing machinery and look for appropriate mechanization to be included in its training courses. The promotion and also support for possible local manufacturing in any given country is the responsibility of that country's national program.

References

Bashford, L. L., Bansal, R. K. and Gharras, O. El. 1990. In: *Rapport d'Activité Année 1989–90*. Centre Regional de la Recherche Agronomique, Settat, Morocco.

Diekmann, J., Papazian, J. and Mutran, H. 1983. *FABIS* 7: 39.

Friedrich, T., Wieneke, F., Diekmann, J., Jegatheeswaran, P. and Erskine, W. 1987. In: *Legume Improvement Program Annual Report 1987*, pp. 82–84. Aleppo, Syria: ICARDA.

KTBL – Taschenbuch der Landwirtschaft. 1984. DLG-Verlag Frankfurt (Main), Germany.

Saxena, M. C. 1987. In: *The Chickpea*, pp. 207–232 (eds. M. C. Saxena and K. B. Singh). Wellington, Oxon: CAB International.

Saxena, M. C., Diekmann, J., Erskine, W. and Singh, K. B. 1987. In: *Proceedings of the IAMFE/ICARDA Conference*, pp. 211–228 (ed. D. Karlsson). Aleppo, Syria: ICARDA.

Silim, S. N., Saxena, M. C. and Erskine, W. 1990. *Agronomy Journal* 82: 927–930.

Zimdahl, R. L., and Brahli, A. El. 1990. Desherbage de la lentille. In: *Rapport d'Activité Année 1989--90*. Centre Regional de la Recherche Agronomique, Settat, Morocco.

Cool season food legume breeding

Potential for wild species in cool season food legume breeding

F.J. MUEHLBAUER[1], W.J. KAISER[2] and C.J. SIMON[1]
[1] United States Department of Agriculture, Agricultural Research Service, 303W Johnson Hall, Washington State University, Pullman, WA 99164–6434, USA, and
[2] United States Department of Agriculture, Agricultural Research Service, 59 Johnson Hall, Washington State University, Pullman, WA 99164–6402, USA

Abstract

Wild species which are crossable to cultivated pea, lentil, and chickpea have been collected and are maintained in major germplasm collections throughout the world. Wild species of *Vicia* crossable to the cultivated faba bean have not been found. The primary, secondary, and tertiary gene pools of the cool season food legumes represent potential genetic diversity that may eventually be exploited in cultivated types to overcome biotic and abiotic stresses. Technical difficulties in obtaining hybrids beyond those within the primary gene pool is a major obstacle. Reproductive isolation, embryo breakdown, hybrid sterility, and limited genetic recombination are major barriers to greater use of wild germplasm. Conventional crossing has been successful in producing interspecific hybrids in *Lens*, *Cicer* and *Pisum* and those hybrids are being evaluated for desired recombinants. *In vitro* culture of hybrid embryos has been successful in overcoming barriers to wider crosses in *Lens*. The successful transfer of genes from wide sources to cultivated types can be assisted by repeated backcrossing and selection designed to leave behind undesired traits while transferring genes of interest. Molecular marker assisted selection may become a valuable tool in the future use of wild species. In general, too little is known about the possible genetic variation available in wild species that could be valuable in developing resistance to biotic and abiotic stresses. Current efforts on the use of wide hybridization in the cool season food legumes are reviewed and discussed.

Introduction

The wild species progenitors of the cool season food legumes have been found and collected in the presumed centers of origin in the Near East and in the nearby areas of southern Europe and central Asia (Ladizinsky et al., 1988; Muehlbauer et al., 1989; Muehlbauer et al., 1990; van der Maesen and Pundir, 1984). Wild relatives closely related to cultivated pea, lentil, and chickpea have been collected. These accessions have the potential to provide the needed

F.J. Muehlbauer and W.J. Kaiser (eds.), Expanding the Production and Use of Cool Season Food Legumes, 531–539.
© 1994 Kluwer Academic Publishers.

genetic variation for the improvement of these crops for resistance to many of the biotic and abiotic stresses that have been particularly troublesome. In addition, the wild relatives may provide a means of expanding the range of these crops to previously unsuitable areas. Genes for tolerance to cold temperatures, tolerance to heat and drought, and resistance to disease and insect pests, may enable cool season food legumes to be grown successfully in previously unsuitable environments.

There is a belief that the collections of the cultivated materials have not been fully utilized in breeding programs (Hawtin *et al.*, 1988), and that breeders should concentrate on cultivated germplasm before looking to the wild relatives. This is especially true where needed genetic variation is known to be present in cultivated material. Nevertheless, the wild species accessions that are available have been subjected to crossability studies, cytological examinations, and limited evaluations for needed genetic variation. The findings have provided an indication of what may be possible by conventional methods and what may require complex procedures.

Important considerations for the exploitation of the wild species to alleviate biotic and abiotic stresses in the food legumes include: the placement of the relevant species into gene pools relative to the cultigen, barriers to interspecific hybridization and how those barriers might be overcome, evaluation of wild species for traits of importance, and techniques of gene transfer if conventional crossing is not possible.

Gene Pools of the Cool Season Food Legumes

Even though species closely related to the cultivated cool season food legumes have been collected and are being maintained in gene banks for use by breeders, crossability barriers have limited their usefulness. A systematic means of categorizing wild species as to their usefulness for improving the cultigen has been formulated by Harlan and De Wet (1971). According to the concept, the *primary* gene pool of a cultivated species is equivalent to a biological species. Species within the primary gene pool are readily intercrossed and produce progenies that are fully fertile or nearly so. Consequently, gene flow between species of the primary gene pool can be accomplished by conventional breeding methods. Any partial fertility that appears is easily overcome by selection among the progenies.

The *secondary* gene pool contains species which are somewhat distant from the cultigen. Hybridization, to obtain gene flow, is more difficult and the progenies have substantial degrees of sterility, usually because of chromosomal rearrangements.

The *tertiary* gene pool contains those species that are related to the cultigen but where hybridization with the cultigen has not been possible or where hybrids have been completely sterile. Species in the primary, secondary and tertiary gene pools of the cool season food legumes are shown in Table 1.

Table 1. Species in the primary, secondary and tertiary gene pools of lentil, chickpea, pea and faba bean

| Crop | Gene pool | | |
	Primary	Secondary	Tertiary
Lentil	*Lens culinaris* ssp. *culinaris* *Lens culinaris* ssp. *orientalis* *Lens culinaris* ssp. *odemensis*	*Lens nigricans* ssp. *nigricans* *Lens nigricans* ssp. *ervoides*	Not known
Chickpea	*Cicer arietinum* *Cicer reticulatum* *Cicer echinospermum*		*C. bijugum* *C. pinnatifidum* *C. judaicum* *C. chorassanicum* *C. montbretii*
Pea	*Pisum sativum* ssp. *sativum* *Pisum sativum* ssp. *elatius* (including ssp. *humile*)	*Pisum fulvum*	Not known
Faba bean	*Vicia faba*	None known	*V. narbonensis* *V. hyaeniscyamus* *V. galilaea* *V. johannis* *V. bithynica*

Lentil

The wild *Lens* species and the cultigen, *L. culinaris* are all diploid (2n = 14 chromosomes) and are predominantly self pollinators. In addition, all *Lens* species have a similar karyotype consisting of one metacentric chromosome with a satellite, three metacentric chromosomes, and three acrocentric chromosomes (Ladizinsky and Sakar, 1982).

All the wild *Lens* species are considered to be crossable to the cultigen (Ladizinsky *et al.*, 1988), but there is difficulty in obtaining certain hybrids (Table 1). Hybridizations are readily obtained between *L. culinaris* ssp. *culinaris*, the cultigen, and ssp. *orientalis* and ssp. *odemensis*; but, embryo rescue is needed to obtain hybrids of the cultigen with *L. nigricans* ssp. *ervoides* or *L. nigricans* ssp. nigricans (Cohen *et al.*, 1984; Ladizinsky *et al.*, 1985). Therefore, the subspecies of *L. nigricans* are considered to be in a secondary gene pool. Progenies have partial fertility but fully fertile segregants can be selected. Accessions of wild *Lens* are available from the US and ICARDA germplasm collections.

Chickpea

The annual wild *Cicer* species, including the cultigen, *C. arietinum*, are all diploid (2n = 16 chromosomes) with similar karyotypes, and are predominantly self pollinators. *Cicer reticulatum* and *C. echinospermum* are included in the primary gene pool of cultivated chickpea (Table 1). Hybridizations of *C. arietinum* with *C. reticulatum* are readily obtained and the progenies are fully fertile. *Cicer echinospermum* is crossable to *C. arietinum* but the hybrids are completely self-sterile due to the presence of a reciprocal translocation (Ladizinsky *et al.*, 1988). Seeds can be obtained by backcrossing to *C. arietinum*. It has been noted that certain accessions of *C. arietinum* produce partially fertile hybrids when crossed to *C. echinospermum* (Muehlbauer, personal observation). Very little work has been done with the perennial *Cicer* species, possibly because of difficulties in producing seeds and maintaining the accessions. *Cicer anatolicum* has shown good resistance to Ascochyta blight in screening trials (Muehlbauer and Kaiser, unpublished data) and could provide an additional source of variation for controlling that disease.

Attempts to cross *C. arietinum* with the remaining annual *Cicer* species (*C. bijugum*, *C. pinnatifidum*, *C. judaicum* and *C. chorassanicum*) have been unsuccessful, and therefore these species are considered to be in the tertiary gene pool.

Pea

All *Pisum* species are diploid (2n = 14 chromosomes), share a similar karyotype, and are predominantly self-pollinators. All the subspecies of *Pisum sativum* are crossable to the cultigen, *P. sativum* ssp. *sativum*, all of which comprise the primary gene pool (Table 1). There is little difficulty in the use of *P. fulvum* in crosses with subspecies of *P. sativum* when *P. fulvum* is used as the pollen parent.

Faba Bean

Vicia faba has 2n = 12 chromosomes and generally exhibits a high percentage of outcrossing. The wild progenitor of faba bean has not been found and there are no wild *Vicia* species which are crossable to the cultigen. Distantly related *Vicia* species might be placed in the tertiary gene pool (Table 1).

Evaluations of Wild Species

Lentil

Collections of wild lentil are available and there has been increased interest in evaluation for resistance to important diseases. Bayaa *et al.* (1991) systematically screened the ICARDA wild lentil collection for resistance to Fusarium wilt caused by *Fusarium oxysporum* f. sp. *lentis* and for resistance to Ascochyta blight caused by *Ascochyta fabae* f. sp. *lentis* (= *A. lentis*). Considerable genetic variation was found for these two diseases even though the majority of the accessions were susceptible. Resistance to Fusarium wilt was identified in 3 of 109 accessions of *L. culinaris* ssp. *orientalis*, 3 of 30 accessions of *L. nigricans* ssp. *nigricans*, and 2 of 63 accessions of *L. nigricans* ssp. *ervoides*. Resistance to Ascochyta blight was identified in 24 of 86 accessions of *L. culinaris* ssp. *orientalis*, 12 of 35 accessions of *L. culinaris* ssp.

odemensis, 3 of 35 accessions of *L. nigricans* ssp. *nigricans* and 39 of 89 accessions of *L. nigricans* ssp. *ervoides*. One accession of *L. nigricans* ssp. *ervoides*, ILWL 138, had combined resistance to both diseases; however, the use of that accession for breeding combined resistance to Fusarium wilt and Ascochyta blight may be difficult because of the need to use embryo culture to obtain hybrids with the cultigen. It would be straight forward to utilize resistant accessions of ssp. *orientalis* resistant to one of the diseases and other accessions with resistance to the other disease even though additional crosses would be required to obtain combined resistance.

Screening of wild *Lens* against *Orobanche* has not revealed accessions with resistance (ICARDA, 1991).

Chickpea

Collections of the wild species of *Cicer* are maintained in genebanks and some have been evaluated for important disease and insect pests. In a limited evaluation of seven *Cicer* spp., one accession each of *C. pinnatifidum*, *C. montbretii* and *C. judaicum* were highly resistant to Ascochyta blight (Singh *et al.*, 1981), while accessions of *C. yamashitae*, *C. bijugum*, *C. cuneatum* and *C. reticulatum* were tolerant to highly susceptible.

Resistance to cyst nematode, *Heterodera ciceri* was identified in 21 accessions of *C. bijugum*, five accessions of *C. pinnatifidum* and, most importantly, in one accession of *C. reticulatum* (Singh *et al.*, 1989; Singh and Reddy, 1991). The finding of resistance in *C. reticulatum* is of particular interest because of the ease of hybridization with the cultigen.

In an evaluation of annual wild *Cicer* species for cold tolerance, most accessions of *C. bijugum*, *C. echinospermum* and *C. reticulatum* were tolerant and were better than the cultigen (Singh *et al.*, 1991). All accessions of *C. chorassanicum*, *C. cuneatum*, *C. yamashitae*, and all but one accession of *C. judaicum* were susceptible. Accessions of *C. pinnatifidum* had both susceptible and tolerant reactions. Of these accessions, those of *C. reticulatum* and *C. echinospermum* are likely to be the most valuable for breeding, at least in the short term, because they are readily crossable to the cultigen. Use of genes present in the other wild *Cicer* species will need to await the development of techniques for obtaining hybrids or more novel means of gene transfer.

Other resistances found in evaluations of wild *Cicer* species include: resistance to Fusarium wilt in *C. judaicum* (Nene and Haware, 1980), and resistance to Botrytis gray mold (Singh, G. *et al.*, 1982).

Pea

All known accessions of wild *Pisum* are readily crossable to the cultigen; however, in using *P. fulvum* it is important to use the wild species as the pollen parent. Recently, there has been increased interest in *P. fulvum* as a possible source of resistance to bruchids (and see Clement *et al.*, this volume).

Faba Bean

There are no reported cases of interspecific hybridization between cultivated faba bean and any of the wild *Vicia* species. Apparently, pollen tubes are able to reach and fertilize the ovules, but post-zygotic barriers prevent hybrid embryo development (Pickersgill *et al.*, 1985). Unfortunately, the wild progenitor of faba bean is yet to be discovered. *Vicia* species in the tertiary gene pool (Table 1) might have the needed variation for such traits as resistance to rust, chocolate spot and possibly *Orobanche* spp. (Ladizinsky *et al.*, 1988). Success in obtaining interspecific crosses between *V. faba* and the other *Vicia* species would provide much needed genetic variation for improving faba bean.

Introgression Using Molecular Markers

Introgression of single genes from wild species into crop species involves the transfer of segments of chromosomes of various sizes into the recipient genome. These segments occur not only proximate to the target gene (linkage drag), but also in other locations throughout the genome (Stam and Zeven, 1981). Donor genome DNA other than the gene of interest may be deleterious and warrant elimination. The recent development of molecular marker systems and linkage maps are important tools for transferring genes of interest while preventing the transfer of genes which may be undesirable but closely linked.

Molecular marker systems, by their very nature, offer direct examination of the introgressed genotype at any given locus. Loci under examination with these systems may or may not have an overt phenotypic effect, but may offer a subtle reversion of plant type toward the wild phenotype due to the infusion of wild species DNA. Because molecular marker systems directly examine the recipient genotype, they have the capacity to detect the presence of donor genome DNA that may have only subtle but additive deleterious effects. Furthermore, markers can analyze the size and location of residual donor genome carried along through gene introgression and provide the breeder additional information for monitoring the elimination of unwanted donor DNA. The most elegant illustration of the use of molecular markers for this purpose can be found in the review article by Tanksley *et al.* (1989), wherein the authors performed computer simulations to show that while traditional backcross breeding may reach 99% recurrent genome in 6.5 generations, marker assisted backcrossing can approach 100% recurrent genome in three generations. They also illustrate that while it would take 100 backcrosses to reduce linkage drag to 1 cM by conventional backcrossing, it could be done in two backcrosses by marker assisted backcrossing in the context of a highly saturated linkage map.

Molecular markers are most useful in the context of a well developed genetic linkage map. Such maps have been developed for a number of cool season food legume genera, including *Pisum* (Weeden and Wolko, 1990; Ellis *et al.*, 1992), *Lens* (Havey and Muehlbauer, 1989; Weeden *et al.*, 1992; Simon *et al.*, 1993), and *Cicer* (Simon and Muehlbauer, 1991). Maps for each of these genera currently consist of approximately 100 loci each (more in pea) and several research groups have indicated that they intend to greatly expand the chickpea map (Muehlbauer and Simon, personal communications).

The maps that have been developed for lentil and chickpea are particularly useful for gene introgression from wild relatives, since these maps have been developed from interspecific crosses of the cultivated species with its putative progenitor species. Havey and Muehlbauer (1989) developed the first RFLP map of lentil by analyzing crosses of *L. culinaris* ssp. *culinaris* with *L. culinaris* ssp. *orientalis*. Weeden *et al.* (1992) expanded upon that by analyzing crosses of *L. culinaris* by *L. ervoides* provided by G. Ladizinsky. Similarly, Simon and Muehlbauer (1991) have developed a map of *Cicer* by analyzing crosses of *C. arietinum* by *C. reticulatum*. In so doing, these authors have already identified numerous molecular polymorphisms between the respective species. In the case of *Cicer*, the resulting map will be especially useful because of the extremely high uniformity of the molecular morphology of the cultivated species, *C. arietinum*.

Molecular marker systems used in the cool season food legumes include isozymes (Zamir and Ladizinsky, 1984; Weeden and Marx, 1987; Ahmad *et al.*, 1992), restriction fragment length polymorphisms (RFLPs) (Ellis *et al.*, 1992; Havey and Muehlbauer 1989; Simon and Muehlbauer, 1991) and most recently, randomly amplified polymorphic DNAs (RAPDs) (Simon and Muehlbauer, 1992; Weeden *et al.*, 1992; Simon *et al.* 1993).

The relative merits of isozymes, RFLPs, and RAPDs for assisting in gene transfer can be debated, but isozymes and RAPDs seem to be the most useful and practical. While isozymes are comparatively simple, the number of polymorphisms available for use is quite limited and linkages sufficiently close are often not found. The RAPD system offers the advantage of simplicity and a sufficiently large number of polymorphisms that close linkages to genes of interest can most often be found. Repeatability of the RAPD system is sometimes called into question and it appears that RAPD loci are often found in clusters within the genome. The RFLP system has the advantage of repeatability, but the technical difficulties of applying the method in practical breeding and the usual need for radioactive ^{32}P has meant that breeders have generally not adopted the method.

Concluding Remarks

Wild species become of interest to plant breeders when desired gene(s) are not available in the cultigen. This seems to be the case with the cool season food legumes, particularly with regard to resistance to Ascochyta blight of chickpea and lentil, rust of lentil and faba bean, nematodes of chickpea, tolerance to heat, cold and drought of chickpea, resistance to *Orobanche* ssp., and resistance to insects. Fortunately, the wild species relatives of the cool season food legumes have been collected and are available for evaluation and breeding purposes. Additional targeted collection of wild species should be undertaken to expand on this genetic resource, but nevertheless it is becoming apparent that useful variation is already available.

Only a limited number of species are crossable to the cultigen; and, where crossing is successful, the resulting progenies must undergo stringent selection and backcrossing to recover useful material with the gene of interest from the wild species source. In most cases, the gene of interest in the wild species is linked to undesireable and deleterious genes and there is difficulty in achieving desired recombinations. These technical difficulties can be formidable and result in limited use of wild species germplasm.

Molecular marker facilitated introgression is an emerging breeding tool which can be used effectively in selection of progenies from interspecific hybridization. Specifically, molecular markers known to be close to the gene(s) of interest can be used to break undesireable linkages and eliminate unwanted or deleterious genes. It is fortunate that the wild species relatives of the cool season food legumes are available for breeding purposes. Techniques of gene transfer using molecular markers should allow breeders to use this genetic resource with the expectation of obtaining useful recombinants within a reasonable period of time.

References

Ahmad, F., Gaur, P. M. and Slinkard, A. E. 1992. *Theoretical and Applied Genetics* 83: 620–627.
Bayaa, B., Erskine, W. and Hamdi, A. 1991. Screening for resistance to vascular wilt and Ascochyta blight diseases. Paper presented at the Fourth Arab Congress of Plant Protection, Arab Society of Plant Protection, 1–5 December 1991. Cairo, Egypt.
Cohen, D., Ladizinsky, G., Ziv, M. and Muehlbauer, F. J. 1984. *Plant Cell Tissue Organ Culture* 3: 343–347.
Ellis, T. H. N., Turner, L., Hellens, R. P., Lee, D., Harker, C. L., Enard, C., Domony, C. and Davies, D. R. 1992. *Genetics* 130: 649–663.
Harlan, J. R. and De Wet, J. M. J. 1971. *Taxon* 20: 509–517.
Havey, M. J. and Muehlbauer, F. J. 1989. *Theoretical and Applied Genetics* 77: 395–401.
Hawtin, G. C., Muehlbauer, F. J., Slinkard, A. E. and Singh, K. B. 1988. In: *World Crops: Cool Season Food Legumes*, pp. 67–80 (ed. R. J. Summerfield). Dordrecht, The Netherlands: Kluwer Academic Publishers.
ICARDA. 1991. *Legume Program Annual Report*, pp. 94–114. Aleppo, Syria: ICARDA.
Ladizinsky, G. and Sakar, D. 1982. *Botanical Journal Linnean Society* 85: 209–212.
Ladizinsky, G., Cohen, D. and Muehlbauer, F. J. 1985. *Theoretical and Applied Genetics* 70: 97–101.
Ladizinsky, G., Pickersgill, B. and Yamamato, K. 1988. In: *World Crops: Cool Season Food Legumes*, pp. 967–978 (ed. R. J. Summerfield). Dordrecht, The Netherlands: Kluwer Academic Publishers.
Muehlbauer, F. J., Kaiser, W. J. and Kutlu, Z. 1989. *International Plant Genetic Resources Newsletter*, Issue 78/79: 33–34.

Muehlbauer, F. J., Kaiser, W. J., Kutlu, Z. and Sperling, C. R. 1990b. *Pisum Newsletter* 22: 97–98.

Nene, Y. L and Haware, M. P. 1980. *Plant Disease* 64: 379–380.

Pickersgill, B., Jones, J. K., Ramsay, G. and Stewart, H. 1985. *Proceedings of the International Workshop on Faba Beans, Kabuli Chickpeas and Lentils in the 1980s*, pp. 57–70 (eds. M. C. Saxena and S. Varma). Aleppo, Syria: ICARDA.

Simon, C. J. and Muehlbauer, F.J. 1991. In: *Agronomy Abstracts*, p. 116. Madison, Wisconsin, USA: American Society of Agronomy.

Simon, C. J. and Muehlbauer, F. J. 1992. In *Agronomy Abstracts*, p. 114. Madison, Wisconsin, USA: American Society of Agronomy.

Simon, C. J., Tahir, M. and Muehlbauer, F. J , 1993. Linkage Map of Lentil (*Lens culinaris*) 2N = 14. In: *Genetic Maps* (ed. S. O'Brien). Cold Spring Harbor: Cold Spring Harbor Press (In press).

Singh, G., Kapoor, S. and Singh, K. 1982. *International Chickpea Newsletter* 7: 13–14.

Singh, K. B., Di Vito, M., Greco, N. and Saxena, M. C. 1989a. *Nematologia Mediterranean* 17: 113–114.

Singh, K. B., Hawtin, G. C., Nene, Y. L. and Reddy, M. V. 1981. *Plant Disease* 65: 586–587.

Singh, K. B., Holly, L. and Bejiga, G. 1991. *Catalog of Kabuli Chickpea Germplasm (an Evaluation Report of Winter-sown Kabuli Chickpea Land Races, Breeding Lines, and Wild Cicer Species)*, pp. 393–397. Aleppo, Syria: ICARDA.

Singh, K. B. and Reddy, M. V. 1991. *Advances in Agronomy* 45: 191–222.

Stam, P. and Zeven, C. 1981. *Euphytica* 30: 227–238.

Tanksley, S. D., Young, N. D., Patterson, A. H. and Bonierbale, M. W. 1989. *Bio/Technology* 7: 257–264.

van der Maesen, L. J. G. and Pundir, R. P. S. 1984. *Plant Genetic Resources Newsletter* 57: 19–24.

Weeden, N. F. and Marx, G. A. 1987. *The Journal of Heredity* 78: 153–159.

Weeden, N. F., Muehlbauer, F. J. and Ladizinsky, G. 1992. *Journal of Heredity* 83: 123–129.

Weeden, N. F. and Wolko, B. 1990. In: *Genetic Maps* (5th edition), pp. 6.106–6.112 (ed. S. J. O'Brien). Cold Spring Harbor: Cold Spring Harbor Press.

Zamir, D. and Ladizinsky, G. 1984. *Euphytica* 33: 329–336.

Current status and future strategy in breeding pea to improve resistance to biotic and abiotic stresses

S.M. ALI[1], B. SHARMA[2] and M.J. AMBROSE[3]

[1] South Australian Research and Development Institute, GPO Box 1671, Adelaide, SA 5001, Australia;
[2] Division of Genetics, Indian Agricultural Research Institute (IARI), New Delhi 110 012, India, and
[3] John Innes Institute, Colney Lane, Norwich, NR4 7NH, UK

Abstract

The economic importance and current progress made in studies of the host-parasite relationship and identification of sources of resistance and breeding strategies of some important biotic diseases of pea are reviewed in this paper. The root rot complex caused by *Rhizoctonia solani*, *Fusarium solani*, *Aphanomyces euteiches*, *Pythium ultimum* and *Fusarium oxysporum* f. sp. *pisi*, race 1 and 2 has been reported from all commercial pea growing areas of the world. Adequate sources of resistance have been identified and there has been impressive success in the control of the Fusarium wilt pathogen following the introduction of wilt-resistant cultivars. Leaf and stem diseases of pea caused by the *Ascochyta* complex, *Peronospora viciae* and *Erysiphe pisi* are prevalent in most temperate pea growing regions of the world. Several sources of resistance are available, some of which are surprisingly durable. The biochemical genetic parameters of phenolic content used for assaying resistance to *Erysiphe pisi* offers an alternative method of evaluating breeding material. Wild relatives of pea (*P. fulvum* and *P. humile*) are valuable additional sources of genetic variability and provide good sources of resistance to pests and diseases. In temperate rainfed pea growing areas of southern Australia, pea seed yield is more closely related to dry matter production than harvest index. Tall and leafy cultivars proved more productive than afila types.

Introduction

Dry pea (*Pisum sativum* L.), the second most important food legume crop in the world, has experienced significant increases in production from 7.5 million hectares planted in 1979–81 to over 10 million hectares in 1989 (FAO, 1989). The increase in area sown has not been accompanied by yield increases which have remained static or have declined in many parts of the world. Diseases are considered the most important causes of reduced biomass production and seed yields. Abiotic stresses caused by adverse environmental conditions such as

F.J. Muehlbauer and W.J. Kaiser (eds.), Expanding the Production and Use of Cool Season Food Legumes, 540–558.
© 1994 *Kluwer Academic Publishers.*

heat, drought and frost are also responsible for heavy production losses.

The economic importance and the current progress made in unravelling host-parasite relationships, identification of sources of resistance and breeding strategies of some important diseases of pea are reviewed. Prospects for using the wild relatives of *Pisum* (*P. fulvum* and *P. humile*) as sources of resistance to insects and diseases, and plant types considered desirable for dryland temperate pea growing areas of southern Australia (similar to Mediterranean climatic zone) are also reviewed.

Biotic Diseases Caused by Fungi

Importance of Soilborne Root Diseases

Common root rot (*Aphanomyces euteiches* Drechs.), Rhizoctonia seedling tip blight, seed rot and stem rot [*Rhizoctonia solani* (Kühn)], Fusarium root rot [*Fusarium solani* (Mart.) Appel & Wr. f. sp. *pisi* (F.R. Jones) Snyd. and Hans.], Pythium seed, seedling and root rot (*Pythium ultimum* Trow), and wilt caused by *Fusarium oxysporum* Schl. f. sp. *pisi* (Van Hall) Snyd. & Hans. races 1 and 2 are important diseases which attack the underground stems and roots of pea (Davis and Shehata, 1985; Kerr, 1963). Common root rot caused by *Aphanomyces euteiches* is probably the most important disease of pea worldwide (Papavizas and Ayers, 1974). Soilborne diseases are widespread in pea growing areas of the world, and are particularly prevalent in the USA, northern Europe, Canada, New Zealand and Japan. In southern Australia pea growing areas, soilborne diseases are not widespread but are localized problems.

Host-Parasite Relationship

The successful control of pea wilt caused by *Fusarium oxysporum* race 1 by incorporation of the specific, dominant, resistant gene *Fw* is a model of pea disease control. Pea wilt was first reported in Wisconsin in 1924 (Kraft *et al.*, 1981) and by 1932 it had spread to all pea growing areas of the United States and seriously affected production of peas for all uses, including processing and seed. Resistance was soon found and was attributed to a single dominant gene, *Fw*, located in linkage group IV (Kraft *et al.*, 1981). Complete disease control was achieved through the introduction of wilt-resistant cultivars. Occurrences of race 1 of *F. oxysporum* have not been eliminated, but have appeared only in cultivars without the resistance gene.

Race 2 of *F. oxysporum* was first reported in 1933 and has been found in most pea-growing areas of the world (Kraft *et al.*, 1981). Resistance to race 2 was also quickly identified and resistance was again attributed to a single dominant gene, *Fnw*, located in linkage group IV and loosely linked with *Fw* (Matthews and Dow, 1973). Races 5 and 6 were identified in 1970 and 1979, respectively

(Haglund and Kraft, 1970, 1979), and resistance was found in Plant Introduction (PI) accessions of US Department of Agriculture and again attributed to a single dominant gene. A total of 11 races of *F. oxysporum* have been described in the literature, but a reappraisal of race classification by Hubbeling (1974) and Kraft and Haglund (1978) suggests strongly that the 11 race classification is based more on virulence differences than on true genetic differences in the host. Both Hubbeling and Kraft and Hagland recommended that the 11 races should be grouped into either race-1 or race-2 types. Kraft and Haglund (1978) recommend that the establishment of a new race of Fusarium wilt should be based on: (1) that the isolate must establish itself in the field and cause the disease; and, (2) that the isolate can be distinguished from other isolates by a recognizable gene difference in the host. Differential lines and their reactions to four races of *F. oxysporum* are given in Table 1 (Haglund and Kraft, 1979).

Table 1. Pea diffential lines to *Fusarium oxysporum* f. sp. *pisi*

Host	Race			
	1	2	5	6
New season (JI 1360*)	R	R	S	R
New Era (JI 1351)	R	R	S	S
Dark Skin Perfection (JI 1362)	R	S	S	S
WSU 28# (JI 1363)	R	S	R	R
WSU 23 (JI 1364)	R	R	R	S
Little Marvel (JI 1365)	S	S	S	S
74SN5°	R	R	R	R

*   John Innes Institute, UK pea accession number

\#   Washington State University, USA pea accession number

°   USDA/ARS, USA pea accession number

Kern and Naef-Roth (1965) identified a positive correlation with virulence of *F. solani* f. sp. *pisi* and production of the toxin isomarticin. However, Marcinkowska *et al.* (1982) found no relationship between degrees of virulence and the phytotoxic properties of cell-free cultural filtrates of 21 isolates of *F. solani* f. sp. *pisi*. They further reported that virulence of *F. solani* f. sp. *pisi* was not correlated with production of isomarticin. Seedling vigor, as measured by electrolyte loss, is linearly correlated with resistance to Fusarium root rot in peas (Kraft, 1986). Ali (unpublished data) in South Australia also observed that breeding lines with poor seed vigor, as measured by electrical conductivity, consistently gave poor seedling emergence in wet seasons. The high leakage of solutes from seed of low vigor attracted and stimulated seed infection mainly by *P. ultimum*. Lines with high seed vigor consisted of both *AA* and *aa* genotypes, but no AA genotype was classified as a low vigor line. Delphinidin, which is an anthocyanin pigment, is fungistatic to *F. solani* f. sp. *pisi* and *P. ultimum* and is present in seedcoats of peas possessing the major *A* gene for anthocyanin production (Kraft, 1978; Stasz *et al.*, 1980). However, sufficient reducing sugars present in the seedling exudates can overcome the fungistatic properties of delphinidin (Kraft, 1977). Consequently, not all pea lines possessing the *A* gene are resistant to Pythium or Fusarium root rot and seedling vigor is a very important requirement for resistance to these two diseases. Muehlbauer and Kraft (1973) reported that resistance to *F. solani* f. sp. *pisi and P. ultimum* is governed by the same gene factors. Kraft and Roberts (1970) identified resistance gene combinations to both *F. solani* f. sp. *pisi* and *P. ultimum* in PI accessions 140165, 183910, 194006, 210587, 223285 and 257593.

Any given soil is usually infested by more than one soilborne pathogen. Kraft (1978) reported that root pruning caused by *P. ultimum*, suppressed Fusarium wilt development in a race 1 wilt-susceptible cultivar. In addition, the effects of *F. solani* f. sp. *pisi* and *P. ultimum* were additive in causing root rot severity. Shehata *et al.* (1982) reported that infection by *R. solani* could break down resistance in peas to *A. euteiches*. An accession (PI 257593) moderately resistant to root rot, was not resistant when inoculated with *R. solani, F. solani* f. sp. *pisi* or with all four soilborne pathogens (*F. solani, R. solani, A. euteiches* and *P. ultimum*).

Identified Sources of Resistance

Probably the most unique source of resistant pea germplasm is Minnesota 494-All (King *et al.*, 1981). It has moderate resistance to *A. euteiches, F. solani* f. sp. *pisi* and *P. ultimum* and is resistant to races 1, 2 and 6 of *F. oxysporum* f. sp. *pisi*. However, this line has pigmented flowers and many other undesirable horticultural traits. Marx *et al.* (1972) reported that tolerance in PI 175227 to *A. euteiches* was associated with dominant, wild type alleles at 3 unlinked marker loci (*Le, A,* and *Pl*). They further reported that substitution of recessive alleles which express horticulturally desirable traits at each of these loci resulted in a reduction in root rot tolerance. However, Kraft (1988) reported that resistance to *A. euteiches* was recovered in breeding lines with desirable horticultural traits. Lewis and Gritton (1988) reported that resistance to *A. euteiches* appears to be quantitatively inherited with low heritability. A recurrent selection program where disease pressure is intense was used to increase resistance to common root rot in horticulturally acceptable types.

"Dark Skin Perfection", "Freezer 76110" and "Wando" are reported to have partial resistance to *R. solani* (Shehata *et al.*, 1981). Kraft (1981) released two F_8 resistant lines (792022 and 792024) which have combined resistance to *F. solani* f. sp. *pisi* and *F. oxysporum* f. sp. *pisi* races 1 and 2 (Kraft, 1981). Later, four F_5 breeding lines (86–638, 86–2197, 86–2231 and 86–2236) were released (Kraft, 1988), which have resistance to *A. euteiches, F. solani* and races 1, 2, and 5 of *F. oxysporum* f. sp. *pisi*.

Mildew Diseases

Importance of Mildew Diseases

Downy mildew of pea, caused by *Peronospora viciae* (Berk.) Casp. is one of the most widespread and damaging foliar diseases of peas (Taylor *et al.*, 1989). Hagedorn (1985) considered it as a disease of moderate importance. However, it is a disease of major importance in the UK and occasionally is important in the Netherlands, Tasmania, Sweden, New Zealand and in the northwestern part of the USA. It is becoming prevalent in South Australia.

Investigations at John Innes Institute have identified that systemic infection by the downy mildew pathogen is largely prevented due to stipules wrapping round the apex (conventional and semileafless peas); whereas, absence of well-developed stipules appears responsible for susceptibility in completely leafless peas. With advance in maturity, pea plants become more resistant to downy mildew (Matthew *et al.*, 1983–84).

Powdery mildew of pea caused by *Erysiphe pisi* Syd. (syn. *E. polygoni* DC) occurs worldwide, including Australia, Canada, France, India, Peru, South Africa, UK, USA and Zimbabwe (Hagedorn, 1985). The disease is favored by warm days and cool nights where dew forms. Powdery mildew often appears in epidemic form turning whole fields white and seriously affecting the photosynthetic activity of the plants. This not only reduces seed yield but also reduces market value of processed green peas and dry seed.

Host Parasite Relationship

Downy mildew

Eight pathotypes of *P. viciae* were identified in the Netherlands and four in Germany (Taylor *et al.*, 1989). Taylor *et al* . (1989) tested a large number of isolates of *P. viciae* collected from a wide range of locations in UK. They were differentiated into 11 pathotypes (UKP 1–11) based on their reaction on four standard differential hosts (JI 411, JI 560, JI 758 and JI 1272). Workers at the John Innes Institute reported that resistance to *P. viciae* may be determined by a three gene system. The line JI 85 has a single dominant gene, JI 411 two complementary recessive genes, while JI 314 has a single recessive gene (Matthews and Dow, 1972). The recessive genes are designated as *rpv-1* and *rpv-2* (Lewis and Matthews, 1985). JI 85 is resistant to all UK pathotypes (Snoad *et al.*, 1981).

Powdery Mildew

Kalia and Sharma (1988) observed that cultivars resistant to powdery mildew contained higher levels of phenolics and phenol-oxidizing enzymes than the susceptible cultivars. Analysis of F_1's, F_2's and backcross progenies suggested a high heritability for all of the biochemical compounds. These compounds may be a useful means of screening segregating populations. Two recessive genes (*er1* and *er2*) reported by Heringa *et al.* (1969) confer resistance to *E. pisi*.

Identified Resistance Sources

Downy Mildew

Sources of resistance to downy mildew include JI 9, 85, 114, 393, 423, 470, 493, 833, 882, and 1181 (John Innes Ann. Rep. 1983–84). Other sources include Orb, CEBECO-212 and Countess (resistant in southern Australia). Reactions of differential hosts to *Peronospora viciae* pathotypes are given in Table 2.

Table 2. Pea differential lines to pathotypes cf *Peronospora viciae*

Hosts	Pathotypes							
	1 #	2 *	3 #	4 *	5 #	6*	7*	8*
Katinka (JI 1272)		4		1		1	4	4
Heralda (JI 1273)		1		1		4	1	4
Clause 50 (JI 560)		1		2		4	3	4
Cobri (JI 411)	R	1	R	2	R	4	3	4
Dark Skin Perfection (JI 540)	S	3	R	4	S	4	3	4
Recette (JI 584)	S	4	S	4	R	4	4	4
Starnairn (JI 758)	R	4	R	4	R	4	4	4
Puget (JI 441)		1		4		4	3	4
Koroza (JI 952)	R	1	R	1	R	4	4	4
Cicero (JI 1215)	R	4		1	R	4	3	4
Wild-type Afghanistan (JI 85)		1		1		1	4	1

*   Hubelling, N. Med. Fac. Landbouww. Rijksuiv. Gent 40, 539–543, 1975.

*   Ester and Gerlagh, M. Zaadbelangen 33, 146–147, 1979

(UK) Taylor, P. Ph.D. Thesis, University of East Anglia, 1984

R & 1: Resistant, 2, 3, 4 and 5: Susceptible

Powdery Mildew

Sources of resistance to powdery mildew are listed in Table 3. The South Australian Research and Development Institute released a tall, semileafless cultivar Glenroy, which is highly resistant to powdery mildew.

Table 3. Sources of resistance to Erysiphe pisi

Hosts	Resistance	gene/s
Mexique 4 (JI 1559) and SVP 750	er1	er2
Stratagem (JI 2302) and SVP 942	er1	Er2
JI 2480, SVP 951 and SVP 952	Er1	er2

Sources of resistance reported by John Innes Institute (Annual Report, 1983-84) include JI 140, 143, 229, 1049, 1064, 1195, 1214. Other available sources of resistance are Wisconsin 7101, 7102, 7103 and 7104.

Source: Heringa et al. (1969)

The Ascochyta Complex

The Ascochyta complex consists of *Ascochyta pisi* Lib. which causes a leaf and pod spot; *A. pinodes* Jones, the conidial state of *Mycosphaerella pinodes* (Berk. and Blox.) Vesterg., which causes blight; and *Phoma medicaginis* var. *pinodella* (Jones) Boerema (Syn. *Ascochyta pinodella* Jones) which causes a foot rot (Ali et al., 1978). These diseases are widespread throughout the world, particularly in the temperate areas of Europe, North America, and New Zealand, and few pea fields are free of some degree of infection (Hagedorn, 1984). In the United States, Canada and Australia, *M. pinodes* is considered to be the most destructive of the three organisms (Jones, 1927; Hare and Walker, 1943; Carter and Moller, 1960; Wallen, 1974; Ali et al., 1982; Ali, 1987). *Mycosphaerella pinodes* is the most important disease encountered in southern Australian dry pea production areas (Ali, 1987). This pathogen has also been reported from subtropical areas in Africa, Central and South America and Haiti (Hagedorn, 1984). Yield losses of up to 50% have been reported with stand reductions of 24 and 14% followed by surface leaf infection and early defoliation and reduction in number and weight of pods (Wallen, 1974).

In South Australia, the organism that had previously been assumed to be *A. pisi* (Ali et al., 1978) was identified during 1980 as *Macrophomina phaseolina* (Tassi) Goid (Personal communication, E. Punithalingam Commonwealth Mycological Institute, Kew, UK). It is now believed that previously reported *Ascochyta pisi* which causes leaf and pod spot disease in peas, does not affect field peas in South Australia. A similar situation may exist in other southern Australian pea growing states as well (Ali and Dennis, 1992).

Host-Parasite Relationship

Both *M. pinodes* and *P. medicaginis* var. *pinodella* form chlamydospores which enable them to survive in the soil for several years. They are also seedborne (Wallen and Jeun, 1968). Soil isolates of *P. medicaginis* var. *pinodella* are weakly virulent until passage through the host by successive inoculations, which explains why this pathogen has not become epiphytotic in nature. In contrast, when *A. pinodes* is isolated from soil it is highly virulent without successive host passage indicating its pathogenic capability (Wallen *et al.*, 1967). *Ascochyta pisi* does not produce chlamydospores and cannot survive in the soil for any length of time (Wallen and Jeun, 1968). Its main means of carryover is through seed (Hagedorn, 1984). Now that seed production has been moved to drier climates and seed treatments have been so widely used, the importance of *A. pisi* in the major pea producing areas has become largely historical (Hagedorn, 1984).

Several sources of resistance to *A. pisi* are available (Darby *et al.*, 1985). Resistance is reportedly (Wark, 1950) controlled by three dominant genes in Austrian winter peas. Lyall and Wallen (1958) in Canada reported resistance in Ottawa-A 100 was due to two duplicate genes while Matthews and Dow (1973) in the UK reported that a dominant and a recessive gene controlled resistance. The field resistance of "Century" in Canada has been surprisingly durable (Darby *et al.*, 1985). More recently, Darby *et al.* (1985) identified five pathotypes of *A. pisi* in the UK, some of which appear to be geographically localized. The British pathotype (BP2, corresponding to Dutch race c), appears to be widespread and is the most aggressive pathotype in Europe. The results of Darby *et al.* (1985) indicate that hypersensitive resistance in JI 423 and symptomless resistance in JI 502 to *A. pisi* (BP2) are controlled by a dominant gene loosely linked to the *i* locus at the far end of the linkage group 1. The gene has been designated as *Rap 2*. It may prove to be highly effective as it prevents the pathogen from developing or surviving.

Mycosphaerella pinodes has exhibited extreme pathogenic variability and five races have been detected in Czechoslovakia (Ali *et al.*, 1978). In South Australia, 26 isolates of *M. pinodes* were differentiated into 15 pathotypes (Ali *et al.*, 1978), while in the UK, 45 isolates from a wide range of geographical locations could be classified, by stem symptoms, into nine pathotypes, or by leaf symptoms into 16 pathotypes (Clulow *et al.*, 1991). Clulow (1989) observed that resistance to *M. pinodes* is qualitative in nature and fungal growth is restricted to a few host cells in the leaves and stems of resistant pea lines but tissues are rapidly colonized in susceptible pea hosts. The most common pathotypes in the UK were stem pathotype 2 and foliar pathotype A. There has been no report of physiologic specialization in *P. medicaginis* var. *pinodella*.

Identified Resistance Sources

Sources of resistance to the *Ascochyta* complex have been reported (Ali *et al.*, 1978; Ann. Rept., 1983–84; Clulow *et al.*, 1991) and are as follows:
a) *Ascochyta pisi*: JI 9, JI 143, JI 370, JI 423, JI 461, JI 470, JI 502, JI 862, JI 1097, JI 1222, JI 181, JI 322.
b) *Mycosphaerella pinodes* JI 64 (*P. elatius*), JI 96, JI 103, JI 174, JI 181, JI 190, JI 251, JI 1089.

c) *Phoma medicaginis* var. *pinodella*. PI 173052, JI 580, JI 776, PI 166159, PI 174922.

Differential hosts and their reaction to British races of *A. pisi* are given in Table 4.

Table 4. Pea differential lines to five *Ascochyta pisi* races

Host	Race				
	1	2	3	4	5
Wyola (JI 320)	S	S	R	R	R
Pisum elatius (JI 198)	S	R	S	R	R
Frazer (JI 403)	S	S	S	S	R
Dwarf Sugar (JI 461)	R	R	R	R	R

Source: Darby *et al.*, 1986

Pea lines with broad-based resistance to *M. pinodes* in glasshouse tests and in seven disease nurseries across southern Australia are listed in Table 5. All lines had disease scores between 0 and 2 indicating good resistance; however, the rankings varied considerably between greenhouse and field tests.

Table 5. Ranking of pea lines with broad based resistance to *M. pinodes* based on mean disease score for field and glasshouse tests (1988–89)

Pea Line	Field Ranking	Glasshouse Ranking#
SA 1161*a	1	8
SA 1161 b	2	2
SA 1164 a	3	19
SA 1150	4	18
SA 1164 b	5	21
SA 1157	6	9
SA 1156	8	10
SA 1149	9	7
SA 597	10	11
SA 943	12	3
SA 506	16	1

\#   Tested against representative isolates collected from field disease nurseries across southern Australia.

*   South Australian pea accession number.

Importance of Macrophomina phaseolina in Field Pea

Macrophomina phaseolina (Tassi) Goid causes charcoal rot and ashy stem blight and is the second most important disease of field peas in South Australia. A similar situation may exist in other states as well (Victoria, New South Wales and Western Australia) (Ali *et al.*, 1982; Ali, 1987; Ali and Dennis, 1992). The fact that *M. phaseolina* is not mentioned in the Compendium of Pea Diseases, indicates the obscurity of *M. phaseolina* as a possible major pea disease. Charcoal rot development is greatest in immature, weak and stressed plants. In South Australia, severe crop losses in field peas have been reported following post-emergence herbicide sprays which stress plants (Ali and Dennis, 1992). This pathogen produces both sclerotia and pycnidia in pea stems and can survive for long periods in soil (Ali and Dennis, 1992). Considering the field pea area of South Australia, 150,000 ha in 1990/91 (ABARE, 1991), it should be considered as a major disease problem of pea.

Host-Parasite Relationship

Alternative hosts were found to include *Vigna mungo, Medicago littoralis, Medicago scutellata* and *Lens culinaris. Macrophomina phaseolina* has exhibited wide variation in pathogenicity and 21 isolates were differentiated into 15 pathotypes. Pathotype 15 is the most pathogenic and pathotype 1 is least pathogenic (Ali and Dennis, 1992).

Identified Resistance Sources

The important sources of resistance to *M. phaseolina* are SA 506, SA 597 and SA 943 (Ali and Dennis, 1992).

Virus Diseases

Importance of Pea Seedborne Mosaic Virus

Virus diseases of peas have been reported worldwide, Hagedorn (1984); however, there are excellent prospects for resistance breeding. Pea seedborne mosaic virus (PSbMV) has been selected for review because of its widespread distribution and its ability to be transmitted by seed.

Host-Parasite Relationship

Pea seedborne mosaic virus (PSbMV) has been disseminated worldwide by infected seed (Khetarpal and Maury, 1987). The disease is also spread by aphids which transmit the virus from plant to plant, including infected peas and other hosts (*Vicia* spp.) (Hagedorn, 1984). Hagedorn and Gritton (1973) identified a single recessive gene in PI 193586 and PI 193835 conditioning resistance to PSbMV. They designated the gene as *sbm*. Certain new pathotypes of PSbMV were discovered on the basis of differential reactions on pea genotypes

(Khetarpal *et al.*, 1990). Studies on inheritance of resistance to the new pathotypes led to the identification of four separate, recessive genes which govern resistance. New symbols have been proposed including *sbm1* (earlier known as *sbm*), *sbm2*, *sbm3* and *sbm4* (Khetarpal *et al.*, 1990). Only *sbm1* located in the *Pisum* linkage group 6 (Hagedorn and Gritton, 1973) has so far not been overcome by any pathotype of PSbMV.

Identified Resistance Sources

Sources of resistance to PSbMV include PI 193586 and PI 193835 which have the recessive gene *sbm1* (Hagedorn and Gritton, 1973). Other sources of resistance to PSbMV include: X78123, X78126 and X78127 which are known to carry *sbm1* (Khetarpal *et al.*, 1990), and PLP40, PLP350, EC15184, EC17451 and PLP564 (Khetarpal *et al.*, 1990).

Seven breeding lines, OSU-547-29, OSU-559-6, OSU-564-3, OSU-584-16, OSU-589-12, OSU-615-15 and OSU-620-1, resistant to PSbMV and pea enation mosaic virus'(PEMV) were released in 1988 (Baggett and Kean, 1988). Two breeding lines, VR-74-410-2 and VR-1492-1, were released which are resistant to PSbMV, race 1 and 2 of *F. oxysporum* f. sp. *pisi* and Fusarium root rot (Kraft and Giles, 1978).

Biotic Diseases Caused by Bacteria

Importance of Pseudomonas in Peas

Two very closely related plant pathogenic bacteria are associated with peas. *Pseudomonas syringae* pv. *pisi* causes significant economic damage especially during spring and summer while the other bacterium, *Pseudomonas syringae* pv. *syringae*, is a less virulent pathogen but can cause severe disease following hail damage or excessively wet conditions (Mazarei and Kerr, 1990). The two organisms (pv. *syringae* and pv. *pisi*) are difficult to distinguish; however, a rapid and convenient serological method was developed to distinguish the two pathovars (Mazarei and Kerr, 1990).

Host-Parasite Relationship

Taylor (1972) presented evidence for two distinct races of pv. *pisi*. Isolates pathogenic to "Early Onward" but non-pathogenic to "Partridge" were designated race 1. Isolates pathogenic to Partridge but non-pathogenic to Early Onward, were designated race 2. The host differentials for *Pseudomonas syringae* pv. *pisi* to seven races are given in Table 6.

Table 6. Pea differentials for *Pseudomonas syringae* pv. *pisi*

Cultivar	JI Number	Race						
		1	2	3	4	5	6	7
Kelvedon Wonder (M)	2430	S	S	S	S	S	S	S
Early Onward (M)	2431	S	R	S	S	R	S	R
Belinda* (M)	2432	R	S	R	S	S	S	R
Shasta*	2433	R	S	R	S	S	S	R
Puget*	2434	R	S	R	S	S	S	R
Hursts Greenshaft (M)	2435	R	S	S	R	R	S	R
Vinco	2436	R	R	R	S	R	S	R
Sleaford Triumph	2437	R	R	S	R	R	S	R
Partridge	2438	R	S	R	R	R	S	R
Fortune	2439	R	R	R	R	R	S	R

(Source: Pers. Comm., M.J. Ambrose, John Innes Institute, UK)

*   These lines all have the same phenotype so any one of these can be used, Belinda being the preferred.

(M) Minimum set of lines required to distinguish the races.

Identified Resistance Sources

Line 3080 and "Virco" were reportedly resistant to both races 1 and 2 (Taylor, 1972).

Pest Resistance

One of the more damaging pests of field pea is the pea weevil (*Bruchus pisorum* [L.]). The larvae of this bruchid reduce yield and quality by feeding on the developing seeds. Only one larva develops to maturity in an individual seed, but consumes a large portion of that seed.

Screening for resistance to bruchids is currently under way at the Waite Agricultural Research Institute, South Australia. The work is based on the preference of the bruchids to feed on the seeds. Some *P. fulvum* lines, under laboratory conditions, were not a preferred host for egg-laying by female weevils (Pers. Comm., D. Hardie, Waite Institute, South Australia).

Breeding programs to develop resistant cultivars also have been reported in the former USSR (Vilkova and Kalesnichenko, 1973; Aleksandrova, 1977; Sokolov, 1977; Verbitskii and Pokazeeva, 1980) and in the USA (Pesho *et al.*, 1977).

Abiotic Stresses

Moisture Stress in Temperate Areas of Australia

The temperate pea growing areas of southern Australia have a typical Mediterranean climate, with most of the rain occurring during the winter-spring growing season. Temperate crops are grown between the 300 mm and 650 mm annual rainfall zones (Richards, 1991). The two major environmental constraints for pea crops in Australia include drought, particularly in the spring, and high temperatures during flowering. In recent years, there have been significant increases in the semi-leafless (afila type) cultivars grown around the world, particularly in Europe. Semi-leafless peas with reduced leaf area, are considered less sensitive to drought than conventional-leafed types (Davies *et al.*, 1985). Based on three years (1988–90) of interstate trial results across southern Australia, the highest yields were obtained from conventional-leafed pea cultivars both in high (475 to 650 mm) and low (300 to 450 mm) rainfall sites (Figures 1a,c). The conventional leafy pea cultivars (Alma and Dundale) are tall, vigorous and have an extended flowering period. The tall semi-leafless cultivar Maitland also has an extended flowering period but has average yields. The semi-dwarf semi-leafless cultivars represented by Bonzer and Dinkum are more determinate and early flowering.

Work by French (Pers. Comm., R. J. French, W. A., Department of Agriculture) in Western Australia indicated that seed yield of peas is strongly correlated with total dry matter production (Figures 2a,b). Conventional tall pea cultivars of Australia have, in general, higher dry matter production than semi-dwarf, semi-leafless cultivars. Work conducted in southeastern Australia (Evans *et al.*, 1989) indicated that larger amounts of biomass correlated directly

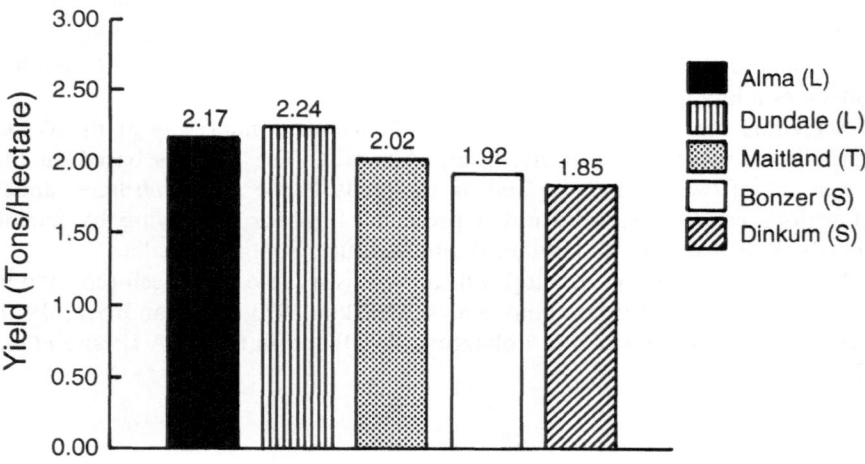

Figure 1a. Seed yield of different plant types.

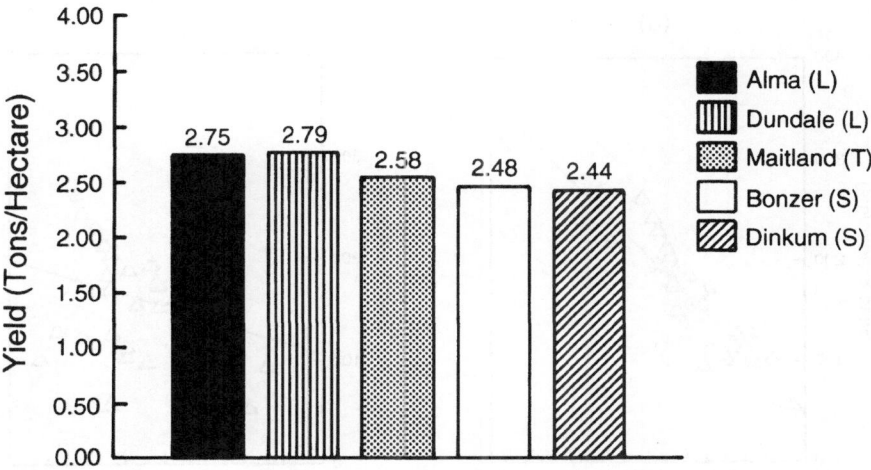

Figure 1b. Seed yield at high rainfall sites (annual rainfall 475–650 mm).

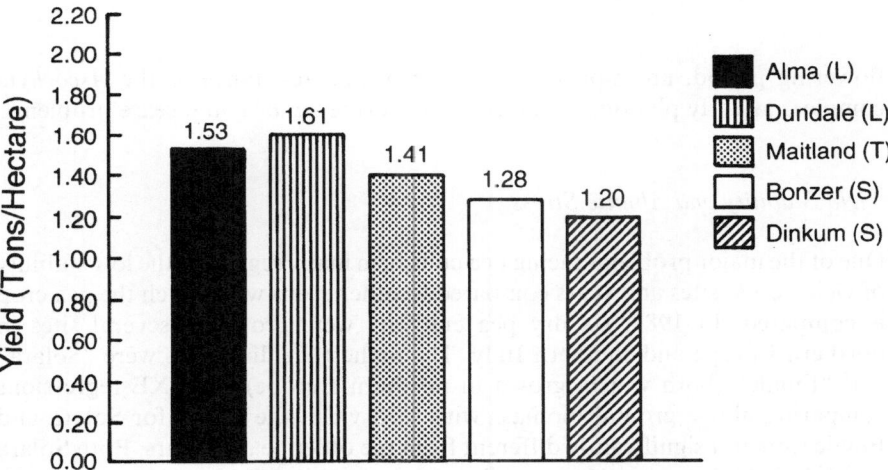

Figure 1c. Seed yield at low rainfall sites (annual rainfall 300–450 mm).

with the amount of fixed N and net N returned to the soil. Evans suggested that increased biomass can be obtained by early planting. Armstrong (1991) in Western Australia found that "Wirrega" used more water than "Dundale", because of deeper and more complete mineral and water extraction due to its larger root system. French (unpublished data) hypothesized that increasing rooting depth and volume can improve tolerance to drought stress. Based on the above experimental data, it is postulated that the most suitable pea type for southern Australia is a conventional pea with vigorous growth, an extended

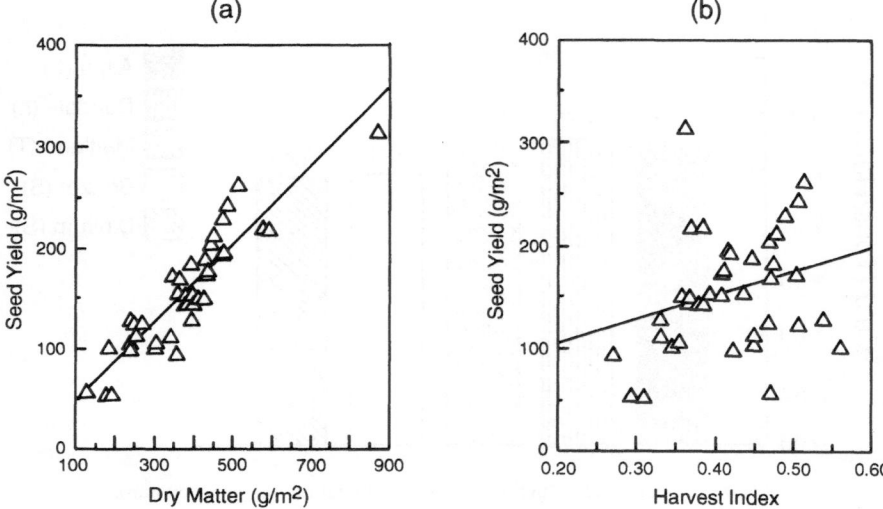

Figure 2. Seed yield of peas correlated to total dry matter production.

flowering period, an improved root system and resistance to the *Ascochyta* complex, as early planting is conducive to severe *Ascochyta* disease problems.

Yield Stability and Abiotic Stress

One of the major problems facing pea culture in many regions is the low stability of yield across sites and years compared to other crops with which the pea crop is compared. In 1984, six dry pea cultivars were grown at several sites in northern Europe and southern Italy. The highest yielding lines were "Solara" and "Finale" (both widely grown in northern Europe). In GXE regressions comparing above ground biomass with seed yield, the slopes for Solara and Finale were not significantly different from the other pea cultivars. Both Solara and Finale had much greater stability of partitioning (harvest index) over the different sites compared to the others. The increase in yield stability was thought to be due to a greater proportion of plants setting an average amount of seed. Interest is growing in selecting for stability of yield as a character in its own right.

Future Breeding Strategy

Breeding to overcome biotic and abiotic stresses depends on economic considerations. To set the priorities of disease-resistance breeding, it is imperative to have a comprehensive survey and quantitative loss assessment data for each disease and information on genetic control based on host-parasite relationships. A good example of comprehensive crop loss assessment work is that of Basu (1978) in Canada for Fusarium root rot, where he has established loss assessments based on disease symptoms.

Significant progress has been realized in identifying sources of resistance to major pea diseases, including soilborne diseases (Kraft, 1989). It is believed that conventional breeding approaches will remain the mainstay in combating them well beyond year 2000. Backcross breeding methods can effectively be used for diseases where durable specific genes have been identified (*F. oxysporum, A. pisi* and *E. pisi*). Resistance to root rot fungi is considered to be multigenetically inherited and suggesting that cyclical recurrent selection will be an appropriate method of breeding (Lewis and Gritton, 1988). The following strategy suggested by Shetata *et al.* (1983) is very appropriate where resistance is quantitatively inherited: (1) use of controlled conditions and proper techniques for screening; (2) clear understanding of the factors that affect disease development and expression of resistance; (3) breeding for a high level of resistance and multiple resistance and not for a single pathogen; (4) carefully choosing parents; (5) developing techniques for multiple disease screening; (6) utilizing breeding methods based on recurrent selection with several cycles of intermating, selfing, and testing; (7) delaying disease screening and selection for resistance until later generations; and (8) intermating among and within the sources of resistance.

The release of G-1000 (Marx's pea breeding line for multiviral resistance) is an approach breeders can follow in developing multiresistant lines based on combining specific genes for resistance. G-1000 combines the genes for resistance to several potyviruses (Provvidenti *et al.*, 1991). The genes for virus resistance in G-1000 are closely linked to either *k* (wings keel like) on chromosome 2 and *wlo* (waxless leaflets) on chromosome 6. Other sources of multivirus resistance have been listed by Provvidenti *et al.* (1991).

Wild accessions from germplasm collections have repeatedly provided important sources of resistance to a wide spectrum of biotic and abiotic stresses. Screening of the test array in current work on *Pseudomonas pisi* has identified a number of rare resistance gene combinations and some low level race nonspecific resistance in wild material from Ethiopia (Pers. Comm., M. J. Ambrose, John Innes Inst., UK).

Cooperation, collaboration and the exchange of information among scientists will lead to standardized differential hosts which are vital for a clearer understanding of race distribution in different geographic locations.

In a multiplication exercise in northern Spain in 1991, a wide range of exotic material from the John Innes collection was grown. The most productive material was a vigorous, deep rooting landrace from Ethiopia (JI 1432) which coped well in a climate where the temperature rose to 40°C in the middle of the day and dropped to 2 to 3°C at night. The growing season had been dry and soil moisture would have only been available deep in the soil.

Pisum fulvum is a possible source of stress tolerance. Its main roots penetrate to greater depths at a rapid rate. Forms of *Pisum elatius* have also been collected from arid regions though their strategy for survival appears reduced water loss. Further observation of material grown in Spain relates to the color of the foliage. The more productive exotic lines under these demanding conditions nearly all had definite layers of wax deposition giving the foliage a blue-green color. This contrasted to the light green types such as *Pisum transcaucasicum* and *Pisum sativum* from India and Afghanistan.

New material is always being collected and some, growing under very extreme conditions, are worthy of further investigation. For example, an accession of *Pisum elatius* (JI 2055) survived temperatures of $-20°C$ and another (JI 1398) was found growing through the snow at 300 m in West China.

Biotechnology techniques of gene mapping, gene cloning and genetic transformation offer opportunities to create new gene combinations to overcome biotic and abiotic stresses. Widespread use of these tools should help breeders in overcoming critical stress problems in peas.

References

ABARE. 1991. *Australian Bureau of Agricultural and Resource Economics, Canberra* 69: 1–15.
Aleksandrova, E. A. 1977. *Selektsiya i Semenovodstvo* 1: 46–47.
Ali, S. M. 1987. Field peas (Market development paper No. 5). Department of Agriculture, South Australia. *AGDEX* 166/10: 1–23.
Ali, S. M. and Dennis, J. 1992. *Australian Journal of Experimental Agriculture* 32: 1121–1125.
Ali, S. M., Nitschke, L. F., Dube, A. J., Krause, M. R. and Cameron, B. 1978. *Australian Journal of Agricultural Research* 29: 841–849.
Ali, S. M., Paterson, J. and Crosby, J. 1982. *Australian Journal of Experimental Agriculture and Animal Husbandry* 22: 348–352.
Annual Report. 1983–84, pp. 9–11. UK: John Innes Institute.
Armstrong, E. L. 1991. *A physiological comparison of morphologically contrasting field pea genotypes.* Ph.D. Thesis, Department of Botany, University of Western Australia, Nedlands, W.A., Australia.
Baggett, J. R. and Kean, D. 1988. *HortScience* 23: 630–631.
Basu, P. K. 1978. *Canadian Plant Disease Survey* 58: 5–9.
Carter, M. V. and Moller, W. J. 1960. *Journal of Agriculture, South Australia* 63: 353–355,363.
Clulow, S. A. 1989. *The resistance of Pisum to Mycosphaerella pinodes.* Ph.D. Thesis, University of East Anglia, UK.
Clulow, S. A., Lewis, D. G. and Matthews, P. 1991. *Journal of Phytopathology* 131: 322–332.
Darby, P., Lewis, B. G. and Matthews, P. 1985. In: *Pea Crop,* pp. 231–236 (eds. P. D. Hebblethwaite, M. C. Heath and T. C. K. Dawkins). London: Butterworths.
Darby, P., Lewis, B. G. and Matthews, P. 1986. *Plant Pathology* 35: 214–233.
Davies, D. R., Berry, G. J., Heath, M. C. and Dawkins, T. C. K. 1985. In: *Grain Legume Crops,* pp. 147–198 (eds. R. J. Summerfield and E. H. Roberts). London: Collins.
Davis, D. W. and Shehata, M. A. 1985. In: *Pea Crop,* pp. 237–245 (eds. P. D. Hebblethwaite, M. C. Heath and T. C. K. Dawkins). London: Butterworths.
Evans, J., O'Connor, G. E., Turner, G. L., Coventry, D. R., Fettell, N., Mahoney, J., Armstrong, E. L. and Walsgolt, D. N. 1989. *N₂ Fixation and its value to soil N increase in Lupin, Field Pea and other Legumes in South-Eastern Australia* 40: 791–805.
FAO. 1989. *Production Yearbook, Vol. 43.* Rome: FAO.
Hagedorn, D. J. (ed.). 1984. In: *Compendium of Pea Diseases,* 57 pp. St. Paul, Minnesota, USA: The American Phytopathological Society.

Hagedorn, D. J. 1985. In: *The Pea Crop*, pp. 205–213 (eds. P. D. Hebblethwaite, M. C. Heath and T. C. K. Dawkins). London: Butterworths.

Hagedorn, D. J. and Gritton, E. T. 1973. *Phytopathology* 63: 1130–1133.

Haglund, W. A. and Kraft, J. M. 1970. *Phytopathology* 60: 1861–1862.

Haglund, W. A. and Kraft, J. M. 1979. *Phytopathology* 69: 818–20.

Hare, W. W. and Walker, J. C. 1944. *Wisconsin Agricultural Experiment Station Bulletin* No. 150.

Heringa, R. J., Van Norel, A. and Tazelaar, M. F. 1969. *Euphytica* 18: 163–169.

Hubbeling, N. 1974. *Overdruk VIT: Mendelingen Fakulteit Landbouwwetenschappen Gent* 39: 991–1000.

Jones, L. K. 1927. *Bulletin New York Agricultural Experiment Station* No. 547: 1–45.

Kalia, P. and Sharma, S. K. 1988. *Theoretical and Applied Genetics* 76: 795–799.

Kern, H. and Naef-Roth, S. 1965. *Phytopathologische Zeitschrift* 53: 45–64.

Kerr, A. 1963. *Australian Journal of Biological Science* 16: 55–59.

Khetarpal, R. K. and Maury, Y. 1987. *Agronomie* 7: 215–224.

Khetarpal, R. K., Maury, Y., Cousin, R., Burghofer, A. and Varma, A. 1990. *Annals of Applied Biology* 116: 297–304.

King, T. H., Davis, D. W., Shehata, M. A. and Pfleger, F. L. 1981. *HortScience* 16: 100.

Kraft, J. M. 1977. *Phytopathology* 67: 1057–1061.

Kraft, J. M. 1978. *Plant Disease Reporter* 62: 216–221.

Kraft, J. M. 1981. *Crop Science* 21: 352–353.

Kraft, J. M. 1986. *Plant Disease* 70: 743–745.

Kraft, J. M. 1988. *Crop Science* 29: 494–495.

Kraft, J. M. 1989. *Pisum Newsletter* 21: 82–85.

Kraft, J. M., Burke, D. W. and Haglund, W. A. 1981. In: *Fusarium Diseases, Biology and Taxonomy*, pp. 142–156 (eds. P. Nelson, T. A. Toussoun and R. J. Cook). University Park, PA: Pennsylvania State University Press.

Kraft, J. M. and Giles, R. A. 1979. *Crop Science* 18: 1098.

Kraft, J. M. and Haglund, W. A. 1978. *Phytopathology* 68: 273–75.

Kraft, J. M. and Roberts, D. D. 1970. *Phytopathology* 60: 1814–1817.

Lewis, B. G. and Matthews, P. 1985. In: *The Pea Crop*, pp. 215–229 (eds. P. D. Hebblethwaite, M. C. Heath and T. C. K. Dawkins). London: Butterworths.

Lewis, M. E. and Gritton, E. T. 1988. *Pisum Newsletter* 20: 20–21.

Lyall, L. H. and Wallen, V. R. 1958. *Canadian Journal of Plant Science* 38: 215–218.

Marcinkowska, J., Kraft, J. M. and Marquis, L. Y. 1982. *Canadian Journal of Plant Science* 62: 1027–1035.

Marx, G. A., Schroeder, W. T., Provvidenti, R. and Mischanec, W. 1972. *Journal of the American Society for Horticultural Science* 97: 619–621.

Matthews, P. and Dow, K. P. 1972. *Annual Report*, 1971, pp. 31–33. UK: John Innes Institute.

Matthews, P. and Dow, K. P. 1973. *Annual Report*, 1972, pp. 36–39. UK: John Innes Institute.

Matthews, P., Dow, K. P., Graves, K., Brittain, M. and Taylor, P. 1983–84. *Annual Report*, pp. 11–12. UK: John Innes Institute.

Mazarei, M. and Kerr, A. 1990. *Plant Pathology* 39: 278–285.

Muehlbauer, F. J. and Kraft, J. M. 1973. *Crop Science* 13: 34–36.

Papavizas, G. C. and Ayers, W. A. 1974. *Aphanomyces species and their root diseases in pea and sugarbeet*. United States Department of Agriculture, Agricultural Research Service, Technical Bulletin 1485, Washington, D.C.

Pesho, G. R., Muehlbauer, F. J. and Harberts, W. W. 1977. *Journal of Economic Entomology* 70: 30–33.

Provvidenti, R., Hampton, R. O. and Muehlbauer, F. J. 1991. *Pisum Genetics* 23: 50–52.

Richards, R. A. 1991. *Field Crop Research* 26: 141–169.

Shehata, M. A., Davis, D. W. and Anderson, N. A. 1981. *Plant Disease* 65: 417–419.

Shehata, M. A., Davis, D. W. and Pfleger, F. L. 1983. *Journal American Society of Horticultural Science* 108: 1080–1085.

Shehata, M. A., Pfleger, F. L. and Davis, D. W. 1982. *HortScience* 17: 33.

558   S.M. Ali et al.

Snoad, B., Dow, K. P., Walkins, P. A. C. and Graves, K. 1981. *Annual Report*, 1980, pp. 27–28. UK: John Innes Institute.

Sokolov, Yu A. 1977. *Zashchita Rastenii* 10: 34.

Stasz, T. E., Harman, G. E. and Marx, G. A. 1980. *Phytopathology* 70: 730–733.

Taylor, J. D. 1972. *The New Zealand Journal of Agricultural Research* 15: 441–447.

Taylor, P. N., Lewis, B. G. and Matthews, P. 1989. *Journal of Phytopathology* 127: 100–106.

Verbitskii, N. M. and Pokazeeva, A. P. 1980. In: *Breeding Pea for Immunity to Bruchus pisorum.* (eds. Sekeltsiya i semenovod zern i kormov kul'tur). USSR: Rostov-on-Don. Vilkova, N. A. and Kolesnichenko, L. I. 1973. *Zashchity Rastenii* 37: 164–171.

Wallen, V. R. 1974. *Canadian Plant Disease Survey* 54: 86–90.

Wallen, V. R. and Jeun, J. 1968. *Canadian Journal of Botany* 46: 1279–1286.

Wallen, V. R., Wong, S. I. and Julie Jeun. 1967. *Canadian Journal of Botany* 45: 2243–2247.

Wark, D. C. 1950. *Australian Journal of Agricultural Research* 1: 382–390.

Current and future strategies in breeding lentil for resistance to biotic and abiotic stresses

W. ERSKINE[1], M. TUFAIL[2], A. RUSSELL[3], M.C. TYAGI[4], M.M. RAHMAN[5] and M.C. SAXENA[1]

[1] *ICARDA, P. O. Box 5466, Aleppo, Syria;*
[2] *Pulses Research Institute, Ayub Agricultural Research Institute, Faisalabad, Pakistan;*
[3] *Department of Scientific and Industrial Research, Private Bag, Christchurch, New Zealand;*
[4] *Genetics Department, Indian Agricultural Research Institute (IARI), New Delhi 110 012, India, and*
[5] *Regional Agricultural Research Station, Ishurdi 6620, Pabna, Bangladesh*

Abstract

Lentil production is limited by lack of moisture and unfavorable temperatures throughout its distribution. Waterlogging and salinity are only locally important. Progress has been made in breeding for tolerance to drought through selection for an appropriate phenology and increased water use efficiency and in breeding for winter hardiness through selection for cold tolerance.

The diseases rust, vascular wilt, and Ascochyta blight, caused by *Uromyces viciae-fabae, Fusarium oxysporum* f. sp. *lentis*, and *Ascochyta fabae* f. sp. *lentis*, respectively, are the key fungal pathogens of lentil. Cultivars with resistance to rust and Ascochyta blight have been released in several countries and resistant sources to vascular wilt are being exploited. Sources of resistance to several other fungal and viral diseases of regional importance are known. In contrast, although the pea leaf weevil (*Sitona* spp.) and the parasitic weed broomrape (*Orobanche* spp.), and to a lesser extent the cyst nematode (*Heterodera ciceri*), are significant yield reducers of lentil, no sources of resistance to these biotic stresses have been found. Directions for future research in lentil on both biotic and abiotic stresses are discussed.

Introduction

In West Asia and North Africa, lentil (*Lens culinaris* Medikus) is winter-sown at elevations below about 850 meters and is usually spring-sown at higher elevations, representing two contrasting agro-ecological regions. Diagrams of the climate at sites representing these two agro-ecological regions are given as Figure 1 (Mueller, 1982). In these areas the crop is grown in drier and colder environments than other food legumes. These agro-ecological regions are characterized by wet winters, springs with rapidly rising temperatures and hot, dry summers. The major limiting factors to crop growth are low moisture availability and high temperature stress in spring, and, at high elevations, cold temperatures in winter. Consequently, selection for tolerance to these abiotic

F.J. Muehlbauer and W.J. Kaiser (eds.), Expanding the Production and Use of Cool Season Food Legumes, 559–571.

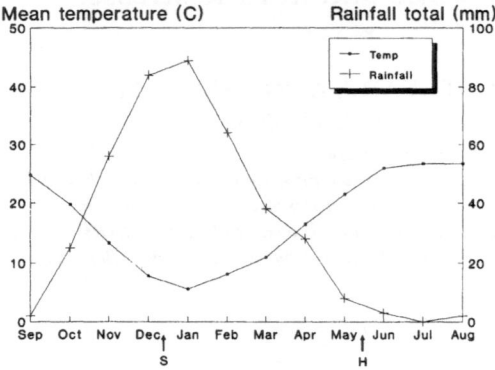

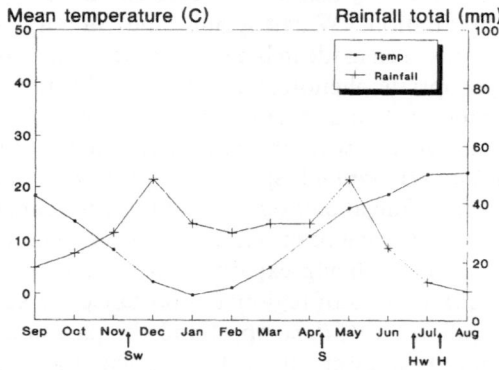

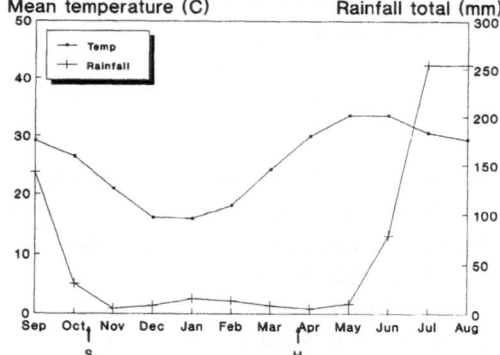

Figure 1. Climato-diagrams showing monthly mean temperatures (°C) and rainfall totals (mm) in Aleppo, Ankara and Kanpur, which are representative of the three major agro-climatic regions of lentil production in the developing world. The elevation above sea level (m) and long-term mean rainfall totals (mm) are given. Typical dates for sowing (S) and harvest (H) are shown; in Ankara dates for winter-sowing (Sw) and harvest (Hw) are also shown.

stresses is of prime concern to lentil breeders in both agro-ecological regions.

By contrast, in South Asia where 50% of lentil is sown (Anonymous, 1990) biotic stresses, particularly diseases such as rust (*Uromyces fabae*), vascular wilt (*Fusarium oxysporum* f. sp. *lentis*), and Ascochyta blight (*Ascochyta fabae* f. sp. *lentis*) are major limiting factors addressable by breeding. Moisture and temperature stresses are also important in limiting yield. A climato-diagram for a representative lentil producing site in South Asia is also given in Figure 1. The relative importance of various biotic and abiotic stresses globally on lentil is indicated in Saxena (1993) and Johansen *et al.* (1994, this volume).

This paper addresses current and future strategies of lentil breeding for resistance to abiotic stresses such as the extremes of temperature, moisture stress and to biotic stresses focusing primarily on fungal pathogens.

Abiotic Stresses

Temperature

Low Temperature

In the agro-ecological region of lowland Mediterranean and south Asia, winters are usually mild. Low temperature is a factor limiting lentil production, but is less important than low moisture availability. For example, at Tel Hadya (280 m.a.s.l.) total seasonal rainfall accounted for 80% of the variance in mean seed yield and the addition of the number of frost nights to the model lifted the variance of seed yield accounted for to 93%, with each frost night reducing seed yield by an average of 15.5 kg ha^{-1} (Erskine and El Ashkar, 1993). Late frost is known to damage early-sown plants more than late-sown material (ICARDA, 1990).

In contrast at higher elevations, where lentil is spring-sown because of the severe winter cold, experiments in Turkey have shown that autumn-sown lentil can yield 50% to 100% more than the traditional spring sowing using cultivars with winter hardiness (Sakar *et al.*, 1988).

Winter survival of lentil often requires tolerance to factors other than cold: e.g., frost-heaving, water-logging, and diseases such as various root pathogens and Ascochyta blight. Cultural practices also strongly affect winter survival, as do frost, and disease and pest attack prior to the onset of winter. In short, a complex of factors is involved in winter survival (Murray *et al.*, 1988), and a myopic preoccupation with cold tolerance is to be avoided. Screening in lentil for winter hardiness has been confined to evaluating survival in the field (Erskine *et al.*, 1981). No recourse has been made to using associations with other morphological, physiological and/or chemical traits or controlled environment facilities. In other species, controlled freeze tests and measurements of plant moisture offer the breeder the best means of predicting cold tolerance, but final evaluation must still be done in the field (Murray *et al.*,

1988). The variability of the field screening environment caused by large differences in winter cold between sites and seasons and local differences in snow cover, soil fertility, etc. make progress slow. Use must be made of many different locations each season. A search for normally snowless, cold, winter areas for screening is warranted.

Despite the problems of field screening, several winter hardy lentil cultivars have been released in Turkey (Sakar *et al.*, 1988) and sources of winter hardiness registered in the USA (Spaeth and Muehlbauer, 1991). A world collection of 3592 lentil accessions was screened for cold tolerance near Ankara, Turkey over a severe winter with temperatures going as low as $-26.8°C$ with 47 days of snow cover (Erskine *et al.*, 1981). A total of 238 accessions were found undamaged by the cold winter with origins mostly from Chile, Greece, Iran, Syria, and Turkey, where natural selection for cold tolerance had occurred. Further confirmation of their cold tolerance was found in joint screening in Italy; and, as a result, a nursery of cold tolerant sources is distributed to cooperators annually in the Food Legume International Testing Program. Pure line selection and mass selection of landraces has, and will continue to have, a role in selecting for winter hardiness.

Additional sources of winter hardiness are being sought at ICARDA within *Lens culinaris* ssp. *orientalis* (see Cubero, 1981), the distribution of which spans areas of severe winter cold. As this species is crossable with the cultigen, utilization of new genetic variation for winter hardiness should be simple.

A major effort is now underway at ICARDA to recombine sources of winter hardiness with other attributes for high elevation areas in simple crosses, and segregating populations with one winter hardy parent are being distributed in the Food Legume International Testing Program. There are no reports on the inheritance of cold tolerance in lentil, but collaborative research between the USA, Turkey, and ICARDA on this topic is in progress. In view of the difficulty in measuring winter hardiness due to environmental variation, it may be useful to study linkages between molecular markers (allozymes, RFLPs and RAPDs) and winter hardiness in order to explore the usefulness of marker-assisted selection. Efforts have begun at ICARDA to recombine various sources to increase the level of winter hardiness using simple recurrent selection facilitated with a new source of cytoplasmic-genetic male sterility (Muehlbauer, pers. comm.).

High Temperature

High temperatures are encountered by lentil in the major production regions mainly during the reproductive stage of growth, usually accompanied by conditions of low moisture availability. Initial efforts to separate the effects of heat and water stresses in the field using supplemental irrigation treatments and heat treatments during the reproductive growth stage by covering with a plastic tunnel have been made at ICARDA (ICARDA, 1989). The paramount importance of moisture stress in a dry season was established and a heat treatment of about 10°C resulted in no change in total biomass but a reduced distribution of dry matter into reproductive growth. Refinement of the technique for providing heat stress is required.

Water

Water Deficit – Drought

In lowland Mediterranean environments lentil is usually sown in the zone with between 300 to 400 mm annual precipitation in the months of December and January. The early stages of vegetative growth are restricted by low radiation and temperature, but this is the time of increasing rainfall and low evaporative demand. From March until maturity in May, the crop experiences increasingly strong sunshine, a rapid rise in maximum temperature (see Figure 1), a fall in rainfall, and high evaporative demand. Drought stress is common during this period, which coincides with the phase of reproductive development in the crop, and as a consequence yields are frequently low. The key importance of moisture in a lowland Mediterranean environment is illustrated by the fact that total seasonal rainfall accounted for 80% of the variance in mean seed yield at a single site over several seasons (Erskine and El Ashkar, 1992). Although lentil is grown in drier environments than other food legumes in West Asia and North Africa, the drought strategy of the species is typically one of drought avoidance with forced senescence and crop maturity induced by conditions of high temperature and/or drought stress.

The breeding program at ICARDA uses simultaneous yield tests at sites spread along a rainfall cline. Selection at these sites has resulted in increases in water use efficiency.

Other approaches to screening for drought tolerance have been tried on lentil at ICARDA with varying degrees of success. Testing performance across sites contrasting in rainfall presupposes that sites differ in moisture availability and not in soil or other factors. To avoid problems, moisture availability may be controlled at a single site. At a dry site water may be added in varying amounts by a line-source sprinkler system, or at a wet site a rain-out shelter may be used to exclude moisture.

Initially at ICARDA, the approach to drought screening was through the exclusion of moisture at a wet site using plastic sheets on the soil in October and November, until a particular rainfall total was passed (e.g., 50, 100 mm), prior to sowing. An additional treatment of supplementary irrigation was applied, creating different levels of moisture supply at a single site (ICARDA, 1983 and 1984). More recently we have used a line-source sprinkler system at a dry site and found that in a dry season (180 mm), 49% of the variation in seed yield among lines was accounted for by variation in flowering time. Drought escape was clearly the key response to drought and, for severely drought-prone areas, selection for early flowering is required (Silim *et al.*, 1992). But, while the line-source sprinkler system is useful in identifying differences between potential parents in response to moisture, it cannot be used for single plant selection.

As another screening method, late sowing to screen for drought tolerance was tried assuming that the crop would experience severe drought and heat stress as it matured late under conditions of low moisture availability and high temperature. However, vegetative and phenological development was abnormal even in wet seasons, because late sown plants were strongly affected by both temperature and photoperiod. Selection under these conditions was clearly for many characters other than drought tolerance, and late sowing as a method of drought screening was abandoned.

In view of the slow progress in other crops in response to direct selection for yield under drought, analytic breeding has been suggested, whereby selection is practiced on another trait, which is strongly associated with yield under drought and has a higher heritability than yield itself (Richards, 1987). Although it is facile to expect any single character to determine yield under drought stress, in view of the variability of stresses and the complexity of yield development, many such characters have been proposed for other crops (Srivastava *et al.*, 1987). In lentil, we have examined traits associated with yield under rainfed conditions and found that early vigor was strongly correlated with biomass and seed yield (Silim *et al.*, 1992). The association of isozyme and RFLP markers with yield under drought conditions and the feasibility of marker-assisted selection for drought is also worth exploring.

Selection for deep rooting has been advocated to increase food legume productivity under moisture limiting conditions (Buddenhagen and Richards, 1988). In the lowland Mediterranean environment the lentil is favored in rotation with wheat over many other legumes because the crop does not utilize all the moisture stored in the soil profile, leaving some residual stored moisture for the succeeding wheat crop (H. Harris, pers. comm.). While an increase in lentil rooting depth may result in increased lentil yield, it might, however, be at the expense of wheat productivity.

In another approach to screening for drought tolerance the differential growth response of genotypes to osmotic stress in a solution containing different levels of polyethylene glycol (PEG) as osmoticum is being investigated and comparisons with field reaction to drought being made (N. Haddad and F.J. Muehlbauer, pers. comm.). The results are not yet available. Investigations are also underway at ICARDA and the University of Saskatoon, Canada of the value of ^{13}C discrimination in selection for drought tolerance in lentil. The naturally occurring isotope of carbon ^{13}C is discriminated against during the fixation of carbon in C3 plants. A correlation between carbon isotope discrimination and grain yield has been found under conditions of non-limited water and selection for carbon isotope discrimination advocated to select for increased yield in wheat (Condon *et al.*, 1987).

Wild lentils, particularly *L. culinaris* ssp. *orientalis*, are often found in habitats receiving low average rainfall. Indeed, selections from crosses of the cultigen with *L. culinaris* ssp. *orientalis* have been distributed as elite lines in the

Lentil International Trials. Wild lentils represent an inadequately explored potential source of drought resistance.

Finally, breeders should be clear whether their aim is to stabilize yields in current production areas and to take advantage of good rains through increased response to moisture availability or to spread lentil into areas currently marginal for production by breeding for drought avoidance and true drought resistance. These aims require different approaches.

Water Excess – Irrigation and Waterlogging

Lentil is sensitive to water-logging and anaerobic conditions causing the low response of the crop to irrigation. In a study of genetic variation in response to irrigation in lentil, large differences were found among genotypes in irrigation response (Hamdi, 1987). An anatomical study, using fluorescent microscopy of genotypes contrasting in their response, revealed that irrigation-responsive genotypes had large aerenchyma or air-spaces in their roots whereas unresponsive genotypes had no such spaces when grown under anaerobic conditions. The value of this anatomical trait in screening for irrigation response requires investigation.

Salinity

Salinity is not considered a major problem in lentil production, consequently there has been limited research on screening for resistance to salinity in the crop. However, a rapid and reliable hydroponic culture technique has been developed for salinity tolerance and compared with field results (Jana and Slinkard, 1979). It has been shown that lentil responds to toxicity of specific ions (e.g., SO_4^-, Cl^-) rather than total salt concentration or osmotic potential.

Biotic Stresses

Fungal Diseases

Rust

Rust, caused by *Uromyces fabae*, is the most important foliar disease of lentil. Complete crop loss is possible from an early infestation with the fungus (Khare and Agrawal, 1978; Sepúlveda, 1989). Epiphytotics of rust are common in Bangladesh, Chile, Ecuador, Ethiopia, India, Morocco, Nepal, and Pakistan; the disease is widespread but unimportant economically elsewhere around the Mediterranean basin.

Field screening has been undertaken at several locations where infection occurs annually, for example, Ishurdi in Bangladesh, Pantnagar in India, and

Akaki, Ethiopia. A method to artificially inoculate rust has been described (Kramm and Tay, 1984). Genetic differences among genotypes and sources of resistance have been reported by many authors (Nene *et al.*, 1975; Khare *et al.*, 1979; Gurha, 1983; Reddy and Khare, 1984; Shukla, 1984; Amin, 1985; Mishra *et al.*, 1985; ICARDA, 1988; Singh and Sandhu, 1988). Resistance to rust has been recently found to be monogenically inherited with resistance dominant to susceptibility (Sinha and Yadav, 1989; Singh and Singh, 1990).

ICARDA is breeding for rust resistance through joint screening with national programs in Ethiopia, Morocco, and Pakistan. As a result, national programs have released rust resistant lines received from ICARDA in Chile, Ecuador, Ethiopia, and Morocco. An international nursery of rust resistant sources was launched in 1990. The nursery will clarify the host-pathogen relationship in different regions and assist in identifying variation in the fungus. Linkage of the rust resistance gene to potential markers is being sought at Washington State University and ICARDA is screening wild lentils for additional sources of resistance.

Ascochyta Blight

Ascochyta blight is caused by *Ascochyta fabae* f. sp. *lentis* (syn. *Ascochyta lentis*) (Gossen *et al.*, 1986). It is economically important in Argentina, Canada, Ethiopia, India, New Zealand, Pakistan, and USSR. Foliar infection has caused yield losses of up to 40% (Gossen and Morrall, 1984), but economic losses from infected seed are much higher in Canada where such seed can only be sold as livestock feed.

Methods of artificial inoculation are known and genotypic differences in foliar reaction following screening for resistance have been reported from Argentina (Mitidieri, 1974), Canada (Slinkard *et al.*, 1983), India (Khatri and Singh, 1975; Gurha, 1983), Pakistan (NARC, 1986), and Syria (ICARDA, 1984). Clearly, resistance to Ascochyta blight is relatively common. Genetic differences in seed damage are also known (Morrall, pers. comm.). Resistance to blight has been found in wild *Lens* (Bayaa *et al.*, 1991).

Breeding for resistance to Ascochyta blight is being conducted in a joint ICARDA/Pakistan breeding program with screening undertaken in Islamabad, a "hot spot" for the disease. As a result the cultivar "Manserha 89" has been released with multiple resistance to Ascochyta blight and rust in Pakistan. Studies of the inheritance of resistance to the disease are being conducted and it is important that the major genes involved are located on the lentil genome to allow the possibility of using marker-assisted selection.

A nursery of Ascochyta resistant sources has been distributed in the International Food Legume Testing Program since 1988. Information on the differential reaction of a common set of lines to indigenous isolates in different countries suggests the presence of variability within the pathogen (Erskine, pers. comm.). Knowledge of the variation in *Ascochyta* species affecting other food legumes suggests that variation in the lentil pathogen should be carefully monitored.

Wilt

The most serious disease of lentil is vascular wilt caused by *Fusarium oxysporum* f. sp. *lentis*, which produces major economic losses in parts of South America, the Mediterranean basin, and South Asia. A screening method for wilt has been developed (Bayaa and Erskine, 1990) and genetic variation in reaction reported (Khare, 1981 and Bayaa and Erskine, 1990). Recently five independently segregating genes for resistance have been found (Kamboj *et al.*, 1990). At ICARDA, breeding for resistance to vascular wilt and a study of the linkage of resistance to isozyme and RFLP markers have been started, the latter in collaboration with Washington State University. Resistance to Fusarium wilt (race 1) in pea is controlled by a locus closely linked to *Est-s*, coding a seed esterase (Hunt and Barnes, 1982). In view of the extent of conserved linkage groups between pea and lentil, this possible linkage is worth investigating in lentil. A recent survey of the wilt reaction of 220 wild lentil accessions revealed resistance in the subspecies *orientalis, nigricans,* and *ervoides* (Bayaa *et al.*, 1991).

Other Fungal Pathogens

Several other fungal pathogens cause economic losses on occasion but their importance has not warranted special attention in lentil breeding (Khare, 1981). Genetic variation in disease reaction has been reported following field infestation for downy mildew (*Peronospora lentis*) (ICARDA, 1982), anthracnose (*Colletotrichum truncatum*) (NARC, 1987), collar rot (*Sclerotium rolfsii*) (Khare *et al.*, 1979), powdery mildew (*Erisyphe polygoni*) (Mishra, 1973), and Botrytis blight (*Botrytis cinerea*) (NARC, 1987).

Viruses

Pea seedborne mosaic virus (PSbMV) is of economic importance in the USA, causing stunting, malformation of leaves and stems, and reproductive abortion. Immunity has been identified and is controlled by a single recessive gene (Haddad *et al.*, 1978). Tolerance to pea enation mosaic virus transmitted by aphids has been identified in the USA (Aydin *et al.*, 1987). Subterranean clover red leaf virus (SCRLV) is endemic to cropping areas in New Zealand and lentil cultivars require a high level of resistance to it. Seven PI lines from the USDA collection have adequate resistance, with one line PI 212610 also resistant to PSbMV and the blue-green aphid (*Acyrthosiphon kondoi*) (Jermyn, 1980).

Insects

In West Asia the major insect pest of lentil is pea leaf weevil (*Sitona crinitus*). Adult weevils eat the leaflets of seedlings and the larvae consume the root nodules of adult plants. Despite finding differences among diverse genotypes in seedling damage by the weevil, no differences in nodule damage have been found (ICARDA, 1981 and 1984). Host-plant resistance to *Sitona* does not have major scope as a control alternative. Through joint research with USA universities, the gene for deltatoxin production from *Bacillus thuringiensis* has been introduced into rhizobial strains as a potential method of Sitona control. This may be seen as a model for the eventual introduction of the same gene into the host plant.

The second most important group of insect pests are the aphids *Aphis craccivora* and *Acyrthosiphon pisum*, which are of particular importance in dry seasons. Genetic differences in susceptibility to *A. craccivora* have been seen in Egypt following natural infestation (Hamdi, pers. comm.) and to *Acyrthosiphon kondoi* in New Zealand (Jermyn, 1980).

In the Indian sub-continent the key insect pest is the pod-borer (*Etiella zinkenella*). Resistance to pod-borer has been found (Chhabra and Kooner, 1980; Sachan, 1991), but no attempt has been made to breed for resistance.

Among seed weevils on lentil *Bruchus lentis* is the most widespread and injurious. Genetic differences in susceptibility have been found (Chopra and Pajni, 1987), but breeding for resistance has not been initiated.

Parasitic Weeds

Broomrape (*Orobanche* spp.) is an important parasite of lentil in the Mediterranean basin. Host plant resistance to *Orobanche* has been exploited in other crops to produce resistant cultivars as a method of control. In lentil, a total of 1774 germplasm accessions have been screened in the field in infected soil and a range in reactions recorded (Erskine and Witcombe, 1984). However, the reaction of the most resistant accessions was examined in petri dishes in the laboratory (Sauerborn *et al.*, 1987) and there were no significant differences among accessions in the number of infections of *Orobanche* per unit length of root. The low incidence of *Orobanche* infection on the roots of "resistant" accessions in the field was probably due to poor root growth. Despite further extensive screening in petri dishes, resistance to *Orobanche* has not been found in the cultigen. Screening has continued within wild *Lens*, but resistance remains elusive.

Nematodes

Among the nematode species that affect lentil, screening for resistance has been undertaken for cyst nematode (*Heterodera ciceri*) as part of an ICARDA/ Institute of Nematology, Bari project. A total of 175 germplasm accessions of the cultigen and a single accession of *Lens orientalis* were screened in Bari, Italy and 75 lines were screened at ICARDA in pot trials with infested soil (ICARDA, 1985). Differences in susceptibility were observed among the lentils, but no resistance found. Although nematodes are not currently major limiting factors of lentil production, it is important to monitor the reaction of promising lines to cyst nematode in West Asia, in order to avoid the release of highly susceptible new cultivars.

Lodging

It is increasingly uneconomic for farmers in West Asia and North Africa to harvest the crop by hand due to the rising cost of labor and research has identified optimum systems for mechanical harvest of the crop (Snobar *et al.*, 1985; Erskine *et al.*, 1991). Genotypes with a high degree of standing ability are desirable in these systems because lodging results in yield losses if it occurs early in the development of the crop (Peñaloza and Mera, 1988) or if a lodged crop is machine harvested (Erskine *et al.*, 1991). Additionally, in wet seasons, lodged crops are more prone to infection from such fungi as *Ascochyta*, *Botrytis*, and *Sclerotinia*. Lentil genotypes vary in standing ability (Mera, 1987); genotypes with thick stems generally have good standing ability (Erskine and Goodrich, 1988). Increasing standing ability is an aim of breeding programs in the Mediterranean region, the USA, and in New Zealand.

Concluding Remarks

Sources of resistance to the key stresses – cold, wilt, rust, Ascochyta blight, and of drought avoidance are now known in lentil and breeding is underway to introduce these resistances into adapted lines. It is important that pathogen variation is monitored in the future. Sources of resistance to several other fungal and viral diseases of regional importance are known. In contrast, although the pea leaf weevil (*Sitona* spp.) and the parasitic weed broomrape (*Orobanche* spp.), and to a lesser extent the cyst nematode (*Heterodera ciceri*), are significant yield reducers of lentil, no sources of resistance to these biotic stresses have yet been found. The real challenge is now to create appropriate combinations of the resistances to different stresses within individual cultivars, as stresses often appear together in the field.

References

Amin, K. S. 1985. Annual Workshop Report on Rabi Pulses on Plant Pathology under the All India Coordinated Pulse Improvement Programme, Kanpur, India.

Anonymous. 1990. *Food production yearbook*. Rome, Italy: FAO.

Aydin, H., Muehlbauer, F. J. and Kaiser, W. J. 1987. *Plant Disease* 71: 635–638.

Bayaa, B. and Erskine, W. 1990. *Arab Journal of Plant Protection* 8: 30–33.

Bayaa, B., Erskine, W. and Hamdi, A. 1991. *Screening wild lentil for resistance to vascular wilt and Ascochyta blight diseases*. Presented at Fourth Arab Congress of Plant Protection, Cairo, December 1991.

Buddenhagen, I. W. and Richards, R. A. 1988. In: *World Crops: Cool Season Food Legumes*, pp. 81–96 (ed. R. J. Summerfield). Dordrecht: Kluwer Academic Publishers.

Chhabra, K. S. and Kooner, B. S. 1980. *LENS* 7: 46–49.

Chopra, N. and Pajni, H. R. 1987. *LENS* 14: 23–27.

Condon, A. G., Richards. R. A. and Farquar, G. D. 1987. *Crop Science* 27: 996–1001.

Cubero, J. I. 1981. In: *Lentils*, pp. 15–38 (eds. C. Webb and G. Hawtin). Slough, UK: Commonwealth Agricultural Bureaux.

Erskine, W. and El Ashkar, F. 1993. Rainfall and temperature effects on lentil seed yield in a Mediterranean environment. *Journal of Agricultural Science, Cambridge* (Submitted).

Erskine, W., Diekmann, J., Jegatheeswaran, P., Salkini, A., Saxena, M. C., Ghanaim, A. and El Ashkar, F. 1991. *Journal of Agricultural Science, Cambridge* 117: 333–338.

Erskine, W. and Goodrich, W. J. 1988. *Canadian Journal of Plant Science* 68: 929–934.

Erskine, W., Myveci, K. and Izgin, N. 1981. *LENS* 8: 5–8.

Erskine, W. and Witcombe, J. R. 1984. *Lentil Germplasm Catalog*, 363 pp. Aleppo, Syria: ICARDA.

Gossen, B. D. and Morrall, R. A. A. 1984. *Canadian Journal of Plant Pathology* 6: 233–237.

Gossen, B. D., Sheard, J. W., Beauchamp, C. J. and Morrall, R. A. A. 1986. *Canadian Journal of Plant Pathology* 8: 154–160.

Gurha, S. N. 1983. Annual Workshop Report on Rabi Pulses on Plant Pathology under the All India Coordinated Pulse Improvement Programme, Kanpur, India.

Haddad, N. I., Muehlbauer, F. J. and Hampton, R. O. 1978. *Crop Science* 18: 613–615.

Hamdi, A. 1987. *Variation in lentil (Lens culinaris Medik.) in response to irrigation*. Ph.D. Thesis, University of Durham, United Kingdom.

Hunt, J. S. and Barnes, M. F. 1982. *Euphytica* 31: 341–348.

ICARDA. 1981. *Annual Report*. Aleppo, Syria: ICARDA.

ICARDA. 1982. *Research in lentil breeding to 1981*. Aleppo, Syria: ICARDA.

ICARDA. 1983. *Annual Report*. Aleppo, Syria: ICARDA.

ICARDA. 1984. *Annual Report*. Aleppo, Syria: ICARDA.

ICARDA. 1985. *Annual Report*. Aleppo, Syria: ICARDA.

ICARDA. 1988. *Food Legume Improvement Program: Annual Report*. Aleppo, Syria: ICARDA.

ICARDA. 1989. *Food Legume Improvement Program Annual Report for 1989*. Aleppo, Syria: ICARDA.

ICARDA. 1990. *Food Legume Improvement Program Annual Report for 1990*. Aleppo, Syria: ICARDA.

Jana, S. K. and Slinkard, A. E. 1979. *LENS* 6: 25–27.

Jermyn, W. 1980. *LENS* 7: 65.

Johansen, C., Baldev, B., Brouwer, J. B., Erskine, W., Jermyn, W. A., Lang Li-Juan, Malik, B. A., Ahad Miah, A. and Silim, S. N. 1994. In: Expanding the Production and Use of Cool Season Food Legumes, pp. 175–194 (eds. F. J. Muehlbauer and W. J. Kaiser). Dordrecht: Kluwer Academic Publishers.

Kamboj, R. K., Pandey, M. P. and Chaube, H. S. 1990. *Eupytica* 50: 113–117.

Khare, M. N. 1981. In: *Lentils*, pp. 163–172 (eds. C. Webb and G. Hawtin). Slough, UK: Commonwealth Agricultural Bureaux.

Khare, M. N. and Agrawal, S. C. 1978. *Lentil rust survey in Madhya Pradesh.* All India Pulse Workshop held at Baroda, Indian Council of Agricultural Research, India.

Khare, M. N., Agrawal, S. C. and Jain, A. C. 1979. *Diseases of lentil and their control.* Technical Bulletin JNKVV, Jabalpur, Madhya Pradesh, India.

Khatri, H. L. and Singh, K. 1975. *Labdev. B* 13: 73–74.

Kramm, V. M. and Tay, J. U. 1984. *LENS* 11: 24.

Mera, M. F. 1987. *Selection for improved standing ability in lentils (Lens culinaris Medic.).* M. S. Thesis, Washington State University, Pullman, Washington, USA.

Mishra, R. P. 1973. *PKV Research Journal* 2: 72–73.

Mishra, R. P., Kotasthane, S. R., Khare, M. N., Gupta, O. and Tiwari, S. P. 1985. *LENS* 12: 25–26.

Mitidieri, I. Z. de. 1974. *Revista de Investigaciones Agropecuarias* 5: 43–45.

Mueller, M. J. 1982. *Selected Climatic Data for a Global Set of Standard Stations for Vegetation Science.* The Hague, The Netherlands: Dr. W. Junk Publishers.

Murray, G., Eser, D., Gusta, L. V. and Eteve, G. 1988. In: *World Crops: Cool Season Food Legumes,* pp. 831–844 (ed. R. J. Summerfield). Dordrecht: Kluwer Academic Publishers.

NARC. 1986. *Pathology of food legume (pulses) crops: Progress Report 1985–86.* Food Legume Improvement Programme, National Agricultural Research Centre, Pakistan Agricultural Research Council, Islamabad, Pakistan.

NARC. 1987. *Pathology of food legume (pulses) crops: Progress Report 1986–87.* Food Legume Improvement Programme, National Agricultural Research Centre, Pakistan Agricultural Research Council, Islamabad, Pakistan.

Nene, Y. L., Kannaiyan, J. and Saxena, G. C. 1975. *Indian Journal of Agricultural Science* 45: 177–178.

Peñaloza, E. and Mera, M. K. 1988. *Agriculture Tecnica* 48: 93–96.

Reddy, R. R. and Khare, M. N. 1984. *LENS* 11: 29–32.

Richards, R. A. 1987. In: *Drought Tolerance in Winter Cereals,* pp. 133–150 (eds. J. P. Srivastava, E. Porceddu, E. Acevedo and S. Varma). UK: John Wiley.

Sachan, J. N. 1991. *Annual Workshop Report on Rabi Pulses on Entomology under the All India Coordinated Pulse Improvement Programme,* Kanpur, India.

Sakar, D., Durutan, N. and Meyveci, K. 1988. In: *World Crops: Cool Season Food Legumes,* pp. 137–146 (ed. R. J. Summerfield). Dordrecht: Kluwer Academic Publishers.

Sauerborn, J., Masri, H., Saxena, M. C. and Erskine, W. 1987. *LENS* 14: 15–16.

Saxena, M. C. 1993. In: *Proceedings for an international conference on Breeding for Stress Tolerance in Cool-Season Food Legumes,* pp. 3–14 (eds. K. B. Singh and M. C. Saxena). Chichester, UK: John Wiley & Sons.

Sepúlveda, P. 1989. *Agricultura Técnica* 45: 335–339.

Shukla, P. 1984. *Indian Journal of Mycology and Plant Pathology* 14: 89–90.

Silim, S. N., Saxena, M. C. and Erskine, W. 1993. *Experimental Agriculture* 29: 9–19.

Sinha, R. P. and Yadav, B. P. 1989. *LENS* 16: 41.

Singh, J. P. and Singh, I. S. 1990. *Indian Journal of Pulses Research* 3: 132–135.

Singh, K. and Sandhu, T. S. 1988. *LENS* 15: 28–29.

Slinkard, A. E., Morrall, R. A. A. and Gossen, B. 1983. *LENS* 10: 31.

Snobar, B. A., Duwayri, M., Haddad, N. I. and Tell, A. M. 1985. *Jordan Dirasat* 12: 7–20.

Spaeth, S. C. and Muehlbauer, F. J. 1991. *Crop Science* 31: 1395.

Srivastava, J. P., Porceddu, E., Acevedo, E. and Varma, S. 1987. *Drought Tolerance in Winter Cereals,* 385 p. UK: John Wiley.

Current status and future strategy in breeding chickpea for resistance to biotic and abiotic stresses

K.B. SINGH[1], R.S. MALHOTRA[1], M.H. HALILA[2], E.J. KNIGHTS[3] and M.M. VERMA[4]

[1] ICARDA, P.O. Box 5466, Aleppo, Syria
[2] Food Legume Laboratory, Institut National de la Recherche Agronomique de Tunisie (INRAT), 2080 Ariana, Tunisia
[3] New South Wales Government, Department of Agriculture, Agriculture Research Centre, RMB 944, Tamworth, NSW 2340, Australia
[4] Department of Plant Breeding, Punjab Agricultural University, Ludhiana 141 004, Punjab, India

Abstract

Chickpea (*Cicer arietinum* L.) production has remained static for the past two decades. One major limiting factor has been susceptibility of cultivars to several biotic and abiotic stresses that adversely affect yield. In recent years, cultivars resistant to Ascochyta blight (*Ascochyta rabiei* [Pass.] Lab.), Fusarium wilt (*Fusarium oxysporum* f. sp. *ciceris*), and cold have been bred and released in many countries. Some progress has been made in breeding for resistance to drought, insects, and cyst nematode, but not for viruses, heat, and salinity. Two or more stresses are of equal importance in most chickpea growing areas. Therefore, future efforts should be directed toward the development of cultivars with multiple-stress resistance. Proper understanding of important stresses in different countries and the genetics of resistance should lead to more systematic approaches to resistance breeding. Wild *Cicer* species hold promise and deserve attention in resistance breeding.

Introduction

Chickpea (*Cicer arietinum* L.) is an important food legume; however, yields tend to be small and unstable. Yield instability is mainly due to the adverse effects of a number of biotic and abiotic stresses. Research has attempted to reduce the effects of stress and stabilize chickpea yields. Research progress towards alleviating stresses affecting chickpea productivity has been reviewed (Saxena and Singh, 1987; Summerfield, 1988; van Rheenen and Saxena, 1990; Singh and Reddy, 1991; Singh and Saxena, 1992).

Breeders have devoted considerable attention to developing cultivars resistant to a single stress to suit the farming systems under which the chickpea crop is grown, but intensified efforts are required to breed cultivars with resistance to multiple stress factors. Such cultivars are expected not only to increase yields in traditional areas but also to widen chickpea cultivation. For

F.J. Muehlbauer and W.J. Kaiser (eds.), Expanding the Production and Use of Cool Season Food Legumes, 572–591.
© 1994 Kluwer Academic Publishers.

example, in moderately low rainfall areas in northerly latitudes, wheat-fallow is a common rotation. Fallow lands could be converted to chickpea production if suitable cultivars with resistance to Ascochyta blight and cold were available. Similarly, in rice-based farming systems of southerly latitudes where rice-fallow is a common rotation, fallow lands could be replaced by short-duration, drought-tolerant cultivars.

In this paper, an effort has been made to review new information and suggest future breeding strategies for implementation in the 1990s with a view to eliminate or at least to reduce the adverse effects of stresses on chickpea yields.

Relative importance of different stress factors

The relative importance, research effort, and success of research programs in alleviating different biotic and abiotic stresses of chickpea are shown in Figure 1. Diseases are the most important stresses and have received about 60% of the total research effort. Fortunately, disease-resistance breeding has been mostly successful. On the other hand, drought is the second most important stress but only modest success has been achieved through breeding. Breeding cold-tolerant chickpeas also has been successful (Singh, 1990). There has been little or no success with all other stresses.

Biotic stresses

Disease

Among 50 diseases affecting chickpea, Ascochyta blight and Fusarium wilt are the most destructive and widespread. Other diseases are either of regional importance or of no significance.

Ascochyta blight

Ascochyta blight, caused by *Ascochyta rabiei* (Pass.) Lab., is the most destructive and widespread disease of chickpea. It has been reported from 29 countries (Nene and Reddy, 1987). Research on this disease began in 1918 in India but the most intensive research has been carried out at ICARDA since 1978 (Singh, 1992). A simple and reliable large-scale field screening technique and a 9-point rating scale were developed. About 20,000 germplasm accessions were evaluated and a number of resistant sources were identified. Five lines, ILC 200, ILC 6482, ICC 4475, ICC 6328, and ICC 12004, were identified through an international testing program as having durable resistance (Reddy and Singh, 1992). Breeding for blight resistance began in 1978; and, with the help of off-season nurseries, more than 1600 lines were bred at ICARDA and

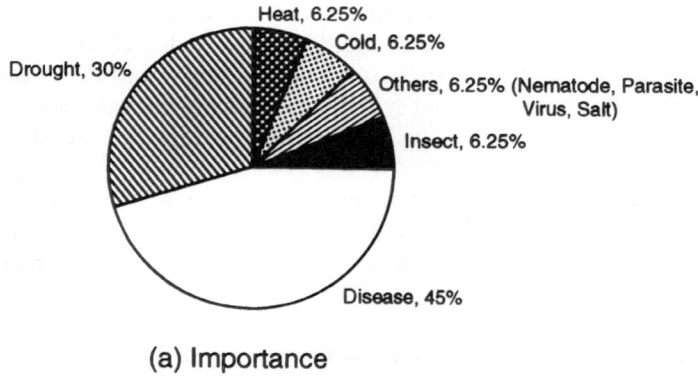

(a) Importance

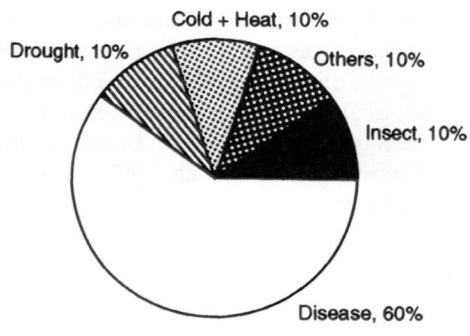

(b) Research effort

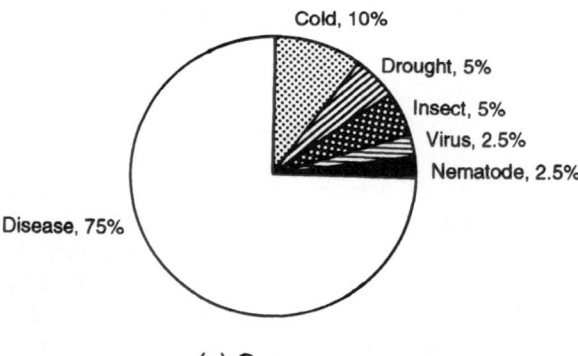

(c) Success

Figure 1. Relative importance, research effort, and success of breeding for different biotic and abiotic stress resistance in chickpea.

shared with national programs. From these breeding lines, 39 cultivars have been released by 12 countries.

In developing cultivars with durable resistance, it is necessary to map races of the pathogen worldwide to study the role of the sexual stage of the pathogen and to understand the epidemiology of the disease.

Fusarium wilt

Fusarium wilt, caused by *Fusarium oxysporum* Schlecht. emnd Snyd. & Hans f. sp. *ciceris* [Padwick] Snyd. & Hans, has been reported from 14 countries (Nene and Reddy, 1987). Research on this disease first began in India and Myanmar (Burma) in the 1920s and later in Mexico (Singh, 1987). Confusion continued on the identity of wilt and other soilborne diseases until clarified by Nene *et al.* (1980). ICRISAT evaluated more than 12,000 accessions and several hundred resistant sources were identified. Four races of the pathogen also were identified. Several hundred resistant lines were bred and shared with national programs, of which a few have been released. Mexico was the first country to breed wilt-resistant cultivars (e.g., 'Surutato 77' and 'Sonora 80'). Tunisia developed 'Amdoun 1' and the USA developed 'UC 15' and 'UC 27' as wilt-resistant cultivars. In general, wilt-resistant cultivars are not widely grown, except in Mexico, because they are susceptible to other soilborne diseases such as Verticillium wilt (*Verticillium albo-atrum* Reinke & Berth.) and dry root rot (*Rhizoctonia bataticola* [Taub.] Butler) that coexist with wilt.

Chickpea wilt is a serious disease, particularly in the southwestern districts of Punjab, India. Four races of this pathogen (races 1, 2, 3 and 4) are known to occur in India. Originally, none of the lines possessed resistance to all races but through hybridization, the Punjab Agricultural University in India developed lines such as GL 87078 and GL 87079 that possess resistance to all races.

Races of blight and wilt pathogens

One of the major obstacles in breeding cultivars with durable resistance to blight and wilt is the presence of races of pathogens. Races of Fusarium wilt have been reported from India, Spain, and USA and these have been reviewed by Jiménez-Díaz *et al.* (1989), whereas races of Ascochyta blight have been reported from India and Syria and these have been reviewed by Singh (1990). A systematic survey of races should be conducted in disease-endemic countries to map the races. This information would facilitate better planning in future cultivar development.

Phytophthora root rot

Phytophthora (*Phytophthora megasperma* Drechs.) is the major disease problem in Australia. No immunity to the disease has been identified in about 1000 accessions tested. Rather, genotypes have been shown to vary in their level of "field resistance" but even the most resistant accessions and the cultivars displaying high mortality have had reduced yields under favorable environmental conditions and/or high inoculum levels (Brinsmead *et al.*, 1985).

Breeding for Phytophthora resistance remains the most important breeding objective of the Australian National Chickpea Breeding Program. In 1992, the first Australian cultivar bred specifically for increased Phytophthora resistance, 'Barwon', will be released. Barwon was developed from a cross between the moderately resistant Russian accession CPI 56564 and the slightly less resistant Iranian accession ICC 2903. Yield data for Barwon, a second moderately resistant selection (243–7), and three susceptible cultivars ('Amethyst', 'Dooen', and 'Tyson') from 12 Phytophthora-affected sites in northern New South Wales and southern Queensland were: Barwon, 1479 kg ha^{-1}; 243–7, 1517 kg ha^{-1}; Amethyst, 950 kg ha^{-1}; Doen, 923 kg ha^{-1}; and Tyson (C 235), 714 kg ha^{-1}.

Given the inadequacy of the resistance of current cultivars (including Barwon) and the apparent mode of inheritance of resistance, the main short-term emphasis will be on a form of recurrent selection. The most resistant genotypes will be intercrossed in an effort to pyramid resistance genes. An alternative approach is to screen other *Cicer* species for improved or additional sources of resistance. Limited screening of *C. reticulatum* has not shown any worthwhile resistance but accessions of *C. echinospermum* have shown markedly superior survival to the best chickpea lines in field and greenhouse tests. Four additional *C. echinospermum* accessions were tested in the same nursery in 1991 at Tamworth, Australia, with disease severity much greater than in 1989, as is evident from the following data:

Accessions	% Survival	% Survival of nearest Barwon check
5–1	60.5	9.4
5–2	77.2	9.6
5–3	57.9	0.0
ILWC 39	75.5	7.4

Identification of lines with multiple-disease resistance

Breeding for resistance to diseases, particularly chickpea blight and wilt complex, is the major objective under many situations. Screening for resistance to chickpea blight at the Punjab Agricultural University, Ludhiana, India has led to the identification of several germplasm lines, including P 1528–1–1, NEC 123, NEC 206, E 100 Y, E 100 Y (M), ICC 76, JM 595, P 919, ILC 2506, ILC 3279, CG 578, Negro, CPI 56566, NEC 138–2, ICC 4075, and ICC 7002. These blight-resistant lines have been extensively utilized in the breeding program to develop wilt-resistant, high-yielding lines. The segregating materials were successively tested in the blight nursery, a wilt-sick plot, and under disease-free conditions. This successive screening has led to the development and identification of several advanced generation breeding lines which combine improved yield and multiple disease resistance. A few of these lines are listed in Table 1.

Table 1. Promising lines of disease-resistant chickpea developed at the Punjab Agricultural University, Ludhiana, India

	Resistant				
Line	Blight	Gray mold	Wilt	Foot rot	Root rot
GL 88341	MR	MR	R	R	R
GL 88355	R	NT	R	R	R
GL 88356	R	MR	R	R	R
GL 87014	R	MR	R	R	R
GL 87047	R	S	R	R	R
GL 87057	R	NT	R	R	R
GL 85107	R	NT	R	R	R
GL 84212	R	NT	R	R	R
GL 84060	R	MR	R	R	R
GL 84099	R	MR	R	R	R
GL 84135	R	R	R	R	R
GL 84210	R	MR	R	R	R

MR = Moderately resistant; R = Resistant; S = Susceptible; NT = Not tested.

Insect pests

Among numerous insect pests attacking chickpea plants and seeds, pod borers (*Helicoverpa* spp. and *Heliothis* spp.), leaf miner (*Liriomyza cicerina* Rond.), aphid (*Aphis craccivora* Koch), and cut worm [*Agrotis epsilon* (Hufnagel) (Noctuidae)] in the field, and seed beetles (*Callosobruchus* spp.) in the store are most damaging. Research on pod borer and leaf miner resistance has been reviewed by Singh *et al.* (1990b) and Weigand and Tahhan (1990). However, additional research carried out on *Helicoverpa* spp. in Australia is reviewed here. Virtually no research has been done on cut worm. Cut worms are active in wet soils at the seedling stage for about a week. Farmers sow 25% more seed to compensate for cut worm damage. The research on stored pests has been reviewed by Weigand and Tahhan (1990) and it seems unlikely that these insects can be controlled through resistant cultivars, at least not in the foreseeable future.

Helicoverpa

Helicoverpa is the most significant insect pest of chickpea, causing variable damage and often meriting chemical control. In Australia, most chickpea crops are sprayed once to control *Helicoverpa*. However, in some situations crops are not sprayed because increased yield from insect control would not offset the cost of control. The major pest is *H. punctigera*, although in Queensland and northern New South Wales *H. armigera* damage can be important in the latter stages of podding. The relative importance of *H. armigera* increases with lower latitude.

The breeding approach to *Helicoverpa* control is an integrated one involving both non-preference/antibiosis and avoidance. Variation among chickpea genotypes with resistance to *Helicoverpa* has been established (Lateef, 1985). Genotypes identified at ICRISAT as being relatively resistant to *Helicoverpa* (*armigera*), presumably due to their high malate concentration, were tested against natural *Helicoverpa* attack at Tamworth in 1987 (the *Helicoverpa* population was likely to have been overwhelmingly *punctigera*). The resistant genotypes (ex ICRISAT) similarly showed less pod damage from the *punctigera*-dominant Australian population. The percentage pod damage in resistant cultivars ICC 506 and CPS-1 was 5.6 and 7.7, respectively. In moderately susceptible cultivars C 235 and Amethyst, pod damage was 12.8 and 13.5%, respectively, whereas in susceptible cultivars 203–3 and ICC 2828, pod damage was 30.9 and 32.4%, respectively.

Given that malate-mediated resistance is most likely to be quantitatively inherited (ICRISAT, 1984) and that sources significantly superior to ICC 506 have yet to be identified, the best prospect for increasing resistance using non-preference/antibiosis is through recurrent selection. The non-preference/antibiosis approach can be complemented by the avoidance approach, i.e., selecting for genotypes with the capacity to set seed under low temperature regimes. Variation among genotypes for ability to set seed under low temperatures has been demonstrated (Savithri *et al.*, 1980).

Nematodes

Among numerous nematodes attacking crop species, three nematodes – cyst (*Heterodera ciceri* Vovlas, Greco et Di Vito), root-knot (*Meloidogyne* spp.), and root lesion (*Pratylenchus thornei* Sher and Allen) – are the most widespread. However, damage done by nematodes is found in pockets and generally is not as damaging as that due to fungi-incited disease. Some recent research has been reviewed here.

Root lesion nematode (Pratylenchus spp.)

Root lesion nematodes are widely distributed in cropping regions throughout the world, and *Pratylenchus thornei* is known to cause crop damage and yield losses in chickpea. In Australia, *P. thornei* is widely distributed in heavy-textured soils throughout the wheat belt and *P. neglectus* has recently been isolated from chickpea. We are not aware of any published work on resistance to root lesion nematode in chickpea and the experience with root-knot nematodes suggests that useful resistant will not readily be found within *C. arietinum*. However, wild *Cicer* species may be a more productive source of resistance, as appears to be the case with cyst nematode (Singh *et al.*, 1990a).

Cyst nematode

Evaluation of nearly 10,000 accessions of cultivated species at ICARDA, Syria has not produced a single resistant line to cyst nematode. However, evaluation of over 200 accessions of 8 wild *Cicer* species at ICARDA has helped in identification of 22 lines of *C. bijugum*, 5 lines of *C. pinnatifidum*, and 1 line of *C. reticulatum* as resistant sources. A scheme to transfer genes for resistance from *C. reticulatum* to the cultigen was initiated in 1990, with promising results from screening F_2 and F_3 progenies. It is proposed to make backcrosses with selected F_3 plants.

Abiotic stresses

Drought

Drought is the second most important abiotic stress in chickpea. The research on drought was reviewed recently by Saxena *et al.* (1990). Major research on drought has been conducted at ICRISAT. Two approaches were used: one included the use of a line-source-sprinkler system while the second approach utilized root biomass to differentiate between resistance and susceptibility. Although both approaches were successful, neither is applicable to large-scale breeding programs.

At ICARDA, a drought screening technique based on delayed sowing has been investigated. Since chickpea plants encounter terminal drought, sowing 20 to 30 days later than normal in the spring has been effective in differentiating those lines that are tolerant to drought and those that are susceptible (Table 2). Tolerant lines produced well both in normal and delayed sowings, whereas the drought-susceptible lines virtually failed to produce any yield in delayed sowing. Likewise, delayed sowing with and without supplemental irrigation to determine yield potential was equally effective in selecting drought-tolerant materials (Table 3). Some attributes, such as early maturity, early plant vigor, fast ground cover, and large seed size, were significantly associated with drought tolerance (K. B. Singh, unpublished data). Tolerant lines produced 40 to 50% of the potential yield, as determined by delayed sowing and comparisons of irrigation regions, whereas susceptible lines produced less than 10% of the

Table 2. Yield performance of selected entries in four dates of sowing at Tel Hadya, Syria during 1990. Data in parentheses indicate rank

Entry	Yield, kg ha^{-1} (Rank)				
	D1	D2	D3	D4	Overall
ILC 72	485(24)	130(25)	11(24)	7(24)	158(24)
ILC 3279	570(23)	365(23)	76(23)	37(232)	262(23)
FLIP 85–142C	339(25)	150(24)	3(25)	6(25)	124(25)
FLIP 86–12C	702(20)	522(22)	98(22)	9(23)	333(22)
ILC 1929	1439(3)	1276(2)	948(3)	737(8)	1100(4)
FLIP 87–59C	1435(4)	1420(1)	935(4)	1004(2)	1199(1)
ILC 6104	1611(1)	1176(6)	893(5)	1007(1)	1172(2)
ILC 6118	1507(2)	1263(3)	789(7)	963(3)	1131(3)
Mean of 25 entries	1065	920	587	551	781
C.V. (%)	19.87	19.01	24.72	18.12	9.43
LSD (P≥0.05)	347.34	287.17	238.29	163.98	120.84

D = sowing date; D1 = 28 Feb, D2 = 10 Mar, D3 = 20 Mar, D4 = 30 Mar.

Table 3. Yield performance (kg ha^{-1}) of selected entries (rank order in parentheses) in response to two dates of planting and two moisture regimes at Tel Hadya, Syria during spring, 1991.

Entry	Dates of planting								
	D1*			D2			Mean		
	R		I	R		I	R		I
	Y(r)	% of I	Y(r)	Y(r)	% of I	Y(r)	Y(r)	% of I	Y(r)
ILC 72	9(25)	1.5	595(22)	1(24)	0.6	166(26)	5(25)	1.3	381(25)
ILC 3279	76(22)	9.7	783(14)	8(21)	3.6	220(25)	42(21)	8.4	502(21)
FLIP 85-142C	5(26)	1.9	265(26)	3(23)	0.6	465(21)	4(26)	1.1	365(26)
FLIP 86-12C	78(21)	12.8	611(21)	4(22)	1.3	318(24)	4(22)	0.9	465(23)
FLIP 87-58C	559(1)	65.0	860(10)	284(2)	47.1	603(17)	422(1)	57.7	732(15)
FLIP 87-59C	492(3)	55.8	882(8)	289(1)	42.8	676(13)	391(2)	5C.1	779(10)
ILC 6104	438(6)	50.1	875(9)	224(4)	33.6	666(15)	331(4)	42.9	771(11)
ILC 6118	474(4)	55.9	848(12)	257(3)	39.2	655(16)	366(3)	48.7	752(14)
Mean of 26 entries	288	36.6	786	102	15.1	676	195	26.7	731
C.V.									46.10
SE of difference between two date means									98.00
SE of difference between two entry means									87.19

*D1 = 28 Feb.; D2 = 20 Mar.; R = rainfed; I = Irrigated; Y = seed yield in kg ha^{-1}; r = rank

potential yield. Hence, both methods seem effective but the first method is somewhat easier to adopt.

Regulation of cellular turgor pressure and hydration through osmotic adjustment have been shown to increase yield potential of chickpeas under water-stress environments in Australia (Morgan *et al.*, 1991). They found that osmoregulation was associated with yield increases of up to 20% in high-water-deficit environments. Interestingly, Tyson (C 235) was one of the best osmoregulators tested. This may account for its persistence as a popular cultivar in India and its relatively better yields under low-yielding situations in Australia. A simple, repeatable seedling assay for osmoregulation, similar to that developed for wheat, would be a useful tool for screening early generations for osmotic adjustment and hence for breeding higher yielding cultivars for water-stress environments.

Sheldrake *et al.* (1978) pointed out that two pods per node were generally found only on the pods formed early, and that double-poddedness increased yield 6 to 11% where the character was well expressed (i.e., under short-season environments). Using inbred sets derived from heterozygous F_3 plants, it was found that for early sowings the mean yield of double-podded lines at Wagga Wagga, Australia was 7.4% less than that of their closely related single-podded lines. However, for late sowings the mean yield of double-podded lines was only 0.8% less (Knights, 1987). More recently, the Knight group in Australia compared the yield of an early maturing Ethiopian accession (DZ-10–11) with its double-podded BC_4F_2-derived "isoline" (unpublished) and found that the double-podded isoline produced slightly more yield than the single-podded original accession (see data below).

Isoline	Yield in kg ha⁻¹ for sowing time		
	23 June	19 July	7 September
Single-podded	2096(63)[a]	1457(44)	782(24)
Double-podded	2202(67)	1491(45)	796(24)

[a] Standard error

Despite these gains, a double-podded character does not seem to have large potential for reducing the effects of drought in chickpea.

Cold

Malhotra and Saxena (1990) reviewed the work on temperature stress in cool season food legumes and indicated that precise information on temperatures which may be stressful at a particular stage in these crops was limited. They concluded that critical temperatures for heat tolerance seem to be higher in chickpea than in lentil, pea, and faba bean and that the inverse order was true for cold tolerance.

The yield advantage of winter-sown chickpea over traditionally spring-sown chickpea has been clearly demonstrated (Singh and Hawtin, 1979; Singh 1990). For winter sowing, cultivars must possess tolerance to cold and Ascochyta blight. Singh et al. (1984) reported evaluation of more than 3000 kabuli chickpea accessions for cold tolerance at Hymana, Turkey during 1979–80, Terbol in the Beqa'a valley in Lebanon and Tel Hadya in Syria during 1978–79. Four lines at Hymana and a large number of lines at Terbol and Tel Hadya were tolerant. They also reported evaluation of chickpea lines for frost tolerance at seedling and at pre-flowering stages at Tel Hadya, Syria. There was no correlation between the tolerance rating at the pre-flowering and seedling stages, but susceptibility to cold was greater at the late vegetative stage than at the seedling stage. Later, Wery (1990) confirmed the finding that cold resistance in chickpea plants tends to decrease from germination to flowering.

Singh et al. (1989) developed a screening technique to evaluate chickpea germplasm and breeding materials for cold tolerance for low to medium elevations in the Mediterranean environments. From evaluations of 3276 germplasm and breeding lines carried out from 1981 to 1987 (Singh et al., 1989), 21 lines were identified as cold tolerant. To date, 10,000 germplasm and breeding lines have been evaluated and additional sources of tolerance have been identified. Two lines, ILC 8262 and ILC 482M, are by far the best sources of resistance.

Wery (1990) used "frost resistance ratio" and grouped the genotypes into three categories: (1) fall type (resistant to frost), (2) winter type (tolerant to frost), and (3) spring type (susceptible to frost).

Genetic studies on cold tolerance in chickpea revealed that cold tolerance is governed by both additive and non-additive gene effects with high narrow sense heritability (Malhotra and Singh, 1990, 1991). They also reported the presence of genic interactions in addition to additive and non-additive gene actions. The cold tolerance in the material was dominant over susceptibility. Thus, in breeding for cold tolerance, selection should be delayed to later generations until the dominance effects are reduced.

In the search for a higher level of cold tolerance, attempts were made to screen the wild species. van der Maesen and Pundir (1984) indicated that the perennial species *C. microphyllum* was tolerant to cold. Singh *et al.* (1990c) evaluated 137 accessions of eight wild annual *Cicer* species for cold tolerance. Most of the accessions of *C. bijugum* K.H. Rech., *C. echinospermum* P.H. Davis, and *C. reticulatum* Ladiz. were tolerant to cold and had significantly higher levels of cold tolerance than the cultivated species. The comparison of cold tolerance reaction of species among themselves revealed that *C. bijugum* had the highest level of tolerance and was closely followed by *C. reticulatum* and *C. echinospermum*. All lines of *C. chorassanicum* (Bunge) M. Popov, *C. cuneatum* Hochst. ex Rich., *C. judaicum* Boiss., and *C. yamashitae* Kitamura were susceptible. Lines of *C. pinnatifidum* Jaub. & Spach showed both susceptible and tolerant reactions. Promising efforts are underway at ICARDA to transfer genes for cold tolerance from *C. reticulatum* and *C. echinospermum* to the cultigen.

Heat

Chickpea crops are sensitive to high temperatures, especially at the full-blooming stage. Temperatures greater than 30°C for 3 to 4 days cause heavy yield losses (Summerfield *et al.*, 1984). Summerfield *et al.* (1984) studied the effects of heat stress in chickpea and found that greater exposure to heat during the reproductive period led to progressively lower yields. Plants exposed to a high temperature of 30°C at 50% flowering were almost barren. All cultivars had reduced yields when plants matured during hot days.

Salinity

Soil salinity occurs in heterogeneous patches (Chandra, 1980) and the ionic composition of the salts vary from place to place. This variability in ionic composition of salts in soils seriously restricts the application of results of any screening of breeding materials. Chandra (1980) also emphasized that because of heterogeneous occurrence of salinity in the field, it is not possible to study responses to soil salinity under field conditons. He reported that chickpea was sensitive to salinity and suggested that in chickpea it might not be possible to make use of yield-based criteria. However, Saxena (1987) suggested that because productivity is a primary concern in the saline environments, it is necessary that only yield-based criteria be used rather than indirect indices of yield performance.

At ICRISAT, field screening is being followed and relatively large rows are used so that genotypes pass through the heterogeneous patches. The scoring is done in two ways; the first is visual, using a 1–9 score (1 = no visible symptoms and 9 = susceptible, plants dead). One genotype was found promising to saline conditions in both field and greenhouse tests. Another method for evaluation of genotypes for salt tolerance is measurement of relative decline in plant biomass and yield, but because of its laborious nature, it may not be practical on a large scale.

Breeding for resistance

Single stress

Success in resistance breeding is mainly confined to disease resistance. Cultivars resistant to different diseases, including Ascochyta blight and Fusarium wilt, have been bred and released in many countries (Singh and Reddy, 1991). *Helicoverpa*-resistant lines have been bred at the ICRISAT Center but due to their susceptibility to Fusarium wilt, none of them could be released as a cultivar. Cold-tolerant lines developed at ICARDA also possess resistance to Ascochyta blight. They have been released in many countries. Lines resistant to Phytophthora root`rot have been bred in Australia and may be released in the near future. To date, no effort has been exclusively devoted to drought-resistance breeding. Breeding efforts for resistance to nematodes, viruses, root rot diseases, leaf miner, and heat are in various stages of development.

Ascochyta blight and Fusarium wilt

Both Ascochyta blight and Fusarium wilt are serious in parts of North Africa and southern Europe. The INRAT, Tunis Program has as its main objective the breeding of blight- and wilt-resistant chickpeas. The program has developed a breeding scheme for alternate generation screening for blight and wilt. A slightly modified version of this scheme is shown in Figure 2.

Two-way crosses are made between blight- and wilt-resistant cultivars. This is followed by three-way crosses, where F_1s are crossed with high-yielding parents. The F_1s of three-way crosses are grown in disease-free conditions. The F_2 bulks are grown in the Ascochyta blight nursery where the disease is artificially created by inoculating plants with diseased debris and spore suspension spray. However, some breeding programs working on blight resistance delay screening until the F_3 to allow for segregation and recombination of recessive genes for resistance. The blight resistant plants are bulk-harvested and grown as F_3 bulks in a wilt-sick plot infested with *Fusarium oxysporum* and *Fusarium solani* at the Beja Station. The wilt-resistant plants are harvested individually and grown as F_4 progenies in the blight nursery. At the end of the growing season, the promising plants from the resistant progenies are individually harvested and grown as F_5 progenies in the wilt-sick plot.

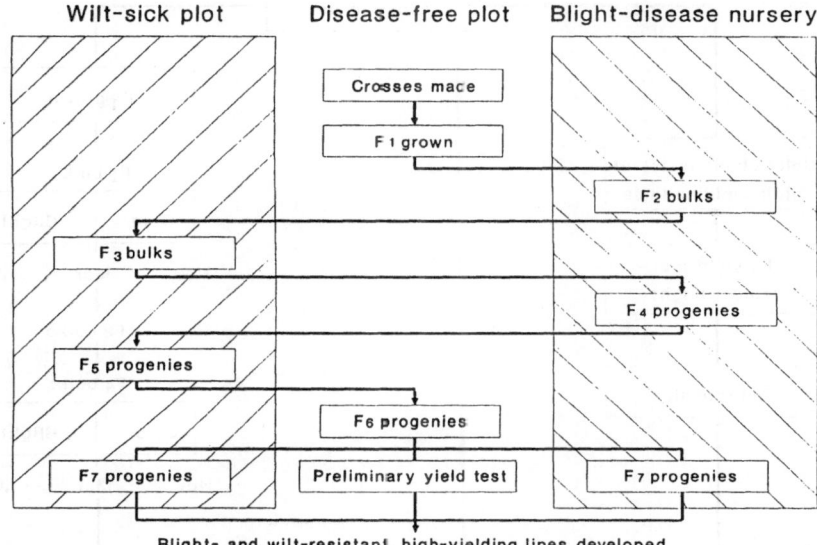

Figure 2. Scheme to develop high-yielding chickpea cultivars with combined resistance to wilt and blight.

Resistant plants from the promising progenies are individually harvested and grown as F_6 progeny rows in disease-free conditions. In this generation, selection of progenies is made for high yield and other desirable attributes. The selected progenies are bulk-harvested and evaluated for yield in preliminary yield trials. Simultaneously, small rows of each entry are grown in the blight nursery and wilt-sick plot for confirmation of resistance. High-yielding lines with resistance to blight and wilt are selected. Following this scheme, lines with combined resistance have been developed such as B91-L1/268, B91-L13/268, B91-L1/274, and B91-L1/247. These lines had a 4.5 rating for Ascochyta blight on a 9-point scale, where 1 = free and 9 = killed, and 0% of wilt incidence. These lines have a 100 seed weight of 35 to 37 g but further improvement is needed to fully satisfy the consumers.

Ascochyta blight and cold tolerance

To introduce winter chickpea in West Asia and North Africa, ICARDA has developed a breeding scheme to breed lines with resistance to cold and Ascochyta blight. This scheme is shown in Figure 3. The crosses between cold- and blight-resistant lines are made the same way as described for blight and wilt resistance. The F_2 bulks are grown from November to June. Negative selection for cold tolerance is made from November to February. The ICARDA Station at Tel Hadya in Syria experiences freezing temperatures from 17 to 56 days in

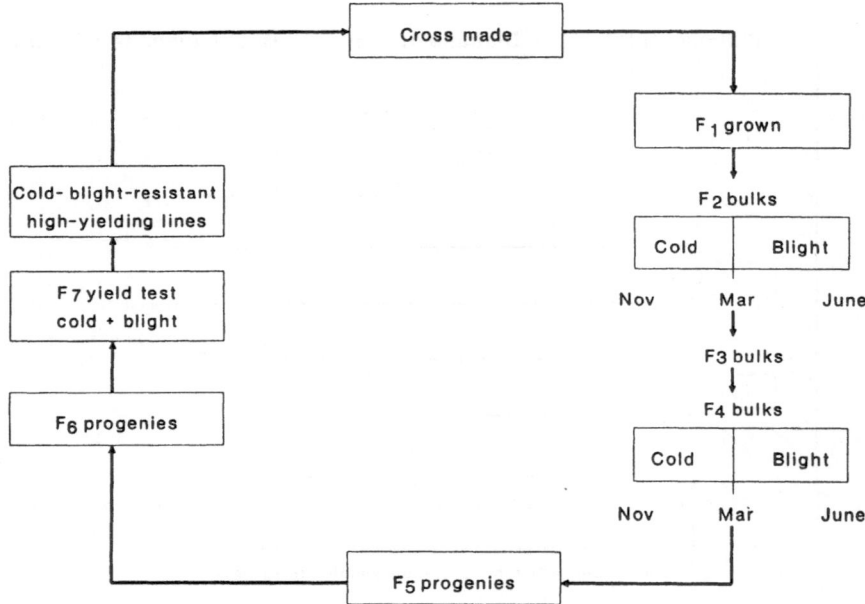

Figure 3. Scheme to develop high-yielding chickpea cultivars with cold and blight resistance.

a year with an absolute minimum of $-10°C$. In the beginning of March, the nursery is inoculated with *Ascochyta rabiei* following the standard procedure as described by Singh *et al.* (1981) and positive selection for disease resistance is made during June. The resistant plants are bulk-harvested and grown as F_3 bulks in disease-free fields. Screening for cold and blight is repeated in the F_4. The F_5 and F_6 progenies also are grown in the disease nursery. The promising F_6 progenies are bulked and evaluated for yield and resistance to cold and blight in the F_7. Following this technique, a large number of lines have been developed. These lines have been shared with national programs and national scientists have released 40 cultivars in 15 countries. Additionally, three lines, FLIP 82–97C, FLIP 82–132C, and FLIP 84–107C, were resistant to cold in the October sowing. Thus, these lines will be useful in breeding programs because they possess a high level of resistance to cold and blight.

Multiple stresses

Nowhere is an effort underway to breed cultivars with resistance to three or more stresses. After having developed cultivars with resistance to two stresses (e.g., blight and wilt or cold and blight), the stage is set to breed cultivars resistant to three stresses. Some examples for multiple-stress resistance are to combine resistance to cold, blight, and leaf miner for the Mediterranean environments and cold, blight, and drought for high-altitude areas of West Asia and North Africa.

Wild *Cicer* species

Multiple stress resistance

While resistance in accessions of cultivated species was generally found for one stress only in an accession, there were several accessions of wild *Cicer* species which had resistance to three to five stresses in ICARDA evaluations (Singh *et al.*, 1989). Ten accessions are listed in Table 4 which possess resistance to four or five stresses. Obviously such accessions would be useful in interspecific hybridization programs. At ICARDA, efforts have been underway to transfer genes for resistance to cold and cyst nematode from *C. echinospermum* and *C. reticulatum*, but the most important resistance sources are the wild species *C. bijugum, C. judaicum,* and *C. pinnatifidum*. It has not been possible to cross these wild species to the cultigen; thus no effort has been underway to transfer genes from these species. However, success has been reported in crossing these three species with the cultigen at Ludhiana, India (Verma *et al.*, 1990, 1991).

Table 4. Sources of multiple resistance in wild *Cicer* species identified at Tel Hadya, Syria

Acc. no.	*Cicer* species	Blight	Wilt[a]	Leaf miner	Bruchid	Cyst nem.	Cold
32	*bijugum*	S	R	S	R	R	R
39	*echinospernum*	S	R	R	R	S	R
46	*judaicum*	S	R	R	R	S	S
62	*bijugum*	R	R	S	R	R	R
73	*bijugum*	R	R	S	R	R	R
79	*bijugum*	S	R	R	R	R	R
81	*reticulatum*	S	R	R	S	S	R
112	*reticulatum*	S	R	S	R	S	R
181	*echinospermum*	S	R	S	R	S	R
236	*pinnatifidum*	S	NE	R	NE	R	R

NE = Not evaluated.
[a] Evaluation carried out at Istituto Sperimentale per la Patologia Vegetale, Rome.

Wide hybridization

Interspecific hybridization between four cultivars (three desi and one kabuli) of cultivated species and five annual wild *Cicer* species (*C. bijugum, C. echinospermum, C. judaicum, C. pinnatifidum,* and *C. reticulatum*) was attempted at the Punjab Agricultural University, Ludhiana, India. Cultivars were used as female parents and wild species as male parents in the crosses (Table 5). A mixture of three growth regulators, in the ratio of 100 ppm GA + 25 ppm NAA + 10 ppm

Kinetin, was applied to the pedicel base just after pollination for two to three days to promote pod setting. A total of 2610 pollinations were made with overall 1.69% success (Verma et al., 1990; 1991). Successful hybridizations were apparently obtained between *C. arietinum* and *C. reticulatum, C. echinospermum, C. judaicum, C. pinnatifidum* and *C. bijugum.*

Table 5. Percentage of success in interspecific crosses of chickpea attempted in 1988–89 and F_1s raised in 1989–90 at Ludhiana, India

Male parents (*Cicer*)	*Cicer arietinum* (Female parents)				
	PBG 1	GL 769	GL 84038	L 550	Total
reticulatum JM 2100	4.3	10.7	0	0	5.0
echinospermum No. 204	0.64	5.0	0	0	2.1
pinnatifidum No. 188 & 199	0	0	8.0	0	1.8
judaicum No. 182 & 185	0	0	8.0	0	1.8
bijugum JM 2113, No. 200 & 201	0	3.6	0	0	0.5
Overall	0.5	2.7	3.0	0	1.7

Success in crossing was dependent on the female genotype and specific combination with the wild species (Table 5). There was no success with the kabuli cultivar as a female parent. Among desis, success was maximum with a green-seeded breeding line GLG 84038 and minimum with PBG 1. Overall success was highest with *C. reticulatum* and lowest with *C. bijugum*. Apparent F_1 hybrids included: (1) *C. arietinum x C. reticulatum*, (2) *C. arietinum x C. echinospermum*, (3) *C. arietinum x C. pinnatifidum*, (4) *C. arietinum x C. judaicum*, and (5) *C. arietinum x C. bijugum*.

Hybrids were recognized by seedling growth habit. The F_1 plants had prostrate growth habits that resembled the male parents, whereas the female parents were either erect or semi-erect. In general, F_1 plants of crosses between cultivated and wild species were intermediate for days to first flower, primary branches, 100-pod weight, and grain yield.

Meiotic behavior and pollen fertility of hybrids and their parents were studied. In certain hybrids, all the pollen mother cells (PMCs) showed eight bivalents at metaphase I/diakinesis and normal disjunction at anaphase I. In others, some PMCs showed 2/4 univalents/quadrivalents at metaphase

I/diakinesis and 9–7 chromosome distribution at anaphase I. Pollen fertility in hybrids ranged from 3.6 to 98.8% in different hybrids.

The F_2 of these crosses was grown in 1990–1991. Data on four characters, namely secondary branches per plant, pods per plant, 100-seed weight, and seed yield per plant from four crosses between *C. arietinum* and *C. bijugum, C. echinospermum, C. judaicum*, and *C. pinnatifidum* are shown in Table 6. A wide range of variability, not found in intervarietal crosses, was observed. No specific data were collected on fertility levels in F_2 plants but the range of seed yield from less than 1 g per plant to 75 g per plant was an indication of wide variation for fertility. The most typical recombinants in the F_2 were plants with many basal secondary branches which resembled the branching pattern of the wild species parent in contrast to the branching pattern of the cultivars where secondary branches arose from the upper half of the primary branches.

The results of this study indicate that it may be possible to cross *C. bijugum, C. judaicum*, and *C. pinnatifidum* with the cultigen.

Table 6. Mean and range of secondary branches per plant, pods per plant, 100-seed weight and seed yield per plant in four populations of interspecific crosses

Cross number	No. of F_2 plants	Mean	Range	S.D.	C.V.	Skewness
	Secondary branches per plant					
1	216	26.3	1–63	13.0	49.6	0.30
2	736	24.0	2–64	11.1	46.3	0.26
3	1793	27.7	2–93	13.7	49.5	0.15
4	101	22.7	5–58	10.9	47.8	0.57
	Pods per plant					
1	216	107.4	1–474	98.1	91.4	1.10
2	736	102.7	1–705	86.0	83.7	0.72
3	1793	137.7	1–679	106.8	77.6	0.72
4	101	143.5	10–563	98.1	68.3	0.56
	100-seed weight (g)					
1	216	16.9	4.0–27.8	3.6	21.0	-0.07
2	736	15.2	4.0–26.6	3.5	23.0	0.19
3	1793	14.9	2.5–26.2	3.4	22.9	-0.06
4	101	15.4	6.0–28.0	3.5	22.6	-0.20
	Seed yield per plant (g)					
1	216	21.0	0.4–77.2	19.5	97.8	1.03
2	736	16.7	0.1–96.5	14.7	88.5	0.93
3	1793	19.4	0.1–85.4	15.0	77.2	0.67
4	101	22.5	1.4–75.2	15.7	70.0	0.85

Cross No. 1 = GL 769 x *C. echinospermum*; 2 = GLG 84038 x *C. pinnatifidum*; 3 = GLG 84038 x *C. judaicum*; 4 = GL 769 x *C. bijugum*.

Future strategies

1. A systematic disease survey should be initiated to identify potentially important diseases in each of the major chickpea-growing countries of the world. The disease survey should be followed by a race survey, especially for the two most important diseases, Ascochyta blight and Fusarium wilt. Attempts should be made to identify sources of resistance to different races and to pyramid the genes involved.
2. Breeding for multiple stress resistance should receive the highest priority because cultivars with multiple-stress resistance are expected to be widely adopted by growers.
3. A major effort should be made to exploit wild *Cicer* species in resistance breeding as they appear to be valuable sources of resistance genes.
4. Develop cultivars which flower two to three weeks earlier in cool temperatures. Such lines will have a longer reproductive period resulting in higher yield. Additionally, they could avoid extreme drought periods during pod-filling.
5. Selection within multiple stress resistant material should be made for two pods per peduncle to take advantage of the benefits of that trait under drought conditions.

References

Brinsmead, R. B., Rettke, M. L., Irwin, J. A. G. and Ryley, M. J. 1985. *Plant Disease* 69:504–506.
Chandra, S. 1980. In: *Proceedings of the International Workshop on Chickpea Improvement*, pp. 97–105 (eds. J. M. Green, Y. L. Nene and J. B. Smithson). Patancheru, Andhra Pradesh, India: ICRISAT.
ICRISAT. 1984. *Annual Report, 1983*. Patancheru, Andhra Pradesh, India: ICRISAT.
Jiménez-Díaz, R. M., Trapero-Casas, A. and Cabrera de la Colina, J. 1989. In: *Vascular Wilt Diseases of Plants*, pp. 515–520 (eds. E. C. Tjamos and C. Beckman). NATO ASI Series, Vol h28. Berlin: Springer-Verlag.
Knights, E. J. 1987. *International Chickpea Newsletter* 17:6–7.
Lateef, S. S. 1985. *Agriculture, Ecosystems and Environment* 14:95–102.
Malhotra, R. S. and Saxena, M. C. 1993. In: *Breeding for Stress Tolerance in Cool-Season Food Legumes*, pp. 227–244 (eds. K. B. Singh and M. C. Saxena). Chichester, UK: John Wiley & Sons.
Malhotra, R. S. and Singh, K. B. 1990. *Journal of Genetics and Breeding* 44:227–230.
Malhotra, R. S. and Singh, K. B. 1991. *Theoretical and Applied Genetics* 82:598–601.
Morgan, J. M., Rodríguez-Maribona, B. and Knights, E. J. 1991. *Field Crops Research* 27:61–70.
Nene, Y. L., Haware, M. P. and Reddy, M. V. 1978. *ICRISAT Information Bulletin 3*. Patancheru, Andhra Pradesh, India: ICRISAT. 44 pp.
Nene, Y. L. and Reddy, M. V. 1987. In: *The Chickpea*, pp. 233–270 (eds. M. C. Saxena and K. B. Singh). Oxon, UK: CAB International.
Reddy, M. V. and Singh, K. B. 1992. *Crop Science* 32:1079–1080.
Savithri, K. S., Ganapathy, P. S. and Sinha, S. K. 1980. *Journal of Experimental Botany* 31:457–481.
Saxena, M. C. 1987. In: *Adaptation of Chickpea and Pigeonpea to Abiotic Stresses*, pp. 135–142 (eds. N. P. Saxena and C. Johansen). Patancheru, Andhra Pradesh, India: ICRISAT.
Saxena, M. C. and Singh, K. B. (eds.). 1987. In: *The Chickpea*. Oxon, UK: CAB International. 409 pp.

Saxena, N. P., Johansen, C., Saxena, M. C. and Silim, S. N. 1993. In: *Breeding for Stress Tolerance in Cool-Season Food Legumes*, pp. 245–270 (eds. K. B. Singh and M. C. Saxena). Chichester, UK: John Wiley & Sons.

Sheldrake, A. R., Saxena, N. P. and Krishnamurthy, L. 1978. *Field Crops Research* 1:243–253.

Singh, G. 1990. *Indian Phytopathology* 43:48–52.

Singh, K. B. 1987. In: *The Chickpea*, pp. 127–162 (eds. M. C. Saxena and K. B. Singh). Oxon, UK: CAB International.

Singh, K. B. 1990. In: *Present Status and Future Prospects of Chickpea Crop Production and Improvement in the Mediterranean Countries*, (eds. M. C. Saxena, J. I. Cubero and J. Wery). Options Mediterraneennes, Serie A: Seminaires Mediterraneens, Numero 9. Zaragoza, Spain: CIHEAM.

Singh, K. B. 1992. In: *Durability of Disease Resistance*, pp. 241–248 (eds. Th. Jacobs and J.E. Parlevliet). Dordrecht, The Netherlands: Kluwer Academic.

Singh, K. B., Di Vito, M., Greco, N. and Saxena, M. C. 1990a. *Nematologia Mediterranea* 17:113–114.

Singh, K. B. and Hawtin, G. C. 1979. *International Chickpea Newsletter* 1:4.

Singh, K. B., Hawtin, G. C., Nene, Y. L. and Reddy, M. V. 1981. *Plant Disease* 65:586–587.

Singh, K. B., Kumar, J., Haware, M. P. and Lateef, S. S. 1990b. In: *Chickpea in the Nineties*, pp. 233--238, (eds. H. A. van Rheenen, M. C. Saxena, B. J. Walby and S. D. Hall). Patancheru, Andhra Pradesh, India: ICRISAT.

Singh, K. B., Malhotra, R. S. and Saxena, M. C. 1989. *Crop Science* 29:282–285.

Singh, K. B., Malhotra, R. S. and Saxena, M. C. 1990c. *Crop Science* 30:1136–1138.

Singh, K. B. and Reddy, M. V. 1991. *Advances in Agronomy* 45:191–222.

Singh, K. B. and Saxena, M. C. (eds.). 1992. In: *Disease Resistance Breeding in Chickpea*. Aleppo, Syria: ICARDA. 196 pp.

Singh, K. B. and Saxena, M. C. (eds.). 1993. In: *Breeding for Stress Tolerance in Cool-Season Food Legumes*. Chichester, UK: John Wiley & Sons.

Singh, K. B., Saxena, M. C. and Griedly, H. E. 1984. In: *Ascochyta Blight and Winter Sowing of Chickpeas*, pp. 159–166 (eds. M. C. Saxena and K. B. Singh). The Hague, The Netherlands: Martinus Nijhoff/Dr. W. Junk Publishers.

Singh, K. B., Weigand, S., Haware, M. P., Di Vito, M., Malhotra, R. S., Tahhan, O., Saxena, M. C. and Holly, L. 1989. Evaluation of wild *Cicer* species to biotic and abiotic stresses 14:13 XII EUCARPIA Congress, February 27 – March 4, Goettingeh, F. G. Germany.

Summerfield, R. J. (ed.). 1988. In: *World Crops: Cool-Season Food Legumes*. Dordrecht: Kluwer Academic Publishers. 1179 pp.

Summerfield, R. J., Hadley, P., Roberts, E. H., Minchin, F. R. and Rawsthorne, S. 1984. *Experimental Agriculture* 20:77–93.

van der Maesen, L. J. G. and Pundir, R. P. S. 1984. *Plant Genetic Resources Newsletter* 57:19–24.

van Rheenen, H. A., Saxena, M. C., B. J. Walby and S. D. Hall (eds.). 1990. In: *Chickpea in the Nineties*. Patancheru, Andhra Pradesh, India: ICRISAT. 403 pp.

Verma, M. M., Sandhu, J. S., Brar, H. S. and Brar, J. S. 1990. *Crop Improvement* 17:179–181.

Verma, M. M., Sandhu, J. S. and Ravi, K. 1991. In: *Proceedings of the Golden Jubilee Symposium on Genetic Research and Education: Current Trends and the Next Fifty Years*, Abstracts Vol. III, pp. 686–687. New Delhi, India: Indian Society of Genetics and Plant Breeding, Indian Agricultural Research Institute.

Weigand, S. and Tahhan, O. 1990. In: *Chickpea in the Nineties*, pp. 169–176 (eds. H. A. van Rheenen, M. C. Saxena, B. J. Walby and S. D. Hall). Patancheru, Andhra Pradesh, India: ICRISAT.

Wery, J. 1990. In: *Present Status and Future Prospects of Chickpea Crop Production and Improvement in the Mediterranean Countries*, pp. 77–86 (eds. M. C. Saxena, J. I. Cubero and J. Wery). Options Mediterraneennes, Serie A: Seminaires Mediterraneens, Numero 9. Zaragoza, Spain: CIHEAM.

Present status and future strategy in breeding faba beans (*Vicia faba* L.) for resistance to biotic and abiotic stresses

D.A. BOND[1], G.J. JELLIS[1], G.G. ROWLAND[2], J. LE GUEN[3], L.D. ROBERTSON[4], S.A. KHALIL[5] and L. LI-JUAN[6]

[1] *Plant Breeding International Cambridge, Maris Lane, Trumpington, Cambridge CB2 2LQ, UK;*
[2] *Crop Development Centre, University of Saskatchewan, Saskatoon, Saskatchewan, S7N 0W0 Canada;*
[3] *Station d'Amélioration des Plantes, INRA, Le Rheu, France;*
[4] *Genetic Resources Unit, ICARDA, P. O. Box 5466, Aleppo, Syria;*
[5] *Field Crop Research Institute, Research Section, Agricultural Research Center, P. O. Box 12619, Giza, Egypt, and*
[6] *Faba Bean Germplasm and Breeding, Zhejiang Academy of Agricultural Sciences, Hangzhou, China*

Abstract

Progress is being made, mainly by ICARDA but also elsewhere, in breeding for resistance to *Botrytis*, *Ascochyta*, *Uromyces*, and *Orobanche*; and some lines have resistance to more than one pathogen. The strategy is to extend multiple resistance but also to seek new and durable forms of resistance. Internationally coordinated programs are needed to maintain the momentum of this work.

Tolerance of abiotic stresses leads to types suited to dry or cold environments rather than broad adaptability, but in this cross-pollinated species, the more hybrid vigor expressed by a cultivar, the more it is likely to tolerate various stresses.

Introduction

Faba bean is a valuable protein-rich food that has sustained large human populations and provides an alternative to soybean meal for animal feed in temperate regions, but the total area in the world is declining (3.7 m ha in 1979–81 to 3.2 m in 1989), partly due to unreliable yields.

It is a plant that is capable of heavy yields but is sensitive to stress, and because of economic pressures, growers are increasingly reluctant to take the risk of poor returns from the crop. The contribution that plant breeding is making towards the alleviation of this problem was recently reviewed by Robertson and Saxena (1993), but in this paper we have taken reports from seven authors covering a wider geographical area from which we hope to provide an overview of the status of breeding for resistance to stress and to discuss how strategies may develop.

The biotic stresses that we examined included the major pests and pathogens;

F.J. Muehlbauer and W.J. Kaiser (eds.), Expanding the Production and Use of Cool Season Food Legumes, 592–616.
© 1994 *Kluwer Academic Publishers.*

also the effect of stress on organisms that are normally beneficial to faba beans. Abiotic stresses were mainly to do with the effect of extremes of temperature and humidity.

The aim was also to consider the impact of changes that are occurring in breeding objectives, such as nutritional quality and in breeding methods, on the plant's ability to tolerate stresses. Though each stress was expected to vary in importance among the regions, a final goal of this review was to try to identify common problems and strategies to tackle them, particularly where international collaboration might lead to a more speedy solution than parochial endeavors.

First, each type of stress was considered for some of the regions where it is known to have importance. Authors have first hand experience of faba beans in UK, France, North Africa, the Middle East, China, North America, and countries in the ICARDA sphere of influence.

Biotic Stresses

Chocolate Spot (Botrytis fabae and B. cinerea)

This has been for many years one of the most widespread and potentially devastating diseases of faba beans. Yield losses can be heavy where luxuriant vegetative growth and a humid microclimate allow the fungus to go into its aggressive stage. Thus, it is most frequently reported from humid regions such as the Nile Delta, Yangtse valley, rainy coastal areas of the Mediterranean, and the more oceanic climate of western France and western UK.

Figure 1. Generally recognised centres of diversity and sources of chocolate spot resistance ● putative centre of origin; ● secondary centres of diversity; ○ sources of chocolate spot resistance ●●●●● ancient region of cultivation of *V. faba*; ◇ region of winter hardy types.

In view of this, it is remarkable that the most consistently resistant accessions (ICARDA lines BPL 710 and BPL 1179) have come from close to a mountainous region in Ecuador (Figure 1) where there have been only about 300 years for resistance to evolve. We attribute this to wide general genetic variation introduced by Spanish settlers, adaptation to extremely varied environments within short distances due to altitude, frequent gene exchanges by insect pollinators and movement of peoples, and natural screening for resistance in the more humid of those environments. Further collections from the same

Table 1. Sources of resistance to major pathogens

Disease or pest	Source (underlined – highly resistant)	Origin
Chocolate spot	BPL 110,112	UK
	BPL 261,266	Greece
	BPL 710, 1179, ILB 3025, 3026, 2282	Equador
	BPL 1196	Spain
	BPL 1278	Syria
	BPL 1821, 1763	Ethiopia
	L83114, L82003, L82009	ICARDA*
	Zhehiang 41 LAO, Qi Dou No. 2, Lu-Xiao-Li-Zhong	China
Ascochyta blight	BPL 74	Iraq
	BPL 230, 365	Morocco
	BPL 460, 465, 471, 472	Lebanon
	BPL 818	Ethiopia
	BPL 266	Greece
	BPL 646, 2485, ILB 752, L83118, 3120, 3124	ICARDA*
	L8 3125, 3127 3129, 3136, 3142, 3149, 3151, 3155	ICARDA*
	L83156, 2001, 31818-1	ICARDA*
	Quasar, Line 224	UK
	29H	France
Rust	BPL 7, 8	Jordan
	BPL 260, 261, 263	Greece
	BPL 309	Turkey
	BPL 406, 417, 427, 490	Spain
	BPL 484	Uruguay
	BPL 524	Japan
	BPL 533	USA
	BPL 539, Qi Dou No. 2	China
	82-15563	Canada
	BPL 552	Iran
	BPL 554, 567, 571, 573, 576, 588, 604, 610	Egypt
	BPL 627	Algeria
	BPL 649, 663, 665, 667, 680	Tunisia
	BPL 640, 643	UK
	BPL 702	Sudan
Stem nematode	BPL 1, 10, 11, 12	Jordan
	BPL 21, 23, 26, 27, 40	Syria
	BPL 63, 88	Iraq
	BPL 183	Afghanistan
	29H	France
	1827, 1698, 1696, 84, 154	ICARDA*
BYMV	BPL 1584, 1567, 2875, 1592, 1530, 1541, 1581, 1597, 1351, 1363, 1366, 1371	ICARDA*
Orobanche	Giza 402, BPL 241	Egypt
	2830, 18105, 2210, 18009, 18035, 18049	N. Africa
	15-2-5, (402 x INIA 06) x 402	N. Africa
	Chiaro TL	Greece
	Baraca	Spain
Aphids	BPL 23	ICARDA*
	Rastatt	Germany
	Line 14	UK

*ICARDA - Lines selected by ICARDA

region have been made more recently and new sources of resistance obtained (Hanounik and Robertson, 1988).

The resistance of the original source (ILB 938 = BPL 1179) (Table 1) was first confirmed in the Nile Delta and, after crossing there to the local cultivar Giza 3, has now been transferred to locally adapted material and released as "Giza 461". Diallel crosses in Egypt indicated that dominant, additive, and reciprocal gene effects for resistance were stable over a number of years. ICARDA has also successfully transferred BPL 710 resistance into other well adapted genetic stocks, including a Moroccan base, and to alternative plant types like determinate habit and independent vascular supply.

However, the transfer of BPL 710 type resistance into the more widely divergent backgrounds of French and English winter hardy faba beans is proving more difficult and a slower process. The genotype BPL 710 is resistant when spring sown in France or England and in all other ICARDA locations, but differences in susceptibility among winter beans in those two countries are much less reproducible (Table 2). Lines that appear resistant as seedlings or when the disease is non-aggressive are often susceptible as adult plants or in the presence of aggressive *Botrytis* (Figure 2). Detached-leaf test data do not always correlate with field observations (Harrison, 1981; Tivoli *et al.*, 1986). Defoliation due to infection is not always related to spread from lesions, extent of crop damage in the field can be more a consequence of the amount of sporulation on fallen and

Table 2. Ranking of five lines, out of a group of 10, for resistance to chocolate spot and Ascochyta at Rennes (1 = resistant)

Disease	29H	Bourdon	Soravi	3.33	48B
CHOCOLATE SPOT					
Detached leaf					
spots after 15 h	10	5	6	8	9
spots per h	9	3	8	10	7
score after 6 days	1	7	2	9	10
Plants in pots					
young plant	9	4	8	4	5
flowering stage Yr 1	2	7	3	10	9
flowering stage Yr 2	6	7	1	9	10
Field assessment					
disease score	9	2	3	10	5
flower damage Yr 1	9	2	3	10	5
flower damage Yr 2	8	1	5	3	9
ASCOCHYTA BLIGHT					
Adult plant					
leaf (Year 1)	3	4	6	8	9
stem (Year 1)	3	4	7	8	10
leaf (Year 2)	3	4	6	8	10
% damaged stems (Yr 2)	2	4	7	6	10
% damaged pods (Yr 2)	1	4	6	7	10
Young plant					
leaves (Year 1)	2	4	6	9	8
pods (Year 1)	2	4	5	9	10
Flowering stages					
leaves (Year 1)	2	4	9	10	8
pods (Year 1)	2	8	9	3	10

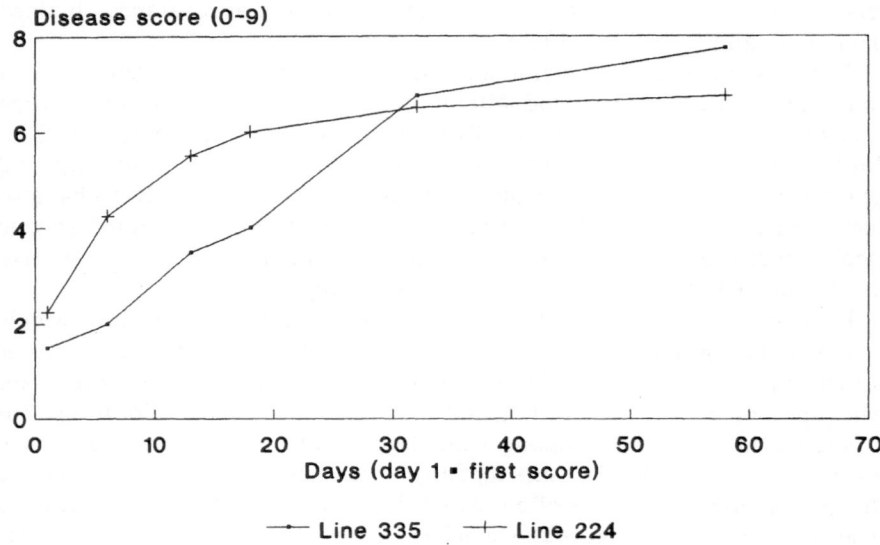

Figure 2. Scores for Botrytis infection after inoculation of two inbred lines at Cambridge, UK.

senesced leaves that re-infect green plants than of rate of disease spread in the live canopy (Gondran, 1975).

The effect of chocolate spot can, however, be ameliorated by breeding for (a) early flowering so that most pods are set before the disease goes aggressive (cv Punch escapes some chocolate spot in UK this way), (b) slow leaf senescence, (c) open canopy or (d) resistance to predisposing factors such as frost or virus damage. Inbreeding often uncovers greater susceptibility than in existing populations, so a high level of heterozygosity should be maintained.

The BPL 710 resistance is thought to be quantitative rather than qualitative (Robertson and Saxena, 1992) and additive with some dominance (El-Hady Mohamed, 1988) so it is not surprising that it is difficult to transfer all the genes concerned to widely different backgrounds. Up to 21 other sources of resistance have been identified by ICARDA and there are three genotypes from China with some resistance (Table 1) but most of these do not have as high a level of resistance or have been shown to be susceptible in certain countries (Halila *et al.*, 1990). Existence of races of *B. fabae* has been proposed on the basis of the reaction of five differentials in seven Mediterranean countries but confirmation is needed from tests of all sources of inoculum on the putative differentials in the same environment because quantitative resistance is often highly influenced by environment.

On limited evidence BPL 710 resistance seems durable, and other sources have incomplete resistance or field resistance so the future strategy should be strict monitoring of durability of resistance, and the maintaining of bean cultivars as mixed populations which rarely challenge the pathogen.

Ascochyta Blight (Ascochyta fabae)

Average yield losses due to *Ascochyta* (5 to 50% in Poland, Zakrweska, 1986; 7.5 to 41% in Czechoslovakia, Ondrej, 1991) are less than for *Botrytis* but *Ascochyta* can spread under cooler and drier conditions, and can be seedborne, so resistance is important not only to reduce biotic stress but also to avoid quarantine and (in the UK) seed certification problems.

Resistance has been found in wider backgrounds, results of tests more positive and reliable (Table 2), and the situation on races clearer than with *Botrytis*. In addition to 24 sources of resistance listed by ICARDA (Hanounik and Robertson, 1988; and Table 1), lines with good resistance have been found in adapted material in France (Line 29H), England ("Quasar"), and in Poland ("Fioletowy" and "Czyzowskich" see Zakrzewska, 1986).

Thus there are fewer places where resistance needs to be transferred between diverse types. However, use of the ICARDA lines, especially BPL 471, 460, 74 and 2485, has already been made in various countries and at ICARDA with the objective of combining resistance with improved plant architecture (Robertson and Hanounik, 1987; Hanounik and Robertson, 1989). Also, as there is evidence of differential interactions between faba bean genotypes and isolates of *A. fabae* (Hanounik and Robertson, 1988, 1989; Halila *et al.*, 1990; Rashid *et al.*, 1991), there is a possibility of a breakdown in resistance and therefore a need for more than one source in each breeding program. Some European breeders, for example, are therefore adopting a strategy of utilizing resistance found in Mediterranean as well as local cultivars. BPL 471 is thought to carry a broad based general resistance compared with a narrow based resistance in BPL 818 (Hanounik and Robertson, 1989). But durability of the present best sources of resistance also needs constant monitoring.

Resistance has been observed in tannin-free as well as tannin-containing cultivars but a suggestion that seed transmission may be more frequent in tannin-free cultivars requires investigation (Jellis *et al.*, 1991).

Rust (Uromyces fabae)

Though previously considered of secondary importance because the disease appeared late in the life of the host plant, recent estimates of yield loss are up to 80% in China and 27 to 32% in UK (Yeoman *et al.*, 1987; Dobson and Giltrap, 1991). Appearances of the disease earlier in the season in northwest Europe have been associated with the introduction there of earlier maturing cultivars.

Race-specific resistance controlled by major genes was established by Conner and Bernier (1982) and Rashid and Bernier (1986), and only one of 65 lines selected by them (Line 82L-15563–1) was rated as resistant in the field in Syria (Hanounik and Robertson, 1988).

The postulate of Conner and Bernier (1982) that a quantitative form of resistance ("slow rusting") would be more durable than race specific immunity has so far been upheld. Most of the resistant lines bred by ICARDA (Table 1) probably have this slow rusting type of resistance. Thus, unless facilities are available for continuous study and prediction of races before they appear in nature, the best strategy is to utilize lines with proven field resistance. BPL 261, for example, is being used in crosses in Europe.

Downy Mildew (Peronospora fabae)

This was not seen to any extent until about 1985, but now occurs in most faba bean trials in northwest Europe. Clear genotypic differences in susceptibility exist with, for example, Maris Bead being resistant and Troy susceptible. Little breeding is being done but a basis for improvement exists.

Foot Rot (Fusarium solani and Other Fusarium spp.) and Wilt (F. oxysporum)

These diseases are soilborne so yield losses may depend on the frequency of faba beans in the rotation. In China, where foot rots and wilts are next in importance after *Botrytis* (Liang, 1989), it was estimated that Fusarium foot rot was responsible for 6% loss in the first year of faba bean but 38% loss with continuous beans. In general, the extent of damage is often influenced by concurrent abiotic stresses like waterlogging, soil compaction, heat, and drought. However in Russia "Burshtyn 56" was reported resistant (Yartiev, 1976), and tolerance as reported in Germany for a homozygous line, KK13, selected from the Austrian cv Kornberger Kleinkornige. KK13 is being used in joint EC-program crosses. Yield of tannin-free genotypes is low in the presence of *Fusarium* infection (Link and Hempel, 1992), and the tannin-free member of some near isogenic pairs is more susceptible to *Fusarium* spp. than the tannin-containing member, especially in stressed soil conditions like low temperature and high moisture (Kantar, 1992). On the other hand, resistance can also be selected in zero-tannin lines (van Loon *et al.*, 1989; Pascual Villalobos and Jellis, 1990).

Inbreeding in populations can often reveal lines that are significantly more susceptible than the mean of the population. Thus a method of maintaining existing tolerance of stress and of *Fusarium* is to breed cultivars with a good level of heterogeneity and which display heterosis.

Stem Rot (Sclerotinia trifoliorum and S. sclerotiorum)

This is a major stress factor only when other susceptible crops occur frequently in the rotation. In the UK, winter faba beans are affected only by *S. trifoliorum* (Jellis *et al.*, 1990) but infections of *S. sclerotiorum* on spring faba beans restrict entry of rapeseed, linseed, and pea in the rotation.

In Greece, damage has been serious enough to promote a breeding program. Resistance has been reported in Lines KU 189, 190, and 191 (Podimatas, Pers. comm. 1990). But more investigations are needed to relate laboratory tests to field observations.

Viruses

There are a large number of viruses contributing to biotic stress and it would be almost impossible to breed for resistance to all of them. Annual average yield losses in Germany were estimated as 8% (Schmidt, 1984). Incidences were: bean yellow mosaic virus (BYMV) 13%, pea enation mosaic virus (PEMV) 5%, bean leaf roll virus (BLRV) 3%, and the seed and weevil transmitted viruses, true broad bean mosaic virus (TBBMV), and broad bean stain virus (BBSV), about 0.5%. Similar incidences were reported in Poland (Blaszczak and Fiedorow, 1979); but, in France and the UK, BLRV is relatively more important.

ICARDA found two accessions from Afghanistan to be resistant to BLRV (BPL 756 and 758), and four lines with resistance to BYMV (BPL 1351, 1363, 1366, and 1371) (Robertson and Saxena, 1993; and Table 1). Two other independent sources of resistance to BYMV were reported by Rohloff and Stulpnagel (1984). Schmidt *et al.* (1986) described a line that combines resistance to 11 strains of BYMV with some resistance to *Aphis fabae*, while Schmidt *et al.* (1989) wrote of a line with resistance to both BYMV and PEMV (Table 3). In the UK, the common cultivars have varied significantly in frequency of plants infected with BLRV, Maris Bead and Wierboon being among the more resistant (Lawes *et al.*, 1983). No resistance to the seed transmitted viruses is known, so care has to be taken with quarantine.

In many countries viruses are not damaging often enough to warrant priority in breeding programs. The main strategy is to attempt to combine resistance to the common viruses with resistance to another major stress factor.

Stem Nematode (Ditylenchus dipsaci)

The giant faba bean race is important in Morocco and could be in other countries if control over crop rotations and checking for seed infection is relaxed. Twelve ICARDA lines have some resistance (Hanounik *et al.*, 1986; Robertson and Saxena, 1993), and this was confirmed for BPL 1696 and 1827 and FLIP 84–154 (Caubel and Leclercq, 1989) but only the French INRA line, 29H, has a high level of resistance (Caubel and Le Guen, 1992). This is controlled by a single gene and is inherited maternally through 29H cytoplasm.

Breeding may commence in France, while Morocco should utilize ICARDA sources of resistance, but otherwise control will mainly be by clean seed (quarantine in countries where the nematode does not exist) and long rotations.

Table 3. Examples of Multiple Resistance

Lines	1	2	3	4	5	6	7	8	9	10	11	12
L82003	R	R										
L82005	R	R	R									
L82006	R	R	R									
L82007	R	R	R									
L82010	R		R									
L82013		R	R									
BPL 261	R		R									
BPL 1179	R		R	R	R							
Qi Dou 2	R		R									
29H		R				R						
BPL 23						R	R	R				
BPL 26						R		R				
88123	R						R					
8810	R							R				
8817	R							R				
Schmidt 1989									R	R		
Schmidt 1986								R	R			
Nadwislanski	R	R									R	

* 1 = *Botrytis*, 2 = *Ascochyta*, 3 = *Uromyces*, 4 = *Alternaria*,
5 = *Peronspora*, 6 = *Ditylenchus*, 7 = *Aphis craccivora*,
8 = *A. fabae*, 9 = BYMV, 10 = PEMV, 11 = *Fusarium*, 12 = *Orobanche*
R = resistant.

Broomrape (*Orobanche crenata*)

Orobanche is a major stress factor in many Mediterranean countries. Resistance of "Giza 402" (Nassib *et al.*, 1982) has proved useful in upper Egypt where *Botrytis* is not a problem, but because Giza 402 is susceptible to *Botrytis* and is unadapted to the western Mediterranean region other sources of resistance to *Orobanche* are being sought. Also resistance of Giza 402 has been partly attributed to late flowering and late stimulation of *O. crenata* to germinate (Kheir *et al.*, 1989), or to low root biomass (Khaled and El-Bastewesy, 1989), so different mechanisms of resistance are required.

Screening in Morocco resulted in BPL 2830 being rated as resistant, but progress is most likely among selections from crosses of the best from the

Spanish (Cubero *et al.*, 1988) and from the Egyptian programs e.g., (402 ×
INIA 06) × 402 (Robertson and Saxena, 1992) or the cross VF 1071 (402
derivative) × "Brocal" which gave the resistant cv Baraca (Cubero *et al.*, 1992).
Several of these are proving to be resistant in Spain, Morocco, Algeria, and
Syria though with lower resistance to *O. foetida* in Tunisia (Hanounik *et al.*,
1992; Kharrat *et al.*, 1992). In Morocco none of the 13 lines tested in 1989/90
and 1990/91 averaged more than 1.2 shoots or 2.2 g of shoots per host plant
compared with 8.9 and 18.7 g, respectively for the control, "Aquadulce". This
difference was reflected in a much higher yield for the selected lines (Figure 3).
Another possible source of resistance is "Chiaro TL", a cultivar that was
significantly less susceptible than 11 others in Greece (Karamanos and
Avgoulas, 1989).

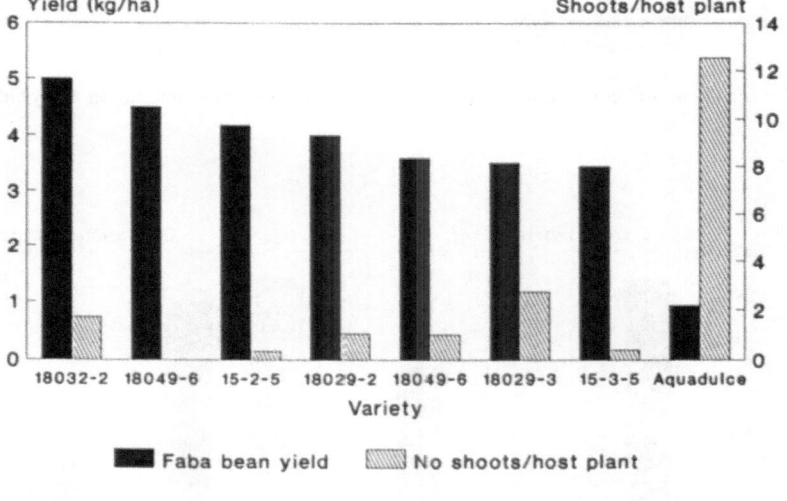

(ICARDA FLIP Annual Report 1990)

Figure 3. Performance of Orobanche-resistant lines in a naturally infested field in Morocco.

Mutation breeding (with gamma rays and EMS) also produced some
resistance which appeared to be under polygenic control (Hussein *et al.*, 1988)
but further single plant selection was necessary in the same way that it was
among populations collected in Egypt. BPL 241 was resistant but other bean
lines were tolerant of only some *Orobanche* accessions (Radwan *et al.*, 1988).

Inheritance of the Giza 402 type of resistance is strongly additive so the most
appropriate breeding method is recurrent selection following intercrosses of
selected F_2 or backcross plants from original resistant x susceptible crosses
(Cubero *et al.*, 1988). Strict control of pollination and continuous selection are
necessary to prevent dilution of the resistance. However, effective use of the

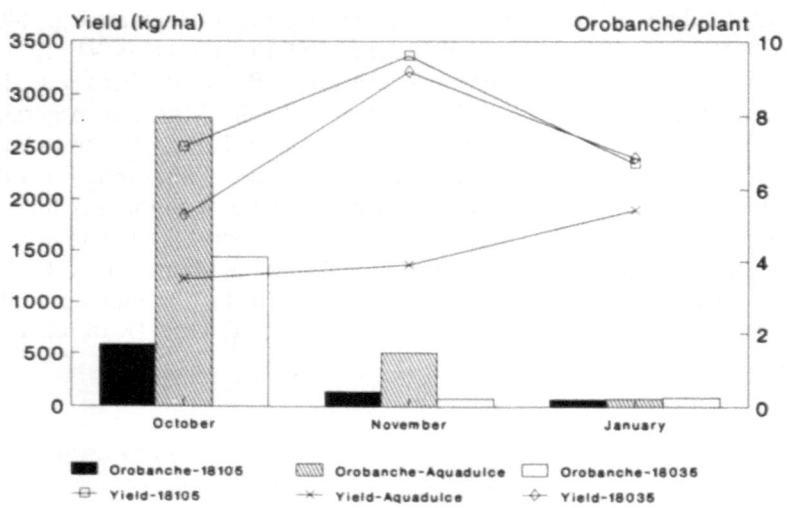

Figure 4. Effect of planting date and genotype on *Orobanche* infestation and yield.

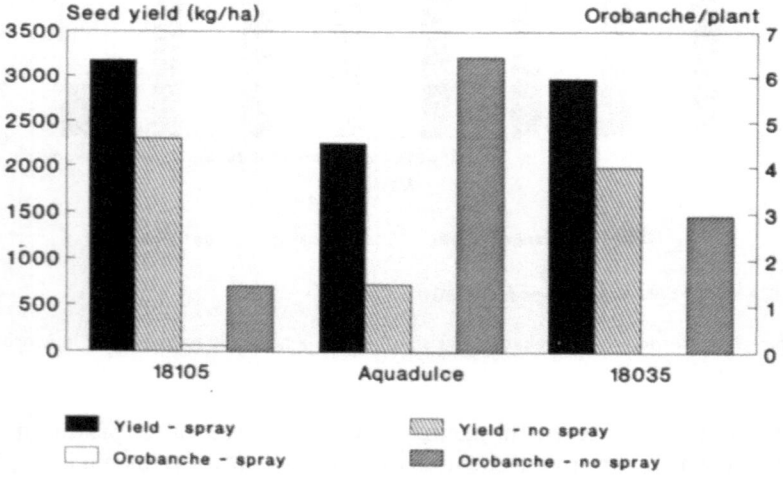

Figure 5. Effect of herbicide and genotype on *Orobanche* infestation and yield.

present resistance to *Orobanche* also involves integration with other control measures like planting date and herbicides. Susceptible cultivars have to be sown late, e.g., January, to avoid *Orobanche*, and thereby suffer a yield loss, but a resistant cultivar could be sown at the normal, early high-yield producing date (Figure 4). In Morocco in 1991, resistant lines had less need of herbicide to give their maximum yield than did the susceptible control (Figure 5). A combination of herbicide, resistant cultivar, early sowing date and appropriate plant density are likely to give the highest yield.

Aphids (*Aphis fabae, A. craccivora*)

Faba bean plants are under considerable stress once sap-sucking aphids start to multiply. Insecticides are relatively inexpensive so resistance breeding has not received priority. However, it may be needed where aphicides are uneconomic or prohibited.

Host-aphid relationships are complex leading to difficulties in assessing resistance (Klinghauf, 1982); for example, "Bolero" was clearly more resistant than "Diana" at 20°C but not at lower temperatures (Zebitz *et al.*, 1988). Nevertheless, the resistance of "Rastatt" to *A. fabae* (Muller, 1968) and also one of its derivatives, Line 14, (Bond and Lowe, 1978) was established. The yield of untreated resistant lines, however, was less than that of susceptible lines treated with aphicides.

In Egypt, over 1000 lines were screened for resistance to *A. craccivora* and 36 classified as resistant. Plastic house screening against *A. fabae* in Syria revealed 5 resistant lines. BPL 23 was resistant to both aphid species and to stem nematode (Table 3). Others were resistant to an aphid and also to *Botrytis* (Bishara *et al.*, 1989). The yielding ability of these lines in the presence and absence of aphids needs to be measured but, if good, the way is open to develop a strategy of multiple resistance that includes aphid resistance.

In the long term, when biotechnology has advanced to the stage of transferring genes from other *Vicia* species, the aim should be to utilize the very high levels of resistance to aphids in wild relatives (Birch, 1985).

Weed Competition

Where herbicides are not applied, or are only partly effective, weed competition can be a major stress that limits yield of faba beans. Breeders can contribute by providing cultivars that are vigorous enough to compete with weeds. In particular, shading by an early-closing canopy can much reduce growth of late germinating weeds when early weed growth has been controlled by pre-sowing cultivations or by non-persistent herbicides.

The most competitive cultivars are usually larger seeded than others within their botanical group and are heterotic due to being in a hybrid or composite population. Weed competition is often worst in low plant densities of pathways and perimeters of fields, so a cultivar needs to tolerate high plant densities without the risk of yield loss. This implies good resistance to diseases, especially chocolate spot, that spread rapidly at high plant densities, and also good autofertility in case bees cannot penetrate the dense canopy at flowering time.

Pollination

Robertson and Saxena (1993) pointed out that stress affects pollinators which in turn affect yield and vigor of bean stocks. The answer is to breed autofertile cultivars which can set seed independently of bee visitation but which, given the opportunity, can also cross pollinate and maintain vigor.

There are a number of sources of autofertility (Hanna and Lawes, 1967; Poulsen, 1980; Robertson and El-Sherbeeny, 1988) and it is well known that most F_1 hybrids are more autofertile than their parents (Drayner, 1959; Holden and Bond, 1960; Salih and Ali, 1989; Link, 1990). The latter type is almost impossible to fix in an inbred line. The process of transferring the fixable types of autofertility, those that are thought to be controlled by relatively few genes, is hampered by the need to distinguish plants in segregating populations that carry the fixable genes from those that are autofertile because of general heterosis. For this reason progress is slow but some breeding programs are now achieving improvements in autofertility.

Symbiosis with Rhizobium

Ineffective strains of *R. leguminosarum* could cause stress to faba bean plants. Fortunately most strains in faba-bean growing areas are effective but there may be stress in soils where *V. faba* has not been grown before. Interaction between faba bean line and Rhizobium strain has been demonstrated (Lawes *et al.*, 1978) and improvement in *Rhizobium* for nitrogen fixation has been induced (Shukkla *et al.*, 1986) but translating these results into practice depends on stability and competitiveness of new strains (even if they are suited to a particular faba bean genotype) in the field.

The importance of *Rhizobium* to *V. faba* was demonstrated by the yield benefits, up to 14%, obtained by insecticidal control of *Sitona* larvae that feed on the Rhizobium nodules (Bardner *et al.*, 1982). Thus this is another way in which stress can be relieved but we know of no resistance to *Sitona*.

Abiotic Stresses

Drought

The faba bean plant requires a large amount of water to support its erect stem and retain turgor in its fleshy leaves. It is therefore prone to drought stress but, where rainfall is fairly predictable, adaptation to a dry climate has been associated with a plant architecture that can minimize this stress. For example, the continental climate of eastern Europe, including Austria, favors tall late maturing types like "Gobo", "Erfano" and "Frinebo". These sometimes lodge and suffer much flower-drop on fertile soils with higher rainfall in western Europe where short, early types are now more popular. However, drought stress can severely limit yields in seasons when rainfall is lacking in regions that are normally wet. Then the short early types suffer most. Care is needed in extrapolating this conclusion to other environments. Whereas van Noel (1985) in Holland concluded that drought tolerance was lower in early lines, Ricciardi (1989b) wrote that early genotypes are better adapted to drought stress in Italy. Determinate habit reduces excessive growth in wet conditions but has not been associated with any drought tolerance. However, as autofertility of moderately autofertile cvs is reduced during drought (Stoddard, 1986), improvements in intrinsic levels of autofertility might ameliorate effects of drought.

Breeding is mainly a matter of testing advanced lines under dry conditions and aiming for maximum water use efficiency at given plant densities and row spacing (Robertson and Saxena, 1993). Water use efficiency is not as good as in pea, so there is need for improvement, and recurrent selection in random mating populations is being practiced in Canada. Drought stress is common in spring faba bean in northern China and any part of the Mediterranean area where it cannot be avoided by early sowing.

In seeking drought resistance, breeders can either aim for drought avoidance or tolerance (van der Wal, 1981). Drought avoidance is not only selecting for rapid early growth but can be approached by extending the root system or by saving water through stomatal control. True drought tolerance is in plants that can endure low internal water potential, e.g., by dehydration tolerance or osmoregulation. Osmotic adjustment was considered to account for drought-tolerance differences at the Centre for Plant Breeding and Reproduction Research at Wageningen in 1987, and Soja *et al.* (1988) found that screening seedlings in nutrient solution with variable osmotic stress correlated with yield under stress in the field.

Direct observations on root growth have been made through transparent tubes (El Shazly and Warboys, 1989) and thick roots found to be associated with rapid wilting (Harrewijn and van Norel, 1986). Other techniques are to control water in field plots and monitor water potential with lysimeters or to construct a bin that gives a gradient for rooting depth and select genotypes that can tolerate shallow rooting (van der Wal, 1981). However, in some soils, rooting volume can be more closely related to drought tolerance than rooting depth.

Stomatal size and frequency vary among genotypes (Nerkar *et al.*, 1981) and stomatal density can be negatively correlated with yield under stress conditions (Ricciardi, 1989a) though not always (Ricciardi and Steduto, 1988). Metabolic markers of drought tolerance being investigated at Rennes, France, include the role of proline accumulation in DOPA (3,4-dehydroxyphenylalanine) degradation. This is estimated from leaf discs in polyethyleneglycol and NaCl solution.

For arid conditions there is a role for fundamental research as well as breeding. Though differences among present genotypes exist, heritability is often low (Nanda *et al.*, 1988) and inbreeding depression results in lower yields in stress conditions (Milewska, 1988). So far as conclusions can be drawn, ILB 1814, 80S43856, and 80L90121 were reported as having good water use efficiency (Robertson and Saxena, 1993), JV-37 performed well in stress conditions (Nanda *et al.*, 1988), while Nadwislanski and Mazur were drought tolerant in Poland (Tomashevski and Yanushevich, 1975).

Waterlogging

Excess water on autumn sown beans is common in southern China and this results in damage to the crop. Elsewhere it is sometimes associated with excessive soil consolidation which can cause up to 48% yield loss (Dawkins and Brereton, 1984). We know of no cultivar tolerance to waterlogging; the problem is best approached through resistance to foot rot (*Fusarium* spp. and *Phythium* spp.) and interpollinating populations that have the vigor to penetrate poorly structured soil.

Low Temperatures

Frost on Seedlings

There are three distinct areas where *V. faba* is autumn sown and which differ in degrees of hardiness for the commonly grown cultivars.
(a) England, northern France and a very small amount in Germany. Cultivars tolerate about −18°C.
(b) Southern and western France, and northern Spain where cultivars are earlier maturing but less hardy, tolerating about −12°C.
(c) Mediterranean countries with a cultivar tolerance of about −6°C, distinguishing them from African cultivars with almost no frost tolerance.
Most spring sown European and Chinese cultivars fall in category (c) or (b). There are also differences in frost resistance among cultivars in the Mediterranean area, Aquadulce, "ILB 3187" and "3188" being among the most hardy (Herzog and Saxena, 1988) but they are not as hardy as the majority of cultivars in areas (a) and (b).

Breeding for improved hardiness in area (a) has been attempted but the very hardy "Cote d'Or" (France) is very susceptible to *Botrytis* and the very hardy "Hiverna" and "Webo" (Germany) are susceptible to *Ascochyta*. Large scale breeding would be needed to combine hardiness with disease resistance.

More promising is the improvement in hardiness of the early maturing cultivars in category (b) by crossing with the hardy late maturing cultivars of category (a). In addition, selection in populations from crosses of French cultivars from area (a) by winter populations from Spain, Greece, and Italy has produced recombinants that have the hardiness of "Soravi" and "Bourdon", yet are 15 to 20 days earlier. These results were obtained under controlled conditions. Some of the effort involved in exposing lines to controlled low temperatures, or in waiting for real winters, could be avoided because 64% of the genotypic variation in frost resistance can be attributed to correlation with easily measured traits like rate of development, water/dry matter ratio and leaf area (Herzog and Saxena, 1988).

Eventually however, survivors of low temperature tests have to experience real winters. Hardiness also involves the ability to harden in the autumn, large reserves in cotyledons to allow rapid root growth into warmer layers of the soil, ability to withstand desiccation from drying winds, and ability to recover in the spring by tillering rapidly. Cultivars that survived best following the 1984/85 winter in England were "Boxer", "Throws MS" and "Webo". Large seeded types survived better than small seeded, and in an F_1 hybrids were less damaged than their mid-parental values (Bond *et al.*, 1986); another instance of heterosis contributing to tolerance of abiotic stress.

Frost on Flowers

In Mediterranean areas early sowing is necessary to avoid spring and summer drought but this can expose faba bean flowers to damage with loss of yield and/or late ripening. Hardiness of northern European seedlings is not related to hardiness at flowering so crosses between these types would not necessarily lead to an improvement but greater general diversity among parents including some from high altitudes should be tried. Flowering at a higher node can easily be bred if frosting of first flowers is inevitable. In Yunnan, China, genotypes with greater numbers of flowers per plant had greater ability for self-regulation and maintenance of yield following low temperatures (Liu *et al.*, 1987).

High Temperatures

Effect on Fertilization

The upper limit of temperature at which fertilization can take place is probably about 30°C; in one experiment there was no ovule fertilization at 35°C and flowers of "Outlook" cultivar collected 36 or 48 h after pollination had better fertilization frequency at 15°C than at 25°C (Graff, 1988). This rather low optimum temperature is probably another reason, in addition to drought, that at low latitudes faba beans are mainly grown in the cool season or at high altitudes. Accessions from hot countries do not tolerate a higher temperature at flowering better than those from temperate countries. However, there is distinct evidence that at 35°C pollen from hybrids germinates better than that from inbreds (Table 4A). Also, fertilization is more successful in hybrids than inbreds even at low temperatures (Table 4B). It may be concluded that F_1 hybrids, or synthetic cultivars which have a high proportion of hybrid plants, would be able to tolerate a wider range of temperatures than inbreds or open-pollinated cultivars.

Table 4. Tolerance of pollen and ovules of inbred and hybrid faba beans to high and low temperatures

(A)   Percent germination of pollen from inbred, hybrid and open pollinated lines after nine hours *in vitro* incubation at 15°C, 25°C and 35°C

Genotype	Temperature (°C)		
	15	25	35
Inbred A	54.5 ± 3.4	61.0 ± 2.8	12.5 ± 1.9
Inbred B	87.8 ± 2.0	79.5 ± 2.6	16.1 ± 2.3
Hybrid AC	66.9 ± 3.1	66.7 ± 3.1	44.6 ± 2.9
Outlook	21.6 ± 2.3	62.9 ± 3.0	10.2 ± 1.9
Chinese	20.3 ± 2.7	59.8 ± 3.1	17.9 ± 2.4

(B)   Effect of temperature on ovule fertilization frequency (%) in inbred and hybrid flowers after 24 h and 36 h incubation

Genotype	Temperature (°C)		
	5	15	25
Inbred 24 h	6.1 ± 1.4	17.6 ± 2.2	25.3 ± 2.5
Hybrid	2.9 ± 1.2	45.0 ± 3.5	35.8 ± 3.5
Inbred 36 h	10.8 ± 1.8	20.5 ± 2.2	47.3 ± 2.9
Hybrid	25.7 ± 3.2	46.3 ± 3.4	72.7 ± 5.5

Effect on Pod Dehiscence

The wrinkled or indehiscent pod is widespread among cultivars in arid regions (Hanelt, 1972). These are normally flowering in the cool season but can tolerate heat without shattering at harvest. More recently, cultivars with indehiscent pods e.g., "Alfred", "Victor" and "Caspar", have been introduced in north-western Europe where they have avoided harvesting losses in hot dry summers there. The genetic control may not be the same as the recessive gene described by Hanelt (1972), or else there are minor modifying genes, because varying degrees of tendency for dehiscence can be identified; e.g., small-seeded long-podded cultivars with a wide angle to the stem are particularly susceptible to shattering.

Salinity

Faba beans are not cultivated much on saline soils but in a few places, especially in Egypt and India, the crop could be extended if salt tolerance were found. El-Karouri (1979) studied the effect of soil salinity on growth and yield of a local Sudanese cultivar and found it to have medium tolerance. This disagreed with the earlier findings of Ayers and Eberhard (1960) who classified faba beans as having low salt tolerance. El-Karouri (1979) suggested that this may demonstrate differences for salt tolerance between genotypes. El-Aal and Waly (1988) tested a 5 × 5 diallel cross for salt tolerance at germination and found good general combining ability effects for the cvs Somaly and Kobrosy. In Germany, "Felix" and Line 49907 developed more root and shoot dry matter in controlled saline conditions than the salt-sensitive and less branched cvs like "Herz Freya" (Melesse, 1988). Dua *et al.* (1989) found variation among faba bean genotypes in threshold salinity in India. There may be a basis for breeding progress.

Discussion of Breeding Strategies

Wild Relatives

Despite many attempts to overcome interspecific barriers (Ramsay *et al.*, 1984), no crosses of *Vicia faba* with any other species have yet been success-ful. This paper cannot, therefore, contribute to the theme "Use of wild species to improve resistance to stress" except to confirm that high levels of resistance to *Botrytis* and to aphids, exist in *V. narbonensis* (Birch, 1985) and to *Orobance* in *V. dasycarpa* (El Moneim *et al.*, 1990); and that resistance to other stresses could also be found in wild *Viciae*. There are good reasons for wanting to utilize genetic variation in wild species once the barriers to crossing are overcome or advances in biotechnology allow gene transfer to *V. faba*.

Multiple Resistance

Sources of resistance to most of the major pathogens exist and the way forward is clearly to combine as many individual resistances in one faba bean cultivar. Some genotypes are already resistant to more than one pathogen without deliberate breeding to combine them. For example, the ICARDA line BPL 1179 has resistance to chocolate spot, rust, *Alternaria* (Khalil *et al.*, 1986) and *Peronospora* (Nassib and Salih, 1983); also INRA line 29H displays resistance to *Ascochyta* and to stem nematodes. These examples suggest close linkage of the resistance genes and/or a common resistance mechanism. Some well-established resistances, however, have limited application unless they can be combined with others. Resistance to *Orobanche* in Giza 402, for example, could be used more widely if combined with resistance to chocolate spot. Extreme resistance to frost in Cote d'Or is unlikely to be employed in a new cultivar unless combined with resistance to lodging and chocolate spot or with earlier maturity.

A number of multiple resistances can be found in ICARDA Annual Reports, notably combinations of resistance to aphids with chocolate spot, aphids with stem nematodes and chocolate spot with rust (Table 3). Bernier (1985) reviewed favorably prospects of multiple disease resistance to most of the fungal pathogens. However, chocolate spot resistance seems to depend on genetic background for full expression so may not be so easy to combine with resistance to other diseases, in all types of faba bean, as will resistance to rust or *Ascochyta*. Resistances to some viruses (BYMV and PEMV) have also been combined (Schmidt *et al.*, 1989). Crosses have already been made with the objective of other combinations particularly resistance to *Orobanche* with resistance to fungal pathogens.

Durable Resistance

None of the resistance genes described in this paper has been used widely enough or over a long enough period to be described as durable *sensu* Johnson (1979). It is known, however, that some sources of resistance are race specific, particularly in the case of rust and *Ascochyta*, and are not likely to be of value in the long term. The fact that most bean cultivars are very heterogeneous, however, provides the opportunity to have a population containing a number of resistances. Such a scheme has been proposed for rust using genes for slow rusting (Bernier and Conner, 1983).

Effect of Breeding Methods

Initial diversity is mainly from local populations or germplasm collections. In general, mutation breeding has produced new plant architecture rather than tolerance to stress, but one exception was greater resistance to chocolate spot and to rust than original cultivars in M2 after use of gamma rays (Abdel-Hak and Mansour, 1980).

After creating new populations by crossing, most faba bean breeders use some form of pedigree or recurrent selection with attempts to preserve identity of the selections in pure lines. But the final cultivar as released may be a pure line, population, F_1 hybrid, or composite of contrasting or of sister lines.

Heterosis of 20% or more is well established in *V. faba*. Composites are heavier and more stable in yield than their inbred components (Bond, 1986) implying the former are on average better able to tolerate yield-limiting factors (= stress); and mixtures of non-inbreds also yield more than their components (Tarhuni and McNeilly, 1990). That greater stability of composites is due to a better tolerance of stresses by the more hybrid plants that they contain is suggested by a number of examples of high yields of hybrids in stress conditions or of inbreeding depression in tolerance of stress.

Due to their autofertility, F_1 hybrids can tolerate lack of pollinating insects better than inbreds and most other pollinated cultivars (Link, 1990). Hybrid flowers fertilized faster at low temperatures and hybrid pollen germinated more at high temperatures (Graff, 1988). F_1s were more resistant than parents to bruchids (Waly *et al.*, 1987); and in winter beans, survival values for hybrids were better than for mid-parents (Bond *et al.*, 1986). In drought conditions, lowest yields were given by lines suffering most from inbreeding depression (Milewska *et al.*, 1988), and Hempel (1983) found no further response to selection for *Ascochyta* resistance after a third cycle owing to inbreeding effects.

Producing hybrid or composite cultivars conflicts with the objective of uniformity, for ease of management and to satisfy registration requirements. Heterogeneity is necessary to provide heterozygosity and heterotic tolerance of stress. (Hybrids are not necessarily uniform because only 3-way crosses are economic in practice). A balance has to be sought between these two opposing pressures, but it is important that registration procedures be formulated in the light of the special case of hybrid tolerance in *V. faba* rather than on models constructed for inbreeding crop species.

Effect of Breeding for Quality

Selection for reduced antinutritional factors, e.g., tannin, trypsin inhibitors, glucopyranosides, most of which have evolved to protect the plant against pathogens, might reduce tolerance to stress. For example, in some near-isogenic pairs of lines that differ in tannin, emergence from the soil is much worse for the tannin-free than for the tannin-containing line at 5°C but not at 15°C (Kantar, 1992). However, this may not be true for all isogenic pairs (van Loon *et al.*, 1989) or for current tannin-free cultivars because they are showing satisfactory emergence and resistance to *Fusarium* in a range of soil conditions. A hypothesis is that alternative plant defenses are being selected as breeding proceeds in tannin-free types, and that a similar scenario could be envisaged for other antinutritional factors. Vicine inhibits *in vitro* fungal growth (Bjerg *et al.*, 1984) but other resistance mechanisms could be selected.

Integration with Other Control Measures

Breeding can make a major contribution to reducing the effect of stress but integration of genetic resistance with agronomic measures can be of greater benefit than either alone. A good example is resistance to *Orobanche* which is likely to allow normal yields if combined with early sowing and herbicide (glyphosate or fosamine) treatment. Aphid resistance could reduce the need for insecticides from two to one application or to borders of fields only. Moderate resistance to chocolate spot could be combined with a reduced rate of fungicide or with higher plant density if resistance reduces the risk of aggressive attack.

Concluding Remarks

Progress has been made in the last few years, mainly by ICARDA coordinated research, though also to a lesser extent in some national programs, in breeding for resistance to chocolate spot, *Ascochyta*, rust, *Orobanche*, and to some viruses. Resistance to chocolate spot is not so clearly defined or so simply transferred into other genetic backgrounds as the others but it can be helped somewhat by breeding for resistance to pre-disposing factors. Improved levels of resistance to *Orobanche* are still required but most can be made of existing sources by integrating them with herbicides and with the appropriate planting dates and densities.

Further multiple resistances will be pursued with the long term aim of using recurrent selection, backcrosses, and eventually gene transformation as a building-block approach to developing the ideal cultivar. However, this will not be easy. Some lines are resistant to one pathogen but very susceptible to another. Also a survey of most of the ICARDA disease-resistant lines showed a skewed distribution toward small seeds and late maturity (Robertson and Saxena, 1993). In other words some linkages may have to be broken.

Another problem is the unknown degree of durability of the present sources of resistance. New races of pathogens may not appear as often in *V. faba* as in inbreeding crops but the trend in faba bean is toward greater uniformity (e.g., where quality is required by the end user or distinctness for Breeders' Rights) so new sources and strict monitoring will be needed.

There is some hope of improving tolerance of abiotic stresses, though within the constraints of the fundamental morphology of the plant. Some cultivars tolerate drought or frost better than others and this gives them good local adaptation. Improvements in these tolerances, in the short term by trials with the stress imposed and in the long term by selection for associated biochemical traits, should allow wider cultivation, but rarely an improved very broad adaptation as in the case of disease resistance. Hybrid vigor is an important factor in allowing faba bean to tolerate various stresses; thus, breeding methods that allow high levels of hybridity in a cultivar should be adopted.

Very few of the above possibilities for breeding would exist if germplasm had not been collected, evaluated and maintained. Because tolerance to some stresses is still lacking and existing resistances may not be durable, there is a continuing need for collection, evaluation, and maintenance of genetic resources in *V. faba*.

The problems and challenges are now too large to be fragmented among national or private companies. Only international collaboration can tackle requirements like multiple resistances, differentiation of pathotypes, the development of strategies to find durable resistance, and the extension of research and breeding needed to improve tolerance of abiotic stresses.

References

Abdel-Hak, T. M. and Mansour, K. 1980. *Agricultural Research Review* 58: 57–63.

Ayers, A. D. and Eberhard, D. L. 1960. *Agronomy Journal* 52: 110–111.

Bardner, R., Fletcher, K. E. and Griffiths, D. C. 1982. *Rothamsted Annual Report for 1991 (Part 1)* pp. 102–103.

Bernier, C. C. 1985. In: *Faba Beans, Kabuli Chickpeas, and Lentils in the 1980s*, pp. 129–136 (eds. M. C. Saxena and S. Varma). Aleppo, Syria: ICARDA.

Bernier, C. C. and Conner, R. L. 1983. In: *Durable Resistance in Crops*, pp. 429–431 (eds. F. Lamberti, J. M. Walker and N. A. van der Graaff). New York: Plenum Press.

Birch, N. 1985. *Annals of Applied Biology* 106: 561–569.

Bishara, S., Defrowy, G., Khalil, S. and Weigand, S. 1989. *Annual Report of the Food Legume Improvement Program for 1989*, pp. 221–223. Aleppo, Syria: ICARDA.

Bjerg, B., Heide, M., Knudsen, J. C. N. and Sorensen, H. 1984. *Zeitschrift für Pflanzenkrankheiten und Pflanzenschutz* 91: 483–487.

Blaszczak, W. and Fiedorow, Z. 1979. *Biuletyn Instytutu Hodowli i Aklimatyzacjii Roslin 1979* 137: 11–22.

Bond, D. A. 1986. *Biologisches Zentralblatt* 105: 129–135.

Bond, D. A., Brown, S. J., Jellis, G. J., Pope, M., Hall, J. A. and Clarke, M. H. E. 1986. *Annual Report of Plant Breeding Institute*, pp. 42–46. Cambridge, UK.

Bond, D. A. and Lowe, H. J. B. 1978. *Annals of Applied Biology* 81: 21–32.

Caubel, F. and Le Guen, J. 1992. In: *1st European Conference on Grain Legumes*, pp. 337–338. 1–3 June 1992. Angers, France.

Caubel, G. and Leclercq, D. 1989. *FABIS* 25: 45–48.

Conner, R. L. and Bernier, C. C. 1982. *Phytopathology* 72: 687–689.

Cubero, J. I., Moreno, M. T. and Hernández, L. 1992. In: *1st European Conference on Grain Legumes*, pp. 41–42. 1–3 June 1992. Angers, France.

Cubero, J. I., Pieterse, A., Saghir, A. R. and Borg, S. 1988. In: *World Crops: Cool Season Food Legumes*, pp. 549–563 (ed. R. J. Summerfield). Dordrecht: Kluwer Academic Publishers.

Dawkins, T. C. K. and Brereton, J. C. 1984. *World Crops* 10: 113–126.

Dobson, S. C. and Giltrap, N. J. 1991. *Aspects of Applied Biology* 27: 111–116.

Drayner, J. M. 1959. *Journal of Agricultural Science, Cambridge* 53: 387–404.

Dua, R. P., Sharma, S. K. and Mishara, B. 1989. *Indian Journal of Agricultural Science* 59: 729–731.

El-Aal, S. A. A. and Waly, E. A. 1988. *FABIS* 22: 9–10.

El-Hady Mohamed, M. M. 1988. *Diallel analysis of resistance to chocolate spot disease (Botrytis fabae Sard.) and other agronomic traits in faba bean (Vicia faba L.).* Ph.D. Thesis, Faculty of Agriculture, Cairo University.

El-Karouri, M. O. H. 1979. *Experimental Agriculture* 15: 59–63.

El Moneim, A. A., Erskine, W., Singh, K. B., Linke K. - H. and Saxena M. C. 1990. *Annual Report of Food Legume Improvement Program 1990*, p. 224. Aleppo, Syria: ICARDA.

El-Shazly, M. S. and Warboys, I. B. 1989. *Experimental Agriculture* 25: 35–37.

Gondran, J. 1975. In: *VII Plant Protection Congress*, pp. 109–123. Reports and information section VI. Moscow, USSR.

Graff, R. 1988. *Factors affecting ovule fertilization frequency and seed set in faba bean (Vicia faba L).* Ph.D. Thesis, University of Saskatchewan.

Halila, H., Kharrat M. and Hanounik, S. B. 1990. In: *Annual Report for 1990 of the Food Legume Improvement Program of ICARDA*, pp. 205–206. Aleppo, Syria: ICARDA.

Hanelt, P. 1972. *Kulturpflanze* 20: 75–128.

Hanna, A. S. and Lawes D. A. 1967. *Annals of Applied Biology* 59: 289–295.

Hanounik, S. B., Halila, H. and Harrabi, M. 1986. *FABIS* 16: 37–39.

Hanounik, S. B., Jellis, G. J. and Hussein, M. M. 1993. In: *Breeding for Stress Tolerance in Cool-Season Food Legumes*, pp. 97–106 (eds. K. B. Singh and M. C. Saxena). Chichester, UK: John Wiley & Sons.

Hanounik, S. B. and Robertson, L. D. 1988. *Plant Disease* 72: 696–698.

Hanounik, S. B. and Robertson, L. D. 1989. *Plant Disease* 73: 202–205.

Harrewijn, J. L. and van Norel, A. 1986. Veldbonen. *Annual Report of Stichting voor Planten-veredeling (SVP) Wageningen for 1985* 41: 74–75.

Harrison, J. G. 1981. *Tests of Agrochemicals and Cultivars* 2: 72–73.

Hempel, K. 1983. *Tagungsbericht, Akademie der Landwirtschaftswissenschaften der Deutschen Demokratischen Republik* 216: II:579–590.

Herzog, H. and Saxena, M. C. 1988. *FABIS* 21: 19–25.

Holden, J. H. W. and Bond, D. A. 1960. *Heredity* 15: 175–192.

Hussein, H. A. S., Youssef, S. S., Hussein, E. H. A. and Hussein, B. A. 1988. In: *Improvement of Grain Legume Productions Using Induced Mutations*, pp. 127–144. Vienna, Austria: IAEA.

Jellis, G. J., Plumb, A. S. and Clarke, M. H. E. 1991. *Aspects of Applied Biology, Production and Protection of Legumes* 27: 63–66.

Jellis, G. J., Smith, D. B. and Scott, E. S. 1990. *Mycological Research* 94: 407–409.

Johnson, R. 1979. *Phytopathology* 69: 198–199.

Kantar, F. 1992. *Studies on the establishment of white flowered (zero tannin) Vicia faba.* Ph.D. Thesis, University of Nottingham.

Karamanos, A. J. and Avgoulas, C. E. 1989. *FABIS* 25: 40–45.

Khaled, A. K. and El-Bastewesy, F. I. 1989. *FABIS* 25: 5–9.

Khalil, S. A., Nassib, A. M. and Abou Zeid, N. M. 1986. *Biologisches Zentralblatt* 105: 155–161.

Kharrat, M., Halila, M. H., Linke, K. H. and Haddar, T. 1992. In: *2nd International Food Legume Research Conference Program and Abstracts*, p. 53. 12–16 April 1992, Cairo, Egypt.

Kheir, N. F., Salem, S. M., Ahmed, A. H. H. and El-Shihy, O. M. A. 1989. *Agriculture University of Cairo* 40: 197–212.

Klinghauf, F. A. J. 1982. In: *Faba Bean Improvement*, pp. 285–295 (eds. G. Hawkin and C. Webb). The Hague, The Netherlands: Martinus Nijhoff/Dr. W. Junk Publishers.

Lawes, D. A., Bond, D. A. and Poulsen, M. H. 1983. In: *The Faba Bean*, pp. 23–76 (ed. P. D. Hebblethwaite). London: Butterworths.

Lawes, D. A., Mytton, L. R., El-Sherbeeny, M. H. and Sorwli, F. K. 1978. *Annals Applied Biology* 88: 466–468.

Liang, X. Y. 1989. *FABIS* 25: 3–4.

Link, W. 1990. *Theoretical and Applied Genetics* 79: 713–717.

Link, W. and Hempel, K. 1993: In *Abstracts of First International Crop Science Congress*, p. 73. (ed. D. R. Buxton) Madison, WI, Crop Science Society of America.

Liu, Z. S., Zhao, Y. Z., Bao, S. Y. and Wen Guang. 1987. *FABIS* 18: 14–17.

Melesse, T. 1988. *Mitteilungen der Gesellschaft für Pflanzenbauwissenschaften* 1: 96.

Milewska, J., Januszewicz, E. and Koczowska, I. 1988. *Biuletyn Instytutu Hodwli i Acklimatiyzacji, Roslin* 165: 107–114.

Muller, H. J. 1968. *Entomologica Experimentalis et Applicata* 11: 355–371.

Nanda, H. C., Yasin, M., Singh, C. B. and Rao, S. K. 1988. *FABIS* 21: 26–30.

Nassib, A. M., Ibrahim, A. A. and Khalil, S. A. 1982. In: *Faba Bean Improvement*, pp. 199–206 (eds. G. Hawkin and C. Webb). The Hague, The Netherlands: Martinus Nijhoff/Dr. W. Junk Publishers.

Nassib, A. and Salih, F. A. 1983. In: *Faba Bean in the Nile Valley*, pp. 59–68 (eds. M. C. Saxena and R. A. Stewart). Aleppo, Syria: ICARDA.

Nerkar, Y. S., Wilson, D. and Lawes, D. A. 1981. *Euphytica* 30: 335–345.

Ondrej, M. 1991. *Ochrana Rostlin* 27: 257–264.

Pascual Villalobos, M. and Jellis, G. J. 1990. *Journal of Agricultural Science* 115: 57–62.

Poulsen, M. H. 1980. *Inbreeding and autofertility in Vicia faba L.* Ph.D. Thesis. University of Cambridge.

Radwan, M. S., Abdalla, M. M. F., Fisbeck, G., Metwally, A. A. and Darwish, D. S. 1988. *Plant Breeding* 101: 208–216.

Ramsay, G., Pickersgill, B., Jones, J. K., Hammond, L. and Stewart, M. H. 1984. In: *World Crops: Production, Utilization, and Description* 10:201–208. (eds. P. D. Hebblethwaite, T. C. K. Dawkins, M. C. Heath and G. Lockwood). The Hague, The Netherlands: Martinus Nijhoff.

Rashid, K. Y., Bernier, C. C. and Conner R. L. 1991. *Plant Disease* 75: 852–855.

Rashid, K. Y. and Bernier, C. C. 1986. *Canadian Journal of Plant Pathology* 8: 317–322.

Ricciardi, L. 1989a. *Agricoltura Mediterranea* 19: 297–308.

Ricciardi, L. 1989b. *Agricoltura Mediterranea* 119: 424–434.

Ricciardi, L. and Steduto, P. 1988. *FABIS* 20: 21–24.

Robertson, L. D. and El Sherbeeny, M. 1988. In: *Faba Bean Germplasm Catalog. Pure line collection ICARDA Aleppo Syria*, 140 pp.

Robertson, L. D. and Hanounik, S. B. 1987. In: *Annual Report for Food Legume Improvement Program for 1987*, pp. 205–212. Aleppo, Syria: ICARDA.

Robertson, L. D. and Saxena, M. C. 1993. In: *Breeding for Stress Tolerance in Cool-Season Food Legumes*, pp. 37–50 (eds. K. B. Singh and M. C. Saxena), Chichester, UK: John Wiley & Sons.

Rohloff, H. and Stulpnagel, R. 1984. *FABIS* 10: 29.

Salih, S. H. and Ali, A. E. 1989. *FABIS* 23: 8–10.

Schmidt, H. E. 1984. *Nachrichtenblatt für den Pflanzenschutz in der DDR* 38: 157–162.

Schmidt, H. E., Geissler, K., Karl, E. and Schmidt, H. B. 1986. *Archiv für Phytopathologie und Pflanzenschutz* 22: 87–99.

Schmidt, H. E., Meyer, U., Haack, I. and Karl, E. 1989. *Archiv für Zuchtungforschung* 19: 193–196.

Shukkla, R. S., Singh, C. B. and Khare, D. 1986. *FABIS* 14: 18.

Soja, G., Soja, A. – M. and Reza, Z. 1988. *FABIS* 22: 20–24.

Stoddard, F. L. 1986. *FABIS* 15: 22–26.

Tarhuni, A. M. and McNeilly, T. 1990. *Journal of Agronomy and Crop Science* 165: 39–46.

Tivoli, B., Berthelem, P., Le Guen, J. and Onfroy, C. 1986. *FABIS* 16: 46–50.

Tomashevski, Z. and Yanushevich, E. 1975. *Biuletyn Instytutu Hodowli i Acklimatyzacjii, Roslin* 128/129: 157–164.

van der Wal, A. F. 1981. In: *Vicia faba: Physiology and Breeding, World Crops* 4, pp. 49–54 (ed. R. Thompson). The Hague, The Netherlands: Martinus Nijhoff.

van Loon, J. J. A., van Norel, A. and Dellaert, L. M. W. 1989. In: *Recent Advances of Research in Antinutritional Factors in Legume Seeds*, pp. 23–25 (eds. J. Huisman, T. F. B van der Poel and I. E. Liener). Lanham, Maryland, USA: Pudoc.

van Noel, A. 1985. *Zaadbelangen* 39: 157–159.

Waly, E. A., El-Aal, S. A. A. and Hussein, M. H. 1987. *FABIS* 17: 3–5.

Yartiev, A. G. 1976. *Sel' Skokhozyaistvennaya Akademiya Imeni Timiryazeva*, pp. 65–68. Moscow, USSR.

Yeoman, D. P., Lapwood, D. H. and McEwen, J. 1987. *Crop Protection* 6: 90–94.
Zakrzewska, E. 1986. In: *Proceedings of XI EUCARPIA Congress*, pp. 311–317. Warsaw, Poland.
Zebitz, C. P. W., Bauru and Tenhumberg, B. 1988. *Mitteilungen mans der Biologischen Bundesanstalt für Landundforstwirtschaft*. Berlin-Dahlem 245: 313.

Current status and future strategy in breeding grasspea (*Lathyrus sativus*)

C.G. CAMPBELL[1], R.B. MEHRA[2], S.K. AGRAWAL[3], Y.Z. CHEN[4],
A.M. ABDEL MONEIM[5], H.I.T. KHAWAJA[6], C.R. YADOV[7], J.U. TAY[8]
and W.A. ARAYA[9]

[1] *Agriculture Canada Research Station, Morden, Manitoba, R0G 1J0 Canada;*
[2] *Indian Agricultural Research Institute (IARI), New Delhi 110 012, India;*
[3] *Indira Gandhi Agricultural University, Raipur, Madhya Pradesh, India;*
[4] *Lanzhou University Lanzhou, Gansu Province, Lanzhou, China;*
[5] *ICARDA, P. O. Box 5466, Aleppo, Syria;*
[6] *National Agricultural Research Centre, Islamabad, Pakistan;*
[7] *Rampur Research Station, Chitwan, Nepal;*
[8] *Instituto de Investigaciones Agropecuarias (INAI), Estación Experimental Quilampu, Casilla 426, Chillán, Chile, and*
[9] *Institute of Agriculture Research, Adet Research Centre, Bahar Dar, Ethiopia*

Abstract

Efforts in grasspea (*Lathyrus sativus*) improvement have increased since the development of lines that are very low in the neurotoxin Beta-N-oxalyl-L-alpha-beta-diamino propionic acid (ODAP); also referred to as Beta-N oxalyl-amino-L-alanine (BOAA). Many programs now address several related aspects of improvement simultaneously. These include reduced ODAP concentrations, insect and disease resistance, nitrogen fixation, agronomic practices, fodder and forage production, and components for increased yielding ability. The coordinated, multidisciplinary approach now being applied to the genetic improvement of grasspea should allow the potential of this largely neglected grain legume to be fully realized.

Introduction

Grasspea (*Lathyrus sativus* L.), also called chickling pea, khesari or sabberi, is produced as a major crop in Bangladesh, China, India, Nepal, and Pakistan, and to a lesser extent in many countries of Europe, the Middle East, northern Africa, and in Chile and Brazil in South America. It is extensively cultivated and naturalized in the Middle East countries of Iraq, Iran, Afghanistan, Syria, and Lebanon, in France and Spain in southern Europe, and in Ethiopia, Egypt, Morocco, Algeria, and Libya in northern Africa.

The origin of *L. sativus* is unknown but it is thought that the natural distribution has been completely obscured by cultivation even in southwest and central Asia, its presumed center of origin (Smartt, 1984). It shows great morphological variation, especially in vegetative characters such as leaf length,

F.J. Muehlbauer and W.J. Kaiser (eds.), Expanding the Production and Use of Cool Season Food Legumes, 617–630.

while floral characters are much less variable (Jackson and Yunus, 1983). They postulated that the forms with blue flowers and speckled seeds are more primitive. The pattern of variation was not unlike that found in lentils and faba beans where primitive forms with small seeds have a distribution to the east of the Mediterranean, while larger seeded forms have been selected in the Mediterranean region. The development of forms with larger leaves may have resulted from selection of forage types. It would appear that there is a large base of germplasm present in many countries that can be utilized by plant breeders in the production of adapted lines for specific or dual purpose requirements.

Grasspea is very tolerant of drought conditions and has been grown successfully in areas with an average annual precipitation of 380 to 650 mm. Despite its tolerance to drought, grasspea is not affected by excessive rainfall and can be grown on land subject to flooding (Sinha, 1977). In India, Bangladesh, Nepal, and Pakistan it is often broadcast into a standing rice crop one to two weeks before the rice is ready for harvest where it flourishes on the residual moisture left after the rice has been harvested. It has a very hardy and penetrating root system and can be grown on a wide range of soil types, including very poor soils and heavy clays. This hardiness and its ability to fix atmospheric nitrogen make the crop one that seems designed to grow under adverse conditions.

Despite its obvious advantages, relatively little effort until recently has been extended in the improvement of this very hardy pulse crop. Indeed, the history of grasspea has been one where it has been banned by many countries due to it containing the neurotoxin Beta-N-oxalyl-L-alpha-beta-diamino propionic acid (ODAP) also referred to as Beta-N-oxalyl-ammo-L-alanine (BOAA). The neurotoxin in grasspea causes irreversible crippling when it is consumed as a major portion of the diet over a 3 to 4 month period. In spite of this, grasspea is still produced in significant quantities in many parts of the world. Improvement of the crop is now being addressed in many countries through germplasm collection and evaluation, as well as in crop improvement programs, many of which include a breeding component.

Crop Improvement by Conventional Breeding Methods

Evaluation

Local and introduced germplasm has been evaluated both for seed and for fodder purposes and to determine their adaptability to various climates. In the littoral zone of Morocco, "Favetta" performed well in comparison to cultivars introduced from Greece, Libya, and Portugal (Villax, 1963). In Cyprus, in comparisons of four local cultivars with 23 introductions from Algeria, Australia, Egypt, Greece, Libya, Portugal, and Turkey of *L. sativus*, *L. cicera*, and *L. ochrus*, an accession of *L. sativus* from Greece outyielded other introductions and was more leafy (Soadou, 1959). In Turkey, nine ecotypes of *L. sativus* were collected from different regions of Turkey from 1965 to 1969 and are now maintained on the west coast of Izmir (Hertzsche, 1970). In Jordan, a cultivar of *L. sativus* from Cyprus performed well (Hopkinson, 1975), while in Syria Van der Veen (1967) stated that two cultivars from Cyprus have proven to be well adapted. In northern Iraq, a local cultivar of *L. sativus* was cultivated extensively in some areas, while a cultivar introduced from Turkey proved to be more productive and cold tolerant than the local cultivar in small adaptability tests at Mosul (Kernick, 1976). In Nepal, a local germplasm collection of 87 accessions was evaluated and compared to 10 exotic lines. The local accessions were higher in yield, had more seeds per pod and were earlier to mature when compared to the imported lines (Yadov, personal communication). Early and final stand counts were also found to be higher in the local material. The exotic lines had larger seed sizes of 10.2 to 11.0 grams per 100 seeds as compared to 4.7 to 5.3 in the local material. A summary of the agronomic characteristics examined has shown a wide range of variability especially in plant height and pods per plant (Table 1).

Table 1. Summary of agronomic characteristics of a collection of *Lathyrus sativus* germplasm evaluated at Rampur, Nepal

Character	Mean	Range
Days to flower	85	68-94
Days to maturity	135	125-139
Plant height(cm)	71	46-106
Pods per plant	36	13-59
Seeds per pod	3.5	2.4-5.0
1000 seed weight(g)	42	30-60

Insect Resistance

There has been little work reported on the resistance of grasspea to various insect pests. The lines JRL 6 and JRL 41 have been found to be tolerant to thrips at Raipur, India (Chitale, personal communication). In Syria, lines resistant to cyst nematodes (*Heterodera ciceri*) and root-knot nematode (*Meloidogyne artiella*) have been identified (Abd El Moneim, personal communication). Certainly more effort is required in this area in the improvement of this very hardy pulse crop.

Disease Resistance

Powdery mildew (*Erysiphe polygoni* DC) and downy mildew (*Peronospora lathipulustris* Gaumann) are the two major diseases which infect grasspea. Losses due to these diseases as well as cultivar reaction have not been critically studied. However, lines showing moderate resistance to powdery mildew have been identified in India (Lal et al., 1985). At Raipur, the lines RPLK 26 and RL 41 have been found to be tolerant to powdery mildew. In addition, 86 lines from a local germplasm collection from around Raipur, India have shown resistance to downy mildew (Agrawal, unpublished). Efforts are underway in India to transfer the powdery mildew resistance to higher yielding, more adapted lines. In Syria, L. sativus lines that were moderately resistant to powdery mildew (Ersiphe pisi) have been identified. Lines in India have also been identified that were free from infection by downy mildew in a three year evaluation under conditions of heavy natural infection (Narsinghani and Kumar, 1979).

Decreased Neurotoxin (ODAP) Concentrations

Most of the reported work on grasspea improvement has involved reducing ODAP concentrations in the seeds. Although lines with reduced ODAP concentrations were identified almost 20 years ago, no concerted effort was placed on this problem until after 1985. The Lathyrus Collogue held that year at Pau, France served as a catalyst in many ways to focus research on many of the different aspects of grasspea improvement.

The ODAP concentrations of grasspea differ widely among accessions and environments (Dahiya and Jeswani, 1975; Ramanujam et al., 1980). The distribution of ODAP concentrations in landraces in several countries was quite similar (Table 2). Although many attempts have been made to find a correlation between ODAP concentration and seedcoat color or flower color the results have either conflicted (Dahiya, 1985; Quader et al., 1985) or have been unsuccessful (Roy and Rao, 1978). A high correlation to a readily identifiable plant characteristic would allow the characteristic to be used as a rapid selection technique for reduced ODAP concentrations. However, a rapid and fairly inexpensive method capable of analyzing over 100 samples per day has been developed for a breeding program for identification of segregants with reduced ODAP concentrations (Campbell, unpublished). This method is a variation of the method of Briggs et al. (1983).

In China, the major result in the past two decades has been the selection of four lines with relatively low toxin concentrations and with good agronomic characteristics. These were selected from 73 lines with ODAP concentrations ranging from 0.075 to 0.993%. Protein content of these lines averaged from 23 to 25%. These low toxin lines have been stable over several years. Toxicological tests of lines from the Wuwei District of Gansu Province have shown that they were safe when fed to animals. Neither acute nor chronic lathyrism was found when donkeys, pigs, and sheep were fed with seeds of these lines constituting 50 to 80% of the daily intake for 180 to 250 days (Chen, unpublished).

Genetic improvement work of L. sativus commenced in 1966 at the Jawaharlal Nehru Agricultural University, Jabalpur (M.P.), India with the

Table 2. Distribution of ODAP content in landraces of *Lathyrus sativus* collected at Raipur, India and Adhet, Ethiopia

Percent ODAP	Number of land races	
	Raipur	Adhet
< 0.200	10	0
0.201-.300	104	0
0.301-.400	391	4
0.401-.600	606	29
0.601-1.00	91	5
> 1.01	21	1
Total	1223	39

collection, maintenance and evaluation of 503 germplasm accessions (Lal *et al.*, 1985). In 1970, over 1500 samples of *Lathyrus* maintained at the Indian Agricultural Research Institute (IARI) at New Delhi were screened for ODAP concentration, with a few lines having concentrations which ranged from 0.15 to 0.30% (Jeswani *et al.*, 1970). In 1971, 1000 samples were screened for ODAP concentration in a research program at IARI and were found to vary from 0.2 to 2.0% (Leakey, 1979). The distribution of ODAP in different plant tissues was also determined (Mehta *et al.*, 1991) (Table 3). A breeding and selection program using recurrent mutagenic treatment was started to produce high-yielding, low-toxin lines of *Lathyrus* (Swaminathan *et al.*, 1971).

The *Lathyrus* improvement program in India was transferred to the Indira Gandhi Agricultural University at Raipur where work is continuing on screening lines for low ODAP concentration as well as developing lines with specific morphological characters. Also, they are using lines from Canada that have been selected for reduced ODAP concentrations in a hybridization and selection program designed to develop adapted high-yielding material with reduced concentrations of the neurotoxin. Recently, the *Lathyrus* breeding program at IARI has been reactivated with emphasis being placed on reducing the ODAP concentration through the use of both hybridization and somaclonal variation.

In Bangladesh, a screening program was initiated in 1979 to explore the possibility of isolating toxin-free lines from germplasm systematically collected from ten districts (Kaul *et al.*, 1985). The screening of the germplasm continued for over 5 years. Lines having as little as 450 mg ODAP per 100 g of sample to as much as 1400 mg were found indicating large variation in ODAP content. Line 3968 was selected as being significantly low in ODAP, as well as the earliest to mature and the largest yielder (Anon, 1980). An ongoing *Lathyrus* improvement program at the Bangladesh Agricultural Research Institute at Joydepur has recently produced a number of lines having ODAP

Table 3. Distribution of ODAP in different tissues of *Lathyrus sativus*

Plant Part	ODAP content
	--mg 100g^{-1}--
Root	14
Stem	64
Leaf	60
Pod	24
Seed coat	81
Embryo	400
Cotyledon	126

concentrations, ranging as low as 0.03% (M.P. Quader, personal communication). These lines, however, are not as high yielding as some of the local accessions.

Canadian evaluation of *Lathyrus* germplasm began in 1967 at the Agriculture Canada Research Station, Morden, Manitoba. A breeding and improvement program was established in 1982 and has resulted in the release of the germplasm LS 8246 (LS82046) having a ODAP concentration of 0.03% in the seed (Campbell, 1987). An additional two lines containing a factor for low ODAP content have been developed. These lines, L900239 and L920278, have been shown to produce low ODAP content in the seeds when used as parents in crosses. They also appear to differ in the content of ODAP in the cotyledons and in the seedlings (Table 4). It is possible that the amount of ODAP found in the seed of grasspea is dependent not only on the amount produced in the plant but also on the amount transferred into the developing seeds. Studies are underway to determine the inheritance and mode of action of the three sources of low ODAP concentration. The present objectives of the program are to transfer the genes for reduced ODAP concentration into high-yielding, adapted lines having good agronomic potential. Some of these lines have been selected for increased seeds per pod and for double pods per node. This involves hybridization of selected lines with screening and evaluation of the resulting progenies.

Grasspea improvement in Nepal commenced with the collection of 17 lines in 1986 and an additional 89 lines in 1987. These are presently being evaluated for agronomic characteristics under the Grain Legumes Improvement Program and are being screened for ODAP concentration at the Morden Research Station. High yielding accessions have been selected as parental material for a hybridization program to develop high yielding, adapted types with very low or zero ODAP concentration in the seed.

Table 4. ODAP analysis of seedlings, cotyledcns, dry seeds and the ratio of seedling concentration to dry seed concentration in selected lines of *Lathyrus sativus*

Line	Seedling	Cotyledons	Dry Seeds	Ratio*
	mg g⁻¹	mg g⁻¹	mg g⁻¹	
Jamalpur	48.04	8.10	2.37	20.3
LS82046	9.32	0.57	0.29	32.1
LS90239	1.52	0.19	0.40	3.8
LS90278	57.18	1.19	0.34	168.2

Data from Lambien, unpublished.

*Ratio of the concentration of ODAP in the seedlings to the ODAP concentration in the dry seeds.

Collection and evaluation of grasspea has also taken place in Ethiopia, with 252 lines of locally collected material being maintained at the Plant Genetic Resource Centre in Addis Ababa. At least 127 of these lines have been evaluated for agronomic characteristics at the Adhet Research Station and genotypes with reasonable yield and lower ODAP levels have been selected.

Male Sterility

Male sterility, due to a condition whereby the plants have short filaments and anthers which are much below the stigmas, has been reported in *L. sativus*. Male fertility was thought to be restored by a pair of complementary genes (Srivastava and Somayajulu, 1981) or possibly by both single and double restorer genes (Quader, 1987). It is possible that male sterility can be useful in population improvement programs and be used to possibly exploit some of the non-additive gene action on characters affecting production.

General and specific combining abilities have been estimated in grasspea for pods per plant, 100 seed-weight, seeds per pod, grain yield per plant and ODAP concentration (Dahiya and Jeswani, 1974). It was generally found that non-additive gene action was predominant in both the F_1 and the F_2.

Improved Lines

Attempts to provide growers with cultivars with low ODAP concentrations led to the selection and release in 1973–74 of "Pusa-24" by IARI (Lal *et al.*, 1985). This cultivar had comparable yields to some local collections in most years. Pusa-24 was a very important breakthrough as it clearly demonstrated that the

624   C.G. Campbell et al.

concentration of ODAP in the seeds could be selected against and that lines reduced in or lacking in ODAP were indeed possible.

In Chile, the cultivar "Quila-blanco" was developed in 1983 (Tay *et al.*, unpublished). It was a bulk of six plants selected from the locally grown heterogeneous population. The principle characteristics of this cultivar are uniform maturity, large white seeds (100 seeds weigh 28.7 g), and a protein concentration of 24.3%.

Cytogenetics

Chromosomal and cytogenetic studies have shown the genus *Lathyrus* to be predominantly diploid with 2n = 2x = 14 chromosomes. The chromosome number of more than 60 species have been reported with only three species having been shown to have more than 14 somatic chromosomes. Two species *L. pratensis* and *L. venosus* are tetraploid with 2n = 28 chromosomes and one species *L. palustris* is hexaploid with 2n = 42 chromosomes. These species have been studied cytologically and have been shown to be autopolyploids. The occurrence of an autohexaploid is among the very few examples of it occurring in the plant kingdom (Khawaja, unpublished). These autopolyploids are, in reality, cytotypes as diploid forms also occur in nature.

Interest in experimental interspecific hybridization in the genus was shown in sweet pea (*L. odoratus*) as early as 1916. Burpee (1916) and others were interested in trying to introduce yellow flower color genes into the cultivated species from wild relatives in the same genus. Successful interspecific hybridization in the genus has been shown to be exceedingly rare as has been found in other genera of the Leguminosae.

Interspecific hybridization between other species in the genus *Lathyrus* has been attempted by many researchers since the report of the successful crossing of *L. hirsutus* × *L. odoratus* by Barker (1916). Most attempts have been failures and even though many thousand cross combinations are theoretically possible, only 16 have been reported as successful. Interspecific hybridization involving *L. sativus* has only been reported as successful in two instances. In 1956, Lwin (1956) succeeded in crossing *L. cicera* with *L. sativus*, however this cross has been unsuccessful in subsequent attempts (Davies 1958; Khawaja 1988). Khawaja (1988) reported that *L. sativus* crosses readily with *L. amphicarpos* when the latter is used as the female. It has also been noted that in certain cross combinations fertilization is successful but the embryo aborts during development. The stage of development at which abortion takes place differs with the cross combination. Cytological studies of the F_1 hybrids between *L. amphicarpos* × *L. sativus*, *L. amphicarpos* × *L. cicera* and *L. odoratus* × *L. chloranthus* were carried out by Khawaja (1988) and showed 50% to 70% chromosome homology and pollen fertility in conformity with the meiotic pairing.

From the information available on crossing, fertility, and chromosome behavior of the hybrids, it may be concluded that breeding strategies involving alien genetic transfer for the improvement of grasspea is possible through the readily crossable species *L. amphicarpos*. There also appears to be a high probability of success in obtaining interspecific crosses between some species that do not cross due to embryo abortion after fertilization through the utilization of embryo rescue techniques.

Plant regeneration techniques have been successful in regenerating plants from explants derived from stem, leaf, and root tissue (Mehta *et al.*, 1991). The resulting plants showed a high amount of somaclonal variation in plant habit. This technique may be successfully exploited in the production of agronomically desirable types in low ODAP lines and thus provide a faster means of improvement than allowed by conventional crossing and backcrossing methods.

Crop Improvement by Mutation Breeding

Genetic Evaluation

Mutation breeding can be a valuable supplement to conventional plant breeding methods. It can be used to create additional genetic variability that may be utilized by the plant breeder in the development of cultivars for specific purposes or with specific adaptabilities. Mutation studies on *L. sativus* have shown that the chemical mutagens EMS (ethyl methane sulphonate) and NMU (N-nitroso-N-methyl urea) are more efficient than radiation in the production of chlorophyll mutations (Nerkar, 1974). However, cultivars have been found to respond differently to exposure to gamma irradiation (Prassad and Das, 1980). A wide spectrum of morphological mutations have been found affecting plant habit, maturity, branching, stem shape, leaf size, stipule shape, flower color and structure, pod size, and seed size and color (Nerkar, 1976). Plant habit mutants such as dwarf and erect, as well as giant forms, have also been induced (Prassad and Das, 1980). There thus appears to be a large selection of both naturally occurring as well as induced morphological characteristics that are available to the plant breeder.

Studies by Singh and Chaturvedi (1987) have shown that at biological comparable doses the mutagenic effectiveness was in the order of N-nitroso-N-methyl urea (NMU), ethyl methane sulphonate (EMS) and gamma-rays and efficiency in the order of Gamma-rays, EMS and NMU.

Decreased Neurotoxin Content

An indication that ODAP content might exhibit simple mendelian inheritance has been reported from variation induced in grasspea through both physical and chemical mutagens (Nerkar, 1972). In the segregating M2 generation in all treatments, the distribution curves showed three distinct peaks which are characteristic of monogenic F_2 segregation. If this proves to indeed be the case, rapid improvement of locally adapted germplasm is feasible through a hybridization and backcrossing program.

Genetic Resources

Characterizing Material with Descriptors

The Germplasm Resources, Crop Improvement and Agronomy Committee of the International Network for the Improvement of *Lathyrus sativus* and the Elimination of Lathyrism (INILSEL) has agreed to use the following descriptors:
1) Days from seeding to 50% plants with flowers.
2) Anthocyanin present in stem.
3) Flower color.
4) Leaf width (narrow = 0.5 cm, medium = 1.0 cm, wide = 1.5 cm).
5) Seedcoat color.
6) Maturity (days from seeding to 90% pods turned brown).
7) Plant height in centimeters.
8) Downy mildew severity (0 = none, 10 = severe).
9) Pod shattering at maturity (0 = none, 9 = 90–100%).
10) Plant type (indeterminant or determinant).
11) Plant habit (1 = erect, 5 = prostrate).
12) 1000 seed weight in grams.
13) Seed density (kg hl^{-1}).
14) Seeds per pod.
15) Insect resistance.
16) ODAP concentration of the seeds.
   Dr. D. Combes, University of Pau, Pau, France coordinates lists of accessions, passport data, and descriptors for all members of the network.

Collections

The collections of grasspea that have taken place in the past have involved large numbers of lines. As an example, the Indian Agricultural Research Institute, Delhi, India analyzed 1500 lines in 1970 and 1000 in 1971. Although these collections have taken place in many different countries and have been accomplished for many different reasons most of them now have one thing in common. The germplasm that was collected has been lost or destroyed and is no longer available to grasspea improvement programs. While this problem area is now starting to be addressed, an organized system of long term storage with backup storage at other locations is desperately needed in the areas where grasspea is now being cultivated. The large amount of variability that has been found in local germplasm must be sampled, described, and stored for future use.

Storage

Collections of *Lathyrus* germplasm are presently being stored in Bangladesh, Chile, China, Canada, Ethiopia, France, Germany, India, Nepal, Pakistan, Russia, Syria, United States, and the United Kingdom. While many of these are relatively small collections and do not represent a true sampling of the possible genetic variability existing in local germplasm, they are a very valuable component of any grasspea improvement program. Coordination of *Lathyrus* germplasm storage for INILSEL has been undertaken by Dr. D. Combes, Pau, France as an initial attempt at a more organized storage system.

Present Crop Improvement Programs

Many crop improvement programs are presently addressing research on the different aspects of grasspea production. Several of these are programs that contain a number of different, but interrelated, fields of specialization. These developments along with increased "networking" and collaboration with other programs has provided a good base for rapid improvement in this formerly badly neglected crop. The various facets that are contained in different programs include not only crop improvement aspects but also those of a nutritional, medical, socio-economic and biotechnical nature.

Some of the major crop improvement programs and their major emphasis on present research are listed below:

The Lathyrus Improvement program at the Indira Gandhi Agricultural University, Raipur, India includes selection and hybridization of low ODAP lines, insect and disease resistance, and plant regeneration techniques.

At the Indian Agricultural Research Institute, New Delhi, India the program includes aspects of selection and hybridization of lines, development of lines with reduced concentrations of ODAP and the study of protein components for quantity and quality.

The *Lathyrus* Improvement program at the National Agricultural Research Institute, Islamabad, Pakistan, is studying forage and fodder aspects of grasspea, nitrogen fixation, inter-specific hybridization through cytogenetical procedures and analytical aspects, as well as the development of high yielding and reduced ODAP adapted lines. A study of the incidence and severity of Lathyrism is also being addressed.

In Bangladesh, the Bangladesh Agricultural Research Institute has emphasized development of lines with reduced concentrations of ODAP combined with disease and insect resistance. At Mymensingh University, drought tolerance and agronomic traits are being studied.

In Chile, grasspea is a minor crop and presently no research is underway on breeding aspects. Research is being conducted on production practices including planting dates and fertilization.

The grasspea improvement project at Lanzhou University, Lanzhou, China is emphasizing agronomic aspects of production, the development of reduced ODAP lines, and mutational breeding.

The Grain legumes Research Program in Nepal is evaluating and selecting grasspea lines for agronomic characteristics. They are also selecting within segregating progenies from crosses made in Canada between reduced ODAP lines and selected local material.

In Ethiopia, grasspea improvement work at the Institute for Agricultural Research includes agronomic studies and the development of reduced ODAP lines together with processing techniques for the removal of ODAP. Nutritional, medical, and socio-economic aspects are also being addressed in a comprehensive study.

Grasspea research in Canada has developed germplasm with very small concentrations of ODAP. Research underway includes the inheritance of reduced ODAP concentration, nutritional aspects both for human consumption and animal feed, and agronomic studies including competition with weeds. Development of components for increased yield are also being addressed.

In addition to the above mentioned programs, many of them also contain a collaborative component. This allows complementation of the strengths of different programs and should result in efficient research. Many of these collaborative agreements encompass two or more countries and several different aspects that are essential to improvement of this hardy pulse crop.

Future Research Areas

Breeding and Evaluation

Many of the crop improvement programs will continue with the main emphasis being on the development of high yielding, adapted cultivars containing very little or zero amounts of the neurotoxin ODAP. Attempts will be made to completely remove the neurotoxin ODAP. Complete elimination of the neurotoxin content is presently being addressed by two main approaches. Identification and elimination of the enzyme responsible for ODAP production, and by transfer of genetic characters through inter-specific hybridization utilizing both cytogenetics and tissue culture techniques. The feasibility of introgression of desirable characteristics from other closely related species will be also be addressed. The emphasis on seed yield will probably remain, and forage aspects of this crop are starting to demand more attention. Enhancing the nutritional value of the crop by reducing antinutritional factors will also recieve more emphasis. Many of the crop improvement programs will add another dimension or direction to their present programs.

In future research at Lanzhou, China, for example, investigations will be conducted along two directions. Breeding of reduced neurotoxin lines with good agronomic characteristics will continue as well as crop improvement by mutational breeding. Special emphasis will be paid to induced mutation caused by Cobalt 64 heavy ion irradiation.

The forage aspects of grasspea production has always been important to producers. In many areas, the fodder value of the crop for grazing or the feed value of the straw is the main reason for production of this crop. These aspects are now increasingly being addressed by several programs.

For example, the long term goals of ICARDA are as follows:

1) Select genotypes with increased herbage production, seed yield, and harvest index.

2) Assess the selected genotypes for tolerance to production constraints (foliar and root diseases, drought, and cold).

3) Develop other important traits (earliness, leaf-retention, erect plant habit, and non-shattering pods).

4) Assist in replacement of fallow with *Lathyrus* species to increase animal production and improve soil fertility.

5) Develop high yielding but low neurotoxin (ODAP) content *Lathyrus* species.

6) Target suitable *Lathyrus* species to specific farming systems.

7) Identify potential insect pests.

8) Develop control measures for *Orobanche* species (broomrape).

Molecular genetics has emerged as an applied research discipline during the past decade and promises to assist in the solution of many agricultural problems. Genetic engineering holds promise for development of specific genotypes since it is based on identification, characterization, and transfer of specific genes into the recipient plant as compared to the mixing of two complete genomes of two parental lines followed by backcrossing for several generations to remove undesirable genes.

Molecular genetic techniques can be employed to characterize the genes and identify important linkages which could facilitate gene transfer to suitable agronomic types, genes for disease or insect resistance, reduced ODAP concentration, increased protein content, and other characters of agronomic importance seem to be adaptable to this approach.

Interdisciplinary research efforts, now underway at several research establishments together with collaboration between institutes, promises to produce needed improvements in grasspea in the near future.

References

Anonymous. 1980. *Annual Report*, p. 197. Pulses Improvement Project, Bangladesh Agricultural Research Institute. Joydebpur, Dhaka, Bangladesh.

Barker, B. T. P. 1916. *Garden Chronicle Series* 3.60: 156–157.

630   C.G. Campbell et al.

Briggs, C. J., Parreno, N. and Campbell, C. G. 1983. *Journal of Medicinal Plant Research* 47: 188–190.

Burpee, D. 1916. Cited by Tayler (Tayler, G. M. 1916. *Garden Chronicle Series* 3.60: 148).

Campbell, C. G. 1987. *Crop Source* 24: 821.

Dahiya, B. S. 1985. In: *Lathyrus and Lathyrism, Proceedings of Collogue Lathyrus*, 234 pp. (eds. A. K. Kaul and D. Combes). New York: Third World Medical Research Foundation.

Dahiya, B. S. and Jeswani, L. M. 1974. *Indian Journal of Agricultural Science* 44: 829–832.

Dahiya, B. S. and Jeswani, L. M. 1975. *Indian Journal of Agricultural Science* 45: 437–439.

Davies, A. J. S. 1957. *Nature* 180: 612.

Gowda, C. L. L. and Kaul, A. K. 1982. In: *Pulses in Bangladesh*, 472 pp. Bangladesh Agricultural Research Institute and Food and Agriculture Organization of the United Nations.

Hertzsh, W. 1970. In: *Technical Report*, Agricultural Research and Introduction Centre, Izmir AGP:SF/TUR 8, p. 71. Rome: FAO.

Hopkinson, D. 1975. *Technical report on the development of crop husbandry in the dry farming areas around Karak*. Jordan 169:518, Rome: FAO.

Jackson, M. T. and Yunus, A. G. 1983. *Euphytica* 33: 549–559.

Jeswani, L. M., Lal, B. M. and Prakesh, Shiv. 1970. *Current Science* 22: 518.

Kaul, A. K., Islam, M. Q. and Hamid, A. 1985. In: *Lathyrus and Lathyrism, Proceedings at Collogue Lathyrus*, 234 pp. (eds. A. K. Kaul and D. Combes). New York: Third World Medical Research Foundation.

Khawaja, H. I. T. 1988. XVIth International Congress of Genetics, Toronto, Canada.

Lal, M. S., Agrawal, I. and Chitale, M. W. 1985. In: *Lathyrus and Lathyrism, Proceedings at Collogue Lathyrus*, 234 pp. (eds. A. K. Kaul and D. Combes). New York: Third World Medical Research Foundation.

Leakey, C. 1979. *Appropriate Technology* 6: 15–16.

Lwin, S. 1956. *Studies in the genus Lathyrus*. M. S. Thesis. Manchester University, UK.

Mehta, S. L., Santha, I. M., Yadav, V. K., Roy, P. P. and Barat, G. K. 1991. In: *Proceedings Golden Jubilee Celebrations Symposium on Grain Legumes*, pp. 325–332. New Delhi, India: Indian Society of Genetics and Plant Breeding, Indian Agricultural Research Institute.

Naringhani, V. G. and Kumar, S. M. 1979. *Indian Journal of Mycology and Plant Pathology* 9: 252–253.

Nerkar, Y. S. 1972. *Indian Journal of Genetics and Plant Breeding* 32: 175–180.

Nerkar, Y. S. 1976. *Indian Journal of Genetics and Plant Breeding* 36: 223–229.

Prassad, A. B. and Das, A. K. 1980. *Indian Journal of Genetics and Plant Breeding* 40: 172–175.

Prassad, A. B. and Das, A. K. 1980. *Journal of Cytology and Genetics* 15: 156–165.

Prassad, A. B. and Das, A. K. 1980. *Journal of Indian Botany* 59: 354–359.

Quader, M. 1987. *Journal of Botany* 16: 9–13.

Quader, M., Ramanujam, S. and Barat, G. K. 1985. In: *Lathyrus and Lathyrism, Proceedings of Collogue Lathyrus*, 234 pp. (eds. A. K. Kaul and D. Combes). New York: Third World Medical Research Foundation.

Ramanujam, K. L., Sethi, K. L. and Rao, S. L. N. 1980. *Indian Journal of Genetics and Plant Breeding* 40: 300–304.

Roy, D. N. and Rao, K. V. 1978. *Journal of Agricultural and Food Chemistry* 26: 687–689.

Singh, M. and Chaturvedi, S. N. 1987. *Indian Journal of Agricultural Science* 57: 503–507.

Sinha, S. K. 1977. In: *Food Legumes: Distribution, Adaptability and Biology of Yield*. Food and Agriculture Organization of the United Nations. Rome, Italy: FAO.

Soadou, A. C. 1959. 25 pp. Department of Agriculture, Cyprus.

Smartt, J. 1984. *Experimental Agriculture* 20: 275–296.

Srivastava, Y. C. and Somayajulu, P. L. N. 1981. *Indian Journal of Genetics and Plant Breeding* 41: 1964–1966.

Swaminathan, M. S., Naik, M. S., Kaul, A. K. and Austin, A. 1971. *Indian Journal of Agricultural Science* 41: 394–406.

Villax, E. J. 1963. La culture des plantes fourrageres dans la region Mediterraneene occidentale. Cahiers Recherche Agronomique No 17, I.N.R.A., 625 pp. Rabat, Morocco.

Management to control biotic and abiotic stress

Crop and soil management practices for mitigating stresses caused by extremes of soil moisture and temperature

M.C. SAXENA[1], A. GIZAW[2], M.A. RIK[3] and M. ALI[4]

[1] ICARDA, P. O. Box 5466, Aleppo, Syria;
[2] Institute of Agriculture Research, Addis Ababa, Ethiopia;
[3] Field Crops Reserarch Institute, Agricultural Research Center, Giza, Egypt, and
[4] Directorate of Pulses Research (ICAR), Kanpur, Uttar Pradesh, India

Abstract

The productivity of cool season food legumes can be seriously limited by excessive soil moisture and poor drainage on heavy clay soils in high rainfall areas and on fields with high water tables. In contrast, for the same crops grown in low rainfall rain-fed farming systems, drought at any stage, and especially during the reproductive phase, is a major constraint, particularly so when the crop relies on conserved soil moisture. In the sub-tropical areas, high temperatures at crop establishment and during the reproductive stages are a constraint to productivity. In more northerly latitudes and especially in high altitude areas, frost can cause severe crop damage either in the early growth stages or later during the reproductive and maturity stages. Using cultivars having tolerance to these stresses and scheduling sowing to avoid weather extremes or match more favorable environmental conditions with particularly sensitive growth stages can reduce the effect of these stresses. The adverse effects of these stresses also might be mitigated by appropriate variations of planting methods, planting density, and soil management.

Introduction

Extremes of soil moisture and temperature are among the most common natural stresses that limit the yield of cool season food legumes (Buddenhagen and Richards, 1988). Although the problem of excessive soil moisture, leading to waterlogging, is not as widespread as is a lack of moisture, it is of considerable regional importance. The cool season food legumes grown for example on heavy vertisols of the highlands of Ethiopia, or in humid sub-tropical lowland areas of southeast Asia, or in coastal areas around the Mediterranean Sea, and in water harvesting catchment areas in the highlands in West Asia and North Africa, and the flood-prone plains around major river systems are often exposed to at least ephemeral waterlogging and their productivity is constrained. Stress associated with drought is, however, more widespread globally than the problem of

F.J. Muehlbauer and W.J. Kaiser (eds.), Expanding the Production and Use of Cool Season Food Legumes, 633–641.
© 1994 Kluwer Academic Publishers.

excessive moisture because a large proportion of cool season food legume production occurs in rainfed areas with erratic rainfall. In many regions these crops are grown on receding soil moisture with little concurrent rain.

Both cold and hot temperatures can adversely affect the productivity of cool season food legumes within the same cropping season; cold and frost stresses often occur during stand establishment and the early vegetative growth stages whereas heat stress, accentuated by soil moisture deficit, are more common constraints during the reproductive and seed-filling stages. The relative importance of these stresses varies considerably depending on the latitude and the elevation of the production site and distance from the sea.

One approach of dealing with the stresses caused by extremes of moisture and temperature is to develop cultivars tolerant of specific stresses in particular target environments. Current efforts in this direction are reviewed elsewhere in this volume. Enhancing, by breeding, the adaptation of cool season legumes to tolerate these abiotic stresses is time consuming. More stable and higher productivity under these adverse environmental conditions can also be addressed by adopting appropriate crop and soil management practices. It is these approaches which are discussed below.

Excessive Soil Moisture

Production of cool season legumes on heavy vertisols is often constrained by excessive moisture in the root zone. Poor drainage leads to crop damage through the development of anaerobic conditions, accumulation of toxic gases, reduction in mineral nutrient availability, inhibition of symbiotic dinitrogen fixation, and increase in susceptibility of plants to root diseases.

In the Ethiopian highlands, where the problem is acute (Hailu, 1988), farmers have devised various methods to reduce the adverse effect of excessive water. These include "burning" of soil (Abebe, 1981) to improve its drainage, delaying the sowing until well beyond the main rainy period, and sowing on ridges. Burning of soil (*Guie* method), using local vegetation, leads to temporary improvement in infiltration rate because of increase in the aggregate size of soil particles and improved availability of some inorganic nutrients, but the practice is not sustainable (Abebe, 1981). Delayed sowing reduces crop duration and so leads to severe yield reductions. Sowing on ridges is only partially successful because under heavy rain the ridges are often washed away (Getachew et al., 1988).

Research has been done by Ethiopian scientists in collaboration with the International Livestock Centre for Africa (ILCA) and the International Crops Research Institute for Semi-Arid Tropics (ICRISAT) and, later, with the International Center for Agricultural Research in the Dry Areas (ICARDA). The work has demonstrated that sowing on broad-bed-and-furrows (BBF) made by animal-drawn implements gives dramatic increases in yield and economic returns with various cool season legumes in contrast to flat-bed or ridge sowing (Table 1). The BBF are 120 cm wide and permit effective surface drainage. They are prepared in advance of the main rainy season and timely sowing of cool season legumes is possible because of good surface drainage.

Table 1. Yield of cool season food legumes as affected by improved soil drainage using broad-bed-and-furrows (BBF) on vertisols in different locations in Ethiopia

Crop	Location	Seasons	Seed yield			Reference
			BBF	Traditional[1]	LSD (P<0.05)	
			- - - t ha⁻¹ - - -			
Chickpea	Akaki	1989-1991	2.021	0.917	0.717**	Tekalign *et al.* (1991a)
	Keteba	1989-1991	2.181	1.125	0.716**	Tekalign *et al.* (1991a)
Lentil	Akaki	1989-1991	2.317	1.458	0.478**	Tekalign *et al.* (1991b)
	Dibandiba	1989-1991	1.268	0.924	0.278**	Tekalign *et al.* (1991b)
	Keteba	1989-1991	1.606	0.806	0.256**	Tekalign *et al.* (1991b)
Faba bean	Inewari	1986	0.810	0.709	+	Getachew *et al.* (1988)
	Wereilu	1986	0.736	0.171	**	Getachew *et al.* (1988)
	Debre Zeit	1985	2.577	1.633	**	Haque *et al.* (1988)

[1] The traditional is system comprised of sowing on flat seedbeds except in case of faba bean at Inewari where broad-bed-and-furrows (BBF) were made by hand and at Wereilu where ridge and furrows were made using a country plow.
+ Non significant
** Significant ($P \leq 0.01$)

Field observations in Egypt have revealed that the broad-bed-and-furrow system also improves the productivity of irrigated lentil on the heavy soils of the Delta region by preventing the transient waterlogging which is a common feature with basin irrigation. There is a need to confirm this observation. Since symbiotic dinitrogen fixation is highly sensitive to waterlogging (Smith, 1987), the practice of top-dressing with small doses of inoraganic nitrogen fertilizer to assist their recovery from transient waterlogging also needs to be evaluated.

Lack of Moisture

Inadequate soil moisture at the depth of seeding reduces or delays emergence. Moisture deficiency in the vegetative stage prevents development of adequate photosynthetic surface and symbiotic dinitrogen fixation and restricts reproductive growth seed yields. Drought stress during the reproductive phase leads to the abortion of flowers and young pods and inhibits seed filling. Cool season food legumes grown on conserved soil moisture can suffer moisture stress at any growth stage. The probability of moisture stress increases as the crop season advances. Even for those legumes grown during the rainy season in the Mediterranean environment, e.g., faba bean, lentil, pea, and grasspea, moisture stress can develop at any stage depending on the intensity and distribution of rainfall. The chances for moisture stress during the reproductive period increases because the probability of rainfall decreases as the evaporative demand increases (Harris, 1979; Smith and Harris, 1981).

Farmers who produce cool season food legumes in northern India use a traditional soil management practice that ensures optimum soil moisture at the seeding depth. The practice consists of cultivating fields late in the evening with a country plow and leaving ridges and furrows on the soil surface. Dew condensed during the night is retained by planking the field early next morning. This practice is repeated several times before sowing seeds in the bottom of the furrow through a tube attached to the country plow. Even though furrows are not flattened, the seeds are covered with about 4 to 5 cm of soil. Farmers in Syria adopt a similar sowing method for spring chickpea. It would be informative if these and other practices were compared for their benefits to crop productivity.

Supplemental irrigation to relieve soil moisture stress during the reproductive phase improves crop productivity of both spring- and winter-sown chickpea in Syria (Saxena et al., 1990), as well as winter-sown chickpea in India (Singh and Das, 1987) and winter-sown lentil in Bangladesh (Hassan and Rahman, 1987). The average seed yield of lentil in two on-farm trials in the Sharkiya Governorate of Egypt increased significantly from 1.36 t ha^{-1} without supplemental irrigation to 2.10 t ha^{-1} with irrigations at 20 and 50 d (vegetative and early reproductive stages) after sowing (Rizk and Hassan, 1991). These experiments were conducted on fields which had been flooded prior to sowing to simulate the effects of a "Nile flood".

Optimum water use efficiency on soils with limited soil moisture holding capacity is absolutely necessary for obtaining satisfactory yields of cool season food legumes. Reducing evaporative losses and regulating transpiration in such a way that the crop does not deplete soil moisture before successfully completing the reproductive phase is critical (Cooper et al., 1987). Recommended management practices for reducing evaporation from bare soil and so improving water use in the Mediterranean-type climates include sowing at narrow row spacings and at high seed densities so that ground cover is rapidly developed (Cooper et al., 1987; Singh and Das, 1987; Silim et al., 1990).

Practices such as mulching to reduce evaporation and the use of antitranspirants and surface reflectants on the crop canopy can also reduce transpiration.

The effects of dust mulching, antitranspirants, and surface reflectants on the productivity of winter-sown faba bean and spring-sown chickpea were studied at Tel Hadya, Syria in 1980/81 when total seasonal rainfall was 372 mm. Dust mulching by hand hoe at 100% flowering and again 14 days later increased chickpea yields significantly whereas no such improvement occurred in faba bean (Table 2). This difference in response to mulching was attributed to the magnitude of ground cover attained by the two crops when the dust mulching treatment was applied. Because of incomplete ground cover in chickpea, the unmulched plots developed large cracks which accentuated soil-moisture loss by evaporation, whereas surfaces of the mulched plots remained intact. On the other hand, because of winter sowing, faba bean developed good ground cover by the time the mulching treatment was applied and hence did not benefit from mulching. Antitranspirant and surface-reflectant treatments did not increase yields, but for reasons which remain obscure. There remains a need to further evaluate the effects of dust mulching or mulching with field-crop residues such as straw on the water use of cool season legumes grown under rainfed conditions with limited rainfall.

Table 2. Effect of antitranspirant, surface reflectant and dust mulching on the yield of rainfed winter-sown faba bean and spring-sown chickpea, Tel Hadya, Syria, 1980/81

Treatment	Seed yield	
	Faba bean (mean of 2 cultivars)	Chickpea
	- - - - kg ha^{-1} - - - -	
Control	2733	575.4
'Vapoguard' spray at 100% flowering	2712	666.8
'Vapoguard' spray at 100% flowering and podding	2737	-
Kaolin (6%) spray at 100% flowering	2643	856.6
Dust mulching at 100% flowering	2579	893.6
Dust mulching at 100% flowering and 14 d later	-	922.2
SED	156.6	160.74
F-test	Not significant	Significant
		(P < 0.05)

Cold Temperature Stress

Cool season food legumes are often exposed to cold stress at sowing and during the early vegetative stages in the Mediterranean lowlands (Harris, 1979; Smith and Harris, 1981). In the highlands (> 1000 m altitude), frost damage can occur at any time prior to flowering. In sub-tropical conditions, where temperatures decline after sowing (Huda and Virmani, 1987), flowering and pod setting are constrained by cold and occasional frost (Porwal and Singh, 1990). Similar to winter cereals, the ability of cool season food legumes to withstand extremes of cold is under genetic control. The precise physiological mechanisms are poorly understood, but the sequence of environmental factors and the stage of exposure to them are important variables (Gusta and Fowler, 1979). Adjusting sowing dates such that the crop has a chance to develop "cold hardiness" by the time cold stress is most likely to occur, is a good agronomic strategy to minimize yield losses (Saxena, 1990; Wery, 1990). This approach, combined with the selection of cultivars with appropriate phenology such that those plant-growth phases most sensitive to cold occur at that time of year when the probability of cold stress is small, can ensure more productive and stable crops in regions where cold stress is common.

Other management practices that may reduce crop damage from cold include depth of sowing and seeding density. In winter wheat, cold damage decreases with increased sowing depth which prevents freezing of the crown (Taylor and McCall, 1936). With winter-sown chickpea at Tel Hadya, Syria during 1986/87, when minimum air temperature at the late vegetative growth stage was $-6.8°C$, there was little difference in the cold-tolerance reaction when sowing depth increased from 5 to 20 cm at 5 cm increments (Malhotra *et al.*, 1990). In the case of autumn-sown faba bean, Pilbeam and Hebblethwaite (1990) found no difference in sowing depths of 5 to 35 cm on the crop productivity. The scope for mitigating against cold stress injury by increasing sowing depth in cool season legumes may therefore be limited.

Field observations in the highlands of Turkey and in Quetta (Pakistan) have indicated that an increase in seeding density of lentil increased survival of seedlings. Studies at Tel Hadya, Syria during the 1991/92 season have also revealed that cold susceptibility ratings of lentil following frost decreased as the plant population increased from 46 to 260 plant m^{-2}.

Use of irrigation and smoke screens to prevent frost damage in chickpea is practiced by some farmers in northern India (Porwal and Singh, 1990), but these techniques await scrutiny by experiment. Porwal and Singh (1990) studied the effects of foliar sprays of a range of chemicals as cryoprotectants on chickpea at Udaipur, India in a 2-year experiment, in which the crop was exposed to freezing temperatures during flowering. A foliar spray of 10^{-3} M aqueous solution of dimethyl sulfoxide at 1000 l ha^{-1} at 50% flowering and again 10 days later significantly increased tolerance to cold and seed yields when compared to the "check" which had received no cryoprotectant. Other chemicals were less effective.

Hot Temperature Stress

The cool season food legumes can experience hot temperature stress at seedling establishment and during the early vegetative stages in the tropics and sub-tropics and at the flowering and podding stages in almost all production regions (Smith and Harris, 1981; Huda and Virmani, 1987; Saxena et al., 1988). The effects are aggravated if soil-moisture supply is limited, a situation common in areas where the crops are grown under rainfed conditions and relying on moisture stored in the soil profile.

Adjusting sowing time and the use of appropriate cultivars which match crop phenology with the occurrence of optimum temperatures is an effective way of avoiding temperature extremes, as shown in chickpea, faba bean, and lentil in Mediterranean environments (Ageeb and Ayoub, 1977; Murinda and Saxena, 1985; Saxena, 1987, 1990; Saxena et al., 1988; Ageeb et al., 1989). In those areas where cool season food legumes are grown in a double-cropping or multiple-cropping system, the scope of adjustment of sowing date may be rather limited.

Losses of seedlings to high temperature stress in Sudan have been reduced by seeding faba bean on the eastern side of north-south oriented ridges, where the maximum temperatures are colder (Saxena et al., 1988). Use of shade crops such as pigeonpea (*Cajanus cajan* L.), sorghum (*Sorghum bicolor* L.), or maize (*Zea mays* L.) to protect faba bean seedlings from heat stress in Sudan did not improve plant stand or crop yield because of excessive competition for light (Salih and Ageeb, 1987; Ageeb et al., 1989). Delayed sowing from the warmer month of October to the relatively cooler month of November increased plant stands and yields significantly.

In areas where irrigation is feasible, the adverse effects of high temperatures can be reduced by adopting a frequent light irrigation schedule as shown in Sudan with faba bean (Serrag-Mohamed et al., 1988; Ageeb et al., 1989; Farah et al., 1990; Salih and Sarrag-Mohamed, 1990), chickpea (Taha et al., 1991; Taha and Ali, 1992), and lentil (Ahmed and Nourai, 1992). However, these studies did not include a precise separation of moisture and temperature effects and so treatment effects are difficult to interpret.

Heat stress also adversely affects symbiotic dinitrogen fixation in cool season legumes (Rawsthorne et al., 1985; Rupela and Kumar Rao, 1987). In areas where early-season symbiosis is constrained, Rawsthorne et al. (1985) have suggested that a small dose of starter nitrogen fertilizer can increase nitrogen assimilation by vegetative chickpea plants and can eventually improve symbiotic dinitrogen fixation when optimum temperature regimes are resumed. There should also be a possibility of identifying strains of *Rhizobium* for different cool season legumes that could develop an effective symbiotic association at higher temperatures (Rupela and Kumar Rao, 1987). This possibility needs further study.

Concluding Remarks

Extremes of moisture supply and temperature, alone or in combination, constrain the productivity and yield stability of cool season food legumes. Use of tolerant cultivars can minimize the harmful effects of these stresses. Development of such cultivars is being attempted but demands long-term commitment and sustained efforts. Reducing the adverse effects of these stresses through agronomic management remains a useful, additional, and complementary strategy. Farmers have developed several management practices to counter the adverse effects of these factors. Very little, however, has been done so far in systematically evaluating the merits of these practices for application under different agroecological conditions. Future efforts must be directed at determining the scientific basis for the "traditional" management practices and to develop more efficient "packages". When combined with appropriate cultivars, these "packages" of agronomic inputs may help to minimize the adverse effects of extremes of temperature and soil moisture in traditional systems of production as well as in their modern alternatives.

References

Abebe, M. 1981. *Ethiopian Journal of Soil Science* III: 57–73.

Ageeb, O. A. A. and Ayoub, A. T. 1977. *Journal of Agricultural Science, Cambridge* 88: 521–527.

Ageeb, O. A. A., Salih, F. A. and Ali, M. A. 1989. *FABIS Newsletter* 24: 8–10.

Ahmed, S. E. K. H. and Nourai, A. H. 1992. In: *Nile Valley Regional Program on Cool-Season Food Legumes and Wheat, 1990/91 Annual Report, Sudan*, pp. 125–127. ICARDA/NVRP-Doc-017. Cairo, Egypt: ICARDA.

Buddenhagen, I. W. and Richards, R. A. 1988. In: *World Crops: Cool Season Food Legumes*, pp. 81–95 (ed. R. J. Summerfield). Dordrecht: Kluwer Academic Publishers.

Cooper, P. J. M., Gregory, P. J., Tully, D. and Harris, H. C. 1987. *Experimental Agriculture* 23: 113–158.

Farah, S. M., Fakki, H. El, Gorashi, A. M. and Ali, A. E. 1990. *FABIS Newsletter* 27: 16–18.

Getachew, A., Jutzi, S. C., Tedla, A. and McIntire, J. 1988. In: *Management of Vertisols in Sub-Saharan Africa*, pp. 263–283. Addis Ababa, Ethiopia: ILCA.

Gusta, L. V. and Fowler, D. B. 1979. In: *Stress Physiology in Crop Plants*, pp. 160–178 (eds. H. Mussell and R. C. Staples). New York: John Wiley & Sons.

Hailu, G. 1988. In: *Management of Vertisols in Sub-Saharan Africa*, pp. 321–334. Addis Ababa, Ethiopia: ILCA.

Haque, I., Jutzi, S. and Nnadi, L. A. 1988. In: *Soil Science Research in Ethiopia. Proceedings of the First Soil Science Research Review Workshop*, pp. 120–127 (ed. Desta Beyene). Addis Abeba, Ethiopia: Institute of Agricultural Research.

Harris, H. C. 1979. In: *Food Legume Improvement and Development*, pp. 7–14 (eds. G. C. Hawtin and G. J. Chancellor). Ottawa, Canada: The International Centre for Agricultural Research in the Dry Areas (ICARDA) and the International Development Research Center (IDRC).

Hassan, A. A. and Rahman, M. A. 1987. *Thai Journal of Agricultural Science* 20: 277–283.

Huda, A. K. S. and Virmani, S. M. 1987. In: *Adaptation of Chickpea and Pigeonpea to Abiotic Stresses*, pp. 15–31 (eds. N. P. Saxena and C. Johansen). Patancheru, Andhra Pradesh, India: ICRISAT.

Malhotra, R. S., Singh, K. B. and Saxena, M. C. 1990. *International Chickpea Newsletter* 22: 19–21.

Mamo, T., Duffera, M., Abebe, M. and Kidanı, S. 1991a. In: *Nile Valley Regional Program on Cool Season Food Legumes, Research Results 1990/91 Crop Season, Ethiopia*, pp. 144–146. Addis Ababa, Ethiopia: Institute of Agricultural Research.

Mamo, T., Duffera, M., Abebe, M. and Kidanu, S. 1991b. In: *Nile Valley Regional Program on Cool Season Food Legumes, Research Results 1990/91 Crop Season, Ethiopia*, pp. 187–189. Addis Ababa, Ethiopia: Institute of Agricultural Research.

Murinda, M. V. and Saxena, M. C. 1985. In: *Faba Bean, Kabuli Chickpeas and Lentils in the 1980s*, pp. 229–244 (eds. M. C. Saxena and S. Varma). Aleppo, Syria: ICARDA.

Pilbeam, C. J. and Hebblethwaite, P. D. 1990. *FABIS Newsletter* 26: 15–18.

Porwal, B. L. and Singh, H. G. 1990. *Legume Research* 13: 169–175.

Rawsthorne, S., Hadley, P., Summerfield, R. J. and Roberts, E. H. 1985. *Plant and Soil* 83: 279–293.

Rizk, M. A. and Hassan, M. W. 1991. In: *Nile Valley Regional Program on Cool-Season Food Legumes and Cereals, 1990/91 Annual Report, Egypt*, pp. 53–54. Cairo, Egypt: ICARDA.

Rupela, O. P. and Kumar Rao, J. V. D. K. 1987. In: *Adaptation of Chickpea and Pigeonpea to Abiotic Stresses*, pp. 123–131 (eds. N. P. Saxena and C. Johansen). Patancheru, Andhra Pradesh, India: ICRISAT.

Salih, F. A. and Ageeb, O. A. A. 1987. *FABIS Newsletter* 18: 18–20.

Salih, F. A. and Sarrag-Mohamed, G. E. 1990. *FABIS Newsletter* 26: 25–26.

Sarrag-Mohamed, G. E., Salih, F. A. and Ageeb, O. A. A. 1988. *FABIS Newsletter* 22: 17–29.

Saxena, M. C. 1987. In: *Adaptation of Chickpea and Pigeonpea to Abiotic Stresses*, pp. 135–141 (eds. N. P. Saxena and C. Johansen). Patancheru, Andhra Pradesh, India: ICRISAT.

Saxena, M. C. 1990. *Options Méditerranéennes Série Séminaires* No. 9: 17–24.

Saxena, M. C., Saxena, N. P. and Mohamed, A. K. 1988. In: *World Crops: Cool Season Food Legumes*, pp. 845–856 (ed. R. J. Summerfield). Dordrecht: Kluwer Academic Publishers.

Saxena, M. C., Silim, S. N. and Singh, K. B. 1990. *Journal of Agricultural Science, Cambridge* 114: 285–293.

Silim, S. N., Saxena, M. C. and Erskine, W. 1990. *Agronomy Journal* 82: 927–930.

Singh, R. P. and Das, S. K. 1987. In: *Adaptation of Chickpea and Pigeonpea to Abiotic Stresses*, pp. 51–61 (eds. N. P. Saxena and C. Johansen). Patancheru, Andhra Pradesh, India: ICRISAT.

Smith, K. A. 1987. In: *Adaptation of Chickpea and Pigeonpea to Abiotic Stresses*, pp. 77–89 (eds. N. P. Saxena and C. Johansen). Patancheru, Andhra Pradesh, India: ICRISAT.

Smith, R. C. G. and Harris, H. C. 1981. *Plant and Soil* 58: 31–57.

Taha, M. B. and Ali, M. E. K. 1992. In: *Nile Valley Regional Program on Cool-Season Food Legumes and Wheat, 1990/91 Annual Report, Sudan*, pp. 105–107. ICARDA/NVRP-Doc-017. Cairo, Egypt: ICARDA.

Taha, M. B., Ali, M. E. K. and Ahmed, S. E. K. H. 1991. In: *Nile Valley Regional Program on Cool-Season Food Legumes and Wheat, 1989/90 Annual Report, Sudan*, pp. 142–144. ICARDA/NVRP-Doc-012. Aleppo, Syria: ICARDA.

Taylor, J. W. and McCall, M. A. 1936. *Journal of Agricultural Research* 52: 557–568.

Wery, J. 1990. *Options Méditerranéennes Série Séminaires* No. 9: 77–85.

Integrated control of diseases of cool season food legumes

S.P.S. BENIWAL[1] and A. TRAPERO-CASAS[2]

[1] *Legume Improvement Program, ICARDA, B.P. 2335, Fes, Morocco, and*
[2] *Departmento de Agronomia, ETSIAM, Universidad de Córdoba, Apdo. 3048, 14080 Córdoba, Spain*

Abstract

Diseases are a major constraint to the successful cultivation of faba bean, chickpea, lentil, and pea. The major diseases include damping-off, wilt and root rots, spots and blights, rusts, and mildews, although viruses and nematodes also affect these crops. These diseases have been controlled through biological, chemical, and physical methods, although biological methods that include host-plant resistance have received most attention. Thus far, only limited efforts have been made on the integrated control of these diseases. The potential methods and their scope for use in the integrated control in both developing and developed countries are highlighted. Integrated control packages that include the use of resistant/tolerant cultivars, healthy seeds, improved cultural practices, seed and foliar treatments, and biological agents are outlined for controlling the economically important diseases. The need and scope of the integrated management of these diseases are emphasized.

Introduction

Of the cool season food legumes cultivated in the world, the four most important in descending order are pea (*Pisum sativum* L.), chickpea (*Cicer arietinum* L.), faba bean (*Vicia faba* L.), and lentil (*Lens culinaris* Medik.) (FAO, 1990). All four crops are far more important in developing than developed countries as indicated by their total hectarage and production. They are primarily cultivated by resource-poor farmers in the developing world as subsistence crops, mainly for self consumption, in low-input agriculture where sustainability is a major issue. Contrarily, in the developed countries they are grown as commercial and "break" crops in the cereal monocropping system for human as well as animal consumption. Over the years, their total production and productivity have almost remained static, especially in the developing world. This is due to various biotic, abiotic, and socio-economic constraints (Saxena and Goldsworthy, 1988). Among the biotic constraints, diseases are

F.J. Muehlbauer and W.J. Kaiser (eds.), Expanding the Production and Use of Cool Season Food Legumes, 642–665.
© 1994 *Kluwer Academic Publishers.*

considered as a major constraint that limits the production and productivity of cool season food legumes.

Thus far, disease control in cool season food legumes has received fairly good attention in both the developed and the developing countries, especially during the last decade. The methods that have been mainly employed include host-plant resistance and chemicals, whereas only limited use has been made of cultural practices and of useful microorganisms. Moreover, attention so far has mainly focused on individual methods of disease control, whereas the integrated control aspects of these diseases have received only very limited attention, and only very recently.

In the present day context of the many problems associated with modern agriculture in the developed world and the issue of sustainable agriculture in the developing world, the integrated approach in disease control undoubtedly has an appeal that must be seriously considered and exploited. Moreover, only integrated control methods can effectively manage certain diseases such as wilt/root rots that usually occur together either in sequence or at the same time in cool season food legume crops. Diseases of cool season food legumes and their control have recently received good reviews (Kraft *et al.*, 1988; Nene *et al.*, 1988). In this article, we present our analysis of the available information on the integrated control of diseases of the four major cool season food legumes and gaps in our knowledge of their control. We also highlight the potential methods and their scope for use in the integrated control of these diseases during the next decade, and finally, outline integrated disease management packages for the major groups of diseases.

Major Diseases of Cool Season Food Legumes

Pea, chickpea, faba bean, and lentil are affected by many diseases caused by fungi, bacteria, viruses, and nematodes. However, only a few are economically important (Kraft *et al.*, 1988; Nene *et al.*, 1988). These are listed with their relative importance in different food legume growing zones of the world in Table 1. They can be grouped into: seed and seedling diseases, wilt and root rots, blights and spots, rusts, mildews, viruses, and nematodes. Here, we mainly focus our discussion on seed and seedling diseases, root rots, and downy and powdery mildews of pea; Fusarium wilt, Ascochyta blight, dry root rot, and Botrytis gray mold of chickpea; chocolate spot and rust of faba bean; and wilt, rust, and Ascochyta blight of lentil. Salient information on these diseases is provided in Table 2.

Present Status of Disease Control

Presently available information on control of the economically important diseases of pea, chickpea, faba bean, and lentil has been placed into three broad

Table 1. Important diseases of cool season food legumes in different food legume growing zones of the world

Crop/disease	Zone (latitude)			
	I (0°–20°)	II (20°–30°)	III (30°–45°)	IV (45°–60°)
Pea				
Seed and seedling diseases	1[a]	3	5–7	7
Root rots	5	5	5–7	7
Downy mildew	1	5	7	5
Powdery mildew	7	8–9	5–7	5
Rust	3	5	5	3
Chickpea				
Fusarium wilt	8–9	8–9	5–7	–[b]
Ascochyta blight	1	3–5	9	9
Dry root rot	7	5–7	3	–
Botrytis gray mold	1	7–8	3	5
Stunt	3–5	5	3–5	–
Faba bean				
Chocolate spot	7–8	8	8	8
Rust	6	6	5	3
Ascochyta blight	3	3	6	7
Wilt and root rot	6	7	1	3
Stem nematode	1	1	5	5
Lentil				
Wilt	7	8–9	3–7	1
Rust	7	7	5–7	1
Ascochyta blight	1–3	5–7	7	7–9
Root rots	7	5–7	3	3

[a] Rated on a 1–9 scale, where 1 = not important; 3 = slightly important; 5 = moderately important; 7 = important; 9 = very important.

[b] Disease not reported.

categories of biological, chemical, and physical control methods as proposed by Gabriel and Cook (1990) and Cook and Veseth (1991). The control methods that are mediated through the pathogen, antagonists and the host are included in biological control. This includes host-plant resistance, use of antagonists, and all cultural practices, excluding tillage (Cook and Baker, 1983).

Single Disease Control by Individual Methods

Until now, a major emphasis has been on the control of individual diseases. Considerable information has been generated on the individual methods of

Table 2. Information on the important diseases of pea, chickpea, faba bean and lentil in the world

Group/name of disease	Crop	Pathogen(s)	Mode of survival	Favorable conditions
Seed & seedling				
Pre- & post-emergence damping-off	Pea	*Pythium ultimum*	Soil(oospores)	Cool temp.
		Rhizoctonia solani	Soil (sclerotia), other hosts	Warm temp. (24-29°C)
Fusarium wilts				
Fusarium wilt	Chickpea	*Fusarium oxysporum* f. sp. *ciceris*	Soil (chlamydospores), seed	Host refuse in soil; moisture stress
	Lentil	*F. oxysporum* f. sp. *lentis*	Soil (chlamydospores), seed (?)	Light soil; moisture stress
Root rots				
Common root rot	Pea	*Aphanomyces euteiches* f. sp. *pisi*	Soil(oospores)	Poorly-drained heavy soils
Fusarium root rot	Pea	*F. solani* f. sp. *pisi*	Soil (chlamydospores)	Both dry & moist soils
Dry root rot temp.	Chickpea	*Macrophomina phaseolina* (*R. bataticola*)	Soil (sclerotia)	Ambient above 30°C
Blights/spots/molds				
Ascochyta blight	Chickpea	*Mycosphaerella rabiei* (*Ascochyta rabiei*)	Crop residues (pycnidia and pseudothecia)	Cool, moist
	Lentil	*A. lentis* (*A. fabae* f. sp. *lentis*)	Crop residue, Seed	Cool, moist
Chocolate spot	Faba bean	*Botrytis fabae* & *B. cinerea*	Soil (sclerotia), seed	Cool (20°C), humid
Botrytis gray mold	Chickpea	*B. cinerea*	Other hosts, seed	Mod. cool temp. (20-30°C);humid
Rusts				
Rust	Faba bean	*Uromyces viciae-fabae*	Soil (teleutospores), seed	Cool temp. (17-25°C)
	Lentil	*U. fabae*	do	do
Mildews				
Downy mildew	Pea	*Peronospora viciae*	Seed (mycelium, oospores)	Cool temp. (14-18°C); high humidity
Powdery mildew	Pea	*Erysiphe polygoni*	Crop residue (cleistothecia), seed (mycelium)	Dry, dewy nights
Viruses				
Chickpea stunt	Chickpea	Bean leaf roll virus	Legume hosts	Viruliferous aphids as vectors
Nematodes				
Stem nematode	Faba bean	*Ditylenchus dipsaci*	Soil, seed	Moist, light soils

Table 3. Inventory of the available methods of controlling economically important diseases of pea and chickpea

Method/ operation of control	Pea			Chickpea	
	Seed/seedling, root rots	Downy mildew	Powdery mildew	Fusarium wilt	Ascochyta blight
I. Biological methods					
1. Host-plant resistance					
Resistant cultivar	Available (Not for CRR)[1]	Avail.	Avail.	Avail.	Avail.
Cultivar maturity & type	Pigmented, smooth- and colored-seeded	--	Early	--	Tall, erect
2. Antagonists Trichoderma Pseudomonas	--	--	--	--	
3. Cultural practices					
Seeding rate	--	Lower	--	--	Avoid higher
Seeding date	Early (CRR)	--	Early	Early	Delay
Crop rotation	Recomm.	Recomm.	--	Recomm.	Recomm.
Intercropping	--	--	--	--	Cereals, mustards
Org. amend.	Org. matter	--	--	--	--
II. Chemical methods					
Seed dressing	Captan, metalaxyl + TBZ	Metala- xyl, fosetyl- aluminum	--	Benomyl + thiram	Tridemo- rph + TBZ, benomyl
Foliar sprays	--	Metalaxyl	Sulphur, benomyl, tridemorph	--	Chloro- thalonil
Soil treatment	Variable results	--	--	Methyl bromide	--
Effect of herbicides	Triflur- alin	--	--	--	--
III. Physical methods					
Land prepar- ation	Level land, improve drainage	Deep plow- ing	Deep plow- ing	Deep plow- ing	Deep plow- ing
Crop refuse destruction	--	Plow- ing- in	Plow- ing- in	Recomm.	Plow- ing-in
Heat	--	--	--	Soil solari- zation	Sun dry- ing of seed
Pathogen- free seed	Clean & healthy	Clean seed	Clean seed	Recomm.	Recomm.
Choice of site	Less- infested, well drained, light texture	--	--	Less infested	Recomm.

[1] CRR = Common root rot.

Table 4. Inventory of the available methods of controlling economically important diseases of faba bean and lentil

Method/ Operation of control	Faba bean		Lentil		
	Chocolate spot	Rust	Wilt	Rust	Ascochyta blight
I. Biological methods					
1. Host-plant resistance					
Resistant cultivar	Avail.	Avail.	Avail.	Avail.	Avail.
Cultivar maturity & type	--	--	Short & less sec.roots	Early	--
2. Antagonists	*Penicillium citrinum P. cylopium*	-- --	*Tricho- derma Arachniatus*	-- --	-- --
3. Hyperpara- site	*Gliocladium catenulatum*	--	--	--	--
3. Cultural methods					
Seeding rate	Lower	Higher	--	Higher	Lower
Seeding date	Early, delay	Early	Delay	Delay	Delay
Crop rotation	Recomm.	--	Rice or sorghum	--	Recomm.
Irrigation	Avoid excess	--	--	--	Avoid excess
Host nutrition	Higher P_2O_5	Higher P_2O_5	Higher P_2O_5 Mn, Zn	Higher ferti- lizer	Higher P_2O_5
II. Chemical control					
Seed dressing	Benomyl + thiram	No effect	Benomyl+ thiram, captan	Diclo- butra- zol	TBZ, benomyl
Foliar sprays	Dithane M-45, vinclozolin, benomyl	Dithane M-45	--	Wett- able sulp- hur	Chloro- thalanil benomyl
Soil treat- ment	--	--	Benomyl+ captan	--	--
Anti- senescence	Delay (Benzyl- adenine)	--	--	--	--
III. Physical methods					
Land prepa- ration	Level land	--	--	Summer plowing	Deep plowing
Crop refuse destriction	Remove crop debris	--	Recomm.	Recomm.	Remove/ burn
Heat	--	--	--	--	Sun dry seed; hot water
Pathogen-free seed	Healthy seed	Clean	Healthy seed	Clean seed	Healthy seed
Choice of site	Well drained, less infested	--	Avoid infested and sandy loam soils	--	Avoid infes- ted field

controlling the major diseases of pea, chickpea, faba bean, and lentil, and is summarized in Tables 3 and 4. The usefulness of different methods is discussed later in this paper.

Integrated Control

Until now, very limited attempts have been made on the integrated control of cool season food legume diseases. Moreover, in the majority of the attempts, mostly a combination of host-plant resistance and fungicides has been employed. The information available on the integrated control of diseases of the four cool season food legumes is briefly discussed below.

Pea

Attempts of integrated control in pea include those on Pythium pre- and post-emergence damping-off (Kraft *et al.*, 1988), root rot complex (Tu, 1987), and Ascochyta diseases (Hagedorn, 1984). However, disease management of the pea root rot complex (*Fusarium solani* (Mart.) Sacc. f. sp. *pisi* (F. R. Jones) W. C. Snyder & H. N. Hans.; *F. oxysporum* Schlechtend: Fr. f. sp. *pisi* (J. C. Hall) W. C. Snyder and Hans; *Aphanomyces euteiches* Drechs. f. sp. *pisi* W. F. Pfender and D. J. Hagedorn; and *Pythium* spp.) offers a good example of control by the integration of several methods. Effective disease management was obtained using moderately resistant cultivars with cultural practices, including crop rotation, tillage to reduce soil compaction, seedbed preparation, the use of dinitroaniline herbicides, and the introduction in the crop rotation of green manure crops together with the treatment of seeds with fungicides, and avoiding sowing in highly infested soil by the soil-indexing method (Tu, 1987). These diseases also were effectively controlled by the use of resistance, *Trichoderma*, and fungicides (Kraft and Papavizas, 1983). Similarly, control of Pythium seed and seedling rot by resistance was improved by planting seed with high vigor ratings, and tillage and crop residue management markedly influenced severity of root rots (Kraft *et al.*, 1988).

A concept of integrating biological and chemical seed treatments for controlling Rhizoctonia root rot of pea has been proposed (Hwang and Chakravarty, 1991). A combination of *Gliocladium virens* J. H. Miller, J. E. Giddens, and A. A. Foster, and Anchor® was found to increase seedling survival and reduce Rhizoctonia root rot compared to Anchor[(R)] alone (Hwang and Chakravarty, 1991). The growth of *R. solani* Kühn was significantly reduced by Anchor® at 50 ppm, whereas that of *G. virens* was not affected at concentrations up to 500 ppm.

The use of a fungicide mixture has been recommended for broad-spectrum disease control of pea seedlings in France (Rossignol, 1988), where a mixture of oxadizyl, cymoxanil, and mancozeb provided good control of downy mildew, foot rot, and leaf spot (*Mycosphaerella pinodes* (Berk. and Bloxam) Vestergr.), and damping-off (*Botrytis cinerea* Pers.: Fr.).

Ascochyta diseases of pea are examples which emphasize the need of integration. Although resistant cultivars exist, most often they are not resistant to all three diseases (Hagedorn, 1985). Also, environmental conditions for disease development are not similar for the three diseases, so a higher level of integration is required for effective disease control. Besides resistant cultivars, the use of pathogen-free seed or seed treated with fungicides is of paramount importance (Hagedorn, 1984). Also, crop rotations and plowing down pea refuse immediately after harvest are important to reduce primary inoculum (Hagedorn, 1984). Foliar application of fungicides may provide some protection to pea crops, but often this practice is not economical.

Disease control based on a single method is effective for some major diseases, including Fusarium wilt controlled with resistant cultivars or Pythium damping-off by seed treatments. However, these methods should not be substituted for good cultural practices, but must be used in combination with them (Hagedorn, 1984).

Chickpea

Compared with pea diseases, efforts on the integrated control of chickpea diseases have been very limited. Ascochyta blight was managed through integration of host-plant tolerance and foliar application of chlorothalonil (Reddy and Singh, 1990). Two sprays, one each at the vegetative (seedling) and reproductive (early podding) stage or both in the reproductive stage in a moderately resistant cultivar, significantly reduced blight severity on leaves, stems, and pods, and increased grain yield. However, the former two sprays were more effective and economical. Similarly, results of the use of a combination of fungicide sprays and the moderately resistant cv. Kasseb in Tunisia showed that a single chlorothalonil spray at the seedling stage significantly reduced pod infection and increased grain yield (2.5 ×) compared with the total kill and zero grain yield in a susceptible cv. Amdoun, under heavy natural disease pressure (H. Halila, unpublished). Two sprays, one at seedling and the second at mid-vegetative stage, provided complete protection of pods and resulted in a higher (3.5 ×) grain yield increase. A third spray at the early podding stage did not provide any additional advantage over the two sprays. In Algeria, seed treatment with carbendazim and two sprays of chlorothalonil before flowering provided significantly better control of Ascochyta blight in a susceptible cv. Sebdou over the untreated check plots (Bouznad *et al.*, 1991). For Botrytis gray mold, a combination of several seed dressings and foliar fungicides was found effective in India (Reddy *et al.*, 1990). However, the disease was better managed and chickpea grain yields improved by integrating two foliar sprays of vinclozolin with a genotype with tall, erect and compact growth habit, and wider inter-row spacings (Reddy *et al.*, 1992).

Integration of host-plant resistance and fungicidal seed and foliar treatments was tested against a wilt and root rot complex of chickpea in Spain (Jiménez-

Díaz and Trapero-Casas, 1985). Although seed dressings with fungicide, singly or in mixtures, significantly increased seedling emergence of the moderately resistant cv. PV-24 and PV-25, and some fungicide seed dressings significantly delayed early development of epidemics for cvs. PV-24 and PV-25, none, except for triadimenol for cv. P-2245, significantly decreased either the rate of disease increase or the final incidence of dead plants. Integration of host-plant resistance with other methods of control has been suggested for managing wilt/root rots of chickpea in Ethiopia (Beniwal et al., 1992).

Faba Bean

A combination of tolerant cultivars and fungicide sprays of mancozeb was successfully employed for the control of chocolate spot, rust, and downy mildew of faba bean in Egypt (Mohamed, 1982). In Sudan, a combination of late sowing (November) and increased plant population (49.9 plants m^{-2}) and a combination of November sowing and a 7-day watering interval significantly decreased Fusarium wilt (*F. oxysporum* f. sp. *fabae* Yu et Fang) and root rot (*F. solani*) in faba bean (Salih and Ageeb, 1987; Ageeb and Salih, 1989). In Syria, increases in grain yield in a moderately resistant cultivar were six-fold compared with two-fold in a susceptible cultivar when nine sprays were applied to control chocolate spot (Hanounik and Maliha, 1985). In the UK, seed dressing with thiabendazole + metalaxyl + thiram coupled with a post emergence spray of CGR 169374 (a new triazole systemic fungicide) + thiabendazole were found very effective for the control of Ascochyta blight (Jellis et al., 1988).

Lentil

Although considerable work on Fusarium wilt of lentil has been done (Kannaiyan, 1974; Khare, 1980; Khare et al., 1993), hardly any attempts have been made on its integrated control. For Ascochyta blight control, recommendations made to lentil growers of western Canada (Morrall and Sheppard, 1981) and Palouse region of eastern Washington and northern Idaho (Kaiser and Hannan, 1986) were based on general pathological principles. These included crop rotation, deep plowing of infested debris, early seeding, use of disease-free seed, agar-plate tests of prospective seed lentils, and use of cv. Laird that was found more resistant than cv. Eston (for western Canada), and seed treatment with a fungicide to control the seedborne phase of the disease (for Pacific Northwest, USA). The use of a resistant cultivar, seed dressing with thiabendazole, sun drying of seed, and foliar fungicides (chlorothalonil and benomyl) have been found effective in Ethiopia (Beniwal et al., 1989; Ahmed and Beniwal, 1991). For rust, several methods of control including resistant cultivars, seed treatment, foliar fungicide sprays, and destruction of crop debris have been suggested (Nene et al., 1975; Khare, 1981). A significant effect of sowing date and cultivars was observed on lentil rust in India; a 2-week delayed sowing (25 November) from the normal decreased rust in both susceptible and moderately resistant cultivars (Singh and Dhingra, 1980).

Multiple Disease Control

Having realized that in real farm situations generally more than one disease affects a food legume crop in one cropping season (Nene, 1988), a number of attempts have been made to redress this situation. However, only host-plant resistance and chemical control have been employed to control multiple diseases in cool season food legumes. Host-plant resistance has been identified for a number of multiple diseases (Table 5), and cultivars with multiple disease resistance have been developed for some situations. For certain crops, fungicides for multiple disease control have been identified – for example, mancozeb for the control of chocolate spot, rust, and downy mildew of faba bean in Egypt (Mohamed, 1982), and rust, Ascochyta blight, and chocolate spot in Syria (Hanounik and Maliha, 1985); maneb + mancozeb for the control of rust and chocolate spot of faba bean (Yeoman *et al.*, 1987); and triadimephon, tridemorph and triforine and mancozeb for powdery mildew and rust of pea (Amin and Ramachander, 1989).

Gaps in Our Knowledge

The gaps in our knowledge about integrated control of cool season food legumes are due mainly to our lack of information about different aspects of the diseases and their control.

Gaps on Disease Aspects

These include either a complete lack or insufficient information on disease epidemiology and forecasting; pathogen variation, development, and geographic distribution; and disease monitoring.

Gaps on Disease Control Aspects

These include gaps in our knowledge of different methods of control. Although use of resistant cultivars is the mainstay of the integrated management of cool season food legume diseases, resistant cultivars are not available for all situations and crop stages (Reddy *et al.*, 1990) or their use is curtailed by development of new pathogenic races (Singh and Reddy, 1990). This emphasizes the need for developing cultivars with stable resistance to multiple races. For this, more information on the genetics of resistance in the major host-

Table 5. Lines/cultivars identified for multiple disease resistance in pea, chickpea, faba bean and lentil[a]

Crop	Diseases	Line/Cultivar
Pea	Powdery mildew, rust	JP, DP 1
Chickpea	Wilt, dry root rot, black root rot	ICC 12237 to 12269
	Wilt, Ascochyta blight, Botrytis gray mold	ICC 1069
	Wilt, dry root rot, stunt	ICC 10466
	Wilt, Sclerotinia stem rot	ICC 858, 959, 4914, 8933, 9001
Faba bean	Chocolate spot, rust	BPL 710-A-1, 1179-1, 1179-2
	Chocolate spot, rust, powdery mildew	BPL 710-A-1, 1179-2
Lentil	Wilt, rust	JL 599, 632, 674,1005, Pant L 406, 639
	Wilt, collar rot, Pythium root rot, one or more viruses	LP 288 PI 431667, 431668, 432018, 435957
	Rust, powdery mildew	Coll. 10066, 10463, 10498, 10509, 10518, 10537
	Ascochyta blight, rust	Laird, ILL 358
	Wilt, collar rot, wet root rot	JL 80, Pusa 3, Pant 234
	Wilt, wet root rot, black root rot	H 5-6-81, 4-4-81

[a] Adapted from Nene (1988), Reddy (1991), and Khare *et al.* (1992).

pathogen systems is required. In some cases (e.g., common root rot of pea) the available resistance, after transfer to horticulturally-acceptable types, is diminished (Kraft *et al.*, 1988). In other cases, the resistant cultivars do not meet the preferred seed quality standards or do not have desirable plant architecture, and thus cannot be included in the recommendations for integrated control. For example, there is a general lack of a suitable level of resistance to Ascochyta blight or to Fusarium wilt in large-seeded chickpea cultivars, which are preferred in some markets in the Mediterranean region (Jiménez-Díaz *et al.*, 1991). Similarly, there is a lack of tall, compact, and early-maturing cultivars of chickpea to ward-off Ascochyta blight and Botrytis gray mold.

Another important aspect where there are sufficient gaps in our knowledge is the use of host-plant resistance for multiple disease control. Although multiple host-plant resistance has been identified in some cases, high yielding cultivars with stable resistance to the major diseases are yet to be developed in the majority of cases (Nene, 1988). This is extremely important in the overall context of successful crop management.

Among cultural practices for disease control, crop rotation is important especially for pathogens that survive in crop refuse. Crop sequence is known to affect efficacy of crop rotations. There is very little research on this aspect and thus should receive increased attention because inclusion of some crops in the rotations may contribute to reduction in some pathogens (Tu, 1987). Also, there is further need of generating specific information on the effect of cultural practices on disease development, singly or in combinations, in cool season food legumes (Allmaras *et al.*, 1988; Cook, 1988). Here, also, the impact of a cultural practice on the multiple disease control in a crop should be kept in mind. Such information will be useful in deciding cultural aspects of crop management. The specific areas that need further attention include rate, method and date of seeding, intercropping, tillage, crop refuse management, soil moisture, host nutrition, effect of weed infestation, etc.

Although considerable information on chemical methods of control is available, there is a lack of information on the minimum number and timing of fungicide sprays needed for use in combination with host-plant resistance. Information on the use of new fungicide mixtures is needed for multiple disease control. Also, more information is required on the effects of herbicide use on disease control, especially of soilborne diseases.

Among the physical methods of disease control, much more specific information is needed in the developing countries on the effect of these methods on disease development. These include the use of clean, vigorous and disease-free certified seed, plant quarantine regulations, and choice of less-infested fields.

Although information on the use of beneficial microorganisms for disease control in cool season food legumes is available (Cook, 1988), there is considerable need for developing methods to improve their application and effectiveness. Similarly, more information is needed on the use of suppressive soils in disease management (Cook, 1988).

Future Strategy

Important Considerations

Problems that Must Be Addressed

The most important considerations in present day agriculture are tackling a number of multifaceted serious problems (Cook and Baker, 1983). These include economics, human and animal health, and sustainability. Thus, any strategy on integrated disease control (IDC) must provide satisfactory answers to these considerations both in the developed and the developing countries. Therefore, we will have to strive to develop IDC strategies that are effective, economical, safe, and compatible with other crop management practices.

A Complex Activity

Although IDC on a single disease basis is complicated enough, its complexity increases many-fold when it is considered as a part of an integrated pest and crop management program, which definitively forms the package to the farmer to optimize productivity with minimal ecological impact. This complexity emphasizes the need for interdisciplinary cooperation as well as essentially all aspects of plant pathology (Bruehl, 1989). Also, it involves an interaction of several partially effective practices influencing its total effect on disease and crop management. This interaction is well exemplified by some cultivars that are resistant to one disease but very susceptible to other disease(s) of the same crop or a cultural practice that reduces one disease but enhances the other. In this context it may be appropriate to reclassify IDC as integrated disease management (IDM).

Situation/Location Specific

Because of the complexity of an IDM package and the different agroecological conditions that favor specific diseases in a crop, its applicability is limited to specific situations or localities. For example, the integrated approach to control major diseases of chickpea will be very different in California or Mexico where Ascochyta blight is absent, or does not have favorable environmental conditions for disease development, and wilt/root rots and virus diseases are prevalent, than in the Pacific Northwest of the USA, where Fusarium wilt does not occur and major diseases are Ascochyta blight and Pythium seed rot and preemergence damping-off, or in Australia where the major disease is Phytophthora root rot. Therefore, integrated approaches of disease management have to be different based on the disease situation, and identification of the major diseases in a particular locality, country, region, or zone will become a prerequisite for developing IDM strategies.

Options Available

As mentioned earlier, options available for IDM will come from biological, physical, and chemical methods. The choice of the available methods for IDM will be predicated by the importance of cool season food legumes as a cash crop in farming systems, cropping systems in practice, stage of technology development and its adoption in agriculture, and the general socio-economic situation and attitudes of farmers in the mandated zone, country, region, or locality. This implies that different IDM packages will have to be thought of for different situations. Factors that are important in deciding IDM packages for the developing and the developed countries are outlined in Table 6. Thus, choices on methods of disease control in an IDM may differ for different situations.

Table 6. Factors influencing the development of an integrated disease management package for cool season food legumes for the developing and developed countries

Factor	Developing countries	Developed countries
Low-input	++	+
Affordable	++	+
Simple and easy to use	++	+
Economic feasibility	++	++
Environment protection	+	++
Ensure sustainable agriculture	+	++

++ = Very important
+ = Important

Another important consideration in deciding options is to determine the most appropriate method of disease management. For example, in the case of host-plant resistance it is important to identify the kind of resistance that would be the most appropriate for each major pathogen(s). Also, options for IDM do not always have to involve several partially effective methods of disease control because in certain circumstances a single method or only a few methods would effectively manage a major disease in a crop. For example, the use of resistant cultivars for the control of Fusarium wilt in pea has been a successful approach and actually is the only economical control measure of this disease (Hagedorn, 1984, 1985). The scope of different methods of disease control for use in IDM of cool season food legumes is discussed below.

Biological Methods

Host-Plant Resistance

Among the biological methods available, the use of host-plant resistance is the most efficient, safe, economical, and easiest method of disease control in cool season food legumes (Meiners, 1981; Kraft *et al.*, 1988; Nene *et al.*, 1988; Haware *et al.*, 1990; Reddy *et al.*, 1990; Reddy, 1991; Khare *et al.*, 1993). However, for real farm situations greater emphasis is required on identification of sources for multiple disease resistance and development of high yielding cultivars with durable multiple-disease and multiple-race resistance with acceptable seed quality and plant architecture. The long-term nature of this project has been emphasized (Nene, 1988). Maintenance of disease resistance to make it last longer is another important consideration that needs attention.

Cultural Practices

The use of cultural practices is another important method that has tremendous scope in IDM of cool season food legumes, especially of the soilborne diseases caused by fungi and nematodes that attack roots. Different components of cultural practices that could be used in IDM are listed in Tables 2 and 3. Of these, important ones are crop rotation and sequence, seeding time, rate and method, management of crop residues after harvest, fertilizer application, intercropping, and eradication of overwintering and/or alternative pathogen hosts (Kaiser, 1981; Papendick *et al.*, 1988; Cook and Veseth, 1991; Reddy, 1991). Although these methods are applicable to both developed and developing country situations, they have greater relevance in the latter situation where certain practices have been followed since time immemorial.

Antagonists and Hyperparasites

Lately, the use of antagonists and hyperparasites has received greater attention, especially in developed countries. Certain encouraging results have been obtained for disease control in chickpea and pea (Kaiser and Hannan, 1984; Kaiser *et al.* 1989; Trapero-Casas *et al.*, 1990; Parke *et al.*, 1991), and are being recommended. But their actual adoption will greatly depend on the development of easier application methods that need further research. It will be some time before these methods would be used in developing countries.

Physical Methods

Of the physical methods available for IDM of cool season food legumes (Tables 2 and 3), use of pathogen-free seed is the most important for all situations (Kaiser, 1987; Mathur *et al.*, 1988). However, seed certification programs do not exist in most developing countries. In the absence of seed certification, the use of seeds produced in dry areas with field inspection is the best way to avoid seedborne pathogens (Kaiser, 1987). Deep tillage to plow down and bury crop refuse immediately after harvest is a very useful practice (Hagedorn, 1984; Kaiser, 1987; Nene and Reddy, 1987). A combination of tillage and refuse management could form an important part of disease management (Allmaras *et al.*, 1988; Kraft *et al.*, 1988). In regions where non-till or minimum tillage practices are used, complimentary measures to reduce inoculum from crop residues are needed, for example, the attempt to eliminate *Ascochyta rabiei* (Pass.) Labrousse. by temporary legal restrictions on chickpea cultivation in northern Idaho (Wiese and Kaiser, 1991). Avoiding heavily infested soils is an effective measure to prevent losses due to soilborne pathogens (Hagedorn, 1984; Tu, 1987) but requires a previous knowledge of the disease potential of the soils.

Soil solarization has been successfully employed for the control of Fusarium wilt of chickpea in experimental plots at ICRISAT (Chauhan *et al.*, 1988). Although its cost may limit its use to only experimental or seed production plots in developing countries, it could be used in cultivation of vegetable peas in plastic houses.

Chemical Methods

Although fungicides have been used as an important method of disease control, economic and environmental conditions will lead to a significant reduction in their use to control foliar and soilborne diseases of cool season food legumes, especially in the developed countries (Kaiser, 1981), and even in developing countries due to economic reasons, and also due to dry conditions prevalent during the growing season in many regions. However, their use will be required in certain situations (Trapero-Casas, 1986; Nene *et al.*, 1988). Similarly, the use of seed dressing fungicides has tremendous scope for controlling seedborne pathogens and seed and seedling diseases (Haware *et al.*, 1986). These include Captan, thiram, mancozeb, and systemic benzimidazoles used singly or the first three in mixtures with fungicides specific to oomycetes and with systemic benzimidazoles.

IDM Packages for Major Groups of Diseases

As emphasized earlier, except for a few cases, complete information needed for the IDM of cool season food legumes is not yet available. However, based on the available information and on general principles of plant pathology, we suggest IDM packages for different major groups of cool season food legume diseases.

Seed and Seedling Diseases

They are known to be important in pea and chickpea in the temperate zones. The IDM package should include the following:
1. Use of resistant cultivars for Pythium pre and post emergence damping-off of pea and chickpea.
2. Use of certified, high vigor seed with tightly adhering seedcoats to reduce susceptibility to pathogens during germination.
3. Seed treatment with *Penicillium oxalicum* Currie & Thom, *Pseudomonas fluorescens* Migula, *Trichoderma harzianum* Rafai, *T. koningii* Oudem., and *T. hamatum* (Bonord.) Bainier for Pythium damping-off and common root

rot, and with *P. oxalicum* and *P. fluorescens* for chickpea in the developed countries. Use of solid mix priming and polysaccharides and polyhydroxy alcohols with these treatments is recommended. For developing countries, seed treatment with fungicides, namely, Captan, thiram, or mancozeb in combination with systemic metalaxyl and thiabendazole fungicides is strongly recommended. For a broad spectrum disease control in pea, oxadizyl, cymoxanil, and mancozeb are recommended for effective control of downy mildew, root rot and leaf spot, and damping-off.

4. Preparation of a well-leveled and well-drained field to avoid waterlogging.

Fusarium Wilts

These are important in chickpea and lentil. The IDM package for them should include the following:

1. Use of resistant cultivars for normal season plantings in chickpea and lentil. These are now available for most chickpea growing areas in the world. In lentil, they are available in India. For spring-planting of chickpea in the Mediterranean region, southern India, and Ethiopia, short duration resistant cultivars would be desirable to avoid terminal drought.
2. Use of disease-free, high quality seed to ensure higher plant stands free from wilt.
3. Seed treatment with fungicides, Benlate®T (30% benomyl + 30% thiram) at 2.5 g kg^{-1} seed.
4. Use of certain cultural practices such as deep plowing in the beginning of summer or autumn for effective crop residue management; early planting of chickpea to avoid high temperatures at maturity while keeping in mind its adverse effect on other major diseases of the crop, and delayed planting to reduce Fusarium wilt of lentil; avoiding planting in heavily infested fields, especially in areas where land availability is not limiting; planting of lentil after rice or sorghum, but rice should be avoided if collar rot becomes a problem; use of higher doses of P_2O_5 and compost, avoiding addition of undecomposed organic matter to soil (cow dung manure) as it makes an ideal substrate for *Pythium* species.
5. Use of soil solarization, especially for seed production plots in both developing and developed countries.
6. Use of pathogen-suppressive soils, especially in developed countries.

Root Rots

Different types of root rots are important in pea, chickpea, faba bean, and lentil. Their IDM package should include the following:

1. Use of resistant cultivars which are available for both dry root rot and black root rot of chickpea, and the root rot complex of pea.

2. Use of early-maturing cultivars to avoid terminal limiting soil moisture and high temperatures, especially for dry root rot of chickpea.
3. Seed treatment with Benlate®T as in chickpea. Seed treatment with Captan + metalaxyl (Ridomil®) + benalaxyl (Galben®) for the pea root rot complex.
4. Use of biological seed treatments, for example, *Streptomyces diastaticus* (Krainsky) Waksman and Henrici and *Serratia marcescens* Bizio for collar rot, and *T. harzianum* for dry root rot, especially in the developed world. Also, *Gliocladium virens* and Anchor® could be used for Rhizoctonia root rot.
5. Use of certain sub-soil tillage practices to maximize the host-plant resistance of pea cultivars to Fusarium root rot, avoiding soil compaction and light tillage for Rhizoctonia and Fusarium root rots, but instead use tine-cultivation to stir soil to encourage ecological successions (Cook, 1988); avoid planting in wet soil and in fields following rice as it can encourage collar rot; use of green manure crops, such as oats, sorghum, or sudan grass to reduce root rot of pea, and crucifers as a preceding crop for reducing common root rot of pea.
6. Use of dinitroaniline and triazine herbicides for the pea root rot complex; planting in fields that have pathogen-suppressive soils for the control of Fusarium root rot of lentil, especially in the developed countries.
7. Use of an on-site soil indexing exercise to determine the level of field infestation as well as cultivar susceptibility, especially in the developed countries.

Ascochyta Blights

These diseases are economically important in chickpea and lentil, although they also cause considerable damage in pea and faba bean in certain areas. The IDM package should include the following:
1. Use of resistant cultivars, wherever available, or slow blighting cultivars (e.g., ILC 3279) in chickpea (Singh and Reddy, unpublished). As far as possible, use tall, compact, and early-maturing cultivars. Alternating resistant cultivars to maintain resistance is recommended, especially for the developed countries.
2. Use of healthy, *Ascochyta*-free certified seed treated with Calixin®M (tridemorph 11% + maneb 36%). In lentil, sun drying of seeds helps in reducing seedborne inoculum. Seed production in drier disease-free areas to produce healthy seed.
3. Limited foliar fungicide sprays (1–2) of chlorothalonil at the early vegetative and early podding stages for chickpea and lentil in a normal planting situation (including the winter-sown crop in the Mediterranean region); at the early vegetative and flowering stages in early spring-planted chickpeas; and at seedling and mid-vegetative stages in normal spring-planted chickpeas. A fungicide spray on the exposed infested debris in the field after tillage should be used.

4. Follow cultural practices to reduce inoculum and disease severity. These include a 2- to 3-year-crop rotation; deep tillage in moist soil to bury debris below 10 cm; physical removal and destruction of crop debris from the soil surface; deep sowing (> 15 cm) to prevent emergence of infected seeds; lower seed rates to avoid a thick crop canopy; intercropping of chickpea with wheat, barley, or mustard; use of higher doses of phosphatic fertilizer; and avoiding pre-threshing storage of harvested material to prevent an increase in seed infection.

Chocolate Spot

An IDM package for chocolate spot of faba bean similar to the one for Ascochyta blight of chickpea and lentil is recommended. Generally, early planting with seed treatment (benomyl + thiram) and foliar sprays of vinclozolin, benomyl, or mancozeb, and raising the crop in well-drained fields is recommended. Also, use of antisenescence compounds (benzyladenine), especially in the developed countries, is recommended.

Botrytis Gray Mold

The IDM strategy should consist of the following:
1. Use of tolerant cultivars, wherever available, and kabuli types that are less susceptible than desi types. Tall and compact plant type cultivars should be used.
2. Treat seed with benomyl (Bavistin® + Thiram®, 3 g kg^{-1} seed). A limited number of sprays (1–2) at flowering and early podding are useful.
3. Use of certain cultural practices, such as delayed sowing, increased interrow spacing from 30 to 60 cm at a constant plant population, and intercropping with linseed.
4. Use of antagonists, *Penicillium citrinum* Thom or *P. cyclopium* Westling, as seed treatments, especially for the developed countries.

Rusts

These are important on faba bean and lentil. The IDM package should consist of the following:
1. Use of resistant cultivars. Early maturing cultivars are suitable for lentil rust.
2. Plant healthy, pathogen-free seed. Treat lentil seed with diclobutrazol.
3. Use of cultural practices such as early planting for faba bean and delayed planting for lentil, higher seeding rate, higher phosphatic fertilizer, crop refuse collection and burning, and summer-plowing for both crops.
4. Limited (1 to 2) fungicide sprays of mancozeb for both crops, and wettable sulphurs for lentil could be used in severe disease outbreaks.

Downy Mildews

These are economically important in pea, and only important in lentil in certain areas. The IDM package should consist of the following:
1. Use of resistant cultivars.
2. Use of clean, disease-free seed. Seed treatment with metalaxyl or fosetyl aluminum is recommended.
3. Limited fungicide sprays (1 to 2) of metalaxyl + mancozeb in areas with a serious disease situation.
4. Cultural practices, like crop rotation, lower seeding rates, and deep plowing, will help reduce disease severity, and aid in better disease management.

Powdery Mildew

It is an economically important disease of pea and can be managed with an IDM package consisting of the following:
1. Use of resistant, early maturing cultivars.
2. Use of clean and certified seed.
3. Limited use of fungicide sprays (1 to 2) with wettable sulphurs, benomyl, or tridemorph.
4. Use of certain cultural practices, like early planting, crop refuse collection and destruction, and deep plowing in summer.

Chickpea Stunt

The IDM strategy for chickpea stunt, the most important virus disease of chickpea, will include the following:
1. Use chickpea cultivars that have field resistance (as in Indian subcontinent where these are available).
2. Adjusting the sowing date so as to avoid peak aphid vector activity. In California, USA, virus incidence dramatically increased when chickpea culture was shifted from summer to winter.
3. Avoid sowing near alfalfa fields that can serve as reservoirs for the causal virus.
4. May be desirable to provide a spray or two of an insecticide to keep the aphid vector activity to a minimum. To be effective, it should be adopted by all the farmers in an area.

Stem Nematode

The IDM strategy for the stem nematode of faba bean will consist of the following:

1. Use of resistant cultivars wherever available.
2. Use of nematode-free, certified seed.
3. Use of cultural practices like crop rotation with non-leguminous crops, avoid planting in heavily-infested fields, and crop refuse collection and destruction.

Action Needed

In order to address the IDM aspects in cool season food legumes, two types of action are needed: (1) direction of efforts towards research needs, and (2) development of an IDM technology for farmer use.

Future Research Needs

The gaps in our knowledge of different methods needed for effective IDM in cool season food legumes have been highlighted earlier. Future research efforts should address these gaps. In addition, efforts should be intensified for the possible use of induced systemic resistance to soilborne and foliar pathogens by the use of plant growth promoting rhizobacteria and vesicular-arbuscular mycorrhizae (VAM), and biotechnology in improving and/or developing biological methods of disease control.

Development of IDM Technology

For an IDM to become effective, suitable components of disease control to be used will have to be developed. Major emphasis needs to be centered on biological and physical methods which take maximum advantage of, or at least not upsetting to nature's own contributions to the health of the crop and the protection of soil and water resources (Cook and Veseth, 1991). As discussed earlier, considerable information has been generated on disease control methods in cool season food legumes. Also, the gaps in their knowledge and future research needs have been highlighted.

As emphasized earlier, IDM packages for the major diseases of a crop will have to be developed for an area, region, or zone. Information on the importance of diseases of cool season food legumes in most areas, regions, or zones, and interim IDM packages for certain crops and areas/regions are already available. So, the first task should be to upgrade these interim IDM packages to make them effective for successful IDM. Then there will be need to demonstrate their applicability through on-farm trials to facilitate their transfer to and adoption by farmers. Infrastructure for this activity exists in some developing countries. For this, "farm adoption approach" as done in the case of wheat in Northwest Europe (Jadoks, 1990) or "village adoption approach" as done in India may have to be adopted.

Future Perspectives

There is a general myth that the yields of cool season food legumes cannot be improved as they, in general, have a limited yield potential. It cannot be true as some breakthroughs in increasing "actual yields" have been achieved, especially in some developed countries. As emphasized earlier, the main reason for their "no breakthrough" status in productivity is the negligible attention they have received in agricultural research compared with temperate cereals. Moreover, they received a further setback due to the "green revolution" in cereal production. It is only recently that a general awareness has been realized about their importance in some developing countries, and as a result they have received increased research attention. In the same context, the two IARC's (ICRISAT and ICARDA) and some other international organizations (FAO, IDRC, etc.) and developed laboratories have played a pivotal role, the gains of which have been and are being utilized by most of the developing countries that grow cool season food legumes. Another reason for the low productivity of cool season food legumes is their susceptibility to biotic and abiotic stresses. Among these, diseases are very important, and without their successful control and management, no real breakthrough in productivity can be achieved. The effective and practical control of diseases of cool season food legume lies in their management through integrated means (IDM) using integration of different disease control methods. In addition, IDM will have to become an integral part of the whole health of cool season food legumes, as emphasized for wheat by Cook and Veseth (1991), and form an important component of an integrated systems approach, like the one adopted for potato crop management (Connell *et al.*, 1991).

The need for technology development has been emphasized earlier in this paper. To achieve this, the national capabilities for research on integrated control aspects will need strengthening. Developing national or regional networks will have to be thought of on the pattern of the "FAO Rice IPM Program in South and Southeast Asia" (Anonymous, 1990). For this, IDM working groups in cool season food legumes will have to be developed, especially in the developing countries. A general awareness will have to be created among farmers and agricultural administrators in each mandated area, country, or region. And, finally, a group action will be required at the national/regional level as was done in the case of "FAO Rice IPM Program in South and Southeast Asia".

Even in the developing countries, general awareness about the need for managing diseases of cool season food legumes is very much there. True, integrated control of diseases alone cannot solve the problem of "low yields" of cool season food legumes in real farm situations, but also yields cannot be significantly improved without greatly expanding research on IDM. So, what is now required is to initiate action to consolidate and evaluate the interim IDM packages so that they can be transferred to the users. To accomplish this, we must not lose any more time.

References

Ageeb, O. A. A. and Salih, F. A. 1989. *FABIS Newsletter* 24: 8–10.
Ahmed, S. and Beniwal, S. P. S. 1991. *Tropical Pest Management* 37: 368–373.
Allmaras, R. R., Kraft, J. M. and Miller, D. E. 1988. *Annual Review of Phytopathology* 26: 219–243.
Amin, K. S. and Ramachander, P. R. 1989. *Indian Journal of Pulses Research* 2: 140–146.
Anonymous. 1990. *FAO Plant Protection Bulletin* 38: 62–64.
Beniwal, S. P. S., Ahmed, S. and Gorfu, D. 1992. *Tropical Pest Management* 37: (In press).
Beniwal, S. P. S., Ahmed, S. and Tadesse, N. 1989. *LENS* 16: 27–28.
Bouznad, Z., Maatougui, M. H., Mouri, N. and Beniwal, S. P. S. 1991. *Afrique Agriculture* 188: 33.
Bruehl, G. W. 1989. *Canadian Journal of Plant Pathology* 11: 153–157.
Chauhan, Y. S., Nene, Y. L., Johansen, C., Haware, M. P., Saxena, N. P., Singh, S., Sharma, S. B., Sahrawat, K. L., Burford, J. R., Rupela, O. P., Kumar Rao, J. V. D. K. and Sithanatham, S. 1988. *Effects of soil solarization on pigeonpea and chickpea*. Research Bulletin No. 11. Patancheru, Andhra Pradesh, India: ICRISAT.
Connell, T. R., Koenig, J. P., Stevenson, W. R., Kelling, K. A., Curwen, D., Wyman, J. A. and Binning, L. K. 1991. *Journal of Production Agriculture* 4: 453–460.
Cook, R. J. 1988. In: *World Crops: Cool Season Food Legumes*, pp. 649–660 (ed. R. J. Summerfield). Dordrecht, Kluwer Academic Publishers.
Cook, R. J. and Baker, K. F. 1983. *The Nature and Practice of Biological Control of Plant Pathogens*, 539 pp. St. Paul, Minnesota, USA: APS Press.
Cook, R. J. and Veseth, R. J. 1991. *Wheat Health Management*, 152 pp. St. Paul, Minnesota, USA: APS Press.
FAO. 1990. *Production Yearbook*, Vol. 44. Rome: FAO.
Gabriel, C. J. and Cook, R. J. 1990. *FAO Plant Protection Bulletin* 38: 95–99.
Hagedorn, D. J. (ed.). 1984. *Compendium of Pea Diseases*, 57 pp. St. Paul, Minnesota, USA: APS Press.
Hagedorn, D. J. 1985. In: *The Pea Crop*, pp. 205–215 (eds. P. D. Hebblethwaite, M. C. Heath and T. C. K. Dawkins). London: Butterworths.
Hanounik, S. B. and Maliha, N. F. 1985. In: *Faba Bean, Kabuli Chickpea, and Lentils in the 1980s*, pp. 107–117 (eds. M. C. Saxena and S. Varma). Aleppo, Syria: ICARDA.
Haware, M. P., Jiménez-Díaz, R. M., Amin, K. S., Philips, J. C. and Halila, H. 1990. In: *Chickpea in the Nineties*, pp. 129–133 (eds. H. A. van Rheenen, M. C. Saxena, B. J. Walby and S. D. Hall). Patancheru, Andhra Pradesh, India: ICRISAT.
Haware, M. P., Nene, Y. L. and Mathur, S. B. 1986. *Seed-borne diseases of chickpea*. Technical Bulletin 1. Copenhagen Denmark: The Danish Government Institute of Seed Pathology for Developing Countries.
Hwang, S. F. and Chakravarty, P. 1991. *Phytopathology* 81: 1152 (Abstract).
Jadoks, J. C. 1990. *Canadian Journal of Plant Pathology* 12: 117–122.
Jellis, G. J., Bolton, N. J. E. and Clarke, M. H. E. 1988. In: *Proceedings of the British Crop Protection Conference on Pests and Diseases V. 3*, pp. 895–900. London: British Crop Protection Council.
Jiménez-Díaz, R. M., Singh, K. B., Trapero-Casas, A. and Trapero-Casas, J. L. 1991. *Plant Disease* 75: 914–918.
Jiménez-Díaz, R. M. and Trapero-Casas, A. 1985. *Plant Disease* 69: 591–595.
Kaiser, W. J. 1981. *Economic Botany* 35: 300–320.
Kaiser, W. J. 1987. *Testing and production of healthy plant germplasm*. Technical Bulletin 2. Copenhagen, Denmark: The Danish Government Institute of Seed Pathology for Developing Countries.
Kaiser, W. J. and Hannan, R. M. 1984. *Plant Disease* 68: 806–811.
Kaiser, W. J. and Hannan, R. M. 1986. *Phytopathology* 76: 355–360.
Kaiser, W. J., Hannan, R. M. and Weller, D. M. 1989. *Soil Biology and Biochemistry* 21: 269–273.

Kannaiyan, J. 1974. *Studies on the Control of Lentil Wilt*. Ph.D. Thesis, G.B. Plant University of Agriculture and Technology, Pantnagar, Uttar Pradesh, India.

Khare, M. N. 1980. *Wilt of Lentil*, Technical Bulletin. Jabalpur, M.P., India: Jawahar Lal Nehru Krishi Vishwa Vidhyalaya.

Khare, M. N. 1981. In: *Lentils*, pp. 163–172 (eds. C. Webb and G. Hawtin). London: ICARDA/CAB.

Khare, M. N., Bayaa, B. and Beniwal, S. P. S. 1993. In: *Proceedings of an international conference on "Breeding for Stress Tolerance in Cool-Season Food Legumes"*, pp. 107–121 (eds. K. B. Singh and M. C. Saxena). Chichester, UK: John Wiley & Sons.

Kraft, J. M., Haware, M. P. and Hussein, M. M. 1988. In: *World Crops: Cool Season Food Legumes*, pp. 351–365 (ed. R. J. Summerfield). Dordrecht: Kluwer Academic Publishers.

Kraft, J. M. and Papavizas, G. C. 1983. *Plant Disease* 67: 1234–1237.

Mathur, S. B., Haware, M. P. and Hampton, R. O. 1988. In: *World Crops: Cool Season Food Legumes*, pp. 351–365 (ed. R. J. Summerfield). Dordrecht: Kluwer Academic Publishers.

Meiners, J. P. 1981. *Annual Review of Phytopathology* 19: 189–209.

Mohamed, H. A. R. 1982. In: *Faba Bean Improvement*, pp. 213–225 (eds. G. Hawtin and C. Webb). The Hague, The Netherlands: Martinus Nijhoff.

Morrall, R. A. A. and Sheppard, J. W. 1981. *Canadian Plant Disease Survey* 61: 7–13.

Nene, Y. L. 1988. *Annual Review of Phytopathology* 26: 203–217.

Nene, Y. L., Hanounik, S. B., Qureshi, S. H. and Sen, B. 1988. In: *World Crops: Cool Season Food Legumes*, pp. 577–589 (ed. R. J. Summerfield). Dordrecht: Kluwer Academic Publishers.

Nene, Y. L., Kannaiyan, J. and Saxena, G. C. 1975. *Indian Journal of Agricultural Science* 45: 177–178.

Nene, Y. L. and Reddy, M. V. 1987. In: *The Chickpea*, pp. 232–270 (eds. M. C. Saxena and K. B. Singh). Oxford, UK: CAB International.

Papendick, R. I., Chowdhury, S. L. and Johansen, C. 1988. In: *World Crops: Cool Season Food Legumes*, pp. 237–255 (ed. R. J. Summerfield). Dordrecht: Kluwer Academic Publishers.

Parke, J. L., Rand, R. E., Joy, A. E. and King, E. B. 1991. *Plant Disease* 75: 987–992.

Reddy, M. V. 1991. Integrated control of pulse diseases. In: *Advances in Pulses Research in Bangladesh: Proceedings of the Second National Workshop on Pulses*, 6–8 June 1989, Joydabpur. Patancheru, Andhra Pradesh, India: ICRISAT.

Reddy, M. V., Ghanekar, A. M., Nene, Y. L. Haware, M. P., Tripathi, H. S. and Rathi, Y. P. S. 1992. Effect of Ronilan[(R)] spray, plant growth habit and inter-row spacing on botrytis grey mold and yield of chickpea. *Indian Phytopathology* 45: (In press).

Reddy, M. V., Nene, Y. L., Singh, G. and Bashir, M. 1990. In: *Chickpea in the Nineties*, pp. 117–127 (eds. H. A. van Rheenen, M. C. Saxena, B. J. Walby and S. D. Hall). Patancheru, A.P., India: ICRISAT.

Reddy, M. V. and Singh, K. B. 1990. *Indian Journal of Plant Protection* 8: 65–69.

Rossignol, Y. 1988. In: *Monograph No. 39*, pp. 99–104. London: British Crop Protection Council.

Salih, F. A. and Ageeb, O. A. A. 1987. *FABIS Newsletter* 18: 18–19.

Saxena, M. C. and Goldsworthy, P. R. 1988. In: *World Crop: Cool Season Food Legumes*, pp. 25–37 (ed. R. J. Summerfield). Dordrecht: Kluwer Academic Publishers.

Singh, G. and Dhingra, K. K. 1980. *Journal of Research Punjab Agricultural University* 7: 233–235.

Singh, K. B. and Reddy, M. V. 1990. *Plant Disease* 74: 127–129.

Trapero-Casas, A. 1986. *II Symposium Nacional de Agroquimicos*. Sevilla, Spain.

Trapero-Casas, A., Kaiser, W. J. and Ingram. D. M. 1990. *Plant Disease* 74: 563–569.

Tu, J. C. 1987. *Plant Disease* 71: 9–13.

Wiese, M. V. and Kaiser, W. J. 1991. *Ascochyta Blight of Chickpea*. University of Idaho, College of Agricultural, Current Information Series 836. Moscow, Idaho, USA: University of Idaho.

Yeoman, D. P., Lapwood, D. H. and Mc Towen, J. 1987. *Crop Protection* 6: 90–94.

Integrated management systems to control biotic and abiotic stresses in cool season food legumes

Y.L. NENE[1] and W. REED[2]

[1] Deputy Director General, ICRISAT, Patancheru P. O., Andhra Pradesh 502 324, India, and
[2] 11 Wilberforce Road, Waterside, Sherborne St., Bourton-on-the-Water, GL54 2BY UK

Abstract

Yield losses in cool season food legumes result from several biotic and abiotic stresses. The most important of these stresses on five cool season food legume crops are listed, and recent progress in research to alleviate some of these is reviewed. Although it is possible to control some stresses by the use of such inputs as agricultural chemicals, economic and environmental concerns limit their use in many farmers' fields. The most desirable means of stress alleviation is through integrated management systems, including host-plant resistance and improved agricultural practices. There is an urgent need to accentuate progress towards cultivars that can withstand various stresses and give large and stable yields. Recent advances in biotechnology may offer us the means by which to do so. Teams of scientists must go beyond their narrow specializations and make the best use of currently available materials and methods within practicable integrated stress management systems.

Introduction

Cool season food legumes are grown over a wide range of environments and are subject to an enormous number of biotic and abiotic stresses. Many of these can completely destroy crops or greatly reduce the quantity and quality of their products. There is great variation in the occurrence of these stresses, both between and within the crop species and across time and space. Stresses that are known to severely reduce yields are given in Table 1. Such a large number of stresses for each of the crops pose a real challenge to scientists to develop effective integrated stress management systems. This paper seeks to review the international progress in developing such components and systems and to discuss the problems and solutions that are apparent at this time.

Chickpea (*Cicer arietinum* L.) is a crop of major importance in South and West Asia, North Africa, and Central America. There are very active research groups in several countries of these regions. Also, it has been selected as a target

F.J. Muehlbauer and W.J. Kaiser (eds.), Expanding the Production and Use of Cool Season Food Legumes, 666–678.
© 1994 Kluwer Academic Publishers.

Table 1. Biotic and abiotic stresses affecting five cool season food legumes

Crop	Biotic stresses	Abiotic stresses
Chickpea	Fusarium wilt, dry root rot, Ascochyta blight, Botrytis gray mold, stunt, *Orobanche*, pod borer, leaf miner, lesion nematode, root knot nematode, reniform nematode, cyst nematode, weeds	Temperature extremes, drought, salinity
Pea	Seedling and root rots, Fusarium wilt, downy and powdery mildews, rust, mosaic, streak, *Orobanche*, pod fly, pea moth, bean fly, lesion nematode, root knot nematode, cyst nematode, weeds	Temperature extremes, drought, waterlogging
Faba bean	Fusarium wilt, root rot, chocolate spot, Ascochyta blight, rust, mosaics, *Orobanche*, aphids, leaf miner, army worm, stubby root nematode, lesion nematode, dagger nematode, root knot nematode, weeds	Temperature extremes, drought
Lentil	Fusarium wilt, Ascochyta blight, rust, stem blight, *Orobanche*, weeds	Temperature extremes, drought
Lathyrus	Downy mildew, weeds	

crop by both the International Crops Research Institute for the Semi-Arid Tropics (ICRISAT) in India and the International Center for Agricultural Research in the Dry Areas (ICARDA) in Syria. Pea (*Pisum sativum* L.), faba bean (*Vicia faba* L.), and lentil (*Lens culinaris* Medik.) are grown extensively in both developed and developing countries, and have undergone extensive research and selection. Consequently, farmers in many areas now have a choice of well-adapted cultivars that have high yield potential, and extensively tested agronomic practices that enable them to produce these crops profitably in most years.

In contrast, *Lathyrus* or grasspea (*Lathyrus sativus* L.) is a crop, grown almost entirely by subsistence farmers, that has been relatively neglected by researchers. The presence of neurotoxin associated with this food legume is well known, and this factor may have discouraged research on its improvement. However, it is a very useful crop in many areas, for it can produce substantial yields under poor agronomic conditions and appears to have relatively few pests and diseases. Recently, cultivars containing low neurotoxin content have been developed in Bangladesh, Canada, and India (Kaul and Combes, 1986; Lal *et al.*, 1986; Islam and Matiur Rahman, 1991).

Components of Integrated Stress Management Systems

The productivity of any crop is dependent upon a large number of components of the system in which it is grown. Most stress management initiatives interact with many of these components, and with each other, and so must eventually be considered and tested as integrated inputs within the crop system, rather than as simple add-on factors.

Plant Resistance

The primary component in any crop is the plant genotype that offers limitless research opportunities to crop improvement scientists. Here there are opportunities, not only for the plant breeders to increase a crop's basic yield potential but also, and perhaps more importantly, to increase the crop's genetic resistance to the many yield-reducing factors that have to be faced in farming situations. As the climatic, edaphic, and biotic threats to cool season legumes vary greatly across the wide geographic range over which they are grown, it is impossible to design a genotype that will be ideal for all environments. Consequently, it must be accepted that the major agro-ecological zones will require differing genotypes of each legume crop.

Each genotype often requires its own, specially designed, agronomic package of practices that will enable it to be profitably exploited within its adaptation range. The type of cultivation, sowing date, and other agronomic practices all interact to influence biotic and abiotic stresses and crop yields.

Ideally, an optimum package of crop genotype and agronomic system should be developed for each location, but this is clearly impossible. Perhaps the best that can be achieved will be the development of widely adapted genotypes, with specific agronomic recommendations for the differing areas over which they will be grown. For example, the optimum sowing date and plant spacing for such a widely adapted cultivar may differ across the range of soil types and climates within its zone of adaptation, and differing crop protection measures may be required to combat the yield-reducing stresses where these are found to be location-specific.

Alternatively, it might be more productive to select a range of genotypes to suit the level of agronomic inputs that can be afforded by the farmers in a specific zone. A high-yielding cultivar that will yield well when provided with purchased inputs can be provided for those farmers who can afford these, while a cultivar with less yield potential but with resistance/moderate resistance to the locally dominant stresses might be more useful to the farmers who cannot utilize such inputs.

Although immunity, or very strong resistance to pests and diseases would be desirable, it is much more likely that lower levels of resistance or tolerance will be more easily found. This seems true in case of faba bean where only moderate degrees of resistance to *Botrytis cinerea* Pers.:Fr. and *B. fabae* Sardina are available, and therefore effective management would involve use of biological and minimal chemical means (Harrison, 1988). Many landrace cultivars are found to have evolved such protection against locally damaging stresses. It is essential to ensure that the improved cultivars developed to replace such landraces are not more susceptible than they are, unless the local farmers have the ability to use alternative means of pest control. For example, no new chickpea cultivar can succeed in peninsular India unless it has better levels of resistance to Fusarium wilt, dry root rot, collar rot, and *Helicoverpa armigera* (Hüb) than the local landraces. For example, the faba bean ILB 1815, an excellent local landrace in Syria and Lebanon, can be crossed with *Botrytis*-resistant BPL 261, to improve its acceptability (Maliha, 1983).

Agronomic Practices

The growth and yield of a crop genotype will be affected by all the components of an agronomic package, i.e., primary cultivation, sowing, spacing, weeding, soil fertility enhancement, and soil moisture control. Changes in any of these components can also have major consequences on the levels of stresses, and their effects on the crop. For example, delaying sowing date and increasing plant population significantly reduced faba bean root rot [*Fusarium solani* (Mart.) Sacc. f. sp. *fabae*] and wilt (*F. oxysporum* Schlechtendahl) incidence (Salih and Ageeb, 1987). In India, delayed sowing of chickpea reduced Ascochyta blight but also reduced yield (Tripathi *et al.*, 1988). Delayed sowing reduced Botrytis gray mold of chickpea in Nepal (Karki *et al.*, 1989).

The crop must not be considered in isolation, for its juxtaposition with neighboring crops can have major effects on the populations of some pests, particularly those that are mobile. When all the farmers in an area sow a crop synchronously, the mobile pests or wind-disseminated pathogens have no opportunity to build up on early sown crops and then invade the later sown; so the available pest populations are diluted across all the fields in the area. The duration and phenology of the crop can also be of great importance in escaping severe pest attacks. For example, a short-duration cultivar may escape severe damage from insect pests that build up their populations in the crop over more than one generation. Short-duration, cold-tolerant chickpeas, if developed, will enable the crop to mature in February, before *H. armigera*, Ascochyta blight, Botrytis gray mold, and hail storms occur in northern India, Pakistan, and Nepal.

It is also obvious that crop rotations have significant effects on some stresses, particularly on the soilborne plant pathogens. Thus, scientists must look well beyond the agronomy of single crops, both in time and space, if they are to develop efficient systems to limit crop losses.

Soil solarization, using polyethylene covers, has been found to be of immense value (Katan, 1981). In field experiments, carried out in Syria from 1985–89, soil solarization for 40 days increased yields of faba bean by 331%, lentils by 441%, and peas by 92%, as a result of controlling *Orobanche crenata* (Forsk.) and the higher soil availability of nitrogen and phosphorus (Linke *et al.*, 1991). Similarly, solarization improved plant growth and yields of chickpea and pigeonpea, reduced wilt incidence, increased mineralization of soil nitrogen to nitrate, reduced populations of *Fusarium* propagules and plant parasitic nematodes, decreased weed infestation, and encouraged fungal antagonists, thus reducing wilt incidence (Chauhan *et al.*, 1988). Solarization also decreases soil salinity (Abdel-Rahim *et al.*, 1988). Cheap biodegradable mulches are required as the present polyethylene mulches are costly and are difficult to dispose of.

Agricultural Chemicals

The agricultural chemicals – fertilizers including micronutrients, herbicides, insecticides, fungicides, and other biocides – have played an important role in increasing and stabilizing the yields of many crops, including the grain legumes. However, as stated earlier, cost and environmental concerns have prompted reaction against the liberal use of these chemicals, and stimulated research on alternatives. The widespread recognition of the need for integrated pest management is relatively recent. Pesticides will continue to have an important role in integrated pest management. The first step in most successful integrated pest management initiatives has been the realization that pesticides should be used according to need rather than as routine, calendar-based applications.

Biological Control

The most attractive alternative to pesticide use is biological control. Although considerable research is being invested in this aspect on other crops, relatively little is directed to the food legumes. However, efforts in biological control are on the increase. For example, Wilding *et al.* (1986) have studied such factors as the form of inoculum and effect of humidity that would allow the entomogenous fungus, *Erynia neoaphidis* Remaudiere et Hennebert, to effectively parasitize *Aphis fabae* Scop. Delfin® (Sandoz Ltd.), a biological insecticide based on a selected strain of *Bacillus thuringiensis* Berliner, is effective against species of *Lepidoptera*, including *Heliothis* spp.

In general, the major successes in biological control have involved perennial crops and the importation of natural enemies to control recently introduced pests. As the cool season food legumes all tend to be short-duration annuals and most of their major pests are long established in the major production areas, there appears to be a greater challenge for classical biological control to play a major role in pest control on these crops.

Parasites, predators, and diseases, which limit pest populations, are normally present in all fields where these legumes are grown, and there are many ways in which these natural enemies can be encouraged. Recent research in the developed countries has revealed the importance of hedgerows and mixed-species fallows as sources for many of these natural control agents. There are many calls for a reversal of the trend towards large areas of monocrops and the elimination of all non-crop species in farming areas. In most areas of the third world, crop species diversity is increased by traditional intercropping. For example, chickpeas are commonly intersown with mustard (*Brassica campestris* L.) in northern India and such traditional intercrops are often found to increase yields and their stability; reduced pod damage by the pod borer *H. armigera* has been reported by Chauhan and Ombir (1987). However, intercrop components must be carefully chosen, for some combinations may result in an increase of pest populations rather than in those of the natural enemies. Also, intercropping reduces the options for rotation and interferes with mechanical harvesting that is widely used for these crops in developed countries.

There is an increasing awareness of the dangers of indiscriminate pesticide use destroying the beneficial fauna; so new chemicals and application techniques are designed to limit such ill-effects. It is strongly recommended that pesticides should only be applied when pest populations threaten to exceed the economic thresholds. The natural enemies can cause substantial suppression of pest populations. Hence, estimates of natural enemy abundance should be included when monitoring pests and calculating economic thresholds.

Pheromones and Other Semiochemicals

There may be considerable potential for using the chemical scents, which play a large part in the host-finding and communication systems of insects, in direct pest control on these crops. Sex pheromones are already in use for monitoring populations of such pests as *H. armigera* (Srivastava *et al.*, 1990). Research is also in progress to determine the chemical scents that attract these moths and larvae to such hosts as chickpea (Rembold *et al.*, 1990). Success in such research may result in the production of attractive traps that would divert this pest from its host.

Research on Stress Management Components and Systems

It is inevitable that most crop improvement research is initially directed towards individual components of a system. For example, many entomologists, pathologists, and physiologists concentrate upon one aspect of the control of a single insect pest, pathogen, or abiotic factor. Such component research is essential, but must be followed up by research towards integration of any advances into practicable systems. Similarly, yield increases and stress management successes, produced from component or systems improvements on small plots in research station fields, must be regarded as only the first step towards crop improvement, rather than as an end result. It is increasingly evident that the success of a new advance on a research farm may not produce similar benefits when transferred to farmers' fields. Part of the problem is that the ecological conditions in research fields are often grossly atypical of those in farmers' fields.

ICRISAT has addressed these problems with several innovations. Most research is carried out in projects that involve multidisciplinary, crop-based teams. A large area of ICRISAT's central research farm is kept free from irrigation and agricultural chemicals, other than small applications of fertilizer. All promising new crop genotypes that are developed and selected in protected, high-fertility pots, plots, and fields are then extensively tested, both in specific disease and abiotic stress nurseries and in large plots in the pesticide-free area, to determine their susceptibility to the stresses that they may encounter in farmers' fields.

Another large area of the research farm is devoted to farming systems research. Here the new crop genotypes, components, and systems are tested in large fields within watersheds. Simultaneously, the new materials and methods are tested on other research stations, usually in cooperation with national research scientists in coordinated networks. Finally, the advances are taken to farmers' fields, where their progress is monitored by both national and ICRISAT research scientists. The whole process, from the preliminary experimentation to confirmation of benefits in farmers' fields takes many years and is very expensive. However, the process ensures that the scientists are encouraged to follow their findings through to farmer utilization, and do not regard the publication of their data as the end product of their research. Many farmers are now benefiting from new crop genotypes, and other innovations that have reached them as a result of this system, and many more major improvements are already in the pipeline.

A good example of integrated management of pea root rots in Canada is the strategy that includes planting resistant cultivars, using fungicides for seed treatment, avoiding phenoxy herbicides, but using dinitroaniline and triazine herbicides, reducing soil compaction, practicing fall chisel plow or fall plow plus spring raised seedbed preparation, using soil indexing to determine both levels of field infestation and cultivar susceptibility, and planting green manure crops between pea crops (Tu, 1987).

The result of another interesting study was published by McEwen and Yeoman (1989). Field experiments compared four spring faba bean cultivars using two pest control programs based on (1) "standard" control of *Ascochyta fabae* Spegazzini, *Ditylenchus dipsaci* (Kühn) Filip., and seedborne viruses and (2) "enhanced" control of *Sitona lineatus*, bean leaf roll virus transmitted by the vector *Acyrthosiphon pisum*, *B. fabae*, and *Uromyces viciae-fabae* in addition to pests and pathogens in the "standard" control. The average increase in yield was 1.1 t ha^{-1} with the "enhanced" treatment.

Most traditional systems of agricultural production have evolved to ensure relative stability in yields, rather than opting for large yields but with greater risk. Chickpea in Syria offers a very good example. When this crop is sown at the beginning of winter, the potential yields are much greater than those produced by the traditional spring sowings. However, the winter-sown crops are frequently destroyed by Ascochyta blight, and the severe frosts that occur in some years can also kill the plants. Consequently, this crop is traditionally sown in the spring. In such circumstances, the obvious approach is to seek genotypes that combined resistance to both blight and low temperatures through the vegetative phase. Chickpea breeders, pathologists, and physiologists from ICARDA and ICRISAT have cooperated in screening the available germplasm with considerable success (Singh *et al.*, 1991). It has been possible to develop breeding lines that resist Ascochyta blight and low temperatures (K. B. Singh, ICARDA, personal communication) and so will add stability to the high-yield potential of the winter-sown crop. Recent efforts of Kamal and Solh (1990) in developing dual-season chickpea, suitable for both winter and spring seasons, are noteworthy.

Drought is a major abiotic stress factor. Irrigation, of course, can eliminate effects of drought, but where it is unavoidable, breeding for drought escape (e.g., short-duration cultivars) and drought resistance should yield useful cultivars. The ICRISAT short-duration kabuli chickpea cultivar ICCV 2, resistant to Fusarium wilt and tolerant to salinity, escapes both terminal drought and heat stress.

The utilization of host-plant resistance to combat both biotic and abiotic stresses is obviously a most attractive option, and considerable research activity is directed towards this component, particularly in the international agricultural research centers. Intensive screening of the available germplasm has revealed that resistance can be found to combat most of the major yield-reducing stresses in a crop such as chickpea (Pundir *et al.*, 1988; Singh *et al.*, 1991). However, the simple identification of a resistant source is only the first, and perhaps the easiest step. The combination of such resistances into a commercially acceptable cultivar can be an extremely difficult and lengthy process.

Multiple Stress Resistance

To breed lines with multiple stress resistance is not easy. There are examples of cultivars with combined resistance to three stresses; but combining resistance to more than three stresses is very difficult. One interesting example of a cultivar of cowpea, however, is worth mentioning. The cultivar Iron has combined resistance to one nematode, four fungal, two bacterial, and three viral diseases (Nene, 1988).

A good example of difficulty in breeding for multiple stress resistance is provided by the research at ICRISAT that is directed towards the selection of useful chickpea genotypes that have resistance to *H. armigera*, the major insect pest of this crop in India. Screening the available germplasm soon revealed several genotypes with considerable resistance to this pest, but these selections were highly susceptible to Fusarium wilt. However, several sources of resistance to this disease were already available, so the breeders, pathologists, and entomologists embarked on an intensive program to combine these resistances into chickpeas that also possessed all the other required characteristics. Recent selections show considerable promise, and it is expected that useful cultivars will be available for widespread use within the next few years. The ICRISAT kabuli cultivar ICCC 32 (ICCV 6), combines high yield with resistance/moderate resistance to Fusarium wilt, some root rots, *Helicoverpa*, and soil salinity and was released in Nepal in 1990.

A major achievement in interspecific hybridization in chickpea was the production of F_2 seeds between the cross *C. arietinum* and *C. echinospermum* P. H. Davis (ICARDA, 1990; Legumes Program, ICRISAT, 1991). Some accessions of *C. echinospermum* are resistant/moderately resistant to cold, Ascochyta blight, bruchids, Fusarium wilt, and leaf miner (Singh *et al.*, 1991). The segregating hybrid progenies will be useful in chickpea improvement and provide an opportunity to transfer stress-resistant genes from *C. echinospermum* to the cultivated species.

Building Integrated Stress Management Systems

Variability, in space and time, of the various factors, including stresses, that determine the growth and yields of crops ensure that there can be no universal blueprint for a system that would ensure the maximum yield of any crop in all fields and years. Similarly, the ideal genotype which possesses all the traits demanded both by consumers and producers, including resistance to all potential stresses, will never be developed. But it is equally certain that the genotypes, and farming systems that are currently in use can be greatly improved.

A wide knowledge of the relative importance of the stresses that occur locally or regionally is essential. It is appropriate to reproduce a table (Table 2) from a recent publication (Van Rheenen, 1991). The information contained in the table clearly identifies the areas of research that should lead to developing integrated stress management systems.

The traditional systems and crop landraces have many unrecognized virtues that have evolved in response to the stresses that have threatened man and his crops over many centuries. Care must be taken to ensure that any changes made through breeding are beneficial, not only in increasing the farmers' profits and yields, but also the crops' stability and sustainability. We have attempted to indicate relationship between research on components and integrated stress management systems (Figure 1).

Table 2. Desirable characters for chickpea in different zones of the world and their priorities for stress management[a]

Desirable characters	Zones (°latitudes)				
	A 0-20	B 20-25	C 25-30	D 30-45	
Stable, high yield	+	+	+	+	
Good seed quality	+	+	+	+	
Resistance to stresses					
Biotic					
Fusarium wilt	2-1	2-1	2-1	3	
Ascochyta blight	-	-	6[b]	1	
Botrytis gray mold	-	5	3	-	
Root rots	3	3	5	4	
Stunt	4	4	4	5	
Helicoverpa	1-2	1-2	1-2	6	
Leaf miner	-	-	-	2	
Nematodes	?	?	7?	7?	
Abiotic				Spring	Winter
Drought	1	1	1	1	-
Salinity	3	3	2	-	-
Excessive moisture	-	4	5	-	-
High temperature	2	2	4	2	-
Low temperature	-	-	3	-	1

a. 1 = highest priority; 9 = lowest priority; + = required;
   - = not required; ? = uncertain.
b. In case of epidemics, the crop damage is severe.

Source: van Rheenen, 1991.

Future Prospects

Two rapidly expanding scientific fields are likely to play a major role in the improvement of stress management systems: simulation models and the advances in gene transfer across specific and high boundaries.

Modern computers can now handle large simulation models containing several variables. Such simulations will not replace field experimentation but they may provide help in testing interactions, and give indications of the likely effects that may reduce the need for complex experimentation in field trials over many seasons (Rabbinge and Bastiaans, 1989). For example, CHICKBUG, a computer-based decision aid for insect pest management in chickpea and other winter grain legumes, provides information about *Helicoverpa* and derives the required advice which can be used to make a sound management decision (McIntyre and Titmarsh, 1989).

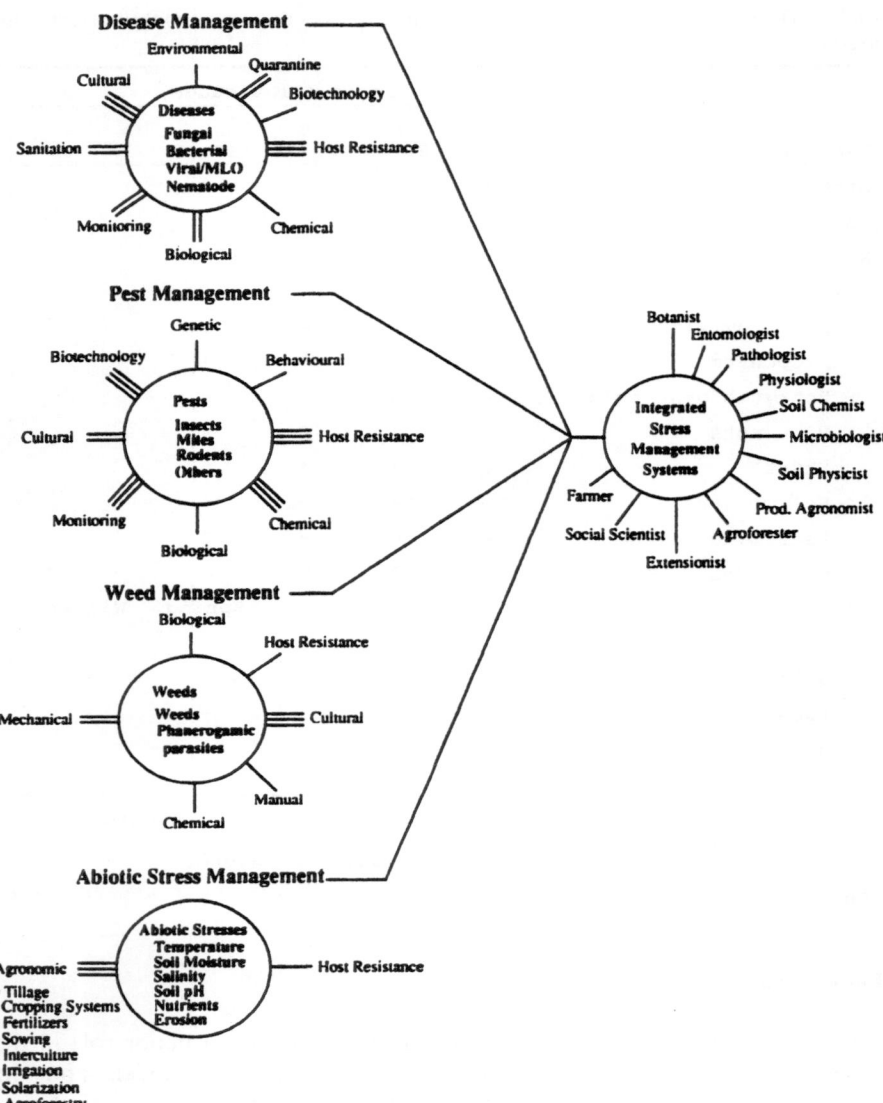

Figure 1. Suggested research emphasis ($\equiv$ high, $=$ moderately high, and — normal) and relationship between components and system of integrated stress management in cool season food legumes.

The potential for biotechnology to enable the rapid transfer of specific resistance genes looks particularly exciting. It is already known that some useful resistances are present in the wild relatives of chickpea and lentil that could not be transferred by conventional plant breeding techniques. Research on wide hybridization and embryo rescue in chickpea is in progress at ICRISAT. At

ICARDA, studies on wide crosses, use of tissue culture techniques, applications of genetic engineering, and molecular marker techniques for the improvement of chickpea and lentil are in progress (ICARDA, 1990, 1991). In addition, desirable genes from unrelated plants could be usefully transferred to these legumes.

Even without such new technologies there are many advances available now, and in the pipeline, that can give farmers substantial increases in profitability from these legumes. It is essential that research scientists should take these advances to the farmers' fields, rather than leave them buried in research journals and conference reports. Scientists should actively seek partnership with extension workers and farmers to finally transfer useful, integrated stress management technologies.

Acknowledgements

Assistance received from Ms. V. K. Sheila and Mr. V. S. Reddy in the preparation of this manuscript is gratefully acknowledged.

References

Abdel-Rahim, M. F., Satour, M. M., Mickail, K. Y., El-Eraki, S. A., Grinstein, A., Chen, Y. and Katan, J. 1988. *Plant Disease* 72: 143–146.

Chauhan, R. and Ombir. 1987. In: *Food Legumes Improvement for Asian Farming Systems: Proceedings of an International Workshop,* pp. 288–289 (eds. E. S. Wallis and D. E. Byth). ACIAR Proceedings No. 18. Canberra: ACIAR.

Chauhan, Y. S., Nene, Y. L., Johansen, C., Haware, M. P., Saxena, N. P., Sardar Singh, Sharma, S. B., Sahrawat, K. L., Burford, J. R., Rupela, O. P., Kumar Rao, J. V. D. K. and Sithanantham, S. 1988. *Effects of Soil Solarization on Pigeonpea and Chickpea.* Research Bulletin no. 11, 16 pp. Patancheru, Andhra Pradesh, India: ICRISAT.

Harrison, J. G. 1988. *Plant Pathology* 37: 168–201.

ICARDA. 1990. Food Legume Improvement Program. *Annual Report for 1989,* 381 pp. Aleppo, Syria: ICARDA.

ICARDA. 1991. Food Legume Improvement Program. *Annual Report for 1990,* 333 pp. Aleppo, Syria: ICARDA.

Islam, M. A. and Matiur Rahman, M. 1991. In: *Advances in Pulses Research in Bangladesh: Proceedings of the Second National Workshop on Pulses,* pp. 1–7. Patancheru, Andhra Pradesh, India: ICRISAT.

Kamal, M. and Solh, M. M. B. 1990. In: *Chickpea in the Nineties: Proceedings of the Second International Workshop on Chickpea Improvement,* pp. 285–286 (eds. H. A. van Rheenen, M. C. Saxena, B. J. Walby and S. D. Hall). Patancheru, Andhra Pradesh, India: ICRISAT.

Karki, P. B., Tiwari, K. R., Singh, O. and Bharati, M. P. 1989. *International Chickpea Newsletter* 21: 21–23.

Katan, J. 1981. *Annual Review of Phytopathology* 19: 211–236.

Kaul, A. K. and Combes, D. (eds.). 1986. *Lathyrus and Lathyrism,* 334 pp. New York: Third World Medical Research Foundation.

Lal, M. S., Agrawal, I. and Chitale, M. W. 1986. In: *Lathyrus and Lathyrism,* pp. 146–160 (eds. A. K. Kaul and D. Combes). New York: Third World Medical Research Foundation.

Legumes Program. ICRISAT. 1991. *Annual Report 1990*, 142 pp. Patancheru, Andhra Pradesh, India: ICRISAT.

Linke, K. H., Saxena, M. C., Sauerborn, J. and Masri, H. 1991. In: *FAO Plant Production and Protection, Paper No. 109, Proceedings of the First International Conference on Soil Solarization*, pp. 139–154. Aleppo, Syria: ICARDA.

Maliha, N. F. 1983. *Resistance of Faba Bean Selections to Different Isolates of Botrytis fabae*. M. S. Thesis, 58 pp. American University of Beirut, Beirut, Lebanon.

McEwen, J. and Yeomen, D. P. 1989. *Journal of Agricultural Science, Cambridge* 113: 365–371.

McIntyre, G. and Titmarsh, I. [Compiler]. 1989. *CHICKBUG: Insect Pest Management in Chickpea. A Computer-based Expert System Using Advisor-2. Recommendations for the Detection and Control of Insect Pests in Chickpea*. Version 1.0, 8 pp., 3.5 and 5.25 in disks, and manual. Brisbane, Australia: Queensland Department of Primary Industries.

Nene, Y. L. 1988. *Annual Review of Phytopathology* 26: 203–217.

Pundir, R. P. S., Reddy, K. N. and Mengesha, M. H. 1988. *ICRISAT Chickpea Germplasm Catalog: Evaluation and Analysis*, 94 pp. Patancheru, Andhra Pradesh, India: ICRISAT.

Rabbinge, R. and Bastiaans, L. 1989. In: *Simulation and Systems Management in Crop Protection*, pp. 217–239 (eds. R. Rabbinge, S. A. Ward and H. H. van Laar). Wageningen: PUDOC.

Rembold, H., Wallner, P., Köhne, A., Lateef, S. S., Grüne, M. and Weigner, Ch. 1990. In: *Chickpea in the Nineties: Proceedings of the Second International Workshop on Chickpea Improvement*, pp. 191–194 (eds. H. A. van Rheenen, M. C. Saxena, B. J. Walby and S. D. Hall). Patancheru, Andhra Pradesh, India: ICRISAT.

Salih, F. A. and Ageeb, O. A. A. 1987. *FABIS Newsletter* 18: 18–19.

Singh, K. B., Holly, L. and Bejiga, G. 1991. *A Catalog of Kabuli Chickpea Germplasm (An Evaluation Report of Winter-sown Kabuli Chickpea Land Races, Breeding Lines and Wild Cicer Species)*, 398 pp. Aleppo, Syria: ICARDA.

Srivastava, C. P., Pimbert, M. P. and Reed, W. 1990. *Insect Science and its Application* 11: 869–876.

Tripathi, H. S., Chaube, H. S. and Singh, R. S. 1988. *International Chickpea Newsletter* 18: 23–24.

Tu, J. C. 1987. *Plant Disease* 71: 9–13.

van Rheenen, H. A. 1991. In: *Uses of Tropical Grain Legumes: Proceedings of a Consultants' Meeting*, pp. 31–35. Patancheru, Andhra Pradesh, India: ICRISAT.

Wilding, N., Mardell, S. K. and Brobyn, P. J. 1986. *Annals of Applied Biology* 108: 373–385.

Integrated control of insect pests of cool season food legumes

S. WEIGAND[1], S.S. LATEEF[2], N. EL-DIN SHARAF EL-DIN[3],
S.F. MAHMOUD[4], K. AHMED[5] and K. ALI[6]

[1] ICARDA, P. O. Box 5466, Aleppo, Syria;
[2] ICRISAT, Patancheru P. O., Andhra Pradesh 502 324. India;
[3] Agricultural Research Center, Wad Medani, Sudan;
[4] Agricultural Research Center, P. O. Box 12619, Giza, Egypt;
[5] National Agricultural Research Center, Islamabad, Pakistan, and
[6] Institute of Agricultural Research, Addis Ababa, Ethiopia

Abstract

Cool season food legumes suffer yield losses from several insects in the field and during storage. The main field insect pests are leafminer and podborer in chickpea; *Sitona crinitus* and aphids in lentil; aphids, leafminer, and *Sitona* weevil in faba bean; *Bruchus pisorum*, aphids and *Sitona* weevil in pea. In view of the relatively high cost of chemical control in these low cash crops and the possible harmful effects on the fragile environment, special emphasis is placed on an integrated pest management strategy utilizing all practical techniques. For control of leafminer and podborer in chickpea, resistance screening, the application of neem extract, and the possible use of parasitoids have given promising results. Podborer damage is also reduced by intercropping, sprays of nuclear polyhedrosis virus, and the microbial insecticide *Bacillus thuringiensis* (Bt). In lentil, seed treatment with Promet effectively controls *Sitona* and is less disruptive to the environment than insecticide sprays. For aphid control in lentil and faba bean, threshold levels for insecticide application and the combination of host plant resistance/tolerance and chemical control are being studied. For control of leafminer in faba bean, dosage and threshold levels of effective chemicals have been identified as well as two main parasitoids. In pea, insecticides are the most widely used technique to reduce densities of aphids and other insect pests.

Introduction

Integrated Pest Management

Integrated pest management (IPM) is the use of a combination of pest control tactics to reduce pest populations to economically satisfactory levels while minimizing disruption to the environment. It is an ecologically based pest control strategy that relies heavily on plant resistance and natural mortality

F.J. Muehlbauer and W.J. Kaiser (eds.), *Expanding the Production and Use of Cool Season Food Legumes*, 679–694.
© 1994 *Kluwer Academic Publishers.*

factors, such as natural enemies; and it seeks to use control methods that are compatible with each other.

While most ideas incorporated in the IPM approach are not new, increased public concern over the extensive use of pesticides, widespread development of insecticide resistance, rise of previously minor pests, difficulties in obtaining insecticides, and other developments have renewed interest in the practice. The concept was formally defined and proposed in 1959 (Stern *et al.*, 1959) and, although it has not been accepted by all practitioners of pest control, its importance as the most effective and balanced approach to pest management has increased over the past 30 years.

Cool season food legumes [chickpea (*Cicer arietinum* L.), lentil (*Lens culinaris* Medik.), faba bean (*Vicia faba* L.), and pea (*Pisum sativum* L.)], are grown worldwide over a wide range of geographic and climatic environments (temperate, subtropical and tropical zones) where they are attacked by a wide range of insect species. Several of the main insect pests attack all food legumes, in addition to other host plants. However, the extent of damage and economic losses vary in the different agroecological regions. In general, the information available on the economic importance and yield losses caused by insect pests of food legumes is limited. The high variability in the occurrence of even the main insects increases the difficulty of accurately assessing the impact of control methods and necessitates the development of nonexpensive control methods. In view of the relatively high cost of chemical control in these low cash crops and the possible harmful effects on a fragile environment, this contribution reviews various approaches to managing major insect pests of food legumes, emphasizing the components of an integrated pest management strategy, namely 1) the identification and use of resistant cultivars, to include the screening of plant germplasm and studies on the extent and nature of resistance of promising material, 2) biological control, involving a survey of naturally occurring parasitoids and predators followed by studies on their biology and effectiveness, 3) chemical control, emphasizing the monitoring of insect populations to ensure precise timing of insecticide applications, and 4) cultural practices, such as the contribution of sowing date, fertilizer application, and plant density to pest control.

Integrated Control in Chickpea

About 60 insect species are known to feed on chickpea (Reed *et al.*, 1987), but relatively few are major pests.

In West Asia and the Mediterranean region the chickpea leafminer (*Liriomyza cicerina* Rondani) is the main pest, occurring in several countries in high densities every year (Cardona, 1983; Reed *et al.*, 1987). The damage is caused by larvae feeding on the leaf mesophyll tissue resulting in desiccation and premature fall of leaves. In India, Pakistan, and Bangladesh, where more than 85% of the world's chickpea is grown and in some regions of West Asia and

North Africa, *Helicoverpa armigera* (Hubner) is the major pest. The larvae damage pods, but also feed on foliage, buds, and flowers. Research has focused on those two pests because other pests, i.e., aphids, cutworms, and other lepidopterous species are of localized importance.

Pest management requires knowledge of the population dynamics of the pest and its natural enemies, as well as knowledge of the cultural practices used to produce the crop. Chickpeas are traditionally grown as a spring crop in West Asia and North Africa. The advancing of sowing date to winter results in substantially higher yields, provided cold tolerant and Ascochyta blight resistant cultivars are available. Such a change in sowing date will also cause some changes in pest incidence. For example, when leafminer populations reach high densities in late April, winter chickpeas are already starting to mature (pod stage), whereas spring chickpeas at that time are just flowering and support leafminer populations for a longer period (Figure 1).

Foliar damage to winter chickpeas does not result in high yield losses; however, spring chickpeas are much more sensitive to losses of photosynthetic area and therefore suffer yield losses. On the other hand, parasitoid populations can build up in spring chickpeas. In Syria, for example, the parasitoids *Diglyphus isaea* Walker (Eulophidae) and *Opius monilicornis* Fischer (Braconidae) are the dominant species, reaching high densities late in the season (Figure 1). These parasitoids could be used as biological control agents if methods can be developed to enhance them early in the season. Without insecticides, biological control agents can be expected to maintain leafminers at subeconomic levels because the economic injury level is very high. Moreover, because other pests (i.e., *H. armigera*) occur at low densities in production areas where *L. cicerina* is the main pest, insecticides are not normally required. The development and application of biological control would help farmers avoid the difficulties experienced by producers in Europe and the USA. That is, the elimination of parasitoids and the rapid development of insecticide resistance in leafminers that has accompanied widespread insecticide use in several vegetable and ornamental crops. Biological control programs have shown promising results in greenhouses, field grown vegetables, and alfalfa (Hendrickson and Plummer, 1983; Trumble and Toscano, 1983; Minkenberg and Lenteren, 1986; Parella *et al.*, 1989).

Insecticides that provide effective leafminer control have been identified and optimal application times determined (Weigand, 1989). Since aqueous solutions of neem seed extract (*Azadirachta indica* A. Juss.) were shown to be quite effective in controlling other leafminer species (Webb *et al.*, 1983; Fagoonee and Toory, 1984; Larew *et al.*, 1985; Stein and Parella, 1985; Schmutterer, 1990), it was tested for chickpea leafminer as an alternative insecticide. Five sprays (500 g neem seeds 10 l^{-1} water; 500 l ha^{-1}) significantly reduced the percentage of mining by leafminers, but yield increases were not significant. At ICARDA, sampling of leafminer parasitoids by D-vac showed that neem sprays had no significant effect on parasitoids, whereas the application of Thiodan (2 cc l^{-1}) significantly reduced the number of parasitoids as compared to the water sprayed plots.

Winter-chickpea

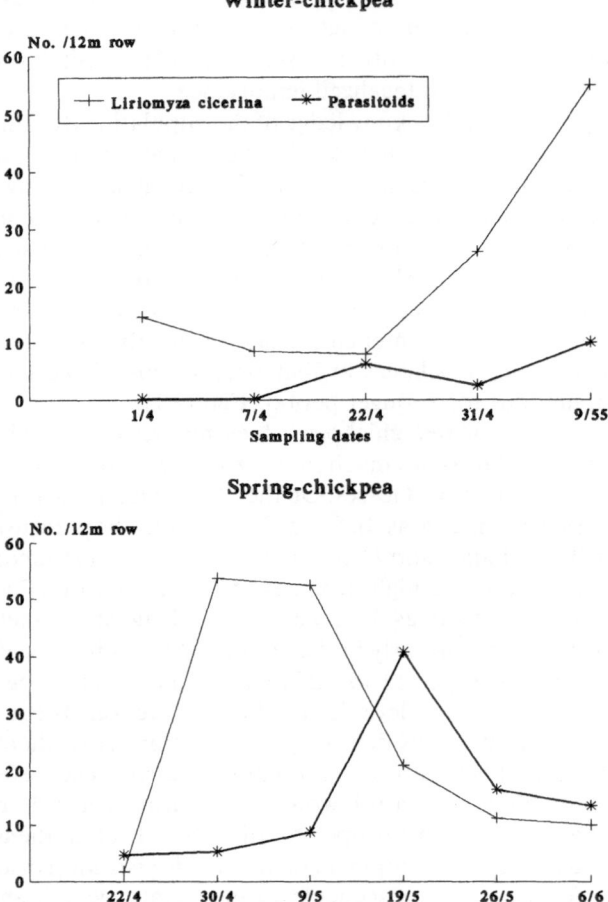

Figure 1. Chickpea leafminer and parasitoid adult population development, as sampled by D-Vac, in winter- and spring-sown chickpea, Syria, 1990/91.

The identification of resistance to chickpea leafminer has been emphasized at ICARDA and several chickpea lines showing low leafminer damage identified. These lines are further studied within an integrated pest management perspective, i.e., the interaction between resistant lines and insecticide application as well as biological control (the effect of resistant lines on parasitoids).

Podborer (*H. armigera*) accounts for high yield losses in some chickpea growing regions of India, Pakistan, and West Asia (Reed *et al*., 1987; Al-Soud *et al*,. 1990; Seghal and Ujagir, 1990). Because the severity of damage varies between regions, pest management research needs are addressed in the context of regional differences (Pimbert and Seghal, 1990). *H. armigera* possesses several biological attributes that make it a serious pest – it is ployphagous, multivoltine, highly fecund, capable of migrating long distances as adults, and larvae are well protected, making it difficult to control with insecticides. In India, many farmers use traditional cultural methods, such as sowing date and intercropping to protect locally adapted cultivars from podborer damage. Early sown (October to mid-November) chickpea have less pod damage than crops sown in December and January (Prased *et al.*, 1985). In contrast, experiments in Syria showed that advancing chickpea sowing date from February/March to November/December increased pod damage from 24 to 45% (Al-Soud, 1992). Yields, however, were still higher in early plantings because plants were able to make full use of limited rainfall.

Under high infestations in India (42 to 90% pod damage), additional control methods are required (Seghal and Ujagir, 1990). Several insecticides, namely cypermethrin, deltamethrin, fenvalerate, endosulfan, and monocrotophos, effectively controlled podborer on chickpea (Seghal and Ujagir, 1990). However, the use of synthetic insecticides is usually not a long term solution because *H. armigera* rapidly develops resistance to insecticides (Wolfenbarger *et al.*, 1981). This is mainly due to heavy insecticide use on cotton and other crops for podborer control. Recently, resistance to cypermethrin and fenvalerate has been recorded in podborer larvae collected from cotton in different areas in India indicating widespread development of resistance to pyrethroids (Singh, 1990). With this development, farmers may be unable to rely solely on synthetic chemical control to control podborer on chickpea. Neem seed extract (5%) was shown to be less effective than insecticides but still significantly better than the untreated check and, therefore, might be an alternative to the synthetic insecticides in the case of medium to severe infestations (Seghal and Ujagir, 1990).

The potential for biological control of *Heliothis* species, including *H. armigera*, was reviewed by King and Coleman (1989). Another contribution summarized the worldwide distribution and abundance of natural enemies of *Heliothis* (King and Jackson, 1989). Yadav (1990) listed the parasitoids and predators attacking *H. armigera* in chickpea. In general, parasitism of podborer eggs is low in chickpea. For example, there was no parasitism of *H. armigera* eggs by *Trichogramma* in chickpea, whereas eggs from cotton, sorghum, and other crops were heavily parasitized by this wasp (Yadav, 1990). Because *Trichogramma* and *Chrysopa* can be mass propagated, augmenative release of these natural enemies is possible for the control of *H. armigera* in cotton (King and Coleman, 1989). On the other hand, because these two natural enemies do not occur in chickpea, the augmentative approach does not appear to be feasible for biocontrol of podborer in chickpea. Perhaps the acidic nature of chickpea leaf exudates interfers or somehow adversely affects the colonization of this crop by these natural enemies. Larval parasitoids have been reared from podborers, of which *Campoletis chloridae* (Uch.) is the dominant species in India. No egg or pupal parasitoids have been found in southern India. No egg or pupal parasitoids have been found in southern Syria; however, two larval parasitoids [C. *chloridae* and *Bracon hebetor* (Say)] were responsible for parasitization rates of 15% to 20% (Al-Soud, 1992). There have been no attempts to use larval parasitoids in a biological control program for *H. armigera* in chickpea. Efficient and cost effective methods for rearing parasitoids must be developed if augmentative releases are to become economically feasible.

Microbes have the advantage of being amenable to commercial production because they can be mass cultured, formulated, packaged, stored, and applied like chemical insecticides. In addition, they can complement the effects of other biological control agents because of their environmental safety and pest selectivity. These positive attributes have prompted researchers to investigate the potential of microbial control of *H. armigera*, with emphasis on the use of a Nuclear Polyhedrosis Virus (NPV) and *Bacillus thuringiensis* (Bt). In one study, Rabindra and Jayara (1988) found that four applications of NPV at a rate of 250 LE (ca. 1.5×10^{12} Polyeder ha^{-1}) gave effective control of *H. armigera* on chickpea, which was almost comparable to podborer control achieved with endosulfan at 350 g a.i. ha^{-1}. Moreover, these researchers reported maximum yields and lowest pod damage with a combination of NPV (250 LE) and a low concentration of endosulfan (175 g a.i. ha^{-1}). Pawar and Thombre (1990) and Yadav (1990) have discussed the role of NPV for control of podborer on chickpea in India. The microbial insecticide *B. thuringiensis* has potential for control of podborer on chickpea, as revealed by studies in Pakistan (Khalique *et al.*, 1989). These studies recorded higher podborer mortality when larvae were exposed to a combination of Bt and malic acid as compared to Bt and malic acid alone. Under field conditions, different concentrations of Bt (Dipel 2X and Dipel ES) with and without 10% molasses significantly reduced pod damage; yields were higher, as well, but not significantly (Table 1). Research to date, however, has shown that Bt is not as effective as insecticides in controlling podborer populations. Moreover, microbial insecticides have the disadvantage of being inactivated by light and temperature, and slow acting as compared to chemical insecticides. And, it usually takes several days before a larva dies after ingesting a lethal dose of a microbial agent. Nevertheless, new formulations may be developed to overcome some of these limitations.

Host plant resistance in chickpea to *H. armigera* has been studied at ICRISAT for many years and several promising lines with resistance have been identified (Lateef, 1990; Lateef and Pimbert, 1990). Host plant resistance has also been considered within an integrated pest management approach (Pimbert, 1990) as well as its compatibility with biological control (Reed *et al.*, 1989). In field studies at ICRISAT, lower rates of parasitism of *H. armigera* larvae were found on resistant than on susceptible chickpea genotypes (Sithanantham *et al.*, 1982). This finding suggests that host plant resistance and biological control may not be compatible in all instances. It also suggests that the compatibility of these two approaches should be evaluated under field conditions.

Table 1. Effect of microbial insecticides with and without adjuvant (10% molasses) on pod damage by *Helicoverpa armigera* and seed yield of chickpea, Islamabad, Pakistan 1988/89

Treatment (kg or l ha⁻¹)	% Pod damage	Yield (kg ha⁻¹)
0.8 Dipel 2x	0.05	2570
1.6 Dipel 2x	5.1	2517
1.5 Dipel ES	16.0	1887
2.5 Dipel ES	7.7	2400
0.8 Dipel 2x*	9.7	3460
1.6 Dipel 2x*	4.4	1945
1.5 Dipel ES*	7.6	3000
2.5 Dipel ES*	4.4	1579
Check	25.7	1075

* Plus 10% molasses

Integrated Control in Lentil

In West Asia, *Sitona crinitus* Herbst is the primary insect pest of lentil (Hariri, 1981; Solh *et al.*, 1986). Adult weevils feed on leaf edges which can severely injure the seedlings. The main damage, however, is caused by larvae feeding on the nodules. This feeding affects the ability of the plant to fix atmospheric nitrogen. In some areas (Egypt, Ethiopia, USA) and in some years, aphids, mainly *Acyrthosiphon pisum* (Harris) and *Aphis craccivora* Koch, cause considerable yield losses.

In Syria, studies have concentrated on control of *Sitona*, especially on the role of cultural practices and chemical control in pest management. For example, planting date affected *S. crinitus* damage and lentil yields (Weigand *et al.*, 1992). On-farm trials over three years showed that early sowing (mid-November) increased *Sitona* infestations. Nodule damage was higher in lentils sown in November, even with a carbofuran treatment (5G, 20 kg ha^{-1}) than in lentils sown in early January. This was attributed to nodule formation in early plantings coinciding with high densities of *Sitona* larvae (Figure 2). However, in spite of higher nodule damage in the early sown lentil, yields were higher than in the late planting. Yield increases due to *Sitona* control were generally higher in the early sown crop. Fertilizer application (50 kg ha^{-1} N and 50 kg ha^{-1} P$_2$O$_5$) was found to have no effect on nodulation and nodule damage (Weigand *et al.*, 1992).

Chemical control studies have focused on the development of effective insecticidal control programs and the evaluation of the economic importance of *S. crinitus*. These studies have considered the fact that applications of insecticides for *Sitona* control must be applied in a timely manner and used only as needed. Granule carbofuran applied at planting is less disrupting to the insect fauna than spray applications and was shown to provide effective control

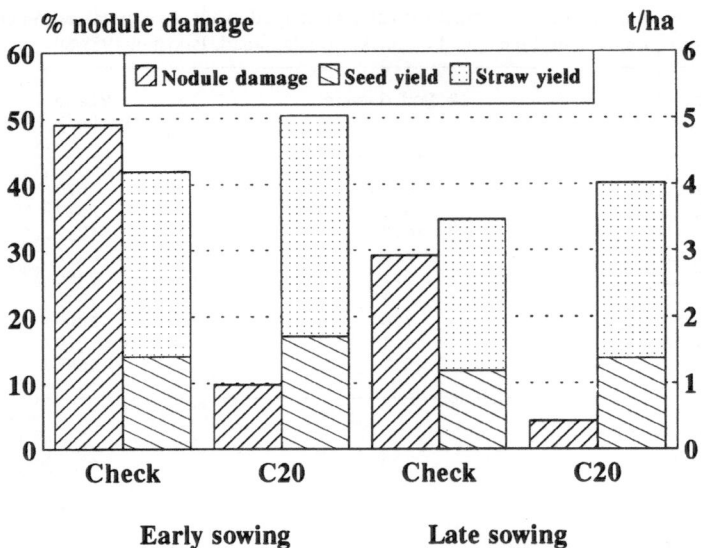

Figure 2. Effect of two sowing dates and carbofuran treatment on *Sitona crinitus* damage and lentil yield (mean of 19 locations and 3 years), Syria 1986–88.

(Figure 2). More recently, seed treatment of lentil with Promet (furathiocarb) proved to be an effective alternative to carbofuran with the advantage of being less toxic and easier to handle. Because chemical seed treatments are preventive in nature, they might often not be economical for *Sitona* control, due to the varying damage levels of this weevil.

With regard to biological control, several parasitoids of *Sitona* spp. were found in surveys of the Mediterranean region. Two species, *Microctonus aethiopoides* Loan parasitizing the adults and *Anaphes diana* Girault parasitizing the eggs, were predominant and most effective (Aeschlimann, 1980, 1986). The egg parasitoid is not restricted to one *Sitona* species, and it successfully completed its development in eggs of *S. crinitus* in preliminary studies at ICARDA. Thus, *A. diana* might have some potential for biological control of *Sitona* in lentil and will be studied further. Bezdicek *et al.* (this volume) discussed the use of engineered Bt in *Rhizobium* bacteria for control of *Sitona*.

Very little information is available on control methods for aphids in lentil. Schotzko and O'Keeffe (1989) developed sampling methods for *A. pisum* and its main predators. Studies in Egypt, Ethiopia, and at ICARDA have focused on chemical control and the establishment of economic thresholds for lentil pests.

Integrated Control in Faba Bean

Aphids, mainly *Aphis fabae* (Scop.) and *A. craccivora*, but also *A. pisum,* are the main pests in most faba bean growing areas of the world. These aphids affect faba bean via direct feeding and by transmitting virus diseases. In addition, the leafminer *Liriomyza trifolii* (Burgess) causes severe damage in Egypt and the Sudan.

With regard to aphids, most studies have been conducted on host plant resistance and chemical control. Only Patriquin *et al.* (1988) studied the effect of cultural practices and reported a higher number of aphids in N-fertilized than in unfertilized plots, in nonweedy than in weedy plots, and in plots without intercrops.

Faba bean lines with some aphid resistance have been identified, but these have not been developed for commercial production. In Egypt, as part of the Nile Valley Regional Program, several breeding lines together with commercial cultivars were tested in the field, with and without insecticide protection, for aphid infestation and yield for several years. One breeding line (30/18/82) is promising as it consistently had low aphid populations. Other lines appear to have some tolerance, as they produced high yields in the presence of aphid infestations (Weigand and Bishara, 1991).

It is unlikely, however, that host plant resistance alone will provide acceptable control of aphids on faba bean. Rather, host plant resistance and chemical control will be required. High aphid infestations often make insecticide treatments necessary and thus reliable and practical recommendations for the proper timing are needed. In experiments in Egypt, critical infestation levels for aphid control were determined using the susceptible "Giza 402" cultivar. The lowest aphid infestation and highest yield was achieved with full protection (weekly sprays of Pirimor) followed by spraying once when 10% of the plants were infested (Figure 3). Applications at 20 and 30% infested plants resulted in higher aphid populations and yield loss. These results illustrate that one properly timed application was sufficient for aphid control and the economic threshold was at 10% infested plants.

Although many predators and parasitoids of aphids are known and naturally occurring these have not been used in a biological control program. Because of the extremely high reproduction of aphids and the short duration of faba bean in the field, the potential for biological control is limited.

In new, nontraditional faba bean growing areas of the Sudan, as well as in some areas of Egypt, the leafminer *L. trifolii* is the main pest. In the Sudan, high infestations require the development of control methods. In a multi-year study, the screening of a number of insecticides and dosages resulted in the recommendation of Danitol S 50% ES (2.4 l ha^{-1}), Tamaron 60% EC (2.5 l ha^{-1}), and Endophos 12/14 EC (2.4 l ha^{-1}). Studies also related infestation levels with yield losses to establish economic thresholds. Insecticide applications of Danitol S were carried out at infestation levels of 15, 25, 35, and 45% infested leaflets resulting in four, three, two, and one application in 1987/88,

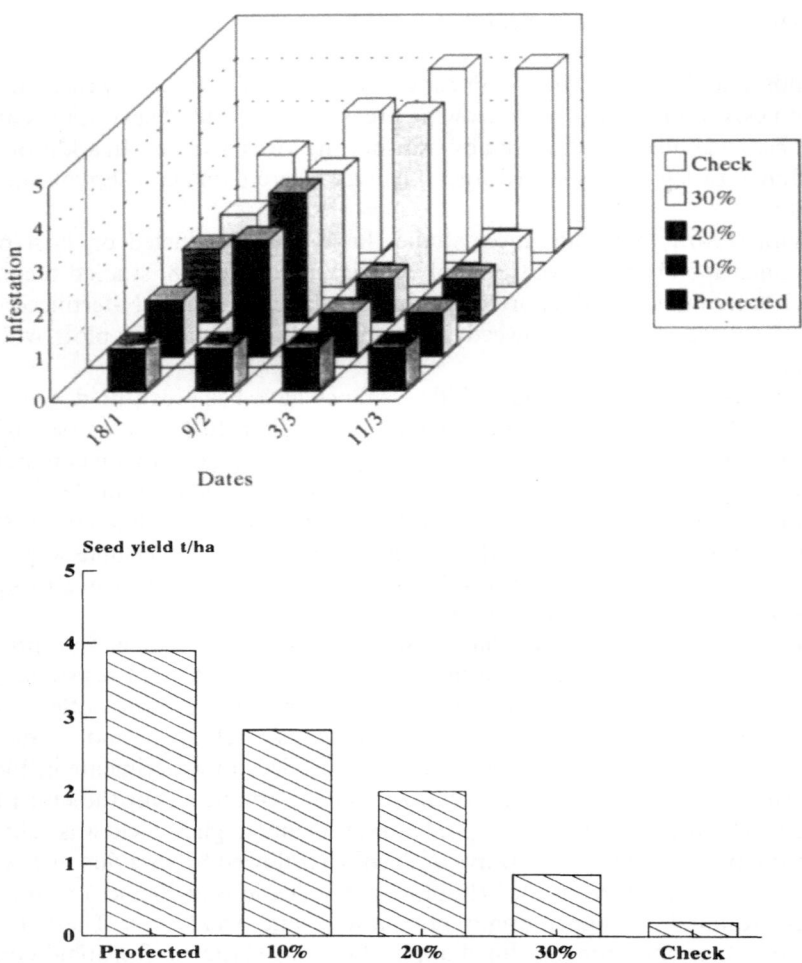

Figure 3. Effect of scheduling insecticidal spray based on percentage plant infestation by aphids on aphid infestation rate (a) and seed yield (b) of faba bean, Egypt, 1990/91.

respectively, as compared with a weekly spray treatment and a nonsprayed check. The optimum threshold level was about 25% infested leaflets coinciding with two or three applications (Table 2). This has been confirmed by further experiments, which also showed that the 25% infestation level has to be reached before pod formation. There was little benefit to applying an insecticide after this stage.

However, chemical control of *L. trifolii* also has limitations. In contrast to the chickpea leafminer this species has a wide host range and is a major pest of vegetables and ornamentals. After resistance to many insecticides was reported

Table 2. Effect of date and number of insecticide applications (Danitol S) on leafminer infestation and seed yield of faba bean, Wad Medani, Sudan 1987/88

% Infested leaflets at spraying	No. of sprays*	% Infested leaflets		Yield
		Nov.-Dec.	Dec.-Feb.	(t/ha)
15	4	14.4	28.8	1.31a
28	3	18.2	32.3	1.14 ab
37	2	23.0	33.3	1.0 b
42	1	27.7	35.7	0.62 c
–	6	13.8	13.5	1.28 a
–	0	27.6	35.7	0.64 c

* Danitol S at 2.4 l ha⁻¹

(Parella *et al.*, 1984; Mason *et al.*, 1987; Keil and Parella, 1990), the use of parasitoids for biological control of *L. trifolii* was studied with promising results in several crops (Minkenberg and Lenteren, 1986; Parella *et al.*, 1987, 1989; Rathman *et al.*, 1991). In the Sudan, two parasitoids, *Chrysonotomyia formosa* (W) and *Hemiptarsenus semi-albicalva* (G) have parasitized high numbers of leafminers (60%) late in the season. Since these naturally occurring parasitoids should not be affected by insecticide treatments, the selectivity of two foliar insecticides was compared with a granular insecticide (Furadan) and sprays of neem extract. In general, no drastic differences were found in the percent parasitism between the treatments (Figure 4). Neem seed extracts and Danitol S were more selective than Evisect and Furadan treatments, but more data are

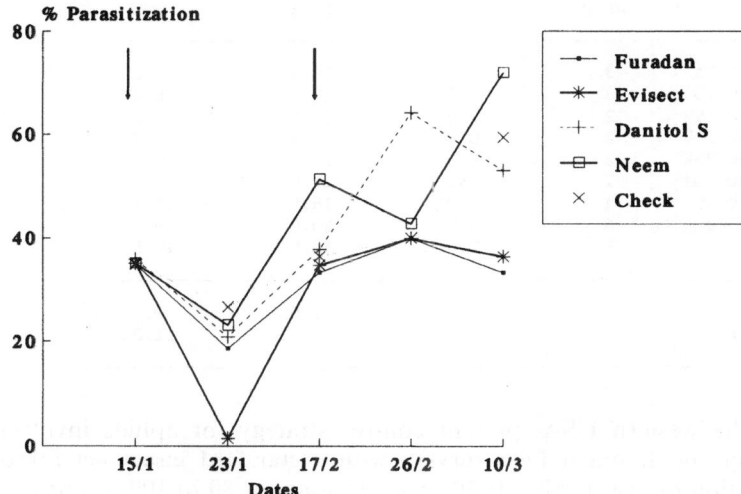

Figure 4. Effect of insecticide treatments on percentage parasitism of larvae of *Liriomyza trifolii* in faba bean, Sudan, 1990/91.

needed to confirm these results. Accounts of the biology of *Liriomyza* species, including *L. trifolii*, are provided by Dimetry (1971) and Parella (1987).

Integrated Control in Field Pea

Yield losses in field pea are caused by many of the same insect pests that attack other food legumes. Most important are aphids (*A. pisum* and *A. craccivora*), pea leaf weevil [*Sitona lineatus* (L.)], the pea moth (*Laspeyresia nigricana* Staph.), and the pea weevil [*Bruchus pisorum* (L.)].

In Ethiopia, where the pea aphid causes high yield losses in some years, chemical control and host plant resistance are being studied as control methods. Since the relationship between varying aphid infestation rates and pea yield is not fully understood, experiments are underway to generate data to establish economic thresholds. In all treatments, yield was significantly higher than in the untreated check (Table 3). The highest yield was achieved with full protection (six sprays), followed equally by two and three sprays. Yield losses were higher in the treatments receiving one spray, but this was more due to the late application rather than the number of sprays. In the economic evaluation, the one spray at 35% infestation resulted in maximum net returns. This clearly shows that only one or two sprays are economical for pea aphid control.

Table 3. Effect of scheduling insecticidal sprays (Pirimor 50% DG, 1 kg ha^{-1}) based on percentage plant infestation by pea aphids on the number of pods per plant and seed yield in field pea, Ethiopia, 1991

Treatment	No. of sprays	Pods Plant^{-1}	Yield (q ha^{-1})	% Yield loss
Spray at 5%	3	9.3	19.5	6.2
Spray at 10%	3	9.2	19.6	5.8
Spray at 15%	2	9.0	19.5	6.2
Spray at 20%	2	8.4	19.4	6.7
Spray at 25%	2	9.0	19.4	6.7
Spray at 30%	1	8.6	17.8	14.4
Spray at 35%	1	9.0	18.5	11.1
Full protection	6	9.8	20.8	-
Check	-	7.6	10.5	49.5
LSD 5%			1.55	0.61
C.V. (%)			7.19	11.85

In the western USA, present control strategy for aphids involves early planting and frequent field surveys with a standard insect net for optimal application of insecticides. If 30 to 40 *A. pisum* or 80 to 100 *A. craccivora* are collected per sweep, the recommendation is to treat the field with a registered insecticide. If aphid numbers are less, the field should be checked frequently until it is known whether the population is increasing or stabilizing (Homan *et al.*, 1984).

There are reports of pea cultivar resistance to *A. pisum*. In Canada, for example, differences in pea aphid densities were observed on different field pea cultivars (Maiteki *et al.*, 1986). Moreover, it was shown that the pea cultivar is an important factor in determination of economic thresholds of pea aphids (Soroka and Mackay, 1990). In another study, Kareiva and Sahakian (1990) found that different cultivars influenced the effectiveness of coccinellid predators of *A. pisum*. This is another example which illustrates that different control tactics cannot be considered separately but must be studied for their interaction with other components.

The pea leaf weevil, *S. lineatus*, is an important pest of peas in Europe and the USA. Studies in the USA have shown that the need for individual control depends on weather, timing of weevil migration, number of pea seedlings, adult weevil numbers, growth stage and rate of seedling development, and seedbed preparation (O'Keeffe *et al.*, 1984). Early sowing is recommended to produce stronger seedlings. Insecticides should be used if one clamshell out of four is damaged, and only up to the two to four leaf stage. Peas past the seedling stage can usually tolerate the feeding damage. The present control strategy for the pea weevil, *B. pisorum*, involves insecticides, but other strategies are being researched (Smith, 1990).

Concluding Remarks

In all four cool season food legume crops different components of pest management have been studied and the state of development is summarized in Table 4. At present, chemical control is the most or only effective technique available to reduce insect pest damage, even though the high variability in occurrence and damage levels of the pests makes the establishment of economic thresholds difficult. Other techniques are under investigation.

Chemical control is the easiest method of control. The use of cultural methods for pest control is limited because the pests are often not so important that the agricultural system should revolve around IPM considerations. Indeed, the environment and climate are the limiting factors in most legume growing areas. The development of biological control systems is difficult because food legume crops are of short duration in the field. They are low cash crops for which the mass rearing and release of natural enemies might not be economical. Finally, breeding for high yield potential and resistance to abiotic stresses and diseases is often of higher priority than insect resistance.

In conclusion, more studies are needed on the population dynamics of the pest species and their natural enemies in relation to weather data and the various components of IPM. Except for chemical control, none of the other pest management options, by themselves, have provided effective control. Nonetheless, all components must be considered more fully within the framework of an overall IPM and crop management strategy.

Table 4. State of development of integrated pest management techniques for food legume insects

Leafminer	XXX	Chemical control
	X	Neem extract
	X	Host plant resistance
	X	Enhancement, use of parasitoids
Podborer	XXX	Chemical control
	XX	NVP virus, Bt spray
	XX	Host plant resistance
	XX	Intercropping
Sitona weevil	XXX	Chemical control
	X	Use of parasitoids
	X	Transfer of Bt toxin gene into *Rhizobium*
Aphids	XXX	Chemical control
	X	Host plant resistance
	X	Use of parasitoids and predators

XXX Most effective technique presently available
XX Moderately effective technique available
X   Technology under investigation

References

Aeschlimann, J. P. 1980. *Entomophaga* 25: 139–153.
Aeschlimann, J. P. 1986. *Entomophaga* 31: 163–172.
Al-Soud, A., Weigand, S. and Tahhan, O. 1990. *International Chickpea Newsletter* 22: 30–32.
Al-Soud, A. 1992. *Studies on some aspects of integrated control of podborer (Helicoverpa armigera and Heliothis spp.) on chickpea in southern Syria.* Ph.D. thesis, Damascus University, Syria.
Bezdicek, D. F., Quinn, M. A., Forse, L., Beck, D. and Weigand, S. 1994. *Expanding the Production and Use of Cool Season Food Legumes,* pp. 740–754 (eds. F. J. Muehlbauer and W. J. Kaiser). Dordrecht: Kluwer Academic Publishers.
Cardona, C. 1983. In: *Proceedings of the International Workshop on Faba Beans, Kabuli Chickpeas and Lentils in the 1980s,* pp. 159–165 (eds. M. C. Saxena and S. Varma). Aleppo, Syria: ICARDA.
Dimetry, N. 1971. *Bulletin Society Entomological Egypt* LV: 55–69.
Fagoonee, I. and Toory, V. 1984. *Insect Science and Application* 5: 23–30.
Hariri, G. 1981. In: *Lentils,* pp. 173–190 (eds. C. Webb and G. Hawtin). Slough, England, UK: CAB International.
Hendrickson, R. M. and Plummer, J. A. 1983. *Journal of Economic Entomology* 76: 757–761.
Homan, H. W., O'Keeffe, L. E. and Stoltz, R. L. 1984. *Aphids on peas and lentils and their control.* University of Idaho, Cooperative Extension Service, Agricultural Experiment Station, Moscow, Idaho. Current Information Series No. 748.
Kareiva, P. and Sahakian, R. 1990. *Nature* 345: 433–434.
Keil, C. B., Parrella, M. P. and Moore, J. G. 1985. *Journal of Economic Entomology* 78: 419–422.
Keil, C. B. and Parella, M. P. 1990. *Journal of Economic Entomology* 83: 18–26.
Khalique, F., Ahmed, K. and Afzal, M. 1989. *Pakistan Journal Science Ind. Research* 32: 114–116.
King, E. G. and Coleman, R. J. 1989. *Annual Review of Entomology* 34: 53–75.
King, E. G. and Jackson, R. D. (eds.) 1989. Increasing the effectiveness of natural enemies. In: *Proceedings of the Workshop on Biological Control of Heliothis.* 11–15 Nov., 1985, New Delhi, India. Far Eastern Regional Research Office, U.S. Department of Agriculture. New Delhi, India.

Larew, H. G., Knodel-Montz, J. J., Webb, R. E. and Warthen, J. D. 1985. *Journal of Economic Entomology* 78: 80–84.
Lateef, S. S. 1990. In: *Proceedings of the First National Workshop on Heliothis Management*, pp. 129--140 (ed. J. N. Sachan) Kanpur, India: Directorate of Pulses Research.
Lateef, S. S. and Pimbert, M. P. 1990. In: *Proceedings of First Consultative Group Meeting on Host Selection Behaviour of Helicoverpa armigera*, pp. 14–18. India: ICRISAT.
Lingren, P. D. 1977. *Environmental Entomology* 6: 72–76.
Maiteki, G. A., Lamb, R. J. and Ali-Khan, S. T. 1986. *Canadian Entomologist* 118: 601–607.
Mason, G. A., Tabashnik, B. E. and Johnson, M. W. 1987. *Journal of Economic Entomology* 80: 1262–1266.
Mason, G. A. and Johnson, M. W. 1988. *Journal of Economic Entomology* 81: 123–126.
Minkenberg, O. P. J. M. and Van Lenteren, J. C. 1986. *The leafminer Liriomyza bryoniae and L. trifolii (Diptera: Agromyzidae), their parasites and host plants: A Review*. Agricultural University, Wageningen, The Netherlands. Papers 86-2.
O'Keeffe, L. E., Homan, H. W. and Schotzko, D. 1984. *The pea leaf weevil*. University of Idaho, Cooperative Extension Service, Agricultural Extension Service, Moscow, Idaho. Current Information Series No. 227.
Parella, M. P. 1987. *Annual Review of Entomology* 32: 201–224.
Parella, M. P., Keil, C. B. and Morse, J. G. 1984. *California Agriculture* 38: 22–23.
Parella, M. P., Jones, V. P. and Christie, G. D. 1987. *Environmental Entomology* 16: 832–837.
Parella, M. P., Yost, J. T., Heinz, K. M. and Ferrentino, G. W. 1989. *Journal of Economic Entomology* 82: 420–425.
Patriquin, D. G., Baines, J., Lewis, J. and Macdougall, A. 1988. *Agriculture, Ecosystems and Environment* 20: 279–288.
Pawar, V. M. and Thombre, U. T. 1990. In: *Proceedings of First National Workshop on Heliothis Management*, pp. 247–258 (ed. J. N. Sachan). Kanpur, India: Directorate of Pulses Research.
Pimbert, M. P. 1990. In: *Chickpea in the Nineties: Proceedings of the Second International Workshop on Chickpea Improvement*, pp. 151–163 (eds. H. A. van Rheenen, M. C. Saxena, B. J. Walby and S. D. Hall). Patancheru, A. P., India: ICRISAT.
Pimbert, M. P. and Seghal, V. K. 1990. *The relative importance of Helicoverpa armigera (Huebner) on chickpea in different agroecological zones of India: Implications for pest management*. Legumes Entomology Discussion Paper No. 1, Patancheru, Andhra Pradesh, India: ICRISAT.
Prasad, D., Premchand and Srivastava, G. P. 1985. *Indian Journal of Entomology* 47: 223–225.
Rabindra, R. J. and Jayaraj, S. 1988. *Tropical Pest Management* 34: 441–444.
Rathman, R. J., Johnson, M. W. and Tabshnik, B. E. 1991. *Biological Control* 1: 256–260.
Reed, W., Cardona, C., Sithanantham, S. and Lateef, S. S. 1987. In: *The Chickpea*, pp. 283–318 (eds. M. C. Saxena and K. B. Singh). Dordrecht: Kluwer Academic Publishers.
Reed, W., Lateef, S. S. and Sithanantham, S. 1989. In: *Proceedings of Workshop on Biological Control of Heliothis*, pp. 529–535 (eds. E. G. King and R. D. Jackson). New Delhi, India: Far Eastern Regional Research Office, US Department of Agriculture.
Schmutterer, H. 1990. *Annual Review of Entomology* 35: 271–297.
Schotzko, D. J. and O'Keeffe, L. E. 1989. *Journal of Economic Entomology* 82: 491–506.
Seghal, V. K. and Ujagir, R. 1990. *Crop Protection* 9: 29–32.
Singh, H. N. 1990. In: *Proceedings of the First National Workshop on Heliothis Management*, pp. 271–282 (ed. J. N. Sachan) Kanpur, India.
Sithanantham, S., Rao, V. R. and Reed, W. 1982. *International Chickpea Newsletter* 6: 21–22.
Smith, A. M. (ed.). 1990. *Proceedings of National Pea Weevil Workshop*. Melbourne, Australia: Department of Agriculture and Rural Affairs.
Solh, M. B., Itani, H. M. and Kawar, N. S. 1986. *Lebanese Science Bulletin* 2: 17–27.
Soroka, J. J. and Mackay, P. A. 1990. *Canadian Entomologist* 122: 503–513.
Stein, U. and Parella, M. 1985. *California Agriculture* 39: 19–20.
Stern, V. M., Smith, R. F., Van den Bosch, R. and Hagen, K. S. 1959. *Hilgardia* 29: 81–101.
Trumble, J. T. and Toscano, N. C. 1983. *Canadian Entomologist* 115: 1415–1420.

Webb, R. E., Hinebaugh, M. A., Lindquist, R. K. and Jacobsen, M. 1983. *Journal of Economic Entomology* 76: 357–362.

Weigand, S. 1989. In: *Proceedings of the International Conference on Integrated Pest Management in Tropical and Subtropical Cropping Systems*, pp. 53–76. Frankfurt, Germany: Deutsche Landwirtschafts-Gesellschaft.

Weigand, S. and Bishara, S. I. 1991. *Serie Seminaires* 10: 67–74.

Weigand, S., Pala, M. and Saxena, M. C. 1992. Effect of sowing date, fertilizer and insecticide on nodule damage by *Sitona crinitus* Herbst (Coleoptera: Curculionidae) and yield of lentil (*Lens culinaris*) in northern Syria. *Journal of Plant Diseases and Protection* (In press).

Wolfenbarger, D. A. and Wolfenbarger, D. O. 1966. *Journal of Economic Entomology* 59: 279–283.

Wolfenbarger, D. A., Bodegas, P. R. and Flores, R. 1981. *Bulletin Entomological Society of America* 27: 181–185.

Yadav, D. N. 1990. In: *Proceedings of the First National Workshop on Heliothis Management*, pp. 259–270 (ed. J. N. Sachan). Kanpur, India: Directorate of Pulses Research.

Integrated control of the parasitic angiosperm *Orobanche* (Broomrape)

A.H. PIETERSE[1], L. GARCÍA-TORRES[2], O.A. AL-MENOUFI[3], K.H. LINKE[4] and S.J. TER BORG[5]

[1] *Royal Tropical Institute, Rural Development Programme, Mauritskade 63, 1092 AD Amsterdam, The Netherlands;*
[2] *Institute of Agronomy and Plant Protection, CSIC, Córdoba, Spain;*
[3] *Alexandria University, Alexandria, Egypt;*
[4] *ICARDA, P. O. Box 5466, Aleppo, Syria, and*
[5] *Department of Vegetation Science, Plant Ecology and Weed Science, Agricultural University, Wageningen, The Netherlands*

Abstract

Various studies have been conducted on integrated control of the parasitic angiosperm *Orobanche* (broomrape). This research, including combinations of cultural, physical, and chemical methods, as well as the use of tolerant hosts, has largely focused on faba bean (*Vicia faba*). The most promising results have been obtained with a combination of slightly delayed sowing and low concentrations of imidazolinone herbicides, such as imazaquin, imazapyr, and imazethapyr, and/or glyphosate. Under certain conditions imazapyr and imazethapyr could also be used in other food legumes, such as lentil (*Lens culinaris*), field pea (*Pisum arvense*), and chickpea (*Cicer arietinum*). In faba bean, lines were recently developed (mainly from the cultivar Giza 402), which are tolerant to *Orobanche*. There is continued interest in the potential of biological *Orobanche* control. However, from a practical point of view, it may be concluded that it is not yet feasible to include biological agents in integrated *Orobanche* control.

Introduction

In general, control of *Orobanche* species (broomrapes) is more difficult than control of other weeds interfering with food legumes. This is largely due to their parasitic mode of life (Pieterse, 1979; Parker and Wilson, 1986). As they are attached to the roots of their hosts, control measures need to be very specific. An additional problem is that *Orobanche* seed will remain viable for many years and only germinate under the influence of a germination stimulant(s) exuded by roots of host plants (and a few non-host plants).

Although effective methods are available, these are generally too expensive to be used in farmers' fields. Examples include soil fumigation (Wilhelm *et al.*, 1958; Zahran, 1970) and solarization (Jacobsohn *et al.*, 1980; Sauerborn and Saxena, 1987). The use of herbicides is restricted, as they should be very specific

F.J. Muehlbauer and W.J. Kaiser (eds.), Expanding the Production and Use of Cool Season Food Legumes, 695–702.
© *1994 Kluwer Academic Publishers.*

for *Orobanche* and not harm the crops. Cultural means of control, such as delayed sowing (Mesa-García and García-Torres, 1986; van Hezewijk *et al.*, 1987; Raaimakers *et al.*, 1988; Linke and Saxena, 1991), rotation (Kasasian, 1971), and fertilizer application (Abu-Irmaileh, 1979; van Hezewijk *et al.*, 1991) have some effect but do not completely control *Orobanche*.

Although manual control does not guarantee a higher crop seed yield, it may be an important technique in reducing the *Orobanche* seed bank in the soil. However, it is very time-consuming and may only be economic if the *Orobanche* infestation is low.

As far as faba bean (*Vicia faba* L.) is concerned, very promising results were recently obtained with low concentrations of glyphosate and the imidazolinones, imazaquin, imazethapyr, and imazapyr (Linke and Saxena, 1991; García-Torres and López-Granados, 1991). The derivation of lines from the faba bean cultivar Giza 402 with a high level of tolerance to *Orobanche* seems to be another breakthrough.

Integration of various methods for *Orobanche* control, already propagated by F. Manoja in 1848 (cit. from Ciccarone and Piglionica, 1979), becomes more and more a focus of attention. In 1984 an integrated control strategy was suggested by Kukula and Masri (1984) implying a combination of tillage, fertilizer, and glyphosate. Recently, Linke and Saxena (1991) presented preliminary results of studies on faba bean, lentil (*Lens culinaris* Medik.), and chickpea (*Cicer arietinum* L.) conducted in Syria before 1989. These studies included early maturing cultivars, slightly delayed sowing, application of low concentrations of glyphosate or imazaquin, hand weeding, and solarization. The results were promising. This also applies to results of studies conducted in Egypt before 1989, including dense planting of Giza 402 in untilled soil, a high level of fertilizer, and application of glyphosate (Nassib *et al.*, 1984; Nassib and Hussein, 1989). Subsequently, integrated control studies have been continued in Syria, Egypt, and Spain.

The present paper gives an overview of the state of affairs with respect to integrated control of *Orobanche* in food legumes. Although the results of biological control have not been very promising so far, prospects for including biological means of control in integrated control programs, have also been discussed.

Studies on Integrated Control in Syria

Since 1985 the combination of different methods for controlling *Orobanche crenata* Forssk. has been studied at the experimental station of ICARDA in Tel Hadya, 30 km south of Aleppo (Linke, 1992). Crops included faba bean, lentil, chickpea, and field pea (*Pisum arvense* L.). As far as faba bean is concerned, most promising results have been obtained with a combination of slightly delayed sowing and low applications of the herbicides glyphosate, imazaquin, or imazethapyr. Although a delay in sowing normally causes yield reduction on soils free of *Orobanche*, this effect is relatively insignificant if the delay is not more than two weeks. Glyphosate (80 g active ingredient [a.i.] ha^{-1}) and imazaquin (40 g a.i. ha^{-1}) were applied twice, postemergence at the *Orobanche* tubercle and bud stage, respectively. Imazethapyr (100 g a.i. ha^{-1}) was applied once, preemergence to the crop. Both imazaquin and imazethapyr, in combination with delayed sowing, resulted in an almost complete control of *Orobanche*. Glyphosate was slightly less effective.

In lentil, a cultivar is available (ILL 8) which is adapted to late sowing. Delayed sowing of ILL 8 in combination with two postemergence applications of 7.5 g a.i. of imazaquin or one preemergence application of 60 g a.i. of imazethapyr resulted in an almost complete control of emerging *O. crenata* shoots. With respect to seed and straw yields of lentil, best results were obtained with imazaquin. As lentil is very susceptible to glyphosate, this herbicide is not recommended.

Generally infestation of chickpea by *Orobanche* is less severe than in faba bean and lentil. The experiments with respect to integrated control conducted so far have indicated that chickpea is even more sensitive to low rates of herbicides than to *Orobanche* attack.

In field pea, combinations have been tested of a genotype that had shown some resistance (accession "290"), delayed sowing and herbicide application. Promising results were obtained with delayed sowing and postemergence application of imazaquin (2 × 20 g a.i. ha^{-1}). In this study differences between *Orobanche* infestation of accession "290" and a local cultivar were minimal.

Delayed sowing and low herbicide treatment were also combined with solarization (Linke and Saxena, 1991). This was accomplished by covering wet soil with transparent polyethylene for a period of 40 days in the months of July and August. Solarization as a third component in the integrated control program substantially increased yields of faba bean and lentil and resulted in 100% control of *Orobanche*. However, the price of polyethylene is more than 500 US dollars ha^{-1} and, as a result, its use in farmers' fields is not economic.

Studies on Integrated Control in Egypt

Research on integrated control in Egypt is mainly conducted on cool season food legumes within the Nile Valley Regional Program (NVRP). Various "control packages" were tested consisting of the faba bean cultivar Giza 402, various amounts of fertilizer, application of various concentrations of the herbicide glyphosate, and different seeding rates (Nassib *et al.*, 1991a,b). Subsequently, the "packages" which yielded the most promising results were included in "demonstration plots" in farmers' fields. In the season of 1990–1991 two "packages" were recommended for faba bean. The first consisted of the cultivar Giza 402, a seeding rate of 184.5 kg ha^{-1} to obtain 27 to 30 plants m^{-2}, 35.7 kg N fertilizer and 71.4 kg P$_2$O$_5$, and 85 g a.i. glyphosate ha^{-1}. In the second package the concentration of glyphosate was reduced to 46 g a.i. ha^{-1}

and fertilizer was applied in the form of a foliar application of a 1% solution of N, P, and K. Generally, the second "package" was the most effective in decreasing *Orobanche* infestation. Imidazolinones also have been tested in Egypt but have not yet been included in integrated control programs (Tewfic *et al.*, 1991; Zahran *et al.*, 1991).

The faba bean line Giza 402 was identified in the early seventies by Egyptian scientists. Initially, it proved markedly tolerant to *Orobanche* (Nassib *et al.*, 1982). However, tolerance proved inconsistent and ultimately a rather limited tolerance (Parker, 1991). Fortunately, new lines have been derived from Giza 402, both in Egypt and Spain, which are tolerant or even partially resistant (S. A. Khalil, pers. comm.; M. T. Moreno, pers. comm.).

Studies on Integrated Control in Spain

In Spain, effective control of *O. crenata* in faba bean could be achieved by combining delayed sowing with preemergence as well as postemergence herbicide application (García-Torres and López-Granados, 1991). Preemergence treatments included imazethapyr at 75 to 100 g active ingredient ha^{-1} or imazapyr at 25 g a.i. ha^{-1}. Postemergence treatments included 60 g a.i. ha^{-1} glyphosate or 40 g a.i. ha^{-1} imazethapyr. When sowing was not delayed *Orobanche* could be effectively controlled when a preemergence application of imazethapyr or imazapyr preceded a single low rate post-emergence application of glyphosate. Pea, chickpea, and lentil appeared tolerant to 100 g a.i. ha^{-1} of imazethapyr and 25 g a.i. ha^{-1} of imazapyr applied preemergence of the crop. Consequently, there is also a potential for a combination of delayed sowing and low herbicide application in these crops.

Potential for Biological Control

The most promising agent for biological control of *Orobanche* is the fly *Phytomyza orobanchia* Kalt. (Agromyzidae) (Linke *et al.*, 1992). The fly is monophagous and its larvae feed in the stem as well as in the fruit capsules of the *Orobanche* species. Most research on the potential of this fly to control *Orobanche* has been conducted in Eastern Europe (Mihajlovic, 1986). In the off-season pupae of *Phytomyza* survive in seed capsules or stems of broomrape. Consequently, the development of *Phytomyza* is affected to a large extent by cultural methods, such as tillage, crop rotation and pesticide application. In this context, it should also be noted that *Phytomyza* is highly parasitized by parasitic insects. Klyueva and Pamukchi (1982) suggested the collection of *Orobanche* shoots infested with *Phytomyza* pupae at the end of the season. After storage under suitable laboratory conditions, shoots could be transferred to fields at the time of *Orobanche* emergence in the following season.

Recently, a survey was conducted in Syria (Linke *et al.*, 1990) to assess the occurrence of *Phytomyza*. The survey covered 31 faba bean fields, 21 of which were infested by *O. crenata*. The fly could be found in 95% of the fields infested with *Orobanche*. Approximately 32% of all capsules examined were parasitized. It was estimated that approximately 30% of the *Orobanche* seed were destroyed due to *Phytomyza* infestation. It was observed that *O. aegyptiaca* was also parasitized. In addition, approximately 5% of the capsules were affected by fungi.

Various fungi have been reported to occur on *Orobanche* of which *Fusarium* spp. and *Ulocladium* spp. are examples (Al-Menoufi, 1986; Linke *et al.*, 1992). Whether or not fungi may be used for controlling *Orobanche* is largely dependent on soil moisture (Kasasian, 1971). In Bulgaria, it appeared that *O. cumana* infestation of sunflower could be substantially increased by artificial inoculation of the soil with *Fusarium* spp. (Bedi and Donchev, 1991).

Concluding Remarks

If looked upon from a practical as well as an environmental point of view, it appears that a combination of slightly delayed sowing and low concentrations of herbicides from the imidazoline family and/or glyphosate are the most promising means for integrated control of *Orobanche* in food legumes, at least as far as things stand at present. If the newly derived tolerant lines of faba bean can be introduced at the farm level, these lines may also be included in the integrated control program.

For several years, glyphosate has been studied extensively. It is generally considered a very safe herbicide, particularly as very low concentrations are used for *Orobanche* control. The fact that it can only be used in faba bean is a disadvantage as the other food legumes are very susceptible to this herbicide. However, also in faba bean there are limitations to the use of glyphosate (García-Torres *et al.*, 1991). Faba bean tolerance is relatively marginal, largely depending on the crop growth stage at the time of application.

Imidazolinones, on the other hand, may also be used for *Orobanche* control in food legumes other than faba bean, both pre-emergence and postemergence. This family of new herbicides is active in the soil and is translocated in the plants through the phloem. They are inhibitors of acetohydroxyacid synthase (AHAS). Consequently, the branched-chain amino acid synthesis is blocked, causing a disruption of protein synthesis which in turn affects DNA synthesis and cell growth. Although the imidazolinones are relatively persistent in soil, environmental problems seem negligible, as concentrations used for *Orobanche* control are very low. In many countries, however, these herbicides have not yet been registered for *Orobanche* control. In this context, it should also be taken into consideration that in the USA resistance to sulfonylureas has rapidly developed in various weed species (Primiani *et al.*, 1990). This group of herbicides also brings about its effect via the branched-chain amino acid synthesis and plants resistant to sulfonylureas appeared cross-resistant to imidazolinone herbicides. It is conceivable that resistance to imidazolinones may also develop in *Orobanche* species. In this respect, the substantial genetic variability in *Orobanche* could play an important role.

Application of fertilizers in combination with chemical and cultural control methods, as practiced in demonstration plots in Egypt, could help to reduce *Orobanche* infestation. However, the effect of fertilizer under field conditions must be investigated in more detail. It may depend on timing, means of application, soil type, and weather conditions. Recently, van Hezewijk *et al.* (1991) reported that application of ammonium sulfate in the field in Syria resulted in a reduction of the number of *Orobanche* shoots at harvest time by 21 to 34% (one application of 28 kg N ha^{-1}) and 34 to 46% (two applications of 28 kg N ha^{-1}). This effect may be connected with an inhibiting effect of ammonium-N on seed germination of *Orobanche*, originally observed *in vitro* by Pieterse in 1981 (cit. from Pieterse, 1991). Penetration of ammonium into the soil, immediately after application, seems to be important. In a comparable field experiment conducted near Alexandria in Egypt, no significant effect of ammonium sulfate on *O. crenata* infestation in faba bean could be observed (van Hezewijk *et al.*, unpublished).

As far as biological control is concerned, it may be concluded that additional research is required. The fly, *Phytomyza orobanchia*, has the best prospects. Although the life cycle has not been studied in detail in the Middle East, it may be assumed in view of the research in Eastern Europe (Mihajlovic, 1986) that the fly will depend on *Orobanche* shoots to survive the summer season. As the main objective of the control programs is to entirely eradicate *Orobanche*, development of shoots in farmers' fields to maintain a *Phytomyza* population seems less practical. This applies in particular to the fact that *Orobanche* produces large numbers of seed and *Phytomyza*, even in dense populations, seems incapable of reducing seed formation by more than 30% (Linke *et al.*, 1990). It will not be very difficult for farmers to collect *Orobanche* shoots, store them in summer, and subsequently hang the shoots in the field in the next season. Contamination with *Orobanche* seed, however, could be risky and if this procedure is followed, special care should be taken to remove and eradicate the seed. From Russian studies (Klyueva and Pamukchi, 1982), it is also conceivable that pupae of *Phytomyza* could be made available to farmers. However, this would require a large scale organization to rear the insect, which at present is not available.

Rotation with trap crops as part of integrated control programs is always advisable as it will reduce the number of seed in the soil. If cultivation of the trap crop is practiced in consecutive seasons, the effect may be substantial. In this respect, Al-Menoufi (1991) reported that infestation of faba bean by *O. crenata* decreased by approximately 95% when faba beans followed three to four successive annual crops of Egyptian clover (*Trifolium alexandrium* L.). Rotation with flooded rice has also been reported to decrease *Orobanche* (Sauerborn and Saxena, 1986). In Egypt, however, cultivation of flooded rice is only permitted in certain areas, due to water distribution regulations.

Hand-pulling of *Orobanche* shoots before seed has set also could be included in integrated control programs. However, hand-pulling is very time-consuming and would only be economic in fields with a very low *Orobanche* infestation.

Acknowledgements

Part of the research discussed in this paper was financed by the European Community (EC) under EUR-contract number TS2-0124-C(GDF).

References

Abu-Irmaileh, B. E. 1979. In: *Proceedings of the 2nd International Symposium on Parasitic Weeds*, pp. 278–283 (eds. L. J. Musselman, A. D. Worsham and R. E. Eplee). Raleigh: North Carolina State University.

Al-Menoufi, O. A. 1986. *Alexandria Journal of Agricultural Research* 31: 297–310.

Al-Menoufi, O. A. 1991. In: *Progress in Orobanche Research*, pp. 241–247 (eds. K. Wegmann and L. J. Musselman). Tübingen: Eberhard-Karls-Universität.

Bedi, J. S. and Donchev, N. 1991. In: *Proceedings of the 5th International Symposium of Parasitic Weeds*, pp. 76–82 (eds. J. K. Ransom, L. J. Musselman, A. D. Worsham and C. Parker). Nairobi: CIMMYT.

Ciccarone, A. and Piglionica, V. 1979. In: *Some Current Research on Vicia faba in Western Europe*, pp. 87–102 (eds. D. A. Bond, G. T. Scarascia-Mugnozza and M. H. Poulsen). Luxembourg: CEC.

García-Torres, L. and López-Granados, F. 1991. *Weed Research* 31: 227–235.

García-Torres, L., López-Granados, F. and Saavedra, M. 1991. In: *Progress in Orobanche Research*, pp. 200–208 (eds. K. Wegmann and L. J. Musselman). Tübingen: Eberhard-Karls-Universität.

Jacobsohn, R., Greenberger, A., Katan, J., Levi, M. and Alon, H. 1980. *Weed Science* 28: 312–316.

Kasasian, L. 1971. *Weed Control in the Tropics*. London: Leonard Hill.

Klyueva, M. P. and Pamukchi, G. V. 1982. *Zaschita Rastenii* 1: 33–34 (in Russian).

Kukula, S. T. and Masri, H. 1984. In: *Proceedings of the 3rd International Symposium on Parasitic Weeds*, pp. 256–261 (eds. C. Parker, L. J. Musselman, R. M. Polhill and A. K. Wilson). Aleppo: ICARDA.

Linke, K. H. 1992. *Biology and control of Orobanche (special project): Final report*. Collaborative research project between ICARDA, Syria and the Department of Agro-ecology, University of Hohenheim, Germany (In press).

Linke, K. H., Sauerborn, J. and Saxena, M. C. 1992. In: *Proceedings of the Eighth International Symposium on Biological Control of Weeds* (eds. E. S. Delfosse and R. R. Scott). Melbourne, Australia: DSIR/CSIRO.

Linke, K. H. and Saxena, M. C. 1991. In: *Progress in Orobanche Research*, pp. 248–256 (eds. K. Wegmann and L. J. Musselman). Tübingen: Eberhard-Karls-Universität.

Linke, K. H., Vorländer, C. and Saxena, M. C. 1990. *Entomophaga* 35: 116–122.

Mesa-García, J. and García-Torres, L. 1986. *Weed Science* 34: 544–550.

Mihajlovic, L. 1986. In: *Proceedings of a Workshop on Biology and Control of Orobanche*, pp. 118–126 (ed. S. J. ter Borg). Wageningen: LH/VPO.

Nassib, A. M. and Hussein, A. H. A. 1989. *FABIS Newsletter* 24: 11–15.

Nassib, A. M., Hussein, A. H. A. and El-Deeb, M. A. 1991a. In: *Nile Valley Regional Program on Cool Season Food Legumes, Proceedings of the Third Annual Regional Meeting*, pp. 9–16. Giza, Egypt: Agricultural Research Center.

702   *A.H. Pieterse* et al.

Nassib, A. M., Hussein, A. H. A. and El-Rayes, F. M. 1984. *FABIS Newsletter* 10: 11–15.
Nassib, A. M., Ibrahim, A. A. and Khalil, S. A. 1982. In: *Faba Bean Improvement*, pp. 189–206 (eds. G. Hawtin and C. Webb). The Hague: Martinus Nijhoff.
Nassib, A. M., Saber, H. A., Farrag, H. M. and Seada, A. Y. 1991b. In: *Nile Valley Regional Program on Cool Season Food Legumes, Proceedings of the Third Annual Regional Meeting*, pp. 45–51. Giza, Egypt: Agricultural Research Center.
Parker, C. 1991. *Crop Protection* 10: 6–22.
Parker, C. and Wilson, A. K. 1986. *FAO Plant Protection Bulletin* 34: 83–98.
Pieterse, A. H. 1979. *Abstracts on Tropical Agriculture* 5: 9–35.
Pieterse, A. H. 1991. In: *Progress in Orobanche Research*, pp. 115–124 (eds. K. Wegmann and L. J. Musselman). Tübingen: Eberhard-Karls-Universität.
Primiani, M. M., Cotterman, J. C. and Saari, L. L. 1990. *Weed Technology* 4: 169–172.
Raaimakers, D., Raaijmakers, J., Ter Borg, S., Nassib, A. M. and Pieterse, A. H. 1988. *FABIS Newsletter* 22: 33–39.
Sauerborn, J. and Saxena, M. C. 1986. In: *Proceedings of a Workshop on Biology and Control of Orobanche*, pp. 160–165 (ed. S. J. ter Borg). Wageningen: LH/VPO.
Sauerborn, J. and Saxena, M. C. 1987. In: *Parasitic Flowering Plants (Proceedings of the 4th International Symposium on Parasitic Flowering Plants)*, pp. 733–744 (eds. H. C. Weber and W. Forstreuter). Marburg: Philipps-Universität.
Tewfic, M. S., Yehia, Z. R., El-Wekil, H. R. and Saber, H. A. 1991. In: *Nile Valley Regional Program on Cool Season Food Legumes, Proceedings of the Third Annual Regional Meeting*, pp. 185–190. Giza, Egypt: Agricultural Research Center.
van Hezewijk, M. J., Linke, K. H., Verkleij, J. A. C. and Pieterse, A. H. 1991. In: *Proceedings of the 5th International Symposium of Parasitic Weeds*, pp. 470–483 (eds. J. K. Ransom, L. J. Musselman, A. D. Worsham and C. Parker). Nairobi: CIMMYT.
van Hezewijk, M. J., Pieterse, A. H., Saxena, M. C. and Ter Borg, S. J. 1987. In: *Parasitic Flowering Plants (Proceedings of the 4th International Symposium on Parasitic Flowering Plants)*, pp. 377–390 (eds. H. C. Weber and W. Forstreuter). Marburg: Philipps-Universität.
Wilhelm, S., Benson, L. C. and Sagan, J. E. 1958. *Plant Disease Reporter* 42: 645–651.
Zahran, M. K. 1970. In: *Proceedings of the Tenth British Weed Control Conference*, pp. 680–684. Oxford: ARC Weed Research Organisation.
Zahran, M. K., Hussein, A. H. A. and El-Deeb, M. A. 1991. In: *Nile Valley Regional Program on Cool Season Food Legumes, Proceedings of the Third Annual Regional Meeting*, pp. 79–85. Giza, Egypt: Agricultural Research Center.

Biotechnology and gene mapping

The potential of gene technology and genome analysis for cool season food legume crops: theory and practice

G. KAHL[1], D. KAEMMER[1], K. WEISING[1], S. KOST[1], F. WEIGAND[2] and M.C. SAXENA[2]

[1] Plant Molecular Biology, University of Frankfurt/Main, 6000 Frankfurt/Main, Germany, and
[2] Legume Program, ICARDA, P. O. Box 5466, Aleppo, Syria

Abstract

The potential of plant gene technology encompasses a multitude of different techniques ranging from the isolation of useful genes, their characterization and *in vitro* manipulation to the reintroduction of the modified constructs into target plants, where they are expressed at a rate that alters the phenotype of the plants. Genome analysis, on the other hand, aims at characterizing the genome architecture and function(s).

Plant gene technology has catalyzed progress in plant breeding, as will be exemplified by a few examples, but has not yet been applied to food legume improvement on a large scale. Genome analysis, however, has a series of practical implications, as is illustrated by the successful introduction of DNA fingerprint and PCR fingerprint techniques to chickpea (*Cicer arietinum* L.) breeding and *Ascochyta rabiei* pathotyping. The present overview addresses both areas of plant molecular biology to illustrate their potential for food legume breeding.

Introduction

This overview portrays the successes on two fields of up-to-date molecular biology of plants: gene technology and genome research. It details some aspects of genetic engineering of plants that promise future application also for the improvement of food legumes, and exemplifies successful applications with other crops that might guide research of legume scientists of whatever convention. A brief survey introduces only one aspect of plant genome analysis, that is presently being studied in a joint venture between the International Centre for Agricultural Research in the Dry Areas (ICARDA, Aleppo, Syria) and the Plant Molecular Biology Laboratory of the University of Frankfurt am Main (Germany).

F.J. Muehlbauer and W.J. Kaiser (eds.), Expanding the Production and Use of Cool Season Food Legumes, 705–725.
© 1994 Kluwer Academic Publishers.

Gene Technology

Basically, plant gene technology aims at isolating a gene, modifying either its regulatory regions or information content *in vitro*, and reintroducing the modified gene into a target plant using either natural vectors or direct gene transfer methodology. It also encompasses getting the expression of the transferred gene, preferably in a regulated mode, in its transgenic environment, and the exploitation of the gene product for pharmaceutical or medical use or, its effect on a desirable trait.

Gene Cloning

The *isolation of a gene* requires the isolation and purification of DNA, usually routine in most laboratories. Since the DNA possesses high molecular weight, and is therefore vulnerable to shearing forces or exo- and endonucleolytic breakdown, it has to be fragmented into sizable pieces by restriction with one or few of a series of about 600 available restriction endonucleases. These enzymes recognize specific DNA sequences and cut the recognition site at a specific position (or, in few cases, also distantly from the recognition sequence). The experimenter may use these fragments in different ways. The conventional start for gene isolation was (and still is) the establishment of a *genomic library* (or gene bank), comprising all the different genomic restriction fragments inserted into a *cloning vector*.

Cloning Vectors

Such cloning vectors have been designed in their hundreds (Winnacker, 1987; Sambrook *et al.*, 1989) and are principally DNA segments that carry a relatively short sequence at which replicative proteins attach and catalyze autoreplication (ori_v). In addition, they contain a number of useful genes (e.g., antibiotic resistance markers, that allow the easy selection of recombinant molecules, i.e., vectors harboring foreign DNA) together with a unique cloning site for the insertion of cloned DNA. Cloning vectors derived from bacterial plasmids usually tolerate insert sizes of up to 6 to 8 kb without loss of vital functions (Bolivar and Backman, 1979; Vieira and Messing, 1982). Such vector plasmids can be engineered to contain desirable functions for the experiment (e.g., express the cloned DNA in *E. coli* by in-frame insertion of a functional promoter sequence located immediately 5' upstream of the cloning site).

For the cloning of DNA fragments larger than 10 kb vectors derived from phage (λ vectors) were designed by simply removing the central stuffer fragment and ligating the foreign DNA onto the resulting arms (e.g., Frischauf and Lehrach, 1983). Basically two derivatives of the wild-type phage were produced that either possess a single target site at which foreign DNA can be inserted (insertional vectors) or a pair of restriction sites defining a fragment that can be removed and replaced by foreign DNA (replacement vectors). Since phage λ can accomodate only 5% more than its normal complement of DNA due to packaging constraints, λ vectors have been streamlined such that they tolerate up to 23 kb, which is close to the theoretical limit. In addition, most λ vectors of the new generation have convenient polylinker sequences (multiple cloning sites) flanking the replaceable fragment (e.g., EMBL 3 and EMBL 4). The presence of cohesive ends (cos sites) at both ends of the λ-DNA allows to concatemerize the DNA into long multimers, which *in vivo* (i.e., the infected bacterial host cell) are split by λ-terminase to yield the packageable λ-DNA.

Since only a small region in the proximity of the cos site is required for recognition by the packaging system, placmid cloning vectors have been designed with both cos sites, thus allowing the insertion of DNA segments of 32 to 47 kb in length (*cosmids*; Hohn and Collins, 1980; Wahl et al., 1987). For the cloning of still larger DNA fragments, other systems recommend themselves as efficient cloning tools (e.g., P1 plasmids and yeast artificial chromosomes, YAC vectors, reviewed by Schlessinger, 1990).

Colony or Plaque Hybridization

The detection of specific genes in such genomic libraries presupposes sequence information, either in the form of a probe (a DNA segment that can be used to screen libraries for homologous, i.e., identical or near-identical sequences, whose base sequence and in most cases identity is known) or a synthetic oligonucleotide that contains a consensus sequence of several genes from different sources encoding the same functional protein. Such probes are radioactively labelled (e.g., by incorporation of ^{32}P from α-labelled dNTP in conventional nick translation or random priming techniques; Sambrook et al., 1989) or non-radioactively (e.g., by the introduction of biotinylated or digoxigenated nucleotides) and hybridized to appropriately blotted genomic libraries (colony or plaque hybridization; Grunstein and Hogness, 1975). Genomic sequences with base homology can easily be found by *hybridization* followed by autoradiography in case of radioactive labelling, or antibody-mediated enzymatic detection in case of most nonradioactive techniques.

Cloning into vectors ensures the identity of each genomic fragment (in most cases) and allows its amplification by either normal plasmid replication or phage infection and phage replication. More recently, cloning techniques have been complemented by non-cloning methods, that presuppose base sequence information of the desired gene (or DNA sequence generally), "sequence-tagged sites" (Olson et al., 1989). Such information can be transformed into the synthesis of primers, short oligodeoxynucleotides recognizing sequences at both ends of the desirable gene and located on opposite strands. These primers hybridize to their target sequences and can be used to start a DNA polymerase-catalyzed primer extension in the *polymerase chain reaction* (PCR; Saiki et al.,

1988). The resulting products again offer primer recognition sites and are amplified the same way, so that after 30 to 35 such cycles the original target sequence has been amplified 10^8 fold and can be detected by simple ethidium-bromide staining procedures. Thermostable DNA polymerases (e.g., from *Thermus aquaticus, Taq* polymerase, *Pyrococcus furiosus, Pfu* polymerase, or a DNA polymerase peptide, the Stoffel fragment) allow to cycle at optimal temperatures for each step (denaturation of the strands: 94°C; annealing of primers: 37°C; primer extension: 72°C) without replacing the enzyme. The amplified fragment can be used for further characterization (e.g., sequencing) directly, or also be cloned.

Chimeric Genes

The ultimate characterization of a gene is by sequencing and by testing its function. *DNA sequencing* is now routine in most laboratories and generates data that can best be handled with the aid of appropriate computer programs that allow e.g., the establishment of restriction maps, the detection of open reading frames as well as the search for regulatory promoter elements (e.g., TATA- or CAAT-boxes, enhancer and silencers elements). The precise sequence information then allows the coding part of a gene to be dissected from the promoter. Both parts can then be used separately. For example, if the characterized gene is strongly expressed (i.e., is driven by a strong promoter,

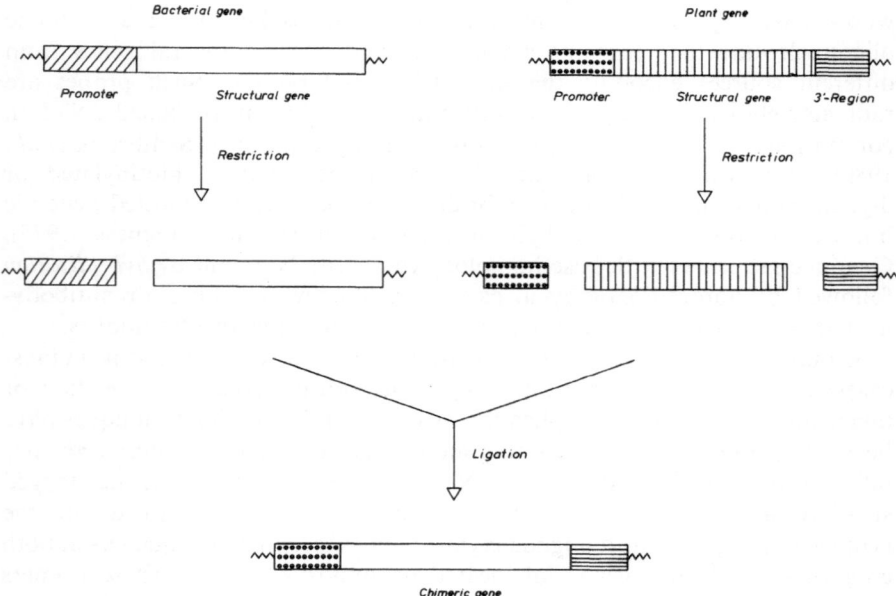

Figure 1. The construction of a chimeric gene (a scheme).

that allows frequent formation of a transcription initiation complex) in the organism of origin, then the promoter can be linked to any foreign gene to ensure its high transcription rate in the original plant. Expression of such genes in a transgenic environment is dependent on two basic steps: first, the construction of functional *chimeric genes*, and the transfer of such genes into a target plant, that becomes a *transgenic plant* by the acquisition of the chimeric gene.

Chimeric genes are hybrid sequences, consisting of e.g., the promoter from one gene, the coding part of another gene, and the 3' terminator sequence from a third gene (Figure 1). Or, alternatively, chimeric genes may be constructed from the promoter of a gene from organism A, the coding sequence of a gene from organism B, and the 3' terminator region from a gene of organism C. The regulatory sequences contained within such chimeric constructs are usually tested for functionality in a *transient expression assay* (Werr and Lörz, 1986; Töpfer *et al.*, 1988), in which the construct is transformed into protoplasts of a favorite plant (suitably the desired host organism) and assayed for expression over several days. Expression assays are facilitated by the inclusion of "reporter" genes which code for enzymes with easily screenable activities (e.g., the widely used β-glucuronidase reporter gene system; GUS system; Jefferson *et al.*, 1987). Usually, the expression is transient, because the chimeric DNA is *not* incorporated into the recipient's genome, but only present in the nucleoplasm where it is accidentally transcribed, but also destroyed exo- and endonucleolytically. If the promoter has proven to be functional in the target plant, it can be ligated to any coding sequence that can then be expressed once transformed into suitable plant systems. *Plant transformation* that leads to the stable incorporation of the foreign DNA into one of the genomes of a cell (e.g., nuclear, plastid or mitochondrial genome) derives from a naturally occurring transfer process, mediated by *Agrobacterium tumefaciens*, or an artifical process designed by the experimenter, coined *direct gene transfer*.

Plant Gene Transfer

Though it has long been known as a plant pathogen, *Agrobacterium tumefaciens* attracted the attention of molecular biologists only after the Ti (tumor-inducing)-plasmid has been identified as the disease-causing agent (Van Larebeke *et al.*, 1974). Part of this plasmid, the T-region, is excised at its border sequences of some 25 bp, packaged with proteins and transferred to plant cells with which the bacterium came into contact. The molecular basis of this natural gene transfer is now well established (reviewed by Kahl and Schell, 1982; Nester *et al.*, 1984; Zambryski, 1988; Zambryski *et al.*, 1989; Kado, 1991) and is outlined in Figure 2.

The bacterium approaches the plant rhizosphere on a gradient of excreted sugar compounds and attaches to wound-exposed cell walls after a positive chemotactic movement that is triggered by plant phenolics and sugars. After its

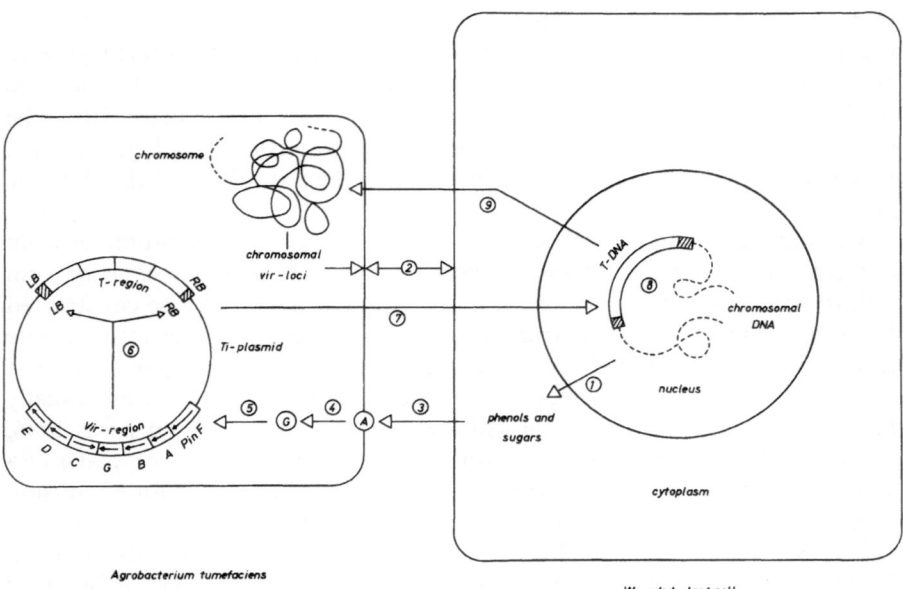

Figure 2. A simplified scheme of *Agrobacterium*-mediated gene transfer. Wounded plant cells synthesize phenolic compounds as part of their defense and wound-healing strategies. These phenols serve as attractants for the ubiquitously occurring soil bacterium *Agrobacterium tumefaciens* (1) that approaches the wound site via positive chemotaxis. The bacterium attaches to the wound-exposed plant cell wall (2). The phenols are recognized by a bacterial membrane protein encoded by gene A of the virulence (vir) region of the Ti-plasmid (3). Recognition involves autophosphorylation and phosphorylation of a second vir protein (VirG), (4) that acts as a DNA-binding protein (5) and activates the promoters of all vir genes. The product of virD is a site-specific endonuclease that nicks the bottom strand of the T-region (6) at the left and right border sequences (LB, RB). The excised T-strand is packaged into virE proteins and piloted by virD proteins into the plant cell nucleus (7). The T-strand is covalently integrated into the host plant genome (8) and expressed. The constitutive transcription of T-DNA genes 1, 2, and 4 leads to the accumulation of auxins and cytokinins that incite permanent proliferative growth (tumor formation). The product of the ocs and nos genes of the T-DNA, opine synthases, catalyze the formation of opines in the host cells. These compounds are secreted (9) and serve as carbon, nitrogen and energy source for the parasitic bacterium.

firm attachment to the plant cell wall, a cascade of chemical reactions takes place. First, the membrane-bound virA protein, encoded by gene A of the virulence (vir) region of the Ti-plasmid, forms a complex with one of the phenolics synthesized by the wounded plant cell. As a consequence, virA phosphorylates virG protein that in turn becomes a gene-activating vir-gene promoter-binding protein. The induction of all vir genes by virG leads to the accumulation of an endonuclease (virD protein) that recognizes the border repeats and introduces a single-stranded nick in the bottom strand and/or double-strand breaks. The bottom strand (T-strand) is covalently attached to virD2 protein, and also binds to single-strand-specific proteins encoded by virE.

The virus-like T-complex is then piloted to the plant nucleus, partly a consequence of nuclear targeting sequences present in the virD2 protein (Herrera-Estrella *et al.*, 1990). The T-strand, or the double-stranded variant of it, is then covalently integrated into the nuclear genome of the infected plant cell. Since all the transferred genes, collectively called T-DNA, are transcribed from promoters with eukaryotic regulatory sequences, their expression within the plant cell leads to the action of enzymes changing its phenotype. The products of gene 1 and 2 are auxin-synthesizing enzymes, and the product of gene 4 is a cytokinin-synthesizing protein. The concerted action of these enzymes raises the auxin/cytokinin level such that a permanent proliferation ensues. The permanent mitotic activity of such transformed cells ultimately leads to a tumor (crown gall). This type of natural gene transfer has several advantages for the bacterium, e.g., the production of opines in the tumor cell that can only be metabolized by the bacterium and thus be exploited as carbon, nitrogen, and energy source.

Plant Transformation Vectors

Though the T-DNA contains genes that are needed for the synthesis of opines and the tumorous phenotype, the transfer process proceeds without them. In fact, only the right and left border sequences are required for precise excision of the T-strand and its integration into the plant cell's genome, provided the presence and action of the vir genes in the bacterium. Consequently, the T-DNA genes can collectively be deleted and replaced by foreign DNA (Zambryski *et al.*, 1983) which can make up some 40 to 50 kb, encoding some 10 genes. A prototype of *plant transformation vectors* (Figure 3) then no longer contains the oncogenes, but rather the border sequences. The removal of oncogenic functions in fact does no longer allow the formation of neoplastic outgrowths that are able to grow on a medium without phytohormones. Detection of the transformed state was made possible by the inclusion, within the borders, bacterial antibiotic resistance genes under the control of a eukaryotic promoter and a polyadenylation signal. Such chimeric resistance genes are dominantly expressed in any eukaryotic genetic background and serve to identify transformants, to prove the presence of closely linked genes, and to guarantee their expression (based on the assumption that the chromatin architecture would be the same for both types of genes).

Many non-oncogenic transformation vectors with different features have been constructed (reviewed by Weising *et al.*, 1988; Corbin and Klee, 1991). They fall into two broad categories, the *cis* and the *trans* vectors. *Cis* vectors contain both the border sequences and the vir region on the same replicon, whereas *trans* vectors are composed of two replicons, one carrying the borders, the other (helper plasmid) carrying the vir function that are supplied *in trans* (binary vector system). Usually such vectors additionally contain an easily screenable reporter gene, again controlled by a eukaryotic promoter that functions in plants.

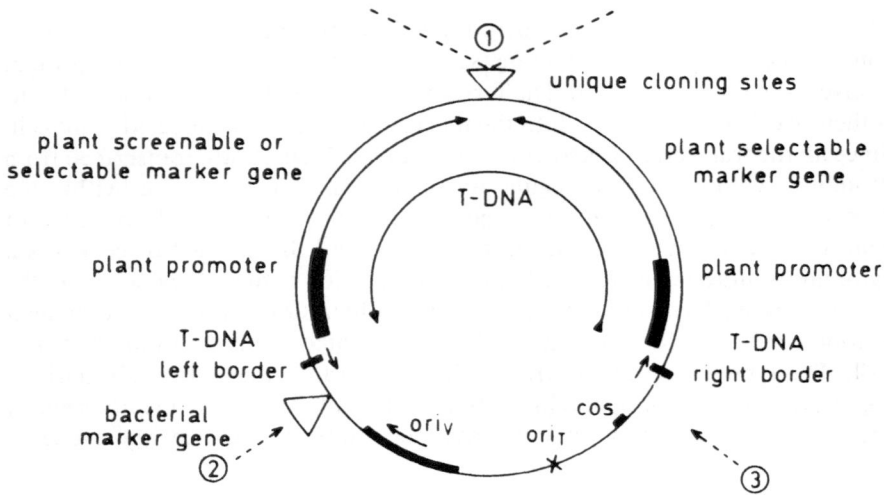

Figure 3. The prototype of a plant transformation and expression vector. Arrows 1, 2, and 3 indicate unique restriction endonuclease cleavage sites for the introduction of additional elements into the cassette. The direction of replication initiated at ori_V sequences, the extent of the modified T-DNA and the polarity of the T-DNA border sequences are denoted by arrows inside the circle. ori_V: origin of vegetative replication; ori_T: origin of conjugative transfer; cos: cohesive ends (cos sites)

The *Agrobacterium*-mediated gene transfer is relatively easy to handle. Usually, leaf disks or other explants of favorite plants are excised, incubated with engineered bacteria that transform some of the wounded cells which in turn have to be regenerated (leaf disk transformation; Horsch *et al.*, 1985). Consequently, the *Agrobacterium* system has successfully been used to introduce new genes into a large variety of plants (Weising *et al.*, 1988; Corbin and Klee, 1991). However, the limited host range of *Agrobacterium tumefaciens* (e.g., monocotyledonous plants such as corn, the cereals, and sugar cane are no natural hosts; De Cleene, 1985) prevents its general applicability for plant transformation.

Direct Gene Transfer

Therefore, a series of alternative techniques collectively called *direct gene transfer* or vectorless gene transfer methods, have been developed which are either based on physical (i.e., mechanical) or chemical means to transform plant cells (reviewed by Potrykus, 1991; Rogers, 1991). The transforming DNA is introduced into target protoplasts, cells, tissues, organs, or whole plants by a series of chemical treatments (e.g., the use of membrane-destabilizing agents such as polyethylene glycol or DNA precipitation by calcium phosphate) or by electrical forces (e.g., by electroporation or electrophoresis), particle bombardment ("biolistics", which uses DNA-coated tungsten or gold particles as vectors, that are driven through target tissues by mechanical force and release their DNA during their penetration of the tissue; reviewed by Christou, 1992), or microinjection (in which the DNA is injected directly into the cell or its organelles, preferrably together with a fluorochrome to monitor the injection process). A series of other vectorless gene transfer techniques have been introduced (e.g., laser microbeam techniques, silicon carbide fiber-mediated transfer, liposome fusion and liposome injection, packaging of the DNA before transfer, sonication loading, simple incubation of cells, tissues and organs in DNA, macroinjection). The availability of such diverse techniques has tremendously catalyzed progress in the transformation of plant cells.

Though very successful, all these techniques do not yet allow to target the incoming DNA to specific regions of the genome, and attempts to introduce foreign DNA into the plant genome via homologous or site-specific recombination are still in their infancy (e.g., Paszkowski *et al.*, 1988; Odell *et al.*, 1990; Halfter *et al.*, 1992). Random integration produces a lot of unexpected problems. One of them, unpredictable expression rates of the transferred genes, is particularly severe. If, for example, the genes are being inserted into genomic sites, where the chromatin structure does not allow transcription, the genes will altogether be silent (position effect). It might also be, that local differences in chromatin architecture permit the expression of a reporter gene, but not of the adjacent gene of interest. Moreover, many copies of integrated genes suffer from mutations, deletions, inversions, truncations, or generally rearrangements, so that a positive proof of the physical presence of transferred sequences by e.g., Southern type experiments does not necessarily prove their integrity.

Despite such obstacles both the *Agrobacterium*-mediated and the vectorless gene transfer techniques have paved the way for a series of successes in plant genetic engineering (see e.g., Gasser and Fraley, 1989). Plants have been transformed with genes conferring resistance (better: tolerance) towards phytopathogenic viruses, bacteria and insects, towards herbicides and, most recently, towards fungi (Broglie *et al.*, 1991; Logemann *et al.*, 1992). Transgenic plants produce an improved complement of proteins and a series of proteins important in medical sciences (e.g., anti-bodies: Hiatt *et al.*, 1989; anti-AIDS drugs). The potential of these techniques will be briefly illustrated with *virus tolerance* and *insect resistance*.

Engineering Virus Resistance

A major goal of the plant breeder and plant pathologist is the production of plants with enhanced resistance towards pathogenic viruses. It has been known for some time that the infection of a host plant by a mild virus strain (the inducer) protects the plant from the superinfection by a second, more virulent but related virus (the challenger). This phenomenon has been coined *cross-protection* (reviewed by Sequeira, 1984; Fulton, 1986), but its mechanism of action has not been deciphered yet. One hypothesis stresses the importance of the presence of inducer RNA (the replicated genome of the mild virus) in the host cell, which hybridizes to the challenger's RNA and prevents either its replication or translation into protein. Cross-protection might also be a consequence of successful competition of the inducer virus for receptors at the plant cell surface, so that the challenger will not recognize the potential host. Yet another theory capitalizes on the abundance of coat proteins in the infected cell, which are synthesized late in the viral replication cycle. These coat proteins, normally used for the assembly of viral RNA into infective particles, now assemble with the incoming RNA of the challenger virus and prevent its replication. As a consequence, the superinfection is impossible or delayed or inefficient.

The first successful demonstration of genetically engineered cross protection towards a plant virus was reported by Powell Abel et al., (1986). The coat protein gene of the model virus TMV (tobacco mosaic virus) was isolated as a cDNA copy, inserted into a plant transformation vector together with appropriate promoter and polyadenylation signals, transformed into tobacco cells and constitutively expressed. Plants regenerated from the cells, otherwise normal, were either not infected or the disease symptoms developed more slowly than in control plants. Since this initial report, genetically engineered cross protection mediated by viral coat proteins has been successfully applied to a variety of crop species such as potato and tomato, and also against a wide spectrum of plant viruses belonging to different virus groups (reviewed by Beachy et al., 1990). The strategy has been shown to be effective under field conditions (Nelson et al., 1988), and against mixed infection by two different viruses (Lawson et al., 1990). Moreover, heterologous protection from related viruses has been repeatedly observed (e.g., Stark and Beachy, 1989; Nejidat and Beachy, 1990; Ling et al., 1991). Though the mechanism of coat protein-mediated cross-protection in transgenic plants is still unclear, and several lines of evidence point to additional mechanisms that work independently of coat protein accumulation (Golemboski et al., 1990; MacKenzie et al., 1991), the strategy certainly has a good perspective for legume crop improvement. It should be kept in mind, however, that the acquired tolerance can in some cases be overcome by an increased concentration of the virus, or the presence of naked viral RNA in the inoculum.

Engineering Insect Resistance

Insect-resistant plants were designed, exploiting plasmid-borne insecticidal genes from the bacterium *Bacillus thuringiensis*. If under environmental stress (e.g., surplus of oxygen, nutrient deficiency) the bacillus is forced to form an endospore, a propagative unit that is able to withstand unfavorable conditions. The bacterial mother cell synthesizes bulks of a specific protein that is necessary for the construction of the endospore cell wall. The surplus is deposited as a paracrystalline protein body. After the endospore is complete, the mother cell decays and releases both endospore and protein crystal. If an insect feeds on leaves of a plant, it also ingests the protein crystal. During the passage through the insect's midgut, the spore protein is cleaved either proteolytically or chemically by the alkaline environment to yield two fragments, one of which obviously forms a complex with specific receptors of the insect's brush border cells. The interaction of this crystal protein (termed δ-endotoxin), which consists of almost all of the amino terminal half of the precursor molecule, with the receptor molecule incites a series of reactions leading to the paralysis of the insect.

The insecticidal properties of *B. thuringiensis* (B.t.) crystal proteins (reviewed by Höfte and Whiteley, 1989) have been exploited for decades, because the lethal effect of the δ-endotoxin is limited to susceptible insects. This makes the toxin safer than chemical insecticides that poison a broad spectrum of organisms indiscriminately. However, insecticides based on B.t. own the disadvantage of being expensive and readily broken down under field conditions. One approach to circumvent these obstacles is the introduction of cloned toxin genes into bacteria that, either living or dead, persist longer in the environment than B.t. itself (reviewed by Feitelson *et al.*, 1992). Another strategy, the transfer of the B.t. toxin gene into the target plant that one wants to protect was first applied by Vaeck *et al.*, (1987). The δ-endotoxin gene, or a truncated version of it, has been cloned into a Ti-plasmid vector, transferred into *Agrobacterium tumefaciens* and subsequently into tobacco plants. These plants appeared to be normal, except that they expressed the B.t. toxin gene. As a consequence, δ-endotoxin proteins were produced that conferred resistance towards larvae of the tobacco hornworm, *Manduca sexta*. In the greenhouse, *Manduca* larvae were released on selected tobacco plants, both wild-type (control) and transgenic. Control plants suffered appreciable damage within a week and were completely consumed after two weeks. On the transgenic plants, however, all the larvae died within few days, and damage of the plants remained minimal.

Using the B.t. endotoxin strategy, a series of transgenic plants with relatively specific insecticidal properties was produced (e.g., tobacco: Barton *et al.*, 1987; Vaeck *et al.*, 1987; tomato: Fischhoff *et al.*, 1987; cotton: Perlak *et al.*, 1990; potato: Chen *et al.*, 1992). With tomato, δ-endotoxin-mediated insect tolerance was also shown to be effective under field conditions (Delannay *et al.*, 1989). An additional advantage of the B.t. system, the availability of may be 1000 different B.t. strains with different toxin genes and different toxins directed towards different spectra of not only insect, but also nematode, trematode, mite, and protozoan species (reviewed by Feitelson *et al.*, 1992) makes it very attractive to engineer pest-resistant crops. Several problems, however, still remain. First, the expression level of endotoxin genes is usually low (Vaeck *et al.*, 1987; Perlak *et al.*, 1990), and insect-tolerant plants will be of value only if they produce sufficient δ-endotoxin to provide consistent plant protection under field conditions. Therefore, modified and/or truncated versions of the endotoxin genes will have to be tested with different crop/pest combinations in the field. Second, the problem of developing resistance of the insect towards the toxin will have to be assessed. To that end, the presence of two or more δ-endotoxin genes from different bacterial strains in the transgenic plant will increase the selection pressure on the insect. Third, the expression of the foreign genes should be only possible after the insect starts feeding on the plant. The use of wound-induced promoters (e.g., Logemann *et al.*, 1992) would therefore be a choice.

Perspectives

All these different strategies, though highly successful with various crops, have not yet been much used for the improvement of food legumes. A major reason for this is the fact that legume species have proven to be less amenable to transformation/regeneration procedures than most other dicotyledonous crop species. Despite these difficulties, however, transgenic plants have been obtained from several leguminous crop species, i.e., soybean (Hinchee *et al.*, 1988; McCabe *et al.*, 1988; Chee *et al.*, 1989), moth bean (Köhler *et al.*, 1987), pea (Puonti-Kaerlas *et al.*, 1990), and alfalfa (Deak *et al.*, 1986; Pezzotti *et al.*, 1991). There is certainly a need to introduce the whole plant gene technology in legume breeding programs, in particular the engineering of virus-, bacteria-, fungus-, and insect-resistance into commercially important crops such as pea, bean, faba bean, lentil, and chickpea.

Genome Analysis

Another set of techniques that have been developed for basic research, have turned out to be highly promising for plant breeding, plant systematics and evolution, plant population biology, plant taxonomy, and plant pathology as well. The techniques altogether were applied to problems of gene and genome structure in the past, but some of them are now increasingly used in the field of agriculture. This chapter briefly reviews two such methods, DNA fingerprinting and DNA amplification fingerprinting, and their potential in chickpea breeding and the control of Ascochyta blight (incited by *Ascochyta rabiei*) disease.

DNA Fingerprinting

DNA fingerprinting exploits the presence of tandem-repetitive DNA within most eukaryotic genomes (see Burke *et al.*, 1991 for reviews). The function of these repetitious sequences is as yet obscure. Tandem-repetitive DNA falls into three broad categories: satellite DNA (repetitive units of more than 50 bp in length), minisatellite DNA (repetitive units of 10–40 bp), and microsatellite DNA with even shorter units. The potential of fingerprinting techniques for various areas of plant breeding will be exemplified with the microsatellite sequences. Microsatellites are ubiquitously present in eukaryotic genomes (Tautz and Renz, 1984; Weising *et al.*, 1991a). They consist of short, tandemly arranged sequence motifs of two to ten base pairs in length ("simple sequences", simple repetitive sequences, SRS) and form more or less monotonous stretches of variable extensions. Such simple sequences represent an extensive pool for genetic variation that is mostly based on different copy numbers of one fundamental core motif (variable number of tandem repeats, VNTR), less so on internal sequence heterogeneity. The genetic variation can be detected by probes complementary to simple sequences. The probes are radioactively or non-radioactively labelled and hybridized to electrophoretically separated fragments of restricted genomic DNA. Restriction fragments carrying simple sequences are then discovered by autoradiography, or by antibody-mediated enzymatic assays. Since the probes are usually short (10 to 15 bp), they can also hybridize to DNA sequences within a dried agarose gel thus circumventing the need for Southern-type capillary blotting procedures.

DNA Fingerprinting of plants

A series of synthetic oligodeoxynucleotides complementary to SRS have been introduced to chickpea DNA fingerprinting (e.g., CA-, CT-, GATA-, GACA-, GAA-, GTG-, GGAT- and TCC-multimers). All these sequences are present in the chickpea genome, most of them are informative, and some even

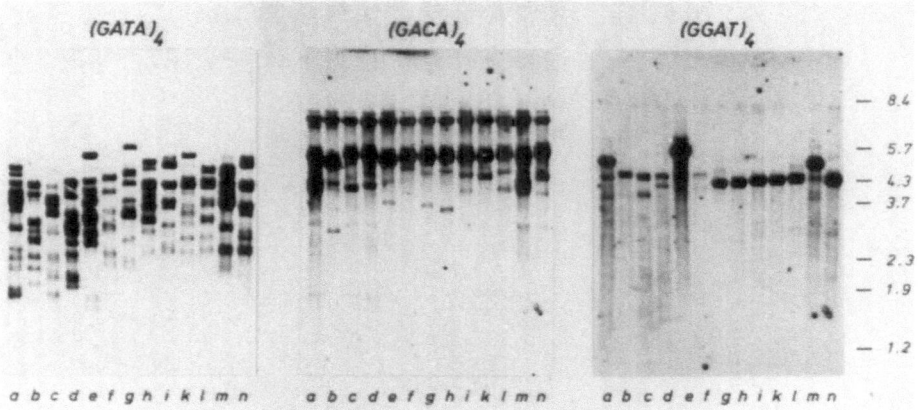

Figure 4. Screening of thirteen different chickpea accessions for genetic polymorphisms with ^{32}P-labeled oligonucleotide probes. Total DNA was purified from leaves of individual plants and digested with Alu I. After electrophoretic separation of the resulting fragments in 1.2% agarose gels (5 ug DNA per lane), the DNA was denatured, neutralized, the gel dried, and consecutively hybridized to the indicated oligonucleotide probes. Molecular weight markers are given in kb.

hypervariable (Weising *et al.*, 1992). For example, hybridization of (GATA)$_4$ and (GACA)$_4$, to Taq I- or Alu I-digested chickpea DNA reveals highly variable intraspecific patterns (Figure 4), whereas (GTG)$_5$ is rather uniformly distributed in the genomes of various accessions. Thus, (GTG)$_5$ creates a species-specific fingerprint: patterns obtained with other species of the genus *Cicer* look different. In essence, different oligonucleotide probes exhibit different levels of informativeness within one and the same plant species, and the usefulness of a particular probe has to be determined empirically.

Not only species, accessions and varieties, but also individuals can sometimes be distinguished by *oligonucleotide fingerprinting*. Individual plants belonging to the same chickpea accession can be readily specified by (GATA)$_4$ and (CA)$_8$, where individual differences are but slight and detected by individual-specific differences in electrophoretic mobility (Figure 5). It seems likely that slight variations in the number of tandemly repeated simple sequence motifs among

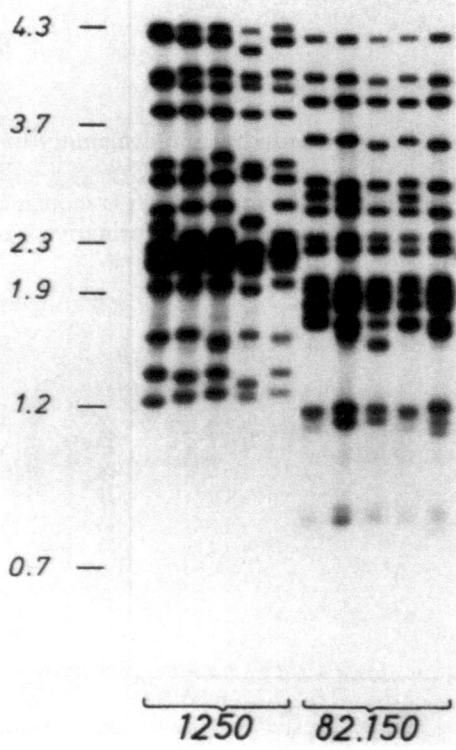

Figure 5. Variability of (GATA)$_4$-fingerprints between individuals derived from the same accession. 1250: Five individuals from accession ILC 1250. 82–150: Five individuals from accession FLIP 82–150. Total DNA was purified from leaves of individual plants and digested with Taq I. For electrophoresis and hybridization conditions see legend of Figure 4 (3 ug of DNA per lane). Molecular weight markers are given in kilobases.

alleles at some loci cause this effect. It is obvious that oligonucleotide fingerprinting can be of use also in backcross breeding programs, by allowing to monitor gene introgression with the aim to select offspring individuals with maximal similarity to the recipient line. We expect to reduce the number of backcross generations by fingerprinting.

In addition to its application for the identification of species, varieties, cultivars, and accessions of *Cicer* (with direct implementation to protect breeder's rights), oligonucleotide fingerprinting is being used for such diverse problems as the molecular characterization of the extent of genetic variation between and within landraces and cultivars, the evaluation of the extent of inbreeding, the detection of genetic relationships among and within populations, the molecular identification of cell hybrids and the characterization of somaclonal variants. We have already applied DNA fingerprinting with microsatellites successfully in all these fields with a series of plants of the temperate regions (e.g., rape, tomatoes), the dry areas (e.g., chickpea, lentil), subtropics and tropics (e.g., banana and plantain: Kaemmer *et al.*, 1992a, yams, cocoa). Presently our laboratories are in the process of isolating multiallelic, single-locus probes from chickpeas. Such sequences will allow linkage analysis and genome mapping, both strategies aiming at the characterization and isolation of genes conferring tolerance (or also susceptibility) towards Ascochyta blight disease. Basically, two techniques are being employed, both finally leading to sequences flanking hypervariable microsatellite loci and their molecular characterization. Once such single-locus sequences are known, they can also be exploited as locus-specific primers for the amplification of adjacent DNA in the polymerase chain reaction.

DNA Fingerprinting of Pathogenic Fungi

The fingerprint approach has also been successfully applied to mycological problems (see Kaemmer *et al.*, 1992b, for a review). For example, the genetic identity of *Ascochyta rabiei* isolates has so far been obscure, although much work has gone into characterizing their virulence. We have genotyped six Syrian isolates of *Ascochyta rabiei*. Four out of the six single-spored isolates were clearly distinguished by oligonucleotide fingerprinting (Weising *et al.*, 1991b; Kaemmer *et al.*, 1992b). The fingerprint patterns are stable during culture of the fungus, and allow discrimination among fungal races. Our future research follows two lines. First, the fungal genome will be characterized in some detail, using such diverse techniques as pulse-field gel electrophoresis (PFGE), *in situ*-hybridization, and linkage studies, which should allow isolation of pathogenesis genes in the future. Second, DNA fingerprinting is already increasingly employed to establish geographic distribution maps of some countries (e.g., Algeria, Morocco, Syria, Tunisia, and Turkey), that detail the occurrence of the diverse fungal races. They will also assist in detecting the genetic drift occurring within *Ascochyta rabiei* populations. To summarize, though at a very preliminary stage, the results obtained so far show that oligonucleotide fingerprinting is a potent technique for exploring the genome of *Ascochyta rabiei*.

Radioactive Versus Non-Radioactive Detection

One of the most severe limitations against a generalized use of autoradiographic detection of DNA fingerprints in developing countries, the absence of isotope laboratories, will certainly be overcome soon. Non-radioactive detection of fingerprint patterns is now routine in our laboratories. The two most prominent techniques are based on enzymatic reactions. The *colorigenic method* visualizes hybridizations between a genomic target sequence and a non-radioactively labelled probe by employing a colorless substrate that is converted to a dye by an enzyme covalently linked to an antibody, specifically raised against the label. Labelling most conveniently is performed by the introduction of an antigenic biotin or digoxigenin molecule by nick translation, random priming or oligo-labeling. The biotin, for example, is coupled to a nucleotide by an allylamine spacer of various length (e.g., varying from 4 carbons in bio-4-dUTP to 17 in bio-17-dATP). This nucleotide is then incorporated into the probe sequence by DNA polymerase instead of dTTP. The detection mechanism is based on a high affinity reaction of avidin or streptavidin towards biotin (Kd = 10^{-15}M). Usually horseradish peroxidase is conjugated to avidin (or streptavidine), and the conversion of its substrate into a colored product allows detection of the probe (i.e., the hybridization event). The antigenic digoxigenin, an alkaloid from *Digitalis purpurea*, in turn is coupled to dUTP (as e.g., digoxigenin-11-dUTP) and as such incorporated into the probe via the Klenow fragment of *E. coli* DNA polymerase. The digoxigenin-labelled probe is used for hybridization and can be detected by the Fab-fragments of a polyclonal antibody conjugated to alkaline phosphatase. This enzyme catalyzes the production of a colored product (e.g., indigo) from a colorless substrate (e.g., bromochloroindolyl phosphate, BCIP). The development of new, luminogenic substrates of alkaline phosphatase (Bronstein and McGrath, 1989) expands the utility of this technique. For example, the spiroadamantan dioxetane AMPPD, or derivatives, if dephosphorylated, lead to the appearance of unstable intermediates that decompose to the stable product adamantan under emission of light at 477 nm. This can be detected by a radiographic film. Such *chemiluminescent techniques* come close to the performance of radioactive detection methods as tested with chickpea and *Ascochyta rabiei* DNA samples in our laboratory (Bierwerth *et al.*, 1992). Figure 6 compares the detection techniques and shows the potential of the chemiluminescent approach.

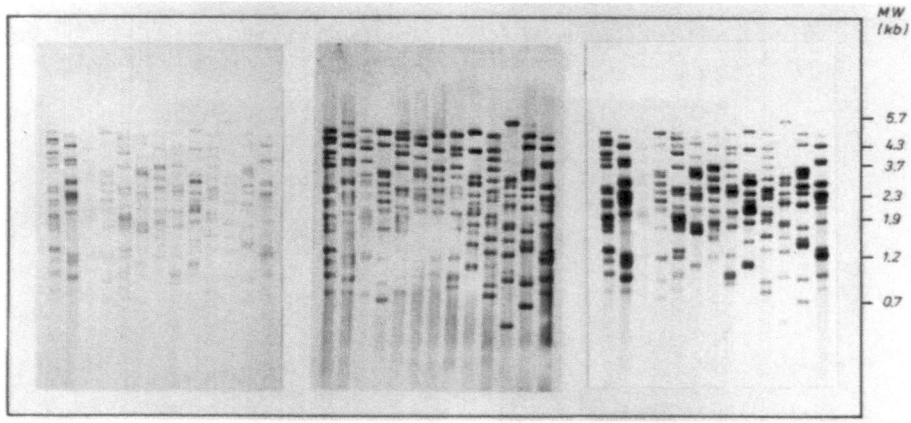

Figure 6. Comparison of nonradioactive and radioactive detection methods for (GATA)₄-fingerprinting of 13 different chickpea (*Cicer arietinum*) accessions. Chickpea DNA was digested with Taq I, separated electrophoretically, blotted onto a nylon membrane and hybridized to DIG-(GATA)₄ or to ³²P-(GATA)₄. Digoxigenin-dependent hybridization signals were detected by an antibody/alkaline phosphatase conjugate followed by treatment with either a colorigenic (NBT/BCIP) or a chemiluminogenic substrate (AMPPD). ³²P-generated signals were detected by autoradiography. The same blot was used for all three experiments. Molecular weight markers are given in kb.

DNA Amplification Fingerprinting

The polymerase chain reaction (PCR) technology can also be added to the repertoire of techniques successfully applied to chickpea and *Ascochyta rabiei*. A variant of the conventional PCR, the *random amplified polymorphic DNA (RAPD) technique* (Williams *et al.*, 1990), also coined arbitrarily primed PCR, AP-PCR (Welsh and McClelland, 1990) or DNA amplification fingerprinting, DAF (Caetano-Anolles *et al.*, 1991) uses either one (simplex) or two (duplex) synthetic oligodeoxynucleotide primers of arbitrary sequence, usually high (i.e., > 60%) GC content and 5 to 15 bp length, to prime DNA polymerase-catalyzed extension and synthesis of a strand complementary to the target strand to which the primer binds. The PCR-driven amplifications of specific (presumed unique) sites in the genome allow to detect amplification fragment length polymorphisms (AFLPs) with high informativeness, but do not require any prior knowledge of DNA sequences, do not depend on cloned probes, are rapidly performed and – if obstacles in the technology itself are overcome – are sufficiently sensitive. The technique has proved to be reliable and the amplification fingerprint patterns to be reproducible with banana and plantain (Kaemmer *et al.*, 1992a) and has been used to further characterize chickpea and *Ascochyta rabiei* (Kaemmer *et al.*, 1992b). In case of *Ascochyta rabiei*, RAPD discrimination of six fungal pathotypes can be obtained, presupposed the

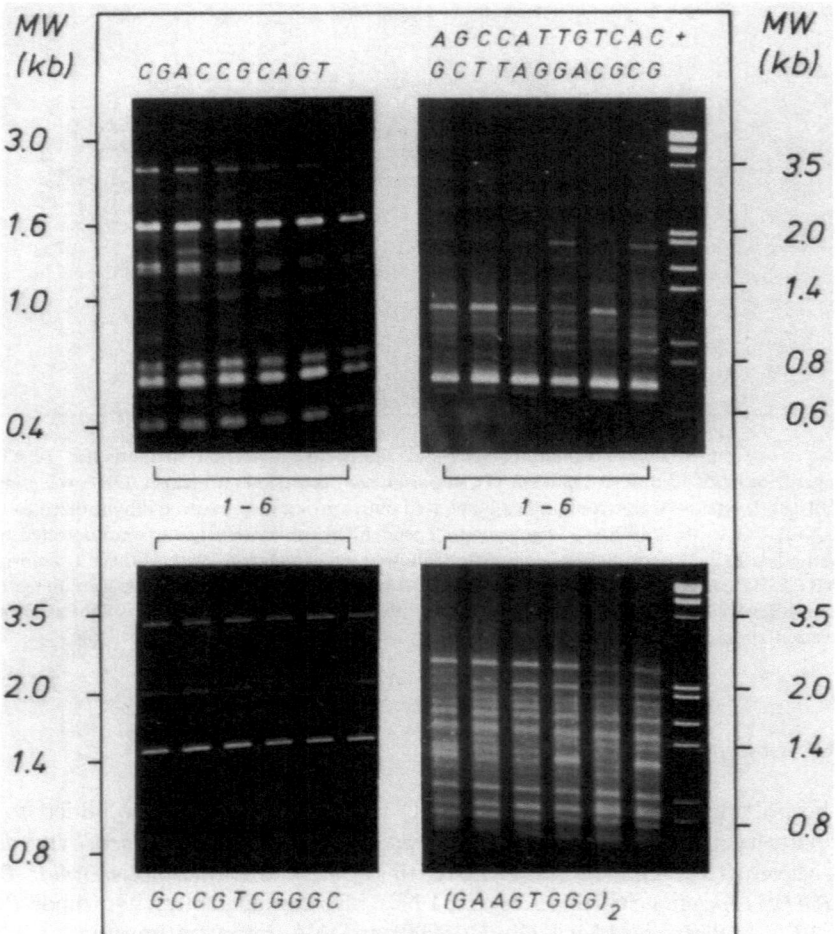

Figure 7. DNA amplification fingerprinting of six single-spored isolates of *Ascochyta rabiei*. PCR amplifications were performed with 150 uM/ml of each dNTP, 10 pmol of each primer (or 20 pmol in single primer experiments), 10 ng of template DNA and 1.25 U Taq polymerase in a total volume of 50 ul. Samples were preincubated for 5 min at 96°C. The reactions were initiated by the addition of the enzyme and carried out for 30 cycles. Each cycle consisted of 1 min at 94°C, 1 min at the annealing temperature (5°C below melting temperature) and 2 min at 72°C. The final elongation step was extended to 10 min. Amplification products were separated on 1.2 % agarose gels and stained with ethidium bromide.

appropriate primer (or, in duplex PCR, primers) are found (Figure 7). In addition to the high resolution of RAPD fingerprints, the technique will also allow to tailor fingerprinting patterns by selecting an appropriate primer. Trial and error-type experiments can result in simple (for e.g., genetic mapping) or complex patterns (for e.g., identification of individuals), as exemplified in Figure 7 for *Ascochyta rabiei*.

It is our hope, that DNA fingerprinting and DNA amplification finger-printing will be used successfully by our colleagues in the developing world, who might then address problems with crops or pathogens specific and relevant to their home country.

Acknowledgements

We appreciate the contributions of scientists from our laboratories who are not co-authors. Research of both laboratories was supported by a grant from Bundesministerium für Technische Zusammenarbeit (BMZ), Bonn, Germany (grant 89.7860.3–01.130).

References

Barton, K. A., Whiteley, H. R. and Yang, N. – S. 1987. *Plant Physiology* 85: 1103–1109.

Beachy, R. N., Loesch-Fries, S. and Tumer, N. E. 1990. *Annual Review of Phytopathology* 28: 451–474.

Bierwerth, S., Kahl, G., Weigand, F. and Weising, K. 1992. *Electrophoresis* 13: 115–122.

Bolivar, F. and Backman, K. 1979. *Methods in Enzymology* 68: 245–267.

Broglie, K., Chet, I., Holliday, M., Cressman, R., Biddle, P., Knowlton, S., Mauvais, C. J. and Broglie, R. 1991. *Science* 254: 1194–1197.

Bronstein, I. and McGrath, P. 1989. *Nature* 338: 599–600.

Burke, T., Dolf, G., Jeffreys, A. J. and Wolff, R. 1991. *DNA Fingerprinting: Approaches and Applications*, 400 pp. Basel, Switzerland: Birkhäuser.

Caetano-Anollés, G., Bassam, B. J. and Gresshoff, P. 1991. *Bio/Technology* 9: 553–557.

Chee, P. P., Fober, K. A. and Slightom, J. L. 1989. *Plant Physiology* 91: 1212–1218.

Chen, J., Bolyard, M. G., Saxena, R. C. and Sticklen, M. B. 1992. *Plant Science* 81: 83–91.

Christou, P. 1992. *Plant Journal* 2: 275–281.

Corbin, D. R. and Klee, H. J. 1991. *Current Opinions in Biotechnology* 2: 147–152.

Deak, M., Kiss, G. B., Koncz, C. and Dudits, D. 1986. *Plant Cell Reports* 5: 97–100.

De Cleene, M. 1985. *Phytopathologische Zeitschrift* 113: 81–89.

Delannay, X., LaVallee, B. J., Proksch, R. K., Fuchs, R. L., Sims, S. R., Greenplate, J. T., Marrone, P. G., Dodson, R. B., Augustine, J. J., Layton, J. G. and Fischhoff, D. A. 1989. *Bio/Technology* 7: 1265–1269.

Feitelson, J. S., Payne, J. and Kim, L. 1992. *Bio/Technology* 10: 271–275.

Fischhoff, D. A., Bowdish, K. S., Perlak, F. J., Marrone, P. G., McCormick, S. M., Niedermeyer, J. G., Dean, D. A., Kusano-Kretzmer, K., Mayer, E. J., Rochester, D. E., Rogers, S. G. and Fraley, R. T. 1987. *Bio/Technology* 5: 807–813.

Frischauf, A. M. and Lehrach, H. 1983. *Journal of Molecular Biology* 170: 827–841.

Fulton, R. W. 1986. *Annual Review of Phytopathology* 24: 67–81.

Gasser, C. S. and Fraley, R. T. 1989. *Science* 244: 1293–1299.

Golemboski, D. B., Lomonossoff, G. P. and Zaitlin, M. 1990. *Proceedings of the National Acaddemy of Science, USA* 87: 6311–6315.

Grunstein, M. and Hogness, D. S. 1975. *Proceedings of the National Academy of Science, USA* 10: 3961–3965.

Halfter, U., Morris, P. – C. and Willmitzer, L. 1992. *Molecular and General Genetics* 231: 186–193.

Herrera-Estrella, A., Van Montagu, M. and Wang, K. 1990. *Proceedings of the National Academy of Science, USA* 87: 9534–9537.

Hiatt, A., Cafferkey, R. and Bowdish, K. 1989. *Nature* 342: 76–78.

Hinchee, M. A. W., Connor-Ward, D. V., Newell, C. A., McDonnell, R. E., Sato, S. J., Gasser, C. S., Fischhoff, D. A., Re, D. A., Fraley, R. T. and Horsch, R. B. 1988. *Bio/Technology* 6: 915–922.

Höfte, H. and Whiteley, H. R. 1989. *Microbiological Review* 53: 242–255.

Hohn, B. and Collins, J. 1980. *Gene* 11: 291–299.

Horsch, R. B., Fry, J. E., Hoffman, N. L., Eichholtz, D. E., Rogers, S. G. and Fraley, R. T. 1985. *Science* 227: 1229–1231.

Jefferson, R. A., Kavanagh, T. A. and Bevan, M. W. 1987. *EMBO Journal* 6: 3901–3907.

Kado, C. I. 1991. *Critical Reviews in Plant Science* 10: 1–32.

Kaemmer, D., Afza, R., Weising, K., Kahl, G. and Novak, F. J. 1992a. *Bio/Technology* 10: 1030–1035.

Kaemmer, D., Ramser, J., Schön, M., Weigand, F., Saxena, M., Driesel, A., Kahl, G. and Weising, K. 1992b. *Advances in Molecular Genetics* (In press).

Kahl, G. and Schell, J. 1982. *Molecular Biology of Plant Tumors*, 615 pp. New York: Academic Press.

Köhler, F., Golz, C., Eapen, S., Kohn, H. and Schieder, O. 1987. *Plant Cell Reports* 6: 313–317.

Lawson, C., Kaniewski, W., Haley, L., Rozman, R., Newell, C., Sanders, P. and Tumer, N. E. 1990. *Bio/Technology* 8: 127–134.

Ling, K., Namba, S., Gonsalves, C., Slightom, J. L. and Gonsalves, D. 1991. *Bio/Technology* 9: 752–758.

Logemann, J., Jach, G., Tommerup, H., Mundy, J. and Schell, J. 1992. *Bio/Technology* 10: 305–308.

MacKenzie, D. J., Tremaine, J. H. and McPherson, J. 1991. *Molecular Plant-Microbe Interactions* 4: 95–102.

McCabe, D. E., Swain, W. F., Martinell, B. J. and Christou, P. 1988. *Bio/Technology* 6: 923–926.

Nejidat, A. and Beachy, R. N. 1990. *Molecular Plant-Microbe Interactions* 3: 247–251.

Nelson, R. S., McCormick, S. M., Delannay, X., Dub, P., Layton, J., Anderson, E. J., Kaniewska, M., Proksch, R. – K., Horsch, R. B., Rogers, S. G., Fraley, R. T. and Beachy, R. N. 1988. *Bio/Technology* 6: 403–409.

Nester, E. W., Gordon, M. P., Amasino, R. M. and Yanofsky, M. F. 1984. *Annual Reviews in Plant Physiology* 35: 387–413.

Odell, J., Caimi, P., Sauer, B. and Russell, S. 1990. *Molecular General Genetics* 223: 369–378.

Olson, M., Hood, L., Cantor, C. and Botstein, D. 1989. *Science* 245: 1434–1435.

Paszkowski, J., Baur, M., Bogucki, A. and Potrykus I. 1988. *EMBO Journal* 7: 4021–4026.

Perlak, F. J., Deaton, R. W., Armstrong, T. A., Fuchs, R. L., Sims, S. R., Greenplate, J. T. and Fischhoff, D. A. 1990. *Bio/Technology* 8: 939–943.

Pezzotti, M., Pupilli, F., Damiani, F. and Arcioni, S. 1991. *Plant Breeding* 106: 39–46.

Potrykus, I. 1991. *Annual Reviews in Plant Physiology Plant Molecular Biology* 42: 205–225.

Powell Abel, P., Nelson, R. S., De, B., Hoffman, N., Rogers, S. G., Fraley, R. T. and Beachy, R. N. 1986. *Science* 232: 738–743.

Puonti-Kaerlas, J., Eriksson, T. and Engström, P. 1990. *Theoretical Applied Genetics* 80: 246–252.

Rogers, S. G. 1991. *Current Opinions in Biotechnology* 2: 153–157.

Saiki, R. K., Gelfand, D. H., Stoffel, S., Scharf, S. J., Higuchi, R., Horn, G. T., Mullis, K. B. and Erlich, H. A. 1988. *Science* 239: 487–491.

Sambrook, J., Fritsch, T. and Maniatis, T. 1989. *Molecular Cloning. A Laboratory Manual. 2nd edition.* Cold Spring Harbor: Cold Spring Harbor Laboratory Press.

Schlessinger, D. 1990. *Trends in Genetics* 6: 248–258.

Sequeira, L. 1984. *Trends in Biotechnology* 2: 25–29.

Stark, D. M. and Beachy, R. N. 1989. *Bio/Technology* 7: 1257–1262.

Tautz, D. and Renz, M. 1984. *Nucleic Acids Research* 12: 4127–4138.

Töpfer, R., Pröls, M., Schell, J. and Steinbiß, H. – H. 1988. *Plant Cell Reports* 7: 225–228.

Vaeck, M., Reynaerts, A., Höfte, H., Jansens, S., De Beuckeleer, M., Dean, C., Zabeau, M., Van Montagu, M. and Leemans, J. 1987. *Nature* 328: 33–37.

van Larebeke, N., Engler, G., Holsters, M., van den Elsacker, S., Zaenen, I., Schilperoort, R. A. and Schell, J. 1974. *Nature* 252: 169–170.

Vieira, J. and Messing, J. 1982. *Gene* 19: 259–268.

Wahl, G. M., Lewis, K. A., Ruiz, J. C., Rothenberg, B., Zhao, J. and Evans, G. A. 1987. *Proceedings of the National Academy of Science, USA* 84: 2160–2164.

Weising, K., Beyermann, B., Ramser, J. and Kahl, G. 1991a. *Electrophoresis* 12: 159–169.

Weising, K., Kaemmer, D., Epplen, J. T., Weigand, F., Saxena, M. and Kahl, G. 1991b. *Current Genetics* 19: 483–489.

Weising, K., Kaemmer, D., Weigand, F., Epplen, J. T. and Kahl, G. 1992. *Genome* 35: 436–442.

Weising, K., Schell, J. and Kahl, G. 1988. *Annual Reviews in Genetics* 22: 421–477.

Welsh, J. and McClelland, M. 1990. *Nucleic Acids Research* 18: 7213–7218.

Werr, W. and Lörz, H. 1986. *Molecular General Genetics* 202: 471–475.

Williams, J. G. K., Kubelik, A. R., Livak, K. J., Rafalski, J. A. and Tingey, S. V. 1990. *Nucleic Acids Research* 18: 6531–6535.

Winnacker, E. 1987. *From Genes to Clones. Introduction to Gene Technology.* Weinheim: Verlag Chemie. 634 pp.

Zambryski, P. 1988. *Annual Reviews in Genetics* 22: 1–30.

Zambryski, P., Joos, H., Genetello, C., Leemans, J., Van Montagu, M. and Schell, J. 1983. *EMBO Journal* 2: 2143–2150.

Zambryski, P., Tempé, J. and Schell, J. 1989. *Cell* 56: 193–201.

Identifying and mapping genes of economic significance

N.F. WEEDEN[1], G.M. TIMMERMAN[2] and J. LU[1*]

[1] *Department of Horticultural Sciences, New York State Agricultural Experiment Station, Cornell University, Geneva, NY, USA, and*
[2] *Crop and Food Research, Cantebury Agriculture and Science Centre, Lincoln, New Zealand*
* Present address: Center for Viticultural Science, Florida Agricultural and Mechanical University, Tallahassee, FL, USA

Abstract

An understanding of the genetic basis of characters of commercial importance is critical if a breeder is attempting to move such characters into breeding material. A number of particularly interesting characters or genes have been identified in cool season food legumes, and in pea many of these have been "tagged" by molecular markers such as allozyme or DNA polymorphisms. This process of mapping and tagging genes has been greatly accelerated by recent developments in molecular biology. It appears that markers will soon be available for many genes in lentil, faba bean, and chickpea and that genetic knowledge developed in one crop will have significant applications in the other cool season food legumes.

Introduction

One of the primary needs of a breeder is to be able to identify those plants in a segregating progeny that possess the appropriate combination of genes desired in a cultivar. Part of this process involves the innate ability of the breeder to recognize appropriate plant types or novel combinations that will represent important innovations. However, as our knowledge of the genetic basis of characters continues to expand, many more genes are being recognized as being a desirable component of the final genotype of a cultivar. If such genes are not easily recognized by morphology or convenient screening procedures, the breeder is faced with tedious backcrossing or complicated assay tests to determine the genotype of the line. In order to avoid lengthy testing procedures, many breeders have been intrigued with the possibility of identifying genetic "tags" for these recalcitrant genes. Ideally, such tags would be quickly and easily scored at the seed or seedling stage and, thus, greatly facilitate screening operations.

The genes *r* (wrinkled seed) and *a* (white flowers, lack of anthocyanin) are examples of important genes in pea (*Pisum sativum*). Their phenotypes can be

F.J. Muehlbauer and W.J. Kaiser (eds.), Expanding the Production and Use of Cool Season Food Legumes, 726–737.
© 1994 *Kluwer Academic Publishers.*

conveniently and accurately determined by simple observation, and the genes have become incorporated into a wide range of germplasm. There is no particular reason to tag these traits with a different marker gene. However, genes such as *er* (conferring resistance to powdery mildew, *Erysiphe pisi*), *Fw* (conferring resistance to Fusarium wilt race 1), or *sn* (conditioning photoperiod sensitivity) are not as easy to score yet are desirable in many breeding lines. Even more challenging are traits such as resistance to Aphanomyces root rot in pea or upright growth habit in lentil (*Lens culinaris*) which appear to be polygenic traits with significant environment effects on their expression. Progress in modifying these phenotypes can be frustratingly slow without some kind of tool for identifying and following the major genes influencing the phenotype.

The feasibility of marker-assisted selection has been demonstrated in many crops, particularly those such as maize (*Zea mays*) and tomato (*Lycopersicon esculentum*) that possess a well established linkage map. A number of markers for economically important traits have been described in pea, but relatively few in the other cool season food legumes. In this paper, we will discuss recent accomplishments in mapping genes and developing linkage maps in cool season food legumes. We will also describe progress in elucidating the genetic basis of several important characters and in tagging these genes with molecular markers.

Recent Advances in Gene Mapping and Tagging

Within the last 10 years there has been a veritable explosion of new markers identified in cool season food legumes. Allozyme and other protein polymorphisms have been described in pea (Casey, 1982; Przybylska, 1986; Murray and Ayre, 1987; Weeden and Marx, 1987), lentil (Zamir and Ladizinsky, 1984; Hoffman *et al.*, 1986; Muehlbauer *et al.*, 1989; Vaillancourt, 1989), chickpea (Gaur and Slinkard, 1990a,b), sweet-pea (Godt and Hamrick, 1991), and faba bean (Mancini *et al.*, 1989). Far more polymorphic loci have been found in *Pisum sativum* than in any other legume species (Weeden, 1988), although significant variation exists for lentil, chickpea, and several other legume crops if one includes both the primary and secondary gene pools. Current research at several locations is emphasizing the identification of DNA polymorphisms in order to supplement the set of markers available to breeders and geneticists in each of these crops.

The current map for pea (outlined in Figure 1) has had many markers added to it since Blixt (1974) assembled a comprehensive linkage map for this crop. The present map confirms, for the most part, that presented by Blixt. Chromosomes 2, 5, and 7 are rearranged, but at least some of these rearrangements may reflect variation in the karyotype within *Pisum sativum*. Much of our DNA marker data is based on a cross between a *P. sativum* ssp. *humile* line (JI1794) collected on the Golan Heights and a dwarf, colored, flowered *P. s.* ssp. *sativum* line we have designated "Slow". It is possible that the linkage shown on chromosome 5 between *Tl* and *Gp* may not be universally present in *P. sativum*. This linkage has been confirmed in studies with *gi* (Murfet, 1990) and *het* (Swiecicki, 1990) but appeared to be present in only some of the crosses examined by Ellis *et al.* (1992). The shift of part of Blixt's chromosome 2 (*Oh* and more distal markers) to the new chromosome 7 has not been confirmed by researchers outside of our program. However, neither our mapping work nor that reported from other laboratories has produced data to clearly associate the *Oh*-lectin segment with any other chromosome.

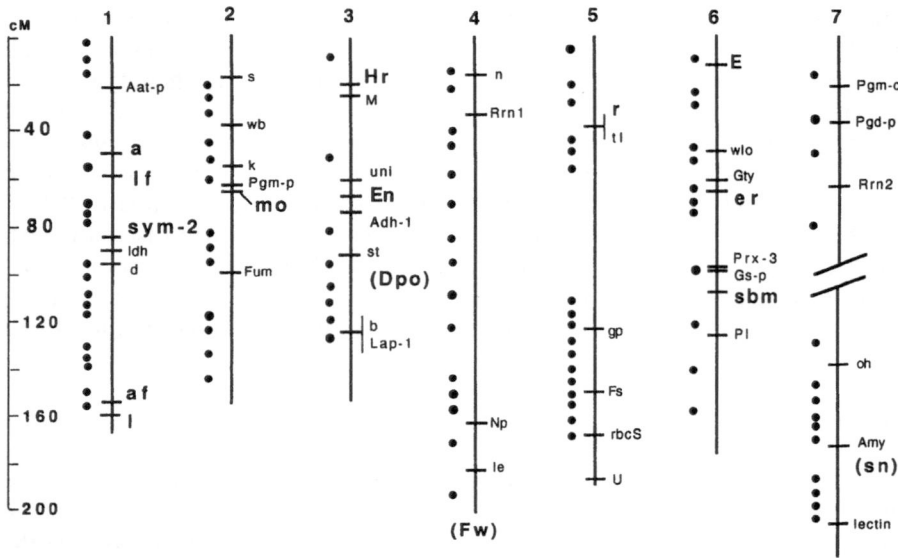

Figure 1. Outline of current linkage map for *Pisum sativum*. Genes shown in bold type are those possessing special commercial importance. Included among these are *a* (absence of anthocyanin), *af* (afila or leafless), *I* (green versus yellow cotyledons), *r* (wrinkled seeds), *Dpo* (dehiscent pod), as well as the genes controlling flowering time (*lf*, *Hr*, *E*, and *sn*) and those conferring disease resistance (*mo*, *En*, *Fw*, *er*, and *sbm*). The dots to the left of each linkage group indicate the position of at least one RAPD marker that has been mapped using recombinant inbred lines from the cross JI1794 × Slow mentioned in the text. The integrity of linkage group 7 remains to be confirmed and is drawn as two sections. The synteny of markers within each segment has been demonstrated many times.

In addition to the progress on the pea linkage map, remarkable progress has been made on the lentil, chickpea, and faba bean maps. Zamir and Ladizinsky (1984) identified many linkage relationships among loci encoding proteins, and Havey and Muehlbauer (1989) published an RFLP map for lentil which also included several morphological and protein markers. The map has been confirmed and extended (Weeden *et al.*, 1992) using an interspecific cross (*L. culinaris* × *L. ervoides*). Comparison of the lentil map with that available for pea revealed that approximately 40% of the loci syntenic in pea also are syntenic in lentil (Weeden *et al.*, 1992). Studies on isozyme loci in chickpea (Gaur and Slinkard, 1990a, b) also indicated a significant similarity with the pea map. The chickpea map is now being extended using RFLP and RAPD markers (Simon and Muehlbauer, 1991), again with the tentative conclusion that about 40% of the linkage associations observed in pea are also present in chickpea.

Pea, lentil, and chickpea constitute the three most divergent taxa among the cool season food legumes. Faba bean is believed to represent an intermediate or perhaps basal group, whereas sweet-pea is placed on the same lineage as pea. Thus, we would anticipate that the linkage maps for these latter crops would be within the range displayed by pea, lentil, and chickpea, albeit faba bean will possess one less linkage group than the others. Indeed, the partial map for faba bean being presented at this conference by A. M. Torres-Romero and coworkers appears to contain at least two regions that are present in both pea and lentil. These results should be encouraging to all working on cool season food legumes because they suggest that molecular tags identified in one crop may be applicable to the other crops as well.

The Tagging of Disease Resistance Genes in Pea

The idea of tagging economically important genes with more conveniently scored genes was undoubtedly conceived soon after linkage was first described in sweet-pea (Bateson and Punnett, 1905) or at least clearly interpreted (Morgan, 1911). In pea, morphological markers have been used for some time as tags for certain virus resistance genes. The flower mutant, k, can be used as a marker for bean yellow mosaic virus resistance conferred by the gene mo (Marx and Provvidenti, 1979), and wlo has been used as a marker for $sbm-1$, a gene conditioning resistance to pea seedborne mosaic virus (Hagedorn and Gritton, 1973). Morphological markers have the disadvantage that they usually are not desired in the final cultivar and must be eliminated in the final steps before the release of a cultivar. Thus, the search for more convenient markers has continued at the molecular level.

The first molecular tag reported for an economically important gene in pea was an esterase polymorphism tightly linked to Fw, the gene bestowing resistance to race 1 of Fusarium wilt (Hunt and Barnes, 1982). Unfortunately, more recent research has been unable to confirm this linkage (M. J. Grajal-Martin, personal communication) and thus, the marker is not currently being used commercially.

Molecular tags that are being used commercially involve isozyme loci linked to three virus resistance genes or gene clusters. Pea enation mosaic virus resistance, conditioned by the gene En on chromosome 3, can be followed in segregating populations by scoring ADH-1 (Weeden and Provvidenti, 1988). Stocks in which the rare "fast" allele at $Adh-1$ has been linked to the resistance gene have proven useful for incorporating the resistance into advanced breeding material.

The mo gene is one of a cluster of genes which in the most common arrangement also provides resistance to clover yellow vein virus, a lentil strain of seedborne mosaic virus, watermelon mosaic virus II, and pea mosaic virus (Provvidenti and Hampton, 1991). Seven years ago we reported that the isozyme locus *Pgm-p* was situated about 2 cM from this cluster and could be used as a very efficient and accurate screening device for the resistance genes (Weeden *et al.*, 1984). The major problem encountered when using *Pgm-p* is that only two alleles at this locus have been identified and both are relatively common in resistant and susceptible types. Hence, a breeder is occasionally faced with the problem where both the resistant and the susceptible lines possess the same allele at *Pgm-p*. Unless the breeder is willing to use a different source of resistance possessing the alternative *Pgm-p* allele, the locus will be monomorphic in the progeny and useless as a marker.

One of the most important virus resistance genes in pea is *sbm-1* on chromosome 6. Again, this gene is part of a cluster which also includes genes conferring resistance to clover yellow vein virus and other pea seedborne mosaic virus pathotypes (Provvidenti and Hampton, 1991). Previous reports have demonstrated that *sbm-1* is linked to chromosome 6 morphological markers, including *wlo* (Hagedorn and Gritton, 1973) and *P* (Skarzynska, 1988). However, neither of these morphological markers were particularly close, nor were they desired in the final cultivar. Many molecular markers have been mapped to this chromosome, and the linkage relationships of some of these markers were examined in a series of segregating populations Weeden *et al.*, 1991; Timmerman *et al.*, 1993). The closest marker to *sbm-1* appears to be the gene coding the plastid-specific glutamine synthetase, *Gs-p*. Timmerman and coworkers used restriction fragment length polymorphisms (RFLPs) generated by the cDNA clone GS185 (Tingey *et al.*, 1987) to show that *Gs-p* is less than 10 cM from *sbm-1*. The potential value of *Gs-p* as a marker was further demonstrated when the resistance/susceptible phenotypes of 32 pea lines were compared with the hybridization pattern generated by GS185. Of the resistant lines examined, most (12 out of 15) displayed a GS185 hybridization pattern which was not observed among the 17 susceptible lines tested (Timmerman *et al.*, 1993). Most of the resistant lines also possessed the PRX-3 allozyme with the slower mobility, although the distinction between the resistant and susceptible groups was not as marked as for the GS185 pattern. Breeders should find GS185 or PRX-3 phenotypes to be more practical tags than the previously used morphological markers.

Another important disease resistance gene is *er*. Marx (1971) reported close linkage (approximately 8% recombination) between *Gty* and *er*. At that time *Gty* was assigned to chromosome 3, but more recently this gene has been shown to reside on chromosome 6 (Wolko and Weeden, 1991), implying that *er* also may be on this chromosome. Recently, the position of *er* on chromosome 6 has been conclusively demonstrated using DNA markers (Timmerman, unpublished). Two RFLPs, each clearly linked to established chromosome 6 loci, flank *er*. The markers map about 15 cM to each side of the gene.

Other important disease resistance genes are lacking genetic tags, and some have yet to be located on the linkage map. The inability to confirm the *Est-Fw* linkage means that none of the wilt resistance genes (*Fw*, *Fnw*, and those bestowing resistance to races 3, 4, 5, and 6) have markers or definite map positions. The gene *lr*, conferring resistance to pea leaf roll virus, also remains unmapped (Drijfhout, 1968; Baggett and Hampton, 1991).

Genetic Dissection and Tagging of Other Traits

A large number of other economically important genes in pea have been identified through traditional genetic studies among which a few stand out as particularly instructive examples. An understanding of the genes influencing time of flowering would be particularly useful for breeders because of the need to adapt morphotypes to local conditions. Studies performed by Marx (1975), Gottshalk (1988), and particularly Murfet and coworkers (Murfet and Reid, 1985; Murfet, 1988) resolved this apparently quantitative character into four major genes (*lf*, *sn*, *hr* and *E*), two rare mutants (*dne* and *veg*) and several secondary "modifier" genes such as *det* and *le*. This work is particularly impressive because it was performed without the aid of molecular markers. All of the genes except *sn* had been mapped to linkage groups before allozyme or DNA polymorphisms became available in pea. As it turned out, *sn* was located in a region of chromosome 7 lacking in classical morphological markers and could be mapped only after allozyme polymorphisms had been found in this region. It is now possible to follow segregation at *Lf*, *Sn*, and *Hr* using molecular tags and predict with reasonable accuracy the node at which a particular plant will flower in a specified environment. Selection for early, late, or photoperiod sensitive types is now possible at the seedling stage.

Another particularly important trait, not only of pea but of most other legumes including all cool season food legumes, is the ability to fix nitrogen. Although considerable advances have been made in our understanding of the genetics of this process in *Rhizobium*, the number and role of host genes in the *Rhizobium*/legume symbiosis and nitrogen fixation remains primitive. Several laboratories have been investigating this problem, and major advances have been made in pea through the identification and analysis of mutants displaying abnormal nodulation phenotypes. At present, mutations at over 30 distinct genes (termed symbiosis or "*sym*" genes) have been described (Jacobsen and Feenstra, 1984; Lie, 1984; Engvild, 1987; Kneen and LaRue, 1988; Duc and Messager, 1989). Most of these genes have been located on the pea linkage map, being widely distributed on the genome (Weeden, *et al.*, 1990).

Knowing the location of the *sym* genes has greatly facilitated our investigations of the nodulation process. For instance, lectin genes are believed to play a role in the host/symbiont recognition (Diaz *et al.*, 1989). We were interested in the possibility that one or more of the *sym* genes, particularly those conditioning strain specificity, may represent a lectin gene. We obtained two pea lectin clones from other laboratories and mapped their location on the pea linkage map using RFLPs (Lu *et al.*, in prep.). Each clone hybridized to a distinct small cluster of sequences. No recombination was observed within the clusters, although only a relatively small progeny (38 plants) was analyzed. Both clusters mapped to one end of linkage group 7, with approximately 15% recombination between the clusters. Only one *sym* gene is located in this region, and this gene is characterized by short root laterals and does not induce strain specificity. We have not tested the closeness of the linkage between this *sym* gene and the lectin clusters, but at present we are assuming that their proximity is coincidental and that the *sym* gene does not involve the lectin cluster. We have established that none of the other *sym* genes reflect mutations at sequences hybridizing to the two pea lectin genes.

A cluster of four *sym* genes map near the leghemoglobin genes on chromosome 1, but none of these genes appears to represent a loss or modification of leghemoglobin expression. Similarly, the one mutant we have examined that forms nodules lacking leghemoglobin maps to chromosome 7 and, thus, cannot be produced by a mutation in the leghemoglobin gene cluster.

Markers in Other Crops

Many traits of interest have been identified in other cool season food legumes; however, relatively few of these have been tagged by markers. Indeed, few monogenic morphological variants have been identified in lentil, and the development of a linkage map for this crop has been feasible only since molecular markers have become available and progeny from interspecific crosses have been generated. A large number of morphological mutants have been described in chickpea (Muehlbauer and Singh, 1987), but their use as markers has been very limited. In both faba bean and sweet-pea the application of markers to tag or dissect traits of commercial interest is just beginning to have an impact.

In lentil several traits of significant commercial interest have been reported and in some cases linkage relationships have been investigated, but none have been tagged by tightly linked markers. Resistance to pea seedborne mosaic virus and to pea enation mosaic virus was reported by Haddad *et al.* (1978) and Aydin *et al.* (1987), respectively. Vaillancourt *et al.* (1986) described a low tannin mutant that has now been incorporated into commercial lines (Slinkard *et al.*, 1992).

Of particular value for assessing the genetic basis of complex traits such as seed size, cold tolerance, or plant habit is the analysis of large (100 to 500 lines) recombinant inbred populations. Each line in the population is derived, by single seed descent, from a different F_2 plant. Thus, the population contains a considerable number of recombinant genotypes but lacks the complicating effects of heterozygosity. Tahir (1990) used recombinant inbred lines generated from the cross *L. culinaris* × *L. orientalis* to investigate the number and location of genes affecting seven quantitative traits (days to flowering and maturity, plant height, biomass, seed yield, harvest index, and seed weight). Each of these traits displayed significant correlations with at least one of the ten marker loci assayed (Tahir, 1990). A similar study on a *L. culinaris* × *L. ervoides* progeny found seed size to correlate well with markers near the ribosomal array on chromosome I (Abbo *et al.*, 1992).

Approaches for the Future

In order to obtain a large number of markers distributed throughout the genome it is clear that a technique will be necessary that directly exposes DNA sequence variation. Morphological markers often interfere with each other and have undesirable pleiotropic effects while protein markers can access only a limited number of genes. RFLPs offer one method for examining DNA sequence variation, and many cloned segments of DNA have now been mapped in both lentil and pea (Havey and Muehlbauer, 1989; Weeden and Wolko, 1990; Ellis *et al.*, 1992; Weeden *et al.*, 1992). However, this technique is expensive and relatively sophisticated, and thus not easily adapted to the needs of breeders.

A recently developed approach that is both easier and more economical than RFLPs also appears to expose much more variability at the DNA sequence level. This technique uses the polymerase chain reaction (Saiki *et al.*, 1988) to amplify small (200 to 2000 base pair) fragments of the genome in sufficient quantity to permit them to be seen after electrophoresis and staining with a fluorescent compound. The specificity of the amplification process is determined by the sequence of the short DNA primer (usually 10 bases long) added to the reaction along with the genomic DNA, nucleotides, and heat stable DNA polymerase. The sequence of the primer is usually selected at random. Hence, one of the names applied to this technique is Random Amplified Polymorphic DNA (RAPD) (Williams *et al.*, 1990).

We have applied RAPD analysis to pea, lentil, chickpea, and faba bean and found all these crops to be amenable to this technique. For pea, we have pursued the mapping of RAPD markers more vigorously than in the other crops and have added many of these markers to the linkage map (Figure 1). RAPD markers appear to be well distributed across the known linkage map, although they exhibit regions of high and low density. The regions of high density may reflect portions of the genome where crossing over is suppressed due to the presence of heterochromatin or it may reflect a non-random distribution of the sequences required for the amplification of the RAPD fragments.

Identification of Gene-Specific Markers

Now that we have the ability to generate markers easily throughout the genome, the feasibility of identifying and tagging traits of commercial interest is much greater. How can we make this process as efficient as possible in order to quickly identify the markers we need? Near isogenic lines (NILs) offer one tool for identifying closely linked markers. Theoretically, such lines are genetically identical except for a small region surrounding the gene differentiating the lines. Markers differing between the NILs presumably are tightly linked to the gene of interest.

Another approach that avoids the lengthy process of developing NILs has recently been applied with RAPD markers by Michelmore *et al.* (1991). In this method two DNA samples are generated from a population segregating for a monogenic trait. One sample contains DNA from 5 to 10 individuals exhibiting the desired trait while the other contains DNA from 5 to 10 individuals lacking the trait. These two samples are screened for differences in their RAPD phenotype for 50 or more primers. Because all other regions of the genome except that containing the gene conditioning the trait of interest should be randomly distributed between the two samples, fragments amplified in one sample but not in the other often are closely linked to the gene. The screening of two samples can be performed much more efficiently than screening each plant in a segregating population individually, and this method permits the investigator to quickly determine which primers may generate appropriate markers for the trait.

We have used both NILs and Michelmore's "batch" method to develop RAPD markers for genes in pea for which we want tightly linked DNA sequences. One of these genes is *sym-2*, the gene originally found in landraces from Afghanistan that alters the specificity of the host/*Rhizobium* interaction. We obtained from Dr. T. A. LaRue (Boyce Thompson Institute, Ithaca, NY) two NILs, one a normal nodulating cultivar ("Sparkle") and the other a backcross 5 line from a Sparkle x "Afghanistan" hybrid. This latter line had been selected for the *sym-2* phenotype in a Sparkle background. Two RAPD fragments missing in Sparkle but present in both Afghanistan and the BC5 line were identified during the screening of 75 arbitrary primers. Joint segregation analysis of these RAPD fragments with *sym-2* using F_2 populations indicated that the markers were approximately 7 and 10 cM from *sym-2* (Lu, unpublished).

The second approach was used for another symbiosis gene, *sym-20*. We used two samples for our screening procedure, one consisting of DNA from five normally nodulating F_2 plants from a cross between Sparkle and Afghanistan (the latter homozygous for *sym-2*) and the other containing DNA from five non-nodulating F_2 plants. In this case, 73 arbitrary primers were screened and one RAPD was found to exhibit polymorphism between the two samples. Linkage analysis demonstrated that the RAPD marker was about 10 cM from *sym-20*. Thus, both approaches were able to identify RAPD markers reasonably close to the target gene. As the number of primers that can be tested is nearly limitless, a sufficiently patient investigator should be able to identify markers very close to a gene of interest. Once such a RAPD fragment is found, that fragment can be sequenced and primers specific for that sequence be constructed.

Concluding Remarks

Molecular techniques currently available for producing genetic markers are so powerful that virtually every character segregating in a progeny can be dissected into its genetic components. RAPD markers appear to be sufficiently abundant to permit the tagging of any economically important gene with gene-specific markers. The question confronting us is how can we most effectively apply this approach to address the problems facing breeders, growers, and processors of cool season food legumes.

Clearly, we must have open and responsive avenues of communication between the various players in the game. International centers such as ICARDA and ICRISAT must play a central role for the crops they handle. More regional organizations (agricultural research centers, the John Innes Institute, university campuses), at which growers, processors, breeders, and molecular biologists can interact and cooperate on specific projects, also will contribute critical information. We need gene specific markers for all genes conferring resistance to specific diseases or races of pathogens. We need recombinant inbred lines to facilitate the study of quantitatively inherited characters. Although much of the above is true for any crop, the cool season food legumes represent a cohesive, relatively small taxonomic group in which genetic knowledge acquired in one crop often will have direct applications in the other crops. Such inter-crop comparisons will become particularly important as biotechnology matures and breeders begin looking for genes in species outside the cross-compatible germplasm.

References

Abbo, S., Ladizinsky, G. and Weeden, N. F. 1992. *Euphytica* 58: 259–266.

Aydin, H., Muehlbauer, F. J. and Kaiser, W. J. 1987. *Plant Disease* 71: 635–638.

Baggett, J. R. and Hampton, R. O. 1991. *Journal of the American Society for Horticultural Science* 116: 728–731.

Bateson, W. and Punnet, R. C. 1905. *Report to the Evolution Committee of the Royal Society of London*, II.

Blixt, S. 1974. In: *Handbook of Genetics*, Vol. 2, pp. 181–221 (ed. R. C. King). New York: Plenum.

Casey, R. 1982. *Qualitas Plantarum Plant Foods for Human Nutrition* 31: 281–295.

Diaz, C. L., Melchers, L. S., Hooykass, P. J. J., Lugtenberg, B. J. J. and Kijne, J. W. 1989. *Nature* 338: 579–581.

Drijfhout, E. 1968. *Euphytica* 17: 224–235.

Duc, G. and Messager, A. 1989. *Plant Science* 60: 207–213.

Ellis, T. H. N., Turner, L., Hellens, R. P., Lee, D., Harker, C. L., Enard, C., Domoney, C. and Davies, D. R. 1992. *Genetics* 130: 649–663.

Engvild, K. J. 1987. *Theoretical and Applied Genetics* 74: 711–713.

Gaur, P. M. and Slinkard, A. E. 1990a. *The Journal of Heredity* 81: 455–461.

Gaur, P. M. and Slinkard, A. E. 1990b. *Theoretical and Applied Genetics* 80: 648–656.

Godt, M. W. and Hamrick, J. L. 1991. *American Journal of Botany* 78: 1163–1171.

Gottshalk, W. 1988. *Theoretical and Applied Genetics* 75: 344–349.

Haddad, N. I., Muehlbauer, F. J. and Hampton, R. O. 1978. *Crop Science* 18: 613–615.

Hagedorn, D. J. and Gritton, E. T. 1973. *Phytopathology* 63: 1130–1133.

Havey, M. J. and Muehlbauer, F. J. 1989. *Theoretical and Applied Genetics* 77: 393–401.

Hoffman, D. L., Soltis, D. E., Muehlbauer, F. J. and Ladizinsky, G. 1986. *Systematic Botany* 11: 392–402.

Hunt, J. S. and Barnes, M. F. 1982. *Euphytica* 31: 341–348.

Jacobsen, E. and Feenstra, W. J. 1984. *Plant Science Letters* 33: 337–344.

Kneen, B. E. and LaRue, T. A. 1988. *Plant Science* 58: 177–182.

Lie, T. A. 1984. *Plant and Soil* 82: 415–425.

Lu, J., Weeden, N. F. and LaRue, T. A. Chromosomal location of lectin genes indicate they are not the basis of *Rhizobium* strain specificity mutations identified in pea (*Pisum sativum*). *The Journal of Heredity* (Submitted).

Mancini, R., De Pace, C., Mugnozza, G. T. S., Delre, V. and Vittori, D. 1989. *Theoretical and Applied Genetics* 77: 657–667.

Marx, G. A. 1971. *Pisum Newsletter* 3: 18–19.

Marx, G. A. 1975. *Pisum Newsletter* 7: 3031.

Marx, G. A. and Provvidenti, R. 1979. *Pisum Newsletter* 11: 28–29.

Michelmore, R. W., Paran, I. and Kesseli, R. V. 1991. *Proceeding of the National Academy of Sciences, USA* 88: 9828–9832.

Morgan, T. H. 1911. *Science* 34: 384.

Muehlbauer, F. J. and Singh, K. B. 1987. Chapter 6. Genetics of Chickpea. In: *The Chickpea*, pp. 99–125 (eds. M. C. Saxena and K. B. Singh). Wallingford, UK: CAB International.

Muehlbauer, F. J., Weeden, N. F. and Hoffman, D. L. 1989. *The Journal of Heredity* 80: 298–303.

Murfet, I. C. 1988. In: *Plant Reproduction: From Floral Induction to Pollination*, pp. 10–18 (eds. E. Lord and G. Bernier). Rockville, MD: American Society of Plant Physiologist.

Murfet, I. C. 1990. *Pisum Newsletter* 22: 38–40.

Murfet, I. C. and Reid, J. 1985. In: *The Pea Crop: A Basis for Improvement*, pp. 67–80 (eds. P. D. Hebblethwaite, M. C. Heath and T. C. K. Dawkins). London: Butterworths.

Murray, D. R. and Ayre, D. J. 1987. *Journal of Plant Physiology* 127: 193–201.

Provvidenti, R. and Hampton, R. O. 1991. *Pisum Genetics* 23: 26–28.

Przybylska, J. 1986. *Seed Science & Technology* 14: 529–543.

Saiki, R. K., Gelfand, D. H., Stoffel, S., Scharf, S., Higuchi, R. H., Horn, G. T., Mullis, K. B. and Erlich, H. A. 1988. *Science* 239: 487–491.

Simon, C. J. and Muehlbauer, F. J. 1991. In: *Agronomy Abstracts*, p. 116. Madison: American Society of Agronomy. Skarzynska, A. 1988. *Pisum Newsletter* 20: 34–36.

Slinkard, A. E., Matus, A. and Vandenberg A. 1992. *2nd International Food Legume Research Conference Program and Abstracts*, p. 73. 12–16 April 1992. Cairo, Egypt.

Swiecicki, W. K. 1990. *Pisum Newsletter* 22: 62–63.

Tahir, M. 1990. *Use of isozymes polymorphisms in lentil (Lens culinaris Medik.) for gene mapping and detection of quantitative trait loci.* Ph.D. Thesis, 115 pp. Washington State University, Pullman, Washington, USA.

Timmerman, G. M., Frew, T. J., Miller, A. L., Weeden, N. F. and Jermyn, W. A. 1993. *Theoretical and Applied Genetics* 85: 609–615.

Tingey, S. V., Walker, E. L. and Coruzzi, G. M. 1987. *The EMBO Journal* 6: 1–9.

Vaillancourt, R. 1989. *Inheritance and linkage of morphological markers and isozymes in lentil.* Ph.D. Thesis, 133 pp. University of Saskatchewan, Saskatoon, Canada.

Vaillancourt, R., Slinkard, A. E. and Reichert, R. D. 1986. *Canadian Journal of Plant Science* 66: 241–246.

Weeden, N. F. 1988. *Pisum Newsletter* 20: 46–48.

Weeden, N. F., Kneen, B. E. and LaRue, T. A. 1990. In: *Nitrogen Fixation: Achievements and Objectives*, pp. 323–330 (eds. P. M. Gresshoff, L. E. Roth, G. Stacey and W. E. Newton). London: Chapman and Hall.

Weeden, N. F. and Marx, G. A. 1987. *The Journal of Heredity* 78: 153–159.

Weeden, N. F., Muehlbauer, F. J. and Ladizinsky, G. 1992. *The Journal of Heredity* 83: 123–129.

Weeden, N. F. and Provvidenti, R. 1988. *The Journal of Heredity* 79: 128–130.

Weeden, N. F., Provvidenti, R. and Marx, G. A. 1984. *The Journal of Heredity* 75: 411–412.
Weeden, N. F., Provvidenti, R. and Wolko, B. 1991. *Pisum Genetics* 23: 42–43.
Weeden, N. F. and Wolko, B. 1990. In: *Genetic Maps*, (5th edition), pp. 6.106–6.112 (ed. S. J. O'Brien). Cold Spring Harbor: Cold Spring Harbor Press.
Williams, J. G. K., Kubelik, A. R., Livak, K. J., Rafalski, J. A. and Tingey, S. V. 1990. *Nucleic Acids Research* 18: 6531–6535.
Wolko, B. and Weeden, N. F. 1991. *Pisum Newsletter* 22: 71–74.
Zamir, D. and Ladizinsky, G. 1984. *Euphytica* 33: 329–336.

Cloning *Bacillus thuringiensis* toxin genes for control of nodule-feeding insects

D.F. BEZDICEK[1], M.A. QUINN[1], L. FORSE[1], D.P. BECK[2] and
S. WEIGAND[2]

[1] *Department of Crop and Soil Sciences, Washington State University, Pullman, Washington 99164–6420, USA, and*
[2] *ICARDA, P. O. Box 5466, Aleppo, Syria*

Abstract

The *cryIII* endotoxin gene from *Bacillus thuringiensis* subsp. *tenebrionis* was cloned into strains of *Rhizobium leguminosarum* and *R. meliloti*. Strains were constructed that used a constitutive *lacZ* promoter or a conditional *nifH* promoter. CryIII toxin expression in nodules resulted in significant reductions in nodule-feeding damage by *Sitona lineatus* on *Pisum sativum* and *S. hispidulus* on *Medicago sativa*. Results from a greenhouse experiment indicated that the genetically engineered rhizobia were competitive with the parent wildtype strain of *R. leguminosarum* and with the same strain containing Tn5 in the chromosome. Biomass of *P. sativum* was reduced when plants were inoculated with single strains of the engineered rhizobia, but were not affected when coinoculated with wildtype rhizobia.

Introduction

Legumes are an integral part of agricultural systems throughout the world, providing an inexpensive source of protein for people and livestock. The value of legumes lies in the symbiotic relationship between rhizobial bacteria and plants that form root nodules and fix nitrogen. Understandably, scientists have placed considerable emphasis on the enhancement of biological nitrogen fixation of plants. In the USA alone, just increasing the efficiency of nitrogen fixation of legumes could result in a benefit of US$ 1,067,000,000 per year and decrease the need for nitrogen fertilizers by 1,547,000 metric tons (Tauer, 1989). However, the current and future value of legumes in agricultural systems is limited, in part, by nodule feeding insects that disrupt nitrogen fixation and reduce crop yields.

Sitona spp. are, perhaps, the most economically important group of nodule-feeding insects because of their world-wide distribution and abundance on numerous forage and cool season legumes. Nodule feeders can have a substantial direct impact on root nodule density, biomass, and nitrogen fixation

F.J. Muehlbauer and W.J. Kaiser (eds.), Expanding the Production and Use of Cool Season Food Legumes, 738–752.

(Quinn and Hower, 1986a; Saxena, 1988). Estimates of nodule destruction from *Sitona* spp. include 80% for *S. cylindricollis* on *Melilotus* sp. (Manglitz *et al.*, 1963), 52 to 62% for *S. lineatus* on *Pisum sativum* (George, 1962), 8 to 12% for *S. hispidulus* on *Medicago sativa* (Quinn and Hower, 1986b), and 46 to 77% for *S. crinitus* on *Lens culinaris* (ICARDA, 1990).

In the Pacific Northwest, *S. lineatus* is a serious threat to pea production (O'Keeffe *et al.*, 1984). In a 3-year study of *S. lineatus* on winter pea, O'Keeffe and Schotzko (unpublished data) found pea yields decreased between 12 and 26% from larval and adult feeding. In the Palouse Region of Idaho and Washington, *S. lineatus* infestations require the treatment of 25 to 75% of all fields each year. *Sitona lineatus*, *S. limosus*, and *S. crinitus* are also important insect pests of cool season legumes world-wide. These insects frequently cause high yield losses of faba bean, pea, and lentil in the Middle East (Cakmakci *et al.*, 1988; ICARDA, 1990). Other important groups of nodule feeders include *S. discoideus* in New Zealand, Australia, and Europe (Goldson *et al.*, 1988), and *S. hispidulus* in the USA (Quinn and Hower, 1986a,b; Hower *et al.*, 1993).

Currently, insecticides are the only option for controlling damage from nodule-feeding insects. Little progress has been made on developing legumes resistant to nodule feeders and there are few significant biological control agents for the insects. One viable strategy for controlling nodule-feeding insects is the genetic engineering of rhizobia to make root nodules insecticidal. Because root nodules are an essential food resource for larvae of *Sitona* spp. (Quinn and Hower, 1986a), nodule-specific toxins may be used to reduce nodule damage and disrupt the insect's life cycle.

Recently, there has been considerable progress in incorporating insecticidal toxin genes (*cry*) from *Bacillus thuringiensis* (Bt) into a broad range of hosts for controlling insect pests. These Gram-positive soil bacteria produce insecticidal proteinaceous crystals during sporulation. Different strains of Bt produce crystal proteins (Cry) which are toxic to different insect species (Höfte and Whiteley, 1989). Most of the described Bt strains are toxic to lepidopteran and dipteran larvae (Aronson *et al.*, 1986). Recently described strains of *B. thuringiensis* subsp. *tenebrionis* (Btt) and subsp. *san diego* affect larvae of some coleopteran species (Krieg *et al.*, 1983; Herrnstadt *et al.*, 1986; Donovan *et al.*, 1988). Microbial formulations of Bt toxins have been used for over 25 years to control a variety of insect pests (primarily lepidopterans) in agriculture and forestry (Klein, 1988). Researchers have investigated a number of novel strategies for deploying various Bt *cry* genes in plants (Barton *et al.*, 1987; Fischhoff *et al.*, 1987; Vaeck *et al.*, 1987; Gasser and Fraley, 1989), root-colonizing pseudomonads (Obukowicz *et al.*, 1986), and in rhizobia of legume nodules (Nambiar *et al.*, 1990; Skot *et al.*, 1990; Bezdicek *et al.*, 1993).

Here we report on the development of a molecular-based strategy for controlling damage to nitrogen-fixing root nodules by larvae of *Sitona* spp. Specific objectives of our study were to: 1) clone the *cryIII* gene from Btt into *Rhizobium*, 2) determine if CryIII protein is expressed at levels sufficient to affect feeding behavior of nodule-feeding insects, 3) determine if the *cryIII* gene is stable in free-living bacteria and in bacteroids within root nodules, 4) determine if the *cryIII* gene affects plant growth and the competitive ability of the engineered rhizobia, and 5) develop a strategy for managing resistance of nodule-feeding insects to Btt endotoxin.

Materials and Methods

Subcloning and Introduction of CryIII into Rhizobium

Source and description of plasmids and bacterial strains used in the study are shown in Table 1. A schematic diagram of the procedure for inserting the *cryIII* gene from Btt into rhizobia is shown in Figure 1. Additional details of the cloning procedure can be found in Bezdicek *et al.* (1993). The *cryIII* gene responsible for the production of the endotoxin resides on plasmids within the bacteria. Several researchers have successfully cloned the *cryIII* gene from Btt into a variety of cloning vehicles (e.g., Sekar *et al.*, 1987; McPherson *et al.*, 1988; Skot *et al.*, 1990; Bezdicek *et al.*, 1993). For our study, we obtained the *cryIII*

Table 1. Plasmids and bacterial strains

Plasmid/strain	Description	Source
Plasmids:		
pBtT-CP-1	5.3 kb clone of Btt plasmid DNA inserted into *BamHI-XhoI* of Bluescript Minus. This Btt fragment contains the *cryIII* gene which encodes a protein (CryIII) toxic to some coleopteran insects.	Whiteley
pRK311	Broad host range vector in Gram negative bacteria; Tet', *lacZ* promoter; with multiple cloning site.	Ditta
pBtt-MK	*HindIII-BglII* fragment from pBtT-CP-1 inserted in *HindIII-BamHI* of pRK311. Whole toxin gene downstream of *lacZ* promoter.	This lab
pBtt-LZ	pRK311 with 2.3 kb *PstI-BglII* subclone of *cryIII* gene derived from pBtT-CP-1, with the 5' flanking sequence and 150 bp of the coding region removed (yields a "truncated" CryIII protein). Transcription relies on the *lacZ* promoter associated with the multiple cloning site of pRK311.	This lab
pBtt-nH	Translational fusion of the *R. meliloti nifH* promoter and 28 amino acid residues of the N-terminus of the NifH protein fused to the truncated *cryIII* gene constructed in pBtt-LZ. This was created by making a *HindIII* cartridge out of the nifH promoter and N-terminus sequences cloned in pVSP9 (Sundaresan *et al.* 1983). The *nifH* promoter is a conditional promoter in *Rhizobium*, and is responsible for the expression of the nifHDK operon which encodes the nitrogenase. This promoter is very actively transcribed by rhizobia within root nodules under conditions of low oxygen and fixed nitrogen.	This lab
Bacterial strain:		
E. coli S17-1		
R. leguminosarum biovar *viceae* 300	wildtype strain; nodulates pea	This lab
R. leguminorsarum biovar *viceae* C1204	wildtype strain; nodulates pea	This lab
R. leguminosarum biovar *viceae* C1204 with IP4	; nodules pea	This lab
R. meliloti 102F51	wildtype strain; nodulates alfalfa	This lab

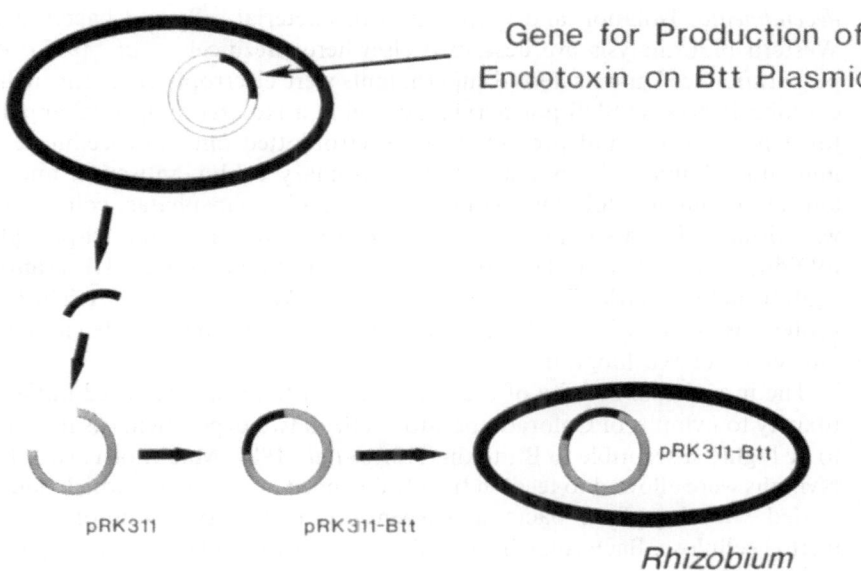

Gene for Production of
Endotoxin on Btt Plasmid

Rhizobium

Insert Btt toxin Gene into
Broad Host Range Plasmid

Figure 1. Schematic diagram of the cloning procedure for incorporating the delta endotoxin gene into *Rhizobium* spp.

gene from Helen Whiteley (University of Washington) as a 5.3-kb clone of Btt plasmid DNA (*pBtT-CP-1*) in Bluescript Minus. A truncated *cryIII* gene was then subcloned into the broad host range plasmid, pRK311, that also contains a tetracycline resistance gene (*tet*) as a selectible marker.

Two different subclones of *cryIII* in pRK311 were developed, *pBtt-LZ* and pBtt-nH. These constructs are similar in that they both have an N-terminus-truncated form of *cryIII* in the broad host range vector pRK311, and differ only in the promoters that drive expression of *cryIII*. In *pBtt-LZ*, expression relies on the constitutive *lacZ* promoter which is expressed *in vitro* and *in planta*. The pBtt-nH was derived from pBtt-LZ, but incorporates a translational fusion of the *R. meliloti nifH* promoter, and is active only when rhizobia are fixing nitrogen inside the nodules. The constructs were conjugated into strains of *R. leguminosarum* biovar *viceae* used to inoculate *P. sativum* cv. Latah, and *R. meliloti* used to inoculate *M. sativa* cv. Ladak.

Detection of CryIII Protein Expression

Two procedures were used initially to detect expression of the endotoxin, Western blot analysis and bioassay with Colorado potato beetle, *Leptinotarsa decemlineata.* Isolation and preparation of bacterial cells and bacteroids for Western blot analysis are described elsewhere (Bezdicek *et al.*, 1993). After preparation, cell and bacteroid supernatants were electrophoresed through a 5-cm long 10% (w:v) SDS-polyacrylamide gel in a Hoeffer midget gel apparatus for 3 h at 30 mA and proteins were electroblotted onto nitrocellulose. The immunoglobulin was obtained using a primary rabbit antiserum and goat antirabbit immunoglobulin G coupled with alkaline phosphatase color reaction was visualized in a solution of BCIP (5-chloro-4-bromo-3-indolyl-phosphate, toluidine salt) and NBT. The primary antiserum (rabbit anti-CryIII serum) was a gift from Dr. Charles Brown, IAREC, Prosser, Washington, USA. The CryIII protein used as an internal standard was a gift from the late Helen Whitely, University of Washington.

The insecticidal activity of the truncated Btt toxin was assessed initially by toxicity to nymphs of Colorado potato beetle in two experiments as it is known to be highly susceptible to Btt toxin (Höfte *et al.*, 1987; MacIntosh *et al.*, 1990). Nymphs were allowed to feed on freshly detached potato leaves which had been coated with log-phase bacterial suspensions in aqueous solutions of 0.5% methyl cellulose. Bacteroids from nodules of 4-wk-old plants were obtained by crushing 1.2 g (fresh weight) of nodules in the methyl cellulose solution, centrifuging 10 sec at low speed to pellet plant debris, then drawing off the supernatant containing the cells.

For the bioassay, leaves were coated with one of the following cell suspensions of *Escherichia coli* and *R. leguminosarum* containing combinations of pBtt-LZ, pRK311, and wildtype controls. Nodule suspensions containing *R. leguminosarum* strain C1204 (hereafter refered to as C1) with pRK311, pBtt-LZ, or pBtt-nH were also exposed to the larvae as were methyl cellulose controls. After coating the leaves thoroughly with the bacterial suspensions or controls, the leaf surfaces were allowed to dry at room temperature before five 1st-instar nymphs were placed on each leaf in six replicated petri dishes. Larvae were allowed to feed at room temperature, and treated leaf tissue was replenished after 24 to 48 h to ensure an adequate supply of food for the larvae. Mortality and number of molts were recorded daily, and feeding damage was estimated visually. Colorado potato beetle eggs were provided by D. Beever, USDA-ARS, Yakima, Washington, USA.

Plasmid Stability

Experiments were conducted to determine if the pRK311 plasmid is stably maintained in *Rhizobium* in the absence of tetracycline selection *in vitro* and *in planta*. For the *in vitro* studies, strains containing pRK311 and pBtt-LZ were grown in 25 ml yeast extract mannitol (YEM) broth at room temperature on a rotary shaker (120 rpm). On the second day after inoculation of the initial culture, and at daily intervals for 3 days thereafter, a 0.1 ml aliquot of the most recent culture was transferred to fresh broth to achieve a total of 40 generations.

At the time of transfer, an additional aliquot was used for a dilution plate series (two plates per dilution) on YEM agar with and without Tet (25 mg l^{-1}). After 2 days incubation of the plates at 28°C, colonies were counted.

For the studies of stability *in planta*, nodules were removed from plants 3 weeks after inoculation with the appropriate strains harboring pRK311 or pBtt-LZ. Five nodules were surface-disinfected and then crushed in sterile water to obtain a suspension of bacteroids. The suspensions were serially diluted in sterile water and aliqouts were plated onto YEM agar. After 2 days incubation of the plates as above, colonies were transferred onto YEM and YEM containing 25 mg Tet l^{-1}. Numbers of resultant colonies on each medium were recorded after an additional 2 days incubation at 28°C.

Sitona spp. Bioassays

Bioassays of CryIII activity against *S. hispidulus* and *S. lineatus* were conducted by allowing neonate larvae to feed on intact root nodules of *M. sativa* or *P. sativum*, respectively. Plants were grown in a soil-based experimental apparatus that is a cone-shaped version of the Leonard jars commonly used in *Rhizobium* research. Plants were grown in cones made from 60-ml autoclavable polypropylene syringes containing a 1:1 mixture of perlite and washed river sand. Nitrogen-free nutrient solution was added to a supply reservoir at the bottom of containers that hold the syringes with plants. Cotton wicks were used to draw the nutrient solution from the reservoir to the soil in the syringe. The apparatus was autoclaved to ensure microbiological control. This system provided for vigorous plant growth and nodulation necessary to assure a sufficient food base for the *Sitona* larvae. Aseptic technique was used throughout the pregermination of seeds and inoculation of seedlings with rhizobia. Surface-disinfected seeds were pregerminated in darkness on moistened, sterile paper towels for 3 to 5 days until the emerging roots were 2 to 4 cm long, at which time the seedlings were inoculated with rhizobia (10^7 cells plant^{-1}) and sown in the cones. Plants were maintained in a growth chamber for a 12 h photoperiod at temperatures of 20°C/15°C (day/night) for *P. sativum* and 25°C/20°C for *M. sativa*. Approximately 21 days after sowing and inoculation of the plants with rhizobia, there was sufficient nodule biomass to support *Sitona* herbivory. *Sitona* eggs were placed on the soil surface at the base of plants (25 eggs plant^{-1}) approximately 21 days after sowing. The timing of egg hatch was estimated by observing the hatching of a subset of eggs placed in a petri dish. To control time of egg hatch, only eggs oviposited on the same day were used. Seven days after egg hatch, the plants and soil were removed to determine numbers of damaged and undamaged nodules per plant. Nine to 15 replicates per treatment were used in a factorial design.

Ecology of CryIII-Containing Rhizobia

A greenhouse experiment was conducted to examine the competitive ability of the *cryIII*-containing *R. leguminosarum* strains with wildtype strains, to determine their effect on plant growth, and to assess the potential for genetic exchange between rhizobial strains within nodules. For the experiments, pea plants were grown in sterile 14-ounce plastic cups that contained a sterilized mixture of perlite and sand (1:1) and that were suspended above a tray that collected excess nutrient solution. The entire system was semi-enclosed to minimize contamination. Sterile nitrogen-free nutrient solution was added through a drip-irrigation system. At planting, surface-sterilized seeds were inoculated with the appropriate strain or strains of rhizobia (a total of 10^7 cells per seed) and the cups sealed with a lid that contained a small hole for the emerging shoot. Inoculant density was confirmed by culturing on YEM agar.

For the competition experiment, the *cryIII*-containing strains were coinoculated (10^7 total cells/plant) with either the parent wildtype strain of *R. leguminosarum* C1 or strain IP4, which is a *Tn-5* mutant of strain C1 carrying the phenotypic marker, kanamycin (Km). This mutant strain has been shown to be equally competitive with the wildtype *R. leguminosarum* C1 (Brockman *et al.*, 1991) and serves as an ideal strain in competition studies. A completely randomized design was used for all experiments, with four replicates per treatment. Six weeks after planting, 24 nodules were removed from each plant, surface sterilized for 3 min in 2.5% Chlorox bleach with 0.04% Tween 80, exhaustively rinsed in sterile water, and placed in individual wells of a microtitre plate containing 100 μL YEM broth. To improve the expression of the inducible *tet* operon of pRK311-containing strains, nodules were crushed in YEM broth containing 50 mg l^{-1} of chlorotetracycline which was autoclaved previously to eliminate the effect of the antibiotic. Nodules were crushed with an aseptic blunt glass rod and suspensions were replica plated with a 48-prong stainless steel device onto three media: 1) plain YEM agar; 2) YEM with 25 mg Tet l^{-1} to detect pRK311-containing strains; and 3) YEM with 100 mg l^{-1} Km to detect expression of strain IP4. At the end of the experiment, plant tissue was harvested, dried at 60°C, and weighed. Analysis of variance and Fisher's protected LSD were used to detect treatment effects and to compare mean biomass of single and double strain-inoculated plants.

The proportion of nodules occupied by more than one strain was determined by positive colony growth from replica plating nodule extracts on separate agar YEM plates containing either 25 mg Tet l^{-1} or 100 mg l^{-1} Km. Frequency of genetic exchange of *cryIII* and *Tn-5* genetic material in the nodule was assessed by replica plating nodule extracts on YEM containing both Tet and Km. Positive colony growth on both Tet and Km from single nodule suspensions derived from double inoculated strains containing either pRK311 or Tn5 would indicate genetic exchange during nodule formation or function.

Results

Detection of CryIII Protein Expression

The presence of CryIII protein as the toxin affecting *Sitona* spp. was verified by Western blot analysis of *Rhizobium* and nodules and by bioassays with Colorado potato beetle nymphs (Bezdicek *et al.*, 1993). Results showed that the CryIII protein was expressed both in free-living rhizobia containing pBtt-LZ, and within root nodules derived from plants inoculated with rhizobia containing either pBtt-nH or pBtt-LZ. These results are expected as pBtt-nH is expressed only during active nitrogen fixation. Bioassays indicated that Colorado potato beetle nymphs were highly susceptible to the endotoxin when coated on potato leaves. Mortality of beetle nymphs ranged from 60 to 83% from rhizobial and *E. coli* cells containing the *cryIII* gene as compared to minimal mortality for potato tissue coated with bacterial cells without the plasmid vector or with methyl cellulose alone (Bezdicek *et al.*, 1993).

Expression of the CryIII protein in nodules significantly reduced nodule damage by nodule-feeding insects (Table 2). Damage to *M. sativa* nodules by *S. hispidulus* larvae was significantly lower in plants inoculated with *R. meliloti* with pBtt-LZ (45%) compared to plants inoculated with the control strain (71%). Similarly, only 30% of *P. sativum* nodules were damaged by *S. lineatus* larvae when plants were inoculated with *R. leguminosarum* strain 300 containing pBtt-nH, compared with 44% of control nodules. Also, percentage of nodules damaged by *S. lineatus* larvae feeding on plants inoculated with *R. leguminosarum* strain C1 with pBtt-LZ and pBtt-nH was 14 and 18%, respectively. In comparison, greater than 25% of nodules were damaged in plants inoculated with either the wildtype or pRK311-containing strains (Table 2). In all experiments, the number of nodules per plant was not affected by the presence of the *cryIII* gene. These results show that the presence of the CryIII protein reduced nodule feeding by both *Sitona* spp. on alfalfa and lentil.

Table 2. Effect of CryIII protein expression in nodules of *M. sativa* and *P. sativum* on nodule feeding insects

Insect	Exp.	*Rhizobium* strain	# nodules per plant ± SEM (n)	% nodules damaged per plant ± SEM (n)
S. hispidulus	1	*R. meliloti* with pBtt-LZ	13.8 ± 3.3 (9)	44.7 ± 9.7 (9)
		R. meliloti with pRK311	8.8 ± 1.7 (10)	70.8 ± 7.2 (10)
S. lineatus	2	*R. leguminosarum* with pRK311	88.1 ± 8.7 (11)	43.7 ± 3.8 (11)
		R. leguminosarum with pBtt-nH	90.4 ± 9.6 (11)	30.0 ± 3.5 (11)
S. lineatus	3	*R. leguminosarum* C1204 (Wt)	108.6 ± 5.8 (15)	25.7 ± 2.8 (15)
		R. leguminosarum with pRK311	80.9 ± 4.5 (15)	30.0 ± 1.7 (15)
		R. leguminosarum with pBtt-LZ	112.8 ± 8.1 (15)	13.6 ± 1.3 (15)
		R. leguminosarum with pBtt-nH	112.9 ± 9.8 (15)	18.2 ± 2.6 (15)

Plasmid Stability

After 40 generations in culture, 95% of the isolates containing pRK311 and pBtt-LZ carried the Tet phenotype (data not shown). Cells cultured from nodules of pea inoculated with rhizobia containing pBtt-LZ also maintained the Tet phenotype from 95% of the isolates replica plated from YEM onto YEM plus Tet. These results suggest that the plasmids were stably maintained in cell culture and in nodules.

Ecology of CryIII-Containing Rhizobia

The *cryIII*-containing rhizobia were competitive with the wildtype *R. leguminosarum* and IP4 strains (Table 3). Nodules from pea inoculated with single strains of Bt containing pRK311, pBtt-LZ, or pBtt-nH expressed the Tet phenotype for 81 to 91% of the isolates. In competition with the wildtype *R. leguminosarum*, the two recombinant strains containing pBtt-LZ or pRK311 were recovered from 63 and 66%, respectively, of nodules on YEM with Tet, which is close to the expected 50% recovery. In contrast, rhizobia containing pBtt-nH was recovered from 40% of nodules.

To determine the competitiveness of a different set of strains and to

Table 3. Recovery of rhizobia from nodules of pea plants inoculated with single or double strains of *R. leguminosarum* (Bezdicek *et al.*, 1993)

Rhizobial strain	Percentage of nodules growing on media[a]				
	YMB	tet	kan	tet→kan[b]	tet + kan[c]
Wildtype	100	0	0	–	–
pRK311	100	87	–	–	–
pBtt-LZ	100	91	–	–	–
pBtt-nH	100	81	–	–	–
IP4	100	0	100	–	0
Wildtype/pRK311	100	66	–	–	–
Wildtype/pBtt-LZ	100	63	–	–	–
Wildtype/pBtt-nH	100	40	–	–	–
IP4/Wildtype	100	0	74	–	0
IP4/pRK311	100	54	64	28	0
IP4/pBtt-LZ	100	97	95	95	0
IP4/pBtt-nH	100	62	76	40	0

[a]96 nodules/4 replicates/treatment.

[b]replica plated on YMB with tet, then YMB with kan to detect double occupancy of nodules.

[c]replica plated on YEM with both tet and kan to detect genetic exchange.

determine the degree of double occupancy of strains in the nodules, two different recombinant strains containing either pRK311 or Tn5 were also co-inoculated on pea seeds in equal proportion and selected based on their respective resistances to Tet and Km. Results suggest that the pRK311-based strain and the Tn5 strain (IP4) were nearly equal in competitiveness, occupying 54 and 64% of the nodules, respectively, which agrees with the results shown above. The IP4 and C1-pBtt-nH strains were recovered from 76 and 62% of nodules, respectively.

Results indicated that at least 40 to 90% of the nodules were double occupied by both *cryIII*-containing and Tn5 containing strains (Table 3). There is no apparent explanation for the high double occupancy (95%) for nodules co-inoculated with IP4 and the strain containing pBtt-LZ. Genetic exchange of the *cryIII* gene between strains within nodules did not apparently occur as none of the isolates from nodules grew on YEM containing both antibiotics.

Total shoot dry weight was significantly affected by strain competition (Table 4). Plants inoculated with single strains of the *cryIII*-containing rhizobia or the pRK311 control (all strains combined) had significantly lower shoot biomass compared with double-inoculated plants (all strains combined)(t = 2.5, P <0.05). There were no significant differences in biomass of plants inoculated with *cryIII*-containing rhizobia and pRK311 (P > 0.05).

Table 4. Shoot dry wts of pea inoculated with different strains of R. *leguminosarum* (Bezdicek et al., 1993)

Rhizobia strain	Shoot dry wt ± SEM[a]
	- - - g - - -
Uninoculated	0.375 ± 0.045
Wildtype	1.585 ± 0.045
pRK311	1.123 ± 0.085
pBtt-LZ	1.180 ± 0.121
pBtt-nH	1.105 ± 0.133
IP4	1.340 ± 0.039
Wildtype/pRK311	1.285 ± 0.093
Wildtype/pBtt-LZ	1.213 ± 0.055
Wildtype/pBtt-nH	1.378 ± 0.058
IP4/Wildtype	1.345 ± 0.112
IP4/pRK311	1.428 ± 0.060
IP4/pBtt-LZ	1.208 ± 0.101
IP4/pBtt-nH	1.220 ± 0.119

[a]n = four plants per treatment.

Concluding Remarks

Several factors must be considered before releasing any genetically engineered microorganism, including the potential for horizontal gene transfer, alteration of competitive abilities, alteration of plant growth and physiology, detrimental interactions with other organisms, and problems with transport and dispersal (NRC, 1989; Levin and Strauss, 1991; Sharples, 1991). Our research has focused on several factors that must be understood before *cryIII*-containing rhizobia can be used to control nodule-feeding damage by *Sitona* spp.: 1) the CryIII protein must be expressed in nodules in sufficient quantities to affect the insects; 2) the gene must be stable in rhizobia and within bacteroids in nodules, and the potential for genetic exchange must be low; 3) engineered rhizobia must be competitive with other rhizobial strains; 4) applications of genetically-engineered rhizobia must not interfere with plant growth and nitrogen fixation; and 5) expression of the Bt endotoxin should not result in the rapid development of resistance by the insect to the toxin.

Our results showed that expression of the Btt endotoxin by rhizobia bacteroids within nodules effectively reduced nodule-feeding damage by *S. lineatus* on *P. sativum* and by *S. hispidulus* on *M. sativa*. The toxin did not totally prevent feeding on nodules and did not cause immediate mortality of the insects. Long-term mortality was not examined. Ideally, the toxin should reduce nodule damage and not totally eliminate the insect. High toxicity would likely increase selective pressure on the insect, resulting in a more rapid development of resistance to the toxin (see below). If nodule damage can be reduced, then the plants may be better able to compensate for the small amount of damage that does occur (Quinn and Hall, 1992).

Skot *et al.* (1990) also reported that CryIII protein produced by genetically engineered *R. leguminosarum* and *R. trifolii* was toxic to larvae of the clover weevil, *S. lepidus*. In their study, endotoxin expression was controlled by the *lacZ* promoter. Our results concur with those of Skot *et al.* (1990) that *Sitona* spp. are among the few coleopterans that are affected by CryIII endotoxin and, in addition, show that expression controlled by the conditional *nifH* promoter also affects *Sitona* spp. Nambiar *et al.* (1990) cloned insecticidal toxin genes of *B. thuringiensis* subsp. *israelensis* into a broad-host range IncQ plasmid vector and transferred this by conjugation into *Bradyrhizobium* spp. which nodulate pigeon pea. Subsequent inoculation and nodulation by the recombinant rhizobia produced nodules that were toxic to the nodule-feeding dipteran, *Rivellia angulata*.

The potential for genetic exchange should be low in genetically engineered microorganisms released into the environment. In our study, transfer of genetic material as determined from phenotypic antibiotic resistance did not occur between the *R. leguminosarum* strains containing the pRK311 plasmid (pRK311, pBtt-LZ, pBtt-nH) and the IP4 strain within nodules. We did not examine genetic exchange between free-living rhizobia. Although pRK311 was stably maintained in free-living rhizobia and in nodules, plasmid vectors are, in general, not as stable as chromosomal insertions of genes. For example, *Rhizobium* are known to exchange genes that regulate important physiological mechanisms such as host compatibility, competitiveness, and nitrogen fixation (Broughton *et al.*, 1987; Schofield *et al.*, 1987; Döhler and Klingmuller, 1988; Van Elsas *et al.*, 1988; Richaume *et al.*, 1989).

Transfer of genes can occur between *Rhizobium* and other bacteria species (Döhler and Klingmuller, 1988; Richaume *et al.*, 1989), but transfer to host plants has not been detected (Shantharam, 1990). Transfer of plasmids between strains of Bt has also been detected (Jarrett and Stephenson, 1990). The transfer of genetic material between bacterial strains is a complex process that is influenced by numerous environmental factors such as soil structure, pH, presence of roots, and energy source (Van Elsas *et al.*, 1988; Kinkle and Schmidt, 1991; Klingmuller, 1991; Stotzky *et al.*, 1991).

Genetically engineered rhizobia should be competitive with other wildtype strains to maximize efficacy of the endotoxin. Results from our study indicated that the *cryIII*-containing rhizobia were competitive with the parent wildtype strain and strain IP4. The rhizobial strain, IP4, was shown previously to be competitive with the wildtype strain (Brockman *et al.*, 1991). Recovery of singly-inoculated rhizobia containing pRK311, pBtt-LZ, and pBtt-nH was similar to recovery of wildtype and IP4 strains. Although recoveries of most inoculated strains did not reach 100%, it is unlikely that this was not due to a loss of the plasmid, because previous experiments indicated that it was very stable *in vitro* and *in planta* (Bezdicek *et al.*, 1993).

Biomass of plants inoculated with single strains containing pRK311, pBtt-LZ, or pBtt-nH was significantly lower than biomass of double-inoculated strains. Thus, it appears that the presence of the pRK311 plasmid or the *cryIII* genes imparts some cost to plant growth which could be reflected by a reduction in the efficiency of N fixation, translocation of fixed N, or other factors. Since we found no significant difference in biomass of plants inoculated with rhizobial strains containing pRK311, pBtt-LZ, or pBtt-nH, the limiting factor may reside in the pRK311 plasmid which is common to all these constructs. Under field conditions, it may be less likely that plants inoculated with *cryIII*-containing rhizobia would exhibit reduced growth, because indigenous soil strains traditionally occupy a significant proportion of the nodules.

The development of resistance to Bt endotoxins in insects is a significant concern among scientists who use Bt in pest management programs. Resistance to endotoxins of *B. thuringiensis* subsp. *kurstaki* has been reported in lab colonies of diamondback moth, *Plutella xylostella*, (Tabashnik *et al.*, 1991), Indian meal moth, *Plodia interpunctella*, (McGaughey, 1985; McGaughey and Beeman, 1988), and tobacco budworm, *Heliothis virescens* (Stone *et al.*, 1989). Rapid resistance to Bt endotoxins in the lab has been attributed to intense selective pressure (Tabashnik *et al.*, 1991). Recently, resistance has also been detected among field populations of diamondback moth exposed to a commercial formulation of *B. thuringiensis* subsp. *kurstaki* (Tabasknik *et al.*, 1990, 1991). The incorporation of endotoxin genes into plants is likely to intensify resistance (Gould, 1988a,b; Raffa, 1989). Once resistance begins to develop in an insect population, it is unlikely that the insects can be managed effectively with Bt (Tabashnik *et al.*, 1991). Thus, it is important to manage crop systems to delay development resistance.

Several strategies have been proposed to manage such resistance in transgenic plants (Gould, 1988a,b,c; Raffa, 1989; Tabashnik *et al.*, 1990, 1991; Van Rie *et al.*, 1990). These strategies include: 1) the use of tissue-specific promoters to target particular insect stages and pests. This would ensure that other secondary pests and nontarget arthropods and non-damaging stages are not exposed to the toxin; 2) the use of tissue specific promoters that protect only damage-sensitive tissue and cause insects to feed on less valuable tissue; 3) the use of inducible promoters that increase the rate of expression as feeding increases. This would allow low pest levels to exist and would decrease the selective pressure acting on the insect population until populations are high; 4) the use of mixtures of resistant and susceptible plants or plant parts to ensure that susceptible genotypes survive. Any such tactic that increases the spatial variation of Bt toxin expression by providing refuges will delay resistance; and 5) the use of genes and/or promoters that produce a sublethal concentration of toxin. Besides reducing the selective pressure on the insects, this would also allow the integration of other biocontrol agents into IPM programs.

The use of *cryIII*-containing rhizobia to control feeding damage from *Sitona* spp. is less likely to cause problems with resistance compared to the incorporation of the toxin gene directly into the plant. The potential for developing resistance with our system is relatively low because the high degree of temporal and spatial variation in toxin expression within nodules and the relatively low recovery of inoculant strains in nodules. This variability is likely to ensure that numerous susceptible individuals are maintained in the population. Temporal variability in toxin expression may be controlled by using different promoters to drive expression of the endotoxin. For example, rhizobial strains containing pBtt-nH rely on the nitrogen-fixing *nifH* promoter that is active only in mature nodules. Thus, it is likely that individual insects would be exposed to a mosaic of toxin-producing nodules.

Spatial variability in toxin expression and exposure to insects should be common under field situations when nodules are occupied by both *cryIII*-containing and wildtype strains of rhizobia. Because of temporal and spatial variability in toxin expression, many insects will not be exposed to toxins, which should greatly slow down the development of resistance to the toxin. The use of a tissue-specific promoter in nodules to drive toxin expression also has additional advantages. For example, inoculation of plants with pBtt-nH that uses the *nifH* promoter to drive toxin expression may protect valuable, nitrogen fixing plant tissue and force insects to feed on less valuable tissue (non nitrogen-fixing), resulting in less plant damage and lower viability of the insects. Toxin that is only expressed in nodules should also ensure that other secondary pests and nontarget insects are not exposed to the toxin.

References

Aronson, A. I., Beckman, W. and Dunn, P. 1986. *Microbiological Reviews* 50: 1–24.

Barton, K. A., Whiteley, H. R. and Yang, N. S. 1987. *Plant Physiology* 85: 1103–1109.

Bezdicek, D. F., Quinn, M. A., Forse, L., Heron, D. and Kahn, M. L. 1993. Ecology and insecticidal activity of *Rhizobium* spp. in nodules containing the *Bacillus thuringiensis* delta-endotoxin gene. *Soil Biology and Biochemistry* (In press).

Brockman, V. K., Forse, L., Bezdicek, D. F. and Fredrickson, J. K. 1991. *Soil Biology and Biochemistry* 23: 861–867.

Broughton, W. J., Samrey, U. and Stanley, J. 1987. *FEMS Microbiology Letters* 40: 251–255.

Cakmakci, M. L., Bezdicek, D. F. and Sakar, D. 1988. In: *World Crops: Cool Season Food Legumes*, pp. 167–174 (ed. R. J. Summerfield). Dordrecht: Kluwer Scientific Publishers.

Döhler, K. and Klingmüller, W. 1988. In: *Risk Assessment for Deliberate Releases. The Possible Impact of Genetically Engineered Microorganisms on the Environment*, pp. 18–28 (ed. W. Klingmüller). New York: Springer-Verlag.

Donovan, W. P., Gonzalez, Jr., J. M., Gilbert, M. P. and Dankocsik, C. 1988. *Molecular and General Genetics* 214: 365–372.

Fischoff, D. A., Bowdish, K. S., Perlak, F. J., Marrone, P. G., McCormick, S. M., Niedermeyer, J. G., Dean, D. A., Kusano-Kretzmer, K., Mayer, E. J., Rochester, D. E., Rogers, S. G. and Fraley, R. T. 1987. *Bio/Technology* 5: 807–813.

Gasser, C. S. and Fraley, R. T. 1989. *Science* 244: 1293–1299.

George, K. S. 1962. *Plant Pathology* 11: 172–176.

Goldson, S. L., Jamieson, P. D. and Bourdot, G. W. 1988. *Annals of Applied Biology* 113: 189–196.

Gould, F. 1988a. *Trends in Ecology and Evolution* 3: 515–518.

Gould, F. 1988b. *BioScience* 38: 26–33.

Gould, F. 1988c. In: *Biotechnology, Biological Pesticides and Novel Plant-Pest Resistance for Insect Pest Management, Proceedings*, pp. 146–151 (eds. D. W. Roberts and R. R. Granados). Ithaca, New York: Boyce Thompson Institute for Plant Research.

Herrnstadt, C., Soares, G. G., Wilcox, E. R. and Edwards, D. L. 1986. *Bio/Technology* 4: 305–308.

Höfte, H., Seurinck, J., van Hootven, A. and Vaeck, M. 1987. *Nucleic Acids Research* 17: 7183.

Höfte, H. and Whiteley, H. R. 1989. *Microbiology Review* 53: 242–255.

Hower, A. A., Quinn, M. A., Leath, K. T. and Alexander, S. D. 1993. *Journal of Economic Entomology* (In press).

ICARDA. 1990. *Annual Report of the Food Legume Improvement Program*. Aleppo, Syria: ICARDA.

Jarrett, P. and Stephenson, M. 1990. *Applied and Environmental Microbiology* 56: 1608–1614.

Kinkle, B. K. and Schmidt, E. L. 1991. *Applied and Environmental Microbiology* 57: 3264–3269.

Klein, M. G. 1988. *Agriculture, Ecosystems and Environment* 24: 337–349.

Klingmuller, W. 1991. *Microbial Ecology* 85: 107–116.

Krieg, V. A., Huger, A. M., Langenbruch, G A. and Schnetter, W. 1983. *Zeitschrift für angewandte Entomologie* 96: 500–508.

Levin, M. and Strauss, H. S. 1991. In: *Risk Assessment in Genetic Engineering*, pp. 1–17 (eds. M. A. Levin and H. S. Strauss). New York: McGraw-Hill.

MacIntosh, S. C., Stone, T. B., Sims, S. R., Hunst, P. L., Greenplate, J. T., Marrone, P. G., Perlak, F. J., Perlak, D. A., Fischoff, D. A. and Fuchs, R. L. 1990. *Journal of Invertebrate Pathology* 56: 258–266.

Manglitz, G. R., Anderson, D. M. and Gorz, H. J. 1963. *Annals of the Entomological Society of America* 56: 831–835.

McGaughey, W. H. 1985. *Science* 229: 193–195.

McGaughey, W. H. and Beeman, R. W. 1988. *Journal of Economic Entomology* 81: 28–33.

McPherson, S. A., Perlak, F. J., Fuchs, R. L., Marrone, P. G., Lavrik, P. B. and Fischhoff, D. A. 1988. *Bio/technology* 6: 61–66.

Nambiar, P. T. C., Ma, S. W. and Iyer, V. N. 1990. *Applied and Environmental Microbiology* 56: 2866–2869.

NRC (National Research Council). 1989. *Field Testing Genetically Modified Organisms: Framework for Decisions.* Washington, D.C.: National.

Obukowicz, M. G., Perlak, F. J., Kusano-Kretzmer, K., Mayer, E. J. and Watrud, L. S. 1986. *Gene* 45: 327–331.

O'Keeffe, L. E., Homan, H. W. and Schotzko, D. 1984. *The pea leaf weevil.* Current Information Series No. 227, University of Idaho, Cooperative Extension Service, Moscow.

Quinn, M. A. and Hower, A. A. 1986a. *Ecological Entomology* 11: 391–400.

Quinn, M. A. and Hower, A. A. 1986b. Alfalfa rhizosphere: the unknown world of clover root curculio. In: *Proceedings of the 16th National Alfalfa Symposium*, pp. 115–133. Fort Wayne, Indiana.

Quinn, M. A. and Hall, M. H. 1992. *Entomologia Experimentalis et Applicata* 64: 167–176.

Raffa, K. 1989. *BioScience* 39: 524–534.

Richaume, A., Angle, J. S. and Sadowsky, M. J. 1989. *Applied and Environmental Microbiology* 55: 1730–1734.

Saxena, M. C. 1988. In: *Nitrogen Fixation by Legumes in Meditterranean Agriculture*, pp. 11–23 (eds. D. P. Beck and L. A. Materon). Dordrecht: Martinus Nijhoff.

Schofield, P. R., Gibson, A. H., Dudman, W. F. and Watson, J. M. 1987. *Applied and Environmental Microbiology* 53: 2942–2947.

Sekar, V., Thompson, D. V., Maroney, M. J., Brookland, R. G. and Adang, M. J. 1987. *Proceedings of the National Acadamy of Science* 84: 7036–7040.

Shantharam, S. 1990. In: *Nitrogen Fixation: Achievements and Objectives. Proceedings of the 8th International Congress on Nitrogen Fixation*, pp. 393–396 (eds. P. M. Gresshoff, L. E. Roth, G. Stacey and W. E. Newton). New York: Chapman and Hall.

Sharples, F. E. 1991. In: *Risk Assessment in Genetic Engineering*, pp. 18–31 (eds. M. A. Levin and H. S. Strauss). New York: McGraw-Hill.

Skot, L., Harrison, S. P., Nath, A., Mytton, L. R. and Clifford, B. C. 1990. *Plant and Soil* 127: 285–295.

Stone, T. B., Simms, S. R. and Marrone, P. G. 1989. *Journal of Invertebrate Pathology* 53: 228–234.

Stotzky, G., Zeph, L. R. and Devanas, M. A. 1991. In: *Assessing Ecological Risks of Biotechnology*, pp. 95–122 (ed. L. R. Ginzburg). Boston: Butterworth-Heinemann.

Sundaresan, V., Ww, D. W. and Ausubel, F. M. 1983. *Proceedings of the National Academy of Sciences USA* 80: 4030–4034.

Tabashnik, B. E., Cushing, N. L., Finson, N. and Johnson, M. W. 1990. *Journal of Economic Entomology* 83: 1671–1676.

Tabashnik, B. E., Finson, N. and Johnson, M. W. 1991. *Journal of Economic Entomology* 84: 49–55.

Tauer, L. W. 1989. *Plant and Soil* 119: 261–270.

Vaeck, M., Raynaerts, A., Hofte, H., Jansens, S., De Beuchelcer, M., Dean, C., Zabeau, M., Van Montagu, M. and Leemans, J. 1987. *Nature* 328: 33–37.

Van Elsas, J. D., Trevors, J. T. and Starodub, M. E. 1988. In: *Risk Assessment for Deliberate Releases. The Possible Impact of Genetically Engineered Microorganisms on the Environment*, pp. 89–99 (ed. W. Klingmüller). New York: Springer-Verlag.

Van Rie, J., McGaughey, W. H., Johnson, D. E., Barnett, B. D. and Van Mellaert, H. 1990. *Science* 247: 72–74.

Crop physiology and productivity

Crop physiology and productivity in the cool season food legumes: recent advances in the measurement and prediction of photothermal effects on flowering

R.J. SUMMERFIELD, E.H. ROBERTS and R.H. ELLIS
University of Reading, Department of Agriculture, Plant Environment Laboratory, Cutbush Lane, Shinfield, Reading, Berkshire RG2 9AD, UK

Abstract

Investigations of the "physiology of yield" in the cool season food legumes have often measured only the end-products of physiological and phenological processes at reproductive maturity. Information on these "static" components-of-yield provides little understanding of why economic yield or any particular component of it varies; neither does it contribute much understanding of the processes by which components of yield are determined, nor how they relate to plant growth or the environment.

Four types of dynamic processes – development (phenology), expansion (e.g., of leaf area and root length), and the assimilation and partition of dry matter – combine to determine economic yield. Environmental factors and agronomic variables have considerable effect on each of these processes, and there can be substantial quantitative differences between genotypes, too. Experience suggests and experiments increasingly confirm that the first category of processes, i.e., phenological events, are of fundamental importance in relation to the capture by crops of environmental resources, notably of solar radiation, water, and nutrients. The first step towards maximizing yield by management or breeding is therefore to ensure that the phenology of the crop is well matched to the resources and constraints of the production environment. Hitherto, empirical approaches to screening germplasm for phenological events such as times to flowering have predominated in food legume breeding programs. These "evaluation descriptors" are often specific to location and season and so are of limited value. However, it is now clear that analyses of the responsiveness of flowering to photothermal conditions are more reliable and more informative when based not on times to flowering (f) but on rates of progress towards flowering ($1/f$). Furthermore, the approach now advocated provides "genetic descriptors" which are independent of environment, but which are capable of predicting flowering responses in any environment, and so are of widespread value.

F.J. Muehlbauer and W.J. Kaiser (eds.), Expanding the Production and Use of Cool Season Food Legumes, 755–770.
© 1994 *Kluwer Academic Publishers.*

Introduction

Growth and reproductive development in plants are to a much larger degree than in animals independent of one another; under strongly inductive conditions, for example, plants can flower precociously when they are still very small whereas in less inductive circumstances they may grow to enormous size but may never flower (Evans, 1975). Between these extremes, timely flowering is of overriding importance in conditioning crop adaptation to the resources and constraints of the production environment (Roberts and Summerfield, 1987) – the more so in those species that are botanically determinate and that have a synchronous pattern of flowering (Richards, 1991).

One aspect of the literature devoted to topics in crop physiology during recent years has been the increasing emphasis given to analysis of economic yield not as "static" components but in terms of four types of dynamic processes, *viz* phenological events (development), expansion (e.g., of leaf area and root length), and the assimilation and subsequent partition of dry matter and nitrogen (Figure 1). A common feature of the seminal contributions on these topics by Lawn (1989), Squire (1990), Richards (1991), and Shorter *et al.* (1991) has been the emphasis given to an appropriate phenology in determining economic yield by influencing the fraction of the environmental resources captured by plant stands (notably of solar radiation, water, and nutrients) and the efficiency with which they use them – as Bunting had also emphasized in 1975. For example, in environments where water is limiting (a common situation with food legumes), the timing of flowering can be crucial; a prolonged vegetative period might reduce seed production when there is terminal drought, but may well increase seed yield when there is a mid-season drought (Squire, 1990). Then again, given that .... *"in morphological terms the present peak of evolutionary development is represented by grain legumes which are compact, erect, free-standing annual herbs with determinate growth patterns and a concentrated flowering period...."*. (Smartt, 1984, 1986, 1990), the timing of flowering is especially critical if modern food legume cultivars seeded at the appropriate density are to have the potential to yield well (see, for example, Erskine *et al.*, 1989; Lawn and Imrie, 1991; Van Rheenen, 1991).

This contribution builds on the previous plenary review of critical physiological traits in the cool season food legumes (Hardwick, 1988). It concentrates on what Hardwick described as "the birth of reproductive nodes and flowers" and briefly summarizes recent progress in the measurement, prediction, and characterization of genetic and environmental effects on the timing of flowering in pea (*Pisum sativum* L.), lentil (*Lens culinaris* Medik.), faba bean (*Vicia faba* L.), and chickpea (*Cicer arietinum* L.). These advances have been described in detail by Summerfield *et al.* (1991).

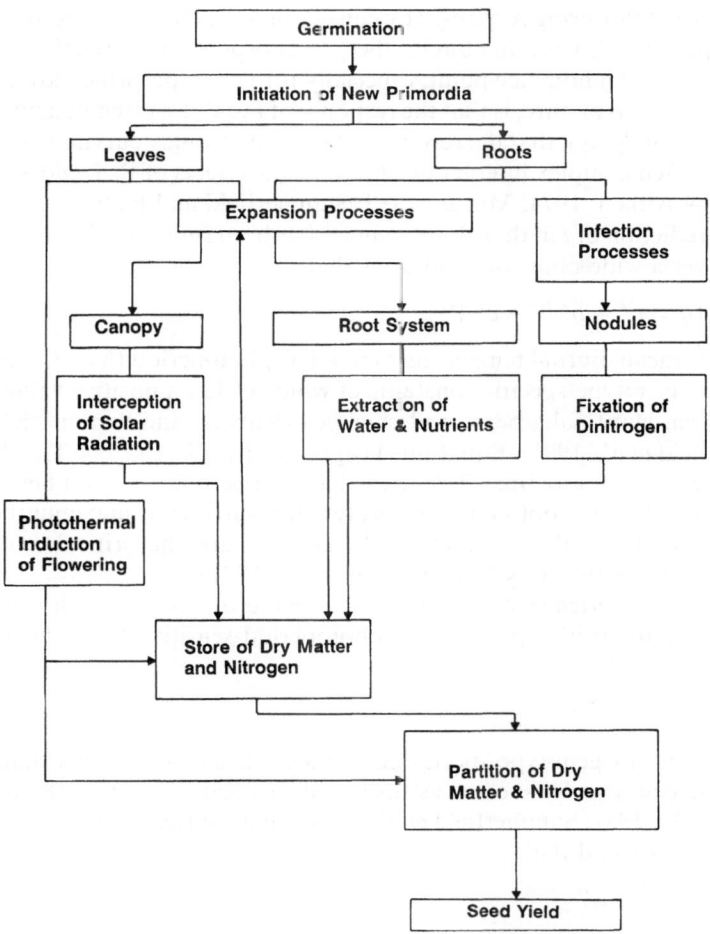

Figure 1. Principal processes in development and growth of a food legume crop (based on the scheme presented as Figure 1.1 in Squire, 1990).

Flowering and Progress Towards It

Most genotypes of all four crops of topical concern are quantitative long-day plants (LDP); temperature (T) also modulates strongly the time from sowing to flowering (*f*), and some genotypes may also respond to vernalization (V) (see Roberts and Summerfield, 1987). Genetic differences and photothermal flowering responses have traditionally been reported in terms of the number of days from sowing to flowering, *f*; genotypes are then described as early-, medium- or late-flowering, but these are "evaluation descriptors" specific to location and season and so are of limited value. This contribution analyzes photothermal flowering responses by considering the effects of photoperiod (P) and temperature (T) not on f but on the rate of progress towards flowering, $1/f$. Hitherto, the common viewpoint, especially for genotypes which are relatively late to flower and mature, is that the response of f to P can often be acute or even qualitative (obligate), that the response of f to T is strongly curvilinear, and that there are often complex interactions between the effects of P, T and V on f (see reviews by Aitken, 1974; Murfet, 1985; Summerfield and Roberts, 1985).

The prediction of f in these four legumes has been simplified because we have shown over a wide range of conditions that:

$$1/f = a' + b' \overline{T} + c' P \tag{1}$$

where $\overline{T}$ is mean diurnal temperature (°C), P is photoperiod (h d^{-1}), and a', b' and c' are genotype-specific constants of which c' has a positive value in LDP such as pea, lentil, faba bean, and chickpea (Roberts and Summerfield, 1987; Summerfield *et al.*, 1991). Equation (1) applies where T $'_b$ < T < T$_o$, where T is the temperature at any time, T $'_b$ is the base temperature (at and below which $1/f = 0$), and T$_o$ is the optimum temperature (at which $1/f$ is maximum) and, for LDP, where P$_c$ > P > P$_{ce}$, where P$_c$ and P$_{ce}$ are the critical and ceiling photoperiods, respectively (Summerfield *et al.*, 1991).

Extensive experience with these and other crops has also shown that in photoperiod-insensitive plants or in photoperiod-sensitive LDP when P > P$_c$, then:

$$1/f = a + b\overline{T} \tag{2}$$

where a and b are genotype-specific constants. When eqn (2) rather than eqn (1) applies, it can also be shown, as discussed in detail elsewhere (Roberts and Summerfield, 1987; Summerfield *et al.*, 1991), that a different base temperature, T$_b$, also applies and that:

$$T_b = - a/b \tag{3}$$

In these circumstances the thermal time, θ, i.e., the day-degrees above T$_b$ which need to be accumulated for flowering to occur, is given by:

$$\theta = 1/b \tag{4}$$

Temperature Responses of Genotypes Insensitive to Photoperiod

We have confirmed the utility of eqn (2) not only for photoperiod-insensitive genotypes over widely different photothermal regimes but also for photoperiod-sensitive cultivars held at a constant daylength (e.g., for faba bean see Ellis *et al.*, 1988a,b). As a further example we reanalyze here the data for field pea published by Aitken (1974), i.e., the flowering responses of "Alaska" when grown in sixteen locations between 71° N and 54° S and from sea level to 3050 m. Flowering in Alaska is insensitive to photoperiod over the range from 8 to 24 hd^{-1} (Berry and Aitken, 1979); in Aitken's field trials f varied between 30 and 120 d.

It is clear from Figure 2a, which adopts Aitken's original format to display the data, that temperatures warmer than 5°C hasten flowering. That said, and notwithstanding the goodness of mathematical fit of the curvilinear model shown, little more can be gleaned from this analysis of an extensive and unique set of data. In contrast, the relation shown and quantified in Figure 2b, i.e., the response of rate of progress to flowering, $1/f$, is the basis of a more informative interpretation of the data. Over the range of mean daily temperatures from 5° to 18°C, the rate of progress to flowering of Alaska is related linearly to temperature, with a slope constant [0.00189 ± 0.00018; the value of b in eqn (2)] which is independent of $\bar{T}$ when $T_b < T < T_o$. The base temperature for progress towards flowering [eqn (3)] is 0.99°C, say 1°C, and the thermal time which needs to be accumulated above T_b for flowering to occur [eqn (4)] is 529°Cd. It is implicit in the thermal time approach (sometimes referred to as "growing degree-days" or by the misnomers "heat sums" or "accumulated heat units") that rate of development is a linear function of temperature (Monteith, 1981).

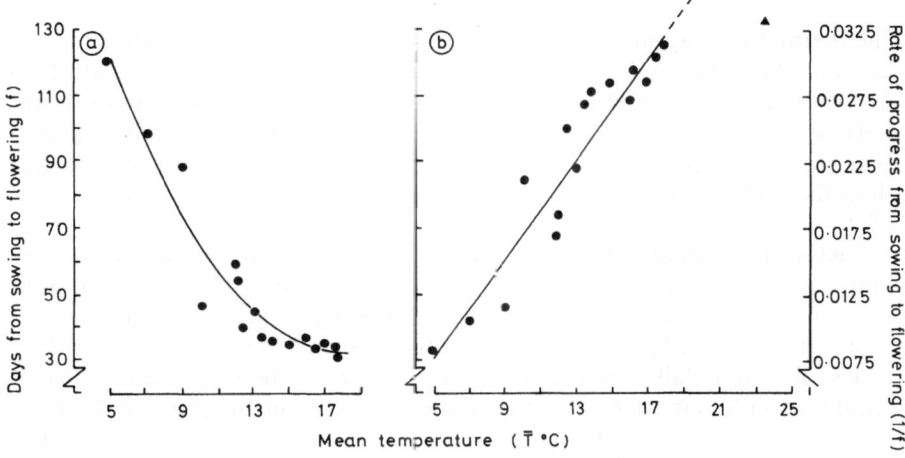

Figure 2. Effect of temperature (T°C) on (a) times from sowing to flowering (f; d) and (b) rates of progress towards flowering ($1/f$) of pea Alaska in diverse field locations (based on the original data of Aitken, 1974). In (a), $f = 209-20.2T + 0.58T^2$ (s.e. 17.6, 3.1 and 0.13, respectively) and in (b) $1/f = -0.00187 + 0.00189T$ (s.e. 0.00024 and 0.00018, respectively). The fitted relations shown have R^2 values of 0.922 (a) and 0.879 (b).

Also clear as a consequence of presenting the data in terms of $1/f$ rather than f is that a mean temperature of 23.5°C (26°C max./21°C min. – the mean values for Aitken's winter sowing in Hawaii – denoted by ▲) is supra-optimal for early flowering in this cultivar.

Table 1. Values of the constants and coefficients of eqns (1) and (2), with s.e. values in parentheses, derived from regression analyses of $1/f$ against P and/or T for faba bean Zeidab Local and for pea Mackay [from Ellis *et al.* (1988a) and calculated from Berry and Aitken (1979), respectively]

Crop and cultivar	Range of photoperiods	Fitted parameter values[+]			
		a'	b'	c'	R^2
Faba bean Zeidab Local	$P_{ce} < P < P_c$	0.0082 (0.0018)	0.00070 (0.00010)	0.00025 (0.00011)	0.96
Pea Mackay	$P > P_c$	0.0000795 (0.00098)	0.001012 (0.000067)		0.98
Pea Mackay	$P_{ce} < P < P_c$	-0.00652 (0.00221)	0.000348 (0.000085)	0.000876 (0.000187)	0.85

[+]When $P > P_c$ for pea Mackay the values given in the body of the table are for a and b in eqn (2).

Photothermal Responses of Genotypes Sensitive to Photoperiod

The thermal-time approach described above takes no account of the typically substantial photoperiodic effects on flowering times in these species, and which are especially important in those genotypes which are typically late to flower and mature. We now illustrate how the approach can be modified to incorporate quantitative photoperiodic responses by way of eqn (1) – the general photothermal response of flowering in annual crops (Roberts and Summerfield, 1987).

Factorial combinations of three photoperiods (10, 13 and 16 hd^{-1}), two day temperatures (18 and 28°C) and two night temperatures (5 and 13°C), were imposed on diverse genotypes of faba bean grown in pots in growth cabinets, and the effects of these 12 photothermal environments on flowering were recorded – as described in detail elsewhere (Ellis *et al.*, 1988a). The response of "Zeidab Local" (from Sudan) over this wide range of environments (but still between T'_b and T_o and between P_c and P_{ce}) is shown in Figure 3 and summarized in Table 1. It is clear that the three-dimensional linear additive rate model based on eqn (1) provides a very adequate explanation of the observations for this cultivar, as it does for faba bean germplasm in general.

For pea, we rely again on data published by Aitken and her colleagues to demonstrate the value of analyses of photothermal flowering responses not in terms of f, the traditional approach, but $1/f$ – in this case for the late maturing "Mackay" grown in the CSIRO phytotron. Berry and Aitken (1979) described the Mackay phenotype as LHR (Late and Highly Responsive to photoperiod) in which strongly curvilinear responses of flowering to $\overline{T}$ (especially in long days) and to P (at both warm and cool temperatures) combined with a marked $P \times \overline{T}$ interaction effect to determine f (Figure 4a).

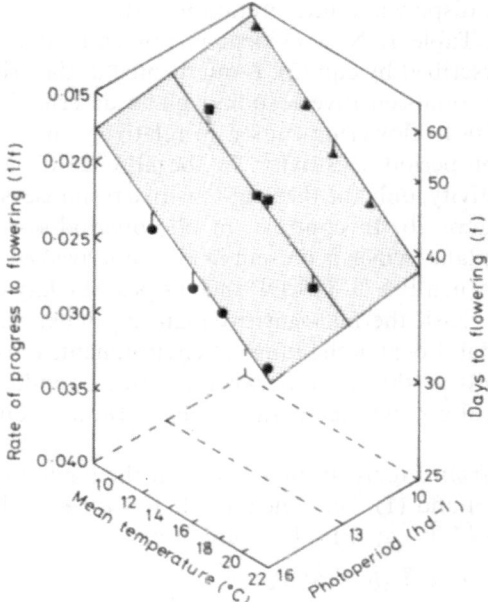

Figure 3. Photothermal responses of $1/f$ in faba bean Zeidab Local (from Ellis *et al.*, 1988a). Plants were grown in 12 different regimes in growth cabinets, comprising mean diurnal temperatures between 9 and 22°C and daylengths of 10 (▲), 13 (■) or 16 hd^{-1} (●). The relation shown is quantified in Table 1.

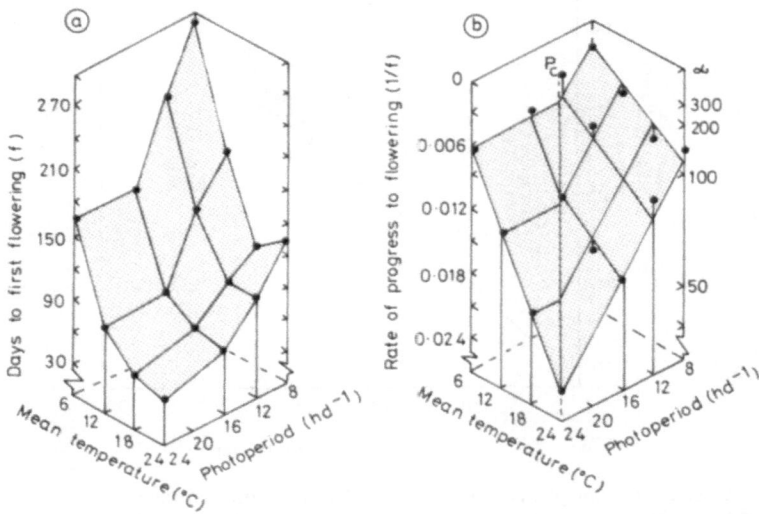

Figure 4. Photothermal flowering responses of pea Mackay: (a) original data (f; d) from Berry and Aitken (1979) and (b) rates of progress towards flowering ($1/f$) based on data in (a) and with fitted response planes as quantified in Table 1.

Our alternative display and interpretation of the data for Mackay are given in Figure 4b and Table 1. Now evident in this cultivar is that the thermal-sensitive plane, described by eqn (2), is intercepted at the critical photoperiod, P_c , by a photothermal-sensitive response plane described by eqn (1) which quantifies the delay to flowering caused in relatively short days by the genes which confer photoperiod sensitivity. In the absence of genes which confer photoperiod sensitivity, only the thermal-sensitive response is evident, as shown for Alaska in Figure 2b. In contrast, in photoperiod-sensitive cultivars the thermal-sensitive plane (which is present in all genotypes) is only exposed when $P < P_c$ in SDP or when $P > P_c$ in LDP such as pea Mackay (Figure 4b). When $P < P_c$ in LDP, the basic thermal-sensitive plane is not exposed. This is the case for faba bean Zeidab Local in the range of environments illustrated in Figure 3 because in this cultivar the range between the two boundary photoperiods is large, *viz.* at typical growing temperatures $P_c > 18$ hd^{-1} and $P_{ce} \approx 7.5$ hd^{-1} (Ellis *et al.*, 1990).

It is also clear from Figure 4b that $P = P_c$ at the junction of the two planes defined by eqns (2) and (1), i.e., when $a + b\,\overline{T} = a' + b'\,\overline{T} + c'\,P$, or when $(a + b\,\overline{T}) - (a' + b'\,\overline{T} + c'\,P) = 0$. Therefore:

$$P_c = [a - a' + \overline{T}\,(b - b')]/c' \tag{5}$$

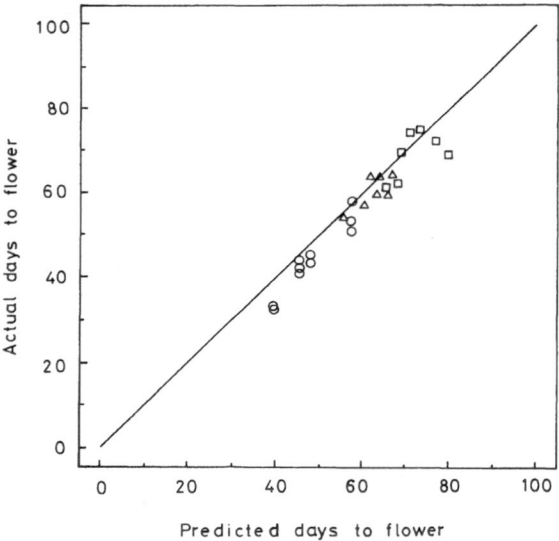

Figure 5. Comparisons for plants of lentil Precoz between observed times from sowing to first flower (d) in four glasshouse environments in UK (°) and to 50% flowering (d) in serial field sowings in Argentina during 1985 (□) and 1986 (△), as reported by Gray and Delgado (1989), with those predicted for *f* by eqn (1) and the values provided for the constants *a' b'* and *c'* by Summerfield *et al.* (1985).

The value of P_c, then, varies with $\overline{T}$: Figure 4b shows that for Mackay it increased from 12 h 5 min at 6°C to 22 h 41 min at 20°C, i.e., by about 45 min°C^{-1}. It is not true, then, as has often been suggested in the literature (see the review by Rees, 1987), that P_c does not vary with temperature.

For lentil, Figure 5 displays how the values of f for "Precoz" grown in wide ranges of P and T in a glasshouse in UK or in the field in Argentina are described well by eqn (1) and estimates of a', b' and c' determined previously and independently (Summerfield *et al.,* 1985). Furthermore, in an analysis of the photothermal flowering responses of 231 lentil accessions from eight major producing countries between about 5° and 50° N, Erskine *et al.* (1990) detected large differences among accessions in their respective values of the constants a', b' and c' of eqn (1). Analysis of the estimates of b' and c' for the 231 accessions showed a weak negative correlation ($r = -0.291$, $P < 0.01$). Nevertheless, the considerable proportion of the variance which remains unexplained suggests that the temperature and photoperiod responses, already shown to be physiologically independent (e.g., Figure 6), are also unlikely to be genetically linked. Appropriate crosses are required to confirm this hypothesis.

Finally, for chickpea, we again rely on independent field data to illustrate the utility of eqn (1) originally developed in controlled environments (Roberts *et al.,* 1985), given that conditions in many field environments where chickpeas are or may come to be grown are likely to be always or mostly within the range $P_{ce} < P < P_c$. We recognize, of course, that simple, linear equations may give absurd values for the dependent variable and for derived entities if the clearly stated and valid limits are transgressed (e.g., Landsberg, 1977). Certainly, more information is needed on the effects on flowering in all four species of transitory excursions into regimes where $T > T_o$ and $T < T_b$ or $T' _b$. In the interim, however, an increasing body of evidence points to the value of eqn (1) as a sound basis for the prediction of field responses over wide ranges of each of P and T – as this fourth example illustrates.

More than 20 years ago, Eshel (1967) sowed two genotypes of chickpea (one from the USA, "California", the other from Bulgaria, "Bulgaria") at successive 3-week intervals between October and August at Rehovot, Israel. Times from emergence to flowering varied between 24 to 30 and 131 to 145 d; average daily temperatures during the pre-flowering period differed by almost a factor of two (from 12.0 to 22.7°C); and average daylength (sunrise to sunset) varied between 10.5 and 14.1 hd^{-1} [and, in passing, times to flowering of both genotypes, we calculate, were strongly correlated ($r = 0.989$ and 0.994, n = 10) with those to reproductive maturity].

When discussing his results, Eshel was unable to "draw firm conclusions concerning the (flowering) response of chickpea to daylength and temperature" but suggested "that the effect of day elongation on shortening the vegetative growth period was stronger than the effect of temperature increase". However, when Eshel's field data on times to flowering, f, are transformed into $1/f$ and eqn (1) is applied, the photothermal model gives R^2 values of 0.84 (Bulgaria) and 0.99 (California) (see Figure 7). Unfortunately, as is common with serial sowing

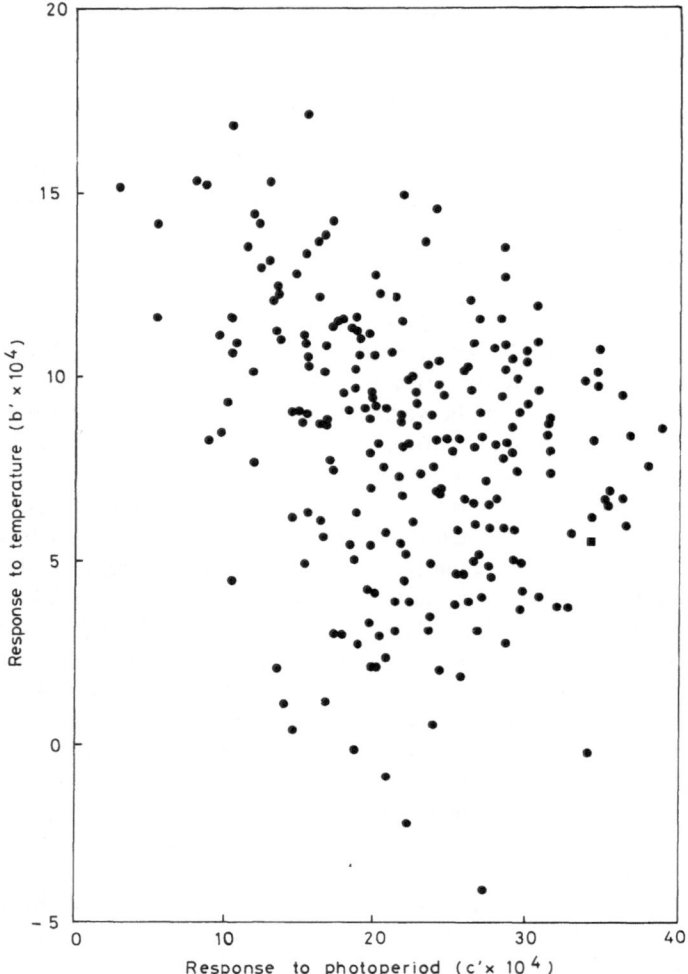

Figure 6. Sensitivity of rate of progress towards flowering to temperature (b') and photoperiod (c') of 230 cultivated lentil accessions (●) and one wild lentil accession from Turkey (■) – reproduced from Erskine *et al.* (1990).

date studies in the field, the seasonal march in temperature and daylength at Rehovot were almost perfectly correlated (i.e., as days lengthened, average temperatures increased). For that reason, we are hesitant to rely, in these circumstances, on the estimated values of the constants a', b' and c'. Nevertheless, application of eqn (1) leaves us in no doubt that California was far more sensitive to P than Bulgaria which, in turn, was far more sensitive to $\overline{T}$. The similar phenological plasticity of these two genotypes was, in fact, a consequence of markedly different responsiveness to photoperiod and

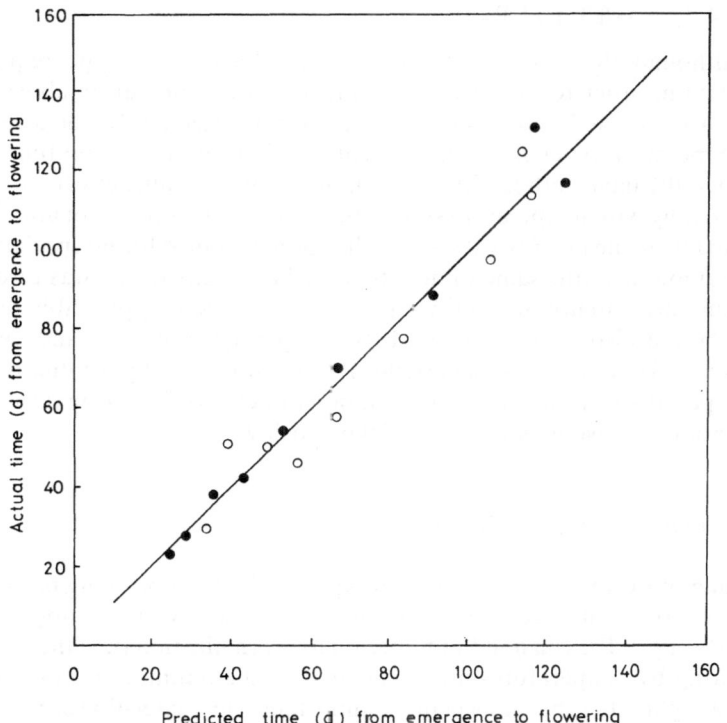

Predicted time (d) from emergence to flowering

Figure 7. Relation between actual times from emergence to first flower (d) and those fitted by eqn (1) for two genotypes of chickpea grown in the field at Rehovot, Israel, one from California (●) and one from Bulgaria (°) — R^2 values of 0.99 and 0.84, respectively (original data of Eshel, 1967). temperature. As with the other crops we have investigated, the inclusion of an interaction term in the photothermal model, i.e.,

$$1/f = a + b\overline{T} + cP + d(P \times \overline{T}) \tag{6}$$

had no significant effect ($P > 0.05$), it increased R^2 compared with those given by eqn (1) by only 0.006 or 0.009. We suggest, then, that as with $1/f$ in lentils, the responsiveness of flowering to daylength and temperature are also under separate genetic control in chickpea.

When conditions in the field are within the photothermal-response plane, as they often are with cool season food legumes (i.e., when $P_{ce} < P < P_c$), then it is feasible to use eqn (1) to devise a photothermal-time concept for predicting the time taken to flower. This is analogous to the thermal-time concept which, while valuable, is only appropriate to photoperiod-insensitive plants (or when $P < P_c$ in SDP or when $P > P_c$ in LDP). In the case of photoperiod-sensitive plants the photothermal time necessary for flowering, θ_Φ, can be measured in similar units, day-degrees (°Cd) above the base temperature. By analogy with thermal time, θ, photothermal time, θ_Φ, is given by $1/b'$, but the base temperature, T $'_b$, above which it is accumulated varies with photoperiod and is given by:

$$T'_b = -(a' + c' \, P)/b' \tag{7}$$

Thus, although the number of day-degrees which a genotype requires to accumulate in order to flower is constant, the base temperature above which they are accumulated depends on photoperiod. Consequently, the same daily mean temperature experienced in different locations or at different times of the year, when the daylength is different, will not have the same effect.

Historically with temperature summations, base values have not always been subtracted from mean daily values, or it has been assumed for no special reason – which amounts to the same thing – that the base temperature was zero. Both approaches are patently invalid; values of T'_b are often appreciably different from zero and also vary markedly between genotypes within- and between-species (e.g., Roberts and Summerfield, 1987). Estimates of photothermal time can be especially helpful in the interpretation of putative flowering responses to vernalization – as we illustrate for chickpea below.

Vernalization in Chickpea: Fact or Artefact?

The timing of flowering in various genotypes of chickpea has long been said to be significantly influenced by quantitative responses to vernalization (e.g., Pal and Murty, 1941; Saxena and Siddique, 1980). Vernalization (i.e., the hastening of flowering by temperatures much cooler than the optimum for growth) has been thought to be especially important in later-flowering, more photoperiodically sensitive genotypes; indeed, in some cases, it has been thought to be capable of substituting for longer photoperiods in this quantitative long-day species (Angus and Moncur, 1980). However, problems arise in the design of experiments which seek to demonstrate vernalization effects.

If a cool pre-treatment is given to imbibed seeds or seedlings before transfer to a subsequent, warmer environment, and pre-treatment effects on f are then compared with a control treatment in which plants have not been exposed to cold, it follows that if the cool pre-treatment had been a regime where T_b or T'_b was exceeded, then some progress towards flowering during pre-treatment would be expected as thermal or photothermal time, respectively, would have accumulated. The essential question is: "Does any reduction in the subsequent time to flowering following a putative vernalization treatment exceed that which is expected due solely to the accumulation of thermal or photothermal time during the pre-treatment period"?

This question has been recently addressed for two genotypes of chickpea known to differ appreciably in their flowering responses to photothermal conditions: kabuli "Rabat", from Morocco, is far more sensitive to P than the desi accession ICC 5810, from northern India, which is far more sensitive to $\overline{T}$ (Roberts *et al.*, 1985). Imbibed seeds of both genotypes were pre-treated in various combinations of cold or cool temperatures (1 to 10°C) and durations (5 to 42 d) before the transfer of seeds or seedlings to each of six growing-on regimes (photoperiods of 11 or 15 hd^{-1} combined with various mean temperatures between 15.6 and 22.7°C). Compared to the non-pretreated controls, these potentially vernalizing treatments hastened subsequent times to flowering by 6% for ICC 5810 and by 12 to 20% for Rabat (Summerfield *et al.*, 1989). One interpretation, then, is that Rabat responds to vernalization whereas ICC 5810 probably does not.

However, knowledge of the values of the constants a', b' and c' in eqn (1) (Roberts *et al.*, 1985) reveals that, at a given value of P, the value of T $'_b$ is considerably cooler in Rabat than in ICC 5810. Consequently, it is not surprising, based only on considerations of photothermal time accumulated during pre-treatment, that exposure to cool temperatures hastened flowering relatively more in Rabat than in ICC 5810. Indeed, Figure 8 shows that when photothermal time was accumulated during both the pre-treatment periods and subsequently, flowering occurred when predicted (for Rabat, $R^2 = 0.910$, n = 162, $P > 0.10$; for ICC 5810, $R^2 = 0.976$, n = 162, $P > 0.25$). There is, therefore,

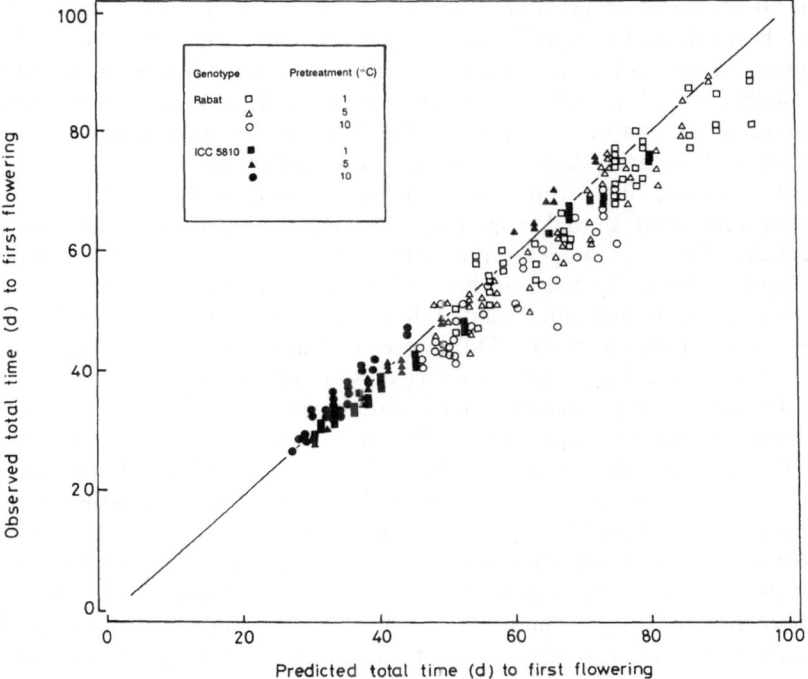

Figure 8. Relations between actual predicted and total times to first flower in plants of chickpea Rabat and accession ICC 5810. The times shown (d) include pretreatment periods in cool environments of 1, 5, or 10°C before transfer to each of six warmer environments in which flowering occurred. Results for 162 plants of each genotype are presented but many points are obscured by coincidence of position. The solid diagonal line (y = x) represents perfect agreement between observed and predicted total times to first flower (for full details see Summerfield *et al.*, 1989).

no evidence of a specific vernalization response in either genotype. Accordingly, a thorough re-evaluation of "responsiveness to vernalization" in the chickpea germplasm might well be useful.

Prospect

It has long been appreciated that the time from sowing to flowering is a principal determinant of adaptiveness to the cropping environment (e.g., Whyte 1946); the cool season food legumes, so often grown in marginal environments, are no exception. An ability to predict phenology in different environments (e.g., in different locations and for different sowing dates) would enable crops to be managed so that environmental resources (e.g., periods during which radiation, water supply, and temperatures were favorable) are fully exploited and the adverse consequences of seasonal constraints (e.g., extremes of temperature and aridity) are minimized. An ability to predict phenological events in crop germplasm would assist in the breeding process, too; it would facilitate more efficient screening of genotypes to characterize their responses and to select those best adapted to specific target environments, and it would enable the range and number of environments over which genotypes need to be tested to be rationally reduced. Equation (1) can be used as the basis and has the advantage of enabling the quantification of genetically controlled flowering responses to photoperiod and temperature to be separated and characterized.

We conclude, then, from our extensive research on the cool season food legumes and other annual crops that a set of robust principles has now been established for examining the photothermal responses of plant development (see Squire, 1990). The basis of these principles is that rate of progress towards flowering, $1/f$, is generally related linearly to P and $\overline{T}$ over wide ranges of agricultural circumstances. Differences in mineral nutrition and even acute water stress (e.g., Singh, 1991) distort these relations only slightly; flowering can, therefore, be predicted reliably on the basis of the genotype-specific constants included in eqns (1) and (2), *viz. a* and *b*, which characterize the underlying temperature response, and a', b' and c', which characterize the delay in flowering caused by photoperiod-sensitivity genes. Furthermore, we conclude that it should be possible to screen large collections of accessions by simple observations and a few sowings at a few carefully selected sites (e.g., see Roberts *et al.*, 1985, Summerfield *et al.*, 1985; Ellis *et al.*, 1990, for recommendations on the minimum number of sites and/or sowings and their climatological characteristics for chickpea, lentil, and faba bean, respectively).

There remains the challenge for all crops to devise an appropriate iterative fitting procedure whereby daily values of P and T are assigned routinely to the appropriate photothermal ranges – i.e., with the cool season food legumes to ranges appropriate for LDP: *viz.* $T < T_b$; or $T_o > T > T_b$ while $P > P_c$; or $T_o > T > T'_b$ while $P_c > P > P_{ce}$; or $T_o > T > T_b$ while $P < P_{ce}$. These matters are the subject of ongoing research. In the interim, we believe that much can be done – and much better and much faster than hitherto.

The analysis of photothermal flowering responses as rates, the expression of development in terms of an integral such as thermal or, more often, photothermal time, and the application of a small number of genotype-specific constants and attributes each with physiological meaning, is extremely valuable in relation to the practical necessity for rapid and inexpensive screening of germplasm (see Ceccarelli *et al.*, 1991) and for modeling purposes. Holistic models which link and describe the physiological nature of a genotype in terms of attributes that govern the processes of development, expansion, production and partition (Figure 1, and see Nanda and Saini, 1990; McKenzie and Hill, 1991; Williams and Saxena, 1991) seem the most appropriate framework in which to integrate crop physiology, modeling and plant breeding (Shorter *et al.*, 1991). Historically, that integration has not always attracted or motivated breeders in their efforts to produce genotypes better adapted to and so more reliably productive in those environments for which they are intended (Simmonds, 1991). The alternative viewpoint, and the one supported here, is that the potential benefits of such an integrated approach are simply too large, and long-term, to be ignored.

Acknowledgements

We thank the UK Overseas Development Administration for supporting the research which led to the development of the main features of the models described here, and for ongoing financial support of our activities at Reading. The contribution to the integrated programs of research in Syria, India, the UK and elsewhere of the dedicated assistance and technical expertise of Mrs. C. Hadley, Mrs. R. Adamson, Mrs. J. Allison and Messrs. D. Dickinson, K. Chivers, S. Gill and A. Pilgrim is gratefully acknowledged. We thank Drs. W. Erskine, M. C. Saxena, J. B. Smithson and C. Johansen for their scientific comments, encouragement and guidance. The pioneering work of Dr. Yvonne Aitken over a period of 40 years (1939–79) has also done much to arouse interest in the way climate affects the flowering of plants. The award of British Council Travel Grants, combined with generous support from the Conference organizers, allowed RJS and EHR to participate in the meeting in Cairo.

References

Aitken, Y. 1974. *Flowering Time, Climate and Genotype*, pp. 193. University Press: Melbourne.
Angus, J. F. and Moncur, M. W. 1980. *International Chickpea Newsletter* 2: 8–9.
Berry, G. J. and Aitken, Y. 1979. *Australian Journal of Plant Physiology* 6: 573–587.
Bunting, A. H. 1975. *Weather* 30: 312–325.
Ceccarelli, S., Acevedo, E. and Grando, S. 1991. *Euphytica* 56: 169–185.
Ellis, R. H., Roberts, E. H. and Summerfield, R. J. 1988a. *Annals of Botany* 61: 73–82.
Ellis, R. H., Roberts, E. H. and Summerfield, R. J. 1988b. *Annals of Botany* 62: 119–126.

770   *R.J. Summerfield* et al.

Ellis, R. H., Summerfield, R. J. and Roberts, E. H. 1990. *Annals of Botany* 65: 129–138.

Erskine, W., Adham, Y. and Holly, L. 1989. *Euphytica* 43: 97–103.

Erskine, W., Ellis, R. H., Summerfield, R. J., Roberts, E. H. and Hussain, A. 1990. *Theoretical and Applied Genetics* 80: 193–199.

Eshel, Y. 1967. *Israel Journal of Agricultural Research* 17: 193–197.

Evans, L. T. 1975. *Daylength and the Flowering of Plants*, pp. 1–15. Menlo Park, California: W. A. Benjamin.

Gray, L. N. and De Delgado, N. C. 1989. *LENS* 16: 19–21.

Hardwick, R. C. 1988. In: *World Crops: Cool Season Food Legumes*, pp. 885–896 (ed. R. J. Summerfield). Dordrecht: Kluwer Academic Publishers.

Landsberg, J. J. 1977. *Experimental Agriculture* 13: 273–286.

Lawn, R. J. 1989. *Experimental Agriculture* 25: 509–528.

Lawn, R. J. and Imrie, B. C. 1991. *Field Crops Research* 26: 113–139.

McKenzie, B. A. and Hill, G. D. 1991. *Journal of Agricultural Science, Cambridge* 117: 339–346.

Monteith, J. L. 1981. *Quarterly Journal of the Royal Meteorological Society* 107: 749–754.

Murfet, I. C. 1985. In: *Handbook of Flowering, Volume IV*, pp. 97–126 (ed. A. H. Halevy). Boca Raton, Florida: CRC Press.

Nanda, R. and Saini, A. D. 1990. *Annals of Agricultural Research* 11: 177–183.

Pal, B. P. and Murty, G. A. 1941. *Indian Journal of Genetics and Plant Breeding* 1: 61–85.

Rees, A. R. 1987. In: *Manipulation of Flowering*, pp. 187–202 (ed. J. G. Atherton). London: Butterworths.

Richards, R. A. 1991. *Field Crops Research* 26: 141–169.

Roberts, E. H., Hadley, P. and Summerfield, R. J. 1985. *Annals of Botany* 55: 881–892.

Roberts, E. H. and Summerfield, R. J. 1987. In: *Manipulation of Flowering*, pp. 17–50 (ed. J. G. Atherton). London: Butterworths.

Saxena, M. C. and Siddique, M. H. 1980. *International Chickpea Newsletter* 2: 7–8.

Shorter, R., Lawn, R. J. and Hammer, G. L. 1991. *Experimental Agriculture* 27: 155–175.

Simmonds, N. W. 1991. *Tropical Agriculture Association Newsletter* 11: 7–10.

Singh, P. 1991. *Field Crops Research* 28: 1–15.

Smartt, J. 1984. *Experimental Agriculture* 20: 275–296.

Smartt, J. 1986. *Experimental Agriculture* 22: 39–58.

Smartt, J. 1990. *Grain Legumes: Evolution and Genetic Resources*, 379 pp. University Press: Cambridge.

Squire, G. R. 1990. *The Physiology of Tropical Crop Production*, 236 pp. Wallingford, UK: CAB International.

Summerfield, R. J., Ellis, R. H. and Roberts, E. H. 1989. *Annals of Botany* 64: 599–603.

Summerfield, R. J. and Roberts, E. H. 1985. In: *Handbook of Flowering, Volume I*, pp. 92–99, 118–124 (with Muehlbauer, F. J.) and pp. 155–164 (ed. A. H. Halevy). Boca Raton, FL, USA: CRC Press.

Summerfield, R. J., Roberts, E. H., Ellis, R. H. and Lawn, R. J. 1991. *Experimental Agriculture* 27: 11–31.

Summerfield, R. J., Roberts, E. H., Erskine, W. and Ellis, R. H. 1985. *Annals of Botany* 56: 659–671.

Van Rheenan, H. A. 1991. *Plant Breeding Abstracts* 61: 987–1009.

Whyte, R. O. 1946. *Crop Production and Environment*, 372 pp. London: Faber and Faber.

Williams, J. H. and Saxena, N. P. 1991. *Annals of Applied Biology* 119: 105–112.

Plant architecture, competitive ability and crop productivity in food legumes with particular emphasis on pea (*Pisum sativum* L.) and faba Bean (*Vicia faba* L.)

M.C. HEATH[1], C.J. PILBEAM[2], B.A. MCKENZIE[3] and P.D. HEBBLETHWAITE[4]

[1] ADAS Arthur Rickwood, Mepal, Ely, Cambridge, CB6 2BA, UK;
[2] Department of Soil Science, University of Reading, Reading RG1 5AQ, UK;
[3] Plant Science Department, Lincoln University, Canterbury, New Zealand, and
[4] Faculty of Agriculture and Food Science, University of Nottingham, Sutton Bonington, Loughborough, Leics., LE12 5RD, UK

Abstract

Plant breeders have developed new plant ideotypes, with modified canopy architecture, in an attempt to increase yield and stability of yield in grain legume crops. Semi-leafless peas, with leaflets replaced by tendrils, have been introduced successfully. Their crop canopy is as efficient photosynthetically as conventional types but far superior in standing ability; the semi-leafless phenotype is a major factor contributing to potentially higher and more stable yields in peas. Completely leafless peas, however, in which stipule size is also greatly reduced, have not proved successful; photosynthetic area is insufficient to produce satisfactory yields at economic planting densities. In conventional indeterminate faba beans, much of the inherent yield variability is attributed to intra-plant competition for photosynthate. It was envisaged that more determinate types would partition photosynthate to yield components with greater efficiency; this has not been realized in practice. Canopies of fully determinate ("topless") faba beans are physiologically and agronomically inferior. Short-strawed spring beans, however, have demonstrated yield improvements. Lentils and chickpeas are relatively new crops on the Canterbury Plains of New Zealand. Lentils appear to be more suited to conditions than chickpeas because when sown early, crops develop larger crop canopies, intercept more radiation, and convert absorbed radiation into dry matter more efficiently.

Introduction

In this review, plant architecture refers to the size, composition, and arrangement of above-ground stems, leaves, and pods; root form and function in relation to crop productivity is covered elsewhere in these proceedings (Gregory *et al.*, this volume).

F.J. Muehlbauer and W.J. Kaiser (eds.), Expanding the Production and Use of Cool Season Food Legumes, 771–790.
© 1994 *Kluwer Academic Publishers.*

Crop yield is largely dependent on plant architecture which in turn is a product of the plant genome, environmental factors (e.g., light, moisture), and the interaction between neighboring plants. Consequently, yield can be affected through modification of plant architecture by altering the genotype (e.g., plant breeding), manipulating the environment (e.g., time of sowing, irrigation), and altering the plant population (plant density and spatial arrangement).

Of the seven subtropical/temperate legumes that can be grown in Europe, only two, combining pea (*Pisum sativum* L.) and faba bean (*Vicia faba* L.) are of major importance in the UK (Heath, 1987). In 1991, 69,000 ha of combining peas and 131,000 ha of faba beans were grown in England and Wales (Anon., 1992a).

In the UK, combining pea and faba bean crops frequently produce relatively low and variable yields (Heath and Hebblethwaite, 1985a). Plant breeders have to some extent attempted to increase yield and improve yield stability by developing new ideotypes with fundamentally changed plant architecture, e.g., "leafless" and "semi-leafless" peas and determinate ("topless"), "semi-determinate" and short-strawed faba beans. Subsequently, agronomists and crop physiologists have sought to identify and account for those crop husbandry practices which optimize the growth and productivity of these new ideotypes.

The first objective of this review is to identify the agronomic benefits of relatively new plant ideotypes by comparing crop canopy performance and productivity of conventional and modified pea and faba bean types grown under variable field conditions and different husbandry practices. Much of the work reviewed was undertaken during the 1980s at Nottingham University, England. Wherever possible, data are compared and contrasted with those from research programs in other countries.

Experience of growing lentils (*Lens culinaris* Medik.) and chickpea (*Cicer arietinum* L.) in the UK is very limited and currently restricted to trial plots (G. Gent, pers. comm.). Some lentils are grown in Europe (e.g., France) and especially Turkey. Lentils are grown mainly in the Indian sub-continent but recently have been grown in non-traditional areas such as the Canadian prairies and on the Canterbury Plains, the most important cropping region in New Zealand (Savage, 1991); similarly, chickpeas are also a recent introduction to New Zealand. Presently, there are ca. 5,000 ha of lentils grown annually in New Zealand; the area sown to chickpeas is much smaller and commercial quantities are not yet grown. Thus, the second objective is to relate crop productivity to canopy performance of lentils and chickpeas by reviewing studies undertaken at Lincoln University, New Zealand.

Canopy Architecture and Productivity of Food Legumes

Canopy Architecture

Peas and faba beans are typically indeterminate in growth habit; seeds develop within pods which form at a variable number of flowering nodes, on both the main stem and branches, following a period of vegetative growth. In pea, the compound leaf is comprised of two stipules and a petiole attached to one or more pairs of leaflets terminating in a tendril or compound tendril (Davies *et al.*, 1985). Of the 33 genes which influence leaf size and shape, the most important are *af* and *st* (Figure 1). *Afila* or semi-leafless types (*afafStSt*) have leaflets replaced by tendrils; additional presence of the *st* gene in completely leafless (*afafstst*) types reduces the size of the stipule (Snoad, 1980). Basal branches are produced at the base of the main stem (from scale nodes) or from nodes immediately below that of the first flower (side branches).

In faba bean, leaves are again compound with stipules and leaflets but tendrils are either absent or rudimentary. Stems are strong, but hollow; one or more branches arise from leaf axils at basal nodes. Conventional indeterminate cultivars of *Vicia faba* minor and equina are typically tall (0.5 to 2.0 m depending on environment) but dwarf, semi-dwarfs, and short-strawed cultivars (e.g., "Alfred", "Troy") are available. In so-called "topless" or determinate types, which possess the *ti* gene (Sjödin, 1971), the number of nodes per stem is reduced to ca. 5 and the stem has a terminal inflorescence (Figure 2); depending on conditions, determinate spring beans will produce a number of compensatory side branches from below the first flowering node. Frauen and Brimo (1983) reported the existence of a true "semi-determinate" growth habit controlled by a single gene (*ti-s*); however, no commercial cultivars have been introduced which contain the *ti-s* gene (D. A. Bond, pers. comm.). In the UK, and elsewhere, short-strawed cultivars such as Alfred and Troy are often described as "semi-determinate". Stem length in faba beans is probably controlled by a complex of "straw-shortening" genes (D. A. Bond., pers. comm.).

Lentils and chickpeas also have an indeterminate growth habit and are of similar height. Lentils are erect and highly branched. The leaves, which are pinnate, consist of three to eight pairs of leaflets terminated by a tendril. The leaflets are small, usually less than 1 cm wide and 2 to 3 cm long and obovate or lanceolate. The cultivars used in New Zealand are small: 40 to 80 cm in height. Both microsperma (small-seeded, 1000 seed wt = 35 g) and macrosperma (1000 seed wt = 70 g) types are grown.

There are no named commercial cultivars of chickpeas grown in New Zealand. However, both desi and kabuli seeds types are grown. Chickpea is very similar in appearance to lentil although the branches are fewer in number and much larger. Leaves are again compound and pinnate with 9 to 17 leaflets. Leaflets are larger than those of lentil and are up to 1.5 cm wide and 2 to 3 cm long. The leaflets are again usually obovate but with a serrated edge.

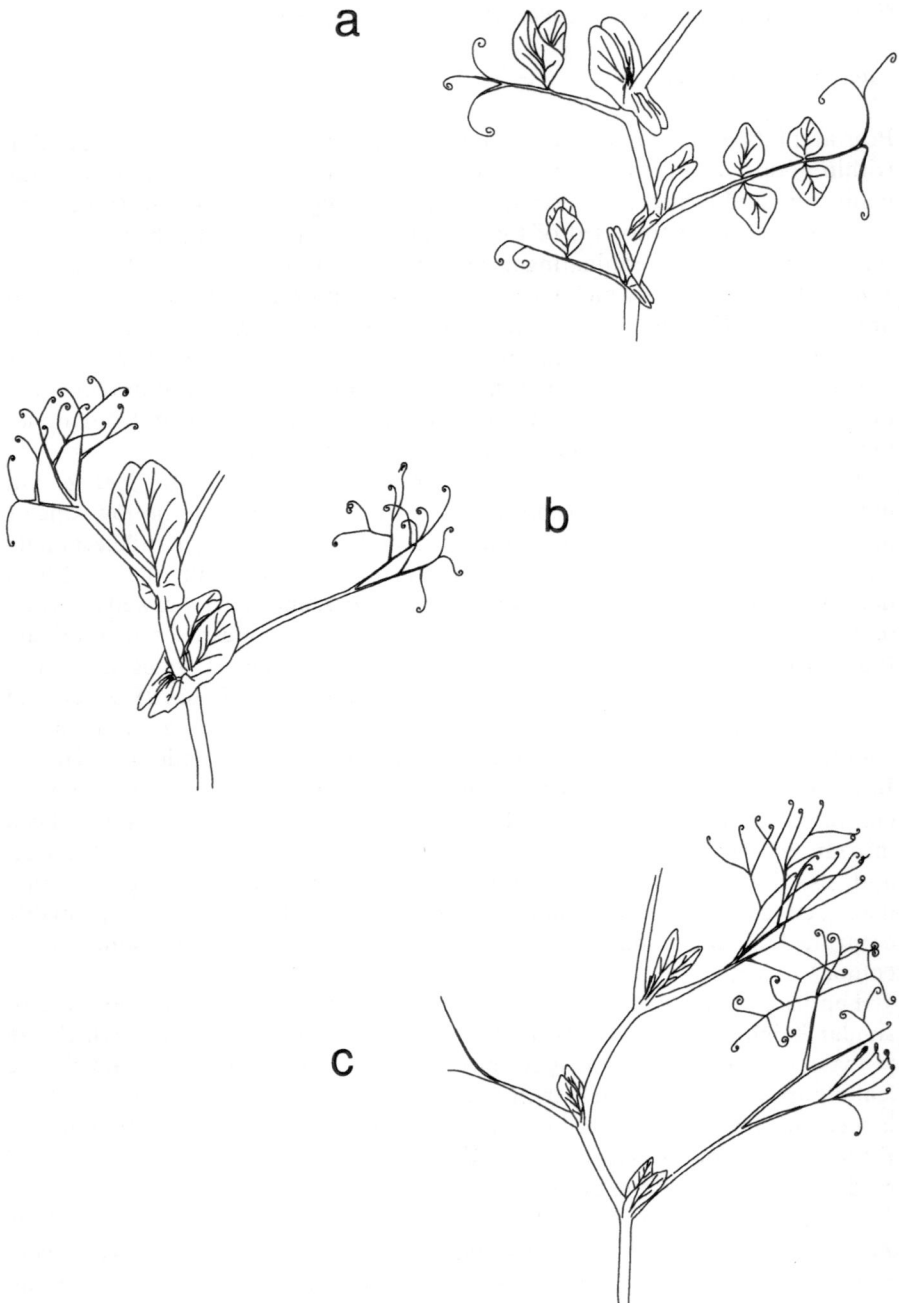

Figure 1. Foliage modification in peas. (a) Conventional leafed (*AfAfStSt*); (b) Semi-leafless (*afafStSt*); (c) Leafless (*afafstst*).

Figure 2. Canopy modification in faba beans. From left to right: (a) Conventional indeterminate ("Maris Bead"); (b) short-strawed (Alfred); (c, d) determinate (Ticol and "PBI composite STI") grown under irrigated and unirrigated conditions (Courtesy of Dr. D. A. Bond, Plant Breeding International, Cambridge, UK).

Productivity

Yields of grain legume species are typically lower and more variable than for cereal species. Average yields for peas in western Canada and the UK were previously 1.6 and 3.0 t ha^{-1}, respectively. However, when productivity was calculated on the basis of yield of dry seed per day of the growing season, it was not vastly different between the two regions (16 and 20 kg ha^{-1} day^{-1}, respectively) (Davies *et al.*, 1985). In the UK, pea yields of 5 t ha^{-1} are becoming more frequent and the average increased to 3.6 t ha^{-1} during the 1980s; average yields were higher in France (4.5 t ha^{-1}) during the same period (Plancquaert, 1991). Faba bean yields range between 0.3 t ha^{-1} in dry areas to 3.5 to 4.5 t ha^{-1} in western Europe where adequate moisture is available (Hawtin and Hebblethwaite, 1983). However, relatively high bean yields of >

4.5 t ha^{-1} are obtained only where the mean daily productivity of seed during the reproductive phase (onset of flowering to maturity) is greater than 50 kg ha^{-1} day^{-1} (Bond et al., 1985). In the UK, bean yields can reach above 5 t ha^{-1} but the average of both spring and winter types is < 4 t ha^{-1}. In 1991, the national average yield for faba beans was 3.3 t ha^{-1} (Anon., 1992a). In NIAB (National Institute of Agricultural Botany) trials over the last 7 years, the average yield of winter and spring faba beans was 4.93 and 4.55 t ha^{-1}, respectively (Anon., 1992b).

Hot and dry weather conditions are required to ripen lentils and chickpeas. In the major lentil-producing countries, seed yields range from 0.5 t ha^{-1} in India to 1.2 t ha^{-1} in Turkey; average yields in the USA and Egypt are 1.1 t ha^{-1} (Muehlbauer et al., 1985). In temperate areas, indeterminancy is a problem and there is a risk that crops will not reach maturity. During the late 1970s and early 1980s, when lentils were grown in the UK on a trial basis, crops produced excessive vegetative growth but very little seed was harvested. Similarly, in New Zealand, McKenzie and Hill (1990) reported that lentils yielded only 1.0 t ha^{-1} in a wet season but up to 3.3 t ha^{-1} in a dry one. In northern India, the world's principal chickpea-producing area, maximum productivity is 30 kg ha^{-1} day^{-1} giving seed yields of up to 5 t ha^{-1} (Smithson et al., 1985). Chickpea yields are usually lower than lentil yields in New Zealand. Hernandez (1986) reported yields ranging from 0 to 2.1 t ha^{-1} while in 1991, McKenzie (unpublished data) obtained yields of up to 4.2 t ha^{-1}.

Yield variability in grain legumes can be attributed to many factors, including the indeterminate growth habit, inherent susceptibility to crop lodging, and sensitivity to extremes of moisture and adverse soil physical conditions. In the UK, crop lodging and associated harvest losses is a major cause of yield variation in peas, particularly in wet seasons; leafless and semi-leafless peas were introduced in an attempt to overcome this problem. Sensitivity to extremes of soil moisture is a cause of yield variability particularly in field beans and peas. In relatively dry conditions, the size of the canopy is restricted, reducing the number of pod-bearing nodes. Poor ground cover also reduces competition against weed invasions (which can become serious due to herbicide failure in very dry springs and early summers). In relatively wet seasons, excessive vegetative growth increases the risk of lodging and encourages foliar diseases. Determinate and short-strawed faba beans have been introduced in an attempt to reduce yield variability (Pilbeam et al., 1990c).

Affect of Canopy Architecture on Crop Performance

Photosynthetic Area Index (PAI) and Radiation Interception

Radiation interception in food legumes is dependent upon leaf orientation, photosynthetic area index (PAI) and incident radiation.

In conventional leafed peas, PAI increases rapidly during the vegetative period, typically reaching peak levels of 8 to 11 shortly before flowering. Optimum PAI, or the PAI at which 95% of incident radiation is intercepted, is ca. 7.5. At constant plant density, PAI values and hence radiation interception levels for leafed peas are higher than for semi-leafless peas, particularly during the post-flowering period. Semi-leafless peas tend to intercept more radiation than leafless peas throughout the entire growing season (Heath and Hebblethwaite, 1985b).

In faba beans, PAI levels are typically higher in winter, compared with spring, types with maximum levels occasionally attaining at least 10 in the former (Pilbeam *et al.*, 1991b). Determinate faba beans have smaller PAIs than indeterminates, a difference which is accentuated markedly post-flowering (Pilbeam *et al.*, 1989b, 1990b).

While morphologically similar, lentils and chickpeas have a very different canopy architecture. Lentils have been shown to produce PAIs as high as 12 (McKenzie and Hill, 1991) while the maximum for chickpeas in Canterbury was 3.4 to 3.5 (Hernandez, 1986; McKenzie unpublished data, 1992). Optimum PAI, at which 95% of incident radiation is intercepted, was ca. 7 for lentils and ca. 3.5 to 4.0 for chickpeas. In Canterbury, lentil crops usually attain the optimum PAI by mid-November. Chickpea crops often do not attain optimum PAIs, even at high sowing rates.

Percentage radiation interception can be related to PAI more precisely by calculating the attenuation or light extinction coefficient (k). In studies at Nottingham University, leafless peas appeared to have a higher k (0.55 to 0.75) than either semi-leafless or leafed peas, both of which had a similar k (0.33 to 0.49). This suggested that leafless canopies intercepted radiation more effectively per unit PAI than either leafed or semi-leafless peas, but this may have been an indirect result of underestimating the extent to which cylindrical foliage structures contribute to the PAI (Heath and Hebblethwaite, 1985b).

In spring faba beans, k was consistently higher in determinate ("Ticol", 0.74 to 0.99) than in indeterminate ("Herz Freya", 0.65 to 0.94) faba beans (Stützel and Aufhammer, 1991). In this study, the "topless" type appeared to produce leaves which were more horizontally orientated and had larger, thicker leaflets than the indeterminate type. Thus, the determinate canopy appeared to be more effective per unit area in intercepting radiation. In winter faba beans, there is no evidence for any difference in k between determinate and indeterminate types (Pilbeam *et al.*, 1991b).

Thus, although modified canopies of leafless peas and determinate "topless" beans attain lower PAI values post-flowering, they both appear to compensate by utilizing foliage more effectively in intercepting radiation.

Lentils have a very low extinction coefficient of ca. 0.26 (indicating erect leaves) (McKenzie and Hill, 1991); chickpeas, however, are less erect, with an extinction coefficient of ca. 0.40.

Table 1. Photosynthetic efficiency (ϵ) and attenuation coefficient (k) for food legumes as reported by various authors

Reference	Crop and cultivar
Stützel and Aufhammer (1991)	Spring beans – indeterminate Herz Freya
	Spring beans – determinate Ticol
Green *et al.* (1985)	Spring beans – short-strawed Alfred
	Spring beans – very short Minica
Heath and Hebblethwaite (1985b)	Peas – leafed Birte
	Peas – semi-leafless BS3
	Peas – leafless Filby
Heath and Hebblethwaite (1987a)	Peas – leafed Birte
	Peas – semi-leafless BS3
	Peas – leafless Filby
Pyke and Hedley (1985)	Peas – leafed JI1196
	Peas – semi-leafless JI1197
	Peas – leafless JI1198
McKenzie and Hill (1991)	Lentils
McKenzie (unpublished)	Chickpeas

Conditions	$\in$	k
	(g MJ^{-1})	
18.5 plants m^{-2}	2.13	0.94
74 plants m^{-2}	2.69	0.84
18.5 plants m^{-2}	1.61	0.99
74 plants m^{-2}	2.18	0.86
High moisture	3.23	–
Dry	2.30	–
High moisture	3.19	–
Dry	2.01	–
High moisture	1.46	0.33
Dry	0.96	0.49
High moisture	1.46	0.33
Dry	0.96	0.49
High moisture	1.46	0.55
Dry	0.96	0.75
Precision drilled	1.65	0.34
Random spacing	1.90	0.34
Micro-plots	1.25	–
Early sowing	1.5	)
		) 0.26
Delayed sowing	1.3	)
Early sowing	1.0	)
Middle sowing	1.2	) 0.40
Delayed sowing	1.1	)

Radiation Interception and Photosynthetic Efficiency

The photosynthetic efficiency (ϵ) of a crop canopy is a measure of its ability to convert intercepted radiation into dry matter. During vegetative growth, all arable crops *inter alia* tend to have a similar ϵ (Monteith and Elston, 1983). It is not surprising, therefore, that no difference in ϵ was found between different pea leaf phenotypes [Heath and Hebblethwaite, 1985b ($\epsilon = 1.46$ g MJ^{-1}); Pyke and Hedley, 1985 ($\epsilon = 1.25$ g MJ^{-1})]. Thus, leafless and semi-leafless peas which are composed largely of cylindrical foliage structures (tendrils, stems, petioles) are as efficient as conventional leafed peas in utilizing solar radiation. No difference in ϵ ($\epsilon = 2.0$ to 3.2 g MJ^{-1}) was found between short-strawed (Alfred) and very short-strawed ("Minica") spring faba beans (Green *et al.*, 1985). However, Stützel and Aufhammer (1991) reported that ϵ was consistently lower in a determinate (Ticol, $\epsilon = 1.6$ to 2.4 g MJ^{-1}) than in an indeterminate (Herz Freya, $\epsilon = 2.7$ to 3.5 g MJ^{-1}) spring faba bean.

Differences in ϵ do tend to occur when crops are grown under various stress conditions and/or when ϵ is calculated on the basis of final seed yield or dry matter production during the post-flowering stage. Photosynthetic efficiency was reduced by moisture stress in both pea (Heath and Hebblethwaite, 1985b, 1987b) and faba bean (Green *et al.*, 1985; Barmouh and Hsiao, 1992) and, in beans, by soil compaction (Assaeed *et al.*, 1990). In peas, crop lodging reduced ϵ by 17% (Heath and Hebblethwaite, 1987b).

In lentils, ϵ was reduced by disease and delayed sowing (McKenzie and Hill, 1991). For early autumn-sown (March to May) lentil crops in New Zealand, ϵ was 1.5 g MJ^{-1}; late-spring sowings used solar radiation less efficiently and ϵ was reduced to 1.3 g MJ^{-1} (McKenzie and Hill, 1991). There was less variability with chickpeas and ϵ values for May, July, and October sowings were 1.0, 1.2, and 1.1, respectively.

The reduction in ϵ for lentil with later sowings in New Zealand may have been due to the relatively greater losses of total carbon in root growth, nodule formation, and nitrogen fixation compared to earlier-sown crops. Early autumn-sown crops were large with a well developed nodule mass and were capable of taking full advantage of good spring growing conditions. Spring-sown lentils, however, were producing nodules and roots during the spring and early summer period; hence ϵ, which is based on above-ground dry matter, was lower. In New Zealand, chickpeas do not nodulate unless inoculated and nodulation is not always successful. Plants in these sowing date experiments were not nodulated, hence there would have been no difference in carbon partitioning response and thus no difference in ϵ between autumn and spring sowing dates.

Coefficients ϵ and k were not affected by changes in plant density or spatial arrangement in peas (Heath and Hebblethwaite, 1987b) but were in spring faba beans (Stützel and Aufhammer, 1991). Data were presented (Barmouh and Hsiao, 1992; Poulain *et al.*, 1992) which further suggest that in faba beans, ϵ and k can be affected by changes in plant density or spatial arrangement. The various values for ϵ and k reviewed here are summarized in Table 1.

Water Use Efficiency

Water use efficiency (WUE) is frequently used to quantify the extent to which crops use water to produce yield (Cooper *et al.*, 1988). In general, WUE of legumes tends to increase with decreasing moisture availability, i.e., crops use water more efficiently when supplies are restricted. Differences in plant architecture might be expected to influence the ability of the crop canopy to utilize available moisture reserves and thus affect WUE.

At Nottingham, the WUE of leafless peas ("Filby") was significantly lower than either leafed ("Birte") or semi-leafless (BS3) peas under relatively dry soil conditions (Heath *et al.*, 1985); however, there were no differences in WUE between near-isogenic phenotypes (JI 1196, JI 1197, JI 1198) under normal soil moisture conditions (Jones, 1989). In New Zealand, two separate studies (Wilson *et al.*, 1981; Zain *et al.*, 1983) showed that there was no difference in WUE between a leafed ("Rovar") and a related semi-leafless (Rovar × semi-leafless) pea when grown in irrigated plots; under dryland conditions, however, there was a tendency for the semi-leafless cultivar to use water more efficiently (Wilson *et al.*, 1981). By contrast, recent studies under dryland conditions in Australia indicated that conventional leafed cultivars had a higher WUE than semi-leafless cultivars (Armstrong, 1992). Similarly, Bailey and Groves (1991) showed that semi-leafless ("Solara") peas were more responsive to irrigation than leafed ("Bohatyr") peas under relatively dry, sandland conditions in England. Thus, it would appear that differences in WUE between pea leaf phenotypes are more likely to occur under conditions of reduced soil moisture availability. Under such conditions, reduced WUE of leafless and possibly semi-leafless peas may be caused by the more open crop canopy. Leafless pea canopies experienced a higher rate of transpiration from the lower parts of the canopy and also a higher rate of evapotranspiration from the soil surface, than either leafed or semi-leafless peas (Jones, 1989).

Plant habit had no effect on WUE in a comparison of indeterminate ("Gobo"), short-strawed (cv. Alfred), and determinate ("Tina") spring beans (Pilbeam, unpublished data). Conversely, Pilbeam *et al.* (1992) showed that a very short-strawed cultivar ("Minica") had a greater WUE than a determinate (Ticol) one.

Thus, there is evidence that "leafless" peas and "topless" beans are not as efficient as conventional types in utilizing water under conditions of limited moisture availability. These modified types are, therefore, unlikely to be high yielding in arid regions or in relatively dry seasons.

Under rain shelters, lentils have been shown to produce 1.29 g dry matter $m^{-2}\,mm^{-1}$ of actual evapotranspiration (McKenzie and Hill, 1990). There have been no detailed studies conducted on water use of chickpeas in Canterbury.

Affect of Plant Architecture on Crop Productivity

Yield

Leafless peas yield less than either semi-leafless or conventional leafed peas at standard planting densities (i.e., 60 to 100 plants m^{-2}). This is because the photosynthetic area per plant in leafless peas is reduced to such an extent that individual plants have inherently lower growth rates than either semi-leafless or leafed peas (Hedley and Ambrose, 1981).

Determinate faba beans yield less than indeterminate types for three reasons. Firstly, determinate cultivars have smaller canopies; equal or greater harvest indices in the non-determinate forms inevitably results in lower yields for the determinates (Pilbeam *et al.*, 1989a, 1990c, 1991a). Secondly, yield production per unit stem is lower on the side branches than on the main stem; in spring faba beans, determinate cultivars have more side branches than their non-determinate counterparts and consequently lower yields; in winter faba beans, where stem numbers are similar, the contribution of the mainstem to total yield is greater in the indeterminate types (Pilbeam *et al.*, 1990b) and again the determinates are at a disadvantage. Thirdly, determinates may be less able to transfer assimilates stored in the stem to the developing pods than indeterminates (Pilbeam *et al.*, 1989a).

Thus, as a result of canopy modification, leafless peas and determinate faba beans are at a physiological disadvantage relative to conventional types. Semi-leafless peas and short-strawed faba beans are not similarly disadvantaged.

Yield-Density Response

Optimum planting densities for the four food legume species are markedly different. For lentils, McKenzie *et al.* (1986) reported little response to yield at densities above 200 plants m^{-2}; the optimum is ca. 150 plants m^{-2}. In England, the optima for peas range from 65 to 95 plants m^{-2}, depending on seed type; these optima are slightly higher in France. For faba beans, optima are ca. 40 to 60 plants m^{-2} for spring types (Pilbeam *et al.*, 1990a) and ca. 10 to 20 plants m^{-2} for winter types (Pilbeam *et al.*, 1991a). Hernandez (1986) and McKenzie (unpublished data) found no positive response with chickpeas to densities above ca. 30 plants m^{-2}.

Canopy architecture affects optimum plant density in both peas and spring faba beans. Leafless peas require higher planting densities than either semi-leafless or conventional peas in order to produce a canopy of sufficient PAI to maximize crop growth rate and attain reasonably high yields (Hedley and Ambrose, 1981; Heath and Hebblethwaite, 1987a). Optimum densities for semi-leafless peas are not necessarily higher than for leafed peas (Heath *et al.*, 1991; Taylor *et al.*, 1991). Similarly, in spring faba beans, optimum plant densities are higher for determinates (Pilbeam *et al.*, 1990a) but there is no difference between short-strawed and indeterminates (Cleal, 1991). By contrast, plant morphology has no effect on the optimum density of winter faba beans; optima for both "Bourdon" (indeterminate) and "858" (determinate) lay between 10 and 20 m^{-2} (Pilbeam *et al.*, 1991a).

Irrespective of pea leaf phenotype or faba bean type, lodging risk increases with increasing plant density (Heath *et al.*, 1991; Taylor *et al.*, 1991); this can be attributed to the taller and weaker stems which are produced by greater inter-plant competition at high densities. Increased lodging will result in reduced yields.

Inter- and Intra-Plant Competition

Grain legumes exhibit considerable compensatory ability or plasticity of yield response. Propensity to branch is a major factor in this respect. In both semi-leafless and leafed peas, yield per unit area is maintained over a relatively wide range of densities (70 to 140 plants m^{-2}) by compensatory increases in branches (Heath *et al.*, 1991). Leafless peas are also capable of branching, yet relatively high densities (e.g., 100 to 150 plants m^{-2}) are required to maximize yields; thus, leafless peas (in contrast to semi-leafless and leafed types) are responsive to increasing density.

Faba beans also make compensatory increases in branch numbers to maintain yield at reduced densities. Determinate cultivars containing the *ti* gene are typically single stemmed but produce numerous side branches at low densities. However, branches are generally not as efficient as the mainstem, irrespective of type. Faba beans are noted for their high losses of flowers and immature pods. This is often attributed to inadequate supplies of assimilate to meet the demands of each reproductive sink. It has been proposed (Gates *et al.*, 1983) that an independent vascular supply to each flower within a raceme may reduce competition and stabilize yield, but this has been rejected by others (Ruckenbauer and Mollenkopf, 1986).

Yield is determined by the number of podding nodes, pods per podding node, seeds per pod, and mean seed weight. Compensatory adjustments between yield components also occur. All food legumes are highly "plastic", particularly with regard to pods per plant. In lentil, the number of pods per plant decreased from 45.7 at 100 plants m^{-2} to 16.5 at 400 plants m^{-2} (McKenzie *et al.*, 1986). With chickpeas, Hernandez (1986) recorded 22.7 pods per plant at 33 plants m^{-2} while at 133 plants m^{-2} there were only 7.1 pods per plant. In the UK, the number of pea pods per plant decreased from 19.7 at 40 plants m^{-2} to 3.3 at 140 plants m^{-2} (Heath *et al.*, 1991).

Spatial Arrangement

Peas and faba beans need to be sown relatively evenly (with regard to spacing and depth) to ensure that crops establish well and optimize leaf area coverage. Conventional establishment methods may not always optimize yield and a more regular plant arrangement could be beneficial.

Uniform spacing (precision drilling) increased radiation interception relative to random spacing (conventional drilling) in leafless and semi-leafless, but not leafed peas, during early vegetative growth when PAI values were low; despite these initial differences, precision drilling did not increase yield (Heath and Hebblethwaite, 1987a). By contrast, McEwen (1978) demonstrated that precision drilling increased yield of faba beans in studies at Rothamsted. Similarly, in determinate (Ticol) and indeterminate (Herz Freya) spring faba beans, both radiation interception and dry matter production increased faster in isometric stands (equivalent distance between plants) than in single or double rows (Stützel and Aufhammer, 1991). However, the determinate Ticol responded to changes in plant distribution similarly to both the short-strawed Alfred (Pilbeam *et al.*, 1989a) and the indeterminate Herz Freya (Stützel and Aufhammer, 1991). Thus, productivity of leafless peas and determinate faba beans cannot necessarily be increased by growing plants in a more uniform plant distribution.

Agronomic Factors Affecting

Standing Ability

Combining peas are inherently susceptible to crop lodging. The major agronomic advantage of *afila* types is improved standing ability; the large profusion of tendrils bind stems together, thereby delaying the onset and eventual severity of lodging (Heath and Hebblethwaite, 1984). Improved standing ability of *afila* types leads to easier harvesting, fewer harvest losses and, hence, higher yields. Determinate and short-strawed faba beans are also of improved standing ability compared to conventional types. However, although determinate faba beans rarely lodge, they can still yield less when compared to lodged indeterminate types.

Disease Considerations

The combination of reduced foliage, a more open canopy structure and improved standing ability would be expected to reduce humidity within the crop microclimate and thereby lead to reduced disease incidence in leafless and semi-leafless peas. Currently, there is no evidence for this. On the contrary, in the UK, *afila* types appear to be as susceptible as conventional types to *Mycosphaerella pinodes* (A. J. Biddle, pers. comm.) and downy mildew (Heath and Hebblethwaite, 1984), both of which are more virulent on tendrils compared to leaves. Similarly, *afila* types are also generally more susceptible to foliar pathogens in the USA (J. M. Kraft, pers. comm.) and Australia (M. Ali, pers. comm.).

Similarly, determinate faba beans, particularly spring-sown cultivars, can be more affected by foliar diseases (e.g., chocolate spot, rust) than conventional types. This is because determinate plants are unable to produce compensatory increases in leaf growth following severe disease infestations; also, old susceptible leaves are not protected by young, more resistant, leaves as in indeterminates.

Sowing Date

For many crops, early sowing tends to increase yield because a greater time is available for crop canopies to develop, intercept radiation, accumulate dry matter and partition into yield components. Yield of both modified and conventional cultivars of spring peas (leafless Filby; leafed "Vedette"), spring faba beans (determinate Ticol; short-strawed Alfred), and winter faba beans (determinate 858, "796"; indeterminate Bourdon) were all increased by early sowing (Silim *et al.*, 1985; Pilbeam *et al.*, 1989b, 1990b). However, there was no interaction between sowing date and canopy architecture.

In New Zealand, lentils have been shown to be most productive when sown in the autumn due to longer leaf area duration and greater interception of photosynthetically active radiation (McKenzie *et al.*, 1985). Chickpeas are usually spring sown in Canterbury, because of the heavy frosts; in exceptionally cold winters, e.g., 1991, plant mortality can reach 50% (McKenzie, unpublished data). However, chickpeas are potentially higher yielding when sown in the autumn under more typical winter conditions (Saxena *et al.*, 1990). In England, autumn sowing of peas is potentially very risky because the crop may fail to survive the winter, and there is no guarantee of higher yields even where winter survival is adequate (Silim *et al.*, 1985). Overwintering of winter faba beans is not a problem, particularly if crops are established by deep plowing. Determinate winter faba beans (e.g., 858) may be less winter hardy than the conventional indeterminates (e.g., Bourdon) (Pilbeam *et al.*, 1990b); however, determinates could be made just as hardy by breeding (D. A. Bond, pers. comm.). Winter faba beans are frequently higher yielding than spring faba beans, except in years of severe chocolate spot attacks or where establishment has been very low and uneven.

Irrigation

Irrigation generally increases plant height, leaf area, and biomass production, radiation interception, rooting depth, number of podding nodes, and hence seed yield in both combining peas (Salter, 1963) and faba beans (Hebblethwaite, 1982); spring-sown faba beans tend to be more responsive to irrigation than winter-sown faba beans (Pilbeam *et al.*, 1990c). Conversely, moisture stress reduces radiation interception and yield by restricting the size of the crop canopy (Pilbeam, unpublished data). It has been suggested (Pilbeam *et al.*, 1990c) that determinate faba beans, because of their lower productivity, show a greater stability of yield over a wide range of soil moisture conditions than indeterminate faba beans.

The extent to which crops respond to additional water supply, or are inhibited by restricted moisture, depends on crop growth stage. In peas, the two moisture-responsive growth stages are flowering and pod fill (Salter, 1962, 1963). Moisture-responsive stages are less well defined in faba beans (Hebblethwaite, 1982) but Pilbeam *et al.* (1990c) suggested that pod fill was the main moisture-sensitive stage in winter faba beans and Green *et al.* (1986) indicated that flowering was the critical stage in spring faba beans. It has been suggested that a period of moisture stress just prior to flowering can be beneficial in both peas (Frohlick and Henkel, 1961) and faba beans (Grashoff, 1990) because it reduces vegetative growth and promotes reproductive development. Some workers have reported that irrigation or excessive moisture availability during the vegetative stage of peas and faba beans can decrease yield by encouraging vegetative growth and lodging (e.g., Green *et al.*, 1986). Bailey and Groves (1991), however, demonstrated that leafed and semi-leafless peas responded to irrigation applied during both the vegetative and post-flowering stages on sandland in two particularly dry seasons.

With lentils in New Zealand, irrigation has given yield increases only when the crop was grown under rain shelters (McKenzie and Hill, 1990). Under rain shelters, and on a soil with a field capacity of 250 mm H_2O m^{-1} of soil, a lentil crop had a limiting potential soil moisture deficit of 132 mm and lost 0.39% of maximum potential yield for each mm increase in deficit beyond the limiting deficit D_1. This loss figure is higher than that reported in New Zealand by Jamieson *et al.* (1984) for field peas (0.22%) and by Husain (1984) for faba beans (0.13%). However, the limiting deficit for lentil was much higher than for either peas (88 mm) or faba beans (65 mm). Lentils are clearly very drought resistant, but when a high D_1 is reached, losses will be severe. Limiting deficits for faba beans have also been defined in England and ranged from 30 to 120 mm at Rothamsted, depending on soil type and season (Day and Legg, 1986); on a coarse sandy loam at Nottingham, D_1 was 65 mm (Hebblethwaite, 1982).

Concluding Remarks

Plant breeders have achieved a major breakthrough in the introduction of semi-leafless peas and short-strawed spring faba beans. These improved crop ideotypes demonstrate yield advantages and other agronomic benefits (e.g., reduced indeterminancy, better standing ability, reduced lodging, more even ripening, easier harvesting) which should lead to greater yield stability between sites and years. Semi-leafless peas (e.g., Solara) and short-strawed spring faba beans (e.g., Troy, Alfred) now dominate the market; they have played a major role in increasing UK production since the mid-1980s and have thus reduced the EC's dependence on imported vegetable protein for compounding into animal feed.

In the future, there is potential for further improvement by incorporating the *afila* and straw-shortening genes into different genetic backgrounds (e.g., improved winter peas and early maturing spring faba beans, respectively). There is also further potential for canopy modification in winter faba beans and for evaluating the semi-determinate *"Frauen ti-s"* gene for agronomic and production potential.

It is well-documented that certain agronomic factors (e.g., early sowing, irrigating under conditions of moisture stress and rapid, even establishment of optimum plant populations) will increase canopy size and usually increase crop productivity. In the future, there will be an R & D need in the UK to identify the optimum husbandry requirements for new improved cultivars: e.g., investigating the feasibility of advancing sowing date by plowing-in spring faba beans and autumn sowing of winter peas; determining the critical growth stages for irrigating spring faba beans; and evaluating different establishment methods for winter faba beans.

Extreme forms of canopy modification (i.e., in completely "leafless" peas and determinate "topless" faba beans) have proved too drastic. Because of their inherently low yield potential they need to be sown at relatively high, uneconomic seed rates. Plant breeders in western Europe are no longer selecting for *st* or *ti* ideotypes. It is highly unlikely, therefore, that completely "leafless" peas (e.g., Filby) or "topless" faba beans (e.g., Ticol) will ever become commercially viable, because of their physiological disadvantages.

Lentils and chickpeas are now becoming established in non-traditional areas such as New Zealand. In Canterbury, lentils are more productive than chickpeas because they can intercept more radiation due to a higher PAI and earlier attainment of optimum PAI; also, intercepted radiation is utilized more efficiently in lentils than in chickpeas. Furthermore, lentils are better adapted to dry New Zealand harvest conditions than chickpeas. In the future, there is scope for undertaking further physiological/agronomic studies, particularly on water use of chickpeas, to identify the limitations to cropping; such studies could point the way for plant breeders to develop new improved crop plant ideotypes for lentils and chickpeas which might extend their cropping ranges further.

Acknowledgements

Financial support from the UK Ministry of Agriculture, Fisheries and Food (MAFF) in the preparation and presentation of this paper is gratefully acknowledged. Much of the R & D on which this Review is based was funded as follows: combining peas, UK Agricultural Food and Research Council; winter field beans, MAFF; spring faba beans, European Economic Community; lentil, New Zealand Department of Scientific and Industrial Research; chickpea, Lincoln University Research Committee. All are gratefully acknowledged. Thanks are also extended to Dr. D. A. Bond, Plant Breeding International, Cambridge for use of Figure 2 and for information supplied during the preparation of this paper and to Mr. J. B. S. Freer for preparing the drawings in Figure 1.

References

Anon. 1992a. Survey of dried pea and field bean production in 1991, England and Wales (including minor holdings). *MAFF Statistics, Ministry of Agriculture, Fisheries and Food, Guildford.*

Anon. 1992b. *NIAB Farmers Leaflet No. 10*, 18 pp. National Institute of Agricultural Botany, Cambridge.

Armstrong, E. L. 1992. *A physiological comparison of morphologically contrasting field pea genotypes.* 189 pp. Ph.D. Thesis. University of Western Australia.

Assaeed, A. M., McGowan, M., Hebblethwaite, P. D. and Brereton, J. C. 1990. *Annals of Applied Biology* 117: 653–666.

Bailey, R. J. and Groves, S. J. 1991. Irrigation of combining peas. *Aspects of Applied Biology 27, Production and Protection of Legumes*, pp. 299–304 (eds. R.J. Froud-Williams et al.) The Association of Applied Biologists, Horticulture Research International, Wellesbourne, Warwick, U.K.

Barmouh, A. and Hsiao, T. C. 1992. In: *2nd International Food Legume Research Conference Program and Abstracts.* p. 45. 12–16 April 1992, Cairo, Egypt.

Bond, D. A., Lawes, D. A., Hawtin, G. C., Saxena, M. C. and Stephens, J. H. 1985. In: *Grain Legume Crops*, Ch. 6 (eds. R. J. Summerfield and E. H. Roberts). Collins: London.

Cleal, R. A. E. 1991. *Aspects of Applied Biology 27, Production and Protection of Legumes*, pp. 89–94. (eds. R.J. Froud-Williams et al.) The Association of Applied Biologists, Horticulture Research International, Wellesbourne, Warwick, U.K.

Cooper, P. J. M., Campbell, G. S., Heath, M. C. and Hebblethwaite, P. D. 1988. In: *World Crops: Cool Season Food Legumes*, pp. 813–829 (ed. R. J. Summerfield). London: Kluwer Academic Publishers.

Davies, D. R., Berry, G. J., Heath, M. C. and Dawkins, T. C. K. 1985. In: *Grain Legume Crops*, Ch. 5 (eds. R. J. Summerfield and E. H. Roberts). London: Collins.

Day, W. and Legg, B. J. 1986. In: *The Faba Bean (Vicia faba* L.): *A Basis for Improvement*, Ch. 10 (ed. P. D. Hebblethwaite). London: Butterworths.

Frauen, M. and Brimo, M. 1983. *Zeitschrift für Pflanzenzüchtung* 91: 261–263.

Frohlick, H. and Henkel, A. 1961. *Archiv für Gartenbau* 9: 405–428.

Gates, P. J., Smith, M. L., White, G. and Boulter, D. 1983. In: *The Physiology, Genetics and Nodulation of Temperature Legumes*, pp. 43–54 (eds. D. R. Davies and D. G. Jones). London: Pitman Books.

Grashoff, C. 1990. *Netherlands Journal of Agricultural Science* 38: 21–44.

Green, C. F., Hebblethwaite, P. D. and Ison, D. A. 1985. *Annals of Applied Biology* 106: 143–155.

Green, C. F., Hebblethwaite, P. D. and Ricketts, H. E. 1986. *Fabis* 15: 6–31.

Gregory, P. J., Saxena, N. P., Arihara, J. and Ito, O. 1994. In: *Expanding the Production and Use of Cool Season Food Legumes*, pp. 809–820 (eds. F. J. Muehlbauer and W. J. Kaiser). Dordrecht: Kluwer Academic Publishers.

Hawtin, G. C. and Hebblethwaite, P. D. 1983. In: *The Faba Bean (Vicia faba L.): A Basis for Improvement*, pp. 3–22 (ed. P. D. Hebblethwaite). London: Butterworths.

Heath, M. C. 1987. *Outlook on Agriculture* 16: 2–7.

Heath, M. C. and Hebblethwaite, P. D. 1984. *Outlook on Agriculture* 13: 195–202.

Heath, M. C. and Hebblethwaite, P. D. 1985a. In: *The Pea Crop – A Basis for Improvement*, Ch. 2 (eds. P. D. Hebblethwaite, M. C. Heath and T. C. K. Dawkins). London: Butterworths.

Heath, M. C. and Hebblethwaite, P. D. 1985b. *Annals of Applied Biology* 107: 309–318.

Heath, M. C. and Hebblethwaite, P. D. 1987a. *Journal of Agricultural Science, Cambridge* 108: 425–430.

Heath, M. C. and Hebblethwaite, P. D. 1987b. *Annals of Applied Biology* 110: 413–420.

Heath, M. C., Jones, C., Hebblethwaite, P. D. and Memar, M. H. 1985. In: *Proceedings, EUCARPIA Meeting on Pea Breeding*. Sorrento, Italy, 10–13 June 1985, pp. 164–176. University of Naples/ENEA, Rome.

Heath, M. C., Knott, C. M., Dyer, C. J. and Rogers-Lewis, D. 1991. *Annals of Applied Biology* 118: 671–688.

Hebblethwaite, P. D. 1982. In: *Faba Bean Improvement*, Ch. 16 (eds. G. Hawtin and C. Webb). The Hague, The Netherlands: Martinus Nijhoff.

Hedley, C. L. and Ambrose, M. J. 1981. *Advances in Agronomy* 34: 225–277.

Hernandez, L. G. 1986. *Study of the agronomy of chickpea (Cicer arietinum) in Canterbury*. Ph.D. thesis, Lincoln College, University of Canterbury, New Zealand.

Husain, M. M. 1984. *The response of field bean (Vicia faba L.) to irrigation and sowing date*. Ph.D. thesis, Lincoln College, University of Canterbury, New Zealand.

Jamieson, P. D., Wilson, D. R. and Hanson, R. 1984. *Proceedings of the Agronomy Society of New Zealand* 14: 75–82.

Jones, C. 1989. *Agronomic and evapotranspiration studies in peas of differing leaf morphology*, 102 pp. Ph.D. thesis. University of Nottingham.

McEwan, J. 1978. In: *Rothamsted Experimental Station Report for 1977, Part I*, p. 127. Harpenden, Herts., England.

McKenzie, B. A. and Hill, G. D. 1990. *Journal of Agricultural Science, Cambridge* 114: 309–320.

McKenzie, B. A. and Hill, G. D. 1991. *Journal of Agricultural Science, Cambridge* 117: 339–346.

McKenzie, B. A., Hill, G. D., White, J. G. H., Meijer, G., Sikken, G., Nieuwenhuyse, A. and Kausar, A. G. 1986. *Proceedings of the Agronomy Society of New Zealand* 16: 29–34.

McKenzie, B. A., Sherrel, C., Gallagher, J. N. and Hill, G. D. 1985. *Proceedings of the Agronomy Society of New Zealand* 15: 47–50.

Monteith, J. L. and Elston, J. 1983. In: *The Growth and Functioning of Leaves*, pp. 499–518 (eds. J. E. Dale and F. L. Milthorpe). London: Cambridge University Press.

Muehlbauer, F. J., Cubero, J. I. and Summerfield, R. J. 1985. In: *Grain Legume Crops*, Ch. 7 (eds. R. J. Summerfield and E. H. Roberts). London: Collins.

Pilbeam, C. J., Akatse, J. K., Hebblethwaite, P. D. and Wright, S. D. 1992. *Field Crops Research* 29: 273–287.

Pilbeam, C. J., Duc, G. and Hebblethwaite, P. D. 1990a. *Journal of Agricultural Science, Cambridge* 114: 19–33.

Pilbeam, C. J., Hebblethwaite, P. D. and Clark, A. S. 1989a. *Field Crops Research* 21: 203–214.

Pilbeam, C. J., Hebblethwaite, P. D., Nyongesa, T. E. and Ricketts, H. E. 1991b. *Journal of Agricultural Science, Cambridge* 116: 385–393.

Pilbeam, C. J., Hebblethwaite, P. D. and Ricketts, H. E. 1989b. *Annals of Applied Biology* 114: 377–390.

Pilbeam, C. J., Hebblethwaite, P. D., Ricketts, H. E. and Hassan, O. A. 1990b. *Journal of Agricultural Science, Cambridge* 114: 339–352.

Pilbeam, C. J., Hebblethwaite, P. D., Ricketts, H. E. and Nyongesa, T. E. 1991a. *Journal of Agricultural Science, Cambridge* 116: 375–383.

Pilbeam, C. J., Hebblethwaite, P. D. and Yusuf, A. A. 1990c. *Journal of Science and Food Agriculture* 53: 443–454.

Plancquaert, P. 1991. *Aspects of Applied Biology 27, Production and Protection of Legumes*, pp. 189–198 (eds. R.J. Froud-Williams et al.) The Association of Applied Biologists, Horticulture Research International, Wellesbourne, Warwick, U.K.

Poulain, D., Boulal, H. and Hateau, B. 1992. In: *2nd International Food Legume Research Conference Program and Abstracts*, p. 45. 12–16 April 1992, Cairo, Egypt.

Pyke, K. A. and Hedley, C. L. 1985. In: *The Pea Crop – a Basis for Improvement*, Ch. 26 (eds. P. D. Hebblethwaite, M. C. Heath and T. C. K. Dawkins). London: Butterworths.

Ruckenbauer, P. and Mollenkopf, K. W. 1986. *Plant Breeding* 97: 264–267.

Salter, P. J. 1962. *Journal of Horticultural Science* 37: 141–149.

Salter, P. J. 1963. *Journal of Horticultural Science* 38: 321–334.

Savage, G. P. 1991. *Outlook on Agriculture* 20: 109–112.

Saxena, M. C., Silim, S. N. and Singh, K. B. 1990. *Journal of Agricultural Science, Cambridge* 114: 285–293.

Silim, S. N., Hebblethwaite, P. D. and Heath, M. C. 1985. *Journal of Agricultural Science, Cambridge* 104: 35–46.

Sjödin, J. 1971. *Hereditas* 67: 155–180.

Smithson, J. B., Thompson, J. A. and Summerfield, R. J. 1985. In: *Grain Legume Crops*, Ch. 8 (eds. R. J. Summerfield and E. H. Roberts). London: Collins.

Snoad, B. 1980. *DAS Quarterly Review* 37: 69–86.

Stützel, H. and Aufhammer, W. 1991. *Journal of Agricultural Science, Cambridge* 116: 395–407.

Taylor, B. R., Richards, M. C., MacKay, J. M. and Cooper, J. 1991. *Aspects of Applied Biology 27, Production and Protection of Legumes*, 309–312 (eds. R.J. Froud-Williams et al.) The Association of Applied Biologists, Horticulture Research International, Wellesbourne, Warwick, U.K.

Wilson, D. R., Hanson, R. and Jermyn, W. A. 1981. *Proceedings of the Agronomy Society of New Zealand* 11: 35–39.

Zain, Z. M., Gallagher, J. N., White, J. G. H. and Reid, J. B. 1983. *Proceedings of the Agronomy Society of New Zealand* 13: 95–102.

Reproductive physiology as a constraint to seed production in cool season food legumes

G. DUC[1], P. GATES[2], B. NEY[3], G.G. ROWLAND[4] and A. TELAYE[5]

[1] INRA, Station de Génétique et d'Amélioration des Plantes, BV 1540, 21034 Dijon Cédex, France;
[2] University of Durham, Department of Biological Sciences, South Road, Durham, DH1 3LE, UK;
[3] INRA, Station d'Agronomie, BV 1540, 21034 Dijon Cédex, France;
[4] University of Saskatchewan, Crop Development Centre, Saskatoon, Saskatchewan, S7N 0W0, Canada, and
[5] Institute of Agricultural Research, P. O. Box 2003, Addis Ababa, Ethiopia

Abstract

Seed legumes are famous for yield instability and related high flower and pod shedding rates. The reproductive physiology post-pollination is considered in this paper.

Factors determining pollen germination, stigmate receptivity, and ovule fertilization are reviewed. After fertilization, the stages of reproductive organ development are described, and their progression on the plant analyzed and modeled.

Abortion events either early or late have been explained either by hormonal or by source-sink competition phenomena. Modified chronologies or morphologies in reproductive organ development, by genetic means or by use of growth regulators, help in the understanding and in the modification of abortion frequencies.

Introduction

Whatever the reproductive regime of food legumes is, autogamous for pea (*Pisum sativum* L.), chickpea (*Cicer arietinum* L.), and lentil (*Lens culinaris* Medik.) or partially allogamous for faba bean (*Vicia faba* L.) and lupin (*Lupinus* sp.), general features of high flower and pod shedding rates are always reported in the literature. Considering this important feature, the reproductive physiology after the pollination step is considered in this paper.

Pollen and Stigmate Aspect

The partially allogamous character of *Vicia faba*, and the problems resulting from this biology, have provided literature with the larger part of illustrations on pollen-stigmate relationship.

F.J. Muehlbauer and W.J. Kaiser (eds.), Expanding the Production and Use of Cool Season Food Legumes, 791–808.
© 1994 *Kluwer Academic Publishers.*

Genetic Incompatibility

Lord and Heslop-Harrison (1984) studied several genotypes of *V. faba* and did not find any evidence suggesting the presence of an effective physiological self incompatibility system. Unilateral incompatibility was not demonstrated in *V. faba*, even though it was observed by Le Guen (1983) that different subspecies or lines did not have the same general ability to pollinate or to be fertilized.

Pollen Fertility

Telaye (1990) described a gradient of pollen fertility, as assayed by the fluorochromatic reaction, in *V. faba* plants. He found that pollen fertility declined progressively from basal to apical racemes. In a wide range of genotypes it was found that the lower nodes bore the most fertile pollen, a factor which may be related to the typical basipetal pattern of pod set in this crop.

Fluctuating temperature is known to adversely affect fruit set in a wide range of crops, while stigma secretion, pollen germination and pollen tube growth are drastically reduced at 35°C in *Acacia* (Marginson *et al.*, 1985). Telaye (1990) showed that interactions between temperature stress and high humidity can greatly reduce pollen germination in *V. faba* and that there was considerable variation for this response between inbred lines drawn from a wide range of geographical origins. The implication was that pollen fertility may be drastically reduced in the field during periods of high temperature and humidity in some genotypes.

Pollen Competition

The partially allogamous breeding system in *V. faba*, whereby the crop is composed of a mixture of autofertile (self pollinating) and autosterile (cross pollinated) plants, in variable proportions, poses many difficulties for breeders. Paul *et al.* (1978) demonstrated that autofertile plants produce a stigmatic exudate which allows self-pollen to germinate and fertilize ovules before flowers are cross pollinated; these plants are functionally protogynous. Conversely, in autosteriles the stigmatic exudate is produced late in development, so cross- and self-pollen are likely to be present simultaneously on the stigma when it is secreted; such plants are functionally protandrous.

This situation raises questions as to whether there are differences between the vigor of cross- and self-pollen tube growth, which might affect the outcome of competition between different pollen genotypes in the style. There are many examples in the literature of pollen competition, leading to so-called gametophytic selection (see Snow and Spira, 1991).

Telaye (1990) investigated competition between male gametophytes in *V. faba* by making mixed pollinations with pollen from a wide range of inbred lines. When self- and cross-pollen were mixed equally in pairwise combinations, he found wide genotype-specific variations in vigor of pollen tube growth, such that some pollen genotypes were more competitive male parents. He also detected a maternal effect, whereby some pollen parents performed better on the stigmas and styles of certain female lines. This pollen/stigma interaction raises important questions with respect to the choice of pollen parents in the production of synthetic cultivars and even F_1 hybrids, since it may be possible to increase the heterozygosity in a population by selecting male parents with rapid rates of pollen tube growth, so that self-pollen is outcompeted in the style of female parents. Conversely, it may be possible to increase autofertility in populations by selecting lines where self-pollen tube growth is exceptionally vigorous, outcompeting foreign pollen introduced by bees.

Duration of Stigmate Receptivity

In *V. faba*, failure of fertilization could occur in any one of the steps leading from pollination to fertilization. Sensitive stages include pollen germination, penetration of the stigmatic surface by pollen tubes, traversing of the stylar canals and cavities and entry to the ovary/micropyle. In this connection, it has been asserted that from anthesis pollen grains can remain viable for 3 to 4 days while the stigmatic surface of the pistle would be receptive for 6 days in *V. faba* (Stoddard, 1986c).

Pollen Storage

Longevity of pollen is of interest to plant breeders for several reasons. The period of viability of pollen under natural conditions may have a significant effect on fertilization of flowers, while storage of pollen might allow breeders to carry pollen over for breeding programs from year to year without growing all lines in a breeding program every season (Bajaj, 1987). *Vicia faba* pollen was shown by Gates (1978) to retain viability for up to 30 months when dried over $CaCl_2$ and stored at $-20°C$ and successful seed set was routinely achieved with pollen that had been stored under these conditions for 12 months. Gupta and Murty (1985) also demonstrated that *V. faba* pollen remained viable for 4 to 5 months when stored at 4°C.

Recently, detailed studies by Telaye *et al.* (1990) showed that temperature during the desiccation process is important in determining pollen longevity and that genotypic-specific variation exists between lines for ability to survive desiccation under unfavorable temperature regimes. Telaye (1990) has described methods for freeze drying and storing *V. faba* pollen in liquid nitrogen so that it will retain greater than 80% viability for at least 9 months.

Ovule Fertilization

Faba Bean

Rowlands (1960) found that only 24% of *V. faba* ovules develop into mature seeds. Prior to 1980, there was no conclusive evidence to point at flower abortion being a pre- or post-fertilization event.

A series of experiments conducted during the 1980s showed that ovule fertilization frequencies can vary tremendously among environments (Table 1). These fertilization levels generally exceeded the number of mature seeds produced in a crop (Rowland and Bond, 1983; Stoddard, 1986a,b) which demonstrated that fertilization of ovules does not directly limit seed yield.

Table 1. Ovule fertilization (%) in faba bean by ovule position in various experiments

Authors	Country	Type	Ovule position			
			1	2	3	4
Rowland et al. (1983)	England	Spring	47.4	39.9	32.4	26.2
Stoddard (1986a)	England	Winter	98.8	95.7	91.0	82.9
Stoddard (1986b)	England	Spring	99.0	96.8	93.4	90.0
Rowland et al. (1986)	France	Spring	94.0	86.0	85.0	74.0
	Canada	Spring	58.0	62.0	56.0	48.0
Graf (1988)	Canada	Spring	79.3	77.1	72.0	66.9
Marcellos and Perryman (1988)	Australia	Winter	91.6	84.4	72.8	56.7

When the data in Table 1 is transformed so that all fertilization is a percentage of ovule 1, the decline in fertilization is fairly consistent among studies from ovule 1 through ovule 4 (Table 2). The variation among studies is greatest at the third and fourth ovule positions. However, as pointed out by Marcellos and Perryman (1990), the failure of fertilization of distal ovules is not likely due to the lack of opportunity as usually more pollen tubes than ovules are observed in the ovary.

Rowland *et al.* (1983) had noted that flower position within a raceme had an influence on ovule fertilization frequency but subsequent studies by Stoddard (1986c) and Marcellos and Perryman (1988) have indicated that this is not likely the case. It was also proposed by Rowland *et al.* (1983) that the apex might have an effect on ovule fertilization. However, Rowland *et al.* (1984, 1986) showed that neither apex excision (topping) nor flower removal influenced ovule fertilization. Sink size and distribution would, therefore, not appear to be a factor in ovule fertilization.

Rowland *et al.* (1986) showed that while ovule fertilization frequencies could differ greatly between environments, this difference did not necessarily lead to

Table 2. Relative frequencies of ovule fertilization as a percentage of those of ovule 1 for the studies in Table 1*

Authors	Ovule position			
	1	2	3	4
Rowland *et al.* (1983)	100	84	68	55
Stoddard (1986a)	100	97	92	84
Stoddard (1986b)	100	98	94	91
Rowland *et al.* (1986)	100	91	90	79
	100	107	97	83
Graf (1988)	100	97	91	84
Marcellos and Perryman (1988)	100	92	80	62

*Adapted from Marcellos and Perryman (1988)

a discrepancy in seed yield. In an experiment conducted with male sterile plants at Dijon, France and Saskatoon, Canada, the Dijon plants had an average ovule fertilization frequency of 83%, while at Saskatoon it was 50%. However, pod retention per flower was only 25% at Dijon but 45% at Saskatoon. This suggests that flower abortions is a post-fertilization event.

Heterosis appears to be a factor involved in the fertilization of ovules of plants within a population. Graf (1988) showed that fertilization of ovules of flowers from hybrid plants occurred more quickly and to a greater degree than flowers from inbred plants (Table 3). Graf (1988) found that inbred pollen was just as likely to fertilize an ovule as hybrid pollen. Therefore, growth rates of inbred and hybrid pollen must be equal but a hybrid pistil allows more rapid ovule fertilization than an inbred pistil.

Table 3. The effect in faba bean of flower hybridity on ovule fertilization frequency (%) determined after 12, 24 and 36 hours of incubation

Flower type	Time interval (h)		
	12	24	36
Inbred	6.5	16.1	26.0
Hybrid	16.7	27.5	41.8

Temperature can have a great influence on fertilization of ovules. Graf (1988) showed in faba bean "Outlook" that the fertilization of ovules was very slow at 5°C (Table 4). Fertilization of ovules was at its highest in the range of 15° to 25°C. This may partly explain Stoddard's (1986d) observation of ovule fertilization being greater in spring than in winter faba bean crops. Winter crops flower earlier than spring crops and thus generally encounter cooler spring temperatures. The observation of Graf (1988) that no ovule fertilization occurred at 35°C, despite pollen germination and growth at this temperature, identifies another factor responsible for *V. faba*'s intolerance of high temperatures.

Table 4. The ovule fertilization frequency (%) of faba bean cv Outlook flowers incubated at 5, 15, 25, and 35°C for 12, 24, 36, and 48 hours

Temperature (°C)	Time interval (h)			
	12	24	36	48
5	0.0	0.0	6.0	7.6
15	9.4	29.9	44.4	55.4
25	13.2	38.5	24.2	48.4
35	0.0	0.0	0.0	0.0

Pea

Link (1961) has shown that most pea ovules are generally fertilized, suggesting flower abortion is a post-fertilization event in pea as in faba bean.

Development of the Legume Seed

The seed number is the most variable component of yield of annual food legume species. Thus, the determination of the period of seed set is of major importance.

Pigeaire *et al.* (1986) and Egli *et al.* (1987) showed that seeds of various legumes do not abort when they have reached a certain dry weight. Kato and Sakaguchi (1954), and more recently Duthion and Pigeaire (1991), concluded that abortions occurred during cotyledon elongation. The latter authors suggested that the length of the seed could be used as a convenient criterion to determine the "final stage in seed abortion". Using nondestructive photographic methods, they showed that seed lengths, which correspond to the critical stage, were 12 mm for two cultivars of soybean [*Glycine max* (L.) Merr.], 8.5 mm for two cultivars of pea, and 13 mm for two cultivars of lupin.

Subsequently, other experiments were carried out to assess if the final stage in seed abortion corresponded to changes in characteristics of the seed. Some parameters of development and growth of the seed were followed in relation to the length of the seed (Duthion and Ney, 1990b). It appeared that the critical length corresponded in pea (Figure 1):

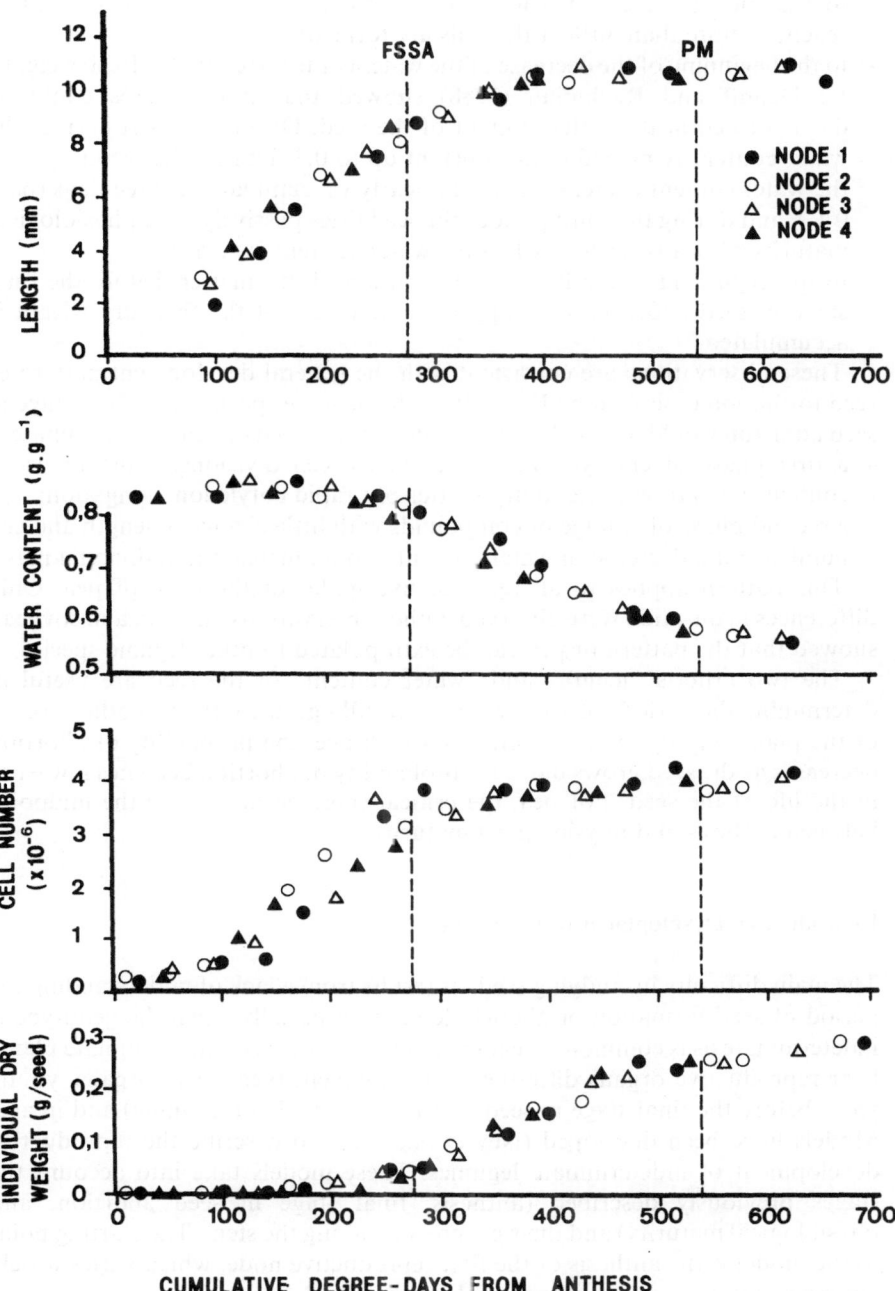

Figure 1. Seed characteristics versus cumulative degree days from flowering for the first four nodes of *Pisum sativum* cv. Solara (FSSA = final stage of seed abortion, PM = physiological maturity).

– to the end of the cell divisions of the embryo when the critical length is reached, more than 80% of the cells are formed;
– to the beginning of the decrease of the water content (relative to fresh weight), Le Deunff and Rachidian (1988) showed that three phases could be distinguished in the water content of the seed. During the first period, the water content remained quite constant up to 0.8. During the second phase, the water content decreases almost linearly vs. cumulated degree-days (base 0°C), and during the third period, the seed dries passively. The physiological maturity of pea is reached when the water content is 0.55;
– to the beginning of the linear accumulation of dry matter, before the final stage in seed abortion, only approximately 15% of the final dry weight is accumulated.

These observations are consistent with the general development pattern of seed formation described by Dure (1975) for legume species. The final stage in seed abortion would divide the development of the seed of pea into two phases:
– a first phase of embryogenesis with active cell divisions, constant water content, low growth rate in dry matter but rapid cotyledon elongation;
– a second phase of storage of compounds with little change in length and cell number but a decrease in water content and an increase in individual mass.

This pattern applies to all reproductive nodes of the stem of pea. Only differences in duration were observed. Other observations on lupin and soybean showed that the pattern of pea can be extrapolated to other legume species.

The two criteria "length" and "water content" of the seed are useful in determining the periods of seed set and seed filling. In contrast to other organs of the plant whose senescence increases with age, the probability of abortion decreases as the seed grows old. This probability of abortion becomes low early in the life of the seed. For pea, the critical stage occurs before the midpoint between anthesis and physiological maturity.

Reproductive Development of the Plant

The main difficulty in studying seed set results from a lack of understanding the period of seed formation on the whole plant, especially when the genotype is indeterminate as is common in legume crops. The stems of indeterminate plants bear reproductive organs differing in development (vegetative organs, young seeds before the final stage in seed abortion, seeds that are filling) and in age. Models have been developed (Ney *et al.*, 1992) to describe the reproductive development of indeterminate legumes. These models take into account the stages previously described (anthesis, final stage in seed abortion, and physiological maturity) and their progression along the stem. The starting point of the model is the anthesis of the first reproductive node, which varies widely among genotypes and environmental conditions.

Flowering

Beginning of Flowering

Flowering depends on temperature and photoperiod. Some models have been developed to assess the influence of these factors on time of flowering.

Roberts *et al.* (1985) for chickpea and Summerfield and Roberts (1988) for pea, lentil, chickpea, and faba bean established the relation between the inverse of the number of days from sowing to first flower (f) and the mean temperature (t) and photoperiod (p): $1/f = a + bt + cp$. The coefficients a, b, and c depend on the genotypes but also on the vernalization of the seed.

Another model allowing a quantification of the effects of vernalization on time of flowering of pea was proposed by Truong and Duthion (1992) based on field trials. This model analyzes the duration of the emergence-flowering period D as a function of two variables R, the leaf appearance rate, and N1F, number of first flower: $D = (N1F - 2) \times (1/R)$. R varies linearly with temperature (base 2°C). Mean photoperiod and temperature prior to floral initiation act additively on N1F within the range of environmental conditions encountered by a genotype.

The flowering time of *Lupinus albus* L. was analyzed in the same way (Duthion and Ney, 1990a; Huyghe, 1991).

These models allow for a more or less precise estimation of the beginning of the flowering time for various genotypes and environmental conditions.

Progression of the Flowering Along the Stem

The progression of flowering time along the stem can be described by linear models based on cumulative degree-days since flowering time of the first node of pea (Ney *et al.*, 1992). Cumulative degree-days (base 0°C) were accurate to describe the flowering progression of pea. When the environmental conditions were favorable (in the glasshouse and in the field at low plant density), the progression rate was almost constant for a given genotype (Figure 2).

When environmental conditions are unfavorable or competition among the plants increases, the growth rate of the stem decreases and flowering progresses more slowly along the stem. The relationship between individual growth rate (IGR) and progression rate of flowering (PRF) can be modelled by the exponential equation: $PRF = a \times (1 - e^{-b \times IGR})$; where "a" is the asymptotic value corresponding to the potential progression rate of flowering of the genotype (Ney and Duthion, 1992).

End of Flowering

The number of flowering nodes varies widely in pea as shown in Figure 2. The factors responsible for the number of flowered nodes are not well understood. Deflowering or depodding treatments cause prolongation of vegetative growth

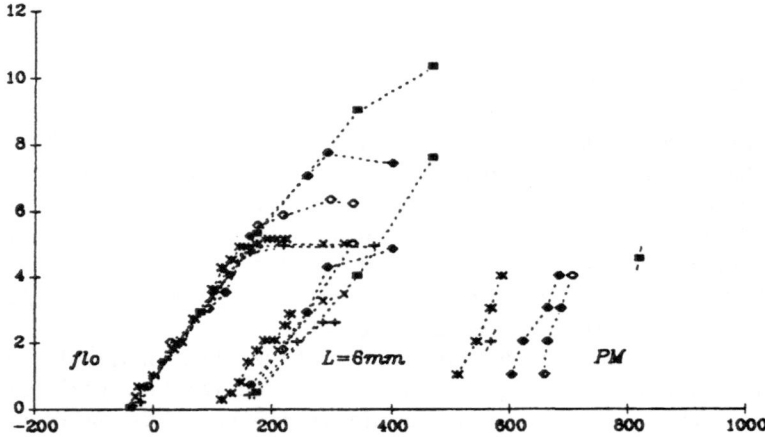

Figure 2. Progression of the stages of development along the stem of pea, starting with the beginning of flowering in various experiments. Flo: flowering; L = 6 mm: final stage in seed abortion (corresponding to a length of 6 mm of the biggest seed of the node); PM: physiological maturity, field 1988 (°), 1989(●), 1990 (■), glasshouse 1988 (×), 1990 (+), 1991 (∗)

and lead to a higher number of reproductive nodes. Many experiments have been conducted to determine whether the termination of flowering was the result of the routing of assimilates to developing pods or to a senescence signal sent by the pods (Hardwick, 1985). Not enough information is available to support one or both of the hypotheses. In all cases, when the amount of assimilate available for the plant increases, the cessation of the flowering occurs later and the number of reproductive nodes is higher. The number of nodes which flower is higher at low than at high plant densities, likely due to the higher amount of assimilates available for each plant.

A water deficit stops the flowering progression along a stem when it occurs before or after the start of anthesis (Turc *et al.*, 1990). The mechanism, lack of assimilates or particular influence of the water deficit on the apex, is not well known. Other research showed that the nitrogen nutrition might be involved in the cessation of the vegetative growth. Brevedan *et al.* (1978) showed that flowering time of soybean grown in the glasshouse was prolonged when fertilized with nitrogen. More recently, Jeuffroy (1992) showed with pea that the number of flowering nodes was related to the amount of nitrogen accumulated in the stem at the beginning of anthesis.

Post Flowering

Final Stage of Seed Abortion

The progression of the final stage of seed abortion was recorded in pea under various conditions (field trials with a wide range of plant densities, glasshouse experiments) (Figure 2). The progression was calculated as the mean of reproductive nodes whose largest seed length exceeded 6 mm, as recommended by Duthion and Pigeaire (1991).

It appears that:
- the progression rates were almost constant among years and type of experiment;
- the progression of the final stage in seed abortion was almost parallel to the progression of flowering;
- the time required to reach the final stage in seed abortion from anthesis was quite constant (about 200 degree-days for pea).

Physiological Maturity

Physiological maturity progresses linearly versus time expressed in temperature sum. The slope of the relationship was greater than the slope of the flowering progression. As reported by Spaeth and Sinclair (1984) for soybean, the time required to reach physiological maturity from flowering time decreased for upper nodes.

The duration of seed growth, from the final stage in seed abortion to physiological maturity, varied widely, unlike the duration from flowering to the final stage in seed abortion. One of the factors involved in the variations in seed filling duration of soybean is the self-destruction of the plant (Sinclair and De Wit, 1976). Nitrogen supplied to seed comes not only from the soil but also from internal reserves in vegetative parts and pod walls. Physiological maturity of the seeds probably occurs when the nitrogen of the vegetative parts and of pod walls is totally removed (Egli *et al.*, 1978). But the loss of nitrogen from the vegetative parts is not the only factor which influences the senescence of the plant. The variability in N content of the vegetative parts at harvest for pea shows that, in some cases, N is not the limiting factor. The individual seed weight is probably limited by the cell number of the cotyledons. The duration of seed filling is probably influenced by both factors: first, the balance between the demand from the seeds and the amount of nitrogen available in the soil and in the vegetative parts, and second, the maximum seed weight of the genotype.

Proposition of a Development Model

As shown in Figure 3, the development of a pea plant can be described by a general model which takes into account the time when the plant enters a new stage (anthesis, final stage in seed abortion, and physiological maturity of the first reproductive node), the rate of progression of the stages along the stem, and the end of these stages.

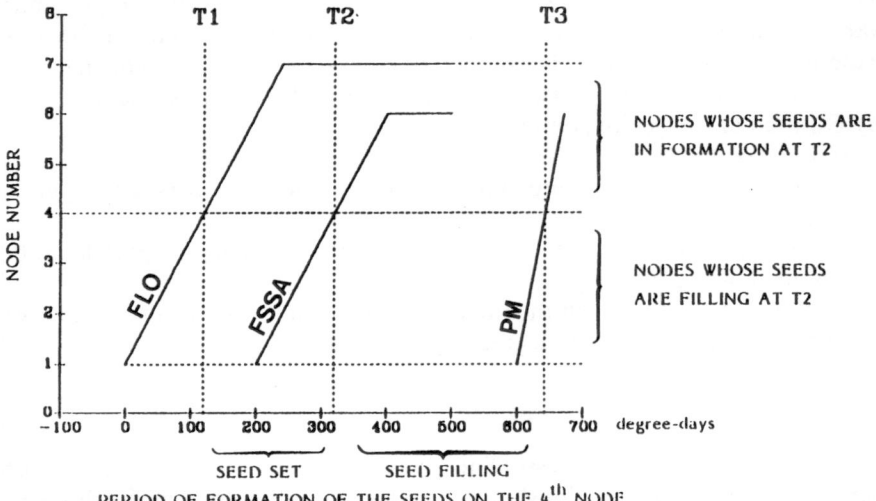

Figure 3. Diagrammatic representation of the development of pea.

This diagrammatic pattern presents some advantages:
− the age and nature of organs present on the stem are known at a given time, and the periods of seed set and seed filling are determined for a given node;
− The main stages, characterized by the presence or absence of competing sinks, are more precisely defined. The first period between emergence and start of flowering corresponds to strictly vegetative growth. Reproductive nodes and pods are formed during the second phase, from start of anthesis to start of seed filling (or final stage in seed abortion of the first reproductive node). During the third period, characteristic of indeterminate species, seeds that are filling at the base of the plant are competing for assimilates with the formation of new nodes and with the growth of the pods in the upper parts of the plant. This period ends when the final stage of seed abortion is reached by the youngest pod on the plant. The last phase is characterized by seed filling and ends with physiological maturity.

This diagrammatic representation for pea can be applied to other legume species with a structure similar to soybean. It has been extrapolated for lupin (Duthion and Ney, 1990).

Plant Factors which Influence Seed Setting

Many agronomic plants produce more reproductive organs, many of which are subsequently aborted. These reproductive losses are very important in seed legumes and consequently were noted by breeders. However, this abortion is not very different in intensity from that of wheat which is less spectacular as wheat flowers are smaller and abort without falling.

As discussed in part 2 of this paper, most studies have reached the conclusion that abortion in seed legumes occurs as a post fertilization event. As a consequence, only the plant factors which influence seed set after fertilization will now be discussed.

Intensity of Reproductive Losses

Among seed legumes, pea is often reported to have the lowest values for flower abortion with a pod per flower ratio close to 0.5, as compared to values of 0.35 to 0.25 frequently obtained for faba bean, lupin, and soybean (Van Stevennink, 1958; Puech *et al.*, 1977; Duc and Picard, 1981).

At the intraspecific level, the literature provides many illustrations of genetic and phenotypic diversity. Table 5 shows for faba bean, different genotype strategies and morphologies with a pod/flower ratio ranging from 0.16 to 0.32 apparently independent of the yield potential.

Stages of Reproductive Abortion

Generally three stages of abortion are observed: a rapid flower fall within 5 to 7 days following anthesis; then an early stage of pod development (pod no longer than 1 to 2 cm) when pod growth can definitely be stopped; finally the pod can stop growing after this young pod stage under the influence of late stresses. By far, flower and young pod abscission are more frequent.

In faba bean (Table 5), abscission rates close to 50% were observed for each of these organs. In this species, surgical topping (Duc and Picard, 1981) and genetic determinant of stem growth (Gates *et al.*, 1983) can stimulate the early stage of pod development reducing flower fall, but are later compensated for by an increased abortion of young pods.

Structuration of Abortion Inside the Plant

As discussed by Hardwick (1985) in pea and by Gates *et al.* (1983) in faba bean, yield appears not as an amorphous "sink" made out of a population of ovules or embryo cells but rather as a structured population. Seed set is frequently higher in bottom podding nodes of a stem and in proximal pods of a raceme.

Table 5. Fertility and yield components measured on six genotypes of faba bean in a field experiment (Dijon, 1981)

Genotypes	Yield	Flowers per plant	Young pods* per flowers	Pods per young pods*	Pods per flowers
	t ha⁻¹		--%--	--%--	--%--
line 319	4.80	55.4	48	67	32
line 370	5.20	39.6	52	62	32
line 247	4.72	47.0	52	46	23
hybrid AxC	6.12	73.7	44	40	18
cv. Ascott	5.41	60.7	42	59	25
cv. Blaze	4.90	80.9	39	41	16
LSD 5%	0.3	7.5	3.3	4.1	4.0

Genotypes	Flowers per flowering node	Podded nodes per plant	Seeds per pod	TGW	HI
				-g-	--%--
line 319	2.4	14.9	1.96	520	44.6
line 370	2.4	8.4	2.54	800	48.5
line 247	2.7	9.1	2.39	650	45.7
hybrid AxC	3.3	9.7	2.65	550	37.8
cv. Ascott	4.3	9.1	3.09	390	44.3
cv. Blaze	4.9	9.3	3.13	410	40.9
LSD 5%	0.2	1.3	0.22	25	3.2

*: young pods are 2×10^{-2} m long.
TGW: thousand grain weight.
HI: harvest index.

Within the pod, abortion is reported to be more frequent at either end of the pod in pea and at the proximal position of the peduncle in *V. faba*. This topographic organization also corresponds to a chronological distribution which generally favors oldest reproductive organs within each level of organization.

Embryo Determination of Abortion

Scant physiological knowledge of seed development in legumes does not help to clearly distinguish what processes are under plant or embryo control. However, some experiments have provided information on an embryo effect. Garcia-Martinez *et al.* (1991) showed in pea that compounds which induce fruit-set (the pod shell) are probably synthesized in the ovules following fertilization. Duc and Rowland (1990) propose in *V. faba* that under the effect of intra- or extra-plant stress, developing ovules containing hybrid embryos may be selectively retained when compared to inbred embryos.

Mechanisms that Contribute to Successful Seed Set

Seed development is dependent upon total sink volume. In soybean, this volume is expressed by an adjusted number of developing ovules, and also by the number of cells in the cotyledons (Egli *et al.*, 1989). Beside this source-sink relationship, the partitioning of resources as previously described must be determined by plant and embryo messages, probably of hormonal nature. These processes must regulate the final stage of seed abortion which was reported in part 2, and also must prevent the development of numerous inviable tiny seeds.

Parts of the regulation processes have been characterized. A comprehensive review by Wang and Sponsel (1985) showed the diversity, complexity, and precise timing of hormones that are contained in the pea fruit and pea seeds or that influence their development. There are auxins, cytokinins, GAs, and abscissic acid but their role and causal relationships are far from being elucidated. Adding to this complexity was the observation in lupin by Van Steveninck (1958) that applied growth substances were not readily transported from the laterals to the main flower stalk unless the carbohydrate stream was induced to run in that direction by defoliating the main stem. This suggested a dependency of hormonal transport on photosynthetic production and transport.

Plant structures which are active in the transportation of resources are also under growth hormone control as shown in soybean by Kuang *et al.* (1991). The benzylaminopurine, which increased pod set, simultaneously caused rachis swelling in positions where pods were present and may have delayed pedical abscission either in the presence or absence of pods.

In vitro pod culture experiments can illustrate the competitive sink regulations. For example, starting with 4-day-old pea pods, Barrett (1986) obtained viable seeds. Basal medium, free of growth hormones, permitted the best seed development and allowed little pod development. Adding GAs switched off this role of major sink in the seed and allowed pod development.

Effect of Modifying Plant Morphology

Determinate Growth

Besides its contribution to increasing lodging risk, excessive stem growth in grain legumes is often considered an additionnal sink, competing with reproductive development. This hypothesis is supported by means of radiotracers, showing the attraction of the apex for assimilates (Jacquiery and Keller, 1978).

Numerous experiments on the application of growth regulators to faba bean have been reported. Chapman and Sadjadi (1981) induced pod set and reduced vegetative growth by treating plants with the antiauxin TIBA. However, to date, no growth regulator has produced a consistent reliable yield benefit.

In *V. faba*, two genes which control determinate growth with a terminal inflorescence were obtained by mutagenesis (Sjödin, 1971; Nagl, 1979). The yield capacity of these genotypes is still lower than indeterminate types. They express a regular harvest index and a lower seed development in pods (Duc et al., 1990), suggesting that the source-sink relationship is not really improved in these types. In soybean, Dayde and Ecochard (1984) suggested that the genetic character of determinate growth could help in improving harvest index. However, they also suggested that this character would introduce a limit in yield potential.

Raceme Structure

In pea, the multipod character which increases flower number and pod set per raceme is later compensated for by higher seed abortion in pods resulting in no yield increase (R. Cousin, personal communication).

In faba bean, Gates et al. (1983) suggested that interactions among pods and flowers in a raceme could be reduced by selecting for synchrony of anthesis within racemes and independent vascular traces to each flower. Most faba bean breeders agree that genetic variability offers the possibility of increasing the number of pods per node whatever the physiological processes turn out to be.

Concluding Remarks

Extra production of reproductive organs is necessary to allow plasticity and compensation processes. It is an adaptative mechanism to match the environmental conditions. As we have no indication that legume plants are paying an expensive insurance, we would not recommend any sink reduction.

Large genetic variability for morphological traits is available in most grain legumes and should help the breeders to approach an ideotype adapted to modern agriculture:
- a moderate top growth should maintain harvest index at a good level and also have good standing ability;

- higher synchrony between reproductive organs seems of interest and can be the result of:
  * synchronous branching;
  * a higher number of synchronous pods per raceme and fewer racemes on a stem;
  * a larger number of ovules per ovary, together with a secure fertilization of ovules far from the stigmate.

Progress has been made in the understanding of abortion events that concern reproductive organs in grain legumes. They occur after fertilization. The stages and kinetics of seed development, common to several species of grain legumes, have now been characterized which aid in the analysis of problems in yield instability.

Based on this knowledge, the next step will be a sharper study of plant physiology (regulation of cell divisions, competition between vegetative and reproductive phases, role of growth hormones, etc.).

References

Bajaj, V. P. S. 1987. In: *Pollen: Cytology and Development* pp. 397–420 (eds. K. Giles and J. Prakesh). London: Academic Press.

Barratt, D. H. P. 1986. *Plant Science* 43: 223–228.

Brevedan, R. E., Egli, D. B. and Leggett, J. E. 1978. *Agronomy Journal* 70: 81–84.

Chapman, E. P. and Sadjadi, A. S. 1981. *Zeitschrift für Pflanzenphysiolische Bodenkunde* 104: 265–273.

Dayde, J. and Ecochard, R. 1984. *Agronomie* 5: 127–134.

Duc, G. and Picard, J. 1981. In: *Vicia faba, Physiology and Breeding*, pp. 283–298 (ed. R. Thompon). The Hague, The Netherlands: Martinus Nijhoff.

Duc, G. and Rowland, G. G. 1990. *Canadian Journal of Plant Science* 70: 79–82.

Duc, G., Berthaut, N., Pelletier, R. and Carteron, A. 1990. *Plant Breeding* 105: 126–136.

Dure, L. S. 1975. *Annual Review of Plant Physiology* 26: 259–278.

Duthion, C. and Ney, B. 1990a. *6th International Lupin Conference*, November 1990. Temuco Pucon, Chile.

Duthion, C. and Ney, B. 1990b. *Proceedings of the First Congress of the European Society of Agronomy*, p. 78, 5–7th December 1990. Paris.

Duthion, C. and Pigeaire, A. 1991. *Crop Science* 31: 1579–1583.

Egli, D. B., Leggett, J. E. and Duncan, W. G. 1978. *Agronomy Journal* 70: 43–47.

Egli, D. B., Wiralaga, R. A. and Ramseur, E. L. 1987. *Agronomy Journal* 79: 463–467.

Egli, D. B., Ramsan, E. L., Zhen-wen, Y. and Sullivan, C. H. 1989. *Crop Science* 89: 732–735.

Garcia-Martinez, J. L., Marti, M., Sabater, T., Maldonado, A. and Vercher, Y. 1991. *Physiologia Plantarum* 83: 411–416.

Gates, P. J. 1978. Ph.D. Thesis, University of Durham, North Carolina, USA.

Gates, P. J., Smith, M. L., White, G. and Boulter, D. 1983. In: *The physiology, genetics and nodulation of temperate legumes*, pp. 43–54 (eds. D. R. Davies and D. G. Jones). London: Pitman.

Graf, R. 1988. Ph.D. Thesis, University of Saskatchewan, Saskatoon, Canada.

Gupta, M. and Murty, Y. S. 1985. *Acta Botanica Indica* 13: 292–294.

Hardwick, R. C. 1985. In: *The Pea Crop*, pp. 317–326 (eds. P. D. Hebblethwaite, M. C. Heath and T. C. K. Dawkins). London: Butterworths.

Huyghe, C. 1991. *Annals of Botany* 67: 429–434.

Jacquiery, R. and Keller, E. R. 1978. *Botanik* 52: 261–276.

Jeuffroy, M. H. 1992. *1st European Conference on Grain Legumes*, pp. 229–230. Angers, France.

Kato I. and Sakaguchi, S. 1954. *Tokaikinki National Agricultural Experiment Station Bulletin* 1: 115–132.

Kuang, A., Peterson, C. M. and Dute, R. R. 1991. *Plant Growth Regulation* 10: 291–303.

Le Deunff, Y. and Rachidian, Z. 1988. *Journal of Experimental Botany* 39: 1221–1230.

Le Guen, J. 1983. *Agronomie* 4: 443–449.

Link, A. J. 1961. *Phytomorphology* 11: 79–84.

Lord, E. M. and Heslop-Harrison, Y. 1984. *Annals of Botany* 54: 827–836.

Marcellos, H. and Perryman, T. 1988. *Australian Journal of Agricultural Research* 39: 579–587.

Marcellos, H. and Perryman, T. 1990. *Euphytica* 49: 5–13.

Marginson, R., Sedgely, M. and Knox, R. B. 1985. *Journal of Experimental Botany* 36: 1660–1668.

Nagl, K. 1979. In: *Some Current Research on Vicia Faba in Western Europe*, 6244: 355–364 (ed. D. A. Bond). European Economic Community, Europe.

Ney, B. and Duthion, C. 1992a. *1st European Conference on Grain Legumes*, pp. 253–254. Angers, France.

Ney, B. and Duthion, C. 1992b. *1st European Conference on Grain Legumes*, pp. 305–306. Angers, France.

Ney, B., Turc, O. and Duthion, C. 1992. *Proceedings of the First Congress of the European Society of Agronomy*, p. 18, Paris, France.

Paul, C., Gates, P. J., Harris, N. and Boulter, D. 1978. *Nature* 275: 54–55.

Pigeaire, A., Duthion, C. and Turc, O. 1986. *Agronomie* 6: 371–378.

Puech, J., Bonnell, J. M. and Hernandez, M. 1977. Observations sur l'importance de l'avortement des organes fructifères de soja placé dans différentes conditions écologiques. C.R. Academy of Science Paris, t 284, D: 2343–2346.

Roberts, E. H., Hadley, P. and Summerfield, R. J. 1985. *Annals of Botany* 55: 881–892.

Rowland, G. G. and Bond, D. A. 1983. *Journal of Agricultural Science Cambridge* 100: 35–41.

Rowland, G. G., Bond, D. A. and Parker, M. L. 1983. *Journal of Agricultural Science Cambridge* 100: 25–33.

Rowland, G. G., Duc, G. and Picard, J. 1984. *Canadian Journal of Plant Science* 64: 95–103.

Rowland, G. G., Duc, G. and Picard, J. 1986. *Canadian Journal of Plant Science* 66: 235–239.

Rowlands, D. G. 1960. *Heredity* 15: 161–173.

Sinclair, T. R. and De Wit, C. T. 1976. *Agronomy Journal* 68: 319–324.

Sjödin, J. 1971. *Hereditas* 67: 155–180.

Snow, A. A. and Spira, T. P. 1991. *Nature* 352: 796–797.

Spaeth, S. C. and Sinclair, T. R. 1984. *Agronomy Journal* 76: 123–127.

Stoddard, F. L. 1986a. *Euphytica* 35: 925–934.

Stoddard, F. L. 1986b. *Plant Breeding* 97: 210–221.

Stoddard, F. L. 1986c. *Journal of Plant Physiology* 123: 249–262.

Stoddard, F. L. 1986d. *Journal of Agricultural Science Cambridge* 106: 89–97.

Summerfield, R. J. and Roberts, E. H. 1988. In: *World Crops: Cool Season Food Legmes*, pp. 911–922 (ed. R. J. Summerfield). Dordrecht: Kluwer Academic Publishers.

Telaye, A. 1990. Ph.D. Thesis. University of Durham, North Carolina, USA.

Telaye, A., Beniwal, S. P. S. and Gates, P. 1990. *Fabis* 26: 6–10.

Truong, H. H. and Duthion, C. 1992. *1st European Conference on Grain Legumes*, pp. 303–304. Angers, France.

Turc, O., Wery, J. and Sao Chan Cheong, G. 1990. *Proceedings of the First Congress of the European Society of Agronomy*, p. 13. Paris, France.

Van Steveninck, R. F. M. 1958. *Journal of Experimental Botany* 9:3 72–382.

Wang, T. L. and Sponsel, V. M. 1985. In: *The Pea Crop*, pp. 339–347 (eds. P. D. Hebblethwaite, M. C. Heath and T. C. K. Dawkins). London: Butterworths.

Root form and function in relation to crop productivity in cool season food legumes

P.J. GREGORY[1], N.P. SAXENA[2], J. ARIHARA[3] and O. ITO[4]

[1] CSIRO Dryland Crops and Soils Research Unit, Private Bag, P. O., Wembley, Western Australia, 6014, Australia;
[2] ICRISAT, Patancheru P. O., Andhra Pradesh 502 324, India;
[3] Hokkaido National Agricultural, Experiment Station, 1 Hitsujigaoka, Toyohira-Ku, Sapporo 062, Japan, and
[4] Legumes Program, ICRISAT, Patancheru P. O., Andhra Pradesh 502 324, India

Abstract

Legumes have a tap root with lateral branches, but there is substantial genetic variation in root characters. The proportion of total plant dry weight found in the root systems of most cool season legumes is typically 0.25 at mid grain-filling, and may be 0.5 for much of the crop's life; this is greater than the proportion found in cereal root systems grown under similar conditions.

Cool season food legumes are frequently grown in conditions of cold winters followed by hot summers so that grain growth occurs during periods when evaporative demand is increasing rapidly. In chickpea crops, water is used initially from the surface soil but as the season progresses, it is used at greater rates from deeper in the profile and, at close to maturity, water at depths > 1 m may supply 40% of the total water use. There are significant differences between genotypes in root length and depth that relate to the yields under drought.

Chickpea has a greater capacity than several other crops to acidify the rhizosphere pH and thereby increase the availability of phosphate on alkaline soils. The potential for increasing crop yields of legumes by selection of root characters that increase water use and nutrient uptake is assessed.

Introduction

Legume root systems have been less intensively studied than cereal roots and as Gregory (1988) has pointed out, most of the studies that have been made were principally designed to investigate nodulation and N fixation by *Rhizobium* so that the roots themselves were incidental to the main study. However, many grain legumes are grown in dryland conditions with low inputs of fertilizer, so that an understanding of how legumes secure supplies of water and nutrients might aid improvement of their yields. In particular, the beneficial effects of legumes grown in rotation with cereals, exploited in western agriculture since Roman times (White, 1970), are difficult to explain wholly in terms of the N fixed (Arihara *et al.*, 1991). While part of the explanation undoubtedly lies with

F.J. Muehlbauer and W.J. Kaiser (eds.), Expanding the Production and Use of Cool Season Food Legumes, 809–820.
© 1994 *Kluwer Academic Publishers.*

the break of disease and insect pests allowed by inclusion of legumes, there is increasing evidence that the root systems of legumes may induce changes in soil chemistry (pH and P solubilization) that give beneficial effects for subsequent crops (Marschner and Römheld, 1983; Arihara *et al.*, 1991).

As with other crop plants, there is increasing interest in exploiting the genetic variation in root form and function that exists within species, particularly in drought-prone environments (Gregory, 1989). Ludlow and Muchow (1990) identified rooting depth as a particularly important drought-avoidance trait for crops growing on stored soil moisture because this allows maximum extraction of limited reserves of water. When this trait is combined with other traits, such as growing large vegetative structures before flowering commences and then partitioning most subsequent growth to seeds, high yields can be obtained compared to genotypes that follow other strategies (Williams and Saxena, 1991). Hamblin and Hamblin (1985) demonstrated significant species and cultivar differences in rooting depth and root distribution among a range of legume species (pea, lupins, medics, vetch and subterranean clover) and speculated that such differences will affect water use and long-term productivity in areas subjected to drought.

The purpose of this brief review is first, to outline the form of legume root systems and detail new knowledge gained since the previous IFLRC conference (Gregory, 1988); second, to demonstrate the role played by legume root systems in the exploitation of soil water reserves; and finally, to examine the role of legume roots in obtaining supplies of phosphate and nitrogen. Particular attention will be given to recent work on the root systems of chickpeas.

The Root System of Legumes

The emerging root radicle of legumes forms a tap root which subsequently branches. In contrast to cereals, growth of the root system appears to continue beyond flowering although the growth rate slows as seed filling starts (Sheldrake and Saxena, 1979). For example, Brown *et al.* (1989) grew kabuli-type chickpeas (*Cicer arietinum* L.) in northern Syria under rainfed conditions sown in either winter or spring and found that root growth of both crops continued beyond flowering. The dry weight of roots of the winter-sown crop increased from 20 to 45 g m^{-2} between flowering and maturity, while that of the spring-sown crop increased from about 20 g m^{-2} at flowering to 42 g m^{-2} at mid-pod filling before decreasing again to 35 g m^{-2} at maturity. Similarly, chickpea (cv. K 850) grown on a Vertisol at ICRISAT Center in India also accumulated about 13% of its root mass during grain-filling (Figure 1) before roots started to senesce. Root growth is highly dependent on the supply of carbon from the shoot. Brown *et al.* (1989) found that root weight as a fraction of total plant weight was about 0.25 during late vegetative growth but fell thereafter to 0.10 and 0.15 at maturity for winter- and spring-sown crops, respectively. In *Vicia faba* L., root weights comprised up to 40% of total plant weight during early growth of two cultivars but this decreased rapidly as the pods grew, to about 8–10% at maturity (Stützel and Aufhammer, 1991). These proportions are greatly affected by environmental conditions so that, for example, when soil water availability is reduced, root growth may remain constant (El Nadi *et al.*, 1969) or increase (Geisler, 1983 cited by Stützel and Aufhammer, 1991), but shoot growth is decreased resulting in increased root:total plant ratio.

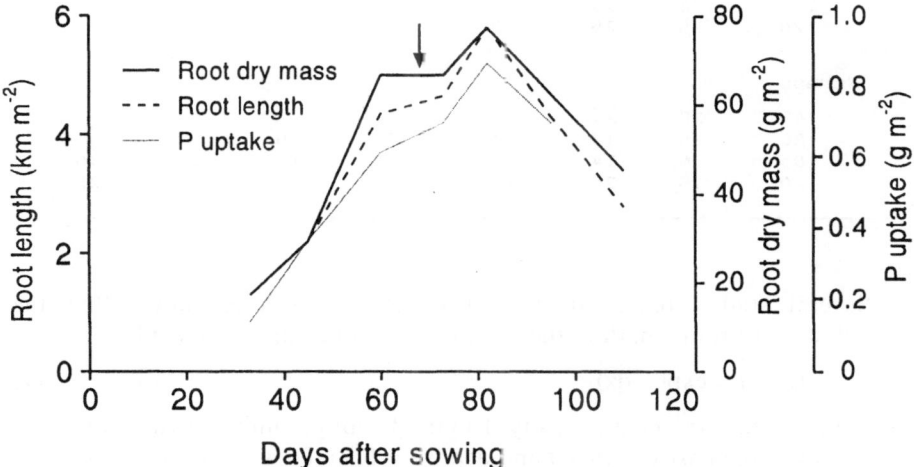

Figure 1. Changes with time in root dry weight (–), root length (--) and phosphorus content of the plant (...) for chickpea (cv. K 850) grown under conditions of receding soil moisture on a Vertisol at ICRISAT, India, 1984/85. The arrow indicates the time of flowering.

Part of the CO_2 fixed in photosynthesis is lost by respiration and in the case of legumes, the respiratory rate is in the order nodules > root > shoot (Hooda *et al.*, 1986 for an example with chickpea). The combined respiratory losses in chickpea total nearly 75% of gross photosynthetic production which is considerably greater than that observed in other legumes; this may result in poor photosynthetic efficiency (Hooda *et al.*, 1986). Studies with $^{14}CO_2$ showed that 24 hours after labelling, about 4–13% of the ^{14}C recovered was in the roots and about twice as much was present in nodules (Table 1). It is noteworthy that the plants grown under conditions of soil water shortage allocated a greater proportion of the ^{14}C to both roots and nodules. In this experiment, the roots respired 80% of the translocated ^{14}C. Although the ratio of maintenance respiration to growth respiration in chickpea (1:21) was similar to that in wheat (1:27), the total root respiration of chickpea was about twice that of wheat (Kallarackal and Milburn, 1985). These results suggest that compared to cereals, chickpeas will obtain less growth for each unit of carbohydrate translocated from the shoot.

Table 1. Changes with time (DAS – days after sowing) in the ^{14}C distribution (%) in various organs of chickpea (cv. H-75–35) 24 hours after $^{14}CO_2$ feeding. The plants were kept at either 60–75% (Control) or 25–35% (Stress) of soil water saturation (from Hooda et al., 1989)

Plant age DAS	Stem	Leaf	Root	Nodule	Bud & Flower	Pod	Seed
Control							
45	21	42	12	25			
85	17	22	13	18	30		
105	8	18	4	10	23	25	11
120	15	25	9	14		15	22
Stress							
45	29	21	14	36			
85	15	18	14	32	20		
105	6	14	5	12	12	15	36
120	15	25	10	6		19	25

Several studies (e.g., Greenwood *et al.*, 1982) have shown that the distribution of root length in the soil profile can be approximated by:

$$L_v = L_{vo} \exp(- qz) \tag{1}$$

where L_v is the root length density (length of root per unit soil volume), L_{vo} is the root length density at time t and depth z = 0, and q is a constant that varies with t. Equation 1 has been applied to chickpea (Brown *et al.*, 1989), pea (*Pisum sativum* L.) and *Vicia faba* root systems (Greenwood *et al.*, 1982) and provides a useful, but site- and season-specific description of root distribution. However, as Figure 2 demonstrates, root length may not always decrease exponentially with depth depending on soil conditions. Siddique and Sedgeley (1987) grew desi-type chickpeas at Merredin in Western Australia and although distributions were approximated by equation 1, q did not decrease uniformly with time (c.f. Brown *et al.*, 1989; Greenwood *et al.*, 1982). Their results demonstrate a substantial turnover of roots particularly in the upper soil layers, a feature that has also been demonstrated in Western Australia with lupin roots (P.J. Gregory, unpublished). Figure 3 shows the daily rate of expansion of the chickpea root system measured by Siddique and Sedgeley (1987); the rates were calculated by fitting equation 1 and then calculating the difference in L_v between time intervals. It is clear that root growth in the upper 20 cm only occurred until flowering (95 days after sowing – DAS) and thereafter there was substantial decay above 20 cm while root proliferation below this continued. Decay of roots occurred throughout the soil profile as maturity was approached (122 DAS).

Chickpea grown in a semi-arid climate has a smaller frequency of branching of both primary and secondary roots compared with other legumes grown under similar conditions (Table 2). Although the specific root length was greatest in chickpea, the reduced branching may place chickpea at a comparative disadvantage in exploiting water and nutrients.

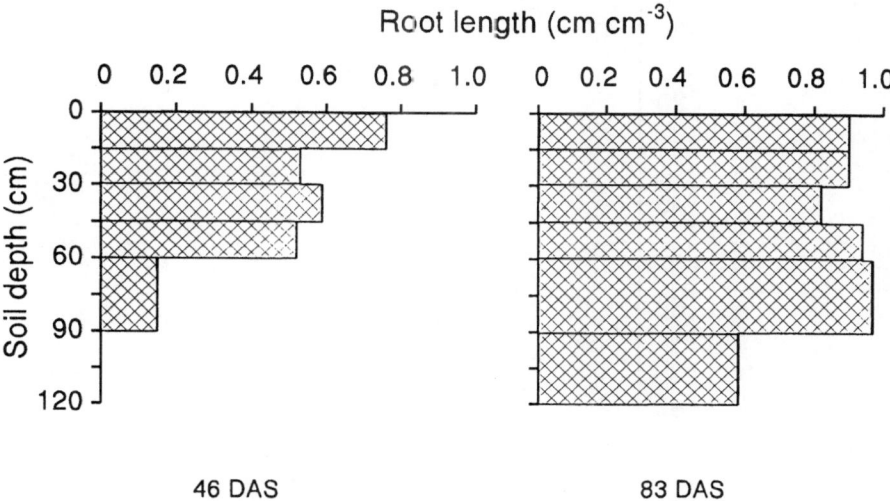

Figure 2. Root distribution of rainfed chickpea (cv. K 850) grown on a Vertisol at ICRISAT, India, 1984/85 (from Arihara *et al.*, 1991).

Table 2. Morphological parameters for the root systems of pigeonpea (ICPL 87), chickpea (K 850) and groundnut (NC 17090) grown in sand culture for a month. Values are the average of five replications with standard errors (O. Ito, unpublished data)

Parameters	Pigeonpea	Chickpea	Groundnut
Branching frequency (cm^{-1})			
Primary root	1.0 ± 0.3	0.6 ± 0.1	1.5 ± 0.3
Secondary root	1.3 ± 0.1	0.7 ± 0.1	2.5 ± 0.1
Specific root length (m g^{-1})	55 ± 10	65 ± 4.7	49 ± 4.1
Water content (%)	78 ± 5.2	91 ± 0.9	92 ± 1.0

Roots and Water Uptake

The importance of root traits in drought-susceptible environments is well recognized but surprisingly little work has been conducted on legumes (for cereals, see Gregory, 1989). Genotypic differences in the length and extent of root systems have been reported in many legume crops, including faba bean

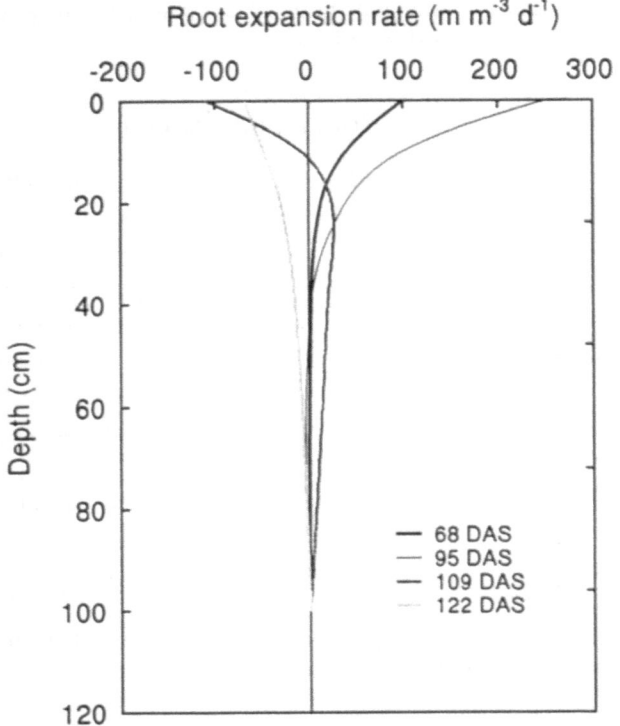

Figure 3. Daily expansion rate of the root system in chickpea (cv. CPI 56288) grown in a xeric alfisol in south-western Australia. The phenology was complete emergence 15 DAS; 50% flowering 92 DAS; start of pod filling 105 DAS; end of flowering 124 DAS; and physiological maturity 131 DAS. (from Siddique and Sedgley, 1987).

(Looker, 1978; ICARDA, 1984), chickpea (Sheldrake and Saxena, 1979; Vincent and Gregory, 1986), and lentil (*Lens culinaris* Medik.) (ICARDA, 1985); a more complete account is provided in the review of O'Toole and Bland (1987). A measure of the likely importance of root traits for the acquisition of water by legumes can be gained from the results of Sponchiado *et al.* (1989). They demonstrated that drought avoidance as a consequence of deeper root growth and greater extraction of soil water was an important mechanism of drought tolerance in common beans (*Phaseolus vulgaris* L.). Their results showed that under conditions of drought, two drought-tolerant lines rooted to 1.3 m, whereas two drought-sensitive lines reached only 0.6 m. This difference in rooting depth was associated with smaller soil water contents and an almost three-fold increase in yields. Concurrent research using grafting techniques demonstrated that the drought tolerance was conferred by genes expressed in the root system and not by the shoot genotype indirectly affecting root characteristics (White and Castillo, 1989).

Vincent and Gregory (1986) identified genotypic differences in root length and the ratio of root length:leaf area during early growth of 4 genotypes of chickpea grown in nutrient solution. They hypothesized that a large ratio might signify that the availability of water in relation to the growth of leaf area would be greater in such genotypes. Further investigation is required to determine whether such differences persist throughout growth and whether they confer an advantage at maturity. Observations at ICRISAT with chickpea (N.P. Saxena, unpublished) show that differences between genotypes in root size are established early in the season and that greater root growth and branching in seedlings is associated with drought resistance (ICRISAT, 1989). This has led to the development of a screening technique, using sand culture, to screen progeny of crosses involving drought-resistant parents.

There have been few field studies of water use by legumes compared with those on cereals. Brown *et al.* (1989) compared the water use of winter- and spring-sown chickpea (ILC 482) and of three genotypes (ILC 482, ILC 1929, ILC 3279) when spring-sown at Jindiress in northern Syria. At the time of maximum water content in the soil profile (late March), both winter- and spring-sown crops had used similar amounts of water despite the absence of a crop canopy on the later sown plots. Thereafter, because of the greater canopy growth, the winter-sown crop used water more rapidly than the spring-sown crop until the soil profile was depleted in late May/early June. By early June, the total water use of both crops was almost the same but the longer duration of the spring-sown crop meant that it used slightly more by maturity. The patterns of water use in the three spring-sown genotypes were similar to each other although the longer growing season of ILC 3279 again resulted in greater water use. Table 3 shows that water was used preferentially from the surface layer (0–30 cm) initially, but as the season progressed, water was used at greater rates

Table 3. Loss of water (mm d^{-1}) from the soil beneath a crop of spring-sown chickpea at Jindiress, northern Syria in 1983 (from Brown *et al.*, 1989)

Depth (cm)	21 April to 5 May	5 May to 19 May	19 May to 29 May	29 May to 8 June	8 June to 15 June
0 – 30	2.06	1.43	1.34	1.19	0.45
30 – 60	0.60	1.28	0.72	0.84	0.27
60 – 90	0.18	0.99	1.12	0.91	0.33
90 – 120	0.11	0.26	0.95	1.29	0.43
120 – 150	0.02	0.10	0.18	0.47	0.49
150 – 180	-0.09	0	-0.01	0.05	0.09
TOTAL	2.88	4.06	4.30	4.75	2.06

from deeper in the soil profile. At close to maturity, water from depths >1 m provided about 40% of the total water use.

Several workers (Keatinge and Cooper, 1983, in Syria; Siddique and Sedgley, 1987, in Western Australia) have shown the benefits of timely sowing in improving the water use efficiency of legume crops. In the study of Brown *et al.* (1989) winter sowing increased water use efficiency from 4.1 kg grain ha^{-1} mm^{-1} in the spring-sown crop to 8.2 kg grain ha^{-1} mm^{-1}. Early sowing, then, not only resulted in higher yields but in greater efficiency of water use principally because water which would otherwise have evaporated from the soil surface was transpired by the crop.

Roots and Nutrient Uptake

Nutrient uptake by legumes, particularly chickpea, has been studied intensively at ICRISAT (Ae *et al.*, 1991) and has revealed several features of legume roots that have rarely been known previously. The results will be described here as a case study and illustration of what may occur in other species.

Chickpea is mostly grown in alkaline soils, such as the Vertisols and Inceptisols of the Gangetic Plain in India and the Aridisols and Vertisols of northern Syria. The available P content of these soils is generally low to very low but despite this, chickpea rarely responds greatly to applications of P fertilizer, especially on the Vertisols at ICRISAT. However, in comparison with other crops, chickpea has a much greater capacity to acidify the pH of the rhizosphere, thereby increasing the availability of phosphate. This was confirmed using a laboratory test in which seven species were grown on agar plates and the amount of acidification assessed (Table 4). The results showed

Table 4. Color change with pH indicator bromcresol green (pH 5.4–3.8) caused by root exudates of various crop species on an agar plate. Standard errors are shown in parentheses (from Ae *et al.*, 1991)

Crop species	Area of color changed (=A) (mm^2)	Area of roots (=B) (mm^2)	A/B
Sorghum	144 (± 36.8)	1240 (± 108)	0.12 (± 0.02)
Pigeonpea	104 (± 11.6)	1112 (± 40)	0.09 (± 0.01)
Chickpea	548 (± 89.2)	2084 (± 192)	0.26 (± 0.02)
Soybean	256 (± 18.8)	1848 (± 300)	0.15 (± 0.03)
Pearl Millet	56 (± 10.8)	1072 (± 228)	0.06 (± 0.01)
Groundnut	516 (± 45.2)	2792 (± 48)	0.19 (± 0.02)
Maize	280 (± 31.6)	2160 (± 288)	0.13 (± 0.01)

that the amount of acid produced per unit surface area of roots was greatest for chickpea; this was later shown to be a consequence of the large amounts of organic acids, particularly citric acid, released from chickpea roots (Ae *et al.*, 1991).

The ability of chickpea to take up soil P efficiently from calcareous soils also was related to its deeper root system compared, for example, with soybean. In a field study using chickpea (cv. K 850) grown on a Vertisol during the post-rainy season, it was found that 50% of shoot dry matter was produced after flowering but that at flowering the roots had attained 87% of their maximum dry mass and that the total P uptake of the plant was 73% of its value at maturity (Figure 1). This was quite different from maize and soybean crops where dry matter and P tended to accumulate in the shoot in tandem.

Most legumes cannot satisfy their nitrogen requirement by N-fixation alone, especially during the early seedling stage and the pod-filling stage when N-fixation decreases markedly due, probably, to the reduced supply of assimilates to nodules (Lawn and Brun, 1974). Supplementary applications of N fertilizer during these periods is often necessary to increase grain production of chickpea, a crop in which the competition between N-fixation and nitrate uptake is low relative to other legume species. Nitrate is known to inhibit nodulation and N-fixation. In soybean, both processes are inhibited by concentrations of 1.5 mM nitrate (Herridge, 1982) but in chickpea, nodule dry weight was stimulated by applications of 3 mM nitrate and not inhibited until 6 mM nitrate was applied (Jessop *et al.*, 1984). Comparative studies show that nodulation of chickpea is less sensitive to nitrate than pea but more sensitive than lupin (Cowie *et al.*, 1990).

Nitrate uptake is governed not only by the structure of the root system but also by the permeability of membranes. The latter can be characterized by analysis using the Michaelis-Menten equation and K_m and V_{max} calculated from a Lineweaver-Burke plot (Figure 4). Chickpea had the lowest V_{max} (0.0019 mol kg^{-1} h^{-1}) among three legumes (0.0037 for pigeonpea and 0.0045 for groundnut). The K_m for chickpea (0.121 mol m^{-3}) was higher than pigeonpea (0.091). These results indicate that the potential rate of uptake and the affinity of root membranes for nitrate in chickpea are low compared with other legumes, especially pigeonpea. The comparative efficiency of nutrient uptake between crop species is usually assessed from the total nutrient uptake of the crops. However, as the preceding study indicates, there may be genetic differences in membrane transport which regulate the uptake process of nitrogen (Rao *et al.*, 1993).

Once taken into the plant, nitrate is first transformed to nitrite by nitrate reductase (NAR). Although there are conflicting results on which organ is the major site for this transformation in chickpea (Sawhney *et al.*, 1985; Sekhon *et al.*, 1987; Wasnik *et al.*, 1988), NAR activity in roots is generally smaller than that in leaves and nodules. It is thought, then, that a large part of the nitrate absorbed by chickpea is translocated to the leaves as nitrate and then transformed to nitrite using the reductive energy provided by photochemical reactions.

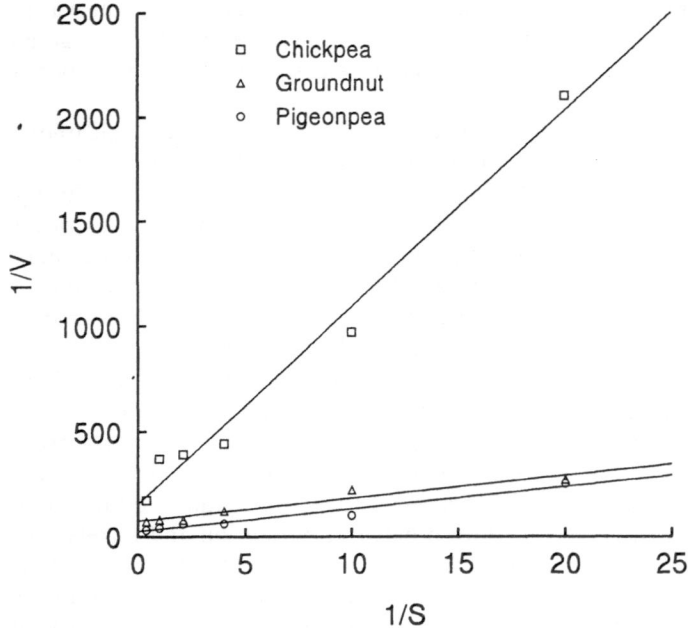

Figure 4. Lineweaver-Burk plot for nitrate concentration in the medium (S; mol m^{-3}) and uptake rate (V; mol kg^{-1} h^{-1}) of detached roots of pigeonpea ° (cv. ICPL 87), chickpea □ (cv. K 850) and groundnut △ (cv. NC 17090) grown in sand culture for a month (O. Ito. unpublished data).

Concluding Remarks

Compared with the review at the first IFLRC (Gregory, 1988), there has been a marked increase in our knowledge of the growth and functioning of chickpea roots. Much of this work has been conducted at ICARDA and ICRISAT. There remains, however, a dearth of information about the other major cool season food legumes. In Western Australia, lupin (*Lupinus angustifolius* L.) has been widely studied and, although not used as a food crop, results from these studies might be useful in understanding the role of legume root systems, particularly in relation to water uptake.

The demonstrable ability of chickpea to change its immediate chemical environment around the roots is of considerable practical importance. Further work in other species is required to determine whether it is a widespread phenomenon and whether the genotypic variation in root growth can be allied with this potential to improve nutrient acquisition and thereby increase growth and yields.

References

Ae, N., Arihara, J. and Okada, K. 1991. In: *Phosphorus Nutrition of Grain Legumes in the Semi-Arid Tropics*, pp. 33–41 (eds. C. Johansen, K. K. Lee and K. L. Sahrawat). Patancheru, Andhra Pradesh, India: ICRISAT.

Arihara, J., Ae, N. and Okada, K. 1991. In: *Phosphorus Nutrition of Grain Legumes in the Semi-Arid Tropics*, pp. 183–194 (eds. C. Johansen, K. K. Lee and K. L. Sahrawat). Patancheru, Andhra Pradesh, India: ICRISAT.

Brown, S. C., Gregory, P. J., Cooper, P. J. M and Keatinge, J. D. H. 1989. *Journal of Agricultural Science, Cambridge* 113: 41–49.

Cowie, A. L., Jessop, R. S. and MacLeod, D. A. 1990. *Australian Journal of Experimental Agriculture* 30: 651–654.

El Nadi, A. H., Brouwer, R. and Locher, J. Th. 1969. *Netherlands Journal of Agricultural Science* 17: 133–142.

Greenwood, D. J., Gerwitz, A., Stone, D. A. and Baines, A. 1982. *Plant and Soil* 68: 75–96.

Gregory, P. J. 1988. In: *World Crops: Cool Season Food Legumes*, pp. 857–867 (ed. R. L. Summerfield). Dordrecht: Kluwer Academic Publishers.

Gregory, P. J. 1989. In: *Drought Resistance in Cereals*, pp. 141–150 (ed. F. W. G. Baker). Wallingford, UK: CAB International.

Hamblin, A. P. and Hamblin, J. 1985. *Australian Journal of Agricultural Research* 36: 63–72.

Herridge, D. F. 1982. *Plant Physiology* 70: 1–6.

Hooda, R. S., Rao, A. S., Luthra, Y. P., Sheoran, I. S. and Singh, R. 1986. *Journal of Experimental Botany* 37: 1492–1502.

Hooda, R. S., Sheoran, I. S. and Singh, R. 1989. *Annals of Applied Biology* 114: 367–376.

ICARDA (International Center for Agricultural Research in Dry Areas). 1984. In: *Annual Report 1983*, pp. 154–155. Aleppo, Syria: ICARDA.

ICARDA (International Center for Agricultural Research in Dry Areas). 1985. In: *Annual Report 1984*, pp. 196–201. Aleppo, Syria: ICARDA.

ICRISAT (International Crops Research Institute for the Semi-Arid Tropics). 1989. In: *Annual Report 1988*, pp. 60–61. Patancheru, Andhra Pradesh, India: ICRISAT.

Jessop, R. S., Hetherington, S. J. and Hoult E. H. 1984. *Plant and Soil* 82: 205–214.

Kallarackal, J. and Milburn, J. A. 1985. *Annals of Botany* 56: 211–218.

Keatinge, J. D. H. and Cooper, P. J. M. 1983. *Journal of Agricultural Science, Cambridge* 100: 667–680.

Lawn, R. J. and Brun, W. A. 1974. *Crop Science* 14: 22–25.

Looker, C. H. 1978. *Studies of the growth and development of roots in Vicia faba L.* Ph.D. Thesis, University of Nottingham, UK.

Ludlow, M. M. and Muchow, R. C. 1990. *Advances in Agronomy* 43: 107–153.

Marschner, H. and Römheld, V. 1983. *Zeitschrift für Pflanzenphysiologie* 111: 241–251.

O'Toole, J. C. and Bland, W. L. 1987. *Advances in Agronomy* 41: 91–145.

Rao, T. P., Ito, O. and Matsunga, R. 1993. In: *Proceedings of the International Colloquium for the Optimization of Plant Nutrition*, September 1992. Lisbon, Portugal.

Sawhney, V., Amarjit and Singh, R. 1985. *Plant and Soil* 86: 233–240.

Sekhon, B., Sandhu, H. S., Dhillon, K. S. and Singh, R. 1987. *International Chickpea Newsletter* 17: 32–34.

Sheldrake, A. R. and Saxena, N. P. 1979. In: *Stress Physiology in Crop Plants*, pp. 466–483 (eds. H. Mussell and R. Staples). New York: John Wiley & Sons.

Siddique, K. H. M. and Sedgley, R. H. 1987. *Australian Journal of Agricultural Research* 37: 599–610.

Sponchiado, B. N., White, J. W., Castillo, J. A. and Jones, P. G. 1989. *Experimental Agriculture* 25: 249–257.

Stützel, H. and Aufhammer, W. 1991. *Annals of Botany* 67: 487–495.

Vincent, C. and Gregory, P. J. 1986. *Experimental Agriculture* 22: 233–242.

Wasnik, K. G., Varade, P. B. and Bagga, A. K. 1988. *Indian Journal of Plant Physiology* 31: 324–327.

White, J. W. and Castillo, J. A. 1989. *Crop Science* 29: 360–362.

White, K. D. 1970. *Agricultural History* 44: 281–290.

Williams, J. H. and Saxena, N. P. 1991. *Annals of Applied Biology* 119: 105–112.

Biological nitrogen fixation: basic advances and persistent agronomic constraints

A. STANFORTH[1], J.I. SPRENT[1], J. BROCKWELL[2], D.P. BECK[3] and H. MOAWAD[4]

[1] *University of Dundee, Department of Biological Sciences, DD1 4HN Scotland, UK;*
[2] *CSIRO, GPO Box 1600, Canberra Act 2601, Australia;*
[3] *ICARDA, P. O. Box 5466, Aleppo, Syria, and*
[4] *National Research Center, Dokki, Cairo, Egypt*

Abstract

Since the last conference there have been some notable advances in understanding legume nodulation and nitrogen fixation, but many gaps remain. For example,

1. Details of signal exchanges between host and rhizobia have been elucidated for *Vicia* and *Pisum*. However, there is some evidence that these may be affected by soil factors and much more work is needed here.
2. There is now a renewed interest in host-rhizobial interactions. Modern molecular methods allow the study of diversity in both partners on a wider scale than was possible earlier.
3. Progress has been made in measurement of N fixation in the field, particularly using ^{15}N methods. These will be assessed.
4. Further studies on environmental factors affecting N fixation have been carried out; with new methodology these should now be more targeted.
5. Interactions between nodulation and mycorrhizas can now be studied in *Vicia* and *Pisum* using a series of *nod-* and *myc*-mutants.
6. The agronomic constraint of turning N fixed by nodules into N in grain remains, although more is now known about N (as opposed to C) nutrition of pods and seeds.

Introduction

Biological nitrogen fixation within agriculture can be split into two distinct categories: firstly, where an effective association takes place between host plant and rhizobial partner and secondly, where no effective partnership develops. These differences are related primarily to availability of effective indigenous strains of rhizobia in the rhizosphere, and then to environmental factors (water stress, temperature extremes, nutrient availability, and pest attack). The following discussion will, therefore, be loosely based around these two categories.

F.J. Muehlbauer and W.J. Kaiser (eds.), Expanding the Production and Use of Cool Season Food Legumes, 821–831.
© 1994 *Kluwer Academic Publishers.*

Where Effective Nodulation Takes Place

The potential for nitrogen fixation and high yields is considerable. However, the high yields often obtained in yield trials are rarely achieved under normal agricultural practice. For example, modern cultivars of faba bean are capable of yielding up to 7 t ha^{-1} of seed. This compares with a world average of 1.1 t ha^{-1} (Hawtin and Hebblethwaite, 1983). With estimated nitrogen yields of up to 600 kg ha^{-1} (Sprent and Bradford, 1977), much of which can be attributed to biological nitrogen fixation (Zapata *et al.*, 1987), the potential of plant/rhizobial symbioses is outstanding.

Why has there been a failure in grain legumes to translate potential into production, in terms of high and stable yields? In cereals, improvements have generally been made through stem shortening and increased diversion of assimilates to grain production. This has not been directly applicable to grain legumes, and high biological nitrogen fixation, in particular, appears to be inextricably linked to high vegetative dry matter production. This was shown in faba bean where, within a single cultivar, increases in both C and N components of seed yield were obtained only by increasing C and N yield of the whole plant, and increases in C and N harvest index were only achieved under conditions which led to drastic reductions in yield (Figures 1 and 2). Reasons for this are not clear, but may be due to plant structure or to a high energy demand from biological nitrogen fixation. Future progress will depend largely on the ability of breeders to maintain high rates of biological nitrogen fixation and high seed sink capacity but with lower plant vegetative dry matter.

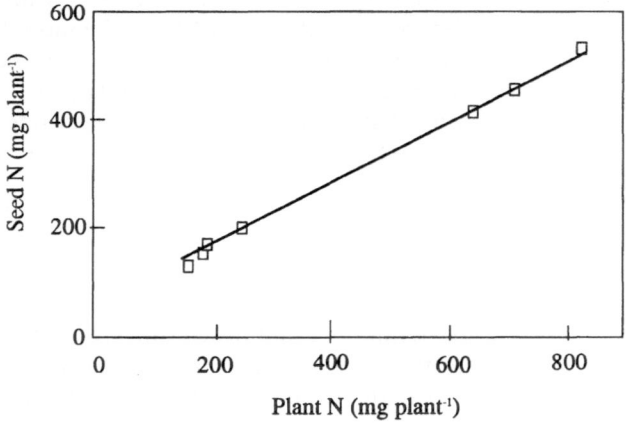

Figure 1. Relationship between seed N yield and total plant N accumulation in an indeterminate cultivar of field grown faba bean. Data taken from two seasons and using different applications of nitrogenous fertilizer.

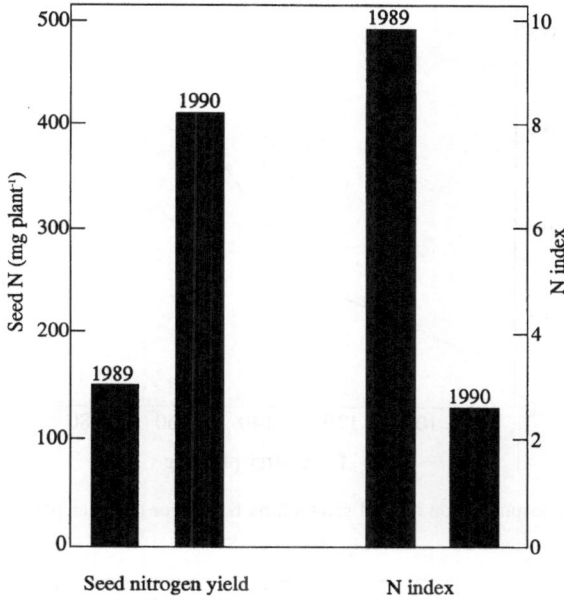

Figure 2. Seed yield and N index in field grown faba bean in a dry season (1989) and a moderately wet season (1990). N index calculated as ratio of N in seed/N in stem at maturity.

Nitrogen Fixation During Reproductive Development

There is a widely held suspicion that grain legumes have limited ability to fix sufficient quantities of nitrogen during reproductive development (e.g., Latimore *et al.*, 1977; Deibert *et al.*, 1979). This has been based on models of competition for carbon assimilates between growing seeds and active nodules (e.g., Sinclair and De Wit, 1975), widely observed senescence, and redistribution of nitrogenous reserves during the seed fill period and measurements of seasonal patterns of acetylene reduction activity (e.g., Alvilio *et al.*, 1979). Recent work on soybeans (Imsande, 1989; Bergersen *et al.*, 1992) utilizing ^{15}N dilution methods has shown that this may not be the case, and high rates of nitrogen fixation continued throughout seed development with virtually all the nitrogen being diverted to the seed. These conclusions were supported by work on faba bean which showed that total nitrogen accumulation rate, and accumulation per unit dry weight of nodule, of field-grown plants remained constant or increased throughout the reproductive period (Figures 3 and 4).

A close relationship was found between seed nitrogen yield and the proportion of plant nitrogen derived from fixation in faba bean (Duc *et al.*, 1988). It has also been suggested that biologically fixed nitrogen was preferentially diverted into seed reserves, compared to combined N, in phaseolus bean (Westerman *et al.*, 1985; Hungria and Neves, 1987) and this may

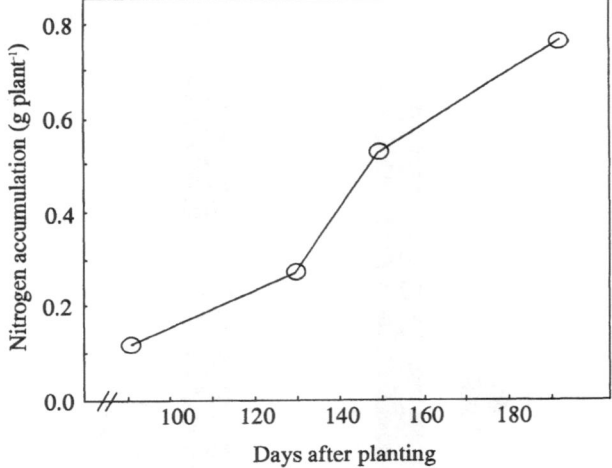

Figure 3. Nitrogen accumulation in field grown faba bean over the later period of development.

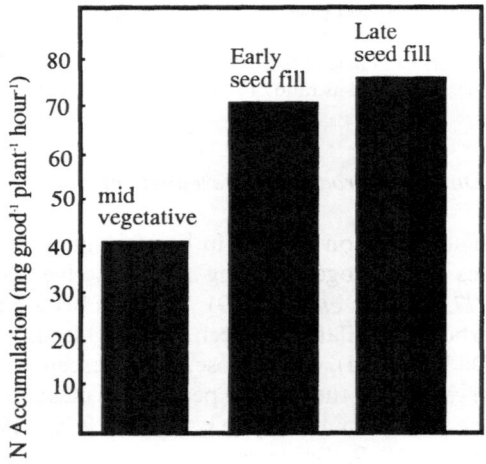

Figure 4. Plant N accumulation, per gram of nodule, during different stages of development in field grown faba bean.

also be the case with faba bean (Figure 5). The concentration of nitrogen within growing embryos was found to be directly proportional to the nitrogen concentration of the assimilate supplied by the parent plant (Figure 6), indicating that low amounts of nitrogen for translocation would decrease seed nitrogen yield. Furthermore, and contrary to expectations, redistribution of nitrogen from senescing leaves became less efficient during the seed filling period (Figure 7) (A. Stanforth, unpublished).

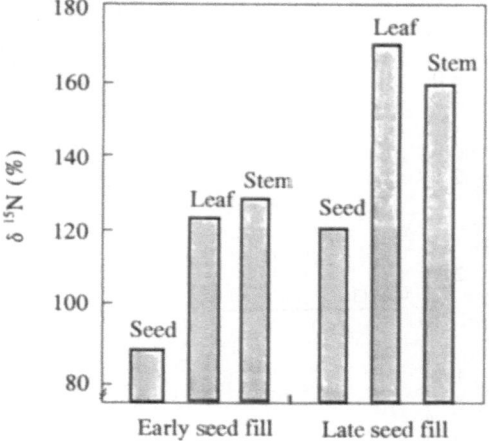

Figure 5. Partitioning of ^{15}N from labelled nitrate into new growth during reproductive development in field grown faba bean.

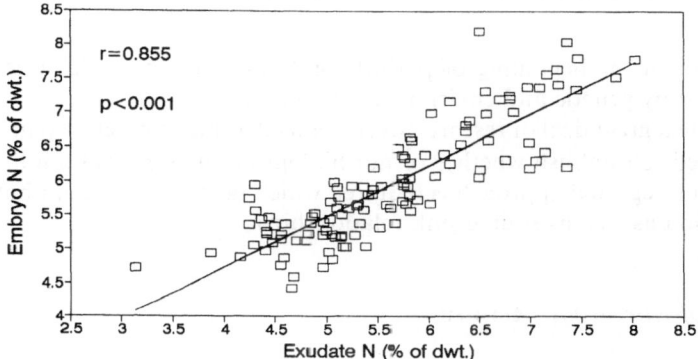

Figure 6. Relationship between the N concentration of an embryo and the N concentration of the assimilates supplying the embryo, sampled at the testa/embryo interface. Samples taken from three cultivars and at different stages of development from field grown faba bean.

Although these results stress the importance of maintaining nitrogen fixation throughout the period of reproductive development, there still remains a great deal of confusion over whether overall plant yield is limited by the ability to produce C and N assimilates (source limited), or is limited by the ability to utilize C and N assimilates (sink limited). The above evidence suggests that insufficient sink capacity exists, but there is also support for the opposite in that available sink capacity is not fully utilized (Aufhammer *et al.*, 1989). A recent hypothesis (Vessey, 1992) proposed that, in pea, maintenance of nitrogen fixation rate throughout pod development was dependent on carbon source to

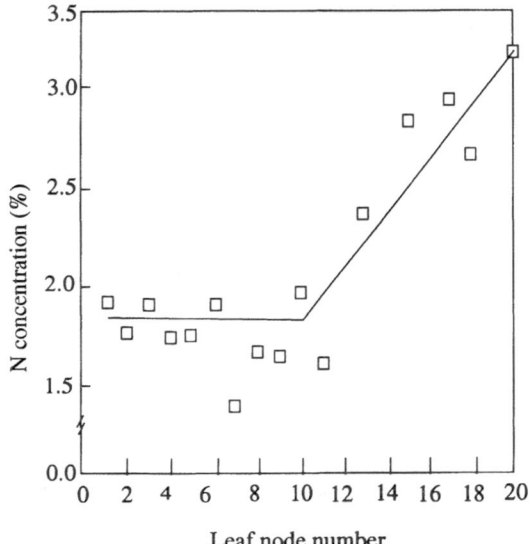

Figure 7. N concentration in abscinded leaves from different nodes of field grown faba bean.

sink ratios at the beginning of pod-filling. Source to sink ratios were in turn influenced by genetic and environmental factors.

Despite a great deal of research over a considerable number of years, there is still a need to establish whether or not biological nitrogen fixation limits crop yield. An integrated approach is required which takes into account both C and N nutrition as well as source sink relationships.

Control of Nitrogenase Activity

Carbon supply to the nodule has often been considered as the major factor limiting rates of nitrogen fixation, especially during reproductive development. This has recently been reviewed (Vance and Heichel, 1991) and it was concluded that changes in carbon utilization within the nodule may be more important than the rates of supply. In soybean nodules, carbon reserves in the form of poly-B-hydroxybutyrate (PHB), may be important in preserving nodule function during reproductive development (Bergersen *et al.*, 1991). At a time when availability of carbon assimilates may undergo transient limitations, these reserves may act as a buffer to maintain bacteroid oxygen consumption and low internal oxygen concentrations, thereby protecting nitrogenase from oxygen damage. This may not apply, however, to temperate grain legumes as there is no evidence of high levels of PHB within the nodule tissue, although starch reserves could play a similar role.

Control of nitrogenase activity within the nodule, therefore, appears to center on processes operating largely within the nodule. The key-controlling mechanism appears to be the variable oxygen diffusion resistance barrier (see Vance and Heichel, 1991). This variable barrier responds to a number of plant stresses and environmental stimuli and has been implicated in the initial response to nitrate application (reviewed by Vessey and Waterer, 1992). By decreasing the availability of oxygen within the nodule-infected tissue, a nodule can limit the production of ATP and reductant and thereby limit nitrogenase activity (Heckman *et al.*, 1989). An oxygen diffusion resistance barrier can, therefore, have both a proactive role, directly limiting nitrogen fixation by limiting oxygen and nitrogen availability, and a reactive role, protecting the nitrogenase enzyme from build-up of oxygen resulting from internally produced decreases in nitrogen fixation (Layzell and Hunt, 1990). It also has been suggested that under normal field conditions, soil oxygen concentration may be a principal factor limiting biological fixation, due to high diffusion resistance of nodules (Hernandez and Drevon, 1991).

The mechanism by which the oxygen diffusion barrier operates is not totally clear. Two hypotheses which are not mutually exclusive have been proposed. The first involves changes of the water-filled diffusion path length due to changes in osmotic potential. The second postulates that oxygen diffusion is regulated by variable occlusion of the intercellular spaces of the diffusion barrier zone with a form of glycoprotein.

It is envisaged that increased understanding of the regulation of the oxygen diffusion resistance barrier, coupled with the ability to manipulate it artificially, could lead to increased nitrogen fixation. This may be particularly important under certain stress conditions where symbiotic nitrogen fixation is generally more vulnerable to environmental stresses than assimilation of combined N. For example, both the total amount of, and the proportion of plant nitrogen derived from, biological nitrogen fixation decreased with increasing drought stress in pea, faba bean, and lentil (Bremer *et al.*, 1988) and in lentil and chickpea (Herridge, *et al.*, this volume). There was also a great deal of variation with cultivars, however, and further efforts are needed to understand the genetic and physiological bases for differing responses.

Although there is a great deal of variation between different cultivars to support biological nitrogen fixation, it has often been difficult to extend potential into productivity. Quantitative traits are difficult to manipulate in classical breeding programs because of effects of environmental factors. Quantitative trait loci (QTL) can be resolved into individual genetic components using molecular polymorphic assays to generate high density linkage maps. This facilitates marker-assisted breeding programs (MABPs), using easily scorable markers (such as flower color) linked to QTL. The logic of this approach has been developed mainly for crop yield, but could equally be applied to nitrogen fixation. Among cool season food legumes, the methodology has been established for faba bean by Van de Ven (1992).

Conditions of Ineffective Nodulation

There are substantial areas of the world where plant/rhizobial associations are ineffective or poorly effective. This may be due either to a failure to form nodules or due to formation of inefficient nodules which fix little or no nitrogen (for a recent review see Bottomley, 1992).

Situations Where Nodulation Does Not Occur

This occurs mainly where there is a lack of indigenous rhizobia or where soil populations are not able to infect a particular host. The essential steps involved in the formation of functional legume nodules have been recently reviewed (Sprent, 1989). A great deal of the progress in understanding of the events leading to infection and formation of effective nodules has been made by modern molecular techniques. In faba bean and pea, for infection to be successful, a number of genes on the symbiotic plasmid of *Rhizobium leguminosarum* bv. *viciae* must first be induced (Long, 1992). A large spectrum of flavones, isoflavones, and related compounds exuded from the plant root into the rhizosphere, which are host-species specific and are involved in the switching on of *nod* genes, have been found. The situation is, however, very complex and a compound which specifically induces one strain may act as an inhibitor in another strain. A further complication is that soil factors may also be involved. Soil pH and aluminum content may affect induction of *nodA* in *R. leguminosarum* bv. *trifolii* (Richardson *et al.*, 1988), and in recent work in our laboratory (Shaw *et al.*, 1993) showed that organic compounds can affect the ability of flavonoids to induce *nodC* in *R. leguminosarum* bv. *viciae*. Despite the advances made in our understanding of the factors involved in nodulation, application to field situations is still some way off, and increased nitrogen fixation will depend more on conventional techniques where inoculation with effective strains of rhizobia has often led to marked increases in yield (e.g., Moawad *et al.*, 1988).

Situations Where Nodulation Occurs But Is Inefficient

In many soils indigenous populations of rhizobia are present which successfully infect and lead to nodulation of the host plant, but are subsequently inefficient at fixing nitrogen. A great deal of variation exists in the genetic pools of both plants and bacteria with some host x strain interactions (Roskothen, 1989). Rhizobial strains have been isolated which show markedly improved nitrogen fixation when compared with indigenous populations (Moawad *et al.*, 1985, 1987). Promising results from inoculations within a controlled situation have, however, generally failed to extend to field situations, often due to an inability of these new strains to compete with indigenous population strains for nodulation sites (Bohlool *et al.*, 1988). Recent work with faba bean has, however, identified an effective strain of *R. leguminosarum* which formed an effective symbiosis, was able to compete with indigenous strains for nodulation sites, and also appeared to be persistent in the soil, forming an effective association in the following year (Moawad *et al.*, 1991). Some concern has recently been expressed, however, (Bottomley, 1992) as new rhizobial strains which are currently being introduced may be difficult to displace at a later date by better nitrogen fixing strains.

Other Factors Affecting Nitrogen Fixation

The soil environment not only affects the ability of rhizobia to infect plants and form nodules, but also the potential for nitrogen fixation. The poor availability of micronutrients, particularly in some soils, may severely limit nitrogen fixation and whole plant growth. Addition of a number of micronutrients was shown to increase biological nitrogen fixation and plant growth (Moawad *et al.*, 1985). The presence of mycorrhizal associations may also be particularly important (Badr El-Din and Moawad, 1988), increasing the uptake of some micronutrients and excluding excess quantities in other cases. In situations where a tripartite association exists, there may be competition for plant-produced carbon assimilates between rhizobia and mycorrhizae. The development of nod^- and myc^- isolines of faba bean and pea (Duc *et al.*, 1989) will facilitate a greater understanding of the contributions made by different symbiotic partnerships and may lead to increases in productivity in some of the least productive agricultural areas.

Assessment of Potential and Tracking Selections

Selection of efficient rhizobial genotypes is likely to provide some of the main advances in increasing biological nitrogen fixation. The problems of assessing both the range of natural variation and following a particular rhizobial strain from isolation through inoculation and back to reisolation have been considerable in the past. Methods of rhizobial identification have involved complex, time consuming, and expensive techniques. Work is progressing in our laboratory on two new techniques to type rhizobial strains. 1) Polymerase Chain Reaction techniques (PCR) that utilize the rapid cloning of DNA fragments which can then be characterized. 2) Capillary Electrophoresis (CE) which utilizes a microbore capillary through which a solution of bacterial proteins is moved by electromotive force. Separation is based on mass/charge ratios as with conventional electrophoretic methods. The advantages over conventional methods are the low running costs, high degree of automation, large scope for changing conditions, and both quantitative and qualitative results.

Field Assessment of Biological Nitrogen Fixation

Whatever advances are made, either in selection of pre-existing high nitrogen fixing partnerships or in genetic engineering of new partnerships, it will be difficult to make clear quantitative judgments of increases in biological nitrogen fixation without accurate field measurement. As legumes in general are efficient scavengers of combined nitrogen in the soil and prefer to utilize combined rather than fixed nitrogen, quantifying the overall contribution from nitrogen fixation can be extremely difficult. The main methods available have been reviewed by Herridge *et al.* (this volume). Although none of the methods appear to be completely accurate, new equipment available for measuring very low levels of nitrogen enrichment have increased the potential use of natural abundance ^{15}N enrichment. This method has already proved to be very useful in soybean (Bergersen *et al.*, 1989, 1992).

Concluding Remarks

Over the 6 years since the last conference, basic advances have been made, with greater knowledge of genetic diversity and nitrogen fixation potential, and attempts to fit together effective partnerships have been made with varying success. Methods to measure nitrogen fixation in the field have been advanced and should help to clarify what contribution biological nitrogen fixation makes to plant yield. Despite these advances, many of the old constraints remain and translating potential into production remains an enigma. With the new methods available to us and with close cooperation between scientists at molecular, cellular, and whole plant levels, there is greatly increased scope for overcoming many of the obstacles and improving biological nitrogen fixation within an agronomic context.

References

Alvilio, A. F., Pereira, J. C. and Neyra, C. A. 1979. *Plant Physiology* 63: 421–424.
Aufhammer, W., Nalborczyk, E., Geyer, B., Gotz, I. A., Mack, C. and Paluch, S. 1989. *Journal of Agricultural Science, Cambridge* 112: 419–424.
Badr El-Din, S. M. S. and Moawad, H. 1988. *Plant and Soil* 108: 117–124.
Bergersen, F. J., Brockwell, J., Gault, R. R., Morthorpe, L., Peoples, M. B. and Turner, G. L. 1989. *Australian Journal of Agricultural Research* 40: 763–780.
Bergersen, F. J., Peoples, M. B. and Turner, G. L. 1991. *Proceedings of the Royal Society of London. B.* 245:59–64. Bergersen, F. J., Turner, G. L., Peoples, M. B., Gault, R. R., Morthorpe, L. J. and Brockwell, J. 1992. *Australian Journal of Agricultural Research* 43: 145–153.
Bohlool, B. B., Bezdicek, D. F., Somasegaran, P. and Moawad, H. 1988. In: *World Crops: Cool Season Food Legumes*, pp. 675–689 (ed. R. J. Summerfield). Dordrecht: Kluwer Academic Publishers.
Bottomley, P. J. 1992. In: *Biological Nitrogen Fixation*, pp. 560–597 (eds. G. Stacey, W. H. Burris and H. J. Evans). New York: Chapman and Hall.

Bremer, E., Rennie, R. J. and Rennie, D. A. 1988. *Canadian Journal of Soil Science* 3: 553–562.

Deibert. E. J., Bijeriego, M. and Olson, R. A. 1979. *Agronomy Journal* 71: 717–723.

Duc, G., Mariotti, A. and Amarger, N. 1988. *Plant and Soil* 106: 269–276.

Duc, G., Trouvelot, A., Gianninazzi-Pearson, V. and Gianninazzi, S. 1989. *Plant Science* 60: 215–222.

Hawtin, G. C. and Hebblethwaite P. D. 1983. In: *The Faba Bean*, pp. 3–22 (ed. P. D. Hebblethwaite). London: Butterworths.

Heckman, M. - O., Drevon, J. - J., Saglio, P. and Salsac, L. 1989. *Plant Physiology* 90: 224–229.

Herridge, D. F., Rupela, O. P., Serraj, R. and Beck, D. P. 1994. In: *Expanding the Production and Use of Cool Season Food Legumes*, pp. 472–492 (eds. F. J. Muehlbauer and W. J. Kaiser). Dordrecht: Kluwer Academic Publishers.

Hernandez, G. and Drevon, J. -J. 1991. *Journal of Plant Physiology* 138: 587–590.

Hungaria, M. and Neves, M. C. P. 1987. *Physiological Plantarum* 69: 53–63.

Imsande, J. 1989. *Agronomy Journal* 81: 549–556.

Latimore, M., Giddens, J. and Ashley, D. A. 1977. *Crop Science* 17: 399–404.

Layzell, D. B. and Hunt, S. 1990. *Physiologia Plantarum* 80: 322–327.

Long, S. R. 1992. In: *Biological Nitrogen Fixation*, pp. 560–597 (eds. G. Stacey, W. H. Burris and H. J. Evans). New York: Chapman and Hall.

Moawad, H., Abd El-Malek, Y., Abd El-Maksoud, H. K. and Gohar, M. 1985. *Zeitschrift für Pflanzenernahrung und Bodenkunde* 148: 584–589.

Moawad, H., Abd El-Malek, Y., Abd El-Maksoud, H. K. and Gohar, M. 1988. *Egyptian Journal of Microbiology* 23: 465–472.

Moawad, H., Badr El-Din, S. M. S. and Khalafallah, M. A. 1987. *Egyptian Journal of Microbiology* 22: 203–212.

Moawad, H., Badr El-Din, S. M. S. and Khalafallah, M. A. 1991. *World Journal of Microbiology and Biotechnology* 7: 191–195.

Richardson, A. E., Simpson, R. J., Djordjevic, M. A. and Rolfe, B. G. 1988. *Applied Environmental Microbiology* 54: 2541–2548.

Roskothen, P. 1989. *Plant Breeding* 102: 122–132.

Shaw, J., Sprent, J. I. and Reynolds, T. 1993. Does the soil environment exert control over root nodulation? (Submitted for publication).

Sinclair, T. R., and De Wit, C. T. 1975. *Science* 189: 565–567.

Sprent, J. I. 1989. *New Phytologist* 111: 129–153.

Sprent, J. I. and Bradford, A. M. 1977. *Journal of Agricultural Science Cambridge* 88: 303–310.

Vance, C. P. and Heichel, G. H. 1991. *Annual Review of Plant Physiology, Plant Molecular Biology* 42: 373–92.

Van de Ven, W. T. G. 1992. *Construction of a genetic linkage map of Vicia faba using molecular, biochemical and morphological markers and molecular analysis of Vicia species relationships.* Ph.D. Thesis, University of Dundee, Scotland.

Vessey, J. K. 1992. *Plant and Soil* 139: 185–194.

Vessey, J. K. and Waterer, J. 1992. *Physiologia Plantarum* 84: 171–176.

Westerman, D. I., Porter, L. K. and O'Deer, W. A. 1985. *Crop Science* 25: 255–229.

Zapata, F., Danso, S. K. A., Hardarson G. and Fried M. 1987. *Agronomy Journal* 79: 505–509.

VA mycorrhiza: benefits to crop plant growth and costs

E. GEORGE[1], S.K. KOTHARI[2], X.-L. Li[3], E. Weber[1] and H. Marschner[4]

[1] Institute of Plant Nutrition, Hohenheim University, P.B. 700562, 7000 Stuttgart 70, Germany and ICARDA, P. O. Box 5466, Aleppo, Syria;
[2] Central Institute of Medicinal and Aromatic Plants, Lucknow, India;
[3] Department of Soil Science and Plant Nutrition, Beijing Agricultural University, Beijing, China, and
[4] Institute of Plant Nutrition, Hohenheim University, P.B. 700562, 7000 Stuttgart 70, Germany

Abstract

Most vascular plant species, including agricultural crops, are commonly colonized by vesicular-arbuscular mycorrhizal (VAM) fungi. VA mycorrhizal colonization of roots occurs in all agroecosystems. The extraradical hyphae of the fungus are able to take up nutrients, such as phosphorus, zinc, and copper, and transport them to the host plant, thereby improving plant nutrition. Thus VAM fungi can be of crucial importance for adequate growth of plant species with a small root surface area when growing in soils low in these nutrients. In addition, roots of individual plants in the field are connected by a common fungal mycelium, allowing for a very limited exchange of nutrients between plants. VAM hyphae can probably not transport large amounts of water, but mycorrhizal plants may be more resistant than non-mycorrhizal plants to drought periods by means of a number of direct and indirect effects of VAM fungi. On the other hand, due to enhanced vegetative growth and lower root/shoot ratios, under rainfed conditions mycorrhizal plants may be more susceptible to drought stress during seed filling. In addition to direct and indirect effects on nutrient and water uptake, VAM fungi can also increase plant resistance to root pathogens.

In exchange for the nutrients taken up, the fungus receives carbohydrates from the host plant to sustain its growth. When the carbohydrate drain of the fungus is higher than the benefit plants derive from their VAM, mycorrhizal colonization of roots may lead to a reduction in plant growth. The carbohydrate demand by VAM fungi may be of particular importance in legumes where symbiotic nitrogen fixation also is a strong sink for photoassimilates.

Introduction

Physiological and ecological implications of vesicular-arbuscular (VA) mycorrhizas have been reviewed in detail lately (for example, Smith and Gianinazzi-Pearson, 1988; Fitter, 1991; Koide, 1991; Read, 1991; Schwab et al.,

F.J. Muehlbauer and W.J. Kaiser (eds.), Expanding the Production and Use of Cool Season Food Legumes, 832–846.

1991) and practical applications by Sieverding (1991). Here, we discuss some recent results on the importance of VA mycorrhizas in an agricultural context.

Possible Contributions of VA Mycorrhizas to Plant Growth

Agricultural crops are commonly colonized by VA mycorrhizal fungi, except in some plant families (*Chenopodiaceae, Cruciferae*). After clearing and staining, plant roots can be checked easily for mycorrhizal colonization under a microscope (Grace and Stribley, 1991).

Vesicular-arbuscular mycorrhizal fungi colonize plant roots by spreading infection in the root cortex from fungal entry points. Hyphae proliferate in the cortex and form branch-like structures (arbuscules) within root cells (Figure 1). Arbuscules are the sites of mineral nutrient (from fungus to plant) and carbohydrate (from plant to fungus) exchange. The fungus also forms storage organs (vesicles) in the root.

In addition to fungal tissue within the root, extraradical hyphae (mycelium) grow into the soil, where spores are also formed (Figure 1). Other plants can be infected from the living mycelium in soil or, in agricultural fields after tillage,

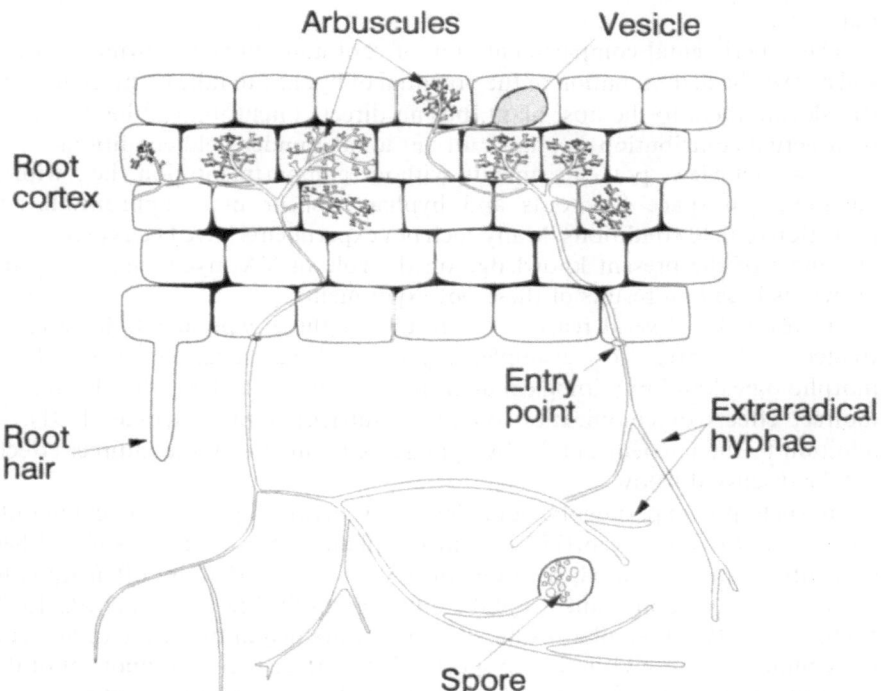

Figure 1. Schematic diagram of vesicular-arbuscular (VA) mycorrhizal colonization of a plant root.

from spores, hyphal fragments, or infected root parts (summarized as "propagules") of preceding crops.

The extraradical hyphae absorb nutrients which are translocated to the arbuscules in the root, and subsequently to the plant. Because extraradical hyphae can extend to several centimetres distance from the root and take up nutrients there (Li *et al.*, 1991a), the spatial exploitation of soil by the plant-fungus association is greatly improved compared to the soil exploitation by the roots alone. This increased spatial availability is of importance for nutrients of low mobility in the soil, especially phosphorus (P), but also ammonium (NH_4^+), zinc (Zn), and copper (Cu).

The nutrient uptake potential of the external hyphae is best examined in pots where root growth is prevented in certain zones of the soil by nylon meshes which can be penetrated by hyphae only, so that nutrient depletion by hyphae can be studied in these soil zones apart from the effects of roots.

Up to approximately 80% of the plant P and 50% of the plant Cu (Li *et al.*, 1991b) and Zn (Li *et al.*, unpublished) were absorbed from soil by the VA mycorrhizal fungus under these conditions. Hyphae also take up considerable quantities of NH_4-N (Ames *et al.*, 1983). These data show that hyphal uptake of nutrients with limited mobility in soil, especially of P, can be very substantial when compared to the uptake of the root alone. Possible contributions of VA mycorrhiza to plant growth are, therefore, large on soils deficient in these nutrients.

The experimental compartmentation of root and hyphae growing zones in soil allows the determination of the potential of hyphae to take up nutrients and translocate them to the host plant, but no direct conclusion can be drawn on their actual contribution to plant nutrient uptake under field conditions.

At a first view, pot experiments without compartmentation, i.e., with a common soil space for roots and hyphae, appear more appropriate for reflection of field conditions. Many such pot experiments have been carried out, and most of the present knowledge on the role of VA mycorrhizas in plant growth is based on results of these pot experiments.

However, for several reasons the results of these experiments have to be treated with care. For example, mycorrhizal colonization also induces morphological and physiological alterations in the host plant, thus leading to indirect effects of colonization on plant nutrient uptake (Koide, 1991), in addition to the nutrient uptake by hyphae. A number of these indirect effects will be discussed below.

In such pot experiments, inoculated mycorrhizal plants are commonly compared to non-mycorrhizal control plants in sterilized soil. When agricultural soils are used, root colonization with VAM fungi often leads to dramatic increases in plant P uptake and growth (Table 1; Raju *et al.*, 1990; Weber *et al.*, 1992). Similar results have been obtained in the last decades for a large number of agricultural crop species. Mycorrhizas are a component of the plant nutrient acquisition strategy, and should always be taken into account when plant nutrient uptake is studied.

Table 1. Plant dry weight and P content, number and specific mass of nodules, acetylene reduction activity and nodule efficiency in seven-week-old soybeans grown on a silt loam in a pot experiment, fertilized with high [high P, −VAM] or low [low P, −VAM] quantity of KH_2PO_4 or colonized by a VA mycorrhizal fungus [low P, +VAM] (Brown *et al.*, 1988)

	Low P −VAM	High P −VAM	Low P +VAM
Total dry weight (g plant^{-1})	4.5	5.7	7.6
Total P content (mg plant^{-1})	2.9	6.0	5.8
Number of nodules (plant^{-1})	33	30	97
Nodule specific mass (mg nodule^{-1})	2.7	7.0	3.5
Acetylene reduction activity (nmol plant^{-1} s^{-1})	1.3	6.3	2.5
Nodule activity/chloroplast activity (nmol C_2H_2 [mmol CO_2]$^{-1}$)	7.3	15.8	4.7
Nodule P use efficiency (mg C_2H_2 [g P]$^{-1}$ s^{-1})	90	140	60

In addition to increased uptake of mineral nutrients, mycorrhizal fungi are assigned a number of other beneficial effects for their host plants. Thus, it is possible that mycorrhizas partly protect plants from heavy metal toxicity. For example, decreased uptake of manganese (Kothari *et al.*, 1991) or increased tissue tolerance to high manganese concentrations (Bethlenfalvay and Franson, 1989) in mycorrhizal plants may increase growth of colonized plants on soils with high manganese concentration.

Mycorrhizas may also increase plant resistance to soilborne pathogenic fungi. As for other observations, this effect is well described only under controlled experimental conditions. It may be related to differences in rhizosphere microorganism populations between mycorrhizal and non-mycorrhizal plants (Linderman, 1988) or differences in the spread of infection in the root.

In a number of experiments, VA mycorrhizal plants proved to be more drought resistant than non-mycorrhizal control plants. This held true both when the water supply was kept at a constant low level (Ibrahim *et al.*, 1990) or when intermittent drought and watering cycles were applied (Bethlenfalvay *et al.*, 1988).

It is still an open question whether this improved drought resistance is related to direct mycorrhizal effects on water uptake and use, or related to improved nutrient (P) uptake of mycorrhizal plants. There is experimental evidence both for (Faber *et al.*, 1991) and against (George *et al.*, 1992) substantial water transfer in hyphae of VA mycorrhizal fungi.

While the small diameter of VA mycorrhizal hyphae may restrict the quantity of water transported within hyphae (Kothari et al., 1990), water movement from soil to root can be improved by the mycorrhizal mycelium connecting the root surface with the surrounding soil (Davies et al., 1992), leading to higher plant water uptake. Also, direct mycorrhizal effects on internal plant water use may occur, for example, via changes in root turgor in mycorrhizal plants (Augé and Stodola, 1990).

Furthermore, because in dry soil root P uptake is specifically impeded (Mackay and Barber, 1985), P uptake by the fungus may become even more beneficial to plants under dry soil conditions (Fitter, 1985). Then, increased plant P uptake leads to enhanced plant water use and better growth (Fitter, 1988). Thus, direct and indirect effects may both act to increase drought resistance of mycorrhizal plants, at least in pot experiments.

However, valid interpretation of pot experiments comparing mycorrhizal and non-mycorrhizal plants is further complicated by a number of other factors. For example, root length densities are higher and fractional VA mycorrhiza colonization is often lower in pots than in the field. This may result in an underestimation of the mycorrhizal contribution in pot studies as compared to the field situation.

On the other hand, in pot experiments using sterilized soil, hyphal growth may be stimulated, while often there is no grazing of hyphae by soil fauna as under field conditions (Fitter, 1985), leading to an overestimation of the mycorrhizal contribution to plant growth.

Some of the shortcomings of pot cultures can be overcome by careful experimentation, such as using control plants inoculated with soil organisms other than VA mycorrhiza. There are, however, other problems inherent to the comparison of mycorrhizal and non-mycorrhizal plants. For example, mycorrhizal colonization may lead to increased root length at very low soil P supply compared to non-mycorrhizal plants, because mycorrhizal colonization increases P uptake, shoot growth and then, in turn, root growth (Figure 2).

However, at higher soil P supply, mycorrhizal colonization may result in a decreased root length (Figure 2; Gnekow and Marschner, 1989; Raju et al., 1990). In addition, colonization also changes the morphology of the root system (Kothari et al., 1990). Thus, there is a number of indirect effects brought about by mycorrhizal colonization which complicate the causal interpretation of quantitative differences between mycorrhizal and non-mycorrhizal plants.

As a further complication, not all VA mycorrhizal fungal species are equally effective in colonization or P uptake. Even within one species, isolates differ in their ability to improve shoot growth of their host plants (Bethlenfalvay et al., 1989).

In summary, as measured in pot experiments, the increased uptake of P is by far the most dominant contribution of mycorrhizas to plant growth (Fitter, 1991). An increased uptake of Cu, Zn, and NH_4-N is also possible. Other effects, such as a protection from heavy metal toxicity or fungal pathogens or

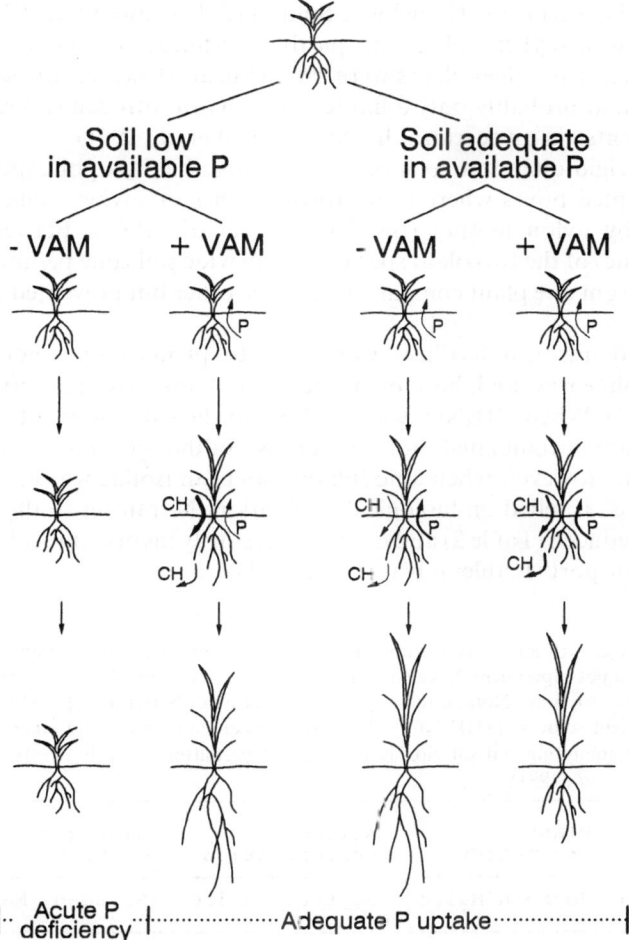

Figure 2. Effect of colonization with VA mycorrhiza and soil P supply on host plant root/shoot ratio (CH = carbohydrates).

improved drought resistance, may be either direct mycorrhizal effects or secondary consequences of increased P uptake, but they occur inconsistently.

Plant-to-Plant Nutrient Transfer by Mycorrhizal Hyphae

In addition to uptake of nutrients, VA mycorrhizal hyphae may also be involved in plant-to-plant exchange of nutrients (Van Kessel *et al.*, 1985). Recent evidence suggests that extraradical hyphae can act as bridges between individual plants, and thus permit nutrient exchange. This has been discussed for sites under natural vegetation, for pastures, but also for arable land.

In a field experiment, Hamel and Smith (1991) showed that ^{15}N applied as $^{15}NH_4NO_3$ to a soybean plant was partly transferred to a maize plant in the adjacent plant row when plants were mycorrhizal. However, the same authors concluded that probably only a limited quantity of nitrogen (N) is transferred between plants via mycorrhizal hyphae (Hamel *et al.*, 1991).

A more rigid quantitative approach was possible in a pot experiment with compartmented boxes where root growing zones of soybean and maize were separated by nylon meshes (Bethlenfalvay *et al.*, 1991). Between the root growing zones of the two plant species, a 6-cm wide soil zone permitted crossing of hyphae from one plant compartment to the other but prevented soil solution N transfer.

When no nitrogen fertilizer was given to plants and a non-nodulating soybean isoline was used, both plant species took up a low quantity of nitrogen (N) (Table 2). When NH_4NO_3 was fertilized to the soybean plant, not only the soybean plant accumulated more N but also (although to a lower extent) the maize plant. However, when a nodulating soybean isoline was used and not N fertilized, i.e., it relied on biological N_2 fixation, N transfer to the maize plant was much reduced (Table 2) and did not necessarily involve direct hyphal plant-to-plant transport (Bethlenfalvay *et al.*, 1991).

Table 2. Plant growth and nitrogen uptake of 65-day-old mycorrhizal soybean and maize on a loamy sand in a pot experiment. Soybeans used were either nodulating [$+N_2$ fix] or non-nodulating isolines of Clark soybean. Non-nodulating soybeans were not N fertilized [− N] or supplied with NH_4NO_3 nutrient solution [$+NH_4NO_3$]. Roots of soybean and maize were separated by a 6-cm-wide soil zone inhibiting soil solution N transfer but permitting growth of mycorrhizal hyphae (Bethlenfalvay *et al.*, 1991)

N supply to soybean	Plant dry matter		Nitrogen concentration		Nitrogen content	
	Soybean	Maize	Soybean	Maize	Soybean	Maize
	g plant^{-1}		g kg^{-1}		mg plant^{-1}	
−N	3.9	7.2	6.9	5.0	30	33
+N_2 fix	25.1	6.9	28.4	5.5	419	40
+NH_4NO_3	21.8	8.6	30.8	8.8	351	55

Therefore, for practical purposes, N transfer from legumes to non-legumes via mycorrhizal hyphae will not be of much significance. It occurs when the donor plant is well supplied with mineral fertilizer. Nutrient transfer in hyphae may be more important in natural ecosystems, where plant nutrient demand is generally smaller, growth duration is often more extended, and hyphae disruption in soil is less than in agricultural fields. Hyphae may also be involved in the nutrient transfer from senescing to young plants in mixed stands (Eason *et al.*, 1991).

Benefits and Costs of Mycorrhizal Colonization

In the above section, we have discussed under which circumstances the plant-fungus association may or may not be of advantage to the plant. We have neglected, however, the costs of mycorrhizas for host plants and the implications of these costs for the role of mycorrhizas in agricultural production.

For example, whether the root-shoot ratio of mycorrhizal plants is different from that of non-mycorrhizal plants depends not only on soil P supply and mycorrhizal infection, but also on the plant carbon (C) assimilation (Figure 2). The mycorrhizal fungi require carbon (assimilates) from the host, with the result that in mycorrhizal plants more C has to be allocated below-ground (Pang and Paul, 1980).

Estimations on this additional carbon demand to the plant differ widely (Paul *et al.*, 1985). Of the total C assimilated by the plant, it may account for between 10% (Fitter, 1991) and up to 20% (Jakobsen and Rosendahl, 1990). Only a smaller part of the C lost to the fungus is used for fungal biomass production, while most is used for fungal respiration (Figure 3).

On low-P soils, VA mycorrhiza colonization leads to higher P uptake, followed by higher leaf area, photosynthesis and plant carbohydrate production, and allocation to the root (Figure 2). In this case, costs are outweighed by the benefits and mycorrhizal plants then continue to grow larger but shoot/root ratios are not changed (Figure 2).

In soils with adequate P supply, however, the additional P taken up via the fungus does not lead to improved shoot growth and C assimilation, while VA mycorrhizal colonization still requires additional carbohydrates from the plant (Figure 2). Root/shoot ratios of VA mycorrhizal plants on soils with adequate P supply are often lower than those of non-mycorrhizal plants. This may then turn out to be disadvantageous for the root uptake of other nutrients and water.

In case root growth is not drastically reduced after colonization, for compensation of the carbohydrate demand of the fungus, mycorrhizal plants have to increase their photosynthetic rate to maintain similar growth rates as non-mycorrhizal plants. Such an increase in photosynthesis induced by mycorrhizal colonization is often observed (Kucey and Paul, 1981) and may not be problematic as long as photosynthetic rate is sink-limited rather than source-limited (Fitter, 1991).

However, in legumes, nodules represent an additional strong sink for photosynthates. Similar to the mycorrhizal fungi, they have a high respiratory activity (Figure 3) and may exhaust C reserves. Mycorrhizal, nodulated legumes may then be well supplied with P and N, but short of C to be used for leaf growth or seed formation.

Internal P use efficiency (shoot dry matter per unit P content) is often lower in mycorrhizal than in non-mycorrhizal plants. The reasons of this decreased P use efficiency have been discussed controversely (Koide, 1991), but they can be taken as expressions of the additional C costs of mycorrhizal colonization. In

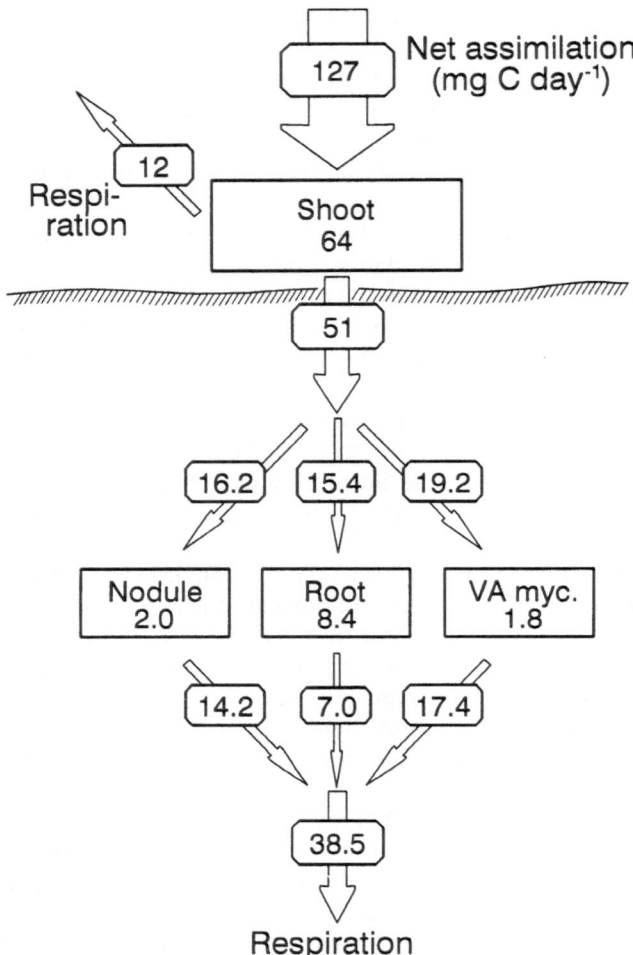

Figure 3. Carbon allocation (mg ^{14}C day^{-1}) in a six-week-old soybean-*Rhizobium*-VA mycorrhiza association measured after supply of ^{14}C to the leaves (Paul *et al.*, 1985).

line with this, mycorrhizal infection can decrease the efficiency of nodule activity in legumes. On the contrary, a higher soil P supply increases nodule efficiency (Table 1; Brown *et al.*, 1988).

Thus, although there are numerous reports on positive interactions in legumes of *Rhizobium* and VA mycorrhiza activity (Piccini *et al.*, 1988; Daft, 1991; Simpson and Daft, 1991), these positive interactions can be expected on low-P soils only.

Raising the P level in the soil may, therefore, be a more economic way to improve legume growth than increase in mycorrhizal colonization (Table 1), both in terms of carbon distribution in the plant and seed yield of legumes.

However, a sufficient mycorrhizal colonization of a crop may often be necessary to efficiently absorb soil and fertilizer P.

Whether benefits of mycorrhizal colonization are higher than the costs will depend on environmental conditions, and the benefit/cost ratio may also change drastically during plant development (Fitter, 1991). The cost/benefit ratios with respect to carbon demand for P acquisition in mycorrhizal plants have also to be compared to other plant strategies for enhanced P acquisition, such as increased root growth (Koide, 1991) or an excretion of organic acids (Dinkelaker *et al.*, 1989).

Occurrence of VA Mycorrhizas

Fractional VA mycorrhiza colonization of plant roots can be high in diverse ecological regions, especially at the lower latitudes of the world (Read, 1991). High colonization has been found in arid, semi-arid, tropical, and temperate pastoral and agricultural systems. Under natural vegetation, the abundance and effectivity of mycorrhizas may be greater on nutrient-poor sites than on nutrient-rich sites (McNaughton and Oesterheld, 1990). In agricultural fields, infection can be decreased following high P and N fertilization.

Mycorrhizal activity is limited by various environmental factors. Low temperatures and also very low soil pH, for example, inhibit the activity of VA mycorrhizal fungi. Below 5°C root colonization can be neglible (Buwalda *et al.*, 1985).

Although environmental factors determine the intensity of mycorrhizal colonization, plant species also differ in their susceptibility to be colonized. Often, in legumes the fractional colonization of roots is higher than in cereals (Jakobsen and Nielsen, 1983; Koide, 1991). Nevertheless, the effect of different crop rotations on spore abundance and infectivity of VA mycorrhizal fungi is mostly rather small, although in general a high mycorrhizal root length density in one season will also support adequate colonization of the following crop (Johnson *et al.*, 1991). Therefore, bare fallow (or growing of non-mycorrhizal crops) can reduce the VA mycorrhiza inoculum potential of a soil (Thompson, 1987).

Soil disturbance prior to planting is also known to reduce plant P uptake and growth (McGonigle *et al.*, 1990). This can at least partly be related to a disruption of the fungal mycelium in soil during disturbance. The effect is most conspicuous in soils which were previously undisturbed for a long period. In practice, this can be important after forest clearing (Jasper *et al.*, 1991). The effect can be pronounced, especially when the soils were under a vegetation dominated by non-VA-mycorrhizal plant species (ectomycorrhizal, ericoid mycorrhizal, or non-mycorrhizal).

Thus, in pasture soils with a higher density of VA mycorrhizal propagules, disturbance does not cause reduced infectivity of VA mycorrhizas (Table 3). In forest or heathland soils, however, where populations of VA mycorrhizal fungi are lower, soil disturbance reduces mycorrhiza infectivity measurably (Table 3). Thus, soils used for surface reclamation after mining often support little mycorrhiza colonization initially (Jasper *et al.*, 1991), and inoculation with VA mycorrhizal fungi may be necessary in this situation to permit vigorous plant growth.

Table 3. Soil characteristics and abundance of VA mycorrhiza under different vegetation types, and effect of disturbing previously undisturbed soil on subsequent VA mycorrhiza formation in four-week-old clover plants in a pot experiment (Jasper *et al.*, 1991)

	Jarrah forest	Heath-land	Annual pasture
VA mycorrhizal plant species (%)	75	55	95
No. of VA mycorrhiza spores (100 g soil)$^{-1}$	19	10	96
Extractable soil P (mg kg^{-1})	4	5	15
Extractable soil N (mg kg^{-1})	13	5	10
VA mycorrhiza colonization (%) of roots of clover plants grown in respective soils:			
Undisturbed soil	28	11	52
Disturbed soil	15	6	50

Agricultural Significance of VA Mycorrhizas

Occurrence of mycorrhizas and results of pot experiments do not provide clear evidence to which extent crop plants in the field depend on their mycorrhizas for nutrient uptake. Fitter (1991) called it paradoxical that most plants are colonized with VA mycorrhizal fungi, yet it is difficult to demonstrate that host plants receive any benefit by this association.

Experimental evidence on the contribution of mycorrhizas to P uptake in the field is contradictory. On the one hand, when early root colonization of maize was reduced in fields to below 15% by rotational or seasonal effects, higher P fertilization was needed to ensure optimal growth (Vivekanandan and Fixen, 1991). On the other hand, intense mycorrhizal colonization sometimes occurs only late in the season while it may be low in the period of rapid P uptake (Figure 4). Also, treatments leading to increased P uptake, such as application of organic matter in the sandy soils of the Sahel, are often not associated with an increase in mycorrhizal colonization (Figure 4).

Thus, there is circumstantial evidence for both greater or lower contributions of mycorrhizas to crop P uptake. Mycorrhizal association between plant and

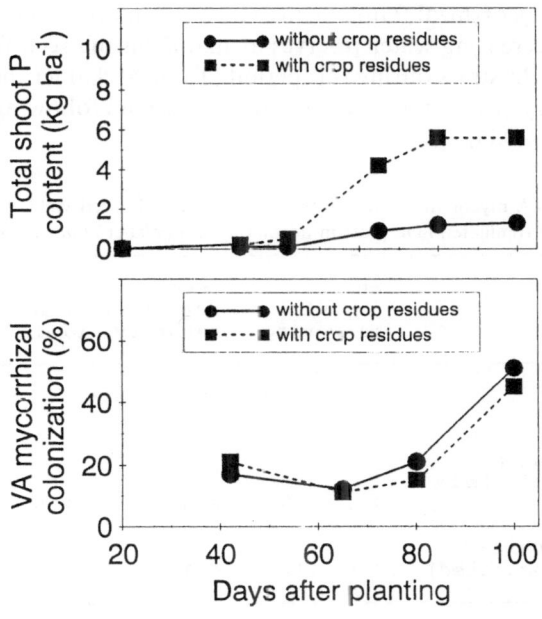

Figure 4. Time-course of P uptake (upper) and fractional mycorrhizal colonization (lower) of a pearl millet crop grown in Niger (West Africa) on an acid sandy soil with or without long-term crop residue (millet straw) application (Hafner *et al.*, unpublished).

fungus may, therefore, at least during large parts of plant life, not function mutualistically (McGonigle and Fitter, 1988).

There is no direct test for the actual mycorrhizal contribution in the field. This is simply because in the field non-mycorrhizal control crops (except non-mycorrhizal plant species) are impossible to obtain experimentally without inducing major microbial and nutritional changes in the soil. Hamel *et al.* (1991) reported a drastic decrease of P uptake by soybean and maize when plants were maintained non-mycorrhizal in a fumigated field. However, other field experiments showed no effect of mycorrhizal inoculation (McGonigle, 1988). Differences in results may be caused by various factors, such as plant species, P status of soil, and infection potential of root pathogens in soil.

At present, there are no suitable research tools and thus there is not sufficient quantitative evidence to determine the role of VA mycorrhizas in agricultural crop production. A last example may illustrate the problems of assessing mycorrhizal contributions to plant growth in the field.

As discussed above, improved plant drought tolerance following mycorrhizal colonization may be a direct mycorrhizal effect on plant water use or related to increased P nutrition. In the field, at least in semi-arid summer-dry environments, however, drought resistance may not be increased by the mycorrhizal association (Weber *et al.*, 1993). Early improvement of shoot growth due to mycorrhizal P uptake may lead to a more rapid consumption of soil water, to decreasing water reserves available during seed filling, and thus enhancing drought stress during this period (Table 4). Under the same climatic conditions, in dry years similar effects can be observed following P fertilization (Keatinge *et al.*, 1985).

Table 4. Effect of VA mycorrhiza inoculation on shoot growth, seed yield and harvest index of chickpeas grown on fumigated field plots in northern Syria (Weber *et al.*, 1993)

	Shoot dry wt. flowering	Shoot dry wt. maturity	Seed yield	Harvest index
	- - - - - (g plant⁻¹) - - - -			(%)
Low VA mycorrhiza infection (fumigated soil)	2.7	6.0	2.5	41
High VA mycorrhiza infection (inoculated)	4.9	6.9	1.9	27

Concluding Remarks

Colonization of roots with VA mycorrhizal fungi strongly increases plant uptake of P. It can also improve uptake of Zn and Cu and can have indirect effects on root/shoot-ratio, plant water use, and resistance to root pathogens. Except after soil sterilization or severe soil disturbance, in most cases field inoculation with VA mycorrhizal fungi does not affect plant growth because field crops are usually sufficiently mycorrhizal. Furthermore, it is not yet possible to consistently replace inefficient fungal strains by introduced efficient strains in the field. At present, there is evidence for and against an important role of VA mycorrhizas in plant growth in the field. A substantial contribution can be predicted when plant growth is limited by P deficiency. When growth factors other than P are limiting yield, a reduced root/shoot ratio in mycorrhizal plants and the fungal carbon consumption may turn the symbiotic association between plant and fungus into a relationship unfavorable for the plant. This may apply especially to legumes where *Rhizobium* nodules and VA mycorrhiza both drain photoassimilates from the host plant and in environments where other soil factors such as water or nutrient supply (other than P) limit plant growth.

References

Ames, R. N., Reid, C. P. P., Porter, L. K. and Cambarcella, C. 1983. *New Phytologist* 95: 381–396.

Augé, R. M. and Stodola, A. J. W. 1990. *New Phytologist* 115: 285–295.

Bethlenfalvay, G. J., Brown, M. S., Ames, R. N. and Thomas, R. S. 1988. *Physiologia Plantarum* 72: 565–571.

Bethlenfalvay, G. J. and Franson, R. L. 1989. *Journal of Plant Nutrition* 12: 953–970.

Bethlenfalvay, G. J., Franson, R. L., Brown, M. S. and Mihara, K. L. 1989. *Physiologia Plantarum* 76: 226–232.

Bethlenfalvay, G. J., Reyes-Solis, M. G., Camel, S. B. and Ferrera-Cerrato, R. 1991. *Physiologia Plantarum* 82: 423–432.

Brown, M. S., Thamsurakul, S. and Bethlenfalvay, G. J. 1988. *Physiologia Plantarum* 74: 159–163.

Buwalda, J. G., Stribley, D. P. and Tinker, P. B. 1985. *Journal of Agricultural Science, Cambridge* 105: 649–657.

Daft, M. J. 1991. *Agriculture, Ecosystems and Environment* 35: 151–169.

Davies Jr., F. T., Potter, J. R. and Linderman, R. G. 1992. *Journal of Plant Physiology* 139: 289–294.

Dinkelaker, B., Römheld, V. and Marschner, H. 1989. *Plant, Cell and Environment* 12: 285–292.

Eason, W. R., Newman, E. I. and Chuba, P. N. 1991. *Plant and Soil* 137: 267–274.

Faber, B., Zasoski, R. J., Munns, D. N. and Shackel, K. 1991. *Canadian Journal of Botany* 69: 87–94.

Fitter, A. H. 1985. *New Phytologist* 99: 257–265.

Fitter, A. H. 1988. *Journal of Experimental Botany* 39: 595–603.

Fitter, A. H. 1991. *Experientia* 47: 350–355.

George, E., Häussler, K., Vetterlein, D., Gorgus, E. and Marschner, H. 1992. *Canadian Journal of Botany* 70: 2130–2137.

Gnekow, M. A. and Marschner, H. 1989. *Plant and Soil* 114: 91–98.

Grace, C. and Stribley, D. P. 1991. *Mycological Research* 95: 1160–1162.

Hamel, C., Furlan, V. and Smith, D. L. 1991. *Plant and Soil* 133: 177–185.

Hamel, C. and Smith, D. L. 1991. *Soil Biology and Biochemistry* 23: 661–665.

Ibrahim, M. A., Campbell, W. F., Rupp, L. A. and Allen, E. B. 1990. *Arid Soil Research and Rehabilitation* 4: 99–107.

Jakobsen, I. and Nielsen, N. E. 1983. *New Phytologist* 93: 401–413.

Jakobsen, I. and Rosendahl, L. 1990. *New Phytologist* 115: 77–83.

Jasper, D. A., Abbott, L. K. and Robson, A. D 1991. *New Phytologist* 118: 471–476.

Johnson, N. C., Pfleger, F. L., Crookston, R. K., Simmons, S. R. and Copeland, P. L. 1991. *New Phytologist* 117: 657–663.

Keatinge, J. D. H., Neate, P. J. H. and Shepherd, K. D. 1985. *Experimental Agriculture* 21: 209–222.

Koide, R. T. 1991. *New Phytologist* 117: 365–386.

Kothari, S. K., Marschner, H. and George, E. 1990. *New Phytologist* 116: 303–311.

Kothari, S. K., Marschner, H. and Römheld, V. 1991. *New Phytologist* 117: 649–655.

Kucey, R. M. N. and Paul, E. A. 1981. *Science* 213: 473–474.

Li, X. – L., George, E. and Marschner, H. 1991a. *Plant and Soil* 136: 41–48.

Li, X. – L., Marschner, H. and George, E. 1991b. *Plant and Soil* 136: 49–57.

Linderman, R. G. 1988. *Phytopathology* 78: 366–371.

Mackay, A. D. and Barber, S. A. 1985. *Agronomy Journal* 77: 519–523.

McGonigle, T. P. 1988. *Functional Ecology* 2: 473–478.

McGonigle, T. P., Evans, D. G. and Miller, M. H. 1990. *New Phytologist* 116: 629–636.

McGonigle, T. P. and Fitter, A. H. 1988. *New Phytologist* 108: 59–65.

McNaughton, S. J. and Oesterheld, M. 1990. *Oikos* 59: 92–96.

Pang, P. C. and Paul, E. A. 1980. *Canadian Journal of Soil Science* 60: 241–250.

Paul, E. A., Harris, D. and Fredeen, A. 1985. In: *Proceedings of the 6th North American Conference on Mycorrhizae*, pp. 165–169 (ed. R. Molina). Corvallis, OR, USA: Forest Research Laboratory, Oregon State University, Corvallis, Oregon.

Piccini, D., Ocampo, J. A. and Bedmar, E. J. 1988. *Biology and Fertility of Soils* 6: 65–67.

Raju, P. S., Clark, R. B., Ellis, J. R., Duncan, R. R. and Maranville, J. W. 1990. *Plant and Soil* 124: 199–204.

Read, D. J. 1991. *Experientia* 47: 376–391.

Schwab, S. M., Menge, J. A. and Tinker, P. B. 1991. *New Phytologist* 117: 387–398.

Sieverding, E. 1991. *Vesicular-arbuscular mycorrhiza management in tropical agrosystems.* Schriftenreihe der GTZ, no. 224. Eschborn, Germany: Technical Cooperation.

Simpson, D. and Daft, M. J. 1991. *Agriculture, Ecosystems and Environment* 35: 47–54.

Smith, S. E. and Gianinazzi-Pearson, V. 1988. *Annual Review of Plant Physiology and Plant Molecular Biology* 39: 221–244.

Thompson, J. P. 1987. *Australian Journal of Agricultural Research* 38: 847–867.

van Kessel, C., Singleton, P. W. and Hoben, H. J. 1985. *Plant Physiology* 79: 562–563.

Vivekanandan, M. and Fixen, P. E. 1991. *Soil Science Society of America Journal* 55: 136–140.

Weber, E., George, E., Beck, D. P., Saxena, M. C. and Marschner, H. 1992. *Experimental Agriculture* 28: 433–442.

Weber, E., Saxena, M. C., George, E. and Marschner, H. 1993. *Field Crops Research* 32: 115–128.

Farmers' constraints and on-farm research

Experience with Ascochyta blight of chickpea in the United States

W.J. KAISER[1], F.J. MUEHLBAUER[2] and R.M. HANNAN[1]

[1] U. S. Department of Agriculture, Agricultural Research Service, Regional Plant Introduction Station, 59 Johnson Hall, Washington State University, Pullman, Washington 99164–6402, USA, and
[2] U. S. Department of Agriculture, Agricultural Research Service, 303W Johnson Hall, Washington State University, Pullman, Washington 99164–6434, USA

Abstract

Ascochyta blight of chickpea (*Cicer arietinum*) incited by *Ascochyta rabiei* was first observed in the United States in chickpea germplasm evaluation trials at Pullman, Washington, in 1983. In 1984, Ascochyta blight was found in over 50% of the commercial fields of "UC-5" chickpea in the Palouse region of eastern Washington and northern Idaho. In 1986, *Didymella rabiei* (syn. *Mycosphaerella rabiei*), the teleomorph (sexual stage) of *A. rabiei*, was discovered on overwintered infested chickpea debris from a field near Genesee, Idaho. The fungus, which is heterothallic, requires the presence of two compatible mating types for the teleomorph to form. Ascospores from the teleomorph are important in the long distance spread of the pathogen. In 1987, over 4,500 hectares of UC-5 and "Surutato 77" chickpea in the Palouse region were devastated by Ascochyta blight. Many kabuli and desi chickpea lines from different countries have been screened for resistance to pathotypes of the fungus that occur in the Palouse region. A chickpea breeding program is incorporating blight resistance into large-seeded kabuli cultivars. An integrated program to control Ascochyta blight is needed in areas where the disease occurs in the US Pacific Northwest if chickpea is to remain a viable crop. The program will need to include the use of blight-free seed, seed treatment fungicides, crop rotation, management of infested crop debris, and resistant cultivars.

Introduction

In the United States, chickpea (garbanzo) (*Cicer arietinum* L.) is grown commercially in California and in the Palouse region of eastern Washington and northern Idaho. In the Palouse region, chickpea is a relatively new alternative cash crop that is grown in rotation with winter wheat, barley, pea, and lentil. The first commercial sowings of large-seeded kabuli chickpea cultivars were made in 1981 on <200 ha. By 1987, over 4,700 ha of kabuli chickpea,

F.J. Muehlbauer and W.J. Kaiser (eds.), Expanding the Production and Use of Cool Season Food Legumes, 849–858.

principally of the cultivars UC-5 and Surutato 77, were being grown commercially in the region (Table 1).

Table 1. Chickpea production from 1983 to 1991 in the Palouse region of eastern Washington and northern Idaho, United States

Year	Area	Mean Yield	Total production
	-ha-	--kg ha^{-1}--	-- kg --
1983	1229	1096	1,346,269
1984	677	1484	1,004,631
1985	2225	1246	2,772,035
1986	3692	1175	4,335,372
1987	4725	581	2,742,594
1988	1817	933	1,056,818
1989	1399	1112	1,554,858
1990	2934	632	1,853,369
1991	1798	1096	1,348,050

Several diseases adversely affect the cultivation of chickpea in the United States. In California, Fusarium wilt, caused by *Fusarium oxysporum* Schlechtend.:Fr. f. sp. *ciceris* (Padwick) Matuo and K. Sato, is the principal disease (Kaiser, 1981). In the Pacific Northwest, seed rot and preemergence damping-off caused by *Pythium ultimum* Trow (Kaiser, 1982; Kaiser and Hannan, 1983) and Ascochyta blight incited by *Ascochyta rabiei* (Pass.) Labrousse [syn. *Phoma rabiei* (Pass.) Khune and J.N. Kapoor] are of major concern (Kaiser and Hannan, 1988; Kaiser and Muehlbauer, 1984, 1988). This paper will address the history and problems caused by Ascochyta blight in the cultivation of chickpea in the Palouse region of the Pacific Northwest.

The Disease

The first report of Ascochyta blight of chickpea in North America was by Morrall and McKenzie (1974) in Canada. The blight disease was first observed in the United States in chickpea germplasm evaluation trials at Pullman, Washington, in 1983 (Kaiser and Muehlbauer, 1984). The pathogen apparently had been imported on seed, but the actual source is unknown. Environmental conditions (cool, wet weather) during June and July 1983 favored spread and development of the disease. Blight affected 77 of 125 lines in eight yield and adaptation trials at Pullman. In 1984, blight was observed in 23 of 30 commercial chickpea plantings in northern Idaho (Derie *et al.*, 1985). In some fields, Ascochyta blight caused a significant reduction in yield. Also in 1984, blight was widespread and damaging in chickpea evaluation trials at Pullman, Washington, but differences were observed in the blight reaction among these lines. The reaction to blight ranged from moderately resistant to highly susceptible (Table 2). None of the lines were immune to infection. The highest yielding lines generally had a higher level of resistance to *A. rabiei*. Two hundred chickpea seeds harvested from each

Table 2. Effect of *Ascochyta rabiei* on incidence of foliar disease, seed infection, and yield of 17 chickpea lines in a dryland adaptation experiment at Pullman, Washington in 1984[a]

Cultivar	Type of seed[b]	Ascochyta blight score[c]	Seed infection[d]	Wt. 100 seeds	Yield
			-%-	-g-	-kg ha⁻¹-
ILC 260	K	3	9.5	38.7	3100
ILC 591	K	3	20.5	36.9	2798
C-235 (PI 462174)	D	2	0.5	13.3	2795
UC 5 (PI 462203)	K	4	46.0	52.3	2384
Garnet (PI 273879)	D	2	11.5	16.9	2359
Lyons (PI 512300)	K	2	9.5	37.1	2329
ILC 171 (PI 462023)	K	4	25.5	38.2	2267
ILC 294 (PI 360207)	K	3	26.0	32.2	2221
ILC 35	K	3	33.5	37.7	2189
ILC 517	K	4	25.0	27.2	1992
Aztec (PI 512301)	D	4	19.0	15.7	1931
Surutato 77 (PI 468948)	K	4	19.5	53.6	1880
ILC 1929	K	5	40.5	32.0	1802
Mammoth (PI 458872)	K	4	27.0	51.5	1576
ICCC 4	D	5	22.0	14.2	1538
ILC 1102	K	5	33.0	23.0	1386
Spanish White (PI 503005)	K	5	58.5	51.2	1194

[a] Trial was planted on 8 May 1984 and harvested on 22 August 1984.

[b] D: Desi lines usually have small angular shaped, dark-colored seeds with thick seedcoats; K: Kabuli lines generally have larger, rounded, light-colored seeds with thin seedcoats.

[c] Ascochyta blight score: (based on foliar lesions): 1 = resistant (no visible lesions on the foliage), 2 = moderately resistant, 3 = tolerant, 4 = susceptible, 5 = highly susceptible (lesions on stems and leaves extensive, many plants killed)

[d] Two hundred seeds were surface disinfested in 0.25% NaOCl for 5 min and cultured on potato-dextrose agar containing antibiotics.

line were tested for seed infection by the blight pathogen by plating surface disinfested seeds on an agar medium. Seed transmission of *A. rabiei* ranged from 0.5% to 58.5%. The desi lines generally had a lower incidence of seed infection.

Life Cycle

The blight pathogen has both anamorphic (asexual) and telemorphic (sexual) stages and two distinct spore types (Figure 1). The anamorph produces pycnidiospores (conidia) in fruiting bodies (pycnidia) that develop in lesions on foliar tissues during the growing season and on infested chickpea debris after harvest. Rain-splash dispersal of pycnidiospores usually occurs over short distances within a field. The teleomorph, *Didymella rabiei* (Kovachevski) v. Arx (syn. *Mycosphaerella rabiei* Kovachevski), produces another spore type (ascospores) on infested plant debris that overwinters on the soil surface. The teleomorph has been reported from several countries (Kaiser and Hannan, 1987).

In 1986, the teleomorph was found on overwintered infested chickpea debris near Genesee, Idaho (Kaiser and Hannan, 1987). The telemorph plays an important role in the epidemiology of the disease in the Palouse region. The

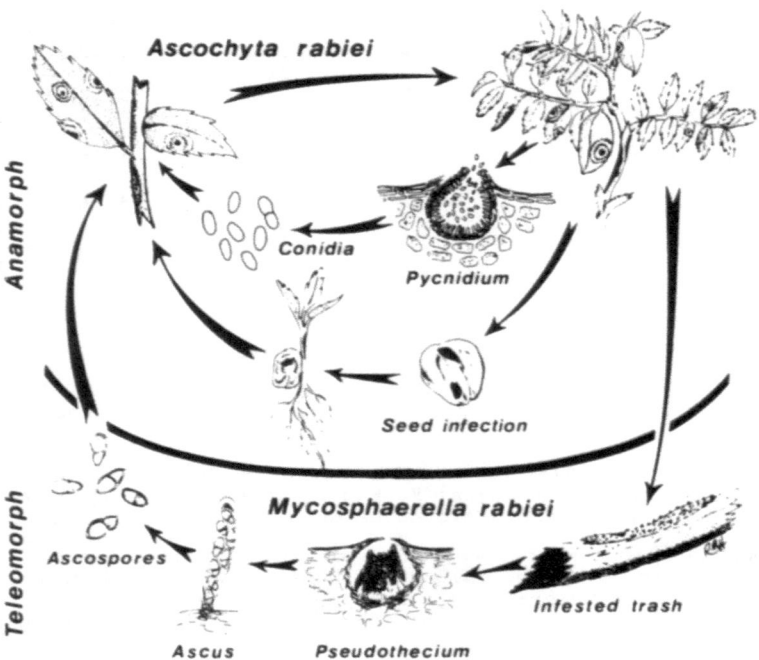

Figure 1. Disease cycle of Ascochyta blight of chickpea caused by *Ascochyta rabiei* in the Pacific Northwest of the United States. Pycnidia can also occur on overwintered chickpea debris.

teleomorph only develops on infested chickpea debris during the fall and winter. In the spring (March to June) when infested debris is wetted by rain, ascospores are forcibly discharged from the fruiting bodies (pseudothecia) and carried by the wind where they can infect susceptible chickpea cultivars in fields located several kilometers downwind. In the Palouse, the teleomorph plays an important role in the long distance spread of the pathogen and in initiating primary infection foci early in the spring. It also provides the possibility of development of new pathotypes of the blight fungus through sexual recombination. The blight pathogen is heterothallic, requiring two compatible isolates (mating types) of the fungus for the telemorph to develop. Compatible isolates of *A. rabiei* occur throughout the Palouse region and the teleomorph is present in most areas surveyed to date. The fungus spreads rapidly from the primary infection foci, particularly during cool, wet weather, by pycnidiospores of the anamorphic stage which also survive on infested chickpea debris from one growing season to the next. Pycnidiospores produced on overwintered debris can provide inoculum for infection of a susceptible chickpea crop in the spring.

Infected Seed

Infected chickpea seed is an efficient method of transporting *A. rabiei* through time (carry over from one season to the next) and space (spread from one area to another). Seedborne inoculum is most likely the means of transporting the blight pathogen into new regions or countries. Infected chickpea seed appears to have been responsible for the introduction of Ascochyta blight into Australia (Cother, 1977), Canada (Morrall and McKenzie, 1974; Tu and Hall, 1984), southwestern Iran (Kaiser, 1972), and the United States (Kaiser and Muehlbauer, 1984).

Seed infected by *A. rabiei* ensures a random distribution of the fungus in a field and provides primary infection foci from which the pathogen can spread. It has been our experience that the blight pathogen can spread very rapidly in a susceptible chickpea crop, even from a few infected seeds, when the weather conditions are cool and wet.

From 1984 to 1987, the incidence of *A. rabiei* in commercial seed lots from the Palouse region was checked. Seed infection in different seed lots varied between years and ranged from 0.5% to 31% (Table 3). Not all seeds showed visible symptoms or signs of infection.

Blight Epidemic

In 1987, much of the chickpea crop in the Palouse region was devastated by Ascochyta blight (Figure 2). Isolated fields were infected by the blight pathogen at distances of > 8 km from the nearest chickpea planting. Many blight-

Table 3. Incidence of *Ascochyta rabiei* in commercial chickpea seed lots from the Palouse region of eastern Washington and northern Idaho, United States

Year	No. seed lots[a]	Seed lots infected with *A. rabiei*		Range in percent seed infection
		-no.-	- %-	
1984	50	4	8	0.5 - 1.5
1985	23	3	13	0.5
1986	16	3	19	1 - 2
1987	38	34	89	1 - 31

[a] The number of seeds plated on potato-dextrose agar per sample in 1984, 1985, 1986 and 1987 were 200-400, 200, 100 and 100, respectively. Seeds were surface disinfested in 0.25% NaOCl for 5 min.

Figure 2. Ascochyta blight of chickpea in a field near Genesee, Idaho, United States. The blight pathogen has spread from primary infection foci (light colored areas) to adjacent plants. Yields in this planting were reduced by > 80%. Photo taken 12 weeks after sowing.

infected fields were located in areas where chickpeas had never been grown. Much of the UC-5 and Surutato 77 chickpea seed used to plant these fields appeared to be free of *A. rabiei* and most of the seed had been treated with

Captan or a combination of Captan and benomyl. These isolated fields appeared to have been infected by airborne ascospores of *D. rabiei.* Once ascopores initiate infection on the foliage of chickpeas, blight spreads quickly to other plants within a field by rain-splashed pycnidiospores. The intermittent periods of cool, wet weather in the Palouse region during June and July 1987 provided ideal conditions for blight development. Symptoms of blight were first observed on the foliage of young chickpea plants in early to mid-May. Blight spread very quickly in some fields and completely devastated the crop by the end of July. Many of the severely damaged chickpea fields were plowed under before harvest resulting in total crop loss.

Chickpea yields in the Palouse region averaged 581 kg on 4,725 ha in 1987, compared with 1175 kg on 3692 ha in 1986 (Table 1). The yield loss was due primarily to Ascochyta blight and resulted in financial losses of over US$ 1 million.

Control

In order to successfully cultivate chickpea in the Palouse region, it will be necessary to implement an integrated program to control Ascochyta blight which will include:
1. Planting of seed free of the Ascochyta blight pathogen.
2. Treatment of seeds for planting with fungicides, such as benomyl or thiabendazole (Kaiser and Hannan, 1988).
3. Practice a three-year crop rotation.
4. Dispose of infested chickpea debris by plowing or other means.
5. Identify blight resistant germplasm and develop resistant cultivars.
6. Encourage the use of blight resistant cultivars by growers when available.

Disease Resistance

In 1987, 330 chickpea lines were screened for resistance to Ascochyta blight under field conditions at Genesee, Idaho. Disease pressure was very high and weather conditions favored infection and spread of the fungus. In 1988, the screening trials were transferred to Pullman, Washington where the plots could be sprinkler irrigated to ensure good disease development and spread, especially during the podding stage. At both sites, sources of resistance were found in several kabuli and desi lines (Figure 3). Unfortunately, much of the resistant germplasm was late maturing and had only medium seed size (30 to 50 g 100^{-1} seeds). However, this germplasm is being used in a breeding program to develop large-seeded (> 55 g 100^{-1} seeds), blight resistant kabuli cultivars for cultivation in the Palouse region. Chickpea lines that have shown lower blight ratings for at least three years are listed in Table 4.

Figure 3. Reaction of chickpea lines to Ascochyta blight in a screening trial near Genesee, Idaho, United States. Most lines were killed within 10 to 12 weeks of sowing. Sources of resistance to blight were found in a few kabuli and desi lines.

Breeding Program

Resistance to the pathotypes of *A. rabiei* present in the Palouse region was found in germplasm lines obtained from ICARDA, Syria. Germplasm lines with good resistance to blight usually were late maturing in the Palouse region and had mostly medium seed size. In order to combine the early maturity and large seed size of the popularly grown Surutato 77 and Tammany cultivars with resistance to blight, crosses were made in 1987 and 1988 (Kusmenoglu *et al.*, 1989). The major objectives of the breeding program were to incorporate blight resistance into early-maturing, large-seeded cultivars; determine the adaptation and yield of blight resistant selections; and release of blight resistant cultivars to growers. Progenies from the crosses were advanced in the greenhouse and screened in the field the following season for blight resistance. Single plant progenies were planted in a single row plots in the blight-screening nursery. A blight susceptible check was planted every fifth row to provide fungal inoculum and for comparison. Supplemental sprinkler irrigation was provided to ensure good disease pressure and spread and to obtain pod infection later in the growth

Table 4. Chickpea germplasm lines with average Ascochyta blight ratings of 5.5 or less for 3 or more years in field screening trials at Genesee, Idaho or Pullman, Washington[a]

Chickpea line	Average disease severity rating[b]				
	1987	1988	1989	1990	1991
FLIP 83-046	5.0	4.0	3.0	4.0	
FLIP 83-047	5.0	4.0	3.0	4.0	
FLIP 83-048	5.0	4.0	3.0	3.0	
FLIP 83-060	5.0	4.0	3.0	3.0	
FLIP 84-012	4.0	4.0	3.0	3.0	
FLIP 85-013	5.0	4.0	2.0	5.0	
FLIP 85-057	5.5	5.0	3.0	4.0	
FLIP 85-058	5.5	5.0	3.0	4.0	
FLIP 85-061	5.5	5.5	4.0	5.0	
FLIP 86-011		5.0	5.5	5.5	
FLIP 86-012		5.0	5.0	5.0	
FLIP 86-013		5.0	5.0	5.0	
FLIP 86-019		4.0	4.0	4.0	
FLIP 86-058		5.5	4.0	5.0	
FLIP 86-060		4.0	3.0	5.0	
FLIP 86-070		4.0	3.0	5.5	
FLIP 86-077		5.0	5.0	5.5	
ICCX 04			4.0	5.0	4.0
ILC 0072	5.0	4.0	3.0	4.0	
ILC 0191	5.0	5.0	3.0	5.5	
ILC 0194	5.5	4.0	4.0	4.0	
ILC 0195	5.5	5.0	3.0	5.0	
ILC 0200	5.0	5.0	3.0	3.0	
ILC 0201	5.0	4.0	3.0	4.0	
ILC 0202	5.0	4.0	2.0	2.0	
ILC 2380	5.0	4.0	2.0	3.0	
ILC 3279 (PI 471915)	5.0	5.0	2.0	3.0	
PI 343019		5.0	2.5	5.5	
PI 343021		5.0	3.0	5.0	
PI 374079	5.5	5.5	4.0	5.0	
PI 451653		5.0	4.0	5.5	
PI 458869	5.5	5.0	3.0	5.0	
PI 471915		4.0	2.0	3.0	

[a] In 1987, the trial was located at S. Evans' farm, Genesee, Idaho. In 1988 to 1991, the trial was located at the Washington State University Spillman Agronomy Farm, Pullman, Washington.
[b] Average disease severity rating based on two replications. Rating scale = 1 to 10; 1 = no disease, 10 = all plants in plot dead.

cycle. Twelve promising blight resistant selections were chosen for winter increase. These selections have large seed size and are earlier maturing than the original blight resistant parents. The resistance, at least to the pathotypes of *A. rabiei* present in the Palouse, appears to be controlled by two recessive genes acting additively (Kusmenoglu *et al.*, 1989), and are not linked to the genes controlling seed size and maturity.

The uniform infection of susceptible chickpea germplasm by the Ascochyta blight pathogen throughout the screening trial was a major factor in the rapid development of promising blight resistant lines. Evaluations in 1992 will include adaptation yield trials in the Palouse region and additional evaluations for blight resistance. Quality evaluations for seed size, shape, color, and cookability are also planned.

References

Cother, E. J. 1977. *Plant Disease Reporter* 61: 736–740.
Derie, M. L., Bowden, R. L., Kephart, K. D. and Kaiser, W. J. 1985. *Plant Disease* 69: 268.
Kaiser, W. J. 1982. In: *Proceedings of the Palouse Symposium on Dry Peas, Lentils, and Chickpeas*, pp. 131–145. Moscow, ID.
Kaiser, W. J. 1981. *Economic Botany* 35: 300–320.
Kaiser, W. J. 1972. *FAO Plant Protection Bulletin* 20: 73–79.
Kaiser, W. J. and Hannan, R. M. 1983. *Plant Disease* 67: 77–81.
Kaiser, W. J. and Hannan, R. M. 1987. *Plant Disease* 71: 192.
Kaiser, W. J. and Hannan, R. M. 1988. *Seed Science and Technology* 16: 625–637.
Kaiser, W. J. and Muehlbauer, F. J. 1988. *International Chickpea Newsletter* 18: 16–17.
Kaiser, W. J. and Muehlbauer, F. J. 1984. *Phytopathology* 74: 1139.
Kusmenoglu, I., Muehlbauer, F. J. and Kaiser, W. J. 1989. *Agronomy Abstracts* p. 89.
Morrall, R. A. A. and McKenzie, D. L. 1974. *Plant Disease Reporter* 58: 342–345.
Tu, J. C. and Hall, R. J. 1984. *Phytopathology* 74: 826.

Lygus bug on lentil in the United States

R.J. SUMMERFIELD[1], R.W. SHORT[2] and F.J. MUEHLBAUER[2]
[1] *University of Reading, Department of Agriculture, Plant Environment Laboratory, Cutbush Lane, Shinfield, Reading RG2 9AD, UK, and*
[2] *U. S. Department of Agriculture, Agricultural Research Service, 303 W Johnson Hall, Washington State University, Pullman, WA 99164-6434, USA*

Abstract

A multidisciplinary, collaborative research effort generated complementary data lending substantial support for the hypothesis that: "Feeding of *Lygus* spp. on immature pods can lead to reductions in quality and losses in the quantity of seeds harvested from crops of lentil (*Lens culinaris*)". Knowledge of research findings on food legumes other than lentil was an important contributory factor in formulating the hypothesis tested. Use was made of the experience and expertise of: farmers, industry administrators, extension specialists, seed graders, food scientists, plant breeders, entomologists, pathologists, botanists, chemists, and physiologists. The element of "novelty" and perceptions of prospects for "success" were important factors in the collaborative research effort. Clearly, specialists cooperating in research directed towards carefully formulated and agreed objectives can make rapid and important advances towards the solving of practical problems in food legume crop production.

Chalky Spot Syndrome (CSS)

Physical deformations combined with localized blemishes and/or a more general discoloration of the testa are sometimes evident in many of the lentil (*Lens culinaris* Medik.) crops harvested in the Palouse (i.e., in eastern Washington and northern Idaho). Farmers have adopted the colloquialism "Chalky Spot Syndrome (CSS)" to describe these symptoms (Plate 1). Lots with CSS on more than 3.5% of seeds are downgraded and so their seed market value is lost. Unknowingly to farmers, seed yields may have also been lessened prior to harvest; CSS may be, in fact, the most obvious manifestation of a more general malaise in particular cropping circumstances with lentil. A recent Current Information Bulletin (O'Keefe *et al.*, 1991) describes the losses due to CSS as follows:

> Crop losses due to chalky spot have fluctuated among production regions and from year to year. In some years, such as in 1982, damage in northern

F.J. Muehlbauer and W.J. Kaiser (eds.), Expanding the Production and Use of Cool Season Food Legumes, 859–876.
© 1994 *Kluwer Academic Publishers.*

Idaho and eastern Washington has been negligible. The more typical situation occurred in 1979 when 17 percent of harvested fields produced sample grade thresher-run lentils. In 1980, more than 30 percent of harvested fields near the Clearwater and Snake River breaks produced sample grade lentils, while fewer than 10 percent of fields in the drier, more westerly areas of the region produced sample grade lentils. A disaster struck in 1983 when more than 51 percent of fields harvested in northern Idaho and eastern

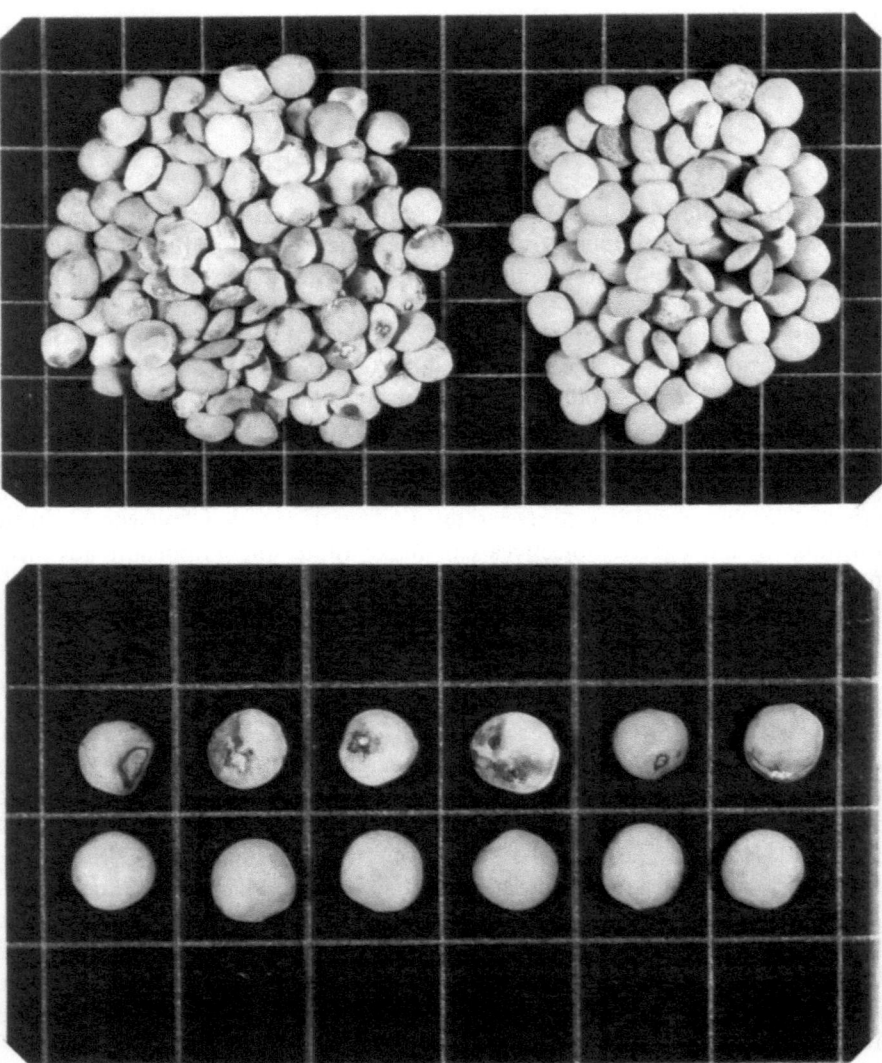

Plate 1. Seeds of lentil, Chilean, damaged (left or above) or not (right or below) by CSS (lines are a 1 cm² grid).

Washington produced sample grade lentils, and some individual fields exceeded 45 percent chalky-spot-damaged seed.

This contribution briefly summarizes a multidisciplinary, collaborative research effort targeted at CSS and which is described in detail elsewhere (Summerfield *et al.*, 1982). At the outset, it is important to remember that *based on local experiences with familiar crops*, farmers differed widely in their opinions as to the most probable cause of CSS; parochial theories apportioned the blame to weather which was either too hot or too cold, or too wet or too dry at one particular growth stage or another. Soils too heavy or too light in texture and various "nutrient disorders" were also implicated. Some producers have insisted that more seeds are damaged if the lentils are harvested as a standing crop at maturity rather than swathed when green.

Synthesis of Information: Pointers for Research

Superficially, CSS is similar to the damage which results from infestations of *L. hesperus* (Knight), *L. elisus* (Van duzee), and *L. lineolaris* (Palisot de Beauvois) on immature fruits of other food legumes grown not only in North America but also in Europe and elsewhere. In lima bean (*Phaseolus lunatus* L.), common bean (*P. vulgaris* L.), and cowpea (*Vigna unguiculata* [L.] Walp.), for example, the consequences of infestation by *Lygus* spp. are similar; damaged seeds of all sizes are misshapen, pitted, discolored, and often with cracked testa or even shrivelled completely (e.g., Elmore, 1955). The symptoms also resemble those which are typical of stink bug (i.e., *Euschistus* spp. and *Acrosternum* spp.) damage to seeds of soybean (*Glycine max* [L.] Merr.) (e.g., Miner, 1966). These insects have in common mouthparts capable of piercing and sucking; they are regarded primarily as pests of meristematic tissues (e.g., stem apices) and especially of young reproductive structures (Gupta *et al.*, 1980). When the topical program of research on lentil was instigated, there were no reports to suggest that *Lygus* spp. were a major or indeed even a minor pest of lentil – either in the USA or worldwide (e.g., Hariri, 1981).

Parochially, CSS in Palouse lentil crops is not only more common in some years than in others (O'Keefe *et al.*, 1991), it is also more prevalent in certain areas (Figure 1). Lentils cropped to the east and southeast of Moscow, Idaho have often been affected whereas those grown in the region of Garfield and Palouse, Washington are seldom damaged. These two general areas of production are only 35 to 75 km apart (Figure 1); they do not experience markedly different climates or weather patterns, nor is the soil appreciably different in physical or chemical composition. Agronomically, however, farming systems differ in the fact that rape crops (*Brassica napus* L.) are more commonly grown east rather than north of Moscow. This, we believe, may be an important difference. *Lygus* bugs are known to be especially attracted to the color yellow (Landis and Fox, 1972); Cruciferous weeds along roadsides and ditches, in orchards, and on fallow and wasteland are major spring hosts of overwintering *Lygus* bug females (Fye, 1980); and rape has been cited (along with carrots, radish, mustard, and collards) as a cultivated host to which *Lygus* will migrate from adjacent weed flora during the late spring and summer in southwestern Washington (Hagel, 1978; and see Auld *et al.*, 1980).

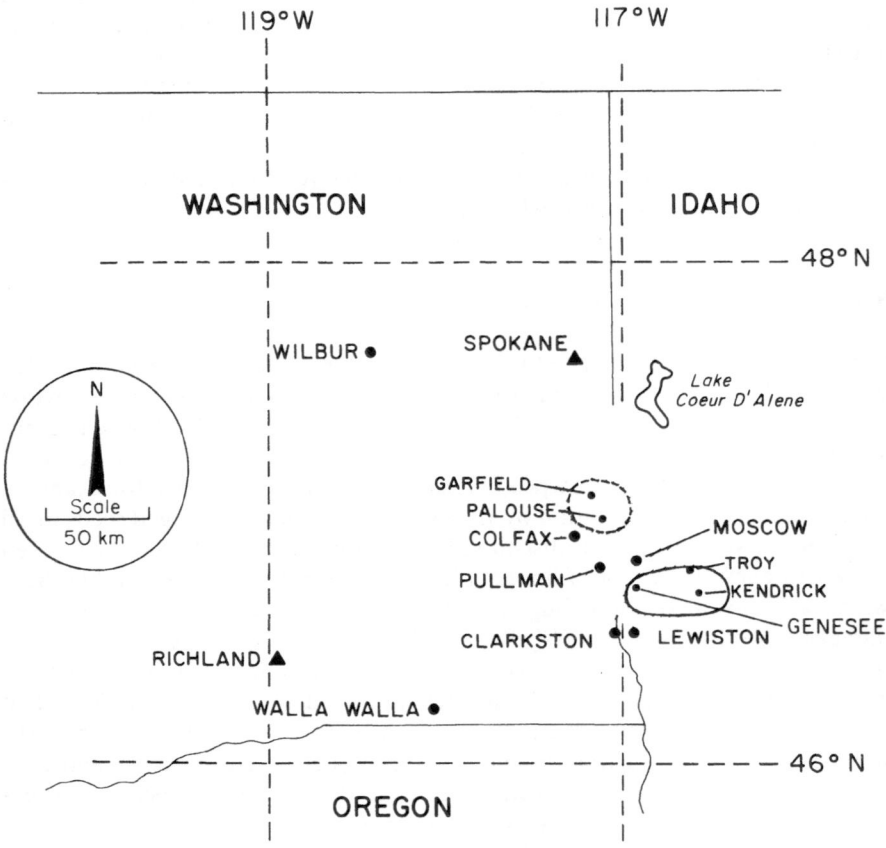

Figure 1. Location of lentil production in the USA (hatched) and areas where crops are often damaged to a greater or lesser extent by CSS (–) or are seldom affected by the problem (– –).

Collectively, this circumstantial evidence suggests that there may be good ecological reasons why populations of *Lygus* have developed and persisted in the area east of Moscow, and that the sequence of growth and death of annual Cruciferous weeds, flowering in rape and then flowering in lentils may provide a prolonged source of nourishment to which the pest is particularly attracted. Then again, lentil fields in the cooler, wetter areas east and southeast of Moscow are more commonly harvested as a standing mature crop (as opposed to swathing), and forage crops (e.g., clovers and alfalfa) are grown more frequently in this area than they are in the region of Garfield and Palouse. Thus, not only are there other crops to which *Lygus* bugs are particularly attracted (and from whence they could migrate to lentils) more prevalent in the area around Moscow, but lentils are also sown later and are present in fields for longer into the calendar year than they are further north.

Caging *Lygus* Bugs Onto Pot-Grown Plants

Replicate plants of three lentil genotypes of diverse origin and habit were grown in a glasshouse using plant culture and husbandry techniques which are described in detail elsewhere (Summerfield and Muehlbauer, 1982). The area available to each pot-grown plant was comparable to that in field crops seeded at reasonable densities (a seeding rate of 67 kg ha^{-1} with 15,400 seeds kg^{-1} to give 1×10^6 plants ha^{-1}; Muehlbauer, 1973).

After 68 d from sowing, when at least some replicates of all three cultivars had initiated flowers [in "Chilean" the first flower had opened (corolla color visible) whereas in "Precoz" and accession LC800028 several flowers had opened fully], 15 replicates of each cultivar were placed into large insect-proof cages. Two cages with wooden frames ($1.02 \times 1.53 \times 1.08$ m, to give a floor area of 1.56 m^2) and nylon mesh (0.5 mm^2 open area) were used. Adults of *L. hesperus* and *L. elisus* (about 90% and 10%, respectively) were introduced into one cage to create an original infestation of 0.63 bugs per plant which, based on experience with other hosts, might reasonably be expected to be injurious if, indeed, lentils proved to be susceptible to the pest (R. E. Fye, personal communication). Plants in the control cage remained insect-free throughout.

All plants were allowed to fruit, mature, senesce, and die. Pods were then collected and counted, and records were taken of the numbers obviously misshapen or malformed and of the respective number and weight of undamaged and damaged seeds.

Adult bugs were frequently observed to feed on the caged plants and in this forced feeding situation they were especially attracted to immature reproductive structures. We also interpret the presence of localized droplets and stains on green pod walls to be evidence of feeding activity compatible with multiple-stylet mouthparts (Plate 2).

That the adult bugs had reproduced during the caging period is beyond doubt; all stages of the *Lygus* bug life cycle were identified at one time or another (G. R. Pesho, personal communication). Thus, we cannot be certain about the proportions of any damage to the lentil plants which had been caused by adults or immature bugs or, indeed, whether male or female adults are likely to have been relatively more destructive. Furthermore, since the original population intentionally comprised both *L. hesperus* and *L. elisus*, there may have been a species component involved too. Data on economic yield and selected components of yield for control plants and those subject to *Lygus* infestation are summarized in Table 1.

None of the seeds harvested from plants in the insect-free cage were damaged

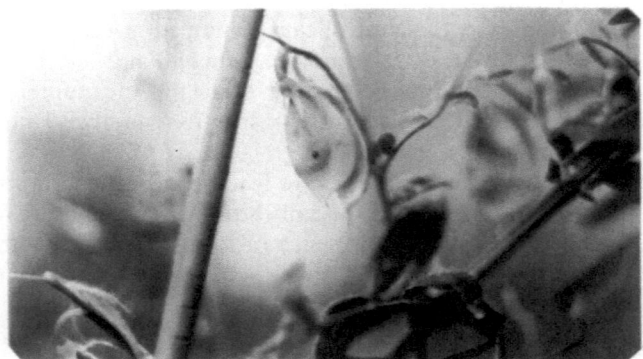

Plate 2. (*Top left*) Scanning electron micrograph of adult *Lygus* bug (× 40) showing the prominent multiple-stylet mouthparts. (*Top right and bottom*) Localized exudates from punctured immature lentil pods.

whereas an average of 26% of those from the plants caged with *Lygus* had "classical" CSS (Table 1). Chilean, which produced the most nodes and flowers (not recorded quantitatively), was seemingly twice as susceptible to damage than the other cultivars (CSS on 40.2% of seeds compared with an average of

Table 1. Effects on seed yield of caging *Lygus* bugs onto pot-grown lentil plants in a glasshouse (mean values of 15 replicates with SE values in parentheses)

Attribute of seed yield measured or derived	Lentil genotype					
	Chilean	Precoz	LC800028	Chilean	Precoz	LC800028
	Lygus-free control			Lygus bugs present		
No. pods plant⁻¹	102.8 (5.8)	40.6 (1.7)	62.4 (3.1)	79.6 (5.3)	40.9 (2.5)	59.9 (3.5)
No. shrivelled pods plant⁻¹	24.3 (3.8)	3.7 (0.8)	4.5 (0.9)	29.4 (4.6)	9.9 (1.1)	12.1 (2.1)
Shrivelled pods (%) plant⁻¹	23.6	9.1	7.2	36.9	24.2	20.2
Increased % CSS due to lygus bug	-	-	-	13.3	15.1	13.0
Estimated pod abortion (%) due to lygus bug	-	-	-	22.6	0	4.0
No. seeds plant⁻¹	96.9 (4.0)	49.7 (2.2)	91.3 (3.7)	71.3 (4.4)	40.2 (3.9)	77.7 (5.9)
No. CSS seeds plant⁻¹	0	0	0	28.7 (2.8)	6.6 (0.6)	16.6 (1.3)
Proportion CSS seeds (%)	0	0	0	40.2	16.4	21.4
No. seeds pod⁻¹	0.94	1.22	1.46	0.90	0.98	1.30
Estimated seed abortion (%) due to lygus bug	-	-	-	4.3	19.7	11.0
Mean seed dry wt (mg):						
a. CSS seeds	-	-	-	30.9	23.0	7.5
b. undamaged seeds	41.7	38.8	16.3	49.2	42.2	16.7
c. overall mean	41.7	38.8	16.3	41.8	39.2	14.7
Estimated change in overall mean seed dry wt (%) due to lygus bug	-	-	-	+0.24	+1.03	-9.8
Seed yield (g) plant⁻¹	4.0 (1.16)	1.9 (0.11)	1.5 (0.09)	3.0 (1.17)	1.6 (0.17)	1.1 (0.10)
Loss in seed yield due to lygus bug (%)	-	-	-	26.2	18.1	23.5

18.9%, respectively). As expected from experience with other grain legumes (Summerfield, 1980), not all fruits on the control plants contained fully formed seeds (Table 1). However, the presence of *Lygus* bugs had increased the prevalence of malformed, more-or-less shrivelled pods by 13% to 15% in all three cultivars and, for Chilean, premature pod abortion was estimated to have been 22.6%.

Estimated seed abortion due to *Lygus* was much more severe (11.0% to 19.7%) in the smaller-seeded accessions than in Chilean (4.3%), as was the loss in individual seed dry weight due to CSS. Compared with individual seeds on control plants, those damaged by CSS were 26.1% lighter in Chilean and 40.7 to 54.0% lighter in the other two cultivars. However, as so often happens in grain legumes when reproductive load is reduced (Summerfield, 1980), the individual undamaged seeds on the poorer-yielding plants were atypically heavy (increases of 2.5 to 18%). Thus, the overall average individual seed dry weight of the populations of seeds from all Chilean and Precoz plants were remarkably invariant. These observations are consistent with findings elsewhere; in general, smaller-seeded species suffering *Lygus* attacks seem likely to experience greater relative losses of individual seed dry matter and of economic yield *per se* than species genetically programmed to produce relatively large seeds (e.g., Gupta *et al.*, 1980).

Plants caged with *Lygus* bugs throughout reproductive growth yielded an average of 22.6% less seed dry weight than those grown in an identical cage but free of insects (Table 1). Thus, in this forced feeding situation, not only the quality but also the quantity of economic yield was reduced. Of course, as we suggested earlier, farmers may not be aware that their seed yields have been reduced, since shrivelled pods and lighter seeds are likely to be lost from the combine at harvest.

Selected Characteristics of Healthy Seeds and Those With CSS

Batches of 1,000 seeds of healthy and CSS-damaged lots were selected at random from each of four accessions from the USDA germplasm collection obtained from breeders' plots at a single location (Genesee, Idaho) in 1980. Using procedures described in detail by Summerfield *et al.*, (1982) the following characteristics were quantified:
 (a) Mean seed dry weight and the distribution of relatively larger and smaller seeds within each lot;
 (b) Mean seed water content (% fwb);
 (c) Relative contributions of testa and cotyledons to the dry weight of whole seeds;
 (d) Germinability of seeds from the respective median class weight range;
 (e) Chemical composition (concentrations of protein-N, P, K, S, Fe, Mg and Ca);
 (f) Leakage of cellular contents into steep water – as reflected by changes in electrical conductivity and pH; and
 (g) Production of normal and abnormal seedlings.
With respect to the characters (a), (b), (c) and (d) listed above, the consequences of CSS in all four accessions were substantially similar quantitatively and so overall average values are presented in Table 2.

The incidence of CSS was not confined to seeds of any particular size (or did

Table 2. Overall average values of selected characters of lentil seeds damaged or not by CSS (means of four accessions with SE values in parentheses)

Attribute	Seeds with CSS	Healthy seeds
Water content (%)[+]	7.67 (0.07)	6.76 (0.06)
1000 seed dry weight (g)*	37.2 (3.9)	50.5 (2.1)
Loss in weight (%) due to CSS	26.9 (5.1)	-
Median seed weight (mg range)[T]	41 - 45	56 - 60
Mean seed dry weight (mg)[S]	48.1 (0.32)	49.9 (0.58)
Mean cotyledon dry weight (mg)	39.9 (0.26)	44.1 (0.46)
Mean testa dry weight (mg)	8.3 (0.30)	5.9 (0.16)
Proportion (%) of whole seed in testa	17.1 (0.56)	11.8 (0.20)
Germination (5 d) (%)	79.8 (4.4)	90.2 (2.9)
Germination (10 d) (%)	81.6 (4.8)	94.7 (2.1)
Seeds without embryos (%)	5.0 (2.2)	0

[+]Expressed as % fresh weight.
*And see Figure 2.
[T]Approximated to the nearest weight class as used in Figure 2.
[S]All seeds from the 51-55 mg fresh weight class.

not result in the formation of seeds of more-or-less the same size) but it virtually precluded the formation of large seeds and greatly increased the prevalence of relatively smaller seeds in the population (Figure 2). For example, damaged seeds heavier than 60 mg and undamaged ones lighter than 35 mg were rare, whereas healthy seeds in excess of 65 mg and damaged ones lighter than 30 mg were quite common. Thus, apart from any effects on seed quality *per se*, CSS also reduced seed dry weight because it limited the ability of seeds to realize their full growth potential.

With *P. vulgaris*, single punctures by *Lygus* bugs have reduced average seed dry weight by 18% compared with undamaged seeds, and three or four punctures per seed are even more deleterious (average reduction in weight of 36%; Scott, 1970). The relative magnitudes of these depredations are remarkably similar to those recorded for lentil (Table 1). Scott (1970) reasoned

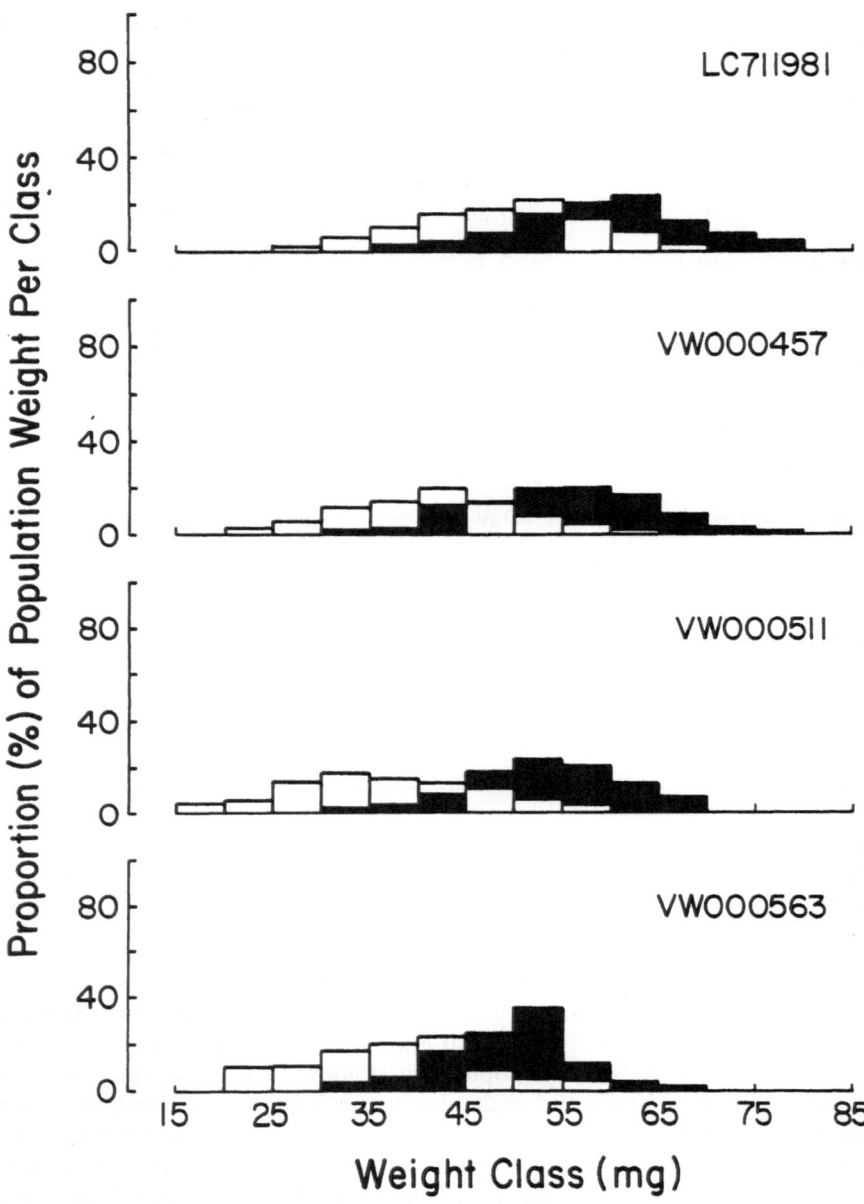

Figure 2. Frequency distributions for lentil seeds of various weights (mg range) within populations damaged (open) or not (shaded) by CSS.

that seed weights were reduced because normal embryogenesis had been prevented ("stunted"), and that lysis of endosperm by salivary secretions was followed by siphoning of partially digested materials by the feeding insects.

Although they were harvested to within one day of each other, and then placed promptly into storage in identical paper sacks at the same temperature (10 to 13°C in the dark), the damaged seeds three months later had larger water contents (7.67 ± 0.7%) than undamaged ones (6.76 ± 0.06%). Certainly, damaged seeds were better able to imbibe water than healthy ones. The few undamaged seeds which had not germinated after 10 d in the germination test (Table 2) had failed to imbibe water (they are described as "hard seeded"), whereas all the seeds with CSS were fully imbibed and swollen. The failure of a larger proportion of CSS-damaged seeds to germinate is attributed to two principal causes: heavy infestations of fungal pathogens and absence of embryos.

The fungal contaminants of non-germinated, CSS-damaged seeds were identified as *Fusarium* spp., *Cladosporium* spp., and *Alternaria* spp. (collectively the most prevalent species) together with *Mucor* spp., *Penicillium* spp., and *Stemphylium* spp. These species of fungi are components of a microflora which might reasonably be expected to colonize seeds either developing on plants in the field, or after the seeds have matured and the plants are either still standing or have been cut and swathed, awaiting threshing. However, we did not record fungal growth on any healthy seeds during the time scale of this experiment. It seems probable that rather than inherent differences in the fungal flora of damaged and healthy seeds, CSS damage allowed exudation or leakage of energy-rich cotyledon reserves during imbibition which then stimulated spore germination and hyphal growth. Indeed, as Table 3 shows, damaged seeds were an average of three times more leaky than undamaged ones. Accession LC711981 ("Brewer") was somewhat atypical in that although damaged seeds leaked more of their storage reserves than healthy ones, the conductivity value of steep water was 38% smaller than the average value for CSS-damaged seeds of the other three accessions. The increase in acidity of steep water (an average decline of 1.44 and 0.84 units of pH for damaged and healthy seeds, respectively) may well reflect leakage of amino, carboxylic, and fatty acids – all of which are known to be readily leached from seeds of other legume species within a time scale similar to the one used here (J. D. Maguire, personal communication).

Lentil embryos typically represent slightly less than 2% of seed dry weight (Singh *et al.*, 1968) and are easily visible with the naked eye when cotyledons are separated. Each of the 120 undamaged seeds examined here contained an apparently healthy embryo, whereas an average of 5.0 ± 2.2% of chalky spot seeds did not (Table 2). The formation of embryoless seeds is symptomatic of *Lygus* bug damage to carrots (*Daucus carota* L.), dill (*Anethum graveolens* L.), and fennel (*Foeniculum vulgare* Miller); as many as 100% of seeds can be affected in forced feeding situations (e.g., Flemion and Olsen, 1950). Hence, although the larger seeds of lentils seem less likely to suffer a loss of embryos than these smaller-seeded Umbelliferous species, the fact that a small proportion of seeds had been killed by CSS because their embryos had been destroyed is further evidence implicating *Lygus* bugs, or an insect with similar mouthparts and feeding niches, as the causal agent.

Table 3. Electrical conductivity (μmhos cm^{-2} g seed^{-1}) and acidity (pH) of steep water from lentil seeds damaged (+) or not ($-$) by CSS (means of four replicates with SE values in parentheses)

Attribute	Control	Seed Stock (USDA Accession No.)							
		LC711981		VW000457		Redchief		VW000563	
		+	-	+	-	+	-	+	-
Weight of seeds used (g)	0	2.93 (1.10)	2.98 (0.04)	1.81 (0.06)	3.25 (0.05)	1.77 (0.06)	3.05 (0.06)	1.79 (0.06)	2.75 (0.07)
Initial pH of steep water	6.28 (0.14)	6.43 (0.08)	6.58 (0.25)	6.39 (0.21)	6.59 (0.19)	6.01 (0.11)	5.91 (0.08)	5.90 (0.08)	6.40 (0.10)
Initial conductivity of steep water[+]	1.37 (0.22)	1.24 (0.13)	2.31 (0.75)	1.62 (0.51)	2.08 (0.40)	0.97 (0.03)	0.93 (0.01)	0.97 (0.04)	1.54 (0.09)
Final pH of steep water after 24h	6.38 (0.09)	4.85 (0.04)	5.40 (0.05)	4.62 (0.03)	5.49 (0.16)	4.74 (0.04)	5.71 (0.10)	4.75 (0.12)	5.53 (0.06)
Final conductivity of steep water after 24h	2.28 (0.27)	123 (6)	68 (5)	129 (7)	62 (5)	120 (12)	51 (2)	112 (4)	50 (1)
Corrected final pH[*]	-	4.75	5.30	4.52	5.39	4.64	4.61	4.65	5.43
Corrected final conductivity[*]	-	122	67	128	61	119	50	111	49
Electrical conductivity per unit wt of seeds (μmhos cm^{2}g^{-1})	-	41.6	22.5	70.7	18.8	67.2	16.4	62.0	17.8

[+] Industrial Instruments Inc., Conductivity bridge Model RC 16B2.
[*] Adjusted values to compensate for drift in control samples.

Compared with the effects of CSS on the characteristics of seeds discussed hitherto, there were no consistent or significant differences in chemical composition between healthy and damaged seeds (Summerfield *et al.*, 1982). In contrast, and as Table 4 and Plate 3 show, seedling emergence from CSS-damaged seed was delayed and dramatically poorer than from healthy seeds (mean values of 20% and 86% emergence, respectively). Furthermore, a large proportion of seedlings which did eventually emerge were abnormal morphologically (Plate 3). Since all seeds were surface sterilized with sodium hypochlorite (1% a.i.) before sowing, the poor performance of those damaged by CSS may have been due, at least in part, to penetration of the sterilant. Nevertheless, since farmers often dress their lentil seeds with fungicides, we might expect the likelihood of seedling emergence from chalky spotted seeds in the field to be similarly poor.

Table 4. Seedling establishment (% emergence) and prevalence of seedlings with abnormal morphology (% of those emerged) for lentil cultivars grown from visibly healthy ($-$) or CSS-damaged ($+$) seeds (Means of 10 replicate pots each of 10 seeds sown with SE values in parentheses)

	Seed Stock (USDA Accession No.)							
	LC711981		VWC00457		Redchief		VW000563	
Attribute	+	-	+	-	+	-	+	-
Emergence (5 d) (%)	2 (1)	48 (3)	12 (4)	68 (5)	4 (3)	64 (5)	6 (2)	71 (4)
Emergence (10 d) (%)	20 (4)	76 (5)	25 (5)	92 (3)	19 (5)	87 (3)	17 (4)	88 (3)
Abnormal seedlings (%)	55 (3)	0	48 (2)	0	79 (4)	0	35 (3)	0

There is an adage often repeated among farmers: "The crop can be no better than the seed sown". Chalky Spot Syndrome certainly has adverse effects on two measures of seed quality – germinability and vigor – as discussed above. However, since lentils are used almost exclusively for food rather than, for example, oil extraction and for animal feed, then other attributes of quality also need to be considered – as we describe below.

Cookability and Taste of Seed

Graders sample lots (2.25 kg) of lentil Chilean harvested in 1980 from farmers' fields close to Troy, Idaho (see Figure 1) were found to contain approximately

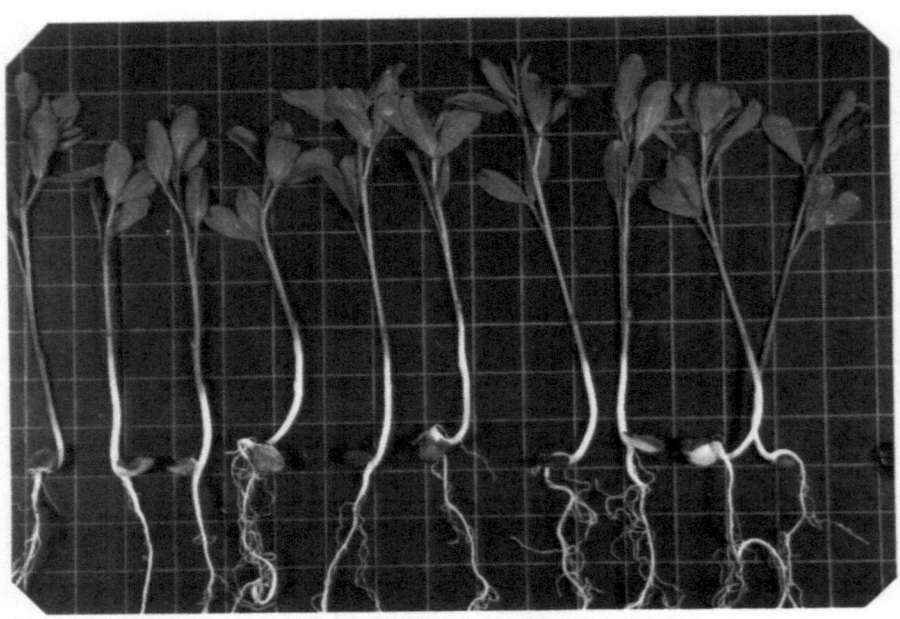

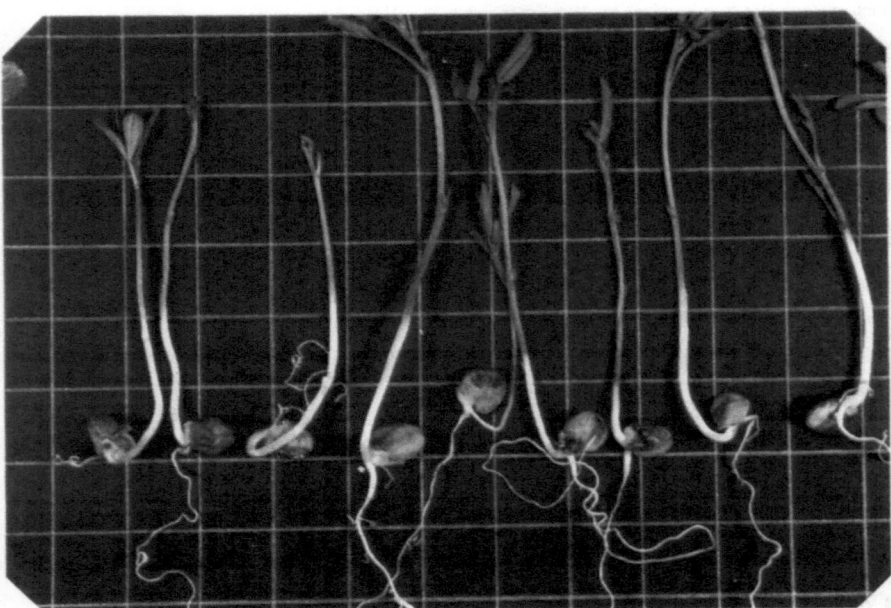

Plate 3. Lentil seedlings harvested 10 d after sowing visibly healthy seeds (above) or ones damaged by CSS (below) in pots in a glasshouse (lines are a 1 cm² grid).

10 to 15% chalky spot seeds. The seed lots were cleaned by hand in order to remove inert matter and seeds of species other than lentils, and samples each of approximately 900 g were then "reconstituted". Three samples comprising (a) visibly undamaged seeds only; (b) 90% undamaged seeds, 10% chalky spot seeds; and (c) 80% undamaged and 20% chalky spot seeds (respective proportions by weight) provided the material for a range of cookability tests and taste panel evaluations. Sensory, texture, and color tests were made on preparations from each of these samples using methods reported in detail by Summerfield *et al.* (1982).

Average values of flavor characteristics of cooked lentil samples containing either 0, 10, or 20% chalky spot seeds as assessed by sensory panelists are presented in Table 5. Descriptive terms used by the panelists indicated that the samples containing CSS-damaged seeds were "slightly less sweet" and a "little more bland" than the undamaged control samples. In texture, the samples containing damaged seeds had "slightly tougher seedcoats" – which is perhaps not surprising since chalky spot seeds are known to have thicker testas (Table 2). Shear press values were slightly smaller for the inferior samples, presumably indicating an increased tendency to become mushy during processing (and note in Table 5 the increased weight of cooking medium for these samples reflecting, perhaps, greater losses of seed integrity during baking).

Table 5. Mean values of sensory characteristics (taste panel), shear force (kg), hunter-color and drained weight (g) of cooked lentil samples comprising various proportions of chalky spot seeds

Attribute of cooked sample	Sample composition [undamaged: damaged seeds (%) by weight]		
	100:0	90:10	80:20
Sensory characteristics[+]			
Flavor	3.4	3.8	3.5
Texture	3.6	3.9	3.7
Appearance	3.8	2.9	3.3
Preference Score	2.68	2.60	3.10
Shear Force Value (kg)*	36.1	35.7	34.9
Hunter-color			
Rd (lightness)	8.7	7.9	8.2
a (redness)	2.0	2.5	2.2
b (yellowness)	14.5	13.5	14.0
Drained Weight (g)[$]			
Cooked lentils	294	293	300
Cooking medium	72	84	80

[+]Mean values of 24 panelists asked to assess the degree of difference from a control sample (100:0) on a 9-point scale (1 = "no difference" to 9 = "extreme difference").
*Mean values of 6 estimates.
[$]Mean values of weights of samples prepared for panels 2 and 3.

Overall, while lentil seed lots comprising either 10% or 20% chalky spot seeds (by weight) were slightly inferior in sensory characteristics and texture to the undamaged samples used in this pilot study (Table 5), differences were not significant. Nevertheless, the presence of CSS detracted from the visual and sensory appeal of the processed product and, if present in proportions greater than 20% by weight, the negative tendencies described above may become correspondingly greater. On the other hand, seed lots containing 10%, or perhaps even 20%, chalky spotted seeds should not be automatically downgraded on the basis of these criteria alone.

Progress During the Decade to 1992

Without reference to the program of research summarized here, subsequent work (undertaken primarily at the University of Idaho) has confirmed that adult *Lygus* bugs usually arrive in Palouse lentil fields as the crop comes into bloom, and that fourth or fifth instar nymphs can combine with the feeding injury to mature pods caused by the adults and so exacerbate CSS damage (O'Keeffe *et al.*, 1991). Sampling by sweepnet during the afternoons can provide reliable estimates of *Lygus* bug densities as a prelude to control actions in individual fields (Schotzko and O'Keeffe, 1986a,b, 1989a,b, 1990b.). Stink bugs (*Thyanta pallidovirens* Stal) usually arrive later in lentil fields and may or may not contribute to the CSS problem (Schotzko and O'Keeffe, 1990a).

Chemical control of *Lygus* spp. (but not of *T. pallidovirens*) is possible, based on sampling estimates of economic thresholds of the numbers of adults and nymphs. Economic thresholds, however, remain to be determined for stink bugs and for the combined populations of *Lygus* spp. and stink bugs. Timing and method of harvesting has been suggested as a means of reducing the incidence of CSS (O'Keeffe *et al.*, 1991). However, that suggestion is somewhat misleading because damage from *Lygus* will already have taken place, and harvesting prior to physiological maturity will measurably reduce crop quality.

A decade after the original and extensive program of research had been completed, then, precise control details for CSS remain to be formulated. Furthermore, it is not yet possible to predict where and when CSS will appear in a particular locality (O'Keeffe *et al.*, 1991). As we argued in 1982, coordinated research in the field will undoubtedly contribute significantly to practical and economically-attractive strategies for minimizing the incidence of CSS in future lentil crops – not only by chemicals but also by cultural practices appropriate for maximum biological control. The passing of time shows that the exploitation in practice of advances in food legume science can be frustratingly difficult. A retrospective investigation of the CSS problem and of the ability of farmers to deal with it may now be timely and informative.

Acknowledgements

We are pleased to recognize the help given to us during the course of this work by many specialists: Drs. R. E. Fye (USDA-ARS, Agricultural Research Laboratory, Yakima) and G. R. Pesho (USDA-ARS-RPIS, Pullman) for advice and guidance with problems on entomology; Dr. W. J. Kaiser (USDA-ARS-RPIS, Pullman) for identifying fungal contaminants of seeds; Ms. C. Kagel (Plant and Soil Analytical Laboratory, University of Idaho) for elemental analyses; Mr. H. Blain (Washington and Idaho Dry Pea and Lentil Commissions) for checking and expanding our synthesis of grower experiences; Dr. S. McCurdy and Ms. D. Halvorson (College of Agriculture and Home Economics, Washington State University) for carrying out the cooking tests and organizing taste panel evaluations; Mr. M. Chestnut (Botany Department, Washington State University) for scanning and transmission electron microscopy investigations; and Dr. W. S. Hawkins (USDA Federal Grain Inspection Service, Standardization Division, Marketing Standards Branch, Grandview, Missouri) for bringing useful publications to our attention. Financial support from the American Dry Pea and Lentil Association, the Washington and Idaho Dry Pea and Lentil Commissions, the USDA-ARS, the University of Reading, and the UK Overseas Development Administration is gratefully acknowledged.

References

Auld, D. L., Murray, G. A., Carnahan, G. F. and others. 1980. *University of Idaho College of Agriculture, Cooperative Extension Service Current Information Series* No. 524, pp. 4.
Elmore, J. C. 1955. *Journal of Economic Entomology* 48: 148–151.
Flemion, F. and Olsen, J. 1950. *Contributions from the Boyce Thompson Institute for Plant Research* 16: 39–46.
Fye, R. E. 1980. *Journal of Economic Entomology* 73: 469–473.
Gupta, R. K., Tamaki, G. and Johansen, C. A. 1980. *WSU-CARC Technical Bulletin* No. 92, pp. 18.
Hagel, G. T. 1978. *Journal of Economic Entomology* 71: 613–615.
Hariri, G. 1981. In *Lentils* (eds. C. Webb and G. Hawtin), pp. 173–190. Commonwealth Agricultural Bureaux, UK: Farnham Royal.
Landis, B. J. and Fox, L. 1972. *Environmental Entomology* 1: 464–465.
Miner, F. D. 1966. *Arkansas Agricultural Experiment Station Bulletin* No. 708, pp. 40.
Muehlbauer, F. J. 1973. *Washington State University Agricultural Experiment Station Circular* No. 565, pp. 3.
O'Keeffe, L. E., Homan, H. W. and Schotzko, D. J. 1991. *University of Idaho CES-AES Current Information Series* No. 894, pp. 4.
Schotzko, D. J. and O'Keeffe, L. E. 1986a. *Journal of Economic Entomology* 79: 224–228.
Schotzko, D. J. and O'Keeffe, L. E. 1986b. *Journal of Economic Entomology* 79: 447–451.
Schotzko, D. J. and O'Keeffe, L. E. 1989a. *Environmental Entomology* 18: 308–314.
Schotzko, D. J. and O'Keeffe, L. E. 1989b. *Journal of Economic Entomology* 82: 1277–1288.
Schotzko, D. J. and O'Keeffe, L. E. 1990a. *Journal of Economic Entomology* 83: 1333–1337.
Schotzko, D. J. and O'Keeffe, L. E. 1990b. *Journal of Economic Entomology* 83: 1888–1900.
Scott, D. R. 1970. *Annals of the Entomological Society of America* 63: 1604–1608.
Singh, S., Singh, H. D. and Sikka, K. C. 1968. *Cereal Chemistry* 45: 13–18.

Summerfield, R. J. 1980. In: *Opportunities for Increasing Crop Yields*, pp. 51–69 (eds. R. G. Hurd, P. V. Biscoe and C. Dennis). London: Pitmans.

Summerfield, R. J. and Muehlbauer, F. J. 1982. *Experimental Agriculture* 18: 3–15.

Summerfield, R. J., Muehlbauer, F. J. and Short, R. W. 1982. *USDA-ARS Agricultural Reviews and Manuals ARM-W-29*, 43 pp.

Addressing farmers' constraints through on-farm research: peas in Western Canada

A.E. SLINKARD[1], C. VAN KESSEL[2], D.E. FEINDEL[2],
S.T. ALI-KHAN[3] and R. PARK[4]

[1] Crop Development Centre, University of Saskatchewan, Saskatoon, Saskatchewan, S7N OWO
Canada;
[2] Department of Soil Science, University of Saskatchewan, Saskatoon, Saskatchewan, S7N OWO
Canada;
[3] Agriculture Canada Research Station, P. O. Box 3001, Morden, Manitoba, ROG 1JO Canada, and
[4] Alberta Agriculture, Field Crops Branch, Bcg Service 47, Lacombe, Alberta, TOC 1SO Canada

Abstract

Dry pea (*Pisum sativum* L.) production in Canada increased from 27,000
hectares in 1976 to over 400,000 hectares in 1993. Most of these pea producers
had never grown peas before and an intensive applied research and
demonstration program was developed to help farmers produce this new crop
successfully. The Alberta, Manitoba, and Saskatchewan Universities,
provincial Departments of Agriculture, and Pulse Growers Associations, in
cooperation with Agriculture Canada and private industry (pulse traders, pea
breeders, *Rhizobium* manufacturers, and herbicide and fertilizer companies),
were involved in rapid development of the pea industry. Registered pea cultivars
increased from 3 to 30, two-thirds of them introduced from western Europe. A
Canadian *Rhizobium* industry was developed, and two feed pea marketing
missions were sent overseas. Many field scale demonstrations were used to show
the merits or lack thereof of new cultivars, inoculation, N and P fertilization,
herbicidal and cultural weed control, various harvesting methods, plus the effect
of seeding rates and dates. Demonstrations were also conducted on the use of
peas in swine and dairy rations. Concurrently, a large private processing and
marketing industry developed. In addition, two plants were constructed to
process peas into dry pea products, such as pea fiber, pea starch, pea protein
concentrate, and pea flour. A plant was established to produce a snack food
similar to potato chips, but with a lower fat and higher protein content. The
expansion of pea production could not have occurred this fast without the
cooperation of all participants.

*F.J. Muehlbauer and W.J. Kaiser (eds.), Expanding the Production and Use of Cool Season Food
Legumes, 877–889.*
© 1994 *Kluwer Academic Publishers.*

Introduction

The Challenge

Agriculture in western Canada has traditionally been based on production of red spring wheat. The resulting one crop economy is characterized by violent fluctuations in net farm income. As a result of the wheat surplus and resulting low wheat prices in the years immediately preceding 1971, the Government of Saskatchewan and the National Research Council of Canada jointly established the Crop Development Centre (CDC) at the University of Saskatchewan in 1971. One of the main research and development programs in the CDC involved development of special crops, especially the dry pea crop, as a means of increasing crop diversification and stabilizing farmer's incomes. The CDC special crops program accepted this challenge and has had major input into many aspects of the development of the dry pea industry in western Canada.

Background and Strategy

Dry pea has been grown on a limited area in western Canada since farmers started plowing the prairies over 100 years ago. Most of the peas were grown in southern Manitoba with its more favorable climate and soil. Peas were grown on about 20,000 hectares annually in the years following WW II until 1976. However, dry pea production in western Canada started increasing rapidly in 1976 (Slinkard and Blain, 1988), based largely on the export market. Saskatchewan became the major pea-producing province in 1986 and pea production has increased markedly in Alberta in recent years. Dry pea production in western Canada increased to over 400,000 hectares in 1993 (Saskatchewan Agriculture and Food, 1991b; Slinkard, personal communication). This increase in pea production occurred with minimal disruption because all facets of the industry worked closely together as a team with full and open communication among all players. The conditions and events contributing to this dramatic and sustained increase in dry pea production are reviewed here.

The Team

Research Agencies

The Crop (Plant) Science and/or Soil Science departments at the Universities of Alberta, Manitoba, and Saskatchewan have some teaching and research responsibilities for pulses (peas and other food legumes). However, the major thrust in pulse research in western Canada is provided by the Special Crops Program of the Crop Development Centre in the Department of Crop Science and Plant Ecology at the University of Saskatchewan. Pea research, development, and demonstration projects involving breeding, management, and technology transfer are largely funded by Saskatchewan Agriculture and Food through the Agriculture Development Fund (ADF).

In addition, Agriculture Canada has 15 research stations and experimental farms scattered throughout western Canada and all do some research on pulses. However, most of their pea breeding, management, pathology, and quality research is concentrated at Morden. MB.

Technology Transfer Agencies

The Alberta, Manitoba, and Saskatchewan Departments of Agriculture each have a Special Crops Specialist, whose primary responsibility is technology transfer for all special crops, including peas. Each province also has regional specialists and local agents involved in technology transfer. In addition, researchers at the provincial universities and Agriculture Canada have some responsibility for technology transfer to pulse growers. Personnel in these agencies cooperate closely, often across provincial boundaries, to effectively facilitate technology transfer.

Provincial Growers Associations

Pea and lentil producers in the province of Saskatchewan were the first to become organized and formed the Saskatchewan Pulse Crop Growers Association in 1976. This association provided an effective medium for coordinating early research, development, and technology transfer. In 1984, it reorganized as the Saskatchewan Pulse Crop Development Board (SPCDB) and under provincial legislation it was empowered to collect a compulsory levy of 0.5% of the initial sale price of all pulse crops (pea, lentil, etc.) sold in Saskatchewan, effective 1 January 1985. The various contracting companies cooperated by deducting the levy from the purchase price they paid the farmer and remitting the levy to SPCDB each month.

In the formative years, most of the levy was used for increased research on breeding and management of pea and lentil. Currently, efforts are made to balance expenditures equally between research and market development. In many cases, Saskatchewan Agriculture and Food matched the research grants through their ADF. Provincial and federal support have been available for market development, both domestic and foreign.

The success of the SPCDB and its levy system prompted pulse crop producers in Alberta to form the Alberta Pulse Growers Association in 1981, which reorganized into the Alberta Pulse Growers Commission under provincial legislation in 1989, and started collecting a levy of 0.5% of the initial selling price of the various pulse crops. Subsequently, the Manitoba Pulse Growers

Association was organized in 1986 and started collecting a levy of 0.5% of the initial selling price of the various pulse crops in 1990. The major difference between the levy in Saskatchewan and the levy in Alberta and Manitoba was that the levy in the latter two provinces was voluntary in that the producer could apply for a refund of his contribution at the end of the year. Fortunately, few refunds were requested and development of the pulse crop industry proceeded rapidly throughout western Canada.

These provincial pulse growers groups are governed by an elected board consisting of pulse growers assisted by several *ex officio* advisors. Each provincial pulse growers group has a regular newsletter with reports on prices, problems, production pointers, and other items of interest to the producers. In addition, an annual meeting is held by each association to hear updates on research, market developments, pesticides, cultivars, other new technologies, and the price outlook.

The need for a regional lobby group was recognized at an early date. This group would lobby federal and provincial governments for concessions to the pulse crop industry as a whole. Accordingly, the Western Canada Pulse Growers Association (WCPGA) was organized in 1984. Directors of the provincial growers groups also serve as directors of the WCPGA. An elected executive coordinates lobby efforts and annually makes a trip to the nation's capital to develop and maintain open communications with different federal agencies that have jurisdiction over various aspects of the pulse industry. In this way, the problems and needs of the pulse industry can be brought to the attention of the regulatory agencies. In addition, they can be advised of any adverse effects on the pulse industry of proposed changes in various regulations. Thus, the pulse producers have input into various regulations that affect them and their industry.

Pulse Contractors and Exporters

The Canadian Seed Trade Association (CSTA) is a national association of private seed companies organized in 1922 for the purpose of coordinating development of the seed industry in Canada. Primary emphasis originally was on forage seeds, then corn, and recently all crops, including pulse crops. In recent years between 20 and 30 companies have been involved in contracting and marketing pulse crops in western Canada. Initially, the CSTA formed a Special Crops Committee to help coordinate development of the commercial aspects of the pulse industry in Canada. In February 1987, the Special Crops Committee of CSTA evolved into a separate organization, the Canadian Special Crops Association (CSCA). It was formed to "establish trade rules and to serve as a forum to deal with the concerns of exporters, dealers, and related business involved in the Canadian special crops industry". It currently has about 50 direct members and 10 associate members among Canadian, American, and European companies. The CSCA publishes four or five newsletters each year and holds an annual convention as part of their efforts in promoting and coordinating development of the commercial aspects of the Canadian pulse industry.

Pulse Processors

As pulse production in western Canada increased, the number of small primary processors increased until now over 300 primary processors are involved in cleaning and bagging pulse crops in western Canada. Initially, equipment in the seed-cleaning plants was simple, but as the industry developed, most seed processors purchased spiral cleaners and gravity machines. The quality of the cleaned and bagged product increased materially as a result. In addition, some lots were shipped bulk in 20-tonne containers. Most primary processors aligned themselves with one of the major pulse contracting and marketing companies and served as their local contracting agent.

The number of pea-splitting operations in western Canada increased from three to nine as pea production increased. The quality of split pea from some of these plants is not the best, but it is improving as the operators gain more experience and receive feed-back on quality requirements from the exporters.

By 1975, Dr. C. Youngs at the National Research Council Prairie Regional Laboratory in Saskatoon, SK had developed two prototype (wet and dry) pilot scale plants for further processing dry pea seeds into pea fiber, pea starch, and pea protein concentrate (40% to 60% protein). The pea fiber was produced from the finely ground pea seedcoat and was added to certain bread mixes to produce a high-fiber bread. The pea starch was used for carbonless paper, adhesives, and as a desliming agent in the potash industry. The pea protein concentrate had potential use in specific human food applications, but had difficulty in competing in the larger, already established industry based on protein concentrates and isolates from soybean meal. One pilot plant was based on a dry process which involved fine grinding and air classification to separate the starch and protein fractions. The other pilot plant was based on a wet process which involved fine grinding, adding water and lime to raise pH to 9, and centrifuging to produce a high protein supernatant and starch solids. The supernatant is spray- or drum-dried to yield a protein concentrate (about 60% protein).

A commercial-scale plant, based on the dry milling process, was established in Saskatoon, SK and operated 1977–79, 1982–85, and since 1989. The company went into receivership twice, primarily because it could not find a high-value food market for the pea protein concentrate and had to sell it to the livestock feed market in competition with soybean meal. A commercial-scale plant, based on the wet milling process, was established in Portage la Prairie, MB in 1977 and has been operating since then. In the wet milling process the pea protein concentrate is in a slurry and must be dried. Various bitter flavor components are volatilized and "flashed off" during drying and this increases the potential food applications for pea protein processed by the wet method. In addition, pea protein isolates are produced for unique food applications.

In 1990, Kelvington Processors in Saskatchewan started producing "Peola Chips" and have since expanded their plant due to increasing demand. Peola chips are a snack food similar to potato chips, but made from wet pea flour formed into chip-sized pieces and deep fat fried in canola oil, and thus the combination name peola chips. This is an innovative approach to value-added processing, originally conceived by Dr. Youngs of the National Research Council of Canada back in 1972 or so and finally effected by a group of farmers and small town businessmen. Hopefully, this new use of peas will expand and provide a continuing market for peas.

Allied Industries

Peas are poor weed competitors and a highly effective chemical weed control program is usually required for profitable yields. Several multinational chemical companies actively promote research, herbicide evaluation, demonstration, and technology transfer for the benefit of the pulse industry in western Canada. They help sponsor field demonstrations of their products, growers field days, growers meetings, and the annual meeting of the provincial pulse growers associations. These efforts are often in cooperation with the provincial departments of agriculture, the universities, and the provincial pulse growers associations and greatly facilitate development of the pulse industry in western Canada.

Soils in western Canada are naturally low in available phosphorus (P) and pulse crops respond well to seed-placed P. In addition, many of the pulse crops responded to nitrogen (N) fertilizer when a pulse was grown for the first time in low N soils. Thus, fertilizer companies readily cooperated in providing research grants and fertilizer to demonstrate the benefits from fertilizer use in research plots and in demonstration plots in farmers' fields. Subsequently, the response to N fertilizer was related to poor and ineffective inoculation with *Rhizobium*. Better adapted strains of *Rhizobium*, improved sticking agents, seeding pulses into the same field a second time in the rotation, and improved application techniques by experienced pulse producers have eliminated the need for N fertilizer on pulses, reducing the cost of production and increasing net return to the producer.

Prior to 1987, all *Rhizobium* inoculant was imported from the United States, primarily from one company. The rapid increase in pulse crop production in western Canada stimulated the development of a *Rhizobium* industry in Saskatchewan. Currently, three new companies, all headquartered in Saskatchewan, are the major producers of *Rhizobium* inoculant for western Canadian pulse growers. All three companies have an effective dealer and retail organization throughout western Canada so that pulse growers can obtain inoculant from their local dealer. These three *Rhizobium* inoculant companies have an active research program and have been involved in farmer research and demonstration plots.

On-Farm Research and Demonstration – Some Examples

The cost of on-farm research and demonstration projects is often reduced by using inputs from various cooperators. Thus, depending on the experiment, the farmer may provide a portion of his field, several pea seed companies may provide seed of their cultivars, a herbicide company may provide herbicide, a fertilizer company may provide fertilizer, or a *Rhizobium* company may provide *Rhizobium*, all at no cost, in an effort to promote both the research project and their individual product. Later in the season, several field days are held so that the local farmers can see if these products provide any benefits. Data are collected on seed yield and other agronomic traits. After the products have been evaluated over a series of environments, the results are released to the public in a manner that emphasizes the value of the experimental results to pea producers and at the same time properly acknowledges the industry contributions.

Pea Cultivars

Until 1984 few pea cultivars were registered in Canada. The cultivar Century, a dry yellow pea, was registered in 1960 and has been the national standard for cooking quality since then. The small-seeded yellow pea cultivar Trapper was registered in 1970. The first feed pea cultivar Tara was registered in 1978. Tara had a 15% yield advantage, but was discounted in the food pea market due to its irregular seed shape, and never attained its full potential.

The increased pea production, starting in 1977, stimulated interest in research on yield and adaptation of European pea cultivars. Subsequently, three pea cultivars from Svalof Seed, Sweden were registered: Victoria (1984), Fortune (1986), and Express (1987). Express still is the highest yielding cultivar in Saskatchewan.

On-farm demonstrations in central Alberta started in 1984 and promptly showed that this area required pea cultivars with a markedly different adaptation than those in other parts of western Canada. The cooler, moister growing season favored earlier maturing and shorter, semi-determinate cultivars. For example, Express, the highest yielding cultivar in Saskatchewan was only about average in central Alberta. Starting in 1988, special cultivar trials have been conducted to facilitate regional registration of pea cultivars for the shorter season areas of western Canada.

In 1985, the first semi-leafless pea cultivar, Tipu, was registered in western Canada. On-farm demonstrations in Manitoba promptly proved that semi-leafless pea cultivars are characterized by reduced harvest loss, easier harvest, and produce a higher quality product (Wall *et al.*, 1991). The end result is that

Tipu and the newer, higher yielding, semi-leafless peas have received prompt farmer acceptance.

Traditionally, western Canada has produced dry yellow peas. Consequently, most of the new pea cultivars that were rapidly registered as pea acreage increased were dry yellow cultivars. By 1992, 30 pea cultivars were registered in western Canada, with several dry green cultivars in recent years (Vaillancourt, 1989).

The Palouse area of eastern Washington and northern Idaho traditionally has grown about 70,000 hectares of dry peas, mostly dry green cultivars. A severe drought in 1988 resulted in a greatly reduced crop and importers turned to western Canada where about 10,000 hectares of dry green peas were being grown. The high premium paid for these dry green peas, relative to dry yellow peas, in a drought year greatly stimulated production of dry green pea cultivars in western Canada. Thus, by 1991 nearly half of the Saskatchewan dry pea crop was seeded to dry green cultivars. Consequently, several of the more recently registered pea cultivars are dry green cultivars. This is but one example of how the western Canada pulse industry is so well organized with open lines of communication that it is able to respond to major changes in the industry in only one or two years. The ability to quickly respond to changes in the industry means greater net returns to all facets of the pulse industry – the producers, processors, exporters, brokers, and the *Rhizobium*, herbicide, and fertilizer dealers. Even the researchers benefit from the collection of increased levies, part of which is allocated for increased research and market development.

Dry green pea seeds may bleach to pale green or light yellow, especially if hot (> 28°C) sunny days follow light showers or heavy dews during the final stages of maturity. The dry green pea market heavily discounts, or even rejects, bleached dry green peas. However, on-farm research and demonstration plots in Manitoba showed that bleaching in dry green peas could be reduced by prompt harvesting at 20% seed moisture and promptly aeration drying the seed to 16% moisture. Alternatively, the near mature plants could be sprayed with a desiccant and harvested after a short drying time. Both measures reduced the probability of severe bleaching losses and helped make dry green pea production feasible in Manitoba. Due to lower temperatures at harvest time in Saskatchewan and Alberta, losses from bleaching are reduced, but prompt early harvest is still practiced.

Rhizobium Inoculation

On-farm research and demonstration experiments have been used to evaluate *Rhizobium* inoculants in Saskatchewan. Initially, inoculants from different companies were compared. Since pea production is relatively new in western Canada, the soils are devoid of native *Rhizobium*, except for isolated low density populations associated with native *Vicia* species. Under these conditions, excellent responses to *Rhizobium* inoculation can be demonstrated, especially on low nitrogen soils following a cereal crop.

However, once inoculated peas have been grown in a field, the resulting indigenous *Rhizobium* becomes competitive and greatly inhibits nodulation by a new strain introduced on the next pea crop. Thus, Saskatchewan researchers have opted to screen for more effective strains in soil with an indigenous *Rhizobium* population since this should provide valuable information about the selected strain when grown under field conditions. As a result, a highly effective strain on pea plants (C-1) has been isolated and is being evaluated under a wide range of field conditions. Results indicate that re-inoculation with strain C-1 in the presence of indigenous *Rhizobium* will result in a good response in some conditions, but certainly should not decrease yield, i.e., re-inoculation is good, low-cost insurance to ensure high levels of N_2 fixation and high yields.

Cooking Quality

On-farm research and demonstration experiments on cooking quality of peas in Manitoba indicated that nitrogen fertilization reduced cooking quality, while phosphorus fertilization increased cooking quality. In addition, harvesting too early (immature pea) or too late (shattering losses of highest quality seed) resulted in reduced cooking quality. This series of on-farm research and demonstration experiments were conducted in cooperation with a commercial pea soup canning company, an excellent example of cooperation with private industry (Gubbels *et al.*, 1982, 1985).

Nitrogen Fertilization

On-farm research and demonstration experiments on effect of nitrogen fertilization of peas over a three-year period indicated no yield response from the application of 50 kg N/ha to properly inoculated peas, even under low soil nitrogen conditions. In several experiments, the pea plants were taller, greener, and matured slightly later, but no differences occurred in seed yield.

Phosphorus Fertilization

Most soils in western Canada are low in available phosphorus. On-farm research and demonstration experiments proved that seed-placed P_2O_5 would result in a "pop-up effect" in that the pea seedling would grow faster, form a ground cover sooner, and, as a result, be slightly more competitive with weeds. Seed yield responses occurred in most on-farm sites that tested low in available phosphorus. Peas are sensitive to seed-placed phosphorus fertilizer and rates should not exceed 20 kg P_2O_5/ha if the fertilizer is placed in a narrow-band in close proximity to the pea seeds. Somewhat higher rates can be used if the phosphorus fertilizer is placed in a wide band with the seed or side-banded.

Herbicides

On-farm research and demonstration experiments are used to familiarize farmers with newly registered herbicides or those that will be registered after the current season. Industry cooperation and leadership are essential in this type of on-farm experiment.

Domestic Market Development

On-farm research and demonstrations on the feeding value of dry peas in hog rations were established in Alberta (1987), Saskatchewan (1991), and Manitoba (1992). Results were consistent in showing that dry peas or dry peas plus canola meal could replace imported soybean meal as the source of protein in a barley-based hog ration. Feed efficiency, average daily gain and cost per kg gain were comparable. In addition, an on-farm research and development experiment in Alberta in 1991 established that peas were competitive with soybean meal as a protein source in dairy rations. These results go a long way in developing a domestic feed market for peas to complement the export market for feed and food peas.

Overseas Missions

While not an on-farm research and development project, several pea producers participated in a Canadian Export Expansion Fund Mission to France, Holland, and Germany in 1989 to explore the export potential for Canadian peas in the EEC. In addition, several Alberta pea producers participated in a Pea Agronomy Mission to England, Sweden, Denmark, Holland, and Germany to evaluate their agronomic practices and cultivars relative to those in Canada. Both of these missions were designed to help bring the Canadian pea producer up to date on European markets and agronomic practices and help make him a more efficient pea producer and marketer.

The Result

The Canadian pea industry has developed rapidly since 1976. All of the above team players have been heavily involved and many on-farm research and demonstration experiments have been conducted over the years. However, this is still not a guarantee of success. The missing ingredient is the proper incentive or price for the peas. If the price is not right, the farmer will not produce the peas.

The Incentive

The major crop in western Canada is red spring wheat. It is easy to grow and the price of red spring wheat is the standard against which all other crops must be compared. In addition, a new crop such as dry peas, needs an extra premium (net return per hectare) in order to stimulate producers to grow it, usually on a small scale at first.

In Saskatchewan, a guideline has been established for technology transfer purposes. This guideline is only appropriate to Saskatchewan conditions with its relative yields per hectare of dry peas and red spring wheat. This guideline states that "Experienced pea producers in the Black soil zone (best pea area) of Saskatchewan will maintain or increase their pea production whenever the farm gate price of pea seed exceeds 125% of the farm gate price of red spring wheat on a weight basis". Since current season price comparisons are not known until after planting, the response in pea production normally lags one year behind changes in the pea price:wheat price ratio.

The Response

The area devoted to dry pea production in western Canada, average pea price, average red spring wheat price, and pea price as percent of red spring wheat price for the years 1968 to 1992 are presented in Table 1. The first time the pea price exceeded 125% of the red spring wheat price was in 1976 and 1977. However, the Saskatchewan Pulse Crop Grower's Association was organized only in late 1976 and growers were not really aware of the potential returns from pea production at that time. Nevertheless, dry pea production increased nearly 40% in 1977 and another 14% in 1978, at which time the pea price dropped markedly.

The second time the pea price exceeded 125% of the red spring wheat price started in 1985 and continued up until at least 1992, except for 1988 and 1989 when the price of red spring wheat increased temporarily (Table 1). By this time the Saskatchewan Pulse Crop Development Board was well organized and was actively promoting research and market development. Members were quickly alerted to the high price of peas, relative to wheat, and promptly responded. As a result, the area devoted to pea production in western Canada, expanded by 76% in 1986, 92% in 1987, and 16% in 1988, at which time the pea price differential had temporarily disappeared due to a two-year increase in the price of red spring wheat. During this time the pea price peaked (1985), the premium relative to wheat peaked (1986), Saskatchewan took over the lead in pea production from Manitoba (1987), and pea production in Canada peaked

Table 1. Dry pea production, dry pea price, red spring wheat price and pea price as percent of red spring wheat price for western Canada, 1968–1992 (Saskatchewan Agriculture and Food, 1991a)

Year	Dry pea (ha)	Dry pea price Can. $ t⁻¹	Red spring wheat price Can. $ t⁻¹	Pea price as % of red spring wheat price
1968	22,000	67	62	108
1969	29,500	67	62	108
1970	33,000	71	61	116
1971	31,500	73	59	124
1972	28,000	93	79	118
1973	26,000	167	168	99
1974	30,000	156	164	95
1975	29,000	171	146	117
1976	27,000	171	117	146*
1977	37,000	231	120	193*
1978	42,000	147	161	91
1979	43,000	140	196	71
1980	48,000	193	222	87
1981	59,000	209	200	105
1982	78,000	198	192	103
1983	64,000	176	194	91
1984	74,000	201	186	108
1985	74,000	209	160	131*
1986	130,000	194	130	149*
1987	250,000	176	134	131*
1988	290,000	198	197	101
1989	176,000	181	172	105
1990	120,000	174	135	129*
1991	197,000	170	125**	136*
1992**	250,000	165	125	132*
1993	450,000***	---	---	---

*Pea price greater than 125% of red spring wheat price.
**Estimated.

(1988). Furthermore, the price premium in 1992 for peas, relative to wheat, resulted in a pea production record for western Canada in 1993 (over 400,000 ha)

Thus, western Canada has become one of the major dry pea producers in less than 15 years. This demonstrates how all individuals involved in an agricultural commodity ranging from the producer to the supplier of inputs, processor, contractor, broker, shipper, and researcher can work together and maintain open lines of communication so that the industry can rapidly respond to favorable prices.

References

Gubbels, G. H., Ali-Khan, S. T., Chubey, B B. and Stauvers, M. 1982. *Canadian Journal of Plant Science* 62: 893–899.

Gubbels, G. H., Chubey, B. B., Ali-Khan, S T. and Stauvers, M. 1985. *Canadian Journal of Plant Science* 65: 55–61.

Saskatchewan Agriculture and Food. 1991a. *Agricultural Statistics 1990*, pp. 84–85. Saskatchewan Agriculture and Food, Regina, SK.

Saskatchewan Agriculture and Food. 1991b. *1991 Specialty Crop Report*, 12 pp. Saskatchewan Agriculture and Food, Regina, SK.

Slinkard, A. E. and Blain, H. L. 1988. In: *World Crops: Cool Season Food Legumes*, pp. 1059–1063 (ed. R.J. Summerfield). Dordrecht: Kluwer Academic Publishers.

Vaillancourt, R. 1989. *Pea Varieties for Saskatchewan*, 2 pp. Saskatchewan Agriculture and Food, Regina, SK.

Wall, D. A., Friesen, G. H. and Bhati, T. K. 1991. *Canadian Journal of Plant Science* 71: 473–480.

Youngs, C. G. 1975. In: *Oilseed and Pulse Crops in Western Canada – A Symposium*, pp. 617–632 (ed. J. T. Harapiak). Calgary, AB, Canada: Western Cooperative Fertilizers Limited.

Addressing production constraints for cool season food legumes in West Asia and North Africa through on-farm research: problems and ways forward

J.D.H. KEATINGE[1], I. KUSMENOGLU[2] and D. SAKAR[3]

[1] ICARDA, PK 39 Emek, Ankara 06511, Turkey;
[2] Ministry of Agriculture and Rural Affairs, P. O. Box 226, Ankara, Turkey, and
[3] Ministry of Agriculture and Rural Affairs, P. O. Box 72, Diyarbakir, 2111 Turkey

Abstract

Overall production levels of food legumes in West Asia and North Africa have not increased substantially in the last decade and certainly not in accordance with their potential. In order to ensure rapid adoption by farmers of new improved technologies, we have examined whether or not scientists should or can have complete confidence in both their identification of critical constraints for on-farm trials, and in the robustness of their technological recommendations. The keys to future success are effective communication and true integration of research and extension efforts.

Introduction

Oram and Belaid (1990) have comprehensively reviewed the status of food legume production in West Asia and North Africa (WANA). They have highlighted the conclusion, disturbing for agricultural scientists, that production levels of food legumes in WANA remain low. This is in relation not only to their realizable potential, but also to that of the production of farmers' main optional crops: wheat and barley. Nevertheless, large increases in production area have occurred in the 1980s in specific countries as a result of fallow replacement policies (e.g., lentil and chickpea in Turkey, Figure 1). However, increases in cropped area have not been mirrored by expected increases in yields per unit area, irrespective of the recent emergence of improved production technologies and new cultivars with enhanced resistance to biotic stresses (Figure 2). Concern arising from the previous statement may be premature, but over the next 20 years can we confidently expect substantial increases in production in WANA as a result of increased use of legume/cereal rotations and from a marked increase in the presently low average yield (Table 1).

We believe a quick adoption of available improved technologies and germplasm will be required if this optimistic hope is to materialize. However, if

F.J. Muehlbauer and W.J. Kaiser (eds.), Expanding the Production and Use of Cool Season Food Legumes, 890–898.

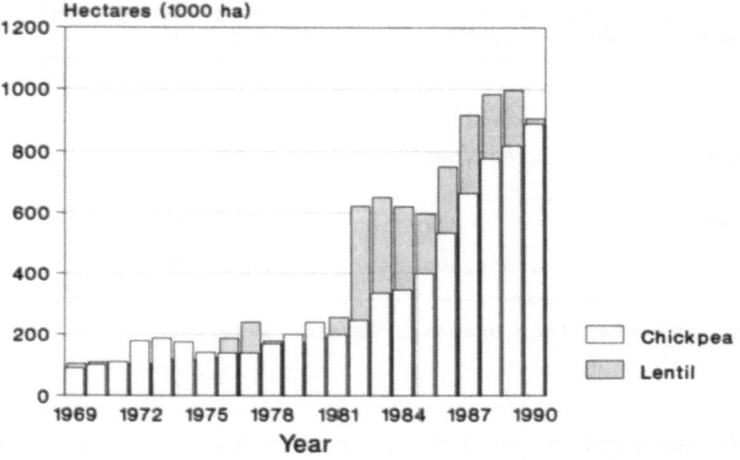

Figure 1. Lentil and chickpea area in Turkey 1969–90.

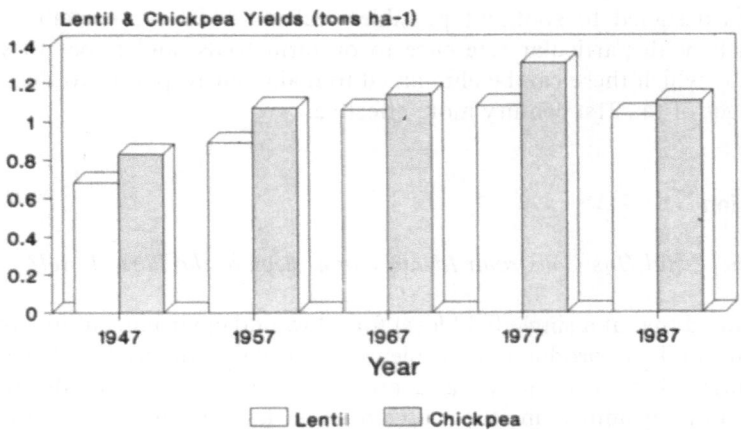

Figure 2. Yields of lentil and chickpea in Turkey 1947–87.

this is to be realized in the rapid manner in which, for example, semi-dwarf wheats were adopted by the farmers of the irrigated areas of the Indian sub-continent [80% between 1966 and 1982, CIMMYT (1989)], then we must be absolutely certain that our technologies are appropriate for farmers to use, economically profitable in both the short and long term, acceptable to cautious farmers, and supported by a strong consensus of the research, extension, and administrative services of the governments and private sectors of the countries in question.

Some rigorous soul-searching and alterations in strategic planning by the present research and extension community to chart a way forward for the next

Table 1. Area, yield and production of all cool season food legumes from the eleven principal producing nations in West Asia and North Africa: 1974/76 and 1985/87

	1974/76	1985/87
Area (ha)	3,685,000	5,069,000
Yield (kg ha^{-1})	831	840
Production (t)	3,062,000	4,259,000

Source: Extracted from Oram and Belaid (1990).

20 to 30 years may be required. It is to be hoped that, within this time period, food legume crops will achieve their potential place in the dryland cropping rotations of WANA and ensure that farmers are provided with robust, sustainable farming systems which will guarantee their future prosperity. This paper is designed to spotlight possible weaknesses in our current research approach, with particular reference to on-farm trials, and hopes to suggest means by which these can be eliminated to make our responses to the coming challenges of the 21st century more effective.

Discussion

How Successful Has Constraint Identification Been at the Farm Level?

We contend that at a single field level for a lowland environment, in any given year, the bulk of production problems has been recognized and improved technologies have been tested and proven which can substantially increase yields. For example, in the case of the kabuli chickpea, agronomic improvements have been comprehensively reported by Saxena (1987); and it is evident from ICARDA (1991) that adapted germplasm resistant to the major biotic stresses exists with adequate quality characteristics and acceptably large seed size. However, at the higher, crop rotational and farm levels, we feel that our assurance of adequate constraint identification and prioritization is considerably eroded. What are the causes of this increased uncertainty? We feel that they are largely related to a recognition of the increasingly complex nature of the problems for which we are seeking solutions. Below we outline four such areas of increased complexity and thus, concern:

1) If increased production levels are to be sustained, scientists must adequately address the increased complexity of decision-making involved in the transition from field to farm level. We are concerned, for example, with such things as the adequacy and means of weed control measures and their

required timing in relation to other farm operations, for crops such as lentil and winter chickpea, and the resistance of farmers to change current operational schedules (Guler, 1990). In addition, we are concerned with weed and herbicide carry-over effects in longer term crop rotations with cereals. Is our knowledge of these factors adequate (ICARDA, 1987)?

2) A further factorial increase in complexity of issues is observed when the complementary, and possibly conflicting, requirements of on-farm livestock are considered. Jones (1990) has highlighted the dual-purpose nature of many "food legumes" such as peas, lentil, and grasspea. This point underscores the need for crop scientists to review constraints in on-farm trials from a farmer perspective, especially for mixed crop and livestock enterprises. This perspective should have a profound effect on experimental trial design, the production variables monitored, and the economic evaluation of the technologies being tested. For example, Cocks and Thomson (1988) have indicated that even small additions of legume straws to the diets of sheep, largely relying on cereal straw and native pasture grazing, could result in higher cereal straw intake and thus substantially improve animal productivity. Therefore, such factors as plot sizes, harvest methods, and timing may need to be re-evaluated in the light of the need for optional grazing treatments for dual-purpose legumes, particularly in dry years, where harvesting costs can thus perhaps be avoided. Although this type of systems thinking in experimental design is reasonably well established at international levels, at the national research agency level it still requires considerable consolidation. This is most obvious in the need for linked efforts between crop and animal scientists. This does not necessarily mean that we recommend the institutionalization of farming systems research departments in national programs, but rather our belief is reflected in the need for a general commitment by all scientists to complex problem-solving with appropriate interdisciplinary planning, execution of experiments and analysis of results.

3) An additional element of complexity in production issues which impinges on both farmers and national production economists, arises from the varied product requirements of consumers in both internal and export markets. These issues should directly affect both on-farm trial management and measurements, as product quality has become of increasingly important concern in marketing and profitability. Blain (1982) states for the USA that "farmers in the Palouse country do a tremendous job of producing, but probably do one of the poorest jobs in the world in marketing their pea and lentil crops". The passage of ten years may have improved this situation in the USA, but we believe this comment on marketing attitudes remains highly relevant for WANA today. As a result, agricultural scientists must build consumer requirements more conscientiously into their products' germplasm whether they concern issues of quality, insect resistance, seed size or taste, and must subsequently test these in on-farm trials to confirm their suitability. For example, in Baluchistan province of Pakistan, lentils imported from

Turkey unnecessarily dominate the local market due to consumer preference. This is on the grounds of either larger seed size, or cleanliness of product, or both, when compared respectively with the locally produced landrace and cultivars from Sind and Punjab provinces (Keatinge *et al.*, 1990). Joint research between the Pakistan Agricultural Research Council and ICARDA is underway to address this problem (Asghar Ali *et al.*, 1991).

4) In consideration of further complexity in constraint identification and prioritization we suggest that national research organizations continue to undervalue the influence of abiotic stresses such as cold, heat, drought, and soil physical and chemical characteristics on yield. There remains a strong and perhaps growing tendency for national organizations to perform "on-farm type and scale" trials on the atypical environments of research stations, or to continue to rely on small plot trials. This may help explain the difficulties in closing the yield gap between farmers and researchers. The logic for retention of this less effective trial methodology is often unchallengeable. It results largely from increasing and perennial budgetary restrictions which limit operational expenditure. Many examples of this harsh reality exist in WANA today.

In the previous section, we raised areas of doubt regarding our identification and ability to prioritize current production constraints for examination in on-farm trials. In the following section we will examine whether our experimental technique in developing solutions to the problems we have correctly identified is as robust as we would desire.

How Confident Are We in our Current Experimental Technique?

The "standard" methodology of crop research – small plot on-station trials, followed by on-farm trials and by extension demonstrations – has a proven track record of success, particularly for cereal crops grown in favorable environments over the last 30 years. However, the record is certainly less clear in the case of crops grown in harsh environments, and especially those which may involve an animal component in the legume phase of the crop rotation. Reappraisal therefore of some of the assumptions underlying our technique could be of value. We examine four further broad areas of concern:

1) We have mentioned in passing the issue of appropriate plot sizes for experimental trials involving livestock. Large plots in the eyes of a crop scientist are often very small plots from the perspective of an animal scientist. In on-farm trials and crop rotational experiments, plot-size factors may be critical in the economic viability and maintenance of the research effort. However, experiments below the minimum plot size to allow animal grazing, which are the norm in WANA today, are also probably too small to permit adequate tillage treatments. Immediately, this partially undermines the robustness of our agronomic recommendations, as factors such as tillage can be basic to determining the responses from other commonly tested variables

such as fertilizer, weed control, sowing depth, and stand density (ICARDA, 1988). Other experimental variables may also be biasing results. Keatinge and Somel (1989) mention the possibility that damage to small plots in rotation trials from faunal agencies such as birds may be overestimated, whereas disease and insect damage effects may be underestimated. Economic and logistical realism may preclude our use of the perfect plot size in on-farm research, but we should at least be aware of potential underlying biases and account for these in our recommendations for new technologies.

2) A similar economic logic may also apply to the ideal time course for experimental trials. However, this is perhaps a more serious problem in on-farm trials than bias due to plot size. In the dry areas of WANA, problems arise for agricultural scientists not only because of drought but also because of the variability in amount of precipitation and other environmental factors experienced. Therefore, the "standard" three-year experiment is only of equivalent robustness to similar work in irrigated systems if scientists have a detailed knowledge of the probability of meteorological events for the specific location at which they are working, and can estimate effectively whether or not their spread of seasonal results represents anything like "average" conditions. Without such detailed agroclimatic information it may be necessary to perform an experiment for 10 years before confidence in the results can be truly obtained. Better awareness of the interactions between crop growth and environmental factors, and better categorization of target environments, seem to us to be primary requirements for possessing confidence in future technological recommendations.

3) Oram and Belaid (1990) state very bluntly that "very few rotation studies include an economic evaluation of the results" for WANA. Underlying this pointed but undeniably valid statement is a host of complex factors such as establishing realistic seed and product prices for introduced crops, choice of inflation rates, choice of which yield variable to measure (grain, hay, feed values, milk yield, lambing percentage, or live-weight gain) (Jones, 1989; Keatinge and Somel, 1989). However, the reality is that most trials in WANA, not just rotation trials, are not getting effective economic analysis. This is probably because it is a very difficult task and requires considerable commitment by economists and agronomists from the first stages of experimental design to the final outcome. This economic support is often either not available or there is inadequate appreciation by crop scientists of its need. Therefore, in our opinion, uncertainty about the soundness of subsequent technological recommendations is valid. We must strive to ensure that for both on-farm trial design and management we account for the needs of a subsequent economic analysis in all cases, and that the objectives we set for our experiment are realizable in a real-world logistical and economic framework.

4) A further complicating factor in determining a confident outcome from on-farm trials is usually the need for determining whether or not a suitable supporting infrastructure exists to render our research results valid. In

Turkey, for example, if a Ministry of Agriculture scientist conducts trials involving new lentil, faba bean, or chickpea germplasm and produces successful results, then, once the registration authority regulations are satisfied, the Turkish State Farm and private sector production system for registered seed takes the cultivar. This system is of such efficiency and size that farmers can get newly released cultivars, freely throughout the country, in as little as two or three years. This is rarely the case in other countries in WANA, and this problem with infrastructural support, particularly in seed supply, availability of agricultural chemicals and credit to resource-poor farmers, is a major problem influencing the "success" of on-farm trials (Erskine *et al.*, 1990). For example, Cocks and Thomson (1988) report a series of ICARDA on-farm fallow replacement trials with grasspea (*Lathyrus sativus* L. and vetch (*Vicia sativa* L.) which were very "successful" in increasing farmers' profits substantially and the introduced technologies were well accepted by farmers. However, in 1991, just three or four years later, no farmers were growing these crops because they claimed they were unable to either obtain or afford the legume seed (Thomson, pers. comm., 1991). An early commitment to supply needs arising from a new technology by the private sector or state agricultural supply enterprises seems to be a precondition for rapid and sustained adoption of new technologies "proven" by on-farm research.

We have attempted to expose possible areas of concern in current efforts on constraint identification and problem-solving methodology in order to ensure that on-farm legume research has every chance to spark a rapid adoption of new technologies by farmers. However, as it is a less challenging exercise to be negative rather than positive, we will attempt in the final section of the paper to make additional constructive recommendations which may help overcome the areas of weakness to which we have referred.

The Way Forward?

Many of the problems of effective constraint identification, prioritization, and their ultimate removal may be overcome by better communication. With a systems perspective this will involve initial discussion, planning, and execution of on-farm trials with the active participation of other scientific disciplines, extensionists, farmers, agricultural suppliers (state and/or private sector), and policy makers. It involves neither a top down nor bottom up "theoretical" approach, but rather a practical understanding of the human, climatic, economic, and policy environment in which the trial is to occur. If this interchange of ideas is comprehensive, then the probability of missing a critical economic, animal management, ecological, climatic, or social constraint to successful food legume crop production is less likely.

The adoption of an integrated team approach, which has proved substantially effective for on-farm faba bean research in the Nile valley, by ensuring an efficient blend of interdisciplinary research, needs to be genuinely – not cosmetically – adopted in other countries. Much more effort needs to be made to ensure the active participation of extension authorities and staff in the research process to ultimately allow them to be truly committed to their job of providing a communication link to and from farmers, and introducing the new technology to the widest possible audience (Talug *et al.*, 1989a,b).

Government and private sector support for infrastructural development, credit and particularly seed supply during the wide-scale testing and subsequent extension of new technologies seems to be vital to ensure broad adoption of technologies. Durutan *et al.* (1990) have described the Turkish Government and World Bank's "Utilization of Fallow Areas" project. This has a major component of on-farm research, and since 1982 has been a major contributor to increased legume production in WANA. This seems to be a concrete role model in which an on-farm systems approach was well integrated with extension efforts, and in which farmers received, in good time, the financial and infrastructural support required from government and private sector agencies.

Finally, in support of good linkages with extension services and policy makers, scientists need to become more aware of, and adept with, modern communication technology. The era of literate farmers and widespread forms of mass communication is now with us. We must educate ourselves to make effective use of video, radio, and television technology to communicate our research results to the farming community and to no longer accept that the end of the job is the publication of a refereed journal article.

In conclusion, we urge all researchers to fully grasp the complexity of problems associated with food legume production in uncertain environments, and to absorb and share the full range of knowledge available to them with their research, extension, and farming colleagues. Next, they must conduct and fully analyze their on-farm trials in the biological, social, and economic dimensions. Finally, in partnership with their extension colleagues, they must produce robust, sustainable technologies which farmers can adopt with full confidence. In this manner, we expect food legume production to become one of the foundations of food security in West Asia and North Africa throughout the 21st century.

References

Asghar Ali, Keatinge, J. D. H., Roidar Khan, B. and Sarfraz Ahmad 1991. *Journal of Agricultural Science, Cambridge* 117: 347–354.

Blain, H. 1982. *Proceedings of the Palouse Symposium on Dry Peas, Lentils and Chickpeas*, pp. 1–2. February 23–24th, 1982, Moscow, Idaho.

CIMMYT. 1989. *1987–88 CIMMYT World Wheat Facts and Trends. The Wheat Revolution Revisited: Recent Trends and Future Challenges*. Mexico, DF:CIMMYT.

Cocks, P. S. and Thomson, E. F. 1988. In: *Increasing Small Ruminant Productivity in Semi-arid Areas*, pp. 51–66 (eds. E. F. Thomson and F. S. Thomson). Dordrecht: Kluwer Academic Publishers.

Durutan, N., Meyveci, K., Karaca, M., Avci, M. and Eyuboglu, H. 1990. In: *The Role of Legumes in the Farming Systems of the Mediterranean Areas*, pp. 239–256 (eds. A. E. Osman, M. H. Ibrahim and M. A. Jones). Dordrecht: Kluwer Academic Publishers.

Erskine, W., Nordblom, T. L., Cocks, P. S., Pala, M. and Thomson, E. F. 1990. In: *The Role of Legumes in the Farming Systems of the Mediterranean Areas*, pp. 273–282 (eds. A. E. Osman, M. H. Ibrahim and M. A. Jones). Dordrecht: Kluwer Academic Publishers.

Guler, M. 1990. In: *The Role of Legumes in the Farming Systems of the Mediterranean Areas*, pp. 131–142 (eds. A. E. Osman, M. H. Ibrahim and M. A. Jones). Dordrecht: Kluwer Academic Publishers.

ICARDA. 1987. *Annual Report for 1986 – Farming Systems Program*. ICARDA 108 En. Aleppo, Syria: ICARDA.

ICARDA. 1988. *Annual Report for 1987 – Farm Resource Management Program*. ICARDA 131 En. Aleppo, Syria: ICARDA.

ICARDA. 1991. *Annual Report for 1990 – Food Legume Improvement Program*. ICARDA 202 En. Aleppo, Syria: ICARDA.

Jones, M. J. 1989. *Barley Rotation Trials at Tel Hadya and Breda Stations*. ICARDA 140 En. Aleppo, Syria: ICARDA.

Jones, M. J. 1990. In: *The Role of Legumes in the Farming Systems of the Mediterranean Areas*, pp. 195–204 (eds. A. E. Osman, M. H. Ibrahim and M. A. Jones). Dordrecht: Kluwer Academic Publishers.

Keatinge, J. D. H. and Somel, K. 1989. *Agronomic Management and Economic Interpretation of Small Plot Crop Rotation Trials in Northern Syria*. ICARDA 150 En. Aleppo, Syria: ICARDA.

Keatinge, J. D. H., Buzdar, N., Farid Sabir, G., Afzal, M., Shah, N. A. and Asghar Ali. 1990. *Lens* 17: 13–15.

Oram, P. and Belaid, A. 1990. *Legumes in Farming Systems*. ICARDA 160 En. Aleppo, Syria: ICARDA.

Saxena, M. C. 1987. In: *The Chickpea*, pp. 207–232 (eds. M. C. Saxena and K. B. Singh). Wallingford: C.A.B. International.

Talug, C., Chowdry, M. B. A., Ali, A. and Aslam, M. 1989a. *MART/AZR Research Report* 38. Quetta: ICARDA.

Talug, C., Aslam, M., Ali, A. and Chowdry, M. B. A. 1989b. The role of extension in FSR. *MART/AZR Research Report* 41. Quetta: ICARDA.

Approaches to overcoming constraints to winter chickpea adoption in Morocco, Syria, and Tunisia

R.N. TUTWILER[1], M. AMINE[2], M.B. SOLH[3], S.P.S. BENIWAL[4] and
M.H. HALILA[5]

[1] ICARDA, P. O. Box 5466, Aleppo, Syria;
[2] DPV, Ministry of Agriculture and Agrarian Reform, Rabat, Morocco;
[3] ICARDA, P. O. Box 2416, Cairo, Egypt;
[4] ICARDA, Legume Improvement Program, B. P. 2335, Fes, Morocco, and
[5] Institut National de la Recherche Agronomique de Tunisie (INRAT), 2080 Ariana, Tunis, Tunisia

Abstract

Following the initial release of winter sown chickpea cultivars in Morocco and Syria, surveys of winter sown producers were done to determine initial acceptance and adoption rates. There were vast differences in acceptance from year to year and place to place, ranging from 80 to 20% in Morocco and from 90 and 29% in Syria. The overall rates were about 50% in both countries. The principal constraints in Morocco are small seed size and poor marketing, although weed control and susceptibility to Ascochyta blight are problems. In Syria adverse climatic conditions and farmers' risk avoidance practices restrict winter chickpea adoption. Tunisia has had less experience in transferring the technology, but all three countries are taking concrete measures to overcome constraints and improve benefits to farmers.

Introduction

Rainfed chickpeas (*Cicer arietinum* L.) grown in the Mediterranean basin are traditionally planted in the spring (February to May) and are largely raised on residual soil moisture since precipitation in this environment generally falls during the winter months. Yields tend to be low. Limited moisture may restrict yields, but spring planting also means that the reproductive growth phase coincides with sharply increasing and possibly limiting temperatures. An obvious strategy for overcoming the moisture constraint is to advance the planting date into the winter rainy season, thereby avoiding the late spring and summer dry periods of high temperature and allowing the crop to take maximum advantage of available moisture.

ICARDA research in the 1970s demonstrated the potential inherent in a winter planting strategy, at least for low to medium elevations where winter cold is not too severe. However, as temperatures rise in the presence of moist conditions, the young plants are susceptible to attack by Ascochyta blight, a disease which left unchecked can destroy the entire crop. Researchers concluded

F.J. Muehlbauer and W.J. Kaiser (eds.), Expanding the Production and Use of Cool Season Food Legumes, 899–910.
© 1994 Kluwer Academic Publishers.

that avoidance of Ascochyta blight, rather than freezing temperatures is probably the principal reason why farmers plant chickpea in the spring rather than winter. Since no clearly effective means of chemical control exists, the development of new cultivars resistant to Ascochyta and also tolerant to cold has been followed as the best method of overcoming chickpea production constraints in the Mediterranean basin. The effort has been successful. National Agricultural Research programs in ten circum-Mediterranean countries released some 23 cultivars of chickpeas for winter sowing during the period 1984 to 1989. Verification trials demonstrated the yield advantages of winter planting with the new cultivars throughout the region.

In 1989 ICARDA began a partnership with a number of national programs for a series of farm-level studies to assess the acceptability of winter chickpea among the region's farmers. The studies are designed to establish a dialogue between the scientists who are developing winter chickpea technology and the farmers who are the intended beneficiaries. The studies have several objectives: 1) identifying the farm-level incentives and constraints regarding adoption of winter sown chickpea, 2) evaluating performance of the new technology in farmers' hands, 3) assessing the benefits farmers derive from winter chickpea, and 4) providing farmers with the means to participate in the further development of the new technology.

Winter Chickpea in Morocco

Technology Transfer Strategy and Acceptability

Total Moroccan chickpea area fell by almost one half between the early 1970s and mid-1980s. Yields during the same period fluctuated wildly between a peak of about 1,000 kg ha^{-1} in 1979 to around 200 kg ha^{-1} in 1981. Yield variations correspond closely with variations in precipitation. The average since 1985 is near 800 kg ha^{-1}. There has been a marked change in the geographical distribution of chickpea in Morocco. Formerly highly productive regions to the south and southwest of Casablanca have considerably reduced chickpea areas, and many farmers have abandoned chickpeas altogether. The retreat has been attributed to various causes: a prolonged drought, the ravages of Ascochyta blight, and the rising costs of hand labor for harvest.

Moroccan authorities have given considerable attention to winter chickpea technology as a possible way of improving and stabilizing national production trends. Beginning with germplasm developed at ICARDA, scientists at the Institut National de la Recherche Agronomique (INRA) started the process of adapting winter chickpea technology to the conditions of Moroccan farmers. INRA followed a dual approach: on-station work concentrated on testing germplasm and cultural practices while on-farm trials aimed at verifying the new technology and identifying the constraints to achieving maximum yields. Two cultivars, ILC 482 and ILC 195 were catalogued and released in 1987/88.

A nation-wide program of farmer-managed demonstrations was begun by the Direction de la Production Vegetale (DPV) in the same year.

The results were promising. The demonstration farmers obtained much better yields with the new winter cultivars as compared to the local spring chickpea. The demonstration program was expanded for the 1988/89 season. However, there had been little in the way of analyses of farmers' reactions to the new technology nor assessment of its adaption potential. The gap was filled with monitoring surveys of farmers participating in the demonstration.

Four categories of farmers were covered in the first survey conducted in 1988/89: 1) participants in trials and demonstrations prior to 1988, 2) spontaneous adopters who had not participated in trials and demonstrations, 3) participants in the demonstrations in 1988/89, and 4) farmers who had knowledge of winter chickpea but had not actually grown the crop. These farmer categories allowed the comparative evaluation of incentives and constraints to adoption according to farmers' levels of experience with the new technology. By including farmers not growing winter chickpea, it was possible to compare the performance of their local spring cultivar with the performance of the winter cultivars.

The sample was further divided into four groups, each representing a different agricultural environment. The northern provinces of Fes-Taounate and Khemisset represent current principal producing areas of chickpea. The two southern provinces, Safi and Settat, are previously important chickpea areas in which chickpea has been largely replaced by cereals, forage maize, and weedy fallow. Safi and Settat are lower rainfall areas, and it is hoped that winter sown cultivars will do well there and facilitate the re-introduction of chickpea into these important agricultural regions.

By surveying pre-1988 trials and demonstrations participants, it was possible to estimate an initial acceptance rating for winter chickpea. Overall, winter chickpea had been adopted by 61% of farmers who had experience of growing the new cultivars prior to the 1988/89 season. For the purposes of measuring acceptability, adoption was defined as the decision to grow the crop in the years following an initial year of production. Acceptance and adoption ratings are based only on the evaluation by those farmers with at least one year's practical experience growing winter chickpea.

Farmer acceptance of winter chickpea prior to 1988/89 was based largely on the significantly higher yields obtained when compared to traditional spring chickpea. However, there had been no major attacks of Ascochyta blight prior to 1988/89. Non-adopters prior to 1988/89 reported dissatisfaction with the smaller seed size of winter sown cultivars and the costs of hand weeding the winter sown crop. Because it is planted after much of the winter rain has fallen, the weeds in spring chickpea fields are mostly controlled by pre-planting tillage operations, but winter sowing usually requires additional weed control operations.

Climatic conditions during the 1988/89 growing season were very poor for chickpea. Although accumulated rainfall was about average, the distribution was uneven. Heavy April rains created conditions conducive to the spread of Ascochyta blight in all areas except Fes province, but here the absence of Ascochyta blight was little consolation to the majority of farmers who suffered from extraordinary weed infestations and hail storms. Spring chickpea suffered equally because of the lateness of the Ascochyta blight development. Overall, the mean yield for winter chickpea was 970 kg ha^{-1}, and spring chickpea was 840 kg ha^{-1}. But in Settat and Khemisset provinces, spring chickpea managed to outyield winter chickpea by some 360 kg ha^{-1} and 180 kg ha^{-1}, respectively.

The ravages of the 1988/89 Ascochyta blight epidemic are reflected in the acceptance rating of first-time winter chickpea producers in that year, which fell to only 40%. The extent of the setback to adoption of winter cultivars is further revealed by the abandonment of previous adopters. Some 26% of pre-1988/89 winter chickpea adopters decided to no longer grow the crop after 1988/89.

Based in part on the results of the 1988/89 survey, the Moroccan national program revised their technology transfer strategy for the 1989/90 season. First of all, the survey showed that ILC 482 had been more affected by Ascochyta blight than had ILC 195, and it was decided to use only ILC 195 in the 1989/90 demonstrations. Second, because Ascochyta blight had been less severe in Safi and Fes provinces, the demonstration program for the next year would put relatively more emphasis on these provinces and adjacent areas away from the Settat and Khemisset hot-spots. Third, the extension agents who were intro-ducing winter chickpea to demonstration participants were given additional training in the various agronomic practices associated with winter sowing so that they could give advice to farmers should problems arise. Fourth, a weed control component was added to the demonstrations should farmers request it. Weed control was intentionally targeted to the demonstrations in the wetter areas.

The success of these changes is amply demonstrated from the results of the 1989/90 demonstrations monitoring survey. The acceptance of winter chickpea among farmers growing the new cultivars for the first time in 1989/90 was 70%. This constitutes a major recovery for efforts to introduce the new technology. It is all the more significant because there was practically no dis-adoption of winter chickpea in the 1989/90 season.

The monitoring surveys conducted in 1988/89 and 1989/90 enabled the Moroccan national program to construct a baseline for farmer acceptance and adoption of winter sown chickpeas that is built upon actual farmer evaluative decisions and farmer perceptions of the incentives and constraints associated with the new technology. Of the 121 farmers with experience growing winter chickpea and interviewed in either one or both years of the survey, some 48% elected to adopt the new cultivars. Moreover, the new cultivars provided them with sufficient incentives to, on the average, significantly increase the annual area they plant to chickpea. Adoption is not uniformly distributed, however. Acceptance is greatest (68%) in Safi province and slightly less in Fes-Taounate region (56%). Winter sown cultivars are not particularly acceptable in the Ascochyta blight hot spots of Settat and Khemisset (only 32 and 28%, respectively).

Lessons Learned

The experience with farmer-managed demonstrations and monitoring surveys in Morocco provides the lesson that introducing winter chickpeas at the farm level is not mechanistic nor is there a fixed prescription which can be followed for all farmers in all locations in all years. Rather, the transfer from researchers to farmers is processual and interactive. Above all, the strategy followed should be flexible and capable of considerable revision. A key part of the process from one year to the next is the dialogue conducted between farmers and researchers through the technology transfer activities. In this case, the dialogue was established through farmer evaluations of new cultivars. By quantifying these evaluations through monitoring surveys, the Moroccan national program was able to either verify or falsify a number of *a priori* assumptions about the acceptability of winter sown chickpea.

An important discovery from the surveys is the importance farmers place on the "information constraint". This was especially revealed in questions about factors limiting adoption of winter chickpea posed to all farmers sampled, including those who knew something about the new technology but had not yet had direct experience growing winter sown cultivars. When assigned a weighted value according to frequency mentioned and priority rank assigned by responding farmers, more information about winter chickpea technology received over twice the value of the next highest constraint, that being the small seed size. This appears to confirm a distinction made by many researchers in the region between simply planting chickpea in the winter rather than spring and understanding that winter planting requires new cultivars which are resistant to Ascochyta. Farmers clearly understand the potential moisture, and therefore yield, advantage of early planting, but what they want is information about the new cultivars and the associated agronomic practices.

Comparisons among different categories of farmers according to experience with the new technology revealed another important dimension to the incentives/constraints matrix. Although diseases and pests were seen as very significant problems encountered in production, their significance diminished considerably when making the decision to adopt the new technology and to expand winter chickpea area. Despite the Ascochyta blight attacks of 1988/89, farmers generally ranked winter chickpea as better than local spring chickpea for resistance to diseases and pests. Winter cultivars also got much better ratings for other characteristics which breeders have carefully selected: higher yield, more stable yield, greater straw production, and plant stand and growth. Also mentioned as distinct economic advantages were the winter cultivars earlier maturity and taller stature allowing mechanical harvesting. Harvesting a month or more earlier than the traditional spring cultivar (which usually coincides with the area harvest) means that poor farmers can take advantage of off-peak lower wage rates, higher chickpea prices before the spring sown harvest glut, and obtaining needed cash to pay off debts and finance the upcoming cereal harvest. For richer farmers with larger fields, winter chickpea's facilitating mechanical harvesting means cheaper costs by reducing labor requirements.

The principal factor constraining future adoption and expansion of winter chickpea technology, other than lack of information about the new technology, is the small seed size of the released cultivars. This has been long suspected by Moroccan scientists. The reason for this is not the technical aspects of production, seed quality, or even food quality, so much as the Moroccan consumers' preference for eating their chickpeas boiled and whole. Large seededness is greatly preferred from an aesthetic point of view. In terms of producers, this consumer preference is usually translated into a range of prices offered by purchases, with large seeded chickpeas such as the local spring cultivar having a distinct advantage over the smaller seeded winter cultivars.

However, the constraint does not apply uniformly to all existing or would be winter chickpea producers. First of all there are those producers who recognize the advantages of economies of scale. By growing large areas mechanically harvested and receiving a 50% or more yield advantage over spring chickpeas, a number of winter sown adopters easily accept a lower selling price because their increased volume of production still results in higher profitability. Some small producers using mostly family labor with a low opportunity cost prefer winter chickpea because of a perceived lower risk of crop failure. Nonetheless, the importance of the aesthetic value placed on large seededness should not be undervalued. A number of farmers disregarded discussions of profit margins and reduced risk and flatly stated that their self-esteem as farmers would not allow them to sell their chickpeas at a lower price than their spring sowing colleagues.

The price disadvantage of winter sown chickpeas in the 1988/89 and 1989/90 seasons, however, varied greatly by location, time, and farmer market position. In general the larger producers and the early sellers experienced the smallest price disadvantage due to seed size. They also tended to have access to better market outlets, such as wholesales and even exporters in large cities. Those most dissatisfied with the seed size and subsequent prices tended to be medium to small producers who relied on local weekly marketplaces and selling in small quantities from time to time. Sales in small quantities in local marketplaces accounted for well over half the producers. Ironically, some of the most vocal advocates of winter chickpea are small farmers (and often sharecroppers) who rely on casual and local marketing. But these farmers sell their chickpeas as seed supplies to their neighbors, receiving good prices but taking the risk of dissatisfied customers should there be problems the following season.

The lesson learned is twofold. First, winter chickpea is evaluated by farmers for its commercial potential, not for subsistence production or its role in a complex dryland farming system. Second, the commercial incentives and constraints for winter chickpea are different for different farmers. This is not simply a question of technical production problems or even calculated net benefits. Rather, circumstances external to the farm, particularly access to market outlets must be considered along with internal circumstances, such as farmer self-evaluations of themselves and the imputed value they assign to unpaid family labor. To illustrate this point, standard partial budgets were calculated using imputed values for family labor (based on prevailing wage rates for individual activities and times) and prices averaged for locations and times. Profitability on this basis could only account for some 60% of actual winter chickpea adoption. Obviously, although farmers overwhelmingly cite price incentives and profitability in their evaluations, other factors, perhaps other than those employed in standard economic analyses, need to be considered.

Actions Taken

Based upon the results of the demonstrations and monitoring surveys, the Moroccan national program has taken a number of positive steps to improve the adoption of winter sown chickpea. The multiplication of the released cultivar which proved susceptible to Ascochyta blight in 1988/89 has been suspended and is not being used in the on-going demonstration program. While continuing to emphasize resistance to blight, the breeding program is paying particular attention to seed size as a criterion for new cultivars. There is increased research attention to crop management, especially weed control. As for market demand and marketing factors, a food legume sector study was conducted in 1991 which specifically examined the question of prices and market outlets. This research is on-going, as is farmer participation in the further improvement of winter chickpea technology in Morocco.

Winter Chickpea in Syria

Technology Transfer Strategy and Acceptability

The first winter sown chickpea cultivars in Syria were "Ghab 1" (ILC 482) and "Ghab 2" (ILC 3279), both released in 1986, one year before winter chickpea was released in Morocco. Average yields (presently about 650 kg ha^{-1}) have tended to decline in Syria since the 1960s, although the annual area sown to chickpea has increased 60% since 1967 despite wild year-to-year fluctuations. The technical problems of producing chickpea in Syria are slightly different from those in Morocco. Disease pressure is somewhat less severe, but problems associated with erratic rainfall and, especially, frequent killing frosts as late as March are of great significance. Winter chickpea technology is appealing in Syria because of its promise to stabilize area planted and reverse the negative trend in yields.

A comparison of the coefficients of variation among chickpea area planted, yield, and precipitation supports the view that rainfall has relatively greater impact on the variability of area planted than on yield variation *per se*. For most rainfed crops in Syria, one would expect yield to vary more from year to year than the variation in area planted, because planting is done at the start of the rainy season according to the farmer's production strategy and resource availability without knowledge of future rainfall. Spring chickpea, however, is anomalous in terms of the statistical relationships between rainfall, area planted, and yield.

Farmers are aware of winter precipitation levels before they plant, and for this reason yields of spring chickpea may be less variable in relation to rainfall because in dry years many farmers may simply choose not to plant, thereby saving the costs of production and avoiding the risk of crop failure. Having decided not to plant chickpea, then they can either fallow the intended spring chickpea land or they will wait to see if enough additional rain falls in the spring to warrant planting a summer crop. Spring chickpea is one of the least risky crops in the rainfed system, not because of its yield level, but because of its yield predictability. The existence of this apparent risk reducing dimension for spring, but not for winter, chickpea has important implications for the adoption of winter sown cultivars, for the substitutability of winter sown for spring sown chickpea, and thus for the goals of reducing annual variations in area planted and increasing national chickpea production.

The principal chickpea production areas fall within two of the rainfall-based agricultural stability zones established by the government. Zone 1 has a mean annual rainfall of over 350 mm and is located along the coastal plain, the coastal mountains, and the Jawlan plateau in the south. It also includes an area to the extreme northeast in the Jazirah near the Tigris River. Zone 2, which has an annual rainfall of 250 to 350 mm with no less than 250 mm falling during two-thirds of the years, lies adjacent to Zone 1 to the east and south behind the western mountains and across the Jazirah. Within the two Zones, there are two geographical areas which together constitute about 95% of the chickpea area. These are the Southwest, mostly Zone 2, which has a median of 60% of the national area, and the Northwest, mostly straddling the line between Zone 1 and Zone 2. Chickpea is only of minor importance in the Jazirah, where lentil is the favored legume. This distribution indicates the significance of the risk aversion factor associated with spring chickpea. Over half the median area lies in Zone 2, in which rainfall is less reliable than in Zone 1. In the northwestern Zone 1 there is greater danger of frost, and apparently farmers utilize their more favorable moisture conditions to grow other, perhaps more profitable, crops.

The Syrian National Program has followed a strategy for transferring winter sown chickpea technology that is very different from the one used in Morocco. Rather than devoting themselves to a modest program of targeted demonstrations in the years following release, the Syrian decision was to undertake a large-scale program of seed multiplication using private farmers under contract to the General Organization for Seed Multiplication (GOSM). Multiplication was done on plots of one to 12 hectares. The results were encouraging. Yields were high, there was no major incidence of diseases or pests, and economic analyses showed high profit margins. Government marketing organizations set attractive purchasing prices for chickpeas. Once GOSM had accumulated sufficient stocks of certified seed, a media campaign and the extension services were used to inform farmers about winter chickpea's advantages and the availability of the new cultivars.

General distribution of the new cultivars began in 1989. In the same year, the Socio-Economic Studies and Training Section of the Syrian Scientific Agricultural Research Directorate, together with ICARDA scientists, organized a farm-level survey to assess the performance of the new technology under farmer conditions and to obtain an evaluation from the farmers themselves of the potential for adoption and positive impact.

The survey was conducted for two successive years, and the sampled farmers were selected from lists of farmers growing winter chickpea in different locations provided by the Ministry of Agriculture and Agrarian Reform and GOSM. Due to limited resources, it was not possible to include in the sampling universe farmers who had obtained seed outside official release channels, such as those who may have received seeds from farmers participating in past on farm trials with the Ministry and ICARDA, although it did include farmers who had been part of the multiplication program. Nonetheless, the lists of farmers purchasing seeds constitutes an appropriate and adequate starting point for establishing a baseline for evaluating the adoption process.

The sample contrasted farmers on the basis of their experience with winter chickpea: those growing for the first time in 1989/90 or 1990/91 (68% of the sample) and those already with a year or more previous experience (32% of the sample). About a third of the farmers were also growing spring chickpea. The sample was distributed over Zone 1 in three provinces in 1989/90: Aleppo and Hama (Zone 1, northwest) and Hassakah (Zone 1, northeast Jazirah). In 1990/91, Daraa province (Zone 2, southwest) was added to the other three locations. Hassakah was included because research trials indicate it has great potential for maximizing winter chickpea performance and impact.

Winter chickpea acceptance and adoption were defined in the same way as in the Moroccan study. Overall, of the 185 farmers surveyed in the 2 years, 55% found the new cultivars acceptable and had adopted them. As in Morocco, the total average masks differences in the year-to-year and location-to-location acceptance rates. In 1989/90, the acceptance rate in the three provinces covered averaged 57%, but it ranged from only 32% in Hama (where there were killing frosts and little rain) to 58% in Aleppo and 78% in Hassakah (where temperatures were mild and rainfall adequate). Acceptance among farmers growing winter chickpea for the first time in 1990/91 averaged 50% over the four provinces covered that year. Again, Hama and Aleppo were disappointing, with acceptance rates of 50% and 29%, respectively. Daraa had a rate of 45%, and Hassakah had a rate of 90%. Acceptance over the 2-year period by province are: Hama 36%, Daraa 46%, Aleppo 53%, and Hassakah 81%. These results indicate that the present Syrian National Program idea to target winter chickpea towards the Jazirah area of Hassakah province is justified.

Lessons Learned

The starting point for assessing adoption decisions is yield performance, since the principal purpose of winter sowing is to raise productivity per hectare. Some 85% of the surveyed farmers said that the higher yields promised by winter sowing was one reason they chose to grow the new cultivars. The potential higher profitability of winter chickpea was cited by only 35% of the farmers. The actual yields reported by farmers correspond closely with their adoption decisions. Those farmers with yields that were disappointing in comparison to anticipated spring chickpea yields in the same year in the same location tended to find the new cultivars unacceptable.

Climatic conditions were cited most frequently as the biggest problem encountered by winter chickpea producers, and not surprisingly this was mentioned more often by non-adopters than those who decided to adopt the new cultivars. It would appear that frost and low rainfall were the most discouraging factors in Hama and Aleppo provinces, even though winter chickpea actually performed as well as or better than local spring chickpea in these locations during the two years. Low rainfall was also a factor in Daraa, were a number of farmers said that they would not have normally planted spring chickpea in 1990/91 because of the disappointing rains. In contrast, mild temperatures and adequate rainfall in Hassakah allowed farmers the full yield advantages of winter sowing.

Economic constraints appeared less important than adverse climatic conditions. Only 24% of farmers cited higher cost to benefit ratios for winter as opposed to spring chickpea. This is in sharp contrast to the Moroccan case where winter chickpea suffers adverse prices and additional weed control requirements raise production costs and lower profit margins for many producers. There was no significant price difference between winter and spring cultivars in Syria, winter cultivars tended to outyield spring chickpeas, and weed control was not reported as a major constraint.

What does show up in the survey results is the relatively greater economic incentives to adopt winter chickpea felt by the larger, commercial-oriented farmers than smaller producers. Large producers contrast with small producers in terms of the mechanization of production. The large producers, who are mostly located in Hassakah, tend to have mechanized all operations, including weed control by spraying herbicides and harvesting winter chickpea by combine, whereas small producers continue with more costly hand weeding and hand harvesting. Large producers are evidently more willing to take the rainfall and frost risks associated with winter planting chickpea, because they cannot afford to leave large areas of land fallow.

Although adopters of winter chickpea have a much larger average farm size than do non-adopters, this figure is distorted by the very large farms and the high rate of adoption in Hassakah. When adoption and non-adoption is divided between farm size categories, some 44% of small farms (i.e., those less than 35 ha) have adopted the new cultivars. The adoption rate among large farms is

68%. Perhaps more indicative of the impact of winter chickpea is the area planted in either type of chickpea in the year following the adoption or non-adoption decision. Winter chickpea adopters, both large and small size farmers together, will increase their chickpea area by 25.2%. Those who do not adopt will increase their (spring) chickpea area by only 1.2%. If this trend continues, then substantial progress will be made towards the objectives of stabilizing year-to-year area differences and increasing yields.

The survey clearly demonstrates that winter chickpea is considered by Syrian farmers as a commercial crop. Only some 1% of production is being withheld for home consumption. Some 6% is kept for next year's seed, and the remaining 93% is being sold either to government agencies or to the private market, and prices do not vary much between these two outlets.

On the whole, there seem to be few technical or economic problems for winter chickpea producers. The principal constraint during the two survey years was the frosts in Hama and Aleppo provinces and the low rainfall in Hama, Aleppo, and Daraa. It should be noted that spring chickpea producers in these areas during the survey years suffered these climatic constraints just as much as did winter chickpea producers. But would-be spring producers had the option of waiting to see if the rains were sufficient to warrant planting in 1989/90 or 1990/91. Winter producers did not, and many of them in Hama and Aleppo suffered very low (and unprofitable) yields. In retrospect they felt that the risk of adverse climatic conditions did not justify planting winter chickpea again. Whether or not they would be more amenable to the new cultivars had the climatic conditions been more favorable in the particular survey years remains an open question.

Actions Taken

The Syrian National Program, partly as a result of the survey findings that farmers felt a need for more information about winter chickpea, has organized training courses for researchers and extension agents. In terms of breeding, emphasis is being placed on cold tolerance.

Adoption monitoring is continuing with a focus on including the more "traditional" spring chickpea growing areas of the southwest in future studies. It is in this Zone 2 area that the as yet unproven hypothesis regarding the risk aversion constraint to winter chickpea adoption can be tested best. The objectives of the new program are to determine the degree to which farmer perceptions of risk have an adverse effect on winter chickpea adoption and to identify for which farm types and locations little adoption of the presently available cultivars can be expected. Preparations are being made for targeted recommendations to the seed multiplication and marketing program.

Winter Chickpea in Tunisia

Although the winter sown cultivars "Chetoui" (ILC 3279) and "Kassab" (FLIP 83–46C) were released in Tunisia in 1986, there has been little in the way of a national technology transfer program. Indeed, it would be fair to say that the national strategy is still in the formulation stage. A number of disease surveys have been undertaken, breeding work has been expanded onto farmers' fields, and field days have been conducted to introduce farmers to the concept of winter sowing. The idea of winter planting appears to be readily understood by Tunisian farmers. What appears to be less understood is that winter planting requires new cultivars which are specially developed to perform well under winter climatic and disease conditions. There is still the genetic constraint of developing cultivars that have a dual tolerance for Ascochyta blight and Fusarium wilt.

As the breeding program proceeds with the objective of cultivars with dual tolerance to these biotic stresses, a baseline survey of chickpea producers is being undertaken to assess the place of chickpea in the farming system and to determine its future potential for improvement.

Concluding Remarks

The experiences of monitoring winter chickpea adoption in Morocco and Syria and the present work in Tunisia point to four important conclusions. First and foremost, it is extremely difficult to construct a predictive model of adoption and impact *a priori* without substantial farmer participation and on-farm experience with the new technology. Second, although national program strategies are necessarily varied to reflect local circumstances, any national technology transfer strategy should be flexible and able to respond to farmer evaluations. Obvious examples are the response of the Moroccan program to the Ascochyta blight epidemic in 1988/89 and the recovery of the acceptance rating among farmers the following year. Similarly, the Syrian programs targeted the Northwest for increased efforts following the results of farmer surveys. A third conclusion is that national policy measures can play an important part in adoption and impact once the technology has achieved improved performance with farmers. The case of commercialized production in Syria with a major seed multiplication program and improved price incentives is illustrative. Above all, researchers, extension agents, and farmers should work together to find ways of overcoming the "information constraint" about winter chickpea which appears to be present in all three countries.

On-farm research addressing lentil farmers' constraints in West Asia

B.A. SNOBAR[1], M. ABI ANTOUN[2], N.I. HADDAD[3], M. TAWIL[4] and A.B. SILKINE[5]

[1] Plant Production Department, Faculty of Agriculture. University of Jordan, Amman, Jordan;
[2] Agricultural Research Organization, Tel Amara, Lebanon;
[3] ICARDA, P. O. Box 950 764, Amman, Jordan;
[4] Directorate of Agriculture, Damascus, Syria, and
[5] ICARDA, P. O.Box 5466, Aleppo, Syria

Abstract

Despite the importance of lentil as a protein source, the crop has received minimal research emphasis. However, recent studies in West Asian countries have identified improved production technologies for lentil. This technology may differ from country to country, but the main problems and constraints facing farmers in growing lentil are similar in West Asia.

Research on farmers' fields has focused on overcoming these constraints. The areas of major importance are:
- Evaluating newly improved lentil cultivars that possess traits suitable for mechanical harvesting;
- Demonstrating in farmers' fields the optimum cultural practices that should be used for lentil production, such as proper seedbed preparation, time of sowing, weed control, fertilization, *Rhizobium* inoculation and mechanical seeding, and;
- Introducing several mechanical harvesting methods that farmers can chose from.

Several of these improved practices are now being used by farmers and have resulted in substantial increases in yields and profits. However, other technical constraints and problems, such as weed control, *Rhizobium* inoculation, efficient harvesting methods, *Sitona* and *Bruchus* control, and residue management still require additional research. Land fragmentation and availability of inputs are two other important constraints that need to be addressed.

Introduction

Lentil in Asia has been an important source of protein since ancient times. This crop is well adapted to the cool season of the area. The area planted to lentil and the per capita production in West Asian countries in the last two decades are presented in Table 1. The area devoted to lentil cultivation doubled in Iran and

F.J. Muehlbauer and W.J. Kaiser (eds.), Expanding the Production and Use of Cool Season Food Legumes, 911–925.
© 1994 *Kluwer Academic Publishers.*

increased seven times in Turkey during this 20 year period. A slight increase in lentil area was observed in Syria; however, in Iraq, Jordan and Lebanon, the area either decreased or did not change. The per capita production decreased in all countries except Lebanon where it doubled and Turkey where it quadrupled.

Table 1. Area of lentil and per capita production in West Asian countries from 1969 to 1989

Country	Area ('000) ha			Per capita production		
	1969-71	1979-81	1989	1969-71	1979-81	1989
Iran	46	46	105	0.98	0.82	0.91
Iraq	9	7	2	0.53	0.45	0.11
Jordan	21	7	9	6.58	2.22	1.89
Lebanon	3	4	4	0.81	1.12	1.41
Syria	121	82	188	11.69	7.04	5.30
Turkey	105	206	715	2.84	5.93	11.38

Source: FAO, 1978 and 1990.

Lentil in most West Asian countries is grown following traditional farmers' practices. Farmers usually consider lentil as a secondary crop in the rotation with cereals, and therefore provide minimum inputs. Farmers grow local landraces, by hand-broadcasting the seed and apply no fertilizers. They delay planting to allow weeds to emerge after the first effective rainfall and then they cultivate the fields and plant the crop. This delay usually results in a serious yield reduction. The lentil crop is susceptible to several damaging insect pests, such as including *Sitona* larvae that feed on the nodules making them ineffective for nitrogen fixation. Also Bruchids cause serious damage to the seed and reduce yields and quality.

One problem facing lentil growers is the unavailability of machines to harvest the crop, labor has become scarce and very expensive. Recently, International Agricultural Research Centers such as ICARDA and funding agencies such as IDRC, have devoted time, money and expertise to increase lentil yields by improving cultivars and crop management practices. Mechanization of the crop has been given major emphasis. The work has been conducted in cooperation with the national programs in the West Asia North African (WANA) countries. After an extensive research effort, some encouraging results were obtained. These include: new and improved cultivars; optimum crop management that includes land preparation and planting time, seeding rate, fertilization, inoculation; and some promising mechanical harvesting options. The results were mainly obtained at research stations and to a lesser extent under farm conditions. The results have recently been made available to farmers.

Problems and Constraints Facing Lentil Growers

As was indicated earlier, farmers consider lentil as a secondary crop. The main crop in rainfed areas of West Asia is wheat. However, with an increase in the price of lentil seed and straw (tibbin), there is a growing interest at the farmer and governmental level to expand lentil cultivation. This expansion has faced several constraints, some of which are difficult to overcome. The problems and constraints that affect improvements in lentil yield are as follows:

Seedbed Preparation and Seeding

This problem is not specific to lentil since it requires the availability of specific equipment which is also needed for cereal cultivation. Needed equipment includes chisel plows, sweep plows, spike tooth harrows and the seed drills. However, in order to mechanically harvest a lentil crop, a roller is used immediately after sowing to provide an even soil surface that allows cutterbars to be used for harvesting.

Improved Lentil Cultivars

Until recently, cultivars used in all West Asian countries were landraces available at the farmer level. However, new cultivars were released in most West Asian countries which have erect and tall plant types, non-shattering pods, and large yield potential. Such new cultivars should facilitate mechanical harvesting.

Weed Control

Weeds are a major constraint in lentil cultivation. The control of weeds in farmers' fields is usually done by hand weeding, either by the farmer and his family or by hired labor. However, with the expansion in lentil cultivation, hand weeding has become a problem and chemical weed control has become necessary. Farmers usually delay the seeding operation until after the first heavy rainfall to allow weeds to emerge. This delay usually results in serious yield reductions. Therefore, attempts at early seeding must address the weed control problem.

Others

There are other constraints to the improvement of lentil yields such as the damage caused by *Sitona* and *Bruchus* spp., and the need for inoculation with Rhizobia.

On-Farm Research Addressing the Constraints

Lentil on-farm research in the countries of West Asia started after adequate information became available from on-station research. The on-farm research aims at verifying the technology under farmers' conditions using research management. After the verification stage is successful, then large demonstration trials are implemented with more involvement of the farmer and less of the researcher. On-farm research is designed to address constraints that farmers are facing and to demonstrate to farmers under their own conditions how these constraints can be overcome.

The constraints facing lentil farmers were discussed earlier. However, the on-farm research did not address each constraint by itself, but instead dealt with them either together or in subgroups which could have a better chance of success under farmers' conditions. For example, in Jordan, two types of on-farm research were conducted:
- One is called the full-package of technology, starting with seedbed preparation, time of sowing, improved cultivars, fertilization, weed control and mechanical harvesting.
- The second is called the minimum input package which includes improved cultivars and fertilization; the remaining practices are performed using farmers' traditional practices.

The first type addresses those farmers who have access to machinery and inputs, whereas the second type addresses farmers who are located far from machinery hiring stations and where it is difficult for them to get a seed drill or have an opportunity to mechanize lentil production. In both demonstrations, a control treatment is included which makes use of local cultivars and the farmers own production practices.

The following are the constraints which were dealt with through on-station and on-farm research:

Cultivars

The most common cultivars of lentil grown in West Asian countries were the local landraces. Although these cultivars are adaptable to the local environment, they are known to have several undesirable traits, such as being small seeded and low yielders, short plant stature and susceptibility to that lodging, pods located low on the plant which shatter easily, susceptiblity to diseases and insects, and hard to cook.

National programs in cooperation with ICARDA and with assistance from funding agencies, such as IDRC, have worked toward the development of new improved lentil cultivars. The new cultivars seem to have desirable traits with high yield potential. The improved cultivars are tall, erect, less susceptible to lodging and shattering, and carry their pods in a higher position on the plant which makes them suitable for mechanical harvesting. Most of these new cultivars are either selections from local material provided to national programs by ICARDA, selections made directly by national programs, or from hybrids that have local material in their background. This makes them adapted to the environment and able to withstand the moisture stress that is usually encountered. The lentil cultivars released in some of the countries of the region are listed in Table 2. The table also includes some of the important traits of these cultivars. Among these cultivars, we found that 78S 26002, a selection made by ICARDA from germplasm that was originally collected in Jordan, has been released in Jordan, Lebanon, Syria and Iraq. This indicates the adaptability of this cultivar to the environmental conditions of this region. It was also observed that only a limited number of improved cultivars have been released. The yields of these new cultivars greatly exceeded the yield of the local landraces (Table 3). However, the question which remains to be answered is how many of the farmers are planting the new improved cultivars. This requires follow up studies.

Table 2. Newly released improved lentil cultivars in some West Asian countries with a description of their important traits

Country	Name	Pedigree	Plant height (cm)	Seed color	100 seed wt. (g)	Protein content (%)	Cooking time (min.)
Jordan*	Jordan 1	UJL176	30–40	Yellow	3.5	-	-
	Jordan 2	81S15	30–40	Red	5.0	-	-
	Jordan 3	78S26002	25–35	Yellow	5.5	-	-
Lebanon**	Talia 2	78S260013	30–36	Red	3.0	28.8	35
Syria***	Edleb 1	78S26002	32	Yellow	4.3	24.6	39

*   University of Jordan, 1984, 1985, and 1986.
**   Abi Antoun *et al.*, 1990, pp. 3 and 4.
***   Ministry of Agriculture, 1989.

Table 3. On-farm grain yield of landraces and improved cultivars of lentil released in some of the West Asian countries

Country	Cultivar	Grain yield
		– kg ha^{-1} –
Jordan*	Local	790
	Jordan 1 (UJL176)	1060
	Jordan 2 (81S15)	980
	Jordan 3 (78S26002)	1030
Lebanon**	Local	1190
	Talia 2 (78S260013)	1500
Syria***	Horani (Local)	890
	Kurdi (Local)	925
	Edleb 1 (78S26002)	1075

*   University of Jordan and Ministry of Agriculture, 1990 and 1991.
**   Abi Antoun *et al.*, 1990, pp. 3 and 4.
*** Ministry of Agriculture, 1989.

Land Preparation and Seeding

The traditional land preparation and seeding practices for lentils was one of several factors contributing to reduced yields. Animal drawn plows, although suitable for most lentil areas, cause delays in land preparation and timely sowing of seed because of the slow nature of such plows. Turning plows, which are most common in the lentil growing areas, leave the soil pulverized and the surface rough, full of stones, and bare which negatively affect the follow up operations, such as seeding, fertilizing and harvesting.

Difficulties in land preparation and seeding are important constraints in lentil cultivation. Therefore, improved land preparation and seeding were developed using the following equipment:

Chisel Plow

This is a primary plow which is used after harvesting the previous crop, and particularly if the preceeding crop is a cereal. Plowing with this implement is done at a depth of 25 to 40 cm. The chisel plow does not invert the soil and therefore leaves the soil less pulverized when compared to the effects of a moldboard plow.

Sweep Plow

The sweep plow which is used after the chisel plow to till the soil to a depth of 8 to 12 cm to create a suitably firm and smooth seedbed to the desired depth of planting.

Spike Toothed Harrow

The spike toothed harrow is another secondary tillage implement which tills the soil to a very shallow depth (less than 5 cm). This harrow cannot be recommended for all uses. It is only recommended when it is necessary to break clods and to smooth the soil surface. It is generally used after mechanical sowing, and is usually pulled behind the seed drill.

Grain Drill

The same grain drills which are used to plant different cereal crops are used to plant lentil. The machine plants the lentil seeds in rows spaced 15–20 cm apart. The machine can also apply fertilizers at the time of seeding. The use of grain drills facilitates harvest mechanization because of the improved uniformity of seeding depth, soil surface, plant emergence and maturity as compared to traditional practices.

Roller

The roller is critically important for lentil production. It is recommended for stony fields and fields full of clods. This implement breaks clods and pushes the stones into the soil to create a smooth surface which facilitates the mechanical harvesting operations. It is used after the seeds are planted with the grain drill.

The use of the newly developed cultural practices significantly increased lentil yields and net income compared to the use of traditional cultural practices (Table 4).

Table 4. Grain yield of lentil produced under traditional and improved land preparation and seeding methods for two seasons 1989/90 and 1990/91 in Jordan

	Grain yield	
Type of cultural practices	1989/90	1990/91
		$-kg\ ha^{-1}-$
Traditional	900	515
Improved	1295	745

Source: University of Jordan and Ministry of Agriculture, 1990 and 1991.

Time of Sowing

Time of sowing is the single most important factor affecting lentil yields. The traditional time of sowing lentil in rainfed areas is after good rainfall which usually occurs in late November or early December. In Syria, for example, 7% of lentils are planted in November, 69% in December, and 24% in January (Mazid, 1989). Planting later than mid-December is considered late (Table 5). The reason for the delay in sowing is to allow for mechanical weeding which is performed before sowing and after weed emergence.

Table 5. The effect of date of sowing on lentil grain yield in Jordan, Syria and Lebanon

Country	Date of sowing	Grain yield
		$-kg\ ha^{-1}-$
Jordan*	29/10–04/11	1385
	20/11–06/12	1535
	05/01–09/01	1030
	31/01–14/02	785
Syria**	14/11–17/11	840
	04/02–08/02	680
Lebanon***	04/12	1190
	24/01	585

+   Average of 1979/80 and 1980/81 seasons for Jordan and 1982–
    1985 seasons for Syria.
*   University of Jordan, 1982.
** Silim *et al.*, 1991, pp. 145–154.
*** Food Legume Improvement Program, Research Highlights
    1979/1980, ICARDA, Aleppo, Syria, pp. 15, October 1980.

Extensive on-station research proved that early sowing will significantly increase yields (Table 5). Nine on-farm trials, covering a wide range of agroecological conditions, conducted in Syria in 1979/80 confirmed that early planting increased grain yields to 1223 kg ha^{-1} compared to 640 kg ha^{-1} for the normal late date of planting (ICARDA, 1980). Therefore, the recommendation was to encourage early sowing during late October, November and early December. Since effective herbicides have become available, this is possible.

Weed Control

Weeds, particularly the broad-leaved ones, are the most serious problem in lentil fields. These weeds cause major yield reductions in the West Asian countries. Therefore, unless lentil fields are hand weeded at least twice, grain and straw yields of lentil are reduced.

Several on-station and on-farm research trials demonstrated the efficacy of herbicides in controlling weeds in lentil fields. Pre- and postemergence herbicides were tested and found to be most effective in controlling both broad- and narrow-leaved weeds. The cost of using herbicides is expected to be cheaper than the cost of manual weeding.

Pronamide (Kerb) for narrow-leaved weeds and prometryne (Gesagard) or cyanazine (Bladex) for broad-leaved weeds proved to be the most effective preemergence herbicides in controlling weeds in lentil fields (Al Thahabi, 1991). Weed control by preemergence herbicides resulted in significantly increased yields (Table 6). Weedy lentil fields in Tel Hadya, Syria, and Terbol, Lebanon, yielded 70 and 63%, respectively, of weed-free lentil fields (ICARDA, 1984).

The use of herbicides at Terbol increased grain yields from 62 to 86% depending on the treatment (Table 7). It should be noted that lentil is susceptible to certain herbicides. Attempts were made to identify safe herbicides which

Table 6. Effect of herbicide treatment on lentil grain yield in Jordan

Treatment	Grain yield
	$-kg\ ha^{-1}-$
Weedy	280
Pronamide + prometryne	1100
Weed free (hand weeding)	1320

Source: Al Thahabi, 1991, pp. 58.

Table 7. Comparison between lentil yields obtained from herbicide treated fields compared to weedy fields at Terbol, Lebanon

Treatment	Increase in grain yield (%) over the weedy check
Prometryne (105 kg a.i. ha⁻¹)	62
Methabenzthiazuron (2 kg a.i. ha⁻¹) + pronamide (0.5 kg a.i. ha⁻¹)	86
Cyanazine (1 kg a.i. ha⁻¹) + pronamide (0.5 kg a.i. ha⁻¹)	73

Source: ICARDA, 1984, pp. 62-64.

would be compatible with lentil Rhizobia. L-28a was the most promising strain of *Rhizobium* for compatibility with Prometryne, which is one of the safer herbicides to use with lentil. *Rhizobium* strain L-28a resulted in the highest yield in the presence of herbicides (ICARDA, 1980).

Fertilization

The belief among many lentil growers was that lentil does not need any fertilization. However, this notion has changed drastically in some major lentil producing regions in West Asia. On-farm survey research in Syria showed that only 13% of the lentil growers did not use fertilizer in the 1988/89 cropping season, compared to 47% a decade ago (Mazid, 1989). The average application rate also has increased substantially from 64 to 145 kg ha^{-1} of triple superphosphate and from 17 to 24 kg ha^{-1} of urea. On-farm research proved that phosphorus is an important fertilizer for lentil which will contribute significantly to increased yields. However, response to phosphate application is variable, depending on the availability of soil moisture (Mater, 1976) and on the phosphorus status of the soil (ICARDA, 1980). Application of phosphorus to lentil resulted in a significant yield increase in the succeeding wheat crop (Saxena and Wassimi, 1980).

The recommended dosages of P should depend on the available P status in the soil. Therefore, an analysis of the soil should be made before recommending the application of P_2O_5. At 5 ppm P in the 0–15 cm layer, 68 kg P_2O_5 ha^{-1} is recommended. At 7.5 to 8 ppm P, no P_2O_5 application is recommended (ICARDA, 1980).

Rhizobium Inoculation

In some West Asian countries inoculation of lentils with *Rhizobium* helped in formation of nodules which improved nitrogen fixation and increased lentil yields. The most effective Rhizobium strains are listed in Table 8. Further on-farm research is needed to test these strains under farmers' conditions.

Constraints Still to Be Addressed

Several constraints still need to be addressed either by on-farm research and/or by decisions to be taken by policy makers. These include the following:

Table 8. Effectiveness of Rhizobium strains in lentil

Cultivar	Stain designations*	SE**	Source
Jordan 1	LE 735	114	Syria
	LE 843	114	Jordan
	LE 715	113	Syria
Jordan 2	LE 898	99	Turkey
	LE 893	97	Turkey
	LE 867	96	Turkey
Jordan 3	LE 804	112	Egypt
	LE 867	108	Turkey
	LE 835	100	Jordan
Cultivar mean	LE 867	100	Turkey
	LE 804	95	Egypt
	LE 893	93	Turkey
	Mean of 21 strains	71	

Source: Beck, 1990.

*   The above strains were the most effective of 21 lentil
    Rhizobium strains selected from the ICARDA collection of
    over 300 lentil *Rhizobium* isolates, based on this high SE in
    previous screening.

**   SE: Symbiotic Effectiveness
    $SE = \dfrac{\text{Shoot dry weight of strain treatment}}{\text{Shoot dry weight of 100 mg N pot}^{-1}\text{ control}} \times 100$

Sitona Control

Sitona is most common pest in lentil fields in West Asia. Damage to nodules
caused by *Sitona* is significant in most West Asian countries. Therefore, the
insect should be controlled to reduce its adverse effects on yield. Some on-farm
research carried out in Jordan, Syria and Lebanon addressed *Sitona* control in
lentil fields. It was found that treating lentil seeds with promet before planting
will reduce *Sitona* damage significantly. Further research is needed in order to
study the interaction of several inputs, such as P and *Rhizobium* inoculum on
Sitona control.

Bruchus Control

Bruchus is an important insect in lentil fields. This insect causes significant
damage to lentil seed, thus reducing yield significantly. Limited on-farm

research and trials were conducted on controlling *Bruchus* in lentil fields in Jordan, Syria and Lebanon.

There are still some unknowns concerning control of *Bruchus* under field conditions. Therefore, further on-farm research is needed to address these issues, such as when is the best time and stage of plant development to apply insecticides, and what are the most effective types of insecticides and dosage rates required.

Mechanical Harvest

Harvesting the lentil crop is the most critical operation. Expenses related to the harvesting operation will determine whether the grower will make or lose money. Presently, the cost of manual harvesting is becoming prohibitive. Therefore, some farmers have ceased to growing lentil because of the high cost of production due to the high cost of harvesting the crop by hand. The high cost of the harvest operation is the main reason for the continuous reduction in the area devoted to lentil production in West Asian countries. In order to encourage farmers to grow more lentil, several mechanical harvesting techniques were tested. These included:
– Tractor side-mounted single knife cutterbar,
– Tractor side-mounted double knife cutterbar,
– Swather with narrow double knife cutterbar, 1.25 m wide,
– Swather with wide single knife cutterbar, 3 m wide,
– Whole crop harvester,
– Plot grain combine,
– Modified field grain combine.

Most the above mentioned mechanical harvesting techniques are cheaper and faster than hand harvesting, but there are some grain and straw losses. However, the swather with a double or single knife cutterbar was the best in terms of low cost, fewer losses and good capacity for harvesting.

Some of the techniques need improvements. Whenever such improvements are made, the puller and the whole crop harvester could be highly recommended because they recover most of the straw and grain.

As it stands now, the cost of hiring a swather in Jordan (from the Jordan Cooperative Organization) is 20 to 30 USD ha^{-1}. If harvesting is performed at the right stage and time of the day, the grain losses will not exceed 15% and the straw losses 30 to 50%, compared to hand losses of 10 and 20% for grain and straw, respectively (University of Jordan and Ministry of Agriculture, 1990 and 1991).

In Syria, on-farm trials showed that the cost savings due to harvest mechanization can compensate for harvest losses which averaged 13% for seed and 24% for straw (Erskine *et al.*, 1991).

Lentil Residue Management

With manual harvesting, little or no lentil residue is left in the field because the plants are pulled up by the roots. Limited studies have been made on the effect of incorporation of lentil residues into the soil. When lentil is rotated with wheat on a two or three year crop rotation system, wheat yields are less after lentil when compared to wheat after other crops or wheat after a fallow period. One possible explanation for reduced wheat yields following a lentil crops is that little or no lentil residue is left in the field when the crop is harvested by pulling. What would happen if the lentil crop was cut instead of being pulled and the lentil stubble and other residues were incorporated into the soil? Will this result in an increase in the yield of the following wheat crop? Would it improve the soil structure?

Land Fragmentation

Not only is land ownership fragmented, but also farmers in West Asia allot only a small percentage of their land for lentil cultivation. A survey in Syria showed that the average lentil producer allots about 18% of the total farm area for lentil. On the average, about 33% of the farmers grow lentil on 0.5 to 1.0 ha and 36% on 1.5 to 2.5 ha (Table 9). In Jordan, the average lentil grower allots about 12% of total farm area for lentil. About 17% of the farmers grow lentil on 0.5 to 1.0 ha, 21% on 1.0 to 1.5 ha and 35.5% on 1.5 to 2.5 ha (Table 9). This land fragmentation creates an obstacle towards progress and use of new technology in lentil production. Fragmentation of land holdings in the region does not favor mechanization, particularly harvest mechanization which is the most pressing constraint in lentil production.

Availability of Inputs

Inputs such as the recommended land preparation equipment, seeders, sprayers, seeds of improved cultivars, fertilizers and herbicides should be made available to lentil farmers at the proper time and place. Different public and private sectors should play an active role in providing farm machinery and custom operation services because farmers rely heavily on such services due to land fragmentation. Seeds of improved cultivars, fertilizers and herbicides and other inputs should be supplied to farmers in different villages in order to assure the use of inputs recommended through on-farm research and demonstrations which will lead to increased yields.

In Jordan, the Jordan Cooperative Organization plays an important role in providing machinery services in three regions of Jordan (North, Central and South), distribution of improved seed through the seed cleaning units it operates in the three regions, and distribution of fertilizers through the cooperatives

Table 9. Frequency distribution of farmers by lentil area in Jordan and Syria

Lentil area (ha)		% of Farmers	
Jordan*	Syria**	Jordan	Syria
0.5-1.0	<0.5	17.0	3.7
> 1.0- 1.5	0.5- 1.0	20.7	33.8
> 1.5- 2.0	1.5- 2.5	18.9	35.6
> 2.0- 2.5	3.0- 3.5	16.6	9.3
> 2.5- 3.0	6.0-10.0	18.3	9.3
> 3.0- 4.0	12.0-17.0	24.3	3.7
> 4.0- 5.0	20.0-30.0	17.2	4.7
> 5.0- 7.5		13.6	
> 7.5-10.0		6.5	
>10.0-15.0		3.6	
>15.0-20.0		2.4	

*   From a survey made in 1990/91 by the Food Legume Improvement and Mechanization Project of Jordan (170 farmers) (unpublished).
**   From a survey made in 1978/79 and 1979/80 in Syria (115 farmers) (Salkini, 1991).

scattered in different agricultural areas, including remote villages. If the recommended inputs, which are part of the technology package, are not made available at the right time and place, then these constraints could become limiting factors in the success of increasing lentil production in the region.

Concluding Remarks

Great efforts are being made to overcome most or all of the constraints facing lentil farmers in the West Asian countries through on-station and on-farm research. Unless these constraints are removed, the area planted to lentil will continue to decline and the most important locally produced source of protein will be lost. If this occurs, it may be necessary to import lentils. It is believed that on-farm research which addressed the constraints facing lentil farmers that were conducted in the last ten years in the West Asian countries have borne fruitful results, thus reversing a decline in lentil production, while at the same time making progress towards increasing the area of lentil under cultivation, and in increasing yields and per capita production to its old records. This can only be done if and when the joint efforts of researchers, extension agents, farmers and policy makers, using the concept of a farming systems research approach, are able to transfer locally developed technology to farmers.

References

Abi Antoun, M., Kiwan, P. and Erskine, W. 1990. *Lens Newsletter* 17: 3–4.

Al Thahabi, S. 1991. Weed control in lentils. MSc thesis. Faculty of Agriculture, University of Jordan, Amman, Jordan.

Beck, D. 1990. *Food Legume Improvement Program, Annual Report for 1990*, pp. 150–152. Aleppo, Syria: ICARDA.

Erskine, W., Dickmann, J., Jeyntheswostan, P. Salkini, A. B., Saxena, M. C., Ghanuim, A. and El Ashkar, F. 1990. *Journal of Agricultural Science, Cambridge* 117: 333–338.

FAO. 1990. *1989 Production Yearbook*. Vol. 43. Rome, Italy: FAO.

FAO. 1978. *FAO Production Yarbook*. Vol. 30. Rome, Italy: FAO.

ICARDA. 1980. *Food Legume Improvement Program, Research Highlights, 1979/1980*, pp. 15–16. Aleppo, Syria: ICARDA.

ICARDA. 1984. *Food Legume Improvement Program, Research Highlights, 1982/1983*. Aleppo, Syria: ICARDA.

Matar, A. E. 1976. Direct and cumulative effects of phosphates in calcareous soil under dry farming agriculture of southern Syria, Soil Science Division, ACSAD/55/P1, Damascus, Syria.

Mazid, A. 1989. *Farm Resource Management Program Annual Report*, pp. 317–327. ICARDA-162 En., Aleppo, Syria: ICARDA

Ministry of Agriculture. 1989. *Annual Report of Agriculture Scientific Research*. Damascus, Syria.

Salkini, A. B. 1991. Training course on lentil harvest mechanization held on 12–23 May, 1991, Aleppo, Syria: ICARDA.

Saxena, M. and N. Wassimi. 1980. *Lens* 7: 52, 53–64.

Silim, S. N., Saxena, M. C. and Erskine, W. 1991. *Experimental Agriculture* 27: 145–154.

University of Jordan and Ministry of Agriculture. 1990. *Food Legume Improvement and Mechanization Project. Annual Report 1989/1990*. Amman, Jordan: Faculty of Agriculture, University of Jordan.

University of Jordan and Ministry of Agriculture. 1991. *Food Legume Improvement and Mechanization Project. Annual Report 1990/1991*. Amman, Jordan: Faculty of Agriculture, University of Jordan.

University of Jordan. 1984. *Food Legume Improvement Project. Annual Report No. 3*. Amman, Jordan: Faculty of Agriculture, University of Jordan.

University of Jordan. 1982. *Food Legume Improvement Project. Annual Report No. 1*. Amman, Jordan: Faculty of Agriculture, University of Jordan.

University of Jordan. 1985. *Food Legume Improvement Project. Annual Report No. 4*. Amman, Jordan: Faculty of Agriculture, University of Jordan.

University of Jordan. 1986. *Food Legume Improvement Project. Annual Report No. 5*. Amman, Jordan: Faculty of Agriculture, University of Jordan.

Addressing farmers' constraints through on-farm research: chickpea in Maharashtra state of India

P.W. AMIN[1], Y.L. NENE[2] and H.A. VAN RHEENEN[2]
[1] Punjabrao Krishi Vidyapeeth, Krishi Nagar, Akola, Maharashtra, India, and
[2] ICRISAT, Patancheru P. O., Andhra Pradesh 502 324, India

Abstract

On-farm research on irrigated and rainfed chickpea in ten states of India (mostly in the state of Maharashtra) was conducted during 1988/89 and 1989/90. Results showed that under residual moisture conditions, extra short-duration chickpeas had a 30% yield advantage over commonly grown long-duration cultivars. Local and improved cultivars produced similar yields when irrigated, but improved management provided a 19% yield advantage when compared to traditional cultivation practices.

Systematic and sustained efforts to train farmers in improved cultivation of chickpea have resulted in a rapid expansion of the area planted.

Needs for further adoption of chickpea include: short-duration cultivars, heat tolerance, bruchid resistance, resistance to *Helicoverpa*, resistance to *Sclerotium rolfsii*, and large white seededness.

Introduction

India produces about 4.5 million tons of chickpea from about 7.0 million ha. The area under chickpea has shown a decline from 7.76 million ha in 1973/74 to 6.78 million ha in 1986/87. Productivity has increased marginally during this period from 528 kg ha^{-1} to 657 kg ha^{-1} (Anon. 1988/89). India is the world's largest producer of chickpea and has exported small quantities estimated at between 0.5 and 3.1 thousand tons per year^{-1}. However, India began to import small quantities of chickpea starting in 1977 (Von Oppen and Rao, 1987).

In 1988, the Indian government asked for ICRISAT's help in increasing the productivity of chickpea. At a meeting held in May 1988, trials were planned in ten Indian states to demonstrate achievable yield potential. The LEGOFTEN Unit which was formed at ICRISAT to coordinate and monitor similar groundnut and pigeonpea trials was given responsibility to carry out the chickpea trials. The trials began in the post-rainy season of 1988 and continued

F.J. Muehlbauer and W.J. Kaiser (eds.), Expanding the Production and Use of Cool Season Food Legumes, 926–937.
© 1994 *Kluwer Academic Publishers.*

through 1989. The ten states represented a wide range of climatic conditions, soils, pests, and diseases (Amin *et al.*, 1988).

Eleven out of 19 irrigated trials and two out of eight non-irrigated trials were conducted under ICRISAT's supervision in the state of Maharashtra. Therefore, this state was chosen for identifying farmers' constraints that limit expansion of chickpea production. The discussion includes experiences on chickpea crop production system, consumer and trade preferences, seed prices, and key biotic constraints to serve as a guide for future research and technology transfer.

Maharashtra ranks fourth among the chickpea growing states of India. The area under chickpea cultivation in the state has remained remarkably stable between 0.4 million ha and 0.46 million ha from 1960/61 to 1982/83 and has shown a steady increase up to 0.55 million ha in 1987/88 (Anon. 1988/89).

Materials and Methods

Before the trials started, a high level planning meeting of the officials of Agricultural Ministry was held. This was followed by a meeting of extension workers responsible for conducting trials that was held at ICRISAT to discuss basic and applied aspects of chickpea research. Soils from trial sites were analyzed, disease and pest histories was discussed, cultivars were identified; and, based on this information, a package of practices was formulated for each location. The government provided funds for conducting the trials, while ICRISAT provided seed of improved, wilt resistant cultivars and funds for monitoring the trials. The trials were monitored by a multidisciplinary team from ICRISAT consisting of a breeder, agronomist, and plant protection specialist who worked jointly with agricultural officials from the different states.

There were four comparisons for each trial: (1) An improved wilt resistant cultivar was grown with improved cultivation practices; (2) the same wilt resistant cultivar was grown with the state-recommended cultivation practices; (3) a local cultivar was grown with improved cultivation practices; and (4) the same local cultivar was grown with the state's recommended cultivation practices. The plot size was 0.2 ha for each treatment.

Two sets of trials, one for irrigated and the other for residual moisture situation, were formulated. For irrigated trials, two cultivars, ICCC 42 and ICCC 37, but mainly the latter; and, for non-irrigated trials, a new extra short-duration cultivar, ICCV 2, were used. Both groups of cultivars have resistance to wilt caused by *Fusarium oxysporum* f. sp. *ciceris*.

The improved package of practices for the irrigated set of trials consisted of raised beds and a furrow irrigation system to obtain proper distribution of water to avoid waterlogging and water deficit patches, proper fertilization based on soil analysis to overcome nutritional deficiencies, controlled irrigation depending on soil texture, and the application of pesticides when needed.

For residual moisture trials, all test conditions were similar to the irrigated trials except that flat seedbeds were used instead of raised beds with furrows.

A total of 43 irrigated and 33 non-irrigated trials were conducted. Of these, 19 irrigated and eight non-irrigated trials were closely monitored by a team consisting of a breeder, agronomist, and entomologist. The approach followed in testing improved chickpea cultivation practices is given in Figure 1.

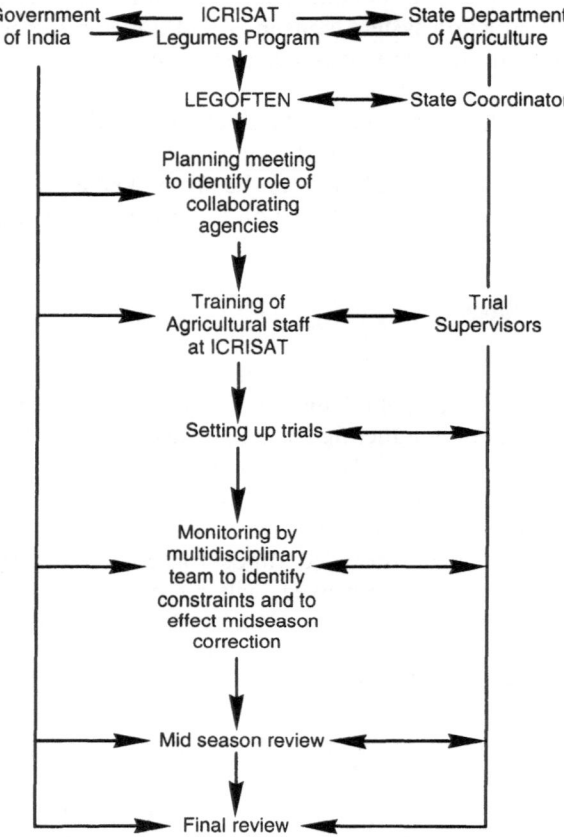

Figure 1. Planning and review process for testing improved chickpea cultivation practices

Results

Improved genotypes and cultivation practices led to increased yields. A yield of 3704 kg ha^{-1} was obtained from ICCC 37 under irrigation which was close to the world record of 4.0 tons ha^{-1} (Saxena, 1984). The average yield for that trial was 1875 kg ha^{-1}. However, local genotypes also responded well to improved management and yields averaged 1735 kg ha^{-1}. Better management practices increased yields for both improved and local cultivars by 16.6 to 22.1%, respectively (Tables 1a,b and 2).

Table 1a. Grain yield (kg ha^{-1}) of irrigated chickpea at different locations, post rainy season, 1988/89

State/location	Improved package		State recommended package	
	Improved cultivar*	State's recommended cultivar	Improved cultivar*	State's recommended cultivar
Andhra Pradesh				
Tangadencha	1554	1278	1085	1087
Gularat				
Hathrol	1426	1699	1221	1444
Karnataka				
Gangavati	920	1275	1275	1465
Madhya Pradesh				
Bhikangaon	810	860	785	800
Narsingpur	1610	1715	1340	1275
Maharasntra				
Chakur	1360	1770	1450	2325
Parbhani	1752	1695	1676	1657
Pokharni	1620	1250	1200	1080
Bori	2951	2815	2197	2951
Niwali	850	780	760	700
Niwali	3704	3605	2963	3111
Jintur	2947	2464	2367	1864
Ridaj	3500	2800	2240	1400
Niwali	2500	2407	2315	2037
Gangakhed	2850	2500	2500	2417
Hingoli	1500	1208	1100	735
Orissa				
Keonjhar	1733	1107	995	448
Rajasthan				
Ajmer	444	227	110	115
Uttar Pradesh				
Amarokh	1590	1510	1595	1370
Average yield	1875	1735	1535	1488

*Mostly ICCC 37, a wilt resistant cultivar.

Table 1b. Two way comparison of average yields

Cultivar	Cultivation package		Average	Percentage increase
	Improved	State recommended		
ICRISAT	1875	1535	1705	5.8
State	1735	1488	1612	---
Average yield	1805	1511.5		
% Increase	19.4			

Table 2. Average yield, income, cost of cultivation and profits of irrigated chickpea, 1988–89

Treatment	Grain yield tons ha[-1]	Income Rs ha[-1]	Cultiva-tion cost Rs ha[-1]	Profit Rs ha[-1]
IRRIGATED (n=19)				
Improved cultivar and methods	1.875	11,250	5,213	6,037
State recommended cultivar and improved methods	1.735	10,410	5,166	5,244
Improved cultivar and State recommended methods	1.535	9,210	4,284	4,926
State recommended cultivar and methods	1.488	8,928	4,242	4,686

Since wilt was not a problem in most trials, the advantage of wilt resistant genotypes was difficult to assess. However, at Tangadencha and Ridaj, wilt killed many plants of the local cultivar and there were large differences in yield between the improved wilt resistant and local wilt susceptible cultivars.

For the residual moisture situation, the change from long duration genotypes (110 to 120 days) to extra short-duration genotypes (80 to 90 days) gave a yield advantage of 30% (Tables 3a,b and 4). This group of genotypes has potential to give stable yields because they escape moisture stresses and high temperatures at the end of the growing season. An important feature of this group of cultivars is that they also respond to irrigation and fertilizer management and produce yields as high as 3.5 tons ha^{-1}. Under irrigation they can be planted as late as early January, while normal long duration cultivars must be planted in November (Tables 3a,b and 4).

Constraints on Chickpea Expansion and Approaches to Overcome Them

Expansion of Chickpea

The recent increase in the cultivation of chickpea in Maharashtra state appears to have taken place in traditional wheat producing area. However, chickpea faces severe competition from safflower and sunflower (Table 5). Expansion of chickpea in non-irrigated areas is limited to sowing dates between 15 October and 15 November for optimum yields. Available water for good crop growth is the major limitation in non-irrigated areas. The area sown to non-irrigated chickpea increases if late rains are received in September or October, as was the case in 1990/91. However, the widely adopted crop sequence is mung bean/urd bean (70 to 90 days, rainfed), followed by safflower sown in September as a second crop (Table 6). On the other hand, chickpea can be sown after kharif

Table 3a. Grain yield (kg ha^{-1}) of non-irrigated chickpea at different locations, post rainy season, 1988/89

State/locations	Improved package		State recommended package	
	ICRISAT cultivar*	State cultivar	ICRISAT cultivar*	State cultivar
Andhra Pradesh				
Tangadencha	822	197	1619	760
Karnataka				
Gangavati	1310	1360	1260	625
Madhya Pradesh				
Bhikangaon	1385	955	1190	845
Narsingpur	1620	1525	1020	1190
Maharashtra				
Chakur	910	1035	485	750
Sawarkhed	1040	950	730	860
Orissa				
Keonjhar	1012	733	817	582
Uttar Pradesh				
Amarokh	1754	875	1380	825
Average	1232	954	1063	804

*ICCV 2, an extra short duration cultivar.

Table 3b. Two way comparison of average yields

Cultivar	Cultivation package		Average	Percentage increase
	Improved	State recommended		
Extra short duration	1232	1063	1147	30.5
Local, long duration	954	804	879	
Average yield	1093	933		
% increase	17.1			

sorghum which is harvested by November. This late sown chickpea crop often suffers from end-of-season drought when rains are not received in November or December. Irrigated chickpea must compete with rabi sunflower.

In non-traditional areas, chickpea can be a profitable crop after rice but establishment of the crop is a problem. Chickpea also is grown extensively in young orange and pomegranate orchards. In old orange orchards, the crop suffers from shading by the fruit trees.

For expansion of chickpea production in traditional wheat producing regions, changes in plant traits are needed. Cultivars need to be developed that

Table 4. Average yield, income, cost of cultivation and profit from non-irrigated chickpea 1988/89

Treatment	Grain yield tons ha^{-1}	Income Rs ha^{-1}	Cultiva-tion cost Rs ha^{-1}	Profits Rs ha^{-1}
Improved genotype and methods	1.232	8624*	4050	4574
State recommend cultivar and improved methods	0.954	5724	4036	1688
Improved cultivar and State recommended methods	1.063	7441*	3684	3757
State recommended cultivar and methods	0.800	4824	3578	1246

*Selling price of ICCV 2 Rs.7000 ton^{-1}
Selling price of other genotypes Rs.6000 ton^{-1}

Table 5. Area (million ha) and production (million tons) of major winter crops in Maharashtra, India from 1960/61 to 1987/88

Year	Chickpea		Wheat Sorghum		Rabi		Safflower		Sunflower	
	A	P	A	P	A	P	A	P	A	P
1960/61	0.40	0.13	0.91	0.40	0.37	2.32	--	--	--	--
1970/71	0.31	0.01	0.81	0.44	0.32	0.67	--	--	--	--
1975/76	0.44	0.18	0.17	0.20	0.33	1.22	--	--	--	--
1979/80	0.43	0.17	1.14	0.98	0.35	1.90	0.49	0.17	--	--
1980/81	0.41	0.14	1.06	0.88	0.35	1.57	0.48	0.17	--	--
1981/82	0.46	0.18	1.13	0.99	0.35	1.62	0.54	0.32	--	--
1982/83	0.44	0.15	1.02	0.80	0.38	1.39	0.55	0.28	--	--
1983/84	0.49	0.21	1.18	1.14	0.36	1.58	0.57	0.37	0.07	0.04
1984/85	0.51	0.20	0.99	0.86	0.37	1.84	0.61	0.36	0.12	0.07
1985/86	0.53	0.18	0.88	0.64	0.37	1.27	0.62	0.25	0.12	0.06
1986/87	0.48	0.13	0.73	0.54	0.35	0.98	0.58	0.20	0.12	0.03
1987/88	0.56	0.23	0.73	0.63	0.37	1.06	0.71	0.33	0.20	0.98

A = Area, P = Production

can be sown in September in order to take full advantage of available soil moisture to produce adequate yields. If such cultivars were available, they could replace safflower in the cropping sequence. Also, cultivars that mature in 70 to 80 days need to be developed. For rice growing areas, short duration cultivars (80 to 100 days) with good seedling vigor may be necessary.

It will not be possible for chickpea to entirely replace safflower or sunflower even when genotypes that are amenable to September sowing are developed. This is because chickpea brings a lower price than safflower and sunflower and requires very high seeding rates and high seed costs (Tables 7 and 8).

Table 6. Major cropping system in Maharashtra

Monsoon Crop	Duration	Winter crops	
		Residual moisture	Irrigated
Hybrid sorghum	125 days	Chickpea	Wheat/chickpea sunflower, groundnut
Cotton	200 days	None	Groundnut
Mungbean	80 days	Safflower	Sunflower/ groundnut wheat
Urd bean	90 days	Safflower	Wheat/chickpea/ sunflower/ groundnut
Pigeonpea	180 days	None	
Pigeonpea	110 days	Chickpea, Castor (relay)	Sunflower
Orange (perennial)	--	Chickpea	Chickpea, wheat
Rice	140 days	Chickpea, *Lathyrus*	Groundut
Soybean	120 days	--	Chickpea, wheat
Fallow	--	Safflower	

Table 7. Seeding rates and seed costs of different rabi crops in Maharashtra, India

Crop	Seeding rates kg ha^{-1}	Cost of seed Rs kg^{-1}	Total seed cost Rs ha^{-1}	Reason for purchase
Chickpea Optimum	65-80	12-15	780-1200	Severe infestations of bruchids in storage
Farmers using	30-40	12-15	360-600	
Safflower	12-15	15-20	180-300	None
Sunflower hybrid cultivar	10-12 10-12	50-200 15-25	500-2400 150-300	Farmers do not produce hybrid seed
Wheat	80-100	4-5	320-500	None
Rabi sorghum Only one cultivar	10-15	10-15	100-225	None
Groundnut (in shell)	180-200	14-20	2520-4000	Poor viability

Table 8. Farm gate price of various rabi crops

Crop	Farm gate price Rs ton[-1]
Chickpea	
'Deshi'	5000 – 8000
White testa	8000 – 16000
Green testa	7000 – 10000
For puffed and roasted	12000 – 15000
Safflower	8000 – 11000
Sunflower	8000 – 12000
Wheat	3000 – 4000
Groundnut	8000 – 12000

To eliminate the price disadvantages of chickpea, four approaches can be followed: (1) development of extra short-duration chickpeas to capture the early market, thus giving more economic benefit to farmers; (2) development of bold, white-seeded (kabuli) cultivars which bring substantially higher prices than brown-seeded cultivars; (3) development of green-seeded cultivars which also bring better prices in local markets; and (4) development of cultivars suited for puffing and roasting (Table 8).

In 1991/92, we introduced medium-bold, cream-white seeded cultivars, ICCV 3 and ICCV 5, and both are becoming popular because of higher market prices, particularly in large cities. Similarly, a cultivar with green testa has been developed. Two extra short-duration cultivars, ICCC 88201 and ICCC 88202, have been introduced in the rice belt and are showing promise. Two additional cultivars, P 18 and P 1329, are being tested at various research farms for September sowing.

Seed availability is another constraint on chickpea expansion. However, this can be overcome if farmers receive training in protection of seed from bruchids. Alternatively, we motivated a large number of farmers to store their seed in warehouses at nominal cost. The results of this endeavor are being assessed.

Training of farmers (Figure 2) in cultivar identification and seed production procedures have resulted in several groups of farmers producing seed at the village level. Preliminary assessment indicates that seed of improved cultivars is becoming increasingly available (Table 9).

Micronutrients

Most soils in Maharashtra have pHs ranging from 7.5 to 8.2. Nutrients, except molybdenum, become less available at these pHs and zinc, iron, and phosphate become deficient. For irrigated chickpea, we recommend 4 kg $ZnSO_4$ ha^{-1} and 1 kg Ferrous sulphate ha^{-1} as soil and foliar applications, respectively. This

Table 9. Seed of improved cultivars multiplied by farmers[1]

Season	Seed multiplied by farmers(t)
1988/89	2
1989/90	45
1990/91	750
1991/92	500[2]

[1] ICCC 37 and ICCV 2.

[2] Area under chickpea (and all rabi crops) declined drastically because of severe drought.

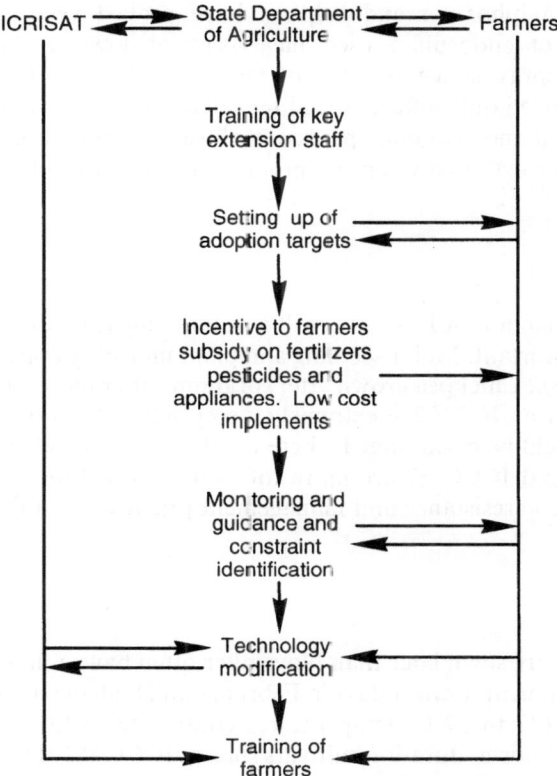

Figure 2. Scheme for training of extension staff and farmers in cultivar identification and seed production procedures.

practice is becoming popular. However, we have no data on its effect on yield, but the crop looks greener, particularly when ferrous sulphate is applied. More research is needed in this area.

For residual moisture situations, farmers are not likely to use costly inputs. We suggest that research should be conducted on seed fortification.

Pests

Occasionally, *Helicoverpa* causes heavy damage to chickpea. This happens when the initial attack on foliage (skeletonization) goes unnoticed. Therefore, training of farmers to recognize these foliar attacks and its symptoms have proved immensely useful. Dust formulations of pesticides such as methyl parathion, quinalphos, or endosulphan have proved very effective. Spray concentration of endosulfan used have been as low as 300 g a.i. ha^{-1}. Alternatively, spray concentrations of 600 g a.i. ha^{-1} (2 liters commercial endosulfan) proved quite effective and repeated sprays were not required. The key to successful and economic pest control was to apply insecticides in early stages of crop growth and when symptoms were first noticed.

Diseases

With the introduction of Fusarium wilt resistant cultivars, the wilt disease has become less important. Major seedling and plant mortality appears to be due to *Sclerotium rolfsii*. Chickpea grown after sorghum suffer more from this disease. The kabuli cultivar, ICCV 2, is extremely susceptible to *S. rolfsii* and cultivation of this line should be discouraged where the disease is prevalent. ICCC 88201, ICCC 88202, and ICCC 37 are more tolerant to *S. rolfsii*. More research is needed on genetic resistance and management practices for this disease.

Heat

Chickpeas that are sown later than November often have a limited cool period for crop development. Longer days in February and high maximum (35 to 37°C) and minimum (22 to 27°C) temperatures contribute to flower abortion and reduced yields. Shorter duration cultivars such as ICCC 88201 and ICCC 88202, overcome this problem to a large extent even when sown late. ICCV 2 sown in the month of February had yields as large as 1.0 tons ha^{-1}.

References

Amin, P. W., Jain, K. C., Kumar Rao, J. V. D. K. and Pawar, C. S. 1988. In: *ICRISAT (International Crops Research Institute for the Semi-Arid Tropics)* pp. 61–65. 1989. Linking Grain Legumes Research in Asia: Summary of Proceedings of the Regional Legumes New Work Coordinators' meeting, 15–17 Dec. 1988. Patancheru, Andhra Pradesh, India: ICRISAT.

Anonymous. 1988–89. District-Wise General Statistical Information of Agriculture Department. *Part II. Epitome of Agriculture in Maharashtra.*

Saxena, N. P. 1984. In: *The Physiology of Tropical Field Crops*, pp. 419–452 (eds. P. R. Goldsworthy and N. M. Fisher). New York: John Wiley and Sons.

von Oppen, M. and Rao, P. P. 1987. In: *The Chickpea*, pp. 383–398 (eds. M. C. Saxena and K. B. Singh). Wallingford, UK: C.A.B. International.

Reports of seven concurrent discussion groups based on geography

Discussion group: North America

CHAIRMAN: W.J. KAISER
RAPPORTEUR: J.M. KRAFT

Abstract

The discussion had representatives from both Canada and the USA. While the USA was the major producer of cool season food legumes in North America two decades ago, Canada now far surpasses the USA in both production and exports. Greatly expanded production of dry peas and lentils has taken place in the Canadian Provinces of Saskatchewan, Alberta and Manitoba.

Current Status of the Crops

Pea

Approximately 120,000 hectares of peas of all types are produced in the USA. About 80,000 ha is devoted to dry peas of which the majority are green smooth-seeded "Alaska" types. The major production area for dry peas in the USA is the Palouse region of eastern Washington and northern Idaho. Pea production in Canada is centered in the provinces of Saskatchewan, Alberta, and Manitoba where approximately 400,000 hectares of the crop are produced.

Lentil

Approximately 60,000 hectares of lentils of all types are produced in the USA. The major portion of the production in the USA is devoted to large-seeded yellow cotyledon cultivars; however, there has been increased interest in small-seeded red cotyledon types. Production of lentil in Canada has shown phenomenal growth over the past decade and currently stands at about 250,000 hectares.

F.J. Muehlbauer and W.J. Kaiser (eds.), Expanding the Production and Use of Cool Season Food Legumes, 941–943.
© *1994 Kluwer Academic Publishers.*

Chickpea

Mexico is by far the largest producer of chickpea in North America where the area sown now stands at over 80,000 hectares annually. The coastal areas of central and southern California and the Palouse region of eastern Washington and northern Idaho are the major production areas in the USA where a total of about 10,000 hectares are produced annually.

Faba Bean

Production of faba bean in the USA is minimal; however, about 2000 hectares are produced in Canada.

Regional Goals by the Year 2000

It is probable that production of dry peas and lentils will continue to increase in Canada from now to the year 2000. Increases in the USA are not expected although there is expanding production is expanding in nontraditional areas of Washington state and also parts of North and South Dakota. Most of the increased production is intended for export either as food or feed. Only a small portion of the production of grain legumes in the USA and Canada is used domestically.

The situation is somewhat different for chickpea. The garbanzo type is preferred in the USA and substantial quantities are imported from Mexico. However, domestic production in the USA is expanding and it is possible that the USA could become a net exporter of chickpea.

Major Production Constraints to Achieving Regional Goals

The major biotic constraints contributing to instability of pea yields include root diseases, such as common root rot caused by *Aphanomyces eutieches*, and Fusarium and Pythium root rot. Major foliar diseases include Ascochyta blight, powdery and downy mildew, Sclerotinia white mold, and occasionally bacterial blight. Significant problems are sometimes encountered with aphid transmitted viruses of which the most prominent are pea enation mosaic and bean leaf roll viruses. Pea seedborne mosaic has been problematic at times but occurrences have nearly always been in breeding nurseries.

Pea weevil (*Sitona lineatus*), seed weevil (*Bruchus pisorum*), and aphids are the serious insect problems associated with dry pea production in the Palouse region. The potential loss of effective insecticides to control insect pests is of concern to producers.

Biotic constraints to lentil production include Ascochyta blight and

Colletotricum blight (*anthracnose*) in Canada and aphid transmitted viruses in the USA. Other than aphids, the only serious insect pest of lentil is lygus bugs which induce a condition in lentil seeds known locally as "chalky spot". The condition reduces yields as well as market quality. Insecticides are effective in the control of aphids and lygus bugs.

Ascochyta blight is the major biotic stress of chickpeas in the USA and Canada. Selections with good resistance have been made that hold promise for alleviating the problem. Other than Ascochyta blight and Pythium seed rot and preemergence damping-off, Fusarium wilt, found mostly in California, is the only other serious disease problem of chickpea.

Faba bean is affected only by chocolate spot and to some extent by aphids and *Sitona* weevils.

The abiotic constraints of grain legumes in North America include susceptibility to cold and frosts, especially in Canada and to heat and drought in the drier areas of the US Pacific Northwest.

Other Major Constraints to Achieving Regional Goals

Accurate forecasting of the commodity and quality requirements of export markets and the timely communication of these requirements to researchers is considered of utmost importance. Breeders must have sufficient lead time to identify useful germplasm and establish critical quality criteria for use in their hybridization and selection schemes.

Improving Collaboration, Coordination and Communication

Close collaboration has developed between grain legume research programs in North America and ICARDA and ICRISAT. These links may be jeopardized by reduced research funding and likely restrictions on travel.

Recommendations

Several areas of research were highlighted by the group as critical to expansion of cool season food legume production in North America. These include: (1) continue research efforts to develop strategies to overcome biotic and abiotic constraints to grain legume productivity, (2) continue genetic mapping of chickpea, lentil, and pea with emphasis on disease resistance genes, (3) use biochemical characterization for the development of new products from grain legumes, and (4) improve domestic and international market information.

A major concern is the seemingly continual reduction in the number of researchers devoted to grain legume crops.

Discussion group: Latin America

CHAIRMAN: J.U. TAY
RAPPORTEUR: G. HERNÁNDEZ-BRAVO

Abstract

The discussion group considered cool season food legume crops grown in the Latin American region. This region extends from north to south and includes: Mexico, Guatemala, Dominican Republic, Jamaica, Venezuela, Colombia, Ecuador, Perú, Bolivia, Chile, Brazil, Paraguay, Argentina, and Uruguay.

The four food legumes, pea, chickpea, faba bean, and lentil, constitute a very important component of the population's diet and are inexpensive sources of protein when compared to meat. These legumes are produced mostly by small farmers who employ traditional farming methods such as crop associations, intercropping, and multiple cropping as well as monocultures with minimal inputs. The production of these legumes is commercialized primarily in local markets and some quantities of pea, chickpea, and lentil are exported by Mexico, Argentina, and Chile.

Current Status of Cool Season Food Legumes in Latin America

Production of cool season food legumes in the Latin American region consists of 90,000 tons of peas, 175,000 tons of chickpeas, 134,000 tons of faba beans, and 72,000 tons of lentils (FAO, 1991). Factors which adversely affect the production and quality of these crops in the region are summarized in Table 1. The major biotic stresses include plant diseases, while drought is the most important abiotic stress factor.

Regional Goals by the Year 2000

Six regional cooperative research projects are recommended (Table 2) to help solve production problems caused by *Ascochyta*, Fusarium wilt, *Botrytis*, and rust as well as drought in this geographic region. The cooperative project number 6 originated from a previous recommendation that was made by the

F.J. Muehlbauer and W.J. Kaiser (eds.), Expanding the Production and Use of Cool Season Food Legumes, 944–946.
© 1994 *Kluwer Academic Publishers.*

Table 1. Major biotic and abiotic constraints to achieving regional goals in cool season food legume production in the Latin American region

	Constraints	
	Biotic	Abiotic
Pea:	Fusarium wilt Ascochyta blight *Liriomyza*	drought frost
Chickpea	Fusarium wilt *Heliothis*	drought low temperature
Faba bean	*Botrytis* *Rhizoctonia* Virus	drought low temperature waterlogging
Lentil	Fusarium wilt *Rhizoctonia* Rust	drought low temperature poor soil fertility
Grasspea		poor soil fertility

Table 2. Recommended regional cooperative research projects between National Research Institutes and International Research Centers

Project No.	Subject	Participants
1	Evaluation and selection of large-seeded peas with *Ascochyta* resistance.	Chile, Colombia, Ecuador, Peru, Argentina, and ICARDA
2	Evaluation and selection of kabuli chickpeas with resistance to Fusarium wilt.	Mexico, Bolivia, Peru, Chile, Argentina, ICARDA, and ICRISAT
3	Evaluation and selection of large-seeded faba beans with resistance to *Botrytis*.	Mexico, Bolivia, Colombia, Ecuador, Peru, Chile, and ICARDA
4	Evaluation and selection of large-seeded lentils with resistance to rust.	Chile, Colombia, Ecuador, Peru, and ICARDA
5	Drought Food Legume Research Network.	Mexico, PROCIANDINO, Chile, Argentina, Brazil, ICRISAT and ICARDA
6	Collection, evaluation, and preservation of food legume germplasm from unexplored areas of South America.	Peru, Bolivia, Chile, Argentina, Brazil, ICARDA, and ICRISAT

participants to the "Training Course on Research and Production of Cool Season Food Legumes in the Andean Region". This training course took place in Ecuador and Colombia in 1988 with the collaboration of ICARDA.

It would be highly desirable to have the participation of the countries shown on Table 2, as well as the International Research Centers (ICARDA and ICRISAT) to ensure the success of these cooperative projects.

Improving Collaboration, Coordination, and Communication

Research collaboration in the region can be improved through the establishment of "Cooperative Research Projects" between the National Agricultural Research Institutes of Mexico and South America. By doing so, the existing Food Legume Research networks, such as PROCIANDINO in the Andean Region should be taken into consideration. The latter is a cooperative Program on Agricultural Research and Technology Transfer for the Andean Region, which has been financed by the governments of five countries and by the Inter American Development Bank, under the administration of the Inter American Institute for Cooperation in Agriculture (IICA). The cooperative program PROCIANDINO is coordinating research activities among the countries of Bolivia (IBTA), Colombia (ICA), Ecuador (INIAP), Peru (INIAA), and Venezuela (FONAIAP) on such crops as pea, faba bean, and lentil. This discussion group encourages regional research collaboration with participation of the National Research Institutes and International Research Centers, ICARDA and ICRISAT.

Discussion group: Europe

CHAIRMAN: J. PICARD
RAPPORTEUR: M.C. HEATH

Current Status of Food Legumes in Europe

Pea and faba bean are grown mainly for animal feed throughout all EC countries. Chickpea and lentil are of lesser importance in Europe and are confined mainly to Spain and southern France. Grasspea remains of academic interest only, confined to small experimental plots, and is likely to remain so unless a market can be recognized.

Imports

The EC currently imports pea (from Canada and the USA), as well as chickpea and lentil. These additional quantities could be produced easily from within the EC but the current price support system discourages increased production.

Exports

The EC exports some food legumes, particularly faba bean to Egypt and the Middle East. However, this market is primarily for human consumption; whereas in the EC, faba bean is grown mainly for animal feed. It is not surprising, therefore, that EC exports are unsuccessful in overseas food markets. Plant breeders expect that the EC could produce faba bean of suitable quality for food export markets; however, current policies favor production for animal feed.

Regional Goals by the Year 2000

It is virtually impossible to set a "regional" goal for the EC for the year 2000 when the price support mechanism for the year 1993 is not yet guaranteed! That is the situation for EC countries. In non-EC European countries, there is room

F.J. Muehlbauer and W.J. Kaiser (eds.), Expanding the Production and Use of Cool Season Food Legumes, 947–950.
© 1994 Kluwer Academic Publishers.

for expansion and the goal should be to maximize production subject to any price constraints.

Within EC countries, there is an increasing likelihood of more and more land being set aside from agricultural production. Attempts should be made to identify agricultural systems in which grain legumes could either improve profitability or provide environmental benefits.

Major Production Constraints to Achieving Regional Goals

The major constraint to goal setting is EC policy. No major biotic or abiotic factors can be identified which are likely to provide a major constraint to yield. It is generally believed within Europe that the technology exists to cope with the most serious limitations to yield. However, increasing environmental pressures are likely to lead to reduced usage of agrochemicals, particularly nematicides and insecticides. As a result, problems with nematodes and insects (e.g., *Sitona* and *Aphis* spp.) could become more serious in some areas.

With regard to new diseases, pea bacterial blight and pea seedborne mosaic virus could become more problematic. Chocolate spot, the most serious disease of faba bean, is likely to remain so. Plant breeders do not envisage any major breakthroughs in breeding for improved resistance under European conditions in the immediate future; chemical control methods also are likely to remain less than fully effective.

Cold and drought are likely to be the major abiotic factors affecting production. Cold-tolerance or frost-hardiness *per se* is not a problem, but winter survival is, particularly under the cold and wet conditions of northern European winters. Drought is likely to cause increasing problems, particularly as farmers are confronted with increasing water restrictions; under such restrictions, water available for irrigation purposes will be used preferentially on horticultural and root crops rather than grain legumes.

Other Major Constraints to Achieving Regional Goals

Policy constraints, imposed under CAP within Europe, are the most important. Other major constraints include technology transfer and utilization.

Technology Transfer

While technology transfer is generally good within EC countries, there is still scope for further improvement, particularly with chickpea and lentil. In the future, the combination of EC policies and environmental concern is likely to create an R&D need for member states to develop rational strategies for integrated use of food legumes within a viable and sustainable agriculture.

Utilization

Feed compounders are the major customers for grain legumes within Europe. This industry is highly sophisticated and demands relatively large quantities of reliable supply and constant quality. In the future, there will remain a need to ensure that grain legume production meets the needs of the compounding industry. Much progress has been made already in stabilizing the supply of peas and faba bean; in the future, attempts to identify and define relevant quality standards would help to achieve this.

Improving Collaboration, Coordination, and Communication

For countries within the EC, collaboration in research endeavors between member States is not a problem; nevertheless, there is a need to maintain, and whenever possible, strengthen these links in order to ensure that dwindling resources are utilized effectively and efficiently.

In an attempt to achieve this, an Association of European Grain Legume Researchers was formed at the European Conference on Grain Legumes held 1–3 June 1992 at Angers, France.

There is also a need to strengthen links between the EC and other European countries; this has been facilitated by the dramatic changes taking place politically in eastern Europe. Already new links are being formed and exchanges of scientists and students are taking place.

At a worldwide level, the main aim should be to maintain, rather than to increase collaboration. European researchers value the links existing with colleagues in North America and wish to see these continue. On faba bean, the research momentum created by ICARDA has been of immense value worldwide; Europe would like to see this momentum maintained.

Recommendations

1. Encourage EC policy makers to consider formulating a new policy within a revised CAP which gives due regard to effective and efficient grain legume production by providing a realistic pricing structure and defined quality standards.
2. Develop, for Europe, rational systems of integrated grain legume production, within a viable and sustainable agriculture, that meet EC policy requirements and minimizes adverse effects on the environment.
3. Set up a European Grain Legume Association, with appropriate links to non-EC countries, which will encourage collaboration between scientists.
4. Encourage the EC to give due consideration to making sufficient funds available for fostering and strengthening international links that will be of ultimate benefit to EC policies.

5. At the 3rd International Food Legume Research Conference to be held in 1997, place more emphasis on quality aspects and identify mechanisms for improving the relationships between plant breeders and the food and feed industry.

6. It is recognized that pea is not covered by the Consultative Group on International Agricultural Research (CGIAR) system and that faba beans are in danger of being removed from it. Against this backcloth, European researchers recommend that the future policy of the CG system should be redefined – particularly with regard to pea and faba bean.

Discussion group: Africa

CHAIRMAN: M.H. HALILA
RAPPORTEUR: S.P.S. BENIWAL

Abstract

The discussion was attended by 29 scientists from eight countries in Africa including Algeria, Libya, Morocco and Tunisia, in North Africa; Egypt, Ethiopia, and Sudan from the Nile Valley Region; and, Burundi in the Great Lakes Region. The discussion proceeded according to the guidelines circulated by the Conference Organizing Committee. Salient points arising out of discussion are summarized below:

Current Status

Of the five major cool season food legumes grown in Africa, faba bean is the most important followed by chickpea, lentil, pea, and grasspea. However, pea is the only cool season food legume cultivated in Burundi; whereas grasspea is mainly cultivated in Ethiopia. In general, the area and production of these crops has been stable in most countries of Africa except for a decline of chickpea and lentil in Ethiopia, of lentil in Egypt, and of faba bean, chickpea, lentil, and pea in Algeria and Morocco. Of special significance is the decline in area of chickpea in Ethiopia, and of lentil in Egypt.

Most countries are self-sufficient in most cool season food legumes except Ethiopia, Sudan, and Egypt in lentil, and Algeria and Libya in all four food legumes.

Goal by Year 2000

Self-sufficiency in cool season food legumes is the major goal of all the countries that cultivate them in Africa, and is expected to be achieved in two ways: (1) increasing the area sown as is the case in Algeria, Libya, and Morocco in North Africa, and Egypt and Ethiopia in Nile Valley Region, (2) increasing productivity through adoption of improved technology resulting from research.

F.J. Muehlbauer and W.J. Kaiser (eds.), Expanding the Production and Use of Cool Season Food Legumes, 951–955.
© 1994 *Kluwer Academic Publishers.*

Major Production Constraints

Biotic and abiotic stresses were highlighted as important constraints to production of cool season food legumes. Biotic stresses in descending order of importance are diseases, insect-pests, weeds, and flowering plant parasites (e.g., *Orobanche*). However, the sequence of importance differs between the North African and Nile Valley Regions. In the former, weeds including *Orobanche* spp. are the most important followed by diseases and insect-pests; whereas, in the latter, diseases were most important followed by insects and weeds.

The diseases in descending order of importance are: chocolate spot, rust, black root rot, Ascochyta blight, and virus diseases in faba bean; Ascochyta blight and wilt/root rots in chickpea; rust and downy mildew in lentil; and, powdery mildew in pea. Important among insect-pests were: aphids and *Sitona* in faba bean and lentil, pod-borer and leaf miner in chickpea, and aphids in pea. *Orobanche* is the most important flowering plant parasite of faba bean and lentil.

Drought was considered the most important abiotic constraint for Africa, followed by cold and high temperatures. Drought was identified as the most important abiotic constraint for cool season food legume production in Africa; others in descending order of importance were poor subsistence farmers and use of traditional agronomic practices. However, the sequence of importance was different for countries in North Africa and the Nile Valley Region. In descending order of importance, drought, poor subsistence farmers, and use of traditional agronomic practices for North Africa; whereas poor subsistence farmers, lack of inputs, drought, and use of traditional agronomic practices are major constraints in the Nile Valley Region.

Other Major Constraints

Other important major constraints identified were poor input availability, low prices, lack of suitable marketing policy, lack of mechanization, lack of or insufficient emphasis on technology transfer, and lack of or underdeveloped processing facilities.

Input Availability

Africa has poor subsistence farmers that lack resources and credit to invest in cool season food legume production. Another important constraint is the nonavailability of inputs to farmers.

Marketing

This includes price instability, low prices, and a lack of a pricing policy for cool season food legumes. These often become disincentives to small subsistence farmers in the region.

Mechanization

In Africa, poor subsistence farmers in the region still follow the traditional methods of cultivating cool season food legumes that are cumbersome compared with production of cereals that has now become more mechanized, and thus, more convenient to farmers. Lack of mechanization especially for planting faba bean, and harvesting and threshing of the four cool season food legumes is a major hindrance to their production, especially in North Africa.

Seed Production

Lack of or insufficient availability of high quality seed of cool season food legumes is a common phenomenon in Africa. The group emphasized its importance as an input in increasing productivity and production of these crops.

Technology Transfer

Another constraint that has affected improved productivity of cool season food legumes is the lack of technology transfer. Newly developed improved technology for production of cool season food legumes has not reached farmers who, as a result, have not realized the potential benefits. This has been clearly demonstrated in the three countries of the Nile Valley Region where increased emphasis on this aspect has already paid dividends.

Processing and Use

Presently, processing of cool season food legumes in Africa is highly underdeveloped indicating a scope for diversification. This has adversely affected its diversified use. Unsuitable small and hard seeds were highlighted as an important hindrance for their use and processing.

Improving Collaboration, Coordination, and Communication

At the National Level

Existence of poor collaboration at the national level is a major problem. Collaboration among different research, teaching, extension, and seed producing institutions, and between research and development institutions needs to be encouraged and developed.

At the Regional Level

Although regional networks have been developed and are operating, a need for strengthening was emphasized. Special mention was made of the necessity for regional collaboration in faba bean research and the need to develop special regional facilities to be used for such things as: screening for resistance to *Orobanche*, aphid and heat tolerance, mechanization, and virus and nematode work. A need for exchange of grasspea germplasm among countries in the region was emphasized.

At the International Level

Important activities mentioned for collaboration at the international level include: (1) exchange of plant material, scientific visits, and information, (2) development of special projects such as on biotechnology (only applied aspects) and Ascochyta blight of chickpea, and (3) training of scientists for higher degrees at advanced institutions dealing with cool season food legumes. The group expressed concern on the withdrawal of support by CGIAR from ICARDA for research on faba bean, the most important cool season food legume in the region, and emphasized the need for CGIAR to consider reversal of this decision.

IARCs and NARs

The group expressed great satisfaction on the excellent rapport that exists between the NARS and ICARDA and ICRISAT, the two IARCs that have mandates for cool season food legumes in the region. A strong satisfaction was expressed on the training imparted to national scientists and technicians by these two centers on these crops. Although attention has been given to certain biotechnology aspects, the group emphasized the need for increased attention on the development of this aspect in the region. Also emphasized was the need for increased attention on improvement of the pea crop that is important in Burundi and Rwanda, and continuity of much-needed support on faba bean research in the region.

Recommendations

The group made the following recommendations:
1. There is an urgent need to increase productivity and production of cool season food legumes in the region. For this, it is important to address the constraints limiting productivity of these crops.
2. There is a definite and urgent need for increased attention on research and development activities on these crops considering their role in sustainable agriculture in the region. In the same connection, CGIAR is requested to reconsider providing much-needed support to ICARDA on faba bean research. Faba bean is the most important cool season food legume crop in the region.
3. Considering the importance of transfer of technology from researcher for its adoption by farmers, efforts on this important aspect needs to be strengthened. Collaboration among different institutions at the national level needs strengthening.
4. There is a definite need for favorable government policies to encourage increased cultivation of cool season food legumes.
5. The existing regional, interregional and international collaboration, coordination, and communication on cool season food legumes in Africa needs to be further strengthened in order to derive mutual benefits.

Discussion group: Near East

CHAIRMAN: I. KUSMENOGLU
RAPPORTEUR: N.I. HADDAD

Abstract

The countries included in the Near East Group are: Turkey, Syria, Lebanon, Cyprus, Jordan, Iraq, Iran, and Baluchestan area. Chickpea, lentil, and faba bean are the major food legumes grown in these countries. Pea production and research is limited, whereas grasspea was recently included in the research programs of some of the countries primarily for use as a forage crop.

Current Status

Chickpea

Kabuli chickpea is produced in all countries and only Iran is producing kabuli and desi types. Except for Turkey, which exports large quantities, other countries are either self-sufficient or importing some of their needs.

Lentil

The region has grown both red and yellow lentils. Turkey is the major producer and exporter, while Syria exports small quantities. The other countries import some fraction for their needs.

Faba Bean

Faba bean major is the principal type grown in the region. Turkey exports some quantity, while other countries are either self-sufficient or import to meet their needs.

F.J. Muehlbauer and W.J. Kaiser (eds.), Expanding the Production and Use of Cool Season Food Legumes, 956–958.
© 1994 *Kluwer Academic Publishers.*

Pea and Grasspea

Very limited area is planted to pea; however, there is a growing interest in production of grasspea. Production in the region is adequate and only small quantities are imported.

Regional Goal by the Year 2000

It is expected that the areas planted to the three major grain legume crops (chickpea, lentil and faba bean) will increase in the coming years to produce sufficient quantities for domestic needs. Turkey, already a major exporter, is working toward increasing exports even further.

Major Biotic and Abiotic Constraints

The major biotic constraint to chickpea production throughout most of the region is Ascochyta blight, while Fusarium wilt is important in Iran. Leaf miner and weeds are also important biotic constraints. Major abiotic constraints include cold, drought, and heat.

The major biotic constraints to lentil production are Fusarium wilt, Sitona weevil, and weeds; while the major abiotic constraints are drought and cold.

The major biotic constraints to faba bean production are *Botrytis*, rust, viruses, and *Orobanche*, while the major abiotic constraint is drought.

Other Major Constraints

The domestic market is adequate in all countries except Baluchestan. Processing of the three major crops is hampered by storage facilities which are either inadequate or nonexistent. There is no problem in crop utilization in the region. Food legume crops are used mainly as food or feed and they are major dietary components.

Technology transfer to farmers is a major constraint for increasing productivity of the three crops. Technology transfer is needed in the area of winter-sown chickpea, weed control, seed production, and crop mechanization.

There is a serious problem in the region regarding the present infrastructure, especially in the areas of seed production of improved cultivars, credit, and availability of inputs. However, the Turkish delegation indicated that they have a satisfactory arrangement in seed production.

With the exception of the major producer (Turkey), inadequate policies for food legume promotion is a major problem in the individual countries of the region.

Land fragmentation was also suggested as a constraint to production improvement in some countries.

Collaboration and Coordination

The present collaboration and coordination at the national, regional, and international level is satisfactory. However, stronger linkage and coordination is required. There is a need to strengthen the West Asia food legume network, and there is an urgent need for regional and international coordination in faba bean, especially after the crop has been phased out from the ICARDA mandate.

Recommendations

Technology transfer to farmers is an important issue that needs more focus by national and international agencies. The areas that need immediate action are: winter chickpea technology, weed control, and mechanical harvesting. However, to achieve satisfactory success in this area, good linkage between researchers, extension agents, and farmers are required. Also, more adaptive research and demonstrations in farmers' fields are needed.

There is a need for research on cold tolerance at high elevation areas for chickpea and lentil. This research should take into consideration resistance to Ascochyta blight and weed control.

There is a need to conduct more basic research on drought tolerance before a regular breeding program can be initiated.

The research on multiple stress resistance needs to be strengthened.

In order to reduce the cost of producing grain legumes, it is essential that a system of legume mechanization from tillage to harvest is transferred to farmers.

National policies on food legumes should support the production of these crops in order to contribute to the sustainability of cereal based farming systems.

Discussion group: Asia

CHAIRMAN: B. SHARMA
RAPPORTEUR: C.L.L. GOWDA

Abstract

Five of the 19 countries comprising the Asia region were represented on the discussion group. Based on climate and farming systems, the regions can be divided into three areas:

South Asia   : Pakistan, India, Nepal, Sri Lanka, Bangladesh, and Maldives
Southeast Asia: Thailand, Laos, Cambodia, Vietnam, Malaysia, Indonesia, Myanmar, and the Philippines
East Asia   : Japan, Korea, Mongolia, Taiwan, and Peoples Republic of China

    The discussion group relied on personal knowledge in their deliberations concerning cool season food legumes in the region.

Current Status of Cool Season Food Legumes in Asia

A summary of the three areas is as follows:

South Asia   : All five cool season food legume crops are grown in South Asia with India and Pakistan being the major producing countries; however, faba bean is a minor crop.
Southeast Asia: Few cool season food legumes are grown.
East Asia   : China is a major producing country for pea and faba bean. Chickpea and lentil are also grown there but to a much lesser extent. Pea and faba bean are produced to a limited extent in the other countries of east Asia.

    Even though Bangladesh, India and Pakistan are major producers of chickpea and lentil, they still need to import those commodities to meet their needs. These three countries are major consumers of pea, lathyrus and faba bean but sufficient quantities are produced to meet their needs. Sri Lanka does not grow any cool season food legumes, but imports lentil from Turkey. Myanmar produces chickpea for export to Bangladesh and India, while Japan imports peas. China also trades pulses with other Southeast Asian countries.

F.J. Muehlbauer and W.J. Kaiser (eds.), Expanding the Production and Use of Cool Season Food Legumes, 959–961.
© 1994 *Kluwer Academic Publishers.*

Regional Goals by the Year 2000

All the countries of the region are striving to increase production of grain legumes, and the minimum goal is to produce adequate quantities for domestic markets. Surplus production would be exported if markets were available. The countries exporting grain legumes at present are also willing to increase their exports within the group of Asian countries or to other parts of the world.

Major Production Constraints to Achieving Regional Goals

Constraints to production in all the grain legumes are weeds and pests in storage. Specific constraints for each crop are listed below:

Crop	Biotic Stresses	Abiotic stresses
Chickpea	Pod borer, wilt and root rots, Ascochyta blight, Botrytis gray mold	Drought, extremes of temperature, salinity
Pea	Powdery mildew, pod borer, stem fly, aphids	Drought, lodging, salinity
Lentil	Wilt, rust, Botrytis, aphids, pod borer	Drought, salinity, shattering, low temperature
Grasspea	Powdery mildew, thrips and aphids, pod borer, downy mildew	Drought, cold, neurotoxin
Faba bean	Chocolate spot, rust, aphids	Drought, temperature variations

Other Major Constraints to Achieving Regional Goals

Important factors needed for the efficient development of grain legume markets in the region include: price structure with incentives to farmers, crop insurance, an efficient procurement and distribution system, and the lifting of the ban on cultivation of certain crops, such as grasspea, which prevails in some parts of the region.

Important considerations for processing grain legumes include: value added products, increased recovery of split pulse (dhal), and detoxification and removal of antinutritional factors. Uses for grain legumes need to be diversified to include uses other than human food, e.g., for starch extraction, cattle feeds, grain fodder.

The most important aspects of technology transfer for increasing grain legume production include: an efficient seed production and distribution system that provides high quality seeds, availability of inputs at reasonable prices, and on-farm demonstrations of improved technology.

The following items were identified as priority areas to facilitate production and processing of grain legumes, leading to higher returns to the farmers: enhancement of research capabilities, creation of an extensive and efficient system of technology transfer, and the manufacture of small scale processing units which can be purchased by individual farmers. Processed commodities can be sold directly by individual farmers at the market place.

It was felt that many countries do not have a defined policy for grain legume research, production and marketing. It is essential that all countries evolve sustainable policies to encourage research and production of grain legumes and to organize their domestic and international trade in harmony with other grain legume producing countries.

Improving Collaboration, Coordination and Communication

Coordinated research programs should be developed for the major grain legumes in each country. Research programs are already operating in some countries.

To enhance these research programs, an Asian Food Legume network was initiated along the lines of the FAO project (FAO-RAS/89/040). The tenure of that project is nearing completion. The group strongly felt that a General Grain Legume Network can be evolved using the facilities and infrastructure already available in the FAO project.

Collaboration with international research centers should be promoted through joint studies in a coordinated manner. A system of working groups can be evolved for each crop or for each specific problem. Group leaders should be identified and a contact person nominated for each research problem.

Recommendations

1. The IFLRC should be expanded to include all legumes (including warm season legumes) that are important in the region.
2. Establish an International Working Group on Ascochyta blight (coordinated by W. J. Kaiser) to coordinate research on this important disease.

Discussion group: Oceania

CHAIRMAN: R.O. REES
RAPPORTEUR: J.B. BROUWER

Abstract

The information presented here refers only to the situation in New Zealand and Australia.

As cool season food legumes are grown in several Australian states and regional differences within Australia very much dominate most industry issues, the group recognizes four geographical regions: (1) New Zealand, (2) northeastern Australia (including southern Queensland and northern New South Wales with their summer rainfall pattern), (3) southeastern Australia (including southern New South Wales, Victoria, South Australia and Tasmania), and (4) Western Australia.

Current Status of the Crops

Pea

The field pea production area in New Zealand of approximately 30,000 ha with an estimated production of 60,000 t annually is largely concentrated on the Canterbury Plains. New Zealand currently exports blue peas, maple peas and marrowfat peas. The current area sown to field pea in Australia is approximately 450,000 ha divided over southeastern Australia (400,000 ha which is expected to produce 430,000 t in 1991/92 and Western Australia (50,000 ha which is expected to produce 40,000 t).

Dun pea (purple flowers) types dominate the Australian pea crop. Approximately 10,000 t of blue peas and 20,000 t of white peas are being produced in southeastern Australia and 20,000 t of white peas in Western Australia which are inadequate to supply exports above domestic requirements. Maple peas produced in southeastern Australia are used domestically, while no marrowfat peas are presently being grown commercially.

F.J. Muehlbauer and W.J. Kaiser (eds.), Expanding the Production and Use of Cool Season Food Legumes, 962–968.
© 1994 *Kluwer Academic Publishers.*

Chickpea

No chickpea industry has yet been established in New Zealand and there is no clear forecast for any future development. Southeastern and northeastern Australia are producing and exporting both desi and kabuli chickpeas. Western Australia has only just started to produce chickpea, and now exports garbanzos produced in the irrigated Ord River Valley.

Faba Bean

Faba bean production is adequate for domestic supply in New Zealand but it is still insufficient in northeastern Australia. Production in southeastern Australia exceeds domestic demand and exports have been made of the small horse bean type "Fiord" with only a very small proportion of exports of the broad bean type "Aquadulce".

Lentil

New Zealand currently has 4,000 ha of red and yellow (brown) lentils used mostly for export. Australian lentil production is still confined to approximately 2,000 ha in the southeastern region, which is unlikely to produce a domestic surplus for export.

Regional Goal by the Year 2000

Field Pea

New Zealand aims to increase exports of all types of field peas. Southeastern Australia is expected to further increase production for export of dun, white, and blue peas, while a domestic surplus is expected to become available of maple and marrowfat peas. The field pea area in northeastern Australia is not expected to increase beyond that required for an adequate supply for the domestic market.

Chickpea

Industry developments for chickpea in New Zealand are uncertain. In eastern Australia an increase in exports of desi chickpeas is expected. In the northeastern region a domestic surplus of kabuli chickpeas is anticipated to be available for export by the year 2000, while a positive drive towards increased production for exports of kabulis is already underway in the southeast. Western Australia is expected to achieve a domestic surplus of desis for export and there is potential for increased production of garbanzos for export.

Faba Bean

Production of faba bean in New Zealand is expected to remain largely for domestic supply. A surplus for export is likely to be achieved in northeastern Australia, but the export of faba beans from southeastern Australia is still expanding. The scope for faba bean production in Western Australia appears limited.

Lentil

Increased exports are expected from New Zealand of both red and green lentils. Lentil production in the eastern Australian states is likely to expand to a domestic surplus level with increasing exports.

Major Production Constraints to Achieving Regional Goals

Field Pea

Biotic Constraints

Foliar fungal pathogens feature high on the list of biotic constraints in field peas in New Zealand, with powdery mildew and downy mildew being very important. Virus diseases caused by pea seedborne mosaic virus and alfalfa mosaic virus are recognized as being serious, but insects are not considered to pose any significant problems here. In northeastern Australia, powdery mildew severely restricts pea production when using current Australian cultivars. The Ascochyta blight complex, predominated by *Mycosphaerella pinodes*, is a serious constraint to pea production in southeastern and Western Australia, followed by downy mildew and by powdery mildew in late-maturing districts. Pea weevil (*Bruchus pisorum*) is a major problem in the two latter regions. *Helicoverpa* spp. are becoming an increasing problem in all regions.

Plant or crop architecture as presented by the current cultivars with associated problems of lodging and poor harvestability is seen as a serious obstacle to acceptance of the crop by growers in northeastern Australia. The combination of biotic and abiotic factors causing a lack of suitable arable land in an increasingly tight crop rotation, is likely to pose a real constraint to expanding pea production in southeastern Australia.

Abiotic Constraints

Limited availability of soil moisture in dryland cropping is regarded as the major constraint to pea production in all regions, including New Zealand in some seasons. Both high and low temperatures create problems in northeastern and southeastern Australia during the reproductive period of the crop in the spring, when spells of hot and dry conditions and radiation night frosts can be experienced. The low fertility and poor structure of the majority of Western Australian soils, with the inherent danger of erosion, is a major obstacle to expansion and improved productivity in that region.

Chickpea

Biotic Constraints

Phytophthora root rot is a major threat to chickpea production in northeastern Australia, with gray mold also a serious constraint. *Helicoverpa* spp. is a serious insect pest in that region. While the lack of reliable and effective control of broadleaf weeds remains a serious problem for chickpea growers in all regions, the increasing pressure to make land available for chickpea in a continuous cropping rotation of cereals and forage legumes has become a serious production constraint for growers in southeastern Australia.

Abiotic Constraints

As with field pea, limited availability of soil moisture is a major constraint to chickpea production in all regions. Unreliability of autumn rains necessary for timely crop planting is regarded as a problem for chickpea growers in northeastern Australia. A narrow optimum window for flowering time exists in Western Australia where cool temperatures prevent successful pod set in early spring, and in northeastern Australia with its likelihood of night frosts at flowering and at podding during late spring. The low pH of the majority of Western Australian soils reduces the availability of land suitable for chickpea growing, although alkaline soils do occur in that region.

Faba Bean

Biotic Constraints

Ascochyta blight and chocolate spot are recognized as major constraints to faba bean production in New Zealand and southeastern Australia, with rust and chocolate spot dominating in northeastern Australia. *Helicoverpa* spp. are regarded as major threats mostly in northeastern Australia. The lack of cultivars with the capacity for an extended podding period is seen as lowering the potential for faba bean in higher rainfall districts of southeastern Australia.

Abiotic Constraints

Inadequate soil moisture and high temperatures during flowering constrain faba bean productivity in most regions. Low yields caused by soil acidity problems in southeastern Australia have been alleviated by selection of more effective *Rhizobium* strains, but the lack of suitable soils is still limiting the potential for faba bean production in Western Australia.

Lentil

Biotic Constraints

Susceptibility to lodging among red lentil types and the lack of cultivars with resistance to Ascochyta blight are major constraints in New Zealand.

The lack of suitable cultivars is considered to be the main factor limiting lentil production in Australia, especially because of the low biomass and inefficient plant architecture making the crop unsuitable for mechanical harvesting. Ascochyta blight and the insect pests, red-legged earth mite and *Helicoverpa* spp., are also seen as important problems to solve when establishing this emerging industry in Australia. Broadleaf weeds remain an important production problem.

Abiotic Constraints

Lentil faces waterlogging problems in New Zealand and southeastern Australia, while insufficient soil moisture is a major constraint in the other regions.

Other Major Constraints to Achieving Regional Goals

Field Pea

The most important determinant for the continued viability of field pea is the profitability of the crop relative to other grain legumes and cereals. Economic returns for this crop in all regions are largely determined by price. In eastern Australia further market development and value-adding processes are regarded as necessary to encourage an expansion of the crop.

Chickpea

Further market development is required for the viability of chickpeas in all regions. The need for some form of processing of the chickpea crop in Australia is seen as important but not as high in priority as the maintenance of economic returns to the growers.

Faba Bean

Relative profitability of the crop is also regarded in the faba bean industry as the most important factor in achieving growth in all regions. Technology transfer problems were indicated for northeastern Australia, while infrastructure, such as costly transport, may be a constraint to developing a faba bean industry in Western Australia.

Lentil

Development of markets is regarded as essential for the further development of the lentil industry in New Zealand. As the lentil industry in Australia is only just emerging, market development and relative profitability, albeit of great significance, are not the only requirements for establishing this crop successfully, and other factors such as the availability of processing facilities in Australia, provision of information on production methodology and market availability to growers and an increased input, will be of equal importance.

Improving Collaboration, Coordination, and Communication

Nationally

The Grains Research and Development Corporation in Australia, which supports grain legume research from grower levies and matching federal funds, has a firm policy of encouraging interaction and collaboration between research groups in Australia. It sponsors inter- and intra-disciplinary workshops involving researchers, traders, growers, marketers, and end users. This approach is greatly improving communication among different industry groups. Recently, the Australian federal government allocated $A12 million to the establishment and operation of a Collaborative Research Center to be based at Perth, Western Australia to develop pasture and grain legumes for Mediterranean environments. This Center will also have a national responsibility of supporting and coordinating research on grain legumes.

Regionally

The discussion group agreed that interaction and better coordination between grain legume programs in New Zealand and Australia should be achieved through more frequent and reciprocal visits and exchange of materials and information.

Internationally

There was a strong feeling that Oceania would benefit most from a strengthening of the International Agricultural Research Centers, *viz.* ICARDA and ICRISAT, and that any move to shift responsibilities to more nationally-oriented programs, as seen in the case of faba bean, would not enhance international access to information, genetic resources, and staff training facilities. The group also agreed on the desirability of ICARDA receiving a world-mandate for the improvement of field pea.

Recommendations

We affirm that Oceania strongly resolves to continue research towards the development of the grain legume industry.

We strongly commend ICARDA and ICRISAT for their activities in past years, and strongly recommend that these activities continue as those of international centers but not as regional centers.

We regret that faba bean is no longer included as a mandate crop at ICARDA.

We note that pea is not included as a mandate crop at any international center at present.

We fully support the continuation of the IFLRC concept.

We recommend that between the successive IFLRC's rapid transfer of information be encouraged by the development of grain legume research networks, and we support the continued publication of the journals *Lens, Fabis* and *International Chickpea Newsletter*.

We recommend that in view of the great importance of lupin in our region, future IFLRC's should include lupin as a food legume crop.

We recommend pilot evaluation studies of grasspea as being highly desirable in our.region, and that economic outlets for this crop be developed.

Continuation of the IFLRC concept

International food legume research conference (IFLRC): concept and continuity

R.J. SUMMERFIELD[1] and F.J. MUEHLBAUER[2]

[1] University of Reading, Department of Agriculture, Plant Environment Laboratory, Cutbush Lane, Shinfield, Reading RG2 9AD, Berkshire, UK and
[2] U. S. Department of Agriculture, Agricultural Research Service, 303 W Johnson Hall, Washington State University, Pullman, WA 99164-6434, USA

Retrospect : the IFLRC Concept

The genesis of the concept of an *International Food Legume Research Conference (IFLRC)* can be traced to an informal meeting in 1983. Those present on that evening in Aleppo, Syria (*viz.* H. L. Blain, G. C. Hawtin, F. J. Muehlbauer, A. E. Slinkard and R. J. Summerfield) were keen to ensure that national crop improvement efforts devoted to the food legumes should not be wasteful of resources by duplicating work ongoing elsewhere. A particular concern was to build on the then unprecedented efforts in food legume research underway at ICRISAT and at ICARDA, and which had begun in 1972 and 1976, respectively. These motives were in time to be articulated as the general objective of the IFLRC, *viz.*:

To build communication linkages in order to promote research collaboration and the interchange of scientific and technical information on a global basis covering all aspects of research and development of cool season food legumes.

The inaugural IFLRC became a reality in July 1986 when, under the guidance of Drs. R. H. Lockerman and D. F. Bezdicek, the combined efforts of an International Advisory Board, an International Observer and an Organizing Committee came to fruition (see Summerfield, 1988).

The four hundred or so delegates from 48 countries who attended IFLRC-1 in Spokane, Washington, USA participated in a series of pre-planned business meetings. By their votes the delegates endorsed both the general and specific objectives of the Conference as recommended by the organizers. Activities of the IFLRC as well as the composition and functions of an International Steering Committee were also agreed with enthusiasm. These details are summarized below.

F.J. Muehlbauer and W.J. Kaiser (eds.), Expanding the Production and Use of Cool Season Food Legumes, 971–980.
© 1994 *Kluwer Academic Publishers.*

Specific Objectives of the IFLRC

1. To promote research collaboration and the integration and dissemination of knowledge on cool season food legumes.
2. To provide an international forum for discussion on priority problem areas requiring research attention.
3. To maximize awareness and use of novel technologies and research methods.
4. To encourage the evaluation and adoption of appropriate new technologies at the farm level.
5. To provide an international forum for promoting links between institutions with the aim of strengthening research manpower development.
6. To promote awareness of the importance and role of cool season food legumes among scientists, policy makers and funding agencies in order to stimulate increased research attention and financial support.

Activities of the IFLRC

The principal activity will be the International Conference to be held approximately every four to five years. Other activities may be undertaken, as agreed upon by the IFLRC.

Functions of the International Steering Committee

1. To ensure continuity and help minimize duplication among other national and international meetings.
2. To assist the host organization with logistic arrangements for the Conference.
3. To formulate the Conference agenda so as to meet the objectives of the IFLRC.
4. To solicit financial and logistic support from donors and other organizations for IFLRC activities.
5. To undertake any other activities approved by the IFLRC in line with Conference objectives.

Proposed Composition of the International Steering Committee

1. Representative from ICARDA
2. Representative from ICRISAT
3. Chairperson of the immediate past IFLRC
4. Program Chairperson of the immediate past IFLRC
5. Host nation Chairperson (to be co-opted, once known)
6–12. Seven representatives from six agro-geographical regions, *viz.*:

Agro-Geographical Regions

I.	North America	(USA and Canada)
II.	Latin America and the Caribbean	(Mexico; C. and S. America; Caribbean)
III.	Europe	(W. and E. Europe; USSR[+]; Israel)
IV.	Africa	(All African countries south of the Sahara)
V.	Near East	(N. Africa; West Asia to Afghanistan)
VI.	Asia and the Pacific (including Pakistan) *[Two representatives]*	(India to China and Japan; Australia; New Zealand and Oceanic isles)

(*+Now the Commonwealth of Independent States*)

April 1992: the Second IFLRC

During the decade since 1983, demands for cool season pulse crops as food and feed have continued to increase, to the extent that there is now a supply bottleneck in most countries in the West Asia and North Africa region (Oram and Belaid, 1990). It is the poorest countries, the urban poor, rural peoples and farm families that suffer most in these circumstances. Given these sorts of impetus and causes for concern, the second IFLRC was in the event to be appropriately timed and appropriately located. The International Steering Committee (ISC) elected in Spokane (Table 1) voted Cairo as the venue for IFLRC-II. Under the guidance of Drs. A. E. Slinkard and M. C. Saxena, the ISC worked *in tandem* with the Organizing Committee to plan the Conference for April 1992. More than 200 delegates came from 38 countries to ensure "another major success" (Roberts, 1994).

Business of the IFLRC was conducted following agreed procedures (see Appendix). A ballot proved heavily in favor of the *Continuation of the IFLRC Concept* (69 votes for, 1 vote against). Thereafter, two, one, four, three, three and six delegates were nominated to represent Agro-Geographical Regions 1, 2, 3, 4, 5 and 6, respectively. Voting was pleasingly heavy and often closely contested.

Continuity : the IFLRC-III

The newly-elected ISC (Table 2) met for the first time on 16 April 1992, when Dr. F. J. Muehlbauer was elected as Conference Chairman. The Committee

Table 1. Membership of the International Steering Committee and the Organizing Committee for IFLRC-II

International Steering Committee

Region	Representative
North America	A. E. Slinkard, Canada
Latin America; Caribbean	M. E. Tapia, Peru
Europe	R. J. Summerfield, UK
Africa (Sub-Sahara)	A. Telaye, Ethiopia
Near East	A. M. Nassib, Egypt
Asia and Pacific	W. A. Jermyn, New Zealand
(two representatives)	B. A. Malik, Pakistan
Representative from ICARDA	M. C. Saxena
Representative from ICRISAT	Y. L. Nene
Chairperson of IFLRC-I	R. H. Lockerman
Program Chairman of IFLRC-I	D. F. Bezdicek

Organizing Committee

Conference Chairman	A. E. Slinkard
Program Chairman	M. C. Saxena
Editors	F. J. Muehlbauer
	W. J. Kaiser
Host Country Representative	A. M. Nassib
Host Country Chairman	Saad Nasser
Local Arrangements	M. B. Solh

agreed that IFLRC-III should take place during 1997 and noted that bids to host the Conference were anticipated from:

■ The European Association of Grain Legume Research (J. Picard);
■ General Directorate of Agricultural Research, Turkey (I. Kusmenoglu);
■ Indian Council of Agricultural Research (P. N. Bhal); and from an
■ Australian Consortium of Food Legume Scientists (R. O. Rees).

As intended, the ISC had completed the voting procedures necessary to decide the location of IFLRC-III by the Spring of 1993: the Conference will be held in Adelaide, Australia, It will take place in 1997 between 13 and 18 October.

In what may well prove to be a seminal contribution, Oram (1994) convincingly argued that sustained improvements in productivity of the order of 28 kg ha^{-1}a^{-1} will be necessary to satisfy demands for grain legumes into the next Century in West Asia and North Africa alone. It will be incumbent on IFLRC-III to focus attention of what by then has been achieved and what is therefore left to do ......

Table 2. Members of the International Steering Committee for IFLRC-III[+]

Region	Representative	Region	Representative
1. North America	Dr. F. J. Muehlbauer (Chair) USDA-ARS 303 Johnson Hall Washington State University Pullman, WA 99164-6434 USA	2. Latin America; Caribbean	Dr. J. U. Tay INIA, Estacion Experimental Quilamapu, Avenida Vicente Mendez 516 Chillan, Casilla 426 Chile
3. Europe	Professor R. J. Summerfield University of Reading Department of Agriculture Plant Environment Laboratory Cutbush Lane, Shinfield Reading RG2 9AD, UK	4. Africa (sub-Sahara)	Dr. G. Bejiga Crop Science Department P. O. Box 32 Debre Zeit Ethiopia
5. Near East	Dr. H. M. Halila National Coordinator Laboratoire des Legumineuses Alimentaires INRAT Ariana 2080 Tunisia	6. Asia & Pacific (inc. Pakistan)	Dr. B. A. Malik National Agricultural Research Center PO NARC National Park Road Islamabad Pakistan
Immediate Past Program Chair	M. C. Saxena ICARDA P. O. Box 5466 Aleppo, Syria	6. Asia & Pacific (inc. Pakistan)	Mr. R. O. Rees Crops Economics Section ABARE Edmund Barton Building GPO Box 1563 Canberra, ACT 2601 Australia
Immediate Past Chair	Dr. A. E. Slinkard Crop Development Centre University of Saskatchewan Saskatoon, Saskatchewan S7N 0W0 Canada	ICARDA Representative	Dr. Willie Erskine ICARDA P. O. Box 5466 Aleppo, Syria
ICRISAT Representative	Dr. Y. L. Nene Deputy Director General ICRISAT Patancheru Andhra Pradesh 502 324 India	Co-opted by invitation	Dr. W. J. Kaiser USDA-ARS-RPIS 59 Johnson Hall Washington State University Pullman, WA 99164-6402 USA

[+]Newly elected members joining Drs. A. E. Slinkard, M. C. Saxena, Y. L. Nene, B. A. Malik
and co-opted member W. J. Kaiser (co-editor of the Proceedings of IFLRC-II)

Relations between the CGIAR Centres and the NARS in West Asia, North Africa and elsewhere are pivotal for the food legumes, for agricultural ambitions, and even for human survival......
His Excellency Professor Youssef Wally, Deputy Prime Minister and Minister of Agriculture and Land Reclamation, Egypt.

Acknowledgements

In addition to those Benefactors listed elsewhere in these Proceedings (pp. xv-xvi) we gratefully acknowledge the technical and secretarial contributions of: in Egypt, Drs. A. M. Nassib, Saad Nasser and M. B. Solh, Mrs. Nagwa Lutfi, Miss Mai Elremisy, Ms. Tahany Ramez and Khaled Genene; in Canada, Dr. B. Vandenberg; in ICARDA, Miss Mary Bogharian; in the USA, Mrs. Joy Barbee, Dr. C. Simon and Mr. D. Hoyt; and in the UK, Mrs. C. Hadley and Mr. A. Pilgrim.

References

Oram, P. and Belaid, A. 1990. *Legumes in Farming Systems*. ICARDA-IFPRI Report, pp. 206.
Oram, P. 1994. *Expanding the Production and Use of Cool Season Food Legumes*, pp. 3–49 (eds. F. J. Muehlbauer and W. J. Kaiser). Dordrecht: Kluwer Academic Publishers.
Roberts, E. H. 1994. *Expanding the Production and Use of Cool Season Food Legumes*, pp. 983–988 (eds. F. J. Muehlbauer and W. J. Kaiser). Dordrecht: Kluwer Academic Publishers.
Summerfield, R. J. (ed.) 1988. *World Crops: Cool Season Food Legumes*, pp. 1179. Dordrecht: Kluwer Academic Publishers.

APPENDIX: IFLRC Administration and Business

■ **Ballot Form (A).** Continuation of the IFLRC Concept

Continuation of the IFLRC Concept

Ballot Form [A]

■ The International Food Legume Research Conference is proposed as a continuing entity devoted to the effective and timely communication of research data on all aspects of the production, quality and utilization of cool season food legumes.

■ It is further proposed that the Conference be held approximately every four to five years to consider: (a) individual and collaborative research progress since the previous Conference; (b) the current status of research; and (c) future opportunities for progress and the exploitation and application of research data.

Please tick appropriate box (√) below

Continuation of the IFLRC	
Approve	Do not approve

■ **Nomination Form (B).** International Steering Committee for IFLRC-III

IFLRC-III : International Steering Committee

NOMINATION FORM [B]

■ Any individual full member of the present Conference (i.e.,
those persons who have paid the full registration fee) may
propose any other individual full member(s) for any of the
proposed agro-geographical regions.

■ **The current officers of the IFLRC will assume that those
persons nominated have indicated their willingness to stand
for election to both their proposer and seconder.**

Please write legibly on the form below

Person Proposed	Regional Number*	Proposed by	Seconded by

*Agro-geographical groupings are:

I	North America	(USA and Canada)
II	Latin America	(Mexico; C. and S. America; and the Caribbean)
III	Europe.	(E. and W. Europe; Israel; Commonwealth of Independent States)
IV	Africa.	(All African countries south of the Sahara)
V	Near East	(N. Africa; West Asia to Afghanistan)
VI	Asia and the Pacific. . . . (*Two representatives*)	(India to China and Japan; Australia; New Zealand and Oceanic Isles)

■ **Election-Voting Slip (C).** International Steering Committee for IFLRC-III

IFLRC - III : International Steering Committee

Election - Voting Slip (C)

Any individual full member of the present Conference (i.e., those persons who have paid the full registration fee) may vote for ONLY ONE CANDIDATE within each region - i.e., you have 6 votes.

NOMINEES

Check one
only (✓)

REGION 1 North America

REGION 2 Latin America,
         Caribbean

REGION 3 Europe

REGION 4 Africa (Sub-Sahara)

REGION 5 Near East

REGION 6 Asia and Pacific
         (incl. Pakistan)

■ **Election Results (D).** The ISC for IFLRC-III

IFLRC - III : International Steering Committee

Election Results [D]

Region	Elected Representative
REGION 1 North America	
REGION 2 Latin America; Caribbean	
REGION 3 Europe	
REGION 4 Africa (Sub-Sahara)	
REGION 5 Near East	
REGION 6 Africa and Pacific (inc. Pakistan)	

Conference summary

Impressions of the Second International Food Legume Research Conference

E.H. ROBERTS

Department of Agriculture, University of Reading, Earley Gate, P. O. Box 236, Reading RG6 2AT, UK

The first conference in this series was held in Spokane, Washington, USA in 1986 (Summerfield, 1988). It was a major event (400 participants from 48 countries) in which 91 papers focused much needed attention on the major cool season food legumes, *viz.* pea (*Pisum sativum*), lentil (*Lens culinaris*), faba bean (*Vicia faba*), and chickpea (*Cicer arietinum*). It was a great success: it brought together scientists working on all aspects of these crops from a wide range of disciplines – and from basic science to marketing and extension. The proceedings (Summerfield, 1988) are still an essential reference for anyone seriously concerned with these crops.

Those who attended the first conference clearly saw that there would be a need for another in about five years. The second International Food Legume Research Conference was held six years later in Cairo and included a fifth crop, the grasspea (*Lathyrus sativus*), of limited importance but of some greater potential. This second conference, which was attended by 230 participants from 38 countries, was also a major success, as I hope readers of these proceedings will have discovered, so that a third conference is now being planned for 1997.

None of the five legume crops which are the subject of these proceedings originated in the New World; all originated in the Eastern Mediterranean or West Asian region (Smartt and Hymowitz, 1985) and much of the early domestication took place in the zone now known by the Food and Agricultural Organization (FAO) and the Consultative Group for International Agricultural Research (CGIAR) as WANA – an acronym for West Asia and North Africa. The origin and domestication of these crops therefore took place in the cradle of Western Civilization, whose early development was made possible by the domestication of wheat, barley, and the cool season food legumes. Cairo is sited at the heart of this region and so, in the year which marked the 500th anniversary of the discovery of the "New World" by the "Old World", it was appropriate perhaps for many of our colleagues from the Americas to return to the "Old World" to attend a conference in an ancient city, in a region of such great significance to their interests.

Since the first International Food Legume Research Conference in Spokane, there have been various international changes which are having or may come to

F.J. Muehlbauer and W.J. Kaiser (eds.), Expanding the Production and Use of Cool Season Food Legumes, 983–988.

have some influence on research on the grain legumes. There has been a sudden and unheralded demise of centrally planned socialist economic systems in Eastern Europe and in what was the Soviet Union. Support for agricultural research in these regions is more likely to be modelled on those in western countries. In the "developed" countries, many of which have agricultural surpluses, there is a tendency to move towards a free market agricultural economy, to reduce subsidies and publicly funded research, to use what public funds available for basic research because, according to this philosophy, it is argued that if applied research is to be done, it should be paid for by the immediate beneficiaries; a consequence has been the withdrawal of public funds from applied research. The outcome in the UK, for example, where this philosophy unclouded by much compromise, has been a major retrenchment in publicly funded agricultural research and advice, and many agricultural research institutes of repute have disappeared.

In the "developing" countries where there tend to be food shortages rather than surpluses, and where, in contrast, the agricultural sector is a major part of the economy engaging the largest proportion of the population, governments take a different view. It is still thought appropriate to use public funds to support agricultural research and development – even though the national resources may be meager. Furthermore the bilateral and international aid agencies also continue to support strategic and applied research. However, many of these agencies have recently been developing a new system of research prioritization. While these systems are not identical, they all use multiple criteria and attempt to be quantitative and objective. Not surprisingly, therefore, the major conclusions are similar: the priority crops are cereals, and research on cool season grain legumes is given a comparatively low priority rating. Taking this trend into account and the fact that, with few exceptions, national and international aid budgets are declining, it would take an optimist of some distinction to imagine the investment in cool season legume research is likely to increase – at least in advance of the third conference.

Let us hope I am wrong because Oram (pp. 3–49) in his keynote address began this conference with a seminal paper. It makes several important points including his analysis which shows that, although the cool season pulses are such traditionally important components of the diets in many developing economies, their per capita consumption in these countries is nevertheless declining. He shows that this trend is the result of decreased supplies, not decreased demand. Furthermore, in order to sustain supplies, even at present levels, requires an increase in production in the 1990s of at least 13 million tons in the developing countries. To achieve this target he argues that the main thrust will have to be to increase yields per unit area through research.

Technically there is indeed much scope for improvement (Johansen *et al.*, pp. 175–194). But if we examine the details of what is required, Oram provides convincing arguments that to meet the demands for grain legumes in WANA, it will be necessary to increase average yields from a current value of 650 kg ha^{-1} to one of 870 kg ha^{-1} by the year 2000, i.e, an increase of about 28 kg ha^{-1} per year. He points out that this does not seem to be an impossible target. Nevertheless if we make a comparison with what has happened elsewhere with another grain legume, we find that it is indeed a formidable task. Summerfield and Lawn (1987) describe how between 1925 and 1992 the average yield of soybean (*Glycine max*) in the United States increased from about 800 kg ha^{-1} to about 2,500 kg ha^{-1} – i.e., by about 22.5 kg ha^{-1} per year. To give some comparative indicator of the scientific effort involved, in 1976 there were an estimated 22 soybean breeders (but more than 400 maize breeders) in the USA. They cite another example: during the period from 1965 to 1990 the average yields of soybean in Brazil increased from about 1000 kg ha^{-1} to 1,700 kg ha^{-1}, i.e, an increase of about 32.5 kg ha^{-1} per year. It is important to remember, however, that in both of these cases we are considering agriculture with *relatively* high inputs on *relatively* good soil in *relatively* stress-free climates. The environmental constraints in the WANA region are much greater, the soils are more fragile and the inputs to legumes are meager.

However, there is one more trend that should, I hope, help to encourage some revision of international research priorities and funding. There has been a growing public concern with environmental issues and sustainability. Some of this is well-intentioned but misinformed – e.g., that nitrate that comes out of a bag is in some way different from nitrate produced through the decomposition of organic matter, or that "natural" and "organically" produced foods are always benign (but what about the antinutritional factors and toxins present in most legumes, for example?). Nevertheless, even though some worries are overemphasized through ignorance, anyone who now has no concern with environmental issues is clearly irresponsible. Sustainability, although ill-defined, is a serious matter and, in the harsh marginal environments of many of the regions where they traditionally grow, these pulses are important components of the agricultural systems – one of their roles arising from their ability to fix nitrogen symbiotically. While there have been important contributions in this conference to considerations of symbiotic nitrogen fixation (e.g., Herridge *et al.*, pp. 471–491), perhaps this is a topic to which more attention should be given in the future – the "capture" and use of nitrogen in drought-prone systems of production.

Discussions on sustainability often center on questions of soil conservation, the maintenance of its physical structure, and the avoidance of irreversible chemical deterioration. All this is clearly important, but Peter Oram also asked at the beginning of this conference: should social scientists be involved more in our researches? I would argue that they should because the first question concerning sustainability in my view is: does the farming system sustain the farmer and his or her dependents, i.e., is it profitable? Otherwise the system clearly is unsustainable.

But, the question of economic sustainability is not a simple one. It has several components, important among which is the characteristic of the low rainfall regions where many of these pulses grow, that the coefficients of variation in annual rainfall and its seasonal distribution are greater than in many other climates. I was told when I was in Queensland, Australia that the large-scale grain farmers there expected a complete crop loss from time to time and could sustain such a loss one year in three. Three successive failures, however, would be terminal. Clearly, most small-scale farmers on low incomes in WANA could not cope with this degree of variation. Such considerations lead to another question which Peter Oram posed in his keynote address: to what extent should we be concerned with systems which aim to maximize yield or, alternatively, improve yield stability? Clearly, one cannot be concerned only with one or the other; a completely stable yield of 100 kg ha^{-1} would not be satisfactory, or therefore, sustainable. Nevertheless, if there is some trade-off between high, maximum, or average yields as against more yield stability, which seems likely, then what would be the compromise at which we should aim?

Is it unreasonable to expect yields in marginal areas to be as stable as in those non-marginal areas? Erskine *et al.* (pp. 559–571) described how, at the ICARDA research station at Tel Hadya, analysis has shown the yield variations for lentil could be attributed as follows: 80% to rainfall, 13% to number of frost nights and 7% to unaccounted factors. This is perhaps not a surprising finding in a dry area where rainfall is a major limiting factor and given its greater variability than in more benign climates. Thus with the best will in the world, there is a limit to how far one can improve yield stability in such a climate. It may be more important – and this is where socio-economic analysis may help – to aim for at least some yield in poor years and, so far as it is consistent with this objective, reasonable average yields. If there remained high variability due to an occasional "bumper year", I expect this would be an embarrassment the farmers would be prepared to tolerate!

Dr. John Peacock, now at ICARDA and who passed through Cairo during the conference, suggested to me that in many marginal areas farmers are faced with multiple physical stresses (a point brought out in many of the contributions to this conference); in addition to perennial problems of drought, one year might be exceptionally cold early on, or in another year a high temperature may occur earlier in spring than usual. In addition, of course, are variations in the direct and interacting effects of various biotic stresses which have received considerable attention in this conference. Furthermore, Peacock suggested it might well be a genetic or a physiological impossibility to include tolerance for all these eventualities in a single genotype. In order to move some way towards the greater yield stability that Oram and many of us feel would be a worthy objective, Peacock suggested we should consider more seriously the use of genetic mixtures. I would add that although experiments on cultivar mixtures have met with limited success in high-input agriculture in benign climates, this is no argument for not considering this approach in marginal environments; in the former circumstances the major concern has been to increase yields rather than yield stability. A special feature of the use of mixtures with regard to stability is that, because of the generally asymptotic nature in the relationship between plant population density and economic yield, the complete loss of any one genotype due to stress from a mixture of several genotypes has relatively little effect on total yield.

Now, although grain legumes in general have some reputation, more so than in other species, for yield instability, paradoxically informed opinion would agree that in many areas they are a vital component of a sustainable, and therefore stable, agricultural system. Although they are not staple foods, and although their earlier perceived importance as protein sources was undoubtedly exaggerated – a point made most vigorously by Bunting (1988) in his commentary on the first conference – it is nevertheless clear that they are important components of food and feed in many cultures. If their role in the agricultural systems and societies where many of the world's poorest live were more fully recognized, then I think the research-priority rating of the grain legumes would indeed be revised upwards.

The Past and the Future

The next conference is planned for 1997; this seems a long way ahead and so we can be optimistic about progress. However, when looking back over the past six years to the first conference in 1986, time has gone quickly. Reflecting on this, it is reasonable to ask is an optimistic view justified? Politicians, the popular press, and some scientists (when seeking grants or promotion) like to talk in terms of breakthroughs. But the historical quantitative analysis of agricultural trends tend to show steady progress e.g., in yield as a function of time – possibly because many interdependent scientific, technical, socioeconomic, and political factors are involved in the process.

My impression of the Cairo conference is that there has been steady and commendable progress in plant breeding, crop protection, crop physiology, the application of on-farm research, and in the understanding of potential utilization of products. Although some have expressed the view that there is room for a little more socioeconomic research, this area has not yet, perhaps, received sufficient attention.

There were two new features in Cairo. One was the inclusion of *Lathyrus sativus* which is not a new crop (it may be as much as 5000 year old) but, to some, a crop with considerable potential because of its reputation of being able to withstand harsh conditions. The main constraint to its further development has clearly been its content of neurotoxin. Now that the toxin problem seems soluble its future does indeed seem more promising. But there are many underexploited crops which have been and continue to be promising, and I suspect many of them will still be both promising and underexploited for many years to come. If *L. sativus* is to fulfill its current promise I would expect there to be more evidence of progress by the next conference. It will be interesting then to see what has happened.

The second new feature of the Cairo conference was the separate session devoted to biotechnology and gene mapping. Most of this was concerned with educating the conference on the nature of the techniques and their potential uses. As these techniques are refined they will no doubt become additional weapons to add to the familiar ones in the plant breeder's armory. At that time there will then be no need for a separate session and we shall be hearing how these techniques are being used routinely in crop improvement programs. It is important to recognize, as most plant breeders and biotechnologists do, that these techniques are not alternatives to plant breeding programs but, if we are lucky, will help to improve the efficiency and productivity of those programs. We hope to see evidence of such progress by the third International Food Legume Research Conference in 1997.

References

Bunting, A. H. 1988. In: *World Crops: Cool Season Food Legumes*, pp. 1155–1167 (ed. R. J. Summerfield). Dordrecht: Kluwer Academic Publishers.

Erskine, W., Tufail, M., Russell, A., Tyagi, M. C., Rahman, M. M. and Saxena, M. C. 1994. In: *Expanding the Production and Use of Cool Season Food Legumes*, pp. 559–571 (eds. F. J. Muehlbauer and W. J. Kaiser). Dordrecht: Kluwer Academic Publishers.

Herridge, D. F., Rupela, O. P., Serraj, R. and Beck. D. P. 1994. In: *Expanding the Production and Use of Cool Season Food Legumes*, pp. 472–492 (eds. F. J. Muehlbauer and W. J. Kaiser). Dordrecht: Kluwer Academic Publishers.

Johansen, C., Baldev, B., Brouwer, J. B., Erskine, W., Jermyn, W. A., Li-Juan, L., Malik, B. A., Ahad, Miah, A., and Silim, S. N. 1994. In: *Expanding the Production and Use of Cool Season Food Legumes*, pp. 175–194 (eds. F. J. Muehlbauer and W. J. Kaiser). Dordrecht: Kluwer Academic Publishers.

Oram, P. A. and Agcaoili, M. 1994. In: *Expanding the Production and Use of Cool Season Food Legumes*, pp. 3–49 (eds. F. J. Muehlbauer and W. J. Kaiser). Dordrecht: Kluwer Academic Publishers.

Smartt, J. and Hymowitz, T. 1985. In: *Grain Legume Crops*, pp. 37–72 (eds. R. J. Summerfield and E. H. Roberts). London: Collins.

Summerfield, R. J. (ed.) 1988. *World Crops: Cool Season food Legumes*, 1179 pp. Dordrecht: Kluwer Academic Publishers.

Summerfield, R. J. and Roberts, E. H. (eds.). 1985. *Grain Legume Crops*, 859 pp. London: Collins.

Summerfield, R. J. and Lawn, R. J. 1987. *Outlook on Agriculture* 16: 189–197.

Author index

Abdel Moneim, A.M. 617
Abdelmagid, S. 367
Abi Antoun, M. 911
Açikgöz, N. 388
Agcaoili, M. 3
Agrawal, S.K. 617
Ahmed, K. 679
Al-Menoufi, O.A. 695
Ali, M. 633
Ali, S.M. 540
Ali, K. 679
Ali-Khan, S.T. 877
Ambrose, M.J. 540
Amin, P.W. 926
Amine, M. 899
Araya, W.A. 144, 617
Arihara, J. 809

Bahl, P.N. 495
Baker, R.J. 429
Baldev, B. 175
Bansal, R.K. 517
Bascur, G. 195, 399
Bayaa, B. 268
Beck, D.P. 472, 738, 821
Beech, D.F. 412
Beniwal, S.P.S. 642, 899, 951
Bernier, C.C. 247
Bezdicek, D.F. 738
Bhatty, R.S. 113
Biddle, A.J. 204
Blain, H.L. 98

Bond, D.A. 592
Borg, S.J. ter 695
Bos, L. 305
Brinsmead, R.B. 412
Brockwell, J. 821
Brouwer, J.B. 175, 412, 962

Campbell, C.G. 617
Caubel, G. 346
Chen, Y.Z. 617
Clement, S.L. 290
Cousin, R. 439
Cubero, J.I. 333

Deshpande, S.S. 113
Dhindsa, K.S. 98
Di Vito, M. 346
Diekmann, J. 517
Duc, G. 791

El-Din Sharaf El-Din, N. 290, 679
El-Fouly, M.M. 457
Ellis, R.H. 755
Er, C. 388
Erskine, W. 175, 559

Faki, H. 367, 495
Faris, D.G. 504
Feindel, D.E. 877
Forse, L. 738

García-Torres, L. 695

Gastel, A.J.G. van 495
Gates, P. 791
Gent, G.P. 361
George, E. 832
Gizaw, A. 633
Gowda, C.L.L. 504, 959
Grashoff, C. 159
Greco, N. 346
Gregory, P.J. 809

Haddad, N.I. 911, 956
Halila, H.M. 219, 572, 899, 951
Hannan, R.M. 849
Harrabi, M. 268
Hashim, A. 367
Haware, M.P. 268
Heath, M.C. 771, 947
Hebblethwaite, P.D. 771
Hernández-Bravo, G. 195, 219, 944
Herridge, D.F. 472
Huisman, J. 53
Hulse, J.H. 77
Hussein, L.A. 98, 113

Ito, O. 809

Jambunathan, R. 98
Jaradat, A.A. 130
Jellis, G.J. 247, 592
Jermyn, W.A. 175
Jiménez-Díaz, R.M. 268
Johansen, C. 175

Kaemmer, D. 705
Kahl, G. 705
Kaiser, W.J. 233, 247, 531, 849,
  941
Karaca, M. 388
Kaul, A. 144
Kearney, J. 144
Keatinge, J.D.H. 890
Kessel, C. van 877
Khalil, S.A. 333, 592
Khawaja, H.I.T. 617
Knights, E.J. 412, 439, 572

Kogure, K. 98
Kost, S. 705
Kothari, S.K. 832
Kraft, J.M. 268, 941
Kusmenoglu, I. 890, 956

Lateef, S.S. 290, 679
Le Guen, J. 592
Li, X.-L. 832
Li-Juan, L. 98, 175, 592
Linke, K.H. 695
Lu, J. 726

Mahmoud, S.F. 679
Mahoney, J.E. 412
Makkouk, K.M. 305
Malhotra, R.S. 439, 572
Malik, B.A. 175, 219
Maniruzzaman, A.F.M. 504
Marschner, H. 832
McKenzie, B.A. 771
Mehra, R.B. 617
Meyveci, K. 388
Miah, A. Ahad 175
Mihov, M.I. 219
Moawad, H. 821
Monroe, G.E. 517
Monti, L. 204
Moreno, M.T. 204
Muehlbauer, F.J. 233, 531, 849,
  859, 971

Nassib, A.M. 495
Nene, Y.L. 666, 926
Ney, B. 791
Nonhebel, S. 159

Oppen, M. von 367
Oram, P.A. 3

Pala, M. 130
Papastylianou, I. 130
Park, R. 877
Peñaloza, E. 399
Picard, J. 947

Pieterse, A. 333
Pieterse, A.H. 695
Pilbeam, C.J. 771
Plancquaert, P. 204, 495
Poel, A.F.B. van der 53
Porta-Puglia, A. 247

Quinn, M.A. 738

Rabbinge, R. 159
Rahman, M.M. 144, 559
Rana, R.S. 457
Reddy, M.V. 247
Reed, W. 666
Rees, R.O. 412, 962
Rheenen, H.A. van 926
Rik, M.A. 633
Roberts, E.H. 755, 983
Robertson, L.D. 592
Rowland, G.G. 592, 791
Ruckenbauer, P. 457
Rupela, O.P. 472
Russell, A. 559

Sadri, B. 219
Sakar, D. 890
Sauerborn, J. 333
Savage, G.P. 113
Saxena, M.C. 130, 457, 559, 633, 705
Saxena, N.P. 457, 809
Serraj, R. 472
Shabana, R. 457
Sharma, S.B. 346
Sharma, B. 540, 959
Short, R.W. 859

Sikora, R.A. 346
Silim, S.N. 175, 439
Silkine, A.B. 911
Simon, C.J. 531
Singh, K.B. 572
Slinkard, A.E. 195, 877
Smartt, J. 144
Snobar, B.A. 495, 911
Solh, M.B. 219, 899
Sprent, J.I. 821
Stanforth, A. 821
Summerfield, R.J. 755, 859, 971

Tawil, M. 911
Tay, J.U. 399, 617, 944
Telaye, A. 791
Timmerman, G.M. 726
Trapero-Casas, A. 642
Tufail, M. 559
Tutwiler, R.N. 899
Tyagi, M.C. 559

Verma, M.M. 572

Walton, G.H. 412
Weber, E. 832
Weeden, N.F. 726
Weigand, F. 705
Weigand, S. 290, 679, 738
Weising, K. 705
Wery, J. 439
Williams, P.C. 113

Yadov, C.R. 617
Youssef, M.M. 98

Current Plant Science and Biotechnology in Agriculture

1. H.J. Evans, P.J. Bottomley and W.E. Newton (eds.): *Nitrogen Fixation Research Progress.* Proceedings of the 6th International Symposium on Nitrogen Fixation (Corvallis, Oregon, 1985). 1985   ISBN 90-247-3255-7

2. R.H. Zimmerman, R.J. Griesbach, F.A. Hammerschlag and R.H. Lawson (eds.): *Tissue Culture as a Plant Production System for Horticultural Crops.* Proceedings of a Conference (Beltsville, Maryland, 1985). 1986   ISBN 90-247-3378-2

3. D.P.S. Verma and N. Brisson (eds.): *Molecular Genetics of Plant-microbe Interactions.* Proceedings of the 3rd International Symposium on this subject (Montréal, Québec, 1986). 1987   ISBN 90-247-3426-6

4. E.L. Civerolo, A. Collmer, R.E. Davis and A.G. Gillaspie (eds.): *Plant Pathogenic Bacteria.* Proceedings of the 6th International Conference on this subject (College Park, Maryland, 1985). 1987   ISBN 90-247-3476-2

5. R.J. Summerfield (ed.): *World Crops: Cool Season Food Legumes.* A Global Perspective of the Problems and Prospects for Crop Improvement in Pea, Lentil, Faba Bean and Chickpea. Proceedings of the International Food Legume Research Conference (Spokane, Washington, 1986). 1988   ISBN 90-247-3641-2

6. P. Gepts (ed.): *Genetic Resources of Phaseolus Beans.* Their Maintenance, Domestication, Evolution, and Utilization. 1988   ISBN 90-247-3685-4

7. K.J. Puite, J.J.M. Dons, H.J. Huizing, A.J. Kool, M. Koorneef and F.A. Krens (eds.): *Progress in Plant Protoplast Research.* Proceedings of the 7th International Protoplast Symposium (Wageningen, The Netherlands, 1987). 1988   ISBN 90-247-3688-9

8. R.S. Sangwan and B.S. Sangwan-Norreel (eds.): *The Impact of Biotechnology in Agriculture.* Proceedings of the International Conference The Meeting Point between Fundamental and Applied in vitro Culture Research (Amiens, France, 1989). 1990.   ISBN 0-7923-0741-0

9. H.J.J. Nijkamp, L.H.W. van der Plas and J. van Aartrijk (eds.): *Progress in Plant Cellular and Molecular Biology.* Proceedings of the 8th International Congress on Plant Tissue and Cell Culture (Amsterdam, The Netherlands, 1990). 1990   ISBN 0-7923-0873-5

10. H. Hennecke and D.P.S. Verma (eds.): *Advances in Molecular Genetics of Plant-Microbe Interactions.* Volume 1. 1991   ISBN 0-7923-1082-9

11. J. Harding, F. Singh and J.N.M. Mol (eds.): *Genetics and Breeding of Ornamental Species.* 1991   ISBN 0-7923-1094-2

12. J. Prakash and R.L.M. Pierik (eds.): *Horticulture – New Technologies and Applications.* Proceedings of the International Seminar on New Frontiers in Horticulture (Bangalore, India, 1990). 1991   ISBN 0-7923-1279-1

13. C.M. Karssen, L.C. van Loon and D. Vreugdenhil (eds.): *Progress in Plant Growth Regulation.* Proceedings of the 14th International Conference on Plant Growth Substances (Amsterdam, The Netherlands, 1991). 1992   ISBN 0-7923-1617-7

14. E.W. Nester and D.P.S. Verma (eds.): *Advances in Molecular Genetics of Plant-Microbe Interactions.* Volume 2. 1993   ISBN 0-7923-2045-X

15. C.B. You, Z.L. Chen and Y. Ding (eds.): *Biotechnology in Agriculture.* Proceedings of the First Asia-Pacific Conference on Agricultural Biotechnology (Beijing, China, 1992). 1993   ISBN 0-7923-2168-5

Current Plant Science and Biotechnology in Agriculture

16. J.C. Pech, A. Latché and C. Balagué (eds.): *Cellular and Molecular Aspects of the Plant Hormone Ethylene*. 1993   ISBN 0-7923-2169-3

17. R. Palacios, J. Mora and W.E. Newton (eds.): *New Horizons in Nitrogen Fixation*. Proceedings of the 9th International Congress on Nitrogen Fixation (Cancún, Mexico, 1992). 1993   ISBN 0-7923-2207-X

18. Th. Jacobs and J.E. Parlevliet (eds.): *Durability of Disease Resistance*. 1993
ISBN 0-7923-2314-9

19. F.J. Muehlbauer and W.J. Kaiser (eds.): *Expanding the Production and Use of Cool Season Food Legumes*. A Global Perspective of Peristent Constraints and of Opportunities and Strategies for Further Increasing the Productivity and Use of Pea, Lentil, Faba Bean, Chickpea, and Grasspea in Different Farming Systems. Proceedings of the Second International Food Legume Research Conference (Cairo, Egypt, 1992). 1994
ISBN 0-7923-2535-4

KLUWER ACADEMIC PUBLISHERS – DORDRECHT / BOSTON / LONDON